AF732320

ADVANCES IN CIVIL ENGINEERING: STRUCTURAL SEISMIC RESISTANCE, MONITORING AND DETECTION

Advances in Civil Engineering: Structural Seismic Resistance, Monitoring and Detection is a collection of papers resulting from the conference on Structural Seismic Resistance, Monitoring and Detection (SSRMD 2022), Harbin, China, 21–23 January, 2022. According to the development of many new seismic theories, technologies and products, the primary goal of this conference is to promote research and developmental activities in structural seismic resistance, monitoring and detection. Moreover, another goal is to promote scientific information interchange between scholars from the top universities, business associations, research centers and high-tech enterprises working all around the world.

The conference conducted in-depth exchanges and discussions on relevant topics such as structural seismic resistance, monitoring and detection, aiming to provide an academic and technical communication platform for scholars and engineers engaged in scientific research and engineering practice in the field of civil engineering, seismic resistance and engineering entity structure testing. By sharing the research status of scientific research achievements and cutting-edge technologies, it helps scholars and engineers all over the world to comprehend the academic development trend and broaden research ideas. So as to strengthen international academic research, academic topics exchange and discussion, and promoting the industrialization cooperation of academic achievements.

PROCEEDINGS OF THE INTERNATIONAL CONFERENCE ON STRUCTURAL SEISMIC RESISTANCE, MONITORING AND DETECTION (SSRMD 2022), HARBIN, CHINA, 21–23 JANUARY 2022

Advances in Civil Engineering: Structural Seismic Resistance, Monitoring and Detection

Edited by

Mohd Johari Mohd Yusof

Faculty of Design and Architecture, Universiti Putra Malaysia

Junwen Zhang

China University of Mining & Technology, Beijing

CRC Press
Taylor & Francis Group
Boca Raton London New York Leiden

CRC Press is an imprint of the
Taylor & Francis Group, an **informa** business
A BALKEMA BOOK

First published 2023
by CRC Press/Balkema
4 Park Square, Milton Park / Abingdon, Oxon OX14 4RN / UK
e-mail: enquiries@taylorandfrancis.com
www.routledge.com – www.taylorandfrancis.com

CRC Press/Balkema is an imprint of the Taylor & Francis Group, an informa business

© 2022 selection and editorial matter, Mohd Johari Mohd Yusof and
Junwen Zhang; individual chapters, the contributors

The right of Mohd Johari Mohd Yusof and Junwen Zhang to be identified as the authors of the editorial material, and of the authors for their individual chapters, has been asserted in accordance with sections 77 and 78 of the Copyright, Designs and Patents Act 1988.

All rights reserved. No part of this book may be reprinted or reproduced or utilised in any form or by any electronic, mechanical, or other means, now known or hereafter invented, including photocopying and recording, or in any information storage or retrieval system, without permission in writing from the publishers.

Although all care is taken to ensure integrity and the quality of this publication and the information herein, no responsibility is assumed by the publishers nor the author for any damage to the property or persons as a result of operation or use of this publication and/or the information contained herein.

ISBN: 978-1-032-31491-4 (hbk)
ISBN: 978-1-032-31684-0 (pbk)
ISBN: 978-1-003-31088-4 (ebk)

DOI: 10.1201/9781003310884

Typeset in Times New Roman
by MPS Limited, Chennai, India

Advances in Civil Engineering: Structural Seismic Resistance, Monitoring and Detection – Mohd Johari Mohd Yusof and Junwen Zhang (Eds)
© 2023 The Editors and Contributors, ISBN 978-1-032-31491-4

Table of contents

Advances in Civil Engineering: Structural Seismic Resistance, Monitoring and Detection – Mohd Johari Mohd Yusof and Junwen Zhang (Eds)
© 2023 The Editors and Contributors, ISBN 978-1-032-31491-4

Preface

It is over two years since COVID-19 broke out. The whole world is still struggling with the virulent COVID-19 pandemic. It is still difficult to undertake international travel. Many conferences are held virtually, for the sake of protecting all the participants and conference staff from getting infected by the virus.

It is uncertain when the COVID-19 will end, so it remains unclear for postponement time, while many scholars and researchers wanted to attend this long-waited conference and have academic exchanges with their peers. Therefore, in order to actively respond to the call of the government, and meet author's request, 2022 International Conference on Structural Seismic Resistance, Monitoring and Detection (SSRMD 2022), which was planned to be held in Harbin, China, January 21–23, 2022 was changed to be held online through Zoom software. This approach not only avoids people gathering, but also meets their communication needs.

The conference is an international conference for the presentation of technological advances and research results in the fields of Structural Seismic Resistance, Monitoring and Detection. The conference brings together leading researchers, engineers and scientists in the domain of interest from around the world. We warmly welcome previous and prospected authors to submit your new research papers to SSRMD 2022 and share the valuable experiences with the scientists and scholars around the world.

The conference brings together about 180 well-known scholars in the field of structural seismic resistance, monitoring and detection at home and abroad. The reports were divided into keynote speeches, oral presentations, and poster presentations to share their latest research results and experiences in related research fields. In the first part, each keynote speakers were allocated 30 minutes to present their talks via Zoom. After the keynote talks, all participants joined in a WeChat communication group to discuss more about the talks and presentations.

We were greatly honored to have welcome five distinguished experts as our keynote speakers. Prof. Junwen Zhang, from China University of Mining and Technology (BEIJING). He has been engaged in research on deep coal mining, mine pressure and ground control, rock mechanics. And then we had Prof. Hongfu Zhou, Shandong University of Science and Technology, China. He presented an insightful speech: *Research on geological hazards effect along the Anninghe fault zone at the eastern edge of the Tibetan Plateau.* Assoc. Prof. Bingxiang Yuan, Guangdong University of Technology. His research area: Geotechnical engineering, Tunnel engineering, Solid waste reuse. Assoc. Prof. Lingkun Chen, Yangzhou University, China. He has studied both static and dynamic modeling of bridges and tunnels, as well as dynamic model effects on wind resistance. Our finale keynote speaker, Assoc. Prof. Ahmad Safuan Bin A Rashid, Universiti Teknologi Malaysia, Malaysia was invited to perform a thought-provoking speech: *Evaluating the Effects of Climate Changes on Landslide Hazard Map using Support Vector Machine (SVM).* Their insightful speeches have triggered heated discussion in the second session of the conference. The WeChat discussion lasted for about 30 minutes. Every participant praised this conference for disseminating useful and insightful knowledge.

We are glad to share with you that we received lots of submissions from the conference and we selected a number of high-quality papers and compiled them into the proceedings after rigorously review. These papers featured the following topics, but were not limited to: Civil Engineering, Structural Seismic Resistance, Monitoring and Detection and other relevant directions. All the papers have been through rigorous review and process to meet the requirements of international publication standards.

Lastly, we would like to warmly thank all the authors who, with their presentations and papers, generously contributed to the lively exchange of scientific information that is so vital to the endurance of scientific conferences of this kind.

The Committee of SSRMD 2022

Advances in Civil Engineering: Structural Seismic Resistance, Monitoring and Detection – Mohd Johari Mohd Yusof and Junwen Zhang (Eds)
© 2023 The Editors and Contributors, ISBN 978-1-032-31491-4

Committee member

Conference Chair
Prof. Junwen Zhang, *China University of Mining and Technology (Beijing), China*
Assoc. Prof. Ts. Gs. Dr. Mohd Johari Mohd Yusof, *Universiti Putra Malaysia, Malaysia*

Technical Committee
Prof. Lei Zhang, *East China Normal University, China*
Prof. Fadi Hage Chenade, *Lebanese University, Lebanon*
Prof. Lishuai Jiang, *Shandong University of Science and Technology, China*
Prof. Mohamed Abdel Kader El Gelany Ismai, *Miami College of Henan University, Henan University, China*
Prof. Sudip Basack, *MAKA University of Technology, Kolkata, India*
Prof. Edén Bojórquez, *Autonomous University of Sinaloa, Mexico*
Prof. Bahram Rezayibana, *Islamic Azad University, Ardabil branch, Ardabil, Iran*
Prof. Giuseppe Failla, *University of Reggio Calabria, Italy*
Dr. Haytham F.A ISLEEM, *Tsinghua University, China*
Dr. Kelvin Kuok King Kuok, *Swinburne University of Technology Sarawak Campus, Malaysia*

Organizing Committee
Prof. Hui Guo, *Southwest University of Science and Technology, China*
Prof. Rashmi Verma, *Mata Sundri College for Women, Delhi University, India*
Prof. Szeto, Wai Yuen, *The University of Hong Kong, China*
Prof. Lishuai Jiang, *Shandong University of Science and Technology, China*
Assoc. Prof. Lingkun Chen, *Yangzhou University, China*
Dr. Kim Hung Mo, *University of Malaya, Malaysia*
Deyu Yin, *Huaiyin Institute of Technology, China*
Dr. Muhd Zaimi Bin Abd Majid, *Universiti Teknologi Malaysia, Malaysia*
Dr. Sharifah Akmam Syed Zakaria, *Universiti Sains Malaysia, Malaysia*
Bo Liu, *Shaanxi Huachun Network Technology Shares Co., Ltd.*

Part 1. Research on geological structure and application of anti-seismic technology

Advances in Civil Engineering: Structural Seismic Resistance, Monitoring and Detection – Mohd Johari Mohd Yusof and Junwen Zhang (Eds)
© 2023 The Authors, ISBN 978-1-032-31491-4

Effects of a deep excavation on its surroundings in Shanghai soft ground

K.Y. Huang, Z.R. Liang & X. Wei
Shanghai Shen Yuan Geotechnical Engineering Co., Ltd., Shanghai, China

ABSTRACT: Deep excavations often cause notable deformations of surroundings in soft soil area. The new Shanghai Children's Library project is introduced in this paper to elaborate the general design and construction measures of deep excavations encircled by complicated surroundings in Shanghai soft ground. In consideration of the shape and the scale of the excavation as well as its unfavorable environment and geology condition, ground improvement, waterproof curtain, retaining walls and struts were adopted to control the displacements of the excavation and its surroundings. The construction sequences such as supporting before excavating block by block and from east to west were carefully arranged and well executed. Field data during excavation and construction is presented to illustrate the performances of retaining walls and the existing surroundings such as bridge, underground pipelines, pump station and embankment, which demonstrates the reliability of the design and construction measures.

1 INSTRUCTION

With rapid development and centralized construction of city, more and more deep excavations located in municipal areas are in close proximity to urban infrastructures, such as metros, tunnels, bridges, roads, pipelines, embankments and buildings.

Deep excavations usually cause significant deformations of surroundings in soft soil area, such as the city of Shanghai. The effect of a deep excavation on an adjacent metro station and shield tunnels was investigated in Shanghai soft clay (Liu 2016). It was revealed that the displacements of shallowly buried pipelines due to a neighboring excavation were predominantly governed by the distances between the pipelines and the excavation rather than the pipeline flexural rigidity or soil bedding stiffness in Shanghai soft ground (Tan 2018). A study analyzed the settlements of several pre-existing buildings with different types of foundation induced by a nearby deep subway station excavated by the covered semi-top-down method in Shanghai (Tan 2016). In addition, some researchers also provided monitoring data of surroundings during excavation in Shanghai, including metro lines, tunnels and buildings (Chen 2018; Huang 2014; Tan 2019).

In this paper, the new Shanghai Children's Library project is taken as an example to introduce the design method of deep excavations surrounded by complicated environment in Shanghai soft soil. A series of construction measures were taken to restrict the deformations of the excavation itself and its surroundings. Field data is minutely presented to show the responses related to the excavation, which illustrates the effectiveness of those measures.

2 PROJECT OVERVIEW

The new Shanghai Children's Library is located on the east of Daduhe Road and south of West Guangfu Road in Putuo District of Shanghai. The land area is 6382 m^2 and the overall floorage is about 16000 m^2, including four-floor aboveground area of 11000 m^2 and one-floor underground area of 5000 m^2.

DOI 10.1201/9781003310884-1

3 CHARACTERISTICS OF THE EXCAVATION

3.1 *Irregular in shape and relatively large in scale*

The excavation of Shanghai Children's Library is irregular in shape as shown in Figure 1 with an area of 5125 m^2, a perimeter of 325 m and a depth of 8.8 m.

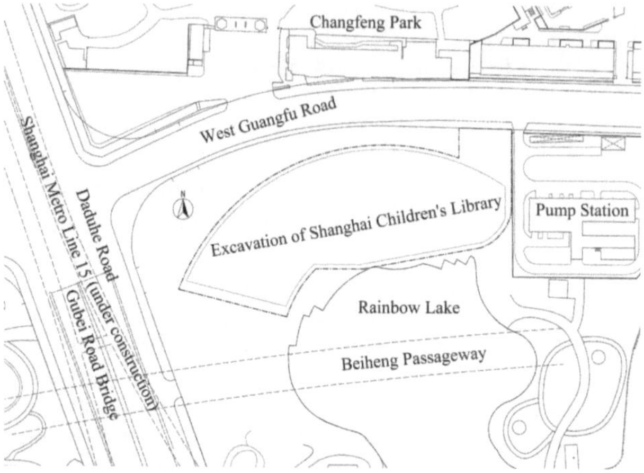

Figure 1. Plan view of the site.

3.2 *Complicated in environment*

The land red line is extremely close to the excavation, and the universal distance between them is merely 1.8 m.

3.2.1 *On the west of the site*

On the west of the site, there lies Daduhe Road, Gubei Road Bridge and Shanghai Metro Line 15 (under construction), whose minimum horizontal distance to the excavation is 22.8 m, 29.9 m and 34.9 m, respectively. The bridge deck is composed of pre-stressed reinforced concrete hollow slabs with a width of 12.75 m, the piles under the bridge are 55 m in depth and 0.8 m in diameter.

3.2.2 *On the north of the site*

The north of the site borders on West Guangfu Road, under which different kinds of pipelines are buried 0.5~0.8 m in the zone $1.5H_e \sim 3H_e$ behind the excavation, H_e denotes the final excavation depth. Details of the underground pipelines are listed in Table 1.

3.2.3 *On the east of the site*

On the east of the site, there is a concrete combined sewer with a diameter of 2 m and a pump station with one-floor superstructure and one-floor basement. The combined sewer is 5.1 m away from the edge of the excavation and buried 3.3 m beneath the ground. The pump station is 10.45 m apart from the excavation and the bottom of its basement is 11 m under the ground without piles.

3.2.4 *On the south of the site*

The south of the site is contiguous to Rainbow Lake, which is connected with Suzhou River and merely 2 m away from the excavation. Under Rainbow Lake lies Beiheng Passageway, a tunnel 20.5 m apart from the excavation on horizontal projection and buried 25 m below the excavation.

Table 1. Underground pipelines of west Guangfu road.

Type	Diameter(mm)	Buried depth (m)	Minimum distance to the edge of the excavation (m)
Power	—	0.6	13.7
Gas	300	0.8	15.2
Telecom	—	0.8	16.7
Telecom	—	0.7	17.7
Storm sewer	1500	0.8	18.9
Water supply	700	0.7	23.9
Water supply	700	0.6	27.4
Water supply	200	0.5	27.9

3.3 *Complex in geological condition*

The site is generally flat with an average ground level of 4.25 m in Wusong Elevation System and belongs to coastal plain in landform. Geotechnical investigations including a battery of laboratory tests and in situ tests were conducted before construction, and the results are presented in Table 2.

Table 2. Physical and mechanical parameters of soils.

Stratum Number	Soil Classification	Thickness (m)	Unit weight γ (kN/m^3)	Peak intensity of consolidated quick shear	
				c* (kPa)	φ** (°)
—	water	2.0~2.5	10	—	—
①$_0$	silt	0.2~0.5	—	—	—
①$_1$	artificial fill	2.0~4.3	—	—	—
②$_3$	clay silt mixed with silty clay	10.5~13.5	18.7	7	28.5
⑤$_1$	clay	8.6~12.5	17.5	17	13.0
⑤$_3$	silty clay	6.4~9.5	18.2	17	17.5
⑤$_4$	silty clay	2.0~3.9	19.6	43	17.0
⑦	sandy silt	6.0~8.4	19.1	5	32.5
⑧$_{1-1}$	silty clay	11.4~13.3	18.3	18	18.0
⑧$_{1-2}$	silty clay	8.1~9.8	18.8	21	20.0

* c denotes cohesion; ** φ denotes angle of internal friction.

3.3.1 *Obvious creek*

Affected by Rainbow Lake, obvious creek lies across the site. The obvious creek consists of water with a depth of 2.0~2.5 m and silt with a thickness of 0.2~0.5 m, which is quite inconvenient for construction.

3.3.2 *Thick fill*

The top soil layer is artificial fill (Layer ①$_1$) with an average thickness of 3.3 m (up to 4.3 m). The thick fill would negatively affect the construction quality of bored piles.

3.3.3 *Silty soil*

The artificial fill is underlain by silty soil (Layer ②$_3$) with a thickness of 10.5~13.5 m. The silty soil is large in permeability and sensitive to flow pressure. Water seepage may occur due to hydrodynamic disturbance.

3.3.4 *Soft clay*

Under the silty soil are clay (Layer ⑤$_1$) and silty clay (Layer ⑤$_2$), which are soft plastic and very wet, therefore unfavorable to restrict the displacements of the excavation.

4 DESIGN AND CONSTRUCTION MEASURES

Due to the obvious creek and thick fill mentioned above, the ground of the site should be treated at first, then other measures were taken as follows to control the deformations of the excavation. Figure 2 shows the plan of the excavation.

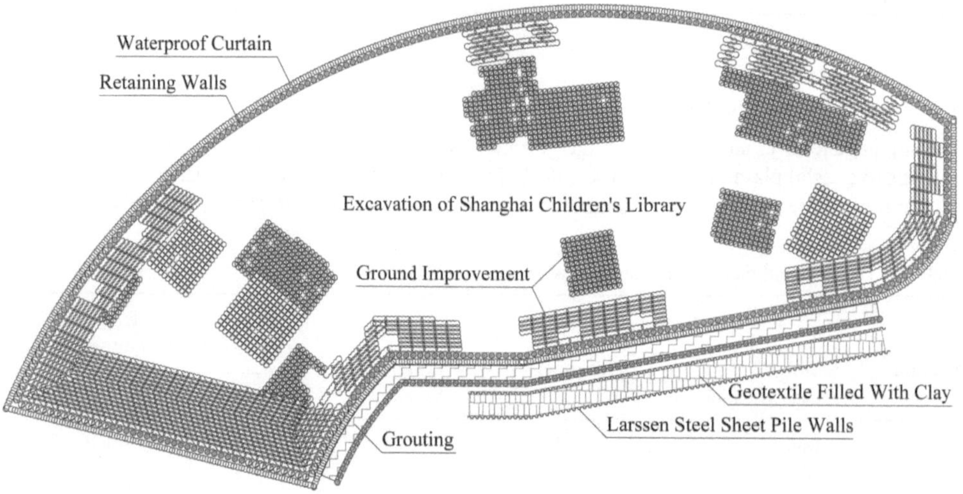

Figure 2. Plan view of the excavation.

4.1 *Ground treatment before construction*

Two rows of Larssen steel sheet pile walls (4# with a length of 12 m shown in Figure 3) were set on the south of the site to separate the obvious creek from Rainbow Lake. Water between the two-row piles was pumped and replaced by geotextile filled with clay with coefficient of compaction no less than 0.95. Water of the obvious creek was drained, and then ground related to the obvious creek was improved by grouting with cement content of 10%.

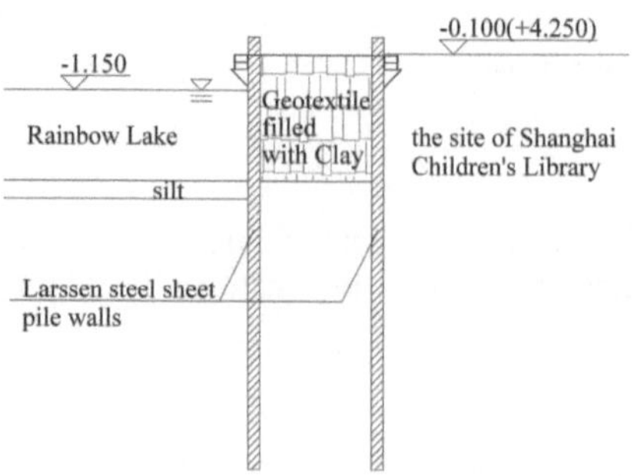

Figure 3. Ground treatment on the south side.

4.2 *Waterproof curtain*

Tri-axial soil mixed walls (3Φ850@1200 with a length of 17.9 m) were adopted as waterproof curtain to abate the hydraulic connections between the excavation and its surroundings, so that the underground water table outside the excavation would not drawdown obviously when dewatering inside.

4.3 *Retaining walls*

Contiguous bored pile walls (Φ850@1050/Φ950@1150 with a length of 20.5~24 m) were employed as retaining structures in consideration of their advantages in deflection control with high stiffness. One row of piles was enough for most areas (Figure 4a) except in the south side (Figure 4b) where the earth pressure was imbalance against the north side owing to Rainbow Lake. Thus one more row was set in the south side and grouting was operated between the two rows of piles.

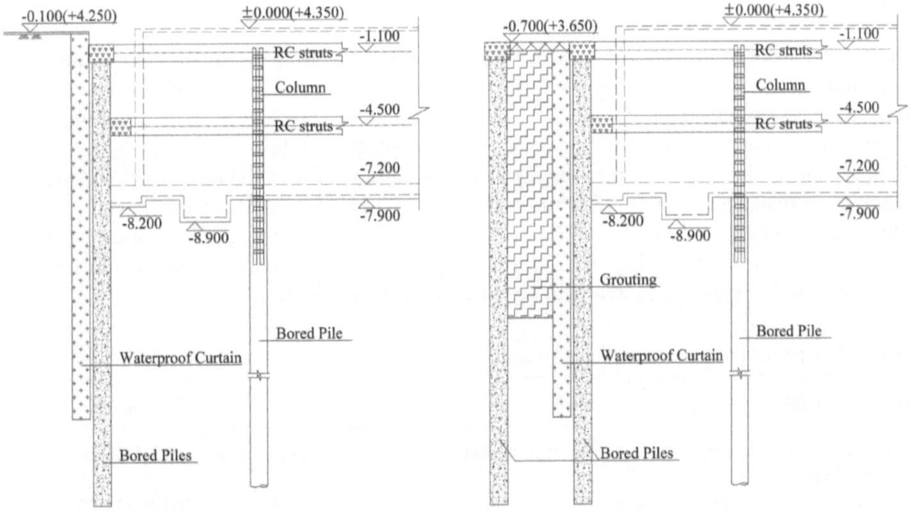

Figure 4. Sections of the excavation.

4.4 *Struts*

Two layers of reinforced concrete struts were installed to increase the stiffness of the support system during excavation. Construction trestles were arranged in combination with the layout of the struts (Figure 5).

4.5 *Ground improvement before excavation*

Tri-axial soil mixed piles (3Φ850@1800) were used to improve the ground along the edge of the excavation.

In most areas, the width of the improvement area was 5 m with a length of 13.3 m. The cement content was 20% below the bottom of the excavation while 13% above.

In the west side, considering the metro line under construction, the improvement width was 10 m with a length of 13.8 m. The cement content was 20% below the second layer of struts while 13% above.

Jet grouting piles (Φ800@700 with a length of 3.7~4.6 m) were adopted in locally descending plate area and the cement content was 25%.

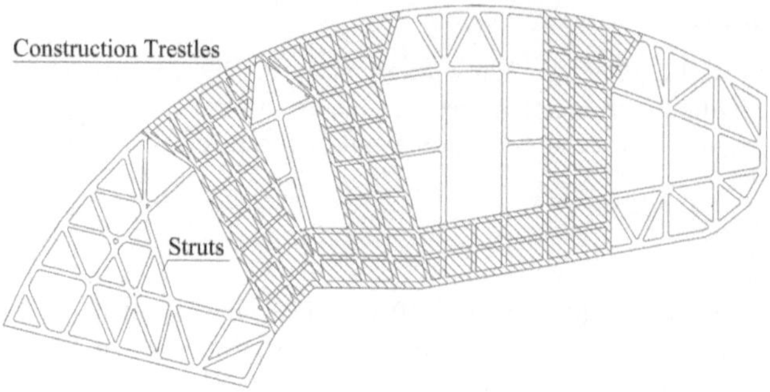

Figure 5. Plan view of the first layer of struts.

5 FIELD DATA

The deformations of retaining structures and the urban infrastructures nearby including Gubei Road Bridge, underground pipelines, pump station and embankment were monitored during excavation and construction from the end of April 2019 to the beginning of December 2019. The time periods corresponding to the main construction stages are presented in Table 3.

Figure 6 shows the lateral displacements of the retaining walls at the end of the construction. The maximum displacement was 13.3 mm and less than 0.18%H_e in the west side while 22.9 mm and less than 0.3%H_e in other areas, H_e denotes the final excavation depth. During excavation and construction, the lateral displacements were generally under control, which illustrates the effectiveness of the retaining system and construction management.

Table 3. Time periods of main construction stages.

Construction stage	Time period
Construction of ground improvement, waterproof curtain, retaining walls and the first layer of struts	2019/4/26-2019/8/19
Excavating to the elevation of the second layer of struts, block by block, from east to west	2019/8/20-2019/8/23
Construction of the second layer of struts	2019/8/24-2019/9/8
Excavating to the bottom of the excavation, block by block, from east to west	2019/9/9-2019/9/28
Construction of baseplate	2019/9/29-2019/10/20
Dismantling of struts	2019/10/21-2019/11/17
Construction of basement	2019/11/18-2019/12/01

The displacements in the west side were significantly smaller than those in other areas, which means ground improvement and construction sequences such as excavating from east to west have important effects on controlling deformations of the retaining walls.

The displacements in the south side were similar to those in the north and east side. It seems that the double row piles played an important role in balancing against the earth pressure transferring through struts from the north side.

Figures 7a-d respectively shows the settlements of Gubei Road Bridge, the pipeline of gas under West Guangfu Road, the pump station and the embankment of Rainbow Lake during excavation.

The maximum settlement of the bridge shown in Figure 7a was merely 1.2 mm, which means the bridge was hardly affected by the excavation.

Figure 7b indicates that the settlements of the pipeline of gas in the north side developed sharply from 0.5 mm to 5.0 mm during excavation above the elevation of the second layer of struts, as if the shallowly buried pipeline is sensitive to the excavation in shallow layer.

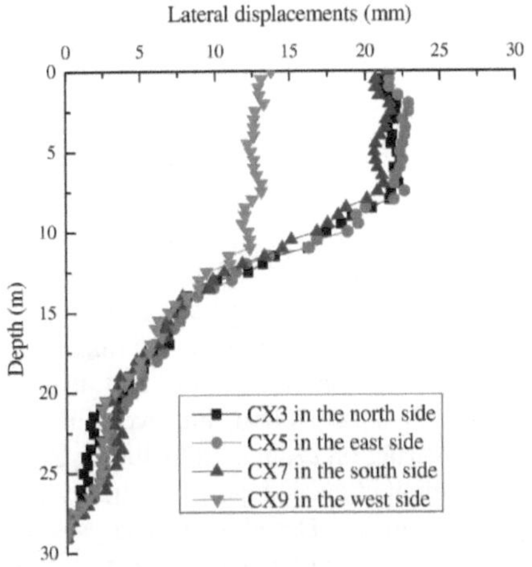

Figure 6. Lateral displacements of retaining walls at the end.

It can be seen in Figure 7c that the settlements of the pump station reached about 8 mm before excavation, which indicates construction of piles would obviously affect buildings on shallow foundation, and excavation aggravated the settlements until the baseplate was constructed.

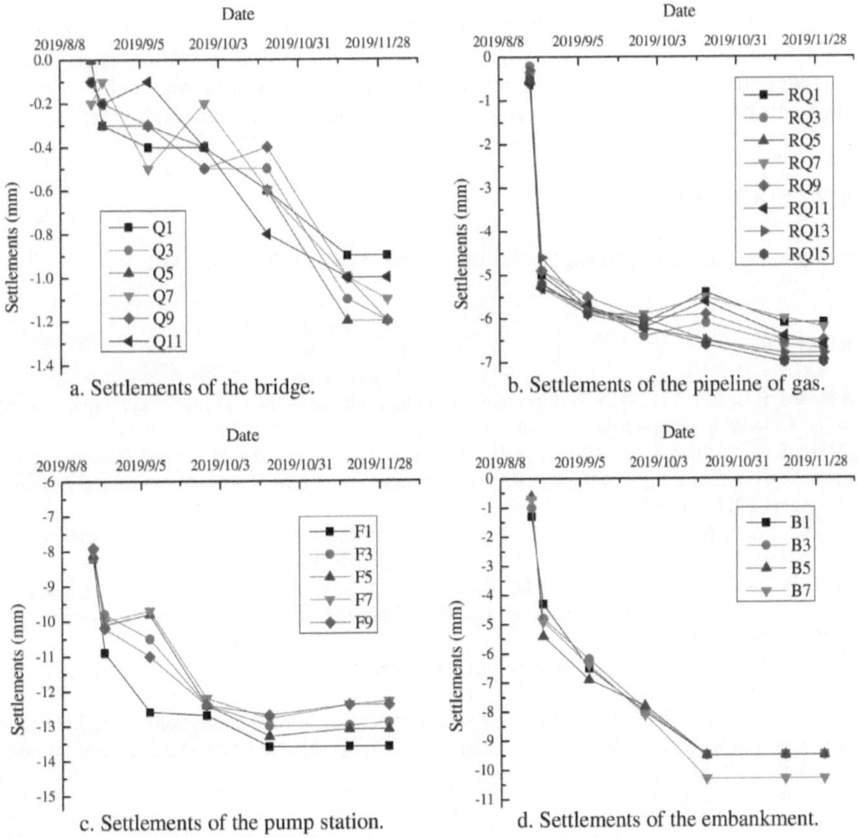

Figure 7. Settlements of the surroundings.

Figure 7d shows that the completion of the baseplate slowed down the increment of the settlements of the embankment which grew rapidly during excavation. The accumulated settlements became steady at 9.5~10.3 mm.

6 CONCLUSIONS

This paper minutely presents the case history of the excavation of Shanghai Children's Library, which was constructed in Shanghai soft ground and surrounded by roads, bridge, metro under construction, underground pipelines, pump station, lake and tunnel.

In order to restrict the deformations of the excavation and its surroundings, several effective design and construction measures such as ground improvement, waterproof curtain, retaining walls and struts were taken. The construction sequences including excavating from east to west and block by block were strictly implemented as arranged. The behaviors of structures related to the excavation were investigated via monitoring. Based on the analyses of the field data, the following major conclusions are obtained. The design and construction measures as well as monitoring data presented in this paper will provide references for similar projects in future.

The maximum lateral displacement of the retaining walls was 22.9 mm and less than 0.3% of the excavation depth, which verifies the effectiveness of the retaining system and construction measures. The deflections in the west side were obviously smaller than those in other areas, which indicates the importance of ground improvement and construction sequences. The significance of double row piles can be seen from the comparison between the deformations in the south and the north side.

The bridge with a long distance to the excavation was hardly influenced. The settlements of the shallowly buried pipeline of gas mainly increased during excavation in shallow layer. The pump station situated on shallow foundation subsided remarkably when piles were being constructed. The settlements of the embankment became steady after the baseplate was completed.

ACKNOWLEDGMENT

This research is sponsored by Shanghai Sailing Program (No. 21YF1432600).

REFERENCES

Chen, H.H. & J. P. Li & L. Li (2018). Performance of a zoned excavation by bottom-up technique in Shanghai soft soils. *J. Geotech. Geoenviron. Eng.* 144, 05018003-1–12.

Huang, Z. H. & X. S. Zhao & J. J. Chen & J. H. Wang (2014). Numerical analysis and field monitoring on deformation of the semi-top-down excavation in Shanghai. *New Frontiers in Geotechnical Engineering*. Geo-Shanghai 2014, 198–207.

Liu, G. B. & P. Huang & J. W. Shi & C. W. W. Ng (2016). Performance of a deep excavation and its effect on adjacent tunnels in Shanghai soft clay. *J. Perform. Constr. Facil.* 30, 04016041-1–14.

Tan, Y. & R. Q. Huang & Z. J. Kang & W. Bin (2016). Covered semi-top-down excavation of subway station surrounded by closely spaced buildings in downtown Shanghai: building response. *J. Perform. Constr. Facil.* 30, 04016040-1–26.

Tan, Y. & Y. Lu (2018). Responses of shallowly buried pipelines to adjacent deep excavations in Shanghai soft ground. *J. Pipeline Syst. Eng.* 9, 05018002-1–14.

Tan, Y. & Y. Lu & D. L. Wang (2019). Practical solutions for concurrent excavation of nerghboring mega basements closely surrounded by utility tunnels in Shanghai Hongqiao CBD. *Pract. Period. Struct. Des. Constr.* 24, 05019005-1–23.

Advances in Civil Engineering: Structural Seismic Resistance, Monitoring and Detection – Mohd Johari Mohd Yusof and Junwen Zhang (Eds)
© 2023 The Authors, ISBN 978-1-032-31491-4

Different structures contributing to wooden bridges' weight capacity

Yiwei Huang
Shenzhen Senior High School, Shenzhen, China

ABSTRACT: To simulate the load-bearing capacity of a real bridge, 3D models built into computers are now widely used. At the same time, old methods such as making wooden bridge models have not been abandoned. Wooden structures are made to study the stress and wear of various structures under real conditions. In this paper, we study the influence of different bridge structures on the maximum bearing weight of the bridge by studying various structural designs and test those designs through constructing them in reality.

1 INTRODUCTION

Wooden bridges are used through out our lives due to their lightness and pleasing to eyes. To test their maximized weight capacity, people tried to construct mini-sized wooden bridges as model. It's believed that different structures, like different numbers of crosses on their side edge, aspect ratio and numbers of beams would contribute the most to their weight capacity other than materials as they have to be solid wooden bridges within restrained weight of wood.

This paper aims to design a bridge within weight as low as possible to weight the most. It focuses on how to construct a bridge with lowest weight and greatest weight capacity. Main aspects contain the type of bridge, mathematical calculation for the design of bridge, and additional structure in the bridge. In addition, this paper will also introduce beam bridges in reality basically.

All the structures used in the upcoming experiment are designed by Xiang-ming Song, who has been studying in this area for years. All the structures are constructed by Yiwei Huang.

2 DESIGNATION IN STRUCTURES AND THEIR CONTRIBUTION

2.1 *Types of bridge*

There are four types of bridge in the list: cable-stayed bridge, beam bridge, circular arch bridge and suspension bridge.

Generally, there is five kinds of forces that may exists in bridges: tension force: compressional force tensile force, bending force, sheering force, torsional force; there is also their conjugate deformations: Tensile deformation, compressional deformation, bending deflection, tortional deformation, sheering deformation.

Due to the special feature of Paulownia (rift grain), the bridge should be designed not to take bending force, sheering force and torsional force, which means, to take compressional force and tensile force as more as possible. As a consequence, we're able to bring out our final choice: beam bridge.

If circular arch bridge: main span taking bending force downward; if cable stayed bridges: main span taking compression force will be transformed to be the tensile force and shear force exerted the surface of the cable connecting the main span and the vertical beam; if suspension bridge: cable connecting beam are taking bending force and the problem exists on the cable stayed bridge also exists on the suspension bridge. After all, the only left type of bridge is beam bridge.

a. Cable stayed bridge [5] b. Beam bridge [6] c. Circular arch bridge [7] d. Suspension bridge [8]

Figure 1. Four types of bridge.

2.1.1 *Number of beams*

Through our examination and calculation, we are able to bring out the design:

Due to $\delta = F/S$, we're able to conclude that $F = \delta \times S$, which means that the force every beam takes is proportionate to the cross section area of the beam. The average δ of Paulownia (rift grain) is between 8.6 mpa and 12 mpa [1]. To facilitate the calculation we use $\delta = 9.8$ mpa. And the beam we use has a cross section with area 9mm^2. Then $F = 8.82$ if the beam is placed vertical.

As it is placed nonvertically, we must calculate it within different situation:

$$\text{Sin}\angle a = \frac{4.5}{\sqrt{4.5^2 + 5.5^2}} = 0.6332 \tag{1}$$

$$\text{Sin}\angle b = \frac{4.5}{\sqrt{4.5^2 + 10.5^2}} = 0.3939 \tag{2}$$

$$\text{Sin}\angle c = \frac{4.5}{\sqrt{4.5^2 + 4.5^2 + 5.5^2}} = 0.535 \tag{3}$$

$F_{ay} = 5.7$ N $F_{by} = 3.55$ N $F_{cy} = 4.8$N

After knowing how much a beam in each position can take, we're able to test that how many beams in each position is the most suitable for our structure.

From the table we are able to conclude that number 5 have the ability to weight most. That is: to make the number of beams on lengthwise to be the most.

2.1.2 *Number of diagonal braces at both sides*

Diagonal braces, or AKA crosses appeared at the diagonal of a tetragon, aim in increase the weight capacity and stability of the square.

In the design, the beams are taking mostly compression force. If the beams are placed vertically, the compression force acting on it will make it more stable. But unfortunately it is not. Due to

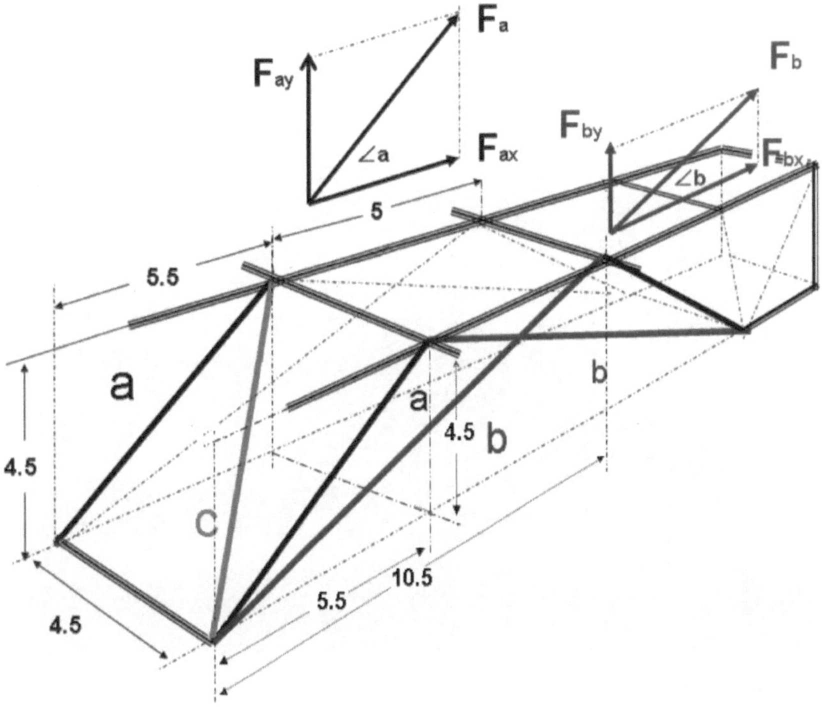

Figure 2. The original design.

Table 1. Tests' data.

Number	Na	Nb	Nc	∑W (kg)
1	4	4	2	46.6
2	8	4	2	69.4
3	12	4	2	92.2
4	12	8	2	106.4
5	16	4	2	115

its inclination angle, the compression force is making it unstable indeed, and both of its ends has become free end. In order to make those free ends stable, we have to introduce the diagonal braces into it.

After reanalysis it, we found that most of the shapes appeared in out structure are triangles and tetragon. As triangle is stable itself, the only issue we have to handle is tetragon. As the author mentioned before, diagonal braces are aimed to stabilize tetragon. But actually, introducing those diagonal braces cannot fix our question about free end even though it does make the structure more stable. The key issue is to reduce the distance between free ends.

After adding a brace into it, four corners of the brace then become free ends other than the original ones. As more and more braces are added, the average distance between free ends is reduced. With shortened distance between them, the entire beam is stabilized.

2.2 *Examination*

2.2.1 *Construct*

With the design listed before, we are able to construct the testing model. Paulownia (rift grain) is used as the material to construct our model, glue (No.502) as the adhesive.

We then cut the beam into the form needed and arrange them together.

Figure 3. Completed model.

2.2.2 *Test its weight capacity*

The structure is placed into a weight horizontally, and is pushed downward until the highest weight capacity is tested (one of the beams broken or the number on the weight stopped increasing.)

3 BEAM BRIDGE IN REALITY

3.1 *Categories of beam bridge*

Beam bridge can be classified by several aspects: geometry, the shape of the cross-section, equilibrium conditions, and type of support.

First, based on geometry, they are classified into straight beam, curved beam and tapered beam. Secondly, based on the shape of the cross-section, they are classified into I-beam, T-beam and C-beam. Thirdly, based on equilibrium conditions, they are classified into statically determinate beam and statically indeterminate beam. Fourthly, based on the type of support, they are classified into supported beam, cantilever beam, overhanging beam, continuous beam and fixed beam. Among them, the bridge of me has the trait of straight beam, statically determinate beam, and fixed beam.

3.2 *Advantages and disadvantages of the beam bridge*

Using beam bridge may have several advantages: beam rests simply on the supports; the effects of thermal expansion and movements of the ground are easily sustained; a beam could also be engineered removed from the ultimate position and raised fleetly into place with the least disruption of traffic or navigation; beam bridges are easy to construct; in comparison to other bridge types, beam bridges are less costly; mostly used widely in urban and rural zones.

However, there are also some disadvantages like: beam Bridge contains forces, which are much larger than the load, and it needs to be relatively massive; beam Bridges have a limited span and do not allow large boats or vehicles to pass underneath; mostly heavy boat traffic or large ships cannot pass underneath; people may not find the design of beam bridges spectacular.

4 OTHER SAMPLES OF BEAM BRIDGE

4.1 *Example of bridge's design: wooden bridge over the Gravina torrent*

The main load-bearing system is constituted by two laminated-timber Pratt-type truss longitudinal beams, thal support the transverse frame (deck), the wooden roof and the steel spatial truss system. The steel space system includes tubular trusses in both longitudinal and transversal directions, with square-hollow or circular cross-sections.

In order to suitably overcome constructional issues. The values of the main mechanical properties of materials. used for numerical simulation, are reported in Table 1. A three-dimensional finite element model of the bridge was built by using RSTAB of the Dlubal software. Only structural elements were included in the model. whereas non-structural elements were considered as extra masses. The model has a total of 928 nodes and 2721 frame elements. The modeling of the glulam truss was carried out by schematizing the joints as rigid nodes [3].

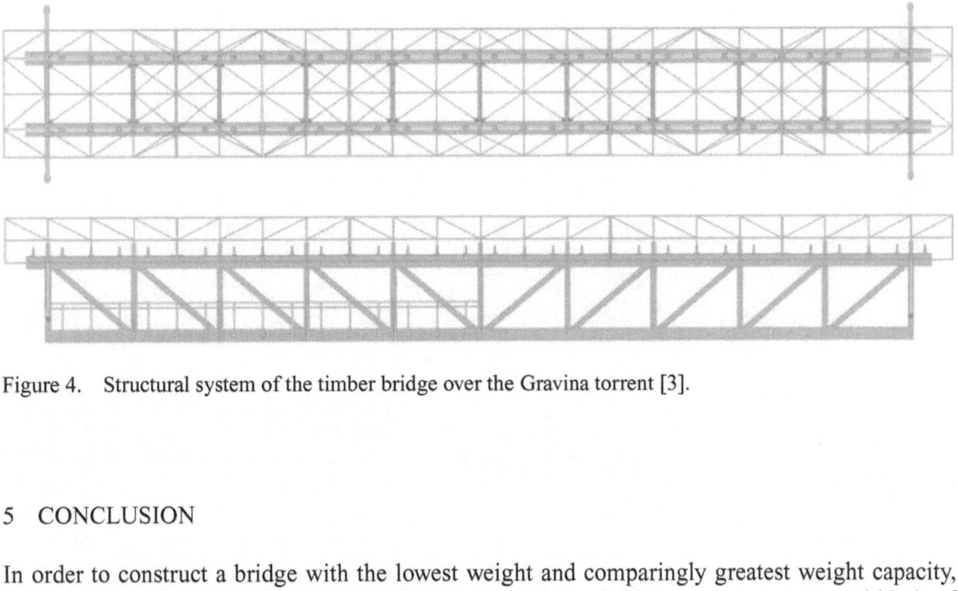

Figure 4. Structural system of the timber bridge over the Gravina torrent [3].

5 CONCLUSION

In order to construct a bridge with the lowest weight and comparingly greatest weight capacity, a bridge should meet the following conditions: 1) use beam bridge to prevent unwanted kinds of deformation; 2) Base on tested data and calculation, the bridge must be under a tested, stable model; 3) In order to increase the inner stability of the bridge, there must be sufficient crosses or AKA diagonal braces.

There is also something that we can improve: this paper included only the bridges that's minimized, and the data of real bridges are cited from other essays. For better and more reliable data, the author will try to include some data from the real world.

ACKNOWLEDGEMENTS

At last, I want to show thankfulness to my teacher Xiangming Song who's the first one to introduce me into this field and guilded me through all the process.

REFERENCES

Greemap, Paulownia: Characteristics Of Wood, https://greemap.es/paulownia-2/paulownia-characteristics-of wood/?lang=en#Physical And_ Mechanical _ Properties_Of_Paulownia_Wood

How Does a Cable-stayed Bridge System Work? Science struck. https://sciencestruck.com/how-does-cable-stayed-bridge-system-work

https://graph.baidu.com/api/proxy?mroute=redirect&sec=1632380206187&seckey=f030f3f669&u=http%3A%2F%2Fwww.changqingfoods.com%2Fnews%2F%E9%8D%9D%E5%A0%9D%E7%9A%B5%E5%A9%8A%2520%E5%A9%8A%E3%84%A5%E7%AA%9E%E9%96%BE%E4%BD%BD%E7%9F%BE%E6%BE%B6%D1%84%CB%89%2F

Martino Antonio Liuzzia, Alessandra Fiorea, RitaGrecob, *Some structural design issues on a timber bridge for pedestrians.*

MIDASIT Co., Ltd. Arch Bridge, https://www.midasbridge.com/en/solutions/arch-bridges

Raqifa Rahman Chowdhury, *Beam Bridge – Definition, Examples, Materials, Types. Civil Engineering.* https://civiltoday.com/construction/bridge/344-beam-bridge

Thousand wonders. *Clifton Suspension BridgeSuspension Bridge in Bristol, England.* https://www.thousand wonders.net/Clifton+Suspension+Bridge

Advances in Civil Engineering: Structural Seismic Resistance,
Monitoring and Detection – Mohd Johari
Mohd Yusof and Junwen Zhang (Eds)
© 2023 The Authors, ISBN 978-1-032-31491-4

Seismic analysis of Jinxing bridge in Yunnan Yangbi earthquake and suggestions for seismic measures

Xingchen Xie, Junqi Lin & Jinlong Liu
Disaster Prevention and Mitigation Engineering, Institute of Engineering Mechanics, China Earthquake Administration, Harbin, Heilongjiang Province, China

ABSTRACT: On May 21, 2021, a earthquake with Ms6.4 occurred in Yangbi County, Dali Prefecture, Yunnan Province, which caused 3 deaths and 34 injuries and direct economic damage of 327,110,000 RMB, resulting in severe impacts on the property and life safety of local people. This paper presents a macroscopic description of the damage of a local prestressed hollow slab bridge, Jinxing Bridge, and analyzes its damage characteristics, and proposes effective seismic measures to provide a reference for the seismic design of similar bridges in the future and to make people more directly aware of earthquakes and their hazards.

1 INTRODUCTION

China locates between two major seismic zones, the Pacific Rim Seismic Zone and the Eurasian Seismic Zone, an earthquake-prone country (Yu 2018). Characteristics of Earthquakes in China are high frequency, high intensity, wide distribution, high casualties ,and severe disasters. Highway bridges are an essential part of the lifeline project and are vulnerable to seismic damage under seismic hazards; In Wenchuan Ms8.0 mega-earthquake, according to statistics, a total of 6140 bridges were damaged to varying degrees in the Wenchuan earthquake (Lin 2017). And once the bridge structure is damaged, it will significantly impact the subsequent rescue mission, so it is vital to research the seismic damage of bridges in China. 6.4 magnitude earthquake occurred in Yangbi County, Yunnan Province (25.67°N, 99.87°E) on May 21, 2012; the main roads in the earthquake area were in good condition and had less impact on the subsequent rescue, Still, some village and town bridges occurred. Some damage occurred, but there was no severe damage such as girder falling or total bridge collapse. The Jinxing Bridge is shown in Figure 1 below.

Figure 1. Jinxing bridge.

DOI 10.1201/9781003310884-3

2 ENGINEERING SITUATION OF THE JINXING BRIDGE

Jinxing Bridge is located in Dali City, Shangguan to North Wuliqiao section of the first-class highway, a prestressed hollow slab bridge, a total of 7 spans, across the Xueshan River, the bridge topography is a U-shaped valley crossing. The span layout is 7×20m, each span consists of 11 prefabricated hollow slabs, the width of the bridge is 12m, 1.5m pavement + 9m traffic lane + 1.5m pavement (as shown in Figure 3). The bridge has 80 expansion joint types, setting principle: one end (Dali bank) has 80 kinds of expansion joint, one end (Lijiang bank), and the bridge deck's back wall makes the continuous bridge structure. The bearing adopts a round plate type rubber bearing. Prestressed hollow slab, hinged joint concrete, now-poured concrete, and slab end head concrete are all C50; 10cm thick asphalt concrete is used for the bridge deck pavement. The general arrangement diagram is shown in Figure 2.

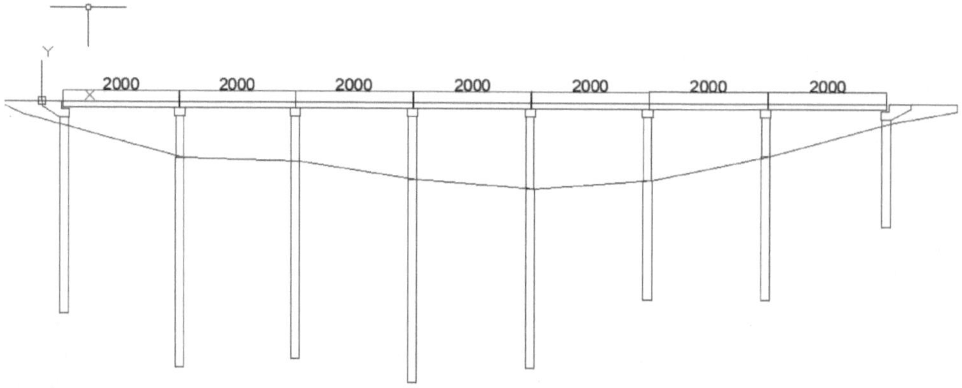

Figure 2. General layout of the Jinxing bridge.

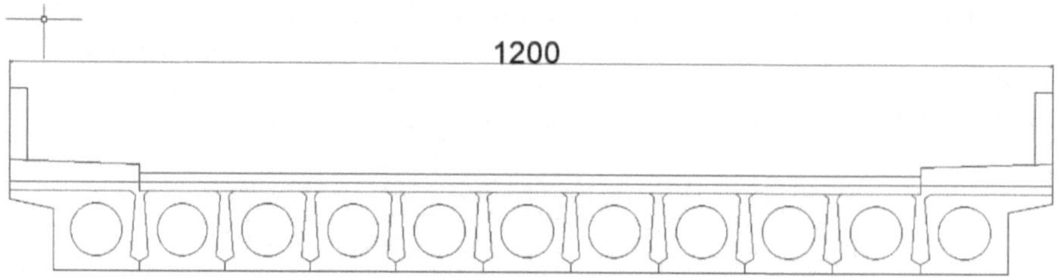

Figure 3. Cross-section of the main girder.

Jinxing Bridge adopts double-column bent frame piers (as shown in Figure 4a) with bored pile foundation, and the abutments on both banks are column-pile abutments, all connected with the roadbed fill. When pier height Li < 7m, cancel the pile top tie beam; when the pier height 12m > Li ≥ 7m, only pile top tie beam; when 20m ≥ Li ≥ 12m on the pier column with a tie beam. The pier heights of each pier are as follows: Pier 1 left, and right 7m; Pier 2 left and exactly 7.5m; Pier 3 left and right 10.5m; Pier 4 left and right 12m; Pier 5 left 8.5m, right 10.5m; Pier 6 left and precisely 6m.

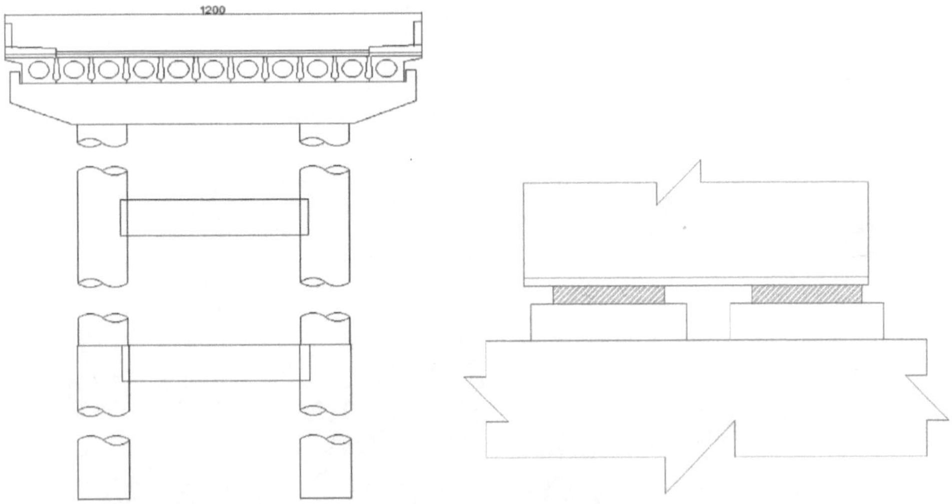

(a) Cross-sectional view of bridge pier

(b) Cross-sectional view of bearing

Figure 4. Cross-sectional view of piers and bearings.

3 ANALYSIS OF SEISMIC DAMAGE AND CAUSES

3.1 *Seismic damage*

Jinxing Bridge is located in the eastern part of Yangbi County city between the urban planning road (Dayang Highway extension) and the western part of Jinxing Village, crossing the Xueshan River along the east-west direction. This bridge is a multi-span prestressed concrete hollow slab girder bridge; each span consists of 11 hollow slabs with seismic anchor bolts set at the end of the middle slab joints. The bridge damage under this earthquake was mainly superstructure damage; the deck pavement at the contact between the hollow slab main girders and the bridge deck was damaged along with the transverse cracking and broke the pavement slab. The retaining blocking on one side of the bridge abutments was cracked, the concrete of the bridge piers and cover beams did not produce obvious cracks and weathering of the protective coating (Liu 2021).

3.1.1 *Seismic damage of superstructure*

Displacement of bridge superstructures is a common occurrence in all earthquakes, with bridge displacement occurring in longitudinal, transverse and torsional directions. In addition to the risk of falling girders, bridge displacements can also lead to cracking of the bridge deck pavement. The damage to the superstructure of the Jinxing Bridge was mainly caused by the shifting of the main girders, which led to the cracking of the concrete of the bridge deck, the cracking of the pavement slab and the cracking of the bridge deck pavement, of which the maximum crack width was over 40mm, as shown in Figure 5 below.

3.1.2 *Seismic damage to the abutment*

As a supporting part of the bridge and both sides of the bank, seismic damage to the abutment will often lead to cracking or even damage to other parts of the bridge. Seismic damage to bridge abutments can be divided into collision damage between the abutment itself and the superstructure, sliding or collapse of the abutment caused by foundation failure, etc. In the Yangbi earthquake Jinxing Bridge is mainly damaged by the block on one side of the abutment, mainly due to the collision of the main girder with the retaining block by producing a large lateral displacement, as shown in Figure 6 below.

(a) Collision of the main girder with abutment

(b) Cracking of deck pavement at the contact between the abutment and main girder

(a) Cracking of bridge deck pavement

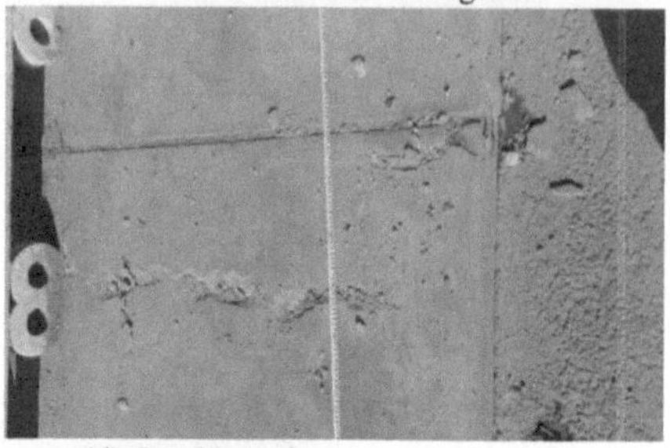

(d) Cracking of pavement concrete

(e) Breakage of pavement slab an abutment

Figure 5. Seismic damage to the superstructure of the Jinxing bridge.

Figure 6. Block cracking on one side of the abutment.

3.2 *Analysis of the causes of seismic damage*

The essential components of simple supported beam bridges and continuous girder bridges are upper main girders, piers, and foundations, and the secondary details are bearings and retaining blocks. According to the investigation of bridge damage in the Wenchuan earthquake, it is known that the deterioration of beam bridges is mainly divided into three kinds of damage: superstructure and support injury, substructure damage, and bridge subsidiary structure damage.

Wenchuan earthquake-stricken area simple supported beam bridges are mostly prestressed concrete hollow slab, T-type, I-type combination beam bridges, set bridge deck continuous, the vast majority of laminated rubber bearings. The significant earthquake damage is (1) main girder displacement, and even the occurrence of girder falling or collapse. (2) pier beam relative displacement resulting in bearing slip or shear deformation, block and expansion joint damage, etc.(3)pier cracking, inclination, collapse, shearing, etc.(4)bridge abutment side wall, conical slope local cracking, filling after the platform sinking, etc. According to the Ministry of Transport Wenchuan earthquake bridge damage report (Liu 2010), Table 1 below (account for Figure 7a).

Table 1. Wenchuan Earthquake Zone VII to XI Seismic damage survey of supported girder bridges in the XI degree area.

Seismic parts of simple supported beam bridges	Main girder displacement (span)	Bearing damage (group)	Block damage (group)	pier damage (group)
quantities	643	1092	720	56
totals	3298	6596	4283	2316
percentage of	19.5%	16.6%	16.8%	2.3%

After the CHICHI earthquake in Taiwan, damage pattern statistics were also done by the relevant authorities, shown in Figure 7b below (Ye 1999).

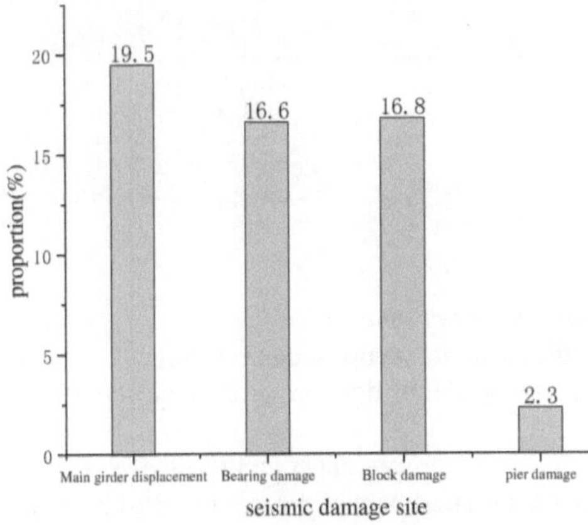

(a) Statistical map of bridge damage sites for the Wenchuan earthquake

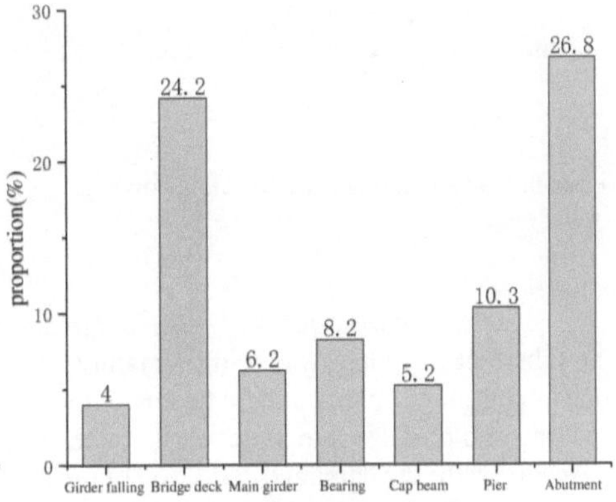

(b) Statistical map of bridge damage sites for the CHICHI earthquake

Figure 7. Statistical map of bridge hazard patterns.

From the Wenchuan and CHICHI earthquakes, the damage of supported girders is mainly concentrated in the superstructure and support damage, while the substructure damage is relatively light. And this phenomenon also appears in this Yangbi earthquake prestressed concrete hollow slab bridge Jinxing Bridge.

3.2.1 *Analysis of the causes of superstructure seismic damage*

Most of the superstructures of bridges adopt the elastic design method with high strength (Ye 2013), and the seismic damage is mainly caused by the displacement of superstructure. From the injury of this bridge, the Lijiang bank and the back wall of the abutment made into a continuous bridge deck cracked, but the deck pavement cracks were not visible around the roadway at the expansion joint of the Dali bank.

Based on the damage mentioned above characteristics of the bridge superstructure, a simple analysis of the damage mechanism of the bridge can be made as follows, as shown in Figure 8. It may be due to the displacement and deflection of the main girder of the hollow slab in the direction of the Dali bank under the earthquake action, which leads to the torsion and longitudinal and transverse displacement of the beam end, thus squeezing and pushing the abutment of the Dali bank and pulling the abutment of the Lijiang bank, so that the continuous deck part of the beam end is subject to large tensile strain, and transverse cracks will be produced when the actual tensile strain exceeds the material limit tensile strain. In addition, the longitudinal displacement of the main girders collided with the abutments causing the concrete of the abutment lap slabs to break up.

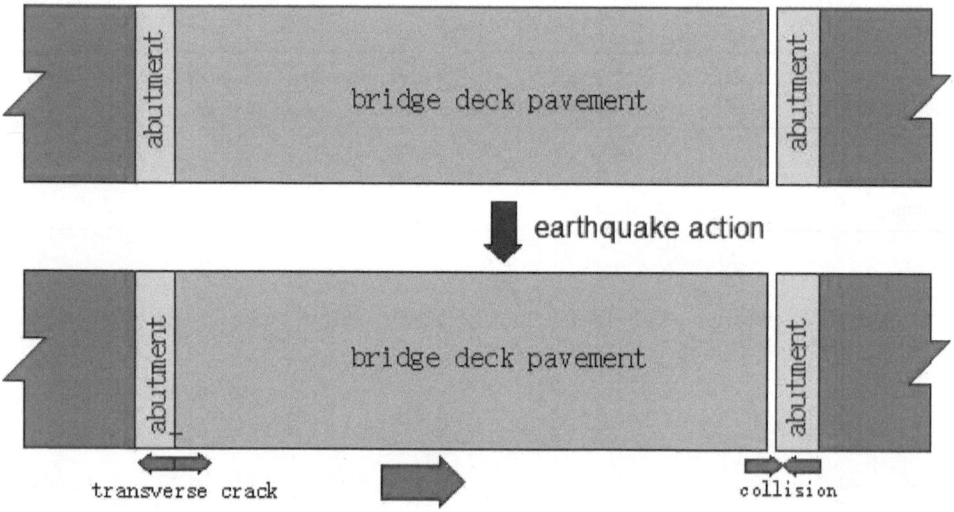

Figure 8. Schematic of superstructure damage.

3.2.2 *Analysis of the causes of seismic damage in the substructure area*

The substructure of highway bridges mainly refers to the structure composed of reinforced concrete such as abutments, piers, and tie beams. Simple-supported girder bridges are often subject to block cracking and pier damage under earthquake action.

According to the damaged condition of the bridge at the site, no evident cracking and weathering of the protective layer were found in the pier, cover beam, and tie-beam of the Jinxing Bridge under the earthquake. The damage mainly occurred in the block cracking on the side of the bridge platform of the Dali bank. We can make the following analysis, analyzing the characteristics and causes of the superstructure earthquake damage in section 3.2.1. Under the action of earthquake, because the Lijiang bank takes the bridge deck pavement and the back wall of the bridge deck continuously, making the bridge deck continuous structure, the integrity is better, while the Dali bank uses 80 type expansion joints to break the whole bridge. Under the effect of severe earthquake, the Dali bank prestressed hollow slab produces a large lateral displacement, while the Lijiang end due to the restraint of the deck pavement, the prestressed hollow slab lateral displacement is smaller, so this also well explains only the main beam at the Dali bank abutment collision with the block, the block is squeezed by the beam, under the action of lateral force, the block first bending, when the force further increases, and finally sheared from the root of the earthquake damage phenomenon.

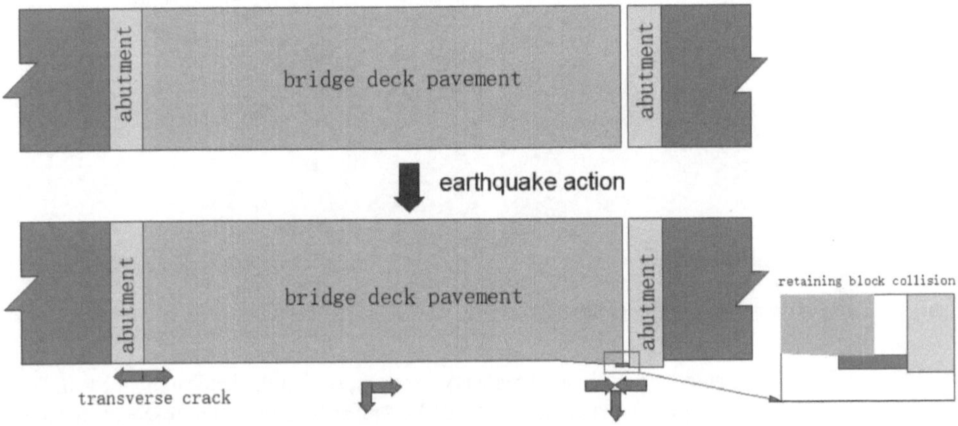

Figure 9. Schematic diagram of damage to abutment retainers.

Pier damage. The bridge pier design should pay attention to its reinforcement rate to ensure that the pier has sufficient strength and flexibility. Jinxing Bridge is a small and medium span girder bridge, and due to the frictional slip of the laminated rubber bearing, shear deformation energy dissipation, reducing the inertia force transferred to the pier, protecting the safety of the pier, so the bridge pier was not found to be obvious cracking and weathering shedding phenomenon.

4 SUGGESTIONS FOR SEISMIC MEASURES FOR THE JINXING BRIDGE

The laminated rubber bearing at the abutment of the Jinxing Bridge was in the form of no connection (as shown in Figure 4b), and it was placed directly on the supporting mat stone without any restraint measures. Because the bearing and the bottom of the beam and the bearing and the mat stone are not connected, the beam and the bearing are easy to slide under the earthquake. Although the anti-seismic anchor bolts against lateral displacement are set at the hinge joints of the two ends of the middle plate, the side plate is not set, so the side plate at the expansion joint is set to have a large lateral displacement and collide with the abutment retaining block in this earthquake. If a larger magnitude earthquake occurs, the anchor bolts may lead to larger girder displacement or even girder falling damage due to insufficient strength. According to the design of Jinxing Bridge and the damage under the crack of the Yangbi Ms 6.4 earthquake, the following seismic recommendations are made for the above seismic damage.

4.1 *Seismic strengthening techniques*

Install reasonable longitudinal and transverse restraint devices to reduce the longitudinal and transverse displacement of the main girder under earthquake action and to avoid severe damage such as a girder falling. Additional limiters (as shown in Figure 10) and Lock-up Device devices can be installed (Wang 1999).

4.2 *Lead rubber bearing*

Lead rubber bearing can be used to replace standard laminated rubber bearing. Lead rubber bearing is made of lead core pressed into RB bearing, in which the lead as a capacity absorbing device has good plastic deformation capacity and energy absorption capacity, and the rubber bearing has

(a) Use of stranded longitudinal limits (b) Use of stranded longitudinal and vertical limits

Figure 10. Use of limiters to prevent girder falling.

good vertical bearing capacity and horizontal recovery capacity, so that the lead rubber bearing has excellent vibration isolation performance (Zhuang 2013). And according to the research of Japan and other countries, the lead rubber bearing has good durability, and its performance remains basically unchanged after 10 years of use. Therefore, as one of the earliest applied seismic isolation bearings, it is widely used in bridge engineering.

However, its seismic isolation design will make the structure softer and more displaced, so there should be enough space between the superstructure and a practical strategy for expansion joints, limiters, etc.

5 CONCLUSION

This paper presents a macroscopic analysis and summary of the seismic damage of Jinxing Bridge in Yangbi, Yunnan Province. It proposes relevant seismic measures to deepen people's knowledge and understanding of bridge seismic damage and prevention techniques and provide some reference and basis for the seismic design of such bridges. in the future.

ACKNOWLEDGMENTS

This work was financially supported by the Natural Science Foundation of Heilongjiang Province Joint Guidance Project (LH2021E123) and Key Special Projects of Key Laboratory of Earthquake Engineering and Engineering Vibration of China Earthquake Administration(2019EEEVL0301).

REFERENCES

Liu, J.L., Ding, Y., Lin, J.Q. (2021) Investigation and analysis of seismic damage to the transportation system of the Yangbi 6.4 magnitude earthquake in Yunnan. *World Earthquake Engineering*, 37(03):31–37.

Liu, L.P., Li, Y.M, Tang, G.W., Zheng, N., H, J., Wang, L.P., Liu, J.W. (2010) Investigation and analysis of seismic damage to the transportation system of the Yangbi 6.4 magnitude earthquake in Yunnan. *World Earthquake Engineering*, 37(03):31–37.

Liu Lping, LI Yingmin, TANG Guangwu, ZHENG Nina, HAN Jun, WANG Liping, LIU Jianwei. Investigation and analysis of seismic damage of bridges in Wenchuan earthquake//Proceedings of the 8th National Conference on Earthquake Engineering (I)., 2010: 29–32.

Lin, Q.L. (2017) Research on vulnerability of highway bridges based on Wenchuan earthquake damage. Institute of Engineering Mechanics, *China Earthquake Administration*.

Wang, K.H. (1999) *Study on control of reducing seismic response for cable-stayed bridge based on mode analysis*. Northern Jiaotong University, Beijing.

Ye, G.Y. (1999) A brief overview of the 1999 9-21 Chi-Chi earthquake in Taiwan. *Building Knowledge.*, (03):16–17.
Ye, A.J. (2013) *Bridge seismic resistance*. People's Traffic Press, Beijing.
Yu, F.L. (2018) Research on bridge seismic damage assessment model based on Wenchuan earthquake. Institute of Engineering Mechanics, *China Earthquake Administration*.
Zhuang, W.L., Chen, L.S., et al. (2013) *Seismic damage analysis of Wenchuan earthquake highways-bridges and tunnels.* People's Traffic Publishing House, Beijing.

Advances in Civil Engineering: Structural Seismic Resistance, Monitoring and Detection – Mohd Johari Mohd Yusof and Junwen Zhang (Eds)
© 2023 The Authors, ISBN 978-1-032-31491-4

Stability reliability analysis of multi-stage slope under earthquake action

Zhao Long, Yi-lei Shi, Wei-li Li & Cheng-ming Cao
Gansu CSCEC Municipal Engineering Investigation and Design Institute Co. Ltd., Gansu Lanzhou, China

ABSTRACT: Seismic action has a significant effect on slope stability, and the adverse effect of seismic action should be considered during slope stability analysis. The intensity and duration of an earthquake are random uncertainties. Therefore, there is a certain deviation in using the single safety factor method in analyzing the slope stability under earthquake action. In this paper, the Monte Carlo method was used to calculate the stability and reliability of a multi-stage high slope under earthquake action. By considering the shear strength parameters and seismic acceleration as random variables, the stability reliability and failure probability of multi-stage high slope were obtained. The results showed that the random uncertainty of seismic action had a significant adverse effect on slope stability. With the change in seismic acceleration, the slope reliability index decreased significantly and the failure probability increased. The stability reliability analysis method can be used to evaluate slope stability more accurately. The research results had a certain guiding effect on the stability analysis of multi-stage high slope under earthquake action.

1 INTRODUCTION

Currently, the method of single dynamic safety factor is used to analyze slope stability under earthquake action. The analysis results of slope stability were obtained by comparing the dynamic safety factor with the stability safety factor of slope. However, most of the factors affecting slope stability have random uncertainty, and the single safety factor method cannot reflect the influence of parameter uncertainty on slope stability. The reliability theory evaluated the stability of slope by considering the uncertainty of various factors as well as by adopting the reliability index or failure probability. Therefore, application of reliability theory in slope stability analysis is developing rapidly.

In recent years, some scholars have begun using reliability theory to analyze the static and dynamic stability of the slope, and had obtained some research results. The influence of the uncertainty of internal friction angle and cohesion on the stability reliability of filled slope was analyzed by Jiang Liao (Jiang 2017). Geo-Studio software was used to carry out sampling simulation calculation using the Monte Carlo method (Gui 2014). An upper-bound numerical method for reliability analysis of soil slope was proposed by Zhang Xiaoyan (Zhang 2018). By considering soil parameters and seismic peak acceleration as random variables, the checking point method was used to analyze the reliability of seismic stability of slope by Zhou Zhijun (Zhou 2016). The limit state equation of rock slope under seismic load was established by Kong Yuyang, and the influence of horizontal and vertical seismic load, rock mass amplification factor, and other factors were analyzed as well (Kong 2017). The reliability of the slope was evaluated by constructing the response surface function and compiling the optimization method and Monte Carlo program based on the optimization principle (Wang 2012). The state function of reliability analysis based on the double wedge method of limit equilibrium principle was established by Sun Shulin (Sun 2012). By using

 DOI 10.1201/9781003310884-4

dynamic finite element time-history analysis and point estimation method, the time-varying reliability of slope under strong earthquake action was solved, and the reliability index and time-history curve of failure probability were obtained (When 2016).

There are many uncertain factors affecting slope stability, such as strength parameters of rock and soil mass, earthquake action, rainfall action, reinforcement structure of frame anchor, and so on. It has been shown that the uncertainty and variability of internal friction angle and cohesion of rock of soil strength parameters have an obvious influence on the stability of slope (Chen 2018; Jiang 2015; Guo 2018; Zhang 2017; Zhu 2019). The acceleration of seismic action fluctuates with seismic action time, and the random uncertainty is obvious, which has a serious adverse effect on slope stability (Dong 2015, 2017). Therefore, during reliability analysis of slope stability, it is necessary to consider the influence of soil strength parameters and the uncertainty of seismic action on slope stability.

Currently, although some progress has been made in the study of the stability reliability of slope, research on calculation of the static reliability of single-stage slope still remains. There are few reports on the stability reliability of multi-stage high slope under seismic action. With developments in infrastructure construction, more multi-stage high slopes are encountered in practical projects. In this paper, selecting multi-stage deep cut high slope as the engineering background, through the analysis of soil shear strength and seismic peak acceleration distribution type, the Monte Carlo method in Geo-Studio software was used to calculate the multi-stage high slope reliability and failure probability. Moreover, according to the calculated results, the stability variation in multi-stage deep cut slope under the influence of random parameters was analyzed. The analysis results can provide a certain reference for the stability reliability analysis and research of multi-stage deep cut slope.

2 MONTE CARLO METHOD FOR SLOPE RELIABILITY ANALYSIS

According to the definition of probability theory, a random sampling test can be carried out on random variables that affect the reliability of the slope, and then the sampling data can be integrated in the slope stability calculation function. The failure probability or reliability index of slope can be obtained by statistical analysis of the calculation results.

The main steps in Monte Carlo method for calculation of failure probability are as follows:

① The sampling values of each random variable were $x_1, x_2, \ldots x_n$

② Substitute the sampling value into the limit state equation to calculate:

$$Z_i = g(X_1, X_2, X_3, \cdots X_n) \tag{1}$$

③ According to the accuracy requirements, the sampling times N can be determined. Repeat the above steps to obtain the limit state value Z_i of each time, and the number of times $Z_i < 0$ was L. If the sampling times were large enough, the slope failure probability can be obtained as follows:

$$P_f = L/N \tag{2}$$

Then, according to the statistical results, the mean μ and standard deviation δ of Z_i were obtained, and the reliable indexes were calculated as follows:

$$\beta = \frac{\mu}{\delta} \tag{3}$$

3 STATISTICAL ANALYSIS OF RANDOM VARIABLE PARAMETERS

3.1 *Statistical analysis of soil strength parameters*

Combined with the actual project, a 30-m high slope was selected as the research object. The internal friction angle and cohesion were set as random variables subject to normal distribution. According to the statistical results, the soil weight was 17.0 kN/m^3. The average value, standard deviation, maximum value, and minimum value of cohesion and internal friction angle are shown in Table 1. Probability density functions of cohesion and internal friction angle are shown in Figures 1 and 2. The correlation coefficient between cohesion and internal friction angle was −0.5.

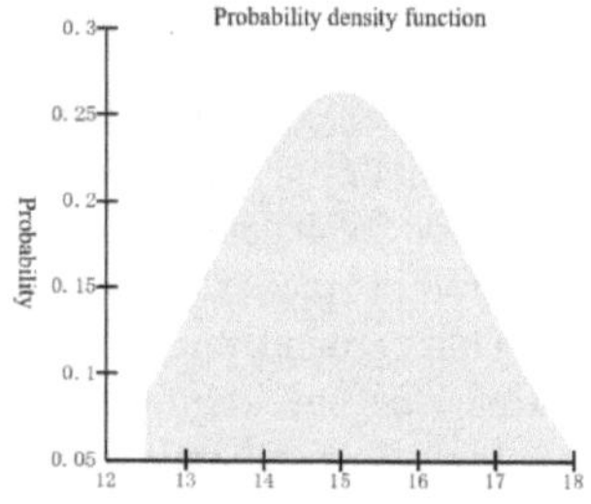

Figure 1. Cohesion.

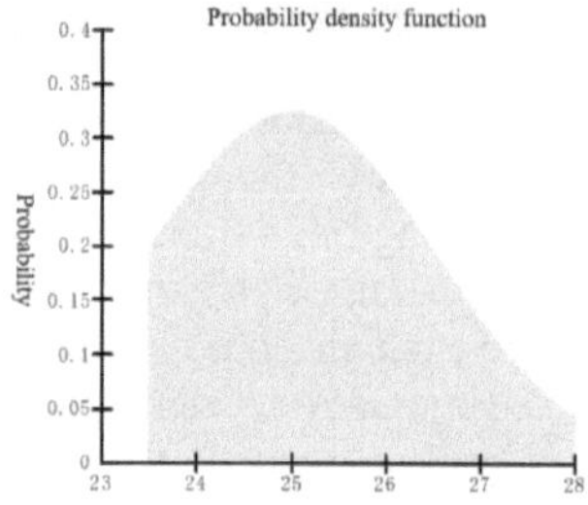

Figure 2. Internal friction angle.

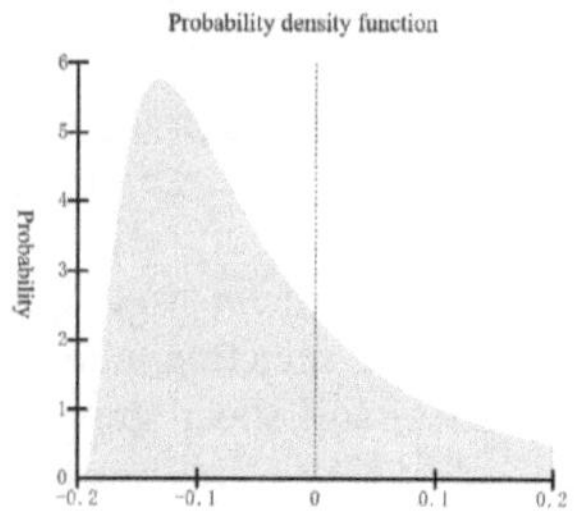

Figure 3. Earthquake peak acceleration.

Table 1. Statistical values of soil mass parameters.

Soil parameters	Distribution type	Mean	Standard deviation	Min	Max
Cohesion c/kPa	Normal distribution	17	1.7	14	19
Internal friction Angle φ/ (°)		26	1.5	23.5	28

3.2 *Statistical analysis of earthquake peak acceleration*

Seismic peak acceleration has obvious time uncertainty. Therefore, it was more reasonable to consider seismic acceleration as a random variable in slope stability analysis. In this paper, the peak acceleration under different fortification intensities was regarded as a random variable and was assumed to follow a lognormal distribution. By querying the earthquake observation database of the National Vibration Network center, the peak acceleration of horizontal earthquake under different seismic intensity was statistically analyzed, and the parameter values of peak acceleration were obtained statistically. Taking horizontal peak acceleration of 0.20g as an example, the mean value of the acceleration was 1.768 m/s^2, the standard deviation was 1.68 m/s^2, and the maximum value and minimum value were 0.2g and −0.2g, respectively. The probability density function curve and sampling function of horizontal seismic peak acceleration were shown in Figure 3, respectively.

Based on the above analysis, it was assumed that soil shear strength parameter (c, φ) and earthquake peak acceleration Sa were continuous random variables. Then, its probability density function can be expressed as: g (c, φ, Sa) = 0. The reliability index and failure probability can be further obtained according to the reliability theory.

4 ESTABLISHMENT OF SLOPE RELIABILITY CALCULATION MODEL

In this paper, taking the multi-stage high slope as the engineering background, a 30-m high typical section was selected to establish the stability reliability calculation model. The slope reliability calculation model was shown in Figure 4.

Slope/W module in Geo-Studio software was used to establish the reliability calculation model of the multi-stage high slope. To ensure sufficient accuracy of the calculation results, the Monte Carlo simulation times in the calculation model were set to 5000 times. The shear strength parameters of soil were set as random variables. The reliability index and failure probability of slope under different earthquake action intensities were calculated by setting different horizontal seismic peak accelerations in the model.

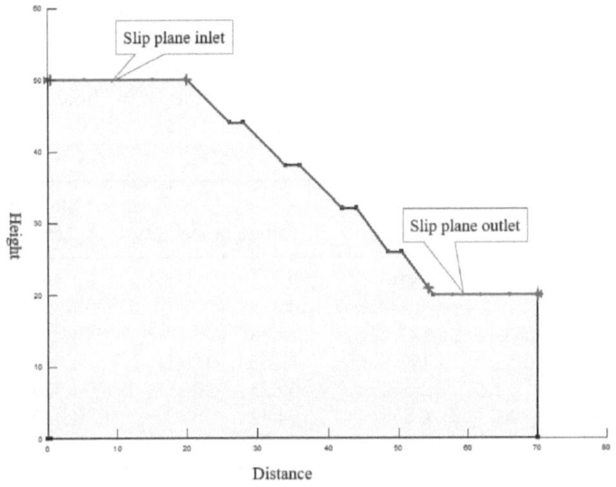

Figure 4. Slope reliability calculation model.

5 ANALYSIS OF CALCULATION RESULTS

5.1 *Calculation and analysis of slope stability*

In this paper, the difference in slope stability calculation results between a single factor of the safety method and the reliability method were compared and analyzed. First, soil parameters and earthquake peak acceleration were set as the determined values, and the stability safety factors of slope under static and 8-degree (0.20g) earthquake were calculated by using the Morgenstern-Price (M-P) method, Bishop method, and Janbu method, respectively. The results obtained using different methods were shown in Table 2.

It can be seen from Table 2 that the static and dynamic safety factors obtained by different stability calculation methods were slightly different. Among them, the Janbu method had the lowest safety factor, the Bishop method had the largest calculation result, and M-P method yielded a value between the other two. Therefore, the M-P method was found to be more reasonable as a safety factor of slope stability in practical engineering calculation. The safety factor of slope decreased obviously under earthquake action, which indicated that the stability of slope was significantly affected by earthquake action and that the seismic design of slope should be strengthened. There

Table 2. Calculation results of safety factor by different methods.

Calculated method	M-P method	Bishop method	Janbu method
Static method	1.521	1.553	1.481
Dynamic method	1.131	1.163	1.117

were differences in safety factors among different calculation methods, which also indicated that reliability theory was more reasonable for slope stability analysis.

5.2 *Reliability calculation of slope stability*

In the calculation model, shear strength parameters of soil were set as random variables. Therefore, input the seismic peak acceleration parameters of different fortifications with lognormal distribution (the peak acceleration values were 0, 0.05g, 0.10g, 0.15g, 0.20g, 0.30g, 0.40g, respectively). The average safety factor, failure probability, and reliability index of slope under different peak acceleration were calculated, and the calculation results are shown in Table 3.

Table 3. Calculation results of slope reliability.

Peak seismic acceleration	Average safety factor	Reliability index	Failure probability	Max safety factor	Min safety factor
0	1.491	10.87	0	1.522	1.416
0.05g	1.456	7.368	0	1.585	1.315
0.1g	1.381	5.133	12.35	1.474	1.123
0.15g	1.127	1.172	36.65	1.331	1.056
0.20g	1.077	−2.502	69.23	1.327	0.928
0.30g	0.982	−6.858	94.72	1.156	0.957
0.40g	0.954	−8.067	99.96	1.024	0.935

According to Table 3, when the seismic peak acceleration was 0.15g, although the minimum safety factor was 1.056, which meets the stability factor required, the slope failure probability reached 36.65%, and the slope failure probability increased sharply. When the peak acceleration was 0.4g, the failure probability had reached 99.96%, but the average safety factor was 0.935, which was slightly lower than 1.05. It can be seen that the reliability theory can evaluate slope stability more reasonably.

(1) Analysis of average safety factor

In Geo-Studio software, the Monte Carlo method was used to calculate the stability reliability of slope, and the probability distribution curve of slope safety factor was obtained, as shown in Figure 5. It can be seen from Figure 5 that when the slope stability calculation with parameter uncertainty was considered, the slope safety factor was close to normal distribution and not the definite value. Therefore, it was not accurate to use a single safety factor to determine slope stability, and the failure probability and reliability index can better reflect the degree of slope stability.

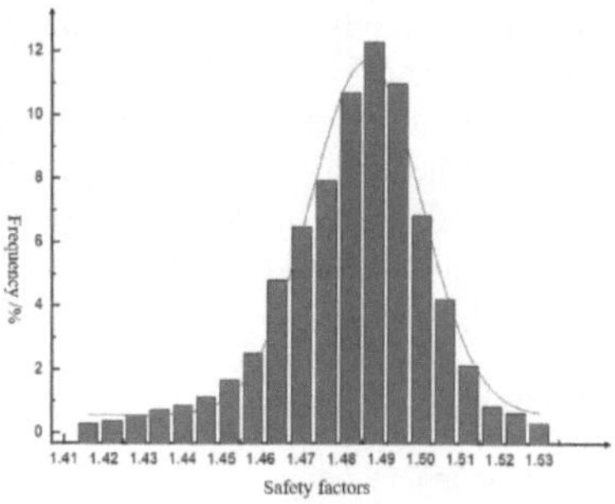

(a) Under static action

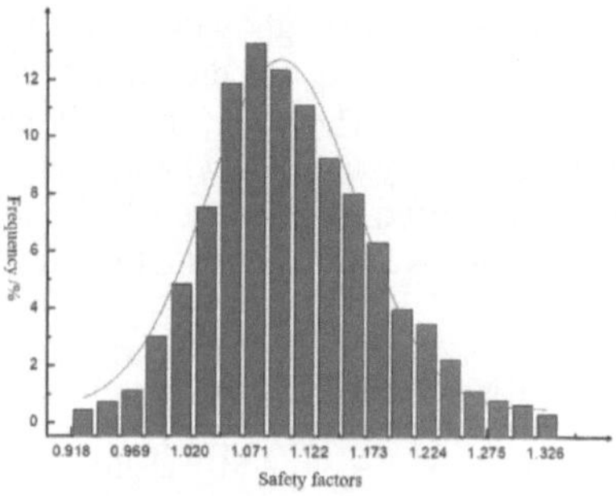

(b) Under seismic action (0.2g)

Figure 5. Probability distribution curve of safety factor.

The average safety factor of slope under different seismic peak acceleration was shown in Figure 6. It can be seen from the variation curve of the safety factor that the safety factor decreased obviously with the increase in seismic peak acceleration, indicating that seismic action had a significant adverse effect on slope stability. When the peak acceleration was 0.05g and 0.1g, the safety factor changed gently. At this point, the slope safety factor was greater than 1.05, which meets the requirements of the slope safety code. With the increase in peak acceleration, the safety factor of the slope decreased sharply, indicating that the slope was more prone to instability caused by strong earthquake. When the seismic peak acceleration reached 0.2g, the slope safety factor was less than 1.0, indicating that the slope had been in a state of instability.

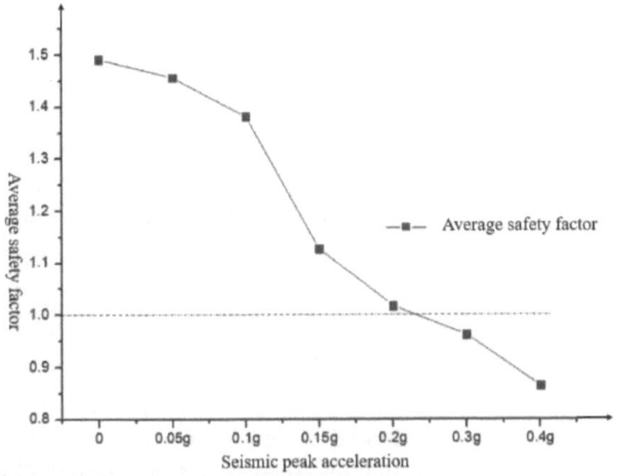

Figure 6. Average safety factor change curve.

(2) Reliability index analysis

The variation of slope reliability index under different seismic peak acceleration is shown in Figure 7. It can be seen from Figure 7 that the slope reliability index decreased with the increase in peak seismic acceleration. With the increase in seismic peak acceleration, the reliability index decreased from 10.87 under normal working conditions to −8.067 when the peak acceleration was 0.4g. The wide range of reliability index also indicated that the strong earthquake had a significant impact on the stability of a slope. When the seismic peak acceleration exceeded 0.15g, the slope reliability index becomes negative.

(3) Comparative analysis of failure probability

The variation of slope failure probability under different seismic peak acceleration was shown in Figure 8. It can be seen from Figure 8 that slope failure probability shows an obvious upward trend with the increase in earthquake peak acceleration. When the peak acceleration exceeded 0.15g, the increasing trend of failure probability was most obvious. This phenomenon showed that strong earthquake had a very obvious adverse effect on the stability of slope. When the acceleration value reached 0.3g, the failure probability of the slope was close to 100%, indicating that the slope had been unstable.

5.3 *Comparative analysis*

In this paper, the safety factor and stability reliability of multi-stage high slope were calculated respectively, and the results were compared and analyzed. When soil parameters and seismic peak acceleration were taken as constant values, the stability safety coefficients of the multi-stage high slope under static and seismic action were 1.521 and 1.131, respectively. When the cohesion, internal friction angle, and seismic peak acceleration were considered as random variables to calculate the stability reliability, the stability safety factor distribution of the multi-stage high

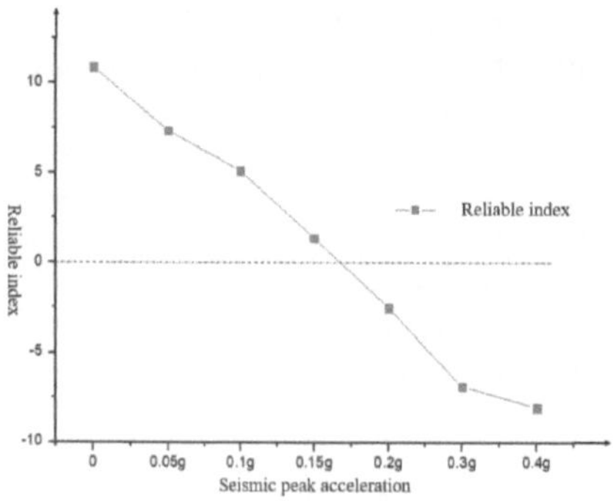

Figure 7. Change curve of reliable index.

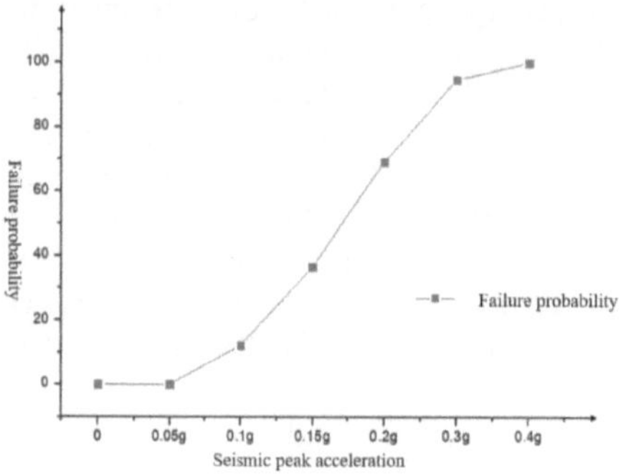

Figure 8. Change curve of failure probability.

slope under static and seismic action was roughly normal distribution, and the minimum safety factors were 1.416 and 0.928, respectively. According to the calculation results, considering the randomness of parameters, the calculation results were small, the design tended to be safer, and the evaluation results were more reasonable.

6 CONCLUSIONS

In this paper, by considering the random uncertainties of internal friction angle, cohesion, and seismic peak acceleration, the slope/W module in Geo-Studio software was used to establish the stability and reliability calculation model of multi-stage high slope. And the following conclusions were drawn:

(1) With the increase in earthquake peak acceleration, the slope safety factor decreased obviously, and the slope safety factor decreased more rapidly under strong earthquake. When the seismic peak acceleration exceeded 0.2g, the slope safety factor derived from the reliability theory decreases to less than 1.0, and the slope was in an unstable state.
(2) With the increase in seismic peak acceleration, the reliability index of slope decreased obviously, indicating that the stability of slope was decreasing continuously. When the seismic peak acceleration reached 0.2g, the reliability index decreased to a negative value, and the slope was in the stage of instability failure.
(3) With the increase of earthquake peak acceleration, the failure probability of slope showed an obvious upward trend, indicating that the possibility of slope failure under earthquakes increased. When the seismic peak acceleration reached 0.3g, the failure probability of the slope reached 94.72%, and the slope was in extreme danger.

ACKNOWLEDGMENTS

This work was financially supported by Construction science and technology project of Gansu Province (JK2019-01); Science and technology project of Gansu Province (20JR10RA570).

REFERENCES

Chen Zhao-hui, Lei Jian, Huang Jing-hua (2018) Finite element limit analysis of slope stability considering spatial variability of soil strengths. *Chinese Journal of Geotechnical Engineering*, 40(6): 985–993.

Dong Jian-hua, Zhang Yuan, Zhu Yan-peng (2015) Random Seismic Response and Dynamic Reliability Analysis of Frame with Prestressed Anchors for Slope Stability. *China Journal of Highway and Transport*, 28(10):26–33.

Dong Jian-hua, Dong Xu-guang, Zhu Yan-peng (2017) Reliability Analysis on Stability of Slope Reinforced by Frame with Pre-stress Anchors under Random Earthquake Action. *China Journal of Highway and Transport*, 30(2):41–47.
Gui Yong, Deng Tong-fa, Luo Si-hai (2014) Establishment of slope stability dual index system based on Monte Carlo simulation and its application. *Rock and Soil Mechanics*, 35(7):1979–1986.
Guo Chong-yang, Li Dian-qing, Cao Zi-jun (2018) Efficient reliability sensitivity analysis for slope stability in spatially variable soils. *Rock and Soil Mechanics*, 39(6): 2203–2210.
Jiang Shui-hua, Li Dian-qing (2015) Reliability analysis of multilayered soil slopes system considering spatial variability of soil properties. *Rock and Soil Mechanics*, 35: 629–633.
Jiang Liao, Yu Xiang, Liu Lin-jie (2017) Reliability Analysis on Filled Slope Based on Monte Carlo Simulation. *Chinese Journal of Underground Space and Engineering*, 201–205.
Kong Yu-yang, Li Shan (2017) Pseudo-dynamic Analysis and Reliability Study of Stability of Rock Slope Under Seimic Load. *Science Technology and Engineering*, 17(22): 169–176.
Sun Shu-lin, Qian Jian, RUAN Xiao-bo (2012) Reliability Analysis of a Municipal Solid Waste Landfill Against Translational Failure Using Monte Carlo Method. *Science Technology and Engineering*, 2: 7–11.
Si-cheng When, Jie Zhang, Hong-wei Huang (2016) Reliability Analysis of Soil High Slope Stability under Strong Seismic Condition. *Chinese Journal of Solid Mechanics*, 37:16–22.
Wang Ling-kuan, Wang Yu, Yuan Wei (2012) Optimization Monte Carlo Method of Slope Reliability Evaluation and Its Application. *Site Investigation Science and Technology*, 2: 7–11.
Zhou Zhi-jun, Yang Zhi-yong, Zou Qun (2016) Reliability Analysis of Slope Under Seismic Load Based on Checking Point Method. *China Journal of Highway and Transport*, 29(11):18–24.
Zhang Wen-sheng, Luo Qiang, Jiang Liang-wei (2017) Variation characterization and type classification of soil strength parameters based on reliability index. *Chinese Journal of Rock Mechanics and Engineering*, 36: 4188–4204.
Zhang Xiao-yan, Zhang Li-xiang, Li Ze (2018) Reliability analysis of soil slope based on upper bound method of limit analysis. *Rock and Soil Mechanics*, 39(5): 1840–1850.
Zhu Yan-peng, Liu Xin, Ma Xiao-rui (2019) Application of reliability theory based on improving the first order second moment method in reinforcement of high fill slope. *Science Technology and Engineering*, 19(27): 351–357.

Advances in Civil Engineering: Structural Seismic Resistance, Monitoring and Detection – Mohd Johari Mohd Yusof and Junwen Zhang (Eds)
© 2023 The Authors, ISBN 978-1-032-31491-4

Stability evaluation and protection of a highly weathered mudstone slope in Yunnan

Weiwei Zhu
School of Architecture and Civil Engineering, Kunming University, Kunming, China

ABSTRACT: To study the stability of a highly weathered mudstone high slope of an expressway in Yunnan, taking the typical section of K57 + 655 as the analysis object, the stability coefficient and potential sliding surface of the slope are obtained using the finite element software ANSYS and shear strength reduction method. Although the stability factor is > 1, considering the particularity of mudstone, the slope still needs to be properly protected. Anchor bolts and prestressed anchor cables are used for slope reinforcement, ~~and the~~ protective measures include shotcrete, six-edge hollow bricks covered with soil and grass planting, and three-dimensional mesh grass planting. The practice shows that the slope has been in a stable state, and the slope protection measures are feasible.

1 GENERAL INSTRUCTIONS

With rapid development of China's transportation infrastructure construction, high-grade highways now extend from the plain to the mountainous areas, with high-grade, wide pavement, and large-scale excavation. There are a large number of high slopes, as well as a large number of high slope deformation and damage, which not only increase the investment but also delay the construction period, and even cause damage to existing projects.

The stability of intense weathering slopes is a common problem in highway engineering. Mudstone is a sedimentary rock that comprises solidified mud and clay; its structure is similar to shale, but it is more durable. It also has a poor degree of consolidation with recrystallization that is not obvious, which leads to its poor weathering resistance and a tendency to soften and disintegrate upon contact with water. Many highly weathered mudstone slopes collapsed during construction in the past, posing considerable threat to the safety of human life and property, as well as the normal operation of roads, also resulting in serious soil erosion. In this paper, the shear strength reduction finite element method is used to calculate the safety coefficient of a highly weathered mudstone high slope of an expressway, and the comprehensive protection measures for the slope are introduced, which can be used as references for other similar slope projects.

2 ENGINEERING GEOLOGY OF THE STUDY AREA

The study area (K57+490~K57+888) mainly exposes the Jurassic and Quaternary strata, that is, from old to new are, Jurassic strata, which are purple-red, brown-red, purple-gray, brown-yellow, and gray-green silty mudstone and argillaceous sandstone, mudstone and argillaceous sandstone, sandstone and thick interbedded mudstone mainly interbedded sandstone and argillaceous sandstone; Quaternary strata, which are extremely complex and changeable. They consist of sub-clay, sub-sandy soil, sand and gravel soil, pebble soil, and block stone soil that are sporadically distributed. Their thickness varies from place to place, and they are in non-conforming contact with the Jurassic strata below.

DOI 10.1201/9781003310884-5

3 SLOPE STABILITY ANALYSIS AND EVALUATION

An analysis was performed based on the K57+655 section (Figure 1). The section had a slope height of 58.8 m and an angle of 52°. The mechanical property index of the rock mass is shown in Table 1.

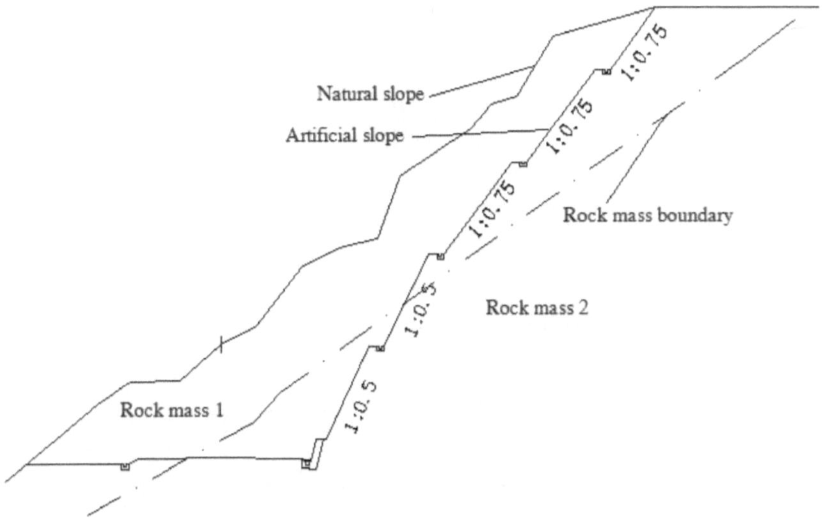

Figure 1. K57 + 655 slope section.

Table 1. Mechanical property index of slope rock mass.

	Elasticity modulus E (MPa)	Poisson's ratio μ	Cohesion c (kPa)	Internal friction angle φ (°)	Weight γ (kN·m^{-3})
Rock mass 1	5000	0.30	95	15.3	19.8
Rock mass 2	6000	0.30	105	17	20.8

Based on the ANSYS software and the finite element method, the slope stability was analyzed, the plane strain problem was resolved, and, in combination with the shear strength reduction method, the safety coefficient of the slope was solved. A planar 8-node element was used in the analysis to determine the following hypotheses: (1) The rock mass corresponds to the Drucker-Prager yield criterion; (2) the right boundary of the slope model X = 0; the left boundary X = 0 and Y = 0; the lower boundary X = 0 and Y = 0; the slope and top are free; (3) the width of the slope model is 150 m (approximately 2.5 times of the slope height), and the lower boundary extends 20 m below the subgrade surface.

The safety coefficient of the slope was solved before protection, with the acceleration of gravity at 9.8 m/s^2. There are 1048 units and 3314 nodes. Under the action of the gravity stress field of the slope, the ANSYS calculation converged, and the equivalent plastic strain was not penetrated, suggesting that the slope was stable, as shown in Figure 2. The safety coefficient of the slope was calculated using the shear strength reduction method, and the material parameters after the strength reduction were input again. Repeated analyses showed that ANSYS reaches critical convergence when the reduction coefficient F = 1.12 and the stability safety coefficient of the slope F = 1.12. The analysis results of the equivalent plastic strain before the critical convergence are shown in

Figure 3. Thus, the penetration of the equivalent plastic strain increases significantly after the shear strength is reduced.

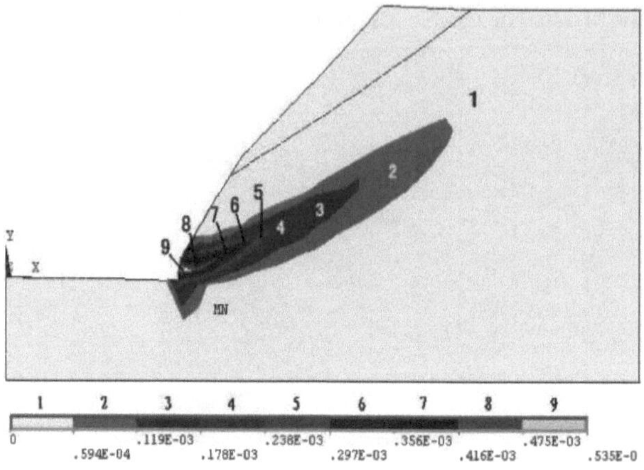

Figure 2. Equivalent plastic strain contour map before strength reduction.

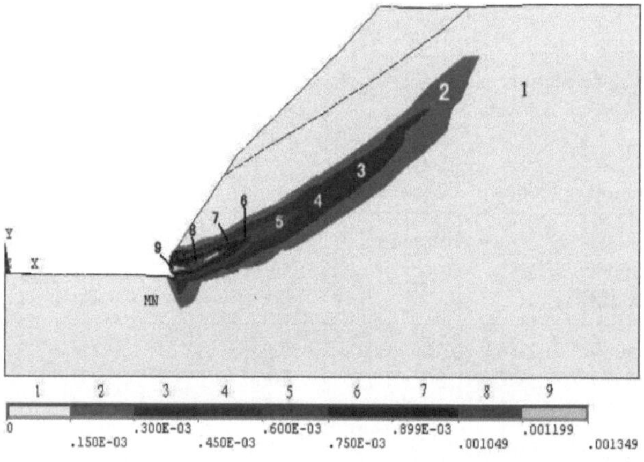

Figure 3. Equivalent plastic strain contour map when the reduction coefficient is 1.12.

Tan Hanhua et al. pointed out that the sliding surface of highly weathered rock slopes is generally circular (Tan 2007). The engineering properties of highly weathered mudstone are closer to coarse-grained soil, and the values of c and ϕ are not high. The finite element analysis showed that the failure surface of the slope is close to a circular arc. After studying 106 slope examples, Zhang et al. (2003) stated that the slope safety coefficient obtained through finite element method for reducing shear strength is, on average, about 5.7% greater than that obtained using simplified Bishop method with a very small dispersion; thus, proving the feasibility of the method. The reason why the safety coefficient obtained with the finite element method is slightly larger is that the rock-soil mass is not regarded as rigid body, but is solved when the rock-soil mass can be deformed. After the stability analysis determined a safety coefficient F greater than 1, a highly weathered mudstone slope outside the study area of the section was protected by sprayed concrete, but the slope still slipped after heavy rains.

A proper understanding of the physical meaning of rock mass failure can ensure a reasonable interpretation of the calculation results. In numerical analysis, the failure of rock mass is generated under a specific calculation model. Because there may be some errors in the calculation model, the calculated area of rock mass failure area may also have errors. However, a large number of calculation results show that the error of the model generally does not lead to significant changes in the failure position of a rock mass. In other words, in numerical simulation, the size of rock mass failure area is relative, and the distribution of failure area is often absolute. When changing the value of rock mass mechanical parameters, the area surrounding the rock failure area changes obviously, but the distribution of the failure area is consistent. Therefore, given the particularity of the mudstone properties and the fact that the mechanical properties of slope rock mass used in the calculation were obtained under normal working conditions, although the results of the finite element analysis showed that the slope is fundamentally stable, it is still necessary to reinforce and protect the slope (Liu 2020, Zhang 2020).

4 COMPREHENSIVE PROTECTION OF HIGHLY WEATHERED MUDSTONE HIGH SLOPE

Due to high and steep natural slope in the section, appropriate measures should be taken depending on the degree of weathering of the rock and the height of the slope. A stepped slope is used, and the slope ratio for a highly weathered stratum is 1:0.5 and the top slope ratio is 1:0.75. The height of the steps is 12 m, and it extends to the top of the natural slope with a 2.0 m long platform reserved for each slope. On certain high and steep slopes, the central platform can be widened to about 4.0 m, and a rectangular intercepting ditch can be set on the platform.

The square cutting process changes the original tension state of the slope, causing the redistribution of the slope body tension and thereby influencing the stability of the slope (Chen 2021). Therefore, the sides of the square cutting should be protected during construction. The first-level slope is reinforced with Φ32 reinforced anchor rod frame, and the length of the anchor rod is 6 m, which is pinned every 3.0 m along the route. The second, third, and fourth-level slopes are reinforced with prestressed anchor cables and anchor frame lattice beams. The anchor rods are reinforced with Φ32 steel bars with a length of 8 m, whereas the prestressed anchor cables are 4Φ15.24 high-strength and low-tension steel strands with a length of 20 m. The frame is pinned every 3.0 m along the route. The fifth-level slope is protected by the C20 concrete hanging net and sprayed concrete to prevent the shallow slope body from sliding and to increase its resistance to weathering and rain erosion.

Intercepting ditches should be available at all levels to drain surface water and groundwater. An intercepting ditch should be set 5 m from the brush side slope line, and the wastewater from the platform intercepting ditch should be drained through the top cutting intercepting ditch. At least one inspection ladder was set up on each slope for inspection and for use during operation. The project is excavated and protected from top to bottom, and the slopes without reinforcement protection are not more than two levels. In addition, during construction, each section is engineered and reinforced before moving on to the next section to avoid slope instability. When building the frame, on the other hand, the vertical ribs of the anchor cable frame are embedded in 30 cm of the slope body. The vertical ribs of the frame are straight up and down, and the beams are on the left and right. To ensure the appearance of the frame, the cavities are filled with mortar and rubble, and there is a 2 cm expansion joint for every 15 m of the frame beam, which is filled with foam board.

During the construction of the anchor cable, a dry drilling method is used to drill the hole with a diameter of 110 mm. When grouting the anchor cable hole, it should be injected from the bottom up all at once. The anchor cable is prestressed twice before the official tensioning. This workstation is divided into three levels of official tensioning, whereby the level load is 200 kN and each tension value is stable for 5 minutes. When the tension reached 600 kN, the deformation was stable and the anchor cable remained intact after 10 minutes of observation. During the construction of the hanging net and sprayed concrete, a rock drill was used to drill the hole with a diameter of 40 mm, controlling the vertical and horizontal differences of the hole within 50 mm. According to the geological conditions of the worksite, a hexagonal drill pipe combined with a high-pressure air cleaning method was used to drill the hole in the rock stratum, and a spiral drill pipe combined with a high-pressure air slag cleaning method was used to drill the hole in the highly weathered rock stratum. After drilling, the incident angle of the drill pipe was checked to ensure that the drill pipe is perpendicular to the slope. Once properly adjusted, drilling should be continued to the bottom of the hole.

5 CONCLUSION

(1) Numerical simulation analysis shows that under normal working conditions, the potential sliding surface of K57 + 655 highly weathered mudstone slope is approximately circular arc, and the stability coefficient F = 1.12.
(2) Considering the special properties of mudstone, when the stability coefficient F of highly weathered mudstone slope is approximately 1, necessary protection needs to be provided.
(3) K57 + 655 slope is comprehensively reinforced by anchor bolt and prestressed anchor cable, and protected with measures such as shotcrete, six-edge hollow brick covering and grass planting, three-dimensional mesh grass planting, intercepting ditch drainage, etc. The slope remains stable all the time, which can be used as a reference for the protection of other similar slope projects in the future.

ACKNOWLEDGMENTS

This work was financially supported by the Basic Research Project of Yunnan Province (202101AT070144).

REFERENCES

Chen, W., Song, B., Wu, W. (2021) Direct and reversal shear behaviors of three kinds of slip zone soil in the Northwest of China. *Bulletin of Engineering Geology and the Environment*, 5: 3939–3952.

Liu, W., Zhang, Z.H. (2020) Experimental characterization and quantitative evaluation of slaking for strongly weathered mudstone under cyclic wetting-drying condition. *Arabian Journal of Geosciences*, 13: 1–8.

Tan, H.H., Luo, Q., Qi, S.W. (2007) Analysis on failure mode and reinforcement effect of fully strongly weathered rock high slope. *Highway* 12: 21–25.

Zhang, L.Y., Zheng, Y., R., Zhao, S.Y. (2003) Study on the accuracy of calculating the safety factor of soil slope stability by finite element strength reduction factor method. *Journal of Hydraulic Engineering*, 1: 21–27.

Zhang, Z., Han, L., Wei, S. (2020) Disintegration law of strongly weathered purple mudstone on the surface of the drawdown area under the conditions of Three Gorges Reservoir operation. *Engineering Geology*, 7: 105584.

Advances in Civil Engineering: Structural Seismic Resistance, Monitoring and Detection – Mohd Johari Mohd Yusof and Junwen Zhang (Eds)
© 2023 The Authors, ISBN 978-1-032-31491-4

Numerical simulation study on dynamic characteristics of high concrete faced sandy gravel dam

Xubin Huang, Shengjie Di, Xi Lu, Zhe Miao & Peng Huang
Northwest Engineering Corporation Limited, Power China, Xi'an, Shaanxi, China
High Slope and Geological Hazard Research & Management Branch, National Energy and Hydropower Engineering Technology R&D Center, Xi'an, Shaanxi, China

ABSTRACT: Based on the equivalent viscoelastic model, the dynamic characteristics of dam, panel and diaphragm wall of concrete faced sandy gravel dam built on deep overburden at Chahanwusu, Xinjiang are calculated and analyzed. The results show that dynamic deformation of the dam along the river increases with the increase of the dam height. The calculated maximum permanent seismic subsidence is about 0.14% of the height of the sandy gravel dam and the thickness of the overburden layer. The maximum stress along slop of panel is located at the middle of the dam bottom, and central of dam crest has the largest deflection. The diaphragm wall will not be damaged under the earthquake. This paper can provide guidance for the construction and design of extra high or high faced sandy gravel dam on deep overburden layer.

1 INTRODUCTION

In recently years, plenty of hydraulic engineering tend to be built in northwest of China, among which concrete faced rockfill dams built on deep overburden cover occupy a considerable proportion (Cao 2013). The concrete face rockfill dam has many advantages, such as strong adaptability, low cost, small construction disturbance, short construction period and good seismic performance, which is the preferred dam type in earth-rockfill dam construction both at home and abroad (Dang et al. 2015; Li & Miao 2012). Therefore, concrete face rockfill dam has become the first choice for building dams on deep overburden.

For the face rockfill dam built on deep overburden, not only the static deformation and stress of the dam body should be analyzed, but also the dynamic characteristics under the condition of occasional earthquake should be analyzed to ensure the long-term safe operation of the dam (Zhang 2021). As for the dynamic characteristics of face rockfill dams on deep overburden, many efforts have been made by predecessors. Such as Wang et al. (2020) discussed the influence of ground motion input method on dynamic response of high face rockfill dam on deep overburden layer, the results show that the acceleration response of the dam obtained by the vibration method is obviously higher than that of the wave method, and it is more significant under the vertical ground motion. Peng et al. (2020) analyzed the acceleration amplification factor of the site, the permanent displacement of the dam body and the stress distribution of the face rockfill dam with different overburden thickness. Yu et al. (2018) studied the dynamic characteristics of earth-rock dam on overburden by taking earth-rock dam, overburden layer and bedrock as the same system. He thought that when the earth-rock dam may suffer from near-field earthquake with large vertical component, it is necessary to use nonlinear wave input method to evaluate the seismic performance of dam.

Although many existing literatures have analyzed concrete faced rockfill dams under earthquake action, however, the filling materials of concrete faced gravel sandy gravel dams are different from those of traditional concrete faced rockfill dams. In order to study the dynamic characteristics of

DOI 10.1201/9781003310884-6

sand gravel dam build on deep overburden layer, this paper adopts equivalent viscoelastic model to calculate the stress and deformation variation law of key parts such as panel, dam and diaphragm wall of high sand gravel dam on deep overburden layer.

2 METHODS AND MATERIALS

2.1 *Project profile*

The crest elevation of the concrete faced sandy gravel dam of Chahanwusu Hydropower Station is 1654.00m, with the maximum dam height of 110m and the crest length of 337.6m. The upstream slope is 1:1.5 and the comprehensive slope of the downstream dam is 1:1.85. The maximum thickness of the overburden layer is 46.7m, which can be divided into three large layers and two rock groups: the upper and lower layer is riverbed sand gravel layer, and the lowest layer is tuffaceous siltstone, as shown in Figure 1.

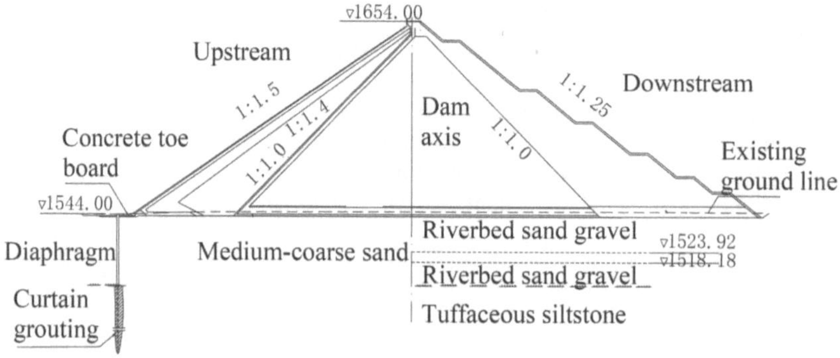

Figure 1. Schematic diagram of concrete faced sandy gravel dam of Chahanwusu Hydropower Station.

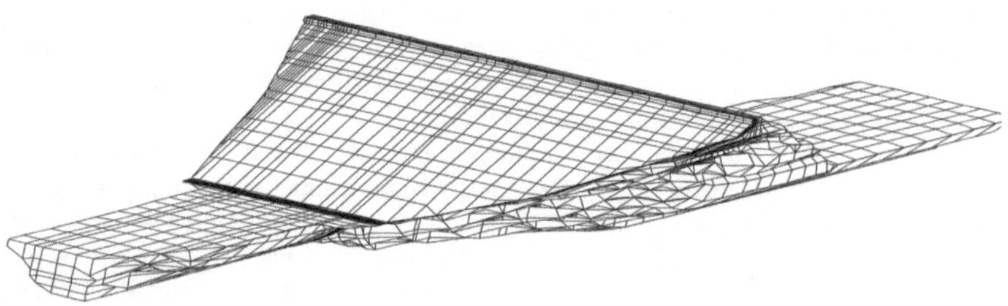

Figure 2. 3D schematic diagram of concrete faced sandy gravel dam.

2.2 *Calculation model and parameters*

A 3D finite element model of concrete faced sandy gravel dam is shown in Figure 2. The calculation model includes 14,465 three-dimensional entity units and 16,323 nodes. When three-dimensional finite element mesh is dissected, the solid element adopts 8-node hexahedral isoparametric element. In order to adapt to the boundary conditions and the variation of dam material partition, some solid models adopt trihedron and tetrahedron as degenerate hexahedron elements. In this model, the equivalent viscoelastic model is adopted in the dam materials and overburden layer materials.

The dynamic parameters of dam and overburden layer are shown Table 1.

Table 1. Dynamic parameters of dam and overburden layer materials.

Materials	k_2	λ_{max}	ν_d	k_1	n	c_1	c_2	c_3	c_4	c_5
Overburden layer	120000	0.10	0.25	0	0	0	0	0	0	0
Sandy gravel	1785.55	0.27	0.45	3.51	0.45	0.0002	0.75	8.9	0.1	1.0
Enrockment	1785.55	0.27	0.45	3.51	0.45	0.0002	0.75	8.9	0.1	1.0
Panel (C25)	156000	0.05	0.167	0	0	0	0	0	0	0
Diaphragm wall (C35)	202500	0.05	0.167	0	0	0	0	0	0	0

2.3 *Input method of ground motion*

The input of ground motion is synthetic seismic wave with the standard design response spectrum (class 1 site) as the target spectrum. Three random synthetic seismic acceleration time-history curves were designed to fit the response spectrum with a 50-year exceedance probability of 10% criterion. Ground motion uses three-way input, and the seismic time-history curve is used as the seismic input along the dam axis and river. The maximum acceleration is 0.145g in the axial direction of dam and along the river direction, and 2/3 of the time history curve (0.097g) of the maximum vertical acceleration is taken as the input of the vertical ground motion.

3 CALCULATION RESULTS AND ANALYSIS

3.1 *Dynamic deformation of dam*

Dynamic deformation distribution of dam under earthquake action is shown in Figure 3, of which Figure 3a is dynamic deformation along the river direction, and Figure 3b is vertical dynamic deformation. It can be seen that because both sides of the dam are embedded in the rock, the dynamic deformation is very small at dam abutment. The dynamic deformation of the dam increases with the increase of the dam height, and the maximum dynamic deformation occurs near the dam crest in the middle of the valley, which mainly because the dam crest is less constrained by the bedrock on both sides and the foundation, hence the dam crest owns the greater swing under the action of earthquake.

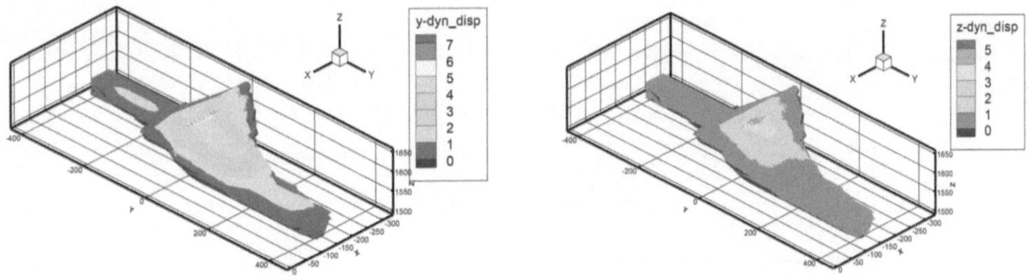

(a) Dynamic deformation along the river direction (b) Vertical deformation displacement

Figure 3. Dynamic deformation of the dam.

Distribution of permanent deformation of dam under earthquake is shown in Figure 4, and Figure 4a is the permanent deformation of dam along the river direction. It can be seen from the figure that the permanent deformation along the river direction is downstream. The maximum value is located in the middle of the dam crest, it decreases along the dam slope and two abutments. Vertical permanent deformation is characterized by seismic subsidence, and its maximum value is located in the middle of dam crest. The permanent deformation of overburden layer is small, which maximum deformation along river direction is about 5.2cm, and the maximum seismic depression is about 1.0cm. In this model, the height of the sandy gravel dam is 110m and the thickness of the overburden layer is about 46m. The calculated maximum permanent seismic subsidence, which is about 0.14% of the height of the sandy gravel dam and the thickness of the overburden layer. According to the seismic damage data of the earth-rock dam, the subsidence is within the normal permanent deformation range.

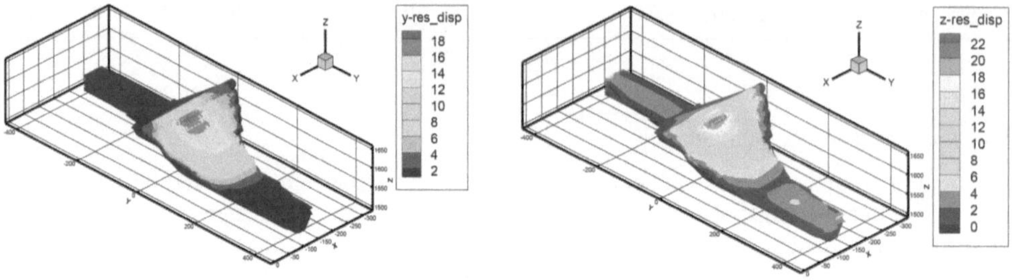

(a) Permanent deformation along the river direction (b) Vertical permanent deformation

Figure 4. Permanent deformation of the dam.

3.2 *Dynamic stress and deformation of panel*

Stress and deformation distribution of panel after the earthquake is shown in Figure 5, of which Figure 5a is the stress distribution of panel after the earthquake. Figure5a shows that as the height of the panel increases, the stress of the panel under the action of earthquake becomes smaller, and the maximum stress along slop of panel is located at the middle of the dam bottom, which is mainly because the bottom of dam is well constrained than dam crest. The results show that the maximum stress on panel is less than maximum compressive strength of concrete, which means that the panel

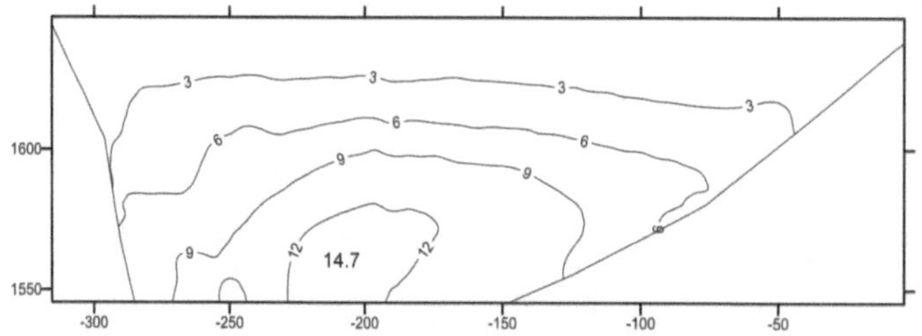

(a) Stress along slope of panel

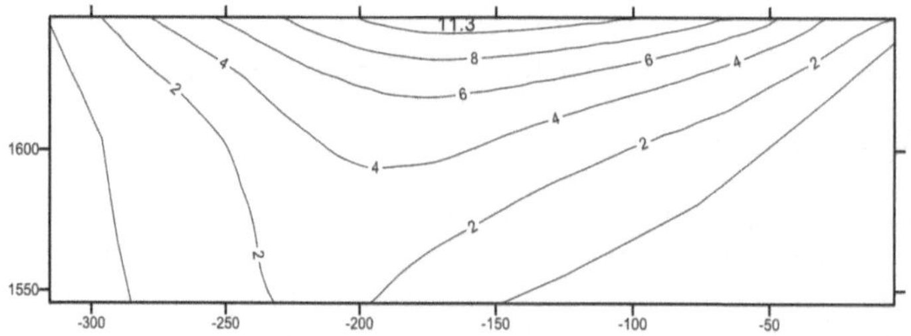

(b) Panel deflection

Figure 5. Stress and deformation distribution of panel after the earthquake.

may be not destroyed under the action of earthquake. Figure 5b is the panel deflection after the earthquake, it can be seen from figure that due to the central position of the dam crest is subject to less constraint than other places, so this position has the largest deflection. And the maximum deflection located on the middle of dam crest.

3.3 *Dynamic stress and deformation of diaphragm wall*

The diaphragm wall is located in deep overburden layer. Due to strength of overburden layer is much less than that of concrete of diaphragm wall, therefore, the deep overburden layer can play a good damping effect under the action of earthquake. Figure 6. shows the dynamic stress distribution of the diaphragm wall under earthquake action, and Figure 6a is the distribution of dynamic compressive stress. It can be seen that the compressive stress on the left side wall is larger than that on the right side wall. Figure 6b shows the distribution of dynamic tensile stress, which is similar to that of dynamic compressive stress. On the whole, the dynamic compressive stress and dynamic tensile stress on the diaphragm wall are less than the maximum compressive and tensile strength of the concrete under the earthquake action, so the diaphragm wall is less likely to be damaged. Figure 6c shows the deflection distribution on the cutoff wall after the earthquake. The results in the figure

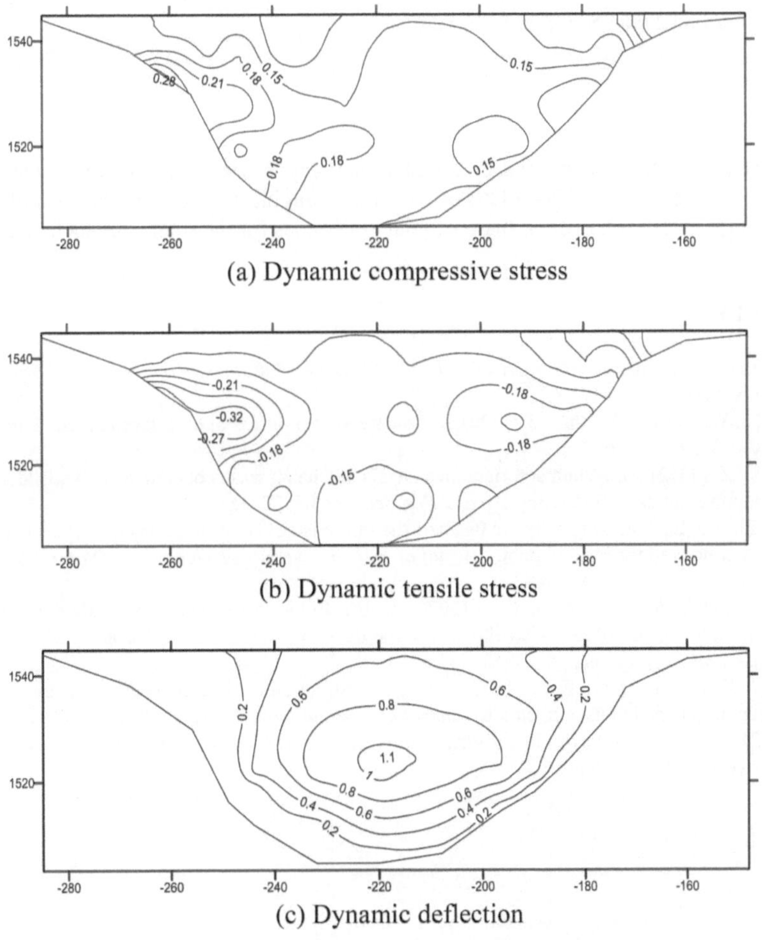

Figure 6. Dynamic stress and deformation distribution of the diaphragm wall.

show that the maximum deflection on the diaphragm wall is located in the center of the cutoff wall and its value is small, which further indicates that the diaphragm wall will not be damaged under the action of earthquake.

4 CONCLUSIONS

Based on the results and discussions presented above, the conclusions are obtained as below:

(1) The dynamic deformation of the dam along the river increases with the increase of the dam height, and the maximum dynamic deformation occurs near the dam crest in the middle of the valley.
(2) The calculated maximum permanent seismic subsidence is about 0.14% of the height of the sandy gravel dam and the thickness of the overburden layer, which is within the normal permanent deformation range.
(3) Because the bottom of dam is well constrained and dam crest is less constrained by basement, hence the maximum stress along slop of panel is located at the middle of the dam bottom, and central of dam crest has the largest deflection.
(4) Due to deep overburden layer can play a good damping effect under the action of earthquake, the stress and deformation on diaphragm wall is very small, which indicate that the diaphragm wall will not be damaged under the earthquake.

ACKNOWLEDGMENTS

This work was financially supported by the National Key Research and Development Program of China (Grant No.2017YFCO404805) of Northwest Engineering Corporation Limited, Power China. In the meantime, we express thanks to our colleagues for their help and technical support.

REFERENCES

Cao, X. X. (2013). *Study on seismic safety of high rockfill dam with earth core on the thick overburden layer*. Wuhan University.

Dang, J. J., Du, W. L., Li, S. P., Zhao, J. F. (2015). Construction progress of high face rockfill dam. *J. Yunnan Water Power*, 31, 71-72+76.

Li, W. & Miao, Z. (2012). Stress and deformation analysis of Chahanwusu concrete face rockfilldam (CFRD) based on monitoring data. *Hydro-Science and Engineering*, 05, 30–35.

Peng, X. F., Zhu, Y. L., Ma, C. (2020). Analysis of the influence of overburden layer thickness on dynamic response of concrete face rockfill dam. *Journal of Hefei University of Technology (Natural Science)*, 43, 98–102.

Wang, Z, J., Liu, H. Y., Meng, T., Zhou, C. G. (2020). Effects of the earthquake input method on the dynamic response of high concrete-faced rockfill dams on the deep overburden layer. *Journal of Water Resources and Architectural Engineering*, 18, 77–81, 111.

Yu, X., Kong, X. J., Zou, D. G., Zhou, C. G. (2018). Construction of time-dependent drought index under changing environment and its application. *Journal of Hydraulic Engineering*, 29, 1378–1395.

Zhang Y. (2021). *Study on seismic fragility of high concrete faced rockfill dam based on support vector machine*. Dalian University of technology.

Advances in Civil Engineering: Structural Seismic Resistance, Monitoring and Detection – Mohd Johari Mohd Yusof and Junwen Zhang (Eds)
© 2023 The Authors, ISBN 978-1-032-31491-4

Influence of bamboo-joint piles to strengthen compact screw pile composite foundation on original piles

Xuemei Wang
China No. 22 MCC Group Corporation, Tangshan, Hebei, China

Jihui Ding* & Zenghui Yu
College of Civil Engineering and Architecture, Hebei University, Baoding, China

Zhipeng Cui
Hebei Zongheng Group Fengnan Iron and Steel Co., Tangshan, Hebei

Zaixing Ma
China No. 22 MCC Group Corporation, Tangshan, Hebei, China

ABSTRACT: In the coastal soft soil area, the material shed compacted threaded pile composite foundation, when the stacking height reaches 5m-6m, the ground subsides, and the grid column appears large deformation, which causes the material shed to be in an over-hazardous state. The slub joints of the slub pile can increase the bearing capacity of the pile side, and the slub pile is used to strengthen the composite foundation of the compacted threaded pile. Design an indoor model test to study the influence of the construction of the bamboo-joint pile on the pile foundation of the shed and the influence of the compacted threaded pile in the composite foundation to guide the construction of the bamboo-joint pile. The test results show that the increase in excess pore water pressure generated during the pile driving process is reduced by 24%–46.3% compared with the pile driving, and the increase in the pressure caused by the pile driving is reduced by 16.26% ~67.37% compared with the pile driving. According to the test, it is determined that the pile is driven by the side of the step flow direction, which reduces the influence of the soil squeezing effect of the bamboo-joint pile on the original compacted threaded pile, side column cap and pile foundation.

1 INTRODUCTION

The engineering properties of the soil in soft soil area are poor, and the pile side resistance provided by the soil is small, which limits the bearing performance of PHC pipe piles. Prestressed high-strength concrete bamboo-joint piles (PHDC piles) have protruding bamboo joints at regular intervals. The bamboo joints can increase the pile side bearing capacity of the pile foundation, thereby increasing the ultimate bearing capacity of the pile foundation (Zhou 2019). The construction of bamboo piles in saturated soft clay will cause soil squeezing effect, causing complex deformations such as settlement, uplift, tilting, and lateral movement of the surrounding ground or buildings, and causing cracks in adjacent building structures or walls. At present, the experimental research on the effect of pile driving and soil squeezing mainly includes two aspects: field test and indoor model test. Compared with the field test, the indoor model test can reduce the interference of many external factors, ensure the accuracy of the test data, and can be repeatedly verified, so it is widely used by scholars at home and abroad.

*Corresponding Author: dingjihui@126.com

DOI 10.1201/9781003310884-7

Li Jingpei, Zhang Shutao, Li Yunong, etc. (Li 2011; Zhang 2009) conducted a pile driving model test of layered soil, and explored the displacement field of the soil near the pile during the process of pile pressing. Li Yunong, Lehane BM et al. (Li 2018) obtained through indoor model tests, the pressure of a single pile in a multi-layer soft clay foundation during the entire process of pile driving, the dynamic end resistance of the pile and the dynamic side resistance of the pile with the penetration depth of the pile The law of change. Chen Wen and Shi Jianyong (Wen 1999) analyzed the spatial distribution of soil displacement and initial excess pore pressure caused by pile driving in saturated clay through indoor model tests. Qian Feng, Liu Ganbin et al. (Feng 2016) conducted static pressure pile driving model tests for open-ended pipe piles, and analyzed the law of lateral pressure, pore pressure, and vertical displacement of the ground surface with the penetration of the pile when the pile was pressed. Xu Jianping, Zhou Jian, etc. (Xu 2000) conducted a pile driving model test in soft clay, and obtained the law of soil displacement with the horizontal and depth directions during pile driving. Jinhua Xia (Xia 2015) et al. studied the influence of the spacing of drainage plates on the dissipation of excess pore pressure when pipe piles were constructed in soft soil foundations. Zhigao Li (Li2008) proved through indoor tests that during pile pressing, the smaller the soil permeability, the more obvious the soil squeezing effect, and the soil squeezing effect is closely related to the construction process. Li Furong and Zhang Yanmei (Zhang 2010) have explored the changes of soil deformation and pore water pressure through the penetration test of static pressure single pile and row of piles. Zhang Jianxin and Lu Qun et al. (Li 2013) analyzed the deformation of the soil near the pile and the variation of the excess pore pressure in the soil with different pile sinking depths through indoor pile group model tests. Existing studies have done more on prestressed pipe piles, but less on special-shaped piles, such as compacted threaded piles and bamboo piles. The effect of composite foundation reinforcement on deep silt sites on the original foundation piles has not been reported. Based on the background of using bamboo-joint piles to reinforce the compacted screw-pile composite foundation of the material shed in coastal soft soil areas, the indoor model test is carried out to compare the pile foundation of the material shed with the original composite foundation during the process of sinking the bamboo-joint pile. The influence of the pile is used to guide the construction of the bamboo joint pile to strengthen the original composite foundation, optimize the construction plan, and reduce the influence on the surrounding pile foundation.

2 MODEL TEST DESIGN

2.1 *Test purposes*

There is a deep silt layer in the storage site, with a thickness of about 10m–13m, in a soft plastic-flowing plastic state, and the stable groundwater level is 1.25m–2.75m. The silos are treated with a composite foundation of compacted threaded piles. The pile spacing is 1.8m×1.8m. The pile body is casted with C20 concrete. The upper part of the pile body is 400mm in diameter and the lower third is 300mm in diameter, and the pile length is 15m.; The scaffolding foundation of the material warehouse adopts the core composite pile foundation, 5 piles and 1 platform, and the buried depth of the platform is 1.6m. The outer periphery is a cement mixing pile with a diameter of 700mm, which is 22m long, and the inner core is a PHC-500AB-125 prestressed pipe pile, which is 20m long. When the material silo was stacked to a height of 5m-6m, the ground subsided, and the maximum subsidence was about 1.30m. The side pillars of the silo had different degrees of axis deviation, and the maximum horizontal displacement reached 680mm. It is proposed to use piled raft foundation for reinforcement, the thickness of the raft board is 0.4m, and C30 concrete is used for pouring. The pile body adopts T-PHC400-B-95 bamboo joint pile, the pile length is 25m~30m. The slub piles are arranged between the original compacted screw pile positions. In order to avoid the influence of large-area bamboo-jointed pile construction on the material shed foundation and the original compacted threaded piles, an indoor model test was designed to study the squeezing stress and excess pore water pressure during the process of pile-driving. There are impacts of compacted piles, undercaps of shed columns and pile foundations. It provides a basis for the construction of bamboo joint piles to reinforce the compacted screw pile composite foundation.

2.2 *Model test chamber*

As shown in Figure 1(a), the model box is made of steel plate with length, width and height of 100cm × 100cm × 100cm, the thickness of the steel plate is 10mm, and the corners are riveted with angle steel. A 30cm sandy soil is placed at the bottom of the test box for drainage and consolidation of the soil layer after the test stratum is set up.

2.3 *Test soil layer*

The test soil was taken from the coastal soft soil area, and the test soil was configured and on-site in the form of soft soil. Stir a certain amount of water and experiment water evenly, pack them in layers, 20cm each, and let each layer stand for 24 hours. From bottom to upper soil layers are: sand layer 20cm, silty clay layer 20cm (water content 45%), silty clay layer 20cm (water content 37%), and soft soil layer 30cm (water content 45%). The soil squeezing experiment was carried out after four weeks of self-weight consolidation.

2.4 *Model pile*

Due to the limitation of the model box size, the vertical similarity ratio is 30 and the horizontal similarity ratio is 16. The model pile is made of pvc pipe. After testing and measurement, the elastic modulus of the pvc pipe is 2410MPa, the Poisson's ratio is 0.3825, and the density is 1.19g/cm^3. Among them, the outer diameter of the compact screw pile (JMP) is 25mm, the pile length is 500mm, and the pile spacing is 4d=112.5mm; the outer diameter of the under-column pile foundation (CJP) is 40mm, and the pile length is 600mm. The bamboo joint pile (ZJP) adopts pvc pipe piles. The bamboo joints are replaced by hard rubber. The length of the protruding part of the bamboo joint is 35mm, the length of the unprotruding part is 30mm, the pile diameter is 20mm, the pile diameter at the bamboo joint is 24mm, and the pile spacing is 112.5mm, the pile length is 750mm.

(a) Model box (b) Model pile (c) Force gauge layout

Figure 1. Indoor test model and layout drawing of earth pressure gauge.

2.5 *Measuring equipment and test elements*

The fiber optic pore water pressure sensor is model HC-25, and the collector is model HCSC-32. The pile side squeezing earth pressure adopts a pressure sensing model, the pressure measurement range is 0–150kPa, and the collector is a static strain gauge of Utech. As shown in Figure 1(c), two fiber optic pore water pressure sensors are arranged in the soil layer of the model box, with buried depths of 35cm and 70cm. Five earth pressure gauges are arranged along different depths on the five pile foundation piles and the original compacted threaded piles, and 4 earth pressure gauges are arranged on the bamboo joint piles.

The construction flow of pile sinking is shown in Figure 2.

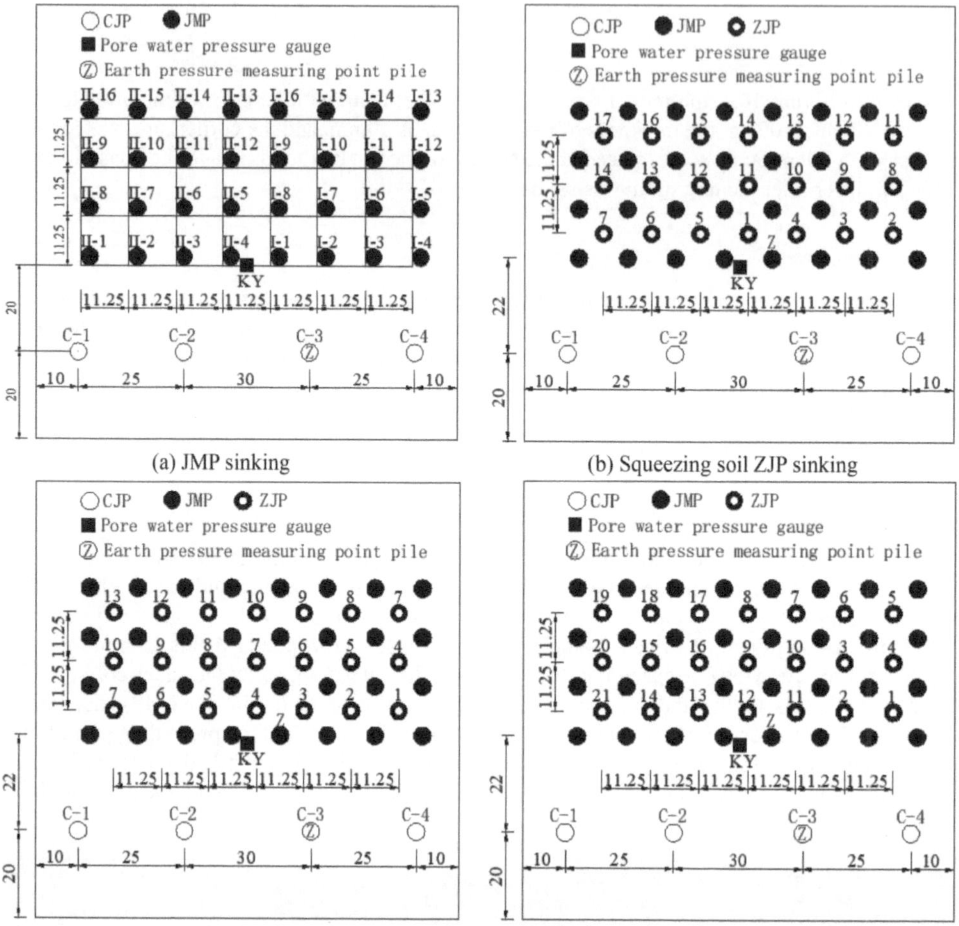

(a) JMP sinking

(b) Squeezing soil ZJP sinking

(c) Non-squeezing soil ZJP step-sequential sinking

(d) Non-squeezed ZJP are sinking from right to left

Figure 2. Construction flow direction of pile driving.

3 TEST RESULTS AND ANALYSIS

3.1 *Influence of sinking bamboo-joint piles on pore water pressure and earth squeezing pressure*

In order to reduce the impact of the sinking of bamboo piles on the original foundation piles of the storage warehouse, the pile sinking, pilot hole and re-sinking tests were carried out in the model box to study the impact of the sinking pile on the original foundation piles. Figure 3 shows the variation of pore water pressure and horizontal squeeze pressure before and after the construction of the pilot hole of the bamboo pile at different positions. The pilot hole depth in this experiment is 60cm. Figure 3 shows the law of pore water pressure change before and after the pilot hole, the pilot hole diameter is 20mm, and the hole depth is 60cm. Figure 3 shows the peak value of excess pore water pressure caused by pile sinking and re-sinking. From Figure 3 and Table 1, it can be seen that the pore water pressure generated during the sinking process of the soil-squeezed bamboo-joint pile after the pilot hole is lower than the pore water pressure of the soil-squeezed bamboo-joint pile. The construction of squeezing soil and sinking piles produces excess pore water pressure and buried depth, and the distance between the bamboo pile and the measuring point is related to the diameter ratio of the bamboo pile. Compared with squeezing bamboo piles, the peak values of excess pore water pressure at the buried depths of *z*=35cm and *z*=70cm are reduced by 24.0–30.0% and 31.0%–46.3%, respectively.

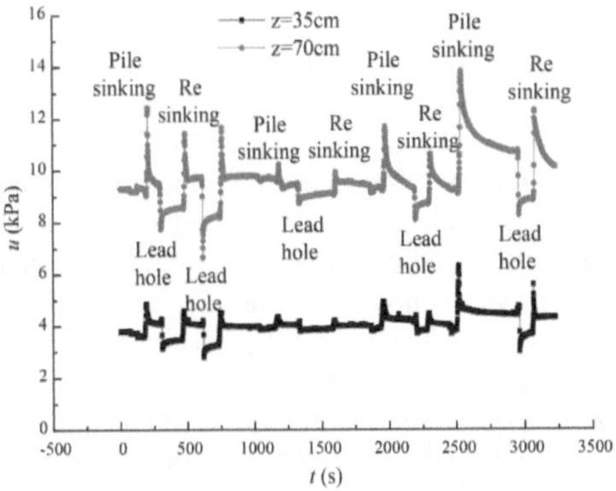

Figure 3. Change in pore water pressure.

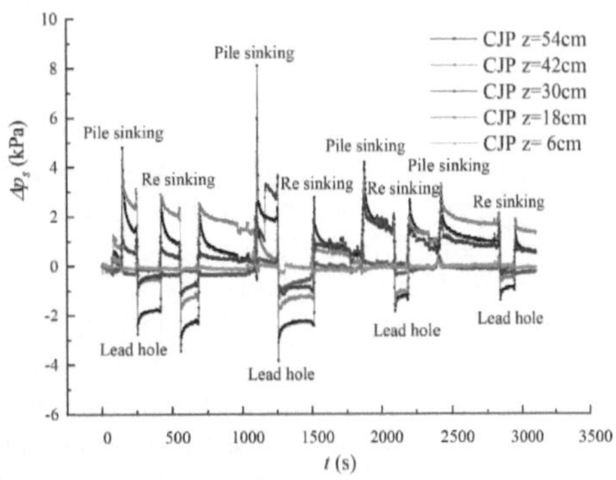

Figure 4. Incremental change in earth pressure.

Table 1. Peak value of excess pore water pressure before and after the lead hole.

	a		B		c	
Diameter ratio	*z*=35cm	*z*=70cm	*z*=35cm	*z*=70cm	*z*=35cm	*z*=70cm
3.75	2.53	4.50	1.80	3.01	28.85	33.11
4.17	1.09	3.14	0.82	2.14	24.77	31.85
6.25	1.15	2.40	0.70	1.29	39.13	46.25
8.33	0.60	0.93	0.45	0.63	25.00	32.26

Note, a: Pile sinking (squeezing soil) excess pore pressure (kPa)
b: Excess pore pressure of pile sinking after lead hole (kPa)
c: Reduced excess pore pressure before and after the pilot hole (%)

Table 2. Peak squeezing pressure increment before and after the lead hole.

Diameter	*z*=54cm			*z*=42cm			*z*=30cm			*z*=18cm		
ratio	d	e	f	d	e	f	d	e	f	d	e	f
4.80	8.02	2.58	67.83	4.56	2.50	45.18	3.26	2.73	16.26	1.98	1.20	39.39
7.20	4.68	2.58	44.87	4.30	2.91	32.33	1.67	1.23	26.35	0.37	0.24	35.14
8.00	4.09	2.78	32.03	3.29	1.98	39.83	2.73	1.67	18.83	0.95	0.37	61.05
12.00	3.10	1.72	44.52	3.29	1.87	43.16	1.93	0.88	54.40	.39	–	–

Note, d: Increment of earth pressure during pile driving (squeezing the soil) (kPa)
e: Increment of soil pressure of pile sinking after pilot hole (kPa)
f: The increase of earth pressure before and after the lead hole decreases (%)

Figure 4 and Table 2 show the influence of the construction of the bamboo-joint pile before and after the pilot hole on the earth pressure at the measuring point on the foundation pile under the column. It can be seen from Figure 4 and Table 2 that when the bamboo-joint pile is constructed, as the horizontal distance between the bamboo-joint pile and the measuring point increases, the horizontal earth pressure caused on the foundation pile is significantly reduced. Table 2 Squeeze piles and lead holes 60cm and then sink piles, so after 60cm, the increase in soil pressure caused by construction is reduced by 16.26% to 67.37% compared with the soil squeezing piles.

3.2 *Influence of sinking pile flow direction on pile foundation of material shed*

Compare the influence of the soil compaction construction of the bamboo pile pile and the non-squeeze soil construction of the pilot hole on the original compacted pile and the foundation pile under the column. Figure 2 (a) uses pipe piles to simulate the original compacted screw pile composite foundation. Figure 2 (b) shows the construction of soil-squeezed bamboo piles. The first row is constructed first, and then the second row of bamboo piles is constructed in a stepwise manner. Figure 2 (c) Three rows of bamboo piles are constructed by leading holes and piles in a ladder form; Figure 2 (d) Three rows of bamboo piles are driven by holes and piles. The construction direction is advancing from right to left. Arrange 1~5 five horizontal earth pressure measuring points on pile I-1, and arrange 6-10 five horizontal earth pressure measuring points on C-3.

Figure 5 shows the pile-sinking sequence of pipe piles (squeezing soil) reinforced with bamboo-jointed piles. According to the pile-sinking sequence of Figure 2(b), the horizontal squeezing earth pressure changes with time when 21 bamboo-jointed piles are driven. No. 1 and No. 4 bamboo joint piles are closest to the 1-1 pipe pile. The maximum horizontal squeezing pressure is 4.58kPa, and the distance is 35cm from the soft soil surface; the maximum squeezing pressure is 0.5kPa~1.82kPa after construction. The maximum squeezing earth pressure at the measuring point C-3 is at z=42cm, and its value is 7.82kPa. After the construction is completed, the earth pressure at the measuring point C-3 is 0.5kPa~2.81kPa.

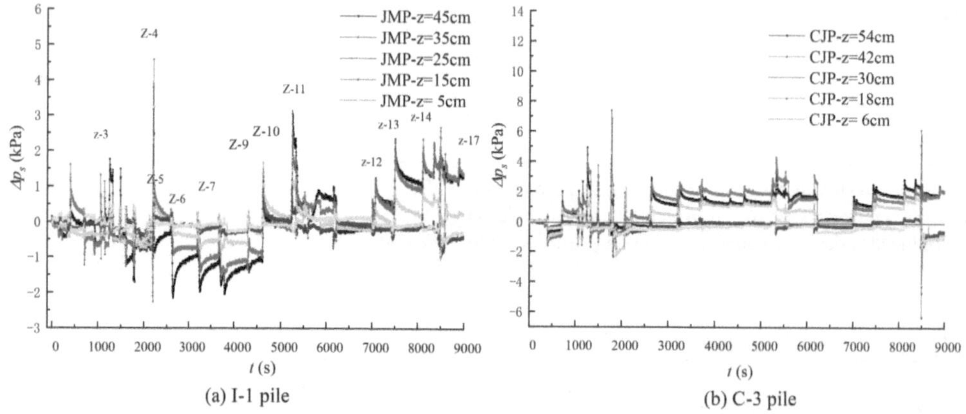

Figure 5. Change of soil pressure increment during soil-squeezing bamboo-joint pile sinking.

Figure 6 shows the sinking of the non-extruded bamboo-joint pile with the lead hole, according to Figure 2 (c) the pile sinking sequence of the lead hole and the pile sinking. After the pilot hole is constructed, the maximum value of the measuring point on the 1-1 pile is 3.36kPa, which is 26.64% less than that of the soil squeezing pile; the maximum soil squeezing pressure after the end of the pile sinking is less than 0.29kPa, which is much smaller than that of the soil squeezing pile. The maximum earth squeezing pressure at the C-3 measuring point is 2.08kPa, which is 73.4% lower than that of squeezing piles. The maximum earth squeezing pressure after construction is 1.29kPa, which is 54.1% lower than that of squeezing piles.

Figure 7 shows the sinking of the non-squeezing bamboo-jointed pile with the lead hole. According to the sequence of Figure 2(d), the pile is driven from right to left. It can be seen from Figure 8 that the horizontal soil squeezing pressure at the measuring point 1-1 gradually increases with

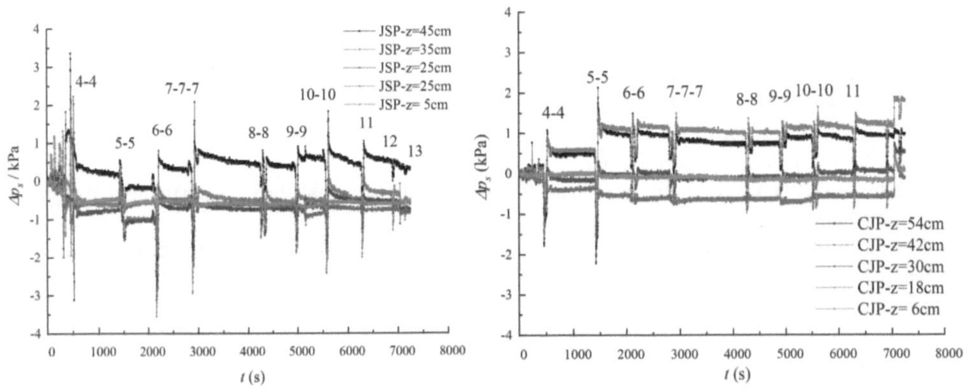

(a) I-1 pile (b) C-3 pile

Figure 6. Incremental change of soil pressure during pile-sinking in step sequence.

time. When the construction reaches 12, the maximum soil squeezing pressure is at the position of z=54cm, and its value is 6.15kPa. The maximum earth squeezing pressure on C-1 is 1.92kPa, which is much higher than the earth squeezing pressure of the stepped construction in working condition 3. The main reason is that the earth squeezing pressure generated by the construction of adjacent bamboo piles has not been dissipated, and the construction of the next pile is started, resulting in The squeezing pressure gradually increased.

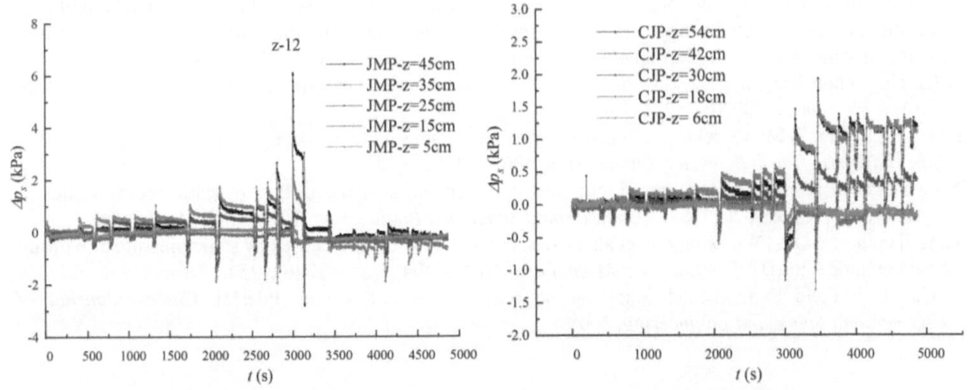

(a) I-1 pile (b) C-3 pile

Figure 7. Incremental changes of soil pressure during pile-sinking in order from there to the left.

Therefore, the lead hole can reduce the excess pore water pressure and the horizontal squeezing earth pressure generated during the pile sinking. The best condition for the lead hole effect is that the adjacent bamboo-joint piles must have a sufficient distance and sufficient time when the pile sinks. Stepped construction just meets these conditions, so that the impact on the original building pile and the original composite foundation is minimal when the pile sinks, and the stepped lead-through pile sinking construction can be considered as a short construction period.

4 CONCLUSION

In this paper, by designing a model test, monitoring the pore water pressure and earth pressure during the model test, the following conclusions are obtained.

(1) The non-extruded bamboo-jointed piles are driven by the lead hole, and the flow direction of the piles adopts a stepped sequence to have the least impact on the original compacted threaded piles and the foundation piles under the column.
(2) Compared with non-introduced holes, the increase in pore water pressure is reduced by 24%–46.3%, and the increase in earth pressure is reduced by 16.26%–67.37% in the process of pile-sinking with lead-through holes.
(3) Indoor model test due to the limitation of the size of the model box, experimental tests are required during on-site construction.
(4) During the on-site construction process, it is necessary to continue to monitor changes in pore water pressure and earth pressure.

REFERENCES

Chen Wen, Shi Jianyong, Gong Youping, et al. Centrifugal model test study on soil squeezing effect of static pressure pile in saturated clay[J]. *Journal of Hohai University (Natural Science Edition)*, 1999(06): 103–109.

Jiajin Zhou, Rihong Zhang. Research on compressive bearing performance of prestressed bamboo pile and pipe pile in soft soil area[J]. *Journal of Tianjin University (Natural Science and Engineering Technology Edition)*, 2019, 52(S1), 9–16.

Jinhua X, Guanghui J, Zhongliang L. Experimental Study on Soil Squeezing Effect of Pipe Pile in Soft Soil Subgrade[J]. *Construction Technology*, 2015.

Li Furong, Zhang Yanmei, Wang Zhaoyu. Model test study on the squeezing effect of static pressure piles in soft soil[J]. *Building Science*, 2013, 29(01): 52–54+19.

Li Jingpei, Li Yunong, Zhang Shutao. Experiment on soil squeezing effect of single static pressure pile in layered foundation[J]. *Journal of Tongji University (Natural Science Edition)*, 2011, 39(06): 824–829.

Qian Feng, Liu Ganbin, Qi Changguang, et al. Model test and numerical simulation of static pressure pile driving in saturated clay[J]. *Hydrogeology and Engineering Geology*, 2016, 43(05): 56–61+69.

Xu Jianping, Zhou Jian, Xu Chaoyang, et al. Model test study on soil squeezing effect of pile driving[J]. *Rock and Soil Mechanics*, 2000(03): 235–238.

Yunong Li, Lehane B M. Model test of pile driving characteristics in double-layered kaolin clay[J]. *Journal of Jilin University (Earth Science Edition)*, 2018, 48(06): 1778–1784.

Zhang Shutao, Li Jingpei, Li Yunong. Model test study on soil squeezing effect of static pressure piles in layered soil[J]. *Chinese Journal of Underground Space and Engineering*, 2009, 5(S2): 1557–1561.

Zhang Jianxin, Lu Qun, Wu Dongyun, et al. Analysis of soil deformation caused by static pressure pile group based on model test[J]. *Rock and Soil Mechanics*, 2010, 31(04): 1243–1246+1252.

Zhi-Gao L I. Field Experimental Study on Squeezing Effect of Mixing Piles[J]. *Chinese Journal of Underground Space and Engineering*, 2008.

Advances in Civil Engineering: Structural Seismic Resistance, Monitoring and Detection – Mohd Johari Mohd Yusof and Junwen Zhang (Eds)
© 2023 The Authors, ISBN 978-1-032-31491-4

Quantitative characterization and research on the genesis of microstructure of loess particles of different strata in LuoChuan, Shaanxi, China

Yupeng Chang*
Power China Northwest Engineering Corporation Limited, Xi'an, China

Shaoqing Yuan
School of Geological Engineering and Geomatics, Chang'an University, Xi'an, China

ABSTRACT: Due to the regionality of geological environmental conditions, climate diversity and near periodicity of climate change in the Loess Plateau, the microstructure of loess in a typical stratum often has regional characteristics in statistics, and the microstructure differences between different loess strata are also very significant. Therefore, this paper carried out micro experiments on loess from different strata in LuoChuan, and established a set of quantitative parameter indexes of loess particle microstructure based on mature and scientific research methods of loess microstructure. Through the quantitative characterization of particle parameter indexes, the results show that probability density of equivalent diameter meets Rational distribution and that probability density of sphericity meets Beta distribution well in each stratum. With the increase of stratum depth, LuoChuan Loess shows the variation characteristics of particle size gradually increasing, sphericity gradually decreasing. The Loess particles in deeper strata are not easy to be deposited vertically, but tend to be deposited gently or horizontally, and most of the particles in each stratum are distributed in the northeast. At the same time, the genesis of LuoChuan loess is discussed, and it is basically concluded that the material source of LuoChuan loess is the same, and the carrying force is the main factor causing great differences.

1 INTRODUCTION

The microstructure characteristics of loess are the most fundamental factors controlling the structural properties of loess, which directly affect the engineering properties and macro-mechanical behavior of loess (Wang 2019; Zhang 2018). For a long time, scholars have gradually noticed the importance of microstructure to solve loess engineering problems (Zhu 1959), and have done a lot of research on the microstructure of loess.

Early scholars such as Zhu Haizhi (1963) observed the microstructure of Malan Loess in 11 places in the middle reaches of the Yellow River by using the "oil immersion light sheet method", and classified the loess in detail according to the particle shape, particle size and content of loess and the relationship between clastic particles and cement. Zhang Zonghu (1964) researched the particle composition, shape, contact relationship, pore characteristics and cementation characteristics of natural loess by slicing and polarizing microscope observation, and analyzed the regional characteristics of the microstructural types of loess soil in China.

With the continuous development of observation techniques and image processing technology, such as the application of scanning electron microscope(SEM), mercury intrusion method and

*Corresponding Author: 2018126092@chd.edu.cn

DOI 10.1201/9781003310884-8

computed tomography(CT), scholars have conducted more specific and detailed research on the microstructure characteristics of loess, including particle morphology, pore distribution, particle contact, connection and arrangement, and achieved remarkable results (DERBYSHIRE 1988; Deng 2010, 2020; Hu 1999; Lei 1987; Tang 2003; Wei 2020; Wei 2019; Yan 2018). At the same time, the research results mainly focused on the two-dimensional microstructure characteristics of loess, and few researchers have studied the three-dimensional microstructure characteristics and quantification of loess.

At present, scanning electron microscope, optical microscope, CT technology and nuclear magnetic resonance are commonly used to study the microstructure of loess. The three-dimensional microstructure of loess can be obtained by high-resolution scanning of optical microscope combined with continuous section and three-dimensional image reconstruction technology; NMR is more suitable for detecting loess pore water; Although scanning electron microscopy and CT scanning can obtain high-resolution images, they are expensive and difficult for most researchers to obtain. Therefore, considering the economic and technical applicability, this paper adopts the optical microscope with low cost and simple operation as the technical means of this study Under comprehensive consideration, this paper selects eight stratum samples of LuoChuan loess section L_1, S_1, L_2, S_2, L_6, L_7, L_8, L_9 and obtains the physical and mechanical parameters of the loess through field experiments and indoor experiments. Based on the continuous slicing and three-dimensional image reconstruction technology of optical microscope, the particle structure distribution characteristics of loess in different stratum are systematically analyzed.

2 SAMPLING AND EXPERIMENT SCHEME

LuoChuan loess section is located on the LuoChuan Loess Plateau in the center of the Loess Plateau. The Loess thickness can reach 138-150m. It has become a typical section in the loess stratum because of its complete stratigraphic sequence and good exposure. In the process of sampling loess from different strata in LuoChuan, due to the destruction of the sections, the sampling difficulty and risk increased, and considering the representativeness of the samples, the loess samples from L_1, S_1, L_2, S_2, L_6, L_7, L_8 and L_9 strata were taken. Indoor physical and mechanical tests were carried out on loess samples from different strata to obtain their physical and mechanical parameters. The density, liquid-plastic limit and shear strength of loess samples in each stratum are shown in Table 1. The cohesion C and internal friction Angle φ were obtained by direct shear t experiment, and the collapsibility coefficient was obtained by double-line experiment.

Table 1. Physical parameters of loess samples from different strata in LuoChuan.

Sampling site	Unit mass γ (g/cm^3)	Liquid limit L_L (%)	Plastic limit P_L (%)	Permeability coefficient k (cm/s)	Cohesion C (kPa)	Internal friction φ (°)	Collapsibility coefficient δ_s
L1	1.32	16.2	11.7	1.43×10^{-4}	63.50	23.53	0.027
S1	1.48	28.4	21.0	4.63×10^{-5}	78.53	15.63	0.017
L2	1.35	30.6	20.8	4.01×10^{-5}	54.20	20.61	0.052
S2	1.48	31.9	21.6	4.34×10^{-5}	101.75	18.36	0.027
L6	1.62	30.7	20.9	3.77×10^{-5}	61.68	31.67	0.015
L7	1.43	31.5	22.1	3.68×10^{-5}	62.29	29.75	0.033
L8	1.54	32.1	22.5	3.49×10^{-5}	37.59	32.07	0.018
L9	1.51	32.6	23.6	3.24×10^{-5}	80.55	30.56	0.026

The continuous slicing method based on optical microscope is used to study the three-dimensional microstructure of loess. The experiment and image reconstruction process (Figure 1) mainly includes the following steps. (1) Sample preparation. The loess sample is cut into a cylinder with a diameter of 1 cm and a height of 1.5 cm, and then saturates with a soaking solution in a vacuum environment (Wang 2010). Finally, sample is prepared using secondary casting molds for polishing and observation. (2) Image acquisition. The sample is polished with Multiprep equipment to make the surface of the sample after secondary casting smooth. The polishing process can be controlled to an accuracy of 0.1μm. The planar image resolution of the sample observed with Leica DM6000M is 0.4μm/pixel. The sample is then repolished to remove about 2μm, which can be controlled using a vertical displacement laser monitor and polishing duration, and then a series of two-dimensional images with a spacing of 2μm are obtained. (3) Three dimensional (3D) structure reconstruction. Use AVZIO software to align all two-dimensional images and select appropriate thresholds by changing the gray level to extract two-dimensional microstructures such as particles and pores, and then construct the morphology of three-dimensional particles and pores. (4) Quantitative analysis (Deng 2021). Basing on the results of experiment and image reconstruction, 3-D microstructure characteristics of loess particle is then obtained, with which many important parameters regard to volume, morphological, orientation and etcetera are then adopted to depict the loess particles characteristics quantificationally. The volume characteristics can be presented using the equivalent

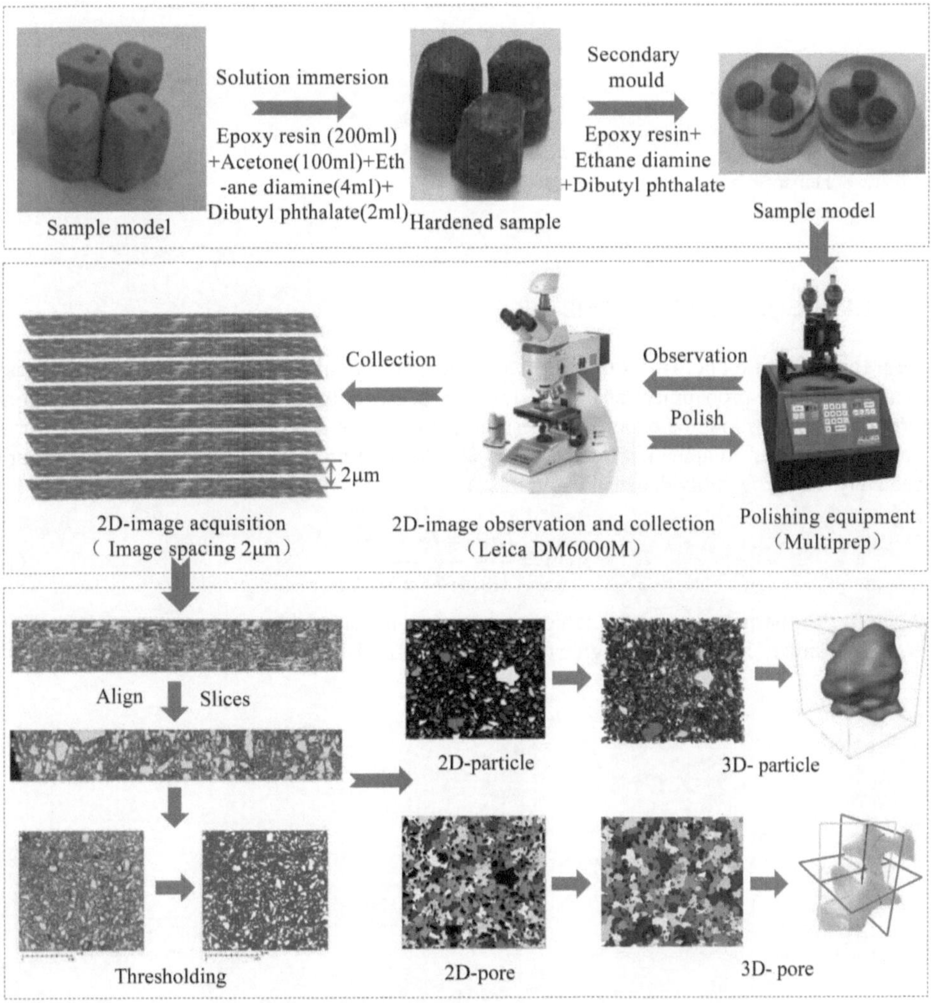

Figure 1. Experiments and model construction process

diameter (Eq-D), which can be defined as Eq-D$=\sqrt[3]{\frac{6\times V}{\pi}}$, in which the parameter V means the volume of loess particle. The morphological characteristics can be depicted using the sphericity, and the sphericity is defined as ratio of the surface area of spheres with the same volume as loess particles to the surface area of loess particles. The orientation characteristics can be depicted using the parameters Phi and Theta which represents the dip angle and strike angle of particle respectively. Definition of the characterization parameters for loess particle is presented in Figure 2.

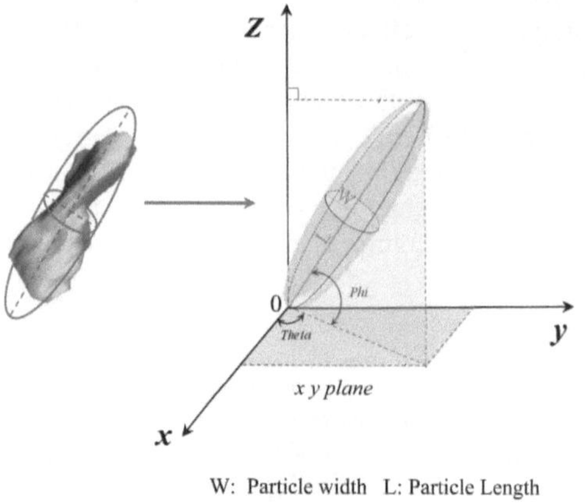

Figure 2. Schematic diagram of particle parameter index.

3 PARTICLE EQUIVALENT DIAMETER

There are obvious differences in particle equivalent diameter composition and content between loess and paleosol, but the probability density of equivalent diameter of loess meets rational distribution, Figure 3(a) and (b) show the experiment results, fitting curve and cumulative probability curve of particle equivalent diameter probability density distribution in L_1 stratum and each stratum respectively. The probability density function is:

$$f(x)=\frac{(ax^2+bx+c)}{(x^2+dx+e)} \tag{1}$$

Where x is the particle equivalent diameter, and a, b, c, d, e are the fitting parameters. According to the fitting situation, the fitting degree of each stratum is above 0.95.

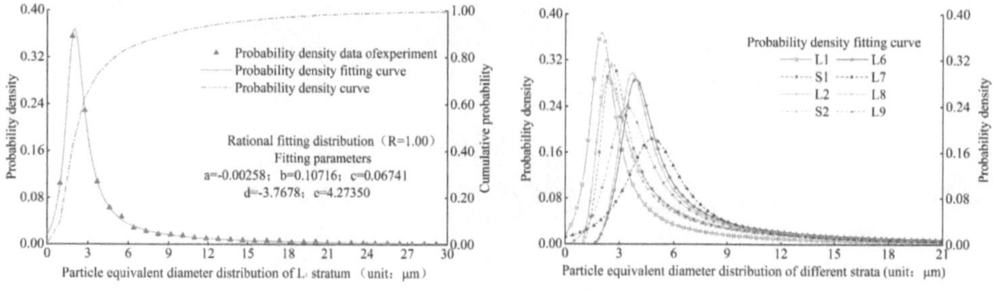

(a) (b)

Figure 3. Probability density distribution of particle equivalent diameter.

During the long deposition process of loess, the particles size of loess in different strata has a certain variation characteristics. Based on the micro scale study of loess particles in the above strata, it can be found that the loess particles tend to coarsen with the increase of stratum depth in Figure 3(b). The change of loess particles size in different strata is not only related to the particles sizes of original loess parent material and the agent of transported material, but also related to the change of climate conditions and weathering during the formation of strata to a certain extent. The main material source of the Loess Plateau comes from the desert of Inner Mongolia next to Mongolia in the northwest. Based on this, it can be inferred that the material source of each stratum is basically the same. Therefore, the material carrying force has become the main reason for the difference of loess particles size in the above strata. The particles size of L_6, L_7, L_8 and L_9 strata is large, so it can be inferred that there was strong carrying wind and dry and cold climate at that time, and L_8 and L_9 strata experienced significant physical weathering, and the particles size is slightly smaller than that of L_6 and L_7 strata; The particles size of L_1, S_1, L_2 and S_2 strata is small, so it can be inferred that the carrying wind force was relatively weakened at that time and suffered different degrees of weathering during carrying process.

4 PARTICLE MORPHOLOGY

According to the distribution characteristics of particles sphericity probability density, the sphericity experiment data of loess particles in L_1-L_9 strata of LuoChuan can be well fitted by using beta distribution. Figure 4 (a) and (b) show the distribution of particles sphericity probability density and fitting curve in L_1 and each stratum. The probability density fitting function is:

$$f\left(x;a,b\right)=\frac{1}{B\left(\alpha,\beta\right)}x^{\alpha-1}(1-x)^{1-\beta} \tag{2}$$

Where x is sphericity, and α and β are fitting parameters. According to the fitting situation of the experiment results at each sampling point, the fitting degree of each stratum is above 0.95.

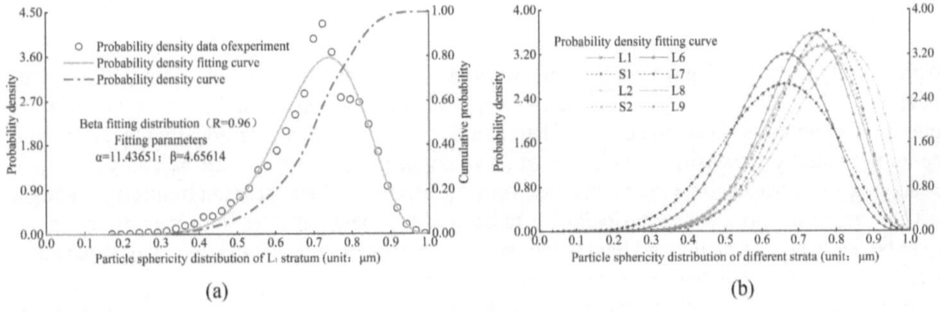

Figure 4. Probability density distribution of particle sphericity.

Comparing the fitting distribution curves in Figure 4(b), it can be concluded that the overall particle ssphericity shows the characteristics of gradual decrease with the increase of stratum depth. It can be inferred that most of loess materials forming L_6, L_7, L_8 and L_9 strata such as silts were suspended in the air during the carrying of strong wind force, resulting in relatively few abrasion times between them. At the same time, medium sand, fine sand and other materials were carried to the vicinity of loess accumulation remotely by leaps and bounds under the carrying of strong wind force. The loess materials forming L_1, S_1 L_2 and S_2 strata such as silts were suspended close to the surface in the air and carried under the carrying of weak wind force, so the materials were constantly abraded. At the same time, some medium sands and fine sands carried from a long distance and near the loess accumulation were carried twice or repeatedly under the carrying of weak wind force, and finally loess particles with high roundness and poor sorting were formed.

5 PARTICLE ORIENTATION

Figure 5 (a) shows the percentage distribution curve of particle Phi angle in L_1-L_9 strata of LuoChuan. The change of particle Phi angles in L_1-L_9 strata shows similar increase and decrease characteristics. Comparing the quantity percentage of Phi angles in the range of 0–50°, it shows that the quantity percentage of deeper stratum is always higher than that of shallow stratum, which indicates that the particles in deeper strata of loess are not easy to be deposited vertically and tend to be deposited gently or horizontally. Figure 5(b) shows the percentage distribution curve of particle Theta angle in L_1-L_9 strata of LuoChuan. The variation trend of Theta angle of particles in L_1-L_9 strata is basically the same. Except for L_6 stratum, the dominant angles of other strata are concentrated between 120°-160°. When the positive direction of Y axis is due south, most of the particles in each stratum are distributed in the northwest direction.

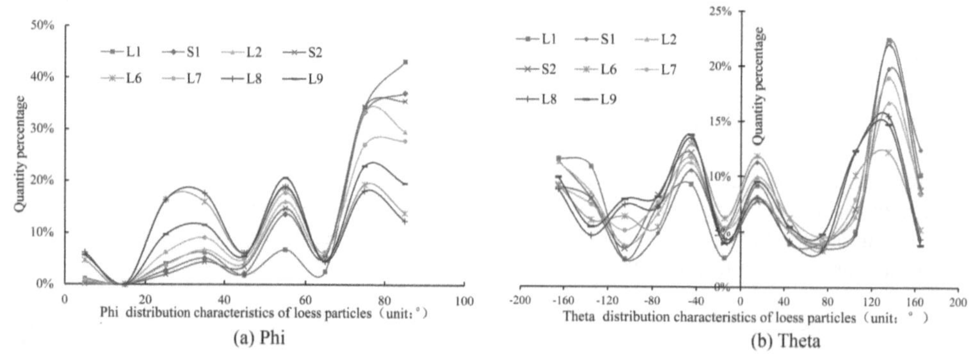

Figure 5. Particle directional angle distribution.

6 CONCLUSION

The probability density of equivalent diameter of LuoChuan loess particles meets Rational distribution, and the probability density of sphericity meets Beta distribution. With the increase of stratum depth, LuoChuan loess shows the change characteristics of particles size gradually increasing and sphericity gradually decreasing, which shows that when the material sources are basically the same, the carrying force has become the main cause for this difference. The variation trends of Phi angle and Theta angle of particles are basically the same, and the loess particles in deeper strata are not easy to be deposited vertically and tend to be deposited gently or horizontally, while most particles in each stratum are distributed in the northeast direction.

The regional distribution characteristics of loess particle microstructure provide a certain scientific basis for the study of the genesis of loess and the evolution of paleogeological environmental conditions. L_6, L_7, L_8 and L_9 strata were formed in the prevailing period of dry and cold climate and strong wind, and L_1, S_1, L_2 and S_2 strata were formed relatively in the period of humid and hot climate and weak wind.

REFERENCES

Deng Jin, Wang Lanmin, Zhang Zhenzhong, et al (2010). Microstructure characteristics and forming environment of late Quaternary Period loess in the Loess Plateau of China. *J. Environmental Earth Sciences*. 59: 1807–1817.

Deng Longsheng, Fan Wen, Liu Shaopeng, et al (2020). Quantitative Research and Characterization of the loess Microstructure in the Bai Lu Tableland, Shaanxi Province, China. *J. Advances in Civil Engineering*. 2020: 1–14.

Deng Longsheng, Fan Wen, Chang Yu-peng, et al (2021). Microstructure Quantification, Characterization, and Regional Variation in the Ma Lan Loess on the Loess Plateau in China. *J. International Journal of Geomechanics.* 21(8): 1–15.

E. Derbyshire, T.W. Mellors (1988). Geological and geotechnical characteristics of some loess and loessic soils from China and Britain: A comparison. *J. Sci. Engineering Geology*. (25): 135–175.

F. Zhang, R. Kong, and J. Peng (2018). Compositional, structural, and physicochemical properties of loess under laboratory conditions. *J. Applied Clay Science*. 152(1): 259–266.

Hu Ruilin, Guan Guolin, Li Xiangquan, et al (1999). Microetructure effect on the subsidence of Loess. *Journal of Engineering Geology*. 7(02): 161–167.

J. Wang, P. Li, Q. Gu, Y. Xu, and T. Gu (2019). Changes in tensile strength and microstructure of loess due to vibration. *Journal of Asian Earth Sciences*. 169(1): 98–307.

Lei Xiangyi, Wang Shujie (1987). Size of Loess Pores in Relation to Collapsibility. *J. Hydrogeology & Engineering Geology*. (5): 15–18.

TANG Liansheng (2003). Synthetic Effect of Microstructure and Uneven Suction on Loess Subsidence. *Journal of Engineering Geology*. 11 (1): 30–35.

Wang Mei (2010). *Study on Structure of Collapsible Loess in China*. Taiyuan University ofTechnology. Taiyuan.

Wei TT, Fan W, Yu NY, et al (2019). Three-dimensional microstructure characterization of loess based on a serial sectioning technique. *J. Engineering Geology*. pp. 261: 256.

Wei Ya-ni, Fan Wen, Yu Bo, et al (2020). Characterization and evolution of three-dimensional microstructure of Malan loess. *J. Catena.* 192: 1–14.

Yan Kai, Gu Tianfeng, Wang Jiading, et al (2018). A study of the micro-configuration of loess based on micro-CT images. *J. Hydrogeology & Engineering Geology*. 45(3):71–77.

Zhang Zonghu (1964). Study on Microstructure of loess like loess in China. *J. Scientia Geologica Sinica*. (03): 113–125.

Zhu Haizhi, Gao Fengying, Pei Jingxian (1959). Analysis of loess particle composition by microscope. *J. Scientia Geologica Sinica*. 3(1): 90–92.

Zhu Haizhi (1963). Some characteristics of particle and structure of Malan Loess in the middle reaches of the Yellow River —Observation results of oil immersion light sheet method. *J. Scientia Geologica Sinica*. (02): 36–48.

Advances in Civil Engineering: Structural Seismic Resistance, Monitoring and Detection – Mohd Johari Mohd Yusof and Junwen Zhang (Eds)
© 2023 The Authors, *ISBN 978-1-032-31491-4*

Analysis on dynamic response simulation and explosion resistance of blast-wall with masonry sandwich steel plate

Zongbo Hu*
Institute of Equipment Management and Support, Engineering University of People's Armed Police Force, Xi'an, China
Postdoctoral Research Station of Civil Engineering College, Xi'an University of Architecture and Technology, Xi'an, China

Juan Zhao
Institute of Equipment Management and Support, Engineering University of People's Armed Police Force, Xi'an, China

ABSTRACT: Due to the low design bearing capacity, large explosion impact surface and poor flexural tensile performance of masonry structure, brittle fracture failure is easy to occur under explosion impact load. For the purpose of anti explosion protection transformation of existing walls, the numerical simulation method is used to study the unreinforced and masonry sandwich steel plate concrete composite reinforced infilled walls, and the dynamic responses of the two walls under explosion impact are qualitatively analyzed and compared. The results show that the protective effect of steel plate sandwich blast-wall is related to the core material. The higher the strength of steel plate and the greater the thickness, the better the protective effect; Compared with reinforced concrete reinforced masonry wall, steel plate sandwich masonry wall has better explosion resistance; When the steel plate thickness of the steel plate sandwich blast-wall is less than 20mm, the maximum plastic displacement of the wall is greater than 0.015m; when the steel plate thickness is greater than 30mm, the maximum plastic displacement of the wall is less than 0.008m; When the thickness of steel plate is less than 10mm, the central speed of explosion point is greater than 58m/s; when the thickness of steel plate is greater than 30mm, the central speed of explosion point is less than 30m/s; Due to the improvement of ductility by sandwich steel plate, the damage of steel plate sandwich masonry wall is obviously better than that of reinforced concrete reinforced masonry wall.

1 INTRODUCTION

For buildings with explosion risk such as ammunition depot and dangerous goods warehouse, explosion-proof wall is usually considered to be set for protection, or laminated plate is used for reinforcement. Laminated slab wall has the advantages of fast construction speed, large overall stiffness and good seismic performance. In recent years, many scholars have done a lot of research on it. The research and experimental results show that the establishment of explosion-proof wall in front of the building can effectively reduce the direct effect of explosion load, so as to reduce the degree of building damage and casualties. Compared with the explosion-proof wall in front of the building, the laminated plate wall has better protective effect and is an important protective measure against explosion (Baylot 2005; Cheng 2010; Mohammed 2011; Johnson 2004). Through experimental research, the passive explosion-proof system should be used to protect the masonry structure in order to improve the explosion-proof performance of the enclosure structure (Galal 2010; Hongrui 2019; Xu 2020; Zhigang 2020). The steel plate, concrete and the original masonry

*Corresponding Author: huzongbo_1985@163.com

DOI 10.1201/9781003310884-9

wall are stacked together to strengthen the structure. Because the reinforcement method has the advantages of steel and concrete at the same time, it has the characteristics of high shear bearing capacity, good ductility and strong energy consumption capacity. It can be used as the transformation direction of masonry explosion-proof wall.

In this paper, for the purpose of anti explosion protection transformation of existing walls, the numerical simulation method is used to study the unreinforced and sandwich steel plate concrete composite reinforced masonry infilled walls, and the dynamic responses of the two walls under explosion impact are qualitatively analyzed and compared.

2 NUMERICAL ANALYSIS MODEL

2.1 *Structural model*

In this paper, the unreinforced masonry infilled wall and composite masonry explosion-proof wall based on steel plate and concrete are studied. The dynamic response analysis of masonry wall under explosion load is carried out by using dynamic nonlinear finite element analysis software ABAQUS. The numerical simulation of steel plate concrete composite masonry explosion-proof wall mainly involves the following components: masonry wall, sandwich steel plate, reinforcement mesh, cast-in-situ concrete, TNT explosive and air medium, as shown in Figure 1.

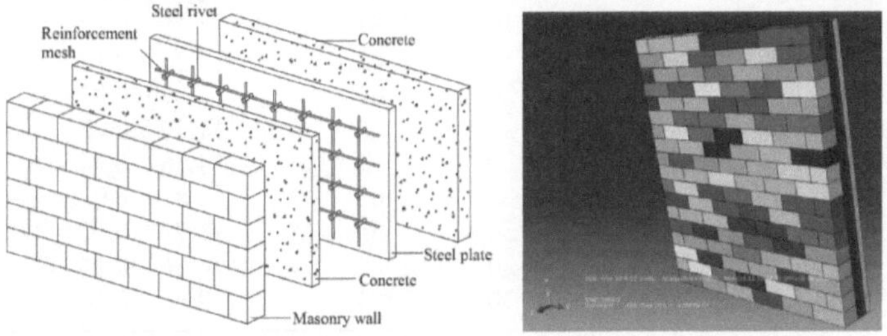

(a) Steel plate sandwich masonry wall (b) Finite element model

Figure 1. Structural model.

2.2 *Explosion load and working condition*

In order to simulate ammunition explosion, TNT explosives of different orders are placed on the ground where the axis of symmetry is located at 0mm, 15mm and 45mm away from the wall surface (the distance from the center of mass of explosives to the surface of masonry wall). The proportional distance Z ($Z = R/W1/3$, where R is the distance/m) and W is the explosive mass,

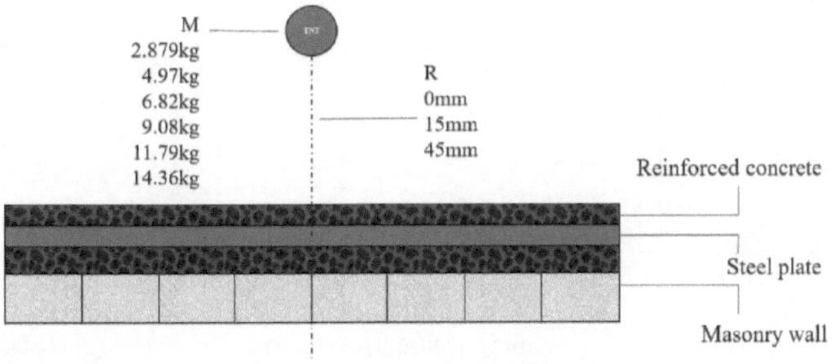

Figure 2. Schematic diagram of TNT explosive and wall position.

which are 2.879kg, 4.97kg, 6.82kg, 9.08kg, 11.79kg and 14.36kg respectively. The relative position relationship of explosive, unreinforced wall and explosion-proof wall is shown in Figure 2. Euler idea gas is used to simulate air around the wall to transmit the shock wave generated by TNT explosive explosion. At the same time, outflow boundary conditions are added to the boundary of Euler grid to simulate the air at infinite boundary. Explosion load and working conditions are shown in Table 1.

Table 1. Specimen parameters and explosion conditions.

Specimen number	Wall panel size/m	Thickness of concrete/ mm	Thickness of Steel/ mm	Thickness of concrete in wall/mm	TNT/ kg	Radius of initiation/m	Explosion point distance/mm
W-N-D15-B1	1.62×1.44×0.24	50	0	65	2.879	0.075	15
W-N-D45-B1	1.62×1.44×0.24	50	0	65	2.879	0.075	45
W-N-D0-B1	1.62×1.44×0.24	50	0	65	2.879	0.075	0
W-N-D0-B2	1.62×1.44×0.24	50	0	65	4.97	0.09	0
W-N-D0-B3	1.62×1.44×0.24	50	0	65	6.82	0.10	0
W-N-D0-B4	1.62×1.44×0.24	50	0	65	9.08	0.11	0
W-N-D0-B5	1.62×1.44×0.24	50	0	65	11.79	0.12	0
W-S10-D15-B1	1.62×1.44×0.24	50	10	65	2.879	0.075	15
W-S10-D45-B1	1.62×1.44×0.24	50	10	65	2.879	0.075	45
W-S10-D0-B1	1.62×1.44×0.24	50	10	65	2.879	0.075	0
W-S10-D0-B5	1.62×1.44×0.24	50	10	65	11.79	0.12	0
W-S10-D0-B6	1.62×1.44×0.24	50	10	65	14.36	0.14	0
W-S20-D0-B1	1.62×1.44×0.24	50	20	65	2.879	0.075	0
W-S20-D0-B5	1.62×1.44×0.24	50	20	65	11.79	0.12	0
W-S20-D0-B6	1.62×1.44×0.24	50	20	65	14.36	0.14	0
W-S30-D0-B1	1.62×1.44×0.24	50	30	65	2.879	0.075	0
W-S30-D0-B5	1.62×1.44×0.24	50	30	65	11.79	0.12	0
W-S30-D0-B6	1.62×1.44×0.24	50	30	65	14.36	0.14	0
W-S30-D0-B7	1.62×1.44×0.24	50	30	65	19.45	0.16	0

Note: W represents the wall, N represents the ordinary wall without sandwich steel plate, and S represents the thickness of sandwich steel plate; D represents the distance from the explosion point to the wall; B represents explosion condition.

3 DYNAMIC RESPONSE SIMULATION AND PERFORMANCE ANALYSIS

3.1 *Analysis of test results*

When simulating the explosion, the explosion energy time history curve in Figure 3 is applied to the inner surface of the wall as a boundary condition, and the calculated wall damage distribution

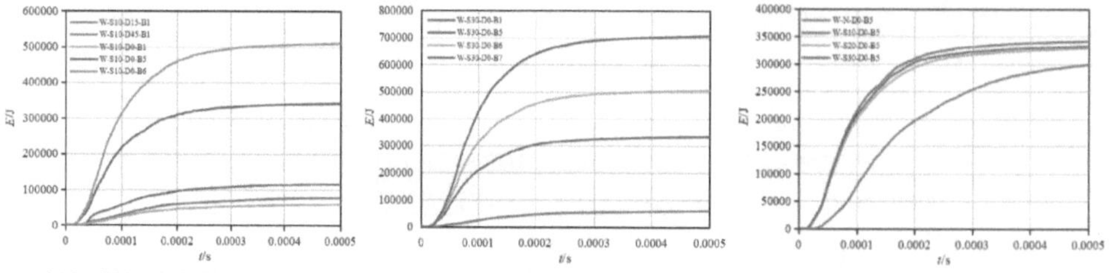

(a) W-S10-D15/D45/D0-B1/B5/B6 (b) W-S30-D0-B1/B5/B6/B7 (c) W-N/S10/S20-D0-B5

Figure 3. Inter-story displacement envelope in the X direction.

is shown in Figures 4 and 5. It can be seen from the figure that with the increase of explosion energy and the proximity of explosion center distance, For the masonry wall strengthened with ordinary reinforced concrete, the bricks and concrete have been seriously damaged. At the position of the explosion contact center line, the blocks fly out, the concrete falls off, the reinforcement is seriously deformed, and finally the bearing capacity is lost.

The strain nephogram is shown in Figure 4 and Figure 5, and the displacement nephogram is shown in Figure 6. It can be seen from the figure that the plastic strain of the reinforcement in the reinforced concrete masonry wall is significantly greater than that in the steel plate sandwich masonry wall, and its strain distribution law is basically the same, and the reinforcement at the explosion center reaches the failure strain first. The strain distribution law of steel plate is consistent with that of reinforcement. Under the same level of explosion load, the steel plate in reinforced concrete masonry wall reaches the failure strain before the steel plate in steel plate sandwich masonry wall. When the steel plate thickness is greater than 30mm, the steel plate strain nephogram in the steel plate sandwich masonry wall shows a uniform diffusion state. When the steel plate thickness is greater than 30mm, the steel plate strain nephogram in the steel plate sandwich masonry wall shows a uniform diffusion state. Under the action of explosion load of the same grade, the displacement of the explosion center of reinforced concrete masonry wall is large, the blocks fall off completely, the concrete denudation is serious, the reinforcement and steel plate reach stress failure, and the center of the wall has been penetrated. At this time, the reinforcement and steel plate at the center of the steel plate sandwich masonry wall are still in the elastic-plastic state, the concrete unit has not been completely penetrated, and the blasting center displacement of the wall is obviously less than that of the reinforced concrete masonry wall. Therefore, the steel plate sandwich masonry wall has good explosion resistance.

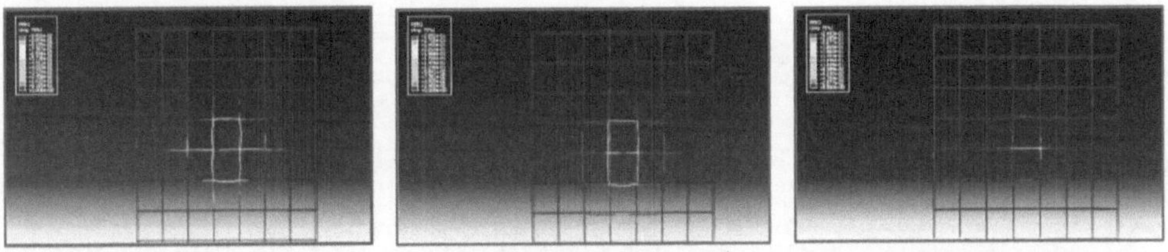

(a) Reinforced concrete reinforced masonry wall

(b) Steel plate sandwich masonry wall/S10

(c) Steel plate sandwich masonry wall/S30

Figure 4. Strain nephogram of reinforcement.

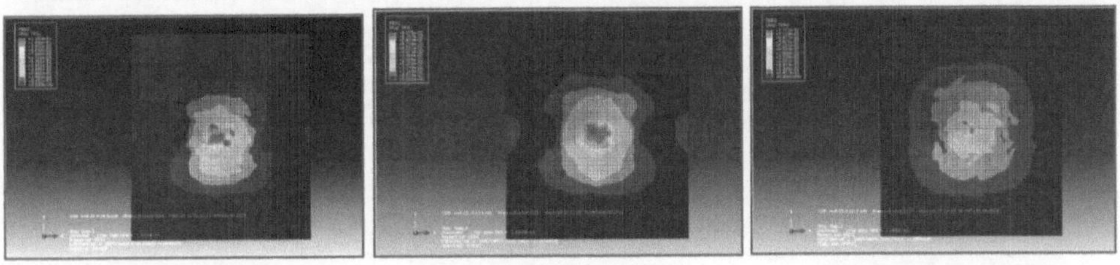

(a) Reinforced concrete reinforced masonry wall

(b) Steel plate sandwich masonry wall/S10

(c) Steel plate sandwich masonry wall/S30

Figure 5. Strain nephogram of steel plate.

3.2 *Dynamic response analysis*

Under the explosion impact load, the maximum displacement of the wall occurs at the center of the explosion, and its time history curve is shown in Figure 7. It can be seen from the figure

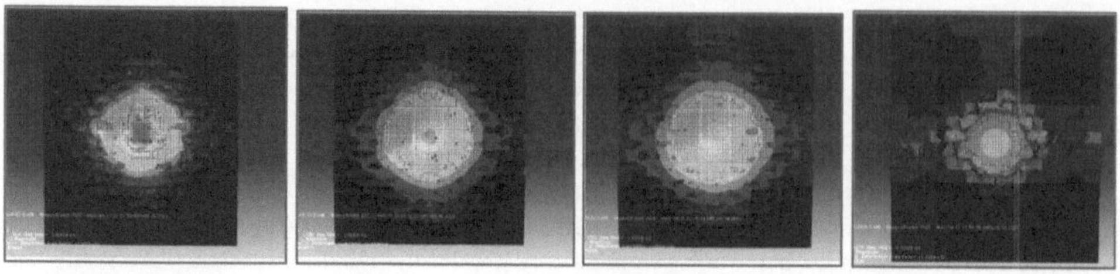

(a) RC masonry wall (b) S10 (c) S20 (d) S30

Figure 6. Strain nephogram of the wall.

that under the action of explosion load, the instantaneous displacement of reinforced steel plate sandwich masonry wall decreases with the increase of steel plate thickness. When the thickness of the steel plate is less than 20mm, the decreasing trend of the blasting center displacement of the wall is not obvious. When the thickness of the steel plate is greater than 30mm, the maximum plastic displacement of the reinforced steel plate sandwich explosion-proof masonry wall decreases obviously. When the thickness of steel plate is less than 20mm, the maximum plastic displacement of wall is greater than 0.015m; when the thickness of steel plate is greater than 30mm, the maximum plastic displacement of wall is less than 0.008m. Therefore, the anti explosion performance of reinforced steel plate sandwich masonry explosion-proof wall is affected by the thickness of steel plate. With the increase of steel plate thickness, the anti explosion performance of original masonry wall and concrete reinforced masonry wall can be significantly improved.

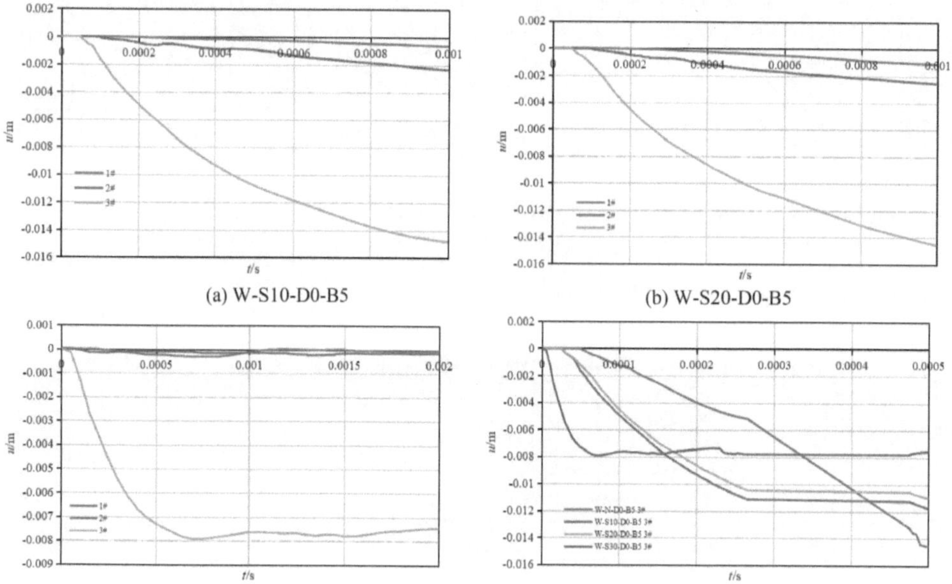

(a) W-S10-D0-B5 (b) W-S20-D0-B5

(c) W-S30-D0-B5 (d) W-N/S10/S20/S30-D0-B5

Figure 7. Displacement time history curve.

Under action of explosion load, the instantaneous speed at the explosion point can reflect the kinetic energy borne by the wall, as well as the instantaneous deformation and recovery capacity of the wall. The speed time history curve at the explosion point is shown in Figure 8. It can be seen from the figure that the maximum instantaneous speed of the original masonry wall, reinforced concrete masonry wall and steel plate sandwich masonry wall appears around 0.0001s under the

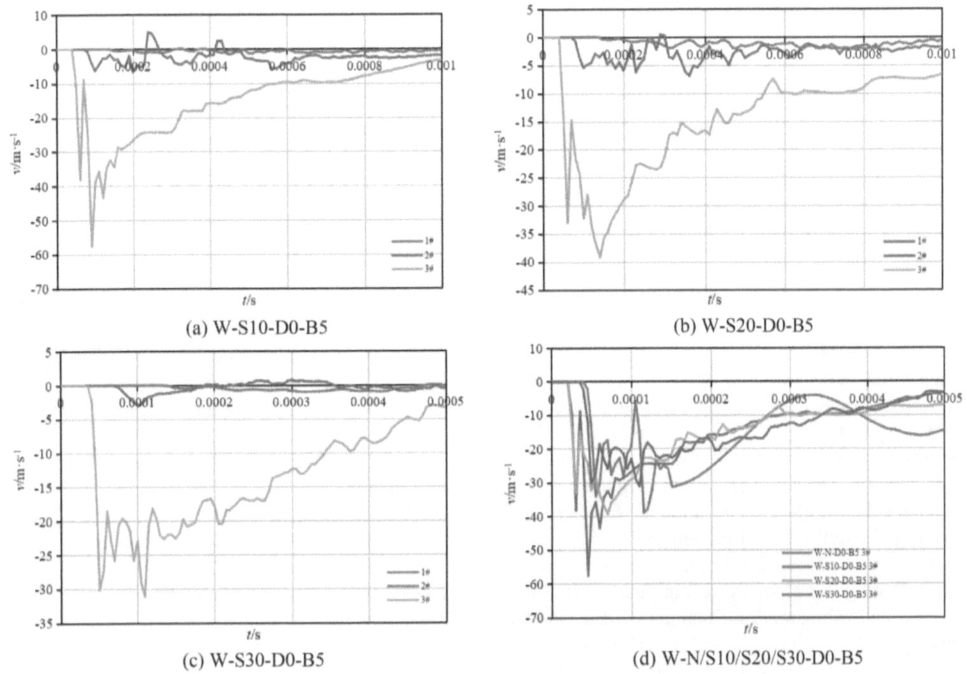

Figure 8. Speed time history curve.

same level of explosion load; When the thickness of steel plate is less than 10mm, the central speed of explosion point is greater than 58m/s; when the thickness of steel plate is greater than 30mm, the central speed of explosion point is less than 30m/s; From the explosion point to the wall edge, the speed of the wall decreases in turn. With the increase of explosion load, the speed change of masonry wall edge is not obvious; Due to the improvement of the ductility of the wall by the sandwich steel plate, the speed of the outer edge of the steel plate sandwich masonry wall is significantly higher than that of the original masonry wall.

4 CONCLUSION

(1) The protective effect of steel plate sandwich blast-wall is related to the core material. The higher the strength of steel plate and the greater the thickness, the better the protective effect; Compared with reinforced concrete masonry wall, steel plate sandwich masonry wall has better explosion resistance.

(2) When the steel plate thickness of the steel plate sandwich blast-wall is less than 20mm, the maximum plastic displacement of the wall is greater than 0.015m; when the steel plate thickness is greater than 30mm, the maximum plastic displacement of the wall is less than 0.008m.

(3) The maximum instantaneous speed of the wall appears around 0.0001s; When the thickness of steel plate is less than 10mm, the central speed of explosion point is greater than 58m/s; when the thickness of steel plate is greater than 30mm, the central speed of explosion point is less than 30m/s; Due to the improvement of the ductility of the wall by sandwich steel plate, the damage of steel plate sandwich masonry wall is obviously better than that of reinforced concrete reinforced masonry wall. The research results can provide scientific basis and technical reference for the protection transformation of envelope structures with anti explosion requirements such as existing ammunition depots.

ACKNOWLEDGMENTS

The financial assistance provided by the Basic Frontier Research Foundation of Engineering University of PAP No. WJY202015, the China Postdoctoral Science Foundation No. 2020M683432, and the Young and middle-aged Scientific Research Backbone Project of Engineering University of PAP No. KYGG202016. These supports are gratefully acknowledged.

REFERENCES

Baylot J. T. & Bullock B. (2005). Blast response of lightly attached concrete masonry unit walls. *Journal of Structural Engineering*, 131(8), 1186–1193.

Cheng L. J. & Mcomb A. M. (2010). Unreinforced concrete masonry walls strengthened with CFRP sheets and strips under pendulum Impact. *Journal of Composites for Construction*, 14(6), 775–783.

Galal K. & Sasanian N. (2010). Out-of-Plane Flexural performance of GFRP reinforced masonry walls. *Journal of Composites for Construction*, 14(2), 162–174.

Johnson C. F. & Slawson T. R. (2004). Concrete masonry unit walls retrofitted with elastomeric Systems for blast loads. *Journal of Structural Engineering*, 130(7), 1120–1128.

Mohammed I., Ahmed A., Alexander H. & Cheng D. (2011). Blast vulnerability evaluationof concrete masonry unit infill walls retrofitted with nano-particle reinforcedpolyurea: modelling and parametric evaluation. Structures Congress 2011. *American Society of Civil Engineers Paper*, 2126–2141.

Xu Z., Baohan H., Cong X. & Yinghua T. (2020). Dynamic response analysis of anti-explosion wall under dynamic loading modes. *Computer Aided Engineering*, 29(2), 39–45.

Zhigang Z., Hongrui C. & Binglin L. (2020). Experimental study on protection effect of anti-blast wall under action of car bomb explosion. *Engineering Blasting*, 26(4), 81–88.

Zhigang Z., Hongrui C. & Tao G. (2019). Experimental on anti-penetration explosion of rapid assembling anti-blast wall. *Engineering Blasting*, 25(5), 1–6.

Advances in Civil Engineering: Structural Seismic Resistance, Monitoring and Detection – Mohd Johari Mohd Yusof and Junwen Zhang (Eds)
© 2023 The Authors, ISBN 978-1-032-31491-4

Study on optimization of supporting structure of carbon slate tunnel based on sensitivity analysis

Chongyang Li, Qingwen Zhang & Yu Xia
School of Civil Engineering, Southwest Forestry University, YunNan, China

ABSTRACT: Sensitivity analysis refers to finding out the sensitive factors that have important influence on tunnel deformation control from many uncertain factors, and analyzing their influence degree and sensitivity degree on tunnel deformation. Sensitivity analysis helps to determine which factors have the greatest potential impact on technology optimization. Common sensitivity analysis methods include orthogonal test method, grey correlation degree method and so on. In this paper, the orthogonal test method combined with MIDAS/GTS, a large-scale finite element analysis software, is used to get the influence rules of various factors on deformation control.

1 INTRODUCTION

At present, most of the research on tunnel construction focuses on the selection of construction methods, including bench method, middle partition method, ring excavation with reserved core soil method, double-side heading method and so on. However, the research on the influence of initial support parameters on tunnel stability is not comprehensive enough. In the initial support structure of tunnel, any project needs to control the cost, and it is impossible to increase the size of steel support or the thickness of shotcrete without limit. Therefore, based on the actual engineering situation of Shanchongqing tunnel, this paper selects steel support, shotcrete thickness, anchor length and grouting effect of advance reinforcement as typical factors of the initial support structure, introduces orthogonal test method to conduct sensitivity analysis, uses numerical simulation to calculate the combination of different factors, and analyzes the general law of the influence of different factors on the stability of initial support of tunnel.

2 TESTING PROGRAM

According to the influencing factors and the number of factor levels, orthogonal table $L_{16}(4^5)$ is selected to arrange the test. The influencing factors include steel support specifications, advanced reinforcement parameters, shotcrete thickness and anchor length. Each factor has four levels, so MIDAS/GTS numerical simulation tests with different combinations of factors and levels need to be conducted for 16 times. Because the tunnel deformation includes two test indexes, vault settlement and horizontal convergence, it will be analyzed in turn below. According to Zhao Jingjing (2015) and others' research, it is found that the advanced grouting reinforcement of the tunnel will effectively improve various physical and mechanical properties of surrounding rock, among which the elastic modulus will be increased by 42% to 56%, the cohesion will be increased by 35% to 51%, and the internal friction angle will be increased by 2° to 3.5°. Therefore, in this paper, when the influencing factors of advanced reinforcement are

DOI 10.1201/9781003310884-10

horizontally divided, the equivalent increase of elastic modulus will be adopted, as shown in Table 1.

Table 1. Four factors and four levels orthogonal test.

	Influencing factor			
level	Steel support specification	Bolt depth	Advance reinforcement effect	Spray mixing thickness
1	20a	2m	Elastic modulus increased by 40%	20
2	20b	3m	Elastic modulus increased by 44%	25
3	22a	4m	Elastic modulus increased by 48%	30
4	22b	5m	Elastic modulus increased by 52%	35

3 NUMERICAL SIMULATION CALCULATION

Simplify the tunnel model into a two-dimensional plane strain model, and establish a two-dimensional calculation model. The analysis area of the model is 10 times of the hole diameter, and the mesh is automatically divided. The buried depth of Shanchongqing tunnel is 140m, which belongs to the deep-buried tunnel. Therefore, the equivalent uniformly distributed stress load is added to the upper part of the model, the size of which is the dead weight of the overlying rock mass, and the periphery of the model is fixed. Moor-Coulomb model and plane strain model are used to simulate surrounding rock of tunnel. The bolt is simulated by 1D embedded beam element; The initial support is simulated by equivalent stiffness method and 2D plate element. Three-stage excavation is adopted, and lining is applied at the same time of excavation. Due to limited space, the first 4 tests are selected for numerical calculation, and the numerical calculation results of each test are shown in Figures 1 to 4.

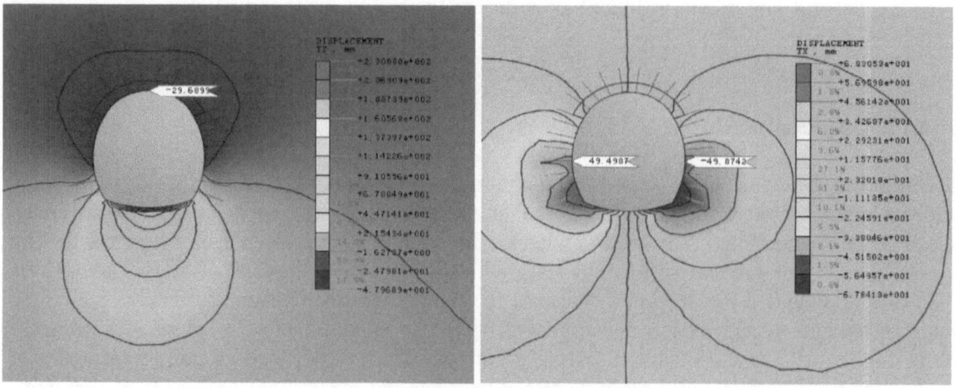

Figure 1. Numerical calculation results of tunnel deformation and displacement (No.1 test).

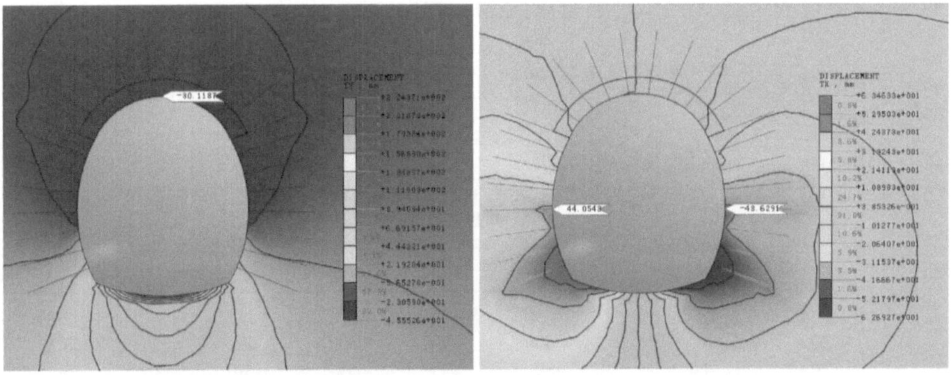

Figure 2. Numerical calculation results of tunnel deformation and displacement (No.2 test).

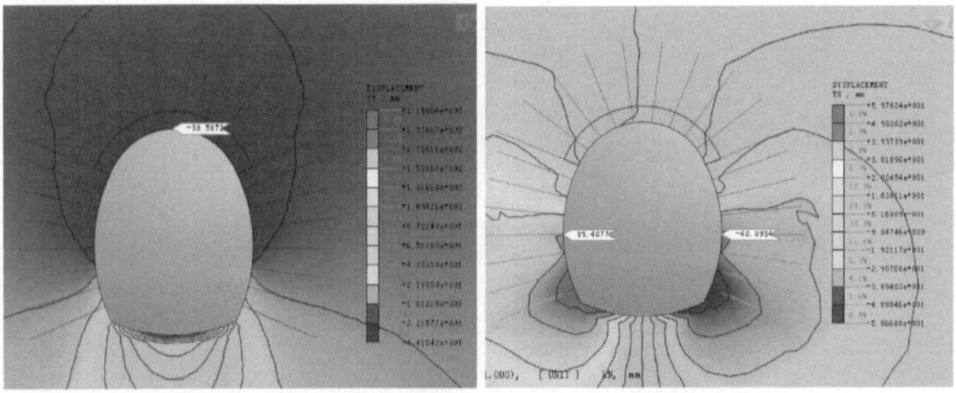

Figure 3. Numerical calculation results of tunnel deformation and displacement (No.3 test)

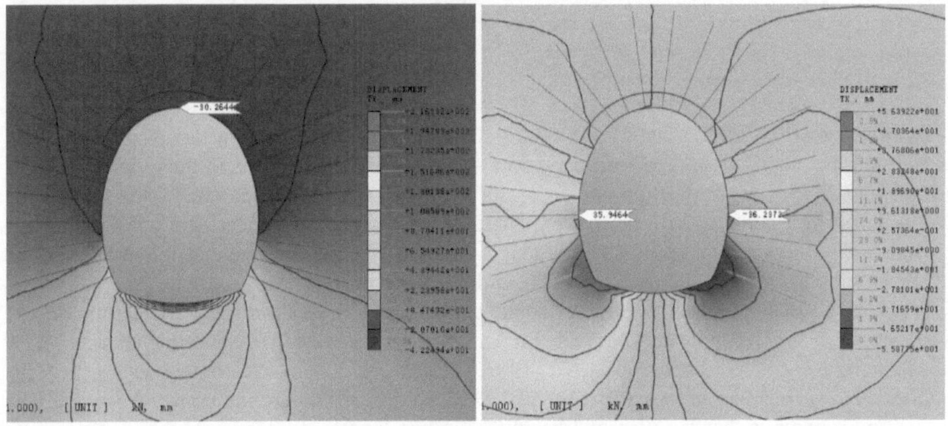

Figure 4. Numerical calculation results of tunnel deformation and displacement (No.4 test).

4 CALCULATION RESULTS AND SORTING ANALYSIS

4.1 *Orthogonal test range analysis method*

Table 2. Analysis table of vault settlement range.

Test number	I-steel model	Roof bolt length	Grouting effect E	Grouting effect C	Grouting effect φ	Spray mixing thickness	Subside
1	20a	2	40%	35%	57.3	20	29.7
2	20a	3	44%	39%	57.7	25	30.1
3	20a	4	48%	43%	58.1	30	30.5
4	20a	5	52%	47%	58.5	35	30.3
5	20b	2	44%	39%	57.7	30	33.7
6	20b	3	40%	35%	57.3	35	33.3
7	20b	4	52%	47%	58.5	20	27.2
8	20b	5	48%	43%	58.1	25	27.6
9	22a	2	48%	43%	58.1	35	35.3
10	22a	3	52%	47%	58.5	30	31.9
11	22a	4	40%	35%	57.3	25	29.1
12	22a	5	44%	39%	57.7	20	25.9
13	22b	2	52%	47%	58.5	25	32.1
14	22b	3	48%	43%	58.1	20	28.4
15	22b	4	44%	39%	57.7	35	32.1
16	22b	5	40%	35%	57.3	30	29.2
K1	120.6	130.8		121.3		111.2	
K2	121.8	123.7		121.8		118.9	
K3	122.2	118.9		121.8		125.3	
K4	121.8	113		121.5		131	
k1	30.15	32.7		30.325		27.8	
k2	30.45	30.925		30.45		29.725	
k3	30.55	29.725		30.45		31.325	
k4	30.45	28.25		30.375		32.75	
R	0.4	4.45		0.125		4.95	

It can be seen from Table 2 that for vault settlement, the range of the four influencing factors such as I-beam type, anchor length, advance support and shotcrete thickness is in the order of 4.95>I-beam length 4.45>I-beam type 0.4>advance support effect. For tunnel settlement, the smaller the displacement and deformation, the better. Therefore, for vault settlement, the optimal support scheme is that the I-beam type is 20a, the length of anchor rod is 5m, the elastic modulus of the advanced support effect is increased by 40%, and the shotcrete thickness is 20cm.

From Table 3, it can be seen that for horizontal convergence, the four influencing factors, I-beam type, anchor length, forepole and shotcrete thickness, are in the order of shotcrete thickness 15.2 > I-beam length 11.675 > I-beam type 1.675 > forepole effect 0.55 For horizontal convergence, the smaller the displacement deformation, the better. Therefore, for horizontal convergence, the best supporting scheme is that the type of I-beam is 22a, the length of anchor rod is 5m, the elastic modulus of the advanced supporting effect is increased by 44%, and the shotcrete thickness is 35cm.

According to previous studies (Wang Zhicai 2021; Wu Yongsheng 2017; Zhao Jianming 2016; Zhang Dehua 2016), advanced grouting reinforcement can effectively improve the stress state of the initial support and reduce the deformation of the initial support of the tunnel. For loose and

Table 3. Analysis table of horizontal convergence range.

Test number	I-steel model	Roof bolt length	Grouting effect E	Grouting effect C	φ	Spray mixing thickness	Subside
1	20a	2	40%	35%	57.3	20	99.3
2	20a	3	44%	39%	57.7	25	87.8
3	20a	4	48%	43%	58.1	30	79.5
4	20a	5	52%	47%	58.5	35	72.1
5	20b	2	44%	39%	57.7	30	86.5
6	20b	3	40%	35%	57.3	35	77.8
7	20b	4	52%	47%	58.5	20	89.5
8	20b	5	48%	43%	58.1	25	79.9
9	22a	2	48%	43%	58.1	35	81.4
10	22a	3	52%	47%	58.5	30	82
11	22a	4	40%	35%	57.3	25	83.7
12	22a	5	44%	39%	57.7	20	84.9
13	22b	2	52%	47%	58.5	25	91.5
14	22b	3	48%	43%	58.1	20	92.9
15	22b	4	44%	39%	57.7	35	74.5
16	22b	5	40%	35%	57.3	30	75.1
K1	338.7	358.7	335.9	366.6			
K2	333.7	340.5	333.7	342.9			
K3	332.0	327.2	333.7	323.1			
K4	334.0	312.0	335.1	305.8			
Q	111963.4	112252.5	111958	112470			
S	6.2	295.3	0.89	513.6			

broken surrounding rock, the surrounding rock pressure after advanced grouting reinforcement can be reduced by 40% compared with that without advanced grouting reinforcement, and the displacement and deformation of the tunnel will also be greatly reduced. Based on the above orthogonal test results, the analysis shows that after the Shanchongqing tunnel is reinforced by advanced grouting, that is, when it reaches level 1, it can meet the engineering requirements, and further increasing the strength of the reinforced area can not further reduce the displacement and deformation of Shanchongqing tunnel.

4.2 *Inspection of construction method optimization scheme*

According to the analysis of the above orthogonal test results, for vault settlement, the optimal supporting scheme is that the I-beam type is 20a, the length of the anchor rod is 5m, the elastic modulus of the advanced supporting effect is increased by 40% and the shotcrete thickness is 20cm, while for horizontal convergence, the optimal supporting scheme is that the I-beam type is 22a, the length of the anchor rod is 5m, the elastic modulus of the advanced supporting effect is increased by 44% and the shotcrete thickness is 35cm; Considering the actual deformation of Shanchongqing tunnel, the horizontal convergence value is far greater than the vault settlement value, so the optimal support scheme of horizontal convergence is the standard, and the test is carried out in two steps. Firstly, three different construction methods, namely, three steps and seven steps (scheme 1), reserving core soil (scheme 2) and adding transverse bracing (scheme 3), are used to test the supporting effects of the three different construction methods after the optimized supporting scheme by MIDAS/GTS. Secondly, select the appropriate supporting scheme and construction method, and return to the engineering practice to test it.

5 CONCLUSIONS AND SUGGESTIONS

In this paper, based on the on-site support structure parameters and construction methods of Shanchongqing Tunnel, through orthogonal test method, combined with finite element analysis software MIDAS/GTS, and through the analysis results of range analysis method, the following conclusions and suggestions are drawn:

(1) For the influencing factors of controlling the deformation of Shanchongqing tunnel, the sensitivity order is as follows: shotcrete thickness > anchor length > I-beam specification > advance reinforcement effect. For controlling horizontal convergence, shotcrete thickness and anchor length are the most significant influencing factors, while for controlling vault settlement, only shotcrete thickness is the most significant influencing factor.

(2) It is the influencing factor of the sensitivity of anchor rod next to the shotcrete thickness. On the premise of controlling the number of existing anchor rods unchanged, properly increasing the length of anchor rod and its embedded depth can give full play to the connecting function of anchor rod, increase the integrity of surrounding rock, and achieve the purpose of effectively controlling the deformation of surrounding rock.

(3) By comparing the three-step seven-step method (scheme 1), three-step reserved core soil (scheme 2) and three-step temporary transverse bracing method (scheme 3), it is concluded that the numerical calculation result of scheme 1 is that the vault settlement is 38.92mm, the optimization effect is 2.2%, the horizontal convergence is 126.33mm, and the optimization effect is 21.24%. The vault settlement of scheme 2 is 32.01mm, the optimization effect is 17.7%, the horizontal convergence is 88.87mm, and the optimization effect is 45.14%. In scheme 3, the vault settlement is 31.73mm, the optimization effect is 20.3%, the horizontal convergence is 62.37mm, and the optimization effect is 61.1%. Based on the comprehensive analysis of the operability of the tunnel construction site, the three-step reserved core soil method (Scheme 2) is finally selected as the final optimization scheme.

REFERENCES

Wang Zhicai, Sun Weifeng, Yang Wanli, Wang Kai, Luo Xin. Study on the Response Law of Supporting Structure in Soft Rock Tunnel with Large Deformation [J]. *Highway Traffic Technology*, 2021, 38(12):91–99+121.

Wu Yongsheng. Study on mechanical properties of surrounding rock of phyllite tunnel and its engineering application [D]. *Beijing Jiaotong University*, 2017.

Zhao Jianming. Study on deformation law and supporting technology of soft rock tunnel with high ground stress in Muzhailing ridge section [D]. *Beijing Jiaotong University*, 2016.

Zhang Dehua, Liu Shihai, Ren Shaoqiang. Mechanical characteristics analysis of supporting structure of high-stress tunnel constructed by three steps and seven steps [J]. *Modern Tunnel Technology*, 2016,53(01):96–102+110.

Zhao Jingjing. Study on equivalent simulation and parameter design of advanced small duct grouting [D]. *South China University of Technology*, 2015.

Advances in Civil Engineering: Structural Seismic Resistance, Monitoring and Detection – Mohd Johari Mohd Yusof and Junwen Zhang (Eds)
© 2023 The Authors, ISBN 978-1-032-31491-4

Study on deformation of weak surrounding rock based on forward inversion analysis

Chongyang Li, Qingwen Zhang & Yu Xia
School of Civil Engineering, Southwest Forestry University, YunNan, China

ABSTRACT: Taking Shanchongqing Tunnel with weak surrounding rock as an example, the elastic modulus and lateral pressure coefficient of the surrounding rock of Dabangwu Tunnel are analyzed and verified by using field measured data and finite element analysis software MIDAS/GTS. The results can meet the engineering precision requirements, and the surrounding rock parameters can be used to guide the subsequent construction design of the tunnel.

1 INTRODUCTION

In tunnel construction projects, soft rock is one of the most engineering geological problems encountered in tunnel construction, and the stress of weak surrounding rock of tunnel has become one of the focuses of tunnel construction problems.

The accuracy of rock mechanics parameters will have a great influence on the numerical results. Therefore, how to obtain more accurate rock mechanics parameters has become one of the important topics in the field of tunnel engineering research. At present, the mechanical parameters of rock mass can be obtained mainly from the following methods: in-situ test, laboratory test, empirical method and displacement back analysis method (Gan Anwu 2018; Zhou Junfa 2019). In these methods, in-situ test can accurately obtain the required mechanical parameters of rock mass, but it often takes a lot of manpower and material resources, and is only applicable in key projects. However, in laboratory, because there are many structural planes in rock mass, such as joints, fissures and faults, the test results are often not representative. On the one hand, the empirical reduction method needs the accumulation of experience; on the other hand, due to the different engineering geological conditions in different regions, the mechanical parameters of rock mass obtained by empirical reduction method often have great errors. The rule of displacement back analysis is to back calculate the mechanical parameters of rock mass by the measured data of surrounding rock deformation. Since H.A.D. Kirsten proposed to back analyze the elastic modulus of rock mass by the measured deformation of rock mass in 1976, many ways of displacement back analysis have been developed up to now, including analytical method (taking the mechanical parameters and initial ground stress of rock mass as unknown quantities to be solved by the process opposite to that of forward analysis), atlas method, forward optimization method, intelligent optimization inversion method, etc (Xi Junjie 2010; Zhou Chuliang 1933; Zhu Junwen 2018; Zhou Xiaojing 2018). Among them, the forward optimization method, also known as the direct method, is essentially built on the basis of iteration. By giving the initial values of parameters, the trial value of unknown functions is gradually corrected by using the minimum error function in the forward iteration process until it approaches the optimal value. The advantage of this method is that it is suitable for the back analysis of most linear or nonlinear problems, and the operation process is clear and clear. Its disadvantages are heavy workload, poor stability of the solution when there are many undetermined parameters, and slow convergence speed. The intelligent optimization inversion algorithm finds out the characteristic relationship (mapping) between input (lithology parameters) and output (displacement) through learning and memory functions. In this paper, forward optimization method

DOI 10.1201/9781003310884-11

is adopted, and the results of different undetermined parameters are calculated by golden section method combined with MIDAS/GTS, a large-scale finite element analysis software, and compared with the measured results, the interval of undetermined parameters is constantly approaching.

2 NUMERICAL SIMULATION ANALYSIS OF TUNNEL BASED ON MEASURED DATA

2.1 *Establishment of tunnel model and selection of parameters*

The calculation adopts M-C criterion. According to the materials collected from the site of Dabangwu Tunnel, the supporting structure parameters are shown in Table 1. In the numerical simulation, the surrounding rock of the tunnel is regarded as homogeneous rock mass, and the mechanical parameters of the surrounding rock selected by the calculation model are shown in Table 2.

When simulating a deep-buried tunnel, the surrounding rock within the range of 3 times the tunnel span is greatly affected by the tunnel excavation, while the surrounding rock outside the range of 3 times the tunnel span is less affected by the tunnel construction. According to the engineering experience, the distance of 3~6 times the tunnel span is generally selected as the boundary of the finite element model. For the function of overlying strata outside the scope of tunnel model, it is simulated by applying uniform load on the upper surface of the model. The uniform load applied on the upper part of the model and the calculation parameters of composite lining are converted according to the principle of equivalent stiffness method, then the elastic modulus of composite lining is:

$$E = E_0 + \frac{S_g \times E_g}{S}$$

Table 1. Tunnel support parameters.

	Supporting measures	Specifications	Remarks
Initial support	mesh reinforcement	20 cm×20 cm	
	roof bolt	Φ25	1.2 m×1.0 m, 3 m
	Longitudinal connecting rib	Φ22	1 m
	I-steel	I20b	
	shotcrete	C25	25 cm
Advance support	Advance small catheter	Φ42	0.4 m×2.4 m, 4m

Regarding the advanced support, the research of Wang Pengfe and Zhao Jingjing of South China University of Technology and Xi'an University of Science and Technology shows that the elastic modulus of the surrounding rock after grouting reinforcement is 42%~56% higher than that of the original surrounding rock, the cohesion is 35%~50% higher, the internal friction angle is 2 ~ 3.5, and the density is slightly higher (Zhao Jingjing 2015).

Table 2. Tunnel modeling parameters.

Geotechnical	Weathering degree	density ρ KN/m^2	compress Es MPa	flexibility E GPa	cohesive strength c KPa	Poisson's ratio
Carbonaceous	Strong	23	6.0	—	25	0.4
Carbonaceous	weak	25	6.0	—	25	0.35
Advance support	—	27	8.1	—	35	0.4
Initial support	—	23	—	28	—	0.2
Roof bolt	—	77	—	200	—	0.3

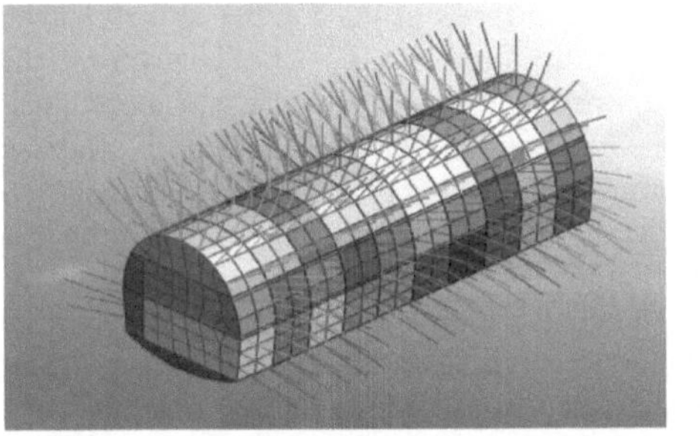

Figure 1. Three-dimensional calculation model.

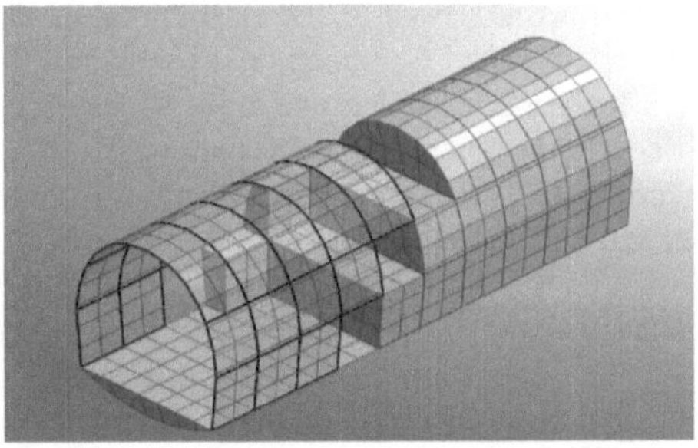

Figure 2. Calculation model of step method.

The three-dimensional tunnel calculation model and the step method calculation model are shown in Figures 1 and 2.

2.2 *Analysis of calculation results*

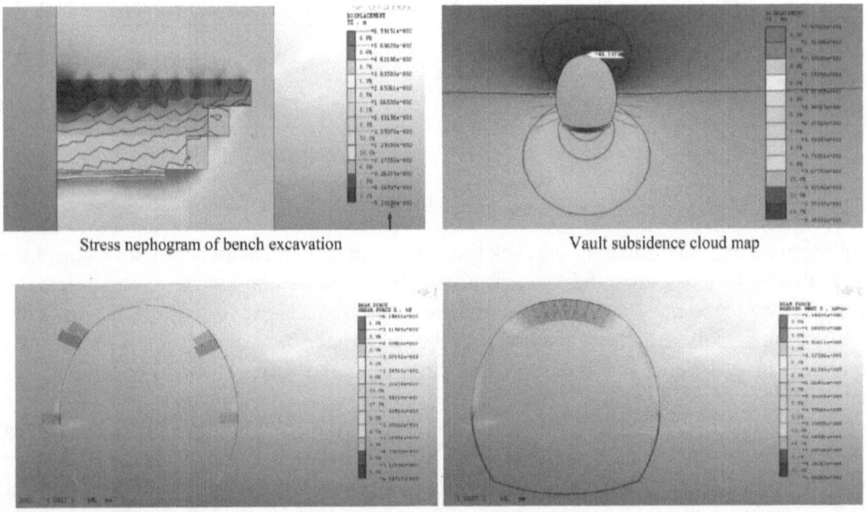

Stress nephogram of bench excavation

Vault subsidence cloud map

Figure 3. Calculation cloud chart.

After the numerical simulation of tunnel excavation, the vault settlement reaches 40.1379 mm, and the horizontal convergence value reaches 135.7214 mm. The numerical calculation displacement results are in good agreement with the field measured deformation. It can be seen from Figure 3 that the stress is mainly concentrated in the tunnel vault and the side walls on both sides, so the tunnel vault and the side walls are most likely to be damaged.

3 INVERSION ANALYSIS OF SURROUNDING ROCK PARAMETERS

3.1 *Selection of inversion parameters*

Mechanical parameters of rock mass include internal friction angle, cohesion C, elastic modulus E and Poisson's ratio. However, in practical engineering, the internal friction angle, cohesion C and Poisson's ratio can be obtained through experiments, and the accuracy can meet the engineering requirements, while the elastic modulus is often converted into elastic modulus by a given deformation modulus or compression modulus, which causes great errors. In the past, people used the empirical formula, that is, the elastic modulus is usually 3–5 times of the compressive modulus. However, no matter what method is used, the elastic modulus of rock mass can not be given, so the elastic modulus is taken as one of the back analysis parameters. Elastic modulus e and lateral pressure coefficient k are selected as parameters of back analysis, and MIDAS/GTS is used to establish the three-dimensional model of Dabangwu tunnel, simulate the whole process of tunnel construction, and calculate the final deformation of Dabangwu tunnel. The buried depth of the tunnel is 140 m, and uniformly distributed load $q = 75kn/m^2$ is applied to the upper part of the model.

The surrounding rock of Shanqing tunnel is in a state of high ground stress or extremely high ground stress. According to Code for Design of Railway Tunnels and engineering geological data, the values of inversion parameters E and K are selected as follows: elastic modulus E = 100–160 MPa, lateral pressure coefficient K = 0.8–1.2. The tunnel construction parameters and surrounding rock parameters are shown in Table 3. The surrounding rock model and tunnel lining model are calculated and analyzed by using the same model in Section 2.

Table 3. Tunnel modeling parameters.

Material	Serious γ kN/m^3	Flexibility E MPa	Cohesive strength C KPa	Internal friction angle $\varphi(°)$	Poisson's ratio μ
Rock	25	100–160	25	55	0.4
Roof bolt	78.5	2e6	—	—	0.2
Initial support	25.5	2.5e6	—	—	0.2

3.2 *Calculation and results of displacement back analysis*

The displacement of tunnel surrounding rock includes vault settlement and horizontal convergence. If back analysis of tunnel surrounding rock parameters is to be conducted, vault settlement and horizontal convergence must be considered at the same time. However, due to the influence of construction, only vault settlement and horizontal convergence of two-step side wall are analyzed and compared here. Inverse analysis and calculation of E and K are carried out, and coordinate rotation method is adopted.

In order to ensure the reliability of the inversion results, the error between the numerical calculation results and the measured results should be controlled within a reasonable range. At present, the research on the control of calculation error in back analysis is not perfect and there is no unified standard, so according to the previous experience and engineering practice, the control of back analysis error within 10% can meet the engineering requirements. The final simulation results of tunnel surrounding rock parameter inversion analysis are shown in Figure 4.

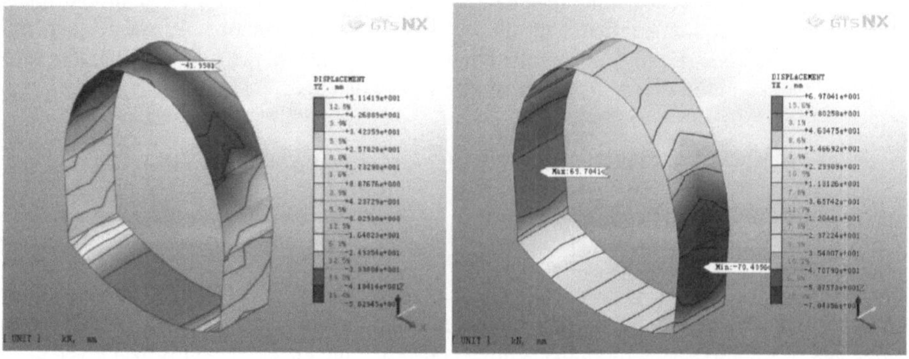

Figure 4. Cloud chart of vault settlement analysis results.

4 CONCLUSION

In this chapter, taking Dabangwu Tunnel as the research background, through regression analysis of the measured data of Dabangwu Tunnel and forward inversion analysis of surrounding rock parameters, the following conclusions are drawn:

(1) According to the measured data of the tunnel site, the tunnel model is established by M-C criterion. After the tunnel excavation, the settlement of the vault reaches 40.1379mm and the horizontal convergence value reaches 135.7214 mm. The displacement calculated by numerical method is in good agreement with the measured deformation on site, and the stress is mainly concentrated near the vault and side wall.
(2) According to the forward inversion analysis theory and the finite element analysis software MIDAS/GTS, when the elastic modulus of surrounding rock of Dabangwu Tunnel is 122.92MPa and the lateral pressure coefficient is 1.048, the numerical calculation results are vault settlement of 41.99mm, horizontal convergence of 140.14mm, measured vault settlement of 39.79mm, measured horizontal convergence of 153.84mm and relative error of vault settlement of 5.5%
(3) The forward optimization method can be used for the inverse analysis of the surrounding rock parameters of the tunnel, and the results can meet the requirements of engineering accuracy. The surrounding rock parameters thus obtained can be used to guide the subsequent construction design of the tunnel.

REFERENCES

Gan Anwu, Long Sichun. Inversion analysis and stability study of tunnel surrounding rock deformation monitoring [J]. *Geodesy and Geodynamics*, 2018, 38(12): 1291–1294+1305.

Gan Anwu. Inversion analysis and stability study of surrounding rock deformation of subway shield tunnel [D]. *Hunan University of Science and Technology*, 2018.

Ministry of Housing and Urban-Rural Development of the People's Republic of China. *GB 50487-2008, Code for Geological Survey of Water Resources and Hydropower Engineering* [S]. Beijing: China Planning Press, 2008.

Ministry of Housing and Urban-Rural Development of the People's Republic of China. *GB/T 50218-2014, Classification Standard of Engineering Rock Mass* [S]. Beijing: China Planning Press, 2014.

National Railway Administration. *TB 10003-2016, Code for Design of Railway Tunnel* [S]. Beijing: China Railway Press, 2016.

Xi Junjie, Li Dewu. Comparative analysis of numerical simulation of three-step and two-step excavation in Zhifang tunnel [J]. *Tunnel Construction*, 2010, 30(02):147–150.

Zhao Jingjing. Study on equivalent simulation and parameter design of advanced small duct grouting [D]. *South China University of Technology*, 2015.

Zhou Xiaojing, Hongxing Luo, Zhong Mingwen, Qin Yuqiao. Research on finite element inversion analysis of soft rock tunnel based on deformation monitoring [J]. *Journal of Water Conservancy and Building Engineering*, 2018, 16(04):79–84+180.

Zhu Junwen, He Yongjun, Huang Haiyan, Yin Zhihao, Jiang Zhaoqiang. Inversion of surrounding rock parameters of a hydraulic tunnel based on numerical simulation [J]. *Henan Science and Technology*, 2018(35):105–107.

Zhou Junfa. Research on safety control technology of highway tunnel construction based on inversion analysis of surrounding rock displacement [D]. *Qingdao University of Technology*, 2019.

Zhou Chuliang, Ding Rixi. Analysis of rock strength attenuation, rheology and surrounding rock stability [J]. *Mine pressure and roof management*, 1993(Z1):7–13+239.

Advances in Civil Engineering: Structural Seismic Resistance, Monitoring and Detection – Mohd Johari Mohd Yusof and Junwen Zhang (Eds)
© 2023 The Authors, ISBN 978-1-032-31491-4

Seismic performance of infilled wall-steel frame structure

Dingding Shi
Interior Design, School of Visual Arts, New York City, NY, USA

ABSTRACT: With the emergence and development of residential steel structures, the research on the influence of infilled wall-steel frame structures on seismic performance has been deepened continuously. This paper systematically expounds the experimental research, the establishment of a simplified calculation model, and the finite element analysis method on the seismic performance of steel frame structures with infilled walls. The key technical problems of seismic performance of steel frame structure with the infilled wall are studied, and the future research direction is proposed. Results show that the infill wall and steel frame can work together to improve seismic performance effectively. The seismic performance enhancement effect and the failure mode of the wall vary with the design.

1 INTRODUCTION

With the rapid development of steel structures, the steel frame-composite wall system develops rapidly. At present, the walls of steel structure residence mainly include light block infilled wall, skeleton wallboard and composite wallboard. Under the environment of implementing green buildings, only composite wallboard can meet the current building energy-saving standards. Light block infilled walls, and skeleton wallboards need to add high-efficiency thermal insulation materials on the outside to reduce energy consumption, which will lead to an increase in wet operation links and reduce construction efficiency and economic benefits. Therefore, considering energy saving and construction factors, choosing composite wallboard in the steel structure system is more appropriate. Compared with ordinary wallboard, composite wallboard has such advantages as high bearing capacity, good thermal insulation effect, and few wet operation links. At the same time, it can realize industrial production and assembly construction, which greatly accelerates the project progress.

In recent years, earthquakes have frequently occurred worldwide, which can seriously threaten people's lives, property, health and safety. Although earthquakes inevitably involve some human factors, there are more uncontrollable natural factors. At the same time, earthquakes are always accompanied by many secondary disasters, such as landslides, mudslides, floods, plagues, fires, explosives, radioactive materials diffusion, which could greatly influence our lives, even endanger our lives, and bring great losses to the national economy. Therefore, for the current situation of frequent earthquakes, the most direct solution is to make great efforts in housing construction, increase the links of seismic design, enhance the safety and seismic performance of houses, and give certain attention.

At present, the frame structure is widely used and widely used in composite wallboard combined with the infilled wall (Figure 1). Although the infilled wall is often regarded as a non-structural member that does not participate in the design stress, the frame itself bears all the horizontal and vertical loads. However, in the actual earthquake process, the infilled wall resists the horizontal earthquake action and cooperates with the frame structure (Dawe 1989; Dhanasekar 1985). Since the 1950s, scholars have carried out a large number of experimental studies on the seismic performance of frame structures with infilled walls. The low cyclic loading tests of frame structures with infilled walls show that steel columns wrapped with bricks and infilled walls can significantly improve

 DOI 10.1201/9781003310884-12

frames' seismic bearing capacity and lateral stiffness. The effect of infilled walls on improving lateral stiffness and bearing capacity of steel frame structures has been widely concerned. According to the research results of seismic performance of infilled steel frame structure by scholars at home and abroad, this paper summarizes the research status of steel frame-infilled wall structure system from three aspects including experimental study, simplified model and finite element analysis, analyzes the key technical problems existing in current research, and puts forward the prospect of future research on seismic performance of steel frame structure system with the infilled wall. It provides a reference for the research, development, and application of infilled wall-steel frame structures.

Figure 1. Typical building steel frame model.

2 THE ROLE AND IMPORTANCE OF SEISMIC DESIGN

Since the 21st century, there has been a high incidence of earthquakes worldwide, and many cities have been seriously affected by earthquakes. These disasters are fatal to the development of the local economy, the safety of people's lives and property, and the threat of earthquakes lies in the collapse of buildings. The continuous development of the construction industry has brought about the completion of buildings. It provides a stable place for people's lives. In construction engineering, strengthening the research on seismic problems in building structures can ensure the safety of people's lives and property. Unfortunately, some construction units have the behavior of cutting corners. As a result, the building can not meet seismic design requirements.

In urban planning, there are many architectural units, so the overall layout should be fully considered in the process of architectural planning. To improve the standardization of municipal engineering, it is necessary to improve the capability of seismic design and promote the process of urban planning. If the building earthquake, the good seismic ability can reduce property losses

Figure 2. Collapsed buildings during the earthquake.

and casualties and the damage caused by the earthquake (Figure 2), improve the service life of the building, and create a good social atmosphere. In a sense, it can promote the continuous progress of society.

Nowadays, construction projects pay more attention to the geographical location and appearance design, ignoring the building itself and reducing resistance to natural disasters. It may threaten people's lives and property safety in front of the earthquake. This requires that the seismic problems be fully considered in the design and construction of civil engineering structures. The emphasis should be paid more attention to ensure the safety performance of building projects.

The function of building seismic design is to ensure that the building can have a certain ability to resist earthquake action in high intensity and will not collapse immediately. And the building will keep a complete state under a certain intensity of vibration for a period of time, thus ensuring the further use of the building and ensuring people's life safety and property safety.

To sum up, seismic design plays an important role in building structure design. It not only affects the safety of our lives but also property. And it is still very important for economic development, social progress and national development. Therefore, the design of the wall-steel is important for seismic resistance (GB 50011, 2010).

3 EXPERIMENTAL STUDIES ON SEISMIC PERFORMANCE OF THE INFILLED WALL-FRAME SYSTEM

The rapid development of steel structures promotes the development of the infilled wall-frame system. As shown in Figure 3, the walls of steel structure residence mainly include light block infilled wall, skeleton wallboard and composite wallboard. A large number of tests show that composite wallboard has better seismic performance than light block infilled wall and skeleton wallboard. Therefore, scholars have used a large number of tests to demonstrate it (Table 1).

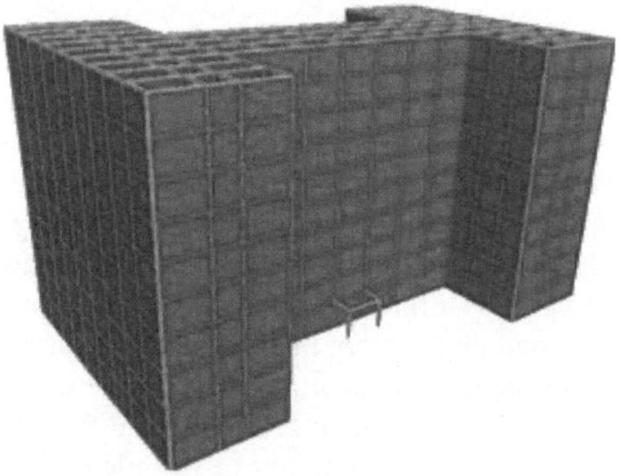

Figure 3. Infill wall-frame system.

The bearing capacity of the structural system was tested and analyzed mainly by adjusting the wall materials and horizontal loading methods to compare the bearing capacity of the structure. They argued that under the action of horizontal force, two diagonals of the infill wall could be separated from the frame, and the other two diagonals will tighten the frame, thus forming a separation area and a contact area. Polyakov proposed that an equivalent root can replace the infilled wall because the contact zone and the steel frame have the same effect as a root, while the separation zone has no interaction with the frame (Polyakov, 1956). The earliest equal-scale experimental study was an equal-scale frame-infilled wall structure (Benjamin, 1958).

In addition, the infilled wall can be simulated by an equivalent compression bar hinged at both ends. The effective width of the equivalent compression bar is 1/3 of the diagonal length of the infilled wall. And the material parameters of the equivalent compression bar are the material parameters of the infilled material. This is the first explicit calculation model of the equivalent compression bar width. The equivalent compression bar calculation model can be used to calculate the deformation capacity and bearing capacity of frame infilled wall structures (Holmes, 1961).

Table 1. Summary of experimental studies on seismic performance.

Experiments	Theories	Advantages	References
Adjusting the wall materials and horizontal loading.	Architectural mechanics	The equivalent root can replace the infilled wall	Polyakov, 1956
Equal-scale frame-infilled wall structure	Architectural mechanics		Benjamin, 1958
Scaled frame infilled wall models	Architectural mechanics	To calculate the deformation capacity and bearing capacity	Holmes, 1961
Nine 1/2 ratio brick infilled wall frame models under unidirectional horizontal loading and horizontal reciprocating loading	Architectural mechanics	To put forward restoring force model and elastic-plastic characteristic parameters of the member	Wu, 1980
Steel frames with external and embedded ALC wallboards	Architectural mechanics	Embedded ALC wallboards have a certain improvement on steel frame stiffness and bearing capacity.	Li, 2005 Qiu, 2011
Low-cycle reciprocating load	Architectural mechanics	The seismic performance of the structure can be improved by increasing the thickness of the wall and strengthening the connection between the wall and the steel frame	Wu,2013
Low-cycle reciprocating load	Architectural mechanics	The ultimate displacement and bearing capacity, ductility and energy dissipation capacity of the structure can be significantly improved by the wall plate with diagonal support	Li,2009
Energy dissipation	Heating and ventilation	The damping effect of the upper structure is better than that of the lower structure	Polyakov, 1960

Wu et al. carried out tests on nine 1/2 ratio brick infilled wall frame models under unidirectional horizontal loading and horizontal reciprocating loading. The corresponding empty frame is tested under the same boundary conditions, and the stiffness, bearing capacity and failure characteristics of each member under load are compared. In addition, the restoring force model and elastic-plastic characteristic parameters of the member are put forward (Wu, 1980).

Many experiments on steel frames with external and embedded ALC wallboards were carried out to study the influence of ALC wallboards on the bearing capacity and lateral stiffness of steel frames. The results show that external ALC wallboards have little influence on steel frame bearing capacity and lateral stiffness. Still, embedded ALC wallboards have a certain improvement on steel frame stiffness and bearing capacity. Therefore, the influence of embedded ALC wallboard on steel frame structure should be considered in structural design (Li 2005; Qiu, 2011).

The low-cycle reciprocating load tests on 8 structural systems of spatial steel wireframe composite wallboard with single-story and single-span steel frames with polystyrene board core, and experimental research and numerical simulation on wallboard thickness, connection mode between wallboard and frame, wallboard position, strong and weak axis direction of steel frame columns, and the joint form of steel frame were carried out. The test results show that the failure position of the wallboard is mainly at the embedded parts connecting the wallboard with the steel frame, but the overall failure mode is better than that of the general wall. Therefore, to ensure that the wall panel and steel framework, it is necessary to make the embedded parts not destroyed at the beginning. Furthermore, the structure's seismic performance can be improved by increasing the thickness of the wallboard and strengthening the connection between the wallboard and the steel frame. Finally, the stiffness reduction factor of the wallboard is 0.07 when the structural system is designed for earthquake resistance (Wu 2013).

The hysteretic performance, ductility, rigid degradation, and failure mode of six scaled steel frame composite wallboard system models were systematically studied through low-cycle reciprocating load tests. The test results show that the system has good seismic performance and lateral stiffness. In the test, the failure mode of "wallboard before frame" accords with the structural design concept, and the wallboard supported by steel plate obliquely can significantly improve the structural system's ultimate displacement and bearing capacity and improve the ductility and energy dissipation capacity of the structure. The initial stiffness of the steel frame is greatly increased by the ceramsite concrete composite wallboard, which is 2.2–2.9 times that of the empty steel frame. The ductility coefficient of the structure system is 3.23–5.45, and the energy consumption coefficient is about 0.65, which makes the structure have good energy dissipation capacity (Li 2009).

Through static and dynamic analysis, the calculation of composite wallboard steel frame connected with energy dissipation by equivalent wall frame method can meet the accuracy requirements of engineering, and the seismic absorption effect under frequent earthquakes is better than that under rare earthquakes; The shock absorption effect on the upper floor of the structure is better than that on the lower floor; However, the composite wallboard with energy dissipation connection reduces the seismic response of the superstructure, and at the same time increases the base shear force and overturning moment of the structure (Polyakov 1960).

4 SIMPLIFIED CALCULATION MODELS OF SEISMIC PERFORMANCE OF INFILLED WALL-STEEL

For the research on the cooperative working condition and seismic performance of infilled wall-steel frame structure, the researchers put forward relevant theoretical theories based on the test phenomena and results and obtained a simplified calculation model of infilled wall-steel frame structure to simplify the tedious calculation process (Figure 4, Table 2).

The three-story and three-span steel frame-infilled walls were studied to investigate the stress of both and put forward the concept of equivalent bracing (Hou 2012; Polyakov 1956). In this model, the infilled wall is regarded as an inclined brace with the same thickness as the wall and hinged with the plane of the frame. The inclined brace is designed as a single compression bar that only bears pressure and does not bear tension and resists horizontal force together with the frame structure (Thiruvengadam 1985). Since then, previous researchers have studied the infilled wall-steel frame system by using the equivalent diagonal brace model and found that the model has good accuracy, but it is difficult to determine the equivalent width of the diagonal brace (Wael 2003; Zhu 1996).

It is argued that the structure can be approximately regarded as a parallel system between the empty frame and infilled wall after cracks appear between the frame and infilled wall. Then, the strength and stiffness of the whole structure are the sums of their superposition, respectively. On this basis, a parallel model of wall and frame is proposed. Although this model can not reflect the formal stress principle accurately, its calculation is slightly simple. Therefore, it has been used in the seismic calculation method of China's seismic design code (Li 2003; Liu 1994).

Liu et al. summarized and put forward the infilled wall element model based on the experimental study. The wallboard element is used to simulate the infilled wall, and rigid beams are set at the upper and lower parts of the wallboard element. The four corner joints only transmit horizontal forces and are hinged with the surrounding beam-column joints. The infilled wall element model can well simulate the actual stress between the infilled wall and the frame structure before cracking. Still, it is difficult to simulate the whole stressful process of the frame infilled wall structure under earthquake action (Liu 2004).

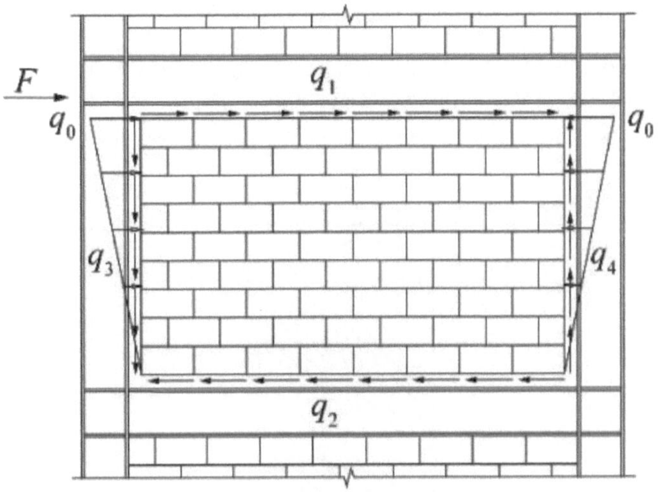

Figure 4. Stress distribution of filled wall and steel frame.

Table 2. Summary of simplified calculation models on seismic performance.

Models	Methods	Advantages	References
Concept of equivalent	Using three-story and three-span steel frame infilled wall to investigate the stress of both	The infilled wall is regarded as an inclined brace with the same thickness as the wall and hinged with the plane of the frame	4, 13–18 Polyakov,1956; Hou, 2012; Thiruvengadam, 1985; Zhu, 1996; Wael, 2003 Li, 2003; Liu, 1994
Infilled wall element model	Wallboard element is used to simulate the infilled wall, and rigid beams are set at the upper and lower parts of the wallboard element	Simulate the actual stress between the infilled wall and the frame structure before cracking	Liu, 2004

5 STUDY ON FINITE ELEMENT ANALYSIS METHOD OF SEISMIC PERFORMANCE OF THE INFILLED WALL-FRAME SYSTEM

With the development of computer technology, The Internet has spread to our daily life. As an effective engineering analysis method, the finite element method (Figure 5) has made fundamental progress in theory, method research, computer program development and application field development. Moreover, it is a good supplement to experiment with. Therefore, scholars often use experiment and numerical analysis methods to analyze infilled wall-frame structures (Table 3).

The experiments and finite element analysis on steel frame filled with cavity block composite wall structure were carried out to analyze the hysteretic behavior, stress distribution, stiffness and internal force distribution. The equivalent three-strut model is adopted as the finite element model, in which the contact length between the wall and the steel frame is calculated by the formula proposed by EI-Dakhakhni Weal W. The results show that the stress of beam-column joints and column feet is very large under the action of horizontal force in the equivalent three-compression bar model, which combines the design concept of "strong joints are like members and strong columns are weak beams". The equivalent three-strut model is more accurate than the equivalent strut model and saves more time than the solid element model. In addition, it can reflect the interaction between the wall and the steel frame more accurately (Dai 2005).

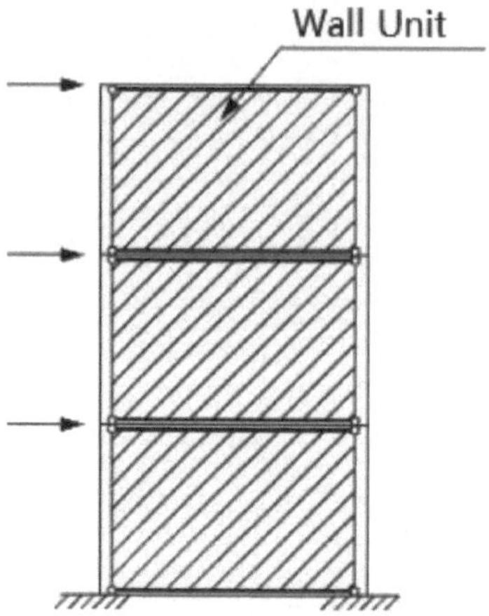

Figure 5. Filled wall model.

Table 3. Summary of FEA method on seismic performance.

Experiments	Models	Advantages	References
Low-cycle reciprocating load	Equivalent three-strut model	Reflect the interaction between the wall and the steel frame more accurately.	Dai 2005
Simulate the contact between the steel frame and wall	3D target 170 and pt-to-surf 175 contact elements	The interaction between the wall and the steel frame can increase the strength of the steel frame by about 20%	Wang 2015
Low-cycle reciprocating load	Nonlinear finite element model	The steel frame with the new CL sandwich composite wallboard has good seismic performance.	Memari 1991
Simulate the element of the wallboard	Filling element model	Convenient to calculate the shear force of each filled wall to get the shear force of the column	Achyutha 1986

The finite element simulation on the aerated concrete wall filled with a steel frame was studied based on experiments. Row comparative analysis 3D target 170 and pt-to-surf 175 contact elements are used to simulate the contact between steel frame and wall. The apex displacement, maximum stress, lateral stiffness and structural stiffness degradation of pure steel frame and aerated concrete wall steel frame under horizontal force are compared. The results show that the interaction between the wall and the steel frame can increase the strength of the steel frame by about 20% and greatly improve the stiffness of the steel frame. At the same time, it can prevent the stiffness of the steel frame from rapidly decreasing due to yield and improve the seismic performance of the structure (Wang 2015).

The low-cycle reciprocating load tests on 8 steel frames were carried out, and ABAQUS established a nonlinear finite element model. Considering material nonlinearity, geometric nonlinearity and boundary nonlinearity, the energy dissipation capacity, ductility, skeleton curve and failure mode were analyzed in detail. The finite element analysis results were in good agreement with the test results. The results show that the steel frame with the new CL sandwich composite wallboard has good seismic performance. The connection mode between composite wallboard and steel frame and the thickness of composite wallboard greatly influences the stiffness and bearing capacity of the structure. The thickness of the composite wallboard is proportional to the structure's stiffness. The closer the wallboard is to the neutral axis of the steel frame, the greater the structure's stiffness. The connection mode of beam and column has little influence on the structure's stiffness (Memari 1991).

A filling element model for embedded infilled walls, which uses filling elements to simulate the element of wallboard, was established. Rigid beams are set at the upper and lower ends of the wallboard, and four nodes of beams and columns are set to be hinged. The simulated wall-frame contact only transmits vertical force but not horizontal shear force. The thickness of the wallboard unit is taken as the actual thickness of the brick infill wall, and the width and length of the wallboard unit are taken as the actual width and length of the infill wall block, respectively. This simulation model deduces the stiffness matrix of the wallboard element. This method is simple and flexible, and it is easy to calculate the shear force of each infilled wall to obtain the shear force of columns. The model is also suitable for programming and computer calculation (Liu 2004).

For finite element analysis methods, researchers have proposed the iterative method (Memar 1991), coupling finite element and boundary element method (Achyutha1986), the nonlinear method (Papia 1988) and plane stress method (May 1991), respectively, for finite element modeling. Using finite element software for actual modeling, they (Riddington 1994) established a brick-filled wall-steel frame structure model using ANSYS based on the three-strut calculation method. Liu et al. (Liu 2005) proposed a finite element model of beam element + solid element + spring element. When the researchers used different finite element analysis methods and software to model the filled wall-steel frame structure, different elements were adopted to simulate the steel frame, the filled wall, and their connection. Although the modeling process was different, the numerical analysis results agreed with the experiment.

6 CONCLUSIONS

Based on the current research results of seismic performance test, simplified calculation model and finite element analysis of steel frame structure system with infilled walls, the following conclusions are made:

(1) The infilled wall and steel frame can work together, which greatly improves the bearing capacity and lateral stiffness compared with pure steel frame, and effectively enhances the seismic performance.
(2) The seismic performance enhancement effect of infilled wall-steel frame structure and the failure mode of the wall is different with the different infilled wall types and embedding methods.
(3) Several simplified calculation models of infilled wall-steel frame structure have their advantages and disadvantages, and more accurate results can be obtained by adopting appropriate models for different studies.
(4) Using finite element analysis method and software, the seismic process of infilled wall-steel frame structure can be accurately simulated, and the lateral stiffness curve, hysteretic curve, and stress nephogram of the structure can be obtained.

REFERENCES

Achyutha H, Jagadish R, Rao P S, et al (1986). Finite element simulation of the elastic behavior of infilled frames with openings. *Computer and Structure*, vol. 23, 685–689.

Benjamin J R, Williams H A (1958). The behavior of one-story shear walls. *Proceedings of the American Society of Civil Engineers*, vol. 84, pp. 308–326.

Dai S B, Yu H, Huang J (2005). Nonlinear Analysis of Cooperative Performance of Filled Wall and Steel Frame. *Earthquake engineering and engineering vibration*, vol. 25, pp. 24–28.

Dawe J L, Seah C K (1989). Behavior of masonry infilled steel frames. *Can. J. Cir. Engrg*, Wtta wa, vol. 16, pp. 1989 865~876.

Dhanasekar M, Page A W, Kleeman P W (1985). The behavior of brick masonry underbiaxial stress with particular reference to infilled frames, *Proc. Seventh In. Brick Masonry Conf.*, pp.814~824.

GB 50011–2010 (2010). *Code for Seismic Design of Buildings*. Beijing: China Architecture and Building Press.

Holmes M (1961). Steel frames with brickwork and concrete infilling. *Proceedings of the Institution of Civil Engineers*, vol. 19, pp. 473–478.

Hou H T, Qiu C X, Li G Q, et al (2012). Experimental Study on Low Cycle Repeated Load of Steel Frame with Energy Saving Composite Wallboard. *Engineering Mechanics*, vol. 29, pp. 177–184+192.

Li G Q, Fang M W, Liu Y J, et al (2005). "Experimental study on seismic performance of aerated concrete exterior wall panels of steel structure residential system". *Journal of Civil Engineering*, vol. 38, pp. 27–31.

Li G Q, Wang C (2005). Experimental Study on Hysteretic Performance of Steel Frame Structure with External hanging and Embedded ALC Wallboard. *Steel Structure*, vol. 20, pp. 52–56.

Li G Q, Li X, Sun F F, et al (2003). Shaking Table Test of Steel Structure Residential System Wallboard and Full Scale Model of Wallboard Joint. *Seismic Engineering and Engineering Vibration*, vol. 23, pp. 63–70.

Li J P (2009). *Study on Energy Dissipation and Shock Absorption Performance of External Wall Panels in industrial residential Buildings*. Heilongjiang: Harbin Institute of Technology.

Liu J X (1994). A new seismic Calculation model of infill wall frame structure. *Earthquake Resistance of Engineering*, pp. 22–25.

Liu Y S, Li G Q (2005). Experimental and Theoretical Study on Lateral Force Resistance of Steel Frame Structure with Infilled Walls. *Journal of Building Structures*, vol. 26, pp. 78–84.

Liu X F, Peng S M, Li S F (2004). Experimental Research and Finite Element Analysis of Steel Frame Masonry Enclosure System. *Journal of Wuhan University of Technology*, pp. 52–55.

May I M, Naji J H (1991). Nonlinear analysis of infilled frames under monotonic and cyclic loading. *Computer and Structure*, vol. 38, pp. 149–160.

Memari A M, Aghakouchak A A, Ghafory A A, et al (1991). Full-scale dynamic testing of a steel frame building during construction. *Engineering structures*, pp. 1115–1127.

Papia M (1988). Analysis of infilled frames using a coupled finite element and boundary element solution scheme. *International Journal for Numerical Methods in Engineering*, vol. 26, pp. 731–742.

Polyakov S V (1956). *Masonry in Framed Buildings, Godsudarstvenoe Is-datel'stvo Literaturyo Stroidal stvui Architecture*. Moscow.

Polyakov S V (1960). On the Interaction Between Masonry Filler Walls and Enclosing Frame When Loading in the Plane of the Wall. *Earthquake Engineering Research Institute*, vol. 25, pp. 36–42.

Qiu C X (2011). *Research on Hysteretic Performance of Steel Frame with Composite Wallboard*. Shandong: Shandong University, 2011.

Riddington J R, Naom N F (1994). Finite element prediction of masonry compressive strength. *Computers&Structures*, vol. 52, 113–119, 1994.

Thiruvengadam V (1985). On the natural frequencies of infilled frames. *Earthquake Engineering'oStructural Dynamis*, vol. 13, pp. 401–419.

Wael W, Dakhakhni E, Mohamed E, et al (2003). Three-Strut model for concrete masonry-infilled steel frame. *Journal of Structural Engineering*, pp. 177–185.

Wang J F, Ye H J, Li J C, et al (2015). Seismic performance of steel frame structure filled with energy-saving composite wall panels. *Progress in building steel structures*, vol. 17, pp. 35–43.

Wu M L (2013). *Research on Hysteretic Performance of Assembled Steel Frame-Energy Saving Composite Wallboard Structure System*. Shandong: Shandong University.

Wu Q Y, Tian J H, Xu X Y (1980). Study on the performance of brick wall filled frame under unidirectional and repeated horizontal load. *Journal of Building Structures*, pp. 38–44.

Zhu R H, Shen J M (1996). Experimental and Theoretical Analysis of pseudo-dynamic Seismic Response of reinforced concrete Frame with brick filled wall. *Journal of Building Structures*, pp. 27–34.

Advances in Civil Engineering: Structural Seismic Resistance, Monitoring and Detection – Mohd Johari Mohd Yusof and Junwen Zhang (Eds)
© 2023 The Authors, ISBN 978-1-032-31491-4

Influence of superstructure on stability of air-raid shelter and reinforcement design

Cheng-ming Cao, Yi-lei Shi, Zhao Long & Wen-jin Yan
Gansu CSCEC Municipal Engineering Investigation and Design Institute Co. Ltd., Gansu Lanzhou, China

ABSTRACT: Currently, most of the underground air-raid shelters are old, which not only present stability problems but also have a serious impact on the deformation of the superstructure. Therefore, it is necessary to analyze the stability of the underground air-raid shelter in the new building area to reduce its influence on the new building. In this paper, for the selection of a new building with an air-raid shelter with different burial depth and direction under the foundation of the main building and the skirt building, the stability of the air-raid shelter was calculated and analyzed by using the unloading arch theory. Then Plaxis 3D finite element software was used to establish the calculation model, and the finite element calculation of the deformation of the air-raid shelter under different working conditions was carried out. The stability of the air-raid shelter was analyzed according to the calculation results. The results exhibited that: the deformation of the air-raid shelter under the main building was far beyond the prescribed critical value after the excavation of the foundation pit and the completion of the superstructure; therefore, it needs to be backfilled. The deformation of the air-raid shelter below the skirt was slightly higher than the critical value, which can meet the stability requirements after taking the corresponding reinforcement measures. The research results can provide some guidance for the treatment of air-raid shelters under new buildings.

1 INTRODUCTION

In recent years, with continuous urbanization, some high-rise buildings are being built above the original underground air-raid shelter. The deformation and stability of an air-raid shelter are directly related to the stability of the superstructure. Therefore, while designing the superstructure, we must consider the influence of excavation unloading and superstructure construction loading on the stability of the underground air-raid shelter, and take corresponding measures to deal with the unstable air-raid shelter after disturbance.

In recent years, a large number of scholars have studied and analyzed the stability of underground air-raid shelters after disturbance, and some research results have been obtained (He 2000; Wang 2017; Yang 2007). Combined with the actual project and according to the specific location, structure size, and integrity of the abandoned air-raid shelter, the disposal method of the abandoned air-raid shelter was proposed by Liu Hailin (Liu 2020). Three theoretical methods for stability analysis of abandoned air-raid shelters were introduced by Huang Zhaozhao (Huang 2016). Structural treatment methods, such as foundation crossing and deep beam support column conversion crossing, to conduct foundation treatment in an air-raid shelter area on the premise were adopted by Liu Changjun (Liu 2013). The influence of soil elastic modulus on the maximum settlement of foundation and the stress of air-raid shelter lining were calculated and analyzed by ANSYS finite element software (Xu 2012). In view of the engineering land filling layer and abandoned air-raid shelter, drilling ramming method combined with grouting method for backfill air-raid shelter was used by Luo Changjun (Luo 2003).

Unloading arch theory was often used in the stability analysis of air-raid shelters, and the analysis results were found to be reliable. Unloading arch theory was used to evaluate the stability

DOI 10.1201/9781003310884-13

of underground air-raid shelters by Jia Shangxing (Jia 2007). The loads acting on the structure of air-raid shelter were calculated according to the conditions of shallow burial and deep burial. According to the nonlinear variation regulation of earth pressure with displacement and the different degrees of deformation in each region, the mechanical characteristics of unloading arch under the action of pressure difference were analyzed, and the axis equation expressions of unloading arch and equilibrium arch was deduced by Song Ming-jian (Song 2008). The bearing capacity of unloading arch based on the basic assumption of reasonable arch axis was calculated and analyzed by Xia Zhi-cheng (Xia 2006).

In this paper, combined with practical engineering, the influence of new construction on the stability of existing underground air-raid shelter was analyzed. First, the unloading arch theory was used to calculate the pressure on the top of the shelter. Then PLAXIS 3D software was used to establish the finite element model, and the finite element calculation was carried out for the deformation of the air-raid shelter under excavation unloading and upper loading conditions. According to the results of the calculation, the stability of air-raid shelter under different working conditions was analyzed, and the influence on the new building was determined. Finally, the corresponding reinforcement design of the air-raid shelter with excessive deformation was proposed to avoid the influence on new buildings. The results can provide some guidance for stability evaluation and reinforcement treatment of similar air-raid shelters engineering.

2 UNLOADING ARCH THEORY

The unloading arch theory was known as the Platts theory. According to this theory, when the upper part of the shelter was covered with thick soil, the natural balance arch was formed, called as the unloading arch. Before the excavation, the stratum was in static equilibrium, but after the excavation, this equilibrium was destroyed. Part of the surrounding rock was out of balance to create an arched collapse circle. After adjusting the displacement of the soil outside the arch, a new equilibrium was finally reached, and the weight of the rock (soil) collapsed inside the arch.

According to the unloading arch theory, for the load-bearing arch structure of underground air-raid shelter, when the soil thickness $H > (2.0\sim2.5)\ Hz$ (unloading arch height) of the top of the air-raid shelter, the unloading arch was formed at the top of the air-raid shelter, as shown in Figure 1. The relationship between unloading arch height Hz, span $2B$, air-raid shelter height H, internal friction angle φ of the top soil layer and hardness coefficient F_k can be calculated according to Formula 1.

$$h_z = \frac{b + h\tan(45^\circ - \varphi/2)}{f_k} \tag{1}$$

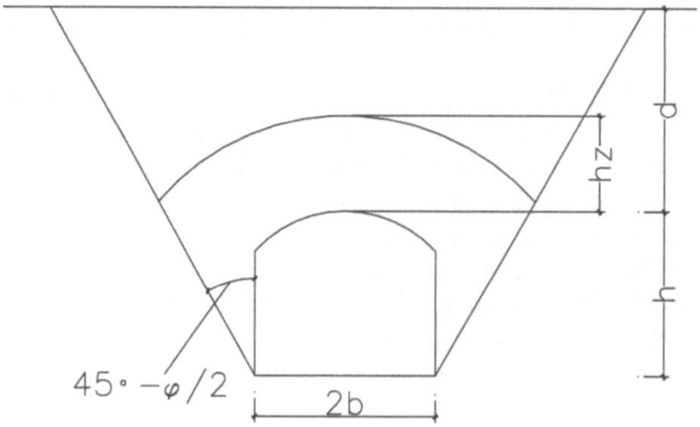

Figure 1. Calculation diagram of unloading arch.

According to the unloading arch theory, the unloading arch can be formed when the air-raid shelter is buried deeply. Its bearing capacity P_x can be calculated by the following formula:

$$P_x = \frac{R_u \tan(45° - \varphi/2)(d - h_z)/k - 2Q_x}{l_x} \tag{2}$$

$$l_x = 2[b + (h + d/2)\tan(45° - \varphi/2)] \tag{3}$$

$$Q_x = \frac{l_x}{2} d\gamma - \frac{2}{3} h_z [b + h \tan(45° - \varphi/2)]/\gamma \tag{4}$$

According to the relationship between the unloading arch bearing capacity P_x and the pressure P at the top of the shelter, the stability of the shelter was judged. When $P > P_x$, the air-raid shelter was in an unstable state, and reinforcement measures are required to be taken for the air-raid shelter. When $P<P_x$, the shelter was in a stable state.

In the above formula, R_u was the ultimate compressive strength of soil; K was safety factor; D was the depth of air raid shelter; L_x was the average span of unloading arch; Q_x was the dead weight of half unloading arch; γ was the weight of the soil.

3 PROJECT OVERVIEW

The proposed site was crisscrossed with air-raid shelters, which is shown in Figure 2. The section of the underground air shelter was horseshoe-shaped. The minimum section clear height was 1.9 m and the clear width was 1.5 m. The depth of the shelter was about 11.2 m from the shallowest part of the current ground. The typical sections of air-raid shelters are shown in Figure 3. The superstructure base was about 78.2 m long and 34.8 m wide. Air raid shelters were distributed under the foundation of the main building and podium building. The shelter was mainly distributed in the pebble layer. The formation strength parameters provided by the survey report were shown in Table 1.

Table 1. Table of soil parameters.

Soil name	Thick/m	Modulus of compression/MPa	r/kN/m^3	c/kPa	φ/°
Fill soil	2.5	3.0	16.5	5.0	15.0
pebble	30	40	21.5		36.0

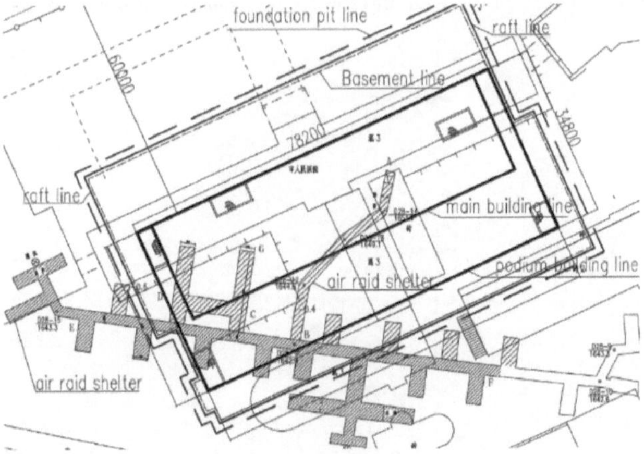

Figure 2. Plane distribution of new building foundation and air-raid shelter.

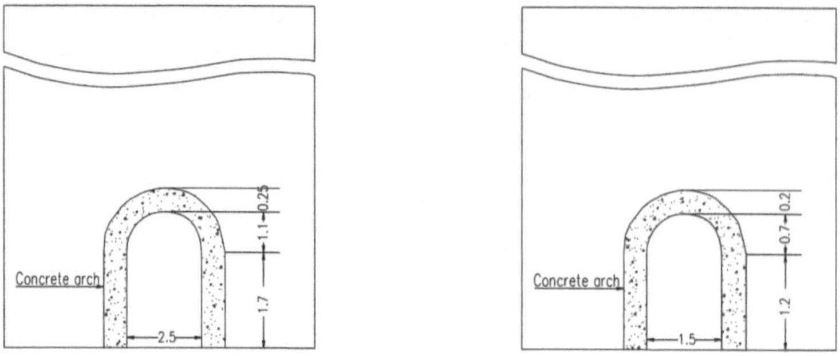

Figure 3. Sectional view of the original structure of the air-raid shelter.

4 FINITE ELEMENT MODEL ESTABLISHMENT

In this study, the PLAXIS 3D finite element software was used to establish the numerical model of air-raid shelter. Finite element calculation was carried out for determination of the current condition, excavation condition, and construction condition of superstructure, respectively. According to the calculation results, the deformation and stability of air-raid shelter under different working conditions were evaluated. In the finite element model, the soil was simulated by a 10-node tetrahedron element. The side wall structure of air-raid shelter was simulated by 6-node plate element converted according to equivalent stiffness. Soil-structure interaction was simulated by 12-node interface element. The interaction between soil and support structure was simulated by setting interface parameters. Based on previous modeling experience, the interface reduction coefficient in the finite element model in this paper was set as 0.75. By setting reduction coefficient, both strength and stiffness were reduced by default in finite element calculation. According to Figure 2, the distribution depth and forms of air-raid shelters under the foundation vary considerably. In this paper, the CD segment in the most disadvantageous position near the double holes and the EF segment under the podium foundation were selected for analysis, respectively. The finite element model and meshing results are shown in Figure 4.

The section height of the air-raid shelter in the CD section is about 2.5 m. The buried depth of the roof was about 20.36 m from the current ground. After excavation, the distance between the roof and the foundation was about 11.65 m. The section height of the EF air raid shelter was about 2.5 m. The buried depth of the roof was about 19.38–20.14 m from the current ground. After excavation, the distance between the roof and the foundation was about 11.35–11.75 m.

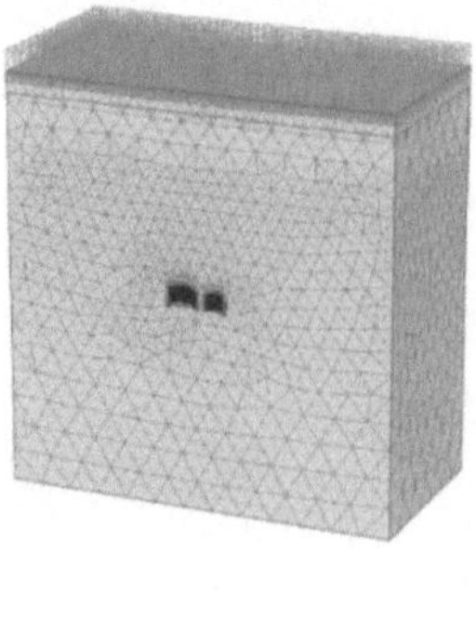

Figure 4. Finite element model.

5 CALCULATION RESULTS ANALYSIS OF AIR-RAID SHELTER DEFORMATION

In this paper, finite element analysis was carried out on the deformation of air-raid shelter under the current condition, excavation condition, and superstructure construction loading condition, respectively.

5.1 *Deformation calculation results analysis of air-raid shelter in CD section*

According to the calculation results, deformation cloud diagrams of the air-raid shelter model in CD section under different working conditions were obtained, as shown in Figure 5.

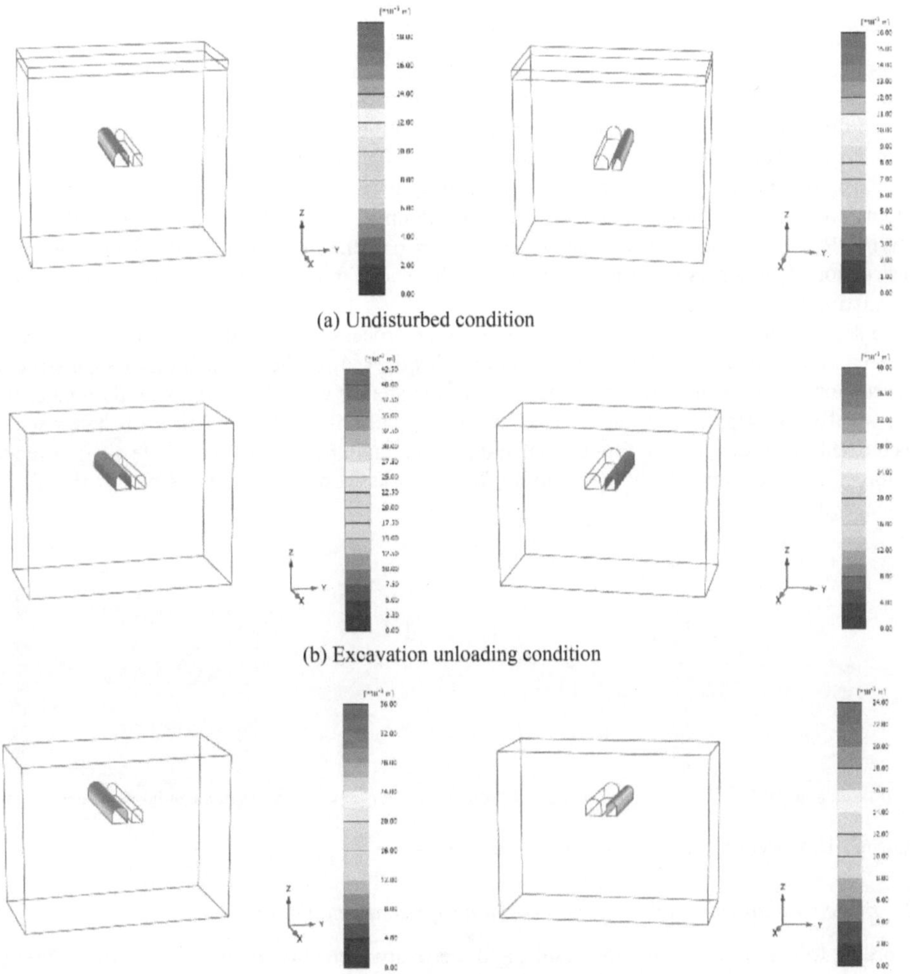

(a) Undisturbed condition

(b) Excavation unloading condition

(c) Construction condition of superstructure

Figure 5. Displacement cloud map of air-raid shelter structure.

It can be seen from Figure 5 that, under undisturbed conditions, the maximum deformation values of the left and right air-raid shelter structures in this section were 15.74 mm and 18.93 mm, respectively. At this time, the shelter was basically stable. In the unloading process of foundation pit excavation, due to decrease in the upper load, large springback deformation occurred, and the maximum deformation reached 39.77 mm and 40.04 mm, respectively. After the construction of the superstructure was completed, the shelter was overloaded, and the maximum deformation value reached 22.17 mm and 34.57 mm, respectively.

In this model, the right cavern was located below the main building, and the deformation value in the process of foundation pit excavation and superstructure construction was far more than the critical value specified in the code of 20 mm. To ensure permanent stability of the superstructure, it was recommended to discard and fill it with plain concrete. In this section of the model, the left cavern was below the skirt, and the load was small in the later stage. The deformation value in the process of foundation pit excavation and superstructure construction was slightly higher than the critical value of 20 mm stipulated in the code. Therefore, this section could be strengthened to meet the requirements of foundation bearing capacity and stability.

5.2 *Calculation results analysis of EF section air-raid shelter deformation*

According to the finite element calculation results, the deformation cloud diagram of EF air-raid shelter model under different working conditions is shown in Figure 6.

According to the calculation results in Figure 6, under the undisturbed condition and the action of ground load and upper soil weight, the maximum deformation value of air-raid shelter structure was found to be 16.78 mm, and the air-raid shelter was in a basically stable state. In the unloading process of foundation pit excavation, due to decrease in the upper load, a large springback deformation occurred together with the soil, and the maximum deformation reached 38.73 mm. After the construction of the superstructure was completed, the load of the shelter increased and the maximum deformation value reached 21.09 mm.

The deformation value of this air-raid shelter in the process of foundation pit excavation was far more than the standard critical value of 20 mm. However, after the completion of superstructure construction, the maximum deformation value of the shelter was 21.09 mm, slightly greater than the critical value stipulated in the code. Considering that only part of the air-raid shelter section was located under the podium foundation, and it was a main passage of the whole air-raid shelter, reinforcement measures can be taken to reinforce, so as to preserve the passage of the air-raid shelter section.

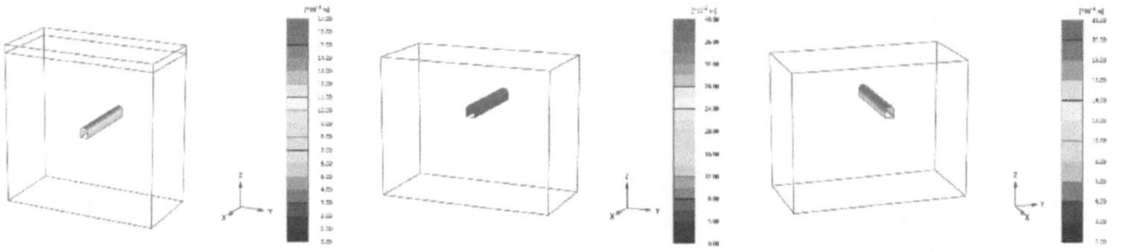

(a) Undisturbed (b) Excavation unloading (c) Construction condition of superstructure

Figure 6. Displacement cloud map of air-raid shelter structure.

5.3 *EF section air raid shelter reinforcement design and calculation*

According to the above calculation results and the relative position relationship between the distribution of air-raid shelters and the superstructure, it can be known that the EF air-raid shelters were located under the foundation of the proposed podium building. And after the upper load was applied, the deformation of the air-raid shelter was slightly greater than the allowable value of the specification. Considering the main passage of the air-raid shelter in this section, reinforcement measures were taken for the air-raid shelter in this section.

According to the calculation results of this paper, the air raid shelter in the influence range of superstructure foundation can be strengthened by hanging steel mesh and spraying concrete. The section of the reinforced air-raid shelter is shown in Figure 7.

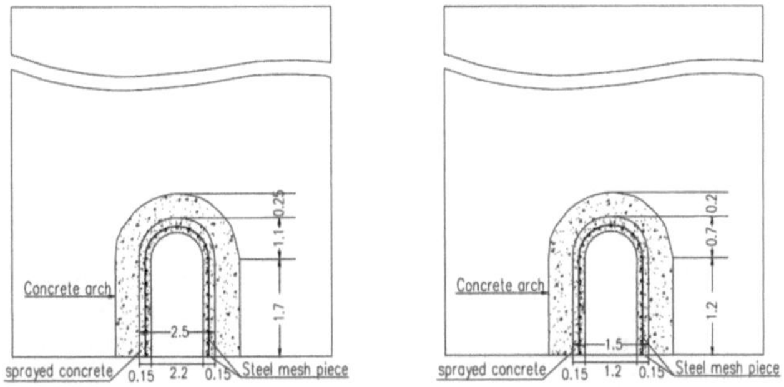

Figure 7. Section view of the reinforced structure of the air-raid shelter.

Finite element calculation was carried out on the reinforced air-raid shelter. According to the calculation results, the deformation cloud diagram of air-raid shelter under excavation conditions and superstructure construction conditions is shown in Figure 8.

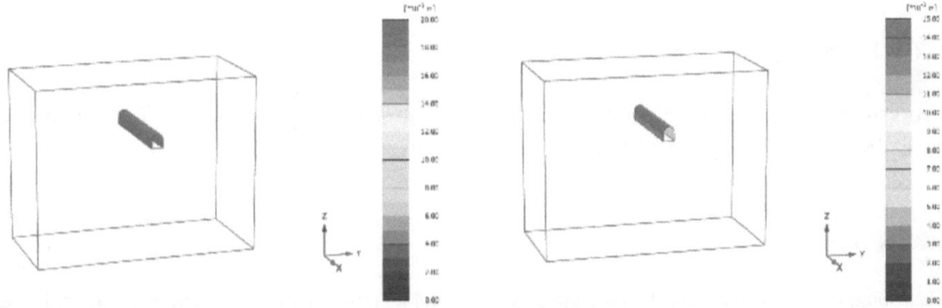

(a) Excavation condition of foundation pit (b) Construction condition of superstructure

Figure 8. Displacement cloud map of air-raid shelter structure after reinforced.

It can be seen from Figure 8 that after reinforcement measures were taken for the air-raid shelter, in the unloading process of foundation pit excavation, large springback deformation occurred together with the soil due to the decrease in upper load. The maximum deformation of the shelter reached 19.31 mm. After the construction of the superstructure was completed, the air-raid shelter was overloaded with a maximum deformation of 14.45 mm. EF section air-raid shelter can meet the deformation control requirements stipulated in the code after reinforcement treatment. The results have shown that the reinforcement effect was good and can ensure prolonged use of air-raid shelter.

6 CONCLUSIONS

In this paper, the influence of foundation pit excavation and superstructure construction of new buildings were mainly considered, and the deformation of the existing air-raid shelter was calculated and analyzed. According to the calculation results, the stability of air-raid shelter under different working conditions was evaluated and reinforcement measures were taken. The main conclusions are as follows:

(1) Under the condition of no disturbance, the deformation of the underground air-raid shelter in this area reached a stable state.
(2) In the process of excavation of the upper foundation pit, with decrease in the upper load, the soil underwent springback deformation. The deformation of air-raid shelter also changed, and its

deformation value increased with the deepening of foundation pit excavation depth. Therefore, the deformation of air-raid shelter should be calculated and analyzed during excavation.

(3) With the construction of the superstructure, the load gradually increased, and the deformation of the air raid shelter under the main building increased sharply. The deformation value had far exceeded the allowable value of the relevant regulations. To avoid the late collapse of the air raid shelter on the superstructure of the serious impact, the air raid shelter under the foundation of the main building should be backfilled.

(4) Due to the small increase in the load on the podium, the deformation of the air-raid shelter under the podium foundation was close to the critical value. Therefore, some reinforcement measures can be taken to deal with the area, to reduce the deformation of the shelter and not affect the safety of the superstructure.

To sum up, because of the existence of air-raid shelters under new buildings, we should first find out the burial depth, structural integrity, and the relative position of air-raid shelters with the main building. Then the deformation of the air-raid shelter after the disturbance was calculated and analyzed. Thus, the stability of air-raid shelters under different conditions can be obtained. Finally, according to the calculation results, the corresponding reinforcement measures should be put forward to ensure long-term stability of the upper structure.

ACKNOWLEDGMENTS

This work was financially supported by Construction Science and Technology Project of Gansu Province (JK2019-01); Science and Technology Project of Gansu Province (20JR10RA570).

REFERENCES

He Xiao-peng, YE Shi-jian (2000) Treatment and implementation plan of abandoned air-raid shelter underground of residential buildings. *West China Exploration Engineering*, 5: 57–58.

Huang Zhao-zhao, Tang Hua-rui (2016) Bomb shelters in the foundation stability analysis. *Sichuan Building Science*, 42(3): 50–53.

Jia Shang-xing, Fu Qiang (2007) Application of Evaluating Air-Raid Shelters Stability Based on Relieving Arch Theory. *Resources Environment and Engineering*, 21(4): 388-390.

Liu Hai-lin, Cui Meng, Fu Xiao (2020) Research on treatment method of abandoned air-raid shelter in deep foundation pit in city downtown. *Journal of Nanchang Institute of Technology*, 39(6): 48–53.

Liu Chang-jun, Meng Fan-ci (2013) Foundation design on air-raid shelter of a project. *Building Structure*, 43: 932–934.

Song Ming-jian, Tang Lian-sheng, Hu Wei-kang (2008) Unloading Arch Considering the Nonlinear Relationship of Displacement and Earth Pressure in Foundation Pit. *Acta Scientiarum Naturalium Universitatis Sunyatseni*, 47(6): 27–31.

Wang Yong-xin (2017) Construction method of foundation pit retaining pile meeting underground air-raid shelter. *Gansu Science and Technology*, 33(20): 112–114.

Xu Jian-hua, Zheng Zhou-lian (2012) On ANSYS Foundation Settlement analyses of Air-Raid Shelters' Lining Stress When Foundation Encountering Underground Air-Raid Shelters. *Chongqing Architecture*, 1: 4–8.

Xia Zhi-cheng, Fan Gan (2006) Study on bearing capacity calculation of unloading arch. *West China Exploration Engineering*, 128: 178–180.

Yang Yong-xin, Dai Xi (2007) Load analysis and treatment methods for the foundation work encountering air-raid shelter. *Journal of Inner Mongolia University of Science and Technology*, 26(1): 70–75.

Advances in Civil Engineering: Structural Seismic Resistance, Monitoring and Detection – Mohd Johari Mohd Yusof and Junwen Zhang (Eds)
© 2023 The Authors, ISBN 978-1-032-31491-4

Study on seismic performance of oil storage tank based on seismic response analysis

Weizhen Cui
Safety Supervision Department, SINOPEC, Beijing, China

Xingyu Xu
Post Doctorate Scientific Research Station Shengli Oil Administration, Dongying, China

Zhengpeng Hou, Feng Zhang & Fuxing Liu
Shengli Oilfield Technology Inspection Center, SINOPEC, Dongying, China

ABSTRACT: Large oil storage tanks—important infrastructure in the oil and gas exploitation and storage process—are bound to cause serious casualties and enormous environmental pollution if damaged under an earthquake. To clarify the seismic performance of large storage tanks under different seismic conditions, the mechanical behaviors of a large oil storage tank under different liquid levels and different earthquake effects were simulated by combining the Adina numerical simulation and the fluid-solid coupling theory. The results revealed that the seismic performance of the storage tank was significantly influenced by the earthquake magnitude and liquid storage height, where the latter was an important index influencing the seismic stability of the tank body. With the increase in the earthquake magnitude and the liquid level, the vertical stress of the storage tank grew while its seismic performance declined.

1 INTRODUCTION

Energy resources, especially oil and gas resources, have been urgently demanded owing to the sustainable high-speed development of the national economy in China. As the important infrastructure in oil and gas exploitation and storage, facilities like oil storage tanks will lead to very serious disasters if damaged in an earthquake, usually accompanied by secondary damages, thus causing severe consequences, imposing enormous losses to the production and national economy and bringing about long-term environmental disruption. The key to ensuring normal oil and gas production lies in studying the seismic fragility of storage tanks and guaranteeing their safety and stability (Qi 2011; Wang 2021; Wu 2021).

All kinds of oil and gas facilities have been extensively investigated by scholars in order to enhance their seismic performance, and certain theories have been formed. Wang S H compared the sloshing situation of different large storage tanks in earthquakes and sought for the optimal structure by simulating the damage conditions of different storage tanks in the event of earthquakes (Li 2020). Wu Y J studied the periodic vibration resistance of large storage tanks, laying a theoretical foundation for the stability design of vertical storage tanks. Jiang G Y et al. analyzed the vibration data of large storage tanks in earthquakes, established a seismic performance rate response library, and completed the aseismatic interactive system design for storage tanks. The abovementioned studies on the seismic performance of oil and gas facilities have been mostly carried out by combining theoretical studies and experiments, while the numerical simulation technology remains to be deeply applied to strengthen the seismic performance of oil facilities. In this

DOI 10.1201/9781003310884-14

study, the dynamic seismic response of oil storage tanks was explored. To be specific, the seismic response characteristics like modal, time history, and nephogram model calculation of equipment and facility structures under different working conditions were analyzed by means of numerical simulation based on the kinetic theory of fluid-solid coupling, thus rendering a scientific basis for pre-earthquake prevention, in-earthquake emergency response, and post-earthquake rehabilitation (Huang 2018).

2 MODELING

The fluid in an oil storage tank sloshes with the tank body in case of an earthquake. In the dynamic seismic response analysis, the influence of fluid in the tank structure is explained as the fluid-solid coupling effect. In this study, the fluid-solid coupling numerical simulation was conducted using the potential fluid elements of Adina finite element software, on basis of the following hypotheses: (1) The medium was free from viscosity, vortex, and heat conduction; (2) The medium could be compressed or slightly compressed; (3) The displacement was relatively small at the fluid boundary. The potential fluid elements were suitable for frequency analysis and transient analysis. The model was constructed with the following considerations: The vibration of the fluid in the storage tank under seismic action was regarded as the linear vibration in a limitedly sealed vessel. φ was set as the potential function of the fluid, which did not permeate through the tank wall, i.e.:

$$\frac{\partial \varphi}{\partial \mathrm{n}} = 0$$

The following kinematical equation is satisfied on the surface of free fluid:

$$\frac{\partial \varphi}{\partial z} = \frac{\partial \zeta}{\partial t}$$

According to the condition of pressure invariance, the gravitational potential is equivalently expressed as below:

$$\frac{\partial \varphi}{\partial t} + gf\zeta = g[X(t)x + Y(t)y]$$

Meanwhile, the kinematic coupling of tank wall-fluid system under seismic action meets the following momentum equation and momentum torque equation:

$$\frac{dQ}{dt} = F + R$$

$$\frac{dG}{dt} = L_F + L_R$$

Where F and FL denote the principal vector and principal moment of fluid (non-liquid level) sloshing force inside the storage tank, respectively. R and LR stand for the principal vector and principal moment of sloshing force at the liquid level, respectively. Q and C represent the momentum and momentum torque, respectively.

In this study, the seismic wave was chosen following the principle of the most disadvantageous ground motion according to the long-periodic structures of three different sites. A total of three seismic waves were chosen, i.e., the artificially synthesized wave comprehensively determined according to the site seismic hazard analysis and the site classification, "Taft (1952) seismic record" and "El-Centro (1940) seismic record". Next, the actual recorded seismic wave was converted based on the corresponding relationship between earthquake magnitude and peak ground acceleration. The converted seismic wave of each magnitude was input into the simulation of dynamic seismic response.

3 SEISMIC PERFORMANCE ANALYSIS OF STORAGE TANK

The seismic performance of a 5000 m^3 vertical doom roof oil tank, which was put into use in an oil union station in 1991, was comprehensively analyzed. This tank was made of A3F rolled steel with a diameter of 23.64 m, a height of 12.518 m, and a safety liquid level of 11.2662 m. The basic technical information of this tank is listed in Table 1.

3.1 *Operation settings*

According to the experience in the historical seismic damage of vertical storage tanks and the site surveys, the height of the storage tank is an important factor influencing the seismic damage prediction results. The liquid level of this doom roof storage tank under normal operation was generally about 90% of the design liquid storage height. When the internal was transported out of the storage tank, the liquid level was generally not lower than the ultimate minimum liquid level. 50% of the design liquid storage height was selected as an abnormal operating condition through the field verification and combining the mechanical property analysis of this storage tank. Therefore, the elastic-plastic time-history analysis of the storage tank was divided into two circumstances: 90% and 50% of design liquid storage height. Subsequently, the seismic time-history response was analyzed using three seismic records under the acceleration amplitude of 0.05, 0.10, 0.15, 0.20, 0.30, 0.40, and 0.80 g, respectively.

Table 1. Margin settings for A4 size paper and letter size paper.

Tank wallboard	Thickness of tank wall (m)	Height of tank wall (m)
The 1st layer	0.010	1.2518
The 2nd layer	0.010	2.5036
The 3rd layer	0.010	3.7554
The 4th layer	0.010	5.0072
The 5th layer	0.010	6.259
The 6th layer	0.010	7.5108
The 7th layer	0.010	8.7626
The 8th layer	0.010	10.0144
The 9th layer	0.010	11.2662
The 10th layer	0.010	12.518

3.2 *Static analysis and modal analysis*

The static analysis and modal analysis of this storage tank were carried out based on 90% and 50% of design liquid storage height, respectively. The static force distribution cloud picture of the fluid and the stress-strain distribution cloud picture of the tank are as shown in Figure 1. The error between the numerical simulation result and the calculated value of hydrostatic pressure was 0.04% and 0.06% under the two circumstances, respectively, reflecting a good simulation effect. The natural vibration frequency of the first 50-order vibration mode under 90% and 50% of design liquid storage height was respectively calculated. The typical structural vibration modes were extracted, as shown in Figure 2.

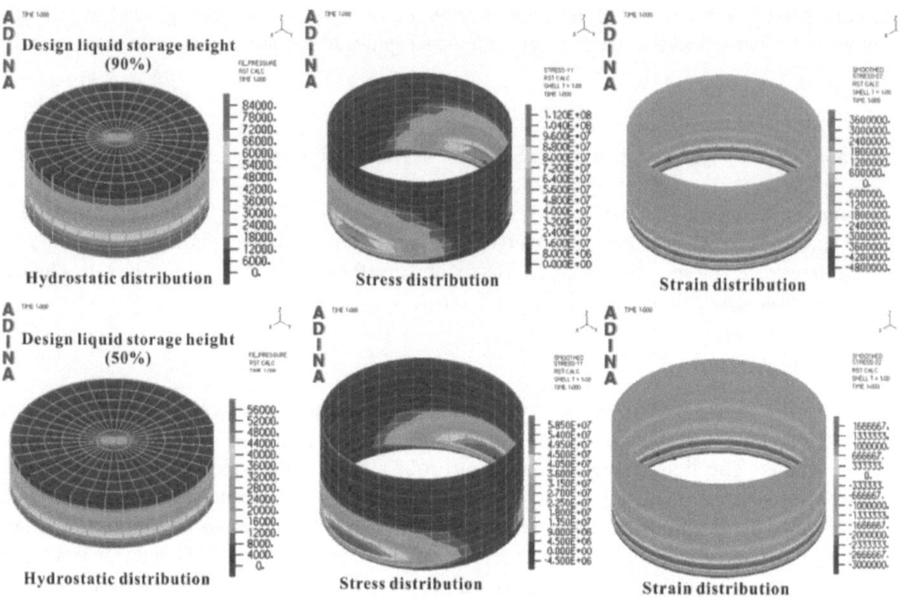

Figure 1. Hydrostatic force and stress-strain cloud pictures of storage tank.

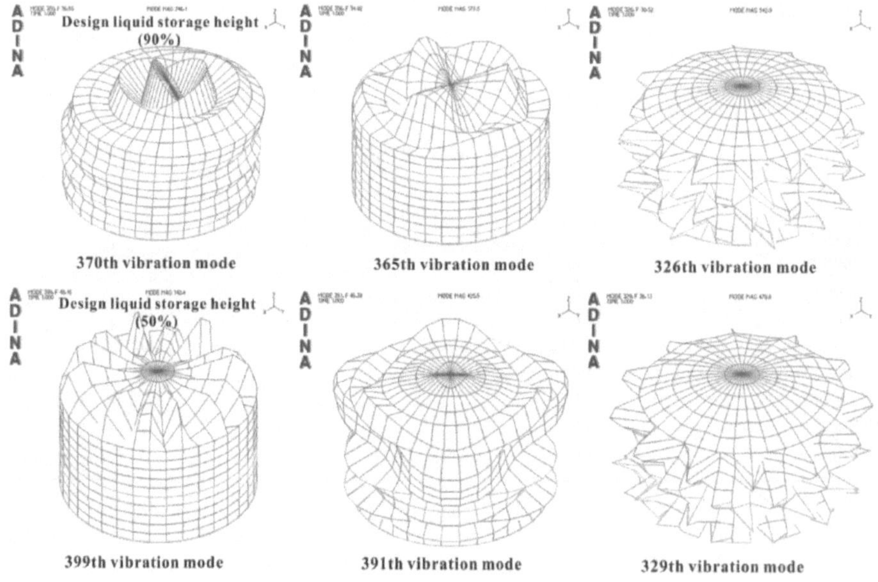

Figure 2. Typical structural vibration modes.

3.3 *Seismic time-history response analysis*

Based on the three seismic records (El-Centro wave, Taft wave, and artificially synthesized wave in the study area) and considering two operating conditions (90% and 50% of design liquid storage height), the seismic time-history response (distribution maps of circumferential tensile stress and vertical compressive stress of the tank wall as well as the time at which their maximum values appeared) of a 5000 m^3 doom roof storage tank in an oil union station built in 1991 was analyzed under the peak ground acceleration of 0.05, 0.10, 0.20, 0.40 and 0.80 g, respectively. Partial analysis results are displayed in Figures 1–2.

The seismic hazard prediction results (Table 2) of various failure modes under different earthquake intensity amplitudes were obtained by calculating the envelope value of each time-history response parameter of the storage tank under different intensity amplitudes (0.05, 0.10, 0.15, 0.20, 0.30, 0.40 and 0.80 g) of three seismic waves and 90% and 50% of design liquid storage height, respectively.

It could be seen from Figures 3–4. that at 50% of design liquid storage height, the failure degree of different failure modes under each earthquake intensity amplitude was generally lower than that at 90% of design liquid storage height. At 50% of design liquid storage height, the structure was intact under the seismic action of 0.05–0.10 g; moderate instability failure occurred at the bottom of the tank wall under the seismic action of 0.20 g; the bottom of the tank wall went through complete instability and the tank was destroyed under the seismic action of 0.40 g. When the liquid level reached 90% of design liquid storage height, the tank wall experienced moderate circumferential

Table 2. Seismic performance prediction results of storage tank.

Seismic amplitude		0.05 g Envelope value	0.05 g Result	0.10 g Envelope value	0.10 g Result	0.20 g Envelope value	0.20 g Result	0.40 g Envelope value	0.40 g Result	0.80 g Envelope value	0.80 g Result
50% of design liquid storage height	Circumferential tensile stress of tank wall (MPa)	71.86	Intact	87.84	Intact	119.81	Intact	183.75	Basically intact	311.62	Seriously damaged
	Vertical compressive stress at the bottom of tank wall (MPa)	1.23	Intact	6.03	Intact	20.32	Moderately damaged	43.56	Destroyed	90.17	Destroyed
90% of design liquid storage height	Circumferential tensile stress of tank wall (MPa)	106.78	Intact	131.44	Intact	186.01	Basically intact	295.14	Moderately damaged	513.40	Destroyed
	Vertical compressive stress at the bottom of tank wall (MPa)	0.68	Intact	10.20	Intact	51.83	Destroyed	91.95	Destroyed	172.18	Destroyed

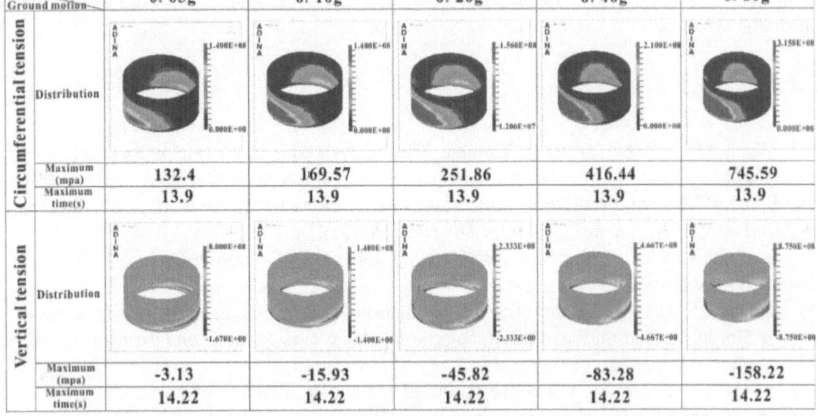

Figure 3. Time-history response analysis of ground motion 1 at 90% of design liquid storage height.

tensile failure under the seismic action of 0.20 g; the tank wall was subjected to circumferential tensile failure and its bottom became completely unstable under the seismic action of 0.80 g.

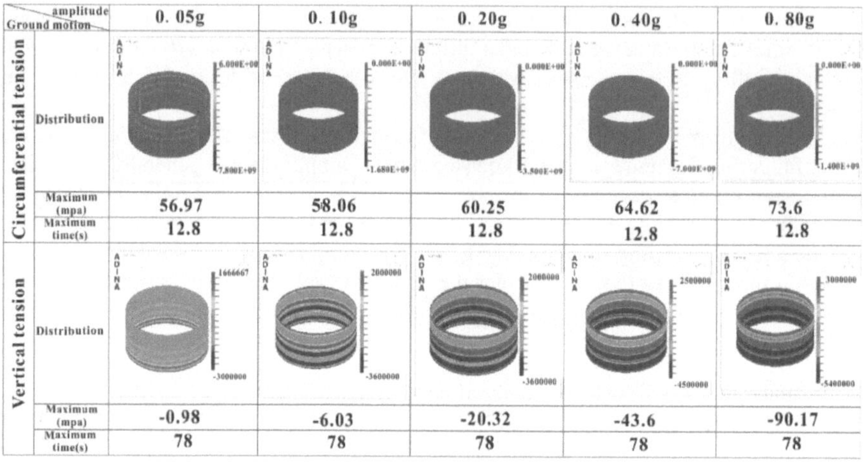

Figure 4. Time-history response analysis of ground motion 3 at 50% of design liquid storage height.

4 CONCLUSIONS

1. At 50% of design liquid storage height: Under the seismic action of 0.05–0.10 g, the storage tank structure was kept intact; under 0.15 g, 8% of storage tanks started experiencing slight damages; under 0.20 g, 38% of storage tanks were damaged to different degrees (moderately damaged-seriously damaged); under 0.30 g, 56% of storage tanks were damaged to different extents (slightly damaged-destroyed); under 0.40 g, the damage degree was aggravated, and 60% of storage tanks were severely damaged or even destroyed; under 0.80 g, 80% of storage tanks were seriously damaged or destroyed.
2. At 90% of design liquid storage height: Under the seismic action of 0.05–0.10 g, the storage tank structure remained intact; under 0.15 g, 58% of storage tanks experienced initial damages to different degrees (slightly damaged-seriously damaged); under 0.20 g, 80% of storage tanks were damaged to different degrees (slightly damaged-seriously damaged); under 0.30 g, 85% of storage tanks were damaged to different degrees (slightly damaged-destroyed); under 0.40–0.80 g, the damage degree was aggravated, and all the storage tanks were subjected to serious damage or even destruction.

REFERENCES

Huang H B, Wang M Z, Gao L. (2021). Comparative analysis and improvement suggestions of typical earthquake damage prediction methods for masonry structures. *Chongqing Architecture*, 17 (02), 47–50.

Li Y Q, Guo E D, Yu T Y et al. Prediction method and software for earthquake damage of water supply pipe network based on VB6.0. *Journal of Natural Disasters*, 29(01), 72–78.

Qi J J. Regressive analysis and application of shear wave velocity of Dongying City area. *Site Investigation Science and Technology*, 02, 23–26.

Wang W Y, Wu Y, You X, Chen L. Seismic fragility evaluation of gas pipelines in the mountainous area based on improved Borda method and attribute recognition. *Oil & Gas Storage and Transportation*, 40 (11), 1265–1271.

Wu H L, Guo E D, Yan P L et al. Predictive study on earthquake damage of horizontal type three-phase separators. *Earthquake Engineering and Engineering Vibration*, 41(05), 106–113.

Advances in Civil Engineering: Structural Seismic Resistance, Monitoring and Detection – Mohd Johari Mohd Yusof and Junwen Zhang (Eds)
© 2023 The Authors,
ISBN 978-1-032-31491-4

Application of finite element strength reduction method in stability analysis of earth-rock dam slope

Jin Quan-hua & Hu Liang
Zhejiang Institute of Hydraulics and Estuary (Zhejiang Institute of Marine Planning and Design), Hangzhou, Zhejiang Province, China
Zhejiang Provincial Key Laboratory of Hydraulic Disaster Prevention and Mitigation, Hangzhou, Zhejiang Province, China

ABSTRACT: Compared with the traditional division method, the finite element strength reduction method takes into account the local plastic failure penetration area of the dam body in the analysis of the slope stability of the earth-rock dam, and comprehensively determines the instability of the dam slope. The calculation results are thus more practical. The calculation results of the single strength reduction coefficient method and the double strength reduction coefficient method are compared with the calculation results obtained using the Bishop method to verify the reliability and rationality of the calculation results of the finite element method as well as further improve the methods and means of the stability analysis of earth-rock dam slopes.

1 INTRODUCTION

Dam slope stability is one of the major concerns of hydraulic engineers. The rigid limit equilibrium method for analysis of the dam slope stability is still the method of choice. Currently, the traditional rigid body limit equilibrium technique is mostly applied using the division method, mainly including the Swedish arc method, the Bishop method, the Morgenstern-Price method, the Spencer method, the Janbu method, and more (Yin 2007).

In recent years, many scholars have proposed to use the finite element strength reduction method to analyze the slope stability (Cheng 2019; Chen 2006; Tang 2007; Wang 2019; Zhu 2017). Currently, the finite element strength reduction method mainly uses a single strength reduction coefficient, whereas some scholars also use the double strength reduction coefficient method. The application involves analytical research to try and discover its regularity. Cheng aims to establish a simple strength reduction strategy to promote the application of the double reduction method. Wang (2019) studies the influence of water content on the change of strength parameters with the reduction coefficient and determines the matching of reduction parameters according to data fitting. Zhu (2017) carried out an application analysis of the reduction supporting mechanism under the linear proportional relationship of the double reduction coefficient. And Tang and Zheng carried out a theoretical analysis of the rationality of the double reduction mechanism.

In calculations using the division method, the range or shape of the sliding surface needs to be assumed, the soil body is regarded as a rigid body, and technical personnel with rich engineering experience are required due to complexity of dam structures. The finite element strength reduction method considers the deformation characteristics of the soil and does not need assumptions about the range of the sliding surface. It only needs to continuously reduce the strength of the soil material and directly analyze the stability of the slope, which is more practical (Cen 2017).

DOI 10.1201/9781003310884-15

2 FINITE ELEMENT STRENGTH REDUCTION METHOD

2.1 *Single strength reduction coefficient method*

Single strength reduction coefficient method was first proposed by Zienkiewicz et al, later widely adopted by many scholars, who proposed a concept of Shear Strength Reduction Factor (Wang 2019). The essence of the strength reduction method is that the cohesion c and the internal friction angle φ of the soil material gradually decrease under the premise of keeping the load constant, so that the shear strength of the soil element cannot match the applied stress. The unbearable stress would gradually be transferred to the middle area of the surrounding soil elements. When a continuous sliding surface or a yield point forms through the surface, the soil will be unstable. The reduced shear strength parameters are expressed as follows.

$$c_i = c/F \tag{1}$$

$$\varphi_i = \arctan(\tan\varphi/F) \tag{2}$$

c_i, φ_i – the reduced soil cohesion and internal friction angle, respectively.
F – Single strength reduction factor.

2.2 *Double strength reduction coefficient method*

Generally, in the process of soil material strength reduction, the reduction coefficients of cohesion and internal friction angle values are not completely synchronized at the same time. Therefore, the double strength reduction coefficients of cohesion and internal friction angle are proposed.

$$c_i = c/F_c \tag{3}$$

$$\varphi_i = \arctan(\tan\varphi/F_\varphi) \tag{4}$$

F_c, F_φ – the cohesion reduction factor and internal friction angle reduction factor, respectively.

3 DAM SLOPE INSTABILITY CRITERIA

The method for determining the instability of the dam slope in this paper: The finite element calculation process does not converge, and the dam body locally forms a plastic failure through the area, which is comprehensively determined.

4 PROJECT OVERVIEW

The earth-rock dam is a homogeneous clay dam with a crest elevation of 198.00 m, a dam height of 20 m, a crest width of 4 m, and a length of 120 m. The upstream slope ratio is 1:2.0, and the downstream slope ratio is 1:1.80.

Table 1. Dam body soil physical and mechanical parameters.

Parameters	Severe (kN/m^3)	Elastic Modulus (MPa)	Poisson's ratio	Cohesion (kPa)	Internal friction angle (°)
Values	21	100	0.35	11	23

5 APPLICATION ANALYSIS

In this paper, ADINA finite element software was used for calculation and analysis of the strength reduction method. The calculation grid model is given in Figure 1. The Mohr-Coulomb constitutive model was used, and the calculation grid model is as follows. The element type is a plane strain 4-node quadrilateral element, with 651 nodes and 600 elements.

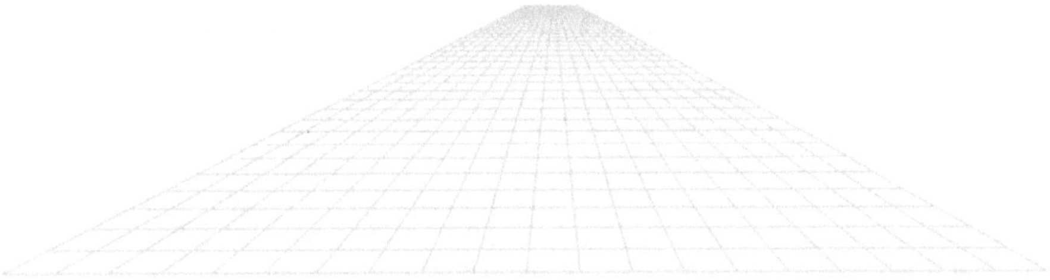

Figure 1. Finite element mesh model.

5.1 *Single strength reduction factor method*

According to the internal friction angle, the cohesion strength index is simultaneously reduced and the strength reduction coefficient F = 1.203, which is the dam slope anti-sliding stability safety factor. The octahedral plastic shear strain distribution is shown in Figure 2.

According to "Design specification for roller compacted earth-rock dam" (SL274-2020), Bishop method is used for analysis, and using the Geo-Studio software to obtain the dam slope anti-sliding stability safety factor of 1.184, the arc position is given in Figure 3.

The dam slope anti-sliding stability safety factor F = 1.203, calculated using the finite element unit strength reduction coefficient method, which is compared with the anti-sliding stability safety factor 1.184 calculated by the Bishop method; giving a difference of 1.6%. Moreover, the distribution of the octahedral plastic shear strain penetration zone calculated by the finite element method is the same as that of the Bishop method.

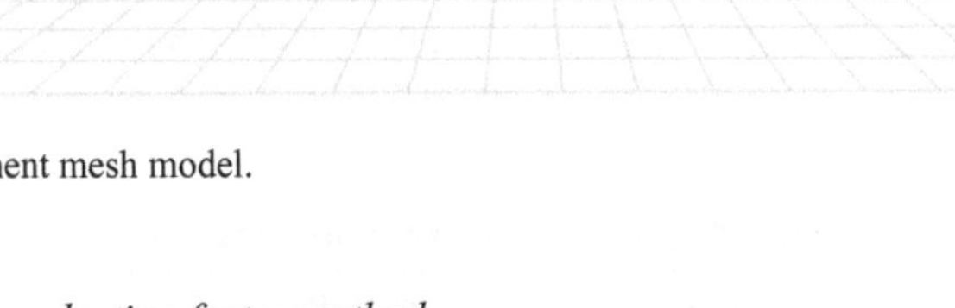

Figure 2. Octahedral plastic shear strain contour.

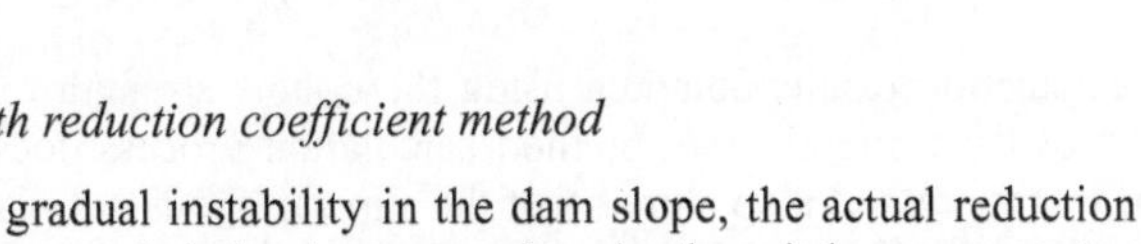

Figure 3. Bishop method – dam slope sliding arc location

5.2 *Double strength reduction coefficient method*

Generally, with the gradual instability in the dam slope, the actual reduction of internal friction angle and cohesion strength index is not synchronized, so it is necessary to use double strength reduction coefficients to simulate the reduction changes of each index. There is no clear or unified understanding of the relationship between them. In this paper, the double reduction coefficient is used to form a linear relationship for application analysis, and the ratio of the internal friction angle and the cohesion double strength reduction coefficient is defined as $R = F\varphi/Fc$.

First, the extreme value analysis of the reduction coefficient of a single strength index is carried out. ① Assuming that the internal friction angle is constant, the single strength index reduction coefficient of cohesion $F_{c0} = 1.800$ is calculated. ② Assuming that the cohesion is unchanged, the single strength of the internal friction angle is calculated. The index reduction coefficient $F_{\varphi 0} = 1.288$.

According to the above assumption of the extreme value of the single strength reduction coefficient, the reduction coefficient of the internal friction angle is obviously smaller than the reduction rate of the cohesion reduction coefficient $R_0 = F_{\varphi 0}/F_{c0}$; while the single strength reduction coefficient method reduction coefficient ratio $R_1 = 1.0$ is the upper limit. This paper considers the ratio of double strength reduction coefficients R = 0.716, 0.75, 0.80, 0.85, 0.90, 0.95, by preliminary analysis, and the results are given in Table 2, and the octahedral plastic shear strain contour in Figures 4(a)–(f).

Table 2. Calculation results corresponding to different double strength reduction factor ratios.

Serial number	Dual strength reduction factor ratio ($R = F_\varphi/F_c$)	Cohesion reduction Factor (F_c)	Internal friction angle reduction factor (F_φ)
1	0.716	1.551	1.111
2	0.75	1.498	1.124
3	0.8	1.424	1.139
4	0.85	1.353	1.15
5	0.9	1.301	1.171
6	0.95	1.248	1.186

According to literature (CHENG2019, Wang2019) and other studies, the calculation formulas (5)–(9) are selected for calculating the comprehensive value of the double strength reduction coefficient. Due to the relatively large proportion of the internal friction angle in the anti-slip effect of the dam slope, the weight of each reduction coefficient is allocated according to the degree of action, and the calculation formula (9) for the comprehensive value of the double-strength reduction coefficient is proposed. The calculation formulas are as follows.

$$\text{Arithmetic mean: } F_s = \frac{F_c + F_\varphi}{2} \tag{5}$$

$$\text{Yuan Wei: } F_s = \frac{\sqrt{2}F_c F_\varphi}{\sqrt{F_c^2 + F_\varphi^2}} \tag{6}$$

$$\text{Geometric mean: } F_s = \sqrt{F_c F_\varphi} \tag{7}$$

$$\text{RMS mean: } F_s = \sqrt{\frac{F_c^2 + F_\varphi^2}{2}} \tag{8}$$

$$\text{This paper: } F_s = \left(1 - \frac{F_{\varphi 0}}{F_{c0}}\right) F_c + \frac{F_{\varphi 0}}{F_{c0}} F_\varphi \tag{9}$$

According to the calculation results obtained using the double strength reduction coefficient method, although part of the finite element method calculation process does not converge, the octahedral plastic shear strain cloud map shows that the slip surface is not plastically penetrated, indicating that the actual instability state has not been reached, such as the reduction coefficient.

Table 3. Calculation results of comprehensive value of double strength reduction coefficient.

Serial number	Comprehensive reduction factor (R)	0.716	0.75	0.8	0.85	0.9	0.95
1	Arithmetic mean	1.331	1.311	1.282	1.252	1.236	1.217
2	Yuan Wei	1.277	1.271	1.258	1.239	1.231	1.216
3	Geometric mean	1.313	1.298	1.274	1.247	1.234	1.216
4	RMS mean	1.349	1.324	1.289	1.256	1.238	1.217
5	This paper	1.236	1.23	1.22	1.208	1.208	1.204

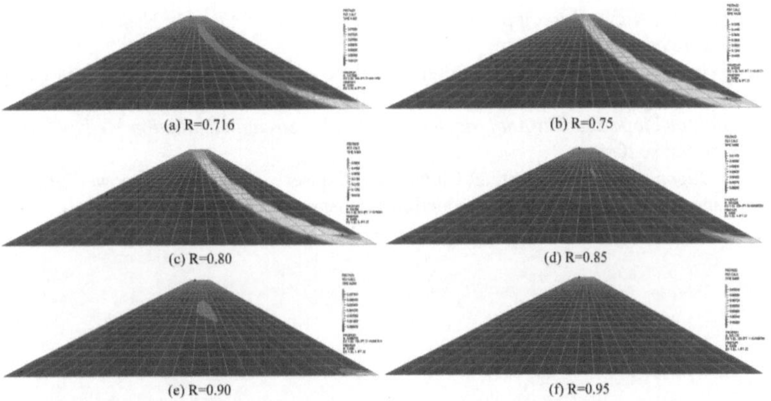

Figure 4. Octahedral plastic shear strain contour.

In several cases when the ratio R = 0.85, 0.90, and 0.95 and the ratio of the reduction coefficient is about to actually reach the unstable state, R = 0.75 and 0.80, it shows that the determination of instability requires calculation non-convergence and combined with the local formation of plastic failure of the dam body through the area to determine.

From the above comprehensive results of the double strength reduction coefficient, it can be seen that the calculation results of this paper are closer to the calculation results of the Bishop method and the single strength reduction coefficient method. When the reduction coefficient ratio R = 0.75 and 0.80, the corresponding comprehensive reduction coefficient values are 1.230 and 1.220, respectively, which are 3.9% and 3.0% different from those calculated by the Bishop method and 1.40% and 2.3% different from those calculated by the single strength reduction coefficient method; overall not much difference.

6 CONCLUSION

Based on the analysis results of this paper, the conclusions are as follows:

(1) The difference between the calculation results of the finite element single strength reduction coefficient method and the Bishop method is small, indicating that the calculation results of the single strength reduction coefficient method are more reliable. In this paper, the calculation method of the comprehensive value of the double strength reduction coefficient is proposed according to the contribution weight of the strength index, which is not much different from the calculation results of the Bishop method and the single strength reduction coefficient method, The results are more reasonable.

(2) Currently, there is no unified understanding or clear mathematical relationship between the double strength reduction coefficients, and the actual calculation results are related to the relationship between the double reduction coefficients, so further in-depth analysis and research are needed.

REFERENCES

Cheng Ziqiao,etc. Slope stability analysis based on a dual factoring strategy[J]. *Journal of Beijing Jiaotong University*, 2019, 43(4):45–51.

Chen Yan. *Stability Analysis of Rock and Soil Slopes Using Finite Element Method.* [D]. Nan Jing. Hohai University. 2006.

Cen Wei-jun, Zhou Tao, Xiong Kun. *Application and Development of Adina in Hydraulic Engineering* [M]. Beijing: Posts & Telecom Press, 2017.

Tang Fen, Zheng Ying-ren. Analysis on Safety Reserve of Slope with Two Strength Reduction Factor[J]. *Journal of Chongqing Jiaotong Unversity (Natural Science)*, 2007, 26(4): 95–100.

Wang Jiqiang. *Study on Slope Stability Analysis Based on Double Strength Reduction Method* [D]. Chongqing. Chongqing University. 2019.

Yin Zong-Ze, ect. *Tugong Yuanli*[M]. Beijing: China Water Conservancy and Hydropower Press, 2007.

Zhu Yan-peng, Yang Xiao-yu, etc. Matching reduction mechanism of double reduction method[J]. *Journal of Lanzhou University of Technology*, 2017, 43(2): 121–126.

Advances in Civil Engineering: Structural Seismic Resistance, Monitoring and Detection – Mohd Johari Mohd Yusof and Junwen Zhang (Eds)
© 2023 The Authors, ISBN 978-1-032-31491-4

Seismic performance analysis of a new layered suspension structure system based on OpenSees

Qingguang He & Lei He
Department of Civil Engineering, Lanzhou University of Technology, Lanzhou, China

ABSTRACT: In recent years, to solve the problem of serious damage caused to the traditional building structures that are difficult to repair after earthquakes, the seismic mitigation and isolation technology have been continuously developed. To reduce the seismic response of the structure, a new type of layered suspension structure system is proposed, and the finite element software OpenSees is used to build a nonlinear model of this structure for numerical simulation; at the same time, material nonlinearity and geometric nonlinearity are also considered. The response of the layered suspension structure and the ordinary frame structure under different seismic waves is compared and analyzed, and the calculation results show that the seismic performance of the layered suspension structure is better than that of the ordinary frame structure.

1 INTRODUCTION

Earthquakes are natural disasters endangering people's lives and property, and causing damage to buildings; how to minimize the damage to buildings caused by earthquakes is an area that people have been committed to research. The seismic design of the traditional structure is mainly to increase the rigidity of the structure by adjusting the performance of the structure itself, such as increasing the cross-section of the component, which is a typical "hard resistance" method and is not economical.

As a new building structure system, the suspended building structure receives more attention, not only because the suspension method has a strong artistic expression, but more importantly, this structural system conforms to the principle of natural force transmission, and can make full use of the mechanical properties of high-strength materials, and improve the seismic and wind resistance of the structure, which has good application prospects. At present, extensive research is being carried out on the suspension structure. Wang (2007) studies dynamic characteristics and parameter optimization of core-tube suspension structure, applied Lagrange equations to derive the motion equation of the single core-tube suspension structure, and based on this, the vibration absorption performance was demonstrated by time history analysis. Zhou (2005) studied natural vibration characteristics of the core-tube suspension structure system, conducted a theoretical analysis on the dynamic characteristics of the core-tube suspension structure, and compiled a calculation program for the suspension structure based on the MATLAB. Liu (1997) studied estimation method of elastic stability of suspended building frame structure, introduced the calculation model of the stability of the suspended building frame structure, used the moment distribution method to derive the critical load estimation formula, and provided a practical curve. Tong (2013) studied the stability of giant suspension structure, studied the elastic buckling of a suspension structure with a single suspended load, and obtained an analytical solution. For the huge suspension structure with multi-layer suspended load, the continuity model was used to derive the stable equilibrium differential equation. However, the current research on suspended structures is concentrated in the field of high-rise buildings and does not reflect the structure's response in three-dimensional space. In

DOI 10.1201/9781003310884-16

summary, this paper proposes a new layered suspension structure system suitable for multi-story buildings, and analyzes its seismic performance.

2 MODEL ESTABLISHMENT OF A NEW LAYERED SUSPENSION STRUCTURE SYSTEM

The new layered suspension structure system proposed in this paper is shown in Figure 1. It keeps direct connection between the roof beam and the column, and setting the hanging point of the suspended floor on the column.

The components of this layered suspension structure are: 1) Prefabricated roof; 2) Concrete columns; 3) Inter-column supports; 4) Steel multifunctional connectors; 5) Steel suspenders; 6) Buffer devices; 7) Prefabricated slab beams. Suspended slab beams are hung on the steel multifunctional connectors through steel suspenders; the suspenders are hinged with the steel multifunctional connectors and the suspended slab beams, respectively, and buffer devices are installed in the middle of the suspended floor beams and the columns to prevent the collision between, and at the same time play a role in energy consumption during the earthquake, the inter-column support is to improve the overall rigidity of the frame structure.

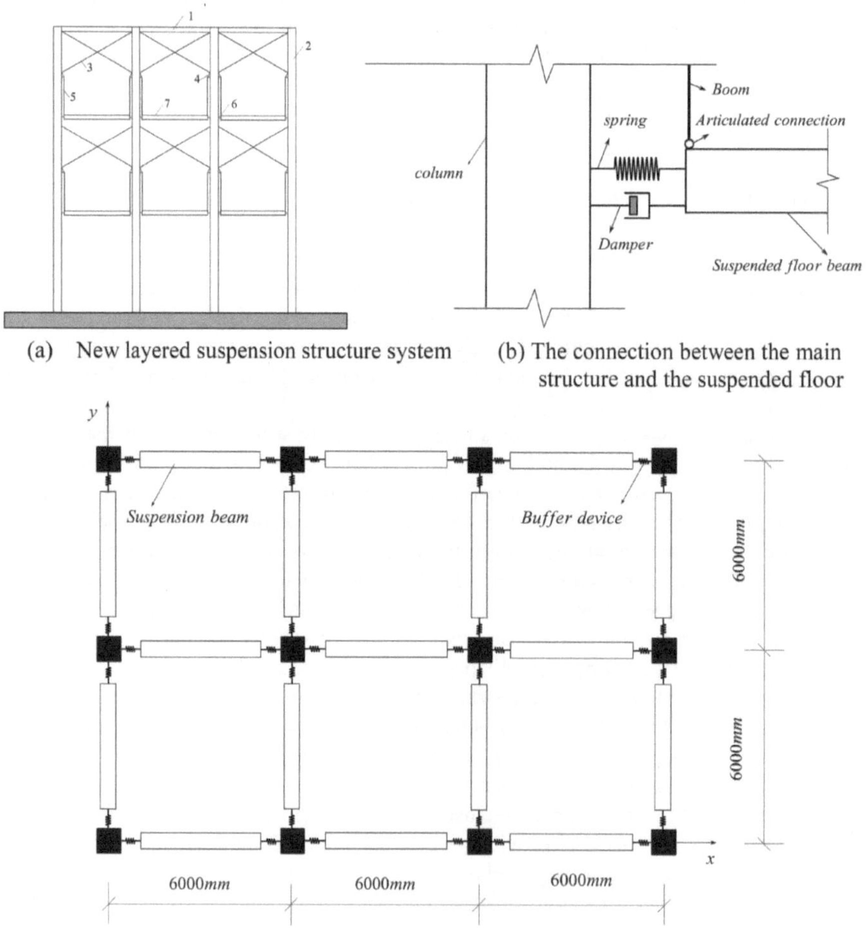

(a) New layered suspension structure system

(b) The connection between the main structure and the suspended floor

(c) Floor plan of suspended floor buffer device

Figure 1. System diagram of a new layered suspension structure.

The basic parameters of the model: 3 spans horizontally, 2 spans vertically, each span is 6 m, the story height is 3.6 m, and there are 3 story's in total. The first and second floors are suspended floors, on the first and second floors, and there are buffer devices at the connection position of each suspended beam and column. Each suspended floor is equipped with 34 buffer devices to provide additional rigidity and damping for the suspended floors, and the stiffness provided by each buffer device is 100,000 N/m, as shown in Figure 1. The beams, columns, and slabs are all made of C35 concrete, the beam section size is 300 mm*500 mm, the column section size is 500 mm*500 mm, the reinforcement is HRB400 grade steel, the thickness of the concrete floor is 120 mm, and the thickness of the concrete protective layer is 50 mm. The suspender adopts a steel suspender with a circular cross-section and a diameter of 20 mm, the length of the suspender is 1.2 m. Because this layered suspension structure only directly connects the top beam and the column, the rest of the beams are suspended on the column by suspenders, which will inevitably cause the lateral rigidity of the structure to decrease; therefore, cross supports are placed between the columns to enhance the stability of the structure and reduce the adverse effects caused by the reduction of beam-column nodes. The cross supports between the columns are made of Q345 steel. The design seismic group is divided into the third group, the site category is category II, the site characteristic period is 0.45s, the seismic fortification intensity is 8 degrees, the design basic seismic acceleration value is 0.2g, and the damping ratio of the structure is 0.05. The standard value of floor and roof live load is 2.0 KN/m^2, and the additional dead load value of floor and roof is 2.0 KN/m^2, with wall weight 5.24 KN/m^2. The wall weight is converted to linear load and evenly distributed on the beam. Wind load does not have a controlling effect on the design of multi-story buildings, so the effect of wind load is not considered. This paper uses OpenSees, a nonlinear finite element analysis program developed by the University of California, Berkeley, to establish a three-dimensional model of this layered suspension structure, at the same time, an ordinary frame structure model is established, the beam and column dimensions and materials of the ordinary frame structure are the same as the layered suspension structure.

3 SELECTION OF SEISMIC WAVES

According to FEMA P695 report of the United States, three far-field seismic waves and near-field seismic waves are selected, respectively. See Table 1 for information on seismic waves.

Table 1. Earthquake record information.

Type	Numbering	Earthquake name	Record station name	PGA_{max} (g)	Magnitude	Portion
Far field	01	San Fernando, USA	LA-Hollywood Sto	0.21	6.6	SFERN/PEL090
	02	Friuli, Italy	Tolmezzo	0.35	6.5	FRIULI/A-TMZ000
	03	Imperial Valley, USA	Delta	0.35	6.5	IMPVALL/H-DLT352
Near field	04	Gazly, USSR	Karakyr	0.71	6.8	GAZLI/GAZ_267
	05	Imperial Valley-06	Bonds Corner	0.76	6.5	IMPVALL/H-BCR_323
	06	Loma Prieta	BRAN	0.64	6.9	LOMAP/BRN_038

4 CALCULATION AND RESULT ANALYSIS OF LAYERED SUSPENSION STRUCTURE

4.1 *Modal analysis*

The period and mode of each order of vibration of the layered suspension structure and the ordinary frame structure are calculated, respectively, and the calculation results are shown in Table 2.

Table 2. Mode shape and period.

Mode shape	Frequency (Hz)		Period(s)	
	Layered suspension structure	Ordinary frame structure	Layered suspension structure	Ordinary frame structure
1	0.388	1.958	2.58	0.51
2	0.394	2.022	2.54	0.49
3	0.407	2.351	2.46	0.43
4	0.413	6.469	2.42	0.15
5	0.469	6.616	2.13	0.13
6	0.476	7.704	2.10	0.09

As can be seen from the above table, due to higher flexibility of the layered suspension structure, it can dissipate energy through the low-amplitude swing of the suspended floor, change the distribution law of vibration mode, and make the natural frequency of the structure smaller than that of the ordinary frame structure.

4.2 *Time history analysis*

To investigate the seismic performance of the layered suspension structure system, the response of the layered suspension structure system and the ordinary frame structure under different seismic waves was compared and analyzed. First, performing modal analysis on the two structures, according to the results of the modal analysis, the y-direction of the building is taken as the main direction of the structural time history calculation. Inputting the above 6 seismic waves in the y-direction of the layered suspension structure and the ordinary frame structure, respectively, we compared the response of the two structures under the action of the earthquake when the amplitude modulation of the seismic wave is 70 gal.

When the amplitude modulation of the seismic wave is 70 gal, the maximum acceleration and maximum displacement of the vertex of the two structures are compared as shown in the figure below:

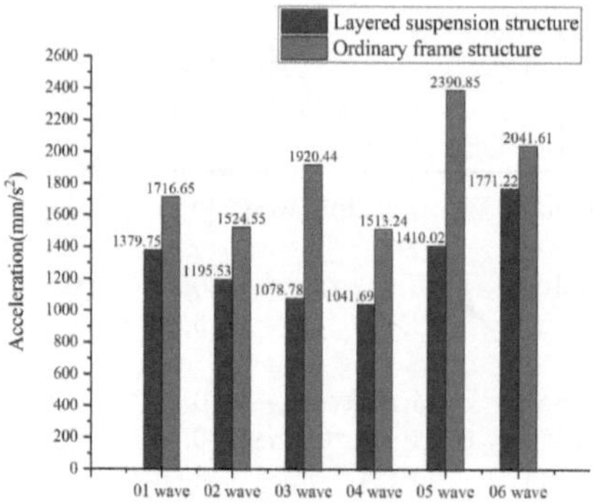

Figure 2. Vertex maximum acceleration comparison chart.

Defining acceleration amplification factor, $\beta =$ maximum acceleration at the top of the structure/peak acceleration of the seismic waves, it can be seen from the above figure that under the effect of frequent earthquakes, the maximum acceleration of the apex of the layered suspension structure is less than that of the ordinary frame structure. The maximum acceleration amplification coefficients of the layered suspension structure and the ordinary frame structure are $\beta_{max} = 2.53$ and 3.42, respectively. It can be determined if the ordinary frame structure is more susceptible to

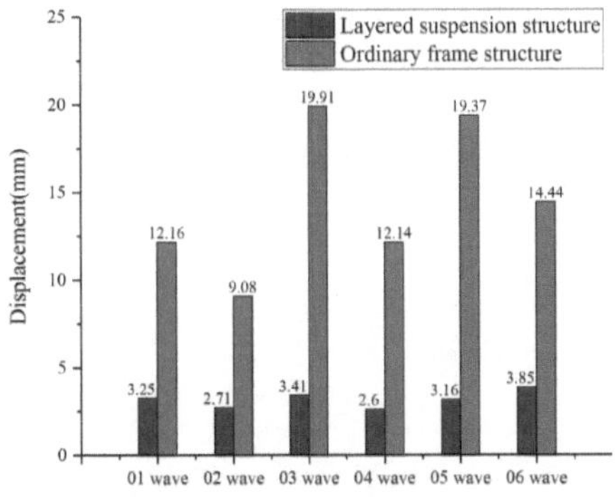

Figure 3. Vertex maximum displacement comparison chart.

earthquakes than the layered suspension structure. The maximum displacement of the apex of the layered suspension structure is smaller than that of the ordinary frame structure. At the same time, under the action of the 05 wave, the maximum displacement of the apex of the layered suspension structure and the ordinary frame structure is 3.16 mm and 19.37 mm, respectively. The maximum displacement of the apex of the layered suspension structure is reduced by 83.7% compared with the ordinary frame structure. Therefore, the layered suspension structure has good seismic performance, at the same time, it shows that due to the existence of suspended floors, the deformation of the main structure of the suspension structure is significantly reduced, which is beneficial in reducing the damage of the structure during the earthquake.

5 CONCLUSION

In this paper, the numerical simulation method is adopted to study the seismic performance of layered suspension structures. The main conclusions can be summarized as follows:

(1) Due to higher flexibility of the layered suspension structure, the natural frequency of the structure is lower than that of the ordinary frame structure.
(2) Under the effects of frequent earthquakes, the maximum displacement of the apex of the layered suspension structure is smaller than that of the ordinary frame structure. Under the action of the 05 wave, the maximum displacement of the apex of the layered suspension structure is reduced by 83.7% compared with the ordinary frame structure, due to the existence of suspended floors, and the deformation of the main structure of the suspension structure is significantly reduced.
(3) The maximum acceleration amplification factor, β_{max}, of the layered suspension structure is smaller than that of the ordinary frame structure, which is more susceptible to earthquakes than the layered suspension structure, and the seismic performance of the layered suspension structure is better than that of the ordinary frame structure.

In terms of future work, such experiments should be continued to enhance persuasiveness.

REFERENCES

C.L. Wang, Z.T. Lv. (2007). Dynamic characteristics and parameter optimization of core-tube suspension structure, *Journal of Southeast University (Natural Science Edition)*. 2,181–185.

G.S. Tong, J. Su. (2013). Research on the stability of giant suspension structure, *Engineering mechanics*. 30, 75–82.

J. Zhou, X.B. Wu. (2005). Research on natural vibration characteristics of core-tube suspension structure system, *Engineering mechanics*. 1,75–81+74.

Y.X. Liu, Z.T. Lv. (1997). Estimation method of elastic stability of suspended building frame structure, *Engineering mechanics*. 14, 29–37.

Advances in Civil Engineering: Structural Seismic Resistance, Monitoring and Detection – Mohd Johari Mohd Yusof and Junwen Zhang (Eds)
© 2023 The Authors, *ISBN 978-1-032-31491-4*

Experimental study on static stability of tailings dam with geotextile tubes

Qiaoyan Li* & Yulin Lu
Institute of Disaster Prevention, Hebei, China

ABSTRACT: Geotextile tubes are one of the emerging and promising technologies that help build fine-grain tailings dam. In this study, static stability tests are conducted to evaluate the effect of the geotextile tubes on the tailings dam. The test results indicate that the geotextile has a very significant effect on the stability of the dam body. The failure mode of the tailings dam with geotextile tubes is the interlayer sliding failure between the geotextile tubes. The fracture surface is deepened to a certain extent compared with the tailings dam without geotextile tubes. The failure pressure is larger and the displacement is smaller than that of the tailing dam without geotextile tubes. The displacement and deformation values of the tailings dam with large geotextile tubes are smaller than that of the tailings dam with small geotextile tubes under the same conditions. The stability of the tailings dam with geotextile tubes is better than that of a tailings dam without geotextile tubes.

1 INTRODUCTION

The safety and stability of tailings dam plays a very important role in ensuring safe production of mines and maintaining good social and economic benefits (Wei et al. 2002). With improvement in the current level of beneficiation technology and recovery rate, the particle size of the tailings entering the tailings pond is getting finer. The characteristics of fine-grained tailings are poor water permeability, long consolidation time, low mechanical strength, and difficulty in dissipating excess pore water pressure after storage. Therefore, using the traditional tailings storage method, tailings accumulation dams often encounter problems such as difficulty in dam construction, poor seepage and drainage through the dam body, the slow slope of the sedimentary beach, and poor stability. In domestic and foreign mines, examples of serious harms are frequent due to the failure of the tailings dam (Xu 2001).

Because of the practical problems existing in the fine-grained tailings dam, some scholars (Xue 2021; Ye 2020) studied the application of geotextile tubes in tailings dam and suggested that geotextile tubes used in tailings dam construction are an effective remedy for all these problems.

The zoomed-in partial view of a prototype tailings dam, which is constructed with geotextile tubes in Yunnan Province, China, is shown in Figure 1, while the schematic of cross-section of such a tailings dam is shown in Figure 2.

The successful application of geotextile tubes in tailings dam solves the problem of fine tailings dam difficulty, which is a major progress in the design and construction of tailings dams. However, this new construction technology still lacks basic theoretical research (Li 2016; Yang 2019). Therefore, it is necessary to carry out systematic research on it, which will be helpful for further popularization and application of the method of tailings accumulation technology.

*Corresponding Author: liqiaoyan@cidp.edu.cn

 DOI 10.1201/9781003310884-17

(a) Interior

(b) Exterior

Figure 1. Zoomed-in partial view of tailings dam: **a)** interior; **(b)** exterior.

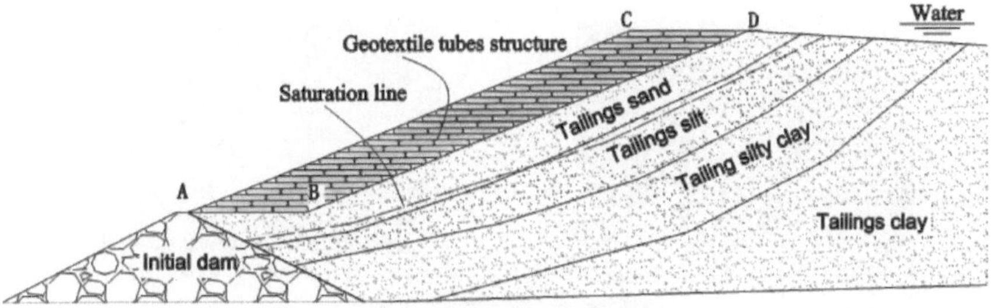

Figure 2. Schematic of cross-section of tailings dam.

In this paper, a physical model similar to the prototype tailings dam is used to conduct an indoor model test on the static stability of the tailings dam with and without geotextile tubes, and it is concluded that the tailings dam with geotextile tubes can improve the stability of fine tailings dam.

2 TAILINGS DAM MODEL TEST

2.1 *Test setting and model design*

The internal size of the model box is 200 cm×60 cm×100 cm (length × width × height) and the front of the model box is made of 20 mm thick transparent plexiglass to observe. To ensure that the model box does not deform during the test, three 5 cm wide ribs are reinforced on the front. The loading system adopts the American MTS hydraulic servo control system, and the size of the loading plate is 30 cm×59 cm. The entire model test device is shown in Figure 3.

Three model tests were carried out in this experiment, and the slope ratios of the three models are all 1:0.5. Scheme 1 is a tailings dam model without geotextile tubes; Scheme 2 is a tailings dam model with the size of 20 cm×20 cm×5 cm of the geotextile tubes; Scheme 3 is a tailings dam model with size 25 cm×25 cm×7.5 cm of the geotextile tubes.

Figure 3. The settings of model test.

2.2 *Test material*

Similar test materials are adopted as those at the construction site of a prototype tailings dam in Yunnan Province (Li 2016).

3 TEST RESULTS AND ANALYSIS

3.1 *Failure mode of tailings dam*

Figure 4 shows failure modes of the tailings dam under the three schemes.

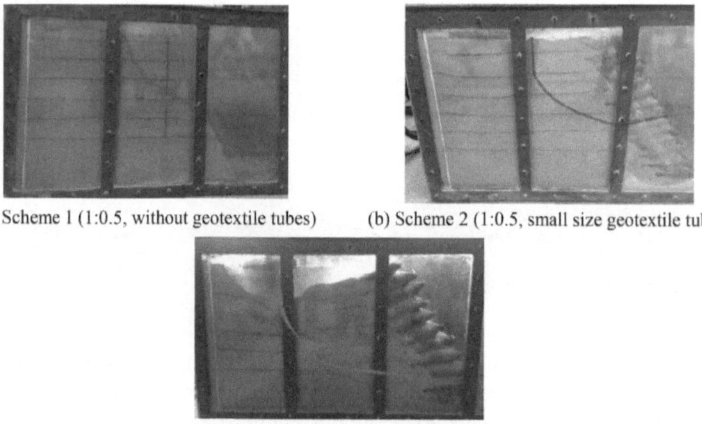

(a) Scheme 1 (1:0.5, without geotextile tubes) (b) Scheme 2 (1:0.5, small size geotextile tubes)

(c) Scheme 3 (1:0.5, large size geotextile bags)

Figure 4. The failure mode of the three-model dam.

According to the failure mode, the following conclusions can be drawn:

(1) The tailings dam with and without geotextile tubes have different failure modes. Figure 3(a) shows the failure of the tailings dam without geotextile tubes. The fracture surface is arc-shaped, which belongs to the arc-type sliding mode. Figures 3(b) and (c) show the damage to the tailings dam with geotextile tubes. Due to the supporting effect of the geotextile tubes on the dam body, the vertical displacement occurs first, and the failure surface is a vertical line. When the load acts to a certain stage, the dam body produces vertical and horizontal displacement, and the failure surface is arc-shaped at this time. As the load continues to increase, when the horizontal action of the upper sliding body on the geotextile tubes is greater than the friction force between the geotextile tubes' interfaces, the entire sliding body slides out along the geotextile tubes layers, and the geotextile tubes are not damaged. The entire failure surface

is divided into three parts—vertical-arc-horizontal, which belongs to the linear-arc-straight sliding mode.

(2) Comparing the three models, the fracture surface gradually deepens, indicating that the stability gradually increases.

3.2 *Stability of tailings dam*

(1) Dam crest pressure and vertical displacement
Figure 5 shows the relationship curve between the vertical displacement of the dam crest and the dam crest pressure under each scheme. It can be seen that with increase in dam crest pressure, the vertical displacement increases. The failure pressure of the tailings dam is 23.65 kN, 29.37 kN, and 34.89kN, respectively. Due to the effect of the geotextile tubes, the failure pressure of the tailings dam increases significantly. Under the same dam crest pressure, the vertical displacement of the geotextile tubes crest decreases; the larger the size of the geotextile tubes, the greater the failure pressure of the dam body.

(2) Dam crest pressure and horizontal displacement of mark point
Figure 6 shows the relationship between the horizontal displacement at the mark point on the geotextile tubes and the pressure on the top of the dam. It shows that under the action of the geotextile tubes when none of the three schemes is damaged, the displacement of the tailings dam without geotextile tubes is larger than others. When the dam failure pressure is reached, the displacement of tailings dams with large-sized geotextile tubes and small-sized geotextile tubes is reduced by 57.9% and 31.5%, respectively.

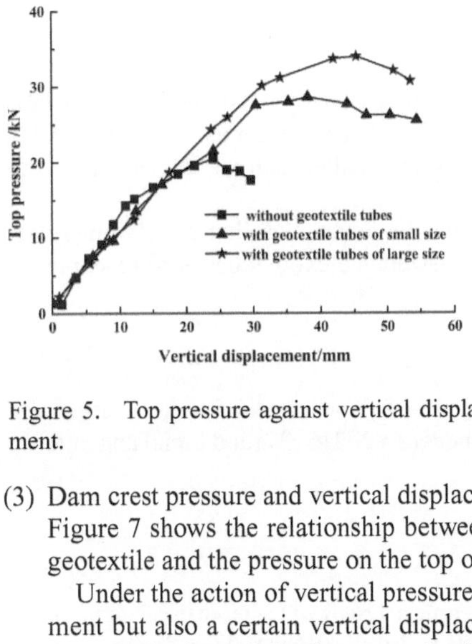

Figure 5. Top pressure against vertical displacement.

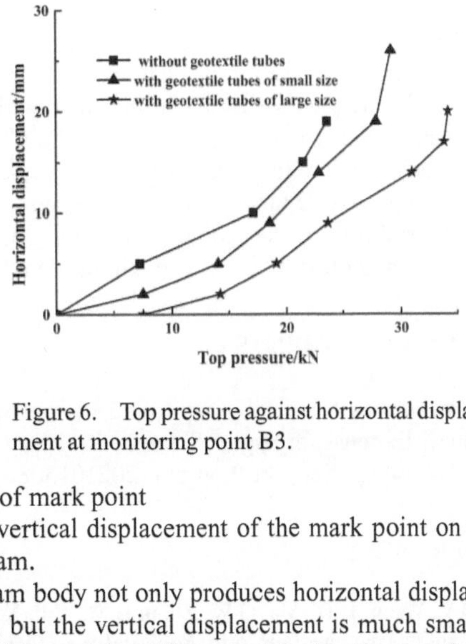

Figure 6. Top pressure against horizontal displacement at monitoring point B3.

(3) Dam crest pressure and vertical displacement of mark point
Figure 7 shows the relationship between the vertical displacement of the mark point on the geotextile and the pressure on the top of the dam.

Under the action of vertical pressure, the dam body not only produces horizontal displacement but also a certain vertical displacement, but the vertical displacement is much smaller than the horizontal displacement. In the initial stage, the tailings sand inside the dam body is squeezed and compacted, and vertical displacement occurs, causing the phenomenon of upturning of the geotextile tubes locally. The geotextile tubes move down, causing vertical displacement. It can be seen from the figure that the displacement of the tailings dam of the large-sized geotextile tubes is 33.3% lower than that of the small-sized geotextile tubes.

(4) Vertical displacement inside the dam body
Figure 8 shows the pressure on the top of the dam and the displacement of the mark point increase due to the tailings sand being compacted under load. The displacement of the tailings dam with the large-sized geotextile tubes is 31.8% lower than that of the small-sized geotextile tubes. The vertical displacement of the tailings dam with the large geotextile tubes is always smaller than others.

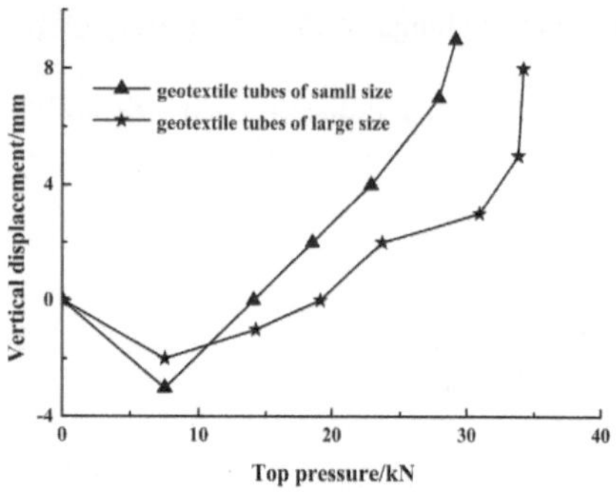

Figure 7. Top pressure against the vertical displacement at monitoring point B3.

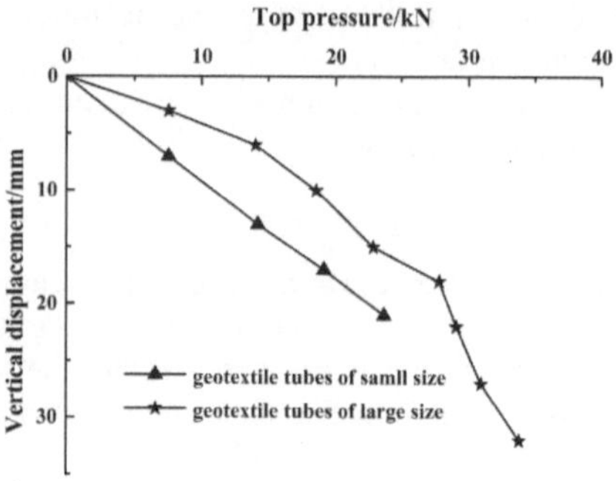

Figure 8. Top pressure against vertical displacement at monitoring point A5.

4 CONCLUSION

Based on the test results, the conclusions are as follows:

(1) The failure mode of the tailings dam with geotextile tubes is different from that of the tailings dam without geotextile tubes. The failure shape of the former is the interlayer sliding failure between the geotextile tubes but a straight line in the upper part and a circular arc in the lower part of the tailings dam without geotextile tubes.
(2) The fracture surface of the tailings dam with geotextile tubes is deepened to a certain extent compared with the tailings dam without geotextile tubes.
(3) The failure pressure of the tailings dam with geotextile tubes is larger and the displacement is smaller than that of the tailing dam without geotextile tubes. The displacement and deformation values of the tailings dam with large geotextile tubes are smaller than the tailings dam with small geotextile tubes under the same conditions.
(4) The stability of the tailings dam with geotextile tubes is better than that of a tailings dam without geotextile tubes. The tailings dam with large-sized geotextile tubes is more stable.

ACKNOWLEDGMENTS

This work was financially supported by the Special Fund of Fundamental Scientific Research Business Expense for Higher School of Central Government (ZY20215127) and LangFang Science and Technology Support Program (2021013086).

REFERENCES

Li, Q.Y., Wang, H.D., Ma, G.W., Zhou, H.M., Cui, X. An experimental study of the mechanical performance of tailings dam geofabriform. *Rock and Soil Mechanics*, Vol. 37, No. 4, 2016, pp.957–964.
Wei Zuoan, Yin Guangzhi, Shen Louyan, et al. Discussing some existing problems on tailings pond design. *Nonferrous Metals (Mining Section)*, Vol.4, No.54, 2002, pp.44–45.
Xu Hongda. Statistical analyses of tailing reservoir disease accidents in China. *Industrial Construction*, Vol.1, No.31, 2001, pp.69–71.
Xue, Z.Y. Feasibility analysis of Xipiao tailings dam with wide top constructed by mold bag method. *Modern Mining*, Vol.37, No.4, 2021, pp.95–97+101.
Yang, Y.H., Wei, Z.A., Cao, G.S., Yang, Y., Wang, H., Zhuang, S.N., Lu, T. A case study on utilizing geotextile tubes for tailings dams construction in China. *Geotext. Geomembranes.* Vol.47, No.2, 2019, pp.187–192.
Ye, E.J., Duo, L.D.X., Zhang, F., Wang X.Y. Application of geofabriform method damming technology in a tailings dam. *Nonferrous Metals (Mining Section),* Vol.72, No.5, 2020, pp. 47–50.

Advances in Civil Engineering: Structural Seismic Resistance, Monitoring and Detection – Mohd Johari Mohd Yusof and Junwen Zhang (Eds)
© 2023 The Authors, ISBN 978-1-032-31491-4

Re-recognition of the characteristics of well-seismic combination in the central and western areas of Lanan

Jinlai Zhang
Oil Production Plant, Daqing Oilfield Co. Ltd., PetroChina, Daqing, Heilongjiang, China

ABSTRACT: The traditional structural interpretation methods only rely on well logging data to interpret and implement well breakpoints, speculate fault morphology, and have no clear understanding of complex structural morphology. Therefore, the fine interpretation technology of well seismic combined with structure is carried out. The method is to introduce seismic information in the process of structural interpretation based on well logging data, based on seismic interpretation fault, combined with variance body and ant body to carry out the spatial homing of breakpoints, the spatial combination of faults and the implementation of fault occurrence. The results show that the accuracy of structural description has been greatly improved, The combination rate of breakpoints after well seismic combination is also higher than that only relying on well breakpoints, and the characteristics of structures and faults are more reliable, which provides a favorable reference basis for exploration and development.

1 INTRODUCTION

Lamadian Oilfield completed high-density 3D seismic acquisition in the south middle block of the oilfield from December 2007 to November 2008. Since 3D seismic acquisition has never been done in the study area before, the fault combination is mainly interpreted and implemented based on logging data. Although the logging data has high vertical resolution, there are still serious deficiencies in crosswell interpretation. Therefore, the fault combination needs to be further improved. At the same time, from the 3D3C seismic structure interpretation results in the north of the oilfield, the structural results of seismic interpretation are indeed quite different from those of logging interpretation. In order to further implement the structural and fault characteristics in the west of the south middle block of the oilfield, it is necessary to organically combine seismic and logging data to form a well seismic combined structural interpretation method, and implement the distribution of structures and faults in the work area, so as to provide a more reliable geological basis for further promoting the fine development and adjustment of Lamadian oilfield.

2 SEISMIC IDENTIFICATION FAULT PRINCIPLE

According to various changes of reflection event axis, the reflection characteristics of fault (Figure 1 and Figure 2) are summarized, which lays a foundation for fault identification. According to these reflection characteristics (Dai 2011), the fault can be judged intuitively on the seismic profile.

DOI 10.1201/9781003310884-18

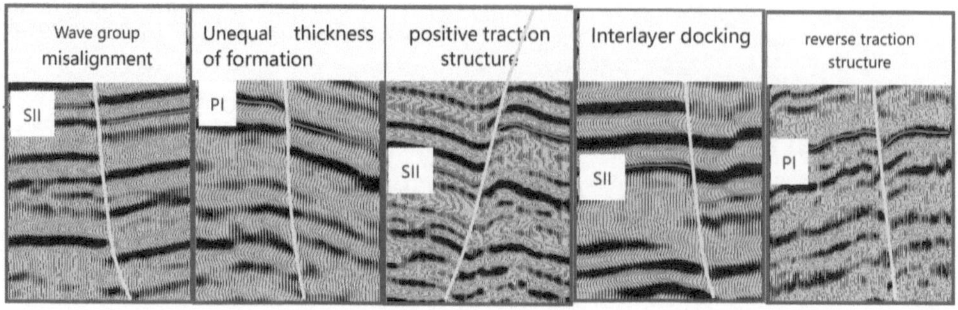

Figure 1. Large fault reflection characteristics.

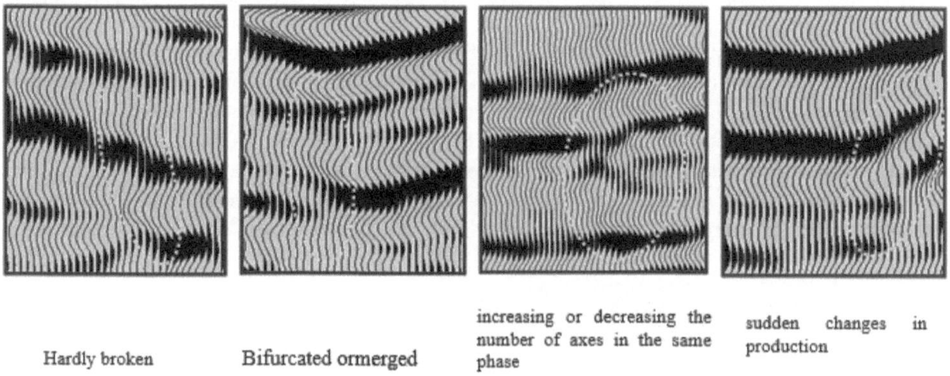

Figure 2. Small fault reflection characteristics.

3 WELL-SEISMIC COMBINED INTERPRETATION FAULT METHOD

Faults with fault distance greater than 8m: well connected seismic profile comparison and interpretation combined with coherent volume, three-dimensional visualization and other technologies are adopted (Figure 3).

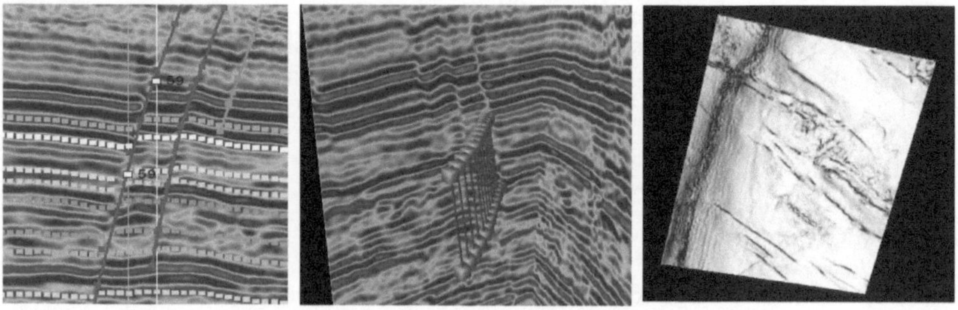

Figure 3. Large fault identification method.

Fault with fault distance less than 8m: the description method of well seismic joint technology leading and breakpoint correction is adopted (Figure 4).

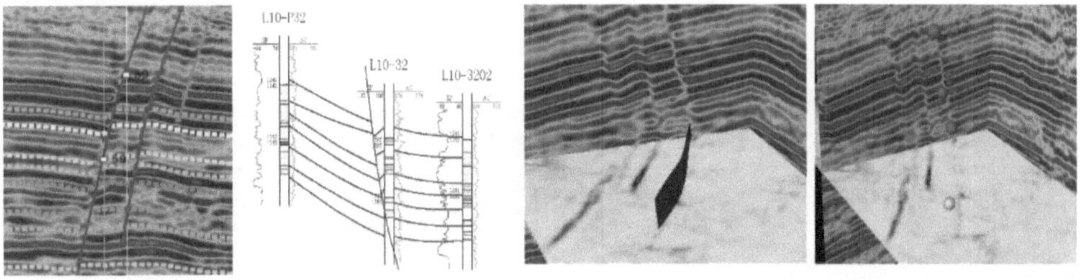

Figure 4. Small fault identification method.

Section identification, using the characteristics of seismic section wave group, for the faults with obvious characteristics, continuous multi section comparison is used for fault identification; Carry out cross well correlation to further determine the accuracy of fault inter well extension; Spatial combination, using variance volume and other means to clarify the spatial distribution of faults; Well seismic interaction, using fault breakpoints to verify breakpoints in three-dimensional space to realize well seismic interaction verification (Ding 2013).

Faults with fault distance less than 3m can be divided into two types.

Identifiable fault: the variable density profile has a slight ability to shake and can combine multiple well breakpoints. This kind of fault is mainly identified by well breakpoint combination.

Unrecognizable fault: the extension length is short, the profile has no obvious characteristics (Zhou 2011), and the breakpoint on the well is an isolated breakpoint.

4 FAULT COMBINATION METHOD

(1) The interactive analysis of main survey line, connecting survey line and arbitrary line explains the spatial closure of the fault and determines the plane extension shape of the fault.
(2) In well seismic comparison, well breakpoints are combined with seismic faults to ensure that there is no contradiction between well breakpoints and earthquakes.
(3) After well seismic comparison, it is implemented on the coherent volume slice for interactive verification to eliminate the interference of human factors.

5 NEW UNDERSTANDING OF THE STRUCTURE AFTER THE COMBINATION OF WELL AND EARTHQUAKE IN THE STUDY AREA

5.1 *61# fault is split into two faults*

Compared with the original 61# fault, two seismic sections are respectively intercepted in the extended direction of the newly recognized 61# fault and the 611# fault, and there are breakpoints in the extension direction (Cai 2008), and the fault features are obvious. The fault on the variance body is clearly displayed (Figure 5).

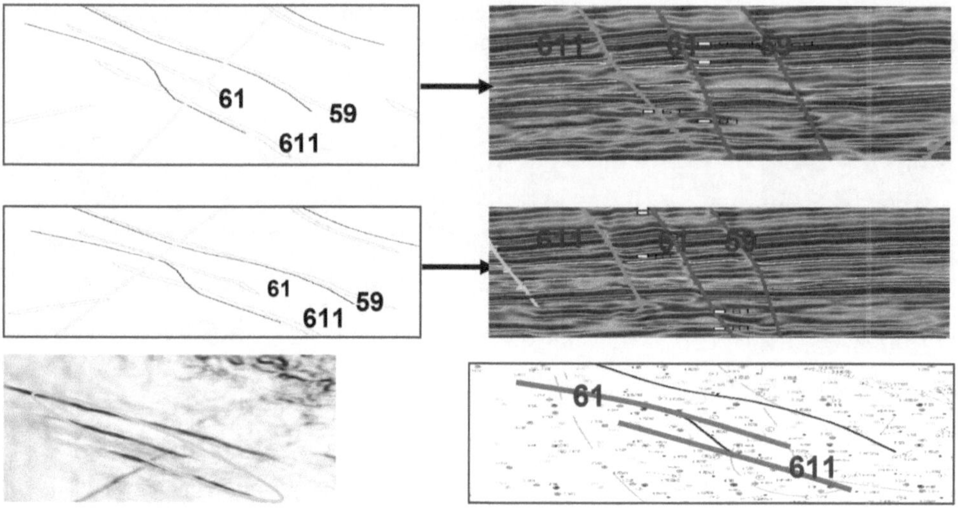

Figure 5. 61# fault split.

5.2 *Recombining three normal faults*

According to the guidance of variance volume, there are new faults near 51#, 541# and 611# faults. The breakpoints of the three faults and the nearby isolated breakpoints are recombined (Jia 2010), which is consistent with the reflection characteristics of seismic profile, and the existence of the three faults is reasonable (Figure 6).

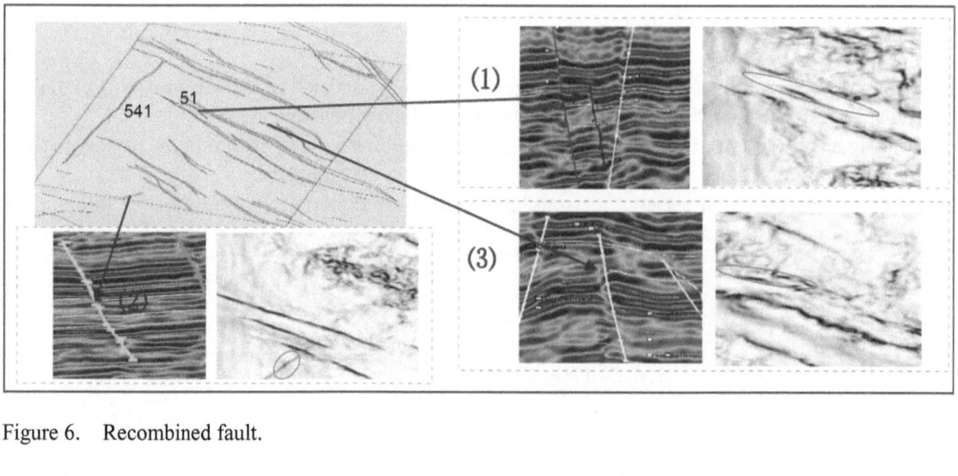

Figure 6. Recombined fault.

5.3 *Breakpoint correction*

In the process of fault interpretation, it was found that the breakpoints belonging to the 53# fault were classified into 541# faults in the breakpoint data (Figure 7).

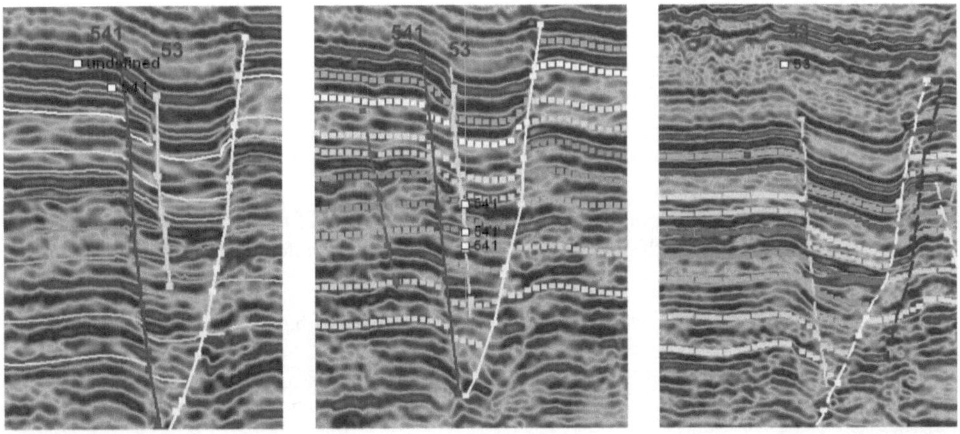

Figure 7. Breakpoint correction profile.

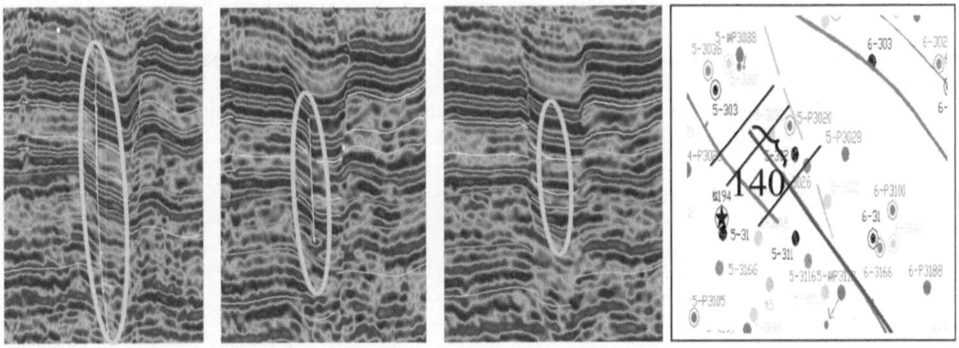

Figure 8. 53# extended length identification.

Further implementation by well-seismic combination (Figure 8), the 53# fault is extended by 140 m.

6 CONCLUSION

In this paper, well-seismic combination method is adopted to study fault characteristics. The main conclusions can be summarized as follows:

(1) Well seismic combined with structural interpretation makes full use of well point information and seismic interpretation results to study faults, guides fault interpretation with well breakpoints, and controls the overall shape and strike of faults with seismic interpretation.
(2) Faults with a fault distance greater than 8m: well connected profile interpretation combined with coherent body and three-dimensional visualization; faults with a fault distance less than 8m: description method dominated by well seismic joint technology and breakpoint correction.
(3) There are 14 interpreted faults with logging data in the study area, and 25 faults are identified with seismic data. After well seismic combination, 18 faults are implemented, 3 faults are

recombined, 1 fault is broken down, and 2 fault through horizons are adjusted. The breakpoint combination rate is increased from 73% to 87%.

(4) The overall structural pattern of the study area is high in the East, low in the west, high in the north and low in the south, which gradually becomes steep from northeast to southwest. On the whole, it is a monoclinal structure, and there are micro amplitude structures in some parts. Horst structures are developed between 47# and 51# faults, and graben structures are developed between 51# and 53# faults.

(5) After the fault changes, the injection production relationship of the well group changes. The accurate understanding of the fault provides a basis for fine potential tapping adjustment in the later stage.

In terms of the future work, well-seismic combination should be carried out to enhance the breakpoint combination rate.

REFERENCES

Cai Dongmei et al. (2008) Application of stochastic modeling technology in a large oilfield abroad. *China National Petroleum Corporation*. 15–17.

Dai Junsheng et al. (2011) *Structural Geology and Geotectonics*. Petroleum Industry Press.

Ding Wenlong et al. (2013) *Analysis of reservoir structure*. Petroleum Industry Press. 10–12.

Jia Ailin. (2010) *Fine Reservoir Description and Geological Modeling Technology*. Petroleum Industry Press. 3–4.

Zhou Huajian. (2011) Application of well-seismic combined fault research method in the well network area. *West China Science and Technology*. Volume 10, Issue 12

Advances in Civil Engineering: Structural Seismic Resistance, Monitoring and Detection – Mohd Johari Mohd Yusof and Junwen Zhang (Eds)
© 2023 The Authors, ISBN 978-1-032-31491-4

Influence of the type of pressure relief hole on explosion-proof performance of blast wall

Erlei Bai
Teaching-Research Office of Airfield and Building Engineering, Air Force Engineering University, Xi'an, China

Xiaorong Li
Transportation Industry Department, Ching International Engineering Consulting Corporation, Beijing, China

Liangxue Nie
Troops of PLA, Xuzhou, China

Xin Luo* & Mengnan Dai
National Defense Engineering Institute, Academy of Military Sciences, Beijing, China

ABSTRACT: In this paper, ANSYS/LS-DYNA is used as the technical means to construct a two-dimensional model of explosive explosion in air, and the influence of two kinds of circular holes and two kinds of specialshaped holes on explosion-proof performance of the blast wall is compared and analyzed, the results show that: the peak reflection pressure in front of the blast wall with YX1 pressure relief hole is the largest, while the overpressure behand the wall is the smallest, and compared with YX2 pressure relief holes, the weakening effect of the former is stronger than the latter on the peak reflected pressure in front of the blast wall, while the effect on the overpressure behind the wall is just the opposite; the setting of special shaped relief holes will reduce the positive pressure on the blast facing surface of the blast wall, but greatly increase the negative pressure strength, and especially the wall with YX2 pressure relief hole is the most obvious; on the whole, the total energy and energy density of all kinds of blast walls increased first and then decreased with the increase of time, the energy and energy density on the blast facing surface of blast wall with the YX2 pressure relief hole increase to the largest extent.

1 INTRODUCTION

Terrorism is a complicated political and social phenomenon (Atran 2021), since the terrorist attacks, the most frightening thing for people is the car bombs (Mahmud 2020), which take vehicles (including cars, trucks, motorcycles, etc.) as the carrier of explosives and explode near buildings, causing extensive damage, leading to the collapse of buildings and the generation of various explosive fragments, resulting in casualties and disastrous consequences. According to the current situation analysis, the set of blast wall is one of the most simple and extremely effective preventive measures, by building anti-explosion and explosion protection walls around the target building, resisting the car rushing into the target building to explode, block the bomb out, and effectively blocking the shock wave generated by the explosion, which plays an effective protective effect on people and buildings.

*Corresponding Author: luoxin_01@163.com

DOI 10.1201/9781003310884-19

According to a large number of research results, the blast wall is mainly used to prevent the explosion of flammable and explosive substances (Das & Weinberg 2012), terrorist explosion (Eytan 2005; Medvedev et al. 2020; Li & Wang 2021), offshore oil and gas explosion (Boh et al. 2005; Chen et al. 2021; Sohn et al. 2015), etc. while the study on the blast wall mainly focuses on the following aspects: firstly, the function of the explosion proof wall, for example, Rose (Rose et al. 1993) carried out a preliminary experimental study on the weakening effect of explosion walls, subsequently, Whitham analyzed the blocking effect of explosion walls; Beyer (Beyer 1986) explored the variation law of parameters of blast wave behind blast walls, so as to study the protective effect of walls on rear protected objects. Secondly is the design of blast walls, for example, Chipley (2003) and Bowles (2003) comprehensively introduce the pressure-load characteristics of blast walls, and analyze the design method; Hedayati (Hedayati et al. 2015) explored a new method for studying the mechanical response of stainless steel shaped walls. Thirdly, the study of the dynamic response of blast walls. Scherbatiuk (Scherbatiuk et al. 2008) carried out the anti-explosion test of blast walls, based on which the dynamic performance of walls under blast wave was studied; Xu W (Xu et al. 2015) and Li W (Li et al. 2015) were both used numerical simulation method to study the dynamic response of autoclaved aerated concrete masonry blast wall under explosion load and the mechanical response of reinforced concrete blast wall under dynamic load respectively.

In general, the current research on blast wall is mainly focused on "resistance", and the effect of its "unloading" is rarely reported. Therefore, in this paper, the finite element model is constructed by means of numerical simulation to preliminarily explore the effect of shock wave on walls with different types of pressure relief holes under the action of explosion.

2 CONSTRUCTION OF FINITE ELEMENT MODEL

For the purpose of numerical simulation, the following assumptions are made: (1) Continuity hypothesis, assuming that the material is a continuum, the effects of creep and relaxation can be ignored if the time is not too long; (2) The assumption that the medium is homogeneous and isotropic, it is assumed that the material is homogeneous and that the physical and mechanical properties of the medium are the same in any direction, that is, isotropic; (3) Because the detonation process of explosive is very short, it is assumed that the detonation process is ideal detonation without considering the influence of thermodynamic parameters. At the same time, the modeling is simplified as follows: the two-dimensional model of explosive exploding in air only considers air and explosive, neglecting soil layer and uses single-layer grid.

When using ANSYS to establish the finite element model, the SOLID164 solid unit is used to establish the finite element model of explosives and air. Euler grid modeling is adopted, and the multi-substance ALE algorithm is used in the unit. The wall is modeled by Lagrange mesh and MAT_ELASTIC material model, and the plane size of wall is sat as 200cm× 10cm.

The explosives were described by a material model of high explosives burning (HIGH_ EXPLOSIVE_BURN) combined with the JWL equation of state, and the $P-V$ relationship is as follows:

$$P = A\left(1 - \frac{\omega}{R_1 V}\right)e^{-R_1 V} + B\left(1 - \frac{\omega}{R_2 V}\right)e^{-R_2 V} + \frac{\omega E}{V} \tag{1}$$

For the TNT explosive model, in the g-cm-μs unit system, the parameters are set as follows: density ρ_0=1.64g/cm^3, detonation velocity D=0.693cm/μs. Champan-Jouget pressure parameters are: PCJ=0.255×10^{11}Pa, A=5.4094, B=0.093726, R_1=4.5, R_2=1.1, ω=0.35.

The MAT_NULL material model is used to establish the finite element model of the air, and described by polynomial state equation EOS_LINEAR_POLYNOMIAL:

$$P = \left[C_0 + C_1\mu + C_2\mu^2 + C_3\mu^3\right] + \left[C_4 + C_5\mu + C_6\mu^2\right]e_{ipv0} \tag{2}$$

For the air model, in the g-cm-μs unit system, the parameters are set as follows: density ρ_0=0.001293g/cm^3, dynamic viscosity coefficient MU=0.001. For the convenience of calculation,

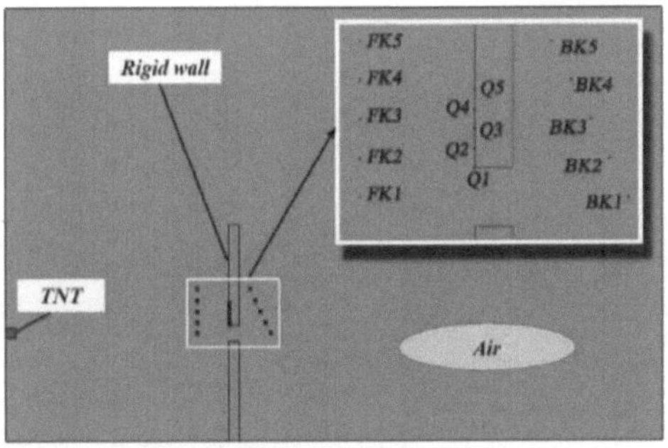

Figure 1. Finite element model and each measurement point.

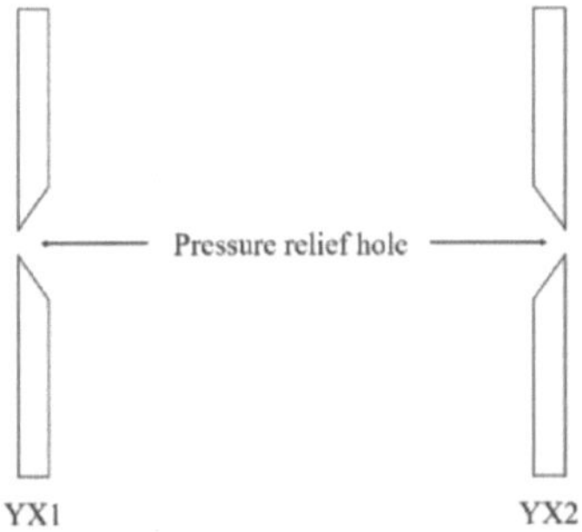

Figure 2. Two different types of pressure relief holes.

the air is regarded as an ideal gas, and the parameters are $C_0 = C_1 = C_2 = C_3 = C_6$=0, $C_4 = C_5$=0.4. In addition, in order to simulate infinite space, a non-reflective boundary condition is applied on the expansion surface of the air model.

In order to effectively record data, the selection principles and specific provisions of measuring points in this paper are as follows: two types of measuring points are included, one is wall unit measuring points, set as group Q, consisting of 5 points, starting from the edge of the relief hole, with an interval of 5 cm; the other is the air unit measurement point, the front of the wall is set as the group FK, and the back of the wall is set as the group BK, both of which are consisted of 5 points. The two initial points are aligned to the horizontal position of the center of the pressure relief hole, and the vertical interval of the measuring points is all 5 cm. The final finite element model and the layout of the measuring points are shown in Figure 1. Two different types of pressure relief holes were designed on the basis of inner diameter $D = 10$ cm and outer diameter $D = 20$ cm, as shown in Figure 2, two groups of blast wall with round pressure relief hole of $D = 10$ cm and $D = 20$ cm were set as the control group at the same time.

3 INFLUENCE LAWS OF THE TYPE OF PRESSURE RELIEF HOLES

3.1 *Influence of relief hole on overpressure in front of the wall*

Figure 3 and Table 1 show the overpressure influence law of the measuring points in group FK in front of the blast wall with different types of relief holes, it can be seen from the figure that the peak reflection pressure in front of the blast wall with YX1 pressure relief hole is on the same level as that of the wall with 10 mm diameter pressure relief hole, and the peak value is the largest, which ups to 4.352 MPa. While the peak reflection pressure in front of the blast wall with 20 mm diameter pressure relief hole is the smallest, which is only 3.801 MPa. The YX2 pressure relief hole can weaken the peak reflection pressure in front of the wall more than the YX1 hole.

Table 1. The peak overpressure at each measuring point in front of the wall (MPa).

Type of pressure relief holes	FK1	FK2	FK3	FK4	FK5
YX1	4.346	3.061	3.045	3.273	3.346
YX2	4.304	3.016	2.807	3.052	3.169
10	4.352	3.057	3.045	3.276	3.343
20	3.801	2.646	2.484	2.818	3.014

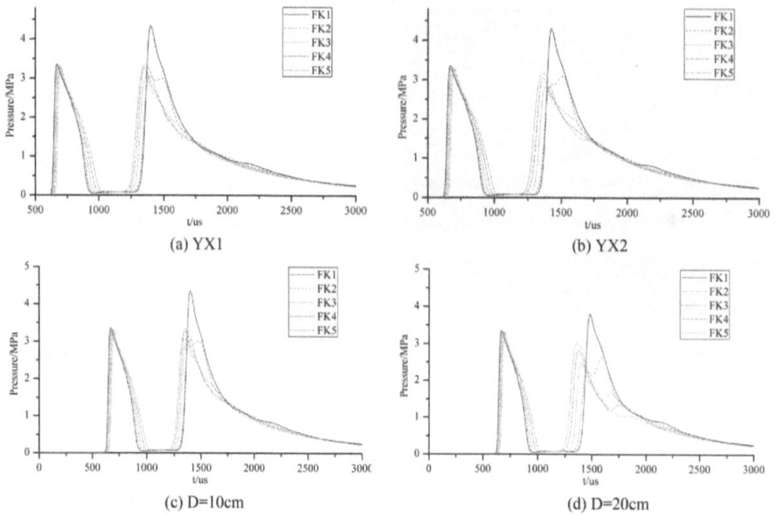

Figure 3. The relationship between overpressure and time of the measured points in group FK.

3.2 *Influence of relief hole on overpressure behind the wall*

The influence law of overpressure on the measuring points in group BK behind the blast wall with two different types of pressure relief holes is shown in Figure 4, and the Table 2 shows the peak overpressure at each measuring point, from which it can be seen that the peak reflection pressure behind the blast wall with YX1 pressure relief hole is on the same level as that of the wall with 10 mm diameter pressure relief hole, and the peak value is the smallest, which is only 1.501 MPa. While the peak reflection pressure in front of the blast wall with 20mm diameter pressure relief hole is the largest, which ups to 1.860 MPa. The YX2 pressure relief hole can increase the peak reflection pressure behind the wall more than the YX1 hole.

Table 2. The peak overpressure at each measuring point behind the wall (MPa).

Type of pressure relief holes	BK1	BK2	BK3	BK4	BK5
YX1	1.501	1.182	0.553	0.304	0.224
YX2	1.585	1.192	0.514	0.349	0.262
10	1.508	1.107	0.486	0.321	0.241
20	1.860	1.794	0.997	0.498	0.332

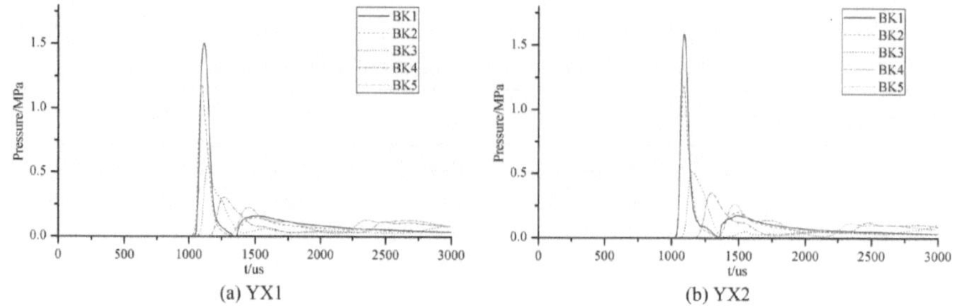

Figure 4. The relationship between overpressure and time of the measured points in group BK.

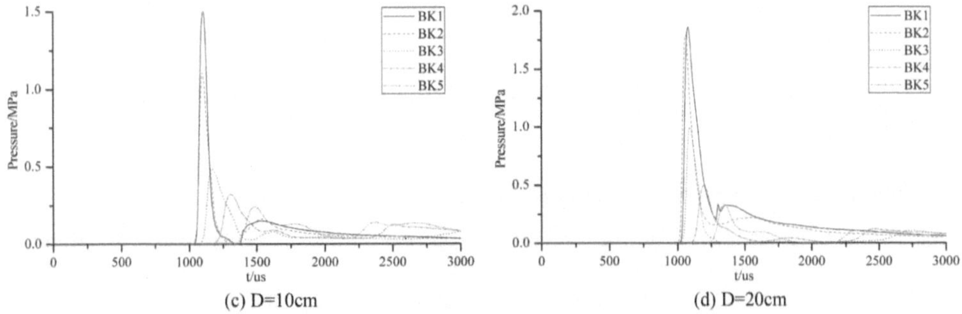

Figure 4. Continued.

3.3 *Influence of relief hole on pressure on the wall*

Figure 5 and Table 3 shows the time-history curve of pressure of measuring points in group Q on the surface of blast wall with different types of pressure relief holes, it can be seen from the

Table 3. The peak overpressure at each measuring point (MPa).

Type of pressure relief holes	Q1-P	Q2-P	Q3-P	Q4-P	Q5-P
YX1	7.995 (−8.793)	7.838 (−52.013)	4.574 (−42.640)	6.713 (−19.030)	7.921 (−9.393)
YX2	0.00 (−98.170)	0.001 (−52.194)	2.804 (−23.276)	6.869 (−16.073)	8.445 (−8.583)
10	10.753	11.106	10.904	10.559	10.420
20	12.037	12.731	11.830	10.853	10.455

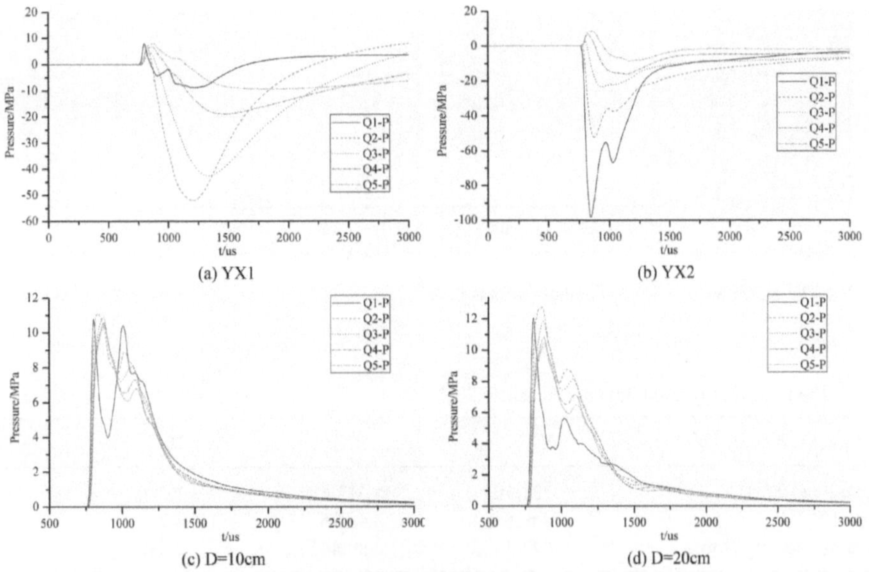

Figure 5. The relationship between pressure and time of the measured points in group Q.

analysis of the figure and table that the setting of special pressure relief holes will reduce the positive pressure on the blasting surface of the wall, but will greatly increase its negative pressure, which is mainly due to the repeated reflection of the surface shock wave caused by the setting of the oblique section, especially the YX2 pressure relief hole, where the maximum negative pressure can reach 98.170 MPa.

3.4 *Influence of relief hole on energy of the wall*

Figures 6 and 7 show the time history curve of energy and energy density of blast wall with different types of pressure relief holes, and the peak values of each curve are given in Table 4, it can be seen from the analysis of the figures and table that the total energy and energy density of all kinds of blast walls increase first and then decrease with the increase of time on the whole, and the energy and energy density on the blasting surface of the wall will be greatly increased by the setting of the special pressure relief hole, especially the setting of YX2 holes, the energy and energy density can reach the maximums of 413.6810^{-3} J and 89.445 J/m^3, and it is 86.59% and 86.60% higher than that of YX1 pressure relief hole, respectively.

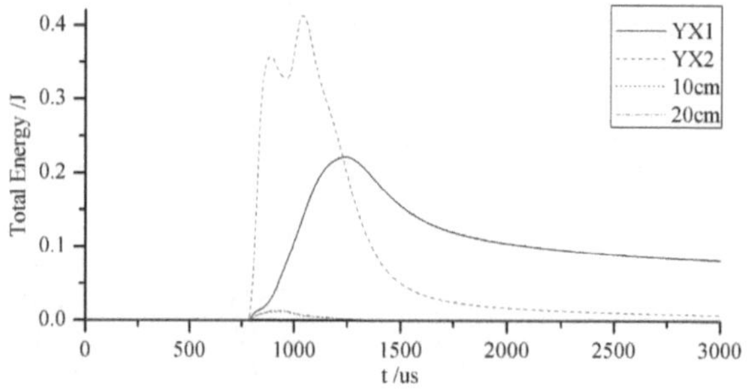

Figure 6. The curve of total energy versus time.

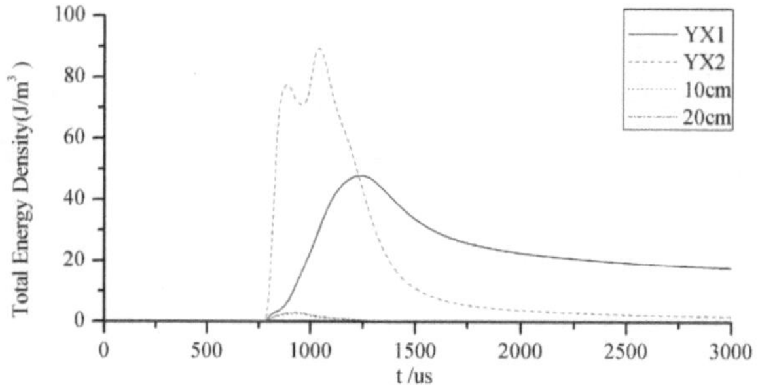

Figure 7. The curve of total energy density versus time.

Table 4. The total energy and total energy density.

Type of pressure relief holes	YX1	YX2	10	20
Total energy (10^{-3}J)	221.70	413.68	12.50	13.42
Volume (10^{-3}V)	4.625	4.625	4.75	4.5
Total energy density (J/m^3)	47.934	89.445	2.632	2.981

It can be seen that the setting of special pressure relief holes will greatly improve the negative pressure, effective stress, energy and energy density of the blast walls on the blasting surface, that is, the stress on the blast wall is more serious, especially the wall with YX2 pressure relief hole.

4 CONCLUSIONS

In this paper, ANSYS/LS-DYNA is used as the technical means to compare and analyze the influence on the explosion-proof property of blast walls with different types of pressure relief holes through numerical simulation. The main conclusions are as follows:

(1) The peak reflection pressure in front of the blast wall with YX1 pressure relief hole is the largest, which ups to 4.352 MPa, while that of the wall with 20mm diameter pressure relief hole is the smallest, which is only 3.801 MPa; the reflection overpressure behind the wall is the smallest, which is only 1.501 MPa, while that of the wall with 20mm diameter pressure relief hole is the largest, which ups to 1.585MPa; compared with YX2 pressure relief holes, YX1 pressure relief holes can weaken the peak reflection pressure in front of the wall more, while the effect on the overpressure behind the wall is just the opposite.
(2) The setting of special shaped relief holes will reduce the positive pressure on the blasting surface of the wall, but greatly increase the negative pressure, especially the setting of YX2 relief holes, which can reach 98.170 MPa.
(3) On the whole, the total energy and energy density of all kinds of blast walls increase first and then decrease with the increase of time, the increase of energy and energy density of the wall with YX2 pressure relief hole are most significantly. Which can reach the maximums of 413.6810^{-3} J and 89.445 J/m^3, respectively.
(4) In this paper, the impact of the type of pressure relief holes on the explosion-proof performance of blast wall is relatively one-sided, and a comprehensive analysis of the interaction law of the number and arrangement of pressure relief holes will be the focus of the next research.

ACKNOWLEDGMENT

This work was sponsored by the key project of Academy of Military Sciences (JK20191A010001).

REFERENCES

Atran S. Psychology of transnational terrorism and extreme political conflict[J]. *Annual review of psychology*, 2021, 72: 471–501.
Beyer M E. *Blast loads behind vertical walls*[R]. 22nd Explosives Safety Seminar, Dept of Defense Explosives Safety Board, Anaheim CA, 1986.
Boh J W, Lnuoa L A, Chooa Y S. Energy ahsnrhing passive impact barrier for profiled blastwalls [J]. *International Journal of Impart Engineering*, 2005, 31: 976–995.
Bowles P K. *Barrier assessment for safe standoff (BASS) test series*[C]//Proceedings of the 11th International Symposium on Interaction of the Effects of Munitions with Structures. 2003.
Chen K, Wei X, Li H, et al. Operational risk analysis of blowout scenario in offshore drilling operation[J]. *Process Safety and Environmental Protection*, 2021, 149: 422–431.
Chipley M. Reference Manual to Mitigate Potential Terrorist Attacks Against Buildings: Providing Protection to People and Building[M]. *Federal Emergency Management Agency*, 2003.
Das B, Weinberg M. Improving flammable mass estimation for vapor cloud explosion modeling in an offshore QRA[J]. *Safety science*, 2012, 50(5): 1218–1227.
Eytan R. Cost effective retrofit of structures against the effects of terrorist attacks–the Israeli experience[C]//Structures Congress 2005: *Metropolis and Beyond*. 2005: 1–12.
Hedayati M H, Sriramula S, Neilson R D. *Reliability of Profiled Blast Wall Structures*[M]//Numerical Methods for Reliability and Safety Assessment. Springer International Publishing, 2015: 387–405.

Li W, Yang H, Zhang B. Dynamic Analysis on Explosion Resistance Performance of Reinforced Concrete Wall[C]//*Advanced Materials Research*. Trans Tech Publications Ltd, 2015, 1078: 162–165.

Li Z, Wang Y. Experiment and Simulation of Critical Parameters for Building Windows under Thermal Explosion[J]. *Case Studies in Thermal Engineering*, 2021, 26: 101014.

Mahmud S F. The Impact of Terrorism on Human Development in Iraq[J]. *Global Journal of Management and Economics*, 2020, 1: 29–41.

Medvedev S P, Anderzhanov E K, Guk I V, et al. Testing of Explosion-Proof Coatings in Cylindrical and Conical Shock Tubes[J]. *Russian Journal of Physical Chemistry B*, 2020, 14(6): 946–950.

Rose T A, Mays G C, Smith P D. Leakage pressures behind blast walls[C]//*Proceedings of the Sixth International Symposium on Interaction of Nonnuclear Munitions with Structures* (6th), Florida, 1993.

Scherbatiuk K, Rattanawangcharoen N, Pope D J, et al. Generation of a pressure–impulse diagram for a temporary soil wall using an analytical rigid-body rotation model[J]. *International journal of impact engineering*, 2008, 35(6): 530–539.

Sohn J M, Kim S J, Seo J K, et al. Strength assessment of stiffened blast walls in offshore installations under explosions[J]. *Ships and Offshore Structures*, 2015, (4): 1–10.

Xu W, Bao Q, Li Z, et al. Numerical Analysis on the Behavior of Autoclaved Aerated Concrete Block Infilled Wall Subjected to Gas Explosion[C]//*Applied Mechanics and Materials*. Trans Tech Publications Ltd, 2015, 723: 259–265.

Advances in Civil Engineering: Structural Seismic Resistance, Monitoring and Detection – Mohd Johari Mohd Yusof and Junwen Zhang (Eds)
© 2023 The Authors, ISBN 978-1-032-31491-4

Microstructural changes of acid polluted laterite under soaking conditions

Yuhang Fan
Kunming University of Science and Technology Oxbridge College, Kunming, China

ABSTRACT: Soil pollution brings changes in the macroscopic mechanical properties of soil. As the pollutants invade the soil and interact with the materials, the soil's original physical composition and chemical composition will change, resulting in the transformation of microstructure. For acid polluted laterite under soaking conditions, acid invades laterite and interacts with substances in laterite, and it reacts with oxides, which play a cementing role in laterite, and free oxides to generate salt. At this time, laterite is in a soaking environment. The generated salt dissolves in the solution with water in pores, which take away the components in laterite, resulting in changes in the linked structure between particles and thus leading to changes in the microstructure of laterite. Under soaking conditions, acid pollutes soil with acid soaking in laterite and out the laterite. Acid invasion results in changes in the particle size, particle properties and pore characteristics of laterite.

1 INTRODUCTION

SEM images of acid-polluted laterite are used to study the microstructure characteristics of acid-polluted laterite under soaking conditions. The representative parameters of microstructure characteristics are extracted by image processing to clarify the characteristics of soil microstructural changes after being polluted by acid.

2 MICROSTRUCTURE TEST OF ACID POLLUTED LATERRITE UNDER SOAKING CONDITIONS

This paper has tested microstructural changes of acid polluted laterite under soaking conditions by using SEM. SEM images can directly reflect the relationship between particles and pores, particles and particles, and pores and pores in acid-polluted laterite. By further digitizing the images, we can get the specific microstructure characteristic parameters of acid-polluted laterite and then study the microstructural change characteristics of acid-polluted laterite from the microscopic point of view.

2.1 *Test equipment and methods*

The prepared samples are sent to the Scanning Electron Microscope Studio of Advanced Analysis and Measurement Center of Yunnan University for microscopic image acquisition tests. Stick the air-dried soil sample on the sample tray and put it into the instrument for conductive treatment. The method is called the metal coating method. Then the well-made sample is put into the instrument for image acquisition. The microstructure picture is taken, and the scanning electron microscope images with a magnification of 500X, 1000X, 2000X, 5000X, 10000X and 20000X are obtained. See Tables 1 and 2 for sample numbers of SEM.

DOI 10.1201/9781003310884-20

Table 1. SEM sample number of acid-soaked laterite.

No.	Concentration (%)	Soaking Time (d)	Ambient Temperature (°C)	Magnification (x)	Number of Images (sheet)	Notes
1#	0	4	20	500, 1000, 2000, 5000, 10000, 20000	4	Before shearing
2#	1	4	20	500, 1000, 2000, 5000, 10000, 20000	4	After shearing
3#	3	4	20	500, 1000, 2000, 5000, 10000, 20000	4	After shearing
4#	8	4	20	500, 1000, 2000, 5000, 10000, 20000	4	After shearing
5#	5	1	20	500, 1000, 2000, 5000, 10000, 20000	4	After shearing
6#	5	7	20	500, 1000, 2000, 5000, 10000, 20000	4	After shearing
7#	5	14	20	500, 1000, 2000, 5000, 10000, 20000	4	After shearing
8#	5	30	20	500, 1000, 2000, 5000, 10000, 20000	4	After shearing
9#	5	7	10	500, 1000, 2000, 5000, 10000, 20000	4	After shearing
10#	5	7	20	500, 1000, 2000, 5000, 10000, 20000	4	After shearing
11#	5	7	40	500, 1000, 2000, 5000, 10000, 20000	4	After shearing

Table 2. SEM sample number of laterite out of acid liquor.

No.	Concentration (%)	Soaking Time (d)	Ambient Temperature (°C)	Magnification (x)	Number of Images (sheet)	Notes
12#	0	4	20	500, 1000, 2000, 5000, 10000, 20000	4	Before shearing
13#	1	4	20	500, 1000, 2000, 5000, 10000, 20000	4	After shearing
14#	3	4	20	500, 1000, 2000, 5000, 10000, 20000	4	After shearing
15#	8	4	20	500, 1000, 2000, 5000, 10000, 20000	4	After shearing
16#	5	4	20	500, 1000, 2000, 5000, 10000, 20000	4	After shearing
17#	5	7	20	500, 1000, 2000, 5000, 10000, 20000	4	After shearing
18#	5	14	20	500, 1000, 2000, 5000, 10000, 20000	4	After shearing
19#	5	30	20	500, 1000, 2000, 5000, 10000, 20000	4	After shearing
20#	5	7	10	500, 1000, 2000, 5000, 10000, 20000	4	After shearing
21#	5	7	20	500, 1000, 2000, 5000, 10000, 20000	4	After shearing
22#	5	7	40	500, 1000, 2000, 5000, 10000, 20000	4	After shearing

2.2 *Preparation of scanning electron microscope samples*

To compare and analyze the microstructure characteristics of acid-soaked laterite and laterite out of acid liquor under different influencing factors, 22 samples were selected for SEM observation considering three factors, including acid concentration, soaking time and ambient temperature. The water content of the samples was 28.2%, and the dry density was 1.45 g/cm^3. The microstructure images with different magnifications were obtained. The SEM samples were taken from the middle part after shearing under corresponding conditions. In the sampling process, in order to avoid artificial shear surfaces, each sample is forcibly broken by hand, and the surface of the soil sample after the break is a natural cross section. The samples were all around 1-3cm^2, and should be as thin as possible, like (b) in Figure 1, which is easy to be air-dried. Figure 1 (c) is the final SEM sample, which is small and thin without cracks on the surface and should be air-dried.

(a) Soil sample after shearing (b) Air-dried soil sample (c) SEM sample

Figure 1. Sample of Scanning Electron Microscopy.

3 COMPARATIVE ANALYSIS OF MIRCROSTRUCTURE IMAGES OF ACID POLLUTED LATERITE UNDER SOAKING CONDITIONS

Before observing and analyzing the microstructure pictures of soil, we should select the magnification of the pictures. The magnification should not be too large; otherwise, there are only 1 to 2 soil particles in each microscopic picture, which cannot represent the general microstructure characteristics of the picture. The magnification should not be too small, either; otherwise, the soil particles and pores on the picture cannot be clearly identified. The magnification of microstructure images, which should be determined by the purpose and subjects of the study, include 500X, 1000X, 2000X, 5000X, 10000X and 20000X. For the conventional microstructure study of clayey soil, the optimum magnification should be within 1500 ± 300. Therefore, this paper selects typical microstructure pictures with a magnification of 2000 for analysis. When the magnification is 20000X, the range of images is too small. Some are soil particles, and some are holes, which cannot accurately show the degree of erosion of laterite. Therefore, microstructure images with a magnification of 2000X and 10000X are selected below. Images with the magnification of 2000X are used to analyze the morphology and pores of soil particles, and images with the magnification of 10000X are used to show the erosion phenomena more clearly.

3.1 *Effect of different magnification on microstructural images of acid-polluted laterite*

Figure 2 shows the microstructural images of acid-polluted laterite under different magnifications and soaking conditions, with an acid concentration of 5% and a soaking duration of 4 days.

Figure 2 shows that the microstructure characteristics presented are different under different magnification. A larger soil range can be seen when the magnification is small, but the more subtle difference cannot be seen. With the increase of magnification, the microstructure image can clearly show the particle morphology and pore characteristics of laterite samples after being eroded by acid. From images (a) to (f), the magnification ranges from 500X to 20000X. It can be gradually seen from the images that after acid erodes laterite, the pores in the soil gradually become clear. Under the soaking condition, cemented iron oxide and alumina oxide in laterite become ionic forms and are washed away into the solution. As can be seen from image (f), there are a large number of fine holes in the soil. The shape of soil particles is irregular, and the aggregate structure of soil particles is loose, which leads to the reduction of compactness and stability of laterite.

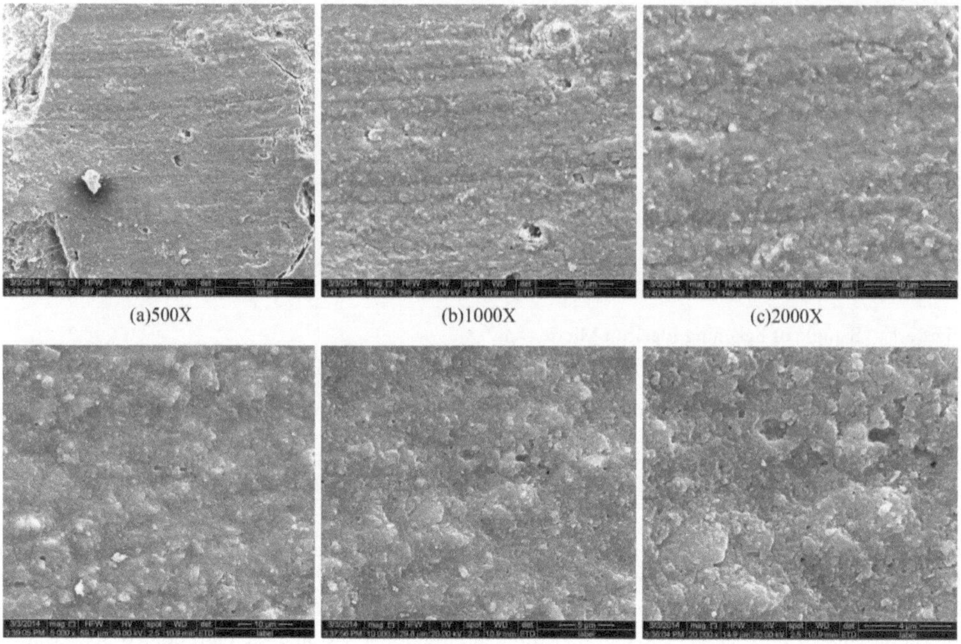

(a)500X (b)1000X (c)2000X

(d) 5000x (e) 10000x (f) 20000x

Figure 2. Microstructural images of acid-soaked laterite under different magnifications.

3.2 *Microstructural images of acid polluted laterite under different acid concentration*

Figure 3 Shows microstructural images of acid-soaked laterite under different acid concentration.

(b)1#,0%,10000X (b)2#,1%,10000X

Figure 3. Microstructural images of acid-soaked laterite under different acid concentration.

Figure 3 shows that:

(1) The microstructural image 1# is laterite before acid pollution, in which the soil particles are closely connected, and the shape of soil particles is mainly round with curved surfaces. The

(a)3[#], 5%,2000X (b)4[#],8%,10000X

Figure 3. Continued.

magnification of picture 1 # (b) is 10000X, so it's clear to see the structure of laterite when the acid concentration is 0%.

(2) The microstructural image 2# shows the microstructural structure of the polluted laterite soaked in 1% acid. The soil's surface becomes rough, pores increase, and there is a small amount of irregular multilateral granules. These irregular multilateral granules are salt crystals formed by the chemical reaction between acid and oxides in laterite, which are attached to the surface of soil particles. With the increase of acid concentration, the granules with curved surfaces in the soil are gradually eroded into irregular granules or flakes, as shown in 2#1%(b). In 3#5%(a), there are many tiny pores, the pores of soil particles slowly increase, and the structure gradually becomes loose. In 4#8%(b), the soil has larger pores, and the soil particles become rougher. It can be seen that acid erodes oxides which play a cementing and bonding role in laterite, which leads to the increase of pores, the decrease of strength and the excellent compressibility of laterite.

3.3 *Microstructural images of acid polluted laterite with different soaking time*

Figure 4 shows microstructural images of acid-soaked laterite with different soaking time

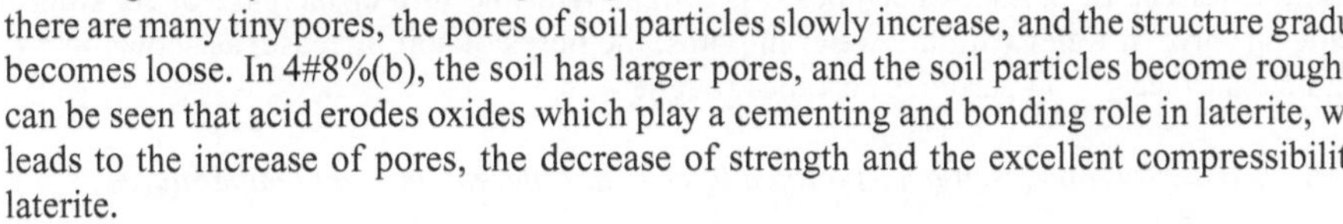

(b)5[#],1d,10000X (b)6[#],7d,10000X

Figure 4. Microstructural images of acid-soaked laterite with different soaking time.

(b)7#,14d,10000X (b)8#,30d,10000X

Figure 4. Continued.

Figure 4 shows that:

(1) Before soaking in acid, the structure of plain laterite is dense and the surface is smooth.
(2) After soaking, the surface of laterite begins to become rough and pores increase. With the extension of soaking time, hydrochloric acid erodes laterite, the pores are larger at 6#7d, the soil surface is rough, and the erosion is severe. With the extension of soaking time, the acid concentration gradually decreases, the erosion weakens, the soluble salts formed to be attached to the surface of soil particles, as shown in 7#14d, the compactness of soil slightly increases, and the surface of particles becomes less rough. With the further increase of soaking time, salts dissolve in water with the water in pores, the pores of soil increase, as shown in 8#30d, and the surface of soil particles becomes rougher.

3.4 *Microstructural images of acid-polluted laterite at different ambient temperatures*

Figure 5 shows microstructural images of acid-soaked laterite at different ambient temperatures

(b)9#,10°C,10000X (b)10#,20°C,10000X (b) 11#, 40 °C, 10000x

Figure 5. Microstructural images of acid-soaked laterite at different ambient temperatures.

Figure 5 shows that with the increase of ambient temperature of soaking, soil pores gradually increase. 9# is the microstructural picture of hydrochloric acid eroding laterite at the temperature of 10°C. As the temperature is low, the molecular activity is lower, the interaction between acid and laterite is not strong, and the erosion is weak. 10# is the microstructural picture of hydrochloric acid eroding laterite at the temperature of 20°C. Compared with the soaking environment at the temperature of 10°C, this temperature is higher, the molecular activity is increased, the acid-soil effect is enhanced, the degree of hydrochloric acid eroding laterite is strengthened, there are obvious holes in the soil and the particle surface becomes rough. 11# is the microstructural image of hydrochloric acid-eroded laterite at the temperature of 40°C. The temperature of soaking environment is further increased, the molecular activity in the solution increases, and the acid-soil effect is enhanced. The oxides of hydrochloric acid in laterite form soluble salts which dissolve in water, thus leading to the increase of soil pores and the roughness of surfaces of soil particles. When the magnification is 10000X, the erosion inside the soil can be seen more clearly.

4 COMPARATIVE ANALYSIS OF MICROSTRUTURAL IMAGES OF POLLUTED LATERITE AFTER ACID SOAKING

4.1 *Microstructural images of acid polluted laterite under different acid concentration*

Figure 6 shows microstructural images of laterite out of acid liquor under different acid concentration

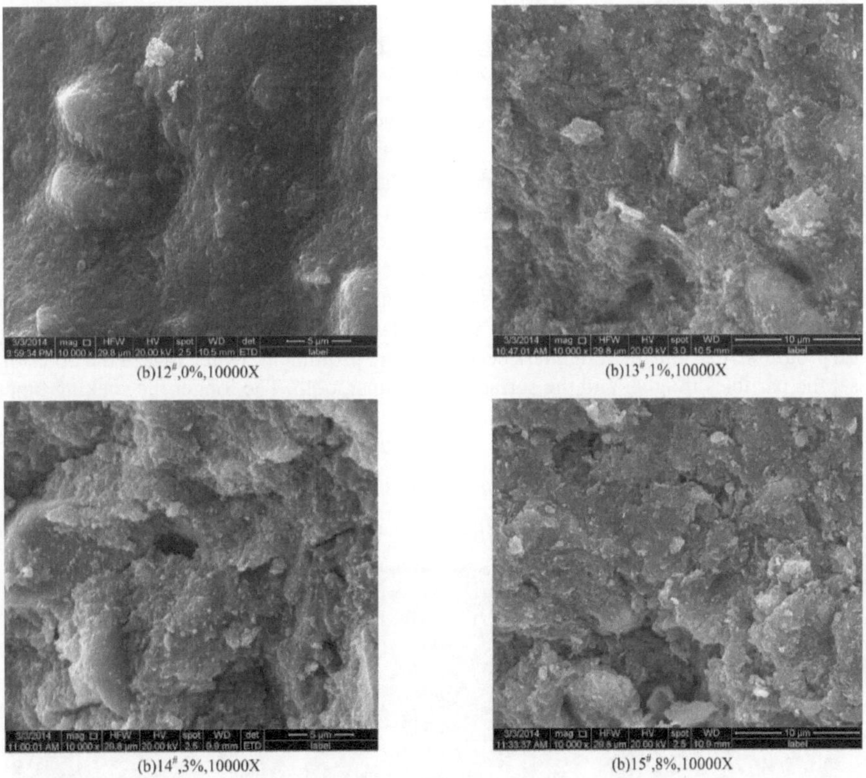

Figure 6. Microstructural images of laterite out of acid liquor under different acid concentration.

Figure 6 shows that:

(1) Before acid pollution, the soil particles are closely connected and arranged in order, and the particles are round with smooth surfaces.
(2) With the increase of acid concentration, the erosion of soil particles becomes more severe. Hydrochloric acid first reacts with oxides in laterite to form salts attached to the surface of soil particles. After being soaked in water, soluble salts gradually dissolve in water with the water in pores. The greater the acid concentration, the more severe the erosion of laterite, the larger the pores of the soil, and the rougher the surface of soil particles. For example, 15# is

the microscopic picture of polluted laterite with an acid concentration of 8%. Because of acid erosion, there are more holes in the soil, and in 15#(b), the morphology of soil particles after erosion can be clearly seen.

4.2 *Microstructural images of acid-polluted laterite with different soaking time*

Figure 7 shows the microstructural images of laterite out of acid liquor with different soaking time

(b)16#,7d,10000X (b)17#,14d,10000X (b)18#,30d,10000X

Figure 7. Microstructural images of laterite out of acid liquor with different soaking time.

Figure 7 shows that for laterite out of acid liquor, the acid reacts with laterite before soaking to form salts attached to the surface of soil particles. With the extension of soaking time, the products of acid eroding laterite gradually dissolve in clear water with the water in pores. The cementing material between soil particles is eroded, which breaks the connection between soil particles. The granular structure formed by soil particles is dispersed, the soil particles become small, the pores between soil particles gradually increase, the salt substances attached to the surface of soil particles decrease progressively, and the roughness becomes more prominent, which makes the pores of soil increase, the particles rougher and the surface of soil unsmooth. The longer the soaking time, the more obvious the erosion.

4.3 *Microstructural images of laterite out of acid liquor at different ambient temperatures*

Figure 8 shows microstructural images of laterite out of acid liquor at different ambient temperatures

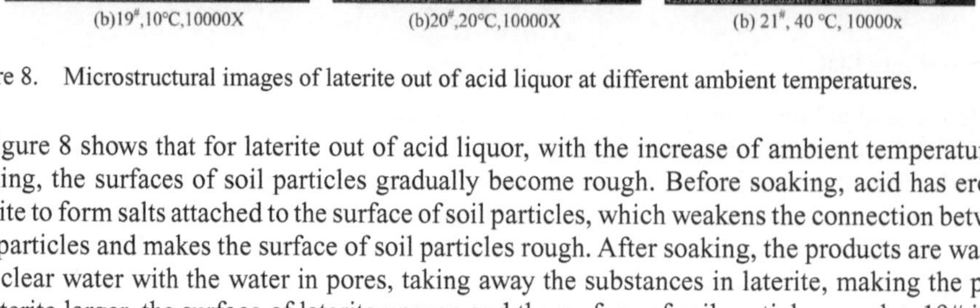

(b)19#,10°C,10000X (b)20#,20°C,10000X (b) 21#, 40 °C, 10000x

Figure 8. Microstructural images of laterite out of acid liquor at different ambient temperatures.

Figure 8 shows that for laterite out of acid liquor, with the increase of ambient temperature of soaking, the surfaces of soil particles gradually become rough. Before soaking, acid has eroded laterite to form salts attached to the surface of soil particles, which weakens the connection between soil particles and makes the surface of soil particles rough. After soaking, the products are washed into clear water with the water in pores, taking away the substances in laterite, making the pores of laterite larger, the surface of laterite uneven and the surface of soil particles rougher.19# is the microstructural picture of acid polluted laterite at the temperature of 10°C. As the temperature is low, the molecular activity is lower, and there are fewer salts washed into the water with the water in pores, and fewer pores in the soil. 20# is the microstructural picture of acid polluted laterite at the temperature of 20°C. Compared with the soaking environment at the temperature of 10°C, this temperature is higher, the molecular activity is increased, the soluble salts dissolve in water with the movement of the water molecule. There are obvious holes in the soil and the particle surface becomes rough. 21# is the microstructural picture of acid polluted laterite at the temperature of 40°C. The temperature of the soaking environment is further increased, the activity of water molecule increases, and more oxides of hydrochloric acid in laterite form soluble salts that dissolve in water, thus leading to the increase of soil pores the roughness of surfaces of soil particles.

The smaller the magnification of the microscopic image of acid-polluted laterite, the larger the structure and the soil range. The larger the magnification, the smaller the structure and the soil range. Through the comparative analysis of the above-mentioned pictures, it can be made clear that for laterite soaking in and out of acid liquor, with the increase of acid concentration, soaking time and environmental temperature, the soil particles become smaller, the pores become larger, the characteristics become more irregular and the surface of the particles become rougher.

The above-mentioned phenomena can be directly observed through images. Also, the characteristic microstructural parameters of acid-polluted laterite can be quantitatively obtained by image processing technology.

5 CONCLUSIONS

This paper mainly analyzed the microscopic images of acid-soaked laterite and acid-polluted laterite after soaking, and obtained the correlation between microscopic parameters and macroscopic mechanical properties.

(1) Through the comparative analysis of the above-mentioned pictures, it can be made clear that with the increase of magnification of microstructural images, the microstructural characteristics are different. When the magnification is small, the microstructural images become a whole structure, while when the magnification is large, the microstructural images show the particles, granules and pores of samples.
(2) The microstructural images show that under different influencing factors, as acid erodes laterite, the pores of the laterite become larger, and there are a greater number of particles, more enormous complexity of soil particles, smaller roundness of soil particles, increased average circumference of particles and decreased fractal dimension of the soil particle distribution.

ACKNOWLEDGMENTS

This work was financially supported by the Fund Project: Scientific Research Fund Project of Yunnan Provincial Department of Education (2016ZZX307).

REFERENCES

Bo, T. Z. (2012). Study on macro and micro characteristics of acid polluted laterite. *D. Kunming University of Science and Technology*, 35–51.

Fan, Y. H. & Huang, Y. & Ren, L.Q. (2014). Influence of Shear Strength of Acid Pollution Laterite by Soaking. *J. Site Investigation Science and Technology* 4, 1–6.

Tang, C. S. & Shi, B. & Wang, B. J. (2008). Factors affecting analysis of soil microstructure using SEM. *Chinese Journal of Geotechnical Engineering* 30, 560–565.

Zhang, X. L. (2007). Experimental study on soil polluted by acid and alkali. *D. Hohai University*.

Zhu, C.P. & Liu, H.L. & Shen, Y. Laboratory tests on shear strength properties of soil polluted by acid and alkali. *D. Chinese Journal of Geotechnical Engineering* 33,1146–1152.

Advances in Civil Engineering: Structural Seismic Resistance, Monitoring and Detection – Mohd Johari Mohd Yusof and Junwen Zhang (Eds)
© 2023 The Authors, ISBN 978-1-032-31491-4

The impact of green space structure in Changsha on the urban heat islands

B. Zheng & J. Liu
School of Architecture and Art, Central South University, Changsha, Hunan, China

ABSTRACT: With Landsat 8 OLI_TIRS as the remote sensing data source, the ground surface temperature of Changsha was inverted. The information of green space within the third ring-road of Changsha extracted based on the remote sensing data source was combined with the inversion data of the ground surface temperature and the correlation between the landscape pattern indexes and ground surface temperature of the selected green space was analyzed. The research found by increasing the percentage of landscape and the length of the boundary of green space within the third ring-road of Changsha, the cold island effect of green space can be enhanced. Few large green patches have a better cooling effect than a lot of small green patches with an equal total area. The aggregated layout of green spaces in a fixed scope within the third ring-road of Changsha has a better cooling effect than the scattered layout.

1 INTRODUCTION

Cities are the inevitable outcomes of social and economic development and they intensively display human civilization (You 2006). According to the report published by the United Nations in 2018, 55% of the population lived in cities around the world in 2018. By 2050, this percentage will increase to 68% (United Nations 2018). In 1833, Howard, a British meteorologist, described the climatic characteristics of the urban heat island effect for the first time and pointed out that the temperature in the downtown of London was higher than that in the surrounding countryside (Howard 1833). The urban heat island (UHI) effect was formally defined by Manley in 1958. Due to the development of cities, buildings, roads, and other infrastructures replaced the open spaces and vegetation in the past. The underlying surface which was permeable and moisturizing in the past became watertight and dry. Besides, the production and lives of urban residents produce a lot of anthropologic heat. These led to the UHI effect. The temperature of a city is higher than that of the surrounding countryside (Nuruzzaman 2015). The urban green spaces have the cold island effect and can slow down the heat island effect. In the cold island effect of urban green space, the impact of type, area, allocation, layout, management mode, and other aspects is focused on (Gherraz et al. 2020; Li et al. 2012; Yu et al. 2015).

The spatial distribution of the landscape affects the radiation flow and energy flow and is critical for determining the ground surface temperature, leading to the heat island. Because the spatial distribution can affect the material flow and energy flow in the vegetation cover, the urban vegetation cover mode in the landscape can affect the ground surface temperature (Song et al. 2014). In different cities, under the premise of the equal total quantity of green spaces, the mean patch area may be positively or negatively correlated with the ground surface temperature. Some researchers believe that they are positively correlated because in these cities a large and continuous green space has a stronger cooling effect than several small green spaces with an equal total area. Some researchers believe that they are negatively correlated because we can enhance the energy flow and exchange between green spaces and provide more shelter for the surrounding area by increasing the number of green patches in the city, thus reducing the ground surface temperature to a greater

DOI 10.1201/9781003310884-21

extent. Besides, the correlation between the degree of aggregation and the cooling effect of green spaces may be different in different cities (Fan et al. 2015; Gherraz et al. 2020; Jafari et al. 2017; Li et al. 2010, 2011, 2012; Masoudi & Tan 2019; Song et al. 2014; Zhou et al. 2011).

The area within the Changsha Ring Expressway was selected as the study area to explore the relationship between the degree of aggregation of green spaces and the ground surface temperature. In the study, data of Landsat8 OLI/TIRS was used to invert the ground surface temperature and analyze the relationship between a series of related landscape pattern indexes and the ground surface temperature. This can help to quantitatively analyze the landscape pattern and the cooling and humidifying function and provide the effective information basis for the structure optimization configuration mode of green spaces in Changsha.

2 MATERIALS AND METHODS

2.1 *Overview of the study area*

Changsha, the provincial capital of Hunan Province, is located at N27°53′ ~ 28°41′ and E111°53′ ~ 114°15′. It is one of the most important metropolises in Central China and is also the central city in the region of Changsha-Zhuzhou-Xiangtan City Cluster. In 2016, the area of land in Changsha was 11,816.0 square kilometers and the urban area was 2,150.90 square kilometers. By the end of 2019, the total permanent resident population of the city was 8.3945 million and the urbanization rate of residents with household registration was 79.56%. The climate type of Changsha is subtropical monsoon climate. It is hot in summer and humid and cold in winter (Changsha Municipal government 2021). Due to human activity, almost all the native vegetation in the urban area of Changsha has disappeared. Almost all the vegetation in the urban area of Changsha is artificial vegetation (Zhang et al. 2009). According to the satellite map, combined with the actual conditions, in this study, Changsha Ring Expressway was the boundary of the study scope. Changsha Ring Expressway is also called the third ring-road.

Figure 1. Study area.

2.2 *Data source*

The data of ground temperature came from the digital product of satellite of Landsat 8 OLI_TIRS, a geospatial data cloud platform (The time was 02:56 GMT on August 17th, 2019. The stripe number was 123 and the administrative number was 40 and 41. The cloud cover was lower than 3%). Among them, the data of the operational land imager (OLI) was used to calculate the ground surface emissivity of the study area; and the data of the thermal infrared sensor (TRIS) was used to invert the ground surface temperature.

2.3 *Inversion of the ground surface temperature*

In this study, the algorithm proposed by Avdan and Jovanovska (2016) was used. ArcGIS10.5 software was used to complete the image preprocessing. The radiance calibration of thermal infrared data and multi-spectral data was made, the ground surface temperature was inverted, and the OLI image was clipped by mask according to the third ring-road of Changsha. In this method, first, spectral radiation on the top of the atmosphere was obtained. And then, the spectral radiation was converted into the brightness temperature. Besides, the normalized differential vegetation index (NDVI) was calculated to speculate the general vegetation conditions, thus obtaining the ground surface emissivity and finally obtaining the ground surface temperature after the correction of radiation.

In this study, Band 4 and Band 5 were used to calculate the NDVI. Band 10 of TIR was used to estimate the brightness temperature. In the first step of the algorithm, Band 10 was entered. After that, at the backstage, the formula from the website of the United States Geological Survey was used to obtain the spectral radiation on the top of the atmosphere.

Table 1. Metadata of the satellite images.

Metadata	Value
Rescaling factor, Band 10 M_L	0.000342
Rescaling factor, Band 10 A_L	0.1
Correction, Band 10 Q_i	0.29
Thermal constant, Band 10 K_1	774.8853
Thermal constant, Band 10 K_2	1321.0789

$$L\lambda = M_L * Q_{cal} + A_L - O_i \quad (1)$$

In which, *ML* represents the band-specific multiplicativity readjustment factor. Q_{cal} represents the image of Band 10. A_L is the band-specific addition readjustment factor. Q_i represents the correction number of Band 10.

Conversion from radiation to the temperature of the sensor. After the digital conversion to reflection, the thermal constant specified in the metadata document should be used to convert the band data of TIRS from spectral radiation to brightness temperature.

$$BT = \frac{K_2}{\ln[(K_1/L\lambda) + 1]} - 273.15 \quad (2)$$

In which, K_1 and K_2 represent the band-specific thermal conversion constants of metadata. To obtain the result with the unit of degree centigrade, by adding the absolute zero (about −273.15°C), the radiation temperature was modified.

Calculation of the NDVI. The visible and near-infrared bands of the earth resource satellite were used to calculate the NDVI. Because the current vegetation quantity is an important factor and the NDVI can be used to speculate the general vegetation conditions, the calculation of the NDVI is very important. Because we should calculate the vegetation coverage subsequently and they are highly correlated with the NDVI, the calculation of the NDVI is very important. And then, we should calculate the emissivity whose value is correlated with the vegetation coverage.

$$NDVI = \frac{NIR\ (band\ 5) - R(band\ 4)}{NIR\ (band\ 5) + R(band\ 4)} \quad (3)$$

In which, NIR represents the near-infrared band (Band 5) and R represents the red band (Band 4).

Calculation of the vegetation coverage. According to Formula (4), the vegetation coverage P_V was calculated. According to a method of the calculation of P_v, it is suggested to use the NDVI value of vegetation and soil ($NDVI_v$=0.5 and $NDVI_s$=0.2) for global conditions. In the formula, the NDVI value ($NDVI_v$=0.5 and $NDVI_s$=0.2) that can be used for vegetation and soil under global conditions according to some researchers is used (Wang et al, 2015).

$$P_V = \left(\frac{NDVI - NDVI_s}{NDVI_v - NDVI_s}\right)^2 \quad (4)$$

However, because the NDVI values of different areas are different, the value of vegetation coverage is 0.5, which may be too low. The global NDVI value can be calculated according to the surface reflectivity. However, if the NDVI is calculated according to the reflectivity of TOA, it is impossible to determine the global value because $NDVI_v$ and $NDVI_s$ depend on the atmosphere conditions.

Calculation of the ground surface emissivity. Because LSE is a scale factor used to measure the black-body radiation and predict the emitted radiation (Planck's law), we must know the ground surface emissivity to estimate LST. The ground surface emissivity is the efficiency of transmitting thermal energy to the atmosphere through the ground surface.

$$\varepsilon_\lambda = \varepsilon_{V\lambda} P_V + \varepsilon_{S\lambda}(1 - P_V) + C_\lambda \quad (5)$$

In which, ε_v and ε_s represent the emissivity of vegetation and soil, respectively. C represents the surface roughness (for a flat and homogeneous surface, C = 0) and is taken a constant value of 0.005. The condition can be represented with the following formula and constant values of emissivity shown in Table 1.

$$\varepsilon_\lambda = \begin{cases} \varepsilon_{S\lambda} & NDVI < NDVI_s \\ \varepsilon_{V\lambda} P_v + \varepsilon_{S\lambda}(1 - P_v) + C & NDVI \leq NDVI \leq NDVI_v \\ \varepsilon_{S\lambda} + C & NDVI > NDVI_v \end{cases} \quad (6)$$

$$\varepsilon_{s\lambda} = 0.996 \qquad \varepsilon_{V\lambda} = 0.973$$

When the NDVI is less than 0, the ground surface is covered by water and the value assigned for the emissivity is 0.991. If the NDVI value is from 0 to 0.2, it is believed that the ground is covered by soil and the value assigned for the emissivity is 0.996. If the NDVI value is from 0.2 to 0.5, it is believed that the ground is covered by soil and vegetation. Formula (6) should be used to obtain the emissivity. Under the first condition, when the NDVI value is greater than 0.5, it is believed that the ground is covered by vegetation and the value assigned for the emissivity is 0.973.

In the last step, the corrected ground surface temperature Ts of ground temperature or emissivity is obtained and calculated as follows.

$$T_s = \frac{BT}{\{1 + [(\lambda BT/\rho)\ln \varepsilon_\lambda]\}} \quad (7)$$

In which, Ts is the ground surface temperature in centigrade degree (°C, (2)), BT is BT in the sensor (°C), λ is the wavelength of emitted radiation (in which, the mean value of peak response and limited wavelength will be used ($\lambda = 10.895$)), ε_λ is the emissivity calculated in Formula (6), and

$$\rho = h\frac{c}{\sigma} = 1.438 \times 10^{-2} mK \quad (8)$$

In which, h is the Planck's constant (6.626×10-34Js), c is the velocity of light (2.998×10^8m/s), and σ is the Boltzmann's constant (1.38×10-23J/K).

2.4 *Information extraction of urban green space*

In this study, ArcGIS10.5 was used for the color composing of satellite images of Bands 1–7 in the satellite digital product of Landsat OLI_TIRS. According to the third ring-road of Changsha, the satellite images were clipped. The maximum likelihood method and NDVI were used to classify the color images. The maximum likelihood method has been widely used for the classification of urban land (Amuladu et al. 2019; Imran & Mehmood 2020; Li et al. 2011). In this study, first, through the image classification tool in ArcGIS10.5, based on artificial recognition, a lot of green space and non-green space training samples were created. The NDVI value generated in the process of inversion of the ground surface temperature was used to judge whether the sample created was accurate. In the vegetation-covered area, the NDVI value may be greater than 0.2. In the non-vegetation-covered area, the NDVI value may be less than 0.2 (Avdan & Jovanovska 2016, USGC 2021, Wang et al. 2013). After the creation of samples, the interactive supervised classification tool in ArcGIS10.5 can be used for classification.

2.5 *The correlation between the degree of aggregation of green spaces and the ground surface temperature*

In this paper, villages and towns (sub-districts) were considered as the analysis units. Villages and towns(sub-districts) are the smallest administrative units in China. Villages and towns(sub-districts) have been widely used as research units to study the relationship between the ground surface temperature and urban green space (Sun 2019; Shi et al. 2019; Xie 2010). In this paper, villages and towns (sub-districts) partially or completely located within the third ring-road of Changsha were selected. Therefore, a total of 83 villages and towns (sub-districts) were selected. With the reference of related studies (Guo et al. 2019; Li et al 2012), in this paper, 5 landscape indexes were selected (Table 2), including the percentage of landscape (PLAND), edge density (ED), mean patch area (AREA_MN), mean Euclidean Nearest Neighbor of patches (ENN_MN), and aggregation index (AI), to reflect the layout characteristics of green spaces of villages and towns (sub-districts). In this paper, Fragstats4.3 was used to obtain the related landscape index of each village and town (sub-districts). Based on the related data of each village and town (sub-districts), in this paper, Origin 2021 was used to generate the scatter diagram and make the Pearson correlation coefficient to discuss the binary relation between mean temperature and each landscape index.

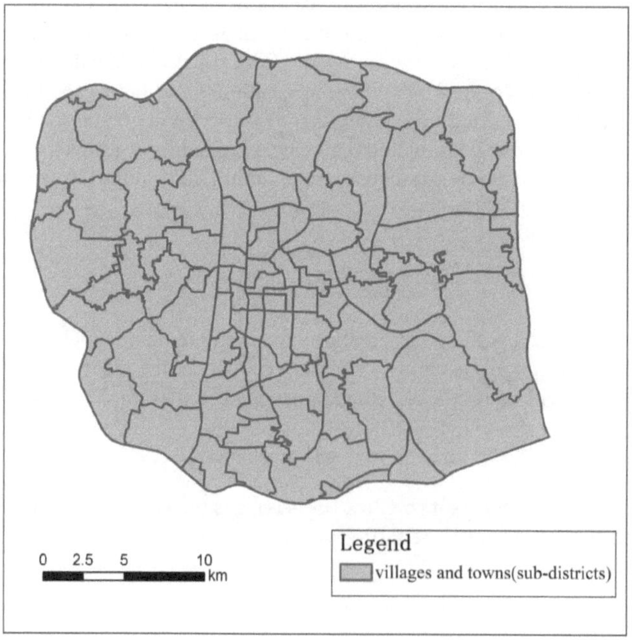

Figure 2. Spatial distribution of villages and towns (sub-districts) within the third ring-road of Changsha.

Table 2. Related information of 5 landscape pattern indexes.

Landscape pattern index	Abbreviation	Unit	Meaning
Percentage of landscape	PLAND	%	The ratio of a type of patches and the total area of landscape.
Edge density	ED	meter/ ha	The length of the boundary per unit area.
Mean patch area	AREA_MN	ha	Mean area of a type of patches.
Mean Euclidean Nearest Neighbor of patches	ENN_MN	m	Reflect the relative intensiveness of the layout of patches. The smaller the value is, the more intensive the layout of patches is.
Aggregation index	AI	%	Dispersion degree of patches in the landscape area. The bigger the value is, the more aggregated the same type of patches in the landscape is.

3 RESULTS AND ANALYSIS

3.1 *Distribution characteristics of the ground surface temperatures and urban green spaces*

Figure 3 illustrates the spatial distribution of the ground surface temperatures within the third ring-road in the urban area of Changsha at 9:56 (Beijing time) (GMT+8) on August 17th, 2019. In the picture, the highest temperature is 40.75 °C, The lowest temperature is 24.23°C, and the mean temperature is 29.28°C. From the picture, we can see that the UHI phenomenon is obvious in the study area. The high-temperature areas spread all over the built-up areas. In the study area, the low temperature is mainly distributed in the north, west, and southeast. As shown in the picture, the eastern high-temperature zone is mainly located in Changsha Economic Development Zone. In this area, factories are intensive. High-temperature areas and sub-high-temperature areas are relatively intensive. In the picture, in the areas in the central part and in the south, the density of population is high and there are many buildings, high-temperature areas and sub-high-temperature areas, but there are fewer high-temperature areas and sub-high-temperature areas than those areas

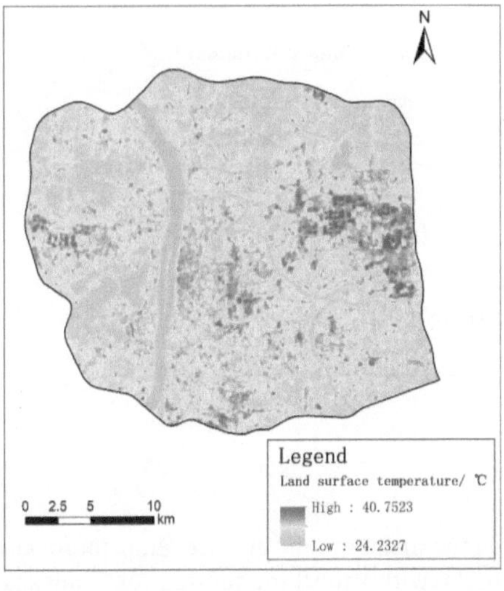

Figure 3. Spatial distribution of ground surface temperatures within the third ring- road in the urban area of Changsha.

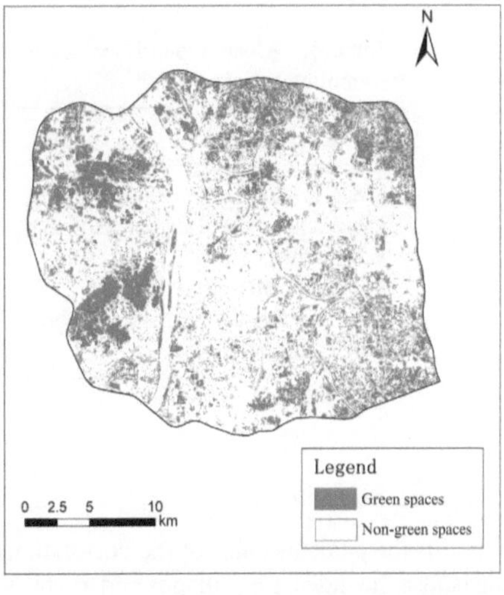

Figure 4. Spatial distribution of green spaces and non-green spaces within the third ring- road in the urban area of Changsha.

shown in the eastern part of the picture. From Figure 4, we can see the layout of green spaces and non-green spaces within the third ring-road of Changsha. In the picture, the layout of green spaces is characterized by "many green spaces in the periphery and few green spaces in the middle". The green spaces are concentrated in the north, west, and southeast. Through the comparison and analysis of Figure 3 and Figure 4, we can know that the ground surface temperature is relatively low in the areas where green spaces gather and the urban green space has an obvious cold island effect on the UHI.

In this paper, villages and towns (sub-districts) were used as the analysis units. The mean temperatures and landscape indexes of 83 villages and towns (sub-districts) partially or completely located within the third ring-road of Changsha were used to explore the impact of the allocation of green spaces within the third ring-road of Changsha on the ground surface temperature. Figure 5 and Figure 6, and Table 3 show the examples of the ground surface temperature map, green space map, mean ground surface temperature, and some landscape indexes of a township (sub-district). This township (sub-district) is Mapoling sub-district in Furong District, Changsha. The PLAND of this street is 24.1252%. At the time, the mean ground surface temperature is 30.03°C.

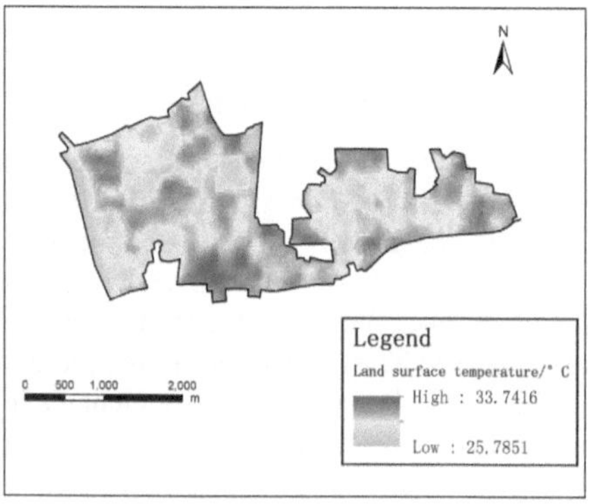

Figure 5. Spatial distribution of ground surface temperatures of Mapoling sub-district.

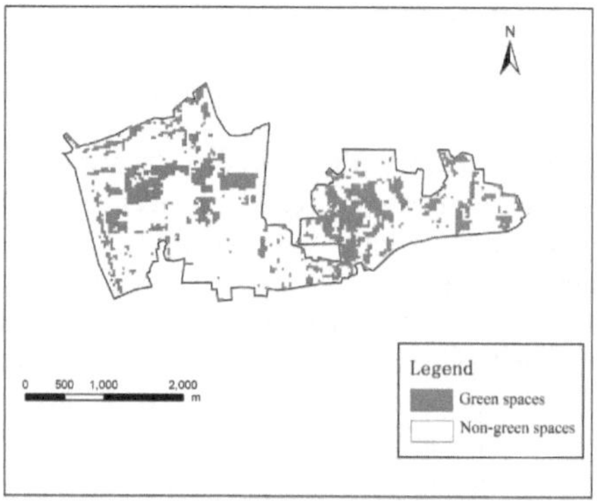

Figure 6. Spatial distribution of green spaces and non-green spaces of Mapoling sub-district.

Table 3. Mean ground surface temperature and some landscape indexes of Mapoling sub-district.

Index	Specific value
Mean ground surface temperature	30.03°C
Percentage of landscape (PLAND)	24.1252%
Edge density (ED)	94.3907 meter/ha
Mean patch area (AREA_MN)	1.4429 ha
Mean Euclidean Nearest Neighbor of patches (ENN_MN)	82.1818 m
Aggregation index (AI)	70.1772%

3.2 *Correlation analysis*

Figure 7 is the analysis chart of the correlation between the mean ground surface temperature and some landscape indexes of villages and towns (sub-districts) within the third ring-road of Changsha. As shown in Figure 7 below, the results of the analysis on the correlation between the percentage of landscape (PLAND) and the ground surface temperature in the research area show that the ground surface temperature decreases as the PLAND increases. This is consistent with the correlation

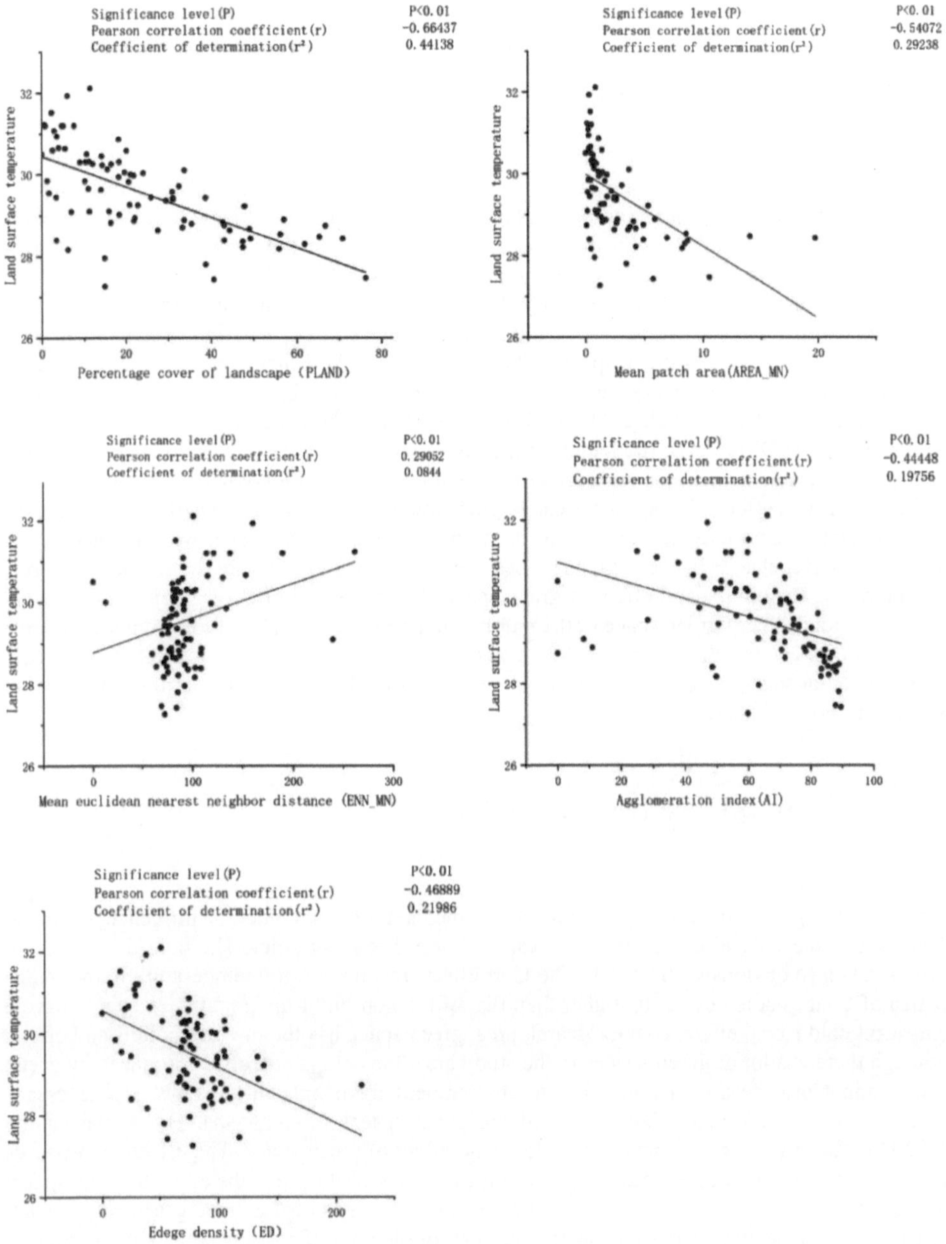

Figure 7. The correlation between the mean ground surface temperature and some landscape pattern indexes of villages and towns (sub-districts) within the third ring-road of Changsha.

study results of previous researchers(Gherraz et al. 2020; Li et al. 2012; Masoudi & Tan 2019). The significant correlation between the percentage of landscape (PLAND) and the ground surface temperature ($r = -0.66437$, $P < 0.01$) also indicates that in the study area relieving the UHI effect by increasing the PLAND has a remarkable effect.

The mean patch area (AREA_MN) is significantly negatively correlated with the ground surface temperature ($r = -0.54072$, $P < 0.01$). This means that the bigger the mean patch area of green space is the more obvious the cold island effect of the green space is without considering the PLAND and other landscape pattern indexes. Therefore, under the premise that the PLAND is definite, we can further slow down the UHI effect by increasing the mean patch area and decreasing the number of patches. The mean Euclidean Nearest Neighbor of patches (ENN_MN) is weakly positively correlated with the ground surface temperature ($r = 0.29052$, $P < 0.01$). The mean Euclidean Nearest Neighbor of patches is not closely correlated with the ground surface temperature, but the cold island effect of the green space reduces with the gradual increase of the patch distance. This also indicates that the more aggregated layout of green spaces has a better cold island effect. Aggregation index (AI) indicates the dispersion degree of patches in the landscape area. The bigger the value is, the more aggregated the same type of patches in the landscape is. The aggregation index in the picture is negatively correlated with the ground surface temperature ($r = -0.44448$, $P < 0.01$), which also indicates that the cold island effect of green space can be improved by improving the degree of aggregation of green spaces. These can support the conclusion of some researchers that a large and continuous green space has a stronger cooling effect than several small green spaces with an equal total area (Li et al. 2011, 2010; Fan et al. 2015; Zhou et al. 2011).

Edge density (ED) refers to the length of the boundary per unit area. In the picture, the ED is negatively correlated with the ground surface temperature ($r = -0.46889$, $P < 0.01$), which indicates that the cold island effect of green space can be improved by increasing the length of the boundary of green patches. By increasing the length of the boundary of green space, we can enhance the energy flow and exchange between green spaces and provide more shelter for the surrounding area, thus improving the cold island effect of green space. To increase the ED of green space, we can increase the total area of green space or the number of green patches. Therefore, from the negative correlation between the ED and the ground surface temperature, we cannot judge that the layout of aggregated or scattered green spaces has a better cooling effect under the premise of the equal total quantity of green spaces.

4 CONCLUSION AND DISCUSSION

4.1 *Discussion*

Climate warming is a global ecological and environmental problem. Under the influence of climate warming and the UHI effect, in many cities, it is very hot in summer. The core urban area of Changsha is also obviously affected by the UHI effect. In summer, the temperature in the build-up area of Changsha is obviously higher than that of the non-build-up area. Green spaces have a significant cold island effect. In the research area, green space has the obvious cold island effect. Although there is a lot of green spaces in the study area, the cold island effect produced by green spaces cannot provide a cool microclimate environment for people in hot days. It's necessary to further improve the cold island effect of the green spaces in Changsha. By increasing the PLAND in the city, we can enhance the cold island effect of green space. The effective measures include planting more trees, encouraging to change the brownfield site in the city into green space, encouraging to increase green spaces with new methods (such as roof garden, green wall, pocket garden, etc.), and having the reasonable requirement for the PLAND of new construction projects.

We should not improve the cold island effect of green space just by increasing the PLAND. By changing the spatial distribution of green spaces, we can change the cooling and humidifying effect. The relationship between the spatial distribution of green spaces and their cooling effects varies by cities. In some cities, the more aggregated layout of green spaces has a better cooling effect. In some cities, the more scattered layout of green spaces has a better cooling effect. The measure of improving the cold island effect of green space by improving the layout of green space should be explored by the specific conditions of the city. In the core urban area of Changsha, we should encourage to plan more green spaces with fewer patches with a large area. This can help to enhance the cold island effect of urban green space. The reason may be that there are many small green spaces in Changsha, which do not have a significant cold island effect. Without changing the area of green patches, by increasing the degree of aggregation of green patches, we can improve the cold island effect of green space. When planning new green spaces, we can pay more attention to the reduction of the distance between green patches, which can help to give full play to the cold island effect of green space. In the planning of green spaces, we should increase the length of the boundary of green space so that the green space can provide more shelter to the surrounding area and promote the flow of energy.

In this paper, villages and towns (sub-districts) within the third ring-road of Changsha were considered as the analysis units. This can help to give suggestions on the optimization of green space in the core urban area of Changsha based on villages and towns (sub-districts) within the third ring- road, especially large-scale green spaces. The area of a village, town, or sub-districts can be tens of square kilometers or more. In this way, we can study the relationship between the urban green space and the UHI effect of Changsha on a large scale. The relationship between the ground surface temperature and allocation of green spaces may be affected by the selection of the scale of the analysis unit. In this paper, it is believed that the results may be different when analysis units of different scales are selected for the study. The study scope of this paper is the area within the third ring-road of Changsha. If the study scope is bigger or smaller, the results of the related correlation analysis may be different.

4.2 *Conclusion*

This paper explores the relationship between the ground surface temperature and some landscape pattern indexes within the third ring-road of Changsha. With the remote sensing information, researchers could invert the ground surface temperature based on the satellite image and related algorithms. The Pearson correlation analysis was used to explore the binary relation between the green space information extracted from the satellite image and the inverted ground surface temperature. In the study, it was found that the composition and layout of green space could affect the ground surface temperature. We should focus on increasing the PLAND and optimizing the layout of green space to mitigate the UHI effect. It is believed that we can improve the sheltering effect of trees and promote the flow of energy by increasing the ED of green space. It is believed that we can improve the ability of green space to mitigate the heat island effect by increasing the patch area and improving the degree of aggregation of green spaces within the third ring-road of Changsha. The correlation between these landscape pattern indexes and the ground surface temperature can be used to guide the urban planning and land use management of Changsha, especially the making of the green space optimization strategy based on villages and towns (sub-districts) in the core urban area of Changsha. Besides, the correlation may vary by the scale of the analysis unit and study scope. In the future, through the comparative study of different scales of analysis units and study scopes, we can explore whether the relationships between the landscape pattern indexes and the ground surface temperatures of some green spaces are obviously different when the scales of analysis units and study scopes are different. To give full play to the cold island effect of urban green space, we should not only increase the urban green spaces and optimize the layout of green spaces, but also give sufficient consideration to the good coordination among green space, buildings, and hydrological characteristics. In this way, we can give full play to the cold island effect of urban green space and hydrological characteristics.

REFERENCES

Amuladu, Xu, B., Li, C., Chen, C., Zhong, C., Li, H. (2019) Spatial-Temporal Variations of Heat Island in Xichang and Its Driving Forces from 2000 to 2014. *J. Hunan Normal Univ. Nat. Sci.*, 48(1): 16–22.

Avdan, U. & Jovanovska, G. (2016) Algorithm for Automated Mapping of Land Surface Temperature Using LANDSAT 8 Satellite Data. *Journal of Sens.*, 10(1): 1–8.

Changsha Municipal government. (2021) Overview of Changsha.

Fan, C. Myint, S., Zheng, B. (2015) Measuring the spatial arrangement of urban vegetation and its impacts on seasonal surface temperature. *Prog. in Phys. Geogr.*, 39(1): 199–219.

Gherraz, Hz., Guechi, I., Alkama, D., (2020) Quantifying the effects of spatial patterns of green spaces on urban climate and urban heat island in a semi-arid climate. *Bulletin de la Société Royale des Sciences de Liège*, 89: 164–185.

Gou, R., Zhao, X., Bu, Y., Gao, T., Zhao, P. (2019) Dynamic Study on the Heat Island Effect and Landscape Pattern in Urban Area of Hangzhou. *Res. water soil conserve.*, 26(1): 316–329.

Howard, L. (1833) *Climate of London Deduced from Metrological Observations.* Harvey and Dorton Press. London. http://www.changsha.gov.cn/xfzs/zjmlzs/zsgl/200907/t20090727_5686409.html.

Imran, M. & Mehmood, A. (2020) Analysis and mapping of present and future drivers of local urban climate using remote sensing: a case of Lahore, *Pakistan. Arab. J. Geosci.*, 13(4): 1–14.

Jafari, E., Soltanifard, H., Aliabadi, K., Karachi, H. (2017) Assessment of the Effect of Neyshabur Green Spatial Configuration on the Temperature of Land Surface and Heat Islands. *Open J. E.*, 7(9): 554–567.

Li, C., Yin,J., Zhao,J. (2011) An extraction algorithm of urban vegetation from remote sensing image based on object-oriented approach [J]. *Surveying and Mapping Science,* 36(3): 113–120.

Li, J., Song, C., Cao, L., Zhu, F., Meng, X., Wu, J. (2011) Impacts of landscape structure on surface urban heat islands: A case study of Shanghai, China. *Remote Sens. Environ.*, 115(12): 3249–3263.

Li, S., Zhao, Z. Xie, M., Wang, Y. (2010) Investigating spatial on stationary and scale-dependent relationships between urban surface temperature and environmental factors using geographically weighted regression. *Environ. Model. Softw.*, 25(12): 1789–1800.

Li, X., Zhou, W., Ouyang, Z., Xu, W., Zheng, H. (2012) Spatial pattern of greenspace affects land surface temperature: evidence from the heavily urbanized Beijing metropolitan area, China. *Landsc. Ecol.*, 27(4): 887–898.

Masoudi, M. & Tan, P.Y. (2019) Multi-year comparison of the effects of spatial pattern of urban green spaces on urban land surface temperature. *Landsc. Urban Plan.*, 184: 44–58.

Nuruzzaman, M. (2015) Urban Heat Island: Causes, Effects and Mitigation Measures- A Review. *Int. J. Environ. Monit. Anal.*, 3(1): 67–73.

Shi, Y., Zhang, B., Gao, J., Tong, Y., Shao, S. (2019) Demand of urban green space cool islands based on heat island pattern: A case study of Chaoyang District of Beijing. *Resource sci.*, 41 (6): 1541–1550.

Song, J., Du, S., Xin, F., Luo, G. (2014) The Relationships between Landscape Compositions and Land Surface Temperature: Quantifying Their Resolution Sensitivity with Spatial Regression. Landsc. *Urban Plan.*, 132: 145–147.

Sun, Z. (2019) Thermal environment characteristic and cooling effect of greenery in Beijing First Green Belt area. *Ecology. Misc.*, 38 (11): 3496–3505.

United Nations. (2018) *2018 Revision of World Urbanization Prospect.* https://www.un.org/zh/desa/2018-revision-world-urbanization-prospects.

USGC. (2021) *NDVI, the Foundation for Remote Sensing Phenology* https://www.usgs.gov/core-science-systems/eros/phenology/science/ndvi-foundation-remote-sensing-phenology?qt-science_center_objects=0#qt-science_center_objects.

Wang, S., Ye, B., Ding, Y., Zhao, C., Qin, J. (2013) The Effect of Water Consumption on Surface Runoff in Oasis in the Aksu River Basin. *Arid zone res.*, 30 (1): 203–210.

Xie, J. (2010) *Research on ecological benefits of urban green space landscape in Wuhua District*, Kunming. Kunming: Southwest Forestry University.

You, J. (2006) Development of city definition. *Shanghai Manage. Sci.*, 28 (2): 67–69.

Yu, Z., Guo, Q., Sun, R., (2015) Impacts of urban cooling effect based on landscape scale: A review. *J. Appl. Ecol.*, 26 (1): 636–642.

Zhang, C., Zhou, F., Sun, G., Liu, H. (2009) Community structure characteristic and optimal structuring of green land plants in Changsha City. *Forestry investing. and plan.*, 34 (2): 44–47.

Zhou, W., Huang, G., Cadenasso, M.L. (2011) Does Spatial configuration matter? Understanding the effects of land cover pattern on land surface temperature in urban landscapes. *Landsc. Urban Plan.*, 102: 54–63.

Advances in Civil Engineering: Structural Seismic Resistance, Monitoring and Detection – Mohd Johari Mohd Yusof and Junwen Zhang (Eds)
© 2023 The Authors, ISBN 978-1-032-31491-4

Analyses on characteristics of microelement and rare-earth element zone of hadamengou gold deposit, Inner Mongolia

Xin Wang, Xiao Bin Dang, Chuan Yun Yue, Yan Wang & Liang Ming
Department of Civil Engineering, Shenyang Urban Construction University, Shenyang, P.R. China

ABSTRACT: The pre-mountain potash belt in Hadamengou Gold Deposit, Inner Mongolia, is located in the south of the ore belt and the north of the pre-mountain Baotou-Hohhot fault. The study of element geochemistry shows that the Piedmont potash belt belongs to the ultrapotassic basalt series. The rocks are characterized by enrichment of LILE (Rb, Ba, K, Sr) and light rare earth elements (LREE), and have obvious negative anomalies of Ti, Ga, and Be, the total rare earth (REE) has the opposite trend with the alteration intensity of potassium. The results show that the rocks of the Piedmont potash belt are structurally fractured and strongly altered, so it is considered that the Piedmont potash belt was originally a tectonically fractured zone, in the later period, the ore-bearing hydrothermal solution with high content of potassium entered the fracture zone and the rocks in the fracture zone underwent hydrothermal alteration.

1 INTRODUCTION

The Hadamengou Gold Deposit is an important part of the Mount Ulla Gold Field and a large-scale gold deposit located in the suburb of Baotou. Based on field observations, sample collection, and isotopic dating of various geological bodies in Hadamengou Gold Deposit, the late variscan formation of the deposit has been determined. However, these studies mainly focus on the geological characteristics (XueJianping 2010), metallogenic conditions, metallogenic epoch (HouWanrong 2011), and ore body geochemical characteristics (YuWanqiang 2010), and the related studies on the Piedmont potassium zone are little inadequate. This paper deals with the relationship between the trace element potassium and the alteration of potassium in the potassium belt in the front of the mountain in the Hadamengou Gold Deposit area through petrography and geochemistry.

2 CHARACTERISTICS OF GEOCHEMICAL ELEMENTS

2.1 *Characteristics of microelement*

The test and analysis results of Rb, Ba, Th, and other trace elements in 7 samples of the Piedmont potassium belt in Hadamengou Gold Deposit, Inner Mongolia are shown in Table 1.

Table 1. Microelement of hadamengou gold mine, Inner Mongolia ($\omega B/10^{-6}$)

Sample No	HD01	HD02	HD04	HD06	HD08	HD10	HD11	Average
Rb	155	163	157	151	154	200	178	165.43
Ba	786	1165	1302	1334	1569	1139	1202	1213.86
Th	9.64	13.9	6.76	10.8	10.2	4.56	4.47	8.62

(*continued*)

DOI 10.1201/9781003310884-22

Table 1. Continued.

Sample No	HD01	HD02	HD04	HD06	HD08	HD10	HD11	Average
U	2.58	0.52	2.18	2.66	1.49	0.97	1.73	1.73
Nb	21.5	6.97	20	27.8	45.8	9.32	23.3	22.10
Ta	1.03	0.17	0.76	0.78	0.86	0.43	0.27	0.61
K	101980.85	86048.94	95093.62	80738.3	113431.91	117248.94	125048.94	102798.79
La	58.4	57	53.4	71.5	65.4	16.3	7.44	47.06
Ce	110	110	107	135	137	37.5	14.5	93.00
Pb	24.7	17.3	17.2	28.7	19.3	52.7	19.4	25.61
Pr	15	12.7	14.1	17.8	18.1	5.15	1.79	12.09
Sr	136	210	185	278	461	296	244	258.57
Nd	58	43.9	57.1	74.8	74.7	21.6	7.35	48.21
Zr	219	95.9	202	275	253	43.9	786	267.83
Hf	5.32	2.53	4.84	7.18	5.51	1.42	21.3	6.87
Sm	10.7	8.07	10.6	12.6	14.5	6.01	2.25	9.25
Eu	2.21	1.41	2.35	2.83	3.35	1.47	0.88	2.07
Ti	5162	840	5490	8480	5654	651	4219	4356.57
Gd	7.71	5.27	7.55	9.68	10.5	6.04	1.68	6.92
Tb	1.09	0.76	1.06	1.31	1.33	1.1	0.3	0.99
Dy	5.29	3.48	5.11	6.12	5.69	6.75	1.86	4.90
Y	29.1	15.8	26.1	32.2	26.1	40	13.3	26.09
Ho	0.99	0.58	0.91	1.1	0.91	1.25	0.41	0.88
Er	2.98	1.73	2.79	3.29	2.63	3.71	1.34	2.64
Tm	0.36	0.18	0.32	0.4	0.28	0.46	0.19	0.31
Yb	2.47	1.09	2.15	2.66	1.77	2.8	1.37	2.04
Lu	0.37	0.15	0.33	0.4	0.25	0.4	0.21	0.30
Ga	21.8	15.4	19.7	20.5	20.1	19.5	24.2	20.17
Be	0.31	0.29	0.34	1.54	0.2	0.26	0.21	0.45

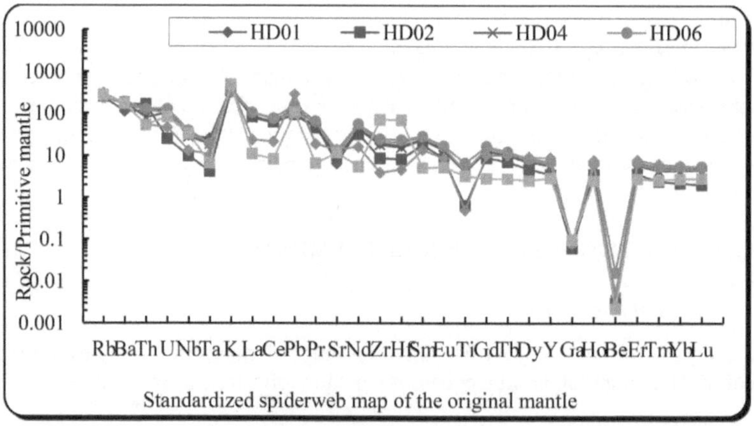

Figure 1. Standardized spiderweb map of the original mantle of the microelement in the Hadamengou gold deposit, Inner Mongolia.

On the normalized micronutrient of the original mantle (Figure 1), the overall features of the samples are consistent, and the large ion lithophile elements are enriched, especially Rb, Th, K, Pb. The high field strength elements are relatively large, and the ionic lithophile elements are relatively deficient, such as Nb, Ta, Zr, Hf, etc. All samples showed negative anomalies of Be and Ga. Except for samples HD11 and HD10, others show obvious negative anomalies of Ti and Sr. It is generally considered that the negative anomaly of Ti may be related to the separation and crystallization of apatite and ilmenite. The Ta-Nb-Ti negative anomaly is similar to the rock characteristics formed in the "island arc" environment. Most scholars believe that the origin of potassic rocks is related to potassium rich and live metasomatic mantle, and this effect is often accompanied by subduction, such as oceanic island arc, post-collision arc environment, and continental arc. Recent studies also show that in Hadamengou Gold Deposit, the ancient Mongolian ocean crust and Angara ancient land block have moved for many times, such as subduction, collision, and docking; therefore, the characteristics of trace elements in the potassic zone have a certain inheritance to the strata in this area. Generally, the negative anomaly of Sr is related to the crystallization differentiation of plagioclase.

On the covariant relationship between trace elements and K_2O content in Hadamengou Piedmont potassic zone, Inner Mongolia (Figure 2), we can see that, in general, the contents of Rb and Ga

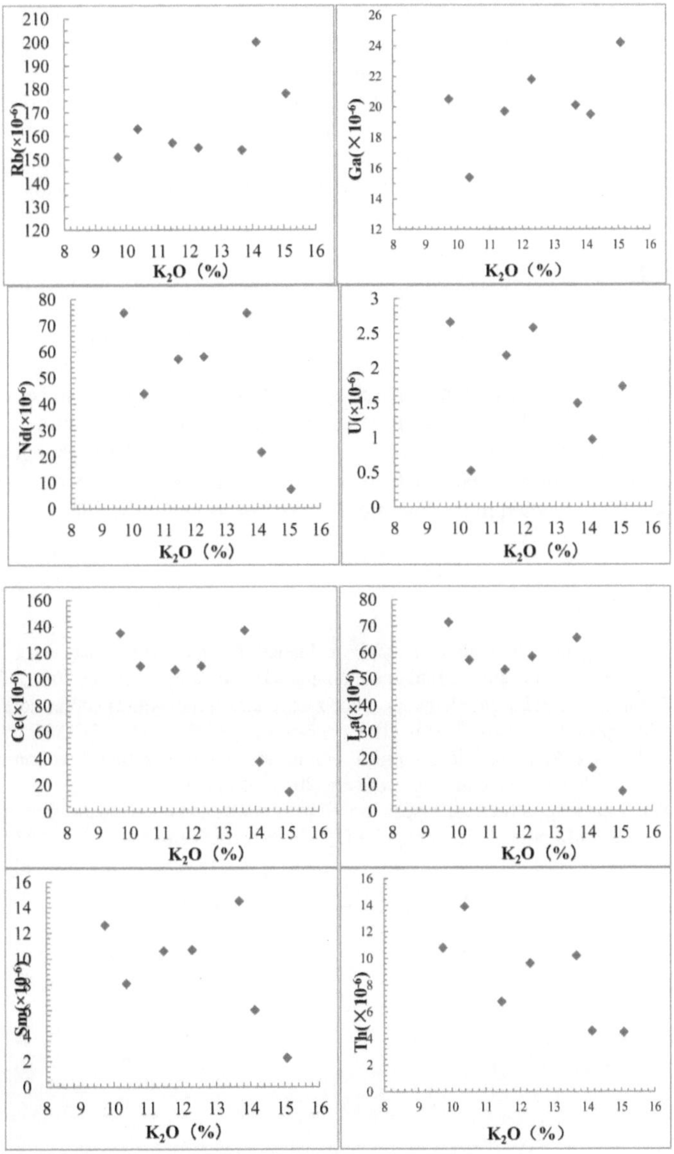

Figure 2. Covariant relationship between trace elements and K2O content in Hadamengou Piedmont potassium zone, Inner Mongolia.

increase with rise in K_2O content. The contents of U, Th, La, Nd, Ce, and Sm decreased with increase in K_2O content. These characteristics are similar to the positive and negative anomalies reflected in the standardized cobweb map of trace element primitive mantle, indicating the relationship between the overall trace element content and the degree of potassium alteration in the Piedmont potassium belt of the mining area.

3 DISCUSSIONS

The REE characteristics of the metamorphic rocks of the Wulashan group reflect that the original rock is a volcanic clastic sedimentary rock combination formed in the Archean active continental margin island arc environment and most of the deposited materials possibly come from the volcanic island arc (RaoWenbo 2002). It can be seen from the standardized trace element spider diagram of the primitive mantle (Figure 1) that the common feature of the trace element distribution pattern of the rocks is that the large ion lithophile elements are enriched, such as Rb, Th, etc., and have the characteristics of obvious Ta, Nb, Ti negative anomaly, which shows that the potassic zone has the characteristics similar to the rocks in the "island arc" environment. This is consistent with the geochemical characteristics of metamorphic rocks in Wulashan Gold Field, suggesting that the Piedmont potassium zone is closely related to the metamorphic rocks of Wulashan group.

4 CONCLUSION

Trace elements are characterized by obvious enrichment of large ion lithophile elements (Rb, Th, K, Pb), the relative loss of high field strength elements (Nb, Ta, Zr, Hf), and obvious TaNbTi negative anomaly, reflecting the rock characteristics formed in the "island arc" environment This characteristic is consistent with the characteristics of trace elements in the strata in this area. Rare earth elements are characterized by the total amount of rare earth Σ REE changes significantly, showing the total amount of rare earth as a whole Σ REE decreased with the increase in K_2O content, and LREE was enriched in the Piedmont potassium zone as a whole.

REFERENCES

Hou Wanrong, Nie Fengjun, Du andao. Isotopic evidence for the determination of Devonian gold (molybdenum) mineralization events in Hadamengou area [J]. *Geological review*, 2011, 57 (4): 584–560.

Rao Wenbo, Gao Zhenmin, Yang zhusen. Element geochemical characteristics and material source of manggang red clay type gold deposit in western Yunnan [J]. *Geochemistry*, 2002, 31 (6): 576–586.

Xue Jianping, Gao Junping. Metallogenic geological conditions and prospecting direction in Hadamengou Gold Deposit Area [J]. *Inner Mongolia coal economy*, 2010, 6: 111–113.

Yu Wanqiang, Li Wei, Liu Gang. Structural superimposed halo model and deep prospecting prediction of vein 13 in Hadamengou Gold Deposit, Inner Mongolia [J]. *Geology and resources*, 2010, 19 (3): 230–236.

Advances in Civil Engineering: Structural Seismic Resistance, Monitoring and Detection – Mohd Johari Mohd Yusof and Junwen Zhang (Eds)
© 2023 The Authors, ISBN 978-1-032-31491-4

Research on optimization of deep foundation pit about excavation support scheme

Yan Wang
Department of Civil Engineering, Shenyang Urban Construction University, Shenyang, China

ABSTRACT: With the development of building technology, the speed of urban development is accelerated, the city scale continues to expand, with the rapid expansion of population, A lot of abuses have emerged. Such as land tension, crowded living space, traffic congestion, backward infrastructure, ecological imbalance, environmental deterioration and other problems, the rational development of underground space can not delay. China has lifted boom of urban underground space development and construction, such as underground mall, underground railway, the comprehensive utility tunnel, underground parking, national defence engineering construction, which inevitably engineering problem is the study of the retaining structure of deep foundation pit engineering, is now a project is critical in the process of conducting a content, need to give a high degree of attention, because of the deep foundation pit supporting structure, when the design is too conservative, produce waste, so is the necessity of optimization. Therefore, this paper gives a detailed analysis on the optimization design of deep foundation pit excavation support structure for subway construction.

1 INTRODUCTION

A subway construction project, geological conditions are mainly silty clay, its obvious feature is diversity and soil water distribution inhomogeneity of surrounding buildings are dense, therefore in the process of construction to the implementation of targeted monitoring real time control of the deep foundation pit design, and the numerical simulation, comparing the related work and then put forward the optimization scheme for supporting pile. In this paper, the excavation and support of deep foundation pit about subway construction in a city as the background of exploration, using on-site monitoring, rectification design and simulation value, so as to effectively optimize the pile distance and corresponding pile diameter of the supporting pile, in order to achieve economic, reasonable and safe construction objectives.

2 STRUCTURAL FORM ANALYSIS OF DEEP FOUNDATION PIT SUPPORT

2.1 *Cement soil retaining wall structure*

Concrete retaining wall supporting structure, the principle is to spray curing agents into the soil, with the help of stirring equipment, the soil and curing agent are forced to mix, thus the overall stability and shear strength of soil are improved. Its advantage is in the process of construction and did not lead to high noise, pollution to surroundings is very small, at the same time,it can save cost, economic effect is obvious, and the technology has matured. The disadvantage is that the construction process is very complex, the use of very professional machinery and equipment, and the construction cycle is long (Chen Dongsheng 2018).

DOI 10.1201/9781003310884-23

2.2 *Underground diaphragm wall*

To be specific, the underground diaphragm wall support structure is that use the equipment to excavate the trench in the mud wall, and put the steel cage into it, and inject concrete to compacting it, then build the continuous underground concrete wall. The outstanding advantage of the underground diaphragm wall is that the structure has very large stiffness, the overall performance is very ideal, reliable security and high bearing capacity, as well as a certain sealing performance, which can effectively control the deformation. The structure can be applied to a wide range of deep foundation pits with mature technology from 10m to 40m.Underground continuous walls can also be used if the construction site is adjacent to important buildings or in a subway station where underground pipelines need to be protected. Disadvantages are relatively high cost, mud will cause pollution to the environment (Kong Junqiang 2019).

2.3 *Soil nail wall*

The structure of soil nailing wall consists of three parts: reinforced soil, soil nailing and concrete panel. It can play an ideal purpose of retaining soil to bear the soil behind the wall. Advantages for the construction process of material saving is very prominent, short construction period, low cost, small vibration, noise, little impact on the surrounding environment. The disadvantage is that the structure will be damaged by rain, which makes the foundation pit and the subsequent partial damage of the structure, or even the whole damage.

2.4 *Pile row support*

The supporting function of pile row supporting structure is soil retaining, waterproof and enclosure. The most prominent advantage of this structure form is that it can carry out good control of deformation, which is very convenient for construction, with higher bearing capacity and better economy. The structure stiffness is large, and the technology is mature, so the support method is widely used. The disadvantage is that there is no ideal integrity and reliability, the sealing performance is not outstanding, can not be carried out at the same time the bored pile and mixing pile operation. Mud will cause pollution to the environment, and it is very difficult to control the perpendicularity of pile body in construction, and it is not easy to have ideal uniformity of large depth water-stopping rotary jet grouting pile (Yang Xiangru 2019).

2.5 *Internal support*

The internal support structure consists of foundation pit retaining structure and support system. Its advantage is that in the support system, the round steel pipe and part of the section steel can be recycled, both installation and disassembly are very convenient, the cost is not high, so it has a very wide range of application, in the pit depth is small, the soil is not ideal, the environment is more suitable for the project. The biggest disadvantage is that in the construction process, the development of machinery will be affected (Zhang Zhilian 2018).

3 ENGINEERING OVERVIEW

The subway station construction is located at the intersection of an economic development zone and the commercial street. It belongs to the underground two-story double-cross-island station, and it is planned to adopt open-cut construction method. There are some municipal pipelines in the station, and the design institute carries out a unified change planning for the pipelines. In the construction process, the municipal route was changed to the safe distance beyond the construction range, and the attached pipelines were constructed by suspension method (DU Guoan 2018).

4 DEEP FOUNDATION PIT RETAINING PILE SCHEME DESIGN AND COST ANALYSIS

In the present stage of construction, there are two problems in foundation pit excavation. First, in the design and construction, some safety accidents often occur in foundation pit engineering, resulting in serious economic losses. Second, too careful and conservative, in the design of supporting structure, caused a waste of resources. Among them, the latter is more likely to attract the attention of the construction side, so it is very critical to improve the economy and safety of foundation pit engineering. In this paper, the supporting structure applied in the standard section and the long distance end of the station is selected to optimize the design.

4.1 *Optimal structural design of supporting pile*

To the design of the supporting pile structure optimization, using the software "reason is the deep foundation pit supporting structure software", the software supporting system to produce the synergy of retaining structure with finite element implementation analysis form, not only can calculate solve all kinds of complex problems of the supporting system of, can also speed up the whole calculation process, intelligent operation, to the user's computing power and speed is not too high requirements. At the same time, the final analysis results are more in line with the actual demand, and the environmental impact program around the foundation pit will be deeply considered (Kong Junqiang 2019).

4.2 *Calculation mode and load of supporting structure*

Support structure stress calculation, the whole construction process is simulated, in accordance with the load "increment method" principle, carry out the relevant structure calculation work, support structure internal force in accordance with the elastic foundation rod finite element method for comprehensive calculation and analysis, and the excavation process, support process, support process simulation. The calculation of the earth pressure outside the foundation pit is carried out according to The active earth pressure of Langken. The theoretical method is that when the permeability coefficient is $k \leq 1m/d$, the combined calculation of soil and water is applied, and when $K > 1m/d$, the soil and water is calculated separately. In the calculation below the excavation surface, a set of springs is used to simulate the horizontal resistance of the stratum. In the process of calculating the support structure, the loads generated mainly include:

(1) The weight of the structure itself: the weight of reinforced concrete itself, about $27kN/m^3$.
(2) Water and soil lateral pressure: in the construction process, the specific calculation will be carried out according to the active earth pressure of Langken. When the permeability coefficient is $K \leq 1m/d$, the combined calculation of soil and water will be applied; when $K > 1m/d$, the calculation of soil and water will be applied.
(3) Ground overload: shield section $30kN/m^2$ The standard section is $21kN/m^2$ considering.

4.3 *Select design calculation parameters*

The most prominent advantage of the deep foundation pit supporting structure is that the specific pile diameter, spacing, crown beam height and width used in the supporting scheme can be optimized based on the previous supporting system. The ultimate goal of optimization is to reduce the overall cost of the deep foundation pit expenditure.

For example, the original standard section 350 is bored pile (Φ1000@1400), and Φ1000@1500 and Φ 900@1500 are selected. Comprehensive comparison and optimization analysis are carried out for the two pile types. The previous Dali section used Φ1000@1200 poured pile type, so Φ1000@1500 is selected for detailed optimization analysis.

The reinforced concrete used in the first bracing system is 800 × 800, and the subsequent three bracing and bracing structures are all applied to the 16th-thickness 609 steel pipe bracing.

Temporary columns are added in the middle of the bracing. The specific information is shown in Table 1.

Table 1. Basic information of pile.

pile position	pile length (m)	depth of pit (m)	embedded depth (m)	Crown beam height (mm)	lateral rigidity (MN/m)
342#	27.55	16.98	12	1.5;1.8	466.7
343#	27.62	17.13	12	1.5;1.8	466.7
349#	26.68	17.72	12	1.5;1.8	466.7
350#	29.34	18.38	12	1.5;1.8	466.7
345#	33.8	20.48	15		1924.8

4.4 *Calculation results and analysis of standard segmental alignment*

In the standard section, four types of piles of different types are applied. This paper only carries out a systematic analysis on the standard section 350#. The main factors for in-depth research on this section are that the pile length is the largest and the foundation pit depth is the deepest. Which using the steel support and supporting pile type and standard of large section of the use of other types of pile type is the same, have very strong similarity, so just pick the representative stronger delve into the standard section of 350 #, 350 # pile before implementation of the optimization, the pile type Φ1000@1400, after the further optimization, application @ 1500, Φ 900 @ 1500 1000.Because Φ 900@ 1400 pile type and Φ1000@ 1500 pile type have different pile distances, comparative analysis is carried out. Because Φ 1000@ 1500 pile type and Φ 900@ 1500 pile type have different pile distances, comparative analysis is carried out.

Through data analysis, it can be seen that the application of Φ 900@ 1500 and Φ1000@ 1400 piles combined with the standard section and large mileage section saves 1.7 million yuan in cost compared with the pile type scheme previously used. Therefore, after comprehensive comparative analysis, it is found that optimization of pile type can realize the goal of safety support and save the total capital invested in the project.

5 SIMULATE THE EXCAVATION VALUE OF DEEP FOUNDATION PIT RETAINING PILE

5.1 *Model building and parameter selection*

(1) Establishing finite element model

Design analysis is carried out based on whether the optimized supporting pile is consistent with the design requirements proposed by ligand optimization, which needs to be verified by ABAQOUS. Among them, the triaxial mixing set outside the retaining pile is very close to the distance generated by the retaining pile, which is beneficial to the implementation of foundation pit excavation. The triaxial mixing pile can be equivalent to the underground continuous wall, and then the simulation is carried out.

Because the pile type of the standard section is similar to the soil quality, there is only a certain difference in length, because the representative 350# is selected as the research object, because the pile is the longest and has a large force. In order to simplify the calculation model, three piles and soils were selected from each model to carry out the calculation. In this study, the model soil is Mohr-Coulomb soil, and the yield criterion is applied. The most significant characteristic is isotropy and homogeneity, and the elastoplasticity is very ideal.

The finite element analysis applied in this study simplifies the supporting structure. The three-dimensional finite element problem is introduced in the analysis process. The pile body is simulated by linear elastic element and the form of embedded assembly is used in the soil.

Select model parameters
Basic information about the model
After the optimization of 350# pile, the pile types used are Φ900@ 1500 and Φ1000@ 1500.
After the optimization of 351# pile, the pile type used is Φ1000@ 1500.
The specific information is shown in Table 2.

Table 2. Basic information of retaining pile before optimization.

pile position	pile length (m)	depth of pit (m)	embedded depth (m)	Crown beam height (mm)	Pile type
350#	27.55	18.480	13	1.5; 0.8	Φ1000@1400
351#	27.62	20.51	126	1.5; 0.8	Φ100@1300

(2) Select calculation parameters, as shown in Table 3.

Table 3. Selecting calculation parameters.

pile position	support the number	horizontal dimension (m)	vertical dimension (m)
350#	1	9.0	2.39
	2	3.0	6.88
	3	3.0	5.37
351 H	1	3.4	2.70
	2	2.3	6.89
	3	2.3	5.43
	4	2.5	1.78

(3) Detailed physical and mechanical parameters of cement earth wall are shown in Table 4.

Table 4. Physical and mechanical parameters of cement earth wall.

structure type	concrete grade	unit weight (kN/m^3)	Poisson's ratio	elasticity modulus (MPa)
cement-soil wall	C35	2300	0.2	32

5.2 *Analysis of calculation results*

After using ABAQOUS software implementation of calculation and analysis for 350 # and 351 # pile diameter and pile distance change implementation analysis, after the optimization, clearly found that the pile supporting force and deformation is reduced, after combination are calculated, the scene monitoring, and ABAQOUS simulation carried out detailed comparison, concluded that effect in the level of support, around the surface subsidence, horizontal displacement results are no more than the control values.

6 CONCLUSION

In short, the structural form of the deep foundation pit supporting pile is detail analyzed this time, and compared with the specific excavation characteristics of engineering foundation pit. After that, the advantages and defects of different supporting structures are explored. Finally, the optimal scheme of support design is selected. After relevant studies, it is found that the on-site monitoring, alignment design and numerical simulation of the mutual comparison and analysis, pile displacement, surface settlement, are basically consistent, fully proved the scientific nature of the optimization scheme.

FUND PROJECT

Road, Bridge and River Crossing Project (No. [2019]58, Urban Construction Institute of China).

REFERENCES

Chen Dongsheng, Numerical experimental study on deep foundation pit excavation in urban subway station [J]., *Engineering and construction*, 2018, 32(04):571–574.

DU Guoan. Common Problems and Countermeasures in design and Construction of Subway deep foundation pit [J]., *Shanxi architecture*, 2018, 44(22):61–62.

Kewen Liu, Yongfen Ruan, Junwei Lin., Zhiliang Wang., *Structural engineer.*, 2017, 33(04):193–199.

Kong Junqiang., Analysis on the influence of deep foundation pit excavation with double row piles on the structure safety of adjacent subway tunnel [J]. *Municipal technology*, 2019, 37(04): 232–236.

Yang Xiangru. Analysis of deep foundation pit supporting construction technology in construction engineering [J]., *Engineering Technology Research*, 2019 (8).

Zhang Zhilian. Analysis of deformation and mechanical characteristics of deep foundation pit excavation in subway station [J]., *Guangdong civil & architecture*, 2018, 25(11):12–16.

Advances in Civil Engineering: Structural Seismic Resistance, Monitoring and Detection – Mohd Johari Mohd Yusof and Junwen Zhang (Eds)
© 2023 The Authors, ISBN 978-1-032-31491-4

Shear strength variation of acid polluted laterite at soaking condition

Yuhang Fan
Kunming University of Science and Technology Oxbridge College, Kunming, China

ABSTRACT: The shear strength of the acid polluted laterite was studied by direct shear test in laboratory under the condition that the laterite was socked in tap water. In the test, the factors, as the variate, included acid concentration, soaking time, ambient temperature, water content and dry density. The results showed that with the increase of acid concentration, the shear strength decreases gradually; the longer soaking time leads to decreased shear strength; with the increase of ambient temperature, the shear strength increases first and then decreases; the shear strength decreases with the increase of water content; the higher dry density causes increasing shear strength.

1 INTRODUCTION

With the constant implementation of urbanization and industrialization in China, as well as the Great Western Development Strategy and West-to-East Power Transmission Project, the local government has accelerated the construction and development in mountainous areas of Yunnan Province, expanded the development scale, and large civil engineering projects have been increasing, such as expressways, water conservancy and hydropower projects. Besides, Yunnan, a paradise of nonferrous metals, where mines, smelters, electroplating plants are over-exploited, and metal processing plants are constantly discharging waste liquid which contains a lot of acid into the river and soil. The laterite serves as foundation soil in Yunnan province, especially Kunming laterite, a widely used building material. Therefore, the acid polluted laterite is bound to cause safety hazards of foundation and upper structures. In addition, Yunnan's backward economic conditions and weak environmental awareness lead to more and more serious problems of contaminated soil in Yunnan. Nonferrous metals are widely distributed in Yunnan, and the wastes from smelters, metal processing plants, chemical plants and paper mills have seriously harmed the structural stability of Yunnan laterite, and also destroyed the sustainable development of local ecology. This causes increasingly severe pollution of laterite. The widespread use of hydrochloric acid leads to the erosion of soil by the contaminated water. The study on how to further define the characteristics of acid polluted laterite is more and more valuable. Therefore, in this paper, a series of tests were conducted on the laterite of Yunnan by selecting hydrochloric acid as the source of pollution and preparing soil samples eroded by hydrochloric acid and socked samples. In this experiment, the laterite samples were taken from Yangzonghai area of Kunming, which were composed of clay particles (55.1%), whose plastic index (22.6) was classified as clay, optimal water content was 32.2%, and the maximum dry density was 1.43 g/cm^3. In this paper, the hydrochloric acid, widely used in hydrometallurgy, bleaching and dyeing industry and metal processing industry, was selected as the source of acid pollution. After the content analysis, the pure hydrochloric acid was 36.5%, including a small amount of burning residue, ferrous sulfate, tin and other percentage content was 0.00~0.05%.

2 TEST SCHEME

As the acid content in acid polluted water (such as the hydrochloric acid used in hydrometallurgy, bleaching and dyeing industry and metal processing industry) varies from less than

DOI 10.1201/9781003310884-24

1.0% to more than 10.0%, the first factor to be considered is changing the acid concentration. The existence time of polluted soil is different. Less polluted soil, not easy to attract attention, but will exist for a long time; Much attention will be paid to those suffering heavy pollution. In this case, suitable measures will be taken to control the polluted soil. So, it's also important to consider the timing of pollution. In terms of temperature, the temperature in Yunnan is generally 0°C to 35°C all the year round, which will affect the degree of molecular activity and the chemical reaction. Therefore, the environment will also affect the pollution. Besides, different moisture content of laterite results in different polluted degree, and the degree of pollutant entering into laterite varies with the dry density of laterite. Therefore, five different factors such as acid concentration, soaking time, ambient temperature, moisture content and dry density of samples were comprehensively considered to carry out tests to obtain the influence of acid pollution on the macroscopic mechanical properties of laterite.

Preparation of Acid Polluted Laterite Samples at Soaking Condition

(1) Preparation of Acid Polluted Laterite
Based on the above experimental scheme, acid solutions at different concentrations were prepared first, and the amount of solution to be used was calculated according to the specified dry density and water content. After the solution was evenly sprinkled on the laterite and stood for 24 hours, acid polluted laterite was well prepared by hitting sample method.
(2) Preparation of Acid Polluted Laterite Samples at Soaking Condition
The prepared acid polluted laterite samples were covered with filter paper and permeable stone, bound with plastic rope (as shown in Figure 1), and then soaked in a beaker containing 300 ml tap water (as shown in Figure 2). After that, all stuff were soaked together in a constant temperature water tank until a preset time. Then the laterite samples were prepared in this way.

Figure 1. Acid polluted laterite sample.

Figure 2. Acid polluted laterite sample at soaking condition.

3 SHEAR STRENGTH VARIATION OF ACID POLLUTED LATERITE AT SOAKING CONDITION

The shear strength of the laterite soaked by the acid solution is the shear strength value under the vertical pressure of 300kPa.

3.1 *Shear strength of acid polluted laterite at different acid concentrations*

At the temperature of 20°C and the moisture content of acid polluted laterite sample at 32.2%, the trend of shear strength τ_f of laterite samples at different acid concentration and soaking days can be obtained by changing the acid concentration and soaking time. The results were shown in Figure 3.

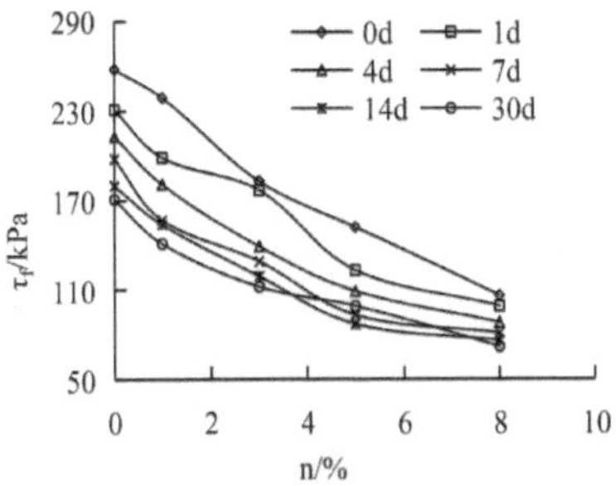

Figure 3. Shear strength of acid polluted laterite at different concentrations.

According to Figure 3, the shear strength of acid polluted laterite under soaking conditions decreased gradually with the increase of acid concentration. The reason was that the cementitious substances, mineral components, and free oxides in the polluted laterite were eroded by acid and then lost with the pore solution, which weakened the connection structure of the acid polluted laterite and reduced its ability to resist deformation. The strength of acid polluted laterite after being soaked was much lower than that of acid polluted laterite.

3.2 *Shear strength of acid polluted laterite at different soaking time*

At the temperature of 20°C and the acid concentration of prepared acid polluted laterite sample at 3%, the shear strength τ_f trend of acid polluted laterite samples after being soaked were drawn by changing the water content and soaking time of acid polluted laterite under the dry density of the prepared sample ρ d = 1.30, 1.35, 1.40 and 1.45 g/cm^3, respectively. The results were shown in Figure 4.

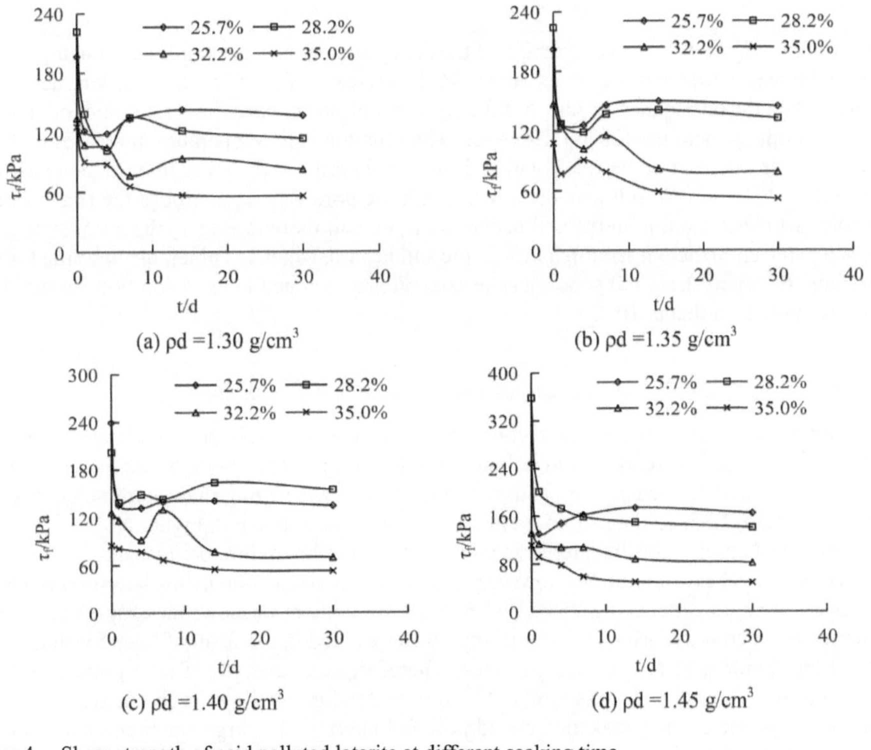

Figure 4. Shear strength of acid polluted laterite at different soaking time.

According to Figure 4, the overall shear strength of acid polluted laterite under soaking conditions decreased gradually with the increase of soaking time. The reason was that after the acid solution pollutes the laterite, the acid solution destroys the cementation of the laterite, and the fine particles in the soil were eroded, and the free oxides in the soil reacted with the acid solution to form soluble salts that dissolve in the water. Therefore, the soil structure became loose, which resulted in the decrease of the strength of the soil.

3.3 *Shear strength of acid polluted laterite at different ambient temperature*

At dry density ρ d = 1.35g/cm^3, the trend of shear strength τ_f of acid polluted laterite samples after being soaked for 7 days with different moisture content and concentration can be obtained by changing acid concentration and ambient temperature. The results were shown in Figure 5.

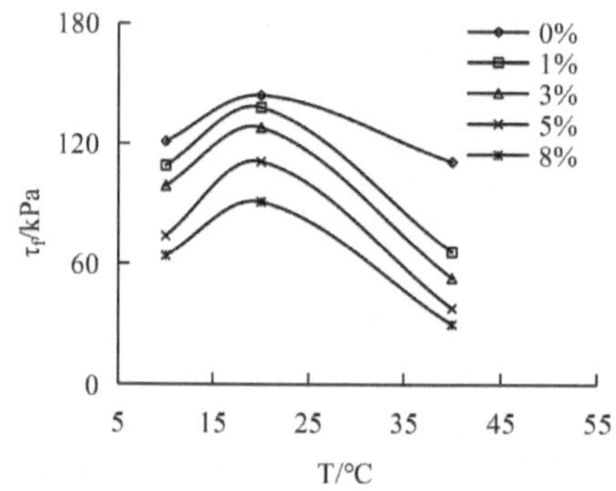

Figure 5. Shear strength of acid polluted laterite at different temperature.

According to Figure 5, the shear strength of the acid polluted laterite samples after being soaked for 7 days increases first and decreases then with the increase of temperature. The laterite soaked in tap water and the one soaked in acid at different concentrations have the least reduction in shear strength at room temperature (20°C). However, when the ambient temperature increases to 40°C, the more intense the reaction between acid and oxide in laterite is, the more intense the erosion of laterite will be. The soluble salt generated loses with the pore water and erodes the fine particles in the soil particles, resulting in the soil becoming more and more loose and the shear strength is lower. When the ambient temperature is 10°C, the soil is hard, which is vulnerable to brittle failure. At this time, the main structural substances in the soil are corroded by acid solution, so the shear strength is lower than that at 10°C.

3.4 *Shear strength of acid polluted laterite at different moisture content*

At the temperature of 20°C and the acid concentration of prepared acid polluted laterite sample at 3%, the shear strength τ_f trend of acid polluted laterite samples after being soaked with different moisture content and dry density, respectively were drawn by changing the water content and dry density ρ d = 1.30, 1.35, 1.40 and 1.45 g/cm^3. The results were shown in Figure 6.

According to Figure 6, for the laterite soaked in acid solution, before being soaked, the laterite has been eroded by the acid solution. However, due to the low water content, the laterite was harder, with stronger strength. After being soaked, when the water content was low, the sample was dry and the laterite absorbed water. Most of the absorbed water existed in the form of bound water, which enhanced the bonding force between particles. Therefore, the strength of acid polluted laterite sample with low water content was larger, which decreased gradually with the increase of water content. Besides, after being soaked, the acid polluted laterite with large water content absorbed water mostly in the form of free water, which resulted in soft laterite samples and reduced strength

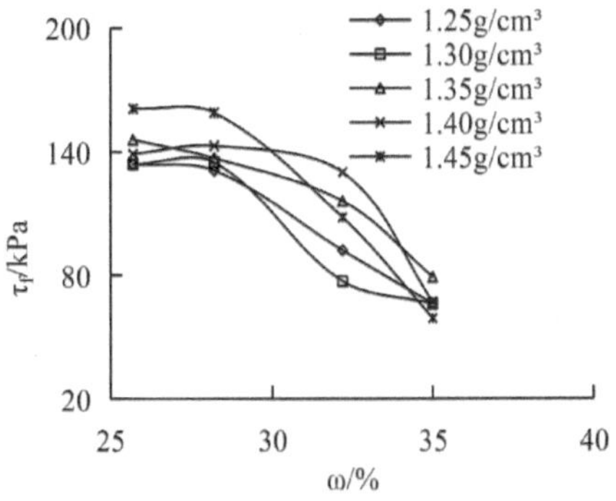

Figure 6. Shear strength of acid polluted laterite at different moisture content.

of acid polluted laterite. For the laterite has been polluted, soaking provides a quick water channel for the acid to enter the laterite. In this case, the laterite samples with higher water content will reach high water content in a short time, while the rapid erosion of acid also happen, thus reducing the shear strength of laterite.

3.5 *Shear strength of acid polluted laterite at different dry density*

At the temperature of 20°C and the acid concentration of prepared acid polluted laterite sample at 3%, the shear strength trend τf of acid polluted laterite samples after being soaked for 7 days were drawn by changing the water content and the dry density of the prepared sample ρ d = 1.30, 1.35, 1.40 and 1.45 g/cm^3, respectively. The results were shown in Figure 7.

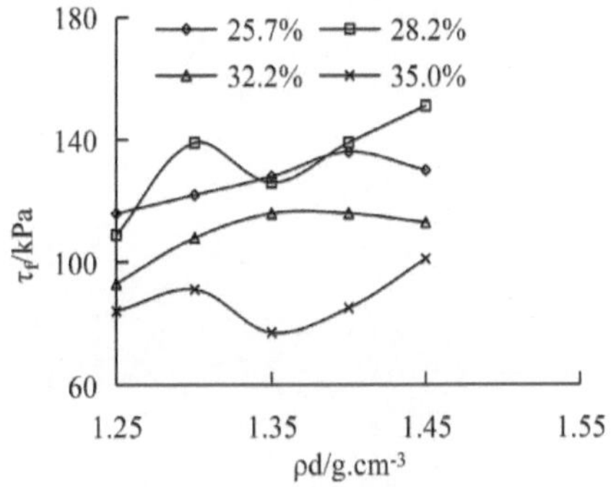

Figure 7. Influence of different dry densities on shear strength.

According to Figure 7, when the soaking time is unchanged, the shear strength of acid polluted laterite under soaking conditions increases gradually with the increase of dry density of samples. The reason is that, increasing dry density allows a denser laterite, which is not easy to be eroded by acid solution, so the shear strength increases with the increase of dry density.

4 CONCLUSIONS

The shear strength of the acid polluted laterite under soaking condition decreases gradually with the increase of acid concentration, which show an overall declining trend with longer soaking time. When the laterite is soaked for 7 days, the shear strength first increases and then decreases with the raising temperature. It shows a decreasing trend with the increase of water content of acid polluted laterite. If the soaking time is unchanged, the shear strength of acid polluted laterite increases gradually with the increase of dry density of sample under soaking condition.

This study can provide a theoretical basis for the remediation of acid polluted soil foundation. With optimal water content and maximum dry density, the shear strength of acid polluted laterite decreases slightly.

ACKNOWLEDGMENTS

This work was financially supported by the Fund Project: Scientific Research Fund Project of Yunnan Provincial Department of Education (2016ZZX307).

REFERENCES

Bo, T. Z. & Hang, Y. & Shi, C. X. & Wang, Y. (2012). Variation characteristic of physical of acid pollution laterite. *J. Hydrogeology & Engineering Geology* 39, 111–115.

Deng, C. Z. (1985). The foundation contaminated by sulfuric acid. *J. Site Investigation Science and Technology* 6, 38–41.

Fan, Y. H. & Huang, Y. & Ren, L.Q. (2014). Influence of Shear Strength of Acid Pollution Laterite by Soaking. *J. Site Investigation Science and Technology* 4, 1–6.

Gu, J. W. (1988). The effect of waste liquid of acid and alkali erosion foundation soil on engineering quality. *Chinese Journal of Geotechnical Engineering* 10, 72–78.

Liu, H. L. & Zhu, C. P. & Zhang, X. L. (1988). Fundamental physical properties of soil polluted by acid and alkali in laboratory. *Chinese Journal of Geotechnical Engineering* 2, 35–38.

Zhang, X. L. (2007). Experimental study on soil polluted by acid and alkali. *D. Hohai University*.

Zhu, C. P. & Liu, H. L. (2007) Study on engineering propefties of polluted soil. *J. Rock and Soil Mechanics* 28, 625–630.

Advances in Civil Engineering: Structural Seismic Resistance, Monitoring and Detection – Mohd Johari Mohd Yusof and Junwen Zhang (Eds)
© 2023 The Authors, ISBN 978-1-032-31491-4

Study on the influence of input ground motion on terrain effect

Minghui Hao & Yushan Zhang
China Earthquake Disaster Prevention Center, Beijing, China

ABSTRACT: In this paper, the finite element finite difference method based on the transmission boundary is used to describe the terrain effect by using the magnification of the response spectrum of the terrain to different control periods. The effects of different input ground motions, including actual ground motion input, artificial ground motion input, simulated ground motion input, white noise input and pulse input, on the ground motion topography effect are studied. The results show that the average spectral ratio curves calculated from the actual ground motion input, the artificial ground motion input and the simulated ground motion input are relatively close; compared with these three types of input, the average spectral ratio curves calculated from the white noise input are larger in some periods; while the spectral ratio curves calculated from the pulse input are quite different from those calculated from the ground motion input. Therefore, in the study of the amplification effect of local topography on response spectrum, it is suggested to use the simulated ground motion and the actual ground motion recording input. In addition, the influence of the input ground motion uncertainty on the regression results should be taken into account when determining the empirical model of amplification coefficient of local convex terrain to seismic response spectrum.

1 INTRODUCTION

The earthquake damage experience, strong earthquake observation records and theoretical analysis results of previous destructive earthquakes show that local terrain significantly affects the characteristics of ground motion, and then significantly affects the degree of earthquake damage and its spatial distribution (Borcherdt 1970; Yuan & Liao 1996). Therefore, the influence of local irregular terrain on ground motion is a popular research topic. Considering the local terrain as a multi-particle system, the seismic response of the system will be determined by both the ground motion characteristics of the input system and the characteristics of the system itself without considering the influence of the source characteristics and propagation paths. Specifically, the characteristics of the system itself mainly include material parameters such as slope angle, slope height, protruding terrain width and medium damping, etc. Scholars have carried out a lot of research work in this field, and achieved certain results (Liu 1996; Hao & Zhang 2014; 2015). As for the characteristics of input ground motion, most researches mainly focus on the amplitude and frequency spectrum characteristics of incident seismic waves. For example, Davis et al. (Davis & West 1973) and Bard et al. (Boore 1973) demonstrated that when the predominant frequency of incident seismic waves is close to the natural frequency of hills, the latter can amplify seismic waves. Nguye et al. (Nguyen 2007) used direct boundary element method to study the influence of complex terrain on SV wave propagation characteristics. The results showed that under a low-frequency input, the ridge top, slope top and canyon edge significantly amplify seismic waves; with the increase of input wave frequency, the amplification effect at the bottom of the canyon and the foot of the ridge becomes greater, but the ridge top is amplified and is not affected by the excitation frequency. Wang Wei (Wang 2011) made a detailed analysis and research on the terrain effect of

DOI 10.1201/9781003310884-25

ground motion by using the mountain ground motion records of the main shock and aftershocks of Wenchuan earthquake, combining the mountain and terrain ground motion records of "Mingdeng No. 1" artificial blasting and the earthquake damage survey data of Wenchuan earthquake related to terrain effect. The results showed that the polarization effect of mountains is significantly related to the frequency components of the input ground motion. Li Yingmin et al. (Zhu et al. 1991) applied dynamic substructure method, with white noise as input, to study the variation law of acceleration and spectral characteristics of each point of the model under different sizes of protruding terrain. In addition, scholars have adopted pulse input to study the amplification effect of local terrain on the peak ground motion. For example, Zhu Yuanqing et al. (Zhou & Chen 2006) studied the influence of viscoelastic terrain on vertical incident SH waves and P waves by using the finite element method, and studied the amplification effect of terrain on peak displacement. Zhou Hong et al. (Li et al. 2019) studied the influence of terrain and source physical characteristics on ground motion by using the spectral element method and taking SH wave propagation excited by point source (pulse) as the object. Rong Mianshui and Li Xiaojun (Rong & Li 2007) applied the explicit finite element-finite difference method to carry out the numerical calculation of viscoelastic site terrain under actual ground motion input, and compared the results of bell-shaped pulse input and actual ground motion input when studying Fourier spectral ratio of slope terrain. They thought that pulse input has the characteristics of small amount calculation and enabling to grasp the main aspects of an issue, but there are some problems in the research of response spectral ratio.

However, the actual ground motion is very complex, and due to the dispersion and interference of seismic waves, there is a complex phase relationship between actual ground motion pulses. Therefore, the amplification factor of terrain to peak ground motion based on pulse input cannot be directly used to evaluate the amplification effect of terrain to actual peak ground motion. When studying the amplification effect of terrain on the ground motion response spectrum, the most reliable method is to directly calculate the seismic response of terrain under the action of input ground motion, and determine the response spectral ratio of local ground motion to horizontal ground motion, and on this basis, the quantitative results of the amplification effect of terrain on ground motion response spectrum can be obtained. Hence, how to determine the input ground motion is the primary problem to reasonably evaluate the terrain effect of ground motion. For this purpose, this paper takes protruding terrain as an example, and studies the influence of input ground motion on the amplification effect of protruding terrain by using the finite element-finite difference method. The input ground motion includes five types: actual ground motion input, artificial ground motion input, simulated ground motion input, white noise input, and pulse input.

2 CALCULATION MODEL

The calculation model of the protruding terrain is shown in Figure 1, and the earth surface includes three parts: the top platform, the slope surface and the slope bottom. The artificial boundary is used to cut the finite element calculation region with local protrusions from infinite region. The vertical distance from the foot to the left/right artificial boundary is 5H/tan α and the vertical distance with the bottom artificial boundary is 5H. It can meet the requirements of calculation accuracy and reduce the influence of artificial boundary. It is scattered with the quadrilateral finite element grid that the size is 1m*1m to obtain reliable numerical simulation results for high frequency components. For the protruding platform, the width is B=50m, the height is H = 100m, and the slope angle is $\alpha = 45$. The medium is assumed to be uniform and viscoelastic. The P-wave velocity is cp, the shear wave velocity is cs, and the damping ratio is ζ. Rock medium is mainly considered in this paper, and its shear wave velocity takes 1400m/s and compressive wave velocity takes 2000m/s, and the wave velocity of medium in the artificial boundary area takes 1200m/s. The incident seismic wave employs a vertically incident SV wave.

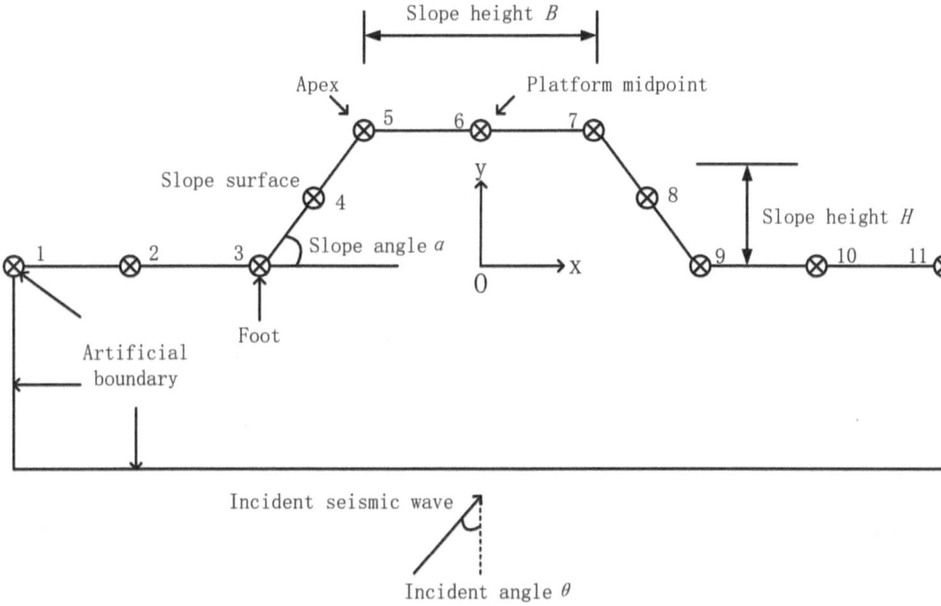

Figure 1. A model with local terrain.

3 CALCULATION METHOD

3.1 *Inner point calculation method for concentrated mass explicit finite element*

The essence of the concentrated mass explicit finite element method is to deduce the motion of the node at the next time from the equation of motion of the node at the current time. It does not need the integration of stiffness, mass and damping matrix. The forming of the right item only needs to be accumulated at the unit level I according to the contribution of each unit to the effective load vector. In this way, the whole calculation is basically carried out at unit level I, so only a very small high-speed storage area is needed, and the calculation efficiency is high. Especially when the stiffness matrix, mass matrix and damping matrix of a series of units are the same, there is no need for double calculation, and the efficiency is higher.

$$M_1\{\ddot{u}_1(t)\} + \sum_{e=1}^{N_e}\sum_{n=1}^{N_n}\left(\alpha M_{in}^{(e)} + \beta K_{in}^{(e)}\right)\left\{\dot{u}_{\underset{u}{\wedge}}(t)\right\} + \sum_{e=1}^{N_e}\sum_{n=1}^{N_n} K_{in}^{(e)}\left\{u_{\underset{n}{\wedge}}(t)\right\} - \{R_1(t)\} = 0 \quad (1)$$

where N_e is the number of all cells (including node 1), N_n is the number of nodes in the eth unit, i is the local number of node 1 in element e, n is the overall number of the nth node in unit e, $M_{in}^{(e)}$ and $K_{in}^{(e)}$ are the mass matrix and stiffness matrix of element e, respectively, α and β are Rayleigh damping coefficients, $\{\ddot{u}\}$, $\{\dot{u}\}$ and $\left\{u_j(t)\right\}$ are the acceleration, velocity and displacement, respectively, $\{R_1(t)\}$ is the equivalent concentrated load of a node, and M_1 is the equivalent concentration quality of node 1, which is calculated as follows:

$$M_1 = \sum_{e=1}^{N_e}\sum_{n=1}^{N_n} M_{in}^{(e)} \quad (2)$$

Using the explicit difference scheme proposed by Li (Li et al. 1992) to solve equation (1), the numerical solution for the motion of node 1 can be obtained:

$$\left\{u_1^{p+1}\right\} = \frac{1}{2}\frac{\Delta t^2}{M_1}\{R_1^p\} + \{u_1^p\} + \Delta t\{\dot{u}_1^p\} - \frac{1}{2}\frac{\Delta t^2}{M_1}\sum_{e=1}^{N_e}\sum_{n=1}^{N_n}\left[K_{in}^{(e)}u_{\hat{n}}^p + \left(\alpha M_{in}^{(e)} + \beta K_{in}^{(e)}\right)\dot{u}_{\hat{n}}^p\right] \quad (3)$$

where $\left\{\ddot{u}_i^p\right\}$, $\left\{\dot{u}_i^p\right\}$, $\{u_i^p\}$ and $\{R_i^p\}$ are the acceleration, velocity, displacement and equivalent concentrated dynamic load, respectively, of node i at time $p\Delta t$ and t is the discrete time interval.

Using equation (3), the dynamic responses of the internal nodes can be solved explicitly.

3.2 *Transmission artificial boundary*

For artificial boundary nodes (nodes 1 and 11 in Figure 1, assuming their overall number is 0), the formula for calculating the dynamic response can be obtained based on the transmission artificial boundary principle:

$$\left\{u_0^{p+1}\right\} = \sum_{j=1}^{N}(-1)^{j+1}C_j^N\left\{u_j^{p+1-j}\right\} \quad (4)$$

Where in, N is the transmission order (second order in this paper), $\left\{u_0^{p+1}\right\}$ is the displacement of artificial boundary node at time $(p+1)\Delta t$; $\left\{u_j^{p+1-j}\right\}$ is the displacement of the calculation point $x = -jc_a\Delta t$ at time $(p+1-j)\Delta t$; c_a is artificial wave velocity value; C_j^N is binomial coefficient.

$$C_j^N = \frac{m!}{(m-j)!j!} \quad (5)$$

3.3 *Program verification*

By using the explicit finite element calculation method based on the transmission artificial boundary, the parallel Fortran numerical calculation program TPG2D is developed and is transplanted to the Beijing industry cloud computing platform to run. In order to verify the correctness of the calculation results of the program, the dynamic response of the elastic half space under the vertical incidence of SV wave and p wave is calculated and compared with the theoretical solution.

A limited scope 6m × 50m is cut from the two-dimensional semi-infinite space. The size of unit grid is 1m × 1m, the elastic modulus of materials is $E = 2.4 \times 10^7$Pa, the Poisson's ratio is $\nu = 0.2$, and the mass density is $\rho = 1000\text{kg/m}^3$.Unit pulse shear displacement wave in vertical upward incident horizontal direction at the bottom and unit pulse compression displacement wave in vertical direction:

$$u(t) = \frac{1}{2}[1 - \cos(2\pi ft)], 0 \le t \le 0.25s \quad (6)$$

Where in, $f = 4Hz$

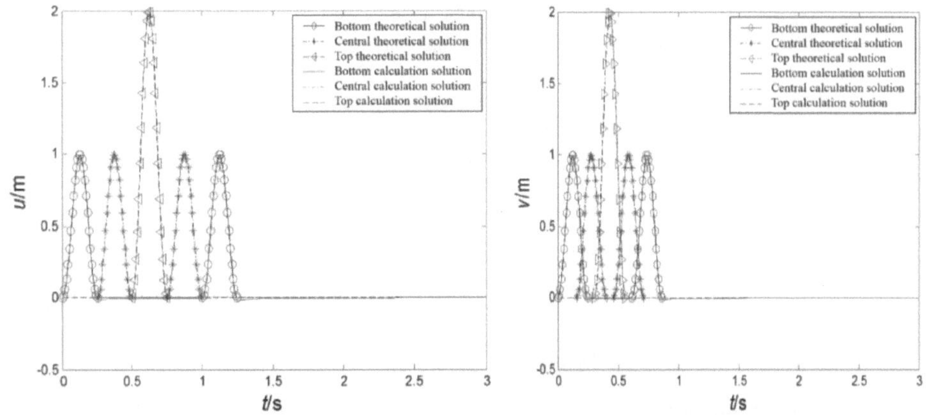

(a) Horizontal (b) Vertical

Figure 2. Comparison of viscous-elastic boundary calculated results with the theoretical solution.

Figure 2 shows the comparison between the horizontal and vertical displacement time-history curves and the theoretical solution of the bottom, middle and top of the finite element model obtained by the program TPG2D. The incident wave is reflected upward from the bottom on the free surface, and the displacement amplitude at the free surface is twice the input wave amplitude. The calculation results of the program TPG2D are in good agreement with the theoretical solution that proves the correctness and validity of the calculation method used in this paper.

4 ANALYSIS METHOD

In order to eliminate the influence of the randomness of the input ground motion on the analysis result, the average value of the spectral ratio calculated by means of multiple groups of input ground motions is used to research the amplification effect of the convex-to-ground motion response spectrum. For the ith input ground motion sample $a_i(t)$, the program TPG2D is used to solve the seismic response protruded in Figure 1 and the acceleration time process $a_i^{\mathrm{r}}(t,\vec{x}$ of seismic response at different output points is obtained. The absolute acceleration response spectrum is $S_{\mathrm{a},i}^{\mathrm{r}}(T,\vec{x})$. $\vec{x}$ is the space coordinates of the output point, $\vec{x}$.

The program TPG2D assumes $a_i(t)$ is the acceleration of incident wave at the bottom left corner of the model shown in Figure 1. On this basis, the seismic response of the free field (that is, there is no protruding elastic half space) can be determined. The acceleration response is denoted as $a_i^{\mathrm{f}}(t,\vec{x})$ and the response spectrum is denoted as $S_{\mathrm{a},i}^{\mathrm{f}}(T,\vec{x})$. Accordingly, the following spectral ratio is defined to describe the amplification effect of local protrusions on ground motion, namely

$$r_i\,(T,\vec{x}) = \frac{S_{\mathrm{a},i}^{\mathrm{r}}(T,\vec{x})}{S_{\mathrm{a},i}^{\mathrm{f}}(T,\vec{x})} \tag{7}$$

5 ANALYSIS OF THE INFLUENCE OF DIFFERENT GROUND MOTION INPUTS ON THE CALCULATION RESULTS

5.1 *Actual ground motion input*

Comparison of viscous-elastic boundary calculated results with the theoretical solution1 shows the basic information of natural ground motions selected in this paper. Figure 3 shows under the natural ground motion input, the response spectrum of the midpoint ground motion and the input ground motion response spectrum (normalized by the peak acceleration of free field ground motion) in the platform section of protruding terrain; Figure 4 shows the spectral ratio curves corresponding to different input ground motions, in which the red dotted line represents the average spectral ratio curve. It can be seen that the randomness of input ground motions has a certain influence on the amplification effect of terrain on the ground motion response spectrum, which is significant in the higher frequency range (less than 0.08s) and the lower frequency range (greater than 2.0s); in the middle frequency band, the spectral ratio curves corresponding to different input ground motions are concentrated.

Table 1. Basic information of natural ground motion.

No.	Earthquake	Magnitude	Station	Epicentral distance (km)	Record component
1	1999 chi-chi earthquake in Taiwan	7.62	TCU101	45.05	EW
2	1999 chi-chi earthquake in Taiwan	7.62	TCU122	21.80	EW
3	1979 Imperial Valley-06 earthquake in USA	6.53	Delta	33.73	262
4	1999 Hector Mine earthquake in USA	7.13	Mill Creek Ranger Station	92.09	270
5	1999 chi-chi earthquake in Taiwan	7.62	TCU051	38.53	EW
6	1987 Superstition Hills-02 earthquake in USA	6.54	Westmorland Fire Sta	19.51	090
7	1979 Imperial Valley-06 earthquake in USA	6.53	El Centro Array #12	31.99	140
8	1999 chi-chi earthquake in Taiwan	7.62	CHY102	70.48	EW
9	1999 Hector Mine earthquake in USA	7.13	Mill Creek Ranger Station	92.09	270
10	1979 Imperial Valley-06 earthquake in USA	6.53	El Centro Array #12	31.99	230

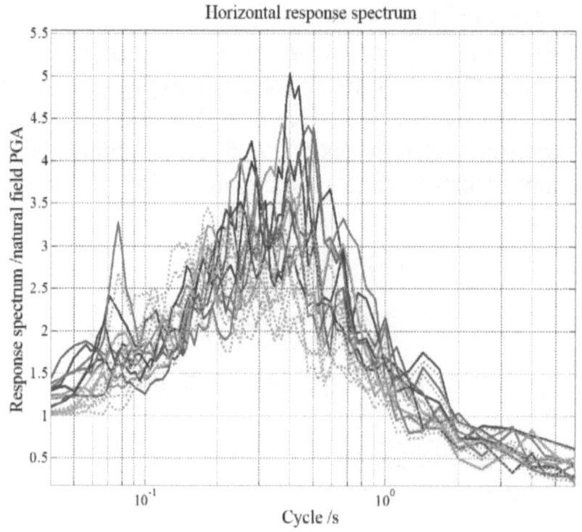

Figure 3. Response spectrum curve of input ground motion and mid point ground motion of platform section in convex terrain.

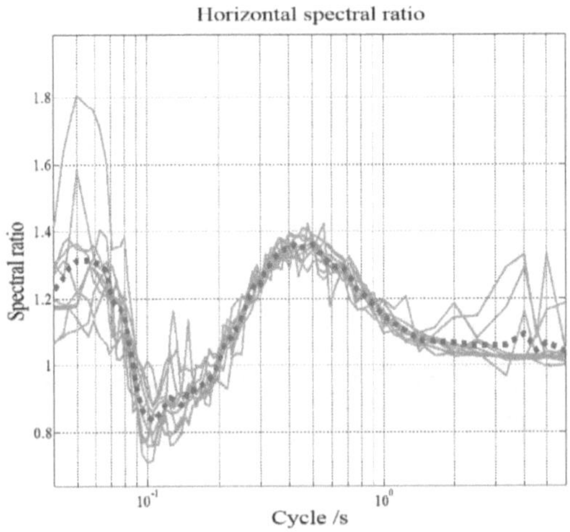

Figure 4. Curve of spectral ratio of raised terrain corresponding to different input natural ground motions.

5.2 *Artificial ground motion input*

The artificial ground motion The response spectrum time-history meeting the target response spectrum shown in Figure 5 is generated by using the response spectrum fitting method based on wavelet function. Totaling 10 samples are generated in this work, and the acceleration, velocity and displacement waveform curves of one sample are shown in Figure 6. The ground motion response spectrum of all 10 samples is shown in the gray curve in Figure 5, and the relative error between the response spectrum of all samples and the target spectrum is within . The response spectrum of midpoint ground motion in the protruding platform section under the above artificial ground motion input is shown in solid line in Figure 7, and the gray dotted line indicates the response spectrum of free field ground motion in elastic half-space under the vertical incidence of SV wave; Figure 8 shows the response spectral ratio of the midpoint ground motion in the protruding platform section to the horizontal free ground motion, and the red dotted line indicates the average value of the spectral ratio curves calculated according to all artificial ground motion samples. It can be seen that compared with natural ground motion input, the spectral ratio curve corresponding to artificial ground motion input is less discrete, and the spectral ratio curve of artificial ground motion with different input fluctuates in a smaller range near the average spectral ratio curve.

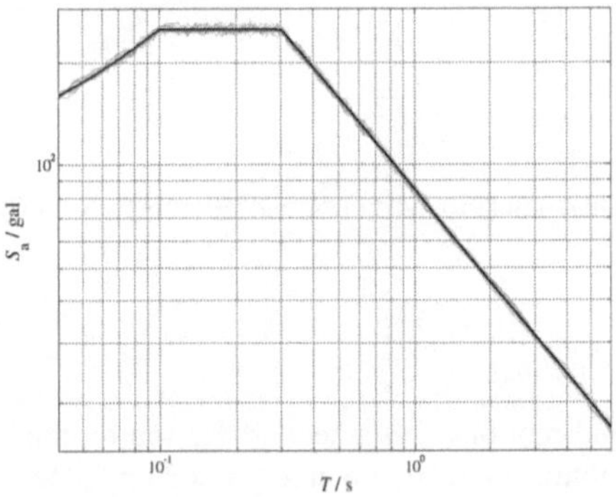

Figure 5. Target response spectrum and response spectrum of artificial ground motion.

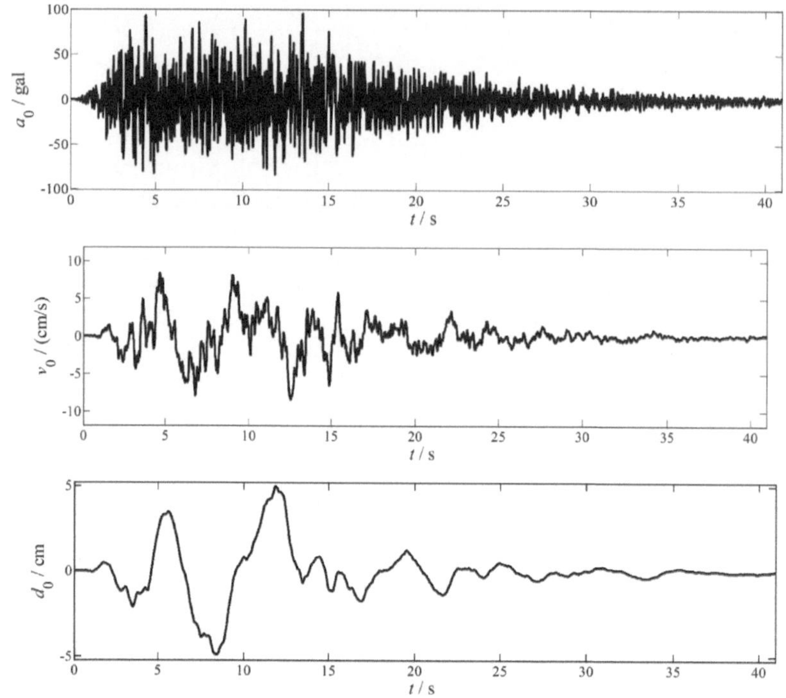

Figure 6. Acceleration, velocity and displacement waveforms of artificial ground motion samples.

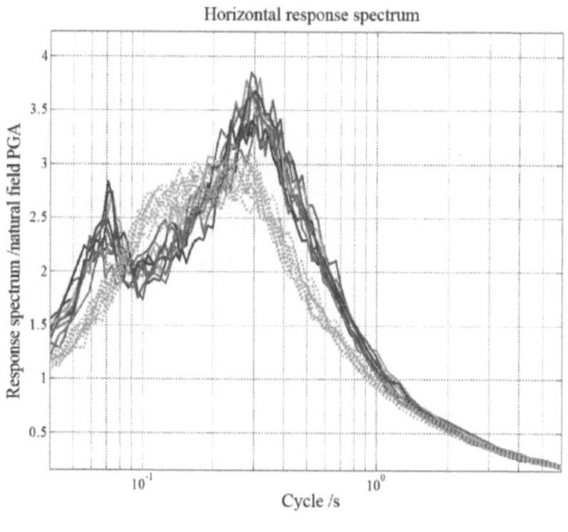

Figure 7. Response spectra of mid point ground motion and horizontal free surface ground motion in platform section of raised terrain.

5.3 *Simulated ground motion input*

Totaling 10 simulated ground motions are generated for the original El Centro wave by using the non-stationary ground motion simulation method based on Hilbert transform (Zhang & Zhao 2014), and the acceleration, velocity and displacement waveform curves of one sample are shown in Figure 9. The gray dotted line in Figure 10 indicates their response spectrum, and the black solid line indicates the response spectrum curve of the original El Centro wave. Figure 11 shows under

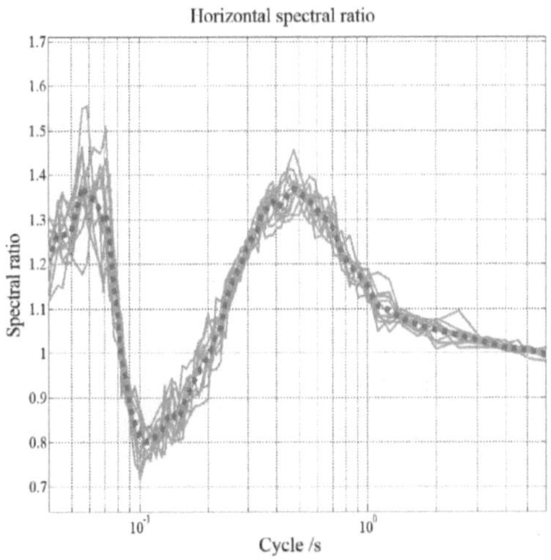

Figure 8. Response spectrum ratio of the mid point of the platform section.

the input of different simulated ground motion samples, the midpoint ground motion response spectrum in the platform section of protruding terrain and the ground motion response spectrum of elastic half space free-surface under the same input, in which the gray dotted line indicates the free field ground motion response spectrum; Figure 12 shows their spectral ratio curves, in which the red dotted line indicates the average curve.

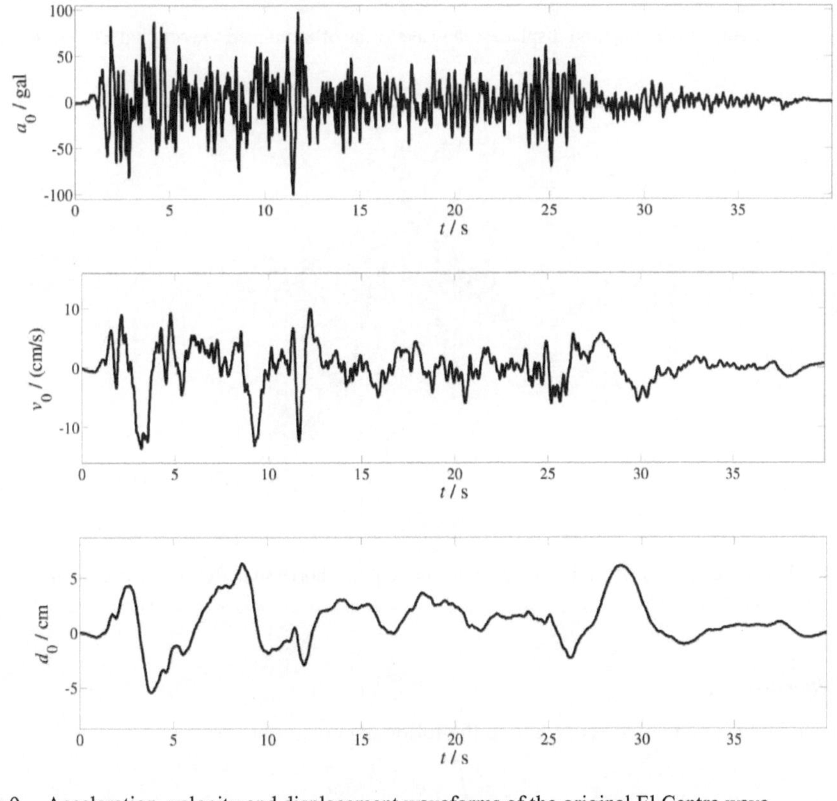

Figure 9. Acceleration, velocity and displacement waveforms of the original El Centro wave.

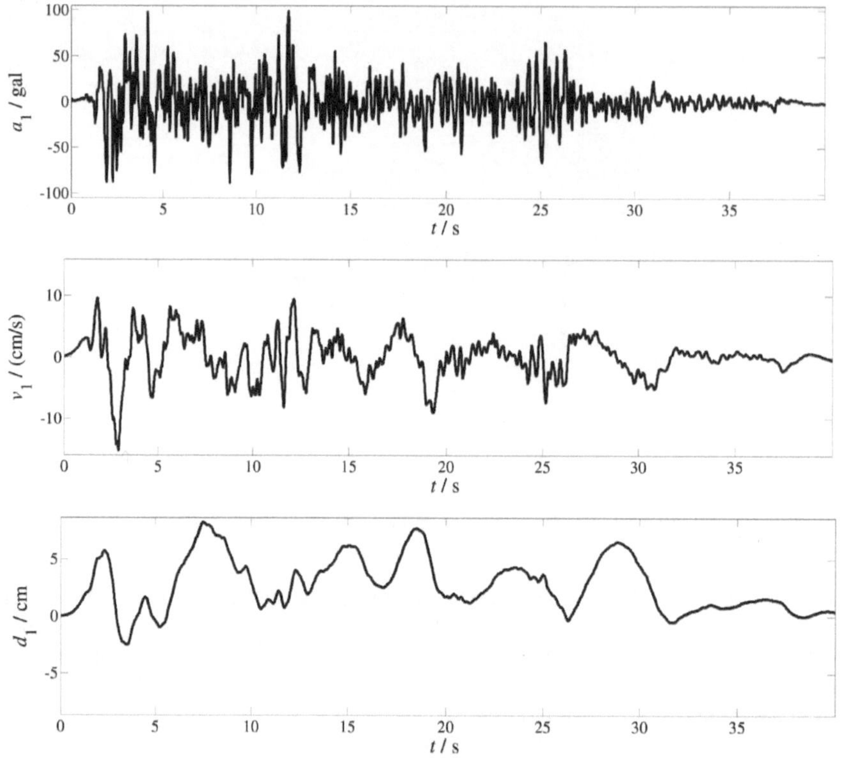

Figure 10. Acceleration, velocity and displacement waveforms of Simulated the original El Centro wave.

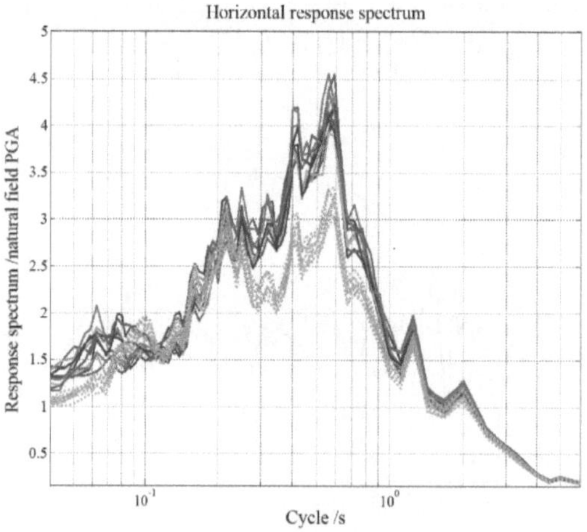

Figure 11. Response spectra of mid point ground motion and horizontal free surface ground motion in platform section of raised terrain.

5.4 *White noise input*

White noise input is generated according to the following equation:

$$a(t) = \int_{-\infty}^{\infty} e^{i(\omega t+\varphi)} \mathrm{d}\omega \tag{8}$$

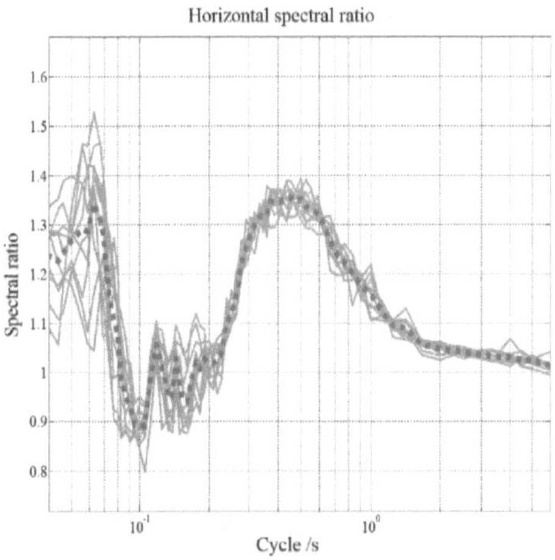

Figure 12. The spectral ratio of the midpoint response of the platform section under the input of simulated ground motion.

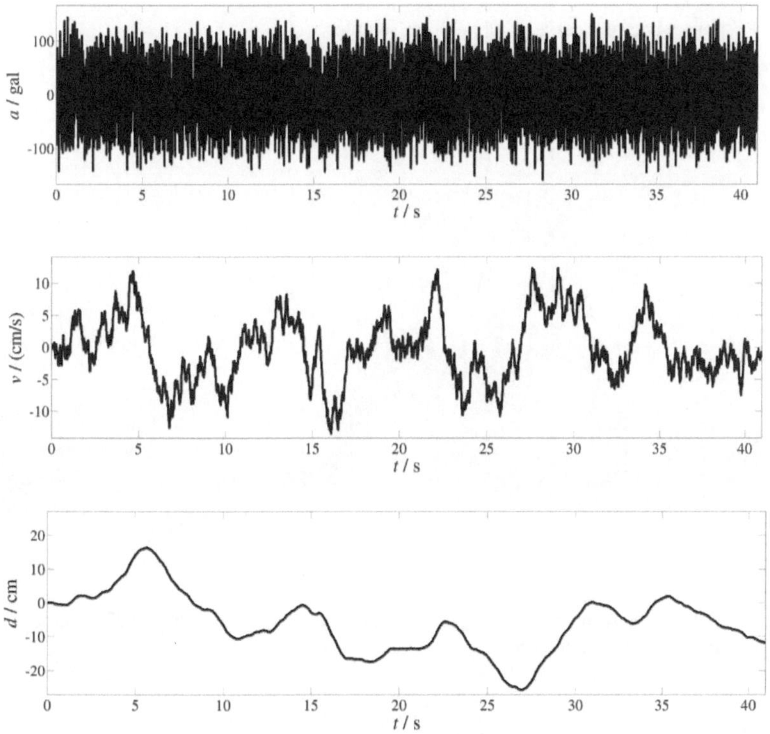

Figure 13. Acceleration, velocity and displacement waveforms of white noise samples.

Wherein, φ is a random variable satisfying the uniform distribution within $[0, 2\pi]$. Totaling 10 white noise samples are generated in this work, and Figure 13 shows the acceleration, velocity and displacement waveform curves of one sample. Due to numerical error, the original ground motion of white noise generated has the problem of baseline drift, and the result shown in Figure 13 undergoes baseline correction. Figure 14 shows the response spectrum curves of all white noise samples. Figure 15 shows the response spectrum curves of midpoint ground motion and free field motion in the platform section of protruding terrain, and Figure 16 shows their spectral ratio curves (the red dotted line indicates the average spectral ratio curve). It can be seen that the spectral ratio curve calculated for most white noise samples is near the average curve, and there is a big difference between the spectral ratio curve calculated for one sample and the average spectral ratio.

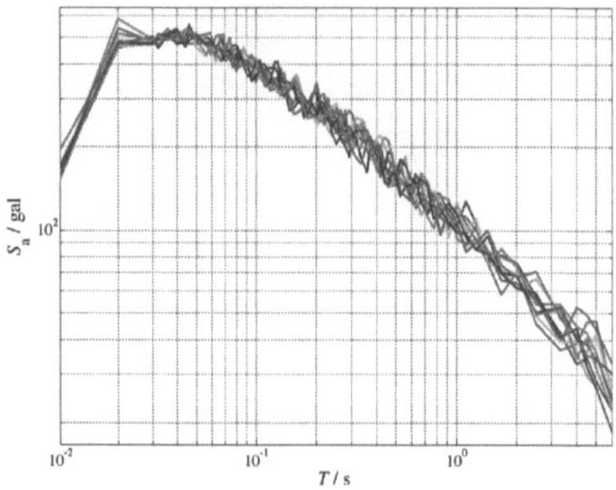

Figure 14. Response spectrum curve of white noise samples.

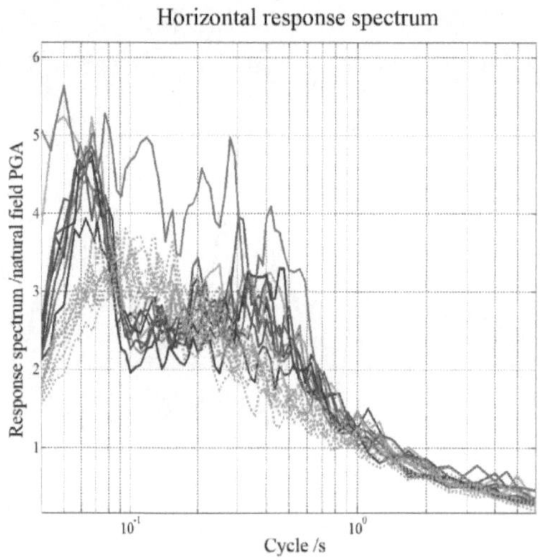

Figure 15. Response spectra of mid point ground motion and horizontal free surface ground motion in platform section of raised terrainunder the input of white noise.

5.5 *Pulse input*

Pulse input adopts the following functional form:

$$d(t) = \psi (t - t_0) \sin [2\pi f (t - t_0)]$$

$$v(t) = \frac{d}{dt} d(t)$$

$$a(t) = \frac{d^2}{dt^2} d(t) \tag{9}$$

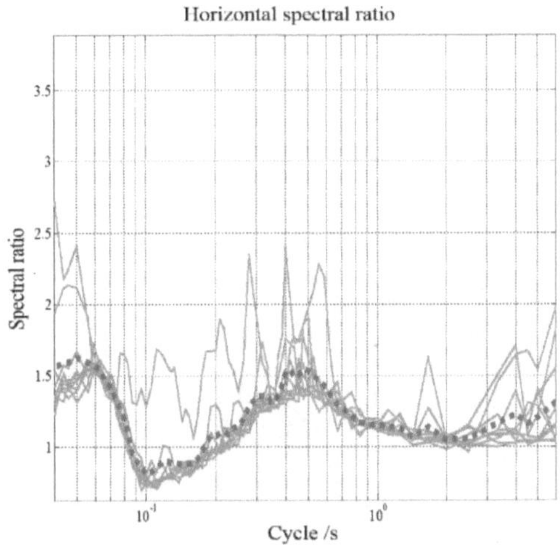

Figure 16. The spectral ratio of the midpoint response in the platform section of the convex terrain under the input of white noise.

Where in, f is the center frequency of the pulse, and t_0 controls the position of the pulse on the time axis; d(t), v(t) and a(t) are the displacement, velocity and acceleration, respectively. In equation (9),

$$\psi(t) = \begin{cases} \exp\left(-\frac{t^2}{c_1}\right), & t \leq 0 \\ \exp\left(-\frac{t^2}{c_2}\right), & t > 0 \end{cases}$$

$$c_1 = -\frac{(t_2 - t_1)^2}{\ln \varepsilon} \tag{10}$$

$$c_2 = -\frac{(t_2 - t_0)^2}{\ln \varepsilon} \tag{11}$$

In the above equation, ε can generally be taken as 10–6, and t1 and t2 define the effective interval of pulse. ε

Totaling 10 pulse inputs are generated according to equation (9), and their control parameters are shown in Table 2.

Table 2. Control parameters of pulse input.

Pulse number No.	t_0	t_2	t_2	ε
1	1.0	0.5	1.5	10^{-6}
2	1.5	0.5	2.5	10^{-6}
3	2.0	0.5	3.5	10^{-6}
4	2.5	0.5	4.5	10^{-6}
5	3.0	0.5	5.5	10^{-6}
6	1.25	0.5	2.0	10^{-6}
7	1.75	0.5	3.0	10^{-6}
8	2.25	0.5	4.0	10^{-6}
9	2.75	0.5	5.0	10^{-6}
10	3.25	0.5	6.0	10^{-6}

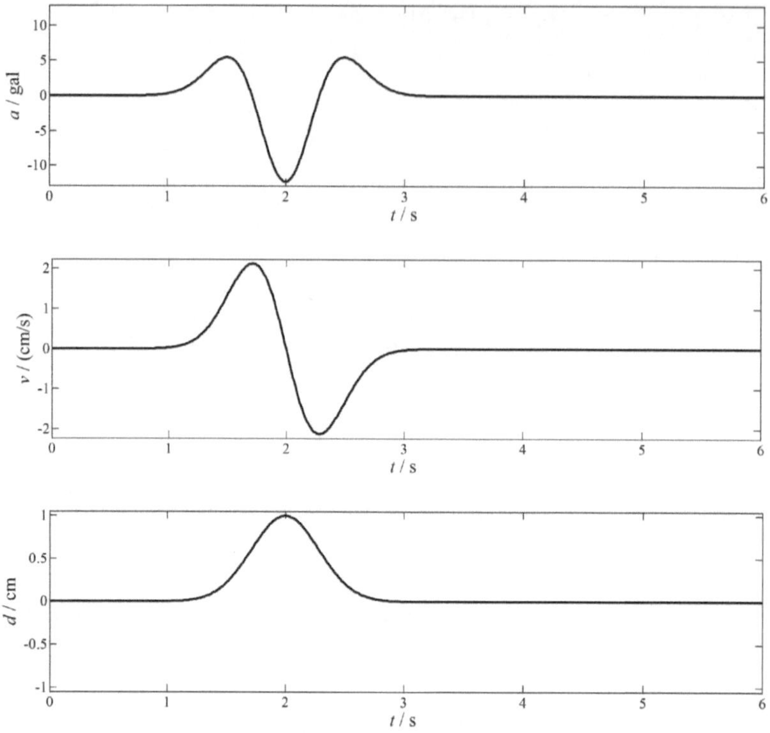

Figure 17. Acceleration, velocity and displacement curves of pulse sample 3.

Figure 17 shows the acceleration, velocity and displacement waveforms of sample No. 3. Figure 18 shows the response spectrum curves of all pulse inputs. Figure 19 shows the response spectrum curves of midpoint ground motion and free field motion in the platform section of protruding terrain, and Figure 20 shows their spectral ratio curves (the red dotted line indicates the average spectral ratio curve). It can be seen that compared with ground motion input and white

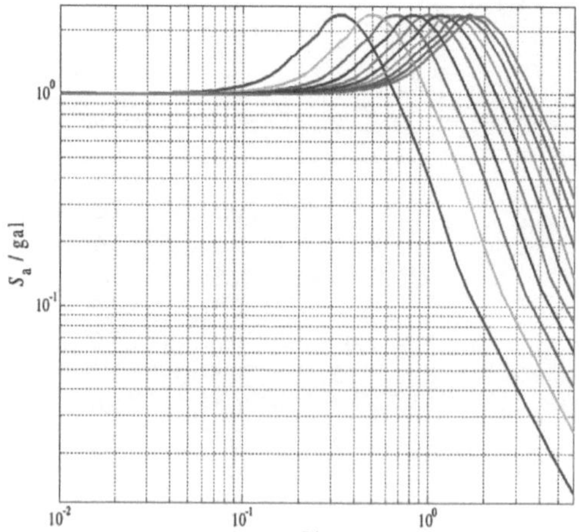

Figure 18. Response spectrum curve of all pulse samples.

noise input, the spectral ratio of response spectrum calculated based on pulse input is quite different. Therefore, pulse input can be well used to study the dynamic characteristics (transfer function) of local terrain, but there is a large deviation when it is used to study the amplification effect of terrain on the ground motion response spectrum.

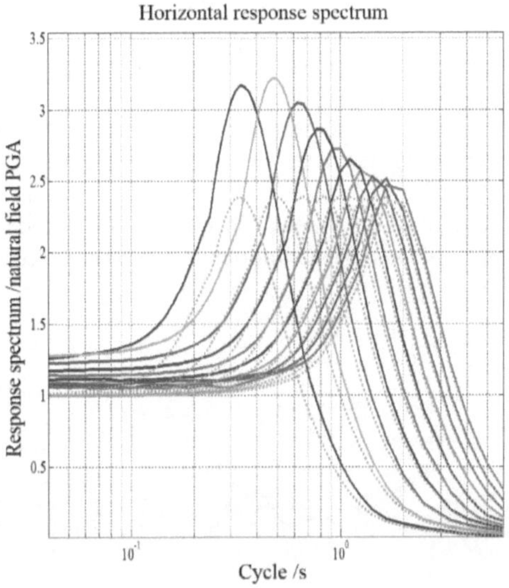

Figure 19. Response spectra of mid point ground motion and horizontal free surface ground motion in platform section under the pulse input.

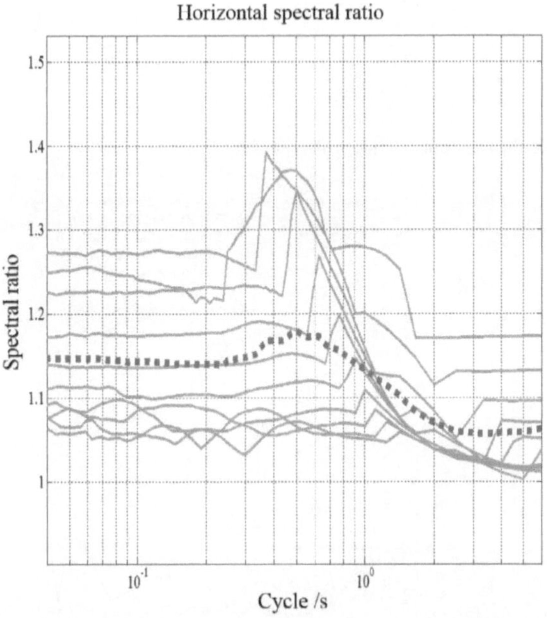

Figure 20. The spectral ratio of the midpoint response in the platform section of the convex terrain under the pulse input.

5.6 *Comprehensive analysis*

Figure 21 shows the average spectral ratio curve of midpoint ground motion in the platform section of protruding terrain under different inputs (i.e., the ratio of the midpoint ground motion response in the platform section of protruding terrain to the free field ground motion response spectrum). The average spectral ratio curves calculated according to actual ground motion input, artificial ground motion input and simulated ground motion input are close to each other; compared with these three inputs, the average spectral ratio curve calculated according to the white noise input is larger in some cycle segments; while the spectral ratio curve calculated according to pulse input is quite different from that under the ground motion input. Hence, when studying the amplification effect of local terrain on response spectrum, pulse input cannot be used as the input ground motion of numerical models.

Figure 22 shows the variation of the variance of spectral ratio with the control period of response spectrum under different inputs. In most frequency bands, the variance corresponding to white noise

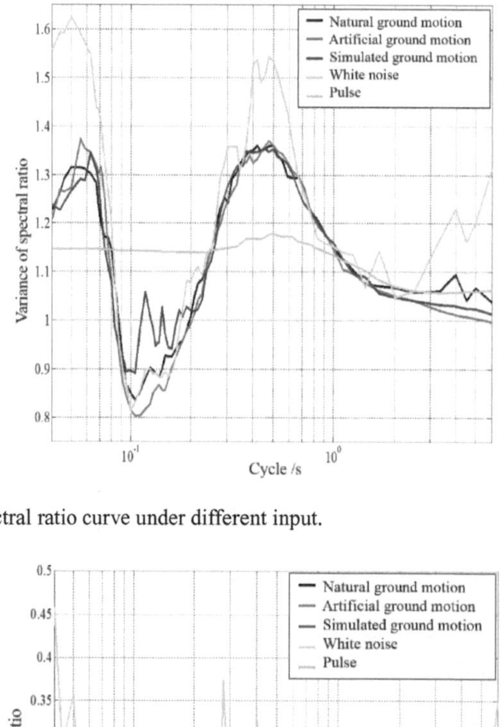

Figure 21. Average spectral ratio curve under different input.

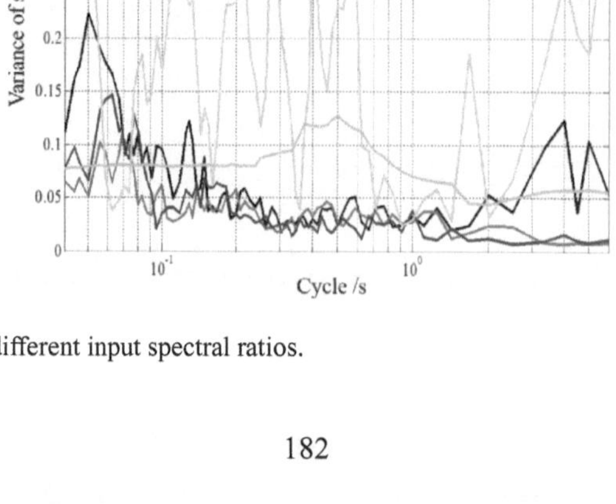

Figure 22. Variance of different input spectral ratios.

input is much larger than ground motion input; the variance of actual ground motion input is slightly greater than that of artificial ground motion input and simulated ground motion input, while the latter has little difference. Therefore, when determining the empirical model of the amplification factor of local protruding terrain to ground motion response spectrum, it is suggested that the artificial ground motion input and actual ground motion input should be mainly used, and the influence of the uncertainty of input ground motion on the regression results should be considered.

6 CONCLUSIONS

This paper takes protruding terrain as an example, and studies the influence of actual ground motion input, artificial ground motion input, simulated ground motion input, white noise input and pulse input on the terrain effect of ground motion by using the finite element-finite difference calculation method based on transmitting boundary. The results show that the average spectral ratio curves calculated according to actual ground motion input, artificial ground motion input and simulated ground motion input are close to each other; compared with these three inputs, the average spectral ratio curve calculated according to the white noise input is larger in some cycle segments; while the spectral ratio curve calculated according to pulse input is quite different from that under the ground motion input. Hence, when studying the amplification effect of local terrain on response spectrum, pulse input is not recommended as the input ground motion of numerical models. In most frequency bands, the variance corresponding to white noise input is much larger than artificial ground motion input; the variance of actual ground motion input is slightly greater than that of artificial ground motion input and simulated ground motion input, while the latter has little difference. Therefore, it is suggested that when determining the empirical model of the amplification factor of local protruding terrain to ground motion response spectrum, the artificial ground motion input and actual ground motion input should be mainly used, and the influence of the uncertainty of input ground motion on the regression results should be considered.

FUNDING

this research is funded by National Key Research and Development Plan of China (2018YFC1504601), Special project of basic scientific research business of central public welfare scientific research institutes (ZDJ2020-09).

AUTHORS

First author: Minghui Hao (1983 –), male, senior Engineer, with research interest in terrain effect and structural earthquake resistance of complex sites. Email: minghuimail@126.com

Corresponding author: Yushan Zhang (1974 –), male, Han nationality, research fellow, with research interest in Ground motion time history synthesis. Email: hyszhang@163.com

REFERENCES

Borcherdt R.D. (1970), Effects of local geology on ground motion near San Francisco Bay[J]. *Bull Seism Soc Am*, 60, 29–61.

Boore D.M. (1973) The effect of simple topography on seismic waves: implications for accelerations recorded at Pacoima Dam[J], San Fernando Valley, California. *BSSA*, 63(4), 1603–1609.

Davis L.L., West L.R. (1973) Observed effects of topography on ground motion[J]. *BSSA*, 63(1), 283–298.

Hao M.H., Zhang Y.S. (2014) Analysis of terrain effect on the properties of ground motion[J]. *Acta Sesimologica Sinica*, 36(4), 883–894.

Hao M.H., Zhang Y.S. (2015) Analysis of the influence of adjacent terrain on ground motion characteristics[J]. *Earthquake research in china*, 2015, 31(3), 656–667.
Liu J.B. (1996) Impact on earthquake ground motion by local irregular terrain[J]. *Journal of Earthquake*, 18(2), 239–245.
Li Y.M., Wang, X.Y., Song W.J. (2019) Application of dynamic substructure method to protruding[J], *Journal of haebin institute of technology*, 51(5), 156–170. topography on characteristics of ground motion.
Li X.J., Liao Z.P., Du X.L. (1992) An explicit finite difference method for viscoelastic dynamic problem[J]. *Earthquake engineering and engineering vibration*, 12(3), 74–80.
Nguyen K.-V. (2007) Gatmiri B.Evaluation of seismic ground motion induced by topographic irregularity. *Soil Dyn. Earthq [J]. Eng*. 27(2), 183–188.
Rong M.S., Li X.J. (2007) The Effect Analysis of Topography on the Spectrum Property of Anti-plane Movement[J]. *Earthquake research in china*, 23(2), 147–156.
Wang W. (2011) Effect of hill topography on ground motion[D].harbin, Institute of engineering mechanics, *China earthquake administration*, 152–156.
Yuan X.M., Liao Z.P. (1996) Scattering of plane SH waves by a cylindrical hill of circular arc cross-section [J]. *Earthquake Engineering and Engineering Vibration*, 16(2), 1–13.
Zhu Y.Q., Hu T.Y., Guo Z.Q. (1991) Propagation of seismic waves in a viscoelastic medium and the effect of topography. *Acta Sesimologica Sinica*, 13(3), 442–449.
Zhou H., Chen X.F. (2006) A new approach to simulate scattering of SH waves by an irregular topography [J]. *GeophysJInt*, 164(2), 449–459.
Zhang Y.S., Zhao F.X. (2014) Matching method of ground motion response spectrum based on the wavelet function[J]. *China civil engineering journal*, 42(1), 70–81.

Advances in Civil Engineering: Structural Seismic Resistance, Monitoring and Detection – Mohd Johari Mohd Yusof and Junwen Zhang (Eds)
© 2023 The Authors, ISBN 978-1-032-31491-4

Stability analysis of bank slope under reservoir water variation and strong earthquake

Xiaodong Fu
State Key Laboratory of Geomechanics and Geotechnical Engineering, Institute of Rock and Soil Mechanics, Chinese Academy of Sciences, Wuhan, Hubei, China
University of Chinese Academy of Sciences, Beijing, China

Liwei Wang
China Nuclear Power Engineering Co., LTD. Zhengzhou Branch, Zhengzhou, Henan, China

Yingwu Wang
Yunnan Institute of Water & Hydropower Engineering Investigation, Design and Research, Kunming, Yunnan, China

Dongge Chen
Central Yunnan Provincial Water Diversion Project Co., Ltd., Kunming, Yunnan, China

Haifeng Ding*
State Key Laboratory of Geomechanics and Geotechnical Engineering, Institute of Rock and Soil Mechanics, Chinese Academy of Sciences, Wuhan, Hubei, China
University of Chinese Academy of Sciences, Beijing, China

ABSTRACT: In many river basins, the deposit slope is widely distributed, it is easy to lose stability under the action of reservoir water variation and strong earthquake, and the induced geological disasters that seriously endanger life and engineering construction. In this paper, the stability of a deposit slope case under the action of reservoir water variation and strong earthquake is studied. The case is located in reservoir bank of Jinjidadan River arch bridge in Yunnan Province, China. Using the Geostudio software, the static and dynamic responses of slope under different reservoir water levels and seismic intensity are simulated. The distribution characteristics of pore water pressure under different working conditions are analyzed, and the evolution law of safety factor of slope with reservoir water variation is studied. Adopting two evaluation indicators including dynamic safety factor and permanent displacement, the dynamic stability of the deposit slope under the combination of water level and seismic intensity is analyzed. The results show that the slope is in a stable state under reservoir water variation, however, it is prone to landslide whether it is a low or high reservoir water level under VIII degree earthquake, which needs to be strengthened.

1 INTRODUCTION

Deposit is a kind of special geological material that mixed rock and soil, which has the characteristics of large variability of material composition and complex spatial structure (Zhang et al. 2019). Its derived geological disasters have the characteristics of multiple occurrences, randomness and recurrence. In China, the areas threatened by geological disasters of deposit slopes are almost all over the provinces. Most of the hillsides are slopes formed by the weathered rocks, slope deposit and eluvium, especially in Western China. In many river basins, these deposit slopes are mostly in the critical stable state without the influence of external load. However, they are easy to lose stability under the action of reservoir water variation and strong earthquake, and the induced

*Corresponding Author: 1807345621@qq.com

DOI 10.1201/9781003310884-26

geological disasters that seriously endanger life and engineering construction (Fu et al. 2020b; Ng et al. 1998; Zhang et al. 2021).

The action of the reservoir water variation reduces the friction and cohesion of the rock and soil. At the same time, the water level variation also affects the seepage field and stability of deposit slope. The rise of reservoir water level increases the groundwater level and the bulk density of landslide body, resulting in the increase of pore water pressure, reverse seepage of groundwater, and the decrease of strength of rock and soil. Because of the influence of permeability coefficient and particle skeleton, the decline rate of groundwater level is lagging behind the decline of reservoir water level when the reservoir water level descends, which will result in the decrease of pore water pressure in the slope and the increase of osmotic pressure difference (Berilgen 2007). Lane and Griffiths (2000) used strength reduction method and limit equilibrium method to compare and analyze the change law of slope safety factor under the condition of sudden and slow drop of reservoir water level. Viratjandr and Michalowski (2006) set the seepage line as a horizontal line, and analyzed the stability safety factor of the slope under the action of sudden rise and fall of water level and unsaturated unstable decline by using the limit equilibrium method.

The analysis of deposit slope under strong earthquake mainly includes the pseudo-static method and the dynamic time history analysis of finite element method. The dynamic time history analysis method is adopted to analyze the influence of the slope under the seismic action, which can better determine the stability of a slope (Kramer 1996). The evaluation criterion mainly includes safety factor and permanent displacement, and the safety factor is widely used as the criterion to evaluate the stability of a slope in China. It is considered that the safety factors that meet the requirements of the specification can ensure the stability of a slope. However, it is found that the safety factor cannot completely guarantee the stability of the slope with in-depth study (Fu et al. 2020a).

So far, there are many research results on the stability of deposit bank slope under reservoir water variation or strong earthquake, but there are few studies that consider both conditions at the same time. To solve this problem, this paper carried out the static and dynamic responses and stability analysis of a typical bank slopes under different reservoir water levels and seisimic intensity based on the reservoir bank of Jinjidadan River arch bridge, a controlling project of Dali-Yongsheng Expressway in Yunnan Province, China.

2 PROJECT OVERVIEW AND NUMERICAL SIMULATION

As shown in Figure 1, Jinjidadan River arch bridge is located in Xialiu Village, Pianjiao Town, Yongsheng County, Lijiang City, Yunnan Province, China. The bridge crosses Ludila reservoir of

Figure 1. Schematic diagram of project site.

Jinsha River Basin from Shaoxiang Island in the east of Xialiu village and connects to provincial highway S220. The total length of the bridge is 478m, the original ground elevation is about 1200 ~ 1297m.

Field investigation and drilling data show that the geological structure of the engineering site of Jinjidadan River arch bridge is very complex, and can be divided into five geological unit layers from top to bottom: 1) Quaternary artificial fill layer; 2) Quaternary Holocene collapse Slope layer (gravel soil); 3) Quaternary Holocene alluvial layer (pebbles, silty sand, medium sand, sand); 4) Quaternary Pleistocene alluvial layer (silty sand, brown coal, silty sand); 5) Quaternary Pleistocene collapsible layer (breccia soil).

The engineering site belongs to two seismic belts, most of the Northeast belongs to the Xianshuihe-East Yunnan seismic belt of the Qinghai-Tibet seismic belt, and the southwest belongs to the Southwest Yunnan seismic belt. The seismic activity of the Xianshuihe-East Yunnan seismic belt is strong, with high frequency and intensity, the largest earthquake level in history is Ms8.0, and the largest earthquake level in the history of Southwest Yunnan seismic belt is Ms7.4. According to the regional potential source area, seismic activity parameters and ground motion attenuation relationship, the basic seismic intensity and fortification intensity of the site are taken as the peak acceleration corresponding to the probability of exceeding 63% and 10% in 50 years.

Selecting a geological profile of reservoir bank slope, the numerical model in the typical section of Jinjidadan River arch bridge is established. As shown in Figure 2, the main strata in the model include medium sand, saturated medium sand, gravel, saturated gravel, artificial fill, pebble, angular gravel. The physical and mechanical parameters of these geomaterials are shown in Table 1. To evaluate the stability of the slope under different reservoir water levels and seismic intensity, the Seep/W, Quake/W, Sigma/W and Slope/W modules of GeoStudio software are used to divide the finite element calculation mesh by combining triangle and quadrilateral.

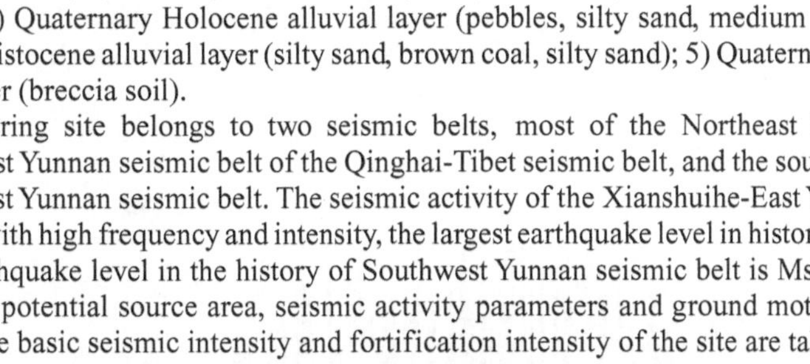

Figure 2. Calculation section of the bank slope along the bridge.

Table 1. Physical and mechanical parameters of the geomaterials.

Material	Elastic modulus E(MPa)	Cohesion (kPa)	Density (kn/m^3)	Friction φ(°)	Water content (%)	Coefficient (m/h)
Medium sand	340000	28	20.5	31	12	0.09
Saturated medium sand	340000	18	21.5	29	14	0.1
Gravel	320000	30	21.5	32	10	0.15
Saturated gravel	330000	18	22.5	29	11	0.16
Artificial fill	320000	15	21	22	12	0.08
Pebble	1800000	0	22	29	10	1.5
Angular gravel	2000000	30	22	29	12	0.036

In terms of reservoir water variation conditions, the lowest reservoir water level of Jinjidadan River is 1212m and the highest reservoir water level is 1224m. The total time for the reservoir water level to rise from the lowest to the highest reservoir water level is divided into five types:

48h, 72h, 96h, 144h, and 288h. The ascent rate is 0.25 m/h, 0.167 m/h, 0.125 m/h, 0.083 m/h and 0.042 m/h. The variation of the reservoir water from the highest level to the lowest reservoir water level uses the same rate as the rise. Observe the change of reservoir bank slope within 720 hours.

In terms of seismic analysis conditions, the siesmic wave detected by seismic stations is processed to obtain the normalized earthquake acceleration time history, as shown in Figure 3. Then use the peak accelerations of 0.043g and 0.151g corresponding to the VI degree and VIII degree earthquakes to obtain the input seismic waves during the analysis of the VI degree and the VIII degree earthquake. The VI and VIII degree earthquakes are brought into the lowest reservoir water level and the highest reservoir water level for action response analysis.

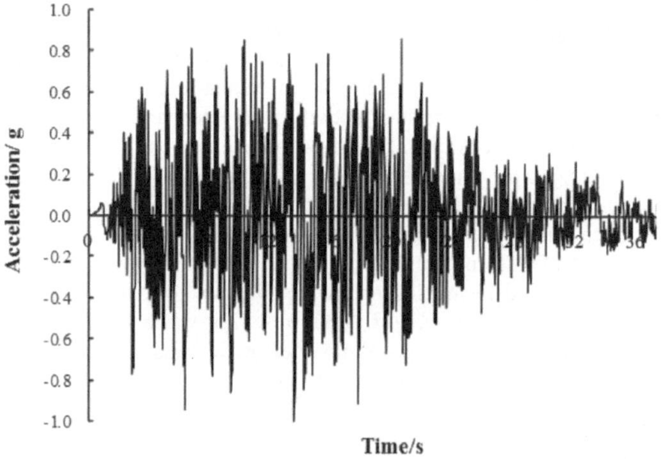

Figure 3. Normalized ground motion acceleration time-history.

3 ANALYSIS ON THE INFLUENCE OF RESERVOIR WATER VARIATION ON THE STABILITY OF RESERVOIR BANK SLOPE

3.1 *Analysis of reservoir water level rising*

Figure 4 shows the time history of safety factor of bank slope under different rising rates of reservoir water level. The overall trend of safety factor under each rising rate is to increase to the maximum value first, then decrease slowly and tend to be stable. The reasons are as follows. In the beginning, the hydrostatic pressure of the reservoir water plays a leading role in the stability of the landslide, and the rising speed of the groundwater level is slower than that of the reservoir water, and the stability coefficient continues to increase. With the infiltration of the reservoir water, while the buoyancy force increases, the physical and mechanical parameters of some rock and soil mass in the sliding zone gradually decrease, the stability coefficient of landslide gradually decreases, then slowly reaches a stable state.

The safety factor of bank slope under the lowest reservoir water level is 1.967. Through the comparison of different rise rates of reservoir water level, it can be seen that when the rising time is 48h, 72h, 96h, 144h, and 288h, the peak of the safety factor is at the end of the reservoir water rise time, and the peaks are 2.463, 2.413, 2.373, 2.319, and 2.248. The shorter the rise time, the longer the time from the maximum value of safety factor to stable state, and gradually stabilized at 2.125. Because the faster the reservoir water rises, the greater the height difference between the reservoir water level and the groundwater level, the greater the hydrostatic pressure, and the greater the slope stability.

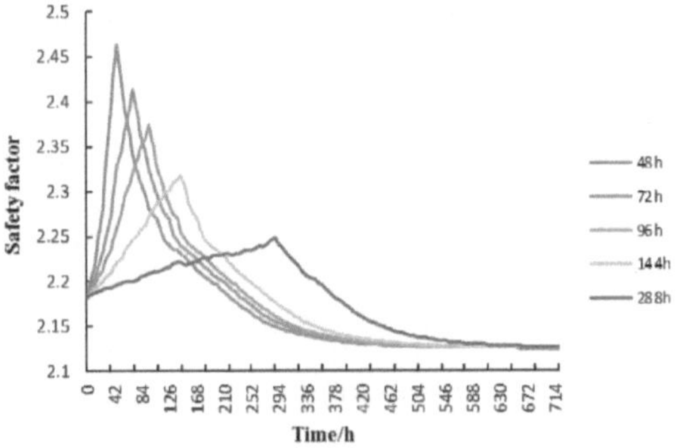

Figure 4. Time-history diagram of safety factor of the bank slope under different reservoir water level rising rate.

3.2 *Analysis of reservoir water level dropping*

Figure 5 shows the time history of safety factor of bank slope under different reservoir water level descending rate. When the reservoir water level drops, the safety factor decreases obviously. The reasons are as follows. There is a relative lag in the decline of groundwater level in the slope, resulting in excess pore water pressure in the slope, which is unfavorable to the stability of the slope. When the reservoir water level suddenly drops, the pore water pressure in the slope cannot dissipate in time, which leads to the lag of the groundwater level decline. When the reservoir water level decreases, the pressure on the slope surface decreases, and the internal pore water pressure field of the slope is too late to adjust, resulting in excess pore water pressure, which is unfavorable to the slope stability, so the safety factor decreases.

The safety factor of bank slope under the highest reservoir water level is 1.967. Through the comparison of different falling time of reservoir water level, it can be seen that when the falling time is 48h, 72h, 96h, 144h, and 288h, the lowest value of safety factor is the end time of reservoir water level falling, and the lowest values are 1.743, 1.754, 1.767, 1.784 and 1.822. The shorter the falling time, the longer the time for the safety factor to rise slowly from the lowest value to the stable state and gradually stabilize to 1.975. When the reservoir water level drops, the hydrostatic pressure of reservoir water plays a leading role in the stability of landslide in the initial stage. As the hydrostatic pressure gradually decreases and the local hydrodynamic pressure increases, the stability continues to decrease to an extreme value. After reaching the extreme value, with the drainage of reservoir water, the physical and mechanical parameters of rock and soil mass in the sliding zone gradually increase, and the increase in shear strength parameters has a greater effect on the stability of the landslide than the hydrostatic pressure and hydrodynamic pressure, the landslide stability coefficient gradually increases until it reaches steady state.

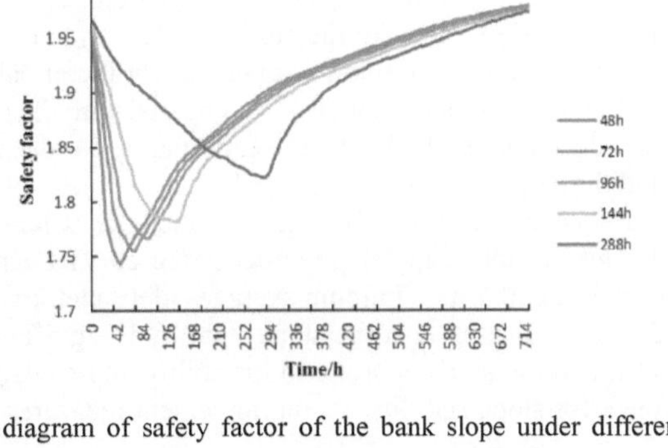

Figure 5. Time-history diagram of safety factor of the bank slope under different reservoir water level descending rate.

4 INFLUENCE OF EARTHQUAKE ON RESERVOIR BANK SLOPE

4.1 *Dynamic stability analysis of slope based on dynamic safety factor*

The dynamic safety factor method is used to calculate the safety factor of reservoir bank slope under VI degree and VIII degree earthquakes at the lowest reservoir water level. The initial safety factor of the slope is 1.672 and the minimum safety factor is 1.202 during the VI degree earthquake. The minimum average safety factor calculated by the above formula is 1.367, which is greater than 1.1 of the slope control index under earthquake action. Compared with the safety factor of 1.672 in the natural state, the decrease of 18.2% indicates that the stability of the slope is reduced under an earthquake of degree VI, but the slope can still meet the stability requirements, and the slope will not be damaged.

Figure 6 shows the time history of the safety factor under a VIII degree earthquake at low reservoir water level. The initial safety factor of the slope under the VIII degree earthquake is 1.672, the minimum safety factor is 0.697, and the minimum average safety factor is calculated as 1.038, which is lower than 1.1 of the slope control index under earthquake action. Compared with the safety factor of 1.672 in the natural state, the decrease of 37.9% indicates that the stability of the slope is significantly reduced under an earthquake of degree VIII, and the slope is easy to instability and damage. It can not meet the requirements of the project for slope stability; therefore, reinforcement measures are needed to ensure the stability of the slope.

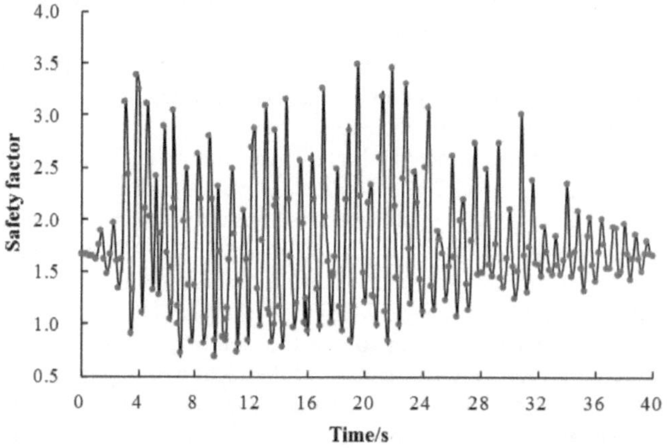

Figure 6. Time chart of safety factor for low reservoir water level VIII degree earthquake.

Similarly, the safety factor of the right reservoir bank slope of the highway bank along the bridge can be calculated under the condition of the highest reservoir water level of VI degree and VIII degree earthquakes. The initial safety factor of the slope is 1.454 and the minimum safety factor is 1.088 during the VI earthquake. The minimum average safety factor calculated by taking into the above formula is 1.216, which is reduced by 16.4% compared with the natural state, indicating that the slope stability is reduced under the VI degree earthquake, but the slope can still meet the stability requirements and the slope will not be damaged.

The time history of safety factor under VIII degree earthquake at high water level is shown in Figure 7. The initial safety factor of the slope under a degree VIII earthquake is 1.454, and the minimum safety factor is 0.650. The minimum average safety factor is 0.931, which is lower than the natural state. 36.0%, indicating that the slope stability is significantly reduced under an earthquake of degree VIII, and the slope is prone to instability and damage. It can not meet the requirements of the project for slope stability, so reinforcement measures need to be adopted to ensure slope stability.

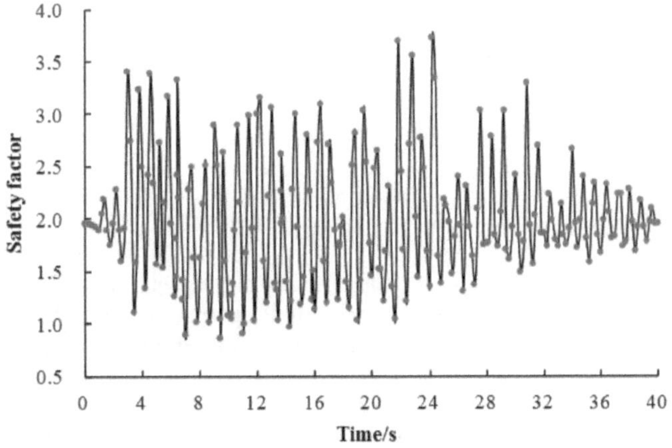

Figure 7. Time-history diagram of safety factor for high reservoir water level VIII degree earthquake.

4.2 *Dynamic stability analysis of slope based on permanent displacement method*

Using the Newmark deformation analysis method in Slope/W module of GeoStudio software, it is calculated that the permanent displacement of the slope under low reservoir water level and high reservoir water level is 0, which means that the slope will not be damaged. The permanent displacement of low reservoir water level is 32.03 cm and that of high reservoir water level is 45.56 cm under the VIII degree earthquake. According to the relationship between permanent display and degree of slope damage given by Jibson and Michael (2009), the damage degree of the slope is ultra-high, which is consistent with the results of the reservoir bank slope calculated by the limit equilibrium method in the previous article. It can be seen that the slope has poor stability and is prone to landslide whether it is a low reservoir water level or a high reservoir water level under a degree VIII earthquake. Therefore, considering the long-term stability of the slope, reinforcement measures should be taken for the slope.

5 CONCLUSION

The deposit slope geological disasters induced by the reservoir water variation and strong earthquake, are widely distributed, and the safety evaluation of slope and the formation mechanism of disaster are very important research topics. Adopting finite element and limit equilibrium analysis, the seepage field and stability of the right selected reservoir water level rising and falling for 48h, 72h, 96h, 144h and 288h on the highway bank of Jinjidadan River Bridge are analyzed. Considered with the impact of VI and VIII earthquakes on high water level and low water level slopes, the conclusions are as follows.

(1) Based on the coupling theory of stress field and seepage field, the evolution law of safety factor of slope under reservoir water variation is studied: During the rise of reservoir water, the safety factor firstly increases to the peak value, then decreases gradually, and the slope stability is improved. When the reservoir water level drops, the safety factor firstly decreases to the minimum value, then gradually increases, and the slope stability decreases. At the same time, the safety factor in the process of reservoir water change meets the specification requirements, and the slope is in a stable state.
(2) Adopting two evaluation indicators including dynamic safety factor and permanent displacement, the dynamic stability of the slope under the coupling action of earthquake and reservoir water is analyzed: During a VI degree earthquake, the slope is in a stable state under both low and high reservoir water levels. Under the action of VIII degree earthquake, the safety factors

of low water level and high water level slopes are 1.038 and 0.931 respectively, and the permanent displacement is 32.03cm and 45.56cm respectively. The corresponding damage degree is ultra-high, the slope is in unstable state, and corresponding reinforcement measures should be taken. In the future research, the mechanics characteristic and reinforcement mechanism of deposit slope under reinforcements must be studied.

ACKNOWLEDGMENT

The work reported in this paper is financially supported by the National Natural Science Foundation of China (No. 52179117), the International Partnership Program of Chinese Academy of Sciences Grant No. 131551KYSB20180042, and the Youth Innovation Promotion Association CAS (No. 2021325).

REFERENCES

Berilgen, M.M., 2007. Investigation of stability of slopes under drawdown conditions. *Computers and Geotechnics* 34(2): 81–91.

Fu, X.D., Sheng, Q., Du, W.J., Mei, H.R., Chen, H., Du, Y.X., 2020a. Evaluation of dynamic stability and analysis of reinforcement measures of a landslide under seismic action: a case study on the Yanyangcun landslide. *Bulletin of Engineering Geology and the Environment* 79: 2847–2862.

Fu, X.D., Sheng, Q., Li, G., Zhang, Z.P., Zhou, Y.Q., Du, Y.X., 2020b. Analysis of landslide stability under seismic action and subsequent rainfall: a case study on the Ganjiazhai giant landslide along the Zhaotong-Qiaojia road during the 2014 Ludian earthquake, Yunnan, China. *Bulletin of Engineering Geology and the Environment* 79 (10): 5229–5248.

Jibson, R.W., Michael, J.A., 2009. Maps showing seismic landslide hazards in anchorage, alaska. *Center for Integrated Data Analytics Wisconsin Science Center* (5): 1-9.

Kramer, S., 1996. *Geotechical Earthquake Engineering*. Upper Saddle River: Prentice Hall.

Lane, P.A., Griffiths, D.V., 2000. Assessment of Stability of Slopes under Drawdown Conditions. *Journal of Geotechnical and Geoenvironmental Engineering* 126(5): 443–450.

Ng, C.W.W., Shi, Q., 1998 A numerical investigation of the stability of unsaturated soil slopes subjected to transient seepage. *Computers and Geotechnics* 22(1): 1–28.

Viratjandr, C., Michalowski, R.L., 2006. Limit analysis of submerged slopes subjected to water drawdown. *Canadian Geotechnical Journal* 43(8): 802–814.

Zhang, Z.P., Sheng, Q., Fu, X.D., Du, Y.X, 2019. An approach to predicting the shear strength of soil-rock mixture based on rock block proportion. *Bulletin of Engineering Geology and the Environment* 79(1): 2423–2437.

Zhang, Z.P., Sheng, Q., Fu, X.D., Du Y.X, 2021. Stability of cracking deposit slope considering parameter deterioration subjected to rainfall. *International Journal of Geomechanics* 21(7). https://doi.org/10.1061/(ASCE)GM.1943-5622.0002045

*Advances in Civil Engineering: Structural Seismic Resistance,
Monitoring and Detection – Mohd Johari
Mohd Yusof and Junwen Zhang (Eds)*
© 2023 The Authors, ISBN 978-1-032-31491-4

Application of well seismic combination in deploying directional wells near faults

Jishuang Xing
Oil Production Plant of Daqing Oilfield Company Ltd, China

ABSTRACT: At present, block A has entered the ultra-high water cut development period, the distribution of remaining oil is highly scattered, only locally enriched, and it is difficult to tap the potential. In the past, when deploying infill well pattern near the fault, in order to avoid losing the target layer, the design well location is as far away from the fault as possible, resulting in low well pattern density near the fault and imperfect injection and production, which is the main remaining oil enrichment area. Since the three-dimensional seismic work in area A, the structural characteristics have been re recognized by comprehensively using logging and seismic data. The number, shape and local structural pattern of faults have changed, and the degree of fault recognition has been further improved, which provides an important basis for tapping the remaining oil near the fault. To this end, since 2019, relying on well seismic and fine fault interpretation results, six large displacement directional wells have been deployed near the fault in block A, which has achieved good development results and broadened the technical path of tapping the potential of oil fields in the later stage of ultra-high water cut.

1 PREFACE

In area A, well seismic combined structural modeling was carried out in 2010. By June 2013, the structural modeling work had been completed, the three-dimensional structural modeling of the whole plant had been fully covered, the structural understanding had been improved, and the structural error was controlled at 0.84 ‰. In order to further tap the remaining oil near the fault in area A, six directional wells were designed in area A in 2019 by combining well seismic research results, and good development results were achieved.

2 RECOGNITION OF GEOLOGICAL STRUCTURE BY WELL SEISMIC COMBINATION

The research adopts the three-dimensional seismic data of Xingshugang taken from Fuyang oil layer in 2002, and the fine processing bin in 2010 is 10m × 10m, the sampling rate is 1ms, and the scope of intercepting the trace in the seismic work area: inline: 3617-4217; Trace:2459-4359. Using this data and well seismic combination, the structure of SP reservoir is finely interpreted and a three-dimensional structure model is established, which further improves the understanding accuracy of SP reservoir structure.

2.1 *Improved construction accuracy*

Through the accurate description of the top surface of SII, SIII and PI oil formation and the bottom structure of PI formation, small faults with a fault distance of more than 5m and an extension length of more than 200m are determined (Figure 1).

DOI 10.1201/9781003310884-27

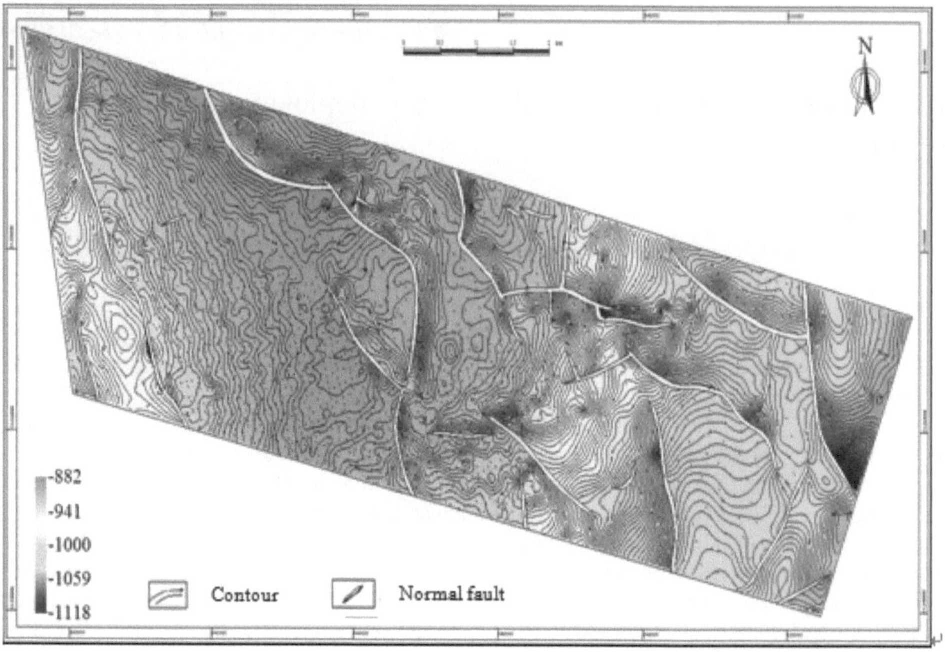

Figure 1. Top structure of PI formation.

2.2 *More detailed description of structural features*

The top surface of SII formation, SIII formation and PI formation in the study area have the same structural characteristics. Both have the characteristics of high in the north, low in the south, low in the East and West and high in the middle. Area A is cut by NW and NE faults, and the local structure is a fault anticline structure formed a PI long the rising wall of the fault.

A total of 45 faults are developed in PI group of area A, all of which are normal faults, mainly NNW and NW trending faults, and NE trending faults are locally developed. Breakpoint data confirm that the number, fault offset, dip angle and shape of faults in each layer are different in the oil-bearing section.

First, the number of faults at all levels is different, and the number of faults gradually decreases from deep to shallow. In the study area, 45 faults are developed in group Portugal I, 28 faults are developed in group SIII and 26 faults are developed in group SII.

Second, the fault displacement is 2 ~ 102m, with a large variation range, which is characterized by large deep fault displacement and small shallow fault displacement. The fault distance of 26 faults in SII group is 2 ~ 84m, with an average fault distance of 40.0m; The fault distance of 28 faults in SIII group is 13 ~ 93m, with an average fault distance of 50.0m; The fault distance of 45 faults in Portuguese PI group I is 10 ~ 102m, with an average fault distance of 53.0m.

Third, the occurrence of large faults on the seismic profile is mostly seat type. From the distribution characteristics of fault profile, most faults are characterized by graben or Horst structure (Figure 2).

2.3 *Improve the accuracy of fault recognition*

The accuracy of fault recognition is further improved through well seismic combined with fine interpretation of structural faults and three-dimensional geological modeling. Compared with the fault results recognized by well data, the distribution direction and extension length of large faults

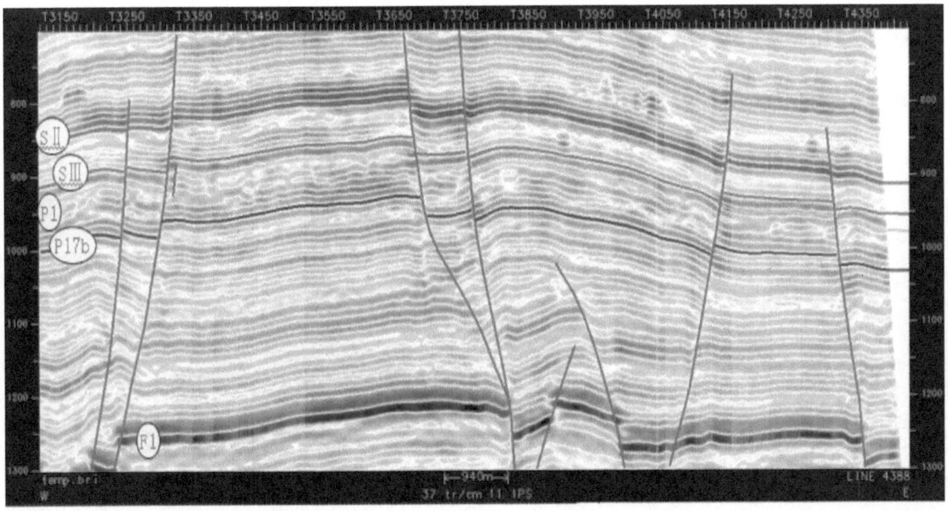

Figure 2. Profile of fault development.

are in good consistency, while small faults are different in breakpoint combination, fault overlapping relationship, extension direction and length, and the number of small faults has increased. On the whole, the fault controlled by more than three breakpoints in the oil layer, regardless of the fault distance, has generally high accuracy, but the tail of individual faults has changed a little; Faults with breakpoint combination of two or less oil layers change in fault strike, tendency and extension length; There are also some changes in the faults controlled by breakpoints in the upper stratum (Figure 3).

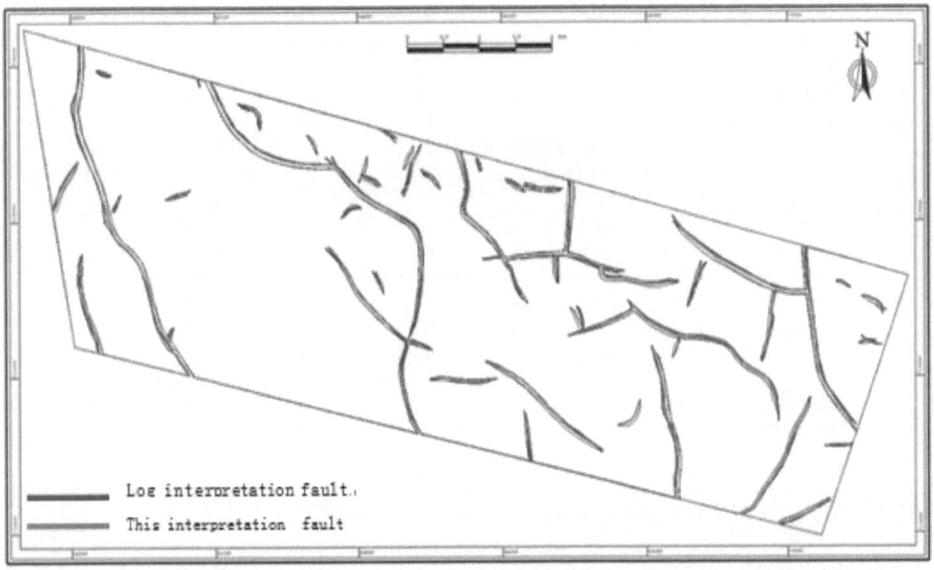

Figure 3. Comparison of well data fault and well seismic combined modeling fault.

The change of fault plane shows the following four characteristics:

First, the extension length changes: including two types of fault extension length extension and shortening. It mainly refers to the fault with good coincidence in the main part of the fault and inconsistent length at both ends (Figures 4 and 5).

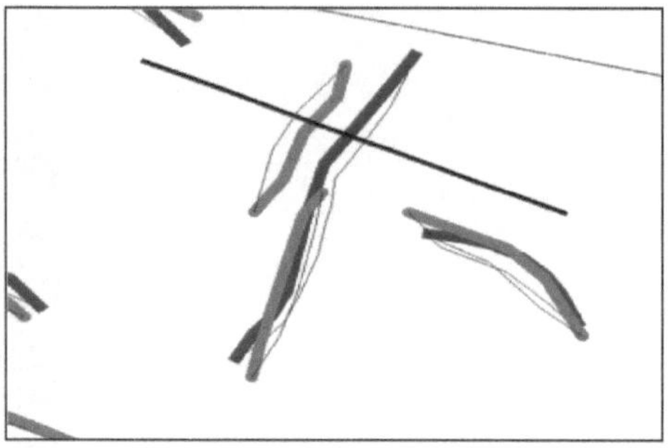

Figure 4. Plane change of fault.

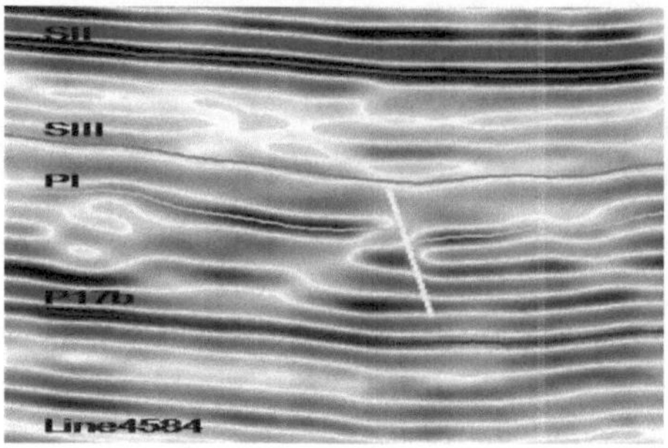

Figure 5. Fault section location.

The second is to reduce faults: mainly faults with short extension length and small fault distance.

Third, multiple faults are combined into one fault: the original fault may be 2 or even more, but it will be one fault after recombination.

Fourth, merging and splitting faults: the fault of the main well combination was originally considered to be one, which became multiple after well seismic combination. It was originally considered to be multiple faults, which became one fault after well seismic combination.

After the combination of well and earthquake, the faults in area A change greatly. According to the change table of top faults in group PI of Portugal, there are 3 newly discovered faults, 4 small faults, 1 extension length change, 17 strike changes, and 19 basically unchanged faults (Table 1).

Table 1. Variation of well seismic combined faults in block A.

Reservoir group fault (piece)		SII group	SIII group	PI group
Well		12	8	20
Well earthquake		26	28	45
Total increase		14	20	25
Quantity change	New discovery	2	2	3
	Write off	2	1	4
	merge			
	split			
	total	4	3	7
Morphological change	Extension length	1	1	1
	Trend change	8	11	17
	total	9	12	18
Basically unchanged		13	13	19

3 ANALYSIS OF PRODUCTION CONDITIONS NEAR FAULTS

Due to the development of faults, in order to prevent drilling from encountering breakpoints and reduce reservoir loss, the fault zone has been deliberately avoided in the process of previous development and adjustment, resulting in imperfect injection production relationship near the fault or no injection production well points in some areas, so that the remaining oil is relatively enriched.

The data of new wells put into operation in recent years show that more than half of the high-yield wells are distributed at the edge of the fault, which proves that there are residual oil enrichment areas at the edge of the fault. However, in the process of water injection development for many years, the edge of the fault has been affected by water injection within a certain range, and there are also areas with relatively good production.

In order to understand the distribution and influencing factors of remaining oil near the fault, an oilfield exploration and Development Research Institute conducted a comprehensive study through conceptual model and practical model. The results show that the distribution of remaining oil at the edge of the fault is controlled by many factors, such as well pattern, well spacing and fault avoidance distance. When oil and water wells are deployed between injection and production at the edge of the fault, the crude oil at the edge of the fault can be used well and the remaining oil potential is small; When the whole row of oil wells are deployed at the edge of the fault, the production condition of the fault avoidance area is poor, which is a potential residual oil enrichment area; The fault avoidance distance is the main factor affecting the enrichment of residual oil at the edge.

4 ANALYSIS AND EVALUATION OF PRODUCTION EFFECT OF DIRECTIONAL WELLS

In 2019, we continued to carry out fine geological research on the structural characteristics, sandstone development and remaining oil distribution near the large fault in area A. On this basis, the fault with large fault distance and long extension is selected as the block for deploying directional wells. Based on the research results of multi-disciplinary reservoir and combined with the injection production relationship of the original well pattern, six large displacement directional wells are designed, which has achieved good results and widened the technical path of tapping the potential of oil fields in the later stage of ultra-high water cut.

5 CONCLUSION

In summary, we can get the conclusion as below,

a) Well seismic combined with fine fault interpretation and the study of fault internal structure provide an important basic basis for tapping the potential of remaining oil near the fault.
b) The combination of well and earthquake is helpful to improve the fine description of geological structure.
c) The combination of well and earthquake is conducive to fine fault interpretation.
d) The deployment of directional wells near the fault is an effective section to tap the potential of remaining oil in the later stage of ultra-high water cut.

REFERENCES

Jia A., Guo J., He D. (2007) Fine reservoir description technology and development direction [J]. *Petroleum Exploration and Development* 34(6):691–695.

Li H. (2017) Application of well seismic combination technology in tapping potential of remaining oil [J] *Chemical Enterprise Management* (23):216–216.

Liang W. (2019) Research on potential tapping of well seismic combined with high efficiency wells in Daqing Changyuan Oilfield [J]. *Journal of Xi'an Petroleum University: Natural Science*, 34(5):63–68

Lv S.(2015)Fault identification technology and its application based on well seismic combination-Taking Xingbei oilfield as an example[J]. *Progress in geophysics*, (5): 2200–2205.

Wang X., Yang X., Chi B.,et al.(2006) Study on fine reservoir description technology of Daqing Peripheral Oilfields [J]. *Acta Petrolei Sinica*, 27(B12): 106–110.

Zhang H (2016) Study on potential tapping technology of high efficiency wells near faults in Changyuan Oilfield [J] *Journal of Changjiang University (Natural Science)* 13(8): 38–40.

Advances in Civil Engineering: Structural Seismic Resistance, Monitoring and Detection – Mohd Johari Mohd Yusof and Junwen Zhang (Eds)
© 2023 The Authors, ISBN 978-1-032-31491-4

Typhoon resistance analysis of a single-story factory building with concrete bent-steel roof truss system

Lijuan Xiao, Yi Jin, Xiaohai Qi & Guo Liu
China Nuclear Power Technology Research Institute, Shenzhen, China

ABSTRACT: The concrete structure workshop has strong lateral stiffness and can resist the general wind load; however, under extreme typhoon, the main structure may still produce large deformation. In this paper, based on the code for a load on building structures (GB50009-2012), the basic wind pressures on the surface of a single-story factory building with concrete bent and steel roof truss along the coast of Guangdong province are calculated, and the whole displacement response and local deformation of the main structure of the workshop under different typhoons are analyzed. The results show that the typhoon action perpendicular to the longitudinal direction of the workshop will cause the large deformation of the main structure, bring certain influence on the use function of the workshop building, and may lead to local crack in the concrete column, but it does not threaten the safety of the main concrete structure. Under the action of a typhoon, the steel frame structure of the parapet wall will appear as a large deformation, which is easy to induce the destruction of the retaining structure.

1 INTRODUCTION

In the past 20 years, our country is experiencing rapid developments in economy and industrial system. Industrial buildings have been built in many industrial development zones all over the country. This type of structure has a large span, wide space, flexible function, and convenient construction. It is often used on the southeast coast of China and is widely used in subway stations, warehouses, and factories (Ma 2015). At present, the single-story factory building structure widely used is plane structure system, mainly with a bent-frame structure and rigid frame structure (Han 2003).

For the single-story industrial plant with a plane structure, the lateral stiffness is mainly provided by the bent column and the wind column with gable wall. Compared with the frame structure and space structure, there is only one anti-overturning line, if the safety of the main structure under the action of a typhoon cannot be satisfied, catastrophic consequences may occur (Jiang 2012). Compared with the concrete bent frame structure, the steel portal frame workshop is more sensitive to the effect of wind load, hence most scholars and designers pay less attention to the wind resistance analysis of the concrete bent frame structure, and most of the research projects are focused on the wind resistance analysis of the lightweight steel portal frame workshop under the common wind pressure (Jang et al. 2002; Fritz 2007). However, in case of larger wind loads, such as extreme typhoon weather, even if the concrete plant has greater lateral stiffness, it will still have larger lateral deformation, which will affect the safety of the structure. It has practical significance in studying the anti-wind performance of concrete workshop in the typhoon-prone area. In this paper, taking a single-story factory building with a composite bent frame of a concrete column and steel roof truss as an example, the overall displacement response of its main structure under typhoon with different wind direction angles above level 12 wind is calculated by using SAP2000 structural finite element analysis software. Based on the analysis of the structural deformation under different typhoons, the safety assessment results of the main structure are presented.

 DOI 10.1201/9781003310884-28

2 DESCRIPTION OF STRUCTURAL SYSTEM

The single-story factory is located in the coastal area of Huizhou, Guangdong province. The longitudinal length of the factory is about 135.1 m, the transverse span is about 19.3 m, and the height of the structure is about 16.6 m. The basic load-bearing structure is a reinforced concrete bent structure with 25 pieces and two wind-resistant columns at the end of the gable wall. The transverse bent frames are connected through the wall beams to provide the longitudinal lateral resistance of the structure, the maximum height of the roof truss is 1.76 m, and the height of the parapet is 4.10 m. The elevation of the concrete bent-steel roof truss is shown in Figure 1.

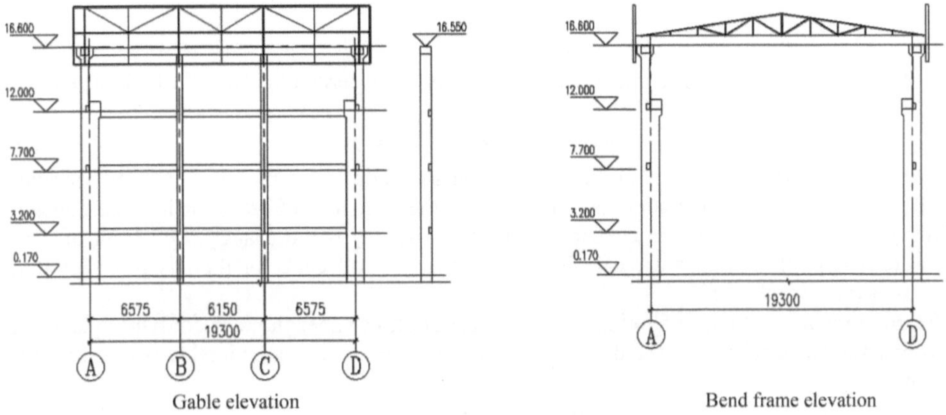

Figure 1. Schematic diagram of horizontal plane bent of the factory building.

The longitudinal bent columns of the workshop are reinforced concrete corbel columns with variable cross-sections, the upper section size is 600 mm × 500 mm, the lower section size is 1300 mm × 500 mm, and the elevation at the variable section is 12.00 m. The triangle steel roof truss of the workshop is made up of European standard section steel, and the strength grade of the steel used is E24.2. The lower chord of the roof truss is made of HEA700 I-section steel and is connected with the top of the concrete bent column by embedded anchor bolts, the upper chord and the web of the roof truss are made of L60x6 double-leg angle steel and the web of the roof truss is welded to the upper and lower chord through the joint plate. Roof purlins are IPE200 I-section steel, located at the node of a triangular roof truss. The steel frame of the parapet is a plane frame structure, which is connected with the lower chord of the roof frame by end-plate bolts. The vertical members of the steel frame adopt the HEA200 I-section steel, the horizontal members adopt the UPN200 channel steel, and the strong axes are located in the plane of the parapet structure.

3 MAIN CALCULATION PROCESS

3.1 *Instructions for load calculation*

According to the national standard of "Wind scale" issued in June 2012, the wind scale is divided into 18 levels according to the wind speed at the height of 10 m in the standard meteorological observation field. In this paper, the anti-wind performance of the structure is analyzed for the extreme typhoon weather above level 12 wind. Because of the basic wind pressure under each level typhoon, according to "Load code for the design of building structures" (GB50009-2012) Appendix E carries on the correlation formula computation, and takes each level wind speed range upper limit value as the basic wind speed, the basic wind pressure under each level of typhoon is shown in Table 1. Typhoon with a wind speed of 65 m/s is defined as level 18 in this paper.

Table 1. Calculation results of basic wind pressure at different wind speed.

Wind level	Wind speed (m/s)	Basic wind speed (m/s)	Basic wind pressure (kN/m^2)
13	37.0~41.4	41.4	1.07
14	41.5~46.1	46.1	1.33
15	46.2~50.9	50.9	1.62
16	51.0~56.0	56	1.96
17	56.1~61.2	61.2	2.34
18	65	65	2.64

For the main members, the characteristic value of wind load perpendicular to the building surface can be calculated as follows:

$$w_k = \beta_z \mu_s \mu_z w_0 \quad (kN/m^2) \tag{1}$$

where β_z is the wind-induced vibration coefficient at height z and the wind-induced vibration coefficient cannot be considered in the calculation of the wind load effect of the low-rise concrete plant, $\beta_z = 1.0$; μ_s is the shape coefficient of wind load; μ_z is the change coefficient of wind pressure height. Both values are calculated according to load code for the design of building structures (GB50009, 2012).

Taking into account the self-weight of the factory roof, consider the roof constant load as 0.5 KN/m^2 uniform surface load, the design value S of the load effect combination is calculated as follows:

$$S = \gamma_G S_{Gk} + \gamma_Q S_{Qk} \tag{2}$$

where γ_G and γ_Q are the partial coefficients of permanent load and variable load respectively; S_{Gk} and S_{Qk} are the effect values of permanent load and variable load, respectively. For the analysis of structural deformation and displacement under the effect of wind load, the standard combination should be adopted, $\gamma_G = \gamma_Q = 1.0$

3.2 *Computational model and material parameters*

SAP2000 V22 finite element analysis software is used for building structure modeling and typhoon resistance analysis. All components are modeled by linear frame elements. The eccentricity of the members is realized by defining the insertion point of the centroid, and the orientation of the bar is achieved by defining a local coordinate axis. The spatial integral model of the main structure is shown in Figure 2. The typhoon action in two directions is considered. The short axis direction is X direction, the wind direction angle of X direction is 90 degrees; the long axis direction is Y direction, and the wind direction angle of Y direction is 0 degrees.

The steel structure material for upper roof truss of structure is European standard material E24.2, and its property is defined as Q235B as per Chinese standards. Concrete materials for lower bent columns and wall beams are defined as C30 concrete. The elastic modulus of C30 concrete is defined as 3.0 × 104 MPa, Poisson's ratio $\mu = 0.2$, compressive strength $fc = 20.1$ MPa; the elastic modulus of Q235B steel is defined as 2.06 × 105 MPa, Poisson's ratio $\mu = 0.3$, and the yield strength is defined as $fy = 235$ MPa.

In the model, the connection between the roof truss web member, purlin, support, and other components and the roof truss chord is defined as the hinged connection. The connection between the concrete wall beam and the bent column is defined as the rigid connection, and the column base joint is connected by a rigid support, that is, relative translation and rotation in any direction is not allowed. According to the detailed construction drawing of the column top joint, there is a 50 mm gap filled by mortar between the concrete column top and the roof beam. Because of the low strength of mortar and the limited filling area, the connection between the top of the concrete column and the lower chord of the roof truss can be regarded as the hinge connection. However,

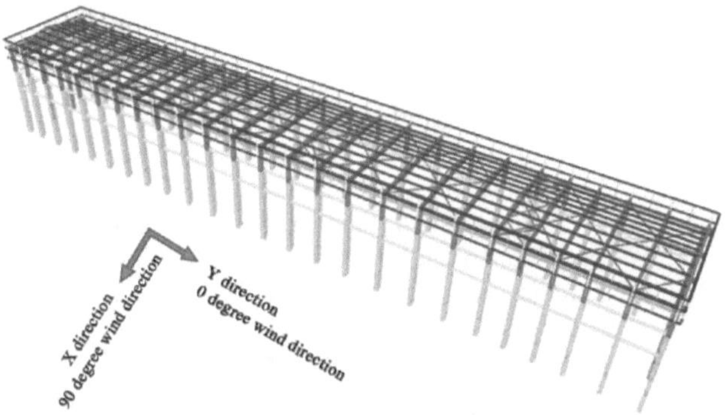

Figure 2. The spatial integral model of the main structure of the factory building.

considering that the bolt spacing at the top of the column is about 1 m and the allowable rotation effect is minimal, the connection between the top of the column and the roof steel beam is regarded as a rigid connection in the SAP2000 overall spatial model.

All the loads in the model are transferred by adding virtual surfaces. The wind load of the lower part of the concrete structure is transferred horizontally to the bent column and the anti-wind column, the wind load of the steel frame of the parapet is transferred vertically to the longitudinal channel steel, and the constant load and wind load of the roof are transferred to the roof purlins.

To analyze the wind load response of the workshop structure under the typhoon with different wind speed grades, it is necessary to calculate and analyze the typhoon above 12 grade under six working conditions. To minimize the modeling and calculation workload, the following simplified methods can be used to calculate and output the results. First, the roof uniform constant load and the wind load standard value under 18 typhoon are applied to the model, respectively, and are transferred to the corresponding load-bearing structural unit. A number of load combinations are established to gradually reduce the proportion coefficient of wind load items, which is equivalent to the reduction coefficient of basic wind pressure relative to typhoon 18 under each wind speed level. The calculation results can reflect the wind load response of each wind speed level.

4 CALCULATION RESULTS AND ANALYSIS

4.1 *Calculation result*

In this paper, in addition to calculating the deformation characteristics of the structure under typhoon by equivalent static wind load according to the code, the natural vibration modal response of the plant structure is also analyzed, it provides a certain reference value for the wind load dynamic response analysis of the follow-up workshop structure.

By using SAP2000, modal analysis of the workshop space model is carried out, and the natural vibration period of the first three modes and deformation parameters in each direction is obtained as shown in Table 2.

Table 2. Modal response parameters of plant main structure under natural vibration.

Mode order	Natural period (s)	UX (N-mm)	Uy (N-mm)	Uz (N-mm)	Rx (N-mm)	Ry (N-mm)	Rz (N-mm)
1	0.897805	0.009713	39.506169	0.000266	158944.2486	51.809744	−182.99113
2	0.755630	−36.671396	−0.012856	1.78E-07	39.629809	−182769.805	80946.5746
3	0.722232	1.805447	−0.004183	−2.40E-05	8.946093	8879.3288	1653274.934

According to the data in the table above, the first period of the main structure of the concrete bent workshop is longitudinal deformation, the second period is transverse deformation, and the third period is torsional deformation. At the same time, the first period of free vibration is about 0.9 s, which indicates that the concrete structure has greater stiffness.

Under the equivalent static wind load of a level 18 typhoon with 90 degree wind direction (X-axis). It can be seen from the displacement nephogram that there is a good synergy between the steel roof truss and the concrete column, the displacement at the top of the bent-frame column is consistent with the lateral displacement of the single roof truss. At the same time, the displacement and deformation of the main structure show the overall spatial effect, that is, the lateral displacement of the roof near the gable wall, the displacement of the roof purlin, and the steel frame of the parapet wall is smaller than the lateral displacement of the middle part of the workshop. The lateral stiffness of the gable wall side is greater than that of the middle-bent column, the displacement at the top of the column and the side of the roof is smaller than that in the middle part of the workshop, and the displacement of the roof purlin and parapet steel frame is more than 154 mm. The maximum lateral displacement of the whole structure occurs at the 13th axis, which is the middle part of the factory building.

Under the equivalent static wind load of level 18 typhoon with 0 degree wind direction (Y-axis). Under the wind load with Y-axis direction, the lateral displacement of the main structure is not significant because of the large rigidity of Y-axis direction, all the displacement of column top and the lateral displacement of the roof are in the range of 20–40 mm. Because the gable parapet steel frame is directly confined to the single side of the roof beam, the torsion constraint is insufficient, and the torsion phenomenon is easy to occur under the action of wind load, details of the structure and deformation are shown in Figure 3. The local maximum deformation of the steel frame is more than 280 mm, far beyond the overall deformation of the main structure.

Figure 3. The deformation nephogram of the parapet steel frame on the side of the gable under level 18 typhoon with 0 degree wind direction.

4.2 *Maximum displacement analysis*

As for the maximum displacement of the main structure under the typhoon with 90 degree wind direction (X-axis), the displacement of the bent column in the middle of the workshop (13axis) is the highest, and the displacement at both ends of the gable decreases gradually; therefore, for each stage of typhoon, the displacement at the gable side and 13 axis is considered for analysis. The maximum displacement analysis and calculation results of each representative node under different wind loads are shown in Table 3, The displacement-wind level correlation curve is shown in Figure 4.

Table 3. The displacement calculation results of representative nodes under typhoon with 90 degree wind direction (X-axis).

Wind speed rating		Level 18	Level 17	Level 16	Level 15	Level 14	Level 13
Maximum column top lateral displacement	Gable side	81.26	72.04	60.33	49.86	40.93	33.06
	13 axis	109.48	96.99	81.22	67.08	55.03	44.4
Maximum displacement of parapet	13 axis	152.89	135.87	114.37	95.11	78.69	64.2

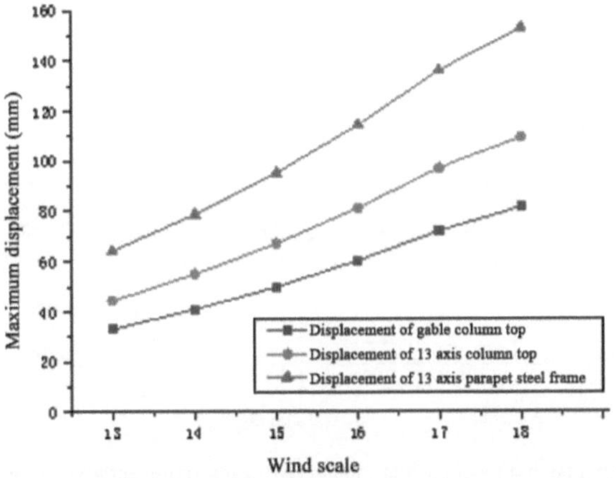

Figure 4. The displacement-wind level correlation curve of each representative node under typhoon with 90 degree wind direction (X axis).

Through the comparison of Figure 4, it can be seen that under the wind load of 90-degree wind direction, the maximum displacement trend of the structure is: the displacement of the parapet wall, the column top displacement in the middle of the workshop, the column top displacement at the side of the gable wall, the displacement increases with the increase of wind speed. The maximum displacement of the main structure in the middle of the workshop is 109.48 mm, i.e., 1/181 of the inter-story displacement angle, while it is 1/373 under the 13-grade typhoon. In the code for seismic design of buildings (GB 50011-2010), the limit value of the elastic-plastic inter-story displacement angle for single-story concrete column bent is 1/30; therefore, for the most single-story concrete factory buildings with good structural functions, the safety of the main structure under the typhoon can be guaranteed. In general, under the premise that the joints are intact and the connection between the roof steel beam and the concrete column can be kept completely rigid, the impact of a level 18 typhoon in the X-axis direction on the main structure is limited and will not pose a serious threat to the safety of the main structure, but will affect the use function of the factory building to a certain extent.

As for the maximum displacement of the main structure of the workshop under the action of Typhoon with 0 degree wind direction (Y axis), the displacement on the windward side is the largest, and the lateral displacement of all bent columns is approximately the same, therefore, the maximum displacement of the anti-wind column and the bent column on the windward side is analyzed. The results of the analysis and calculation of the maximum displacement of each representative node under the action of typhoon with different wind stages are shown in Table 4.

Table 4. Displacement of each representative node under 0 degree wind direction (Y axis) typhoon.

Wind speed rating		Level 18	Level 17	Level 16	Level 15	Level 14	Level 13
Maximum column top lateral displacement	Anti-wind column	37.9	33.6	28.16	23.3	19.15	15.49
	Bent column	24.17	21.41	17.93	14.8	12.14	9.8
Maximum displacement of parapet		275.74	243.01	201.66	164.62	133.04	105.19

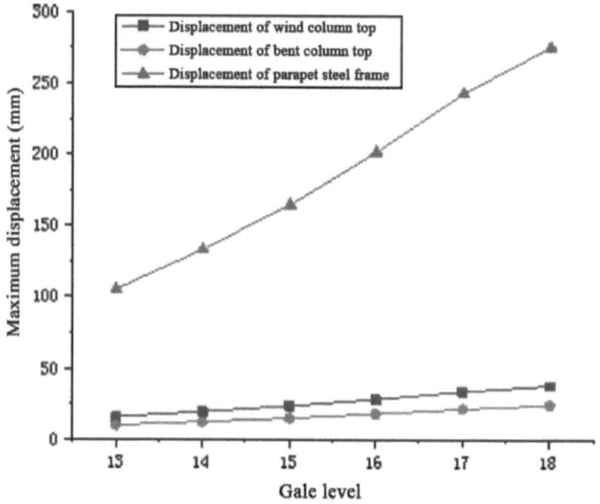

Figure 5. The displacement-wind level correlation curve of each representative node of the workshop under 0 degree wind direction (Y-axis) typhoon.

As shown in Figure 5, the lateral displacement of the main structure is small under the wind load in the y direction, and only 37.9 mm of the lateral displacement is produced at the anti-wind column under level 18 typhoon, and the corresponding displacement angle between the layers is only 1/438. However, for the enclosure structure, due to the particularity of the connection between the gable-side parapet and the roof steel beam, this part is easy to produce torsional deformation, which aggravates the deformation effect of gable parapet steel frame. The maximum displacement of this part is 275.74 mm and the relative displacement to the top of the column is 237.84 mm when the typhoon is level 18. The envelope structure in this part may fail due to excessive deformation, which should be the focus of structural wind resistance.

5 CONCLUSION

In this paper, based on the actual project of a single-story factory building with a concrete bent column-steel roof truss structure system, through SAP2000 modeling calculation and analysis, the following main conclusions and research prospects are obtained.

(1) The maximum displacement angle of the column top of the main structure is found to be between 1/373 and 1/181 under the action of typhoon, which is perpendicular to the longitudinal direction of the factory building. This degree of structural deformation does not threaten the safety of the main concrete structure, but it will affect the use function of part of the factory building, and cause the concrete column local crack.

(2) Under the action of typhoon, which is perpendicular to the direction of the gable wall of the workshop, the main structure will not deform greatly because of the large longitudinal stiffness of the workshop, and the maximum inter-story displacement will not exceed 1/400 under the action of level 18 typhoon. However, because of the special connection between gable parapet wall and the roof steel beam, the parapet wall may fail due to its large deformation.
(3) Compared with the lightweight steel portal frame workshop, the concrete bent frame workshop has larger lateral stiffness and higher resistance to wind load, but under the extreme typhoon, the wind load response of the main structure should not be neglected, especially the displacement response of the main structure under wind load should be studied when evaluating the use function of the factory building or the safety of the enclosure structure.

REFERENCES

DD uthinh, Fritz W P. Safety Evaluation of Low-Rise Steel Structures under Wind Loads by Nonlinear Database-Assisted Technique[J]. *Journal of Structural Engineering*, 2007, 133(4): 587–594.

Han Shu qiu. Status and development trend of industrial building structure at home and abroad in recent years. *Heilongjiang technology information*, 2003(03): 152.

Jang S, Lu LW, Sadek F, et al. Database-Assisted Wind Load Capacity Estimates for Low-Rise Steel Frames[J]. *Journal of Structural Engineering*, 2002, 128(12):1594–1603.

Jiang Shiyong, Fei Wei, Ye Yong, etc. [C]/China Association for Standardization of Engineering Construction; National Technical Committee for appraisal and reinforcement of buildings. China Association for Standardization of Engineering Construction; National Technical Committee for building appraisal and reinforcement, 2012.

Ma Jun. Wind field numerical simulation and wind resistance optimization of industrial workshop [J]. *Journal of Water Conservancy and Building Engineering*, 2015.

Advances in Civil Engineering: Structural Seismic Resistance, Monitoring and Detection – Mohd Johari Mohd Yusof and Junwen Zhang (Eds)
© 2023 The Authors, ISBN 978-1-032-31491-4

Intelligent comparison and selection of structural schemes of a 400m high-rise building

Jie Wang, Shen Zhang & Pengfei Yin
Central-South China Architectural Design Institute Co., Ltd., Wuhan, China

ABSTRACT: In early architectural schematic design phase of a 400m super high-rise building, the architect provides a variety of schemes with different facade inclination angles, from which structural engineers need to select the optimal scheme to meet the requirements of building safety and economy. However, due to the tight time and heavy workload of structural scheme selection, structural engineers are facing great challenges to find the optimal scheme in the expected time. Therefore, an integrated scheme comparison- and-selection method based on parametric modeling technology and multi-level fuzzy comprehensive evaluation is proposed, which realized the rapid modeling, analysis and scheme evaluation of the 400m super high-rise building automatically. The effects of different inclination angles on the overall performance and economy of the structure are accurately investigated. According to the analysis results, with the increase of facade inclination, the minimum shear weight ratio and stiffness weight ratio of structure first increase and then decrease, and the basic period and maximum story drift rotation first decrease and then increase. The proposed method greatly reduces the workload of structural scheme comparison and selection in the preliminary schematic design phase, improves the work efficiency, and has significant engineering application value.

1 INTRODUCTION

The project is a super high-rise building with a square architectural plane, 400m height, 87 floors, 4.3m standard floor height, and an above-ground total floor area of 198000m^2. The typical building plane is shown in Figure 1. The design reference period of the structure is 50 years, the fortification earthquake intensity is the 6th degree, the site category is class II, the design earthquake group is group II, the ground roughness category of the proposed site is class B, and the basic wind pressure with a return period of 50 years is 0.35kN/m^2.

In early phase of architectural schematic design, due to the need of architectural facade effect, the architect gives a variety of architectural schemes with different vertical inclination angles, as shown in Figure 2. The structural engineer needs to cooperate with the architect to select the best scheme. For different architectural schemes, the inclination angles of the facade vary between 0^0 and 2^0. In order to ensure the total floor area of the building unchanged, the heights of the inclined facades vary with the inclination angles. The larger the inclination angle is, the smaller the inclination height is. The facade parameters of each scheme are shown in Table 1.

 DOI 10.1201/9781003310884-29

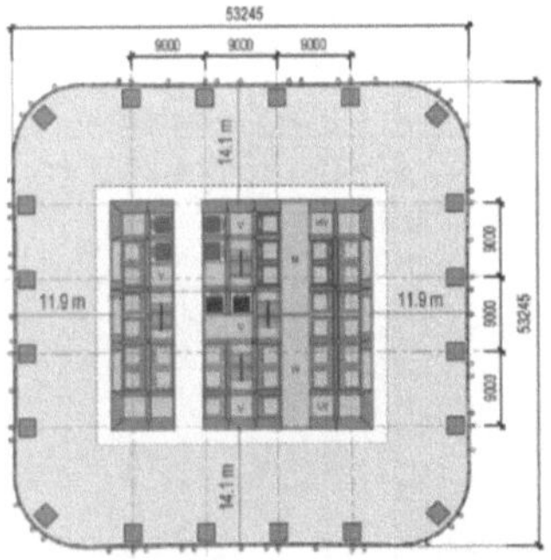

Figure 1. Typical building plane.

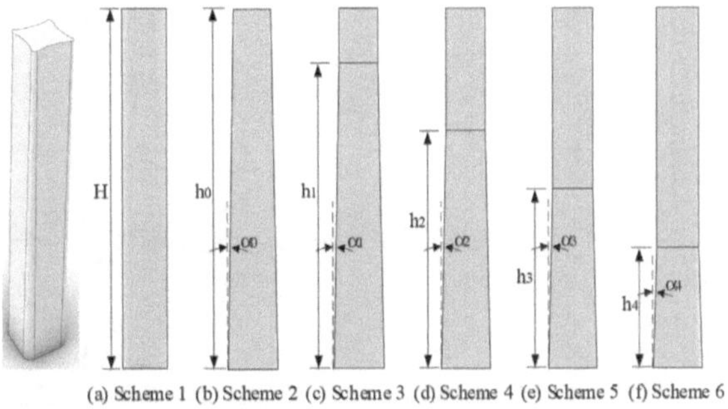

Figure 2. Architectural models of different schemes.

Table 1. Facade parameters of each scheme.

Architectural schemes	Scheme1	Scheme2	Scheme3	Scheme4	Scheme5	Scheme6
Inclination angle/°	0	1.0	1.1	1.2	1.4	1.8
Height of the inclined facade/m	0	h0 = 400	h1 = 339	h2 = 247	h3 = 200	h4 = 134

In order to obtain a reasonable structural scheme quickly, experts and scholars have carried out a lot of research. Ma Lingyong introduced neural network into structural scheme selection and developed an expert system for structural scheme selection based on MATALB neural network to assist designers in decision-making(Ma 2010); Lv Dagang applied the case-based reasoning method based on neural recognition to structural scheme intelligent selection, and developed the corresponding intelligent decision system to realize the structural selection decision in the conceptual phase (Lv 2003, 2004); Zhang JinFang selected a structural concept scheme of a super high-rise building based on analytic hierarchy process (Zhang 2021). These methods use statistics or regression algorithm to select the optimal scheme of structural form, without considering the specific layout form of the structure and analysis results, which is almost impossible to persuade the owner to accept the scheme in the actual project.

Therefore, in order to quickly investigate the impact of different inclination angles on the overall performance of the structure, an intelligent integrated scheme comparison-and-selection method based on parametric design technology and multi-level fuzzy comprehensive evaluation is proposed in this project. The specific steps of this method are as follows:

(1) By analyzing the characteristics of architectural schemes, sort out the structural modeling parameters, and use the parameter modeling technology to establish different structural analysis models;
(2) By secondary development technology, different structural models are analyzed by Etabs software automatically;
(3) By analyzing the calculation results of each structural scheme, the multi-level fuzzy comprehensive evaluation method is used to evaluate each structural scheme, and the optimal scheme is selected.

The whole process of modeling, calculation and scheme evaluation from architectural scheme to structural scheme is realized automatically by the proposed method. The whole process hardly needs human intervention, which greatly improves the efficiency of structural scheme comparison and selection. The process of intelligent comparison and selection of structural schemes is shown in Figure 3.

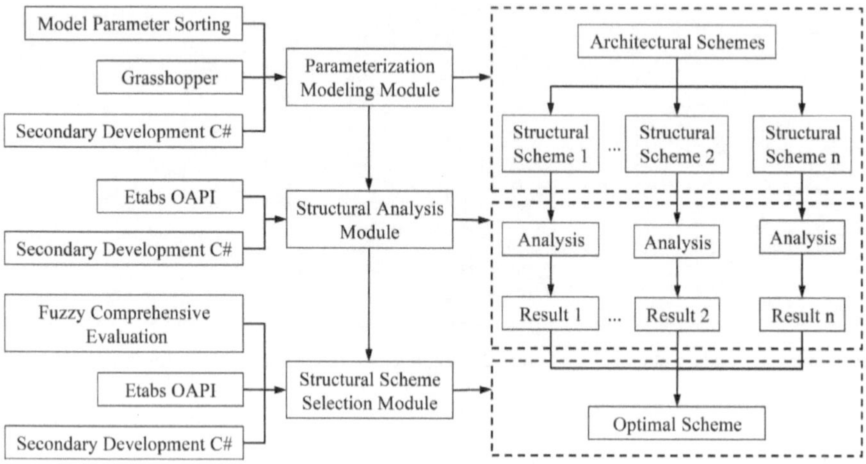

Figure 3. Flow chart of intelligent comparison and selection of structural schemes.

2 PARAMETRIC DESIGN

2.1 *Introduction of structural parametric design*

Parametric design is to convert the factors affecting the design results into parameters and form a parametric model with a certain logical relationship. The design methods of different design schemes can be quickly generated by modifying the design parameters (Chen 2018). For example, for projects with strong logic such as high-rise structures, the building surface, story heights, core tube parameters and the number of columns can be selected as input parameters, and the structural geometric model can be automatically generated after logical calculation.

Grasshopper is used as the parametric modeling software to create the parametric modeling program of the super high-rise building to generate the geometric models of different structural schemes. In addition, the structural attributes such as floor attributes, load attributes and section attributes are defined by the secondary development tool, which are given to the structural geometric model and assembled into the structural analysis model. Then the structural analysis software Etabs is called for calculation, and the structural parametric design flow is shown in Figure 4.

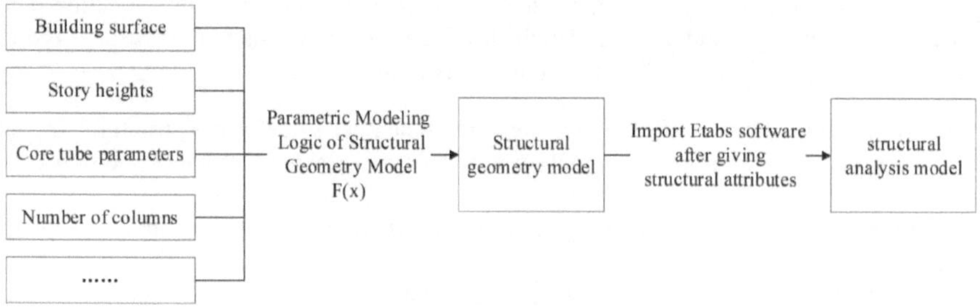

Figure 4. Structural parametric design flow.

2.2 *Structural geometric logic and parametric design*

The building facade of the project is an irregular curved surface. The bottom plane is a square with arc chamfer, while the top plane is a rounded curve with concave edges. The plane dimension is gradually retracted upward along the height, as shown in Figure 2. Considering the architectural requirements, characteristics and structural composition logic of the building, the parametric modeling will be divided into three parts: the core tube, the outer frame, and the floor beams.

(1) Create the core tube. The core tube generally has various plane forms and regular vertical layouts. During modeling, first draw the structural layout plan of the core tube manually, and distinguish the positioning lines of shear walls, coupling beams and frame beams, and then provide these plane positioning lines as input parameters to the core tube parametric modeling tool to generate the three-dimensional model of the core tube, as shown in Figure 5. Due to the reduction of the number of elevators and equipment rooms in the high area of the building, the core cylinder of the building has been retracted twice at the 33rd floor and 55th floor respectively, as shown in Figure 6.

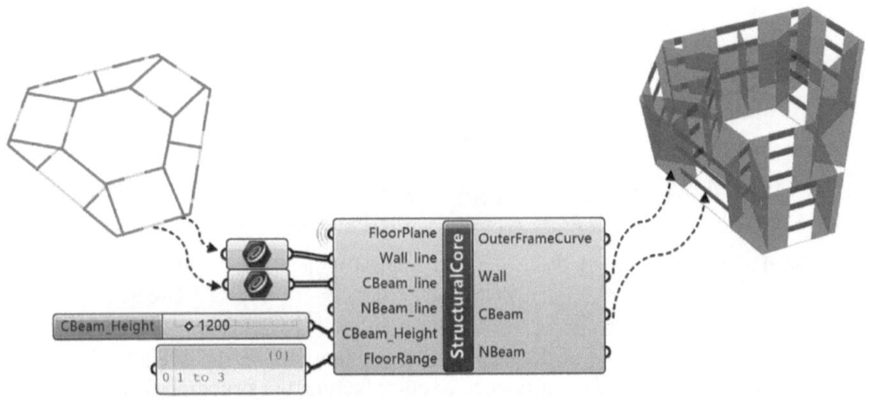

Figure 5. Core tube parametric modeling tool.

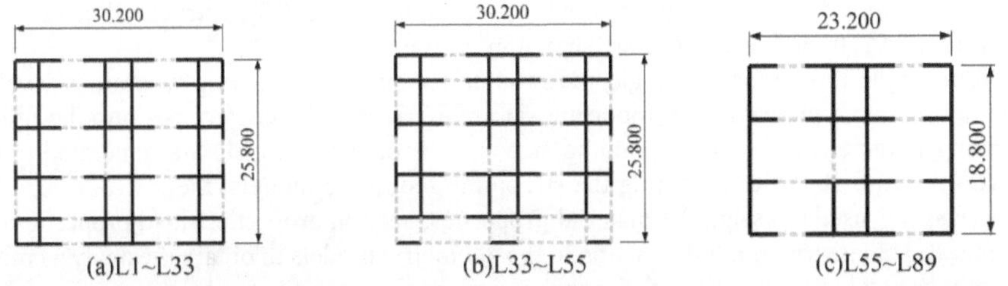

Figure 6. Layouts of core tube.

By inputting the shear wall positioning lines, coupling beam heights and corresponding floor ranges in the core tube plane of each partition (shown in Figure 6) as parameters to the core tube parametric modeling tool, the core tube geometric models of low, medium and high areas can be automatically generated, as shown in Figure 7.

(2) Create the outer frame. The outer frame is mainly composed of outer frame columns, outer frame beams and waist trusses. In order to ensure the outer frame fitting the building surface as much as possible, the building surface, the number of columns and the floors where the belt trusses are located are selected as input parameters, and the corresponding parametric modeling program is compiled in combination with the layout logic of the outer frame. Through the developed modeling program, the outer frame model under different design parameters can be quickly generated. As shown in Figure 8, frame1 is a 20 columns scheme with a belt truss in the middle floor, frame2 is a 16 columns scheme with a belt truss in the upper floor, and frame3 is a scheme obtained by changing the middle columns at bottom floors of frame2 into inclined columns.

(3) Create floor beams and slabs. According to the structural layout logic, create the floor beams between the core tube and the outer frame, and then create the slabs by core tube edge lines, floor beam lines and opening edge lines.

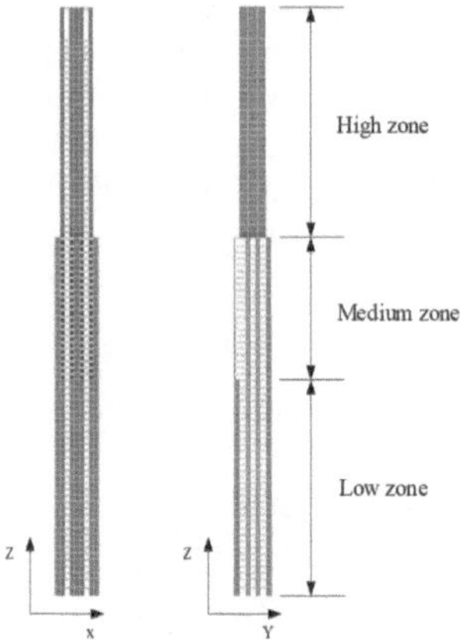

Figure 7. Model of the core tube.

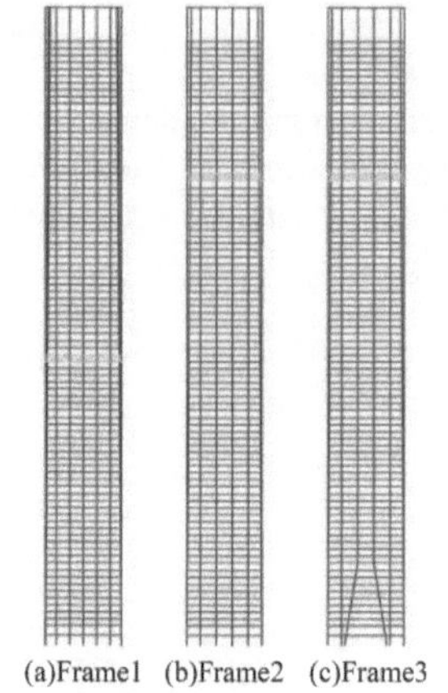

Figure 8. Models of the outer frame.

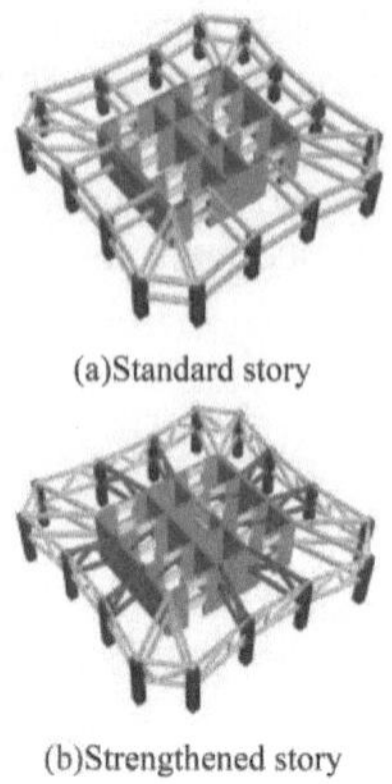

Figure 9. Structural floor layouts.

In order to study the influence of different facade inclinations on the overall performance of the structure accurately and avoid the influence of other factors, the structural form of all schemes adopts RC frame-core structure. In order to increase the overall stiffness of the structure, a structural strengthened story is set at the refuge floor (66F) according to preliminary trial calculation. The outer frame of the structure has 16 columns, of which the middle columns are inclined to both sides below the 12th floor due to building functions, as shown in Figure 8. The structural layouts of the standard story and strengthened story are shown in Figure 9.

According to the above modeling logic, the structural parametric modeling program is developed by using grasshopper platform. By importing six architectural scheme surfaces into the program as modeling parameters, the corresponding structural geometric models are generated quickly, as shown in Figure 10. After obtaining the structural geometric models, the self-developed tool swallow (ESD) is used to assign the material properties, section properties, load properties, node constraints and other structural information to the geometric models to obtain the analysis models, which are then send to Etabs software for calculation, as shown in Figure 11.

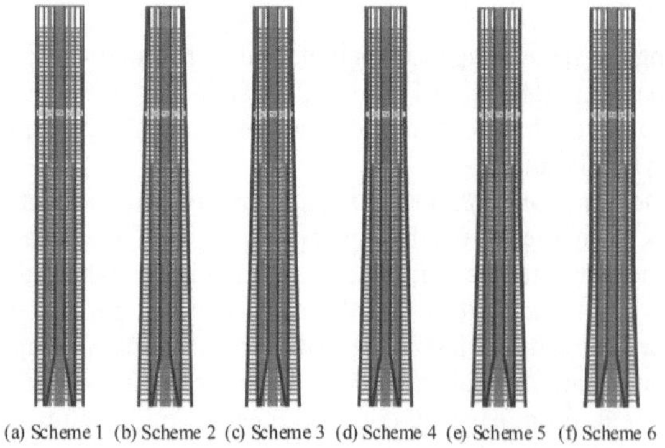

Figure 10. Structural models.

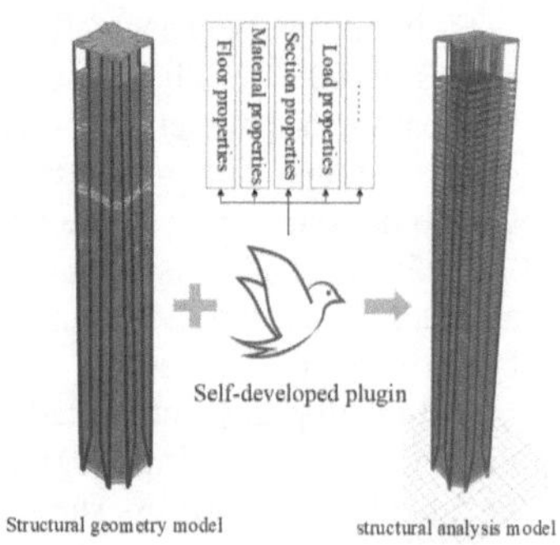

Figure 11. Data interaction between GP and Etabs.

3 THE EVALUATION STRATEGY OF THE STRUCTURAL SCHEMES

The comparison and selection of structural schemes is not a single structural design problem, but a multi-disciplinary, multi-objective and uncertain comprehensive decision-making problem. The most scientific method to solve this kind of problem is fuzzy mathematics (Yang 2011). For this project, the process of the optimal structure scheme selection by fuzzy mathematics is shown as Figure 12. And the basic steps are as follows: (1) Determine the factors affecting the optimal scheme selection, and establish the corresponding hierarchical relationships; (2) Finite element analysis of the structural models; (3) Confirm the membership and weight of factors based on the analysis results; (4) Select the optimal structural scheme by multi-level fuzzy comprehensive evaluation method.

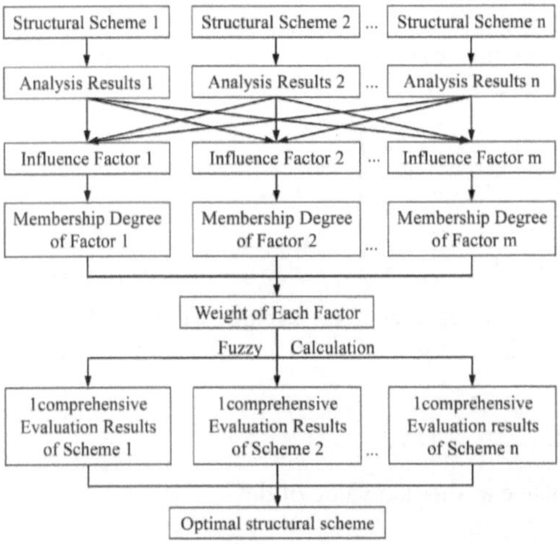

Figure 12. Process of the optimal structure scheme selection.

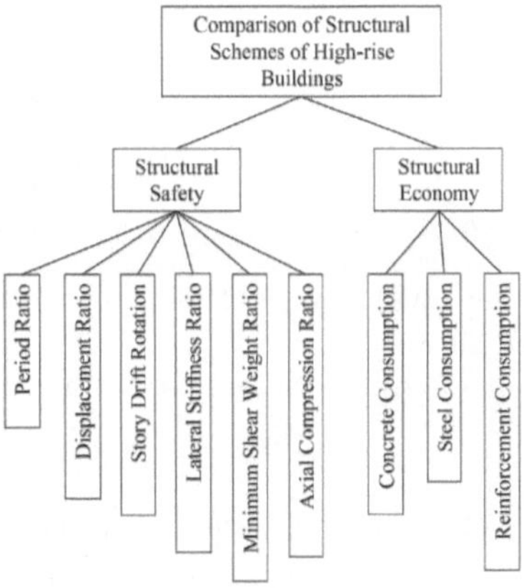

Figure 13. Scheme influence factors.

3.1 *Selection of influence factors and determination of their membership degrees*

There are many factors affecting the structural scheme selection, which can be divided into two categories: structural safety and economy (Du 2019; Feng 2013; Ma 2020; Wang 2020). Among them, structural safety refers to the overall performance indexes of the structure, including structural period ratio, displacement ratio, story drift rotation, minimum shear weight ratio, stiffness weight ratio, etc.; economic indexes refer to the cost of structural engineering, including the amount of concrete, reinforcement and steel, as shown in Figure 13.

There are many methods to calculate the membership degrees of influencing factors (Li 1994). For the index values of structural safety factors, China's specifications have clear limit requirements. The membership degree of each factor can be determined by comparing the factor's index value with its specification limit. If the index value meets the specification limit, the membership degree is 1.0. If the index value exceeds the specification, the membership degree is calculated according to the degree of exceeding the specification.

For economic indexes, the following formula can be used to calculate their membership degrees:

$$\mu_{ij} = 1 - \frac{x_{ij}}{x_{i\max} + x_{i\min}} \tag{1}$$

Where, μ_{ij} represents the membership degree of the j-th index of the i-th scheme, and $x_{i\min}$ and $x_{i\max}$ respectively represent the minimum and maximum values of the i-th index in all schemes.

3.2 *Confirm the weights of the influence factors*

In order to reflect the importance of each influence factor to the evaluation object, it is necessary to give corresponding weight to each influence factor. The greater the weight, the higher the importance. In practical, commonly used weight calculation methods include expert investigation method, Delphi method and analytic hierarchy process (AHP) (Saaty 1980). For this project, the weight of each factor is determined by AHP, which is mainly divided into the following four steps: (1) Determine the factor hierarchy system according to the hierarchical relationship of factors; (2) Establish the judgment matrix by comparing factors with each other; (3) Check the consistency of the judgment matrix; (4) Calculate the weights of the factors by the judgment matrix.

3.3 *Multi-level fuzzy comprehensive evaluation method*

3.3.1 *Single-level fuzzy evaluation*

For a single level factor set $X=(x_1,x_2,\cdots,x_n)$, suppose the evaluation matrix of the factor x_i is $R_i=\{r_{i1},r_{i2},\cdots,r_{im}\}$, where represents the membership degree of the i-th factor to the j-th scheme, and so on, the total evaluation matrix R of n factors is:

$$R=\begin{Bmatrix} R_1 \\ R_2 \\ \vdots \\ R_n \end{Bmatrix}=\begin{Bmatrix} r_{11} & r_{12} & \cdots & r_{1m} \\ r_{21} & r_{22} & \cdots & r_{2m} \\ \vdots & \vdots & \cdots & \vdots \\ r_{n1} & r_{n2} & \cdots & r_{nm} \end{Bmatrix} \tag{2}$$

Set weight matrix $V=(v_1,v_2,\cdots,v_n)$, Where v_i represents the weight of the i-th factor, and all weights meet $\sum_{i=1}^{n} v_i=1$. According to the fuzzy mathematical relation equation, the comprehensive evaluation results are obtained:

$$B=V\cdot R=(b_1,b_2,\cdots,b_m) \tag{3}$$

Each member b_j in matrix B is the comprehensive evaluation value of the j-th scheme.

3.3.2 *Multi-level fuzzy evaluation*

For this project, there are more than one level of evaluation factors. Therefore, the factor set of each layer will be evaluated from the bottom level to the top level. Taking two-level factor set $X=(X_1,X_2,\cdots,X_n)$ as an example, for factor subset x_i, the corresponding evaluation result matrix $B_i=(b_{i1},b_{i2},\cdots,b_{im})$ can be obtained by single-level fuzzy evaluation, and all B_i constitute the final comprehensive evaluation matrix $\tilde{R}$ of the factor set X:

$$\tilde{R}=\begin{Bmatrix} B_1 \\ B_2 \\ \vdots \\ B_n \end{Bmatrix}=\begin{Bmatrix} b_{11} & b_{12} & \cdots & b_{1m} \\ b_{21} & b_{22} & \cdots & b_{2m} \\ \vdots & \vdots & \cdots & \vdots \\ b_{n1} & b_{n2} & \cdots & b_{nm} \end{Bmatrix} \tag{4}$$

Suppose the weight matrix $\tilde{V}=(\tilde{v}_1,\tilde{v}_2,\cdots,\tilde{v}_n)$, where $\tilde{v}_i$ represents the weight of the factor subset X_i. According to the fuzzy mathematical relation equation, the comprehensive evaluation result matrix $\tilde{B}$ of factor set X is as follows:

$$\tilde{B}=\tilde{V}\cdot\tilde{R}=\left(\tilde{b}_1,\tilde{b}_2,\cdots,\tilde{b}_m\right) \tag{5}$$

And so on, more levels of fuzzy evaluation can be carried out.

3.4 *Development of structural scheme comparison program*

In order to realize the automatic comparison and selection of the structural schemes and improve work efficiency, a structure scheme comparison and selection program based on multi-level fuzzy comprehensive evaluation is developed by object-oriented development technology and Etabs OAPI. The overall design framework of the program is shown as Figure 14, which can be divided into two parts:

(1) Extraction of the calculation results. By reading the Etabs analysis results of the structural models, the overall performance indexes of the structures are extracted, and then written into SQLite database file.
(2) Comparison and selection of the structural schemes. Read the results of each structure from the SQLite database file, and then select the optimal structural scheme by the multi-level fuzzy comprehensive evaluation method.

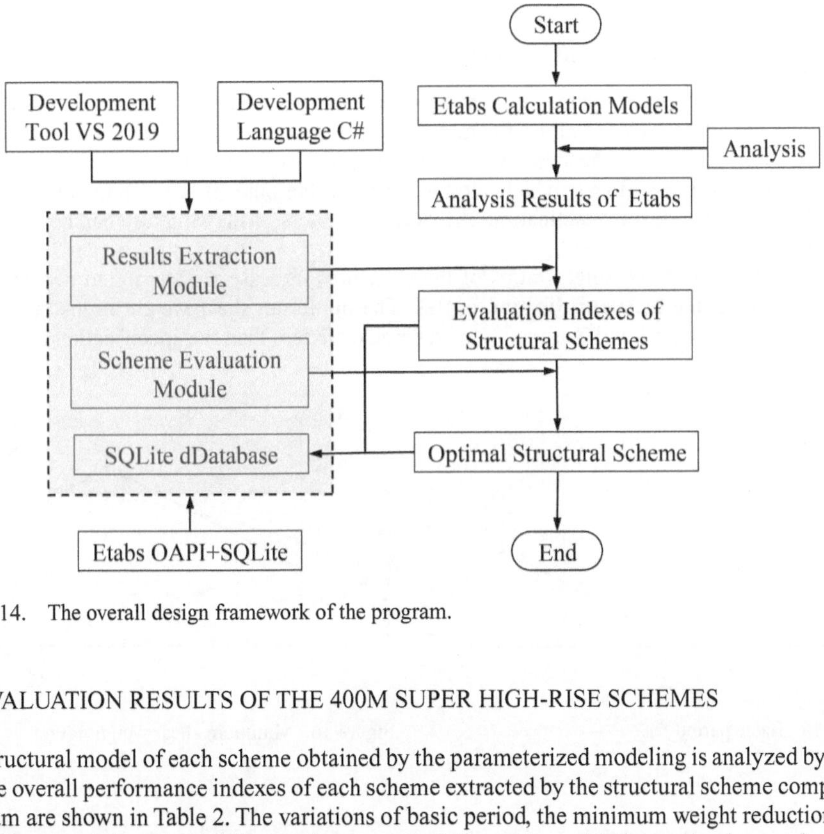

Figure 14. The overall design framework of the program.

4 EVALUATION RESULTS OF THE 400M SUPER HIGH-RISE SCHEMES

The structural model of each scheme obtained by the parameterized modeling is analyzed by Etabs, and the overall performance indexes of each scheme extracted by the structural scheme comparison program are shown in Table 2. The variations of basic period, the minimum weight reduction ratio, the stiffness weight ratio and the maximum story drift rotation with the inclination angles of the building facade are shown in Figure 15~Figure 18.

Table 2. Statistical results of performance indexes of each scheme.

Overall performance indexes		Scheme1	Scheme2	Scheme3	Scheme4	Scheme5	Scheme6
Period/s	T1	6.844	6.164	6.120	6.170	6.258	6.422
	T2	6.288	5.696	5.657	5.701	5.775	5.914
	T3	1.883	1.776	1.783	1.799	1.802	1.81
Minimum shear weight ratio	X	0.467	0.488	0.489	0.487	0.484	0.478
	Y	0.473	0.483	0.484	0.484	0.483	0.481
Structural stiffness	X	1.776	1.947	1.954	1.959	1.953	1.917

(*continued*)

Table 2. Continued.

Overall performance indexes		Scheme1	Scheme2	Scheme3	Scheme4	Scheme5	Scheme6
weight ratio	Y	1.494	1.662	1.668	1.669	1.658	1.622
Maximum story drift rotation	X	1/2519	1/2898	1/2898	1/2832	1/2762	1/2703
under earthquake	Y	1/2100	1/2577	1/2583	1/2492	1/2398	1/2298
Maximum story drift	X	1/1216	1/1545	1/1536	1/1464	1/1398	1/1331
rotation under wind	Y	1/1011	1/1315	1/1307	1/1239	1/1175	1/1114
Displacement ratio	X	1.15	1.08	1.08	1.08	1.08	1.09
	Y	1.16	1.09	1.09	1.09	1.09	1.09

According to the analysis results, the following conclusions can be drawn:

(1) The basic periods of the structures first decrease and then increase with the increase of the facade-inclination angles. Among them, scheme3 (i.e. facade inclination angle of 1.1°) has the smallest period and the best overall stiffness, and scheme1 (i.e. vertical scheme) has the smallest overall stiffness, which shows that when the outer frame columns of the structure have a certain inward inclination, the overall stiffness of the structure can be effectively improved.

(2) The minimum stiffness weight ratios of the structures increase first and then decrease with the increase of the facade-inclination angles. The minimum shear weight ratios in X and Y directions of scheme 1 and X direction of scheme 6 are less than the specification limit $0.8 \times 0.006 = 0.48\%$.

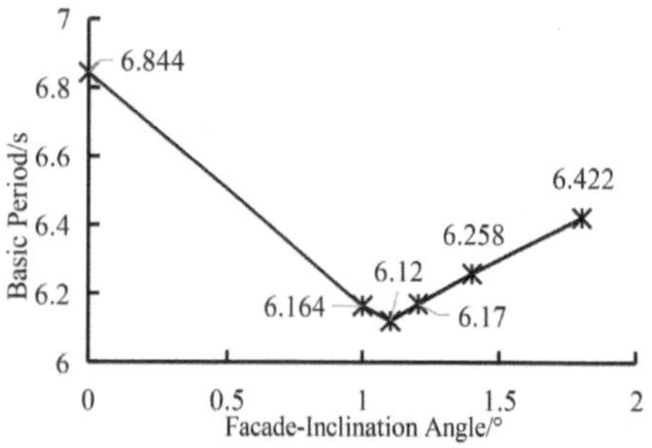

Figure 15. Basic period T1.

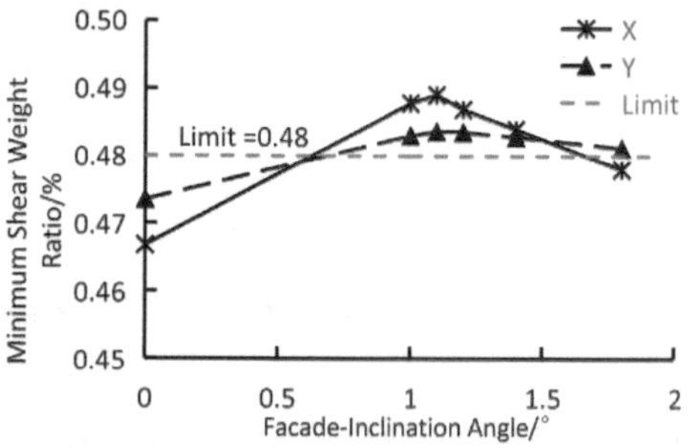

Figure 16. Minimum shear weight ratio.

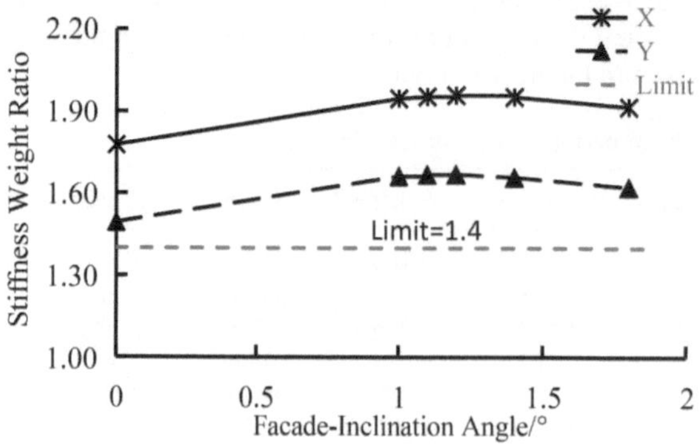

Figure 17. Structural stiffness weight ratio.

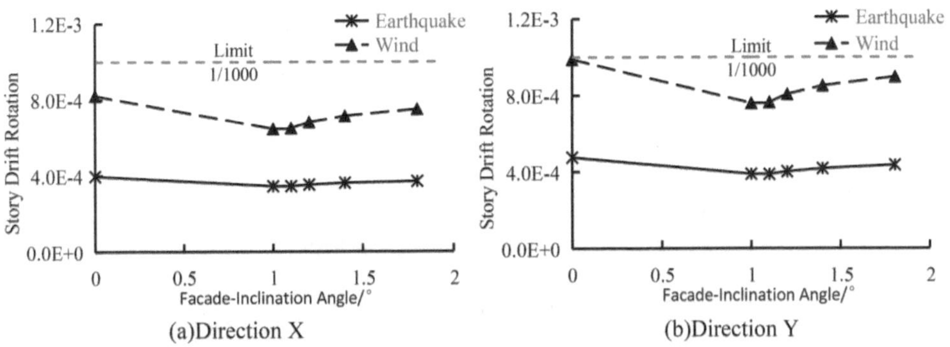

Figure 18. Maximum story drift rotation.

(3) The structural stiffness to weight ratios first increase and then decrease with the increase of the facade-inclination angles, and all structural schemes meet the code requirement 1.4.
(4) Under the earthquake and wind, the maximum story drift rotation of each scheme is less than the code limit 1/1000, and the lateral displacements of the structures are controlled by wind. The story drift rotation angle curve of each scheme under the wind is shown in Figure 19.

The consumption of concrete and reinforcement of each scheme calculated by the developed program are shown in Table 3, and the project cost of each structural scheme is shown in Figure 20. According to the results, the project cost of scheme 3 is the lowest and that of scheme 6 is the highest, with a difference of 10.41 million yuan.

By the developed program, the evaluation score of each structural scheme is calculated by comprehensive evaluating the overall performance indexes and economic indexes, as shown in Figure 21. According to the results, scheme3 has a better overall performance and economy, and its comprehensive score is the highest; the evaluation scores of scheme2, scheme4 and scheme5 are similar; scheme6 has lower score for its poor overall performance and economy; scheme1 has the lowest score, for its bad minimum shear weight ratio, which does not meet the code requirement. Therefore, scheme3 is the optimal scheme.

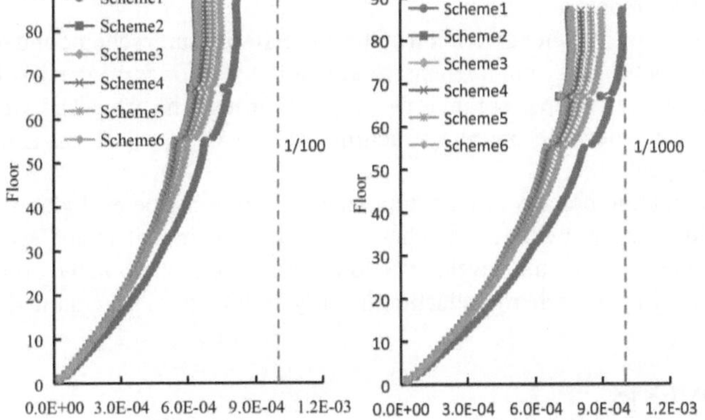

Figure 19. The story drift rotation angle curves under wind.

Table 3. Statistical results of material consumption of different schemes.

Material consumption	Concrete/ m^3	Reinforcement/t
Scheme1	1228e5	2216e4
Scheme2	1214e5	2241e4
Scheme3	1213e5	2147e4
Scheme4	1209e5	2163e4
Scheme5	1206e5	2165e4
Scheme6	1206e5	2313e4

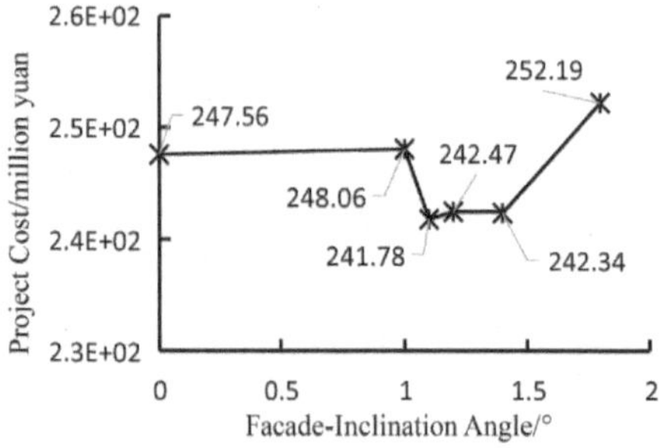

Figure 20. Project cost of each scheme.

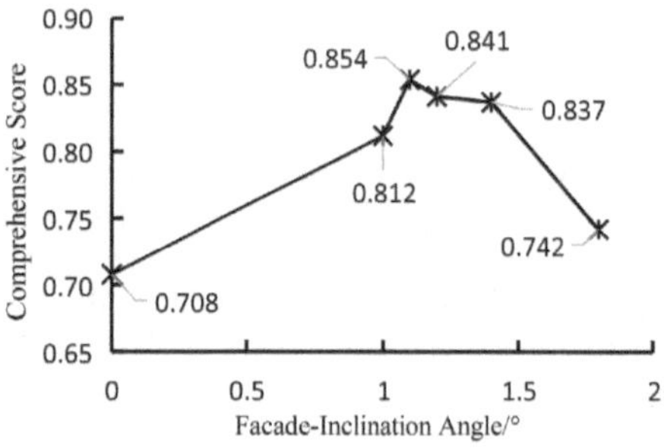

Figure 21. Comprehensive evaluation Score.

5 CONCLUSION

Based on the results and discussions presented above, the conclusions are obtained as below:

(1) According to the building facade schemes, the variations of the overall performance and project cost of the building with the facade-inclination angles are studied. According to the results, when the outer frame columns of the structure have a certain inward inclination, the overall stiffness of the structure can be effectively improved.

(2) According to different building facade schemes, different structural analysis models are quickly established by parametric modeling technology, which greatly improves the efficiency of structural modeling and realizes the seamless connection between architectural scheme and structural analysis model.

(3) In order to realize the intelligent selection of the optimal structural scheme and improve the work efficiency, multi-level fuzzy comprehensive evaluation method is adopted, and corresponding structural scheme evaluation program is developed. Through the project application, the developed program can select the optimal structural scheme well, which has a wide engineering application prospect.

(4) The whole automation process of modeling, analysis and scheme evaluation of a 400m super high-rise building from architecture scheme to structural model is achieved by parametric modeling technology and multi-level fuzzy comprehensive evaluation method, which provides a way for rapid structural scheme selection in early architectural schematic design phase.

ACKNOWLEDGMENTS

This work was financially supported by Wuhan Science and Technology Planning Project of Hubei Province (2019010702011282).

REFERENCES

Dagang Lv, Fangliang Xing, Shihai Zhang, Guangyuan Wang (2003). Scheme generation system for structural form selection based on neural pattern recognition. *Journal of Harbin Institute of Technology*: 1418–1421.

Dagang Lv, Fangliang Xing, Shihai Zhang, Guangyuan Wang (2004). Intelligent decision support system for high-rise building structural form selection. *Journal of Harbin Institute of Technology*: 205–208.

Fangxi Wang (2020). Comparison and selection of strengthening layer setting schemes for a super high-rise building—Taking a 54 story commercial and residential building as an example. *Fujian Architecture & Construction*, 266: 72–76.

Hongxing Li, Peizhuang Wang (1994). *Fuzzy Mathematics*. National Defense Industry Press.

Jikun Chen, Zhu Wang and Yuansheng Zhang (2018). The Parametric from design method of high-rise building. *Urbanism and Architecture*: 106–109.

Jinfang Zhang, Longli Yu, Zhijiong Zhou, Hongdou Chen (2021). Structural selection of super high-rise building based on analytic hierarchy process. *Building Structure*, 51:358–362.

Jingwen Feng, Bo Yu (2013). Comparison and selection for structural scheme of a super high-building in Guiyang. *Building Structure*, 43:226–230.

Lingyong Ma, Gongwei Liu, Wei Jiang (2010). A Study of the Structural Form Selection of High-rise Building with MATLAB Neural Network. *Journal of Information Technology in Civil Engineering and Architecture*: 14–19.

T.L. Saaty (1980). The Analytic Hierarchy Process (AHP), McGraw-Hill Company.

Xiaoyu Ma, Weifeng Wang (2020). Structural system selection for a super high-rise building. *Building Structure*, 50: 18–21.

Yang Lunbiao (2011). *Principles and applications of Fuzzy Mathematics* (5th Edition). South China University of Technology Press.

Zhenyu Du, Zuqiang Liu, Jianyang Xue, Zheng Luo (2019). Structure scheme comparison of a high-rise hotel in eight seismic degree area. *Building Structure*, 49:68–72.

Advances in Civil Engineering: Structural Seismic Resistance, Monitoring and Detection – Mohd Johari Mohd Yusof and Junwen Zhang (Eds)
© 2023 The Authors, ISBN 978-1-032-31491-4

Research on the stress and deformation of the support in the foundation pit

J.P. Liu, W.K. Zhang, S. Yang, D. Wang, Z.J. Hu, H.X. Song & H.Y. You
Shandong University of Science and Technology, Qingdao City, Shandong, China

ABSTRACT: In supporting the deep foundation pit of the sand-filled artificial foundation, it is difficult to describe its characteristics by the force and deformation law of the support in the conventional foundation pit. In this paper, numerical simulation and on-site monitoring are combined to analyze the stress and deformation characteristics of the internal support during the excavation of the foundation pit under the conditions of different support layout schemes. The change in the support axial force increases linearly in the short term. (2) At the same monitoring point, the horizontal displacement of the underground diaphragm wall is positively correlated with the change in the inner support axial force. (3) Adjusting the design scheme of the foundation pit support structure will lead to maximum axial force of the inner support and the deformation of the diaphragm wall.

1 INTRODUCTION

With continuous growth in China's economy, the connection between cities has become increasingly close. However, the dense urban layout has made the construction of intercity transportation a problem, restricting the regional economic development of China, especially in the developed coastal areas, and the urban construction has been nearly saturated. To give full play to the stratigraphic and lithologic advantages of coastal cities, promote the construction of urban transportation networks, and solve the difficulty of land use, it is urgent to develop intercity rail construction. The deep foundation pit is the key issue of the intercity track. Especially, under complex geological conditions, how to select the safest, economic, and reasonable deep foundation pit support scheme is the core issue required to ensure safe construction of the project, which needs to be studied in depth (Liang et al. 2018; Liu et al. 2018; Su et al. 2021; Peng et al. 2020).

Rezvani et al. (Chen et al. 2017; Luo et al. 2011; Rezvani et al. 2021) better controlled the deformation of the foundation pit by selecting a reasonable internal support design scheme and real-time monitoring of the site. Yang et al. (Yang et al. 2013) proposed a new type of support structure in the foundation pit composed of prefabricated steel structural units, from the support structure to the ring on-site monitoring tests of stress, soil deformation, and lateral soil pressure, etc. Analysis showed that the support structure in the new foundation pit is safer, more convenient, and reliable. He et al. (He et al. 2013) took a foundation pit project in Beijing Olympic Park as an example, with limited rods, based on the calculation results of the element method, the force and deformation characteristics of the retaining structure during the excavation of the foundation pit were analyzed.

It can be seen that many scholars at home and abroad have conducted in-depth studies on the influencing factors of foundation pit deformation under different supporting schemes and excavation methods, and have optimized the supporting design schemes of foundation pits to different degrees according to the research results. But, at present, about the steel support in foundation pit supporting, the research for the global stability of foundation pit support position

 DOI 10.1201/9781003310884-30

is less concrete. This paper takes Xiamen Intercity Railway Line R1 airport engineering for deep foundation pit engineering, combined with the field monitoring measurement, support for steel and concrete under the condition of different position of deep foundation pit excavation are numerically simulated analysis. The variation characteristics of internal force and displacement of the internal bracing under various bracing schemes are revealed, and the optimization measures of the bracing schemes are developed according to the research results, which provides a reference for similar deep foundation pit supporting projects.

2 PROJECT OVERVIEW AND GEOLOGICAL CONDITIONS

2.1 *Project overview*

The deep foundation pit project of the airport section of Xiamen Intercity Railway Line R1 is located in the Dadeng Reclamation Area to the east of Dadeng Island in Xiang'an District, Xiamen City, Fujian Province, China, with Nan'an City in Quanzhou in the north, Jinmen Island in the south, Xiamen Island in the west, and Jiaoyu Island in the east. The geographic location map of this project is shown in Figure 1.

The corresponding mileage of the open-cut foundation pit project of the airport section of the R1 line of the intercity railway is DK93+050-DK94+050. The object in this paper is the DK93+130-DK93+450 section that has been excavated.

Figure 1. Engineering geographical location diagram.

2.2 *Subsection*

The project is located under the T1 terminal building and GTC and commercial development plots of China Xiamen Xiang'an International Airport. The original landform unit belongs to the coastal accumulation area and is an intertidal zone. Later, due to the need for airport construction, the site was filled and transformed into the current situation. The current landform is a reclamation area of about 1.9 km in the center, with an elevation of 5 m–15 m; the two ends are 500 m and 800 m, respectively, which have not yet been reclaimed but are still intertidal zones, and the seabed elevation is -0.7 m–3 m. The reclaimed section of the site is mainly composed of artificial fill (backfilled sand) layers, forming an artificial reclamation area 10 m–15 m higher than the original ground. The site is undergoing preloading with heap loading, which has large elevation changes and complex geological conditions. The geotechnical physical and mechanical parameters are shown in Table 1.

Table 1. Geotechnical physical and mechanical parameters.

Geotechnical name	Thickness	Bulk density	Cohesion	Internal friction angle	Deformation modulus	Poisson's ratio
	m	kN/m^3	kPa	°	MPa	
Backfill sand	3.70	17.5	8	7.5	5	0.30
Silty sand	2.30	17.5	10	8	5	0.30
Silty clay	7.61	18.5	15	15	10	0.30
Cohesive soil	9.30	19.0	18	18	15	0.30
Granite common rock (fully weathered)	6.00	25.0	35	70	20	0.30
Granite common rock (fully weathered granular)	5.00	21.0	27	40	40	0.30
Granite common rock (medium weathering)	5.60	25.0	22	70	70	0.28
Diabase (strongly weathered)	8.49	24.0	20	65	60	0.25

3 SIMULATION CALCULATION OF INTERNAL SUPPORT AXIAL FORCE

3.1 *Calculation model and parameters*

In this paper, finite element software is used to simulate the construction conditions of the foundation pit, and the modified Mohr-Coulomb constitutive relationship is adopted for the soil. The soil parameters in the model fully consider the indicators provided by the geological survey report, as shown in Table 1. To simulate the actual working conditions more accurately, a three-dimensional model is used for simulation. Since the foundation pit is regular and symmetrical, 1/4 of the foundation pit is used for modeling as shown in Figure 2.

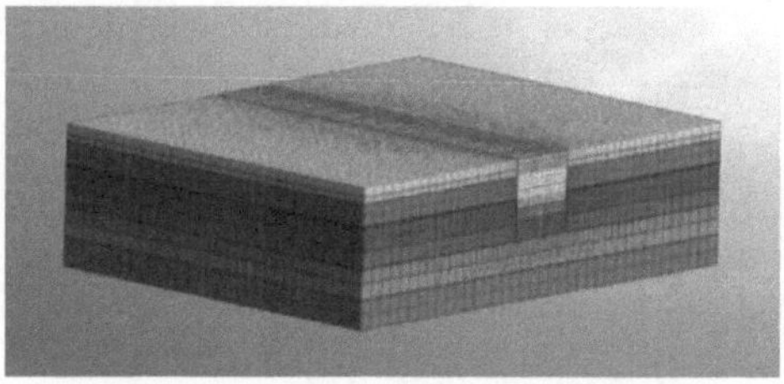

Figure 2. Three-dimensional model of foundation pit retaining structure.

The length, width, and height of the foundation pit model are 88 m, 25 m, and 21.5 m, respectively. Materials such as internal supports and underground continuous walls are regarded as continuous and average. The internal supports and underground continuous walls are simulated by beam elements and slab elements, respectively, and elastic models are selected for constitutive relations. The material parameters of the supporting structure are shown in Table 2.

Table 2. Parameters of foundation pit enclosure structure.

Structure name Steel support	Elastic Modulus	Poisson's ratio	Sectional form	Layout
Reinforced concrete support	30	0.2	Rectangle, 800*1000 mm	Spacing 6 m
Steel bracing	200	0.25	Round tube, diameter 800 mm, wall thickness 20 mm	Spacing 6 m
Diaphragm Wall	30	0.2	Plate, thickness 1000 mm	

3.2 *The design scheme of the foundation pit supporting structure*

The deep foundation pit of the airport section of Xiamen Intercity Railway R1 has a width of 25 m and an excavation depth of 21.5 m. It is located in coastal beach geology and has complex stratum structure conditions. The excavation depth is below the confined water level, which is prone to dangers such as water seepage, soil liquefaction, and collapse, and the construction risk coefficient is high. Safe and stable foundation pit supporting structure has become particularly important, hence the deep foundation pit engineering project department of the airport section of Xiamen Intercity Railway R1 line proposed two kinds of foundation pit supporting structure design schemes (see Figure 3). Both schemes adopt the underground continuous wall design, the thickness is 1000 mm, the wall depth is about 42 m, the wall is embedded with diabase (strongly weathered) not less than 1 m, and the crown beam is poured on the top of the underground continuous wall. The number of supports in the two kinds of foundation pit supporting structure design schemes is both: 3 reinforced concrete supports and 2 steel supports.

In the first kind of foundation pit supporting structure design scheme (Scheme 1), the first, second, fourth roads are supported by reinforced concrete with a length of 1000 mm and a width of 800 mm, and the horizontal spacing of each support is 6 m. The third channel and the fifth channel are supported by steel with a wall thickness of 20 mm and a diameter of 800 mm, and the horizontal spacing of each channel is 6 m. The cross-sectional view of the foundation pit supporting structure is shown in Figure 3(a).

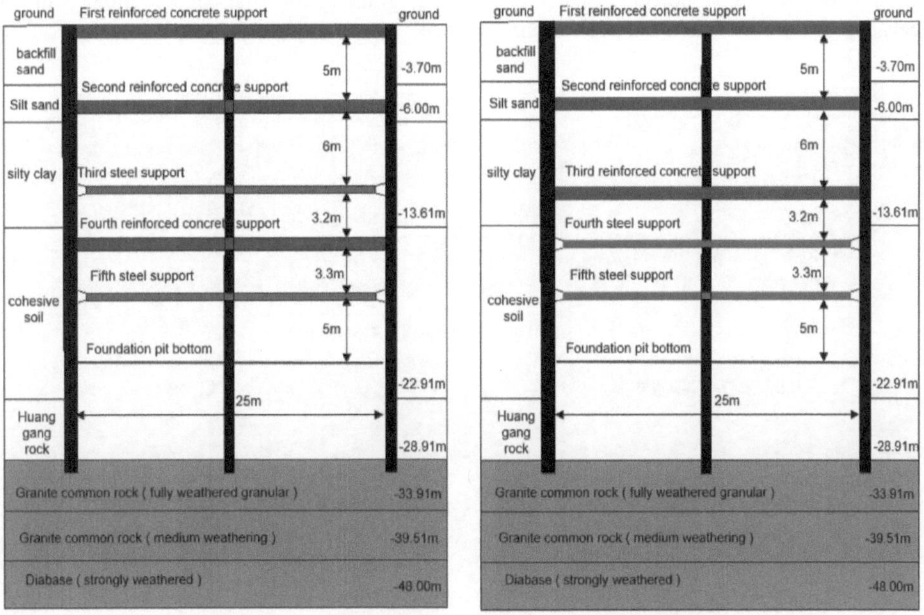

(a) Scheme1 (b) Scheme 2

Figure 3. Cross-sectional view of foundation pit supporting structure.

Scheme 1: first calculate the initial ground stress of the soil to be excavated, and carry out the construction of the diaphragm wall, and then carry out the excavation of the foundation pit five times:

(1) Excavate the first layer of the foundation pit to a height of 5 m, and pour the first reinforced concrete support.
(2) Excavate the second layer of the foundation pit to a height of 11 m, and pour the second reinforced concrete support.

(3) Excavate the third layer of the foundation pit to 14.2 m, and erect the third steel bracing;
(4) Excavate the fourth layer of the foundation pit to a height of 17.5 m, and pour the fourth reinforced concrete support.
(5) Excavate the fifth layer of soil of the foundation pit to 21.5 m from the bottom of the pit, and erect the fifth steel bracing, then clear the foundation pit, carry out waterproof cushion construction, steel banding, and foundation slab pouring.

With other conditions unchanged, we obtained the second type of foundation pit supporting structure design scheme (Scheme 2) by adjusting the positions of the third steel support and the fourth reinforced concrete support. The cross-sectional view of the foundation pit supporting structure is shown in Figure 3(b).

In accordance with the excavation principle of "layered excavation, support first and then dig", this model will excavate five times, and the excavation depth of each stage is consistent with the actual working conditions. Before the excavation of each foundation pit, the support is activated first. The excavation stage of the two options remains the same.

3.3 *Scheme 1 support axial force simulation results and analysis*

According to the design of the internal support structure of the first scheme, the simulation software is used to obtain the supporting axial force cloud diagram of the first scheme along with the excavation of the foundation pit, as shown in Figure 4.

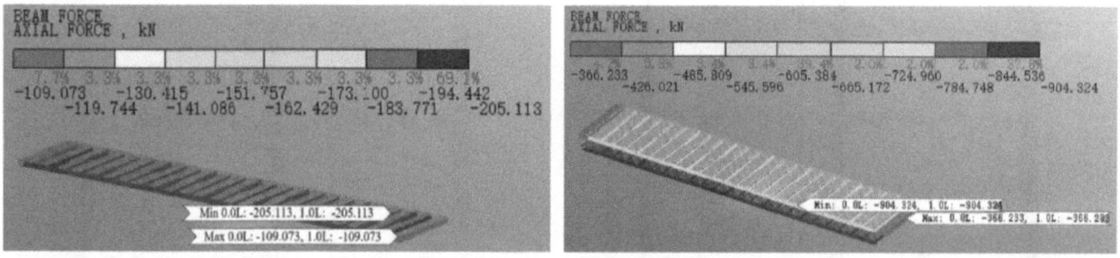

(a) First excavation (b) Second excavation

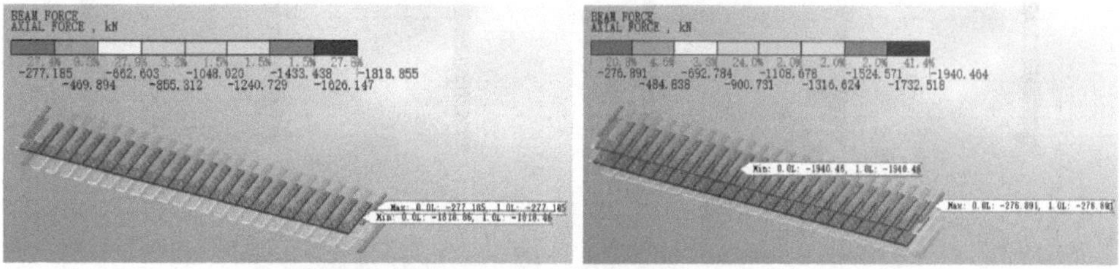

(c) Third excavation (d) Fourth excavation

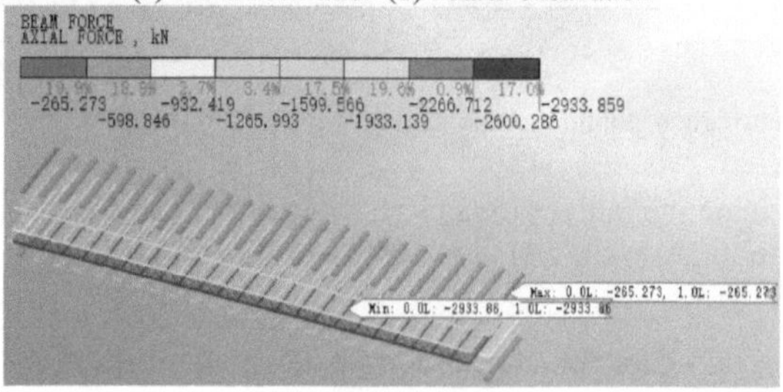

(e) Fifth excavation

Figure 4. The axial force cloud image of scheme one.

Through the axial force simulation of the foundation pit excavation process of Scheme 1, it was found that the axial force value increased significantly when the reinforced concrete support and the steel support were installed at the same monitoring point. However, after the excavation to the bottom of the pit, the internal supports have been fully installed, and the axial force has slowly increased and stabilized. Under the influence of the space-time effect of the foundation pit, the deformation of the underground continuous wall is the largest at the supporting position in the third channel of the foundation pit. Therefore, among the five excavation internal support axial forces calculated by simulation, the third steel support axial force is the largest, reaching 2933.86 kN, which has seriously exceeded the warning value, (the steel support warning value is 1995 kN, which is 70% of the design value). It can be seen that Scheme 1 is not an ideal foundation pit support structure design.

3.4 *Scheme 2 support axial force simulation results and analysis*

We also carried out a numerical simulation on scheme two and obtained the axial force cloud chart of scheme two along with the excavation of the foundation pit as shown in Figure 5. Because only the position of the third steel support and the fourth reinforced concrete support is adjusted, while other conditions remain unchanged, the first design of the two supporting structures of the foundation pit is consistent with the second axial force cloud map. Therefore, only the axial force cloud map of the last three foundation pit excavations is shown in Figure 5. By analyzing the data of the axial force cloud chart, we can easily find that the maximum axial force in Scheme 2 also occurs in the third internal support during the excavation of the foundation pit. The maximum value is 2807.73 kN, which is only 56.15% of the design value, 80.22% of the warning value (the warning value of reinforced concrete support is 3500 kN, which is 70% of the design value), and the overall internal support axial force changes more smoothly and reasonably.

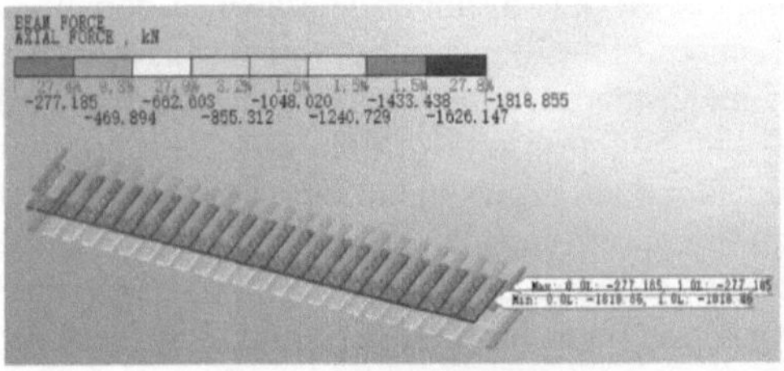

(a) Third excavation

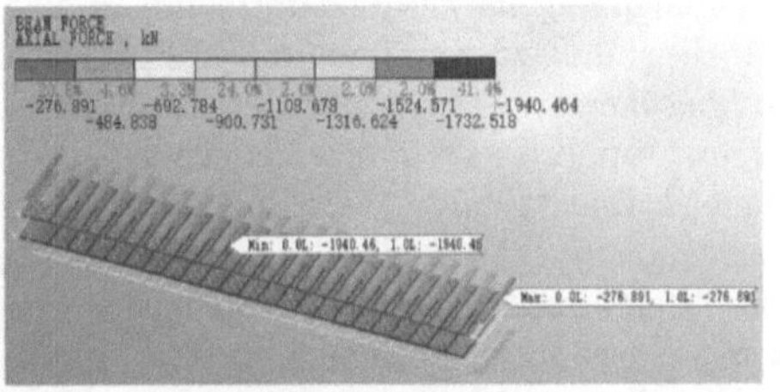

(b) Fourth excavation

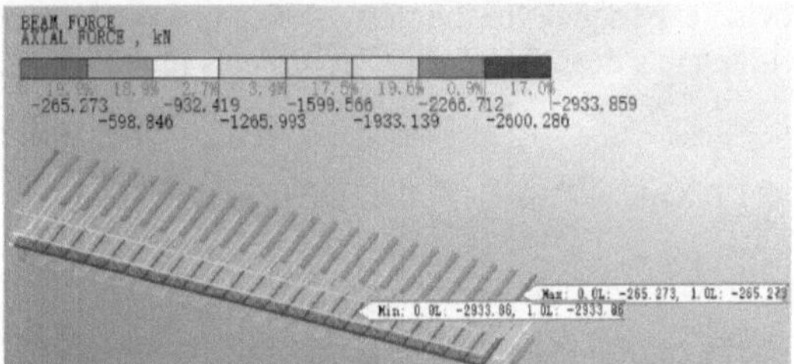

(c) Fifth excavation

Figure 5. The axial force cloud image of scheme one.

3.5 *Comparative analysis on deformation of diaphragm wall of two supporting schemes*

The author then simulates the deformation of the diaphragm wall of the two design schemes of foundation pit supporting structure, and it was seen that after the excavation of the foundation pit to the bottom of the pit, the deformation of the diaphragm wall of the two supporting schemes is at the support position in the third internal support. The deformation reaches the maximum at the excavation depth of 10 m and gradually decreases. Near the bottom of the pit, the deformation stabilizes. The deformation displacement curve of the diaphragm wall is a large arc in the middle and small at both ends. The deformation and displacement curve of the underground diaphragm wall is an arch curve with large middle and small ends. The maximum deformation and displacement of the underground diaphragm wall in Scheme 1 and Scheme 2 are 19.39 mm and 18.34 mm, respectively, which are far less than the warning value (the warning value is 40 mm). The deformation and displacement of the diaphragm wall in the two design schemes of foundation pit supporting structures are shown in Figure 6.

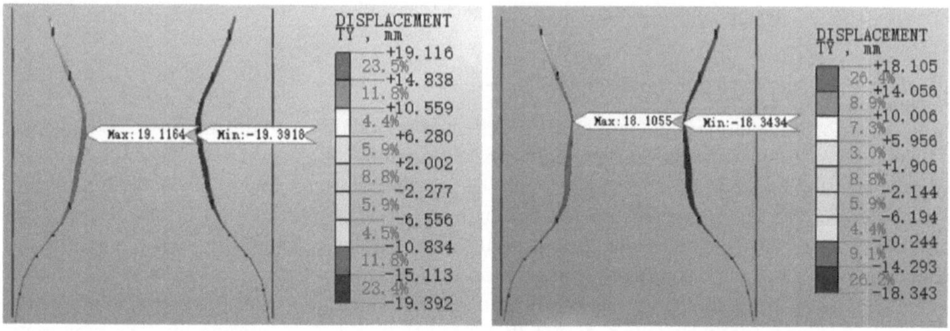

(a) Scheme 1 (b) Scheme 2

Figure 6. Deformation and displacement diagram of diaphragm wall of two schemes.

Through the analysis on deformation of diaphragm wall of two supporting schemes, combined with the simulated axial force analysis of the two sections of the foundation pit supporting structure design scheme, we finally suggest choosing Scheme 2, that is, using the first, second, and third internal supports as reinforced concrete supports, and the fourth and fifth The internal support of the road is steel support as the final design scheme of foundation pit supporting structure for actual foundation pit construction. This proves the feasibility of replacing the third steel support and the fourth reinforced concrete support position between Scheme 2 of the foundation pit supporting structure design scheme and Scheme 1 of the foundation pit supporting structure design scheme, which will make the maximum axial force of the internal support. The deformation of the diaphragm wall is significantly reduced, which effectively enhances the integrity and stability of the supporting structure, which can provide experience for the design of the internal supporting and supporting design of similar open-cut tunnels deep foundation pit projects in the future.

4 INTERNAL SUPPORT AXIAL FORCE MONITORING ON SITE

4.1 *Axial force monitoring data analysis*

The design axial force of the reinforced concrete support of this project is 5000 kN, the design axial force of the steel support is 2850 kN, and the warning value is 70% of the design value. The fourth and fifth steel supports have pre-added axial forces of 600 kN.

This paper selects the internal support axial force of two representative monitoring points, ZCL17 and ZCL22, for research, and in the diaphragm wall at the two monitoring points, some inclinometers can monitor the deep horizontal displacement of the wall. The change curve of the internal support axial force with the excavation stage is shown in Figure 7.

According to analysis of on-site axial force monitoring data in Figure 7, it can be concluded that: for ZCL17 and ZCL22, two monitoring points at the end of each layer of soil excavation are installed promptly according to the requirements of the construction support, although the strut axial forces

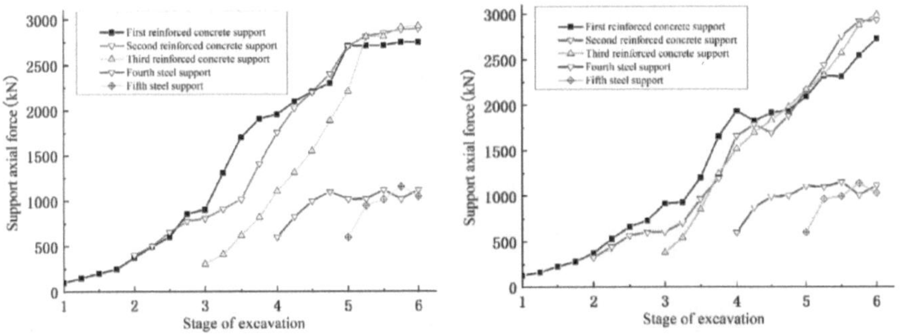

(a) ZCL17 monitoring point (b) ZCL17 monitoring point

Figure 7. Variation curve of internal support axial force with excavation stage.

in increasing, but were not close to early warning value, because the two monitoring points have similar hydrogeological conditions and construction conditions. Hence, they monitor the change in axial force along with the excavation phase curve that is also similar. For reinforced concrete support, as the frequency of steel meter components is significantly affected by temperature, the data of the two monitoring points are significantly different, but also within a reasonable range. In addition, the maximum axial force at the ZCL17 monitoring point occurred on the third reinforced concrete support, and the axial force value was 2928.1 kN, accounting for 83.66% of the warning value (the warning value was 350 kN). In the ZCL22 monitoring point, the maximum axial force was also on the third reinforced concrete support, which was 2996.2 kN. Accounted for 85.61% of the early warning value (the early warning value was 3500 kN), all of which did not exceed the early warning value of axial force. The design scheme of the foundation pit supporting structure is reasonable.

Through field monitoring and comparison of daily field data, it was found that the deep horizontal displacement of the diaphragm wall increases with the rise in excavation depth of the foundation pit. Therefore, during the fifth excavation of the foundation pit, the deep horizontal displacement of the diaphragm wall reaches the maximum. The maximum value of the cumulative change is 18.00 mm, which is far less than the warning value of 40 mm, which shows that Scheme 2 has a strong foundation pit protection and stability, which enhances safety during the construction process. Figure 8 shows the change curve of the deep horizontal displacement of the diaphragm wall in different excavation stages.

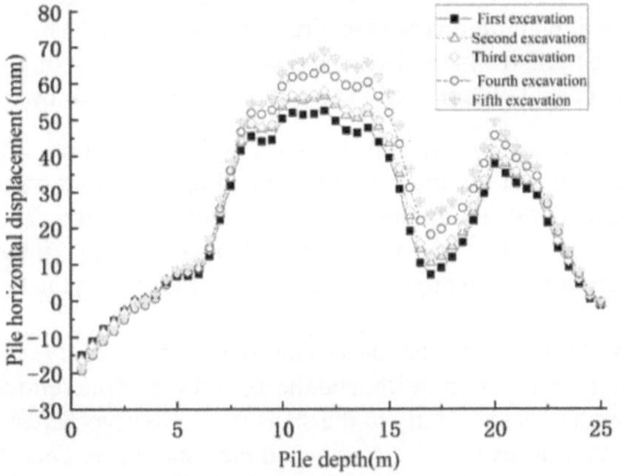

Figure 8. Deep horizontal displacement of diaphragm wall in different excavation stages.

4.2 *Comparative analysis of support axial force simulation results and monitoring results*

During the excavation of each layer of soil in the foundation pit, the comparison between the maximum value of the on-site axial force of the internal support at the monitoring points of ZCL17 and ZCL22 and the maximum value of the numerical simulation are shown in Figure 9.

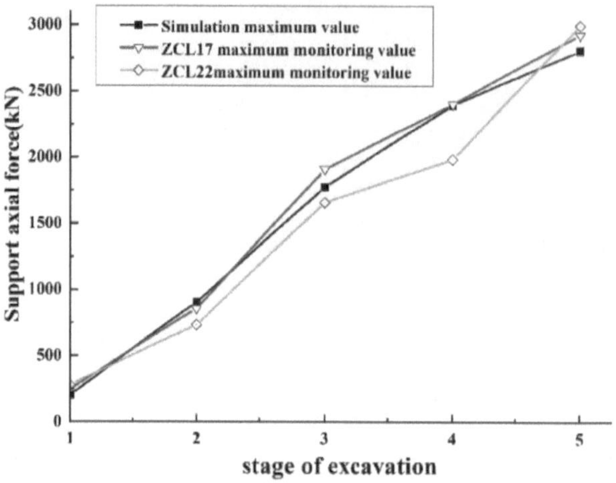

Figure 9. Comparison of the maximum internal support axial force simulation with the monitored value.

From the analysis of Figure 9, it can be seen that there is a slight deviation between the numerical simulation results of the internal support axial force and the on-site monitoring results. This is mainly due to the lack of consideration of influencing factors, such as on-site mechanical construction vibration and temperature changes during the numerical simulation calculation. However, the magnitude and change curve of the axial force of the simulation results and the monitoring results are the same, which further proves that the second choice of the foundation pit supporting structure is reasonable and can reflect the variation of axial force of support in the different excavation stages of the foundation pit and provide a reference for the optimization analysis of the supporting structure in the foundation pit.

5 CONCLUSION

Based on the deep foundation pit project of the airport section of the Xiamen Intercity Railway R1 line, this paper conducts numerical simulation and on-site monitoring of two types of foundation pit supporting structure design schemes, and the conclusions are as follows:

(1) By adjusting the positions of the third steel bracing and the fourth reinforced concrete support, the maximum axial force of the internal support and the deformation of the diaphragm wall is significantly reduced, thereby optimizing the mechanical performance and stability of the supporting structure. Both the maximum displacement and the maximum axial force occur at the third internal support, which proves the rationality of replacing the third and fourth internal support.

(2) Under the condition of soft sand fill in the coastal zone, the difficulty in foundation pit excavation increases with the excavation depth, and the axial force of the reinforced concrete support and the steel bracing increase linearly in the short term. However, after excavation to the bottom of the pit, the axial force changes slowly and then stabilizes. Therefore, daily monitoring should be strengthened before the internal support axial force changes stably, and timely communication with the construction unit should be carried out to ensure the safety of the project.

(3) At the same monitoring point, numerical simulation and on-site axial force monitoring results are mutually verified. The results show that the changing trend of the deep horizontal displacement of the diaphragm wall and the internal support axial force is the same, and they are positively correlated, which proves the correctness of the finite element simulation calculation results. It is recommended to further study the relationship between changes in groundwater level, the ground settlement around the foundation pit, etc., and changes in the axial force of the internal support to make it closer to the engineering reality and provide a reference for the optimization of the internal support structure of the foundation pit.

REFERENCES

Chen, K. et al. (2017) Analysis of the influence of different internal support systems on the deformation of deep foundation pit excavation. *Journal of Tianjin University*. 50 (S1), 1–6.

He, M. D. et al. (2008) Analysis of force and deformation characteristics of a foundation pit retaining structure in the Olympic Park. *Chinese Journal of Geotechnical Engineering*. 30(S1), 582–587.

Liang, F. Y. et al. (2018). Analysis of the force and deformation characteristics of the double ring support system of deep and large foundation pits in soft soil area. *Journal of Hunan University*. 45(S1), 97–143.

Liu, X. W. et al. (2018) Stability analysis of composite steel supporting beams for deep foundation pits. *Engineering Mechanics*. 35(04), 200–207+218.

Luo, Y. W. et al. (2011) Analysis of force and deformation characteristics of ring-shaped ultra-deep foundation pit retaining structure. *Rock and Soil Mechanics*. 32(02), 617–622.

Peng, T. et al. (2020) Investigation and assessment of construction risks of deep foundation pits near subway lines. *Journal of Civil Engineering and Management*. 37(06), 113–117.

Rezvani, R. & Tutunchian, M. A. (2021) Horizontal Displacement of Urban Deep Excavated Walls Supported by Multistrands Anchors, Steel Piles, and In Situ Concrete Piles: Case Study. *International Journal of Geomechanics*. 21(1), 05020008.

Su, X. T. et al. (2021) Lateral movement law of soft soil deep foundation pit support based on im-proved MSD method. *Science Technology and Engineering*. 21(05), 2002–2010.

Yang, Y. Y. et al. (2013) Sensor monitoring of a newly designed foundation pit supporting structure. *Journal of Central South University*. 20(04), 1064–1070.

Advances in Civil Engineering: Structural Seismic Resistance, Monitoring and Detection – Mohd Johari Mohd Yusof and Junwen Zhang (Eds)
© 2023 The Authors, ISBN 978-1-032-31491-4

Application analysis of isolation technology for multi-story frame structure in an upper isolation layer

Jian Fu, Bingying Xie, Liyun Zeng & Qian Zhang
Department of Civil Engineering and Architecture, Panzhihua University, Panzhihua, China

ABSTRACT: At present, in domestic seismic isolation buildings, seismic isolation bearings are usually set on the top of the foundation. At the same time, the domestic seismic isolation technology related regulations only have clear provisions for basic seismic isolation buildings. In this paper, for the construction project where the seismic isolation support is located on the top of the basement, using the CSI ETABS 2013 Nonlinear C tool, the seismic isolation model establishment, seismic isolation calculation and analysis are carried out. The research in this paper shows that the use of suitable isolation supports, reasonable isolation analysis models and seismic waves, as well as correct isolation calculation and analysis methods can reduce the inter-story shear and inter-story displacement of the upper isolation layer building project ; The seismic response of the structure above the isolation layer can be reduced by 1.5 degrees, and the seismic measures can be reduced by 1 degree; it can reduce the seismic response of the upper isolation layer construction project, and effectively improve the seismic performance of the structure above the isolation layer.

1 INTRODUCTION

At present, all countries have current specifications for building seismic design and isolation technology. Through research and analysis, it can be seen that the isolation design in the current national standards adopts the base isolation design based on rubber isolation bearings. For such basic isolation design, the isolation scheme, the basic requirements of the isolation support, the basic model of the isolation analysis, the calculation method of the isolation analysis, and the structural requirements of the isolation layer are all strictly regulated. Therefore, the basic isolation theory and rubber isolation bearing technology have been very mature and widely used (Li 2006, 2011, Zhu et al. 2009). However, the research and application of the isolation technology in the field of building construction engineering by the majority of scholars in China and out of China mainly focus on the research and application of the isolation design of the isolation layer on the foundation top surface, and the base isolation system for the sliding friction, and The layered isolation system in which the isolation layer is placed at other locations requires further study.

The research work in this paper is mainly based on the research of scholars in China and out of China, and in-depth study and analysis of the theory of basic isolation technology. Based on the mature theory of rubber isolation bearing technology, this paper does not repeat the theoretical knowledge of basic isolation technology, mainly based on the specific engineering examples, and proposes the isolation design idea of setting the isolation layer on top of the basement. That is, an isolation space is formed between one floor below ground level and one floor above ground level, and one or more isolation bearings are arranged at positions corresponding to the frame columns. Through the research work, the reasonable isolation bearing type, the seismic isolation calculation model and the seismic wave are selected. The ETABS Software is used to analyze the time-separation of the isolated and non-isolated models of the project under earthquake action, and the calculation results are carried out. Comparative analysis, the two parameters of the horizontal direction damping coefficient and the isolation layer displacement of the isolation model are finally

 DOI 10.1201/9781003310884-31

determined. Through the application of engineering application research results, in the fortification area with seismic fortification intensity of 7.0 degrees and seismic acceleration of 0.15g in China (equivalent to 2A area in UBC, B area in IBC/ASCE7), such up-position is adopted. The multi-layer frame structure of the isolation layer can reduce the inter-layer shear force and interlayer displacement of the structure above the isolation layer. The seismic response of the upper structure can be reduced by 1.5 degrees, and the seismic measures can be reduced by 1.0 degree.

Through the research work and results of this paper, firstly, it can prove the seismic isolation effect of the multi-layer frame structure system with the upper isolation layer, and it can also increase the diversity of the multi-layer frame structure's seismic isolation design. By adopting corresponding seismic isolation structure measures, the seismic response of such buildings can be reduced. Second, it can provide a reliable reference and basis for researchers at home and abroad to further study the seismic isolation technology of buildings. Third, it can promote the adjustment and update of the current seismic isolation technical specifications, and be more able to guide the design and construction of seismic isolation buildings. Fourth, it can promote the promotion and application of domestic seismic isolation technology, promote the rapid development of seismic isolation buildings, and reduce the damage of earthquakes.

2 ENGINEERING SURVEY

The project has 1 underground, 6 above ground. The length of the building is 75.6 metres. The width of the building is 27.6 metres. The height of the building is 21.9 metres. The total area of structure is 24185 square metres. The first basement of this project is usually an underground garage and civil air defense basement in wartime. The height of the 1 layer of underground is 3.90 m. On the ground, there are two six-story dormitories. It's a symmetrical layout. The height of the floor is 3.6 metres. The effect of the building is shown in Figure 1.

Figure 1. Architectural renderings.

This project uses a fully cast-in-place concrete structure. The seismic fortification intensity is 7 degrees. The acceleration of the earthquake is 0.15 g. The earthquake is grouped into the third group. The site type of the project is Class II with characteristic period Tg = 0.45s. Acceleration value of frequently occurred earthquake is 55cm/s^2, fortification intensity 150 cm/s^2, rarely occurring earthquake 310 cm/s^2, this design uses basic wind pressure of 0.30 kn/m^2 with Surface roughness B. The type of seismic fortification in this project is class B. Safety classes of structure are level second. If the project structure design is carried out under the fortification intensity of the region with larger size of the cross section of frame column, affecting the functional arrangement and daily use of basements and student dormitories, it is determined that the project adopts seismic isolation technology.

3 ISOLATION SCHEME

By comparing and analyzing the seismic codes of buildings in China, the United States, Japan and other countries, it can be seen that the norms of various countries stipulate or suggest that the isolation layer should be set below the first layer or below the basement layer. The plane layout stress of the isolation layer should have good symmetry. The center of gravity of the isolation layer is consistent with the center of gravity of the upper structure to avoid the torsion of the isolation layer. The plane layout of the isolation bearing should generally be set according to the position of the vertical force member of the superstructure, thereby improving the reliability of the analytical calculation results (China Association for Engineering Construction Standardization 2001; Kilar & Koren 2009; Yang et al. 2003; Zhao et al. 2006).

By analyzing and studying the design drawing of the construction scheme of the project, isolation space has been formed in the first underground and the first over ground. One or more isolation supports are arranged at the location corresponding to the frame column to form an isolation layer, which can isolate the basement and superstructure. Then the isolation of seismic energy is realized.

Four types of isolation bearings have been used in this project: LNR600, CLRB600, 'CLNR700 and LRB700. The parameters of the rubber support are shown in Table 1. The size of the seismic isolation support is shown in Figure 2.

Table 1. Performance parameters of rubber bearings.

Isolation support type	LRB600	LNR600	LRB700	LNR700
Total rubber thickness (mm)	90	90	108	108
First shape coefficient S_1	≥15	≥15	≥15	≥15
First shape coefficient S_2	≥5.0	≥5.0	≥5.0	≥5.0
Vertical compressive stress (Mpa)	12	12	12	12
Vertical bearing capacity (kN)	3391	3391	4616	4616
Vertical stiffness (kN/mm)	2988.5	2719.6	3368.2	3132.4
Just bend degree (kN/mm)	17.3	/	20.7	/
Post-flexion stiffness (kN/mm)	1.758	/	2.037	/
Yield force (kN)	93.8	/	127.7	/
Equivalent stiffness (r100%) (kN/mm)	2.625	1.696	2.981	1.923
Equivalent damping ratio (r100%) %	23.0	5.0	23.0	5.0
Total number of seismic isolation bearings	18	30	22	70

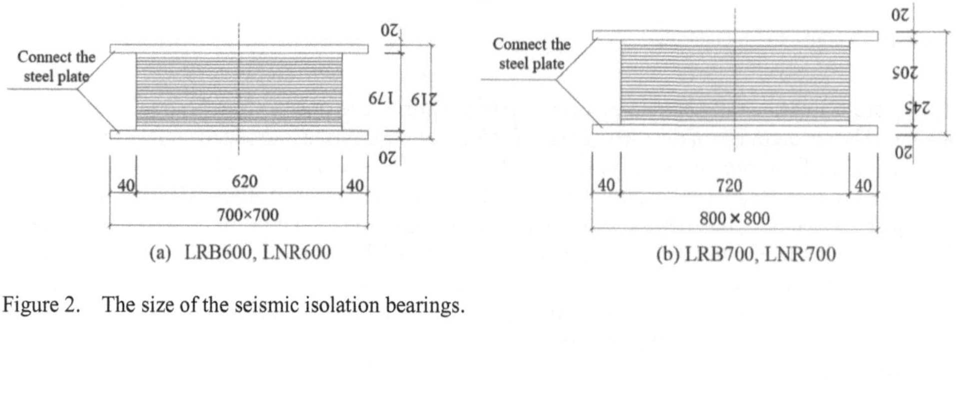

(a) LRB600, LNR600 (b) LRB700, LNR700

Figure 2. The size of the seismic isolation bearings.

4 STRUCTURAL ANALYSIS

4.1 *Verification of the correctness of structural calculation model*

Adopting CSI ETABS2013 Nonlinear C, this project has built non-isolated and isolated structure model. The mode decomposition response spectrum analysis and time history analysis have carried out either (Zhao 2017; Zhang 2011). By the comparison between ETABS non-isolation model and PKPM model, verify the accuracy of ETABS model, the content of the comparison is (Liu et al. 2010; Su 2001): Total mass of non-isolated structures, first three order period and interlayer shear. The results are shown in Tables 2 to 4, where the difference=(PKPM-ETABS)/PKPM × 100%.

As can be seen from the results in Tables 2 to 4, there is very little difference between the total mass, the first three periods and the interlayer shear force between the ETABS non-isolation model and the PKPM model. Therefore, the finite element model of isolation analysis of this project is accurate and reliable by using ETABS model. It can ensure the design of isolation is safe and effective.

Table 2. Comparison of the total mass of non-isolated structure.

PKPM (Ton)	ETABS (Ton)	D-value (%)
17787.0	17773.8	0.07

Table 3. Comparison of the first 3 orders of the non-isolated structure.

Order	PKPM (s)	ETABS (s)	D-value (%)
1	1.0814	1.0785	0.3
2	1.0062	1.0025	0.4
3	0.8717	0.8695	0.3

Table 4. Comparison of the shear force between the layers of the non-isolated structure.

	PKPM (kN)		ETABS (kN)		D-value (%)	
Storey	X	Y	X	Y	X	Y
7	377.15	433.39	380.94	434.93	−1.0	−0.4
6	2841.34	3042.33	2881.12	3068.98	−1.4	−0.9
5	5055.29	5420.08	5138.42	5470.92	−1.6	−0.9
4	6818.05	7312.45	6940.46	7381.46	−1.8	−0.9
3	8225.37	8804.24	8374.94	8886.76	−1.8	−0.9
2	9289.58	9900.49	9462.45	9982.05	−1.9	−0.8
1	9942.35	10543.96	10123.78	10626.58	−1.8	−0.8
Isolation layer	10029.94	10631.93	10212.00	10715.21	−1.8	−0.8

4.2 *Selection of time-history seismic wave*

With regard to the selection of seismic waves, according to the relevant provisions of the current code for seismic design, two sets of actual strong earthquake records (Artificial wave) and a set of simulated acceleration time of history curves are used in this project (Natural wave). The results are safer, more reasonable and reliable (Construction Dept. of PRC 2010, Tan 2008).

4.2.1 *Natural wave selection*

When analyzing natural waves, we should analyze the three factors of seismic waves: Amplitude-Seismic wave peak; Time-sustaining and seismic wave energy; Spectrum characteristics-spectral shape, peak value and remarkable acceleration (Ceccoli & Savoia 1999, Li & Zhou 2002, Sun & Pan 2008). This project mainly analyzes the steps of natural seismic wave as shown in Figure 3. According to the selection step, we select the Taft wave (Figure 3) and Zone wave (Figure 4) as the natural wave for the project. Details of ground vibration are shown in Table 5. The time curve is shown in Figures 4 and 5.

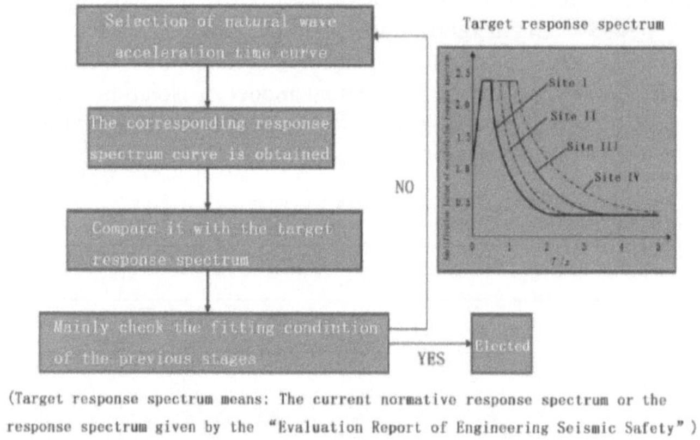

Figure 3. Select the seismic wave step.

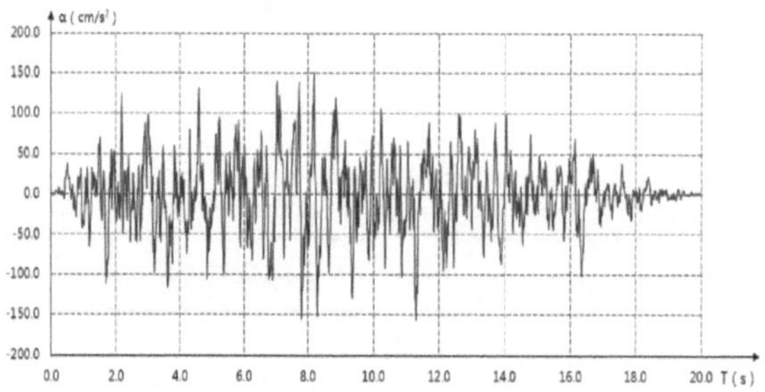

Figure 4. Taft seismic wave time course.

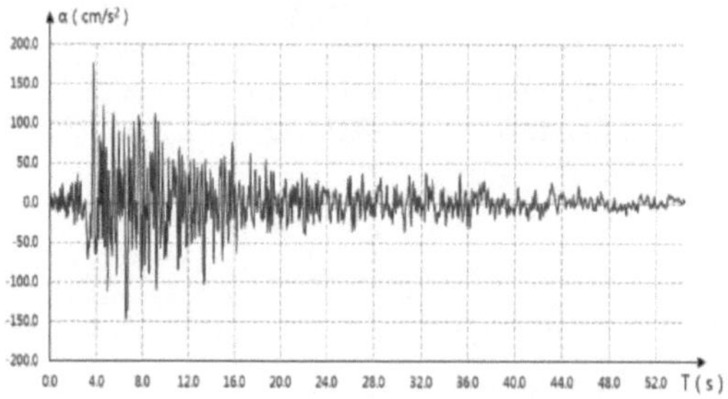

Figure 5. Zone seismic wave time course.

Table 5. Earthquake dynamic detail information.

Name	Source of earthquake motion	PGA
ACC1	Taft	152.7 cm/s^2
ACC2	Zone	156.3 cm/s^2
ACC3	Artificial simulated in seismic acceleration recording	183.9cm/s^2

4.2.2 *Simulation of artificial waves*

The simulation of artificial wave mainly uses iterative fitting method, mainly using seismic wave records; its phase angle is used as the initial phase angle. Then the response spectrum earthquake is used as an artificial seismic wave by iterative synthesis method and the response spectrum of the gauge design. Then the artificial seismic wave is used as an artificial seismic wave through the iterative synthesis method and the normalized response spectrum earthquake. The simulation flow of artificial seismic waves is shown in Figure 6. A synthetic artificial seismic wave curve is obtained, as shown in Figure 7. The results of time history analysis and response spectrum of three kinds of seismic waves are shown in Table 6.

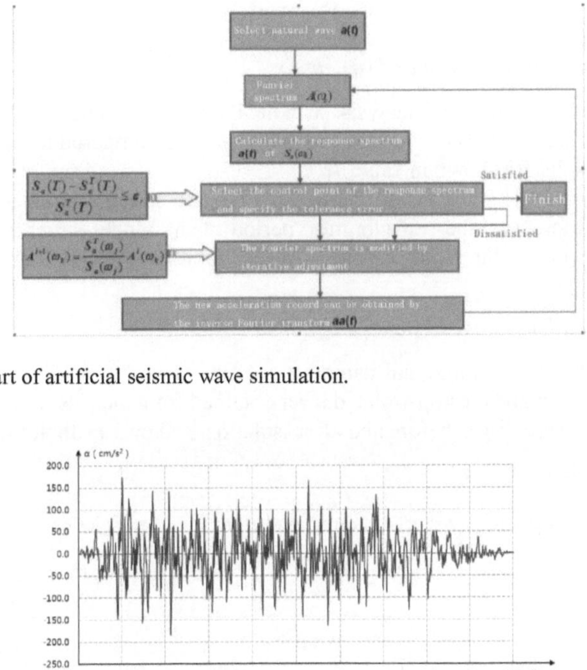

Figure 6. Flow chart of artificial seismic wave simulation.

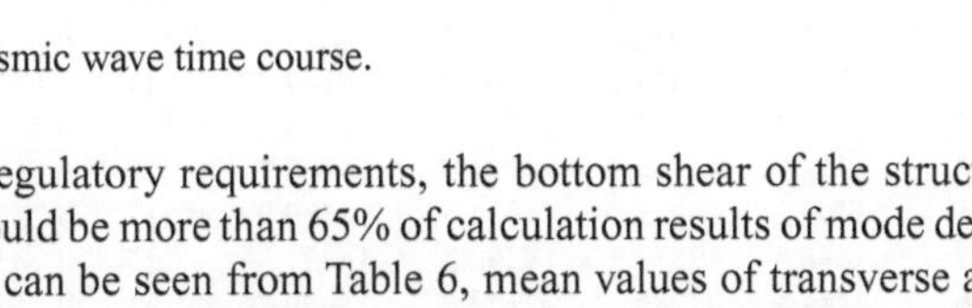

Figure 7. Artificial seismic wave time course.

Based on current regulatory requirements, the bottom shear of the structure calculated by the time history curve should be more than 65% of calculation results of mode decomposition response spectrum method. As can be seen from Table 6, mean values of transverse and longitudinal shear forces calculated by three time-history curves are 8681kN and 8680kN, respectively, which cannot be higher than 80% of the resolution analysis results. It meets the requirements of (GB50011-2010) 5.1.2, paragraph 3 in the current Code for Seismic Design of buildings (Construction Dept. of PRC 2010).

Table 6. The bottom shear force of non-isolated structure under multiple earthquake.

	Seismic direction	Taft	Artificial wave	Zone
Broadwise	Time history analysis(KN)	8687	8996	8361
	Response spectrum (KN)		10212	
	Time history/response spectrum	0.85	0.88	0.82
Direction	Time history analysis (KN)	7853	7872	9114
	Response spectrum (KN)		10715	
	Time history/response spectrum	0.73	0.73	0.85

Table 7. Comparison of the period before and after isolation.

Mode of vibration	Structural cycle before isolation (s)	Structural cycle after isolation (s)
1	1.0785	2.4628
2	1.0025	2.4171
3	0.8695	2.1922

4.3 *Vibration characteristics of isolated structures*

Through time history calculation and analysis, with the ETABS Model before and after isolation of the project, the first three models of vibration are analyzed and compared for the vibration period of the model. The results are shown in Table 7.

As can be seen from Table 7, the vibration period of ETABS structural model is different before and after isolation. Moreover, the self-vibration period of the model structure after isolation is generally longer than that of the model before isolation, approximately 2.30 times longer.

4.4 *Determination of horizontal damping coefficient*

For multistory buildings, the horizontal damping coefficient is the maximum shear force ratio between the isolated layer and the non-isolated layer obtained from the elastic calculation results[8]. The ratio of X and Y shear forces before and after isolation is shown in Tables 8 and 9.

Table 8. Comparison of shear force between X direction before and after isolation.

	Interlayer shear force after isolation			Interlayer shear force before isolation			Comparison of interlayer shear forces before and after isolation			
Storey	Taft	Artificial wave	Zone	Taft	Artificial wave	Zone	Taft	Artificial wave	Zone	Maximum
7	120.5	114.0	117.5	1036.1	895.3	1015.4	0.116	0.127	0.116	0.127
6	1154.9	1137.9	1157.0	8024.2	6602.7	6187.5	0.144	0.172	0.187	0.187
5	2153.3	2143.5	2171.7	11943.6	10524.8	11274.1	0.180	0.204	0.193	0.204
4	3005.3	2955.8	2768.3	14794.1	13724.7	13220.7	0.203	0.215	0.209	0.215
3	3795.4	3583.5	3409.2	16032.1	15934.6	14489.9	0.237	0.225	0.235	0.237
2	4397.4	4016.7	3836.6	19572.2	19390.3	16821.3	0.225	0.207	0.228	0.228
1	5436.1	4853.6	4784.5	23286.7	22123.2	18999.7	0.233	0.219	0.252	0.252
Isolation layer	4324.0	3453.0	3909.2	23690.6	22448.4	19339.7	0.183	0.154	0.202	0.202

Table 9. Comparison of shear force between Y direction before and after isolation.

Storey	Interlayer shear force after isolation			Interlayer shear force before isolation			Comparison of interlayer shear forces before and after isolation			
	Taft	Artificial wave	Zone	Taft	Artificial wave	Zone	Taft	Artificial wave	Zone	Maximum
7	114.7	107.2	121.2	1056.2	1198.9	1138.5	0.109	0.089	0.106	0.109
6	1172.6	1080.7	1091.3	6902.6	7181.7	7522.4	0.170	0.150	0.145	0.170
5	2003.5	2020.4	2254.7	11016.9	12461.5	14796.4	0.182	0.162	0.152	0.182
4	3083.8	3074.1	2907.1	14992.7	15942.9	16213.9	0.206	0.193	0.179	0.206
3	3598.4	3422.1	3815.4	15984.9	18023.5	20736.6	0.225	0.190	0.184	0.225
2	4547.5	4235.2	4097.1	17424.7	19013.4	19594.4	0.261	0.223	0.209	0.261
1	5393.3	4869.0	4781.7	20124.4	21074.4	19495.7	0.268	0.231	0.245	0.268
Isolation layer	4511.1	3654.2	3929.4	20536.4	21367.2	19501.9	0.220	0.171	0.201	0.220

As we know from Tables 8 and 9, the interlaminar shear stress of the structure decreased obviously after the isolation of the project. The maximum ratio of interlaminar shear stress is 0.268. According to the provisions of the Code for Seismic Design of Buildings (GB50011-2010), the seismic response of its upper structure can be reduced by 1.5 degrees.

4.5 *Checking calculation of interlayer displacement angle of isolation model*

Three-way input is used (two natural waves and one man-made wave) for the analysis and calculation of the maximum interlayer displacement angle of the seismic isolation model. The results are shown in Table 10.

Table 10. The inter story drift angle under the action of rare earthquake.

Storey	Interlayer displacement angle in X direction	Interlayer displacement angle in Y direction
7	0.000462	0.000597
6	0.000718	0.000762
5	0.001168	0.001147
4	0.001566	0.001539
3	0.001824	0.001868
2	0.001981	0.001886
1	0.002169	0.002000
Isolation layer	0.003083	0.003435

According to current norms and regulations, for concrete frame structures, the maximum displacement angle between Elastic-plastic layers under rare earthquakes should be smaller than 0.01. Through Table 10, we find that the interlayer displacement angle meets the demand.

5 CONCLUSION

Through the design of isolation and the analysis of isolation technology, we draw the following conclusions:

(1) After the use of isolation technology, the self - shock period of the superstructure will be extended by 2.3 times. The seismic action is obviously reduced. And the horizontal displacement of the structure is concentrated in the isolation layer. The interlaminar acceleration and shear force of the structure are decreasing.

(2) After the use of isolation technology, the maximum of inter - layer shear ratio of the isolated structure under the fortification earthquake is 0.268. According to the provisions of the Code for Seismic Design of Buildings (GB50011-2010), the seismic response of the upper structure can be reduced by 1.5 degrees. Seismic measures can reduce 1.0 degree.
(3) For a multi-tier frame structure, in the area of fortified intensity of 7 degrees/0.15 g, the reduction of seismic response of superstructure can be realized and effectively improve the seismic performance of superstructure using upper isolation layer technology and the technical treatment of isolation layer and substructure.

REFERENCES

C. Ceccoli, M. Savoia. (1999). Non—Linear Seismie AnalysiS Of Base—Isolated RC Frame Structures. *Earthquake Engineering and Structural Dynamics*, vol.28, no.2, pp. 633–653.

China Association for Engineering Construction Standardization. (2001). *Technical Specification for Seismic-isolation with Laminated Rubber Bearing*: CECS126:2001, Beijing: China Association for Engineering Construction Standardization.

Construction Dept. of PRC. (2010). *Code for Seismic Design of Buildings*: GB50011-2010, Beijing: China Building Industry Press.

Li Hui. (2011). Comparative study on seismic behavior of seismic codes for buildings in China, the United States. Europe and Japan, Harbin: *Harbin Institute of Technology*.

Li Wei. (2006). Comparative study on seismic response spectrum and isolation design methods of China. Japan and the United States, Guangzhou: *Guangzhou University*.

Li Zhongxi, Zhou Xiyuan. (2002). Self-vibration characteristics and seismic response analysis methods of regular isolation houses. *Earthquake Engineering and Engineering Vibration*, vol.22, no.2, pp. 33–41.

Liu Qing-zhi, He Xiao-gang, Zhao Zuo-zhou, Qian Jia-ru. (2010). Discussion on Rational Model of RC Frame Structures Without Basement. *Building Structure*, vol.40, no.10, pp.36–40.

Su Jing-yu, Zeng De-min. (2001). Research and Application of Seismic Isolation Technology Building Structure. *Earthquake Engineering and Engineering Vibration*, vol.24, no.4, pp.94–101.

Sun Baifeng, Pan Wen. (2008). Two-stage design method for multi-layer isolation structure. *World Earthquake Engineering*, vol.9, no.3, pp.151—153.

Tan Ping, Zhou Fu-lin. (2008). Isolation Technology Research and Engineering Applications. *Construction Technology*, vol.37, no.10, pp.5–8.

Vojko Kilar, David Koren. (2009). Seismic behaviour of asymmetric base isolated structures with various distributions of isolators. *Engineering Structures*, vol.13, no.1, pp.910–921.

Yang Di-xiong. et al. (2003). The general introduction and mjor problem of isolation structure. *Advances in Mechanics*, vol.33, no.3, pp.302–312.

Zhang Jian. (2011). Research on Issuses of Elastic-plastic Dynamic Time-history Analysis. *Earthquake Resistant Engineering and Retrofitting*, vol.33, no.05 pp.75–78.

Zhao Wansong. (2017). Study on characteristic parameters of response spectrum in seismic design codes. Hefei: *Hefei University of Technology*.

Zhao Yamin, Su Jingyu, Zhang Yumin. (2006). Comparative Analysis of Design Standards for Building Isolation at Home and Abroad. *Earthquake Research*, vol.10, no.4, pp.396–400.

Zhu Hongping, Mei Shilong, Li Li, Ye Kun, Yuan Yong. (2009). Research and developm ent of seismic base isolation technique for civil engineering structures. *Engineering Science*, Received 9 July. pp. 16–20.

Advances in Civil Engineering: Structural Seismic Resistance, Monitoring and Detection – Mohd Johari Mohd Yusof and Junwen Zhang (Eds)
© 2023 The Authors, ISBN 978-1-032-31491-4

The analytical solutions of the maximum horizontal displacement of the pile head in pile-anchor-support composite retaining structures for deep and narrow foundation excavation

Haolan Wang
Urban and Rural Construction Institute, Hebei Agricultural University, Baoding, China

Ying Wang*
China Hebei Construction & Geotechnical Investigation Group Ltd., Shijiazhuang, China

Jinglin You & Shijie Wang
Urban and Rural Construction Institute, Hebei Agricultural University, Baoding, China

ABSTRACT: In engineering construction, due to the large excavation depth of the foundation pit and the close distance between the edge of the foundation pit and the adjacent existing buildings, a single supporting form is often difficult to meet the design requirements. The supporting structure composed of more than two supporting forms is often a better choice. At present, pile-anchor-bracing composite retaining structure has been successfully applied in many deep foundation pit projects. According to the analysis of the measured data of pile displacement in the site, and the expression of the potential energy is derived for the pile-anchor-support composite retaining structures. Based on the principle of minimum potential energy, deriving the analytical solution of the maximum horizontal displacement of the pile head; Using the formula for the calculation of the engineering example in this paper, the relative error between the theoretical calculation value of maximum displacement of the supporting pile head and measured value is only 12.9%, and the calculation result can meet the requirements of engineering precision.

1 INTRODUCTION

The pile-anchor or pile-supported composite support structure of deep foundation pit has its own advantages and disadvantages (Xiao 2021). For some deep and narrow foundation excavation with large excavation depth and small width, the pile-anchor or pile-supported composite support structure may still not meet the deformation and stability requirements of foundation pit support structure, or cause construction inconvenience and high cost. In recent years, pile anchor (pile anchor + a horizontal support at the top of the foundation pit) combination of this new type of support structure has been successfully applied to many deep and narrow foundation pit engineering support. For this complex foundation pit combination support form, due to its many influencing factors, there is still a lack of mature design theory (Li 2020; Liu 2020; Xiao 2021; Zhao 2020; Yin 2018), which makes the theoretical calculation value and the measured value of the deformation of the support structure often have large differences, and the theoretical research is far behind the needs of engineering practice. How to accurately predict the horizontal displacement of this combined foundation pit supporting structure is crucial. Using the principle of minimum potential energy, Qin L.K. et al (2009) gave the deformation expression of top supporting pile supporting

*Corresponding Author: 497971066@qq.com

DOI 10.1201/9781003310884-32

system; Xu X.C. (2006) derived the analytical solutions of the maximum displacement of pile top in the foundation pit of cantilever pile and pile anchor supporting structure. In this paper, the pile-anchor supporting structure is taken as the research object. Based on the principle of minimum potential energy, the analytical solution of the maximum displacement of the pile top of the internal supporting pile-anchor supporting structure is derived, in order to be used for the deformation prediction of such foundation pit supporting structure.

2 DEFORMATION AND STRESS MODE OF SUPPORTING STRUCTURE

2.1 *Fundamental assumption*

Rectangular deep and narrow foundation excavation adopts pile anchor brace combination support, the design calculation diagram is shown in Figure 1. Considering the symmetry of the supporting system, two internal support sections on both sides of the midpoint of the edge of the foundation pit are taken as the calculation unit, see Figure 2.

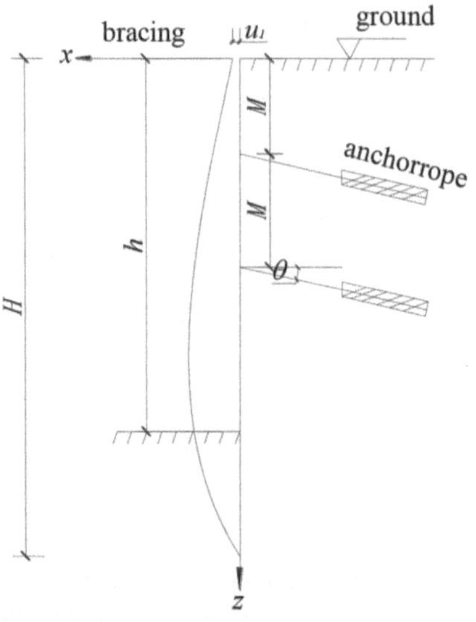

Figure 1. Sketch of deformation mode for supporting piles.

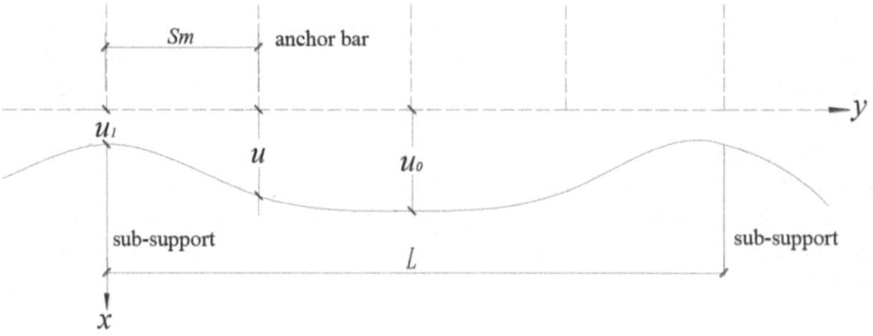

Figure 2. Deformation curve of the top structure.

The coordinate system is established as shown in Figure 1 and Figure 2, and the following assumptions are made: (1) the supporting pile, crown beam, purlin, anchor cable and inner support are all linear elastomer; (2) Only bending deformation of supporting piles and crown beams is considered; (3) The compression deformation of the horizontal support is only considered; (4) Only tensile stress is considered for anchor cables.

2.2 *Crown beam deformation*

According to the engineering measurement in this paper and the existing research results (Fu 2019; Shu 2003) the deformation mode of pile top crown beam in the calculation unit is shown in Figure 2. The lateral displacement of the connection between the crown beam and the support is u_1, Maximum lateral displacement in middle of supporting unit, Set as u_0, then the crown beam deformation curve can be approximately expressed as:

$$u = (u_0 - u_1)\sin\left(\frac{y}{L}\pi\right) + u_1 \tag{1}$$

In the formula: u is the lateral displacement of crown beam at y away from steel support in support unit; L is the spacing between two steel supports. u_1 is the displacement of the pile head before the internal support is set can be calculated by the equivalent beam method.

2.3 *Supporting pile deformation*

The support of the inner support and the application of the prestress of the anchor cable can effectively control the lateral displacement of the supporting pile, and the maximum lateral horizontal displacement appears roughly in the middle and lower part of the supporting pile (Li 2020; Shi 2021). The coordinate system shown in Figure 1 is established. Assuming that there is no displacement at the end of the pile, the lateral displacement curve of the supporting pile can be approximately expressed as a parabola.

Total excavation depth of foundation pit $h = \lambda H$, H is the length of supporting pile; Excavation depth coefficient $\lambda$is the ratio of excavation depth to supporting pile length; the lateral displacement of the pile at the bottom of the pit is u_2, $u_2 = \xi u_1$, ξ is the shape coefficient of deformation curve, which can be calculated by reference [8], then the lateral displacement of supporting pile can be expressed as:

$$u_{(x,z)} = \left[(u_0 - u_1)\sin\left(\frac{\pi}{L}x\right) + u_1\right] \cdot \left(az^2 + bz + c\right) \tag{2}$$

Replace $h = \lambda H$, $u_2 = \xi u_1$ and boundary conditions: $z = H$, $u = 0$ with (2):

$$a = \frac{\lambda^2 + \xi - 1}{\lambda^2 H^2 - hH} - \frac{1}{H^2}, \quad b = \frac{\lambda^2 + \xi - 1}{h - \lambda^2 H}, \quad c = 1$$

3 EARTH PRESSURE DISTRIBUTION

The foundation pit with pile-anchor-support composite retaining structures has strict restrictions on the lateral horizontal displacement of pile top. In order to consider the deformation of supporting structure in the limit state, the distribution assumption of soil pressure along the pile is shown in Figure 3 [6].

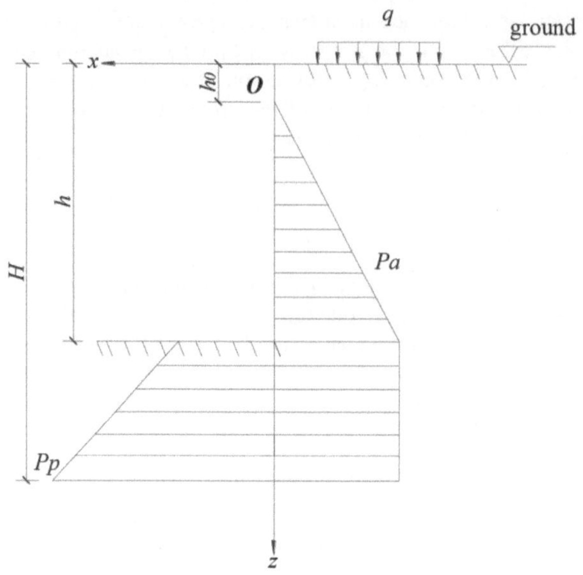

Figure 3. Soil pressure distribution pattern.

3.1 *Active earth pressure*

Active earth pressure is calculated by Rankine's earth pressure theory:

$$p_a = \begin{cases} 0 & z \leq h_0 \\ (q+\gamma z)\ k_a - 2c\sqrt{k_a} & h_0 \leq z \leq h \\ (q+\gamma h)\ k_a - 2c\sqrt{k_a} & h \leq z \leq H \end{cases} \tag{3}$$

3.2 *Passive earth pressure*

$$p_p = \gamma\ (z-h)\ k_p + 2c\sqrt{k_p} \tag{4}$$

In the formula: γ, c and ϕ are the weighted average weight, cohesion and internal friction angle of each soil layer thickness within the pile length range; q is overload at the top of foundation pit; h_0 is the depth of tensile stress zone; K_a, K_p are the main, passive earth pressure coefficient.

4 ENERGY ANALYSIS

The potential energy of support unit includes the strain energy of crown beam bending, support compression, support pile bending, purlin bending, bolt stretching, and the external potential energy of active and passive earth pressure.

It is assumed that the total number of bolts is j, the vertical spacing is M and β is the relative spacing coefficient, which is defined as the ratio of the vertical spacing M from the first bolt to the top internal support to the excavation depth of foundation pit, namely, $M = \beta h$. The bending, tensile and compressive strain energy of each supporting member can be obtained according to material mechanics. In addition, according to the integration of soil pressure in horizontal and vertical [6~8], the potential energy of active and passive soil pressure can be obtained. The specific expressions are as follows:

(1) Bending strain energy of crown beam

$$V_{\text{top beam}} = \frac{E_g I_g}{2} \int_0^L w''^2 dx = \frac{\pi^4 E_g I_g}{4L^3} (u_0 - u_1)^2 \tag{5}$$

In the formula: $E_g I_g$ is the bending stiffness of the crown beam.

(2) Compressive strain energy of support

$$V_{\text{steel bracing}} = \frac{EA}{B} u_1 \tag{6}$$

In the formula: EA is the horizontal support compression stiffness; B is the inner support length.

(3) Bending strain energy of supporting pile

$$\begin{aligned} V_{\text{tangent pile}} &= \sum_{j_0=1}^{n} \frac{E_z I_z}{2} \int_0^H \left\{ \left[(u_0 - u_1) \sin\left(\frac{\pi}{n} j_0\right) + u_0 \right] \cdot \left(az^2 + bz + c\right) \right\}''^2 dxdz \\ &= E_z I_z (u_0 - u_1)^2 a^2 H \frac{l}{S_z} \end{aligned} \tag{7}$$

In the formula: $E_z I_z$ is the bending stiffness of the supporting pile; j_0 is the root number of supporting piles.

(4) Tensile strain energy of bolt

Assuming that the inclination angle of the bolt is θ, the horizontal spacing is Sm, and the displacement increment of the bolt in the deformation process is equal to $\Delta_{\text{tangent pile}} \cos\theta$, the total energy is:

$$\begin{aligned} V_{\text{m}} &= \sum_{i=0}^{j} \frac{K_{\text{m}}}{2} \left\{ \left[(u_0 - u_1) \sin\left(\frac{\pi x}{L}\right) + u_1 \right] \left(a\,(iM)^2 + b\,(iM) + 1 \right) \cos\theta \right\}^2 \\ &= \sum_{i=0}^{j} \frac{L\cos\theta^2 K_{\text{m}}}{2S_{\text{m}}} \cdot \left(\frac{2u_0 + (\pi - 2)u_1}{\pi} \right)^2 \cdot \left(a\,(iM)^2 + b\,(iM) + 1 \right)^2 \end{aligned} \tag{8}$$

In the formula: K_m is the linear stiffness of 1.5×10^5 $textkN/m$; i is the number of anchor bolts.

(5) Bending strain energy of purlin

$$\begin{aligned} V_{\text{anchor purlin}} &= \sum_{i=1}^{j} \frac{E_{\text{wm}} I_{\text{wm}}}{2} \int_0^L \left\{ \left[(u_0 - u_1) \sin\left(\frac{\pi}{L} x\right) + u_1 \right] \left(a\,(iM)^2 + b\,(iM) + c \right) \right\}''^2 dx \\ &= \sum_{i=0}^{j} \frac{\pi^4 E_{\text{w}} I_{\text{w}}}{4L^3} (u_0 - u_1)^2 \cdot \left(a\,(iM)^2 + b\,(iM) + c \right)^2 \end{aligned} \tag{9}$$

In the formula: $E_{\text{wm}} I_{wm}$ is the bending stiffness of purlin.

(6) Active earth pressure potential energy

$$W_{pa} = -\int_0^L \int_0^H p_a\left\{\left[(u_0 - u_1)\,sin\left(\frac{\pi}{L}x\right) + u_1\left(az^2 + bz + c\right)\right]\right\}dxdz$$
$$= \left(-\frac{2L}{\pi}u_0 + \left(\frac{2L}{\pi} - L\right)u_1\right)\cdot(A_1 + A_2 + A_3) \tag{10}$$

In the formula:

$$\left.\begin{aligned} A_1 &= \left(qk_a - 2c\sqrt{k_a}\right)\cdot\left(\frac{1}{3}aH^3 + \frac{1}{2}bH^2 + H - \frac{1}{3}ah_0^3 - \frac{1}{2}bh_0^2 - h_0\right)\\ A_2 &= \gamma k_a\left[\frac{a}{4}\left(h^4 - h_0^4\right) + \frac{b}{3}\left(h^3 - h_0^3\right) + \frac{c}{2}\left(h^2 - h_0^2\right)\right]\\ A_3 &= \gamma hk_a\left[\frac{a}{3}\left(H^3 - h^3\right) + \frac{b}{2}\left(H^2 - h^2\right) + c\left(H - h\right)\right] \end{aligned}\right\} \tag{11}$$

(7) Passive earth pressure potential energy

In foundation pit engineering, the displacement of retaining structure required to achieve stable passive earth pressure is much larger than that required to achieve active earth pressure. Therefore, the passive earth pressure should be reduced, and the reduction coefficient is set as 0.75 according to Reference [7].

$$W_{pp} = \int_0^L \int_h^H 0.75p_p\left[(u_0 - u_1)\sin\frac{\pi}{L}x + u_1\right]\left(az^2 + bz + c\right)dxdz$$
$$= \left[\frac{1.5L}{\pi}(u_0 - u_1) + 0.75u_1L\right]\cdot(A_4 + A_5) \tag{12}$$

In the formula:

$$\left.\begin{aligned} A_4 &= (2c\sqrt{k_p} - \gamma hk_p)\cdot\left[\frac{1}{3}a(H^3 - h^3) + \frac{1}{2}b(H^2 - h^2) + \right.\\ &\quad c(H - h)]\\ A_5 &= \gamma k_p\left[\frac{1}{4}a\left(H^4 - h^4\right) + \frac{1}{3}b\left(H^3 - h^3\right) + \frac{1}{2}c\left(H^2 - h^2\right)\right] \end{aligned}\right\} \tag{13}$$

(8) Total potential energy of the system

$$\Pi = \frac{\pi^4 E_g I_g}{4L^3}(u_0 - u_1)^2 + \frac{EA}{B}u_1 + E_z I_z\left(u_0 - u_1\right)^2 a^2 H\frac{L}{s_z}$$
$$+ \left[\frac{\pi^4 E_w I_w}{4L^3}\left(u_0 - u_1\right)^2 + \frac{L\cos\theta^2 K_m}{2S_m}\times(\frac{2u_0 + (\pi\text{ -}2)u_1}{\pi})^2\right]$$
$$\times G_1 + \left[\frac{1.5L}{\pi}(u_0 - u_1) + 0.75u_1L\right]\cdot(A_4 + A_5)$$
$$+ \left(-\frac{2L}{\pi}u_0 + (\frac{2L}{\pi} - L)u_1\right)\times(A_1 + A_2 + A_3) \tag{14}$$

According to the minimum potential energy theorem, when $\frac{\partial \Pi}{\partial \delta_0} = 0$ is, the potential energy Π takes the minimum value, so as to obtain:

$$u_0 = \left[\left(E_g I_g + \frac{4E_z I_z H a^2 L^4}{\pi^4 S_z} + E_{wm} I_{wm} G_1\right) \times \pi^6 S_m u_1 - 3\pi S_m L^4 (A_4 + A_5) \right.$$
$$\left. -4(\pi - 2) L^4 \cos\theta^2 K_m G_1 u_1 + 4\pi S_m L^4 (A_1 + A_2 + A_3)\right]$$
$$\times \left[\left(E_g I_g + \frac{4E_z I_z H a^2 L^4}{\pi^4 S_z} + E_{wm} I_{wm} G_1\right) \times \pi^6 S_m + 8L^4 \cos\theta^2 K_m G_1\right]^{-1} \quad (15)$$

In the formula:

$$G_1 = \left(aM^2 + bM + c\right)^2 + \left(4aM^2 + 2bM) + c\right)^2 \quad (16)$$

5 VERIFIED BY AN ENGINEERING EXAMPLE

ChiPing 2# dumper project, the foundation pit width 19.5m, depth 14.3m, using layered basin excavation, north and south sides of the horizontal steel pipe top support + three anchor support (Figure 4), supporting pile length range according to the thickness of each soil layer weighted average heavy $\gamma = 19.66$ kN/m^3, cohesion $c = 33.50$ kPa, internal friction angle $\phi = 20.86°$, overload $q = 100$ *kPa*, supporting structure design parameters are shown in Table 1. The displacement monitoring points are arranged in the middle of the crown beam. The maximum displacement of the top of the supporting pile calculated by the formula is 19.3 mm, the measured displacement is 22.17 mm, and the relative error is 12.9 %.

Figure 4. The scene photos.

Table 1. Design parameters.

supporting structure	Diameter (model)	material	specification	remark
tangent pile	ϕ 800 mm	concrete	C30	Spacing 1.2 m the length of the piles 24m
top beam	1.0×0.6 m	concrete	C30	
purlin	25#a channel steel	rolled steel	Q235	Double welding
steel bracing	ϕ 609 mm	rolled steel	Q235	wall thickness 16 mm

6 CONCLUSION

In this paper, the analytical solution of the maximum lateral horizontal displacement of the pile top of the pile-anchor support structure for deep and narrow foundation pits is given. Theoretically, in the given formula, the corresponding constraints such as anchor support, anchor or support are removed, which can also be used to determine the maximum horizontal displacement of pile top in cantilever pile, pile support or pile-anchor supporting structure.

Limited to the determination of the maximum horizontal displacement of the pile top of such supporting structure at present, there is no more mature and reliable method, so the calculation method proposed in this paper can provide a reference for engineering practice.

In the internal bracing pile-anchor supporting structure, the stress mechanism of pile-anchor-bracing composite supporting structure is complex and has many influencing factors, but it is not considered in the formula derivation. At the same time, several assumptions are made in the formula derivation process, hoping to further study in the future.

REFERENCES

Fu H.M., ZHANG Y.Y. (2019) Review on spatial effect of deep foundation pit[J]. *Soil engineering and foundation*, 33(04):465–470.

Li M., CHEN Q.C., ZHAO X. (2020) Application of pile anchor support technology in deep foundation pit engineering [J]. *Building structure*, 50(S2):789–792.

Liu H.L., Fu X., Cui M. (2020) Application of multi-process combined support in complex deep foundation pit [J]. *Journal of Henan University Of Urban Construction*, 29(05):22–28+36.

Qin L.K., Wang X.M., Li Y.Z. et al. (2009) Spatial deformation of top braced pile retaining system for stabilizing long strip foundation pit [J]. *Journal of engineering geology*, 17(01):105–110.

Shi K., Sun Y.S., Zhang W. et al. (2021) Deformation law of pile-anchor support and deformation analysis around foundation pit [J]. *Geotechnical Investigation & Surveying*, 49(10):6–14.

Shu W.C. (2003) Analysis of spatial deformation of cement mixing pile-roof beam support structures [J]. *Rock and soil mechanics*, 24(4):557–560.

Xiao Z. (2021) Technology and application of deep foundation pit support in geotechnical engineering [J]. *Building technique development*, 48(12):153–154.

Xu X.C., Chen S.X., Xu H.B. (2006) Spatial deformation analysis of cantilever soldier pile retaining structure in deep foundation pit [J]. *Rock and soil mechanics*, 2006(2):184–188.

Xu X.C., Ge X.R. (2006) Application of principle of minimum potential energy to spatial deformation analysis for deep foundation pit with anchored pile supporting [J]. *Rock and soil mechanics*, 2006(05):705–710.

Yin Q., Fu H.L., Liu Y.S. et al. (2018) Theoretical model for calculating anchor tension of ribbed beam-pile-anchor supporting structure and its application [J]. *Journal of Central South University (science and technology)*, 49(09):2301–2307.

Zhao F.M., Huang Z.G., Wen P.P. (2020) Influence of different factors on deformation of pile-anchor retaining structure of foundation pit [J]. *Journal of Nanchang University (engineering & technology)*, 42(02):163–168+204.

Advances in Civil Engineering: Structural Seismic Resistance, Monitoring and Detection – Mohd Johari Mohd Yusof and Junwen Zhang (Eds)
© 2023 The Authors, ISBN 978-1-032-31491-4

Review and prospect of seismic isolation and energy dissipation systems

Haohuan Xu
School of Civil and Environmental Engineering, Georgia Tech, Atlanta, GA, USA

ABSTRACT: In the motivation of bringing tons of benefits and positive influences to civil engineering field and society, seismic isolation and energy dissipation systems are of extremely significance in discussion and developing. By researching tons of literature associated with seismic isolation and energy dissipation systems, a newly developed earthquake-resistant structure, from the beginning of the 21st century, this paper summarizes the development, progresses and applications of this technology. Compared with the conventional seismic design, the structural isolation system is a more reasonable, effective, safe and economical damping system. It has significant advantages in the effect of damping controlling, disaster prevention and mitigation, and benefits in terms of society and economy (Xiao et al. 2021). This paper emphatically introduces a variety of commonly used isolation devices and the development status of isolation technology in the world. In addition, this paper compares the conventional seismic design with the isolation design, and briefly introduces what are the advantages of the isolation structure. A model simulation has been performed in SAP2000 to analyse the damping effect of lead rubber bearing isolation and compare the performances of structures with isolators and structures without isolators during an earthquake. Finally, by taking an example of a newly constructed training hall on the north side of the China National Stadium, this paper does a case study and briefly talks about the effective utilization of seismic isolation and energy dissipation systems in real life. Based on the summary of the present situation of structural isolation systems, this paper discusses the future development direction of seismic isolation technology research and its positive influence on the field of civil engineering. Seismic isolation and energy dissipation system is the future trend of earthquake-resistant structures due to several advantages, and the widespread using of this technology is suggested and recommended in contemporary society.

1 INTRODUCTION

Seismic isolation is a seismic protection strategy designed to separate the movement of a structure from the ground vibration, thereby reducing structural forces. Seismic isolation, one of the most effective and successful seismic protection technologies, is now a mature and viable alternative to traditional capability design and has been implemented in many Bridges, buildings and other special structures around the world. Through the study of this system, great progress and breakthrough will be made in the field of civil engineering earthquake resistance. Using this technology in housing construction can effectively maintain social stability and protect residents' property. Therefore, it is extremely important in researching this area.

Isolation technology is a new technology that appeared in the second half of the last century. This technology changes the conventional structure to adopt the "resistance" based thinking. Because of the nature of the seismic wave itself, there are often long period components in the process of an earthquake, so the effect of vibration control by isolation technology alone is not ideal. Therefore, rubber isolation bearings are often used in conjunction with dampers to reduce the response of structures to earthquake effects. First, the combination of isolation technology and energy dissipation technology can make the period of the structure longer and reduce the seismic

DOI 10.1201/9781003310884-33

action (Jiang et al. 2021). Secondly, the seismic energy is absorbed by the damper to reduce the possible displacement of the base (rubber isolation bearing). They complement each other and provide important guarantees for the safety of structures in an earthquake.

2 LITERATURE REVIEW

Nowadays, the technology of seismic isolation and energy dissipation system has gradually become mature after half a century of development, and researches of scholars tend to improve rather than developing.

Energy-dissipating systems can provide increased seismic protection to both old and new buildings. This maul hardware has been used in either matrix or Chevron brace configurations (Constantinou et al 2001). Constantinou et al. (2001) proposed three new configurations that greatly amplify the effects of damping devices using toggle-brace mechanisms for small structural drift. Shaking table tests were performed to analyze the utility of these configurations. The test results show that the damping ratio is significantly increased, and the seismic response of the rigid structure is also significantly reduced.

Lee et al (2010) developed a new type of rolling isolation bearing for Highway Bridges. The new bearing uses cylindrical rollers to roll over a V-shaped inclined surface to achieve shock isolation. It is characterized by constant spectral acceleration and self-centering ability under horizontal ground motion. Constant spectral acceleration reduces the possibility of resonance between the bearing and horizontal earthquake, and the self-centering ability ensures that the bridge superstructure can move from the center to its original position after an earthquake (Lee et al 2010). The built-in sliding friction mechanism reduces seismic response by dissipating seismic energy.

3 AN EXPOSITION OF ISOLATION STRUCTURE AND ENERGY DISSIPATION SYSTEM

3.1 *Working principles of seismic isolation and energy dissipation systems*

The conventional aseismic design uses the strength of materials and the plastic deformation ability of structural components to resist earthquake action, so that the building cannot be damaged irreparably. Seismically isolated technology is a specific measure adopted, generally adding base isolators to isolate the impact of the earthquake reaction on the superstructure, so that buildings in the earthquake will only produce a small vibration, will not cause too much damage. This approach can effectively reduce the stiffness of the building to prolong the period of natural vibration of the structure, thereby reducing the seismic response of the structure. The initial stiffness of the isolator is sufficient to allow the house to "stand still" during minor earthquakes or gusts of wind. In a large earthquake, the horizontal deformation of the isolation is very effective to separate the seismic vibration and significantly reduce the seismic response. When the earthquake is very large, the isolator also acts as a lock to limit the excessive displacement of the structure (Jiang et al. 2021).

The technology of energy dissipation and shock absorption refers to the design of some structures as energy dissipation components or the installation of dampers at the nodes or joints of structures. Under the action of a small earthquake and design wind load, these energy dissipation components are in elastic state, and the structural system has enough lateral stiffness to meet the requirements of normal use (Lin & Huang 2016). In the case of medium and large earthquakes, the dampers first enter the inelastic state, which produces large damping, dissipates a large amount of energy input into the structure from the earthquake reaction, and rapidly attenuates the dynamic response of the structure, so that the main structure does not appear obvious elastic or plastic deformation to ensure the safety of the main structure under the action of earthquakes. In other words, the main structure can be intact with the existence of energy dissipation components.

3.2 *Advantages of seismic isolation and energy dissipation systems*

As stated above, seismic isolation and energy dissipation systems adopt a new approach of "softening" the structure (Pan et al. 2012). This approach effectively reduces the shear wall setting, reduces the section structure, reduces reinforcement and saves the structure fraud. Since the seismic action of the building is greatly reduced and the inter-layer deformation of the building is small, the design of isolation structures no longer focused on increasing material strength of the superstructure and improve the deformation ability, but through the design of each town device from the overall control of the deformation of the structure and the total seismic action of the building (Zhou et al 2020). Experimental research and practical cases both prove that, compared with the buildings with five installations, the seismic action and inter-story deformation of one real building are significantly reduced, which brings great convenience to the design of non-structural components, indoor and outdoor decoration, and equipment piping, which is another obvious advantage of the building with isolators. To be more specific and clearer, Table 1 shows the advantages of seismic isolation and energy dissipation systems over conventional aseismic design.

Table 1. Comparison of traditional seismic structure and isolation & energy dissipation structure.

Method	Conventional seismic structure	Isolation & energy dissipation structure
Principle	Traditional concept of hard resistance, strengthening structure, thickening section	Adjust the dynamic characteristics of the structure with a new concept of softness over hardness
Structural response	Amplified ground response	Reduced structural response
Practice	The structure is fixed to the foundation	The structure is placed on the isolation layer or separated by isolation supports
Performance in Earthquake	Violent vibration, inelastic deformation	Slow translational, elastic deformation
Design basis	According to the predetermined earthquake intensity	Consider sudden super – intensity large earthquakes
Protection object	Only think about the structure itself	Protect both the structure itself and the equipment and instruments within the structure
Scope of application	Newly designed building structure	Both applicable to the new and old general structure, also applicable to important structures, instruments and equipment
Economy	Increase the cost by 3% to 20%	Save the cost by 3%–15%

Admittedly, energy dissipation and shock isolation technology still have many shortcomings. First, this system does not work ideally in responding to impulsive loading. Second, this approach introduces excessive additional force in structural frame because of the installation of dampers and isolators. Also, dampers are costlier. In other words, they are less affordable in residential buildings. However, these disadvantages can be ignored in relation to the huge benefits it brings to society and its advantages. And, in the near future, these shortcomings can be solved by the development of materials technology.

3.3 *Advantages of seismic isolation and energy dissipation systems*

At present, the widely used isolators can be divided into two categories: multilayer rubber bearings and various series of dampers. The multilayer rubber bearing isolation device is installed between the building and the foundation. Standard laminated rubber bearing, high damping laminated rubber bearing, laminated rubber bearing with lead core are three more mature rubber bearings (Zhang et al. 2013). In all these devices, the most widely used is lead-core laminated rubber bearing, which is characterized by: large vertical bearing capacity; the isolation effect is obvious and stable. With stable elastic displacement function, simple structure, convenient installation; Good durability; isolators can be installed in different elevation positions, and the influence of uneven settlement of building foundation is not obvious.

As for the damper series, the main role is to absorb seismic energy and limit the overall deformation of the structure, unlike the isolation bearings which need to bear vertical loads. At present, dampers widely used in engineering field include viscous dampers, viscoelastic dampers, friction dampers, and vibration dampers. Viscous dampers consist of a piston-cylinder arrangement in which silicon-based gel or fluid passes. As the silicon-based gel passes through the piston, the seismic energy is absorbed. Viscoelastic dampers are the type of dampers that consist of an arrangement of steel plates between which viscoelastic materials are placed. Friction dampers are the type of seismic damper that consists of an arrangement of inclined steel plates sliding against each other in an inclined position. Vibration dampers also commonly known as vibration absorbers are the type of seismic dampers that consist of a half tunnel-like passive control device mounted on the structure itself.

4 SAP2000 MODELLING

4.1 *Isolator parameters*

This part of report mainly analyses the damping effect of lead rubber bearing isolation. The aim of this part is showing the effectiveness of isolators and comparing the performances of structures with isolators and structures without isolators during an earthquake. The damping effect of isolated structures is analysed from the natural vibration period and the maximum shear force. Compared with the traditional seismic structure, the corresponding isolation measures (LRB800 isolator) can prolong the natural vibration period of the structure, reduce the maximum inter-story shear force, and reduce the maximum inter-story displacement and acceleration. The isolator used in model is LRB800 fabricated by Wuxi Fuyo Tech Co., Ltd. The parameter of this isolator is shown in Table 2.

Table 2. Parameter of LRB800.

Product Model	DP	Long Duration Load	Vertical Stiffness	Effective Stiffness	Effective Damping
LRB800	160	5026	3114	1795	0.265

4.2 *Periodic and frequency contrast*

The first 3 vibration modes are selected for comparison, and the results are shown in Table 3.

Table 3. Comparison of periodic and frequency of isolator exclusive and isolator inclusive structures.

	Isolator Exclusive		Isolator Inclusive		
Mode #	T	f	T	f	Period Difference in Percentage (%)
1	0.93276	1.07209	2.86585	0.34894	207
2	0.93025	1.07498	2.85265	0.35055	207
3	0.89009	1.12348	2.48535	0.40236	179
4	0.29606	3.37769	0.48459	2.06362	64

As shown in the table above, with the addition of isolator LRB800, the natural vibration periods of the first five modes all increased, and the largest one increased by 207%. Considering that the natural vibration period of the response spectrum, it can be known from the acceleration response spectrum theory that the natural vibration period of the structure is prolonged, and its seismic response decreases, to avoid the dominant period of the site and reduce the seismic effect. Therefore, the isolation effect of lead core rubber base isolation structure is obvious.

Figure 1 to Figure 3 shows the comparison of deformation modes between structure with isolator and structure without isolator.

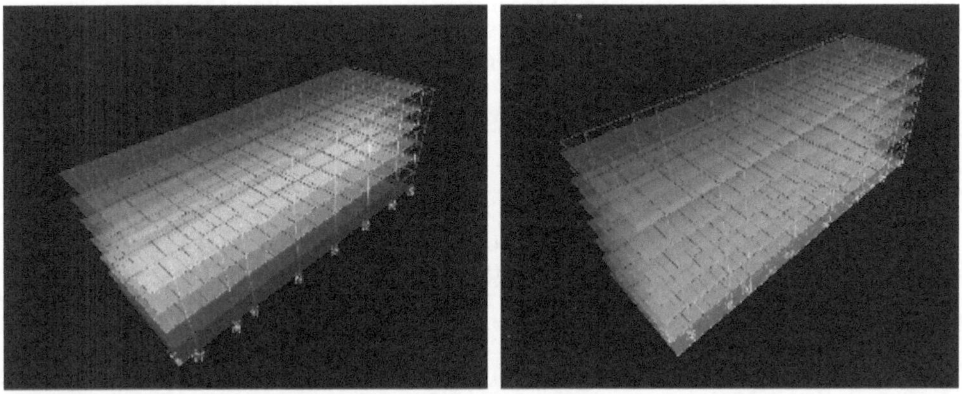

Figure 1. Mode 1 Comparison of deformation modes.

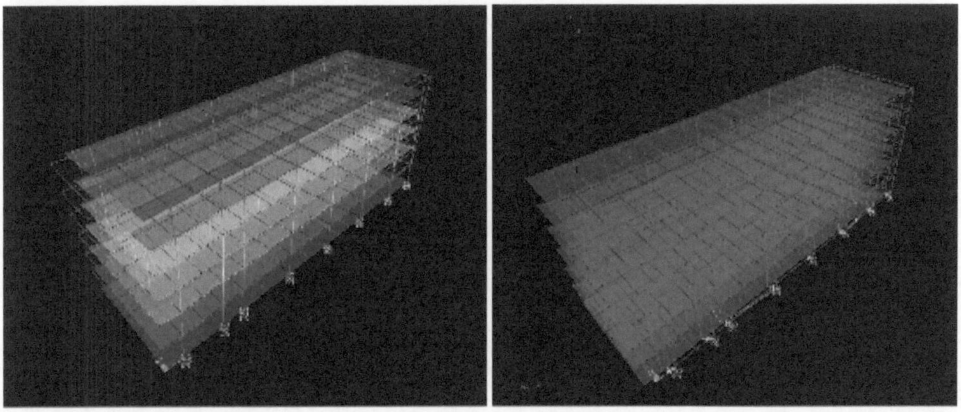

Figure 2. Mode 2 Comparison of deformation modes.

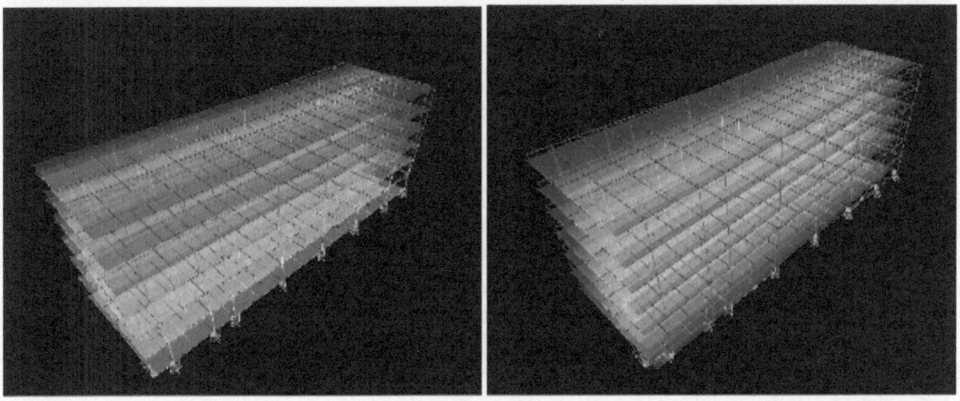

Figure 3. Mode 3 Comparison of deformation modes.

4.3 *Maximum inter-story shear ratio*

In this report, the column shear force at X=42 and Y=0 was taken for comparison. The comparison results are shown in Table 4. The value of shear force of the bottom floor of the frame structure decreases from 816.385 KN to 199.395 KN by 75.58% after adopting the foundation isolation, and the value of the shear forces of the other floors above the foundation isolation layer changes evenly, which is less than that of the traditional seismic frame structure. Therefore, it can be said that the seismic effect is significantly improved after the isolation.

Table 4. Shear Force at X=42 and Y=0 in Y-axis.

Story #	Isolator Exclusive (KN)	Isolator Inclusive (KN)	Difference of Inter-story shear force (KN)
1	816.385	199.395	616.99
2	812.744	248.07	564.674
3	449.919	132.285	317.634
4	413.795	116.842	296.953
5	290.942	74.17	216.772
6	178.89	46.805	132.085

According to the simulation results, the structure with isolators will perform better in earthquake than the structure without isolators. It can be concluded as that compared with the traditional seismic structure, the corresponding isolation measures (LRB800 isolator in this paper) can prolong the natural vibration period of the structure, reduce the maximum inter-story shear force, and reduce the maximum inter-story displacement and acceleration.

5 APPLICATIONS IN CIVIL ENGINEERING FIELD

Seismic isolation and energy dissipation design has been widely used in various fields all over the world. The Toggle-Brace-Damper system mentioned above was proposed by Constantinou and Panos Tsopelas et al. It has been used in practical applications. In the construction of the 37-story Yerba Buena Tower in San Francisco, The 37- Story Millennium Place in Boston and the 38- Story 111 Huntington Avenue in Boston, the proposed Toggle-Brace-Damper system has been adopted (Constantinou et al 2001).

The China National Stadium is in the south of the central area of Beijing Olympic Park. The men's and women's ice hockey semifinals of the 2022 Winter Olympics will be host there. The new training facility will be built on the north side of the National Stadium to meet the training requirements of ice hockey, figure skating and short track speed skating for the 2022 Winter Olympics. The structure is a large-span floor structure is supported by a concrete frame, and the substructure is a concrete frame. The roof is a long-span truss that supported on the concrete column by friction pendulum isolation bearing and equipped with an eddy current damper for energy dissipation and shock absorption (Zhang et al. 2013). The large-span and heavy-duty structure like the training hall has a high demand for seismic resistance and fire resistance, and the problem can be solved well by using seismic isolation technology. Because of the large rotation angle at the bearing of the long-span structure and the limited rotation ability of the classical rubber isolation bearing, it cannot be used in the long-span structure ideally (Zhou et al 2020). Friction pendulum bearing is fabricated by frictional materials with high compressive strength, so the vertical bearing capacity of the bearing is high. Therefore, it can adapt to the rotation of the upper structure and resist fire as necessary (Zhou et al 2020). Thus, Friction pendulum bearing is suitable for large span and heavy-duty structure of vibration isolation. Compared with traditional passive energy dissipation damping devices such as friction damper and viscous damper, eddy current damper does not depend on friction. Also, due to the absence of working fluid, no worries about oil leakage are needed (Zhou et al 2020). It has the advantages of simple structure, high reliability, good durability, easy adjustment of damping coefficient and so on. The isolation design of training hall has reference value because of its characteristics.

6 PROSPECT

With the development of seismic isolation and energy dissipation systems, they have been applied in more and more building structures. The application of the system undoubtedly brings remarkable benefits to the society. However, with the continuous development of energy dissipation and seismic isolation technology, some new problems have been gradually exposed in engineering application. Solving these appeared problems would be the future research trend. First, in terms of design and innovation of dampers and isolators, manufactures should be more active in producing more advanced dampers and isolation bearings to meet the application of hundreds of building structures. At the same time, it is urgent to perfect the shortcomings of the existing isolation bearings and dampers. Second, the use of seismic isolators and dampers has not been standardized. Standardization can promote the application of seismic isolation and energy dissipation technology and reduce property losses caused by improper applications and maintain social stability.

7 CONCLUSION

Isolation and energy dissipation structures are undoubtedly the future trends. Its maturing technology also safeguards the viability and reliability in the practical application. The future trend will be the integration of various technologies, such as restorable functional structures, composite materials, buckling and bracing systems, and high-performance structures, to improve the performance of isolation and energy dissipation technology. By integrating these concepts into existing structures, we are sure to build better, less costly, and more reliable seismic designs.

In this paper, much literature on isolation and energy dissipation damping is briefly reviewed. This paper briefly introduces the working principle of isolation and energy dissipation structure and its advantages over conventional seismic structure. This paper introduces the widely used vibration isolators and dampers in existence. The advantages and disadvantages of these isolation devices are introduced. In addition, some isolating devices or isolating concepts developed or improved by some scholars are also introduced. This paper introduces the application of seismic isolation and energy dissipation structure in practical engineering by taking the new training hall of national stadium as an example. Finally, the future development direction of isolation and energy dissipation design is planned and prospected. After presenting the structure and weighing its advantages and disadvantages, we can conclude that the system will bring great benefits to human society.

Energy dissipation and isolation technology will become an important way to realize the seismic recovery of cities and structures in the 21st century. It has broad application space and development prospect and will make great contribution to reducing earthquake disaster (Zhou et al 2020).

ACKNOWLEDGEMENT

This paper is a review and summary of some literature on seismic isolation and energy dissipation. The information in this paper is the result of the round-the-clock efforts of professors and scholars in many universities around the world. This paper serves as a summary, and thanks to those scholars who have contributed to this field.

REFERENCES

Chen, Huating, and Lingyun Peng. (2020). Dynamic Analysis of Heavily Damped Structures with Supplemental Energy Dissipation Devices." *Journal of Earthquake Engineering*, pp. 1–32. Crossref, doi:10.1080/13632469.2019.1698479.

Constantinou, M. C., Tsopelas, P., Hammel, W., & Sigaher, A. N..(2001). Toggle-Brace-Damper Seismic Energy Dissipation Systems. *Journal of Structural Engineering*, vol. 127, no. 2, pp. 105–12. Crossref, doi:10.1061/(asce)0733-9445(2001)127:2(105).

Falborski, Tomasz, and Robert Jankowski. (2013). Polymeric Bearings – A New Base Isolation System to Reduce Structural Damage during Earthquakes. *Key Engineering Materials*, vol. 569–570, pp. 143–50. Crossref, doi: 10.4028/www.scientific.net/kem.569-570.143.

Jiang, S. J., Yao, S. Z., & Liu, D. W. (2021). Comparison and analysis of seismic isolation, energy dissipation structure and traditional seismic structure. *E3S Web of Conferences*, 248, 01001. https://doi.org/10.1051/e3sconf/202124801001

Lee, G. C., Ou, Y. C., Niu, T., Song, J., & Liang, Z. (2010). Characterization of a Roller Seismic Isolation Bearing with Supplemental Energy Dissipation for Highway Bridges. *Journal of Structural Engineering*, vol. 136, no. 5, pp. 502–10. Crossref, doi:10.1061/(asce)st.1943-541x.0000136.

Lin, S., & Huang, Z. (2016). *Comparative Design of Structures: Concepts and Methodologies*, SpringerVerlag, Berlin Heidelberg.

Pan, P., Ye, L., Shi, W., & Cao, H. (2012). Engineering Practice of Seismic Isolation and Energy Dissipation Structures in China. *Science China Technological Sciences*, vol. 55, no. 11, pp. 3036–46. Crossref, doi:10.1007/s11431-012-4922-6.

Xiao, Y., Shang, S., & Wang, Z. (2021). Shaking Table Experiment and Energy Analysis of a New Rebar Isolation Pier. *Advances in Civil Engineering*, 1–15. https://doi.org/10.1155/2021/8895586

Zhang, Y., Zhang, Y., & Jia, L. G. (2013). The Application of Seismic Isolation Technology in Civil Engineering. *Applied Mechanics and Materials*, 372, 247–252. https://doi.org/10.4028/www.scientific.net/amm.372.247

Zhou Zhongfa, Zhu Zhongyi, Zhou Sun, Li Pei, and Cai Chunjie. (2020). *Design of friction pendulum vibration isolation for the new training hall of the National Stadium for 2022 Winter Olympics. Building Structure.* http://www.buildingstructure.com.cn/Item/25167.aspx.

Advances in Civil Engineering: Structural Seismic Resistance, Monitoring and Detection – Mohd Johari Mohd Yusof and Junwen Zhang (Eds)
© 2023 The Authors, ISBN 978-1-032-31491-4

Research on optimization of seismic design of continuous rigid frame bridges with high and low piers based on parameters of main piers and tie beams

Yuanlu Feng
China Railway Siyuan Survey and Design Group Co. Ltd, Wuhan, China

ABSTRACT: With the vigorous development of bridge construction in the western region, there are more and more bridges with special structures such as high and low piers and large spans. Especially for areas with complex terrain, high pier structures of unequal height are often used. For continuous rigid frame bridges with high piers, in order to reduce the secondary internal forces caused by temperature and concrete shrinkage and creep, two-limbed thin-walled piers with lower longitudinal rigidity are usually used. When the pier height is high, in order to ensure the longitudinal stability of the structure, longitudinal transverse beams are usually added between the two-limbed piers to improve the overall longitudinal rigidity of the bridge. The arrangement of longitudinal transverse beams not only improves the static stress state of the bridge structure, but also affects the dynamic characteristics of the entire bridge. Especially under the action of seismic loads, the seismic response of the structure will change significantly. In order to study the influence of main pier and transverse beam parameters on the seismic performance of high and low pier continuous rigid frame bridge, the finite element Midas civil model for dynamic analysis of the whole bridge is established based on the Xialong Wushui super large bridge of Zhang-Ji-Huai high speed railway (prestressed concrete double thin-walled high and low pier continuous rigid frame bridge with a span of (94+168+94) m). Artificial seismic wave, which is acquired on the base of the design seismic response spectrum, has been performed time-procedure analysis, and the variation law of bridge natural vibration characteristics and seismic response influenced by parameters such as main pier wall thickness, pier height ratio and the number of transverse beams have been obtained. Moreover, the fitting response surface function and objective optimization function has been deduced according to the different distribution positions of the three transverse beams along the pier height, so as to optimize the seismic performance of the bridge. The results show that with the increase of the main pier wall thickness and pier height ratio, the natural vibration period of the first few orders of the bridge gradually decreases, while the high-order mode natural vibration period basically tends to be the same. When the wall thickness of the main pier is 2.75m and the pier height ratio is 0.67, the seismic response of the key section is significantly reduced compared with the original structure. The installation of tie beams has a greater impact on the first mode of the structure, while the other modes of the bridge are less affected by the tie beams. Increasing the number of tie beams can significantly reduce the seismic response of the bridge. With the location combination of three tie beams (down/middle/up = 0.2181H/0.5H/0.85H, respectively), the seismic performance of the bridge is optimal.

1 INTRODUCTION

The main girder of continuous rigid frame bridge has more reasonable stress and larger span capacity because of the rigid frame system formed between main girder and pier. The pier plays an important role in structural force and has more advantages than continuous girder bridge (Zhang 2018). With the continuous advancement of bridge construction, large-span and high-low pier continuous rigid frame bridges are more and more common in complex engineering sites with

DOI 10.1201/9781003310884-34

complicated terrain conditions (Xu 2017). At present, the main pier of continuous rigid frame bridge generally adopts the structure form of double-legged thin-walled pier, which reduces the peak negative bending moment at the pier top of the main girder and make the distribution of internal force of the main girder more reasonable; Besides, the double-thin-walled pier has strong flexural and transverse torsional resistance along the bridge, which can ensure the safety and stability during the construction of the bridge. Although the pushing rigidity along the bridge is relatively small, the flexible structure system of high pier can effectively curtail the influence of concrete shrinkage and creep, temperature change and earthquake action (Xu 2017).

The slenderness ratio and axial compression ratio of high pier structures are both large, and plastic deformation dominated by bending often occurs in strong earthquakes, and the maximum curvature distribution range is not necessarily concentrated at the bottom of the pier. This is in line with the design concept of the current bridge seismic design code. In and out. In fact, there is currently a lack of effective measures and in-depth research on the seismic resistance of high-pier bridges at home and abroad. In the design of double thin-walled high piers, in order to reduce the calculated length and ensure the stability of the flexible high pier, sometimes a transverse beam is set between the thin-walled limbs. In addition, the cross beam has a certain effect on the improvement of the static performance and stability of the high pier, but the influence on the dynamic performance is often ignored.

Zhang, Y.L. et al. (Zhang 2017) studied the influence of wall thickness of double thin-walled piers on seismic response of low-pier continuous rigid frame bridge. It has been found that the internal force of pier control section increases significantly as the wall becomes thicker, while the longitudinal displacement and plastic angle of pier top decrease continuously at the same time. Zhang, Y. et al. (Zhang 2020) studied the influence of pier height ratio on seismic response of continuous rigid frame bridge in high earthquake intensity area and concluded that the bending moment of mid-span and transverse bridge can increase with the pier height ratio, up to 28%. You, H.M. (You 2018) et al. took the number, location of longitudinal tie beams and stiffness ratio of tie beams to pier section, etc. into consideration to study the influence on seismic performance of double-thin-walled high-low-low pier continuous rigid frame bridge. However, the effects of pier and cross-beam on structural seismic performance have not been taken into account at the same time. Based on the Zhang-Ji-Huai high-speed railway—Xialong Wushui Bridge, this paper comprehensively analyzes the influence of the parameters of the main pier and the transverse beam on the natural vibration characteristics and seismic response of the bridge. In the final optimization, the optimal position parameters of the cross beam are obtained.

2 RESEARCH METHODS

2.1 *Analysis theory of bridge dynamic characteristics*

Under the action of dynamic load, the response of the bridge will also change with time. The dynamic response of the bridge structure is closely related to its own natural vibration characteristics, not only related to driving comfort, but also related to the structural safety design of the bridge. Therefore, the analysis of bridge dynamic characteristics is an important content of the study of bridge seismic performance. For most structures, structural dynamic response characteristics can be characterized by mode shape, natural frequency and damping, which are called structural dynamic characteristics. Assuming that the structure has a finite number of degrees of freedom, the infinite displacement components of the actual structure can be discretized into a finite number, and then some analytical calculation methods can be used to obtain the expression of the structural displacement.

According to D'Alembert principle, the dynamic balance equation is established by applying seismic load to the discrete multi-degree-of-freedom structure system, and formula Eq. (1) is obtained:

$$[M]\{\ddot{\delta}\} + [C]\{\dot{\delta}\} + [K]\{\delta\} = -[M]\{I\}\{\ddot{\delta}_g(t)\} = P \tag{1}$$

Where: $[M]$, $[C]$, $[K]$ are mass matrix, damping matrix and stiffness matrix respectively; $\{\ddot{\delta}\}$ is the acceleration vector; $\{\dot{\delta}\}$ is the velocity vector; $\{\delta\}$ is the structural displacement; $\{\ddot{\delta}_g(t)\}$ is the horizontal ground acceleration time history.

When $[C]$ =0, the Eq. (1) is called undamped vibration. when P=0, it is called free vibration. Instead, when $P \neq 0$, it is called forced vibration.

2.2 *Calculation method of natural vibration characteristics*

When there is no load on the structure, the undamped motion equation of the structure is shown in Eq.(2).

$$[M]\{\ddot{\delta}\} + [K]\{\delta\} = 0 \tag{2}$$

Solving the natural frequency and mode shape of the structure is to solve the eigenvalues and eigenvectors of Eq.(2). Assuming that the free vibration is a simple harmonic vibration, the solution form is as shown in Eq.(3):

$$\{\delta\} = \{\varphi\} \sin(\omega t + \theta) \tag{3}$$

In the formula: $\{\varphi\}$ is the characteristic vector; ω is the natural frequency; $\{\varphi\}$ represents the magnitude of the vibration amplitude of each particle, which is a vector that does not change with time. Calculate the second derivative of Eq.(3) with respect to t as shown in Eq.(4):

$$\{\ddot{\delta}\} = \{\varphi\}\omega^2 \sin(\omega t + \theta) \tag{4}$$

Substituting Eq.(3) and Eq.(4) into Eq.(2), the vibration mode equation can be obtained, as shown in Eq.(5):

$$\left([K] - \omega^2[M]\right)\{\delta\} = 0 \tag{5}$$

Where: The real coefficient equation of ω^2 is the problem of finding generalized eigenvalues.

The multiple Ritz vector method treats the dynamic response of the structure as a function of the spatial load distribution. Considering the spatial distribution of the dynamic load, it can avoid missing and introducing un-excited modes, which can significantly improve calculation efficiency. When calculating the dynamic response of the structure, the same number of modes are used for superposition. The Ritz vector direct superposition method can have higher accuracy than the subspace iteration method, because the latter contains modes that are not actually excited. Therefore, this paper chooses the multiple Ritz vector method to analyze the dynamic characteristic value of the bridge.

2.3 *Time history analysis method*

The time history analysis method is a method that uses the direct integration method to obtain the seismic response time history of the structure by inputting actual seismic records or synthetic seismic waves to the structure. Although compared with the static method and the response spectrum method, the amount of calculation is greatly increased. However, this method can draw a good seismic response, especially the peak time and response sign can be obtained. Therefore, the time history analysis method is of great significance in structural seismic analysis.

The time history analysis method can be divided into three steps: First, select the appropriate seismic wave. Secondly, establish the numerical model and dynamic equation of the multi-degree-of-freedom structure. Finally, use a suitable integration method to solve the dynamic equation, and calculate the displacement, velocity and acceleration of the structure at each moment. Since the time history method is complicated to calculate and can only be calculated by a computer, it is necessary to select an appropriate numerical integration method before calculation. The current numerical integration methods generally used are *Wilson-θ* method and *Newmark-β* method.

Time history analysis is mainly divided into linear time history analysis and nonlinear time history analysis. This paper considers the linear elastic seismic action of the bridge, and the form of its motion equation is shown in Eq.(6):

$$[M]\{\ddot{u}\} + [C]\{\dot{u}\} + [K]\{u\} = -[M]\{I\}\ddot{u}_g(t) \tag{6}$$

In the equation: $[M]$ is the mass matrix, $[C]$ is the damping matrix, $[K]$ is the stiffness matrix; $\{u\}$ represents the relative displacement vector of the particle of the structure, and the independent variable is time; $\{I\}$ is the unit array, used to represent ground motion Direction; $\ddot{u}_g(t)$ is the time course of ground motion. The direct integration method and the formation superposition method can be used to solve the equation of motion.

3 ENGINEERING BACKGROUND AND ESTABLISHMENT OF BRIDGE FINITE ELEMENT MODEL

Zhang-Ji-Huai High-Speed Railway is a high-speed railway connecting Zhangjiajie City, Jishou City and Huaihua City in Hunan Province. It is the regional railway connection line of the *eight vertical and eight horizontal* high-speed railway network in the *Medium and Long-Term Railway Network Planning* (2016 Revision). With a total length of 246.6 km and a design speed of 350 km/h, Zhang-Ji-Huai high-speed railway starts from Zhangjiajie City in the north, passes through Furong Town Station, Guzhang Station, Jishou Station, Fenghuang Station, and connects to Huaihua City in the south. Zhang-Ji-Huai High-speed Railway is a milestone project in the history of West Hunan and acts a central nerve and national transportation hub in West Hunan. Its construction will promote the revitalization of rural areas and the development of regional tourism in China. It is of great significance to strengthen regional advantages and expand opening up to the outside world.

The research is based on the project of Zhang-Ji-Huai High-Speed Railway Lower Ridge Wushui Bridge, whose structural form is (94+168+94)m continuous rigid frame bridge with double-thin-walled high and low piers. Its length and width are 357.6m and 12.6m, respectively. The main girder is a pre-stressed C55 concrete box girder with variable section, while the height of middle pivot girder is 10.5 m and that of side pivot and midspan of main girder is 4.5 m., and the center line of the side fulcrum is 0.8 m away from the end of the girder. The main pier is a double-thin-walled circular End-shaped C40 reinforced concrete pier with a distance of 6.3m, the transverse straight section of the pier body is 9.2m long, the half-circle diameter is 3.0m, and the heights of No. 44 and No. 45 main piers are 41.5m and 49.5m respectively. The pier to height ratio is 0.84. The engineering category of the bridge is B, the peak ground motion acceleration is 0.05g, the site category is II, the characteristic period of ground motion response spectrum is 0.35s, and the damping ratio is 0.05. The bridge type layout diagram is shown in Figure 1.

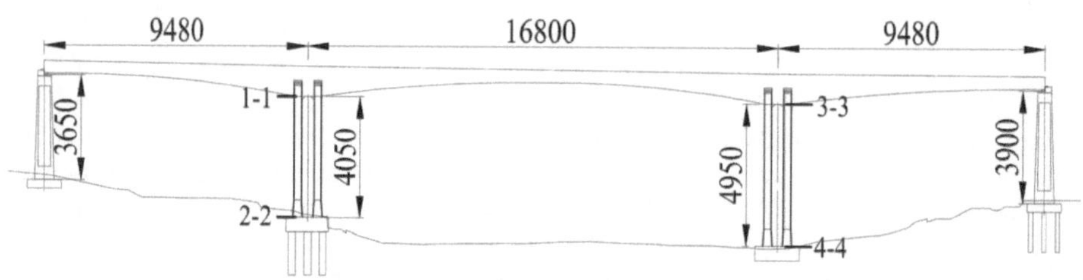

Figure 1. Layout of continuous rigid frame bridge with high and low piers (unit: cm).

Midas Civil software has been used to establish the finite element model for dynamic analysis of the whole bridge. Beam element is used for main girder and thin-walled pier simulation. The weight of pre-stressed steel bundle, reinforcement and second stage pavement is simulated by modifying element capacity. There are 167 joints and 162 beam elements in the whole bridge.

No.44 and No.45 double-thin-walled piers are coupled by rigid connection between the top of each limb and the joint of the main girder, the side fulcrum the main girder is simulated by general support, and the bottom of the main pier is simulated by consolidation. The finite element Midas Civil model for dynamic analysis of the whole bridge is shown in Figure 2.

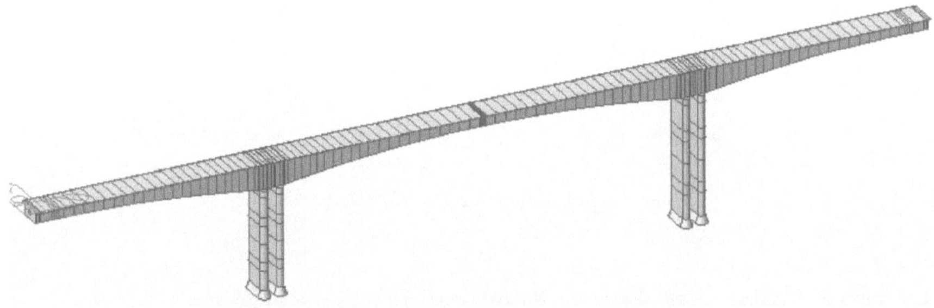

Figure 2. Finite element Midas civil model for dynamic analysis of the whole bridge.

4 INFLUENCE OF MAIN PIER AND TRANSVERSE BEAM PARAMETERS ON NATURAL VIBRATION CHARACTERISTICS OF BRIDGES

Reasonable adjustment of parameters such as wall thickness of main pier, ratio of pier height and number of transverse tie beams can improve the pushing and bending rigidity of pier and the seismic performance of structure as a whole (Zhou 2016). In this paper, the self-vibration characteristics of bridges under different parameters are calculated and analyzed, and the variation rule of each order modes of bridges with the parameters of main piers and crossbeam is compared. The Ritz vector method is used to analyze the self-vibration characteristics of bridges. Considering the three directions of X, Y and Z, the total 180-order modes are taken into account. It is concluded that the cumulative mode mass participation coefficients of X, Y and Z are more than 90% (Tan 2016), as shown in Figure 3. It indicates that the first 180 modes included in the original structural model are effective and can be used for subsequent seismic response analysis.

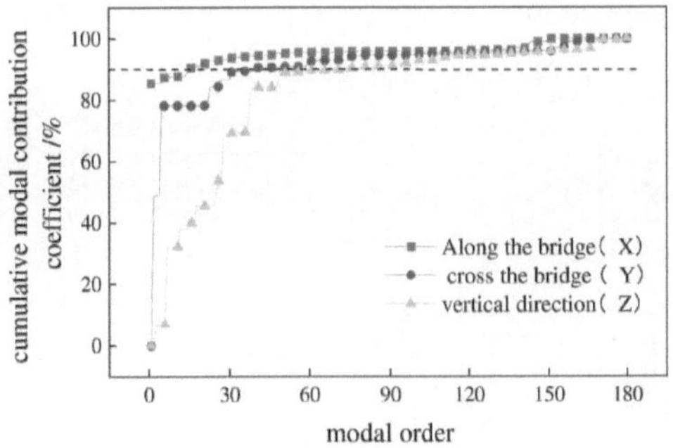

Figure 3. Cumulative contribution of the first 180 modes of the bridge.

4.1 *Natural vibration characteristics of bridges with different wall thicknesses of main piers*

This section aims to study the influence of the wall thickness of the main pier on the natural vibration characteristics of the bridge (Wang 2016), the wall thickness of the main pier of the original bridge is 3.00m. Then adjust the wall thickness of the main pier to four different structures: 2.50m, 2.75m, 3.00m and 3.25m, while keep the height ratio of the main pier unchanged at 0.84 and cancel the transverse tie beam. Calculate the natural vibration period of the bridge. Only the first 10 steps are listed and shown in Figure 4.

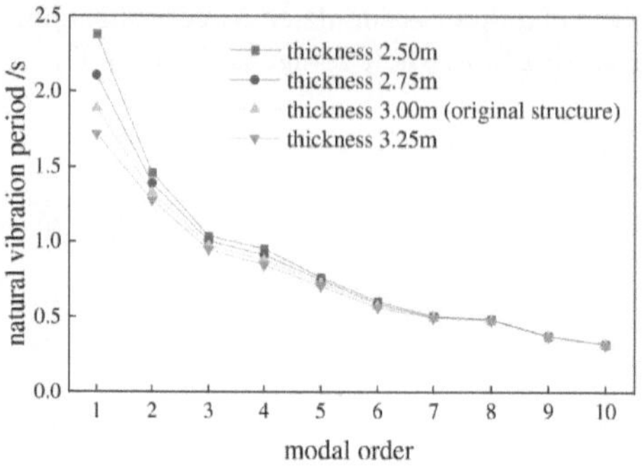

Figure 4. Natural vibration period of bridge with different wall thickness of main pier.

It can be seen from Figure 4 that the first natural vibration period of the bridge decreases significantly with the increase of the wall thickness of the main pier. When the wall thickness of the main pier gradually increases from 2.50m to 2.75m, 3.00m and 3.25m, the first-order natural vibration period of the bridge decreases from 2.380s to 2.106s, 1.891s and 1.718s, reducing 11.49%, 20.54% and 27.82% respectively. The 2nd to 6th natural vibration periods decrease slightly with the increase of wall thickness, and the reduction range is basically kept within 10%. From the 7th mode, the increase of the wall thickness of the main pier has no obvious effect on the natural vibration period, and the period basically tends to be the same.

4.2 *Natural vibration characteristics of bridges with different main pier height ratio*

In order to study the influence of the height ratio of main piers on the natural vibration characteristics of the bridge, the heights of main piers No.44 and No.45 of the original bridge are 41.50m and 49.50m respectively, and the pier height ratio is 0.84. First, keep the wall thickness of the main pier 3.00m and the height of the main pier No.44 main pier unchanged at 41.5m and do not set the transverse tie beam of the main pier. Increase the height of the main pier No.45 by four values: 0.50, 0.67, 0.84 and 1.00. Calculate the natural vibration period of the bridge. Only the first 10 steps are listed and shown in Figure 5.

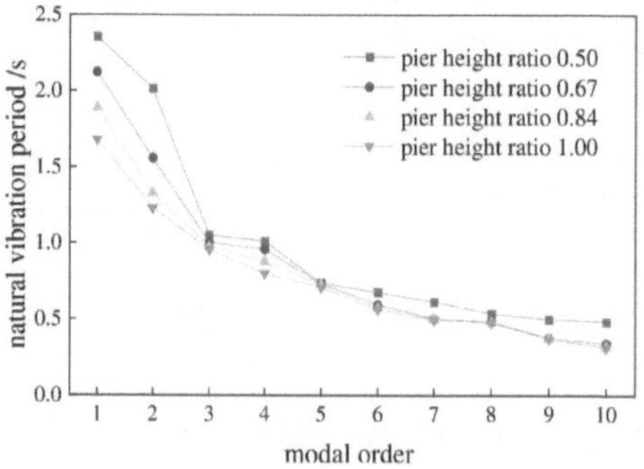

Figure 5. Natural vibration period of bridge with different main pier height ratio.

It can be seen from Figure 5 that the first natural vibration period of the bridge decreases significantly with the increase of the height ratio of the main pier (Li 2018). When the height ratio of main pier gradually increases from 0.5 to 0.67, 0.84 and 1.00, the first-order natural vibration period of the bridge decreases from 2.354s to 2.122s, 1.891s and 1.679s, respectively, by 9.85%, 19.69% and 28.68%. With the increase of pier height ratio approaching 1, the natural vibration period of the bridge in high-order modals gradually approaches. Starting from the fifth mode, the natural vibration periods of the three bridges with pier height ratios of 0.67, 0.84 and 1.00 are basically the same, and the difference is kept within 1%.

4.3 *Natural vibration characteristics of bridges with different number of transverse tie beams of Main Piers*

Transverse tie beam is not adopted in the main pier of the original structure. In order to study the influence of the number of transverse tie beams on the natural vibration characteristics of the bridge, keep the wall thickness of the main pier in 3.00m and the pier height ratio in 0.84 unchanged, and employ different number of transverse tie beam of 0, 1, 2, 3, respectively. The transverse tie beam section size is 9.2m×1.0m, and the transverse tie beam is equally arranged along the pier height, as shown in Figure 6. Calculate the natural vibration period of the bridge, only the first 10 modes are listed as shown in Table 1.

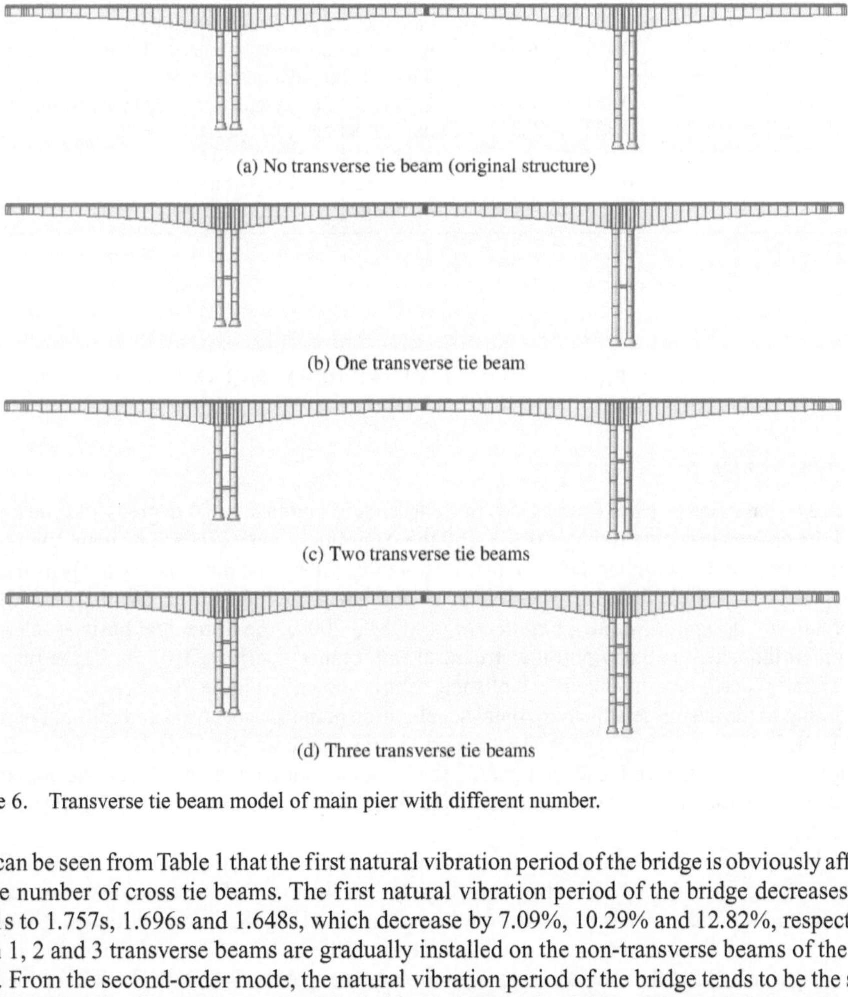

(a) No transverse tie beam (original structure)

(b) One transverse tie beam

(c) Two transverse tie beams

(d) Three transverse tie beams

Figure 6. Transverse tie beam model of main pier with different number.

It can be seen from Table 1 that the first natural vibration period of the bridge is obviously affected by the number of cross tie beams. The first natural vibration period of the bridge decreases from 1.891s to 1.757s, 1.696s and 1.648s, which decrease by 7.09%, 10.29% and 12.82%, respectively, when 1, 2 and 3 transverse beams are gradually installed on the non-transverse beams of the main piers. From the second-order mode, the natural vibration period of the bridge tends to be the same, indicating that the first-order mode of the bridge is longitudinal drift of the main beam + bending of the main pier. Increasing the number of transverse beams increases the bending stiffness of the main pier. Therefore, the natural vibration period of the first-order mode is more sensitive to the number of transverse beams, while the high-order mode is dominated by the side bending and vertical bending of the main beam, and the natural vibration period is almost not affected by the number of transverse beams.

Table 1. Natural vibration periods of bridge with different number of main pier transverse tie beams.

	Natural vibration period (s)				
Modal order	no tie beam	1 tie beam	2 tie beams	3 tie beams	Modal shape
1	1.891	1.757	1.696	1.648	longitudinal drift of main girder
2	1.329	1.329	1.329	1.329	main girder positive symmetrically bent across the bridge
3	0.976	0.964	0.959	0.953	main girder positive symmetrically vertical bent + main pier curved along the bridge
4	0.878	0.878	0.879	0.879	main girder anti-symmetrically bent across the bridge
5	0.727	0.727	0.727	0.727	main girder positive symmetrically bent across the
6	0.576	0.570	0.567	0.564	bridge main girder anti-symmetrically vertically bent + main pier curved along the bridge
7	0.500	0.499	0.499	0.498	main girder positive symmetrically vertically bent + main pier curved along the bridge
8	0.481	0.481	0.481	0.480	main girder anti-symmetrically bent across the bridge
9	0.373	0.373	0.372	0.372	main girder anti-symmetrically vertically bent + main pier curved along the bridge
10	0.320	0.320	0.321	0.321	main girder positive symmetrically bent across the bridge

5 INFLUENCE OF MAIN PIER AND TRANSVERSE BEAM PARAMETERS ON SEISMIC RESPONSE OF BRIDGE

5.1 *Ground motion time history input*

The bridge is classified as class B with a seismic fortification intensity of 6 degrees and site type of class III. Its characteristic period T_g is 0.35s, and the damping ratio ξ is 0.05. The main span of the bridge is 168m and the height of the main pier is more than 40m. The time history analysis method has been used to analyze the seismic response of the bridge. According to *GB 50111-2009 the code for seismic design of Railway Engineering* (AQSIQ 2009), the horizontal basic acceleration of design earthquake $\alpha = 0.05g$ and the structural importance coefficient $C_i = 1.0$, the response spectrum under design earthquake is established, as it is shown in Figure 7(a).

According to the design seismic horizontal acceleration response spectrum and relevant parameters, SIMQKE_GR is adopted to generate three artificial seismic waves. After the seismic response of bridge has been analyzed and compared, the ground motion that produces the maximum response of the structure has been selected as the ground motion input of the final time history analysis.

It can be seen from the natural vibration characteristics of the bridge that the thrust stiffness of the pier along the bridge direction is significantly less than that of the transverse and vertical directions, and the main pier is prone to bending vibration along the bridge direction. Since the seismic response of the main pier is an important factor in the seismic design of the bridge, the left pier top (section 1-1) and the left pier bottom (section 2-2) of pier No.44, the right pier top (section 3-3) and the right pier bottom (section 4-4) of pier No.45 are selected as the key sections for seismic analysis. Considering the three-dimensional seismic effect, and the three-dimensional seismic input mode is: cis (x): transverse (y) : vertical (z) = 1:1:0.65, and the spatial combination mode of seismic response is: cis (x) + transverse (y) + 0.5 vertical (z). After calculation, the seismic response of the bridge under the excitation of artificial wave No.3 is the largest, so artificial wave No.3 is selected as the ground motion input, and the acceleration time history is shown in Figure 7(b).

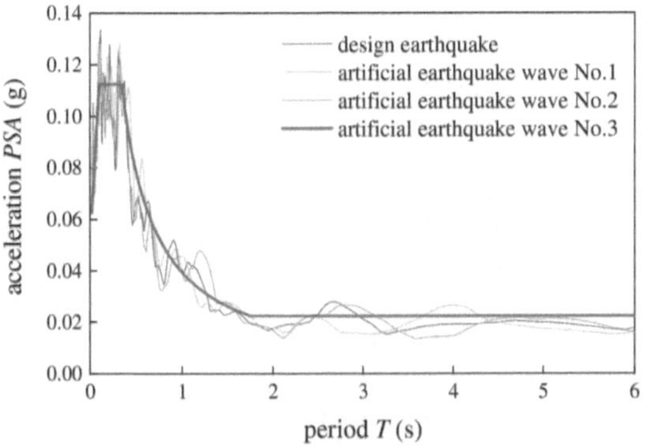

(a) Response spectrum of horizontal acceleration corresponding to design earthquake and artificial wave

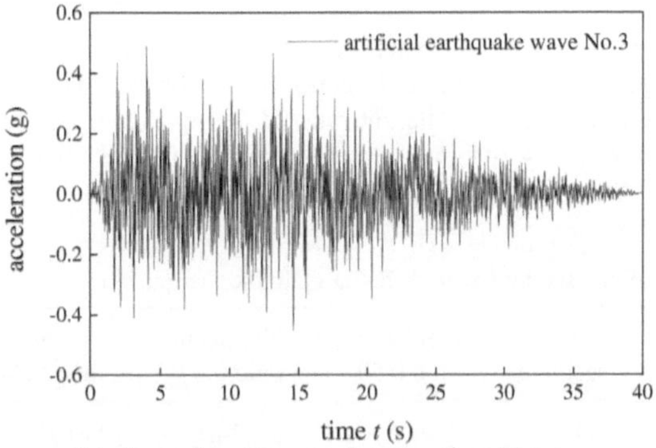

(b) Ground motion acceleration time history

Figure 7. Design acceleration response spectrum and time history analysis of ground motion.

5.2 *Influence of wall thickness of main pier on seismic response of bridge*

Since the first mode of the bridge plays a controlling role in the dynamic response of the whole bridge, in order to study the influence of the wall thickness of the main pier on the displacement D_x and bending moment M_y along the bridge direction of the key section of the bridge, the other parameters have been kept unchanged, While the wall thickness of the main pier has been set as four different structures: 2.50m, 2.75m, 3.00m (original structure) and 3.25m, and the maximum seismic response of the key section of the bridge has been calculated, as it is shown in Table 2.

Table 2. Maximum response of key sections with different wall thickness of main pier.

Earthquake response	Key sections	Thickness			
		2.50m	2.75m	3.00m	3.25m
Mending moment M_y/(kN·m)	1-1	68198.9	68868.6	86105.3	95274.2
	2-2	65528.7	65289.9	83470.4	94409.3
	3-3	76104.9	72399.5	79225.8	94619.5
	4-4	77566.8	69389.2	75457.7	95718.3
Displacement D_x/mm	1-1	30.9	15.0	16.0	14.1
	3-3	31.9	15.4	16.4	14.4

It can be seen from Table 2 that the wall thickness of the main pier has a great impact on the bending moment M_y and displacement D_x response of the key section of the pier. As the wall thickness of the main pier increases from 2.75m, the section bending stiffness and bending moment M_y of the main pier grows, while the displacement D_x of the pier top decreases. When the wall thickness of the main pier is 2.50m, the displacement of the main pier along the bridge direction is doubled compared with the original structure, indicating that the cross-section size of the pier is small at this time, resulting in insufficient flexural stiffness. Therefore, when the wall thickness of the main pier is 2.75m, the seismic response of the key section of the pier is reduced compared with the original structure, which shows the best seismic performance. Taking No.44 pier as an example, the bending moment M_y of pier top and pier bottom sections are reduced by 20.02% and 21.78% respectively compared with the original structure, and the displacement D_x of pier top section is reduced by 6.26% compared with the original structure.

5.3 *Influence of height ratio of main pier on seismic response of bridge*

In order to study the influence of the height ratio of the main pier on the displacement D_x and bending moment M_y along the bridge direction of the key section of the bridge, The other parameters of the original structure remain unchanged, the height ratio of the main pier has been set to the four different structures of 0.50, 0.67, 0.84 (original structure) and 1.00, and calculates the maximum seismic response of the key section of the bridge, as it is shown in Table 3.

Table 3. Maximum response of key sections with different main pier height ratios.

Earthquake response	Key sections	Pier height ratio 0.50	Pier height ratio 0.67	Pier height ratio 0.84	Pier height ratio 1.00
Bending moment M_y/(kN·m)	1-1	72983.6	69114.1	86105.3	89898.7
	2-2	88283.1	62944.1	83470.4	89217.7
	3-3	67545.8	75344.9	79225.8	88754.3
	4-4	67998.2	73490.4	75457.7	89758.2
Displacement D_x/mm	1-1	28.6	15.3	16.0	13.7
	3-3	31.7	16.3	16.4	13.5

It can be seen from Table 3 that the height ratio of the main pier has a great impact on the moment M_y and displacement D_x response of the key section of the pier. With the increase of the height ratio of the main pier from 0.67, the overall flexibility of the bridge is diminishing, resulting in the increase of the overall thrust stiffness along the bridge and main pier bending moment M_y, while the pier top displacement D_x is decreasing. When the height ratio of the main pier is 0.50, the displacement of the main pier along the bridge is nearly twice that of the original structure, indicating that increasing the pier height is not conducive to restraining the displacement response of the structure along the bridge. Therefore, when the height ratio of the main pier is 0.67, the seismic response of the key section of the pier is reduced compared with the original structure, which shows the best seismic performance. Besides, the bending moment M_y of the section at the top and bottom of pier No.44 is reduced by 19.73% and 24.59% respectively compared with the original structure, and the displacement D_x of the section at the top of pier is reduced by 4.39% compared with the original structure.

5.4 *Influence of the number of main pier transverse tie beams on the seismic response of the bridge*

In order to study the influence of the number of cross tie beams of the main pier on the displacement D_x and bending moment M_y along the bridge direction of the key section of the bridge, the other parameters are kept unchanged. Four different structures are set between the two limbs of the main pier, including no cross tie beam (original structure), 1 cross tie beam, 2 cross tie beams and 3 cross tie beams, and the cross tie beams are equally arranged along the pier height. Calculate the maximum seismic response of key sections of the bridge, as it is shown in Table 4.

Table 4. Maximum response of key sections of tie beams of different numbers of main piers.

Earthquake response	Key sections	No tie beam	1 tie beam	2 tie beams	3 tie beams
Bending moment M_y/(kN·m)	1-1	86105.3	74702.8	57435.7	41223.1
	2-2	83470.4	99404.7	90182.0	88479.0
	3-3	79225.8	73671.8	60878.6	49926.1
	4-4	75457.7	94357.4	89621.4	92505.3
Displacement D_x/mm	1-1	16.0	14.7	13.7	13.4
	3-3	16.4	15.1	14.0	14.2

It can be seen from Table 4 that the number of cross tie beams of the main pier has a great impact on the moment M_y and displacement D_x response of the key section of the pier. With the increase of the number of transverse tie beams of the main pier, the displacement D_x and bending moment M_y of the pier top decreases, while the bending moment M_y of the pier bottom increases (Teng 2017). Increasing the number of cross tie beams is conducive to improving the bending stiffness of the main pier and the overall longitudinal thrust stiffness of the bridge (Yang 2018). When three cross tie beams are set for the main pier, the bending moment M_y and displacement D_x of the pier top section are significantly reduced compared with the original structure, and the bending moment M_y of the pier bottom section is slightly increased. Moreover, the bending moment M_y at the top section of pier No.44 is 52.13% lower than that of the original structure, the bending moment M_y at the bottom section of pier is 6.00% higher than that of the original structure, and the displacement D_x at the top section of pier is 16.26% lower than that of the original structure. Therefore, the addition of transverse beams is conducive to suppress the seismic response of the bridge. When three transverse beams are set, the seismic performance of the bridge is the best.

6 SEISMIC PERFORMANCE OPTIMIZATION OF POSITION PARAMETERS OF UPPER AND LOWER CROSS TIE BEAMS

6.1 *Seismic response of key sections at different positions of upper and lower cross tie beams*

Although the seismic performance of the structure is the best when three transverse beams are set at the main pier, the influence of different transverse beam positions on the seismic response of the structure is different. In order to analyze the reasonable arrangement of transverse tie beams along the beam height direction, The height of the main pier is taken as H and the position of the middle transverse tie beam is kept unchanged at 0.5H, and only adjusts the arrangement of the upper and lower transverse tie beams along the pier height to combine the relative positions of the transverse tie beams. Taking the pier bottom as the starting point of the relative position of the cross tie beam, the setting range of the position A of the lower cross tie beam is 0.15H~0.35H, and the position B of the upper cross tie beam is 0.65H~0.85H, with steps of 0.10H. A total of 9 cross tie beam position combinations are formed (You 2018), and the seismic response of the key section of the bridge corresponding to the combination is calculated and compared with that without cross tie beam, as it is shown in Table 5.

It can be seen from Table 5 that the arrangement of three cross tie beams has a significant impact on the moment bending response M_y of key sections and has a certain inhibitory effect on the

Table 5. Maximum response of key sections at different cross tie beam layout positions

Relative position of tie beam (down/up)	Bending moment M_y/(kN·m)				Displacement D_x/mm	
	pier No.44 top	pier No.44 bottom	pier No.45 top	pier No.45 bottom	pier No.44 top	pier No.45 top
0.15H/0.65H	56949.6	88800.2	63011.5	91363.2	13.3	14.1
0.15H/0.75H	41660.4	89646.6	46885.2	89411.1	13.4	13.8
0.15H/0.85H	47999.4	88854.6	26826.5	87325.5	13.9	14.2
0.25H/0.65H	55485.9	87336.7	66037.3	94038.2	13.6	14.5
0.25H/0.75H	41181.7	88478.6	49767.4	92524.9	13.3	14.2
0.25H/0.85H	47155.9	88872.3	26225.8	88964.1	13.4	13.8
0.35H/0.65H	56179.5	91621.4	67654.2	98037.7	13.9	14.8
0.35H/0.75H	39855.3	89682.6	51609.0	96876.2	13.7	14.5
0.35H/0.85H	47544.8	89778.9	26092.5	93555.7	13.3	14.2
No tie beam	86105.3	83470.4	79225.8	75457.7	16.0	16.4

displacement response D_x. The upper transverse tie beam plays a decisive role in the pier top M_y of the main pier. As the upper transverse tie beam gradually approaches the pier top, M_y of the pier top becomes smaller, and the influence of pier No.45 is more obvious than that of pier No.44; The lower transverse tie beam plays a decisive role in the pier bottom M_y of the main pier. As the lower transverse tie beam gradually approaches the pier bottom, M_y of the pier bottom gradually becomes smaller (Song 2017).

6.2 *Establish the optimization objective function of position parameters of upper and lower transverse tie beams*

Based on the combined seismic response results of the above three cross tie beam positions, the following cross tie beam position a and the upper cross tie beam position b are taken as optimization variables, the top of No.44 pier M_y, the bottom of No.44 pier M_y, the top of No.45 pier M_y, the bottom of No.45 pier M_y, the top of No.44 pier D_x and the top of No.44 pier D_x are selected as the optimization functions and being set as $y_1, y_2, y_3, y_4, y_5, y_6$ respectively, and the seismic response without cross tie beam is taken as the benchmark, The ratio of the response values of the three cross tie beams and the non cross tie beams is adopted as the optimization target value, and the overall optimization objective function is established, as it is shown in Eq.(7) and Eq.(8).

$$y_i(a,b) = y_{i,\,tie}/y_{i,\,no} \tag{7}$$

$$y(a,b) = y_1 + y_2 + y_3 + y_4 + y_5 + y_6 \tag{8}$$

6.3 *Construction of optimized response surface function*

In this paper, the lower tie beam position a and the upper tie beam position b are the optimization variables, and the binary quadratic polynomials of y_1, y_2, y_3, y_4, y_5 and y_6 are constructed as their response surface functions, as shown in Eq.(9).

$$y_i(a,b) = p_{00} + \sum_{j=0}^{1} p_{j,1-j} a^j b^{1-j} + \sum_{j=0}^{2} p_{j,2-j} a^j b^{2-j} \tag{9}$$

According to the sample values of functions y_1, y_2, y_3, y_4, y_5 and y_6 under the parameter combination, the optimized function is fitted with binary quadric surface by MATLAB program. The response surface function term coefficients are shown in Table 6, and the response surface function fitting surface is shown in Figure 8.

Table 6. Coefficient of fitting response surface function term.

Function term coefficient	y_1	y_2	y_3	y_4	y_5	y_6
p_{00}	8.088	1.009	−0.3548	0.8455	0.9017	1.095
p_{10}	−0.3734	−0.3978	1.663	−0.04872	0.9686	0.6367
p_{01}	−19.66	0.2423	4.475	1.175	−0.4895	−0.7686
p_{20}	0.492	1.799	−0.417	1.218	0.9374	0.6088
p_{11}	0.09162	−0.5681	−1.697	−0.1472	−1.875	−1.065
p_{02}	12.76	−0.0702	−4.349	−0.9586	0.6249	0.6088

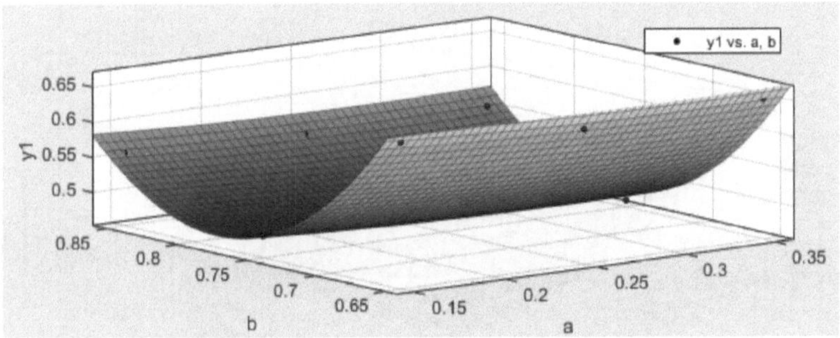

(a) Bending moment y1 response surface of pier No.44 top

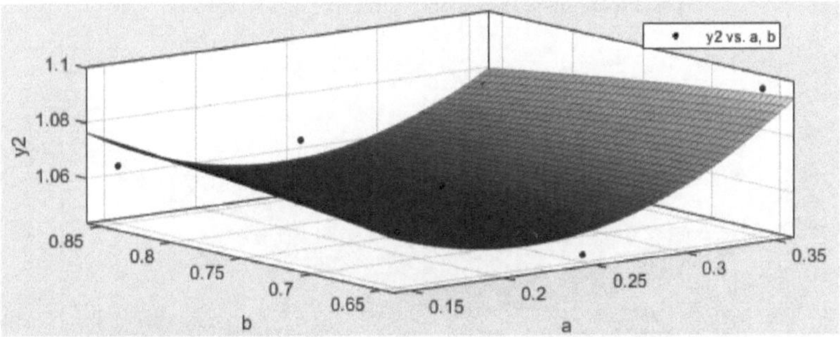

(b) Bending moment y2 response surface of pier No.44 bottom

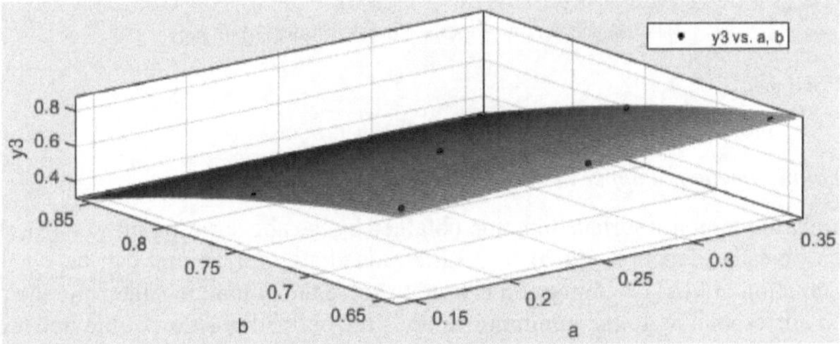

(c) Bending moment y3 response surface of pier No.45 top

Figure 8. Moment and displacement response surface of key section of pier.

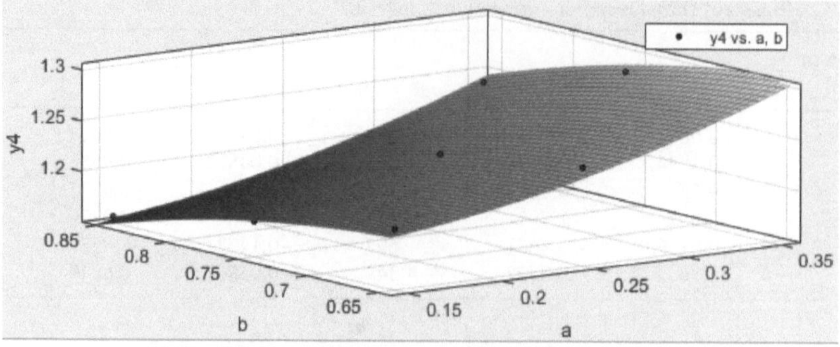

(d) Bending moment y4 response surface of pier No.45 bottom

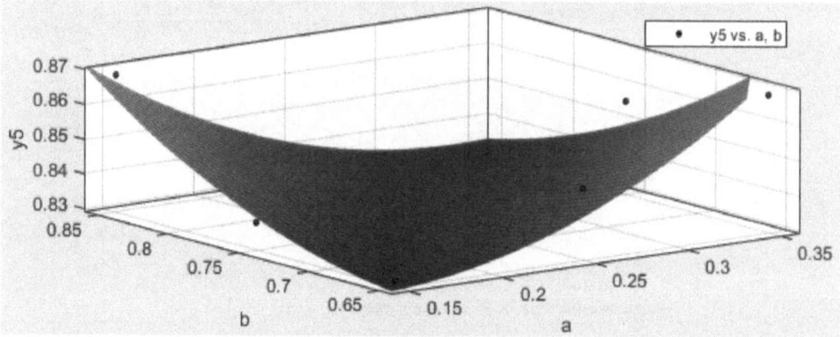

(e) Displacement y5 response surface of pier No.44 top

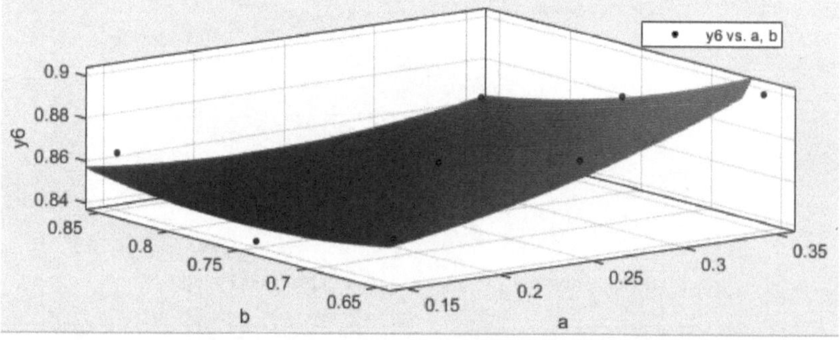

(f) Displacement y_6 response surface of pier No.45 top

Figure 8. Continued.

6.4 *Optimal parameters of upper and lower transverse tie beams*

By substituting the response surface function obtained by the above fitting into Eq.(8), the overall optimization objective function $y(a, b)$ of bivariate quadratic polynomial can be obtained. The fmincon( ) function in MATLAB program is used to solve the minimum value, and the parameter combination corresponding to the minimum value is the optimal position of the upper and lower transverse beams. After calculation, the optimal parameters of the lower and upper transverse tie beams are: $a = 0.2181$H, $b = 0.85$H. At this time, the function values of each response surface are $y_1 = 0.555$, $y_2 = 1.058$, $y_3 = 0.335$, $y_4 = 1.172$, $y_5 = 0.845$ and $y_6 = 0.852$ respectively. The seismic response comparison with the key section of the original structure under the optimal parameter combination of the three cross tie beams is obtained, as shown in Table 7.

Table 7. Comparison of seismic response between the optimal position of 3 cross tie beams and the original structure.

Earthquake response of key sections	Original structure (no tie beam)	Optimal position combination of 3 tie beams (down/middle/up=0.2181H/0.5H/0.85H)
Bending moment of pier No.44 top /(kN·m)	86105.3	47792.7
Bending moment of pier No.44 bottom /(kN·m)	83470.4	88289.3
Bending moment of pier No.45 top/(kN·m)	79225.8	26545.7
Bending moment of pier No.45 bottom /(kN·m)	75457.7	88412.6
Displacement of pier No.44 top/mm	16.0	13.5
Displacement of pier No.45 top/mm	16.4	14.0

It can be seen from Table 7 that three transverse tie beams of the main pier are arranged according to the optimal parameter combination (down/middle/up = 0.2181H/0.5H/0.85H). It is obtained that the moment response at the top of No.44 pier is 44.5% lower than the original structure, while the moment response at the bottom of No.44 pier is 5.8% higher than the original structure, and the moment response at the top of No.45 pier is 66.5% lower than the original structure, and the moment response at the bottom of No.45 pier is 17.2% higher than the original structure, and the displacement response of No.44 pier top is 15.5% lower than that of the original structure, and the displacement response of No.45 pier top is 14.8% lower than that of the original structure. Therefore, the down/middle/up tie beam position combination of0.2181H/0.5H/0.85H can optimize the seismic performance of the bridge compared with other combinations.

7 CONCLUSION

Based on the results and discussions presented above, the conclusions are obtained as below:

(1) The wall thickness of main pier, pier height ratio and the number of tie beams have significant effects on the first few modes of the bridge. With the increase of the wall thickness and pier height ratio of the main pier, the natural vibration period of the bridge gradually decreases. After reaching the high-order mode, the difference between the natural vibration period of the structure is small and basically tends to be the same.

(2) The first mode of the bridge is longitudinal drift of the main beam and bending of the main pier, mainly along the displacement of the bridge. The number of transverse tie beams of the main pier has a great impact on the first-order mode of the structure, while the other modes of the bridge are mainly side bending and vertical bending, which are less affected by the transverse tie beams.

(3) The wall thickness of the main pier increases the flexural stiffness of the pier. When the pier height ratio gradually approaches 1, the flexibility of the bridge is reduced and the thrust stiffness of the structure is enhanced. When the wall thickness of the main pier is 2.75m and the pier height ratio is 0.67, the seismic response of the key section of the pier, which is the best, is reduced compared with the original structure.

(4) Increasing the number of cross tie beams is conducive to restraining the seismic response of the bridge. The number of three cross tie beam could help the structure possess the best seismic

effect. Moreover, the bending moment M_y at the top section of pier No.44 is 52.13% lower than that of the original structure, the bending moment M_y at the bottom section of pier No.44 is 6.00% higher than that of the original structure, and the displacement D_x at the top section of pier No.44 is 16.26% lower than that of the original structure.

(5) Three transverse tie beams (down/middle/up = 0.2181H/0.5H/0.85H) are arranged along the pier height of the bridge according to the transverse tie beams of the main pier. The bending moment and displacement response at the top of the main pier decrease significantly, while the bending moment at the bottom of the pier increases slightly, which shows the best seismic performance of the structure.

ACKNOWLEDGMENTS

This work was financially supported by the Research on Design and Construction Technology of Railway High Pier and Long-span Integral Bridge in Wide and Deep Valley Area (2019K006).

REFERENCES

AQSIQ. (2009). *GB 50111-2009 Code for seismic design of railway engineering*. In: China Planning Press, Beijing.

Li, L.G. (2018). Numerical analysis of seismic resistance of high pier large span concrete continuous rigid frame bridge. *J. Highway engineering*, 43(05): 248–251+283.

Song, L.M. & Xu, Y.Q. & Yang, H.Y (2017). Study on reasonable seismic system of double leg thin-walled high pier rigid-frame bridge in high intensity earthquake area. *J. Highway*, 62(07): 110–115.

Tan, G.S. (2016). Earthquake response analysis of long-span continuous rigid frame bridge with high piers subjected to rare earthquake shock. *J. Railway Engineering*, (06): 27–30.

Teng, J. & Xu, L.Q. & Liu, J.C. & Liu, Y (2017). Effect of link beam between slender double-wall piers on the seismic responses of continuous rigid frame bridges. *Journal of Disaster Prevention and Mitigation Engineering*, 37(02): 187–193.

Wang, S.W. (2016). Key techniques for the design of double-leg and thin-wall piers of railway high-pier and large-span continuous rigid frame bridges. *J. Railway Standard Design*, 60(02): 81–84.

Xu, L.Q. & Peng, Y.C. & Zhou, S.X (2017). Study of economic span length and pier-top girder depth for open-web continuous rigid-frame bridge. *J. World Bridges*, 45(03): 55–59.

Yang, X.H. (2018). Influence of cross tie beam on anti-seismic performance of long-span continuous rigid-frame double-deck bridge with very high piers. *J. World Bridges*, 46(04): 78–82.

You, H.M. & Feng, Z.R. & Chen, B.B (2018). Influence of longitudinal tie beam on seismic performance of continuous rigid frame bridge with slender double-wall high and low piers. *J. Highway*, 63(05): 101–108.

Zhang, Y. & Xue, L. & Chen, S. & Zhou, M (2020). Parameter sensitivity analysis of continuous rigid-frame bridges in high seismic intensity regions under seismic action. *J. China Earthquake Engineering Journal*, 42(02): 311–317.

Zhang, Y.L. & Lu, X.S. & Chen, X.C (2017). Seismic Design of the Continuous Rigid Frame Bridge with Double Thin-walled Short Piers in High Intensity Earthquake area. *Journal of Railway Engineering Society*, 34(11): 45–50.

Zhang, Y.L. & Wang, Y. & Chen, X.C. & Liu, Z.W (2018). Analysis of parameters having influence on seismic responses of continuous rigid-frame bridges with double thin-wall piers. *J. Bridge Construction*, 48(04): 17–21.

Zhou, Y.J. & Yu, M.C. & Chen, Q. & Zhao, Y (2016). Uniform calculation pattern for fundamental vibration frequency of continuous rigid frame bridge with double-thin-wall piers. *Chinese Journal of Applied Mechanics*, 33(06): 1009–1015+1119.

Advances in Civil Engineering: Structural Seismic Resistance, Monitoring and Detection – Mohd Johari Mohd Yusof and Junwen Zhang (Eds)
© 2023 The Authors, ISBN 978-1-032-31491-4

Spatial variability analysis of physical and mechanical indexes of loess

Xiqi Chen, Yanjie Zhang, Xu Wang, Daijun Jiang & Jiandong Li
School of Civil Engineering, Lanzhou Jiaotong University, Lanzhou, China

ABSTRACT: Loess is the product of natural history. The spatial variability of soil parameters is complex and has strong regional characteristics, which is difficult to be evaluated accurately by the traditional random variable model. In view of this, this paper based on the Huayin test site of Zhengzhou Xi'an passenger dedicated railway, the correlation distance and spatial variability of loess soil parameters are statistically analyzed by using random field theory and semi-variogram model. The semi-variogram method is used to solve the correlation distance and variance reduction function. With the help of variance reduction function, the point variation coefficient is transformed into spatial variation coefficient, and the spatial variation coefficient is used as the basis for spatial variability evaluation. According to the semi-variogram, Kriging interpolation is used to obtain the predicted value of coefficient of collapsibility and the calculated collapse value by Kriging method. The results show that spherical model and exponential model are more suitable for the random field model of loess soil parameters. The correlation distance of soil parameters varies due to the influence of error disturbance. From small to large, they are collapsibility index, consistency index, mechanical index and physical index, and the spatial variability of mechanical index is generally greater than physical index. The correlation distance of soil parameters varies due to the influence of error disturbance, from small to large, they are collapsibility index, physical state index, compressibility index and physical property index. The spatial variability of mechanical index is generally greater than physical index. The calculated value of collapse value considering the spatial variability of soil parameters is increased compared with the standard method.

1 INTRODUCTION

Loess is a special quaternary sediment in arid and semi-arid areas. Due to its specific generation environment and storage environment, it has experienced long-term and multi cycle geological processes in a long geological age, which makes the loess soil parameters have significant spatial variability.

In the past, the modeling methods for the probability characteristics of soil parameters are random variable theory, such as parameter estimation theory of mathematical statistics and probability density function of geotechnical random parameters, but there are two major problems: first, the soil parameters of the same soil layer are treated as independent random variables, ignoring the autocorrelation between different points of soil parameters; Second, the sampling at any depth in the same soil layer is regarded as an independent test sequence at the same depth, which is essentially an analysis of the "point characteristics" of soil parameters, rather than the "spatial average characteristics" of soil parameters in the soil layer. The objective evaluation of the spatial variability of soil parameters is the theoretical basis for the reliability analysis of geotechnical engineering. Therefore, how to analyze the overall "spatial average characteristics" in the soil layer should seek a method to reasonably consider the variability of soil parameters in the spatial range.

Vanmarcke first proposed the random field model of soil profile in 1977 (Vanmarcke 1977), and then scholars at home and abroad widely studied the spatial variability of soil parameters based on

DOI 10.1201/9781003310884-35

the random field theory. Yan verified the theoretical equivalence of recursive method and correlation function method in random field theory (Yan 2007). Ni calculated the correlation distance according to the physical and mechanical indexes and static cone penetration test data of loess in Tongguan, Jiangzhang and Yan'an, and pointed out that there is a certain potential correlation between the physical state index (plasticity index) and cone tip resistance, and the correlation distance between them is basically the same (Ni 2002). Yang[4] pointed out that the reliability based on shear strength index considering spatial variability is closer to reality than conventional analysis (Yang 2007). Zhu[5] analyzed the influence of sampling spacing on stationarity and ergodicity of states (Zhu 2003). Li[6] analyzed the reasonable determination of sampling spacing and correlation distance by using the static cone penetration test data of multiple holes in a loess site in Xi'an (Li 2011). Deng and Dou analyzed the influence of spatial variability of soil parameters on slope reliability by numerical simulation method (Deng 2019; Dou 2017). Jiang verified the feasibility of using random inversion method to analyze the spatial variability of geotechnical parameters (Jiang 2020). Liu believed that with the decrease of the correlation distance of undrained shear strength, the failure probability of slope is small (Liu 2021).

For the solution of correlation distance, in addition to the recursive method of random field theory and correlation function method, another method for analyzing the randomness of soil parameters is the semi-variogram method of geostatistics. Geostatistics can be defined as the science of studying spatial natural phenomena with both structure and randomness on the basis of regionalized variable theory and with the help of semivariogram. It was originally used in the mining of mineral resources. With the development of understanding of engineering discipline, it is now widely used in all kinds of Engineering with the needs of spatial variability analysis and spatial prediction. Soulie first put forward the application of geostatistics theory to the study of spatial variability of soil parameters, and studied the spatial variability of undrained shear strength of soil by using semi-variogram (Soulie 1990). Zhang[12] quantitatively evaluated the variation law of rock and soil parameters by using semi variation function and comprehensive variation index of structural direction (Zhang 1996). Xue analyzed the influence of spatial variation characteristics on landslide deformation by using semi-variogram (Xue 2020). Tan[14] pointed out the theoretical relationship between the semi-variogram method and the correlation function method, and discussed the effects of whether the data is de trended, the initial value of fitting parameters and the number of participating curve fitting points on the correlation distance solved by the semi-variogram method (Tan 2020).

The spatial variability of loess is closely related to its region, and has strong regional characteristics. Based on the random field theory and geostatistics theory, this paper uses the semi-variogram model to calculate the correlation distance, analyzes the spatial variability of the physical and mechanical property indexes of loess at the dk354 + 150 test site of Zhengxi high speed railway in Huayin City, Shaanxi Province, and uses Kriging interpolation to calculate the collapse value of loess site. It provides valuable spatial variability parameters and collapse value calculation values for similar projects in loess area.

2 RANDOM FIELD THEORY AND SEMI-VARIOGRAM

The random field model proposed by Vanmarcke essentially studies the spatial variability of soil parameters by using homogeneous normal random field, and transitions the "point characteristic" variance of soil parameters to "spatial average characteristic" variance with the help of variance reduction function, so as to establish the relationship between starting point variability and spatial variability.

For the soil parameter point characteristic $X(t)$, if $\{X(t)\}$ is assumed to be a wide stationary random field, its "spatial average characteristic" (mean square integral) in the depth direction $(t, t+h)$ can be characterized as:

$$X_h(t) = \frac{1}{h}\int_t^{t+h} X(t)\,dt \tag{1}$$

For the soil parameter point characteristic $X(t)$ and its spatial average characteristic $X_h(t)$, its mathematical characteristics are:

$$\mu = E[X(t)] = E[X_h(t)] = \mu_s \tag{2}$$

$$\sigma^2 \Gamma^2(\mathrm{h}) = \sigma_s^2 \tag{3}$$

$$CV \cdot \Gamma(h) = CV_s \tag{4}$$

In the formula, μ, σ^2 and CV are the average value, variance and coefficient of variation of soil parameter "point characteristics"; μ_s, σ_s^2 and CV_s are the average value, variance and coefficient of variation of soil parameter "spatial average characteristics" ; $\Gamma^2(h)$ and $\Gamma(h)$ are variance reduction function and root mean square attenuation factor; H is the average thickness of soil layer.

For soil parameters, within a certain distance range, the soil parameters at each point have strong autocorrelation. Beyond this distance range, it can be considered that the soil parameters at each point are basically irrelevant, and this range is the correlation distance δ. Vanmarcke suggests that formula (5) can be used to solve the correlation distance δ:

$$\delta = \lim_{h \to \infty} h \cdot \Gamma^2(h) \tag{5}$$

The key to solve the variance reduction function is to obtain the correlation distance δ. The correlation distance of loess soil parameters is solved by using the semi-variogram method of geostatistics theory.

The definition of variogram $2\gamma(t, t+h)$ is the variance of the difference between the values of $X(t)$ at positions $t+h$ and t, and its half is recorded as the semi-variogram $\gamma(t, t+h)$ (Nie 2019):

$$\gamma(t, t+h) = \gamma_{t,t+h} = \frac{1}{2} Var[X(t+h) - X(t)] \tag{6}$$

If $\gamma(h, t+h)$ is independent of the spatial position t and only related to the distance h, it can be abbreviated as $\gamma(h)$.

According to the relationship between variance and expectation:

$$\gamma(h) = \frac{1}{2} E\left\{[X(t+h) - X(t)]^2\right\} \tag{7}$$

The steps of solving the correlation distance δ by the semi-variogram method are as follows:

(1) Select the data with continuous and uniform sampling sample size n and sampling spacing Δt;
(2) Take i =0,1,2,…,n-1 and solve $\gamma(h)$ according to formula (8):

$$\gamma(h = i\Delta z) = \frac{1}{2(n-i)} \sum_{j=1}^{n-i} \left[X(t_{j+i}) - X(t_j)\right]^2 \tag{8}$$

(3) Draw the curve of $\gamma(h) \sim h$, select three semi-variogram models of Spherical model, Exponential model and Gaussian model in Table 1 to fit the curve, select the function model with the largest coefficient of determination R^2 to obtain the corresponding parameters, and look up Table 1 to obtain the correlation distance of the corresponding model.

3 SPATIAL VARIABILITY ANALYSIS OF LOESS

Based on the test data of physical and mechanical properties of loess from two exploratory wells in the test site, a random field model is established by using random field theory to analyze its spatial variability.

Table 1. Semi-variogram models.

Semi-variogram model	Formula	Correlation distance
Spherical model	$\gamma(h) = C_0 + C\left(\frac{3h}{2\delta} - \frac{h^3}{2\delta^3}\right)$	δ
Exponential model	$\gamma(h) = C_0 + C\left(1 - \exp\left(-\frac{3h}{\delta}\right)\right)$	δ
Gaussian model	$\gamma(h) = C_0 + C\left(1 - \exp\left(-\frac{3h^2}{\delta^2}\right)\right)$	δ

3.1 *Detrended processing of test data*

Soil parameter $X(t)$ can be expressed as:

$$X(t) = T(t) + F(t) \tag{9}$$

In the formula, t is the depth coordinate, $T(t)$ is the trend component, and $F(t)$ is the fluctuation component.

The premise of applying random field theory to obtain spatial variability parameters is to process the original test data to obtain stationary or weak stationary data with zero mean value. The most commonly used method to obtain stable random field data is de trending. Linear function is usually used to fit the trend component (Lin 2015). The difference between the original data and the linear function value is the fluctuation component. Considering the fluctuation of soil parameter sample value with the trend component, formula (10) is used to process the fluctuation component, and the results are shown in Figure 1.

$$F(t)' = [X(t) - T(t)]/\sigma[X(t)] \tag{10}$$

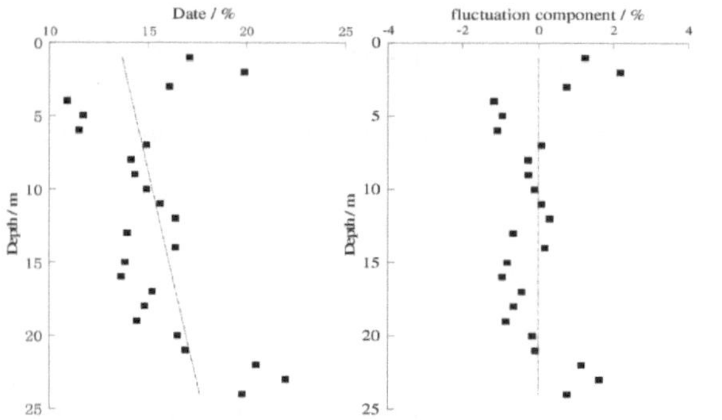

Figure 1. Detrending of date.

3.2 *Parameters of random field model*

It can be seen from formula (8) that when the value of i is larger, the data capacity of $[X(t_{i+j})$-$X(t_j)]$ is smaller, that is, when the spacing h increases, the error of $\gamma(h)$ is also increased accordingly. The calculation of correlation distance in formula (8) has accuracy only for the data within the first quarter of the sample length of the test data, and the accuracy of the data within the last three quarters of the sample length will be reduced. Therefore, the semi-variogram method is used to calculate the correlation distance, which is Only the first quarter of the $\gamma(h) \sim h$ curve is fitted.

Taking moisture content of the first exploration well as an example, its sampling spacing is 1m. Figure 2 and Table 2 show the fitting results of fluctuation components of three semi-variogram on test data and the corresponding correlation distance. The coefficient of determination of curve fitting with spherical model, exponential model and Gaussian model are 99.89%, 98.32% and 98.40% respectively. It can be seen that spherical model is the best fitting model, and the corresponding correlation distance is 3.74m. Similarly, the optimal fitting model and its correlation distance of each soil parameter of exploration wells can be obtained, as shown in Table 3.

Table 2. Correlation distance of semi-variogram fitting.

Semi-variogram model	Coefficient of determination	Correlation distance /m
Spherical model	99.89%	4.24
Exponential model	98.32%	6.90
Gaussian model	98.40%	3.03

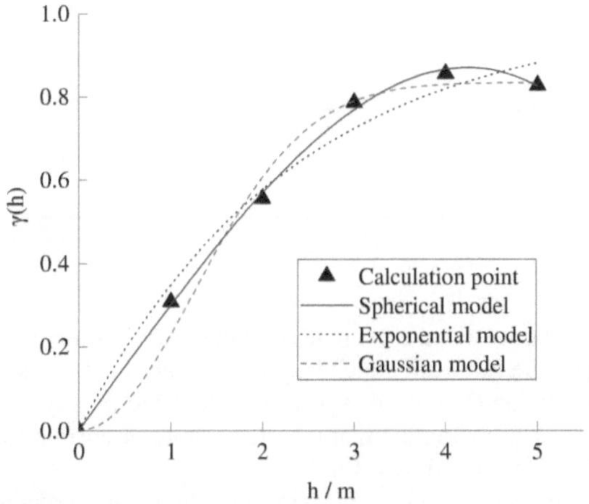

Figure 2. Semi-variogram fitting results.

The coefficient of variation is used to evaluate the variability of soil parameters. The magnitude of the coefficient of variation reflects the structure and randomness of soil parameters. Corresponding to "point characteristic" and "spatial average characteristic", there are point variation coefficient CV and spatial variation coefficient CV_s respectively. For the test data, the point coefficient of variation is:

$$CV = \sigma / \mu \tag{11}$$

Combining formula (4), formula (5) and formula (11), the spatial variation coefficient CV_s can be obtained as follows:

$$CV_s = \sigma / \mu \cdot \sqrt{\delta / h} \tag{12}$$

The correlation distance and variation coefficient of soil parameters of the two exploration wells are summarized in Table 4, and the evaluation criteria of variability are shown in Table 5.

It can be seen from Table 3 that the probability of spherical model, exponential model and Gaussian model as the optimal model is 50%, 46% and 4% respectively. Although there is no unified semi-variogram model that can be used as the random field model of loess soil parameters,

Table 3. Optimal semi-variogram model and correlation distance of loess parameters.

	Optimal semi-variogram model		Coefficient of determination		Optimal correlation distance/m		
Parameters	Exploratory well 1	Exploratory well 2	Exploratory well 1	Exploratory well 2	Exploratory well 1	Exploratory well 2	Average value
moisture content	SPM	SPM	99.89%	83.41%	4.24	4.33	4.28
natural density	SPM	EXM	99.23%	91.90%	4.85	4.70	4.77
dry density	SPM	EXM	96.54%	96.98%	4.46	3.81	4.14
specific gravity of soil particles	SPM	EXM	96.92%	93.24%	3.68	4.72	4.20
void ratio	SPM	EXM	96.14%	95.87%	5.36	3.37	4.37
saturation	SPM	EXM	99.26%	88.69%	5.50	3.95	4.73
liquid limit	EXM	SPM	98.43%	99.96%	1.52	3.03	2.28
plastic limit	EXM	SPM	98.44%	99.92%	1.84	3.12	2.48
plasticity index	EXM	SPM	98.51%	99.43%	1.17	2.94	2.05
liquid index	SPM	EXM	99.75%	53.51%	2.33	1.21	1.77
coefficient of collapsibility	EXM	EXM	94.80%	68.61%	1.19	1.01	1.10
coefficient of collapsibility under overburden pressure	GAM	SPM	97.06%	99.78%	1.18	1.51	1.34
compression modulus	EXM	SPM	96.63%	80.95%	2.99	3.50	3.25
coefficient of compressibility	SPM	EXM	72.62%	96.10%	5.19	1.97	3.58

the spherical model and exponential model are used as the optimal function. The probability of the number model is basically the same and much larger than that of the Gaussian model.

Theoretically, the correlation distance of soil layer calculated by any soil parameter should be the same, but it can be seen from Table 4 that the mean value of correlation distance of different physical and mechanical property indexes is not the same.

The first is the physical property index of loess, including water content, natural density, dry density, specific gravity of soil particles, void ratio and saturation. The correlation distance range is 4.14 ~ 4.77m, with an average value of 4.41m. Moisture content, natural density and specific gravity of soil particles are the measured indicators, while dry density, void ratio and saturation are the derived indicators. The test methods of the measured indicators are simple and similar, and will hardly disturb the soil sample. The derived indicators are converted according to the measured indicators, so the correlation distance is basically consistent with the measured indicators. The maximum difference between the correlation distance and the mean value of each parameter is only 15.22%, indicating that the random field model for this site is applicable and reasonable. Except for saturation, the spatial variability of other parameters is weak, the spatial variability of water content is 12.8%, the spatial variability is relatively strong, and the pore ratio has a positive correlation with water content, so the spatial variability of pore ratio is greater than dry density, and the spatial variability of saturation is also affected by pore ratio and water content, so the spatial variability is greater, The spatial variability was medium.

The second is the consistency of loess, including liquid limit, plastic limit, plastic index and liquid index. The correlation distance range is 1.77 ~ 2.48m, with an average value of 2.15m. The determination of liquid plastic limit adopts remolded loess, and the material structure of soil is damaged. The plastic index and liquid index are the derived indicators of liquid plastic limit, and the correlation distance is basically the same and generally small, which is 51.2% less than the physical property index. The liquid plastic limit reflects the characteristics of the physical state

Table 4. Parameters of random field model.

Parameters	Correlation distance/m			Point coefficient of variation/%			Spatial variation coefficient/%		
	1	2	Average value	1	2	Average value	1	2	Average value
moisture content	4.24	4.33	4.28	17.5	12.9	15.2	14.7	10.9	12.8
natural density	4.85	4.70	4.77	6.5	5.4	6.0	5.8	4.8	5.3
dry density	4.46	3.81	4.14	4.8	4.4	4.6	4.1	3.5	3.8
specific gravity of soil particles	3.68	4.72	4.20	0.1	0.2	0.2	0.1	0.2	0.1
void ratio	5.36	3.37	4.37	9.0	8.2	8.6	8.5	6.1	7.3
saturation	5.5	3.95	4.73	26.3	19.5	22.9	25.2	15.8	20.5
liquid limit	1.52	3.03	2.28	4.4	4.4	3.7	1.5	3.1	2.3
plastic limit	1.84	3.12	2.48	2.9	3.5	2.9	1.3	2.5	1.9
plasticity index	1.17	2.94	2.05	2.3	5.9	4.9	1.7	4.1	2.9
liquid index	2.33	1.21	1.77	−157.2	−469.5	−313.4	−98.0	−210.8	−154.4
coefficient of collapsibility	1.19	1.01	1.10	31.6	47.2	39.4	14.1	19.3	16.7
coefficient of collapsibility under overburden pressure	1.18	1.51	1.34	60.8	63.6	62.2	27.0	31.9	29.4
compression modulus	2.99	3.50	3.25	70.5	79.1	74.8	49.8	60.4	55.1
coefficient of compressibility	5.19	1.97	3.58	70.0	77.8	73.9	65.0	44.6	54.8

Table 5. Evaluation of variability.

Coefficient of variation	0%–15%	16%–35%	>36%
Degree of variation	weak variation	moderate variation	strong variation

of the soil sample when the water content changes. It changes in a small range along the depth direction, and the difference between the upper and lower soil is small. The plastic index is the difference between the liquid limit and the plastic limit, so the spatial variability of these three parameters is weak. The liquid index is not only related to the liquid plastic limit, but also related to the water content, which reflects the difference between the soft and hard of soil. The variation of any parameter of liquid limit, plastic limit and water content will rapidly aggravate the variation of liquid index, so its spatial variability is strong variation.

The third is the mechanical index of loess, including compression modulus and compression coefficient. The correlation distance is 3.25m and 3.58m, and the average value is 3.41m. It is measured by consolidation test. In the test loading process, graded loading will change the material structure of undisturbed loess, so the correlation distance is small, which is reduced by 22.7% compared with the physical property index. Because the self weight pressure of the overlying soil of the natural soil sample increases along the depth direction, the compression modulus and compression coefficient show the law of changing with the depth, so the spatial variability is strong.

The fourth is a special index representing collapsibility of loess, including coefficient of collapsibility and coefficient of collapsibility under overburden pressure. The correlation distance is 1.10m, 1.34m and the average value is 1.22m. During the test, not only compression deformation is caused by loading, but also the change of water content caused by water injection. Therefore, the correlation distance is the smallest, which is reduced by 72.3% compared with the physical property index. The point variability of coefficient of collapsibility and coefficient of collapsibility under overburden pressure is strong variation, but because its correlation distance is the smallest, the root mean square attenuation factor is the smallest, so the spatial variability is medium variation.

4 CALCULATION OF COLLAPSE VALUE BASED ON SPATIAL VARIABILITY

The conventional calculation of soil collapsibility is based on $\Delta t = i$m the coefficient of collapsibility obtained by sampling at a uniform interval or uneven interval replaces the coefficient of collapsibility of all soil within the whole section. In fact, the coefficient of collapsibility of soil should be different from each point but interrelated.

The sampling interval of exploratory well 1 is 1m, and the total sampling depth is 24.50m. Subdivide the 1m interval into 0.1m interval, and uses the semi-variogram to carry out ordinary Kriging interpolation to obtain the predicted value of coefficient of collapsibility with an interval of 0.1m, as shown in Figure 3. The predicted value at 1m is basically consistent with the measured value. The mean value of the measured value of the coefficient of collapsibility under overburden pressure is 0.0262, the standard deviation is 0.0163, the coefficient of variation is 62.1%, the mean value of the predicted value is 0.0262, the standard deviation is 0.0092, and the coefficient of variation is 35.2%; The mean value of the measured value of the coefficient of collapsibility is 0.0399, the standard deviation is 0.0126, the coefficient of variation is 31.6%, the mean value of the predicted value is 0.0399, the standard deviation is 0.0071, and the coefficient of variation is 17.9%. The predicted value is the same as the measured value, which verifies that the Kriging interpolation model is well fitted. The standard deviation and coefficient of variation of the predicted value are far less than the measured value, or even differ by an order of magnitude, which indicates that there is indeed autocorrelation between the points of soil parameters, and this autocorrelation will significantly reduce the variability, indicating that the coefficient of collapsibility interpolated considering the spatial variability of soil parameters is more in line with the actual distribution law of coefficient of collapsibility.

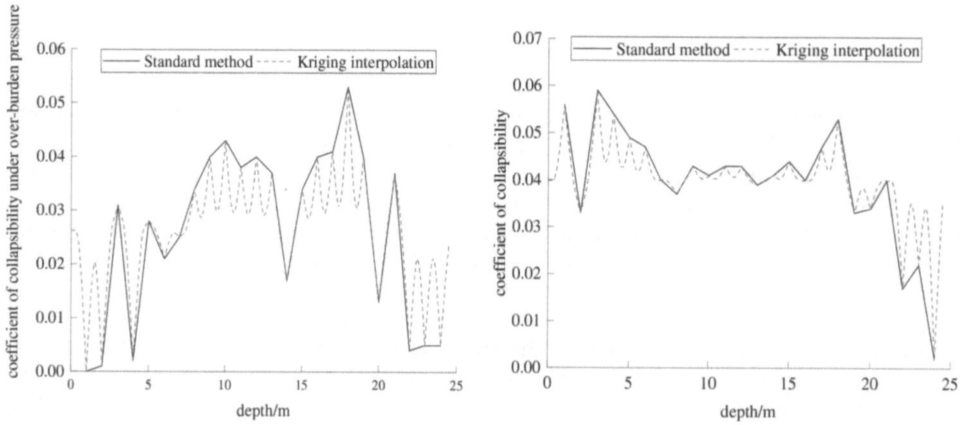

(a) Coefficient of collapsibility under overburden pressure (b) coefficient of collapsibility

Figure 3. Coefficient of collapsibility under over-burden pressure and coefficient of collapsibility.

The collapse value under over-burden pressure and collapse value are calculated according to 'Standard for building construction in collapsible loess regions (GB 50025-2018)'(China Architecture & Building Press Publishing 2018), and the results are shown in Table 6. According to the standard, the calculated collapse value under over-burden pressure of exploration wells 1 and 2 are 531mm and 396mm respectively, and the collapse value under over-burden pressure obtained by Kriging interpolation method is 556mm and 431mm respectively, an increase of 4.71% and 8.84%, with an average increase of 6.78%. The collapse value is 860mm and 690mm respectively. The collapse value obtained by Kriging interpolation method is 976mm and 726mm respectively, an increase of 13.49% and 5.22%, with an average increase of 9.36%. It can be seen that the spatial variability of soil parameters can represent the uncertainty of formation soil parameters and reveal the real uncertainty of the objective world.

Table 6. Collapse value under over-burden pressure and collapse value.

Number of exploratory well	Collapse value under over-burden pressure/mm		Collapse value/mm	
	standard method	Kriging interpolation	standard method	Kriging interpolation
1	531	556	860	976
2	396	431	690	726

5 CONCLUSION

Based on the physical and mechanical property indexes of loess, the random field theory and geostatistics are used to solve the correlation distance, analyze its spatial variability and calculate the site collapsibility. The following conclusions are drawn:

(1) Compared with Gaussian model, spherical model and exponential model are more suitable as the optimal function model of loess random field model.
(2) The correlation distance of physical and mechanical property indexes of loess varies according to the influence of test disturbance. From small to large, they are collapsibility index, consistency index, mechanical index and physical property index, and the corresponding mean correlation distance is 1.22m, 2.39m, 3.41m and 4.41m.
(3) The spatial variability of loess physical indexes is basically weak variation, and the saturation and liquid index show medium variation and strong variation due to the influence of multiple parameters. The spatial variability of mechanical indexes is strong. The variability of collapsibility index points is strong variation, but because its correlation distance is the smallest, the root mean square attenuation factor is the smallest, the spatial variation coefficient decreases, and the spatial variability presents medium variation.
(4) Compared with the standard method, the theoretical self weight collapsibility and total collapsibility deformation considering the spatial variability of soil parameters are increased by 6.78% and 9.36%. The spatial variability of soil should not be ignored in the calculation of loess collapse value.
(5) Although this paper has made a systematic study on the spatial variability of loess through random field theory and geostatistics theory, compares the difference of collapse value obtained by considering spatial variability with standard methods, and has achieved some results, it still needs more field investigation data and test data to verify. In addition to the spatial variability of soil parameters, there is also stratum variability in reality, which needs to be considered in the follow-up research.

REFERENCES

Deng Z.P., Niu J.T., Pan Min, et al. (2019). Full Probabilistic Design Method for Slopes Considering Geological Uncertainty and Spatial Variability of Soil Parameters. *Chinese Journal of Geotechnical Engineering*, 41(06): 1082–1090.

Dou H.Q., Wang H. (2017). Probabilistic Analysis of Spatial Variability of Saturated Hydraulic Conductivity on Infinite Slope Based on the Non-Stationary Random Field. *China Civil Engineering Journal*, 50(08): 105–113+128.

Jiang S.H., Liu Y., Zhang X.B., et al. (2020). Stochastic Back Analysis and Comparison of Spatially Varying Geotechnical Mechanical Parameters Based on Limited Data. *Chinese Journal of Rock Mechanics and Engineering*, 39(06): 1265–1276.

Li X.S., Liu C. (2011). Analysis of Correlation Distance and Reasonable Determination of Sampling Distance of Survey Boreholes. *Hydrogeology & Engineering Geology*, 38(04): 74–77

Lin J., Cai G.J., Zou H.F., et al. (2015). Assessment of Spatial Variability of Jiangsu Marine Clay Based on Random Field Theory. *Chinese Journal of Geotechnical Engineering*, 37(07): 1278–1287.

Liu H., Zheng J.J., Zhang R.J. (2021). System Failure Probability Analysis of Cohesive Slope Considering the Spatial Variability of Undrained Shear Strength. *Rock and Soil Mechanics*, 42(06): 1529–1539.

Ministry of Housing and Urban-Rural Development of the People's Republic of China. (2018). *GB 50025—2018 Standard for building construction in collapsible loess regions*. China Architecture & Building Press Publishing, Beijing.

Ni W.K., Niu F.J., Liu D.Y. (2002). A Study on the Spatial Self-correlation of Loess Parameters. *Journal of Xi'an Engineering University*. 24(02): 4–8+12.

Nie Z.H., Gan C.Z., Xie Y. (2019). Identification and processing method of single outlier of continuous compaction quality measured value. *Journal of Harbin Institute of Technology*, 51(03): 150–157.

Soulie, M., Montes, P., Silvestri, V. (1990). Modelling spatial variability of soil parameters. *Canadian Geotechnical Journal*, 27: 617–630.

Tan X.H., Du L.F., Fei S.Z., et al. (2020). Calculation Method and Influence Factors for Fluctuation Scale of Soil Parameters. *Transactions of the Chinese Society of Agricultural Engineering*, 36(24): 91–98.

Vanmarcke, E.H. (1997). Probabilistic modeling of soil profiles. *Journal of the Geotechnical Engineering Division*, 103(11): 1227–1246.

Xue Y., Wu Y.P., Miao F.S., et al. (2020). Seepage and Deformation Analysis of Baishuihe Landslide Considering Spatial Variability of Saturated Hydraulic Conductivity Under Reservoir Water Level Fluctuation. *Rock and Soil Mechanics*, 41(05): 1709–1720.

Yan S.W., Zhu H.X., Liu R., et al. (2007). Study on Methods for Estimating Correlation Distance of Soil Layers. *Rock and Soil Mechanics*, 28(08): 1581–1586.

Yang J.H., Liu H.D., Qin S.Q., et al. (2007). Reliability Analysis of Slope Stability Taking into Consideration of Spatial Variability of Soil Parameters. *Journal of Engineering Geology*, 15(02): 205–211.

Zhang Zheng, Liu S.C., Ju S.H. (1996). Analysis Principle and Optimal Estimation Model of Spatial Variability of Geotechnical Parameters. *Chinese Journal of Geotechnical Engineering*, 18(04): 43–50.

Zhu D.F., Gao D.Z. (2003). Space Statistical Characteristics Analysis of Smooth Stochastic Field of Soil Characteristics. *Rock and Soil Mechanics*, 24(03): 455–457+462.

Advances in Civil Engineering: Structural Seismic Resistance, Monitoring and Detection – Mohd Johari Mohd Yusof and Junwen Zhang (Eds)
© 2023 The Authors, ISBN 978-1-032-31491-4

Seismic design and analysis of twin towers with vibration absorbers

Yining Liu
Department of Chang'an Dublin International College, Chang'an University, Xi'an, China

ABSTRACT: Earthquakes are one of the most frequent and most destructive natural disasters. How to control production costs while reducing the risk of damage has been a problem in recent years. This article provides solution for this problem, which discusses the structural system of super high-rise symmetrical twin tower conjoined buildings and the seismic design of such buildings in combination with existing architectural design examples. Through model-making, we observe the role of vibration absorber and the twin-tower structure in energy absorption and shock absorption, and improve the vibration absorber by adding replacement limiter to summarize the advantages of reducing earthquake damage to the building by designing and using the vibration absorber. The conclusion is that putting the vibration absorber in the super high-rise building has advantages of better shock absorption effect and low cost compared with other shock absorption measures.

1 INTRODUCTION

Earthquakes are natural calamities that can lead to serious destruction. Approximately 5 million earthquakes occur worldwide every year, which cause varying degrees of damage to natural environment and artificial facilities, threatening human life and property. Meanwhile, in recent years, with the development of economy and architecture, it is seemed that the number of super high-rise buildings has gradually increased. Due to the beautiful appearance and efficient use of the space, one of the building types, twin towers, has gained popularity. For high-rise buildings, the threat of earthquakes is much higher than that of ordinary buildings. The reason is as follows. Above all, as for high-rise buildings, it is hard to disperse the crowd. Moreover, the higher the building, the lower is its natural vibration frequency. In such case, the seismic force and overturning moment received by the building will be higher. Finally, damage caused by earthquake will be worse.

The purpose of this paper is to reduce the vibration response of this type of building structure during earthquake and improve its seismic performance by designing a super high-rise symmetrical twin-tower conjoined building structure system, including constructing a twin-tower building model, designing a vibration absorber, and applying its energy dissipation and shock absorption effect to the model. In the long term, using the structural characteristics of the building and the energy-consuming and shock-absorbing equipment for seismic design to reduce the damage caused by the earthquake can not only limit the damage degree of the structure under the action of earthquake to make it within the range that people can expect, but also minimize the loss of people's lives, government economy, and people's property (Lu et al. 2018).

In recent years, in addition to the improvements in the building structure, energy-consuming and shock-absorbing equipment are often used in building design. For example, to mitigate the vibration damage caused by the earthquake to the main structure of the building, vibration absorbers are often introduced into the design of super high-rise buildings. Vibration absorber is also named as tuned mass damper (TMD). Since Frahm researched dynamic shock absorbers in 1909, the idea of using TMD shock absorption system was born (Yin & Wang 2019). TMD is a passive energy-consuming device, with simple structure (Li 2020). It does not need additional energy, and only needs to be installed within the structure to achieve the purpose of absorbing and consuming

DOI 10.1201/9781003310884-36

seismic energy during an earthquake (Yin & Wang 2019). The principle of TMD is also relatively easy to understand; it is actually a second order mass damping system composed of free masses, springs, and dampers attached to the column structure. During vibration, due to the difference between the main structure and the quality of the TMD, the oscillation effect is also different. In that case, through the spring between the two, an inertial force that opposes the mass inertia of the main structure is applied; therefore, the energy of the main structure can transfer to the TMD. The function of the damper increases the bandwidth of vibration control, thereby inhibiting the wider frequency band movement of the main structure and realizing shock absorption (Yin & Wang 2019). In today's building structure design, TMD has been widely used in the field of vibration reduction and earthquake resistance in civil engineering. Meanwhile, on the base of different engineering backgrounds, the number and style of vibration absorbers must be improved according to the structural characteristics of the project itself (Li 2020). For domestic, one of the most famous instances is the design of wind and earthquake resistance of Taipei 101. This is a derivative version of TMD, which is used to reduce the vibration of the tower due to wind and earthquake. Consisting of a spherical mass, high-strength steel cable, hydraulic damper, and other structures, this type of vibration absorber is similar to a simple pendulum (Li 2020). When the building is affected by acceleration, the pendulum-type tuned mass damper reduces the lateral acceleration of the building; thereby, significantly suppressing structural vibration (Li 2020). There are also many examples abroad. For instance, in Tokyo Bay Bridge, there are some specific spans, which will vibrate due to vortices, so TMD is required to add to absorb energy and reduce shock. In such a case, 8 TMDs are installed inside the box girder. There is a large frame and a small frame in the box girder. The mass of TMD is located at the lower part of the big frame and the spring and oil damping is located at the lower part of the small frame. Both parts vibrate up and down with the frames connected to each other. This improvement not only reduces the size of the TMD and saves space, but also effectively consumes energy and reduces shock (Zhang 2005).

This article first conducts theoretical analysis and designs a simplified model using AutoCAD and MATLAB, then it selects the materials needed for the physical model and desgins the physical model. Ultimately, it conducts static load test and shaking table test on the model.

2 OVERALL DESIGN

2.1 *Structural test design*

The structural design of this project is aimed at designing a symmetrical twin-tower conjoined building. The simplified structure model can highlight the essence and basic characteristics of the structure, which can be drawn by AutoCAD software. First, the overall building height of the structural design model is set to 600 mm, the number of stories is three, and each story is 200 mm divided into left and right columns, the connecting beam is located on the second floor, the span is 150 mm, and it is constrained by the horizontal load (Fx) caused by the seismic wave, the simplified structure model is shown in Figure 1.

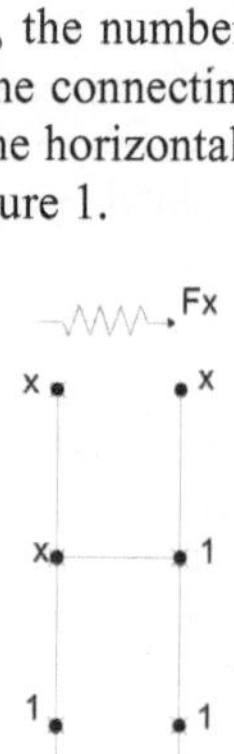

Figure 1. Simplified structure model.

To resist vibrations, there are generally two options for the design of the structural system. One is a system with higher stiffness and the other is a system with better flexibility. In a system with higher stiffness, the increase in stiffness leads to increase in elastic coefficient k. Here, the displacement decreases, as a result of the structure's ability to resist elastic deformation under the action of seismic forces strengthens. In a system with better flexibility, when the flexibility increases, the elastic coefficient k decreases as well, therefore, the natural vibration period T also increases. Combined with the acceleration response spectrum curve of El Centero, it can be seen that the increasing of period T will cause the acceleration Sa to decrease. In such a case, the seismic response gradually decreases and levels off. However, a system with higher stiffness is more expensive to manufacture, and very high stiffness renders the structure unstable. Therefore, the production of this model is based on a relatively rigid system after comprehensive consideration of economical and seismic performance and introduces the vibration absorber. To verify the correctness of the structural design methods and parameters, load tests must be carried out on the solid building structure, and there two parts of the experiment. One part is to detect the resistance and deflection of the structure, which can be achieved by applying a static load to the structure by a heavy object; the other part is to test the seismic performance of the structure, which can be achieved by applying a dynamic load to the structure through a vibrating table.

2.2 *Material performance analysis*

When carrying out structural load test, it is necessary to make a physical model corresponding to the symmetrical twin-tower conjoined building by referring to the structural model. The materials needed for the model must be tested through material performance tests to determine the tensile strength, compressive strength, and elastic modulus and other performance parameters.

2.3 *Cylinder design analysis*

To ensure the overall stability of the structure, the column is designed as a lattice system component with crossed strips in this scheme, and the lattice column is composed of two parts: limbs and embellishment.

When bearing the seismic load, the column part bears the greatest force, and the limb part mainly bears the axial force. The structural design of the building uses a relatively rigid system, so the column needs to have a higher bearing capacity. However, the column cannot have a heavy burden. In such case, vibration absorbers are placed in four positions:the second and third floors of the two towers to absorb the energy generated by the earthquake. After many comparisons, the largest paulownia pole was finally chosen to make the load-bearing limb, as shown in Figure 2.

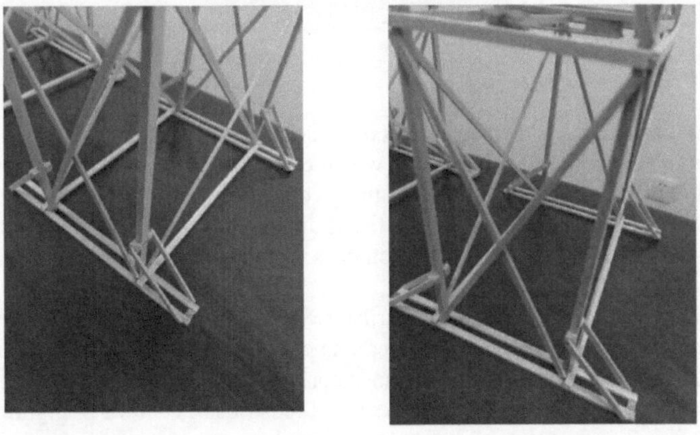

Figure 2. Cylinder design.

The embellishment forms mainly include embellishment strips and embellishment plates, which mainly resist lateral force, relative to the axial direction of the limb. There are many advantages to the installation of the garnishes. First, to prevent the structure from twisting during vibration, the garnishes on one side must be tense; in such case, setting the garnishes on both sides can prevent the structure from twisting in two directions at the same time. In addition, the setting of the strips also reduces the effective length of the unit column and saves materials. In short, the strips improve the stability of the structure, and under the same axial resistance conditions, the bending resistance of the structure is improved, and the possibility of an unbalanced state due to excessive bearing capacity is reduced

2.4 *Beam design analysis*

In this structural design, there are two types of beams, namely frame beams and connecting beams. The beam body mainly bears the vertical load, so it is less affected when it vibrates. After comprehensive analysis, the frame beams in this structure include box beams and solid web beams. To facilitate the observation of the vibration absorber, there are no strips on the inner side of the second and third layers, so the bearing capacity of the second and third layers is weak. Therefore, box beams are used on the inner side of the second and third layers of the structure, and solid web beams are used for the rest. Compared with other types of frame beams, box girder has the advantages of being lightweight, less consumable, and has good bending resistance. On the other hand, structurally, the top and bottom plates of the box beams can provide strong bending resistance; in such case, under the action of load, the vertical deformation is small, which can enhance the integrity of the structure. Therefore, box beams are used on the inner side of the second and third floors, as shown in Figure 3.

Figure 3. Beam design.

Coupling beams are beams that connect structural members. In the twin-tower building structure, it is mainly used to connect two main towers, which can enhance the structural integrity and thus the overall rigidity of the structure. When designing connecting beams, there are two options. One is to link the connecting beams on one side of the structure, and the other side is not connected, so that the two parts vibrate independently; the other is to connect both parts and design multilayer connecting beams.

For convenience of calculation, the symmetrical twin-tower conjoined structure building model will be simplified, and can draw the single-layer connecting beam and twin-layer connecting beam structure with CAD. The nodes in the figure have concentrated masses, the connecting beams are equivalent to rods, and the connecting beams connect the nodes as shown in Figure 4(a)–(b).

Then use MATLAB software to program and solve the frequency and main mode to find the response of the structure to the earthquake load. According to the results, it can be seen that the

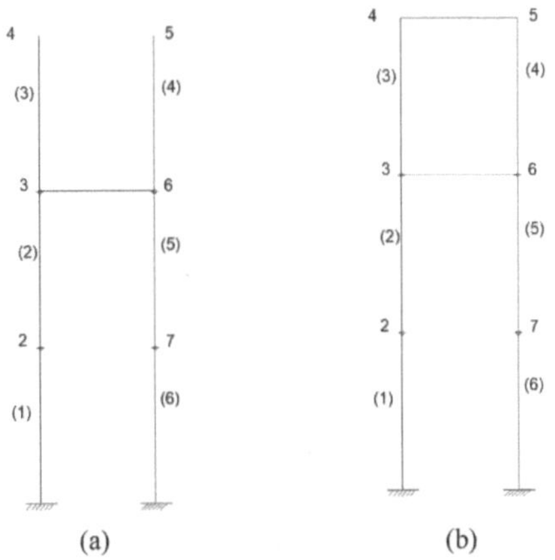

Figure 4. Structural calculation diagram under two design schemes of connecting beams.

position of the connecting beam and the stiffness of the connecting beam mainly affect the structural vibration, and the influence is as follows:

The basic natural vibration period of the structure: the conjoined position has little influence on it, and the natural vibration period of the entire structure is between the natural vibration periods of adjacent towers.

Top displacement of structure: the position of the connecting beam has a great influence on it. The connecting beam can reduce the displacement of the top layer. The higher the position of the connecting beam, the smaller the displacement of the top layer.

Displacement between tower floors: the position of the Siamese has a greater impact on it. For asymmetric Siamese structures, the connecting beams should be placed at the top level as much as possible. As shown in Figure 5.

Figure 5. The connecting beam physical model.

In summary, two layers of connecting beams should be set. Moreover, to strengthen the connection between the two main towers, improve the overall rigidity of the structure, and reduce the displacement of the topmost structure, struts and braces of the same cross-sectional size must be installed at the top of the second floor. At the same time, the two should be tightly connected, so that the integrity of the structure can be enhanced, the vibration absorber can give full play to its advantages, and reduce the loss of the structure.

3 VIBRATION ABSORBER STRUCTURE DESIGN

3.1 *Simplified structure model*

To reflect the energy absorption and shock absorption effect of the vibration absorber, two simplified structural models were established using AUTOCAD software, one with a vibration absorber and the other without a vibration absorber, as shown in Figure 6.

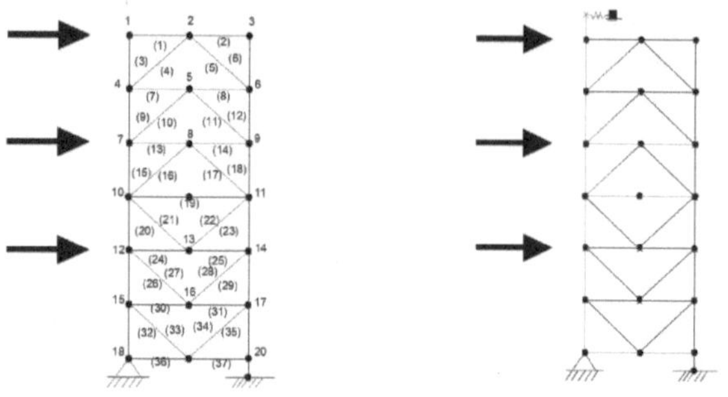

(a) Without vibration absorber (b) Node 1 plus vibration absorber

Figure 6. Simplified structure model with vibration absorber.

Use MATLAB software programming to assemble the overall stiffness matrix and overall mass matrix of the structure. It can be seen from the node displacement diagram that the vibration absorber has good shock absorption effect and bears the vertical load. After adding one or two vibration absorbers to the structure, it can be seen that the displacement of the structure under dynamic load is significantly reduced, but the displacement reduction effect of adding three vibration absorbers is smaller than that of adding two vibration absorbers. Therefore, considering the damping effect and economic factors, one or two shock absorbers can be designed for this structure.

3.2 *Final plan design*

During model-making, a physical model of a swinging vibration absorber was made using paulownia strips and other materials. Among them, the design of the vibration absorber also includes the design of the cable, the platform, and the displacement limiter.

The material used for the cable of vibration absorber is wood board. To ensure that the structure has a good tensile capacity, the wood board needs to have good toughness. Each vibration absorber has four cables, and the top section is close to the inner side of the beam ends of the second and third layers.

Because the weight blocks placed on the left and right sides of each floor are different, the load-bearing capacity of the shock absorber where 5 kg weight is loaded is stronger than that of 1 kg. Therefore, the vibration absorber where the 5 kg weight is loaded needs to use two layers of thick wood to make the cable, and the 1 kg weight uses one layer of thick wood. After shaking table test, the cables of the first and second layers fully meet the requirements of bearing capacity.

The platform of the vibration absorber is made of wooden poles that are closed and overlapped all around. To ensure that the platform has a good pressure bearing capacity, the cross-sectional area of the rod should be appropriately increased. The specific method is to use four wooden strips 100 mm long, 5 mm wide, and 6 mm high to clamp the end of the cable in pairs, so that the platform is level and the height is slightly higher than the nearby beams. The two above the wooden strips and the inside of the cable should also be overlapped. A wooden strip with a length of 100 mm, a width of 5 mm, and a height of 6 mm is used as a platform for the vibration absorber.

The design of the vibration absorber platform is also different for the platform at the 5 kg weight and the platform at the 1 kg weight. The rigidity of the platform at the 5 kg weight is greater.

Therefore, the cross-sectional area of the wooden pole required for the production of the vibration absorber platform at the 5 kg weight is larger than that at the 1 kg weight, so that the structure can be under the action of external loads. The deformation is small. After shaking table tests, the platforms on the first and second floors fully meet the requirements of bearing capacity.

In the design of this absorber, the displacement limiter is one of the highlights. In the shaking table test, it can be found that in case of a given seismic wave, the vibration absorber has a larger swing amplitude, which will cause a larger eccentric load and reduce the bearing capacity of the structure. In response to this phenomenon, a displacement limiter was designed.

The displacement limiter is located under the platform, and the structure is divided into two parts. One part is composed of four small wooden blocks with a width of 5 mm, a height of 6 mm, and a length of 18 mm, and two pieces of wood with a width of 9 mm, a length of 54 mm, and a thickness of 1 mm. Two pieces of wood are bonded under each piece of wood, and two pieces of wood are set aside. The 18 mm distance allows the vibration absorber to swing, and the two small structures are respectively attached to the underside of the two overlapping wooden strips on the platform. The other part requires a wooden strip with a width of 3 mm, a height of 5 mm, and a length of 200 mm, four small wooden blocks with a width of 3 mm, a height of 5 mm, and a length of 7 mm, and two small wooden boards with a length of 15 mm, a width of 5 mm, and a thickness of 1 mm. The wooden strip runs through the gap. Between and overlapped on the vertical beam, the ends of each wooden strip were tightened with wooden blocks, and then covered with two wooden boards. These two parts together constitute the displacement limiter of the vibration absorber.

When the vibration absorber swings, once the limit is exceeded, the small wooden blocks on both sides will limit the displacement of the wooden strips under the platform. When the vibration absorber swings left and right, when the limit is exceeded, the wooden strips under the platform can effectively restrict it, thereby absorbing vibration resilience. The size of the restricted displacement is based on the better limit size measured in the experiment, which can not only make the vibration absorber exert its effect but also reduce the structural bearing capacity due to the vibration of the vibration absorber. After several tests, it can be clearly observed that the seismic wave causes less loss to the structure, and the vibration absorber exerts a good energy consumption and shock absorption effect as shown in Figure 7.

Figure 7. Physical model of vibration absorber structure.

4 DISCUSSION

In this design of vibration absorber, one of the highlights is the adding of displacement limiters. The illustration about this is as follows.

When the vibration absorber swings, once the limit is exceeded, the small wooden blocks on both sides will limit the displacement of the wooden strips under the platform. When the vibration absorber swings left and right, when the limit is exceeded, the wooden strips under the platform can effectively restrict it, thereby absorbing vibration resilience. The size of the restricted displacement is based on the better limit size measured in the experiment, which can not only make the vibration absorber exert its effect, but also reduce the structural bearing capacity due to the vibration of the vibration absorber. After several tests, it can be clearly observed that the seismic wave has less loss to the structure, and the vibration absorber exerts a good energy consumption and shock absorption effect.

5 CONCLUSION

In this paper, a symmetrical twin tower conjoined structure and a vibration absorber structure were designed for the seismic design of super high-rise buildings, and the two structures are analyzed. On the one hand, it is necessary to analyze whether the bearing capacity of a certain line meets the requirements through static load test and shaking table test. On the other hand, CAD, MATLAB, and other software are used to draw simplified models and numerical simulations of the structure. Finally, it is verified that the swingable vibration absorber can play a good role in energy consumption and shock absorption in super high-rise two-tower conjoined structures Besides, reasonably arranging the position of the vibration absorber in the building structure and improve the vibration absorber by adding displacement limiters is also necessary.

In the seismic design of this building, the vibration absorber is one of the highlights. Compared with other anti-seismic means, the advantages of introducing the vibration absorber into the main structure are the following three points.

(1) Economy. Compared with the traditional method of increasing the rigidity of the structure, putting a vibration absorber in the structure will not cause problems such as structural instability and increased cost due to excessive mass. On the contrary, the mass of the vibration absorber is smaller, the cost is lower, and the energy consumption and shock absorption effect are more obvious, which can reduce the construction cost while effectively resisting earthquakes.
(2) Universality. Only for each specific building type under the super high-rise twin-tower conjoined structure building, the application of vibration absorbers is highly universal. In addition to the application of vibration absorbers in the design and construction phases to improve seismic resistance, they can also be used in the later reinforcement and reconstruction phase of the building.
(3) Fundamental. Different from increasing structural rigidity, the vibration absorber fundamentally reduces the vibration amplitude of the main structure to reduces the damage to the building during vibration. The vibration absorber can reduce the mass of the building itself through energy consumption while ensuring the bearing capacity, thereby improving the seismic resistance.

REFERENCES

Li Zheyan. (2020). Research on the application of TMD in domestic civil engineering [J]. *Shanxi architecture*, 1009–6825 (2020) 24-0041-03

Lu Yilin, Yin Fengxia, Guo Xianzhou. (2018). Strengthening measures and suggestions for seismic design of transfinite high-rise buildings – Taking an transfinite high-rise building as an example [J]. *Journal of seismic engineering*, 2018, 40 (06): 1252–1258.

Yin Xinsheng, Wang Qi. (2019). On the application of TMD system in Architecture [J]. *Henan building materials*, 2019 (03): 67–68.

Zhang Wenwen, Chen Wei, Li Wenping.(2005). *Wind induced vibration and control of Tokyo Bay Bridge* [J], 1671–7767 (2005) 04-0060-05

Advances in Civil Engineering: Structural Seismic Resistance, Monitoring and Detection – Mohd Johari Mohd Yusof and Junwen Zhang (Eds)
© 2023 The Authors, ISBN 978-1-032-31491-4

Numerical analysis of immersion influence on bearing capacity of pile foundation in loess area

Guo Jian* & Sun Wen
School of Civil Engineering, Lanzhou Jiao Tong University, Lanzhou China
Key Laboratory of Road & Bridge and Geotechnical Underground Engineering of Gansu Province Lanzhou, China

ABSTRACT: In order to study the mechanical behavior of pile foundation in collapsible loess area under different immersion states, the lateral friction, bearing capacity, axial force of single pile and settlement of soil around pile in collapsible loess foundation in Lanzhou area under three conditions (non-immersion state, loading after soaking state, loading until the soil approached saturation point) are studied by numerical analysis. The analysis results show that the bearing capacity of single pile foundation is significantly reduced after saturated loess foundation is immersed. The immersion causes obvious collapsible deformation of loess foundation and negative friction resistance occurs on the side of pile. The axial force distribution of pile body also changes with the amount of immersion. In natural state, the side friction resistance of the pile is positive, the friction resistance above the middle of the pile body is small, and the friction resistance below the middle of the pile body gradually increases. The negative friction resistance is not only related to the water-bearing state of the foundation soil, but also to the sequence of immersion and loading and the vertical load on the top of the pile. The bearing capacity of single pile decreases with the increase of water immersion. Axial force of pile body redistributes with the increase of immersion volume of pile and soil; Settlement deformation of the soil layer around the pile is caused by both collapsible deformation of loess and negative friction resistance on the side of the pile, which increases the collapsible deformation. The research results have important theoretical value and guiding significance for the design and construction of pile foundation in the loess area of Lanzhou.

1 INTRODUCTION

In recent years, the foundation engineering in collapsible loess is increasing, and the influence of loess collapsible deformation on structural foundation also increases gradually, especially in the structure with pile foundation. It is of great significance to study the load transfer law of pile foundation in collapsible loess foundation, the distribution of friction resistance on the surface of pile and the collapsibility of loess, which is of great significance to the design and treatment of pile foundation in collapsible loess area (Liu 1996; Qian 1985). As the engineering experience on loess collapsibility is not perfect and the understanding of loess collapsibility mechanism is not comprehensive, how to correctly grasp the collapsible deformation characteristics of pile foundation in loess area and eliminate the engineering problems caused by collapsible deformation has become a difficult problem for engineering technicians (Guan 1990; Xie 2001). The numerical analysis method simulates the impact of rainfall on the bearing capacity of pile foundation in collapsible loess, which is more simple, economical and effective than the immersion test. At the same time, it can well reflect the collapsible characteristics of loess and the change of bearing capacity of pile foundation (Feng 2000; Huang 2007; Wang 2017). In order to study the influence of rainwater

*Corresponding Author: gjhealth@163.com

DOI 10.1201/9781003310884-37

on the lateral friction and bearing capacity of pile foundation in collapsible loess area, the action mechanism of pile-soil interface and the difference of vertical bearing capacity of single pile under three conditions, natural state, soaking after loading and loading after soaking of loess foundation in Lanzhou area is analyzed by numerical method. The influence of the collapsible deformation of loess caused by rain on the settlement, negative friction and axial force of pile body is studied. The stress characteristics of pile foundation under the action of rain are discussed. The research results are valuable for improving the design and construction methods of pile foundation in loess area, and studying the settlement and deformation law of pile foundation under the action of rain water and pile top load.

2 SITE ENGINEERING CONDITIONS AND ANALYSIS CONDITIONS

The distribution of foundation soil layers of the analysis site is shown in Figure1. The soil layers from the ground to the pile bottom are Holocene loess, Aeolian Loess, upper Pleistocene loess and Miocene loess respectively. The pile foundation is constructed by dry manual hole digging process. After construction, the concrete quality of the pile body is good, there are no obvious defects, the pile body is complete, and the slag at the bottom of the hole is basically clean.

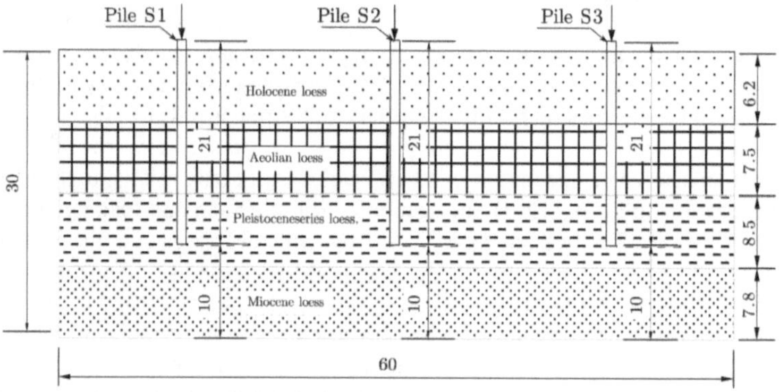

Figure 1. Pile and soil layer layout.

In order to study the influence of different rainfall conditions on the bearing capacity of pile foundation in loess foundation, the numerical analysis is divided into the following three working conditions: first, natural state. The bearing capacity characteristics of pile foundation constructed without rainfall and collapsible deformation in practical engineering is studied in his working condition. Second, the pile foundation shall be constructed first, and then the vertical load on the pile top shall be applied. After the pile deformation is stable, the foundation shall be soaked to the saturated state, and the pile top load shall be applied after the soil collapse deformation is stable. This working condition mainly studies the stress characteristics of the pile in case of rainfall and vertical load on the pile top after pile foundation construction. Third, the foundation soil is saturated by continuous immersion, and then the load is applied to the pile top after the soil collapsible deformation is stable. This working condition mainly studies the mechanical characteristics of the pile after continuous rainfall and serious collapsible deformation of the foundation soil.

3 NUMERICAL ANALYSIS METHOD

3.1 *Overview of numerical mode*

According to the pile length, pile diameter and soil layer thickness, the length and width of the numerical model are 10 times of the pile diameter, the thickness of foundation soil is 40 times of the pile diameter, and the thickness of soil layer at the bottom of the pile is 10 times of the pile diameter. The dimensions of the numerical model are: 20m long, 20m wide, 30m high, in which 1.0m is reserved at the pile head for loading. Both soil and pile foundation are simulated by solid element, and pile foundation is simulated by isotropic elastic constitutive model. The total number of grid elements is 51244, including 55496 soil elements and 56 pile elements. The bottom of the model is fixed, the horizontal displacement is constrained, and the upper boundary is a free. Considering the symmetry, half of the soil is taken for analysis. The finite element model is shown in Figure 2.

Considering the obvious plastic deformation characteristics of loess, the loess is simulated by Mohr Coulomb constitutive model. At present, Mohr Coulomb criterion is widely used in theoretical analysis of granular materials under monotonic load, and it is more in line with the geotechnical test results (Cao 1986; Li 2004; Liu 2008). However, due to the sharp angle of the yield surface, the continuous smooth elliptic surface is generally used as the plastic potential surface [8-9]. It is assumed that the pile and soil are homogeneous and isotropic materials, and the depth of foundation soil in horizontal and vertical directions is infinite. The stress characteristics of pile-soil interface are simulated by contact element, and the interaction between pile and soil is simulated by Coulomb sliding friction model.

Figure 2. Finite element model.

3.2 *Pile soil interaction*

3.2.1 *Calculation of bearing capacity of friction pile*

According to the actual project, the pile foundation is a friction pile. Without considering drainage, the bearing capacity of the friction pile is mainly determined by the pile lateral friction and pile end friction. The ultimate lateral friction around the pile is generally calculated by formula (1):

$$f_s = \alpha c_u \tag{1}$$

Where α is the empirical coefficient. c_u is the undrained strength of soil.

Different scholars have done a lot of research on α and have many empirical formulas. In this paper, the calculation formula given in API standard is adopted, as shown in formula (2):

$$\alpha = \begin{cases} 1 - \dfrac{c_u - 25}{90}, & 25kPa < c_u < 70kPa \\ 1.0 & 25kPa \leq c_u \\ 0.5 & c_u \geq 70kPa \end{cases} \tag{2}$$

The ultimate friction of pile end is generally calculated by formula (3):

$$f_b = (c_u)_b N_c \tag{3}$$

Where $(c_u)_b$ is the drainage strength of the pile bottom soil. N_c is the bearing capacity coefficient, generally taken as 9.0.

Since loess belongs to cohesive soil, the effect of groundwater should be considered in the analysis, so the effective stress is generally used for calculation, and the ultimate lateral friction can be calculated by formula (4):

$$f_s = \mu \sigma_h' \tag{4}$$

Where σ_h' is the horizontal effective earth pressure. μ is the friction coefficient between pile and soil. According to the theory of soil mechanics, the relationship between horizontal effective earth pressure and vertical earth pressure is $\sigma_h' = k_0 \sigma_v'$, then the ultimate lateral friction around the pile can be expressed in the form of formula (5):

$$f_s = \mu k_0 \sigma_v' \tag{5}$$

Where k_0 is the horizontal earth pressure coefficient.

For normally consolidated soil, $k_0 = 1 - \sin\varphi'$, φ' is the effective friction angle of soil. The friction angle of the soil around the pile can be taken as $(0.75 \sim 1)\varphi'$, generally, the friction angle of soil changes little, therefore k_0 is also changed little. Therefore, when the effective stress method is used for the analysis, the ultimate end friction at the bottom of the pile can be formulated as:

$$f_b = (c')_b N_c + (\sigma_v')_b N_b \tag{6}$$

Where $(c')_b$ is the effective cohesion of the soil under the pile. $(\sigma_v')_b$ is the vertical effective stress of the soil under the pile. N_c and N_b is the bearing capacity coefficient, it is generally calculated by the formula (7, 8), which is proposed by Janu.

$$N_b = (\tan\beta + \sqrt{1 + (\tan\beta)^2} e^{2\eta \tan\beta} \tag{7}$$

$$N_c = (N_q - 1)\cot\beta \tag{8}$$

Where β is the angle that controls the shape of the failure surface at the pile end, according to the different properties of the soil, it varies from 0.33 (clay) to 0.58 (sand), and can be taken according to experience in calculation.

3.2.2 *Simulation of pile-soil interaction*

The normal behavior of the contact surfaces of two materials in numerical analysis is generally simulated by the contact pressure interference model. This model stipulates that normal pressure can only be transferred when the contact surface between two objects is compressed. Normal pressure cannot be transferred when there is a gap between two objects and there is no limit on the normal pressure. When the contact pressure between the pile-soil contact surface becomes zero or negative, the contact surface separates and the normal pressure is no longer transferred, so the contact constraint disappears. This behavior is called "hard" contact. The normal force is transferred when the contact surfaces contact each other and the tangential force, friction, is also transferred between the contact surfaces. The tangential behavior of the contact surfaces of two materials is usually simulated by friction model. When the contact surfaces are closed (under contact pressure), the tangential behavior between the contact surfaces is called friction.

When the friction force is less than the limit friction force τ_{cret}, the contact surfaces are bonded together. When the friction force is greater than the limit friction force τ_{cret}, the contact surfaces are in a slip state.

In numerical analysis, Coulomb's law is generally used to calculate the friction force, i.e., $\tau = \mu p$, where μ is the coefficient of friction, p is the normal pressure of the contact surface.

When bonding function is considered, $\tau_f = \mu^* p$, and $\mu^* = \mu + c/p$. Where μ^* is the equivalent coefficient of friction considering the interaction of bonding and friction.

3.3 *Numerical simulation of pile soil seepage stress coupling*

3.3.1 *Numerical analysis method of loess collapsibility*

Collapse is a special deformation that occurs when water immerses into loess. Due to the combined action of water and load, the internal composition and structure of the soil body have changed, which shows that the loess has undergone significant collapsible deformation on the macro level. The deformation process of loess under load-water coupling is essentially a process of structural damage evolution and structural strength reduction of loess. The experimental research shows that with the increase of water content, the structural strength of loess decreases gradually and the collapsibility of loess increases accordingly. The structural strength is the main factor influencing the collapsibility deformation of Loess [10]. When analyzing the collapsible deformation of loess under the coupling action of load and immersion by numerical method, the method of reducing the mechanical properties of loess, such as elastic modulus, internal friction angle and cohesion, is mainly adopted according to the measured mechanical properties of loess (Hang 2006; Hu 2002; Li 1993; Wang 2012; Zhu 2004).

3.3.2 *Analysis of water soil fluid structure coupling*

Seepage of soil is related to pore ratio. When analyzing seepage of soil, it is necessary to define initial stress state, initial pore ratio, initial pore pressure, initial effective pressure, etc. Pore fluid in initial state is in hydrostatic pressure equilibrium state, at which time equilibrium condition should be calculated by Formula (9):

$$\frac{du_w}{dz} = -\gamma_w \tag{9}$$

Where γ_w is the weight of water.

It is generally considered that γ_w is independent of the vertical coordinate Z and integrals equation (10),

$$\frac{d}{dz} = -u_w = \gamma_w(z_w^0 - z) = \gamma_w \Delta z \tag{10}$$

Where z_w^0 is the height of the free water surface, here $u_w = 0$, above this height is the unsaturated region ($u_w < 0$), If shear stress is ignored, the balance condition of vertical total stress can be formulated as:

$$\frac{d\sigma_{zz}}{dz} = \rho_g g + S_r n^0 \gamma_w \tag{11}$$

Where ρ_g is the dry density of soil. g is the acceleration of gravity. n^0 is the initial porosity. The relationship between initial porosity and initial void ratio can be formulated as $n^0 = e_0/(1 - e_0)$. S_r is saturation.

According to the relationship between the effective stress and the total stress can be formulated as the following formula:

$$\overline{\sigma} = \sigma + S_r u_w I \tag{12}$$

By deriving equation (13), the result can be express as follows:

$$\frac{d\bar{\sigma}_{zz}}{dz} = \rho_g g + \gamma_w \left[S_r \left(1 - n^0\right) - \frac{dS_r}{dz}\left(z_w^0 - z\right) \right], z < z_0^1 \tag{13}$$

$$\frac{d\bar{\sigma}_{zz}}{dz} = \rho_g g, \quad z \geqslant z_0^1 \tag{14}$$

Where z_0^1 is the elevation at the boundary between dry soil and saturated soil. When $z > z_0^1$, it is dry soil, $S_r = 0$. when $z_0^1 > z > z_0^w$, it is partially saturated soils, the saturation needs to be measured. when $z < z_0^w$, it is completely saturated soil, $S_r = 1$.

For partially saturated soils, the saturation needs to be measured. After rainfall on the Loess foundation, when rainwater seeps into the soil, the shear strength of the loess will be reduced, and the groundwater level will also be raised, which will increase the pore water pressure in the soil. In addition, a short-saturated area will appear above the groundwater level due to high intensity rainfall for a long time, which will increase the pore water pressure. The saturation degree of loess has a significant influence on collapsible deformation, while the saturation degree of loess is related to the matrix suction of soil. For pore water in unsaturated soil, the water pressure is $u_w < 0$, $-u_w$ which reflects the capillary suction of material (matrix suction). Considering the possible hygroscopic and dewatering characteristics of loess, the saturation degree of soil under certain matrix suction should be within a range. The relationship between soil saturation and matrix suction is shown in Figure 3. The equation of soil moisture absorption curve can be expressed as follows (Khalilin 2008, Fei 2010):

$$\left.\begin{aligned} u_w &= \frac{1}{B}\ln\left[\frac{S_r - S_{r0}}{(1-) + A(1 - S_{r0})}\right], \quad S_{r1} < S_r < 1 \\ u_w &= u_w|_{S_{r1}} - \left.\frac{du_w}{dS_r}\right|_{S_{r1}} (S_{r1} - S_r), \quad S_{r0} < S_r < S_{r1} \end{aligned}\right\} \tag{15}$$

Where A, B is the material parameter. S_{r0} and S_{r1} is determined by the moisture absorption and dehydration curve of soil.

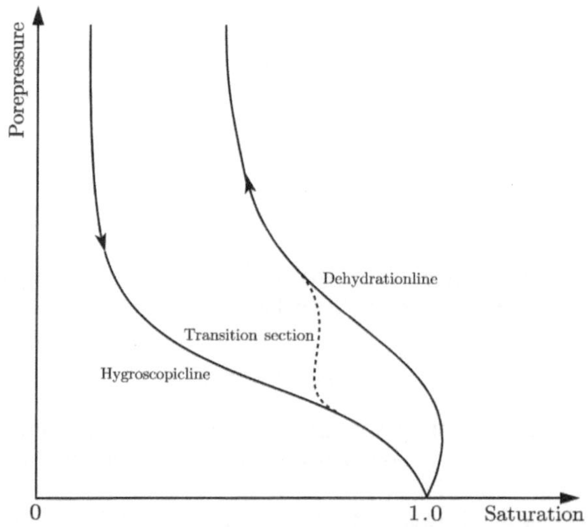

Figure 3. Relationship between soil saturation and matric suction.

3.3.3 *Fluid solid coupling boundary conditions*

In numerical analysis of fluid-solid coupling, pore flow related to pore pressure is generally associated with the flow rate and pore pressure of pore fluid, which can be expressed as

$$u_n = k_s(u_w - u_w^{\infty}) \tag{16}$$

Where v_n is the velocity in the normal direction of the boundary. k_s is the seepage coefficient. u_w is the pore water pressure on the boundary. u_w^{∞} is the reference pore pressure.

Pore pressure is defined as positive and the flow rate of pore fluid is proportional to the pore pressure on the boundary allowing free flow of water. When the pore pressure is negative, the flow rate is 0, and the fluid will not enter the region.

Here, $v_n = k_s u_w$, $u_w > 0 v_n = 0$, $u_w \leq 0$, When the value of seepage coefficient k_s is large, the pore pressure at the boundary can be approximated to 0.

It is also assumed that the surface runoff will not occur during the analysis, that the rainfall will all infiltrate and the water infiltration rate will remain unchanged, and that the rainfall intensity is less than the saturated permeability coefficient of the soil. In the calculation, the rainfall boundary is taken as the flow boundary. When the rainfall intensity is less than the saturated permeability coefficient of the soil, this treatment is appropriate (Li, 2015; Santoso 2011; Zhu 2006).

In order to study the change of pile lateral friction resistance in collapsible loess foundation under the combined effect of continuous rainfall and load. The numerical simulation method of rainfall is to add seepage water with unit flow rate to the top surface of foundation soil. In order to simulate the most unfavorable rainfall, according to meteorological data, when the maximum rainfall in Lanzhou area is 0.04m/h, the seepage coefficient of soil is calculated by formula (17).

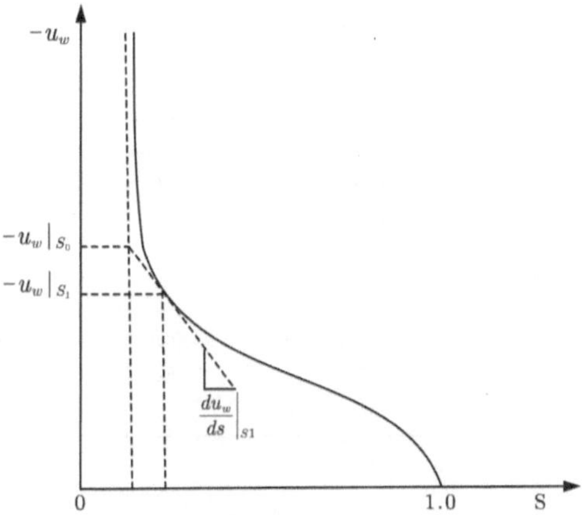

Figure 4. Moisture absorption and dehydration curve of soil.

$$K_w = \frac{a_w K_{ws}}{\left[a_w + (b_w \times (u_a - u_b))^{cw}\right]} \tag{17}$$

Where K_w is the permeability coefficient of saturated soil, according to the experiment it is 0.0020 m/h. u_a, u_w is the air pressure and water pressure in the soil, and the top of the actual foundation soil is in contact with the atmosphere, therefore $u_a = 0 \cdot a_w, b_w, c_w$. is the material coefficient of soil, the values can be 1000, 0.01, 1.65 according to engineering experience. The relationship between soil saturation and soil suction can be expressed by equation (18),

$$S_r = \frac{S_i + (S_n - S_i)\, a_s}{\left[a_s + (b_s \times (u_a - u_b))^{c_s}\right]} \tag{18}$$

Where S_r is the saturation of the soil. is the residual saturation of soil. S_n is the maximum saturation of the soil, according to the experiment it is 1. is a_s, b_s, c_s the material coefficient of soil, it is 1, 0.000005, 3.5, according to engineering experience.

According to the laboratory test results, the physical and mechanical parameters of loess in the experimental site are shown in Table 1 and those of loess and piles are shown in Table 2.

Table 1. Physical properties of soil.

	Physical properties of soil			Limiting moisture content	
Soil sample	Moisture content (W)%	Dry density (g/cm^3)	Void ratio (e)	Liquid limit (W_L)%	Plastic limit (W_P)%
Holocene loess	12.6	1.35	1.09	30.4	19.6
eolian loess	6.6	1.42	1.0	30.6	20.4
upper Pleistocene loess	9.7	1.44	1.0	29.6	19.1
Miocene loess	9.0	1.56	1.0	29.7	19.0

Table 2. Physical and mechanical properties.

Material	Elastic modulus/ E (MPa)	Poisson's ratio/v	Density/γ (kN/m^3)	Cohesion/c (kN/m^2)	Friction angle (°)
Holocene loess	210	0.35	1450	16.9	21.1
Aeolian loess	275	0.30	1430	27.6	22.9
Upper Pleistocene loess	326	0.33	1420	21.2	24.4
Miocene loess	486	0.25	1400	25.9	23.3
Reinforced concrete pile	31000	0.35	2500	–	–

4 RESULT ANALYSIS

4.1 *Q-S curve of pile and plastic zone of soil*

4.1.1 *Natural state*

Figure 5 shows the Q-S distribution curve of pile when the foundation soil is in natural state. It can be seen from the figure that when the load is less than 1000kN, the Q-S curve changes approximately linearly with the increase of the load, and there is no obvious inflection point. When the load increases from 1720kN to 2150kN, the first inflection point of Q-S curve appears; When the load increases to 3870kN, the Q-S curve shows a sharp drop, which indicates that the resistance at the end of the pile reaches the limit, the soil produces plastic deformation, and the calculated vertical displacement at the top of the pile reaches 85mm. According to the Q-S curve, the ultimate bearing capacity of the pile is 3870kN.

4.1.2 *Loading before immersion state*

Firstly, the vertical load is applied to the pile top to 1936kN, and then the foundation soil is soaked to saturation before loading. In this state, the Q-S curve of the pile is shown in Figure 6. It can be seen from the figure that before the foundation soil is soaked, when the load is less than 1720kN, the Q-S curve changes approximately linearly without obvious inflection point. When the load increases from 1720kN to 1936kN, the first inflection point of Q-S curve appears, and the pile top settlement begins to increase, reaching 27mm. When the load increases from 2200 kN to 3400kN, the Q-S curve decreases obviously. According to the Q-S curve, the ultimate bearing capacity of the pile is 3400kN.

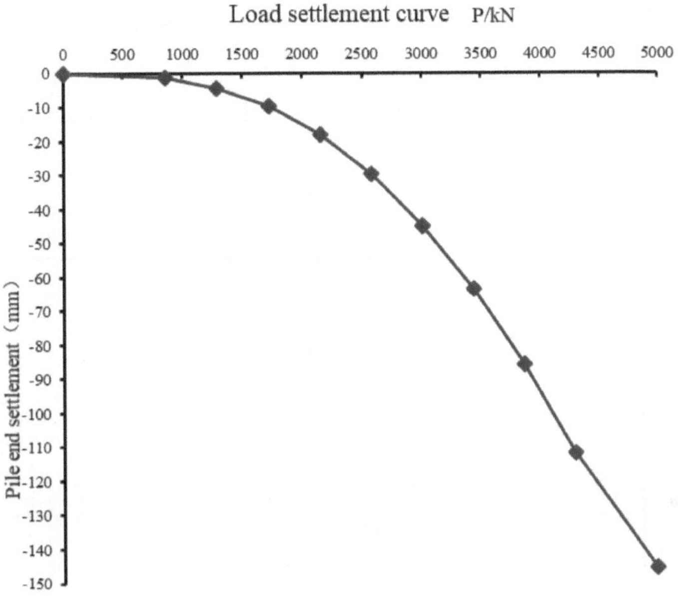

Figure 5. Q-S curve of pile (Natural state).

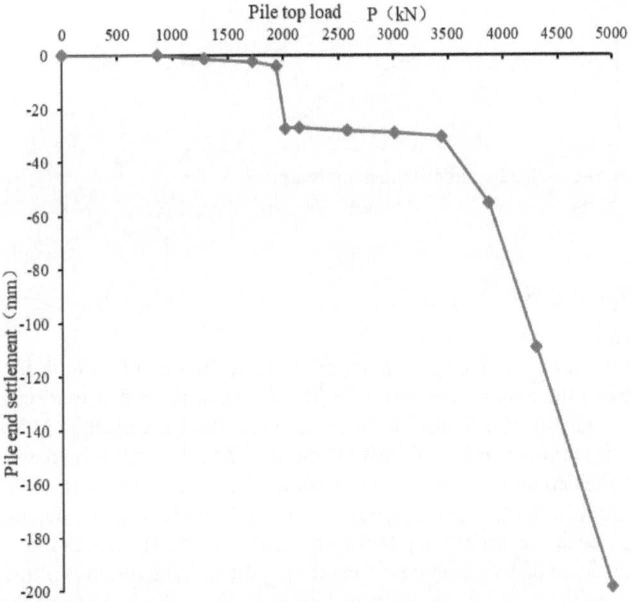

Figure 6. Q-S curve of pile (loading before immersion state).

4.1.3 *Loading until the soil approached saturation*

This condition simulates the mechanical characteristics of the pile after the foundation soil is saturated with water after rainfall and when the pile foundation is constructed later. The Q-S distribution curve of the pile is shown in Figure 7. It can be seen from the figure that when the load is less than 3000kN, the Q-S curve has no obvious inflection point. When the load increases to 3440kN, the first inflection point of Q-S curve appears, and then the curve begins to drop sharply. When the load increases to 4300kN, the vertical displacement has reached 80mm, but it is impossible in practice. According to the Q-S curve, the ultimate bearing capacity of the pile is 3580kN.

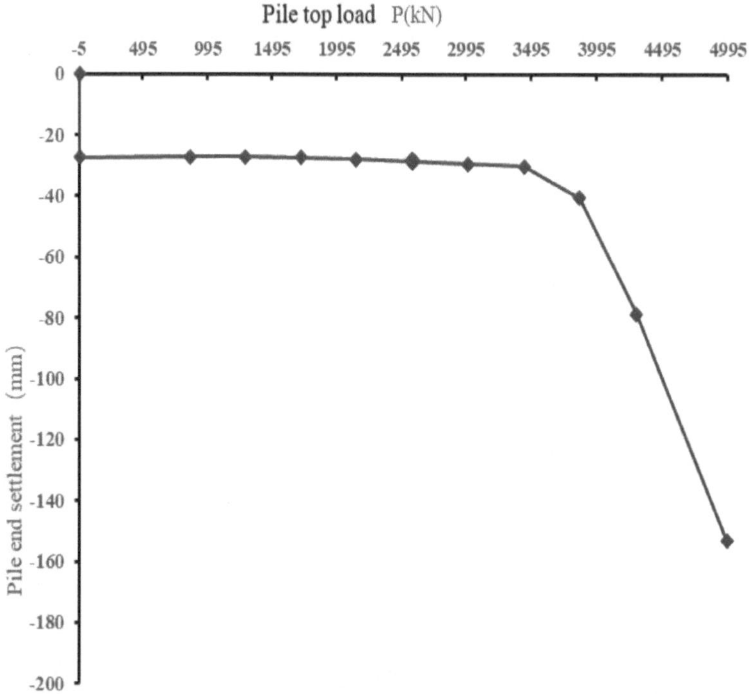

Figure 7. Q-S curve of pile (loading after immersion state).

4.2 *Load transfer law of pile*

4.2.1 *Natural state*

The axial force distribution curve of pile in natural state is shown in Figure 8. It can be seen from the Figure 8 that when the load is less than 1720kN, the axial force distribution is more uniform, and the axial force near the pile bottom is smaller. When the load reaches 2150kN, the bending point of axial force is approached 1/2 of the pile length; When the load is increased to 2580kN, the lateral friction is balanced and the anti-bending point of axial force moves downward. When the load increases to 4300kN, the anti-bending point of the axial force diagram moves to the bottom of the pile, the axial force at the top of the pile increases, and the axial force at the bottom of the pile decreases. The axial force distribution conforms to the force characteristics of the friction pile.

In the natural state, the distribution curve of lateral friction of pile body is shown in Figure 9. It can be known from the figure that when the load is less than 1290kN, the distribution of lateral friction is more uniform. When the load is more than 2150kN, the lateral friction is smaller in the upper part of the pile, and gradually increases in the middle and lower part, which is opposite to the

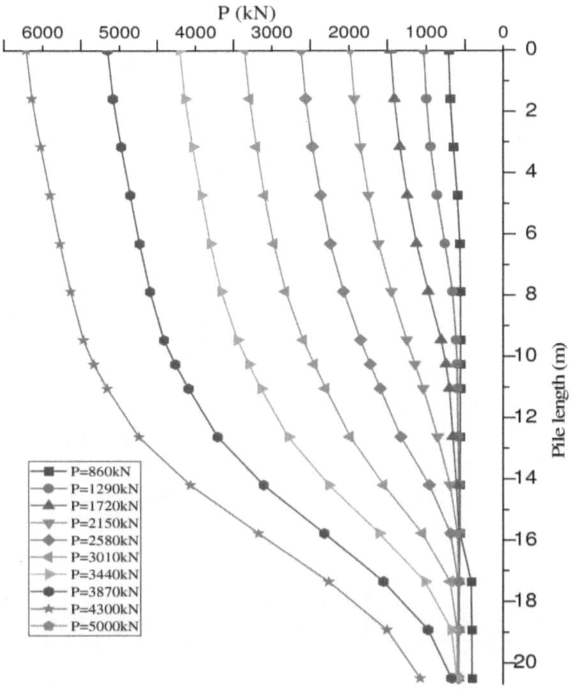

Figure 8. Axial force distribution curves (natural state).

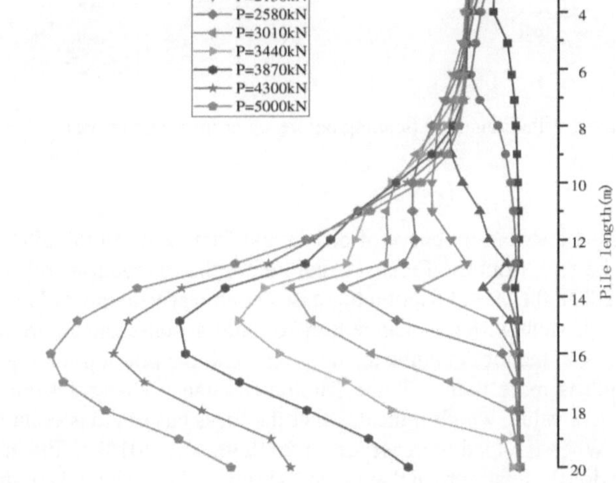

Figure 9. Friction distribution curve (natural state).

change trend of the pile axial force. This is because the vertical earth pressure gradually increases, the horizontal earth pressure also increases, and the horizontal earth pressure on the upper part of the pile body is small, so the pile side pressure is small, and the horizontal earth pressure on the lower part of the pile body gradually increases, so the lateral friction resistance also increases. When the load is greater than 3010kN, the friction of the lower part of the pile is much greater than that of the upper part, and this trend is more obvious with the increase of the load.

4.2.2 *Loading before water immersion state*

Figure 10 shows the axial force distribution curve of the pile when it is loaded first and then soaked. It can be seen from the Figure 10 that before the foundation soil is soaked, when the load is less than 1720kN, the change trend and value of the axial force are close to the natural state. When the pile continues to load after immersion, the axial force decay rate of the lower part of the pile decreases gradually. When the load is increased to 1930kN, as the soil is soaked and the pile top load remains unchanged, the axial force at the upper part of the pile body becomes significantly smaller and that at the lower part of the pile body becomes significantly larger. Then, when the load continues to increase, the axial force does not change significantly.

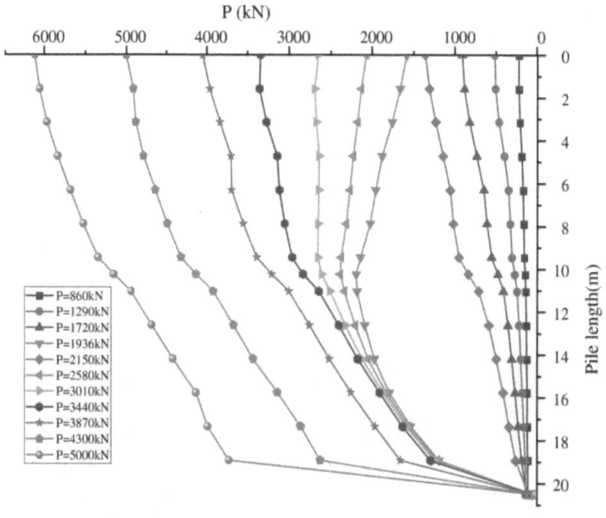

Figure 10. Axial force distribution curve (loading before water immersion state).

Figure 11 shows the distribution curve of pile lateral friction when the pile is soaked first and then loaded. It can be seen from the Figure11 that before the foundation soil is soaked, when the load is less than 1720kN, the lateral friction increases nonlinearly downward along the pile length, which is basically consistent with the change trend of natural state, and all are positive. When the load is equal to 1936kN, the loess begins to submerge, and the pile top load remains unchanged. When the pile length is more than 1/2, the friction resistance changes from negative value in natural state to positive value, which indicates that the loess has obvious collapsible deformation after submergence. When the load increases from P=1936kN to 3010kN, the increase of negative lateral friction is obviously smaller than that in natural state; When the load continues to increase to 3400kN, the negative lateral friction area of the pile decreases; When the load continues to increase to 3870kN, all of them become positive friction along the whole length of the pile.

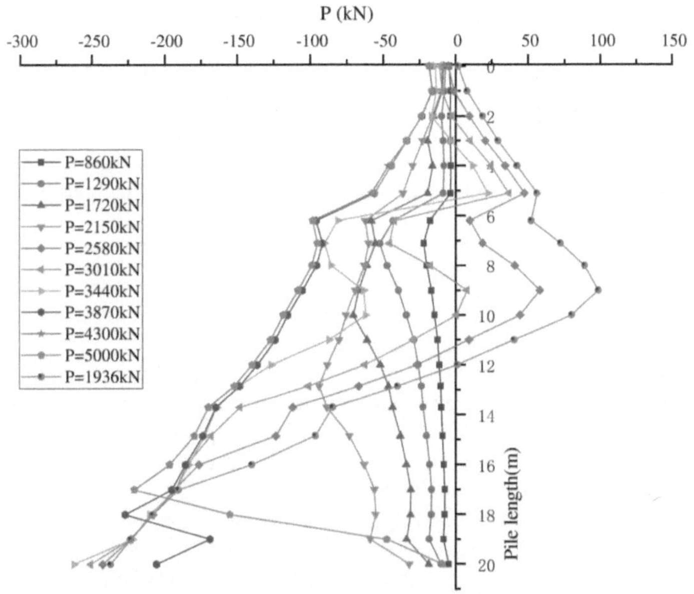

Figure 11. Friction distribution curve (loading before water immersion state).

4.2.3 *Loading after water immersion state*

The axial force distribution curve of the saturated foundation soil under loading is shown in Figure 12. It can be seen from the Figure 12 that when the load is less than 2150kN, the axial force at the upper part of the pile is small, while that at the lower part is large, which is completely different from the natural state; When the load is greater than 2580kN, the axial force of the upper part of the pile body gradually increases, and the axial force of the lower part gradually decreases. The change trend of the axial force is the same as that of the natural state.

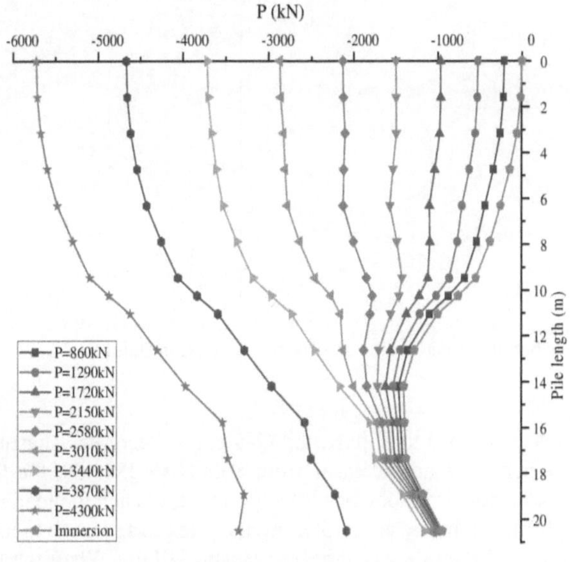

Figure 12. Axial force distribution curve(loading after water immersion state).

Figure 13 shows the distribution curve of pile lateral friction after the foundation soil is saturated and reloaded. It can be seen from the figure that when the pile is not loaded after immersion, most of the pile body is negative friction only under the self-weight of the pile, and only positive friction appears at the bottom of the pile. When the load is less than 1290kN and the pile length is more than 2/3, the lateral friction is negative, which indicates that the loess has collapsible deformation after immersion, so the negative friction appears around the pile; When the load is less than 2580kN, the positive and negative friction appears alternately. When the load is greater than 30100kN, the friction will be positive again.

4.2.4 *Comparison of load transfer law in three cases*

The Q-S curves of three condition are shown in Figure 14. In natural state (no rainfall), when the load is less than 1720kN, the Q-S curve is approximately linear and there is no obvious inflection point. When the load increases to 2150kN, the first inflection point of Q-S curve appears; When the load increases to 3870kN, the Q-S curve shows a sharp drop. According to the Q-S curve, the ultimate bearing capacity of the pile is 4300kN.

When the load is less than 1720kN, the Q-S curve changes linearly, and there is no obvious inflection point. When the load is increased from 1720 to 1936kN and the load remains unchanged, the Q-S curve decreases sharply and the settlement displacement of pile top is 28mm. When the load increases from 1936kN to 2050kN, the Q-S curve changes little; When the load continues to increase to 3440kN, the settlement of pile top increases sharply, which indicates that the soil around the pile has appeared plastic failure; When the load continues to increase to 3870kN, the vertical displacement of the pile top increases to 55.3mm, and the pile reaches the ultimate bearing capacity.

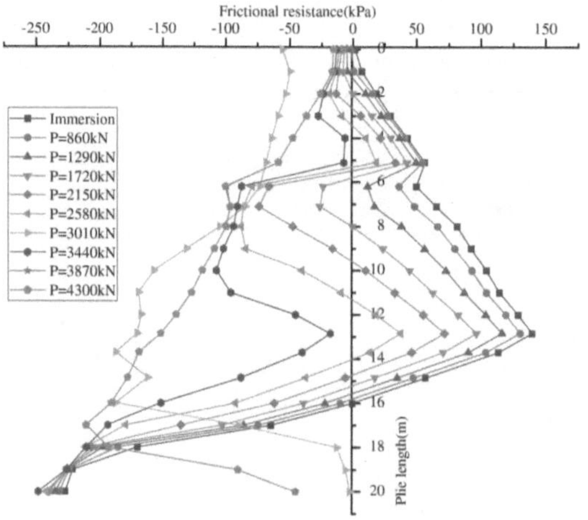

Figure 13. Friction distribution curve(loading after water immersion state).

When the load increases from 0 to 860kN, the Q-S curve decreases sharply, and the pile top settlement is 30mm; When the load increases from 860kN to 2580kN, the Q-S curve changes approximately linearly and there is no obvious inflection point. When the load is equal to 3440kN, the pile top settlement begins to increase sharply, and the pile top settlement reaches 40mm, which indicates that the soil around the pile has appeared plastic failure; When the load continues, the vertical displacement of the pile increases sharply, and the pile reaches the ultimate bearing capacity.

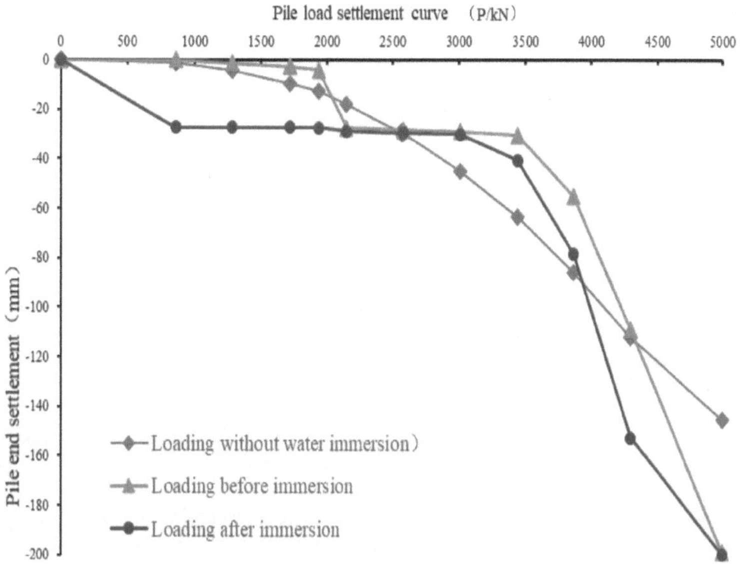

Figure 14. Q-S curve of three state.

5 CONCLUSION

Through the numerical analysis of the force characteristics of pile foundation in collapsible loess of Lanzhou area under three different immersion states, the main research achievements are as follows:

(1) The ultimate bearing capacity of piles before and after immersion is obviously different. The bearing capacity of single pile after immersion is significantly lower than that of natural state, but the bearing capacity of single pile changes greatly when the amount of immersion, loading time and loading sequence are different. According to the analysis results, when the pile top load is the same, the larger the water content of soil, the smaller the bearing capacity of pile.
(2) Axial force of the pile body decreases gradually along the length of the pile when it is not immersed. Axial force decreases slightly or even increases in the range of soil collapsibility after immersion. However, when the range of soil collapsibility is exceeded, the distribution law of axial force is basically the same as that when it is not immersed. When the amount and depth of soil immersion are different, the maximum value of axial force gradually decreases.
(3) In the natural state, the lateral friction resistance is positive. Above the middle of the pile body, the friction resistance is small and gradually increases below the middle of the pile body. Negative friction resistance occurs on the pile body after the soil is immersed in water. The magnitude of negative friction resistance is related to the water-bearing state of the foundation soil and also to the loading sequence and the vertical load on the pile top. When the pile load is fixed, the negative friction resistance changes greatly when the immersion and loading sequence are different.
(4) Settlement deformation of the soil layer around the pile is mainly caused by loess collapsible deformation, which is caused by both loess collapsible deformation and negative friction resistance on the side of the pile. When the soil is immersed, the negative friction resistance on the side of the pile increases the settlement deformation of the pile.

REFERENCES

Cao Han-zhi. (1986). Numerical calculation method of axial load transfer and load settlement curve of pile. *Chinese Journal of geotechnical engineering.* 8(6): 37–49.

Fei Kang, Zhang Jian-wei. (2010). *Application of ABAQUS in geotechnical engineering.* M. Beijing: China water resources and Hydropower Press.

Feng Zhong-ju, Huang An-lu. (2000). *Study on bearing capacity of large diameter bored piles. A. Proceedings of the Ninth National Conference on Structural Engineering.* C. Beijing: Tsinghua University Press, 449–453.

Guang Wen-zhang. (1990). New chapter on engineering performance of collapsible loess. M. Xi'an: Xi'an Jiao tong University Press.

Hu Qing-li, Zhang Ke-xu. (2002) Research on axial bearing behavior of large diameter piles. *Chinese Journal of Geotechnical Engineering.* 24(4): 491–495.

Huang Xue-feng, Chen Zheng-han, Fang Xiang-wei et al. (2007). Study on foundation treatment thickness and treatment method for collapse loess with large thickness. *Chinese Journal of Rock Mechanics and Engineering.* 26(Supp.2): 4332–4338.

Huang Xue-feng, Chen Zheng-han, Ha Shuang, et al. (2006).Large area field immersion test on characteristics of deformation of self-weight collapse loess under over burden pressure. *Chinese Journal of Geotechnical Engineering.* 28(3): 382–389.

Khalili N, habte M A, ZARGARBASHI S. (2008). A fully coupled flow deformation model for cyclic analysis of unsaturated soils including hydraulic and mechanical hysteresis. *J. Computers and Geotechnics.* 35(6): 872.

Li Da-zhan, He Yi-hua, Sui Guo-xiu. (1993) Study and test on immersion of Q2 loess in large area. *Chinese Journal of Geotechnical Engineering.* 15(2): 1–11.

Li Ning, Xu Jian-cong. (2015). Development and verification of slope rainfall infiltration boundary based on ABAQUS. *Journal of water resources and water engineering* 26(1): 198–199.

Li Su-hua, Zhou Jian, Yin Jian-hua, et al. (2004). Study on bearing performance of friction pile foundation in complicated soil-layers. *Chinese Journal of Rock Mechanics and Engineering.* c, 23(15): 2659–2664.

Liu Hai-song. (2008). Preliminary study on the relationship between structural strength and collapsibility of loess. *J. Geotechnical mechanics.* 29: 722–725.

Liu Zu-dian. (1996). *Loess Mechanics and engineering.* M. Xi'an: Shan xi science and Technology Press.

Qian Hong-jin, Luo Yu-sheng, et al. (1985). *Collapsible loess foundation.* M. Beijing: China Construction Industry Press.

Santoso A M, Phoon K.K, Quek S T.(2011).Effects of soil spatial variability on rainfall induced landslides. *J. Computers – Structures.* 89(11–12) : 893–900.

Wang Fu-chun, Yao Xian-hua, et al. (2017). Numerical simulation study on influence of salt marsh corrosion on vertical ultimate bearing capacity of highway bridge pile foundation. *J. Highway.* 1.1: (60–63).

Wang Xiao-jun, Mi Wei-jun, Xiong Zhi-wen, et al. (2012). Water immersion field tests of collapsibility of loess foundation of loess foundation of Zhengzhou-Xi′ an passenger dedicated line. *Journal of the China Railway Society.* 34(1): 83–90.

Xie Ding-yi. (2001). Exploration of some new tendencies in research on loess mechanics. *Chinese Journal of Geotechnical Engineering.* 23(1): 3–13.

Zhu Wei, Cheng Nan-jun, Chen Xue-dong, et al. (2006).Discussion on some basic problems of unsaturated seepage. *Journal of geotechnical engineering.* 28 (2): 235.240.

Zhu Xiang-rong, Wang Jin-chang. (2004).Brief introduction and engineering application of some soil models in ABAQUS software. *J. Rock and Soil Mechanics.* 25(Supp2):145–147.

Advances in Civil Engineering: Structural Seismic Resistance, Monitoring and Detection – Mohd Johari Mohd Yusof and Junwen Zhang (Eds)
© 2023 The Authors, ISBN 978-1-032-31491-4

Experimental study on dynamic consolidation method to improve saturated soft soil foundation

Haolan Wang
Urban and Rural Construction Institute, Hebei Agricultural University, Baoding, China

Ying Wang*
China Hebei Construction & Geotechnical Investigation Group Ltd., Shijiazhuang, China

Jinglin You & Shijie Wang
Urban and Rural Construction Institute, Hebei Agricultural University, Baoding, China

ABSTRACT: Among many geological types, the soft soil foundation has the characteristics of high water content and low shear strength. If the pre-treatment is not well done, the stability of the project will be affected. Therefore, it is necessary to carry out reasonable reinforcement. The dynamic compaction method has the characteristics of simple operation, low cost and obvious treatment effect. Therefore, reinforcement of saturated soft soil foundation by dynamic compaction method. The field monitoring laws of lateral displacement of the foundation after dynamic compaction, the increasing and dissipation regularity of the excess pore water pressure are analyzed and discussed. The improvement effect is verified by static cone penetration test and loading test. The determination methods of effective reinforcement depth after dynamic compaction are put forward. The practice shows that, by field-testing methods, such as static cone penetration test and pore water pressure test, to determine the effective reinforcement depth after dynamic compaction and other relevant design parameters is reliable.

1 INTRODUCTION

Dynamic compaction method is a foundation reinforcement method which was first created by French Menard Company in 1969. Because of its simple equipment, convenient construction, wide range of application, economical and easy, remarkable effect and saving material, it spreads to all over the world quickly. China introduced this method in the late 1970s, and then quickly promoted and applied it nationwide. At present, the scope of foundation treatment by dynamic compaction method includes industrial and civil buildings, warehouses, oil tanks, storage bins, highways, railway subgrades, airport runways, port terminals and revetment projects. Although dynamic compaction technology has been widely used in the treatment of gravel soil, sand, collapsible loess, miscellaneous fill and low saturation clay or silt foundation reinforcement (Liu 2020; WU; Xu 2020; Zhang 2019), but for saturated soft clay, especially silt, mucky soil foundation reinforcement experience is still little. Practice shows that whether the excess hydrostatic pressure caused by dynamic compaction can dissipate rapidly is a key factor to determine the success or failure of dynamic compaction in such foundation reinforcement. The current standard dynamic compaction process is characterized by large energy, heavy compaction first, multiple strikes without drainage, which seriously damages the structure of soft clay. The pore water pressure remains high, and "rubber soil" appears, which leads to the failure of dynamic compaction. In view of this situation,

*Corresponding Author: 497971066@qq.com

DOI 10.1201/9781003310884-38

Zheng Y.R. (Zheng 2000) proposed a new technology for treating soft clay foundation by dynamic compaction. The main features of the process is "first light and then heavy, gradually add energy, less hit many times, layer by layer reinforcement"; Changing the hammer standard does not destroy the soil structure, and artificial drainage (such as setting plastic drainage plate) is used to increase the reinforcement effect of dynamic compaction.

Based on the engineering background, this study analyzes the effect of dynamic compaction by designing experimental scheme. The new technology combined with a variety of monitoring and testing methods is used to conduct experimental research on the reinforcement effect of dynamic compaction method for soft soil foundation. The mechanism of dynamic compaction reinforcement is further analyzed and discussed, and the parameters in the design and construction of dynamic compaction are analyzed, which are verified by practical engineering and further applied.

2 DYNAMIC COMPACTION TEST DESIGN

2.1 *Engineering geological condition*

In the expansion project of an industrial plant, the construction site was originally a shrimp pond, which was backfilled by plain soil (undercompaction and partial construction waste). The backfill time has been more than 1 year. Groundwater depth is 2.2 m. The physical and mechanical properties of each soil layer from top to bottom in the test site are shown in Table 1. The design requires that the effective reinforcement depth of foundation is 8 m, and the characteristic value of foundation bearing capacity after treatment is not less than 120 kPa.

Table 1. The Physical Mechanics Properties Index of Foundation Soil.

soil name	thickness/m	moisture content/%	void ratio	liquid limit/%	compression modulus/MPa	bearing capacity/kPa
plain fill	2.0				3.50	75
mucky soil	5.0	49.2	1.348	43.21	2.60	70
silt	1.6	35.4	0.956	33.69	4.01	100
silty clay	>6.0	27.0	0.737	28.05	4.72	115

2.2 *Dynamic compaction test scheme*

The crane used in this project is a domestic 50t crawler crane. The rammer has a diameter of 2.15 m, a bottom area of 3.63 m2, a hammer weight of 10t, and an automatic decoupling device. During the trial ramming, according to the principle of "first light and then heavy, step by step, less hit multiple times, layer by layer reinforcement". First, the surface fill is leveled, and a 0.6m thick medium sand cushion is laid on it. Then, the plastic drainage plate is inserted, with a plane spacing of 1.5m and a plum-shaped arrangement of 11m in length. Lay soft permeable pipe, around the test area are open trenches for drainage. Dynamic tamping construction is divided into five tamping, the first four times each time 2, the fifth time 1, each tamping can be: 600, 800, 1000, 1200, 600 kN·m. The spacing of the first four tamping points is 3.5m, and the arrangement of the tamping points is plum blossom type. The fifth time is full ramming, hammer printing lap 1/4, single point total ramming subsidence control within 0.6m. The interval time of two tamping was 4d.

3 EFFECT AND ANALYSIS OF DYNAMIC COMPACTION

3.1 *Lateral displacement of soil*

Horizontal displacement tests were carried out during and at the end of dynamic compaction. Sliding inclinometer was used with the accuracy of 0.1 mm/0.5m and the guide wheel spacing of

500mm. The distance between the measuring point and the center of the tamping point is 2.5 m, and the depth of the inclined pipe is 9.0m. The horizontal displacement value of each 50cm is tested, and Figure1 is the deep horizontal displacement curve at the end of dynamic compaction. The graphic monitoring results show that the maximum lateral displacement of soil occurs at about 3.5m below the original ground, and decreases rapidly under this depth. The displacement below 8.5m is small or basically no lateral displacement occurs. The observation of surface uplift outside the area shows that there is basically no uplift phenomenon, indicating that the selected parameters not only achieve the reinforcement effect, but also do not destroy the soil structure.

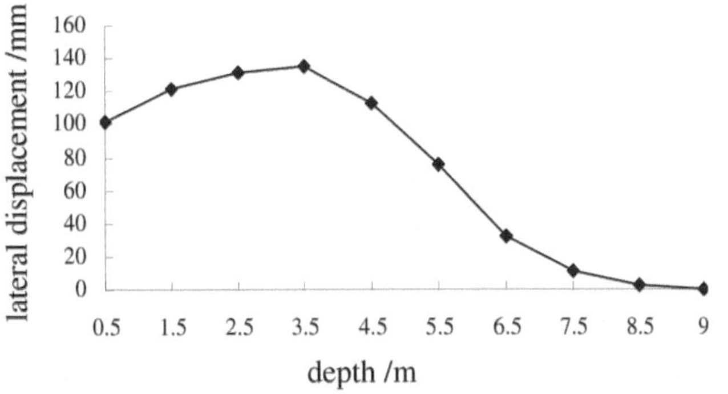

Figure 1. Horizontal displacement curve of deep soil.

3.2 *Excess hydrostatic pressure monitoring results analysis*

Before dynamic compaction construction, a group of pore water pressure gauges were buried in the central position of the test compaction area, with a total of 6 measuring points. The buried depths of each measuring point were 2.3m for KY-1, 3.8m for KY-2, 5.3m for KY-3, 6.8m for KY-4, 7.8m for KY-5 and 8.8m for KY-6, respectively. The measuring points KY-1 ~ KY-4 are set in the silty soil layer, KY-5 is located in the middle of the silty soil layer, and KY-6 is located at 0.2m below the top of the silty clay layer. Figure 2 shows the variation curve of pore water pressure with depth at the end of each compaction. Figure3 is the dissipation process curve of excess hydrostatic pressure.

It can be seen from Figure 2 and Figure 3 that the growth and dissipation of pore water pressure after each tamping are basically consistent. The maximum pore water pressure after each tamping

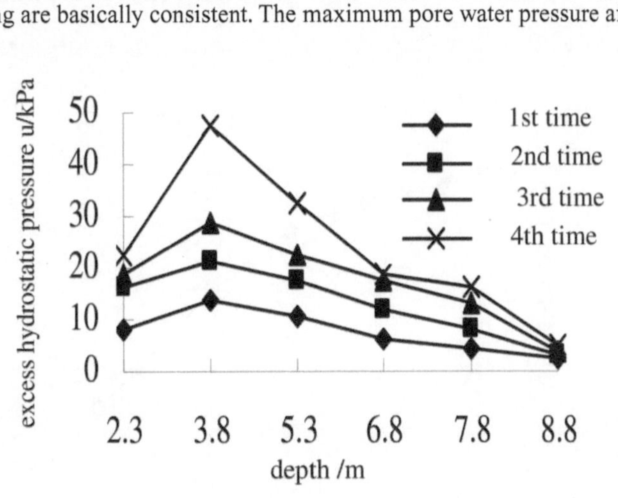

Figure 2. Variation of pore water pressure.

occurs at about 3.8m below the original ground. The first day of excess hydrostatic pressure dissipation is the largest, and then gradually slow down. The pore water pressure dissipates 60% – 66% within 24h, and the dissipation rate reaches 90% after 4d, indicating that the installation of plastic drainage plate provides a good drainage channel, and it is reasonable to control the interval time of two times of dynamic compaction for 4d in large area construction.

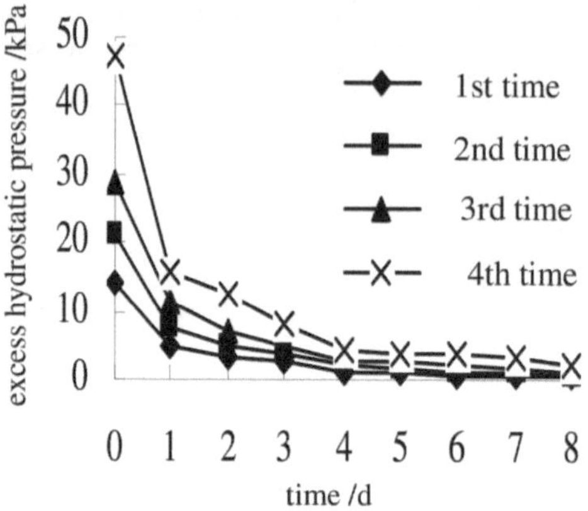

Figure 3. Dissipation curves of excess pore water pressure.

3.3 *Detection of foundation bearing capacity*

Considering that the soft soil layer has a long thixotropic solidification stage after dynamic compaction, the detection work is carried out 15 days after the end of dynamic compaction. The size of the bearing plate in the load test is 1m × 1m, with 10-stage loading and the last stage loading of 250kPa. Figure 4 is the p-s curve of static load test. A total of 2 static cone penetration holes, and try to keep in the same position. Figure 5 is the CPT-1 hole dynamic compaction reinforcement before and after the static cone penetration test results comparison chart. It can be seen from Figure 5 that the p_s value of plain fill and muddy soil layer increased by 1.1 times and 2 times respectively. Based on the test results of load test and static cone penetration test, the characteristic value of

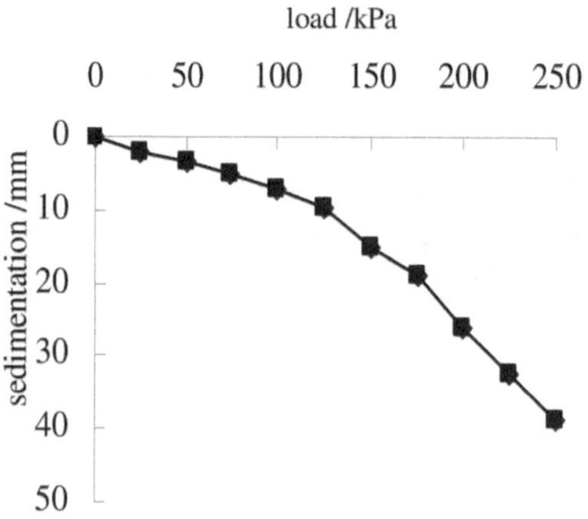

Figure 4. Loading-settlement curve.

foundation bearing capacity after dynamic consolidation has reached more than 120kPa, which meets the design requirements.

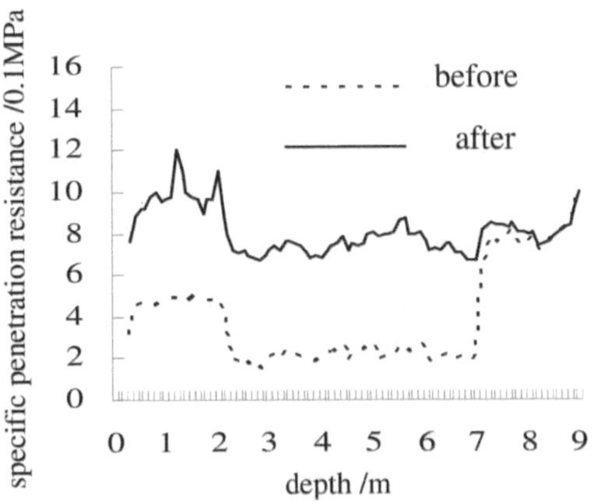

Figure 5. Comparison of the static cone penetration.

3.4 *Determination of effective reinforcement depth*

The reinforcement depth of dynamic compaction is one of the most critical issues in the design and construction of dynamic compaction, which is related to the economy and applicability of dynamic compaction. Many scholars have carried out a series of studies and discussions on this issue, but the definition of effective depth of foundation reinforcement by dynamic compaction is still unclear (Zhou 2006). Usually for different engineering purposes. Generally speaking, the reinforcement depth refers to the strength, that is, the strength of soil in the reinforcement depth is improved (Lu 2002). It can be seen from Figure 5 that the reinforcement depth determined by the comparison of static cone penetration test results before and after reinforcement is about 8.1m. If according to Wang H.C.'s method (Wang 1991), the reinforcement depth is the place where the excess hydrostatic pressure reaches 0.2 times the self-weight stress of soil. According to the monitoring results of pore water pressure in Figure 2, it can be seen that the excess hydrostatic pressure at the depth of 7.8m after the fourth tamping is 16.2kPa, which is very close to the 0.2 times the self-weight stress of the depth. Therefore, the effective reinforcement depth determined by this method is approximately 8m. The above results are basically consistent with the laws revealed by the monitoring results of soil lateral displacement.

4 CONCLUSION

It is feasible to treat saturated soft soil foundation by adopting dynamic compaction method combined with the construction technology of inserting plastic drainage plate according to the principle of "light first and heavy later, gradually adding energy, less hitting more times and layer by layer strengthening". This new technology can greatly improve the bearing capacity of soft soil foundation, reduce the settlement, shorten the construction period, and achieve remarkable economic benefits.

It is feasible to treat saturated soft soil foundation by adopting dynamic compaction method combined with the construction technology of inserting plastic drainage plate according to the principle of "light first and heavy later, gradually adding energy, less hitting more times and layer by layer strengthening". This new technology can greatly improve the bearing capacity of soft soil foundation, reduce the settlement, shorten the construction period, and achieve remarkable economic benefits.

REFERENCES

Liu T., Liu D.S., Zhang P.S. et al. (2020) The application of dynamic compaction + CFG pile composite foundation in corrosive collapsible loess area[J]. *Building structure*, 50(23):114–117+50.

Lu X., Zheng Y.R., Zhou L.Z. (2002) Reinforcement mechanism, calculation and application of dynamic compaction in soft clay foundation[J]. *Sichuan building science*, 28(1): 50–52.

Wang C.H. (1991) Review on estimation method of reinforcement depth of dynamic compaction foundation[J]. *Foundation treatment*, 2(1): 20–24.

Wu S.F., Du J.F., Wei R. et al. Theoretical method and applied research on dynamic compaction reinforcement scope of high fill soil and stone[J]. *Chinese journal of geotechnical engineering*, 42(S2):43–49.

Xu Y.J. (2020) Optimum design of foundation reinforcement by dynamic compaction[J]. *Heilongjiang hydraulic science and technology*, 48(11):74–75.

Zhang H.B., Wang K., Liu Y. et al. (2019) Field test of dynamic consolidation of saturated silty clay foundation[J]. *Road Machinery & Construction Mechanization*, 36(03):86–90+96.

Zheng Y.R., Lu X., Li X.Z. et al. (2000) Study on theory and technology of dynamic compaction for soft clay foundation[J]. *Chinese journal of geotechnical engineering*, 22(1): 18–22.

Zhou J., Zhang S.F., Jia M.C. et al. (2006) Research status and latest technology progress of dynamic compaction theory[J]. *Chinese journal of underground space and enginee*, 2(3): 510–516.

Advances in Civil Engineering: Structural Seismic Resistance, Monitoring and Detection – Mohd Johari Mohd Yusof and Junwen Zhang (Eds)
© 2023 The Authors, ISBN 978-1-032-31491-4

Study on the reaction mechanism and application trend of geopolymer concrete

Hui Xing*, YinBo Li & ZhuangLong Lin
School of Civil Engineering, Guangdong Construction Polytechnic, Guangzhou, Guangdong, China

ABSTRACT: Silicate cement is a cementitious material with high energy consumption, high pollution and huge resource consumption. During the cement production process, a large amount of carbon dioxide is emitted, causing serious climate anomalies. In order to reduce the amount of cement, a large number of industrial byproducts are widely used in commercial concrete, partially or completely replacing cement. The geopolymer concrete, which replaces cement with industrial by-products such as mineral powder and fly ash as concrete cementing material, has advantages in terms of low emissions and high strength, so cement-free concrete with geopolymer as the substrate was born. In this paper, based on the analysis of the current status of the research on geopolymer concrete, the main reaction mechanism of geopolymer concrete is further investigated, and the application and development trend of geopolymer concrete is studied with this guidance. The aim is to promote the research and development of green concrete as a contribution to the global environmental protection cause.

1 INTRODUCTION

While cement manufacturing brings economic benefits, it is also accompanied by serious problems. According to statistics from China's National Bureau of Statistics, China's cement production was as high as 2.31 billion tons in 2019 alone. The production of large amounts of cement clinker causes high energy and resource consumption, and the production process of two grinding and one burning of cement is often accompanied by the production of large amounts of carbon dioxide gas, which releases about 0.73~0.99 kg of CO2 for every 1 kg of cement produced. In addition, the cement production process also emits a large amount of harmful gases NOx, SO2 and dust, causing serious environmental pollution, which has caused great pressure on environmental protection and rational use of resources in China. The process of cement clinker firing needs the support of large amount of electric power resources, while China's electric power production mainly uses coal as the basic fuel, and a large amount of fly ash and slag can be obtained after coal burning and collected by dust collector, usually every two tons of coal consumed can produce one ton of fly ash. Undisposed fly ash accumulates in large quantities in urban industrial areas, aggravating dust and air pollution; if discharged into water systems, it may cause siltation of rivers, and the toxic chemicals in it may also be harmful to humans or organisms. The rational use of fly ash and other waste solid resources has become one of the urgent environmental problems in China, bringing a serious challenge to the sustainable development of China's power industry. Based on this, the pollution control of cement dust can not be delayed. To solve the above problems, some researchers began to think about how to use fly ash and other potentially active and cementitious materials to partially or completely replace cement, reduce the amount of cement in a single square of concrete, thus reducing the environmental pollution caused by the production of cement clinker, and promote

*Corresponding Author: xinghuiedu2020@163.com

DOI 10.1201/9781003310884-39

the consumption of fly ash and other industrial waste to achieve the sustainable development of the concrete industry, without reducing the total demand for concrete. And with the development of alkali-inspired cementitious materials in recent years, the application of geopolymers in the field of green concrete has been gradually explored.

2 ANALYSIS OF THE CURRENT STATUS OF RESEARCH ON GEOPOLYMER CONCRETE

2.1 *Progress of research on geopolymer concrete*

In the 1980s, French scientist Davidovits realized that the solidification and hardening process of geopolymer materials is a reaction process in which the silicon-oxygen and aluminum-oxygen bonds in the raw material are broken under the action of alkaline catalysts and then reorganized. In addition to the geopolymers named by Davidovits, scientists and research institutions in various countries have also carried out many fruitful studies on geopolymer concrete (Zhao et al. 2020). At the beginning of the research, metakaolinite was mainly used as a raw material, but metakaolinite is not a renewable resource and is not conducive to application. Jarsel et al. used kaolinite and fly ash to prepare geopolymer concrete, studying not only its mechanical properties, but also the effect of heavy metal ions on its performance. raw material (Zhou et al. 2019). A large number of scholars have shown that when mineral powder reaches a certain fineness, it has good properties and can play an active role, which can improve the performance of cement and can also benefit environmental protection.Swanepoel et al. used a certain ratio of mixed mineral powder, $Na2SiO3$, NaOH and water in geopolymer materials to prepare geopolymer concrete, and studied the effect of both temperature variables and maintenance condition variables on its Nyalel studied the properties of fly ash geopolymer concrete and found that the fly ash is surrounded by a gel layer, which is probably the silica alumina produced in alkaline environment (Yan et al. 2019). The performance of geopolymer concrete in the marine environment was studied and found that the strength of geopolymer concrete was higher than that of silicate cement concrete by cyclic testing; moreover, the durability of geopolymer concrete was also better than that of silicate concrete. Comprehensive above we can find, through the scholars of various countries for the study of geopolymer concrete, it does have good mechanical properties. With the progress of the research, the geopolymer concrete is not only limited to theoretical research, but will be more applied in the practical engineering.

2.2 *Research on geopolymer concrete combined with recycled concrete*

By reviewing the data, we found that the global literature on the performance of geopolymer concrete and recycled concrete is more extensive and the research is more systematic. Regarding geopolymer concrete, which originated in Ukraine, it has received a lot of attention worldwide since the 1970s. Most of the geopolymer concrete is based on mineral powder as well as fly ash, and utilizes materials such as sodium hydroxide, water glass and potassium hydroxide, sodium carbonate and potassium carbonate as alkaline excitants (Mao et al. 2019). Geopolymer concrete has high strength, high corrosion resistance and high freeze-thaw resistance; however, it has characteristics such as too fast setting, shrinkage and rising carbonation rate compared to ordinary concrete. The early setting properties of geopolymer concrete make it more difficult to use in commercial concrete, so the manufacture of prefabricated components and other products for the use of geopolymer concrete more appropriate application. In recent years, research on geopolymer concrete has delved into the effects of the chemical content of mineral dust and fly ash on the performance of concrete. kathivel et al. used fly ash as a substrate and used recycled aggregates instead of natural aggregates to study the mechanical and partial durability of concrete (Zhou 2017). The study showed that the use of recycled aggregates can make a small increase in the mechanical and durability performance of the components.

3 THE MAIN REACTION MECHANISM OF GEOPOLYMER CONCRETE

The reaction mechanism of geopolymer is currently the key issue studied by researchers, and it is also an urgent problem to be solved when studying geopolymer cementing materials. Today, the global excitation mechanism of geopolymer is basically based on alkali excitation, i.e., the silica-aluminate containing minerals mixed with alkaline exciter to cause polymerization. However, with the development process of geopolymer mechanism research, the following reaction mechanisms are currently represented.

3.1 *Alkali-excited geopolymer formation process*

For fly ash based polymers concrete, the following reactions occur in the process of strength generation: Unlike ordinary silicate cement concrete, in the preparation of geopolymers, fly ash as a raw material component must first be excited by a high alkali environment to destroy the glass phase structure of fly ash, so that the silica-alumina phase active material in fly ash is depolymerized. In this process, the preparation of alkali excitation solution has a great influence on the reaction rate of geopolymer. From the geopolymerization process proposed by various scholars, it can be seen that the most important part of the geopolymerization reaction is based on the reconfiguration and self-organization of silica-alumina and other monomers in the silica-alumina containing active materials under the activation of medium to high alkaline environment, which eventually condenses and hardens into geopolymers (Zhang et al. 2016). In this process, it can be seen that, unlike silicate cement concrete, the strength of geopolymers originates mainly from their unique three-dimensional network structure. The geopolymerization reaction causes the silica-alumina active monomers in the raw material to be rebridged through ionic and covalent bonds, and the silica-oxy tetrahedra and alumina-oxy tetrahedra are connected by bridge oxygen, but since no condensation occurs between any two alumina-oxy tetrahedra appearing to share the same bridge oxygen, the alumina-oxy tetrahedra are often linked to multiple silica-oxy tetrahedra in a single form by bridge oxygen, thus forming a spatial three-dimensional mesh structure.

3.2 *CO2 transport in geopolymer slurry*

The carbonation of concrete is the process of CO2 gradually invading from the concrete surface to the concrete interior by diffusion or adsorption, and carbonating with alkaline substances such as calcium hydroxide within the concrete slurry, thus causing the neutralization of concrete. Research shows that the transport process of CO2 and other external erosion factors in the slurry is mainly influenced by the porosity of concrete itself, and its transport in the concrete slurry is mainly through continuous, discontinuous and obstructive transport.CO2 and other external erosion factors mainly invade the concrete slurry through the continuous path (CCP) composed of pores and cracks, while the load on the sparseness of the concrete pore structure, micro-crack The closure, development or penetration of cracks has an important influence (as shown in Table 1).

Table 1. Formatting sections, subsections and subsubsections.

Compressive stress ratio	Performance
0–0.3	Almost no tendency to expand, even some microcracks appear to be closed
0.3–0.5	Microcrack area starts to increase steadily
0.5–0.7	Gradual cracking of the concrete matrix
>0.7	Microcracks begin to develop rapidly, and with the increase in stress ratio, the concrete is sharply deformed to destruction

It is worth noting that the CO_2 transport process in geopolymer slurry is not only influenced by external loads, but also the maintenance conditions of geopolymer preparation, maintenance age, CO_2 concentration on the surface of geopolymer during carbonation erosion, relative humidity of environment, etc. will have some influence on the CO_2 transport in geopolymer slurry. However, there is still some debate on the interrelationship between different influencing factors and carbonation, but it is certain that the carbonation process is controlled by CO_2 diffusion and good maintenance can effectively reduce the carbonation rate of geopolymers.

3.3 *Carbonation reaction process of geopolymers*

For geopolymers prepared from alkali-excited inorganic silica-aluminates, the carbonation mechanism is different from that of ordinary silicate cement concrete because of the difference in strength generation reaction and hardening mechanism (Zhang et al. 2016). For slag fly ash based geopolymer concrete, the main physicochemical processes occurring at the end of the polymerization reaction when faced with carbon dioxide erosion are as follows:

- CO_2 from the air gradually invades the geopolymer slurry from the geopolymer surface through continuous paths such as geopolymer pores and microcracks;
- Dissolution of CO_2 in the pore solution of the geopolymer slurry and reaction with carbonizable Substances such as $Ca(OH)_2$;
- Further dissolution of $Ca(OH)_2$ in the pore solution;
- Slight decrease of porosity with carbonation products and gradual decrease of pore solution pH;
- As water vapor concentrates on the geopolymer pore wall, a new equilibrium is reached at the external relative humidity and temperature.

It is worth noting that because the polymerization product $Ca(OH)_2$ content in the compounded slag fly ash base polymer is small, and the N-A-S-H gel as the main polymerization product does not decalcify, therefore, when it is subjected to carbonation erosion, CO_2 reacts directly with C-A-S-H gel to produce CaCO, and the main change in the carbonation process is that the geopolymer pore solution changes from high alkalinity to high sodium carbonate concentration. Different types of concrete and even the same material different ratio preparation of geopolymer concrete, often due to the different microstructure of the concrete itself and the role of external factors, resulting in different types of geopolymer concrete carbonation durability between a large difference.

4 APPLICATION AND DEVELOPMENT TREND OF GEOPOLYMER CONCRETE

4.1 *Application of geopolymer concrete*

Compared with ordinary silicate cement, geopolymer has many excellent properties, and it belongs to inorganic polymer material (Mao et al. 2016), which has the advantages of ceramic, cement and concrete. At present, geopolymer is mainly used in the following fields:

(1) Using the characteristics of fast hard and early strength of geopolymer, it is used as fast repair material and reinforcing material in engineering. The Pyrament brand alkali-inspired volcanic ash cementing material developed in the United States has been widely used in the fields of road repair and communication facilities repair.
(2) Using the characteristics of geopolymer resistance to acid and alkali corrosion, it is applied to corrosive environments (Huang et al. 2015). The FJK concrete base material developed by Suzhou Concrete Cement Products Research Institute in China, which is made of blast furnace slag, has been used in the production of corrosion-resistant pipes.
(3) The use of geopolymer can solidify the characteristics of heavy metal ions, it will be applied to the solid waste with strong pollution characteristics of the curing. The use of geopolymer curing contaminants has the advantages of both the simplicity of the cement method and the stability of the ceramic method.

(4) Using the characteristics of the high bond strength of geopolymer, it is applied to the reinforcement of concrete structures. Researcher Wu Bo of South China University of Technology used geopolymer instead of epoxy-type adhesive for carbon fiber cloth reinforcement of concrete beams (Zhu et al. 2014). The study showed that the load capacity of the concrete beams reinforced with single-layer carbon fiber cloth by geopolymer at room temperature was not much different from that by epoxy adhesive, but the high temperature resistance of geopolymer was significantly better than that of epoxy adhesive.

4.2 *Development trend of geopolymer concrete*

The excellent mechanical properties and physical characteristics of geopolymer concrete determine its very broad application prospects, its specific development trend in the following specific aspects.

(1) The use of geopolymer as a base material for interior and exterior coatings
Compared with ordinary cement materials, the polymerization reaction in the early stage will make it have a dense structure and excellent fireproof and waterproof performance. By using calcined kaolin as the basic material, adjusting a certain modulus of water glass as an alkaline exciter, and adding a certain amount of admixtures, the geopolymer-based coating is made, which has excellent corrosion and erosion resistance. Therefore, the geopolymer base material as coating has a broader development prospect.
(2) Application of geopolymer-based materials for solidification of toxic waste and nuclear waste
The final product of geopolymer cementing material is zeolite-like and has a three-dimensional network of hydrated aluminum silicate structure. The zeolite structure is capable of sequestering and adsorbing toxic chemicals, nuclear radioactive elements, and industrial waste materials. Ordinary silicate cement concrete, on the other hand, does not have the effect of sequestering highly corrosive substances because it contains lime, which can react with both chemical substances containing alkali metals and strongly acidic metallic ores (Long et al. 2013). On the contrary, geopolymer is an effective cementitious material for curing these toxic substances because it has a special structure and does not contain lime like substances.
(3) Application of geopolymer as a base material for composite materials
Research shows that geopolymer has high flexural strength, high early strength, corrosion resistance and good plasticity. Therefore, its excellent characteristics can be used in composite panels such as block materials in construction. As far as the maintenance system is concerned, geopolymer materials do not require water bath maintenance compared to cement materials, and the maintenance period is short, green, pollution-free and low-cost. At the same time geopolymer is easy to process, and the appearance is similar to natural stone, is a good performance characteristics, high durability of decorative materials.
(4) Application of geopolymer as the base material for prefabricated parts
Through the study found that geopolymer concrete has early and strong mechanical properties under the high temperature maintenance of 40-80C. With the development of society, the development of bridges and highways is increasing in demand at the same time, the requirements for construction progress are becoming increasingly demanding. The early strength characteristics of ground polymer concrete can be applied to engineering components, so that the building, bridge and sorghum in terms of large beams, slabs can be built in the factory, and then transported to the site assembly can be. This will not only improve the processing speed of the components in the factory and the construction schedule, but also the application of green, environmentally friendly materials will reduce the pressure on the social environment.

5 CONCLUSION

With the gradual development of concrete materials to green, environmentally friendly, good durability and other properties, friendly concrete materials are gradually being paid attention to by researchers. Geopolymer concrete as a new composite non-polluting concrete materials are also being more and more attention of scholars. The ground polymer concrete material is different from ordinary silicate concrete, which does not require high temperature calcination like the production of cement to produce a large amount of CO2, and this cementitious material can also be industrial waste (such as fly ash) as a base material, which can save energy, but also to protect the environment. Nowadays, many scholars have systematically studied the preparation, formation mechanism, proportioning, and application of geopolymers, and the research results show that their superior performance can not only be a substitute for ordinary cement concrete in the construction field, but also have good performance in various engineering applications. Therefore, vigorously promoting the application of geopolymer concrete will have a profound impact on both socio-economic development and environmental protection.

ACKNOWLEDGMENTS

This work was supported by Guangdong Science and Technology Innovation Strategy Special Fund (Science and Technology Innovation Cultivation for University Students, No. pdjh2020b0947.

REFERENCES

Huang Q, Shi Qingshuang, Wang Qingyuan, Tang Ling, Zhang Hongen. Effect of recycled coarse aggregates on the carbonation properties of fly ash-based aggregates concrete[J]. *Silicate Bulletin*,2015,34(05):1264–1269+1281.

Long T, Shi Qingshuang, Wang Qingyuan, Li Lang. Mechanical properties and microstructure of fly ash-based aggregates recycled concrete[J]. *Journal of Sichuan University (Engineering Science Edition)*,2013,45(S1):43–47.

Mao MJ, Ren Jinyang, Zhang WB, Chen W. Study on mechanical properties of fly ash ground aggregate concrete[J]. *Concrete*,2016(05):78–80.

Mao MJ, Wei XP, Han PF, Zhou ST. Effect of maintenance conditions on the early shrinkage performance of fly ash ground aggregate concrete[J]. *Concrete*,2019(03):90–93.

Yan Jia, Zhang Haiyan, Wu Bo. Experimental beam study on the adhesion performance of steel-reinforced-geopolymer concrete[J]. *Journal of Building Structures*,2019,40(12):178–186.

Zhang H Y, Yan J J, Wu B. Study on the adhesion performance of geopolymer concrete with steel reinforcement[J]. *Journal of Civil Engineering*,2016,49(07):107–115.

Zhang H Y, Yuan Z S, Yan Jia. Study on the mechanical properties of metakaolin-fly ash geopolymer concrete after high temperature[J]. Journal of Disaster Prevention and Mitigation *Engineering*,2016,36(03):373–379.

Zhao Rinda, Wang Yongbao, Yuan Yuan, Zhao Chenggong, Qin Pengju. Review of research on shrinkage of geopolymer concrete[J]. *Silicate Bulletin*,2020,39(06):1695–1702.

Zhou Yanhua. Effect of recycled aggregates on the mechanical and durability performance of high-calcium fly ash-based aggregates concrete[J]. *Science Technology and Engineering*,2017,17(11):295–300.

Zhou Zhongyi, Wang Tao, Teng Rui, Yue Zhemeng, Wang Jin. Seismic performance of lightweight assembled frame-deformation controllable geopolymer walls[J]. *Journal of Harbin Institute of Technology*,2019,51(12):46–54.

Zhu Jingsei, Xu Jinyu, Luo Xin, Li Weimin, Bai Erlei, Gao Zhigang. A comparative study on the evaluation indexes of carbon fiber reinforced geopolymer concrete toughness[J]. *Journal of Construction Materials*,2014,17(02):303–308.

Part 2. Construction technology optimization and intelligent detection technology

Advances in Civil Engineering: Structural Seismic Resistance, Monitoring and Detection – Mohd Johari Mohd Yusof and Junwen Zhang (Eds)
© 2023 The Authors, ISBN 978-1-032-31491-4

Application and scheme optimization of new technology for reconstruction of the existing abutment

Yaming Wang
Jiaxing Jiatong Expressway Management Co., Ltd., Jiaxing, Zhejiang, China

Jialin Niu
Beijing Gongke Guqiao Technology Co., Ltd., Beijing, China

ABSTRACT: Taking the reconstruction of the abutment of a separated overpass as an example, this paper expounds the application of a new technology and the scheme optimization in the reconstruction of an existing abutment. Then the comparative analysis of the scheme optimization results is made, and then a safe, stable and economical scheme is chosen as the basis of the reconstruction. Moreover, the construction process and controlling essentials are explained, which can be regarded as reference for reconstruction of similar bridges.

1 INTRODUCTION

Many existing bridges can no longer meet the needs of the current transportation with the development of economy and social progress, so they need to be reconstructed. How to reconstruct the existing bridges and reduce the cost of building new bridges has become the priority that will benefit the country and the people. Therefore, bridge builders should constantly innovate to apply more new technologies in the construction of bridges. More economical and improved schemes can be selected by the comparison and optimization of schemes.

Taking the reconstruction of the abutment of a separated overpass as an example, this paper explains the application of a new technology. Compared with the multiple schemes, a new reconstruction scheme is chosen as the final plan, which not only makes full use of the existing structure of the bridge but also saves abundant investment for the project. Furthermore, it both promotes the new technology and provides reference for the reconstruction for similar bridges.

2 OVERVIEW OF THE BRIDGE

The first span of the separated overpass crosses a national highway while the second and third spans cross an expressway. The total length of the bridge is 127.2 m. The layout of the spans of the bridge is 4×30 m, and the bridge deck width is $8+2\times0.5$ m. The intersection angle of the bridge deck, and the national highway and the expressway is 65°. The horizontal curve of the bridge is in the straight section while the vertical curve is in the convex curve whose uphill is 1%, downhill 4% and R = 2000m. Since the slopes of the abutment, ribbed plates, and bearing platforms of the first span occupy the space under the bridge, which can't be suitable for the net width of the national highway, it needs to be reconstructed. The vertical view of the bridge can be seen as Figure 1.

DOI 10.1201/9781003310884-40

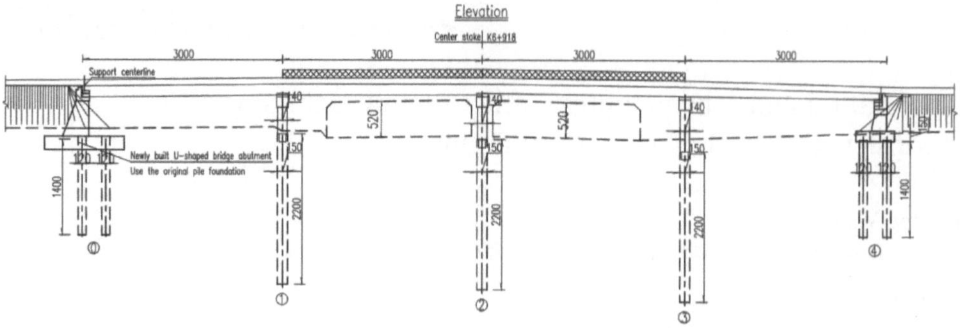

Figure 1. Vertical view of the existing bridge and its ribbed abutment.

3 APPLICATION OF THE NEW TECHNOLOGY FOR ABUTMENT RECONSTRUCTION

The original plan of the construction of the bridge is to demolish the original separated overpass and build a new one. The newly-built separated overpass will have a span of 20+3×30+20 m, and the width of the bridge deck won't be changed. However, the expressway under the separated overpass is an important national highway where there is heavy traffic. The demolition and reconstruction of the separated overpass would seriously affect the transportation of the expressway. And temporary bypass should be built and there would cause huge cost and difficulties, as well as risks. Based on the survey of the topography and physiognomy around the overpass and the force analysis of the structure of the separated overpass, a new reconstruction scheme is proposed in order to minimize the influence of the reconstruction of the separated overpass on the running of the expressway, or even not to affect the running of the expressway.

Based on the economic and technical demonstration, a temporary supporting superstructure is used and a U-shaped abundant is built. It means a temporary supporting structure will be set on the section of the small box girder near the abutment. When the No. 0 jack is used to lift the beam body and transfer the support force to the temporary supporting structure, the structure except the pile foundation of the original abutment will be dismantled. Then a U-shaped abundant will be built. After the jack is used again to lift the beam body, the temporary supporting force will be transferred to the abutment support. In this way, the abutment is successfully changed.

4 COMPARISON AND SELECTION OF THE SCHEMES

4.1 *Comparison of the schemes of the temporary support*

As for this scheme, the relative relationship between the temporary support, the original abutment and the newly-built U-shaped abutment should be considered. The temporary supporting points should be set as close as possible to the original support to ensure the safety the beam during the reconstruction of the abutment. Meanwhile, the location where the temporary support is set shouldn't affect the construction of the new abutment. On the basis of what is mentioned above, the schemes are as follow:

Scheme 1: the temporary supporting column is set in the gap of the bearing platform along the bridge.

The middle supporting steel pipe of the temporary support is set 20cm away from the inner side of the original H-shaped bearing platform, while the temporary supporting line is 3.00m away from the supporting line of the abutment support. The temporary support can enlarge the foundation of the second layer. The size of the top layer is 10.876m*2.7m*1m while the bottom layer 11.876m*3.7m*1m. The relative position of the temporary support and the base plane of the original abutment will be shown in Figure 2, while the lateral view of the temporary support is shown in Figure 3.

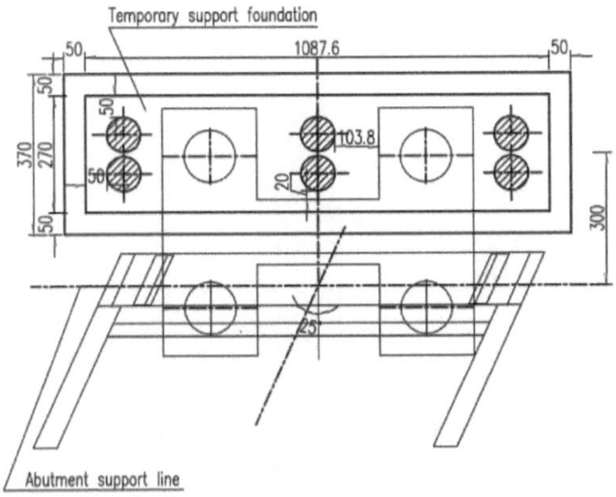

Figure 2. Plan graph of the relative position of the temporary support and the base plane of the original abutment.

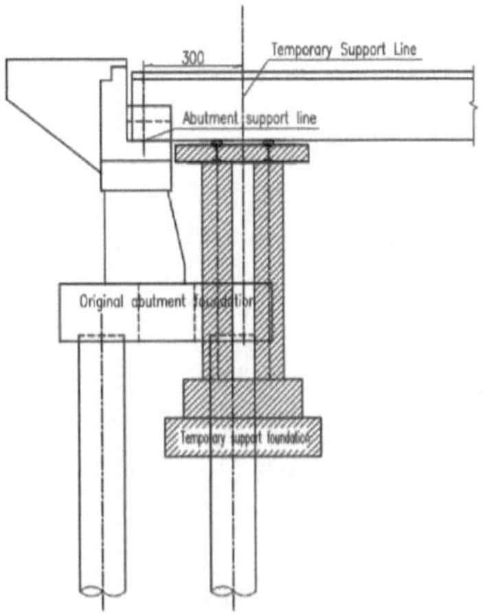

Figure 3. Lateral view of the temporary

When the temporary support is constructed, the flexural capacity of the small box girder and its stress in this scheme are calculated. And the results are shown in Table 1.

Table 1. Checking calculation of the temporary support section. (length of the cantilever: 3.375m)

Checking content	Moment effect (kN•m)	Bending strength (kN•m)	Flexural strength/ Moment effect
Flexural capacity of the box girder	329.6	638.6	1.94
Ultimate stress state when normally used	Maximum stress (MPa)	Stress limit (MPa)	Meet the requirement or not
Upper edge of the box girder	−1.67	−1.86	Yes
Bottom edge of the box girder	16.73	22.40	Yes

The results of the checking calculation show that the flexural capacity of the small girder and its stress when the temporary support is being constructed can meet the requirements, which mean the stress of both the upper and bottom edges is eligible. However, the tensile stress of the upper edge of the box girder is 1.67 MPa, which is relatively close to the limit value, 1.86 MPa. It means the upper edge of the box girder will be prone to crack, which should be paid more attention to.

Scheme 2: Temporary supporting steel tubes should be set on both sides of the temporary support, and they should be at the both sides of the original abutment. And the middle supporting steel tubes should be arranged in the gap between the original bearing platforms.

Compared with Scheme 1, the temporary support is moved to the original support. And the supporting points of the temporary support should be at both sides of the bearing platforms of the original abutment. The distance between the vertical steel pipes and the side of the bearing platform of the original abutment, and the leading edge of the abutment cap should be 30cm. The temporary support line is 1.955m away from the support line of the abutment support. The temporary support is considered to be integrated with the newly built U-shaped abutment foundation, so its foundation is made into one layer of 2m thick reinforced concrete slab. The relative position of the temporary support and the base plane of the original abutment will be shown in Figure 4, while the lateral view of the temporary support is shown in Figure 5.

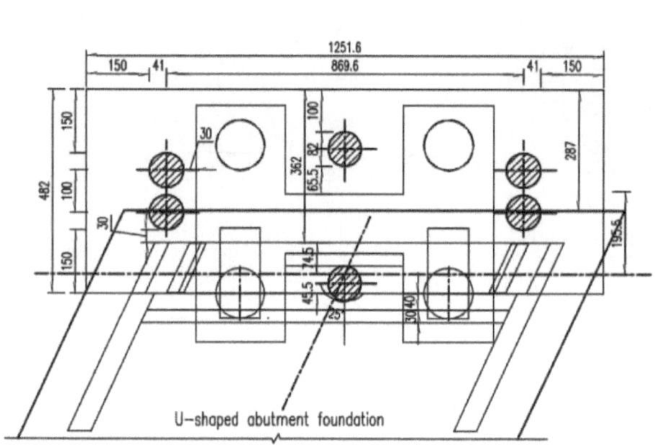

Figure 4. Plan graph of the relative position of the temporary support and the base plane of the original abutment.

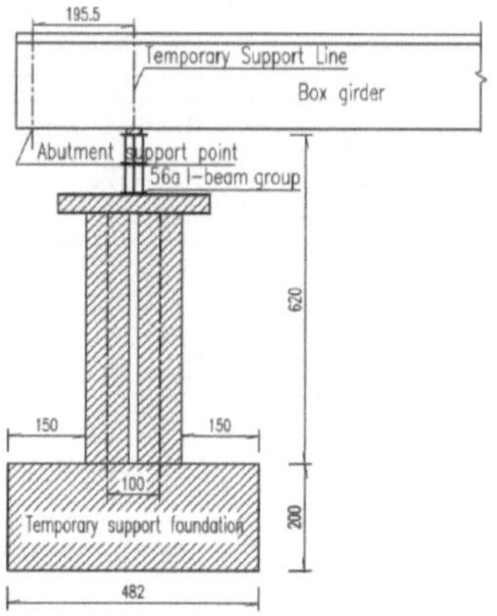

Figure 5. Lateral view of the temporary support.

When the temporary support is constructed, the flexural capacity of the small box girder and its stress in this scheme are calculated. And the results are shown in Table 2.

Table 2. Checking calculation of the temporary support section. (length of the cantilever: 2.33m)

Checking Content	Moment effect (kN•m)	Bending strength (kN•m)	Flexural strength/ Moment effect
Flexural capacity of the box girder	189.5	638.6	3.37
Ultimate stress state when normally used	Maximum stress (MPa)	Stress limit (MPa)	Meet the requirement or not
Upper edge of the box girder	−0.61	−1.86	Yes
Bottom edge of the box girder	14.91	22.40	Yes

The results of the checking calculation show that the flexural capacity of the small girder and its stress when the temporary support is being constructed can meet the requirements, which mean the stress of both the upper and bottom edges is eligible. Moreover, the results are far away from the limitation, so this scheme is feasible.

4.2 *Comparison of the Schemes of the Newly-built U-shaped Abutment*

Scheme 1: The U-shaped abutment is designed without considering the temporary support.

The U-shaped abutment cap is a reinforced concrete structure, while its front wall, side wall and foundation are a flake concrete structure. The lateral and plane views of the U-shaped abutment are shown in Figures 6 and 7.

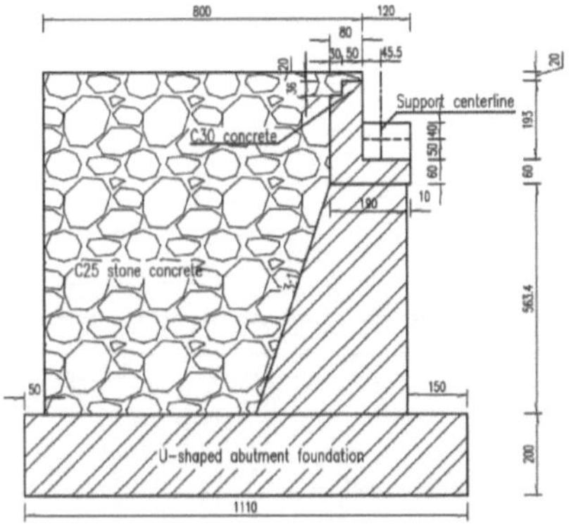

Figure 6. Plan graph of the relative position of the temporary support and the base plane of the original abutment.

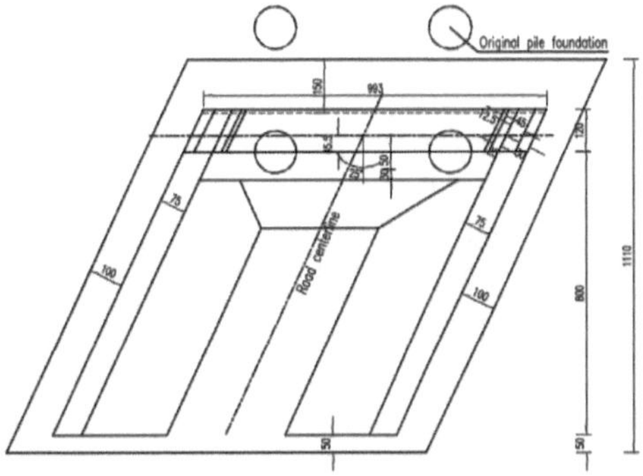

Figure 7. Lateral view of the temporary support.

The parameters in this scheme are calculated, and the results of the base stress are shown in Table 3, while the stability of the base is shown in Table 4.

Table 3. Checking results of the base stress.

Position	Maximum stress at the side of the span of the bridge (MPa)	Minimum stress at the side of the span of the bridge (MPa)	Maximum stress deviating from the side of the span of the bridge (MPa)	Minimum stress deviating from the side of the span of the bridge (MPa)
Base	281	200	207	118

The results show that the most unfavorable condition of the maximum stress of the base happens when there are full vehicles. The maximum stress of the base is 281MPa, and the bearing capacity of the base is 360MPa, which can both meet the requirements.

Table 4. Checking results of the base stability.

Checking content	Calculation factor	Coefficient limit	Meet the requirement or not
Checking calculation of overturning stability	8.33	1.3	Yes
Checking calculation of sliding stability	2.91	1.3	Yes

The most unfavorable conditions of the overturning stability and sliding stability of the base both occur at the back of the abutment. When there's no vehicle on the bridge, the condition is the most disadvantageous. Based on the vehicle loading, the minimum values of the overturning stability and the sliding stability are 8.33 and 2.91 respectively, which are both larger than the limit value, 1.3, and can meet the requirements.

Scheme 2: The temporary support is used to connect the newly built U-shaped abutment as a whole.

The temporary support is connected with the newly-built U-shaped abutment to form an integral foundation. The lateral and plane views of the U-shaped abutment are shown in Figures 8 and 9 respectively. The relative position of the newly-built foundation and the original abutment is shown in Figure 10.

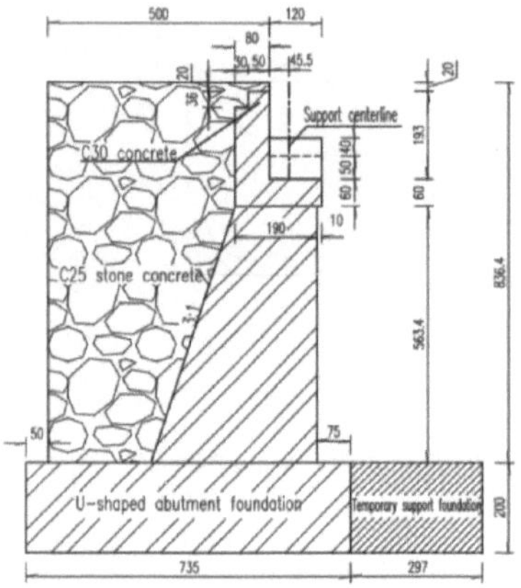

Figure 8. Lateral view of the U-shaped abutment.

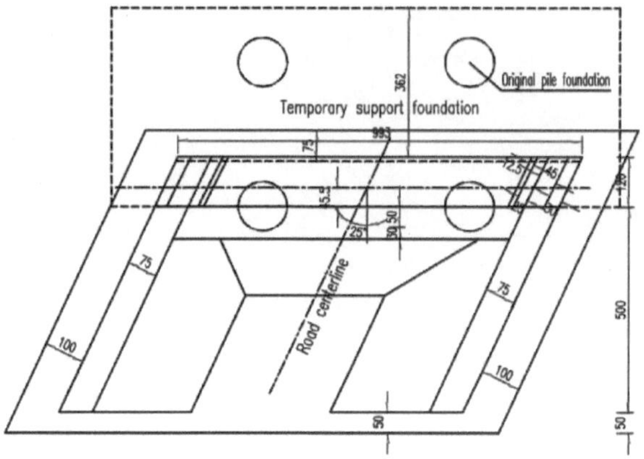

Figure 9. Plan graph of the U-shaped abutment.

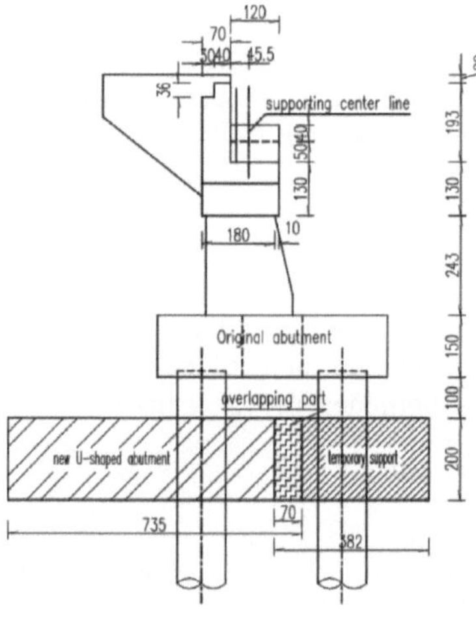

Figure 10. Plan graph of the relative position of the newly-built foundation and the original abutment.

The parameters of the U-shaped abutment in this scheme are calculated, and the results of the base stress are shown in Table 5, while the stability of the base is shown in Table 6.

Table 5. Checking results of the base stress.

Position	Maximum stress at the side of the span of the bridge (MPa)	Minimum stress at the side of the span of the bridge (MPa)	Maximum stress deviating from the side of the span of the bridge (MPa)	Minimum stress deviating from the side of the span of the bridge (MPa)
Base	170	104	243	156

According to Table 5, it can be seen that the most unfavorable condition of the maximum stress of the base happens when there are full vehicles. The maximum stress of the base is 243MPa, and the bearing capacity of the base is 360MPa, which can both meet the requirements.

Table 6. Checking results of the base stability.

Checking content	Calculation factor	Coefficient limit	Meet the requirement or not
Checking calculation of overturning stability	8.56	1.3	Yes
Checking calculation of sliding stability	2.38	1.3	Yes

From Table 6, it can be seen that the most unfavorable condition of the overturning stability occurs when there're vehicles on the bridge. Based on the vehicle loading, the minimum value of the overturning stability is 8.56, which is larger than the limit value, 1.3, and can meet the requirement. Moreover, the most unfavorable condition of the sliding stability occurs when there're vehicles on the back of the platform. Based on the vehicle loading, the minimum value of the sliding stability is 2.38, which is larger than the limit value, 1.3, and can meet the requirement.

4.3 *Result*

According to the analysis, the scheme that the temporary support is set 1.955m away from the supporting line of the original abutment is chosen. Meanwhile, the newly-built U-shaped abutment is connected to the temporary support as a whole.

5 CONSTRUCTION FLOW END CONTROLLING ESSENTIALS

5.1 *Construction Flow*

The construction process is shown in Figure 11.

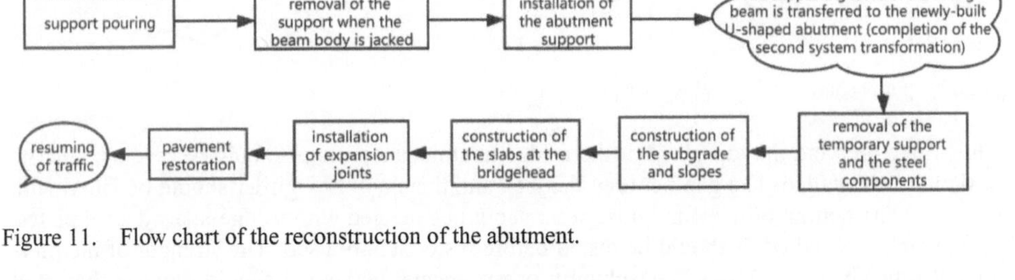

Figure 11. Flow chart of the reconstruction of the abutment.

5.2 *Controlling Essentials*

5.2.1 *Monitoring and Measurement*

When the integral foundation is excavated to the designed depth, bearing capacity of the foundation should be tested. At this time, the cap, ribbed plates, bearing platforms and part of the pile foundation are exposed, and the end force of the small box girder of the superstructure is still borne by the abutment support. During the foundation dredging, removal of the components of the original abutment as well as the whole construction, the elevation and the plane coordinates of the controlling points should be measured, and the shape of the bridge needs to be monitored. If something unexpected happens, the construction should be stopped at once, and it shouldn't be started again until the reason is found. Since the temporary support is relatively close to the abutment, the removal of the original abutment and the reconstruction of the new U-shaped abutment should be carried out carefully and deliberately to avoid the scratches and damages to the temporary support that will lead to the shift or even the collapse of the small box girder of the superstructure.

5.2.2 *Planting Steel Bars*

The planting process of steel bars is shown in Figure 12.

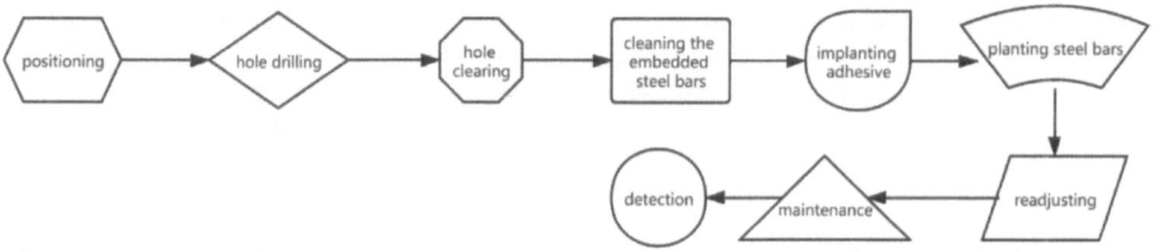

Figure 12. Flow chart of the planting of steel bars.

According to the construction drawing, the location where steel bars are planted should be a distance away from the main steel bar of the wall, which can avoid the damages to the steel bars of the original structure as well as hidden safety hazards. The holes should be dug to the required depth by electric drill, and the diameter of them should be larger than that of the steel bars, 2mm. Air pumps or brushes can be used to clean the dust inside the holes. Steel bars should be slowly planted into the holes with adhesive by rotating until the depth can meet the requirement of the anchorage length. When the steel bars are planted, they can't be moved in 24 hours. The angle of the steel bars should be adjusted to be perpendicular to the implanting surface after 24 hours of consolidation. The pulling test will be carried out by a jack after that, and the test results should meet the related planting specifications.

5.2.3 *Synchronous Jacking*

The flow chart of the synchronous jacking process is shown in Figure 13.

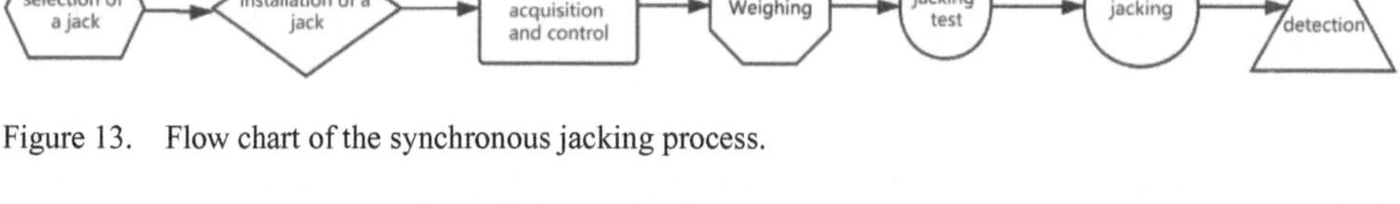

Figure 13. Flow chart of the synchronous jacking process.

The bridge pier and the top plane of the bridge abutment top plane should be leveled, and the jack should be installed. The gap between the jack and the small box girder should be filled with steel plates. The bottom of the small box girder should be leveled with wedge-shaped steel plates. After the jack is installed, it should be tested before it's officially used. The strength of the piers should also be checked. During the jacking process, special personnel should monitor the small box girder, the pier and the support. When some abnormal situations occur, the jacking process should be stopped immediately, and it shouldn't be resumed until the reasons are found. And this is the way to ensure the safety of the construction.

6 CONCLUSION

Based on the optimization and comparison of the reconstruction scheme of the separated overpass abutment, the original scheme is finally adjusted to a new one that reconstruction will only be carried out on part of the abutment. It can both shorten the construction period and save the cost of the project. Moreover, it can even ensure the running of the expressway under the overpass. In this paper, the technological process and controlling essentials are described in detail, which will provide reference for the reconstruction of similar bridges. Besides, the scheme can also be motivation for bridge builders to further study new technologies.

REFERENCES

Haiqing, Yan. (2019) Utilization and Rebuilding Strategies of Old Bridges in Highway Reconstruction Projects [J]. *Traffic world*, Issue 15 (below May):104–105.

Hong, Yuan. (2018) *Specifications for Design of Highway Reinforced Concrete and Prestressed Concrete Bridges and Culverts* [S]. People's Communications Publishing House Co., Ltd., Beijing.

Min, Li. (2020) Comparison and Analysis of Old Bridge Reconstruction Schemes [J]. *Shandong Transportation Technology*, Issue 4:88-89.

Wei, Shi. (2019) Application Analysis of Synchronous Lifting Technology in Old Bridge Reconstruction Project [J]. *Analytical research and discussion*, 10:233.

Xiang, Tang. (2017) Discussion on Design Ideas for Old Bridge Reconstruction [J]. *China Water Transport*, 199-201.

Xigang, Zhang. (2015) *General Code for Design of Highway Bridges and Culverts* [S]. People's Communications Publishing House Co., Ltd., Beijing.

Advances in Civil Engineering: Structural Seismic Resistance, Monitoring and Detection – Mohd Johari Mohd Yusof and Junwen Zhang (Eds)
© 2023 The Authors, ISBN 978-1-032-31491-4

Experimental study on flexural performance of reinforced concrete beams with shape memory alloy

XuYang Che
Yanbian University, Civil Engineering, Yanji, China

ABSTRACT: In order to explore the influence of shape memory alloy on the flexural performance of reinforced concrete beams, the bending test was carried out by embedding shape memory alloy (SMA) instead of tensile reinforcement in reinforced concrete beams. The section size of shape memory alloy and the influence of alloy types on the flexural performance of reinforced concrete beams were studied. The experimental results show that the shape memory alloy (SMA) has a negative effect on the ultimate load of rc beams, but greatly improves the ductility of rc beams, and the rc beams with two properties are better than those with only one.

1 INTRODCTION

Shape Memory Alloy (SMA or Alloy for short) is a new kind of functional material. The main characteristics of the Alloy include Shape Memory property, superelasticity and elastic modulus changing with temperature, etc. Widely used in aerospace, medical devices, mechanical and electrical fields, and in the field of civil engineering, one of the applications of SMA is: Buried SMA in reinforced concrete beams, this can not only improve the bearing capacity of reinforced concrete beams, and because of the shape memory alloy shape memory effect will produce recovery stress, so heating treatment or electrifying excitation can also repair the resulting cracks to a certain extent. Kuang Ya-chuan, OU Jin-ping et al (Kuang & Ou 2008) The effect of SMA on the repair of polymer cracks was studied by heating the alloy in the beam, and the results show that the repair condition is good. Sun Li, Zhou Jianhong et al (Sun et al. 2013) The embedded SMA wire polymer beam was fabricated by three-point bending test. The results show that the shape memory alloy wire can be used to repair the beam. Deng Zongcai, Li Qingbin (Deng & Li 2002) The influence of eccentric shape memory alloy on beam and its influencing factors are verified by experiments. Li Shuangbei, Liang Qingguo et al (Li et al. 2014) The self-healing ability of SMA concrete beams was analyzed by double-spline QR method. The results show that increasing the initial strain and diameter of SMA improves the self-healing ability of SMA concrete beams. Cui Di, Li Hongnan et al (Cui et al. 2010) The concrete beam was made with SMA wire instead of the main reinforcement, and the mechanical properties of the beam were tested by kuazhong single point loading test. The experimental results show that SMA material can obviously improve the mechanical properties of concrete beams and self-repair ability. Wang Wei, Shao Hongliang (Wang & Shao 2014) The mechanical properties of NiTi shape memory alloy bars with diameters of 6mm, 12mm and 19mm were studied experimentally at room temperature. The experimental results show that there is still a gap in the superelasticity level of alloy bars. The 12mm diameter bar has a maximum recoverable elastic strain of about 5% and can be applied to steel structure joints to obtain self-recovery properties. Strain amplitude and cycle times have important effects on mechanical properties of NiTi shape memory alloy bars. Wang Wenwei, Zhou Chang et al (Wang et al. 2021) The flexural properties of CFRP/SMA reinforced beams were tested. For CFRP/SMA composites, the recovery stress increases gradually with the increase of temperature, showing an “S” shape change law.

 DOI 10.1201/9781003310884-41

When the temperature reaches the end temperature of austenite transformation, the recovery stress tends to be stable.

2 SPECIMEN PRODUCTION

The mixing ratio of stone, water, Portland cement and sand is 1.88:0.36:1:1.14. Gravel with particle size between 5~10mm gravel, continuous gradation. The sand is medium sand and the cement is 325# ordinary Portland cement. The shape memory alloy material in the test was purchased from the Ni-Ti shape memory alloy manufactured by Yongshengda Special Steel Co., LTD., Dongwan City. The main parameters of this material are shown in Tables 1 and 2:

Table 1. Metal content of SMA unit: wt%.

Ni	Ti	Co	Cr	Nb	Fe	Cu
55.8	44.1051	0.003	0.03	0.003	0.004	0.003

Table 2. Related parameters of SMA.

Tensile strength/MPa	Yield strength/MPa	Elongation/%	Phase change pseudo-elasticity/%
850	195-690.	25 to 50	8%

The geometrical dimensions of the test beam are shown in Figure 1 below:

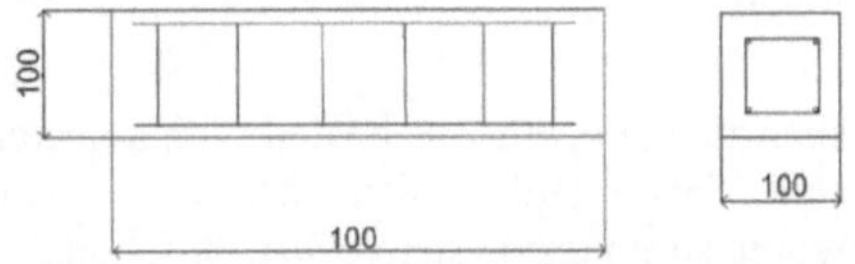

Figure 1. Schematic diagram of specimen form and shape memory alloy layout.

The diameter and types of shape memory alloy used in each beam are given in Table 3. The length of shape memory alloy rod in the specimen is 390mm, and the thickness of protective layer is 5mm.Three days after pouring, the specimen was removed and placed under the same conditions for maintenance. The cross section size of the alloy, different kinds of matching, the influence of the alloy and other factors were studied respectively.

3 TEST BEAM FLEXURAL PERFORMANCE TEST

After the installation of the test beam, the mechanical properties of the test beam were tested by the three-point bending experiment method. Due to the small size of the specimen and the low bearing capacity, the instrument was selected as the PWS-500 hydraulic fatigue tester. The test load was small and the plastic deformation was small, so that the mechanical properties of the beam could be accurately reflected. This experiment is divided into three stages:

In the first stage, the test beam is loaded. With increase of experimental force, when the test force reaches 10KN, the loading is stopped.

Table 3. Performance parameters of the test beam.

specimen Serial number	Failure mode	Alloy diameter (mm), number of roots, type	Diameter (mm) and number of reinforcing bars at the bottom	Ultimate load (KN)	Yield load (KN)
D1	The bendig failure	Four, two, memory	0	14.38	10.08
D2	Shear failure	Six, two, memory	0	27.57	15.14
D3	The bendig failure	Eight, two, memory	0	29.78	22.05
C1	The bendig failure	Four, two, super bullet	0	13.74	11.59
C2	Shear failure	Six, two, super bullet	0	26.38	12.45
C3	The bendig failure	Eight, two, super bullet	0	24.69	13.59
Z1	Bendig failure (Figure 2)	4.1. memory.Four, one, super bullet	0	18.32	15.96
Z2	The bendig failure	6,1. Memory.Six, one, super bullet	0	20.88	13.09
Z3	Shear failure (Figure 3)	8,1, memory.Eight, one, super bullet	0	31.46	16.54
G1	The bendig failure	0	4, 2	21.14	17.53
G2	Shear failure	0	6,	34.05	23.63
G3	Shear failure	0	8, 2	42.38	30.24

Figure 2. Z1 broken ring by bending

Figure 3. Z3 shear failure

In the second stage, the experimental beam was heated with a heater for 10 minutes to realize the self-repair of the experimental beam, and the data changes in this time period were recorded by the DH2817 dynamic and static strain test system every 30 seconds.

In the third stage, the heating beam was damaged by loading and the data were recorded. In the whole process, the changes of specimens were carefully observed, and the whole loading process was recorded by computer for later experimental discussion and summary.

The experimental phenomena are as follows:

4 EXPERIMENTAL RESULTSS AND ANALYSIS

Through processing and analysis of the third stage data recorded in the PWS-500 hydraulic servo fatigue tester, the following graph is obtained.

In Figure 4, it can be seen from the picture that the cracking load of D1 is 4.16kN at the minimum and cracks appear at the earliest. Z1's deflection increases with the increase of experimental load. After the experiment, its performance is better than other experimental beams with the same diameter. It follows that heterozygosis is superior to monodirection.On the contrary, C1 and D1 have little difference in performance, which may be due to the fact that the material diameter is too small and the effect on the beam is the same.

In Figure 5, the maximum yield load of D2 is 15.14kN, and the maximum ultimate load of G2 in the experiment is 34.05kN. However, in the later stage, the change is relatively fast, and the graph line decreases rapidly, reflecting the poor ductility of G2. In contrast, C2 has the best ductility.

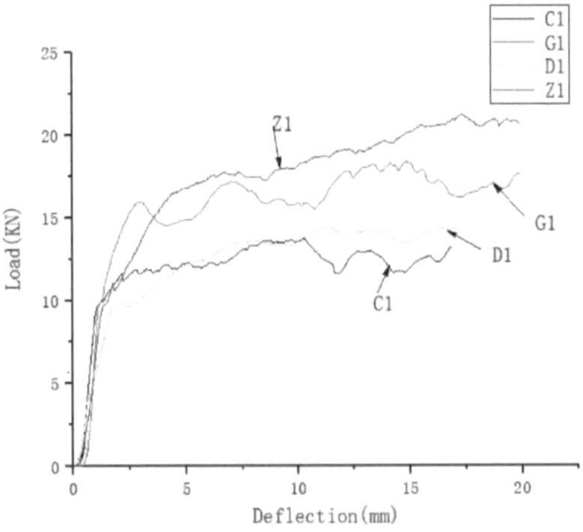

Figure 4. Load-displacement diagram.

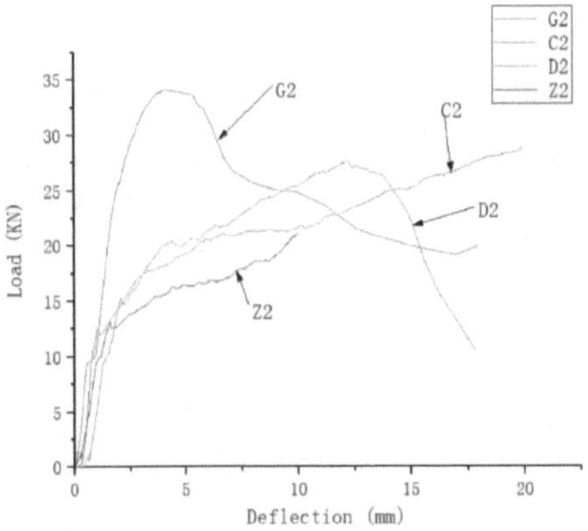

Figure 5. Load-displacement diagram.

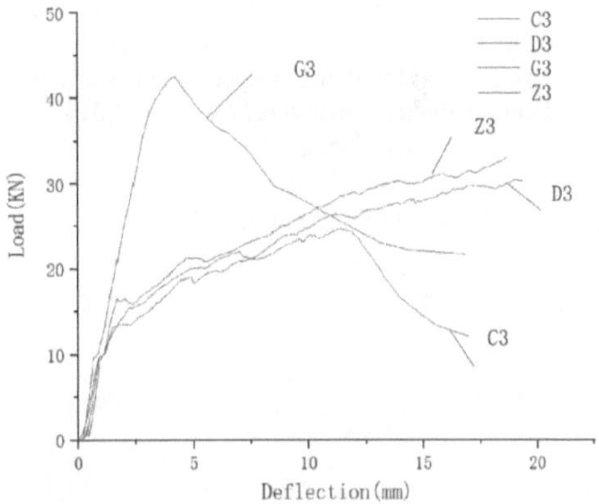

Figure 6. Load-displacement diagram.

Until the end of the experiment. However, in this group of data, Z2 is weaker than D2 and C2 in ultimate load, yield load and ductility, reflecting that the test beam with single performance is better than the test beam with superelastic and memory performance. In Figure 6, G3 is subjected to shear damage, so the figure line comes out when rising, but drops suddenly now, and its ultimate load reaches 42.38kN.Z3 is similar to D3, but the ultimate load, yield load and ductility of Z3 are better than D3. At this time, the image reflects that the experimental beam with memory and hyperelastic performance is better than the experimental beam with single performance. The decrease of C3 in the final stage is due to shear failure.

5 CONCLUSION

By using shape memory alloy to replace the main reinforcement, made of SMA concrete beam, using three-point bending test method to test to explore the impact of concrete beam bending performance. The following conclusions are drawn from the experimental study:

(1) In the case of the same diameter, it can be seen from the above figure, the ultimate load of reinforced members are too large, and with SMA instead of reinforcement, the ultimate load of the test beam will be negatively affected, but its ductility will be well improved.

(2) In the exploration of beams with two properties and only one property, in the three groups of experiments, two groups of experimental beams with two properties are better than experimental beams with one property in bending resistance.

(3) Due to the limitation of the experimental scale, a strong conclusion can not be drawn here, but it may be suggested that, with the increase of the diameter of the material, the test beam with two properties will be due to only one of the test beam? In a test beam with both properties, which property plays a major role?

REFERENCES

Cui Di, Li Hongnan, Song Gangbing. Experimental study on mechanical properties of Shape memory alloy concrete Beams [J] *engineering Mechanics*, 2010(2): 117-123

Deng Zongcai, Li Qingbin. Analysis of driving effect of shape memory alloy on concrete beam. *Journal of Civil Engineering*, 2002.

Kuang Ya-chuan, Ou Jin-ping. Research on Deformation Characteristics of Shape memory Alloy intelligent Concrete Beam [J]. *China Railway Science*, 2008.

Li Shuangbei, Liang Qingguo, Jiang Linjie, Mo Duyi. Double spline OR method for analysis of self-healing performance of SMA concrete beams. *Journal of Guangxi University*, 2014.

Sun Li, Zhou Jianhong, Gao Qianqian. Experimental study on self-repairing ability of embedded SMA wire polymer beam. *Journal of Fuzhou University*, 2013.

Wang Wei, Shao Hongliang. Experimental study on superelasticity of NiTi shape memory alloy bars with different diameters. *Structural Engineer*, 2014

Wang Wen-wei, Zhou Chang, Zhang Ya-fei, Xue Yan-jie, Yin Shi-ping, Li Shan. Experimental study on flexural performance of REINFORCED concrete beams reinforced with CFRP/SMA composite sheets. *Journal of Applied Basic and Engineering Science*, 2021.

Advances in Civil Engineering: Structural Seismic Resistance, Monitoring and Detection – Mohd Johari Mohd Yusof and Junwen Zhang (Eds)
© 2023 The Authors, ISBN 978-1-032-31491-4

Research on production technology of lightweight anticorrosive partially prestressed RPC transmission tower

Shichuan Chen
State Grid Fujian Electric Power Co., Ltd., Fuzhou, China

Xinmin Yu
State GRID Fujian Economic Research Institute, China

ABSTRACT: This paper introduces and analyzes the production and processing process of the light-duty anti-corrosion partially prestressed RPC transmission towers in detail, and analyzes the mechanical performance tests in detail with the test data. The production of RPC (Reactive Powder Concrete) transmission towers is divided into raw material preparation, steel bar processing and skeleton forming, reactive powder concrete preparation, prestressing, centrifugal forming and other processes. Through the analysis of the mechanical properties, it is concluded that the force of the RPC transmission tower is divided into the elastic phase, the working phase with cracks, and the failure phase. The force performance of the transmission tower conforms to the theoretical calculation results, and the production and processing of the product is reasonable, which meets the conditions of promotion and application.

1 INSTRUCTIONS

In recent years, the global wind disasters have been particularly serious, as the frequency and intensity of typhoons have shown high values. Meteorological experts have also admitted that human society's response to extreme convective weather has always been a worldwide problem. Typhoon disasters frequently occur in many areas in my country, causing serious accidents such as damage to the fittings of transmission and distribution lines, broken wires, broken transmission towers, which pose a great threat to the safe operation of the power grid (Cheng 1996). The direct and indirect economy caused by typhoons Losses can be hundreds of millions of yuan.

Fujian Province is in a key area for typhoon intensity zoning, and most of the coastal transmission and distribution lines are in areas with high temperature, high humidity and high salinity. Typhoon Moranti struck Fujian Province in September 2016 and brought a huge disaster to the Fujian Power Grid (Tian 2009; Xiong 2009). The distribution lines of the Fujian Power Grid of 10kV and below were seriously damaged. A total of 12,167 10kV feeders in the province have been damaged to varying degrees. The length of the affected 10kV overhead transmission lines reached 105197.69km, and the length of the affected 0.4kV overhead lines reached 39844.23km; the number of power cutoffs was 5640, and the transmission towers were damaged. There are broken rods, inverted rods, and inclined rods. The typhoon accident caused huge losses to the power grid.

At present, some of Fujian's transmission networks and most of the overhead lines of distribution networks mainly use non-prestressed concrete transmission towers. This project intends to promote reactive powder concrete (RPC) materials to make high-strength and high-durability transmission towers in areas with strong typhoons in the zoning map, and solve the problem of broken transmission towers caused by insufficient bending moment, pitted surfaces, honeycombs, falling blocks, and leaking tendons caused by salt rot (Sui 2002). As well as technical problems such as cracks; and according to its characteristics, a matching basic type is designed to further

DOI 10.1201/9781003310884-42

improve the overall horizontal resistance and solve the problems of inverted rods and inclined rods. The results of the project will fundamentally change the current situation of poor wind resistance of concrete transmission towers in Fujian, and ensure the strong power grid under extreme weather conditions.

After the promotion and application of the technical results of this project, the wind resistance of the transmission tower lines can be significantly enhanced, transmission tower breakage, oblique transmission tower and transmission tower down accidents can be greatly reduced, the operation safety and power supply reliability of the power grid can be greatly improved, and the loss of power outage can be reduced. Considerable direct and indirect economic benefits (Wang 2011).

- The high-strength and high-durability transmission towers and supporting foundations promoted by this project can significantly enhance the wind resistance of the transmission lines, improve the line's disaster prevention capabilities, ensure its safe and stable operation, avoid transmission tower breakage accidents, and greatly reduce transmission tower failures. The probability of occurrence of oblique transmission towers will produce greater direct economic benefits for power companies. Avoid combining SI and CGS units, such as current in amperes and magnetic field in oersteds. This often leads to confusion because equations do not balance dimensionally. If you must use mixed units, clearly state the units for each quantity that you use in an equation (Wang 2002).
- The promotion and application of the results of this project can avoid large-scale power outages in the power grid and provide power guarantee for post-disaster reconstruction. Avoiding the social impact and indirect losses caused by power emergencies to the greatest extent, maintaining social stability and the safety of people's lives and properties, has significant social benefits.

2 THE PRODUCTION PROCESS OF RPC TRANSMISSION TOWER

2.1 *Raw materials*

Copy the template file B1ProcA4.dot to the template directory. This directory can be found by selecting the Tools menu, Options and then by tabbing the File Locations. When the Word programme has been started, open the File menu and choose New. Now select the template B1PROCA4.dot and start by renaming the document after clicking Save As in the Files menu.

Wherever Times is specified, Times Roman or Times New Roman may be used. If neither is available on your word processor, please use the font closest in appearance to Times. Avoid using bit-mapped fonts if possible. True-Type 1 or Open Type fonts are preferred. Please embed symbol fonts, as well, for math, etc.

1) Rebar
   a) The types, specifications and technical conditions of steel bars should meet the design requirements or the regulations in Table 3-3.
   b) All steel bars shall be checked and accepted in accordance with GBJ204 "Code for Construction and Acceptance of Reinforced Concrete Engineering".
   c) The cold-drawn low-carbon steel bars shall be checked and accepted one by one.
   d) The steel products that have passed the inspection must be stacked in batches and sorted to avoid confusion and avoid rust and contamination.
2) Reactive powder concrete material
   a) For cement, see Appendix A.1 of GB4623-1984. For areas with erosion, different types of cement should be used according to the level of erosion (see Table I)
   b) Standard sand (without coarse aggregate) according to design requirements.
   c) Drinking water should be used for mixing concrete, and the water quality should conform to JGJ632.
   d) The performance of admixtures mixed into concrete should meet the requirements of GB8076. It is strictly forbidden to mix chlorine salt admixtures.

Table 1. Cement selection table for erosion areas.

Erosion level	No Erosion	Weak Erosion	Medium Erosion	Strong erosion
Suitable cement varieties	Portland cement	Ordinary Portland	Portland slag cement	Sulfate resistant cement
	Ordinary Portland Cement	Cement Portland slag cement	Sulfate resistant cement	Super sulfate resistant cement

2.2 *Rebar processing and skeleton forming*

1) Cold-drawn grade IV steel bars should adopt dual control, the control stress is 750MPa, and the elongation rate should not be greater than 4%. The cold drawing speed should not be too fast, after drawing to the specified value, continue for 1 minute and then relax.
2) Floating rust and oil stains should be removed before the steel bar is cut. Steel bars with oxidized iron skin, honeycomb rust, heavy skin and cracks shall not be used.
3) The joints of cold-drawn steel bars are welded first and then pulled. The joint strength is not less than 95% of the base material strength.
4) Rebar processing technology and the maximum allowable deviation:
   a) After the steel bar is cut, there should be no local bending, the end surface should be flat, and the length deviation should not be greater than 1.5/10000 of the full length of the steel bar. Take a group (not less than 10) every day for length inspection.
   b) When the diameter of V-grade smooth steel bars and carbon steel wires is greater than 5mm, compulsory anchoring measures should be taken.
   c) When the main reinforcement adopts quenched and tempered grade V steel bar, carbon steel wire and cold drawn low carbon steel wire, welded joints are not allowed.
   d) Rebar heading:

Quenched and tempered grade V steel bars, carbon steel wires, and cold drawn low-carbon steel bars should be cold headed.

Rebar heading should be operated by a dedicated person. At least one set of samples is tested every day.

The strength of the upsetting head of the assembled rod steel bar is not less than 95% of the standard strength of the raw material (if it fails to reach it, the strength can be reduced for use).

Reinforcement heads should be formed at one time, no heavy heading, no cracks, and the eccentricity should not be greater than 0.12 times the diameter of the rebar. In the same transmission tower, the relative error of the effective length of the steel bars after upsetting should not be greater than 2/10000.

5) Continuous reinforcement:
   a) When using continuous reinforcement technology, the relative length deviation between a single steel wire (from one hanging point to another) shall not be greater than 2/10000.
   b) The total breaking force of the two steel wires around the hanging point should not be less than 95% of twice the total breaking force of one steel wire.
   c) The anchoring strength of the start and end of the continuous steel wire shall not be less than 95% of the strength of the steel wire.
   d) The shear strength of the hanging point rivet shall not be less than 1.5 times the tensile strength of the two steel wires, and the diameter of the hanging point rivet shall not be less than 4 times the diameter of the steel wire.
6) Frame forming:
   a) The deviation of the main rib spacing shall not exceed 5mm.
   b) In order to control the position of the main bar, when the strength loss of the main bar does not exceed 5% after the test, seam welding can be used, or the main bar control ring can be set. When the length of the skeleton is less than 9m, a main reinforcement control loop

should be set in the middle; when the length of the skeleton is equal to or greater than 9m, two main reinforcement control loops can be set equally. The main ribs should be tied point by point with the claws of the control ring (as shown in Figure 3-3).

c) In order to ensure the protective layer of the main reinforcement, for the transmission tower section with a length equal to or greater than 6m, the skeleton should be provided with reinforcement pads.

7) For segmented transmission towers, the height of the joint steel plate ring should not be less than 140mm.
8) The non-prestressed steel bars in some prestressed transmission towers should be firmly welded to the steel plate ring or flange.

2.3 *Preparation of reactive powder concrete*

The design of concrete mix ratio shall be determined through experiments. Manufacturers are required to adjust the mix ratio in the laboratory, and select the best mix ratio for the production of transmission towers.

2.4 *Apply prestress*

1) The center of the tensioning machine head is aligned with the axis of the steel mold and then tensioned. Tensioning procedure: 0→initial stress (10%)→105% (holding load for 2min)→0; or 0→103%.
2) For the prestressed tendons of limited prestressed transmission towers and some prestressed transmission towers (two types of steel bars are used at the same time), the maximum control stress shall not exceed the following values: hard steel 0.7, mild steel 0.9, cold drawn low carbon steel wire 0.75. The pre-compression stress of concrete is less than 0.4 (the demoulding strength of concrete).
3) For the IV and V grade steel bars of prestressed transmission towers with low prestress degree, the minimum control stress of carbon steel wire shall not be less than 0.4.
4) The application of prestress should adopt dual control of stress and elongation, but stress control should be the main method.
5) All kinds of machinery, equipment and instruments for applying prestress should be operated by special personnel.
6) The accuracy of the testing machine or dynamometer should not be less than ±2%, and the accuracy of the pressure gauge should not be less than 1.5. The calibration period of tensioning equipment should generally not exceed half a year, and the calibration period of pressure gauges should generally not exceed two months. When the tensioning equipment is abnormal during use or the jack is overhauled, it should be re-calibrated.

2.5 *Centrifugal molding*

1) Before centrifugation, the steel mold running wheel and the centrifuge supporting wheel should be stable.
2) The centrifugal speed generally adopts the values in Table 2 and Table 3.
3) During the centrifugation process, you should pay close attention to the operation of each part. If you find that the slurry is running, the steel mold is violently jumping, the screw is loose or the screw is broken, the machine should be stopped immediately.
4) The centrifugation must be completed before the initial setting of the concrete. After centrifugation, the remaining slurry in the rod should be poured clean and the centrifugation record should be made.

Table 2. Cemtrifugal speedometer.

Steel mold speed(r/min)		Time(min)
Slow	8.01~20	1.5~3
Speed up		2~3
Fast	Rotation speed see TABLE III	Lasts for 6~10
Total centrifugation time		Less than 12

Table 3. Cemtrifugal speedometer.

The tip diameter and diameter of the Transmission Tower(mm)	Fast speed range (r/min)
Less than φ 190	400~500
ϕ190~ φ 310	350~400
More than φ 310	330~370

3 ANALYSIS OF MECHANICAL PROPERTIES OF RPC TRANSMISSION TOWER

When the RPC transmission tower is properly reinforced, the transmission tower will undergo "fit-reinforcement failure". The characteristics of this failure are: the tensioned steel bar first reaches the yield strength, and when the stress remains unchanged, significant plastic deformation occurs until it is subjected to stress. When the edge strain of the concrete in the compression zone reaches the ultimate compressive strain of the concrete bending compression, the compressed concrete is crushed, and the cross section is destroyed. Before the transmission tower is completely destroyed, the steel bar will undergo greater plastic elongation, which will cause cracks. The rapid development and the sharp increase in transmission tower deflection will give people obvious signs of damage. The force process is divided into three stages:

3.1 *The first stage: the elastic stage*

When the load is small, the bending moment generated in the section of the transmission tower is very small, so the strain on the section is also very small. The concrete is basically in the elastic working stage, and the section stress becomes proportional to the strain. In addition, the deflection of the transmission tower and the bending moment also maintain a linear relationship.

When the load increases, the bending moment and strain of the transmission tower section also increase. Since the tensile capacity of concrete is much smaller than its compressive capacity, the strain at the edge of the tension zone first appears to grow faster than the stress. The plastic properties. As the bending moment continues to increase, until the section bending moment reaches its cracking moment, the strain value of the fiber at the edge of the tension zone will reach the ultimate tensile strain of the concrete under bending, and the section is about to crack, which is called the first At the end of the first stage, the crack resistance of the transmission tower is reflected.

3.2 *The second stage: working stage with cracks*

The moment the RPC transmission tower reaches its cracked state, as long as the load is slightly increased, the first batch of cracks will appear on the section with the weakest concrete tensile strength in the pure bending section. The stress of the steel bar suddenly increases a lot, so the deflection and section curvature of the RPC transmission tower suddenly increase when cracks appear. The stress on the section will be redistributed, the concrete at the crack will no longer bear the tensile stress, and the concrete in the compression zone will appear obvious plastic deformation.

3.3 *The third stage: the destruction stage*

As the tensile steel bar yielded, cracks developed rapidly, and the curvature of the section and the deflection of the transmission tower also suddenly increased, forming a sign before failure. As the neutral axis continues to move to the compression zone on the other side of the transmission tower, the height of the compression zone is further reduced, the compression stress of the concrete in the compression zone increases rapidly, the edge strain of the concrete in the compression zone also increases rapidly, and the plastic characteristics are also about to behave better. For full.

When the bending moment continues to increase to the ultimate bending moment, the concrete at the edge of the compression zone will reach its ultimate compressive strain (usually 0.0033), the concrete at the edge of the compression zone will be crushed and bulge outward, and the RPC transmission tower will soon be destroyed. At this time, under the condition that the load remained almost unchanged, the cracks further developed sharply, the concrete was completely crushed, and the cross section was destroyed.

4 CONCLUSION

This paper introduces and analyzes the production and processing process of the light-duty anti-corrosion partially prestressed RPC transmission towers in detail, and analyzes the mechanical performance tests in detail with the test data. The production of RPC transmission towers is divided into raw material preparation, steel bar processing and skeleton forming, reactive powder concrete preparation, prestressing, centrifugal forming and other processes. Through the analysis of the mechanical properties, it is concluded that the force of the RPC transmission tower is divided into the elastic phase, the working phase with cracks, and the failure phase. The force performance of the transmission tower conforms to the theoretical calculation results, and the production and processing of the product is reasonable, which meets the conditions of promotion and application.

REFERENCES

Cheng, Z. (1996). Calculation and setting out according to the control size of iron tower. *Electric power construction*, vol. 2, 53–54.

Sui, Y. & Shao J. (2002). Rod length adjustment method to improve the bearing capacity of statically indeterminate truss. *Applied Mathematics and mechanics*, vol. 23, 269–272.

Tian, H. & Xiong H. (2009). Assembly stress analysis of spatial grid steel structure model. *Enterprise technology development (Academic Edition)*, vol. 28, 53–54.

Wang J. (2002). Discussion on bending resistance and manufacturing accuracy of tower web members. *Shanxi electric power*, vol. 22, 65–68.

Wang J. (2011). Discussion on bending resistance and manufacturing accuracy of tower web members. *Shanxi electric power*, vol. 22, 65-68.

Yang F.& Zhu B. & Xing H. (2002). Study on slip characteristics and model parameters of bolted joints of transmission tower. *Engineering mechanics*, vol. 34, 116–127.

Advances in Civil Engineering: Structural Seismic Resistance, Monitoring and Detection – Mohd Johari Mohd Yusof and Junwen Zhang (Eds)
© 2023 The Authors, ISBN 978-1-032-31491-4

Internal grounding scheme of concrete pole in Fujian Province

Xinmin Yu & Xiaogang Li
State GRID Fujian Economic Research Institute, Fuzhou, China

ABSTRACT: Most of the lightning arresters of distribution lines in Fujian Province are naturally grounded by concrete poles, which can easily lead to concrete burst and fall in case of lightning overvoltage, resulting in the exposure of steel bars within the poles to the air, thus accelerating corrosion. In this paper, an internal grounding scheme of electric pole is proposed, which directly grounded the low-voltage side of lightning arrester. The thermal stability calculation of grounding wire, connection resistance calculation, and experiment of grounding scheme meet the requirements of relevant specifications.

1 INTRODUCTION GENERAL

Fujian Province is located on the southeast coast, with strong and multi-thunder areas accounting for 97% of the total area of the province. The coverage of distribution line arresters exceeds 27%. Most of the line arresters are not equipped with independent grounding devices and are naturally grounded by rods. The lightning strike on the 10 kV overhead line activates the lightning arrester, because the metal cross arm at the tower head is usually not equipotentially connected to the internal structural reinforcement of the reinforced concrete pole (Gu 2016). The lightning overvoltage is easy to discharge the structural reinforcement of the pole at the cross arm position, resulting in the bursting and falling off of concrete on the surface of the pole and exposing the internal structural reinforcement, as shown in Figure 1. Due to erosion from rain and moisture, the exposed structural reinforcement of the pole gradually rusts, affecting the mechanical strength of the pole (Li 2017).

Figure 1. Structure corrosion diagram of exposed pole.

2 CURRENT POLE GROUNDING FORM

The devices that require grounding of the current distribution network poles mainly include lightning arrester, overvoltage protection (Sheng 2019), integrated distribution box shell (Wang 2018), transformer shell (Xu 2000), transformer working grounding, etc. Taking the 12-m pole as an example, the working grounding positions of the shell of the comprehensive distribution box and the shell of the transformer are shown in Figure 2. The shell of the comprehensive distribution box

DOI 10.1201/9781003310884-43

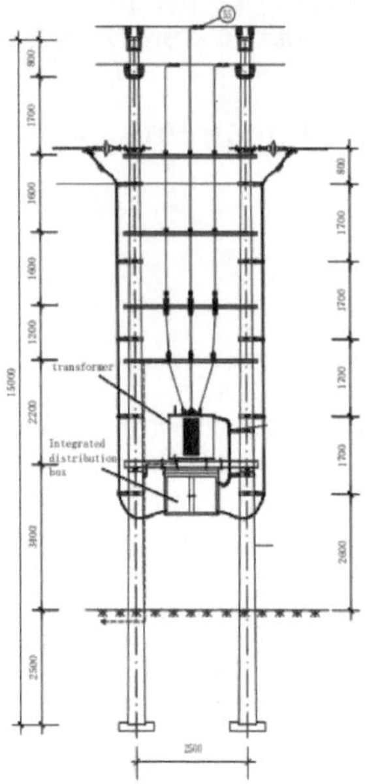

Figure 2. Grounding diagram of integrated distribution box shell.

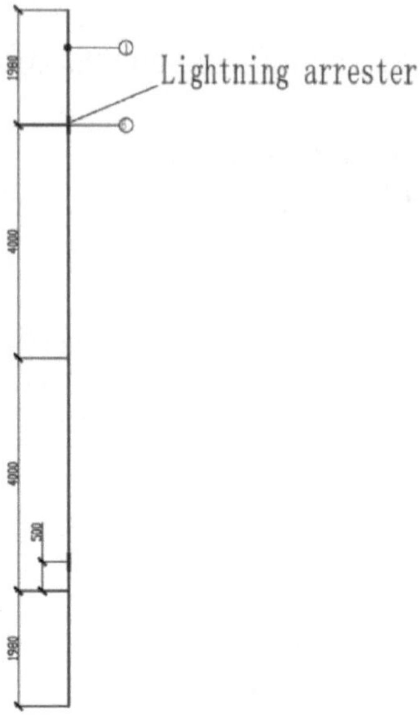

Figure 3. Location diagram of pole arrester.

is located at about 3 m above the ground, the transformer is located at about 4 m above the ground, and the single circuit pole arrester is located at about 3 m above the pole from the top of the tip (Figure 3). The connection between the grounding round steel at the bottom of the pole and the grounding grid is located 0.8 m below the ground plane (Figure 4).

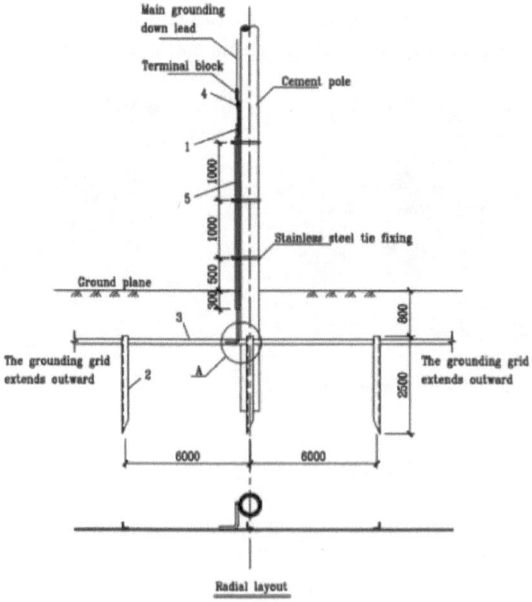

Figure 4. Schematic diagram of grounding at the bottom of pole.

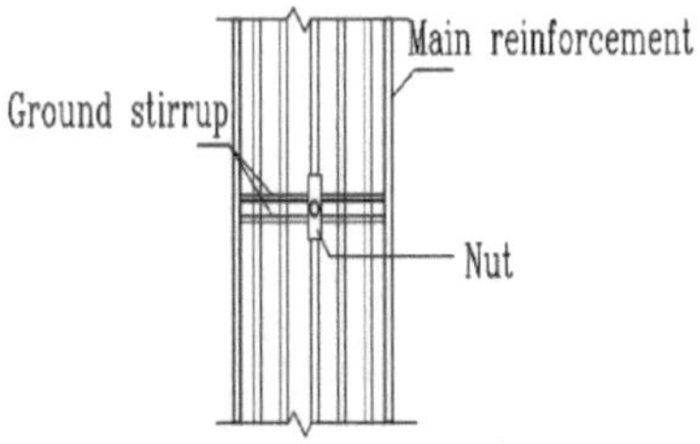

Figure 5. Position diagram of inner stirrup and main reinforcement.

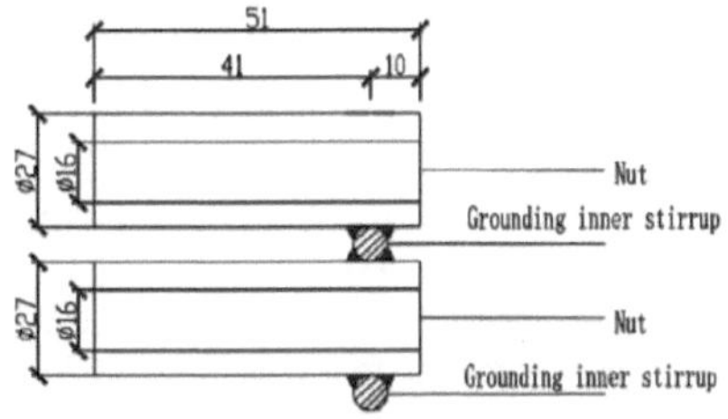

Figure 6. Schematic diagram of inner stirrup and nut.

3 INTERNAL GROUNDING SCHEME OF POLE

3.1 *Setting scheme of rod body grounding terminal*

According to technical guidelines for distribution network (Q/GDW 10370-2016, 5.10.7) for lightning protection of conductors of newly built or reconstructed overhead insulated lines, the reinforcement of annular concrete poles should be used for natural grounding, and the grounding resistance should not be greater than 30 Ω. Considering the standardized design of distribution network, pole production, and bidding, the internal grounding scheme of pole should be simplified as much as possible. Two access points should be designed at the upper and lower sides of pole, respectively. To ensure the grounding stability, two nuts are designed at each access point. The specific scheme is as follows: considering the position difference of single and double circuit line arresters, the access point of pole at the top of pole should be set 2 m away from the top of the tip; pole and the grounding grid access point is set 0.5 m above the buried depth line of the pole. The embedded nut is welded with the grounding inner stirrup, welded on both sides. The outside of the nut is hot-dip galvanized, the nut is welded with the grounding inner stirrup on both sides, the grounding inner stirrup is cross bound with the main reinforcement, the bolt is pressed with various types of grounding terminals through gaskets, and the bolt and nut are fastened and engaged to ensure internal grounding stability of the pole (Figures 5, 6, and 7).

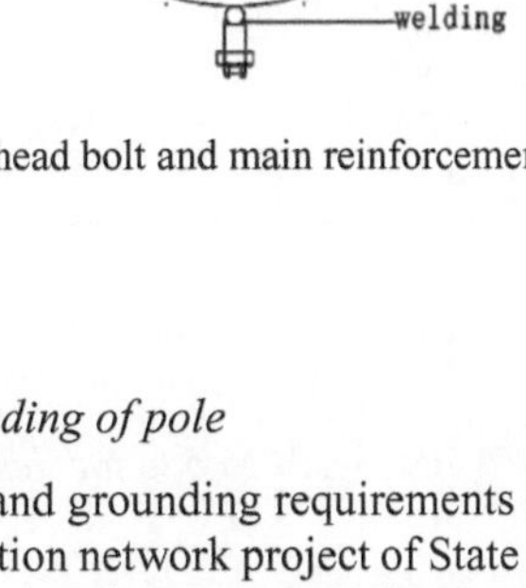

Figure 7. Position diagram of single head bolt and main reinforcement.

3.2 *Calculation of internal grounding of pole*

According to lightning protection and grounding requirements in Section 14.1.3 of typical design of 10kV overhead lines for distribution network project of State Grid Corporation of China (2016), two methods, namely gap arrester or overhead ground wire, can be adopted for lightning protection of 10 kV bare conductor line.

Table 1. *Theoretical calculated value of resistance.

Connection type	Section (mm^2)	Resistance of this section (μΩ)	Total resistance (μΩ)
2× (2) Φ16 bolt + nut section)	486.86	3.62	314.60
8Main reinforcement (7.6 m)	1231.50	0.00	
2Short reinforcement (7.6 m)	307.87	63.53	
2Short reinforcement (6.6 m)	307.87	52.94	
2Short reinforcement (5.1 m)	307.87	45.38	
2Short reinforcement (4.1 m)	307.87	39.71	
2Short reinforcement (3.1 m)	307.87	109.42	

*The total length of the pole is 12 m, and the buried section is 1.9 m. The access point of the pole at the top of the pole is set 2 m away from the top of the tip; the access point of the pole and the grounding grid is set 0.5 m above the ground line, so the distance between the upper and lower bolts is about 7.6 m.

4 CURRENT GROUNDING CALCULATION INTERNAL

The calculation of grounding type resistance in typical design of 10 kV overhead lines of distribution network project of State Grid Corporation of China (2016) is given in Table 2 below.

Table 2. * 12 m pole grounding body resistance.

	$\rho(\mu\Omega$ cm)	l (m)	s (mm^2)	R ($\mu\Omega$)
Round steel	9.78	7.9	50.24	15378.58
Copperclad round steel	4.31	7.9	50.24	6777.27
Flat steel	9.78	7.9	48.00	16096.25
Copperclad flat steel	4.31	7.9	48.00	7093.54
Copper bar	1.75	7.9	50.24	2751.79
Flat copper	1.75	7.9	50.00	2765.00
Copperclad steel strand	4.31	7.9	50.24	6777.27
Copper strand	1.75	7.9	50.00	2765.00
Galvanized steel strand	9.78	7.9	25.00	30904.80

* Copper: 1.75$\mu\Omega$ cm; steel: 9.78$\mu\Omega$ cm

Resistivity of different types of copper-clad steel (the resistivity adopted this time is 4.31 $\mu\Omega$ cm)

Table 3. Copper-clad steel resistance.

Model	Resistivity ($\mu\Omega$ cm)	Conductivity
CCSXX20	8.62	20%
CCSXX25	6.89	25%
CCSXX30	5.74	30%
CCSXX40	4.31	40%

5 THERMAL STABILITY ANALYSIS

According to appendix e.0.1 of *GB/T 50065-2011 code for design of grounding of AC electrical installations*:

$$Sg \geq \frac{\mathrm{Ig}}{\mathrm{C}}\sqrt{\mathrm{te}}$$

The value of C is 70 (400°C) for steel, 249 (700°C) for copper, and 167 (700°C, 40% conductivity) for copper-clad steel. The cross-sectional area of each material in Table 4-1 of typical design 10KV overhead line volume of distribution network project of State Grid Corporation of China (2016) is converted to the cross-sectional area of steel (mm^2) as follows:

Table 4. Material cross-sectional area converted to steel cross-sectional area.

Type	Sectional area	Converted to steel material	Cross-sectional area of this scheme
Round steel	50.24	50.24	360 mm^2 (in this scheme, M16 hexagon nut is adopted, the opposite side length is s = 24 mm, the thickness is m = 30.0 mm, one side of the nut is welded, and the thickness of the contact surface is not less than 15 mm)
Copper-clad round steel	50.24	119.86	
Flat steel	48	48	
Copper-clad flat steel	48	114.51	
Copper bar	50.24	178.71	
Flat copper	50	177.86	
Copper-clad steel strand	50.24	119.86	
Copper strand	50	177.86	
Galvanized steel strand	25	25	

6 THERMAL STABILITY ANALYSIS

Without considering the solder joint resistance and the binding of stirrup and main reinforcement in the pole grounding, the resistance of 12 m pole in the pole internal grounding scheme is 314.60 $\mu\Omega$, which is very less than the current grounding mode (common mode of current grounding 4), the resistance of 48 flat steel is 4828 $\mu\Omega\phi$, the resistance of 12 round steel is 6834.7$\mu\Omega$, and the material cross-sectional area of the internal grounding scheme of the pole in the thermal stability analysis is greater than that of the current scheme, so the internal grounding scheme of the pole meets the design grounding requirements.

Table 5. Experimental data

Test date	2021.06.04	Weather	Sunny
Temperature	32°C	Humidity	83%
Test object	Internal grounding resistance ($\mu\Omega$)	External round steel resistance ($\mu\Omega$)	Remarks
Sample 1	2541.6	12230	PlusΦ16 reinforcement
Sample 2	1499.5	12780	NothingΦ16 reinforcement
Sample 3	2078.3	/	NothingΦ16 reinforcement

7 SUMMARY

Although sample 1 is one more than samples 2 and 3 16 grounding reinforcement, considering that the welding process is adopted for the nut, grounding reinforcement, and inner stirrup, and the binding process is adopted for the inner stirrup and main reinforcement, that is, the nut is welded with the grounding reinforcement and bound with the main reinforcement. The difference in welding and binding conductivity leads to incomplete parallel connection between the grounding reinforcement and other main reinforcement, so the resistance of sample 1 is greater than that of samples 2 and 3 (this test requires the pole processing plant to produce according to the standard production process without special reinforcement treatment).

Because of the contact resistance caused by the binding of the main reinforcement of the pole and the grounding inner stirrup, and the welding resistance between the nut and the inner stirrup, the measured resistance is quite different from the theoretical calculation, but the trend that the internal grounding resistance is obviously less than the external round steel resistance remains unchanged.

REFERENCES

Gu Ming, *Analysis, calculation and design of direct lightning protection for compact 220kV intelligent substation*; JiNan: ShanDong university, April 2016.

Li Xun, Huang Ronghui, Wu Guoxing, *Study on impact of lightning strike grounding model of transmission tower on operation safety of nearby pipeline*; 2017(10), 25–30.

Sheng Kuang, "*Research and design of installing lightning conductor for 10kV overhead distribution line in vulnerable section in strong lightning area,*" Wuhan University of Technology, pp. 40–43, April 2019.

Wang Hui, Liu Jun, He Jinliang, *Influence of transmission line tower on lightning current*, April 2018.

Xu Naixin, *Corrosion and protection of reinforced concrete pole, Corrosion and protection*, 2000, pp 509–510.

Advances in Civil Engineering: Structural Seismic Resistance, Monitoring and Detection – Mohd Johari Mohd Yusof and Junwen Zhang (Eds)
© 2023 The Authors, ISBN 978-1-032-31491-4

Research on dust suppression of bulk cargo operating areas of a domestic port group

Xiaomeng Liu, Yaohongling Chen & Chunyi Zhang*
Environmental Technology Development of TIWTE(TIANJIN) CO., Ltd. Tianjin Research Institute of Water Transport Engineering, Ministry of Transport, Tianjin, China

ABSTRACT: In view of the practical issues in dust pollution prevention and control in bulk cargo operating areas of a domestic port group, with the aim of improving the efficiency and capability of air pollution prevention and control and optimizing the structure and effectiveness of environmental protection investment, we developed an effective and feasible optimization plan by region, priority, and stage through data analysis, field survey, on-site monitoring, numerical simulation, system architecture, and others. The paper further evaluates its economic, environmental, and social benefits comprehensively.

1 BACKGROUND

With increasingly stringent national environmental protection policies, bulk cargo ports with particulate matter as the main pollutant have gradually become the center of social attention. Considering four main operating port areas, the total bulk cargo throughput of the group in 2019 was close to 180 million tons. With characteristics of large throughput, various types of cargos, complex operation techniques, different construction, and management levels, environmental management is much more difficult. Therefore, this research intends to use various methods to figure out existing environmental problems, pollution situations, and governance priorities. It then forms a targeted dust pollution prevention and control management system from the perspective of the group. In addition, it analyzes the environmental protection investment in a semi-quantitative manner. The results of this research are of great significance for guiding the group to enhance air pollution control capabilities, reduce dust emissions, and improve the efficiency of environmental investment.

2 RELATED RESEARCH STATUS

2.1 *Application of port dust pollution control measures*

Dust pollution control technologies for dry bulk cargo ports at home and abroad are relatively mature. They can be divided into dust prevention and dust removal. Specifically, these mainly include wind barrier, wet spraying, dry fog dust suppression, dry dust removal, road dust suppression, and other auxiliary technologies. Wind barrier dust suppression is mostly used in bulk cargo yards, and controls the dust by decreasing the influence of wind or other meteorological factors (Gao 2018). The wet spray technology increases the moisture content, adhesion strength, and surface tension to increase the starting wind speed of the dust. It is currently the most economical method for bulk cargo terminals with the characteristics of high dust removal efficiency, simple operation, and wide range of applications (Chang 2013, Guo 2006). In recent years, dry fog dust suppression method has been widely used for partial operation dust removal with the advantages of significant dust removal effect and water conversation compared with traditional water spraying

DOI 10.1201/9781003310884-44

method. Dry dust removal method is mainly used in relatively closed facilities of bulk cargo terminals. The common forms are bag and electrostatic dust removal (Xu 2011). However, due to high cost and the maturity of alternative technologies, the application of dry dust removal technology decreases. Road dust suppression is committed to reducing the secondary dust pollution caused by vehicle transportation and is particularly effective in port areas where automobile transportation is the mainstay. In addition, auxiliary technologies such as belt conveyor closure, ship loader slide, and road isolation are also widely used in various bulk cargo operation terminals (Li 2020; Tian 2011).

2.2 *Benefit analysis of environmental protection investment*

Environmental protection investment benefit is an important basis for determining the rationality of investment. Through benefit analysis, it is possible to evaluate the economic, social, and environmental benefits brought about by project operations, as well as predict the investment, construction, and expected effects. Overall, it plays an important guiding role for enterprises to make efficient environmental protection investments.

At present, the research on environmental protection investment benefits mostly starts from the perspective of economic or environmental benefits, and there are relatively few comprehensive benefit analyses based on multiple factors and multiple levels. Some scholars have conducted researches on comprehensive benefits, but most of them are qualitative analysis. Yuan (2004) proposed an index evaluation method for the investment benefit of environmental protection projects. After sub-evaluating the completion index, investment completion index, and project quality index of the project, the weighting method was introduced to determine the performance index. Yang et al. (2014) analyzed the economic benefits of an irrigation area project in Nanyang from various aspects, and calculated the payback period through dynamic analysis. Gu (1995) analyzed the environmental and economic benefits of the investment in a cement plant from the aspects of direct and indirect benefits. Shen et al. (2015) took a paper-making company in Jiangsu as an example to compare the economic benefits of environmental protection investment from both horizontal and vertical aspects. Li (2002) discussed the principle of cost-benefit analysis of environmental protection investment, and introduced the steps and evaluation techniques of cost-benefit analysis. Liu (2004) constructed a quantitative evaluation method for social benefits through Analytic Hierarchy Process (AHP). Luo (2020) established an investment efficiency evaluation system of environmental protection through AHP with environmental efficiency, economic efficiency, and social efficiency as the criteria, and studied the annual and industry distribution characteristics based on disclosed data. Peng (2011) took seven coal companies under China Coal as examples to study the relationship between the cost of input and the benefits of economic and environment through Data Envelopment Analysis (DEA), and evaluated the ecological efficiency of the group.

Environmental investment benefit analysis methods have been applied to cement manufacturing, papermaking, coal, and other industries, but the existing analysis methods are not applicable to the port industry. Therefore, based on the actual situation of the port, to establish a benefit analysis method that suits port industry well is of great significance to improving the effectiveness as well as efficiency of port environmental protection investment.

3 CURRENT SITUATION OF DUST POLLUTION PREVENTION AND CONTROL

3.1 *Dust-generating nodes*

Dust in bulk cargo ports is mainly generated from the operation process of loading and unloading, storage, and transportation of bulk cargos. Due to differences in the regions, level of infrastructure, and trans-shipment methods, the properties of dust generation are different. Divided by operation process, the dust-generating nodes of specialized and general bulk cargo terminals are roughly the same, and the biggest difference lies in the number. The dust-generating nodes of specialized and general bulk cargo terminals are shown in Figure 1.

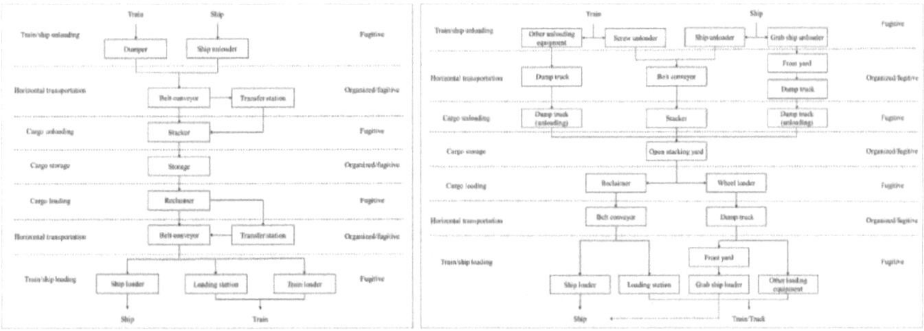

Figure 1. Dust-generating nodes of specialized (left) and general(right) bulk cargo terminal.

The cause of dust generation in each bulk cargo operating port area can be roughly divided into static wind erosion and dynamic operating disturbance. Static dust generation is mainly caused by wind erosion during the period of storage, which has great relevance to external wind speed and the moisture content of the pile surface. Dust in dynamic operation is mainly caused by machinery and manual disturbance and the operating height difference. Meanwhile, it is also considerably affected by external wind speed.

3.2 *Current situation of environmental protection facilities and management*

To control dust pollution during the operation process, each bulk cargo operating area has been equipped with corresponding prevention and control facilities, and a relatively complete environmental management system has been formulated, which helps control dust pollution to a certain extent. Due to differences in infrastructure, environmental management, and resource allocation, the level and effectiveness of dust pollution control in each operating area is also different. As a result of the higher approval requirements in the initial construction phase of the specialized bulk cargo terminal project, modern large-scale facilities were used for operations, which guarantees few dust-generating nodes and higher operating efficiency. However, except for the larger number of dust-generating nodes, general bulk cargo terminal also has the feature of scattered operation locations, which makes it more difficult to form a complete system of protection measures and further increases the difficulty of dust control.

Overall, there are still some problems in the application of environmental protection measures and management system. The first is the incomplete implementation of laws and policies. The lack of dust control measures in key operating processes in some operating areas has greatly reduced the efficiency of dust control. Second, the establishment of pollution prevention and control system is imperfect, which restricts the full use of the effectiveness of control measures. Although the environmental management system has been formulated in each operating area, it is too general and untargeted. The incomplete implementation of the system and the lack of effective on-site feedback mechanism in actual operation process aggravates the passivity and hysteresis of its internal environmental management. Moreover, the vitality of technological research and innovation is insufficient. Since most of the bulk cargo terminals have been in operation for many years, it is difficult to break its operation mode in a short period of time. Currently, the operating areas, to a great extent, rely on traditional technologies for dust prevention and control. Few of them start to combine its own characteristics of operation and dust emission to develop dust removal or dust suppression technologies that are more suitable.

3.3 *Dust emission and diffusion situation*

Based on the layout and the distribution of pollution sources in the operating areas, manual, vehicle navigation, drone navigation monitoring, and other technical means have been applied.

And monitoring of particulate concentration in the operating areas and surrounding sensitive areas, as well as that in key dust-generating nodes and transport roads, was accomplished. In addition, the drones were flown for the distribution characteristics of particulate concentration in specific areas. Finally, we analyzed the comprehensive dust pollution situation based on monitoring data of each port area in a multi-dimensional space. The scope includes the four main operating areas, among which, the D operating area has been divided into four parts for monitoring separately due to its dispersed locations. To evaluate the pollution situation of the storage yards and sensitive areas visually, statistical analysis has been carried out under the data of vehicle navigation monitoring. Introducing the ambient air quality composite index through the method specified in the "Technical regulation for ambient air quality assessment (on trial)" (HJ633-2013) to compare the impact of dust pollution of operating areas. The higher the comprehensive index value, the more serious the pollution. And the individual index indicates the contribution rate of the factor to the comprehensive index. The statistics of ambient air quality composite index and individual index in each operating area are shown in Figure 2.

It can be seen that the primary atmospheric particulate pollution indicators in different operating areas are not completely consistent. And the main reason should be the differences in the types of cargo and operating methods. Therefore, it is necessary to conduct targeted measures based on actual pollution characteristics to control dust pollution.

4 DUST CONTROL OPTIMIZATION

Based on the analysis of current dust control situation, further combining relevant research results and governance experience, and taking the actual situation of each operating area as the starting point, a dust control optimization plan has been put forward. The plan is divided into three phases that are basic control, enhanced control, and optimized control, with the implementation period from 2021 to 2025.

To compare the contribution of bulk cargo operations to the particulate concentration in surrounding areas before and after the implementation of the plan, AERMOD system was used for predictive analysis. It has been divided into five parts according to the location of the operating areas. The results show that after the optimization plan is implemented, the contribution of the operating area to the particulate concentration of surrounding area is effectively reduced. Since the overall diffusion trend of TSP, PM_{10}, and $PM_{2.5}$ is roughly the same, take the diffusion curve of TSP in one of the operating areas as an example, as shown in Figure 3.

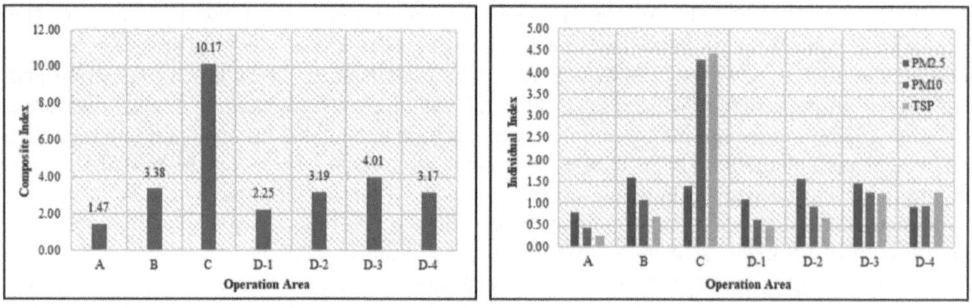

Figure 2. Histogram of composite index and individual index.

As shown in Figure 3, the dust diffusion range after optimization is significantly smaller than current situation, indicating an effective dust control effect. Comparing the diffusion simulation results of different operation areas, it is believed that the dust suppression effect of the plan is related to the operation methods and cargo types. However, it is necessary to start with the upgrading and transformation of the operating process to control dust pollution fundamentally.

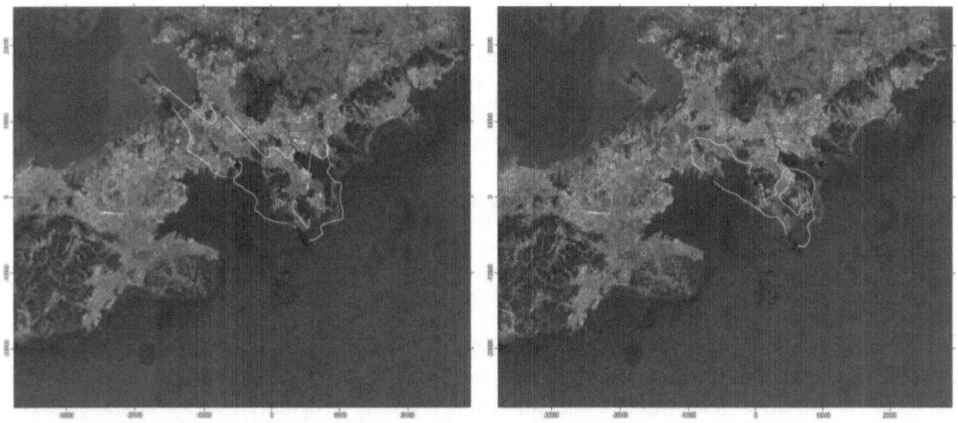

Figure 3. Contribution value of TSP daily(left) and annual(right) mean concentration.

5 BENEFIT ANALYSIS OF OPTIMIZATION PLAN

5.1 *Environmental benefit*

The comprehensive benefits of environmental protection investment refer to the multiple benefits that port companies invest in environmental protection with the aim of reducing the impact of pollutant emissions during port operations. It can be divided into three parts: environmental benefit, economic benefit, and social benefit.

Environmental benefit is the greatest benefit brought by the investment, which is mainly manifested in the improvement of environmental quality by the reduction of total amount of pollutants generated. The annual dust pollutant emissions of the port group's bulk cargo operations will be reduced from 53,293 tons to 41,392 tons after the implementation of the plan under current throughput. Indicating that its implementation has effectively reduced the discharge of dust pollutants on the basis of the same amount of work. Meanwhile, it played a good role in promoting the improvement of the environmental quality of the port area and its surroundings, which shows a positive environmental benefit.

5.2 *Economic Benefit*

The economic benefit generated by the increase in the amount of work during the implementation period was considered as the direct economic benefit while the reduction of environmental taxes was the indirect economic benefit. And the total economic benefit is the sum of the direct and indirect economic benefit. Considering the timeliness of the market and policies, forecasting the bulk cargo throughput from 2021 to 2030 based on the actual throughput in the past 5 years under the limitation of emission standards. Referring to the operating costs and benefits of each bulk cargo operating area at this stage, the main business revenue for bulk cargo operations of the group in 2030 will increase by 1064.34 million yuan compared with current status. With the phased implementation of the plan and the increase in throughput, four of the operation areas can make a reduction of environmental taxes of 0.031, 0.586, 2.882, and 1.206 million yuan, respectively, by 2025. While one of the operation areas will still maintain its environmental capacity after the plan is implemented in 2025. The tax reduction should increase as the throughput, and by 2030, the dust pollution emissions will be reduced by 10,412 tons compared with the condition that the plan has not been implemented in the same year. And the annual environmental tax can be reduced by 3.124 million yuan.

5.3 *Social benefit*

With regard to social benefits, the implementation of the optimization plan can improve the ambient air quality near the operation areas, thereby improving the living comfort and environmental security of residents. Also, it is particularly important for a port company to maintain and enhance the local natural welfare and establish a good role model. In the long run, the implementation of the plan will not only reduce the pollution, but also improve the competitiveness of the company. Overall, it is believed that the plan has multiple positive social benefits.

6 CONCLUSION

This paper takes the bulk cargo operating areas of a port group as the research object, starting from the dust pollution of bulk cargo operation that has attracted much attention of the industry and society at this stage. It analyzed the pollution situation of each operating area comprehensively through the point, line, and surface on-site monitoring. Based on the field survey of the whole process, systematically sorting out the issues in prevention and control of dust pollution of each operating area. Finally, it proposed a feasible optimization plan from the perspective of the group. Moreover, it analyzed the benefits of the plan from the aspects of economy, environment, and society, to prove detailed theoretical and data support for the differentiated environmental management of the group. It is recommended to conduct quantitative analysis based on the actual implementation situation, such as the return in investment, to improve the benefit analysis of environmental protection investment of port enterprises and improve the efficiency of environmental protection investment.

REFERENCES

Chang Hong et al. (2013). The analysis of impact factor of harbor dust and pollution controlling measures. *Northern Environment*. v.25; No.102(06):98–102.

Gao Sheng & Huang Lu (2018). Analysis of dust pollution prevention technology in bulk cargo port. *Constration Materials Decoration*. No.533(24):141–142.

Gu Runnan (1995). Analysis of environmental and economic benefits of dust prevention and control in cement plants. *Chongqing Environmental Science* (6):35–38.

Guo Shan et al. (2006). Review of dedust technique in coal harbor. *Journal of Waterway and Harbor* (01):40–44.

Li Yehui et al. (2020). Influencing factors and control measures of dust pollution in Qinhuangdao Port. *Mechanical and Electrical Information* No.623(17):143–144.

Li Zhendong (2002). Cost-Benefit Analysis for the Investment of Environmental Protection. *China Environmental Protection Industry* 000(002):42–43.

Liu Xijian et al. (2004). Quantitative comprehensive evaluation of social benefits of petroleum engineering projects. *Science & Technology Progress and Policy* 021(012):141–142.

Luo Jing (2020). Evaluation of Enterprise Environmental Protection Investment Efficiency. *Statistics & Decision* (3).

Peng Yi & Nie Guihua (2011). Ecological Efficiency Evaluation of Chinese Coal Industry Based on DEA. *Coal Mining Technology* 16(002):110–113.

Shen Juqin et al. (2015). Research on economic benefit of environment protection in China's papermaking enterprise: based on a certain papermaking enterprise in Jiangsu province. *Journal of Chongqing University of Technology (Natural Science)* 06(No.620):58–62.

Tian Zhentao & Yu Jia (2011). Strengthen dust control, Create a green port. *Port Economy* No.87(01):39–40.

Xu Ping et al. (2011). Dust prevention and control of conveyor system in Caofeidian port. *Hoisting and Conveying Machinery* (S1):70–73.

Yang Likui et al. (2014). The economic efficiency analysis of an irrigation project in Nanyang. *Journal of Shanxi Agricultural University (Natural Science Edition)* v.34(05):473–476.

Yuan Ming (2004). Index evaluation method of investment benefit of environmental protection project. *China Audit* (17):40-40.

Advances in Civil Engineering: Structural Seismic Resistance, Monitoring and Detection – Mohd Johari Mohd Yusof and Junwen Zhang (Eds)
© 2023 The Authors, ISBN 978-1-032-31491-4

Study on the influence of the test environment on geotextile test result

Airong Zheng
Key Laboratory of Geotechnical Engineering of Ministry of Communications; Key Laboratory of Geotechnical Engineering of Tianjin; CCCC Tianjin Port Engineering Institute Co., Ltd.; CCCC First Harbor Engineering Company Ltd., Tianjin, China

Jiahuan Xu & Panpan Song
College of Civil Engineering Hebei University Baoding, Baoding, China

ABSTRACT: As an important geosynthetic material, geotextile is widely used in the fields of water conservancy, harbor and navigation, roads and bridges, etc. The quality of the geotextile is of great concern as a prerequisite for ensuring the safety of the project. To determine the influence of the test environment on the geotextile, the geotextile was tested by changing the state regulation, temperature, and relative humidity during within a certain range. The test results were analyzed using mathematical and statistical methods to obtain discrete characteristics of the test results along with the characteristics of the change in the numerical magnitude with temperature and humidity. The study results show that: the influence of temperature and humidity on the test results is manifested in the change of numerical value and data dispersion (18°C–25°C, 50%–70%), range of temperature and humidity changes, the geotextile unit area mass changes less with temperature and humidity, the coefficient of variation dispersion is smaller; geotextile tensile strength is affected by temperature and humidity changes, the coefficient of variation has a certain dispersion, and the dispersion of elongation is smaller than the tensile strength. The coefficient of variation is affected by changes in temperature and humidity, and increases with temperature.

1 INTRODUCTION

As the most commonly used geosynthetic material, geotextiles play an important role in water conservancy projects, port and waterway projects, roads and bridges, etc. Geotextiles possess the functions of filtration, drainage, reinforcement, and protection, can be in optimum condition under long-term loading, can resist external damage to buildings, have fewer creeps and certain degree of permeability and water permeability, and are widely used in engineering (Bian 2021).

The quality of geotextiles is the key to ensuring the quality and safety of the project. Many factors affect the test and inspection results, including test personnel, instrumentation, test methods, environmental influences, management systems (He 2018; Yang 2005), and the material on its own due to sample inhomogeneity resulting from the manufacturing process. The test method is more objective, and its influence on the test results is more researched (Su 2016; Shi 2007; Wu 2012; Zhu2011). Test personnel, instruments and equipment, environmental impact and management system belong to the management of the test chamber and are less researched. Environmental influence mainly refers to the influence of temperature and humidity on the test results, especially the influence on mechanical parameters has been in much focus, but less researched. There are two main views on the influence of temperature and humidity on test results: first, the influence on test results is minimal and second, there is certain influence on test results, and this influence cannot be ignored.

DOI 10.1201/9781003310884-45

Affected by differences in raw materials and manufacturing processes, geotextiles have different performances under different temperature and humidity conditions. Huang Xin (Huang 2020) conducted a comparative test by changing the temperature and humidity and concluded that temperature would change the material properties of geotextiles and humidity had no significant effect on the tensile properties of geotextiles.

He Tianwen (He 2011) concluded through comparative tests on the effect of temperature and humidity regulation on material properties before testing was related to the type of material and had a non-negligible effect on the testing of mechanical property parameters of two cloths one film filament woven geotextiles, while there was almost no effect on the testing of mechanical property parameters of two cloths one film plastic flat wire woven geotextiles. The reason is that the main force structure of two cloths one film geotextiles is the two layers of geotextiles on both sides of the membrane, while the cloth of filament woven geotextiles is woven from silk threads spun from polyester, polypropylene, ethylene, nylon, and other fibers. The size of the fibers affects the ability of the threads to withstand tension, with wet threads having a stronger ability to withstand tension and dry threads having a weaker ability to withstand tension (Raumann 1984). In contrast, the two cloths one film plastic flat filament woven geotextiles is insensitive to small changes in temperature and humidity due to the single non-woven structure of the substrate and the plastic.

Chen Ting (Chen 2017) and Li Xin (Li 2016) studied the effects of state conditioning time and temperature and humidity (0°C–50°C, 25%–85%), which change on tensile test results for staple needle punching nonwoven geotextiles whose main component is polyester. They found that the effects of state conditioning time and humidity are less regular and significantly affected by temperature, with the state conditioning temperature increasing, the tensile strength decreases and the elongation increases. This is because the raw material is thermoplastic, and high temperature will make irreversible changes in the raw material, resulting in changes in the mechanical properties of the material, while polyester is poorly hygroscopic and does not interact with water, so humidity regulation has less effect on its tensile properties (Xiong 2013).

Zhang Xiaoqiang (Zhang 2021) et al. studied the effect of stability of geomembrane test results under general indoor temperature and humidity conditions. The mass per unit area has good temperature and humidity stability. The thickness decreases with increasing temperature; maximum tensile strength value, elongations at break, and right-angle tear strength are less stable, and elongation at break increases with increasing temperature; puncture strength decreases with increasing temperature.

In this paper, the physical and mechanical properties of staple-punched nonwoven geotextiles were tested under a range of temperature and humidity conditions and the effect of temperature and humidity on the dispersion of the data and the magnitude of the test results were analyzed.

2 EXPERIMENT OVERVIEW

2.1 *Materials*

Geotextile is a cloth-like product made through production processes such as opening and loosening, combining, messing, webbing, and needling of nonwoven threads. The test uses staple/needle-punched nonwoven geotextiles, the main component of which is polyester, and the basic constituent is polyethylene terephthalate, also known as polyester fiber (PET).

2.2 *Experimental design*

Temperature and humidity regulation in geotextile testing generally consists of two aspects: first, the state of temperature and humidity regulation before the test, and second, the control of environmental temperature and humidity during the test. The temperature and humidity adjustment before the test is to allow the material to reach equilibrium in a specific temperature and humidity environment; the temperature and humidity of the test environment are the temperature and relative humidity of the environment in which the sample is tested.

Considering the temperature and humidity range of the general indoor environment, the test set temperature and humidity range at (18°C–25°C, 30%–70%), and adopted the representative conditions of 18°C, 21°C, 25°C and relative humidity of 30%, 50%, and 70% in the range, and carried out 9 sets of orthogonal tests under different temperature and humidity, respectively, and ensured that the temperature and humidity of the test environment were the same as the conditioned state. The tests were divided into unit area mass and tensile tests [14], with 10 specimens per group for the unit area mass test in squares of 100 mm side length, and 7 specimens per group for the tensile test in squares of 200 mm side with length.

To ensure reliability and comparability of test results, several measures are taken to avoid variations in the test results caused by other factors, such as the selection with stable quality samples, uniform sample cutting in the sample area, fixed test equipment, and testers.

2.3 *Statistical analysis methods*

The accuracy of the test results closely depends on the influence of external factors, in the form of trends in concentration, dispersion, the shape of the distribution, and the influence of correlations in the distribution of the series of results. The influence of changes in external conditions is only reflected through changes in the size of the values, the correlation between different parameters, and the dispersion of the data. In this paper, the following indicators are selected for analysis at different temperatures and humidity levels and their average values.

Arithmetic average: Reflecting the average level and concentration trend of the data, reflecting the inevitable characteristics of the data, the average value is influenced by extreme conditions.

Standard deviation: The arithmetic square root of the variance, reflecting the degree of dispersion among a group of data individuals, is a measure of the dispersion of the mean of a group of data, and is also an indicator of the accuracy and stability of the data.

Coefficient of variation: The coefficient of variation, also known as the standard deviation rate or unit risk, is the ratio of the standard deviation of the data to the arithmetic mean. The coefficient of variation is dimensionless and eliminates the effect of different units or means of the comparison of the degree of variation of two or more data.

(4) Relative deviation of extremes: A measure of the width and dispersion of the distribution of the test values, as a ratio of the absolute deviation of the extremes to the arithmetic mean.

3 RESULTS AND ANALYSIS

3.1 *Mass per unit area*

Table 1 shows the results of the statistical analysis of the unit area mass test values and their mean values for 9 different temperature and humidity states of staple non-woven geotextiles. As can be seen from Table 1, the standard deviation of the unit area mass test values ranged from 14.3 $g.m^{-2}$ to 16.4 $g.m^{-2}$ with certain dispersion, but the coefficient of variation and the relative deviation of the extreme values tended to be the same, and the dispersion did not change much with temperature and relative humidity, so the temperature and humidity variations in the range of 18°C–25°C, 50%–70% had no effect on the data of the unit area mass determination of staple felted nonwoven geotextiles. The dispersion of the data is mainly influenced by the inhomogeneity of the material and the level of sample preparation.

The mean value of the mass per unit area under 9 different temperature and humidity conditions was 224.6 $g.m^{-2}$ with a standard deviation of 0.8 $g.m^{-2}$. The dispersion of the niggardly value was found to be very small, with a coefficient of variation of only 0.4%, and the relative deviation of the extreme value was also very small, at 0.45%, with a strong tendency to converge, so the test results of the mass per unit area of staple non-woven geotextiles have good stability. The influence of indoor temperature and humidity on the size of the value is very small, and the influence on the test results can be ignored.

Table 1. Statistical analysis of mass per unit area of geotextile.

Environmental conditions	Average/ (gm^{-2})	Standard deviation/ (gm^{-2})	Coefficient of variation deviation (%)	Minimum relative (%)	Maximum relative deviation (%)
18°C,30%	225.2	16.0	7.1	−9.4	8.7
18°C,50%	224.9	15.6	6.9	−9.3	8.7
18°C,70%	225.4	15.7	7.0	−9.2	8.9
21°C,30%	224.4	15.0	6.7	−9.2	9.2
21°C,50%	224.0	15.9	7.1	−9.4	8.6
21°C,70%	225.8	16.2	7.2	−10.0	8.6
25°C,30%	224.8	16.4	7.3	−10.1	8.8
25°C,50%	223.7	14.6	6.5	−8.6	9.5
25°C,70%	223.5	14.3	6.4	−8.6	9.4

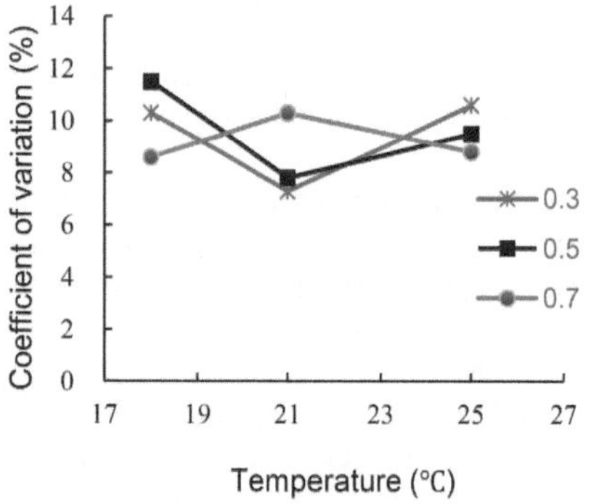

Figure 1. Variation coefficient of test results of geotextile.

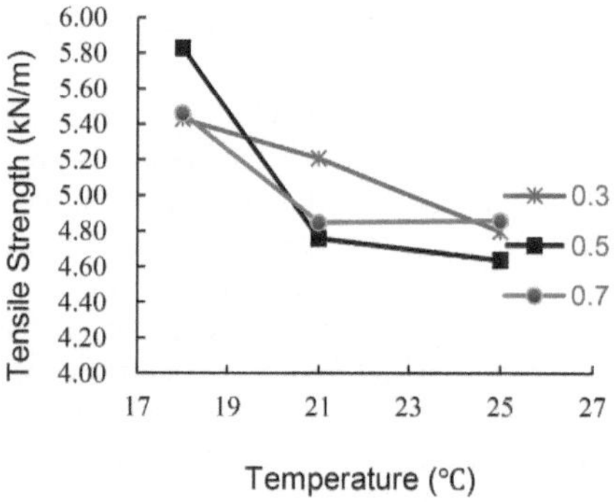

Figure 2. Tensile strength of geotextile.

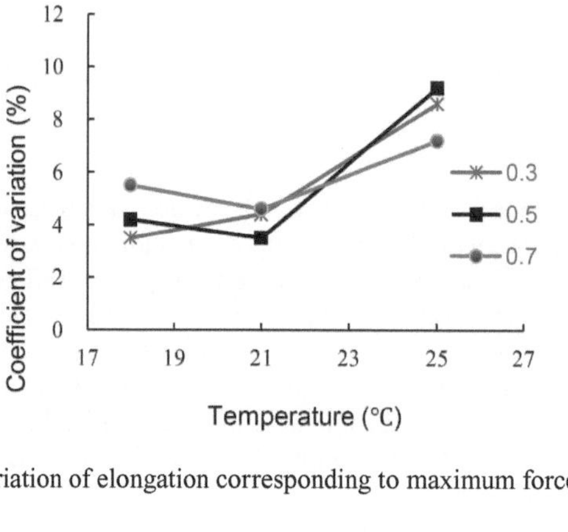

Figure 3. Coefficient of variation of elongation corresponding to maximum force in geotextiles.

3.2 *Stretching*

3.2.1 *Tension*

The tensile strength T_{max} is the maximum tensile strength that the geotextile can withstand per unit width. The results of the statistical analysis of the tensile strength test values, and their average values are shown in Table 2. This is related to the unevenness of the tensile properties caused by

Table 2. Statistical analysis of tensile strength of geotextile.

Environmental conditions	Average/ (kNm^{-1})	Standard deviation/ (kNm^{-1})	Coefficient of variation (%)	Minimum relative Deviation (%)	Maximum relative Deviation (%)
18°C,30%	5.43	0.56	10.3	−12.5	14.0
18°C,50%	5.83	0.67	11.5	−15.3	17.3
18°C,70%	5.46	0.47	8.6	−13.2	12.1
21°C,30%	5.21	0.38	7.3	−12.9	6.7
21°C,50%	4.76	0.37	7.8	−13.9	9.7
21°C,70%	4.85	0.50	10.3	−15.9	11.5
25°C,30%	4.80	0.51	10.6	−16.9	12.3
25°C,50%	4.64	0.44	9.5	−8.4	15.3
25°C,70%	4.86	0.43	8.8	−11.1	13.2

Table 3. Statistical analysis of tensile strain of geotextile.

Environmental conditions	Average (%)	Standard deviation (%)	Coefficient of variation (%)	Minimum relative Deviation (%)	Maximum relative Deviation (%)
18°C,30%	65.8	2.3	3.5	−6.8	3.8
18°C,50%	61.5	2.6	4.2	−5.2	5.7
18°C,70%	65.1	3.6	5.5	−7.1	9.2
21°C,30%	68.2	3.0	4.4	−5.7	7.5
21°C,50%	74.4	2.6	3.5	−4.8	5.2
21°C,70%	65.0	3.0	4.6	−8.8	6.6
25°C,30%	68.8	5.9	8.6	−12.9	14.8
25°C,50%	66.0	6.1	9.2	−10.2	13.2
25°C,70%	65.4	4.7	7.2	−8.0	9.3

the manufacturing process. The coefficient of variation of tensile strength with the temperature at different relative humidity levels is shown in Figure 1, with the tensile strength at 70% relative humidity showing an inverted shallow "V" shape with temperature. At 21°C, the coefficient of variation of the tensile strength is relatively small, which means that the dispersion is small.

The variation curve of tensile strength with temperature and humidity is shown in Figure 2. The tensile strength decreases with increasing temperature and does not change significantly with humidity. As can be seen from Table 2, the coefficient of variation of the average tensile strength under different temperature and humidity conditions is 8.1%, with certain dispersion. The deviation between the minimum and maximum values is 20.4%, with a wide range of data distribution, and the average tensile strength is 5.09 $kN.m^{-1}$ with a standard deviation of 0.41 $kN.m^{-1}$, with certain dispersion.

Therefore, changes in temperature and humidity conditions have certain influence on the magnitude of the tensile strength values of staple-punched nonwoven geotextiles, especially the influence of temperature cannot be ignored.

3.2.2 *Elongation*

The elongation ε_P is the strain corresponding to the maximum tensile strength value, and the results of the statistical analysis of the 9 sets of test values and their average values are shown in Table 3. The coefficient of variation of elongation with temperature was slightly different from that of tensile strength, with the coefficient of variation of elongation increasing with increasing temperature.

As can be seen from Table 3, the mean value of elongation at different temperatures and humidity is 66.7%, with a standard deviation of 3.6% and a coefficient of variation of 5.4%, which has a certain degree of dispersion. And the minimum value deviates from the maximum value by 17.3%, with a wide distribution of data, indicating that the variation of temperature and humidity conditions in the range of 18°C–25°C, 50%–70% has certain influence on the magnitude of the values of the elongation results of staple-punched non-woven geotextiles. The pattern of change in the mean elongation values with temperature and relative humidity is not obvious.

4 CONCLUSION

1) Geotextiles have a certain dispersion in their mass per unit area test values, and the dispersion and size of the test values are not significantly affected by changes in temperature and humidity within the general indoor temperature and humidity range.
2) The coefficient of variation of elongation shows a general upward trend with temperature, and the data has a certain dispersion, with a relatively high room temperature having greater influence on the coefficient of variation corresponding to elongation.
3) The tensile test value of geotextile has some dispersion and tends to increase with the increase in temperature. The tensile strength decreased with increasing temperature and varied insignificantly with humidity.

In summary, the tensile strength and elongation of geotextiles are easily affected by changes in temperature and humidity, while the mass per unit area is less affected. Hence, the influence of the test environment on the geotextile should be fully considered in the test inspection to provide a strong basis for further ensuring the accuracy of the test.

REFERENCES

Bian Huarui, Wang Chao, Zuo Wenjing. Mechanical and photo-oxidative ageing resistance of polypropylene spun bond needle-punched geotextiles[J]. *Technical Textiles*, 2021(3), pp. 39–44.

Chen Ting. Analysis of factors affecting the tensile performance test results of geotextiles[J]. *Fujian traffic science and technology*, 2017(3), pp: 41–44.

He Qianqian, Pei Sheng, Yang Guangchao. Performance comparison and application analysis of non-woven geotextiles[J]. *Technical Textiles*, 2018(2), pp. 30–34.

He Tianwen. Exploration of geotextile testing in practical work[J]. *Transportation Science and Technology*, 2011(5), pp. 86–88.

Huang Xin. Analysis of factors influencing tensile testing of staple-punched non-woven geotextiles[J]. *Engineering Equipment and Materials*, 2020(24), pp. 121–122.

Li Xin. Study on the factors influencing the tensile test of short-fiber needled nonwoven geotextiles[J]. *Water Resources and Hydropower Construction*, 2016(32), pp. 153–154.

Raumann.G, Jiang Shigu. Some tensile test methods for geotextiles[J]. *Textile Specialties Technology*, 1984(21), pp. 56–63.

Shi Dongmei, Xing Jin, Sun Jiawen, Ding Jinhai. Study of carbon black content of polyethylene geomembranes[J] *Chemical Building Materials*, 2007, 23(4), pp. 27–29.

Su Shuqing, Wu Weijun. Discussion on the method of determining effective pore size of geotextiles[J]. *Journal of Geotechnical Engineering*, 2016, 38(S1), pp. 156–159.

Wu Xiaomeng, Ni Bingxuan, Zhang Peng. Comparative analysis of geotextile pore size testing methods[J]. *China Fiber Inspection*, 2012(8), pp. 45–47.

Xiong Rongjun, Li Heng, Sun Aiguo, Tang Zhengtao. Study on the factors influencing the tensile test of short-fiber needle-punched non-woven geotextiles[J]. *Water Transport Engineering*, 2013(9), pp. 6–16.

Yang Mingchang. Testing errors in geosynthetic material laboratories[J]. *Journal of hydraulic engineering*, 2005(3), pp. 63–70.

Zhang Xiaoqiang, Zheng Airong. Influence of temperature and humidity on the stability of geomembrane test results [J]. *China Harbor Construction*, 2021, (8), pp. 38–41.

Zhu Tiange, Liu Chang, Ding Jinhai, Zer Dongmei. Differences in tensile properties of high-density polyethylene geomembranes by test method standards for plastics[J]. *Plastics*, 2011, 40(5), pp. 106–109.

Advances in Civil Engineering: Structural Seismic Resistance, Monitoring and Detection – Mohd Johari Mohd Yusof and Junwen Zhang (Eds)
© 2023 The Authors, ISBN 978-1-032-31491-4

Optimization of node structure of transmission tower

Xinmin Yu & Xingyun Chen
State GRID Fujian Economic Research Institute, Fuzhou, China

Xin Ye, Jingfeng Guo & Bijian Chen
Fujian Yongfu Power Engineering Co., Ltd, Fuzhou, China

ABSTRACT: As in the previous test analysis, relevant specification requirements, and internal force analysis, the structural connection between the main material node and large-size angle steel of transmission tower is analyzed, and the node optimization measures of transmission line angle steel tower are elaborated to provide safe level design support for relevant projects.

1 INTRODUCTION

Currently, a tower structure is widely used in power grid construction at home and abroad, but the node application analysis of tower structure is not perfect. Many structural measures are mainly selected based on engineering experience, which makes the design conservative or have unknown dangers (Bai 2019; Zhao 2016). Node structure is an important part of tower structure design, which is directly related to whether the bearing capacity of components is consistent with the actual bearing capacity. It is very important for the safe and reliable operation of tower, and also affects the weight of tower. More attention should be paid to node design. The research shows that the stronger joints can improve the bearing capacity of connecting members to a certain extent.

On the other hand, with increase in power consumption, long-span and long-distance transmission poles and towers are widely used (Huang 2010). Therefore, as the main material of the tower body, conventional angle steel is difficult to meet the requirements of tall iron towers, and double (multiple) combined angle steel or large-size angle steel needs to be used. Compared with combined angle steel, large-size angle steel has clear force transmission and simple construction and installation. It is widely used in UHV Tower (Chen 2013; Yuan 2021).

In view of the importance of the joint in the angle steel tower, this paper mainly analyzes the optimization of the main material of the tower and the joint structure of large-size angle steel, to provide a reference basis for the relevant design in the future.

2 NODE STRUCTURE OF MAIN MATERIAL

2.1 *Joints of main materials*

In the diagonal steel tower scheme, the joint eccentricity of main materials of different specifications lead to additional bending moment, resulting in the smaller stress capacity of materials than calculated. In the past, to solve the eccentricity problem between the main materials of the tower body, the form of lap joint between the upper and lower main materials was adopted. According to the technical code for design of tower structure of overhead power transmission line (DL/T 5154-2012), the bearing capacity of double-sided connection of main materials can be increased by 20%–30% compared with single-sided connection. Therefore, when arranging the head form and body structure form of a linear tower, double-sided connection should be adopted for main material connection.

DOI 10.1201/9781003310884-46

Generally, when the specifications of the main materials of the tower body are increased regularly and step-by-step from top to bottom, the eccentricity of the main materials between sections is about 10 mm, and the resulting additional bending moment can be ignored. However, when the specifications of the main materials of the upper and lower sections of the tower body differ considerably, the center of gravity lines of the main materials of the sections are connected together. For slope change of the tower body, node strength should be strengthened.

2.2 *Connection between main material and inclined material*

In the mid-1980s, China studied the connection between the main material and the inclined material of 500 kV iron tower. The alignment of the inclined material crosses the skin of the main material or the outer core line of the double row core line of the main material. The test and application results of the real tower are ideal, and they have minimal influence on the force transfer of the rod system, the establishment of mechanical model, and internal force analysis.

According to the current technical code for design of overhead transmission lines in China, the calculated slenderness ratio of components is related to the end constraints of components. See Table 1 below.

Table 1. Correction coefficient of slenderness ratio.

Slenderness ratio λ	End constraints	Correction coefficient of slenderness ratio K
$120 < L/r < 220$	Unconstrained at both ends	1
	Unconstrained at one end and constrained at the other	0.90+11.89/(L/r)
	There are constraints at both ends	0.82+21.64/(L/r)

It can be seen from Table 1 that the worse the constraint conditions, the greater is the slenderness ratio correction coefficient *K*, and the slenderness ratio (λ) can be calculated. The larger the, the worse the stability of the member. Table 2 compares the common connection methods of the important inclined materials.

Table 2. Correction coefficient of slenderness ratio.

	Scheme I	Scheme II
Node detail	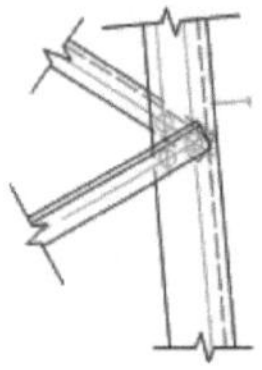	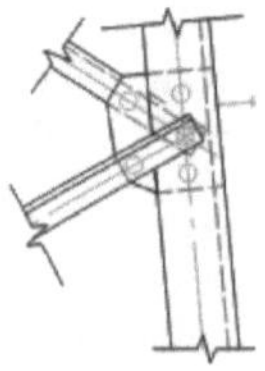
Advantage	The inclined material is intersected with the "one core" of the main material, and the bolts are directly connected with the main material. The structure is compact, and the main material has good restraint on the end of the inclined material	The inclined material is connected with the "one core" of the main material, and the bolt must be connected with the main material through the connecting plate
Shortcoming	Angle cutting is required for inclined materials, and double rows of holes are drilled for main materials, so the processing is relatively complex	To increase the connecting plate, increase the steel consumption weight

When the limb width of the main material and the angle with the inclined material are appropriate, the constraint of the main material on the inclined material in scheme I is better than that in scheme II, and the gusset plate is omitted. It is recommended to be widely used in the tower design of the project, to achieve the purpose of safe, reliable, and economic tower design.

2.3 *Connection at crank arm*

The force transfer system at the inflection point of the curved arm of the linear iron tower is relatively complex. During the design, the joint structure should be consistent with the mechanical model, and the connection mode between the inner and outer inclined materials and the curved arm should be consistent, to match the joint stiffness. According to the calculation and analysis and the test results of real towers in previous projects, reinforcement measures such as setting stiffening plates and pasting angle steel inside the joints at the connection of upper and lower curved arms and the connection between curved arms and tower neck are considered, as shown in Figure 1.

Figure 1. Connection diagram at the crank arm.

2.4 *Analysis of welded connections*

Most of the connection types of angle steel tower are bolt connection, but a few joints are still welded. Because the welding process of high-strength steel is not easy, it is recommended to optimize the joints in the tower design.

1) The welding connection position adopts Q345 steel design, and the bolt connection method is used instead.
2) Z-direction steel plate to prevent lamellar tearing of steel shall be adopted.
3) New design schemes are proposed, such as using double-layer plates and arranging new stiffeners to reduce the plate thickness.
4) The main materials shall be embedded in the foundation concrete, and the thick tower foot plate shall be avoided.

3 OPTIMIZATION OF JOINT STRUCTURE OF LARGE-SIZE ANGLE STEEL

The general structural requirements of large-size angle steel joints are basically the same as those of ordinary size angle steel towers. Due to increase in limb width and limb thickness, the structure needs to be optimized.

(1) For single wrapped joint of ∠220–∠250 angle steel, three rows of alignment lines should be adopted to reduce the connecting plate, such as cross arm root, tower foot, slope change point without inner wrapped angle steel, etc., as shown in Figure 2.
(2) Structural scheme of "K" node of main material of tower body
  a) For the main material of large-size angle steel and the "K" node of the main material of the tower body, the inclined material should be extended into the main material as far as possible, to cancel the gusset plate. This node tower has the lightest weight and the best stress.

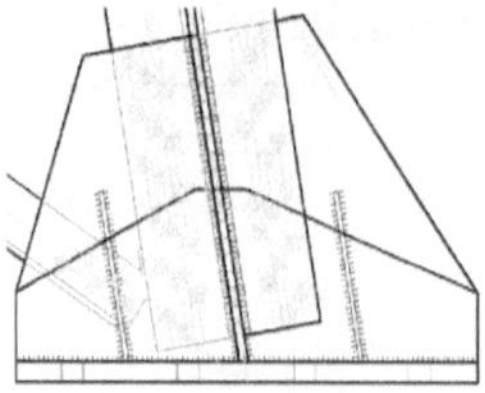

Figure 2. Schematic diagram of three rows of holes at tower foot.

b) When the gusset plate cannot be cancelled, one inclined material should be taken out. At this time, although the gusset plate is slightly larger than that when two inclined materials are extended at the same time, it can prevent the tower from being installed by mistake or missing due to too many filling plates, which facilitates the construction and installation.

(3) The cross joints are connected by two bolts
The limb width of large-size angle steel is large. On the premise of not adding plates, make full use of its limb width, and use two bolts to connect the cross joints to increase the overall stiffness of the joints, as shown in Figure 3.

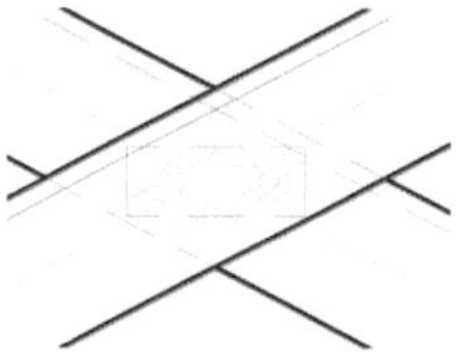

Figure 3. Cross node.

(4) Variable slope node
Mutual welding of the front and side panels of the variable slope point increases the joint stiffness, and the upper and lower main materials are under large stress and there are many bolts. Double wrapped joints should be used, that is, inner wrapped angle steel should be used to reduce the size of the small point plate. However, at this time, the inner wrapped angle steel and outer sticking plate need fire bending, and Q345 steel should be used, as shown in Figure 4:

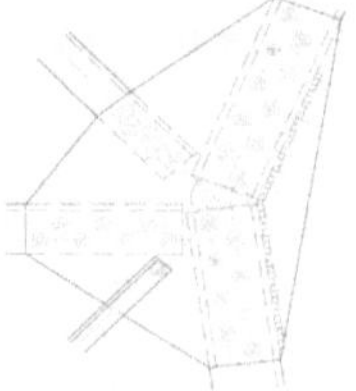

Figure 4. Slope change node.

(5) Cross arm connection node
The main material of cross arm and the main material of spacer face are subject to large stress and are connected through a single package. Three rows of bolts should be used, as shown in Figure 5.

(6) Single leg large-size angle steel tower foot plate
As the large-size angle steel is under great stress, 4 anchor bolts can meet the requirements for some specifications, but the larger specifications ($\angle 250\times26$ and above) 4 conventional 35#

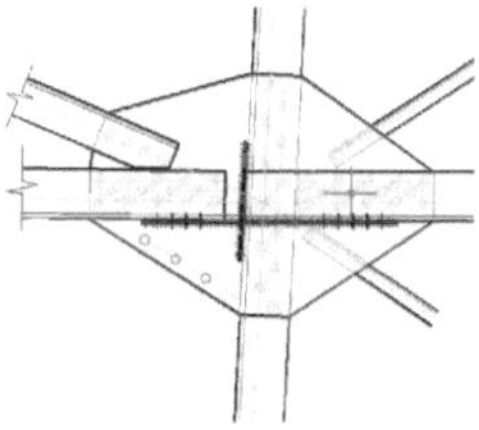

Figure 5. Connecting node of cross arm.

steel anchor bolts cannot meet the requirements, and therefore 8 anchor bolts must be used. The layout of single leg large-size angle steel tower foot plate is shown in Figure 6.

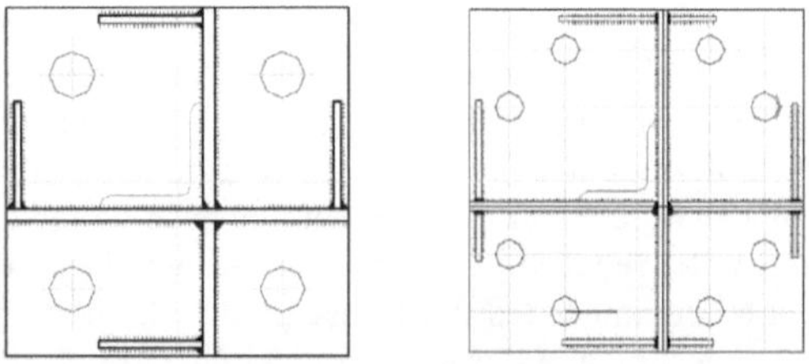

(a) 4 anchor bolts for connection (b) 8 anchor bolts for connection

Figure 6. Single leg large angle steel tower foot plate.

4 APPLICATION OF HIGH-STRENGTH BOLTS

In connection design, bolt connection is preferred for the structural type of transmission tower, followed by welding (insert plate welding or intersecting welding). Except that grade 6.8 ordinary bolts can be used for inclined materials and auxiliary materials with small stress, it is necessary to use high-strength bolts for main materials and inclined materials with large stress. Currently, high-strength bolts above grade 8.8 have been widely used in long-span transmission line towers. Their material and mechanical properties shall comply with the provisions of the current specifications mechanical properties of fasteners bolts, screws, and studs GB/T 3098.1 and mechanical properties of fasteners nuts coarse thread GB/T 3098.2, respectively. However, due to some fracture phenomena in the process of use in recent years, the design institute has some concerns about the application of high-strength bolts, but according to the application experience, most of them are due to material defects, heat treatment, and other processing problems. High-strength bolts above grade 8.8 are widely used in large-span iron towers and steel pipe towers of 500 kV lines. Table 3 shows the use of high-strength bolts in some line projects.

Table 3. Brief table of high-strength bolts above grade 8.8 used in power transmission project.

Project name	Specification of high-strength bolt	Bolt grade	Operation
Steel pipe tower of 500kV Yangdou line	M24 ~ M36 flange bolts	Level 8.8	Good
Steel pipe tower of 500kV Yangyang line	M24 ~ M36 flange bolts	Level 8.8	Good
Section tower of 500kV Jiangyin large span combination	M20 ~ M27 shear bolt	Level 8.8	Good
Large span steel pipe tower in Wuhu of ± 500kV Longzheng line	M24 ~ M45 flange bolts	Level 10.9	Good

Due to the large load of 1000 kV AC UHV line, the component bears large stress and there are many connecting bolts. The use of high-strength bolts can effectively reduce the number of connecting bolts at the end of the component and the size of the connecting plate, to reduce the tower weight. In this paper, the number of bolts and the size of gusset plate are analyzed according to the strength level of bolts (Table 4).

Table 4. Shear strength of bolts of different grades.

Strength level	Level 4.8	Level 6.8	Level 8.8	Level 10.9
Shear strength (N/mm^2)	170	240	300	380
Shear strength ratio	0.71	1	1.25	1.58
Tensile strength (N/mm^2)	200	300	400	500
Tensile strength ratio	0.67	1	1.33	1.67

It can be seen from the above table that shear strength and tensile strength of different bolt grades are different. The diagonal steel tower generally adopts the bolt shear connection mode. According to the ratio of shear strength of grade 6.8 and grade 8.8 bolts, when grade 8.8 bolts are used, the number of bolts can be reduced by 20%, and the gusset plate can be reduced by about 10%, that is, the weight of the plate can be reduced by about 10%. Although the use of grade 10.9 bolts can lead to more saving, it is recommended not to use grade 10.9 bolts because they are easily prone to hydrogen embrittlement fracture during acid pickling and rust removal and lack mature operation experience. Therefore, grade 6.8 (M16, M20) and grade 8.8 (M24) galvanized bolts are recommended for the tower of the project to reduce the gusset plate to optimize the nodes and reduce the tower weight.

5 CONCLUSION

Based on the previous test analysis, relevant specification requirements and internal force analysis, this paper analyzes the structural connection between the main material node and large-size angle steel, and obtains the following conclusions: when arranging the tower head form and tower body structure form of linear tower, the main material connection should adopt double-sided connection. The alignment of the inclined material is crossed with the skin of the main material or with the outer core line of the double core line of the main material, and the stress effect obtained is thus optimum. For the single wrapped joint of ∠220–∠250 angle steel, three rows of alignment lines should be adopted to reduce the connecting plate.

REFERENCES

Yuan Shiyan & Jiang Yuting & Yang Lidong (2021). Study on bolt joint strength of large-size angle steel axial compression members., *Jilin Electric Power*, vol. 49(02),25–27.

Zhao Hongru & Yang Lidong & Liu Fuhai (2016). Damage identification of transmission tower structural members and joints. *Jilin Electric Power*, vol. 44(02),14–16+20.

Chen Xi & Guo Yaojie & Wu Hanqing (2013). Study on reducing the number of holes in large angle steel bolted joints under compression. *Journal of Wuhan University (Engineering Edition)*, vol. 46(S1). 188–191.

Huang Huang & Li Qinghua & Meng Xianqiao (2010). Application of Q420 large size angle steel in ± 800kV UHV Tower., *Power construction*, vol. 31(06),65–69.

Bai Yinghua & Shen Kang & Chen Bo (2019). Study on bolt stress of transmission tower connection joints. *Steel structure*, 34(02):46–52.

Advances in Civil Engineering: Structural Seismic Resistance, Monitoring and Detection – Mohd Johari Mohd Yusof and Junwen Zhang (Eds)
© 2023 The Authors, ISBN 978-1-032-31491-4

Research on geogauge instrument for testing the quality of subgrade filling process

Jianyou Yu
Yanchong Construction Office of Hebei Expressway Group Limited, Zhangjiakou, Hebei, China

Yunfei Zhao*
School of Urban Geology and Engineering, Hebei GEO University, Shijiazhuang, Hebei, China

Weichao Liu
Department of Civil Engineering, Shijiazhuang Tiedao University, Shijiazhuang, China

Zhizhong Liu
Yanchong Construction Office of Hebei Expressway Group Limited, Zhangjiakou, Hebei, China

Guangqing Yang
Department of Civil Engineering, Shijiazhuang Tiedao University, Shijiazhuang, China

ABSTRACT: In the construction of highways in the mountainous areas of northern China, large-sized fillers are often used for roadbed construction, and the quality inspection of the roadbed often becomes a difficult problem. In actual construction, the settlement difference method is often used to control the quality of subgrade, and this empirical method cannot accurately mark the quality of concrete subgrade filling. In order to solve this problem, the GeoGauge method was introduced to conduct comparative experiments to obtain accurate construction quality data. The regression analysis method is used to study the quantitative correlation between the Young's modulus and the settlement difference. And it finally came out with the standard value for process quality.

1 INTRODUCTION

In the process of road construction, it has always been the goal of road designers and builders to maintain good service status for a long time. The factors that affect the service state run through the entire process from road design to material selection to construction and final operation, of which construction is the most critical factor. To control the quality of road construction, it requires quality inspection and control. As the most basic structure of the road, the quality inspection of the roadbed is very important. Generally, road damage is caused by local minor diseases, which spread to the entire road. Most of the diseases are caused by uneven compaction of filling materials during subgrade construction. If the unevenness during subgrade construction could be reduced, the chances of minor diseases can be reduced and the service life of the road can be extended. The current method to solve the unevenness compaction is to control the quality during construction process.

Currently, roadbed detection methods all over the world are divided into two categories: physical index detection methods and mechanical index detection methods. The index of the physical index detection method is generally the degree of compaction. The calculation of the degree of compaction is actually a physical quantity obtained by calculating the dry density (Liang 2003). The better

*Corresponding Author: 89zyf@163.com

DOI 10.1201/9781003310884-47

the material integrity is, the greater the dry density and the greater the degree of compaction would be (Li 2007). Compaction testing methods are divided into destructive testing methods and non-destructive testing methods (Tatsuoka & Correia 2016). Sand replacement method, irrigation method and cutting ring method are destructive testing methods; nuclear densitometer (RI), nuclear density meter (GeoGauge, EDG, PT, PQI, etc.), soil densitometer (SDG) and settlement difference method are non-destructive testing. There are many detection methods, such as Beckman beam method to measure static deflection, drop hammer deflection instrument to measure dynamic deflection, and soil modulus/stiffness meter to measure dynamic stiffness and Young's modulus. The rapid detection method of active subgrade in hot and humid areas has been studied (Zhang et al 2019 & Tatsuoka 2021). The variability of soil modulus/stiffness meter to control the quality of subgrade has been explored, and obtained the "2σ" principle of mathematical statistics to obtain the "soil modulus" principle, and obtained the soil modulus "principle" (Ren 2011). In actual engineering, the differential settlement method is generally used to control the quality of the subgrade filling process, and the Beckman beam method is used to control the quality of the subgrade after the filling is completed, but both methods have defects. The differential settlement method requires a fixed observation point. Once the observation point changes, the measured data will be inaccurate, which will affect the on-site personnel to judge the compaction of the filling layer. As for the Beckman beam method, it is complicated to operate, requires a lot of manpower and time, and affects project progress.

The soil modulus/stiffness meter is a new type of roadbed and pavement testing instrument that has been used on-site in the last five years. It has been applied on the construction site overseas and there was no application in actual projects in China. The testing indicators of this testing instrument are Young's modulus and dynamic stiffness, which respectively reflect the elastic characteristics and dynamic anti-deformation characteristics of roadbed and pavement materials, and can describe more comprehensively than other simple methods of testing compaction, as well as mechanical properties of roadbed materials. Therefore, the H-4140 soil modulus/stiffness tester was used to perform joint tests with the differential settlement method in the actual roadbed compaction process, and uses mathematical statistics to fit the data obtained by the two methods to obtain the soil modulus and the control index of the stiffness meter in the actual roadbed detection has been proposed.

2 RESEARCH ON THE DETECTION METHOD OF GEOGAUGE SOIL MODULUS METER

The research was based on the subgrade project of Yanqing-Chongli Expressway. The expressway is an important passage connecting the Beijing competition area and the Zhangjiakou competition area for the 2022 Beijing Winter Olympics. The total length of the route is 113.684 kilometers. The highway project is large in scale and has a short annual effective construction period. Climate in this area is very harsh and construction condition is poor, especially for the construction of roadbed. However, the quality of roadbed construction is the basic guarantee to ensure the entire project. Subgrade inspection is an important way to inspect the construction quality. In the subgrade process quality inspection of the Yanqing-Chongli Project, the settlement difference method is used, that is, the elevations of the respective measuring points are measured and subtracted from each layer of the subgrade filling during the last two passes of rolling. The settlement difference of each point is obtained, but this method requires two people to cooperate and set up fixed observation points. Not only would it take a lot of time, but also requires a lot of work. The soil modulus/stiffness meter is now used in this project, without the need for observation points, it can be directly placed at the measuring point for detection, and the detection time period can be completed by only one person, and there is no soil modulus/stiffness meter testing at present. To set the standard value, it is necessary to set up a test section and use the soil modulus/stiffness instrument method and the settlement difference method to determine the detection standard value of the instrument.

2.1 *Test section setting and joint testing*

The test section of the joint test was K65+135 to K65+387 that is undergoing subgrade filling at the time, and each test point was taken every 40 m in the area, and the test points were located at the center of the left and right half of the subgrade. Marked as A_i and B_i (i=1~6) respectively, 12 points are detected in total, and the distribution of the measuring points is shown in Figure 1.

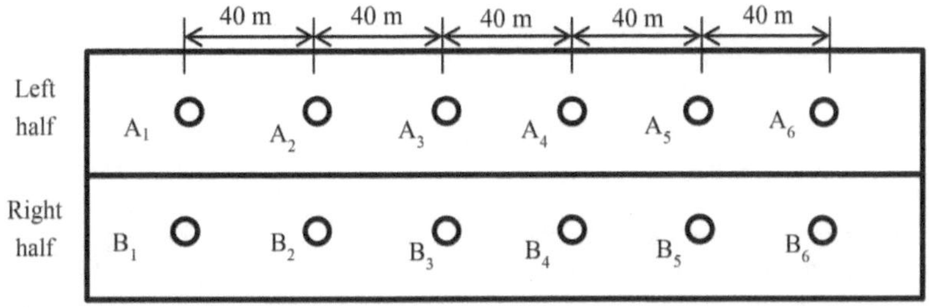

Figure 1. Location distribution of test points.

After the test section and measuring points are arranged, the joint inspection is carried out. The joint inspection steps are as follows:

(1) Differential settlement method. Before the compaction of the iron tailings roadbed, select a suitable observation point, and select stone or iron plate as the reference point for measurement and mark it with red paint. Place a precision level on the observation point, perform 5 times of compaction, have the elevations of A_i and B_i measured and recorded, then continue for another two compaction. When the elevation difference (that is, the settlement difference) is not more than 2.0 mm, the compaction quality was determined to be qualified; otherwise the compaction will continue until it was qualified.
(2) Soil modulus/stiffness meter method. First, flatten the surface of the two sets of measuring points of A_i and B_i to remove the influence of uneven surface factors on the side measurement results, and then place the equipment on the surface of the corresponding roadbed measuring point, slightly rotate the instrument to make the instrument base and the roadbed surface contact well , Turn on the instrument, click the "MEASURE" button on the screen to start the test, wait for about 1 minute, the test result appears on the display and recorded.

2.2 *Determination of detection standard value*

After the completion of the test, the test results obtained by the two test methods are analyzed. As the soil modulus/stiffness meter has two test quantities, in practice, the modulus is often used as the research standard value, so Young's modulus was used as the standard amount of research. The data fitting method was used to study the test results. Figures 4 and 5 show the regression curves of the differential settlement ΔH and Young's modulus E_Y fitted by the A_i and B_i groups, respectively.

As shown in Figure 2 and Figure 3, the correlation R2 between the differential settlement ΔH and Young's modulus EY of the two groups A and B is greater than 0.8, so they all have a good logarithmic relationship. The correlation coefficient R2=0.8683 is the largest, so the regression curve fitted by group A is used as the relationship curve between ΔH and EY, as shown in Equation 1.

$$E_Y = -15.58 \ln(x) + 115.59 \quad (1)$$

According to the general differential settlement ΔH, the control range is 2.0 mm to 6.0 mm. According to the Equation 1, the control range of Young's modulus EY can be determined, so that the quality of the iron tailings roadbed filling process can be reflected by EY. In order to make EY more in line with actual engineering needs, the safety factor of steep slope and high slope subgrade

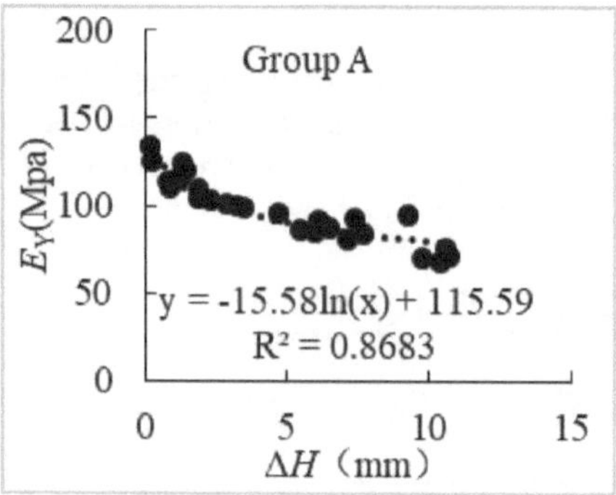

Figure 2. Regression curve of settlement difference ΔH and Young's modulus E_Y of Group A.

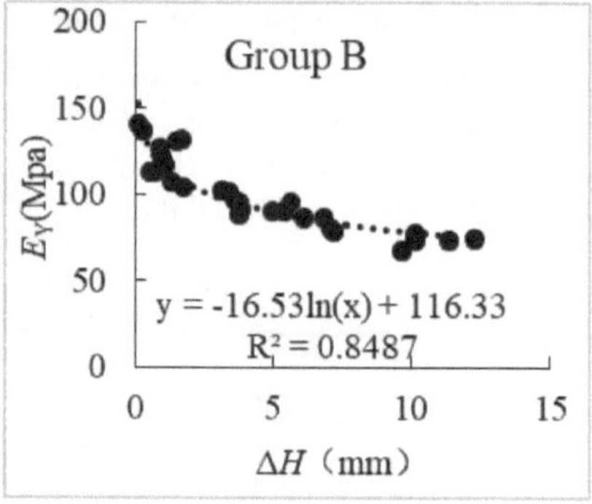

Figure 3. Regression curve of settlement difference ΔH and Young's modulus E_Y of Group B.

in the specification is also referred to, and the control index value of Young's modulus EY adopts a safety factor of 1.25. The control index of EY is calculated as shown in the Table 1.

Table 1. Control index of Young's modulus EY.

ΔH (mm)	E_Y (MPa)
6	109.6
5	113.1
4	117.5
3	123.1
2	131.0
1	144.5

The calculation shows that when the quality control settlement difference ΔH in the process of filling the roadbed with iron tailings in the project is not greater than 2mm, the Young's modulus EY needs to be no less than 131.0 MPa, and the roadbed can meet the design requirements.

3 CONCLUSIONS

To study the applicability of the GeoGauge in highway construction in China, a series of tests have been conducted on-site. The regression analysis method is used to establish a correlation between the two detection methods, the results are as follows.

The joint test of soil modulus/stiffness meter method and the differential settlement method showed that the control index of the Young's modulus EY can be obtained when the instrument detects the subgrade filling process. When the settlement difference is not more than 2.0 mm, Young's modulus should not be less than 130.1 MPa. When the test reaches this standard value, the subgrade filling quality meets the requirements of the specification. The results indicated that the GeoGauge meter could be used, and performed well in the detection process of highway construction.

REFERENCES

Fumio Tatsuoka, Takeshi Hashimoto, Kazuyoshi Tateyama. (2021). Soil stiffness as a function of dry density and the degree of saturation for compaction control. *Soils and Foundations*. 61 (4): 989–1002.

Li S.B. (2007). *Research on the Theory and Testing Technology of Road Structure Nondestructive Testing*. Tianjin: Tianjin University.

Liang Q.Y, (2003) Factors Affecting Subgrade Compaction, *Shanxi Architecture*, 6 (1): 257–258.

Ren S. J. (2011). *Research on the application technology of soil modulus stiffness meter to control the variability of roadbed construction*. Jinan: Shandong University.

Tatsuoka F, Correia A G. (2016). Importance of Controlling the Degree of Saturation in Soil Compaction. *Procedia Engineering*, 143: 556–565.

Zhang J.H, Deng Z.H, Liu J, etc. (2019). Research on the rapid detection method of existing roadbed in the hot and humid area of the south. *Journal of China & Foreign Highway*. 39 (06): 6–10.

Advances in Civil Engineering: Structural Seismic Resistance, Monitoring and Detection – Mohd Johari Mohd Yusof and Junwen Zhang (Eds)
© 2023 The Authors, ISBN 978-1-032-31491-4

Comparison and selection method of temporary hoisting system for suspender replacement of long-span arch bridge

Wei Li
Taizhou Highway and Transportation Management Centre, Taizhou, China

Guotao Shao*
Collage of Civil Engineering and Architecture, Taizhou University, Taizhou, Zhejiang, China

Hui Jin
School of Civil Engineering and Architecture, Zhejiang University of Science and Technology, Hangzhou, Zhejiang, China

Minglei He
Taizhou Highway and Transportation Management Centre, Taizhou, China

ABSTRACT: To study the comparison and selection method of the temporary hoisting system for suspenders replacement of long-span arch bridges, a half though arch bridge and a tied-arch bridge were selected. Two replacement schemes, the arch rib hoisting scheme and the deck hoisting scheme, were designed for each bridge. Based on the suspender replacement works that have been completed, the same replacement process, which with twenty-one construction steps, was adopted in each replacement scheme. The finite element model of each replacement scheme was built, and the deck displacement, arch rib bending moment, and adjacent suspender cable forces were compared. It is found that the most unfavourable effects caused by different hoisting methods are different under the same bridge type, while the structural response of different bridge types under the same hoisting method is also different. According to the results, the application scope and scheme comparison parameters of the arch rib hoisting method and deck hoisting method are recommended, which provides a reference for the suspender replacement of the long-span arch bridge.

1 INTRODUCTION

The temporary hoisting system is the key structure to transfer the old suspender cable force to the new suspender in suspender replacement. The commonly used temporary hoisting methods are divided into the arch rib hoisting method and the deck hoisting method. The arch rib hoisting method transfers the cable force of the old suspender to the arch rib, while the deck hoisting method transfers the cable force of the old suspender to the adjacent suspenders.

Most of the existing studies focus on construction monitoring and control, such as Wang put forward a displacement control method to the replacement control process of a long-span arch bridge (Wang et al. 2021). For the convenience of construction, Yan pointed out that the bridge deck hoisting method is more convenient and will not damage the arch rib, but the local force of the arch ribs and adjacent hangers will change a lot (Yan et al. 2013). The arch rib hoisting method does not change the original force system, but the installation of temporary hoisting equipment on the arch rib may cause damage to the arch rib. Based on the construction monitoring of the cable replacement of a long-span half through arch bridge, Shao verified the feasibility and effectiveness

*Corresponding Author: eaglexshao@chd.edu.cn

DOI 10.1201/9781003310884-48

of the deck hoisting scheme (Shao et al. 2014). In the study of hoisting process, Li pointed out that the bridge deck deformation of variable tension step length is smaller than that of equal tension step length (Li et al. 2018; Li et al. 2017). Aiming at replacing the suspenders under uninterrupted traffic, the process of using temporary stay cables to replace stay cables without interrupting traffic was introduced (Mehrabi 2016). Under considering the impact of traffic vibration, Su pointed out that the arch rib hoisting method is better than the bridge deck hoisting method from the angle of stress distribution (Su et al. 2019).

To sum up, most of the existing researches on suspender replacement methods is aimed at specific bridges, lacking universal rules and appropriate comparison and selection principles. This study intends to study the applicability of the arch rib hoisting and deck hoisting methods in the replacement of arch bridge suspenders and to find out the control factors of the comparison methods.

2 COMPARATIVE EXPERIMENTAL DESIGN

2.1 *Introduction to comparative bridges*

For the sake of generality, this paper selects two different types of arch bridges for comparison. Arch bridge A is a half-through concrete-filled steel tube (CFST) arch bridge with a main span of 192m, the rise height is 38.4m and rise span ratio is 1/5. Four main steel pipes are used for the two arch ribs, which are filled with C50 micro expansive concrete, and the transverse spacing between the arch rib axes is 15.8m. The suspender beam adopts prestressed concrete beam, with height of 1.4m, width of 0.8m and spacing of 8m. The suspender consists of 73 parallel steel wires.

Arch bridge B is a CFST tied-arch bridge with a main span of 80m. The calculated rise height is 16m and the rise span ratio is 1/5. The tie rod, arch end node and middle beam are poured as a whole first, and the bridge deck is poured as a whole. A special cable suspender for finished arch bridge is set between the tie bar and the arch rib.

The elevations of the two bridges are shown in Figures 1 and 2.

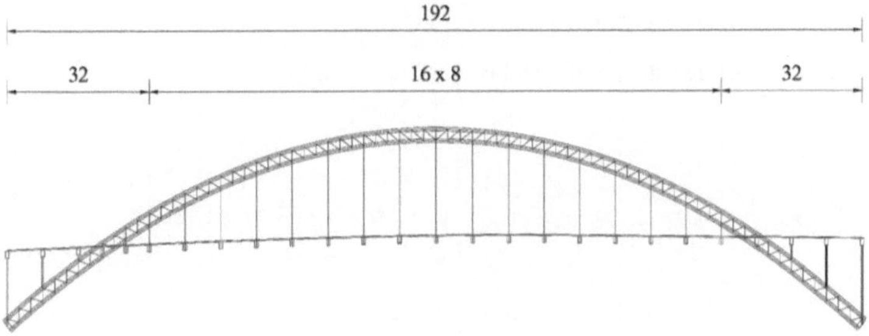

Figure 1. Arch bridge A (Unit: m).

The deck of arch bridge A is a typical large transverse beam and small longitudinal beam system. The suspender hangs the transverse beam, and the longitudinal beam is placed on the transverse beam. It represents a long-span arch bridge with thrust at the arch feet. Arch Bridge B is a tied-arch bridge, with transverse beams and longitudinal beams consolidated, representing a tied arch bridge with no thrust at the arch feet.

2.2 *Hoisting system scheme*

The arch rib hoisting scheme has the problem of how the temporary hoisting device is fixed to the arch rib. For the steel tube concrete arch rib, welding the temporary device will cause damage to the arch rib (Sun et al. 2017). The non-damaged scheme is to uses the hoop method or the

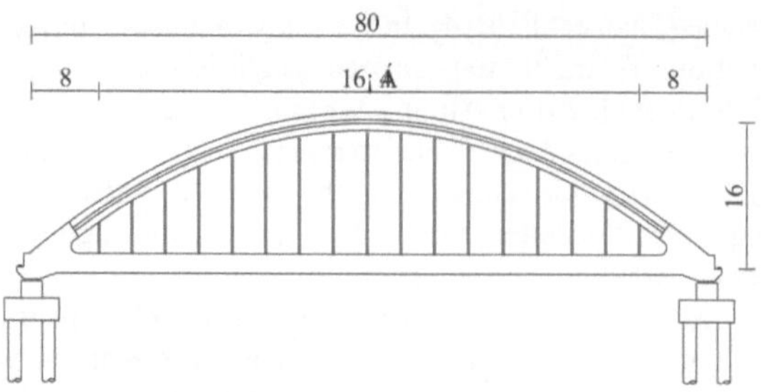

Figure 2. Arch bridge B (Unit: m).

wire rope hoisting method (Feng et al. 2018). The bridge deck hoisting schemes require temporary suspension rods to be drilled under the bridge deck, which may cause local damage to the deck (Wang 2014). The deck hoisting beams usually use truss beams with sufficient rigidity, such as Bailey beams. Regarding the hoisting system used in the replacement of the two bridge suspenders, the deck hoisting and arch rib hoisting schemes were designed for the two arch bridges. The four schemes are shown in Figures 3 to 6.

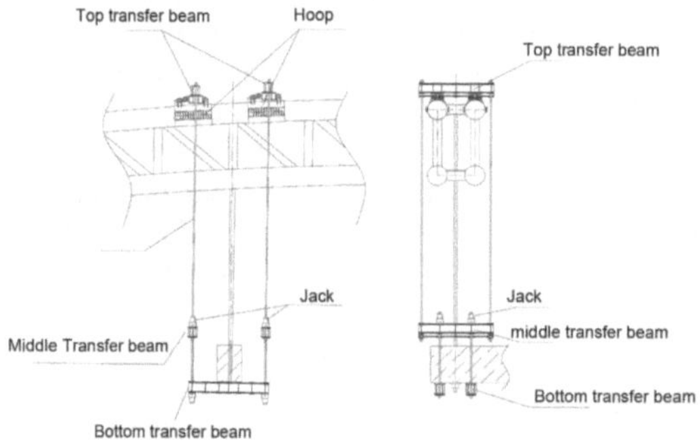

Figure 3. The arch rib hoisting scheme of arch bridge A (scheme 1).

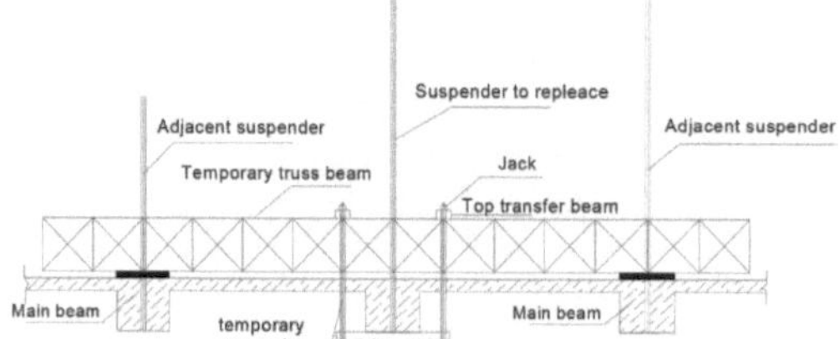

Figure 4. The deck hoisting scheme of arch bridge A (scheme 2).

3 CONTROL OF THE REPLACEMENT PROCESS

3.1 *Temporary hoisting force*

During the suspender replacement process, the tension of the temporary hoisting system is the key indicator. The selection of the tensile force of the hoisting system is mainly determined according to

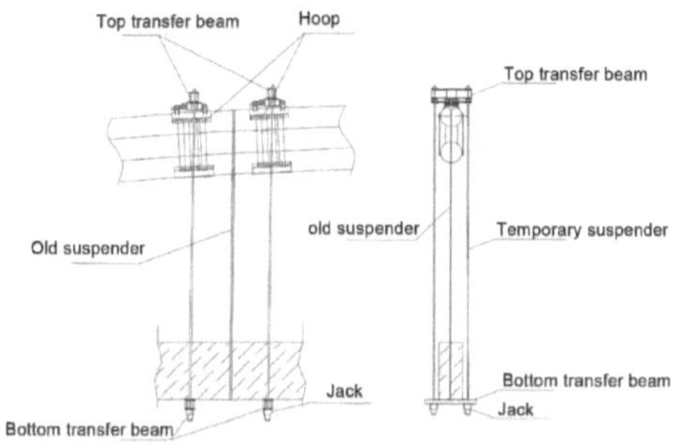

Figure 5. The arch rib hoisting scheme of arch bridge B (scheme 3).

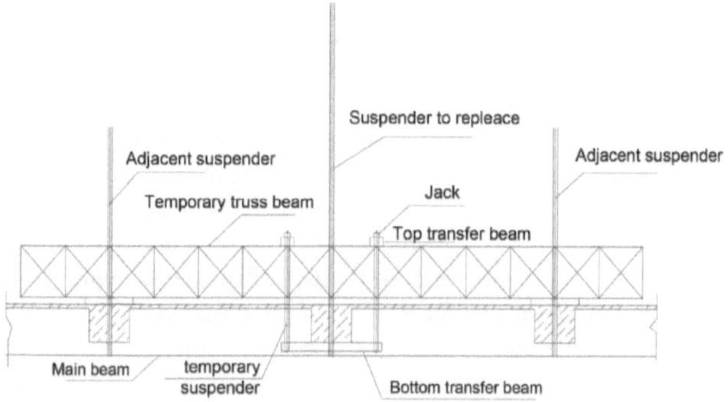

Figure 6. The deck hoisting scheme of arch bridge B (scheme 4).

the stress distribution and the deck deformation. Similar to the bridge deck structure of the transverse main girder of arch bridge A, the bridge deck displacement is more sensitive to cable forces. Similar to the strong longitudinal beam deck structure of arch bridge B, the deck displacement is less sensitive to cable forces (Fu et al. 2015).

(1) Suspender replacement scheme 1:

This scheme is a combination of the strong transverse beam deck and the arch rib hoisting method. Because the longitudinal beams are weak, the bridge deck displacement is more sensitive to the cable force changes. The temporary hoisting force of this scheme can be determined according to the principle of equivalent replacement, that is, the temporary hoisting stress should be the same as the original suspender stress.

(2) Suspender replacement scheme 2:

This scheme belongs to the combination of a strong transverse beam deck system and the deck hoisting method. During the replacement of the old suspender, the deck displacement at the adjacent suspender anchor point changes, and since the hoisting position is inside the anchor point of the original suspender, the temporary hoisting force here should be controlled according to the bridge deck displacement to ensure that the deck displacement is within the allowable range.

(3) Suspender replacement scheme 3:

This scheme is a combination of a strong longitudinal beam deck system and the arch rib hoisting method. The longitudinal beam was hoisted at the place where the suspender is replaced. Since this scheme is a local replacement with less stress distribution, the control target of the temporary suspender tension should be equivalent to the original suspender cable force.

(4) Suspender replacement scheme 4:

This scheme belongs to the combination of the strong longitudinal beam deck system and the bridge deck hoisting method. Due to the stiffness of the longitudinal beams being greater than that of the temporary support beams, the cable force of the old suspenders will be partly distributed.

The temporary hoisting force should be comprehensively determined based on the bridge deck displacement of the replaced suspender and its adjacent suspenders.

For the convenience of comparison, for the scheme of using bridge deck hoisting, this study controls the temporary hoisting force according to the feasible range of the deck displacement within 10mm.

3.2 *Replacement Steps*

According to the existing research, this study adopts the variable step length method to design the replacement process, and the replacement steps are as follows:

Step 1: Stretch the temporary hoisting force to 20% of the target value;
Step 2: Unload the old suspender to 70% of the original cable force;
Step 3: Tension the temporary suspender force to 40% of the target value;
Step 4: Unload the old suspender to 50% of the original cable force;
Step 5: Tension the temporary lifting force to 60% of the target value;
Step 6: Unload the old suspender to 30% of the original cable force;
Step 7: Tension the temporary suspender to 80% of the target value;
Step 8: Unload the old suspender to 10% of the original cable force;
Step 9: Tension the temporary suspender to 100% of the target value;
Step 10: Unload the old suspender to 0, remove the old suspender;
Step 11: Install a new suspender and stretch it to 20% of the design value;
Step 12: Unload the temporary suspender to 70% of the target value;
Step 13: Tension the new suspender to 40% of the design value;
Step 14: Temporarily hoist and unload 50% of the target value;
Step 15: Pull the new suspender to 60% of the design value;
Step 16: Unload the temporary suspender to 30% of the target value;
Step 17: Pull the new suspender to 80% of the design value;
Step 18: Unload the temporary suspender to 10% of the target value;
Step 19: Stretch the new suspender to 95% of the design value;
Step 20: Unload the temporary suspender to the target value of 0%;
Step 21: Pull the new suspender to 100% of the design value, and remove the temporary hoisting device.

3.3 *FE model*

According to the design plan of suspender replacement, four replacement models were established using finite element analysis software Midas/civil. Simulate the process of replacing the long boom in mid-span respectively. As shown in Figures 7 to 10.

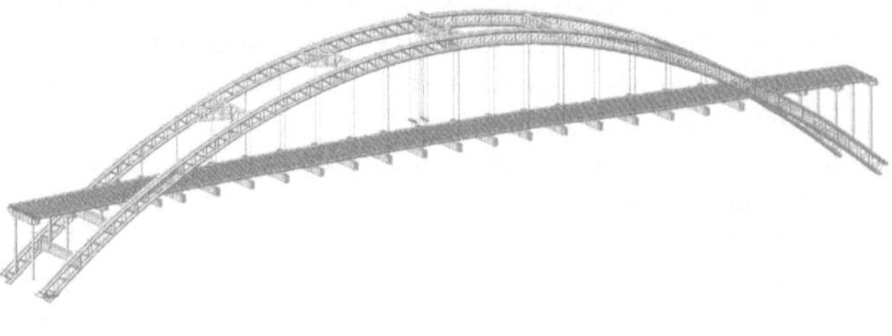

Figure 7. FE model of scheme 1.

After the analysis, the calculated dead-load cable force of arch bridge A is 860kN and arch bridge B is 660kN, which is consistent with the measured value on-site. So the accuracy of the FE models was verified, and then the suspender replacement simulation analysis was performed.

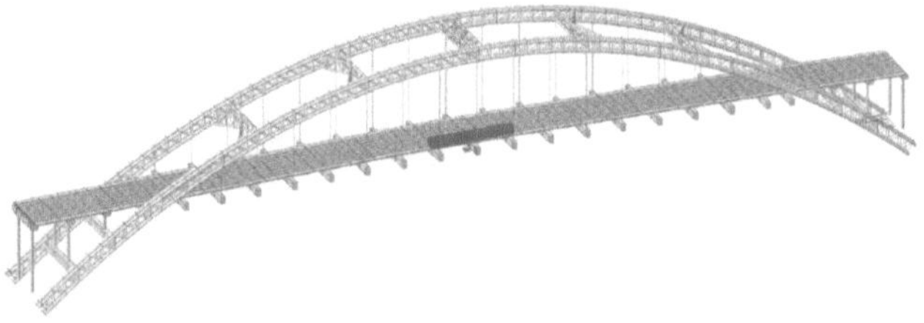

Figure 8. FE model of scheme 2.

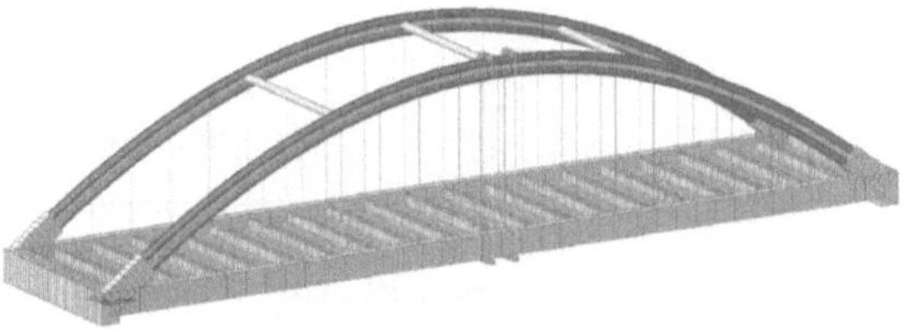

Figure 9. FE model of scheme 3.

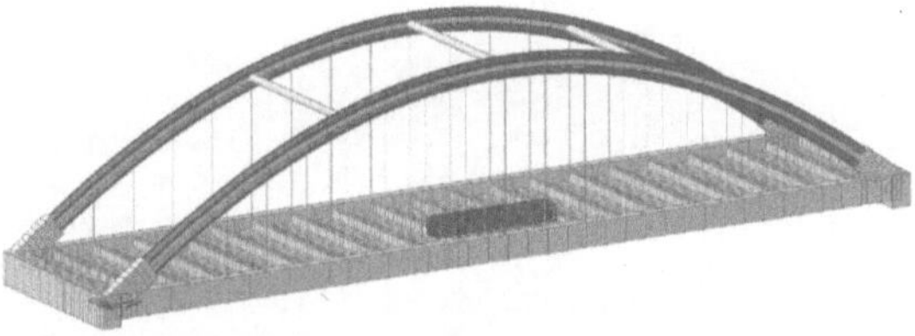

Figure 10. FE model of scheme 4.

4 RESULTS

4.1 *Deck displacement*

According to the design scheme and the control principle of temporary suspender tension, the deck displacement at the anchor point of the suspender that needs to be replaced is shown in Figure 11.

It can be seen from Figure 11 that the deck displacements of the four schemes are all within 10mm, indicating that the temporary hoisting tension meets the control target. Scheme 2 has the largest deck displacement change, which indicates that the use of the deck hoisting on the deck of the strong transverse beam will cause a large deck displacement. The deck displacements of Scheme 1 and 2 are significantly larger than Schemes 3 and 4, which indicates that the deck system with a strong transverse beam is more sensitive to cable force changes. Under the same bridge type, the deck displacement of the deck hoisting scheme is greater than that of the arch rib hoisting scheme, which indicates the arch rib hoisting method has little effect on the deck displacement.

4.2 *Bending moment of arch rib*

The relative bending moment changes of the arch rib caused by the four suspender replacement schemes at each stage are shown in Figure 12.

It can be seen from Figure 12 that the bending moment caused by the deck hoisting scheme of arch bridge B represented by scheme 4 is the largest, followed by the arch rib hoisting scheme of arch bridge B in scheme 3. The bending moment of the arch rib caused by the deck hoisting method of arch bridge A is greater than that of the arch rib hoisting method. The arch rib hoisting scheme of arch bridge A causes the smallest change in the arch rib bending moment.

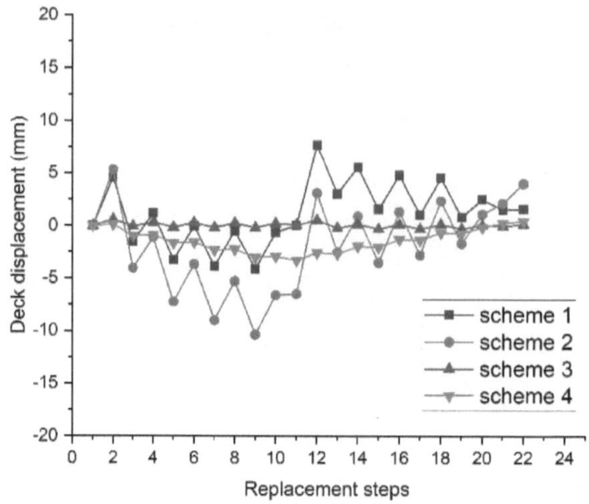

Figure 11. Deck displacement of each replacement scheme.

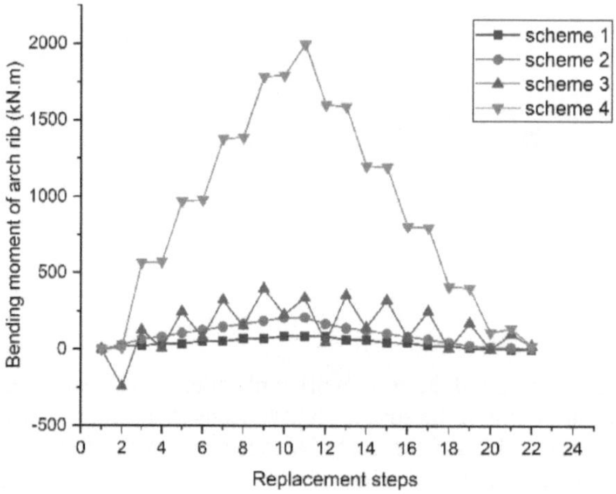

Figure 12. The influence on the bending moment of the arch rib.

4.3 *Bending moment of the main beam*

The relative change values of the bending moment of the main girder (the transverse beam of arch bridge A and the longitude beam of arch bridge B) at each stage caused by the four boom replacement schemes are shown in Figure 13.

It can be seen from Figure 13 that the main girder bending moment caused by the deck hoisting scheme of arch bridge A represented by scheme 2 is the largest, which indicates that the deck hoisting method on the strong transverse beam bridge has a greater change in the stress state of the main beam. The bending moment of the longitudinal beam caused by the deck hoisting scheme of arch bridge B is greater than that of the arch rib hoisting scheme. The arch rib hoisting scheme of arch bridge A causes the smallest bending moment change.

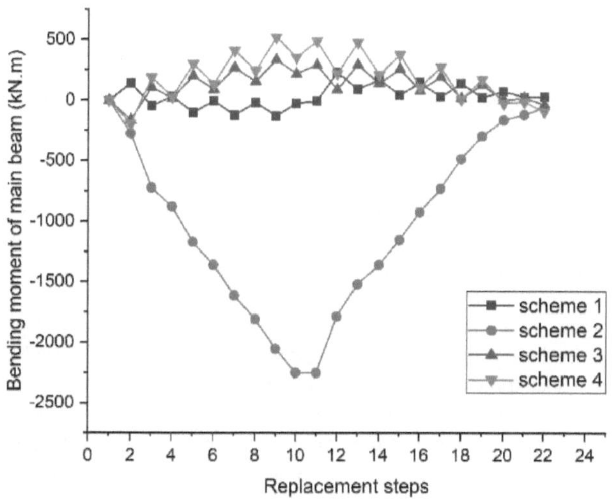

Figure 13. The influence on the bending moment of the main beam.

4.4 *Cable Force of the Adjacent Suspender*

The relative cable force change of the adjacent suspenders caused by the four suspender replacement schemes at each stage is shown in Figure 14.

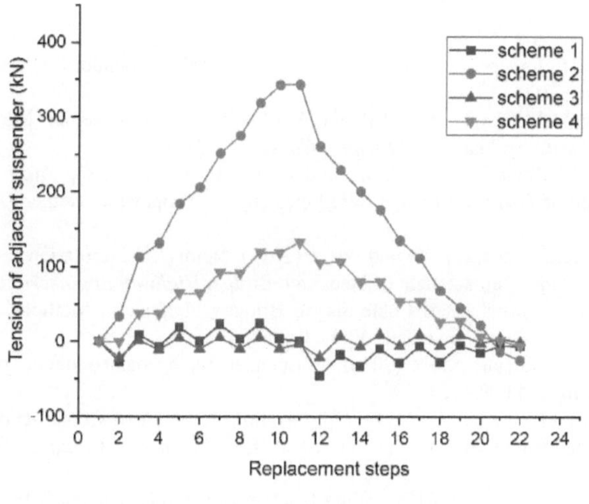

Figure 14. The influence on the cable force of adjacent suspenders.

It can be seen from Figure 14 that the cable force changes caused by scheme 2 is the largest, which indicates that the deck hoisting method of the strong transverse beam bridge distributes more stress from the original suspender to the adjacent suspenders. The change of the adjacent suspender cable force in scheme 4 is greater than that of scheme 3, but the increase ratio is much smaller than that of arch bridge A. The main reason is that the strong longitudinal beam of arch bridge B distributes part of the old suspender cable force. The cable force changes of adjacent suspenders caused by the arch rib hoisting scheme of arch bridge A and arch bridge B are both small.

5 CONCLUSION

In view of the comparison and selection of temporary hoisting method for the suspender replacement of the arch bridge, two representative bridges were selected and four replacement schemes were designed. Through the comparative analysis, the main conclusions are as follows:

(1) The temporary hoisting tension of the arch rib hoisting is easy to control. Regardless of the deck structure of strong longitudinal beams or strong transverse beams, the impact of arch rib hoisting on the deck displacement, banding moment of the arch rib and the main girder, and the cable force of adjacent suspenders are minimal. The arch rib hoisting method should be the first choice for suspender replacement.
(2) The deck hoisting method will cause a large bending moment change in the arch rib of the tied-arch bridge. In an ordinary arch bridge with strong transverse beams, the bridge deck hoisting will change the stress state of the transverse beams, and the original suspender anchoring part will become a cantilever beam, which will cause a large negative bending moment at the temporary hoisting point section. At the same time, hoisting on the bridge deck will greatly increase the cable force of the adjacent suspenders, especially in the strong transverse beam deck system.
(3) If deck hoisting is used in a tied arch bridge, attention should be paid to the bearing capacity of the arch ribs and adjacent suspenders; if deck hoisting is used in arch bridges with strong transverse beams, attention should be paid to the bearing capacity of the transverse beam and the adjacent old suspenders.

In this study, suspender replacement of two conventional bridges is considered, and further research is needed for other types of bridges.

REFERENCES

Feng, D. M., Mauch, C., Summerville, S., and Fernandez, O. (2018). Suspender Replacement for a Signature Bridge. *J Bridge Eng*, 23(11).

Fu, Z. Q., Ji, B. H., Yang, M. Y., Sun, H. B., and Maeno, H. (2015). Cable Replacement Method for Cable-Stayed Bridges Based on Sensitivity Analysis. *J Perform Constr Fac*, 29(3).

Li, H. S., Huang, P. M., Wang, T., Cai, C. W., and Zhang, L. Y. (2018). Optimization Study of Hanger Replacement Scheme of Tied Arch Bridge Based on Different Proportion Method. *Railway Engineering*, 58, 49–52.

Li, Z. J., Sun, G. J., Li, H. J., Zhang, S. X., and Wu, J. (2017). Numerical Simulation and Working Procedure Optimization of Suspender Replacement of Tied Arch Bridge. *Highway Engineering*, 42, 48–53.

Mehrabi, A. B. (2016). Performance of Cable-Stayed Bridges: Evaluation Methods, Observations, and a Rehabilitation Case. *J Perform Constr Fac*, 30(1), C4014007.

Shao, G. T., Jin, H., Guo, F. B., and Xu, Y. (2014). "Suspender Replacement Technology of Jiantiao Bridge." *Construction Technology*, 43, 69–72.

Su, X. M., Xu, H. S., and Yan, D. H. (2019). Optimization Research of Construction Process in Hanger Replacement of Arch Bridge without Interrupting Traffic. *Journal of China & Foreign Highway*, 03, 115–118.

Sun, Z., Ning, S. W., and Shen, Y. F. (2017). Failure Investigation and Replacement Implementation of Short Suspenders in a Suspension Bridge. *J Bridge Eng*, 22(8).

Wang, B. G. (2014). Replacement Construction Technology Analysis of Half-through Reinforced Pipe Concrete Bridge Derrick. *Advanced Materials Research*, 1065–1069, 949–955.

Wang, H., Wang, L. L., Zhuo, X. L., Huang, K. N., Wang, X. R., and Wang, W. S. (2021). Study on the Precise Displacement Controlling Method for a Suspended Deck in the Hanger Replacement Process of an Arch Bridge. *Appl Sci-Basel*, 11(20), 9607.

Yan, G. T., Wang, S. X., and Zhao, D. K. (2013). Cables Repairing Method for Arch-bridges and Temporary Cable System. *Construction Technology*, S1, 445–449.

Advances in Civil Engineering: Structural Seismic Resistance, Monitoring and Detection – Mohd Johari Mohd Yusof and Junwen Zhang (Eds)
© 2023 The Authors, ISBN 978-1-032-31491-4

Simulation study on the influence of thermal performance of energy-saving doors and windows on building energy consumption

Ming Cao, Shanshan Xu, Yunli Zhang & Leixin Yu
Shenyang Urban Construction University, Shenyang, Liaoning, China

Bingxiang Zhao*
Liaoning JIMEI Decorative Engineering Co., Ltd, Shenyang, Liaoning, China

ABSTRACT: In this research activity, we simulated the thermal parameters of energy-saving doors and windows of several different frame materials, and in the process of building energy consumption simulation, we also discussed the influence of energy-saving doors and windows according to the climatic conditions of different regions. According to the simulation data, we can know that energy-saving doors and windows made of different glass materials and different frame materials have different influences on the energy consumption of buildings in summer and winter, and the actual selection of energy-saving doors and windows should be specifically considered according to the local climate conditions. In addition, this paper also discusses the future development direction of energy-saving doors and windows, which also provides an effective reference for the sustainable development of the industry.

1 INTRODUCTION

As the opening part of the building's outer protective structure, doors and windows are the basic barriers to block the external climate intrusion. According to statistical data, the average heat transfer coefficient of exterior doors and windows is about 4.70W/ $(m^2 \cdot k)$, the average heat transfer coefficient of wall structure is about 1.40W/ $(m^2 \cdot k)$, the average heat transfer coefficient of roof structure is about 0.80W/ $(m^2 \cdot k)$, the average heat transfer coefficient of ground structure is about 0.52W/ $(m^2 \cdot k)$. The overall area of doors and windows accounts for 12%-15% of the surface area of the whole external enclosure structure, and the energy consumption level accounts for 40%-50% of the whole external enclosure structure. Therefore, it can be seen that the glass energy saving design needs to pay attention to in the building energy saving design.

2 MAIN PARAMETERS OF BUILDING GLASS ENERGY-SAVING EVALUATION

Among many performance indexes of architectural glass, heat transfer coefficient and solar thermal coefficient can be used to judge its energy-saving characteristics. Heat transfer coefficient (U or K, hereinafter referred to as K value) refers to the amount of heat transfer through $1m^2$ glass per unit time with the air temperature difference on both sides of the window of 1°C under the condition of stable heat transfer, expressed in $W/(m^2 \cdot K)$. Heat transfer coefficient is a general parameter of enclosure structure. Generally speaking, the lower its value, the better its thermal insulation performance. The thermal coefficient of the sun (SHGC) refers to the ratio of the amount of solar radiation energy entering the room through the window glass and the amount of solar heat entering

*Corresponding Author: 2003408665@dlvtc.edu.cn

DOI 10.1201/9781003310884-49

the room through the same size but no glass opening under the same conditions. For K value and SHGC value, the former mainly measures the heat transfer caused by temperature difference, while the latter mainly measures the heat transfer ratio caused by solar radiation, and the two kinds of influences exist at the same time in real life environment. However, for different climatic zones in China, K and SHGC values have different influences on energy saving effect. In this paper, we use DeST software to simulate and calculate the energy consumption of buildings located in different climatic zones, and discuss the influence of K and SHGC values of door and window glass on building energy consumption in different climatic zones.

3 SIMULATION KEY POINTS OF ENERGY-SAVING DOORS AND WINDOWS THERMAL PERFORMANCE ON BUILDING ENERGY CONSUMPTION

3.1 *Building model and environmental conditions*

Build a building model with DeST software, and calculate the building energy consumption of installing different doors and windows. The building model is 12 m×8 m in size and 3.5 m in height. There are 3 houses and 1 staircase on each floor, with 2 floors in total, with a total construction area of 192 m^2. There are 1 outer door and 11 outer windows in the peripheral protection structure. The components and materials used in the model are shown in Table 1.

Table 1. Data of building model components.

Component name	External wall	Internal wall	Floor	Roof	Ground
Thickness/mm	240	200	100	300	40
Materials used	Brick + polyplastic board	Brick	Reinforced concrete	Reinforced concrete	Reinforced concrete

The main function of the building model is an ordinary office, using the central air-conditioning system. In summer, the indoor temperature will be controlled at 24~26°C, and the use time of air conditioner will last from mid-May to the end of September. After November 1st, heating will start, and heating will last until March 15th. DeST simulation software is used to simulate the influence of different types of doors and windows on building energy consumption.

3.2 *Thermal performance simulation*

3.2.1 *Door and window glass*

According to statistical data, the area of glass accounts for more than 70% of the total area of doors and windows, and the heat dissipated from the glass of doors and windows accounts for more than 65% of the heat dissipated from the whole door and window system. In the process of thermal performance simulation, it is necessary to analyze the performance of door and window glass. Judging from the glass materials currently circulating in the market, insulating glass has good application value. In the initial case, the air in the interlayer is mainly dry air. Some interlayers will also be filled with inert gas, and the gap width will be controlled above 6mm. In Windows 6.0 software, the thermal performance parameters of six common doors and Windows are calculated as shown in Table 2:

Low-E and commom in the above table respectively represent one layer of low-emissivity glass and two layers of ordinary glass. Take (5+12+5) Low-E insulating glass series as an example. The glass consists of two layers of 5mm glass with a gap of 12mm in the middle, one of which is low-emissivity glass. In use, the SC value of this series of glasses is 0.695%, SHGC value is 0.620, and the heat transfer coefficient is 1.863 W/(m^2·k). Compared with the parameter of (5+12+5) comm, its performance has been effectively improved. At the same time, by comparing (5+12+5)

Table 2. Calculation results statistical table of thermal performance parameters.

Type	W/(m^2·K)	SC/%	SHGC/%
(6+12+6) Low-E	1.963	0.487	0.416
(5+12+5) Low-E	1.863	0.695	0.620
(5+9+5) Low-E	2.235	0.753	0.633
(6+12+6) commom	2.369	0.862	0.721
(5+12+5) commom	2.715	0.886	0.766
(5+9+5) commom	2.933	0.891	0.779

Low-E with (6+12+6) Low-E, it can be found that the increase of glass thickness will not have too much influence on the system parameters. However, with the increase of the hollow gap, the heat transfer coefficient will be reduced to a certain extent, which has good popularization value.

3.2.2 *Door and window profiles*

In the analysis of door and window profiles, several common manufacturing materials are mainly analyzed, including pure wood materials, broken bridge aluminum alloy materials, aluminum-wood composite profiles and so on. The energy-saving doors and windows used this time are made of aluminum-wood composite profiles. Considering the differences in position and function, we will use THERM6.0 software to complete the U values of different parts of doors and windows in this simulation analysis. When determining the boundary conditions, we will follow the contents in the Code for Thermal Calculation of Glass Curtain Walls of Buildings, Doors and Windows (JJ/T151-2008), that is, the temperature difference between indoor temperature (20°C) and outdoor temperature (−20°C) is 40°C, while the indoor convective heat transfer coefficient is kept at 3.0w/(m^2·K), the outdoor convective heat transfer coefficient is kept at 12, and the solar irradiance is 0. We can refer to this standard to calculate the U value of different parts of different doors and windows. Take 1500mm×1500mm aluminum-wood composite profiles doors and windows as an example. Its parts include the upper edge, left edge, left lower edge of the left window, the upper edge, left edge, left lower edge of the right window and the butt joint between the left and right windows. According to the obtained values, it can be known that the heat transfer coefficients all meet the specification requirements, and the heat dissipation can be controlled in a small range (Jin 2021).

3.2.3 *Whole window system*

In addition to the analysis types mentioned above, it is also necessary to analyze the whole window system. In the specific simulation analysis, we will directly import the calculation data obtained by THERM6.0 software into Window6.0 software for analysis. The calculated whole window system includes 86 series pure wooden windows, 86 series aluminum-clad wooden doors and windows, 60 broken bridge aluminum alloy doors and windows, and 90 broken bridge aluminum alloy doors and windows. The calculated data are shown in Table 3.

Table 3. Calculation results statistical table of thermal performance parameters of whole window system.

Type	W/(m^2·K)	VT/%	SHGC/%
86 series pure wooden windows	1.923	0.313	0.276
86 series aluminum-clad wooden doors and windows	1.915	0.312	0.283
60 broken bridge aluminum alloy doors and windows	2.735	0.316	0.291
90 broken bridge aluminum alloy doors and windows	2.569	0.366	0.310

According to the above calculation data, it can be known that the parameters such as heat transfer coefficient and light transmittance of 86 series pure wood windows and 86 series aluminum-clad wood doors and windows are in a low state in application. However, in terms of durability, 86 series aluminum-clad wood doors and windows have stronger durability and are more suitable for door and window systems (Wang 2019).

3.3 *Building energy consumption simulation*

3.3.1 *Severe cold areas*

Northeast China (such as Shenyang, Harbin, etc.) can be selected as the research object when simulating building energy consumption in severe cold areas. According to statistical experimental data, in winter, if 86 series pure wood windows and 86 series aluminum-clad wood doors and windows use ordinary glass as the main material, the heating energy consumption is relatively low, while the heating energy consumption generated by aluminum alloy doors and windows with broken bridges is relatively high when they are assembled with Low-E glass. At the same time, compared with ordinary glass, assembled Low-E glass can have good thermal insulation performance, but low-emissivity glass will reduce the thermal coefficient of the sun, making the indoor solar heat gain capacity lower, thus increasing the heating energy consumption. In summer, if 86 series pure wood windows and 86 series aluminum-clad wood doors and windows use ordinary glass as the main material, the refrigeration energy consumption at this time is relatively high, while when Low-E glass is assembled, the refrigeration energy consumption is relatively low. Compared with the energy consumption data of the whole year, although doors and windows equipped with Low-E glass will increase the heating energy consumption in winter, its overall energy consumption is less and it has good use value (Zheng 2019).

3.3.2 *Cold areas*

When simulating building energy consumption in cold areas, we can choose the central region (such as Zhengzhou, Shangqiu) as the research object. According to the statistical experimental data, in winter, if ordinary glass is used as the main material, the heating energy consumption is relatively low, while the heating energy consumption generated by Low-E glass is relatively high. When assembling Low-E glass in summer, the refrigeration energy consumption is at a low level, and it has remarkable effect in reducing energy consumption. According to the data, using Low-E glass can reduce energy consumption by 12% (Zhang 2018). Compared with the energy consumption data of the whole year, doors and windows equipped with Low-E glass can reduce the energy consumption by 7.5% compared with ordinary doors and windows, but the influence of the whole window system type is not obvious, and the fluctuation range is less than 2%, so the emphasis will be placed on the glass types when constructing in such areas.

3.3.3 *Warm area*

When simulating building energy consumption in warm areas, we can choose the southern areas (such as Kunming, Yuxi, etc.) as the research object. According to the statistical experimental data, window frame selection and glass types will have obvious influence on energy consumption, among which 86 series pure wood windows and 86 series aluminum-clad wood doors and windows have relatively low overall energy consumption. However, when Low-E glass is assembled, the refrigeration energy consumption is at a low level, and it has remarkable effect in reducing energy consumption. According to the data, the refrigeration energy consumption in mild areas accounts for more than 75% of the total energy consumption, while the use of Low-E glass can reduce the energy consumption by 10%. Compared with the energy consumption data of the whole year, doors and windows equipped with Low-E glass have better energy-saving performance than ordinary doors and windows, so when constructing in such areas, the emphasis will be placed on glass types and window frame selection (Wang 2016).

3.3.4 *Hot summer and warm winter areas*

When simulating the building energy consumption in hot summer and warm winter areas, we can choose the southern coastal areas (such as Guangzhou and Shenzhen) as the research object. According to the statistical experimental data, the selection of window frames and glass types will have obvious influence on the energy consumption, and the situation is basically consistent with that in warm areas. However, the refrigeration energy consumption in such areas accounts for more than 90% of the total energy consumption, and the use of Low-E glass can effectively reduce the solar transmittance, and the use of broken bridge aluminum alloy profiles can effectively reduce the light transmission area, which can play a good role in reducing the refrigeration energy consumption in summer. The thermal insulation performance of window frames will have a certain impact on refrigeration, but from a comprehensive point of view, the effect of reducing annual energy consumption is not significant. Therefore, when constructing in such areas, the emphasis will be placed on the types of glass, door and window selection profiles, and there is no need to pay too much attention to the window frames selection (Yi 2015).

4 FUTURE DEVELOPMENT DIRECTION OF ENERGY-SAVING DOORS AND WINDOWS

4.1 *Energy-saving performance is further improved*

In the future development process, the energy-saving performance of energy-saving doors and windows will be further improved, thus reducing the building energy consumption. Its contents are reflected in the following aspects: First, the energy saving of materials is further improved. In the future research and development, the application proportion of recyclable materials and new polymer materials is also increasing, thus improving the application value of energy-saving materials (Wang 2015). Second, the energy saving of door and window construction technology will continue to improve. In addition to the currently used construction technology, in the future development, the construction technology will continue to be enriched, thus effectively improving the integrity of doors, windows and exterior walls and laying the foundation for building energy consumption reduction.

4.2 *Durability is constantly improving*

In the future development process, the durability of energy-saving doors and windows will also be continuously improved, thus playing a role in prolonging the service life of buildings. Its contents are reflected in the following aspects: First, the durability of door and window materials. In the future research and development, the application popularity of some new polymer materials will continue to increase, and this kind of materials can also improve the durability of materials on the basis of ensuring their quality. Second, the durability of other auxiliary materials, such as sealant and external fixing structures, will also be continuously improved. Furthermore, it will promote the continuous improvement of the overall durability of the building structure, which can also effectively reduce the energy loss and improve the energy saving of the building (Yang 2015).

4.3 *Production costs are constantly decreasing*

In addition to the above two points, in the future development process, the production cost of energy-saving doors and windows will be continuously reduced, thus playing a role in reducing the comprehensive cost of buildings. Its contents are reflected in the following aspects: First, the construction cost of materials is constantly decreasing. In the future research and development, the production technology of new energy-saving materials is also constantly maturing, so the production cost of materials will also continue to decrease at this time, which also provides an effective reference for the promotion of new materials and reduces the construction cost of construction projects.

Second, the economy of door and window construction technology will continue to improve. In the long-term practice process, the maturity of construction technology is also constantly improving, so the application cost of construction technology will continue to decrease at this time, which can also effectively reduce the building project construction cost and improve the economy of project construction (Guo 2014).

5 CONCLUSION

To sum up, in the use of energy-saving doors and Windows, its thermal performance will bring a certain impact on building energy consumption. It is mainly reflected in door and window glass, door and window profiles, window system and so on. By establishing reliable application model and analyzing the influence of thermal performance on building energy consumption, we can accumulate valuable application data, which has positive significance for the continuous improvement of application system and sustainable development of industry economy.

REFERENCES

Guo Xingzhong, Yang Chuang, Yang Hui, Wu Heping.(2014)Simulation Study on the Influence of Thermal Performance of Doors and Windows on Building Energy Consumption. *New Building Materials*, 41(S1):54–57+60.

Jin Xi. (2021) Study on the Application of Green Energy Saving Technology of Building Door and Window Curtain Wall in Practice. *Chinese Construction Metal Structure*, (01): 126–127.

Wang Shixiao, Meng Qinglin, Chen Zhechao. (2015) Simulation Study on Energy-saving Performance of Aluminum-plastic Co-extruded Doors and Windows. *Heating Ventilating & Air Conditioning*, 45(09): 76–79.

Wang Xiaopeng. (2019)Latest Requirements for Thermal Performance of Doors and Windows in EU Member Countries. *Doors & Windows*, (19):4–6.

Wang Yi, Tong Yanping. (2016)Simulation Analysis the Influence of Thermal Performance of Energy-saving Doors and Windows on Building Energy Consumption. *Doors & Windows*, (08):34–35.

Yang Bao. (2015) Application Status and Development Trend of Energy-saving Doors and Windows in Green Buildings. *Journal of Green Science and Technology*, (05):250–252.

Yi Bing. (2015) Application Research on Energy Saving Technology of Typical Building Doors, Windows and Curtain Walls. Hunan University.

Zhang Meng, Chen Xu, Qin Song, Wang Yong. (2018) Analysis on the Present Situation of Popularization and Application of Low-E Energy-saving glass in China. *China Building Materials*, (11):143–145.

Zheng Wenjie. (2019) Brief Analysis on the Energy Saving Design Points of the System Doors and Windows. *Wall Materials Innovation & Energy Saving in Buildings*, (07):63–66.

Advances in Civil Engineering: Structural Seismic Resistance, Monitoring and Detection – Mohd Johari Mohd Yusof and Junwen Zhang (Eds)
© 2023 The Authors, ISBN 978-1-032-31491-4

Research on mechanical properties of optimized metal buckling restrained shear panel damper

Zijian Xun, Lingxin Zhang & Baijie Zhu*
Key Laboratory of Earthquake Engineering and Engineering Vibration, Institute of Engineering Mechanics, China Earthquake Administration, Harbin, China

ABSTRACT: With a view to improving the problem of stress concentration and out-of-plane buckling at the end of the rectangular energy dissipation plate of traditional metal shear damper, the end shape of the energy dissipation plate was optimized by using chamfers with different radii, and an optimized metal buckling restrained shear panel damper was proposed. The stiffness calculation formula is given, the numerical model of this type of damper is established by finite element software, and the quasi-static test is carried out. The test results are used to verify the accuracy of the numerical model and the stiffness calculation formula. Accordingly, the verified numerical model was used for parametric analysis, the optimization effects of five different radius chamfers on the end of the energy dissipation plate were compared, and the mechanical properties of five kinds of span-to-height ratio dampers were studied. Using the simulation results, the calculation formula of the bearing capacity of this type of damper is revised, and the simplified constitutive model calculation formula of this type of damper is regressed, which lays a foundation for the engineering application of this type of damper.

1 INTRODUCTION

Metal shear panel dampers are often used in high-intensity areas or important buildings in cities with relatively low cost and stable performance (Tsai 1993). In the advent of earthquakes, this type of damper can absorb most of the seismic energy and reduce the damage to the main structure. After the earthquake, the damaged damper can be replaced, and the building function can be restored (Whittaker 1991; Xing 2003). Scholars from various countries have carried out a lot of research on this type of damper (Ji 2014; Okazaki 2005). The results show that under the low-cycle reciprocating load, the failure mode is mainly concentrated at the end weld, and the web has obvious out-of-plane buckling. In order to solve the problem of out-of-plane buckling and end stress concentration, Ge and Chen (Chen 2008; Ge 2006) proposed a hysteretic model of stiffened shear panel dampers, Koike and Ma (Koike 2008; Ma 2018) also proposed methods such as welding stiffeners, which can prevent out-of-plane buckling and increase the influence of thermal effects. Zhang et al. (Zhang 2012) proposed a low yield strength steel shear panel damper with high deformation capacity, Liu et al. (Liu 2013) measured the strain distribution of a shear panel damper, Chang et al. (Chang 2016) proposed a hourglass shaped strip damper, Deng (Deng 2014) et al. optimized the shape of the damper web to relieve stress concentration but weakened the bearing capacity of the damper. Under the premise of ensuring the bearing capacity of the damper, this paper uses chamfering to optimize the shape of the end of the energy dissipation plate to relieve stress concentration and adopts a new type of anti-buckling structure to restrain the external buckling of the energy dissipation plate. In this regard, the assembly idea is further combined to design an optimized metal buckling restrained shear panel damper (RSPD for short), this structure can avoid the influence of welding thermal stress, and is easy to disassemble after an earthquake.

*Corresponding Author: baijie_zhu@126.com

DOI 10.1201/9781003310884-50

2 SREUCTURE OF OPTIMIZED BUCKLING RESTRAINED SHEAR PANEL DAMPER

The damper proposed in this paper is mainly composed of an energy dissipation plate, a bending bond, buckling restrained plates (one long and one short on one side), shim, bolt, nuts, etc., as shown in Figure 1. The bending bond is used to provide the flexural bearing capacity of the damper, and the shear bolts are embedded in the wall to provide the shear bearing capacity of the damper. The structure has a clear force transmission path and enables the damper to work stably. In order to avoid the influence of welding thermal effect, the damper adopts a fully bolted assembly connection, which is easy to disassemble and assemble, and the part except the energy dissipation plate can be used repeatedly after the earthquake. The ultimate bearing capacity on the energy dissipation plate of the damper should not exceed the yield bearing capacity of the connector. During an earthquake, all the connection parts of the damper except the energy dissipation plate are in an elastic state. In order to prevent out-of-plane deformation of the energy dissipation plate, out-of-plane restraint is provided by a buckling restrained plate integrally formed with the support. A shim is used to isolate the buckling restrained plates on both sides, thus providing enough working space for the energy dissipation plate.

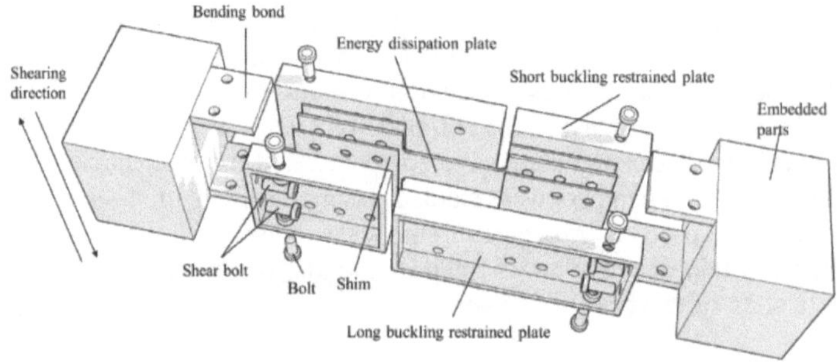

Figure 1. Configuration of dampers.

The end of the energy dissipation plate of the damper is optimized by chamfering to alleviate the phenomenon of stress concentration. The design of the energy dissipation plate of the damper is shown in Figure 2. The end is provided with a chamfer with a radius of R. The rectangular energy dissipation plate in the middle of the energy dissipation plate is h in length and b in height, and an appropriate number of bolt holes are provided in the clamping section.

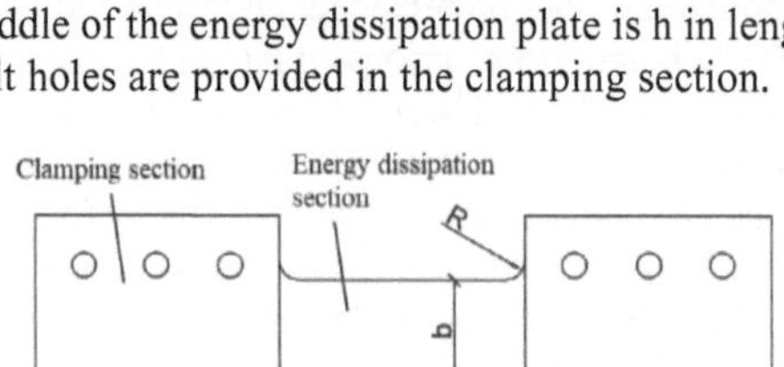

Figure 2. Section design diagram of energy dissipation plate.

3 ESTABLISHMENT AND VERIFICATION OF THE FINITE ELEMENT MODEL OF THE OPTIMIZED METAL BUCKLING RESTRAINED SHEAR DAMPER

Taking a damper of a given size as an example, a finite element model of the example is established for numerical simulation, and a quasi-static test is carried out for the example, the test results of which are used to verify the accuracy of the finite element model.

3.1 *Example design and establishment of finite element model*

A damper is designed with a thickness of 1.3 mm, a height of 20 mm, a length of 50 mm, an end chamfering radius of 5 mm, and a material selection of Q235-B grade steel. According to the stress characteristics of the energy dissipation plate, the yield bearing capacity can be calculated according to Formula (1).

$$V = f_y t b^2 / 3h \tag{1}$$

Where f_y refers to the yield strength of the material; h represents the length of the damper; b denotes the height of the damper; t means the thickness of the damper.

Referring to the suggestion of Bao Shihua et al. (Bao 2005) for the effective calculation of the span of the coupling beam, the effective calculation length of the damper is taken $h + 2^*(b/2)$ according to the same principle, and the calculation formula of the initial stiffness k of the damper is obtained as follows:

$$k = 1/(\frac{h^3}{Etb^3} + \frac{\mu(h + 2 * \frac{b}{2})}{Gtb}) \tag{2}$$

In the formula, E is the elastic modulus of the material; μ denotes the shear section coefficient; it is recommended to take 1.2 (Kong 2015) for the rectangle; G signifies the shear modulus of the material.

The general finite element software ABAQUS is used to establish the finite element model of the above examples, and the specific settings are shown in Figure 3. All parts of the damper are modeled with C3D8R solid elements, and the material properties of the steel are given according to the material properties test data in Section 3.2, the elastic modulus of the material is $2 \times 10^5 MPa$ the Poisson's ratio is 0.3. Hardening model is kinematic hardening. The energy dissipation plates, connection parts and bolts use the hard contact. The boundary condition of the damper is that one side is completely fixed, and the point-surface rigid body coupling is used to constrain it at point 1, and the other side is rigidly connected to point 2. At point 2, only the axial constraint and the up and down translation constraint are released. The target control displacement is applied to point 2.

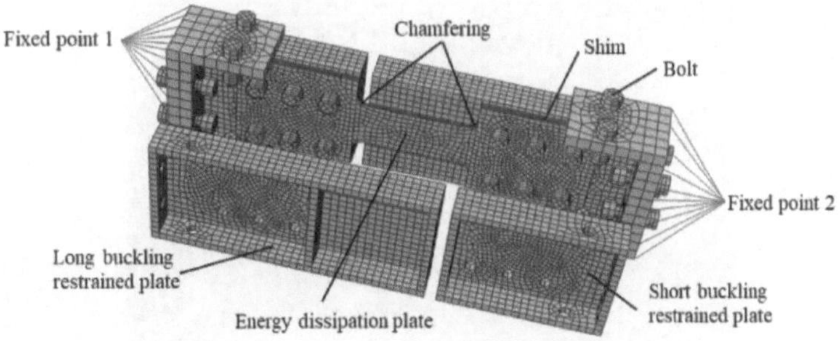

Figure 3. Design of dampers.

3.2 *Verification of finite element model*

In order to test the accuracy of the above numerical model, in this section, the numerical simulation analysis of the above example is carried out according to the loading system in Table 1, and the test body is made according to the example to carry out the pseudo-static reciprocating loading test, and the test and simulation results are compared.

According to GB 50017-2003 *Code for Design of Steel Structures* (2003), the energy dissipation plate test piece of the quasi-static test body is designed, and the material property test results are listed in Table 2. In order to avoid the influence of welding thermal effect, the remaining connecting members are milled out of 13.5 mm and 30 mm thick steel plates respectively.

A 60t horizontal electro-hydraulic servo actuator is used, one end is connected to the ground beam, the other end is connected to the loading beam and the top of the damper through a steel

Table 1. Loading protocol.

Test Number	Amplitude/mm	Cycle-index
1	0.5	1
2	1	1
3	1.5	1
4	2	1
5	2.5	1
⋮	⋮	⋮

Table 2. Material test results.

Material test piece	Q235-1	Q235-2	Q235-3	Mean
Yield strength (MPa)	361.7	399.4	377.6	379.6
Tensile strength (MPa)	502.8	528.8	522.6	518.0
Yield strength	1.39	1.32	1.38	1.36

conversion head, and the bottom end of the damper is connected to the ground beam. The steel conversion head connected to the actuator is connected to the ground beam through the slide rail, which can limit the out-of-plane displacement of the loading device and balance the self-weight of the loading device, as shown in Figure 4. The center of the actuator and the center of the damper are on the same level. The horizontal displacement of the damper was measured by a guyed displacement meter, and the bearing capacity of the damper was measured by a force sensor on the head of the actuator.

Figure 4. Diagram of the test setup.

Figure 5 is the comparison of the failure mode of experimental and numerical results. It can be seen that the damper presented the out-of-plane buckling deformation, which is similar to the deformation of finite element result. The stress contour of finite element is basically consistent with the experimental result.

The test restrained the out-of-plane rotation and displacement, which is consistent with the boundary condition of the finite element model. Figure 6-(a) is the hysteresis curve of the relationship between the top shear force and the loading displacement of the damper obtained from the test. Figure 6-(b) is a comparison diagram of the skeleton curves obtained by numerical simulation and test. It can be seen from the figure that the shapes of the skeleton curves are basically similar, and for the bearing strength, the numerical simulation results are slightly larger than the experimental

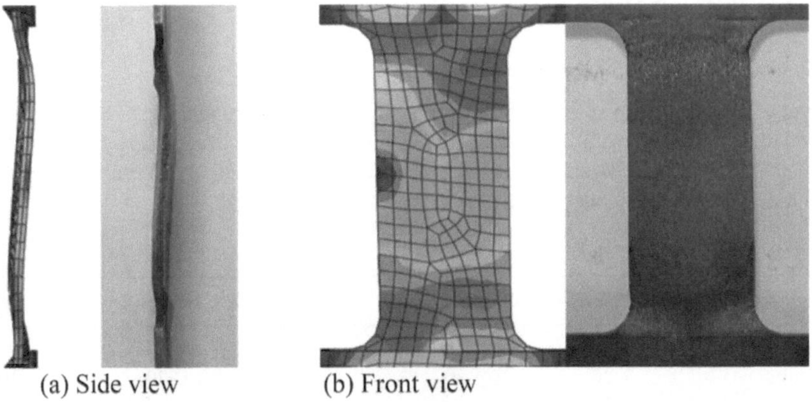

Figure 5. Failure mode of experimental and numerical results.

results, which may be due to the fact that the boundary constraints of the test are not as strict as those of the numerical simulation. The key mechanical property indicators of the damper obtained from the hysteresis curve are listed in Table 3. It can be seen that chamfering improves the yield capacity of the damper by about 25%. The errors between the test results of the initial stiffness of the dampers and the simulation results as well as the results calculated by the formulas do not exceed 3.5%. Therefore, the test results verify the accuracy of the finite element model and the stiffness calculation formula in the previous section, and the model will be used for parametric analysis in the following.

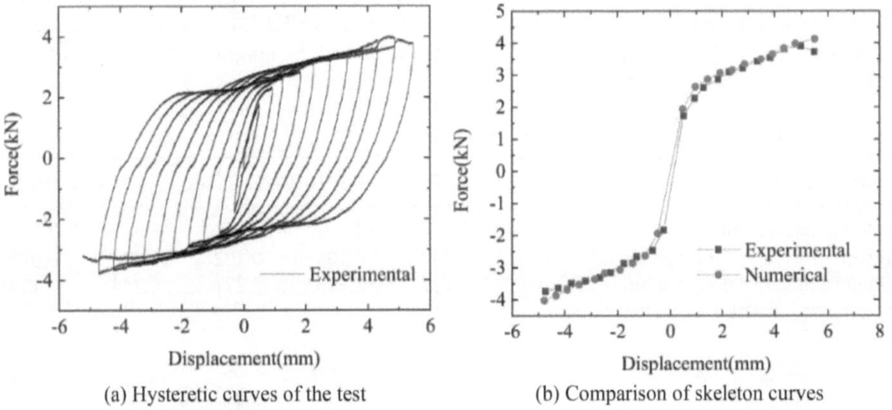

Figure 6. Hysteretic curves and skeleton curves.

Table 3. Comparison of mechanical parameters.

Calculation methods	Initial stiffness (kN/mm)	Yield bearing capacity (kN)	Yield displacement (mm)	Ultimate strength (kN)	Ultimate displacement (mm)	Ductility coefficient
Test results	4.326	1.619	0.374	4.017	5.470	14.6
Simulation results	4.395	1.653	0.376	4.162	5.485	14.6
Formula calculation	4.478	1.317	0.294	–	–	–

4 PARAMETRIC ANALYSIS

4.1 *Optimization analysis of the end shape of the energy dissipation plate*

In order to alleviate the stress concentration distribution phenomenon at the end of the energy dissipation plate, chamfers with five radii were designed for the end of the energy dissipation plate. The specific parameters are listed in Table 4. Numerical simulation analysis is carried out for each model in the same way as in Section 3.

Table 4. Design parameters of dampers.

Design parameters	RSPD-R3	RSPD-R4	RSPD-R5	RSPD-R6	RSPD-R7
Length(mm)	50	50	50	50	50
Height(mm)	20	20	20	20	20
Thickness(mm)	1.3	1.3	1.3	1.3	1.3
Chamfering radius (mm)	3	4	5	6	7

Figure 7 shows the hysteresis curves of the five models obtained by numerical simulation. The energy dissipation of the damper is calculated by the hysteresis curve and divided by the corresponding damper volume to obtain the energy dissipation per unit volume, which is listed in Table 5. The maximum value of cumulative plastic strain (PEEQ) at the final time is extracted from the results of time history analysis and summarized in Table 5. In order to visually show the changing trend of each parameter with the chamfering radius, the two key parameters in Table 5 are drawn as a scatter plot (Figure 8). It can be seen from Table 5 and Figure 8 that with the increase of the chamfering radius, the energy dissipation per unit volume gradually decreases, and it tends to be stable from specimens RSPD-R5 to the RSPD-R7. The maximum equivalent plastic strain value also decreases with the increase of the chamfering radius, with a large decrease from specimen RSPD-R3 to specimen RSPD-R4, and relatively stable from RSPD-R4 onward. By and large, the damper with an end chamfering radius of 5 mm (that is, 1/4 of the height of the energy dissipation plate) has better mechanical properties.

Table 5. Mechanical parameters of numerical simulation.

Key parameter	RSPD-R3	RSPD-R4	RSPD-R5	RSPD-R6	RSPD-R7
Chamfering radius (mm)	3	4	5	6	7
Energy dissipation per unit volume (kN mm/mm^3)	0.114	0.106	0.102	0.098	0.097
Maximum equivalent plastic strain (PEEQmax)	2.732	2.318	2.271	2.244	2.138

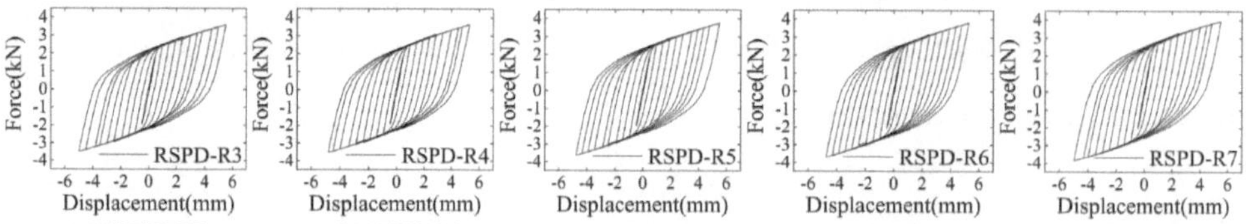

(a)RSPD-R3 (b) RSPD-R4 (c) RSPD-R5 (d) RSPD-R6 (e) RSPD-R7

Figure 7. Hysteretic curves.

4.2 *Mechanical performance analysis of dampers with different span-to-height ratios*

In order to more comprehensively analyze the mechanical properties of dampers of different sizes, the scheme with better optimization results above is adopted, i.e., the optimized energy dissipation plate scheme with 1/4 height chamfering of the energy dissipation plate. Four additional dampers

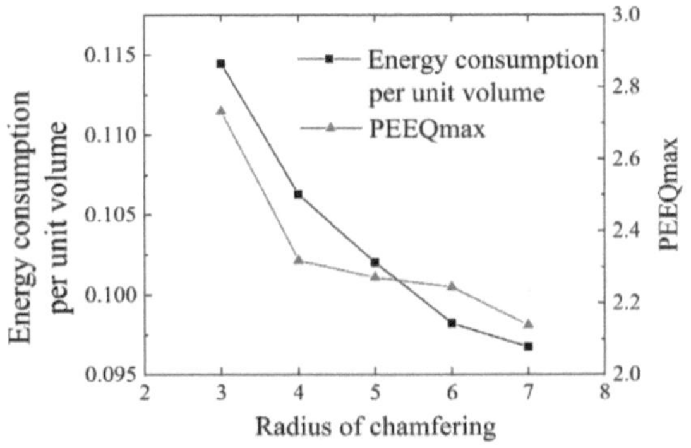

Figure 8. Mechanical parameters of numerical simulation.

were modeled for numerical simulation according to different span-height ratios (height to length ratio of the energy dissipation plate) of 2.63, 2.57, 2.44, and 2.38, respectively. The hysteresis curves are shown in Figure 9, and the results of the key mechanical parameters of the dampers obtained from the hysteresis curves are listed in Table 6. It can be seen from Table 6 that as the span-to-height ratio decreases from 2.63 to 2.38, the initial stiffness of the damper increases by 13.0%, the yield bearing capacity increases by 12.1%, the energy dissipation per unit volume also increases correspondingly by 22.8%, and the bearing capacity overstrength coefficient was basically stabilized between 2.501 and 2.538.

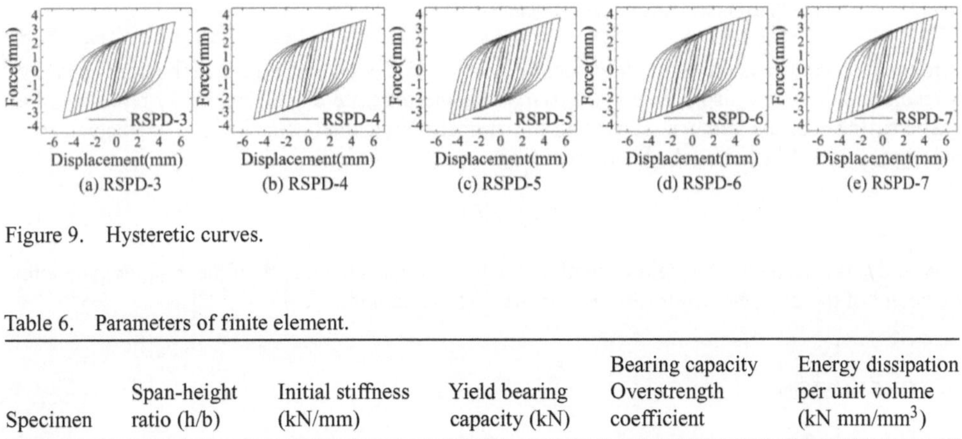

Figure 9. Hysteretic curves.

Table 6. Parameters of finite element.

Specimen	Span-height ratio (h/b)	Initial stiffness (kN/mm)	Yield bearing capacity (kN)	Bearing capacity Overstrength coefficient	Energy dissipation per unit volume (kN mm/mm^3)
RSPD-3	2.63	4.116	1.573	2.524	0.092
RSPD-4	2.57	4.254	1.606	2.537	0.099
RSPD-5	2.50	4.395	1.653	2.518	0.102
RSPD-6	2.44	4.527	1.694	2.538	0.109
RSPD-7	2.38	4.652	1.764	2.501	0.113

4.3 *The elastic-plastic constitutive model of the optimized metal buckling restrained shear panel damper*

In order to apply the damper to practical engineering, the constitutive model of the damper is simplified as a double-polyline, as shown in Figure 10. According to the results of finite element simulation in Section 4.2, the mean of the absolute value of the extreme value point of each load circle on the hysteresis curve of the five kinds of span-height ratio dampers is taken as the sample point, the yield point is taken as the starting point, and the minimum sum of the distance from each sample point to the second fold is sought as the optimal solution to obtain the equation of the intrinsic model of the dampers as follows.

$$F = 2.364V' \tag{3}$$

$$K = \begin{cases} k_d & (elastic) \\ 0.09467k_d & (plastic) \end{cases} \tag{4}$$

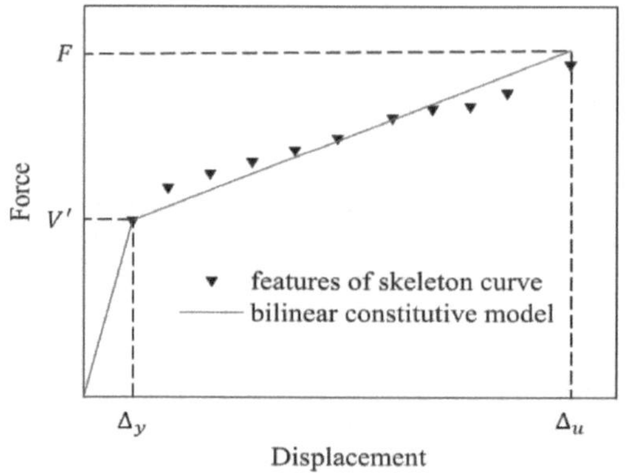

Figure 10. Double linear constitutive model of damper.

In the formula, V' is the actual yield bearing capacity; F means the ultimate bearing capacity; Δ_y indicates the yield displacement; Δ_u stands for the ultimate displacement; k_d represents the elastic stiffness of the damper. According to the analysis results in Section 3.2, the actual yield bearing capacity V' can be calculated according to the following formula:

$$V' = 1.25f_y tb^2/3h \tag{5}$$

Where f_y is referred to the yield strength of the material; h is the length of the damper; b denotes the height of the damper; t signifies the thickness of the damper.

5 CONCLUSION

In this paper, an optimized metal buckling restrained shear panel damper is proposed, the stiffness calculation formula is given, and the quasi-static test and numerical simulation study are carried out. In this regard, the conclusions are drawn as follows:

(1) The quasi-static test results of the damper reciprocating loading are basically consistent with the finite element simulation results, verifying that the numerical model in this paper is accurate. Furthermore, the test results are used to verify the accuracy of the given formula for calculating the stiffness of the dampers, and the formula for calculating the yielding bearing capacity is modified.
(2) The end shape of the energy dissipation plate of the damper is optimized by using chamfering to avoid premature cracking of the energy dissipation plate caused by stress concentration and to improve the energy dissipation capacity of the damper. The finite element simulation results show that when the chamfering radius of the end of the energy dissipation plate is 1/4 of the height of the energy dissipation plate, the mechanical properties of the damper are better.

(3) The numerical simulation results of five kinds of span-height ratio dampers are obtained through the finite element parameterization analysis, and the simplified elastic-plastic constitutive model of the optimized metal buckling restrained shear panel damper is obtained by further regressing the data, providing a scientific basis for the engineering application promotion of this kind of damper.
(4) The out-of-plane buckling of the energy dissipation plate is effectively limited by the buckling restrained plate. However, the effect of the set location of buckling restrained plates on the mechanical properties of the damper is uncertain. In the following research, this problem will continue to be accurately analyzed through experiments and finite element simulation.

ACKNOWLEDGMENTS

This work was financially supported by the: Scientific Research Fund of Institute of Engineering Mechanics, China Earthquake Administration (Grant No. 2021C04, 2019B03), the Natural Science Foundation of Heilongjiang Province of China (Grant No. LH2021E120), and the National Natural Science Foundation of China (Grant No. 52008381).

REFERENCES

Bao S, Zhang T. *Structural design and calculation of high-rise buildings* [M]. Tsinghua University Press, 2005.

Chang Hwan Lee, Seung Hee Lho, Do Hyun Kim, Jintak Oh, Young K. Ju. Hourglass-shaped strip damper subjected to monotonic and cyclic loadings. *Engineering Structures*, 2016, 122–134.

Chen Z, Ge H, Usami T, Yuan Y. Research on hysteretic performance parameters of shear panel damper [J]. *China Civil Engineering Journal*, 2008, {4} (11): 13–17.

Deng K, Pan P, Sun J, et al. Shape optimization design of steel shear panel dampers [J]. *Journal of Constructional Steel Research*, 2014, 99(8): 187–193.

GB 50017-2003 Code for design of steel structures[S]. Beijing: China Planning Press, 2003.

Ge H, Chen Z, Usami T. Hysteretic Model of Stiffened Shear Panel Dampers [J]. *Journal of Structural Engineering*, 2006, 132(3): 478–483.

JI X, Ma Q, WANG Y et al. Cyclic tests of replaceable shear links in steel coupling beams [J]. *Journal of Building Structures*, 2014,35(06): 1–11.

Koike Yohei, Yanaka, Tsutomu Usami, akihisa. et al. An experimental study on developing high-performance stiffened shear panel dampers [J]. *Journal of Structural Engineering*, 2008, 54: 372–381.

Kong Z, Wang T, Shi W. Experimental study on mechanical properties of steel dampers with slits [J]. *China Civil Engineering Journal*, 2015, 48(09): 11-22.DOI: 10.15951/j.tmgcxb.2015.09.002.

Liu Y, Aoki T, Shimoda M. Strain distribution measurement of a shear panel damper developed for bridge structure [J]. *J Struct*, 2013, 1–11.

Ma N, Su L. Experimental study on hysteretic behavior of shear buckling restrained steel panel damper [J]. *Journal of Building Structures*, 2018, 39 (11): 148-157

Okazaki T, Arce G, Ryu H C, Engelhardt M D. Experimental study of local buckling, overstrength and fracture of links in eccentrically braced frames [J]. *Journal of Structural Engineering*, ASCE, 2005, 131(10): 1526–1535.

Tsai, K. C. and Chen, H. W. et al. Design of Steel Triangular Plate Energy Absorbers for Seismic Resistant Construction, *Earthquake Spectra*, Vol. 9, No. 3, 1993.

Whittaker, A. S., Bertero. Seismic Testing of Steel Plate Energy Dissipation Devices, *Earthquake Spectra*, Vol. 7. No. 4, 1991.

Xing S, Guo X. Study on mechanical behavior and effectiveness of a new type of mild steel damper[J]. *Earthquake Engineering and Engineering Vibration*, 2003, 2(6): 179–186.

Zhang C, Zhang Z, Shi J. Development of high deformation capacity low yield strength steel shear panel damper [J]. *Journal of Constructional Steel Research*, 2012, 75(7): 116–130.

Advances in Civil Engineering: Structural Seismic Resistance, Monitoring and Detection – Mohd Johari Mohd Yusof and Junwen Zhang (Eds)
© 2023 The Authors, ISBN 978-1-032-31491-4

Experimental study on dynamic shear modulus and damping ratio of lignin-cement modified expansive soil

Yuguo Zhang & Tai Guo
School of Civil Engineering and Architecture, Zhongyuan University of Technology, Zhengzhou, Henan, China

Weijie Zhang
Central China Branch of China Railway Construction Group Co. LTD, Zhengzhou, Henan, China

Zhenghao Chen
School of Civil Engineering and Architecture, Zhongyuan University of Technology, Zhengzhou, Henan, China

ABSTRACT: In order to ameliorate the adverse effects of expansive soil on engineering, lignin and cement were used as modifiers to study the dynamic characteristics of modified expansive soil under different lignin content and moisture content based on resonance column test. The experimental results show that the dynamic shear modulus of modified expansive soil is larger than that of plain soil, and the damping ratio is smaller than that of plain soil. The dynamic shear modulus of modified soil reaches the maximum when 4% cement + 1% lignin is added. Under the same mixing content and confining pressure, the dynamic shear modulus and damping ratio of the improved soil decrease with the increase of moisture content.

1 INTRODUCTION

Expansive soil exists in many areas of China. Expansive soil contains strong hydrophilic clay mineral composition, which has strong swelling and contraction characteristics, multi-fissure and strength attenuation. Its particles are highly dispersed, and are very sensitive to the change of humidity and heat in the environment. When water expands, deformation increases rapidly, which is very unfavorable to engineering construction (Tan 2021).

Lignin, as a non-toxic and pollution-free industrial by-product, has been used in soil improvement for many years and has obtained abundant achievements in related fields. For example, lignin can improve the strength of soil, increase the anti-deformation ability, improve the anti-erosion characteristics. Liu Songyu summarized a large number of literature on lignin soil improvement and found that lignin has a good improvement effect on soil mechanical properties and anti-erosion properties (Liu 2014). Through experiments, Zhang Jianwei found that lignin effectively improved the structure of silt and enhanced the shear strength and unconfined compressive strength of silt (Zhang 2021; Zhang 2020).

Some scholars also studied the dynamic characteristics of expansive soil, Considering the factors such as confining pressure, vibration frequency, type and dosage of modifier, the variation law of expansive soil in the above factors is obtained. Zhuang Xinshan considered the influence of confining pressure and vibration frequency on dynamic strength, dynamic elastic modulus and damping ratio of expansive soil through dynamic triaxial test (Zhuang 2018). Zhou Xiaosheng considered the influence of confining pressure, moisture content and vibration times on the dynamic characteristics of expansive soil through dynamic triaxial test, and obtained that the dynamic elastic modulus of expansive soil increases with the increase of confining pressure, which is greatly

 DOI 10.1201/9781003310884-51

affected by moisture content, and tends to a certain value with the increase of vibration times (Zhou 2013). Li Xing studied the dynamic characteristics of cement-improved expansive soil and found that cement significantly enhanced the ability of expansive soil to resist dry-wet cycles (Li 2016). Luo Fei used RCA resonance column instrument to study the dynamic shear modulus and damping ratio of silty clay in Dujiangyan area under different confining pressures, and established an empirical formula for the change of relevant dynamic parameters with confining pressures (Luo 2014). Lv Xiaofei studied the dynamic elastic modulus and damping ratio of Hangzhou Bay silty clay by resonance column test, and analyzed their variation trend with the increase of strain and the attenuation law of dynamic modulus (Lv 2010).

At present, there are few researches on the application of lignin in the improvement of expansive soil. The composite of lignin and cement was used to improve the expansive soil in Nanyang. The dynamic characteristics of expansive soil under different lignin content, moisture content and confining pressure were studied by resonance column test.

2 TEST MATERIALS AND PRINCIPLES

2.1 *Test materials*

The test soil sample was taken from Shigang Town, Neixiang County, Nanyang City, Henan Province. The basic physical property indexes are shown in Table 1. Lignin was obtained from a material factory in Beishankou Town, Gongyi City, Henan Province. Its related parameters are shown in Table 2. Cement is ordinary Portland cement, taken from Tianrui Brand, Zhengzhou city, Henan Province, the main chemical composition content is shown in Table 3.

Table 1. Basic physical indexes of expansive soil.

Moisture content (%)	Dry density (g/cm^3)	Liquid limit (%)	Plastic limit (%)	optimum moisture content (%)	maximum dry density (%)	free expansion rate (%)
23.31	1.38	48.8	26.33	18.2	1.75	58

Table 2. Lignin related parameters.

Name of index	Index parameters
Substances not dissolved in water	<1.5%
Heat stability	>130
moisture content	<5%
Inorganic salt content	<5%
Water reducing rate	12

Table 3. Main chemical composition and content of cement.

Chemical component	CaO	SiO_2	Al_2O_3	MgO	SO_3	Fe_2O_3	Na_2O
Content (%)	62.69	23.62	5.09	4.6	3.1	0.76	0.14

2.2 *The experimental principles*

The test instrument is GDS-RCA resonance column tester. Resonant column test is by changing the voltage to change the vibration frequency, the electromagnetic drive for cylindrical specimens under different vibration frequency exerts a longitudinal or torsional vibration, thus can get under different voltage sample the resonant frequency when the resonance occurs, according to the resonance frequency and the sample size can be calculated shear wave velocity Vs.

$$V_s = 2\pi f l \beta^{-1} \tag{1}$$

$$I = I_0 \beta \tan \beta \tag{2}$$

$$8I = md^2 \tag{3}$$

f represents the resonance frequency (Hz) measured in the test, l represents the height of the sample (m), I represents the moment of inertia of the sample, and I_0 represents the moment of inertia of the resonant column drive system, which needs to be calibrated to obtain. $I_0 = 0.0037$ of the resonance column instrument in this test, m (kg) represents the mass of the sample, and d (m) represents the outer diameter of the sample.

The shear wave velocity V_s and bulk density ρ were used to obtain the corresponding dynamic shear modulus G.

$$G = \rho V_s^2 \tag{4}$$

The damping ratio can be directly measured by the instrument corresponding data, without calculation.

3 TEST RESULTS AND ANALYSIS

3.1 *Effects of lignin content on dynamic shear modulus and damping ratio of improved expansive soil*

The dynamic characteristics of improved expansive soil have similar trends under different confining pressures. In order to simplify the content of this paper, the dynamic shear modulus G and damping ratio D of improved soil with different moisture content under 50kPa confining pressures are listed and explained.

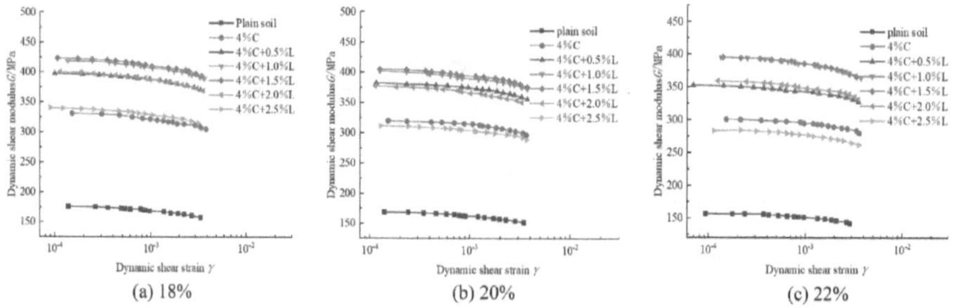

(a) 18% (b) 20% (c) 22%

Figure 1. Variation of dynamic shear modulus of improved soil under 50 kPa confining pressure.

Figure 1 shows that the dynamic shear modulus of expansive soil improved by cement and lignin is larger than that of plain soil, and the increase is larger. With the increase of lignin content, the dynamic shear modulus firstly increased and then decreased. The dynamic shear modulus of both plain soil and improved soil decreases with the increase of dynamic shear strain. The reasons are analyzed as follows. after 4% cement is added to expansive soil, gel materials are generated by hydration and gelation reaction of cement, and gel materials are connected with soil particles to form soil-cement skeleton, which improves the strength of soil, thus showing a significant increase in dynamic shear modulus. With the increase of lignin content, the appropriate amount of lignin can fill the skeleton pores of soil-cement, strengthen the connection between soil particles, and increase the bearing capacity and strength of soil. When the content of lignin is too high, the excess lignin not only increases the fine particles inside the soil, but also attracts more water molecules around the lignin particles, making the soil particles more prone to dislocation during vibration, which is not conducive to the improvement of soil strength.

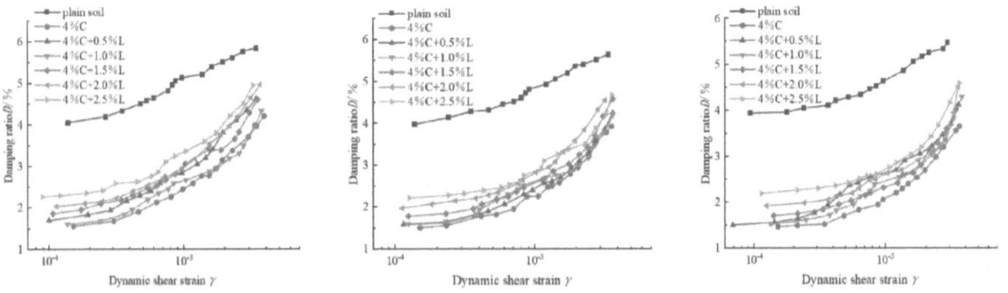

(a) 18% (b) 20% (c) 22%

Figure 2. Variation of damping ratio of improved soil under 50 kPa confining pressure.

It can be seen from Figure 2 that the damping ratios of improved soil under different moisture contents are all smaller than those of plain soil, and the decrease is slightly larger. This is because the soil-cement skeleton formed by the combination of cement and soil particles increases the strength of soil, but the internal pores become larger, reducing the energy consumed by stress waves in the propagation process, the damping ratio of cement-improved soil drops significantly. On the basis of 4% of cement quantity, with the increase of the lignin content, the damping ratio increased slightly, the reason is that the addition of lignin filled the porosity of the soil-cement skeleton, the connection of the soil particles closer to hinder the spread of stress wave in the soil, in the process of transmission of energy increases, the damping ratio increased with the increase of lignin content increased.

3.2 *Effects of moisture content on dynamic shear modulus and damping ratio of improved expansive soil*

The moisture content of this test is designed to be 18%, 20% and 22%. The dynamic shear modulus and damping ratio of the improved expansive soil have similar changes under different confining pressures. This section only lists the influence rules of moisture content on the dynamic characteristics of the improved expansive soil under different mixing amounts when the confining pressure is 50kPa.

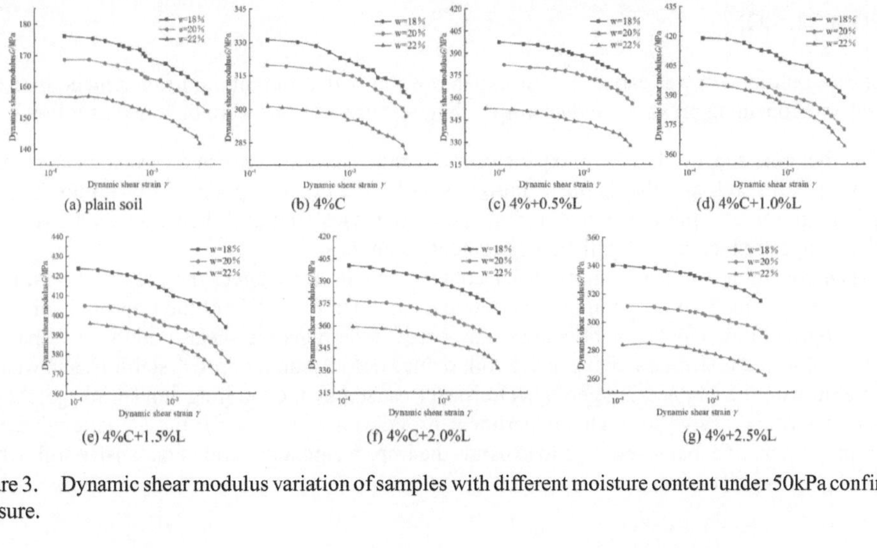

(a) plain soil (b) 4%C (c) 4%+0.5%L (d) 4%C+1.0%L

(e) 4%C+1.5%L (f) 4%C+2.0%L (g) 4%+2.5%L

Figure 3. Dynamic shear modulus variation of samples with different moisture content under 50kPa confining pressure.

Figure 3 shows that the dynamic shear modulus of the improved soil decreases with the increase of the dynamic shear strain, while the dynamic shear modulus decreases with the increase of moisture content. Because expansive soil contains more clay minerals, expansive soil is sensitive to water. With the increase of moisture content, more water molecules will be attracted around soil particles. The lubrication of water molecules makes the friction between soil particles smaller, and the strength of soil decreases.

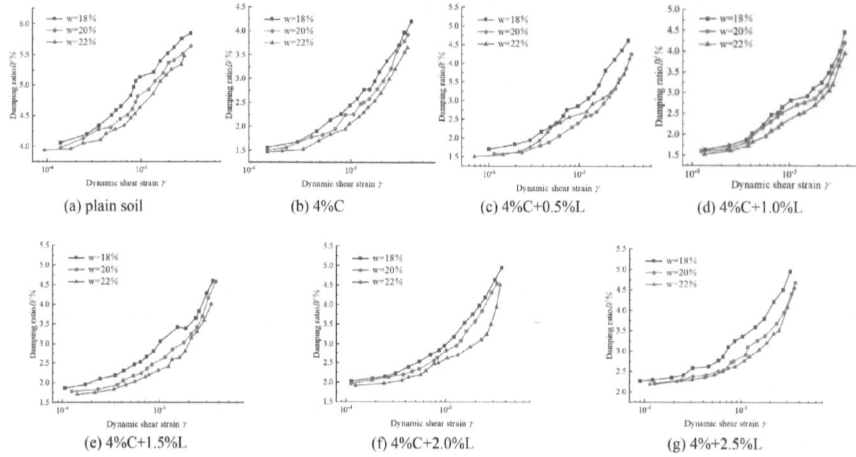

Figure 4. Variation of damping ratio of specimens with different moisture content under 50 kPa confining pressure.

Figure 4 shows that the damping ratio increases with the increase of dynamic shear strain, and the growth rate is gradually accelerated. The damping ratio decreases with the increase of moisture content. Because of the expansive soil containing hydrophilic mineral and water reducing effect of lignin, the expansive soil are more sensitive to the change of moisture content, with the increase of moisture content, soil particle will attract more water molecules around, the lubrication of water molecules contribute to the spread of stress wave in the soil, reduces the consumption of energy, the damping ratio.

4 CONCLUSION

Resonance column test was carried out on expansive soil with different modifier content, moisture content and confining pressure, and corresponding test conclusions were obtained as follows.

(1) The dynamic shear modulus of expansive soil modified with lignin and cement is larger than that of plain soil, and the damping ratio is smaller than that of plain soil. With the increase of lignin content, the dynamic shear modulus increases first and then decreases, while the damping ratio increases slightly at 4% cement content.
(2) When confining pressure and modifier content remain unchanged, the greater the moisture content, the smaller the dynamic shear modulus and damping ratio of the improved soil.
(3) Appropriate lignin and cement content can increase the dynamic shear modulus of expansive soil, enhance the stiffness of expansive soil, reduce deformation, improve stability, and weaken or eliminate the engineering geological disaster caused by the swelling and shrinkage characteristics of expansive soil. This experiment provides a reference for the actual engineering improvement of expansive soil, and expands the improvement method of expansive soil, which is of great significance.

(4) The test considers the changes of modifier content, confining pressure and modifier content on the dynamic characteristics of the improved expansive soil, but does not consider the influence of the curing age on the dynamic characteristics of the improved expansive soil. If the curing age is considered, the study on the dynamic characteristics of the improved expansive soil will be further improved and the understanding of the dynamic characteristics will be more comprehensive.

FOUNDATION ITEMS

National Natural Science Foundation of China(U1204511); Key scientific research project of henan Province (19A560027).

AUTHORS

About author: Yuguo Zhang (1973-), male, born in Tengzhou, Shandong province, professor, ph. D. Mainly engaged in geotechnical engineering and underground engineering teaching and research work.

Corresponding author: Tai Guo (1995-), male, born in Shangqiu, Henan, master candidate, mainly engaged in geotechnical engineering research.

REFERENCES

Liu, S.Y., Zhang, T., Cai, G.J., Li, H.J., Jie, D.B. (2014) Research progress of bioenergy by-product lignin in soil reinforcement. *China Journal of Highway and Transport*. 27(08).130–137.

Li, X., Cheng, Q.G., Zhang, J.C., Yao, Y. (2016) Experimental study on dynamic characteristics of expansive soil improved by high railway cement under dry-wet cycle. *Railway Engineering*, 6.99–103.

Luo, F., Zhu, Z.Y., Tian, Y.D., Cai, C., Mo, Q.C. (2014) Effect of confining pressure on dynamic parameters of silty clay by resonance column test. *Journal of Disaster Prevention and Mitigation Engineering*, 34(01).119–123.

Lv, X.F., Chen, P.X., Yang, Y.J., Chen, X.L., Li, D. (2010) Experimental study on resonance column of Marine silty clay. *World Earthquake Engineering*, 26(S1).1–5.

Tan, W., Liu G.K., Wang, Q.S. (2021) Study on the evolution law of the long - term performance of cement improved expansive soil. *Highway engineering*. 1–9.

Zhang, J.W., Wang, X.J., Li, B.B., Han, Y., Bian, H.L. (2021) Experimental study on EICP-lignin combined solidified silt. *Journal of Civil and Environmental Engineering*. 43(02).201–202.

Zhang, J.W., Kang, F.X., Bian, H.L., Yu, K. (2020) Experimental study on unconfined compressive strength of silt improved by lignin under freezing-thawing cycle. *Rock and Soil Mechanics*. 2020(S2).1–6.

Zhuang, X.S., Wang, J.X., Wang, K., Li, K., Hu, Z. (2018) Study on dynamic characteristics of expansive soil improved by weathered sand. *Rock and Soil Mechanics*. 39(S2).149–156.

Zhou, X.S., Kong, L.W., Guo, A.G. (2013) Experimental study on dynamic characteristics of expansive soil under bidirectional cyclic loading. *Journal of Highway and Transportation Research and Development*. 9(04).63–67.

Advances in Civil Engineering: Structural Seismic Resistance, Monitoring and Detection – Mohd Johari Mohd Yusof and Junwen Zhang (Eds)
© 2023 The Authors, ISBN 978-1-032-31491-4

Dynamic displacement measurement method of bridge structure based on photographic image

Peijun Liu & Qingxin Guo*
Inner Mongolia Road and Bridge Group Co., Ltd., Hohhot, China

ABSTRACT: During the construction and operation of the bridge, structural damage may be caused due to the influence of load and environmental factors, which may lead to safety problems. This paper proposes a new bridge displacement monitoring method based on photographic image. Firstly, the image feature module and marking module are introduced, and two different marking modules are proposed, namely one target circle and two target circles. In order to select the marking module with better stability and accuracy, experiments were carried out under different light intensities, that is, static monitoring was carried out at different time points. The stability is evaluated by the scale factor and the accuracy of the abscissa and ordinate of the target circle is evaluated. The test results show that the performance of the marking modules of the two target circles is better in terms of stability and accuracy. The new monitoring method studied in this paper can accurately monitor the displacement of bridge structure in real time. It has the advantages of high efficiency, low cost and convenient installation. It can play an ideal performance during bridge construction, operation and demolition.

1 INTRODUCTION

With the rapid development of social economy, the rapid improvement of the scientific and technological level of civil engineering and the gradual advancement of China's urbanization process, there are more and more long-span bridges, span bridges and urban viaducts to meet the increasing traffic demand caused by social progress, material transportation and population flow (Kromanis 2021).

During the construction of bridge structure, there are three influencing factors that promote the vertical displacement of bridge structure: concrete creep and shrinkage, prestressed loading and structural self weight. Creep and shrinkage of concrete is one of the basic characteristics of concrete. It has a great impact on bridge structure and lasts for a long time. Creep of concrete refers to the phenomenon that the deformation of concrete increases with time under long-term load; the shrinkage of concrete refers to the dry shrinkage phenomenon caused by the condensation of cement slurry and the drying of the environment (Kim 2020). The influence of concrete shrinkage and creep on bridge structure is mainly reflected in the construction stage. It is mainly divided into cantilever stage and cantilever closure stage, which is shown in Figure 1.

The creep of concrete makes the bridge produce vertical displacement and rotation angle. The creep makes the structure produce vertical deflection, and the rotation angle at this stage is 0. Creep increases the deformation of the compression zone and the initial eccentricity of the biased structure.

Leroy et al. monitored the real-time horizontal displacement of Normandy bridge and achieved the measurement accuracy of centimeter level (Khuc 2017). Kashima et al. realized the deformation

*Corresponding Author: liupeijun@126.com

 DOI 10.1201/9781003310884-52

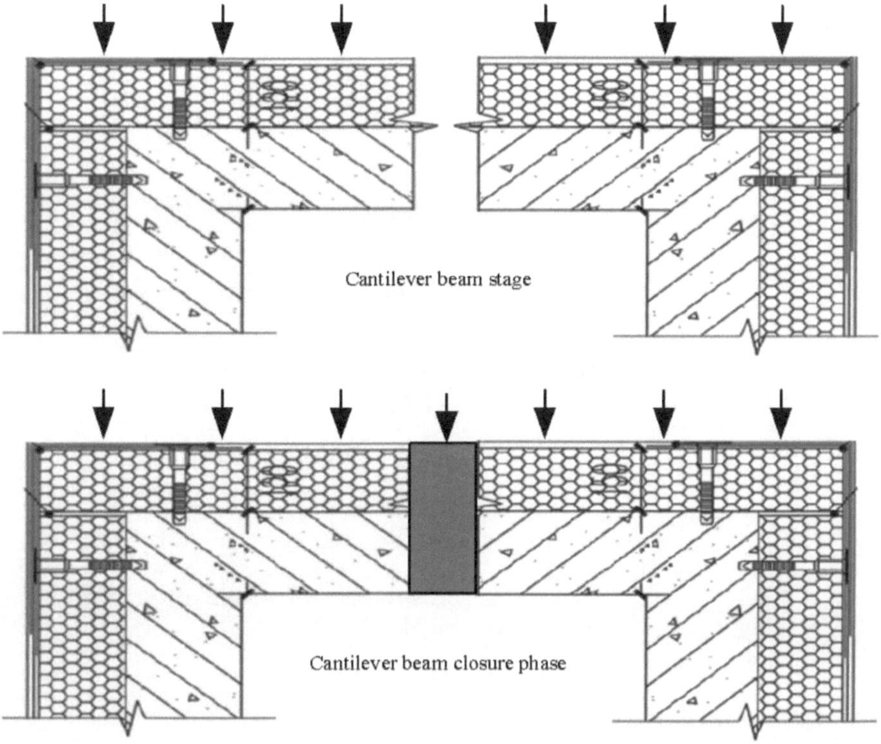

Figure 1. Deformation drawing of bridge construction stage.

monitoring of Akash kaikyo bridge under wind load and temperature load (Lee 2020). Nakamura monitored the deformation of a bridge with a span of about 720m and concluded that there was a strong correlation between the transverse displacement of the main beam and the wind load. The monitoring results are basically consistent with the numerical simulation results (Liu 2019). Qiao Yan et al. monitored the deformation of Runyang Bridge under traffic load, temperature load and wind load (Liu 2018). Guo Jing et al. used RTK technology to monitor the real-time displacement of Humen Bridge. A bridge monitoring system composed of GPS reference station, GPS monitoring station, optical fiber communication link and GPS monitoring center is established (Bai 2020).

Firstly, the resolution of this method is measured by stability test and displacement loading test, which proves that this method has high stability and resolution of displacement monitoring. Then, when the camera module is far away from the marking module, the stability test is carried out. Through the long-distance test, the influence on the displacement monitoring results and how to select the application scenario are analyzed.

2 TARGET RECOGNITION AND COORDINATE CALCULATION METHOD

The main purpose of target recognition is to obtain the coordinates of the center point of the target circle. After threshold binarization and binary "not" operation, the statistical function of pixel gray value is $f(x,y)$. The value of the function is shown in formula (1).

$$f(x,y)=\begin{cases} 0 & dst(x,y)=0 \\ 1 & ds(x,y)=255 \end{cases} \tag{1}$$

Figure 2 lists the statistical function values of the gray value of a pixel.

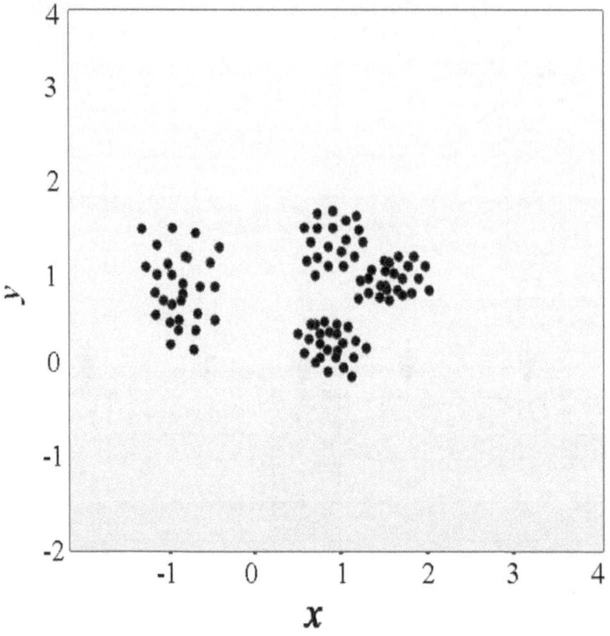

Figure 2. Statistical function value of pixel gray value.

The centroid method [87,88] is used to calculate the center coordinates of the target circle, and the formula is shown in (2).

$$(x_0, y_0) = \frac{\sum f(x_i y_i) x_i}{\sum f_i(x_i, y_i)}, \frac{\sum f(x_i, y_i) y_i}{\sum f(x_{i,j})} \tag{2}$$

In it, x_0 and y_0 are the abscissa and ordinate of the pixel of the centroid point respectively. x_i and y_i are the abscissa and ordinate of the *ith* pixel point respectively (Liu 2018).

3 CONVERSION METHOD BETWEEN PIXEL COORDINATES AND ACTUAL COORDINATES

In order to obtain the structural displacement from the captured image, the relationship between pixel coordinates and actual coordinates needs to be established. The system adopts a practical scale factor, that is, the scale factor (Bai 2020) is obtained based on the known size of the reference target and its corresponding pixel size, as shown in formula (3).

$$I = \frac{D_r}{D_p} \tag{3}$$

The pixel size of the unknown target is obtained from the image, and then the real size of the structure is obtained according to formula (4).

$$d_{ir} = I d_{ip} \tag{4}$$

In it, D_r and D_p are the real size and pixel size of the known target, and d_{ir} and d_{ip} are the real size and pixel size of the unknown target i.

4 MARKING MODULE TEST

The fundamental principle of displacement recognition method based on image features is to extract the displacement of image feature points according to image features. Therefore, the basis and key of this monitoring method is the appropriate image features. Different marking modules have different effects on the stability and accuracy of displacement monitoring (Heliani 2020).

4.1 *Test environment*

The experiment was carried out under different light intensities, i.e. 12 noon and 18 PM. Fix different marking modules on the precision displacement meter. Figure 3 shows two different marking modules.

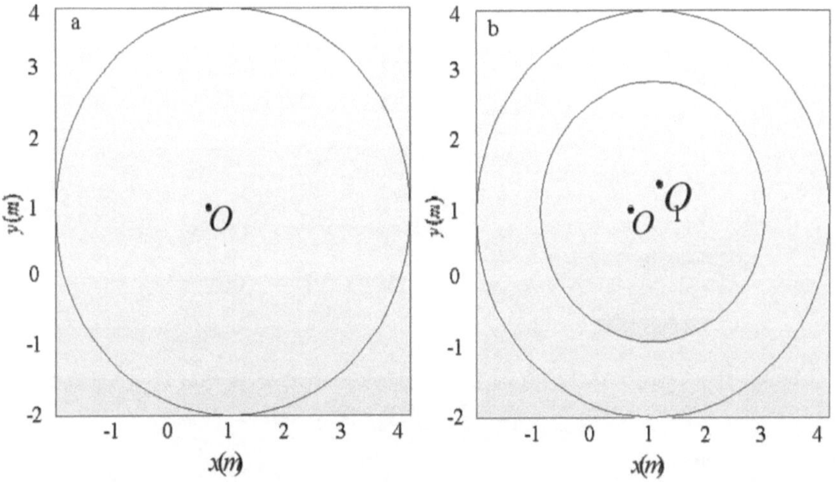

Figure 3. Marking modules for different target circles.

Similarly, on the background, the target in Figure 2 (a) is a solid circle with a radius of 15mm. The target in Figure 2 (b) is a solid circle with a radius of 10mm and a radius of 15mm, with a distance between the centers of the two circles of 50mm. In each test, the telephoto lens is used to place the marking module at the position of 10m of the camera module so that the marking module can be completely and clearly displayed in the field of view of the camera module. Then fix the positions of the two. Place the equipment at rest for 1 minute, and then conduct static monitoring for minutes (Alamdari 2019).

4.2 *Results and analysis*

When the light is strong, the light should be blocked appropriately in order to obtain the image with obvious target characteristics in the experiment. Through the proposed method for image recognition, the recognition results of the two marking modules at different times of the day are obtained.

The stability of the marking module is described by the stability of the scale factor. Determine the scale factor. A marking module of the target circle obtains the pixel radius of the target circle. And the scale factor is the ratio of the real radius to the pixel radius; the marking module of two target circles obtains the pixel distance between the center points of the two target circles, and the scale factor is the ratio of the real distance to the pixel distance. The comparison results of scale factors are shown in Figure 4.

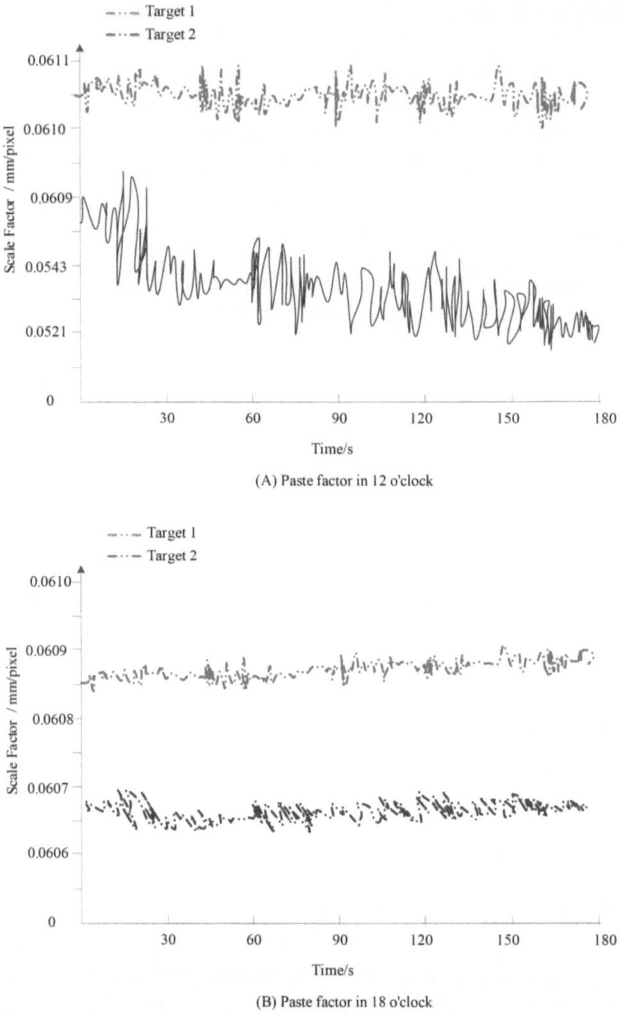

Figure 4. Scale factor of two marking modules at different times.

It can be seen from the results in Figure 4 that at the two time points, except that the number of target circles of the marking module is different, other conditions are under the same conditions.

At the same time, take one of the images of the two marking modules, and count the impact of different thresholds on the scale factor. Table 1 shows the results at 12 o'clock and Table 2 shows the results at 18 o'clock.

Table 1. Effect of threshold on scale factor of marking module at 12 o'clock.

threshold	A target circle (mm/pixel)	Two target circles (mm/pixel)
[110,120,120]	0.05439	0.06102
[120,120,120]	0.05528	0.06097
[130,120,120]	0.05601	0.06089

Table 2. Effect of threshold on scale factor of marking module at 18 o'clock.

threshold	A target circle (mm/pixel)	Two target circles (mm/pixel)
[110,120,120]	0.06087	0.06065
[120,120,120]	0.06096	0.06060
[130,120,120]	0.06103	0.06067

At 12 noon, the scale factors of two target circles are relatively stable, with a range of 0.000102mm/pixel, and the scale factor sub range of one circle is 0.00023mm/pixel; at 18:00 p.m., the range of the scale factor of one target circle is 0.0000673mm/pixel. The scale factor of the two target circles is also more stable, and the distance between the two center points is also more stable. Therefore, in the two marking modules, the stability and accuracy of the two target circles are better.

According to the analysis of data results, when the change of threshold is the same, the scale factor obtained by software will be compared. At 12 o'clock, when the threshold changes from [110120120] to [130120120], the scale factor of one target circle changes by 0.00162mm/pixel. The scale factor of two target circles changes by 0.00013mm/pixel. At 18 o'clock, the threshold changes from [110120120] to [130120120]. The scale factor of one target circle changes by 0.00016mm/pixel, and the scale factor of two target circles changes by 0.00002mm/pixel. Therefore, it can be concluded that the scale factor of the marking module of two target circles is less affected by the threshold. When the light is weak, the scale factor is less affected by the threshold. In addition, when the light intensity is strong, there is a large difference in the scale factor between one target circle and two target circles.

5 CONCLUSION

Aiming at the structural damage caused by load and environmental factors during the construction and operation of the bridge, which may lead to safety problems, this paper puts forward a new method for bridge displacement monitoring. The bridge displacement monitoring method based on machine vision selects the marking module with better stability and accuracy under different light intensities. Static monitoring tests are carried out on the two marking modules. The marking module is evaluated by the stability of scale factor and coordinate, and it is verified that the marking module with two target circles has better stability. This method overcomes the shortcomings of the traditional displacement monitoring system, such as difficult installation, high price and unable to monitor in real time.

This method only obtains good test results in the laboratory and model, which verifies that this method has good long-term monitoring stability, but it has not been used for long-term displacement monitoring in the real bridge structure. Therefore, in the future, it is necessary to install the system in the bridge structure to test the performance of this method.

REFERENCES

Alamdari M M, Ge L, Kildashti K, et al. (2019). Non-contact structural health monitoring of a cable-stayed bridge: case study[J]. *Structure and Infrastructure Engineering*, 15(8): 1119–1136.

Bai X, Yang M, Ajmera B. (2020). An Advanced Edge-Detection Method for Noncontact Structural Displacement Monitoring[J]. *Sensors*, 20(17): 1–16.

Heliani L S, Pratama C, Danardono, et al. (2020). Spatiotemporal variation of vertical displacement driven by seasonal hydrological water storage changes in Kalimantan, Indonesia from GPS observation[J]. *Geodesy and Geodynamics*, 11(5): 350–357.

Kromanis R, Kripakaran P. (2021). A multiple camera position approach for accurate displacement measurement using computer vision[J]. *Journal of Civil Structural Health Monitoring*, 1–18.

Kim J, Jeong Y, Lee H, et al. (2020). Marker-Based Structural Displacement Measurement Models with Camera Movement Error Correction Using Image Matching and Anomaly Detection[J]. *Sensors*, 20(19): 1–24.

Khuc T, Catbas F N. (2017). Completely contactless structural health monitoring of real-life structures using cameras and computer vision[J]. *Structural Control & Health Monitoring*, 24(1): 1–17.

Lee J, Lee K-C, Jeong S, et al. (2020). Long-term displacement measurement of full-scale bridges using camera ego-motion compensation[J]. *Mechanical Systems and Signal Processing*, 140: 1–18.

Liu P, Liu C, Zhang L, et al. (2019). Displacement monitoring method based on laser projection to eliminate the effect of rotation angle[J]. *Advances in Structural Engineering*, 22(15): 3319–3327.

Liu P, Xie S, Zhou G, et al. (2018). Horizontal displacement monitoring method of deep foundation pit based on laser image recognition technology[J]. *Review of Scientific Instruments*, 89(12): 1–10.

Xu Y, Brownjohn J M W, Hester D, et al. (2017). Long-span bridges: Enhanced data fusion of GPS displacement and deck accelerations[J]. *Engineering Structures*, 147: 639–651.

Advances in Civil Engineering: Structural Seismic Resistance, Monitoring and Detection – Mohd Johari Mohd Yusof and Junwen Zhang (Eds)
© 2023 The Authors, ISBN 978-1-032-31491-4

Experimental study on compressive strength of recycled aggregate concrete with artificial sand

Li Chen*
School of Urban Construction, Hubei Business College, Wuhan, China

Jinhua Xu*
Department of Urban Construction, Wenhua College, Wuhan, China

Yue Zhang* & Yafen Song*
School of Urban Construction, Hubei Business College, Wuhan, China

ABSTRACT: In order to study the influence of different artificial sand replacement rates on the compressive strength of RACAS (recycled aggregate concrete with artificial sand), 36 standard test blocks were designed in the experiment. Two strength grades of concrete (C35 and C45) and different replacement rates of artificial sand (0%, 30%, 50%, 70%, and 100%) were considered. The results show that the cubic compressive strength of RACAS is higher than that of natural concrete with the same strength grade, and the increase range is between 7% and 20%. Replacement rates of artificial sand had an effect on the compressive strength. When the replacement rate of artificial sand is about 70%, its cubic compressive strength is lower than those of other replacement rates. Like natural concrete, the compressive strength of RACAS decreases with the increase of water cement ratio.

1 GENERAL INSTRUCTIONS

The research on recycled aggregate concrete in developed countries is mature, and the recycling utilization rate of waste concrete can reach 90%. However, due to the late start of research on recycled aggregate concrete in China and the lack of relevant technical specifications and standards, the recycling utilization rate of waste concrete is only 5% (Wang 2021; Zhao 2021). Considering that recycled concrete has good environmental and economic benefits (Wang 2020), In recent years, domestic researchers pay more attention to the performance research of recycled aggregate concrete. For example, Jina Wang through experimental comparative research shows that under the condition of CO_2 curing, the compressive strength of modified recycled aggregate concrete is higher than that of recycled aggregate concrete, but lower than that of ordinary concrete (Wang 2016). Hongxia Qiao through experimental comparison found that when the additive amount of recycled coarse aggregate is 15%, the variation of concrete splitting tensile strength and compressive strength will be relatively stable, and the compressive strength of concrete is relatively higher than that of other cases (Qiao 2018). Zhanfeng Wang's test results show that with the increase of the replacement rate of recycled coarse aggregate, the cube compressive strength and prism compressive strength of recycled aggregate concrete gradually increase, the splitting tensile strength decreases to a certain extent, and the modulus gradually decreases (Wang 2012). Because of the compressive strength is very important for the application of concrete, this paper prepared RACAS, and studied

*Corresponding Authors: 251076308@qq.com, 165688344@qq.com, 24400959@qq.com and 791294427@qq.com

DOI 10.1201/9781003310884-53

the influence of different artificial sand replacement rates on the cube compressive strength of recycled aggregate concrete.

2 MATERIAL COMPOSITION AND TEST METHOD

2.1 *RACAS definition and material composition*

RACAS as a new type of concrete, its fine aggregate is made from crushed limestone waste, and the coarse aggregate is obtained by crushing the abandoned pavement concrete of the expressway. In the test, the influences of two concrete strength grades (C35 and C45) and five replacement rates of artificial sand (0%, 30%, 50%, 70%, and 100%) were considered. And the selection of concrete materials is as follows: 1) Cement selects conch P O 42.5 ordinary Portland cement. 2) As shown in Figure 1 (a), the fine aggregate respectively adopts the artificial sand (RS) produced by Wuming and Yong River sand (NS). The fineness modulus of artificial sand is 2.9 and that of river sand is 2.3, both of which are medium sand. 3) Coarse aggregate, as shown in Figure 1(b), includes natural coarse aggregate (NA) and recycled coarse aggregate (RA). The particle size of coarse aggregate is 5~20mm. Natural coarse aggregate is crushed stone aggregate produced by a quarry in Nanning, and the recycled coarse aggregate is made from the waste concrete of an expressway pavement in Guangxi after crushing, manual screening, cleaning and drying.

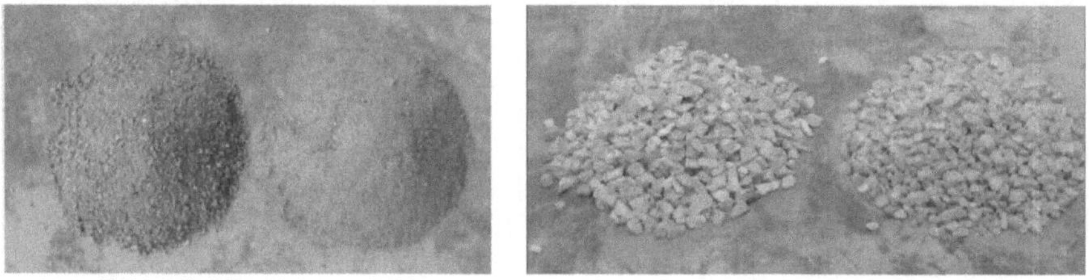

(a) Artificial sand (left) and river sand (right)

(b) Natural coarse aggregate (left) and recycled coarse aggregate (right)

Figure 1. Comparison photo of artificial aggregate and natural aggregate.

2.2 *Mix proportion design of artificial sand reclaimed concrete and natural concrete*

The mix proportion design of specimen concrete is adjusted through trial mixing on the basis of the preliminary research results of the research group (Zhu 2018). The concrete strength grades are C35 and C45. Five replacement rates of artificial sand are designed for each strength grade specimen and compared with natural concrete, and three cube specimens are designed for each group of concrete. Because the particle size of coarse aggregate in this test is smaller than that in the previous test. If the concrete is mixed according to the preliminary test results, the mortar wrapping condition is poor. Therefore, the sand ratio in this test is increased by 1% ~ 3% and the water cement ratio is appropriately reduced. The mix proportion technical parameters of RACAS are shown in Table 1 below, and table 2 for technical parameters of natural concrete mix proportion.

Table 1. Mix proportion of recycled aggregate concrete with artificial sand.

Number	Strength grade	Artificial sand replacement rate (%)	Water cement ratio	Unit water consumption (*kg*)	Sand rate (%)	cement content (*kg*)	Recycled coarse aggregate (*kg*)	Artificial sand (*kg*)	River sand (*kg*)	Additional water (*kg*)
C3500	C35	0	0.4	199.5	44	498.8	1010.8	0	794.2	27.3
C3530	C35	30	0.4	199.5	44	498.8	1010.8	238.3	555.9	27.3
C3550	C35	50	0.4	199.5	44	498.8	1010.8	397.1	397.1	27.3

(*continued*)

Table 1. Continued.

Number	Strength grade	Artificial sand replacement rate (%)	Water cement ratio	Unit water consumption (*kg*)	Sand rate (%)	cement content (*kg*)	Recycled coarse aggregate (*kg*)	Artificial sand (*kg*)	River sand (*kg*)	Additional water (*kg*)
C3570	C35	70	0.4	199.5	44	498.8	1010.8	555.9	238.3	27.3
C35100	C35	100	0.4	199.5	44	498.8	1010.8	794.2	0	27.3
C4500	C45	0	0.35	199.5	40	570	1022.4	0	681.6	27.6
C4530	C45	30	0.35	199.5	40	570	1022.4	204.5	477.1	27.6
C4550	C45	50	0.35	199.5	40	570	1022.4	340.8	340.8	27.6
C4570	C45	70	0.35	199.5	40	570	1022.4	477.1	204.5	27.6
C45100	C45	100	0.35	199.5	40	570	1022.4	681.6	0	27.6

The amount of materials in the table is the amount of materials required to prepare a cubic of concrete. Among them, the specimen number is represented by letter C plus two groups of numbers. The number C represents concrete, the two numbers after C represent strength grade, and the third and fourth numbers after C represent artificial sand replacement rate. For example, C3530 indicates that the design strength grade of RACAS is C35, and the replacement rate of artificial sand is 30%.

Table 2. Mix proportion of natural concrete.

Number	Strength grade	Water cement ratio	Unit water consumption (*kg*)	Sand rate (%)	cement content (*kg*)	Natural stone (*kg*)	River sand (*kg*)	Additional water (*kg*)
C35	C35	0.4	194.25	40	485.6	1092	728	12
C45	C45	0.35	194.25	38	555	1067.6	654.4	11.7

2.3 *Mechanical deformation and failure process of cube specimen*

The test was carried out on the RMT-201 rock and concrete loading system of the Structural Laboratory of Guangxi University. The cubic standard compression test function was selected to load in a bit-controlled manner (in order to obtain the descending section of the curve), and the continuous loading was carried out until the concrete test block is damaged. Figure 2 shows the crushing fracture plane of each cubic specimen.

Figure 2. Fracture characteristics of the cube specimen.

The Figure 2 shows that the final failure mode of RACAS is close to that of natural concrete, The cracking surfaces of all cube specimens are not only split from the joint surface of old and new mortar, but also directly split from the interior of coarse aggregate particles. It shows that recycled aggregate has sufficient strength as natural aggregate. However, in the process of damage, the damage sound of RACAS is clearer and more sudden than that of natural concrete.

It can be seen from the change of physical characteristics of each specimen during the loading process. In the process of continuous loading, artificial sand reclaimed concrete cube test block, like natural concrete, will undergo three stress stages of elasticity, elastic-plastic and final failure with the increase of load F. In elastic stage, the load and deformation develop linearly, and no cracks appear on the concrete surface. In the elastic-plastic stage, the load and deformation curves show bending characteristics. When the load F is close to the ultimate compressive capacity F_{cu} of the cube, the vertical cracks appear on the surface of the specimen, located in the center of the test block. With the increase of load, there is a slight cracking sound of concrete inside the specimen. Once there is a large crack, the bearing capacity of the concrete specimen decreases rapidly, and the failure process is rapid and sudden.

3 RESULTS AND ANALYSIS OF CUBE BLOCK TEST

The test results of each group of natural concrete and RACAS are shown in Table 3.

Table 3. Test results of compressive strength of concrete cube.

Number	Compressive strength of of specimen (MPa) I	II	III	Concrete Average value f_{cu}(MPa)	preparation strength $f_{cu,o}$(MPa)	Ratio of concrete strength	$f_{cu}/f_{cu,k}$	$f_{cu}/f_{cu,o}$
C35	42.4	45.7	46.8	45	43.2	1	1.28	1.04
C3500	50.44	48.48	57.57	52.2	43.2	1.16	1.49	1.21
C3530	55.89	55.31	50.97	54.1	43.2	1.20	1.54	1.25
C3550	56.17	50.05	50.66	52.3	43.2	1.16	1.49	1.21
C3570	47.70	55.31	52.75	51.9	43.2	1.15	1.48	1.20
C35100	50.99	49.70	59.33	53.3	43.2	1.19	1.52	1.23
C45	51.5	54.1	46.9	50.8	53.2	1	1.13	0.96
C4500	60.6	53.2	56.9	56.9	53.2	1.12	1.26	1.07
C4530	53.4	61.8	58.7	57.9	53.2	1.14	1.28	1.09
C4550	57.9	58.9	52.3	56.4	53.2	1.11	1.25	1.06
C4570	57.2	50.2	56.2	54.5	53.2	1.07	1.21	1.02
C45100	62.9	52.8	58.6	58.1	53.2	1.14	1.29	1.09

Table 3 shows that: 1) The average compressive strength of RACAS is higher than that of natural concrete. For C35 strength grade, the strength increase rate is 15%–20%, and for C45 strength grade, the strength increase rate is 7%–14%. 2) The cube compressive strength (f_{cu}) of the concrete specimens with the design strength grade of C35 and C45 is higher than the design strength ($f_{cu,k}$). 3) ($f_{cu}/f_{cu,k}$) measured in C45 is less than the strength ratio measured in C35, indicating that the increment of cube compressive strength relative to design strength decreases with the decrease of water cement ratio. 4) Except for natural concrete in C45 strength grade, the cube compressive strength (f_{cu}) measured is lower than the prepared strength ($f_{cu,\ o}$), and the cube compressive strength (f_{cu}) prepared by all other specimens is greater than the configured strength ($f_{cu,o}$). The experimental results meet the requirements

3.1 *Effect of replacement ratio of artificial sand on compressive strength of recycled aggregate concrete cube*

Figure 3 shows the relationship between compressive strength (f_{cu}) of RACAS and the replacement rates (ρ) of artificial sand

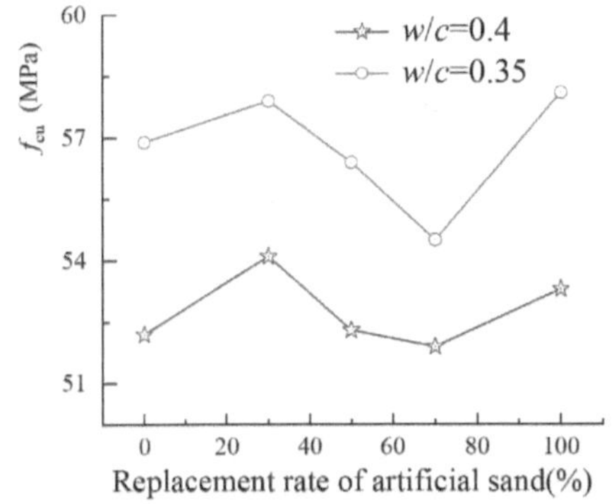

Figure 3. Relationship between f_{cu} and ρ.

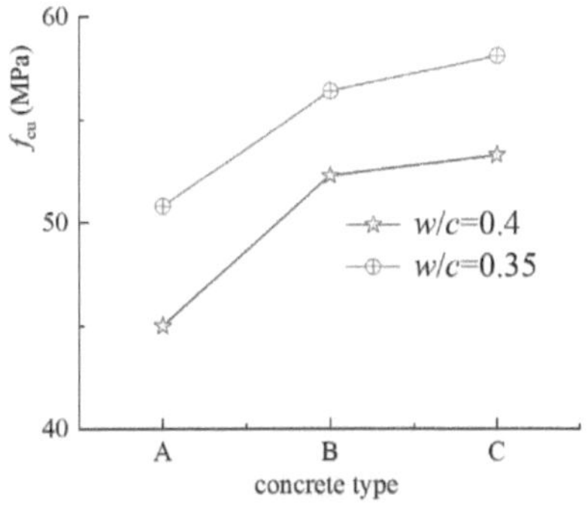

Figure 4. The f_{cu} under different concrete types.

As can be seen from Figure 3: 1) The replacement rate of artificial sand has a great influence on the strength of RACAS. When the replacement ratio of artificial sand is less than 30%, the strength of RACAS increases with the increase of replacement ratio of artificial sand. When the replacement ratio of artificial sand is between 30% and 70%, the strength of RACAS decreases with the increase of replacement ratio of artificial sand. When the replacement ratio of artificial sand is between 70% and 100%, the strength of RACAS increases with the increase of replacement ratio of artificial sand. 2) The strength of artificial sand is the lowest when the replacement rate is around 70%. The main reasons for this phenomenon are as follows: when the replacement rate of artificial sand is different, the grading of fine aggregate will change. When the replacement rate of artificial sand is 70%, the gradation of fine aggregate is relatively poor, which affects the compactness inside the prepared concrete. It can be seen that when the replacement rate of artificial sand is in a certain range, the strength of RACAS is low.

3.2 *Influence of different aggregate types and compositions on cube compressive strength of concrete*

Different aggregate composition has great influence on the strength of concrete. This test includes three types of concrete: Class A – natural concrete with natural aggregate; Class B – recycled aggregate concrete with river sand as fine aggregate and reclaimed aggregate as coarse aggregate; Class C – RACAS with fine aggregate as artificial sand and coarse aggregate as recycled aggregate. The influence of these three aggregate compositions on the compressive strength of concrete is shown in Figure 4:

Through the analysis of Figure 4 and Table 3, when w/c=0.4, the strength ratio between C35, C3500 and C35100 is 1:1.16:1.19; When w/c=0.35, the strength ratio between C45, C4500 and C45100 is 1:1.12:1.14. It can be seen that the concrete with strength grades of C35 and C45 made of recycled coarse aggregate is higher than that of natural concrete, which meets the test requirements. When the replacement rate of artificial sand is 100%, the strength grade of RACAS is the highest, which further demonstrates the feasibility of using artificial sand to make concrete.

3.3 *Effect of water cement ratio on cube compressive strength of concrete*

The water cement ratio has a decisive influence on the compressive strength of RACAS. When other conditions are the same, the strength of concrete increases with the decrease of water cement ratio. The compressive strength of RACAS in this paper also meets this law. Figure 5 shows the variation law of concrete compressive strength of each specimen with water cement ratio.

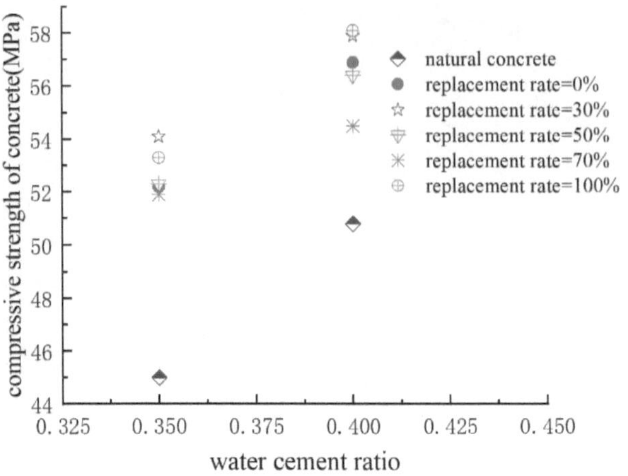

Figure 5. Relationship between compressive strength of concrete and water cement ratio.

It can be seen from Figure 5 that the compressive strength of RACAS decreases with the increase of water cement ratio, just like that of natural concrete, but the reduction range is less than that of natural concrete. It shows that the effect of changing the same water cement ratio on the strength of RACAS with higher strength grade is weaker than that of natural concrete.

4 CONCLUSION

Cube compressive performance is one of the basic mechanical properties of concrete materials. By studying the effects of different artificial sand replacement rates and different concrete water cement ratio on the compressive strength of concrete, this paper draws the following conclusions:

1) Under the aggregate conditions in this paper, when the mix proportion is the same, the cube compressive strength of RACAS is higher than that of natural concrete under the same conditions. For C35 strength grade concrete, the strength is higher by 15% ~ 20%. For C45 strength grade concrete, the strength is higher by 7% ~ 14%. However, the failure of RACAS is more sudden and shows stronger brittleness.
2) The cubic compressive strength of specimens changes with the replacement rate of artificial sand. When the replacement rate is less than 30% or close to 100%, the strength is higher. And when the replacement rate is about 70%, the strength is the lowest, only 0.958 ~ 0.994 times when the replacement rate is 0%. and the compressive strength measured When the replacement rate is 100%, the measured compressive strength is basically the largest, about 1.02 times that when the replacement rate is 0%.
3) Like natural concrete, the compressive strength of recycled aggregate concrete decreases with the increase of water cement ratio. When the water cement ratio decreases, the strength increment of RACAS is less than that of natural concrete, indicating that the sensitivity to the change of low water cement ratio is lower than that of natural concrete.

This experiment only studies the compressive strength of RACAS with design strength of C35 and C45, and the research on other strength grades needs to be further carried out.

REFERENCES

Qiao Hongxia, Guan Lijuan, Cao Hui, et al (2018). Mechanical properties of recycled aggregate concrete with different proportions. *Journal of Lanzhou University of technology*. 44(03), 142–145.

Wang Shuxin, Xiong Hui, Xie Xinghua (2021). Effect of recycled aggregate substitution on mechanical properties of concrete, *China harbour construction*. 41(04), 24–28.

Wang Xueqing, Shen Zhengyan, Pan Guodong, et al. (2020). Experimental study on performance of artificial sand concrete, *Mining and Metallurgy Engineering*. 40(02), 28–32.

Wang Jina, Xu Kaidong, Ma Xianwei, et al. (2016). Effect of CO2 curing on mechanical properties of recycled aggregate concrete, *Concrete*. 2016 (12), 12–14.

Wang Zhanfeng, Wang Shiliang, Weng Guangyuan. (2012). Experimental study on mechanical properties of recycled aggregate concrete with different coarse aggregate substitution rate, *Journal of Zhengzhou University (Engineering Edition)*. 33 (4), 32–35.

Zhao Hua, Gao Yikang, Tian Qian, et al. (2021). Research and development status and review of recycled aggregate concrete, *Qinghai communications technology*. 33(03), 1–11+38.

Zhu Helong. (2018). *Experimental study on mechanical properties of recycled aggregate concrete with stone powder artificial sand*. Nanning: Guangxi University.

Advances in Civil Engineering: Structural Seismic Resistance, Monitoring and Detection – Mohd Johari Mohd Yusof and Junwen Zhang (Eds)
© 2023 The Authors, ISBN 978-1-032-31491-4

Damage detection of space truss structure based on wavelet analysis

Yue Zhang, Li Chen*, Xiong Yin*, Yafen Song* & Guihong Pei*
School of Urban Construction, Hubei Business College, Hubei, China

ABSTRACT: The space truss structure is composed of elements and nodes. The node damage information cannot reflect the accurate element damage location, so it is necessary to use the element damage information to better determine the damage location. To solve this problem, wavelet transform and mode analysis methods are used to detect the damage of space truss structure. Based on curvature mode difference of structure as a space truss structure damage judge index of recognition, combined with the wavelet transform, single and more damage locations of space truss structure are determined, and the preliminary damage degree is determined. The results show that the method can not only be used to effectively determine the single and more damage locations but also can be used to preliminary determine the local damage degree of the structure. It has a certain guiding significance for practical engineering.

1 GENERAL INSTRUCTIONS

With wide applications of the space structure, some engineering accidents like the collapse of space structure also took place (Zhang 2013). To accurately find the hidden danger of the structure in time and reduce the cost of maintenance, it is very important to detect the structure. Many methods of space structure damage detection have been studied. For example, neural network (Zhang 2013), dynamic analysis (Liu 2014), and wavelet transform (Chen 2016; Ghasemi 2018; Liu 2015; Shahsavari 2017; Yu 2020). Damage shows changes in local characteristics of the signal, while wavelet analysis can reflect local damage traits of the signal in time domain and frequency domain. Therefore, wavelet transform is a recent area of research in structural health monitoring (Cui 2018; Li 2007).

The element curvature mode difference is sensitive to structural damage (Zhang 2010), the low mode can be obtained to determine damage information. Therefore, the methods of wavelet transform and mode analysis are used to detect the damage in space truss structure, depicted in this paper. With the curvature mode difference as a space truss structure damage judge index, combined with the wavelet transform, single and more damage locations of space truss structure can be defined, and the local damage degree of the structure can be preliminary determined. The results show that the method can be used to determine the element location of the damage in the truss structure.

2 THEORY OF DAMAGE IDENTIFICATION

2.1 *Element strain mode*

Spatial pole elements mainly load axial force and any element strain gain from the node displacement. In addition, it assumes that element strains are small. The strain of the k element can be shown:

$$\varepsilon_k = \left[(u_j - u_i)(x_j - x_i) + (v_j - v_i)(y_j - y_i) + (w_j - w_i)(z_j - z_i)\right]/L^2 \tag{1}$$

*Corresponding Authors: 251076308@qq.com, 329570186@qq.com, yfsong@hbc.edu.cn and 570769297@qq.com

 DOI 10.1201/9781003310884-54

Where x_i, y_i, z_i and x_j, y_j, z_j are coordinates of node i, j; u_i, v_i, w_i, and u_j, v_j, w_j are displacements of node i, j, by Eq. (1), element strain mode will be obtained by nodal displacement mode.

2.2 *Element curvature mode*

In practical application, strain mode and curvature mode can be approximated by displacement mode and second order central difference method, and the expression is:

$$\frac{\Phi_{r(k-1)} - 2\Phi_{rk} + \Phi_{r(k+1)}}{\Delta^2} = \frac{M_m}{E_m I_m} \tag{2}$$

Where f_m is the bending deflection of the beam at section m, f_{m-1}, f_{m+1} is the bending deflection at the measuring points on two adjacent beams whose distance from section m is Δ.

The bending deformation of the beam corresponds to the strain ε, and the strain can be expressed as:

$$\varepsilon = -\frac{h}{\rho} = -h\frac{\Phi_{r(k-1)} - 2\Phi_{rk} + \Phi_{r(k+1)}}{\Delta^2} \tag{3}$$

Where h is the distance between the measuring point on the beam and the neutral layer.

According to the relationship between curvature mode and strain mode shown in Eq. (3), the curvature mode of the pole element can be obtained by substituting the strain mode of rod element calculated in Eq. (2) into Eq. (3).

2.3 *Curvature mode difference*

The absolute value of curvature mode difference is sensitive to structural damage and can be used to identify structural damage. Therefore, the variation of curvature mode before and after the damage is taken as damage index in this paper.

The definition expression of damage identification index is:

$$\Delta\varphi_m(k) = \left|\varphi_m^u(k) - \varphi_m^d(k)\right| (k = 1, 2, \cdots N) \tag{4}$$

Where $\varphi_m^u(k)$, $\varphi_m^d(k)$ are the curvature mode value of the first mode before and after structural damage, respectively.

3 THE THEORY OF THE WAVELET ANALYSIS

Discrete wavelet transform $\psi_{(a,b)}(x)$ can be given by:

$$\psi_{(l,m)}(x) = a_0^{-1/2}\psi(a^{-1} - mb_0) \tag{5}$$

Where a is the scale factor, b is the translation factor, l is the integer dilation index, and m is the integer position index. Selecting $a_0 = 2$, $b_0 = 1$, the wavelets can be shown as:

$$\psi_{(l,m)}(x) = 2^{-1/2}\psi(2^{-l}x - m) \tag{6}$$

Discrete wavelet coefficients transform is expressed as:

$$C_{l,m} = \int_{-\infty}^{+\infty} f(x)\psi_{l,m}(x)dt = \langle f, \psi_{l,m}\rangle \tag{7}$$

Structural damage $f(x)$ can be identified by wavelet transform coefficients:

$$f(x) = \left|\varphi_m^u(k) - \varphi_m^d(k)\right| \tag{8}$$

After substituting Eq. (8) into Eq. (7), the wavelet transform coefficient of element curvature mode difference can be obtained as:

$$C_{l,m} = \int_{-\infty}^{+\infty} |\varphi_m^u(k) - \varphi_m^d(k)| \psi_{l,m}(x) dt = \int_{-\infty}^{+\infty} f(x) \psi_{l,m}(x) dt \tag{9}$$

4 NUMERICAL ANALYSIS OF SPACE TRUSS STRUCTURE

The space truss structure is shown in Figure 1. The structure has 61 node sand 200 elements, which have 36 nodes in the upper chord and 25 nodes in the lower chord. The geometric properties of the space truss structure are the following: element section area A = 500 mm^2, element length ex = 2.06e^{11} Pa, mass density $\rho = 7800 kg/m^3$, and Poisson ratio is 0.3.

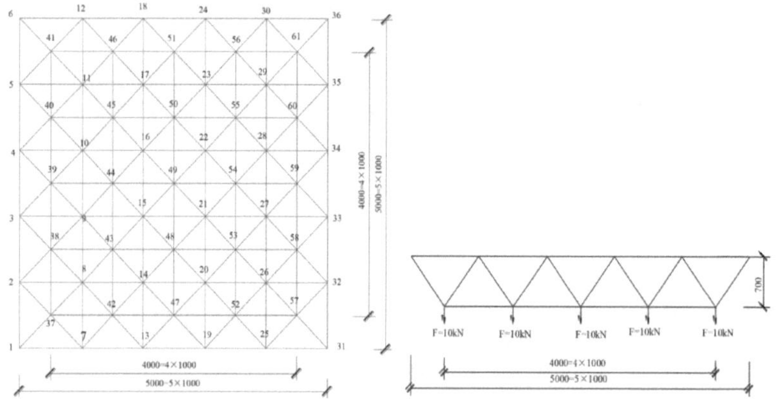

Figure 1. The space truss model.

To verify the accuracy of the damage identification method based on the curvature model and wavelet transform for the different damage degrees of the same damage location and more damage locations, four damage cases (see Table 1) are simulated in this paper. Different damage degrees of the same damage location is simulated in damage case 1 to damage case 3 and three different damage locations are simulated in damage case 4.

Table 1. Damage cases.

Damage cases	1	2	3	4
Damage element (connected code)	37(26~27)	37	37	37, 60(26~32), 100(53~58)
Damage degree	10%	30%	50%	30%

4.1 *Identification of the first to third damage case by wavelet method*

Different damage degrees of element curvature mode differences are shown in Figure 2. It can be seen from the figure that it cannot determine the specific location of the damage. Wavelet transform coefficient of element curvature mode difference in different degree damage is shown in Figure 3. From the figure, the damage position (element 37) can be identified accurately.

It can be seen from Figure 2 that according to the increasing element curvature mode difference, its damage degree could be determined preliminarily. When the damage degree is 10%, the peak value of element curvature mode difference is 2 × 10^{-3}. When the damage degree is 30%, the peak value of element curvature mode difference is 3 × 10^{-3}. When the damage degree is 50%, the peak value of element curvature mode difference is 4×10^{-3}. With the increase of the degree of damage, the element curvature mode difference also follows to increase.

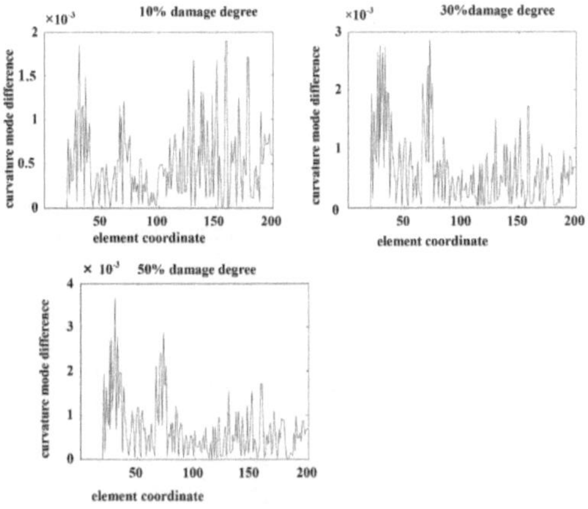

Figure 2. Different degree damage of element curvature mode difference.

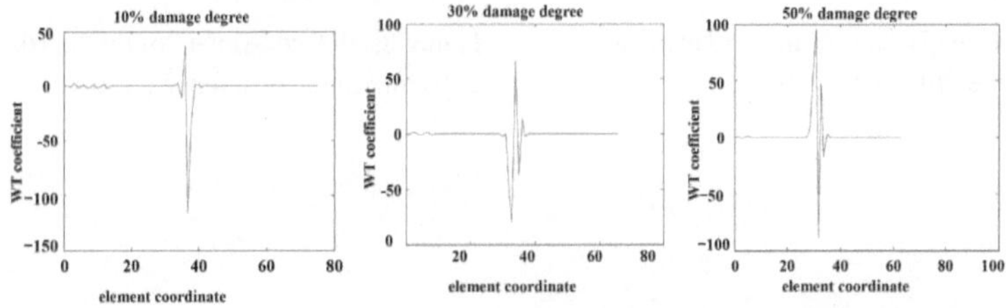

Figure 3. WT coefficient of element curvature mode difference in different degree damage.

4.2 *Identification of the fourth damage case by wavelet method*

The former five-element curvature mode difference of the fourth damage case is shown in Figure 4. It can be seen from the figure that it is difficult to find the damage location by element curvature mode difference.

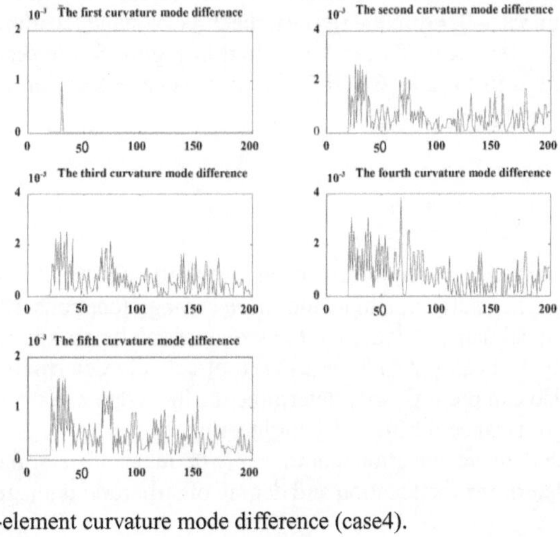

Figure 4. Former five-element curvature mode difference (case4).

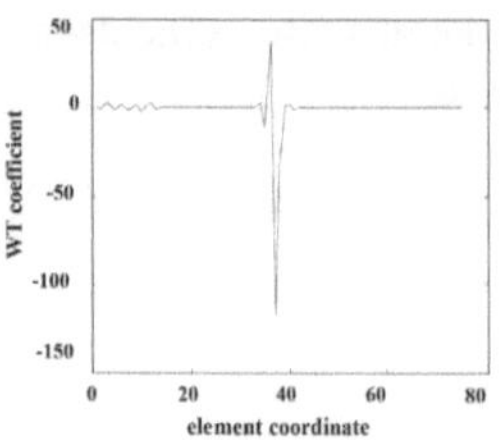

Figure 5. WT coefficient of the first element curvature mode difference (case 4).

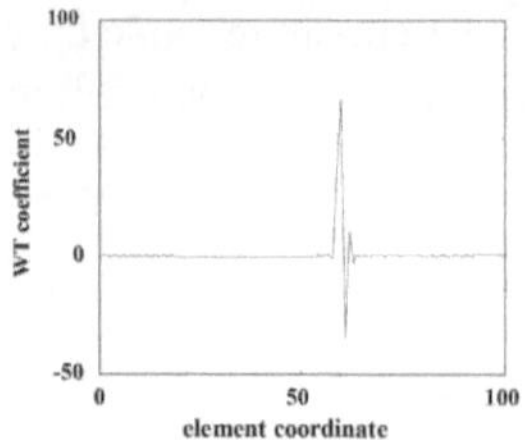

Figure 6. WT coefficient of the second element curvature mode difference (case 4).

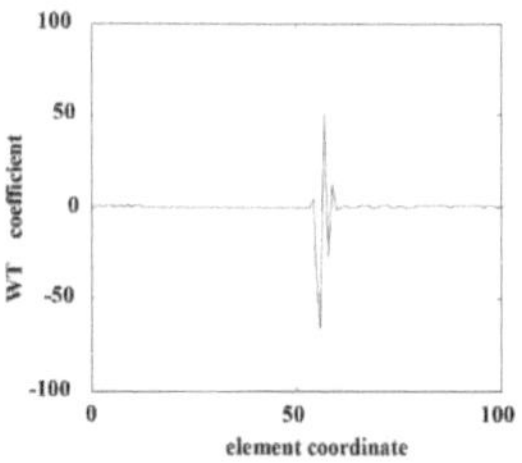

Figure 7. WT coefficient of the third element curvature mode difference (case 4).

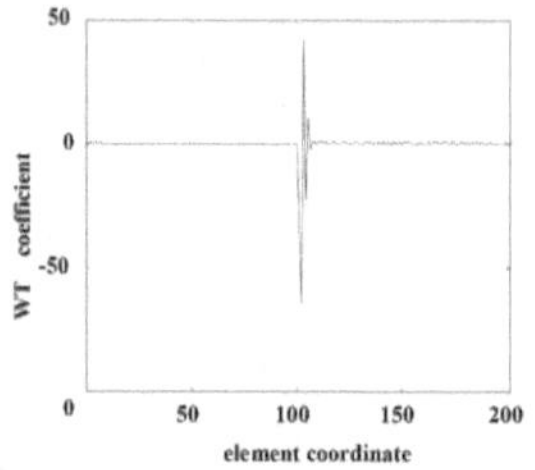

Figure 8. WT coefficient of the fourth element curvature mode difference (case 4).

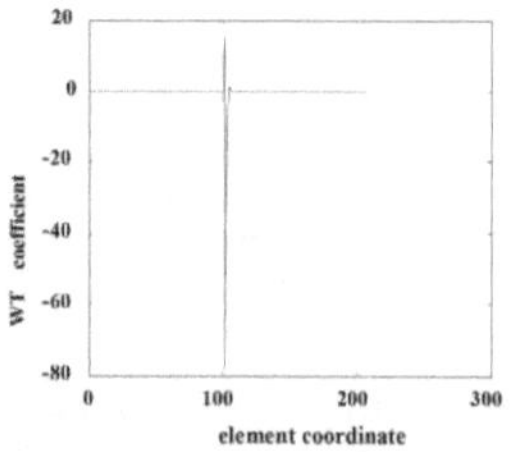

Figure 9. WT coefficient of the fifth element curvature mode difference (case 4).

WT coefficients of the five-element curvature mode difference in different damage locations (case 4) are shown in Figures 5–9. From the figures, the three damage locations will be determined. The first damage location (element 37) can be located in Figure 5; the second damage location (element 60) can be found in Figures 6 and 7; the last damage location (element 100) can be identified by Figures 8 and 9.

5 CONCLUSION

In this paper, with the curvature mode difference as a space truss structure damage judge index, combined with the wavelet transform, single and more damage locations of space truss structure can be defined, and the local damage degree of the structure can be preliminarily determined. The results show that the method can not only be used to effectively determine the single and more damage locations but also can preliminarily determine the local damage degree of the structure. It has a certain guiding significance for practical engineering.

Currently, most of the damage detection methods rely on intact structural original data, but it is difficult to accurately determine the location and degree of structural damage due to many factors in practical engineering. Therefore, a new parameter can be found, which can be used to determine the structural damage location without relying on the original data of the structure.

ACKNOWLEDGMENTS

This work was financially supported by Scientific research planning project of Department of Education of Hubei Province (B2021381) and Key projects of Educational Science Planning of Hubei Province (2019GA067).

REFERENCES

Chen Yijin, Xie Shilin, Zhang Xinong. (2016) Damage identification based on wavelet packet analysis method. *J. International journal of applied electromagnetic and mechanics*, 52(1/2): 407–414.

Cui Hongyu, Xu Xin, Peng Weiqiang, et al. (2018) A damage detection method based on strain modes for structures under ambient excitation. *J. Measurement*, 125:438–446.

Ghasemi M R, Nobahari M, Shabakhtyn. (2018) Enhanced optimization-based structural damage detection method using modal strain energy and modal frequencies. *J. Engineering with computers*, 34(3): 637–647.

Liu Yang.(2014) The development and application of safety detection based on structural dynamic characteristic. *J. Changchun Inst. Tech. (Nat. Sci.Edi.)*

Li Jun, Yu Dedong, Bai Huiren. (2007) Damage localization of changes based on the strain modes for simply supported beam. *J. World earthquake engineering*, 23(1): 104–109.

Liu Xijun, Xiang Linjie, Zhang Suxia. (2015) Damage identification of simply supported beam bridges based on wavelet analysis. *Journal of vibration,measurement & diagnosis*, 35(5): 866–872.

Shahsavari V,Chouinard L,Bastien J. (2017) Wavelet-based analysis of mode shapes for statistical detection and localization of damage in beams using likelihood ratio test. *J. Engineering structures*, 132: 494–507.

Yu Xianying,Lin Chenxu,Wu Jinfu, Liu Jingliang. (2020) Damage localization method of piles using complex continuous wavelet transform. *Journal of Hua qiao University (Natural science)*, 41(2): 177–184.

Zhang Limei,Liu Weiran,Zhang Liwei,Du Shoujun. (2013) Damage testing method of space truss based on elemental strain mode difference and RBF neural network. *Journal of Hebei University of Science and Technology*.

Zhang Yong,Jing Hang, Yuan Hai-qing. (2010) Structural Damage Detection Based on Change Rate of Curvature Mode. *J. of HUST. (Urban Science Edition)*, 27(2): 82–85.

Advances in Civil Engineering: Structural Seismic Resistance, Monitoring and Detection – Mohd Johari Mohd Yusof and Junwen Zhang (Eds)
© 2023 The Authors, ISBN 978-1-032-31491-4

Experimental study on unidirectional tension of grouting sleeve with defects

Yan Wang
School of Architectural Engineering, Nanjing Institute of Technology, Nanjing, Jiangsu, China

Tongliang Xiao*
School of Architectural Engineering, Nanjing Institute of Technology, Nanjing, Jiangsu, China
Nanjing Chixia Construction Co., Ltd., Nanjing, Jiangsu, China

Chuang Li, Shilin Liang, Xiang Shen & Qin Hao
School of Architectural Engineering, Nanjing Institute of Technology, Nanjing, Jiangsu, China

Pu Xun & Chengfang Wang
Nanjing Chixia Construction Co., Ltd., Nanjing, Jiangsu, China

ABSTRACT: Compared with traditional cast-in-place concrete structures, prefabricated concrete structures are more widely used because they have shorter construction period and are more friendly to the environment. Usually, longitudinal reinforcement at member joints is connected by the sleeve grouting method. However, various types of construction defects often occur during the construction. According to the types of defects that often occur in actual construction, three types of specimens were designed in this paper, namely anchorage length defects, longitudinal eccentricity defects and longitudinal grouting defects. Uniaxial tensile tests were conducted on eight specimens of the three types, test data such as ultimate load and corresponding displacement were obtained. The test results showed that the damage modes were divided into reinforcement tensile failure and reinforcement pull-out failure, most of the specimens with defects were reinforcement pull-out failure. The anchorage length defects, longitudinal eccentric defects and longitudinal grouting defects have some influence on the bearing capacity of the specimens, among which the eccentric defects have the greatest influence on the bearing capacity. Defects can cause an increase in bond stress when the specimen reaches the yield point. The bond stress of the anchorage length defect specimen is significantly greater than that of other defect specimens. The bond stresses for eccentric defect specimens and longitudinal grouting defect specimens are slightly higher than those for non-defective specimens. Moreover, different defects can lead to increased strains in the reinforcement. The results of the study provide a reference for future construction plan improvement and numerical simulation of the mechanical properties of grouted sleeves with defects.

1 INSTRUCTIONS

Prefabricated building can reduce construction period and pollution, and its performance is the same as that of cast-in-place concrete structure building China's prefabricated structure started late and lacks scientific research results, but in recent years the country has been vigorously promoting the development of prefabricated buildings, and many excellent scientific research results and innovative inventions have emerged in the industry.

*Corresponding Author: xtl@njit.edu.cn

DOI 10.1201/9781003310884-55

Among the prefabricated components, shear walls have better wind and seismic stiffness (Han 2020) and are widely used, while the main connection methods of prefabricated components are grouting sleeves. Grouting sleeves are divided into two categories, which are full grouting sleeves and half grouting sleeves (Xiao 2021). In the actual use of the sleeve, different degrees and types of defects may arise between the sleeve and other parts, and Linqing Zheng (2017) et al. classified defects as central defects, horizontal defects to test. The study shows that: compared to end defects and vertical grouting defects, central defects and horizontal grouting defects have a greater impact on the performance; specimens with eccentric defects are more likely to produce slip.

Liu (2020) et al. conducted uniaxial tension tests on grouted sleeve connection specimens of different sizes and obtained the effects of some parameters on the connection performance, the tests showed that the strength was fully satisfied when the anchorage length reached 7 more than times the diameter of the steel bar. Xu Ma (2020) et al. found that when the anchorage length is less than 4 times the diameter of the steel bar, the steel bar is pulled out and the bearing capacity of the steel bar can not be brought into full play. Haider (Hamad 2020) et al. found that when the anchorage length is not more than 6 times of the steel bar, it can not resist the seismic load. A lot of test reports (Hu 2021; Liu 2018; Poluraju 2013) show that the anchorage length of 7 times steel bar is safe anchorage length. Most of the tests were conducted only on single anchor length defects and not on other types of defects under the same conditions.

For eccentric defects, Chen (2020) et al. studied the influence of vertical eccentricity defect and oblique eccentricity defect of sleeve, and carried out uniaxial tension test, which showed that the steel bar located at the outlet of the grout had the highest strain and was the weakest location, while the grout damage depth caused by the oblique eccentric sleeve connection was the greatest. The eccentric steel bar in uniaxial tension will produce additional moment on the structure and lateral pressure on the grout, resulting in concentrated stress on the grout and uneven stress on the surface of the steel bar, which eventually leads to a deeper grout damage depth after damage than other kinds of defects.

In addition, the grouting process is susceptible to air bubbles and impurities, which can have an effect (Zeng 2016) on the saturation of the grout, thereby reducing the bond between the steel bar and the grout. When the sleeve is connected horizontally, if the grout is not sufficient, chord height defect (Zheng 2017) will occur, reducing the strength of the entire sleeve grouting system.

Despite the emergence of research results on sleeve defects, studies on chord height defects have been few in recent years. In this paper, we study anchorage length defects, eccentric defects, and chord height defects, and by controlling the variable method, we ensure that other conditions are the same while changing only the defect type and defect degree, and can fully compare the performance of different defects, which can better analyze the coupling between various defects and explore the influence of defect type and degree on the mechanical properties, bonding performance, and damage form of the specimen compared with the test of studying a single defect. The types of defects, damage forms, bonding capacity, and critical slip of the sleeve connections are related to provide a basis for analyzing the structural performance of assembled structures, safety design of assembled buildings, and risk prediction.

2 TEST INTRODUCTION

Three types of sleeve defects were designed, including anchorage length defects (MG type), longitudinal eccentricity defects (PX type) and longitudinal grouting defects (XG type). As shown in Figure 1, the grouting volume was reduced to simulate the anchorage length defect; the reinforcement was kept vertical and inserted against the inner wall of the sleeve to simulate the eccentric defect; a stopper was set at the end of the horizontal grouting sleeve to reserve 3 mm cavity area with insufficient grouting volume to simulate the longitudinal grouting defect of the horizontal grouting specimen. At the same time, a full grouting set was added to compare the effect of defects on the performance of the specimens.

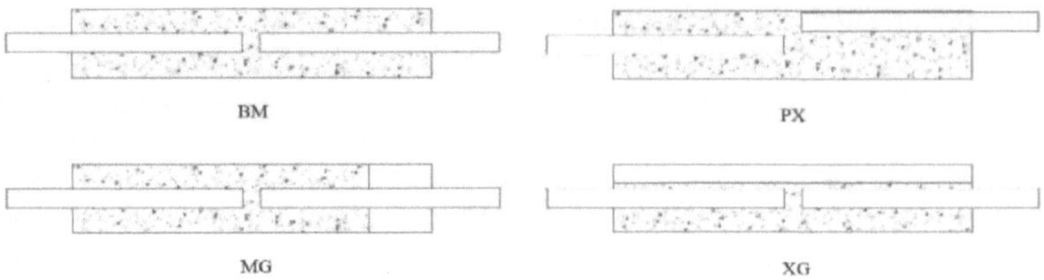

Figure 1. Schematic diagram of test piece.

3 TEST RESULTS AND ANALYSIS

3.1 *Load-displacement curve analysis*

The strain of the reinforcement in the non-defect specimens increased with the increase of load. The load-strain curve of the specimens grows slowly at the beginning due to the adequate bonding of the reinforcement grouting material, and then the reinforcement enters the yielding stage and the strain increase becomes larger. As the load continues to rise, the reinforcement enters the strengthening phase and the curve enters the plateau section. Finally, the reinforcement material could not withstand the load and tensile fracture occurred. the average ultimate load reached 73.12 kN, and The bond stress of the reinforcement when the yield point is reached is approximately 16 MPa. All strain gauges performed well, and the strain data were measured effectively.

For the specimens with anchor length defects, the strains on both sides of the bars were approximately the same, and the load-strain curves of most of the specimens were consistent with those of the specimens without defects. At the later stage of the tensile test, the slope of the curve was greater than that of the specimens without defects. Under the same load, the reinforcement lacking grouting material produced greater strains. Finally, due to insufficient anchorage length, the bonded bearing capacity was less than the instrumented tensile load, the stress on the bonded surface was too high, resulting in premature debonding at the interface of the reinforcement grouting material and failure of the anchorage of the reinforcement in the lower part of the specimen. The maximum ultimate load was 72.84kN, which was lower than that of the specimen without defects. Due to the reduction of anchorage area, the bond stress of reinforcement when yield point is reached increased significantly, which was 28.98 MPa.

The load-strain curves of the eccentrically defective specimens are consistent with those of the non-defective specimens in the early stage. After effective stretching, the strengthening stage is entered. In the later stage, the bond force is less than the instrument load and fractures prematurely. During the stretching of the specimen, the instrumented collet will be extruded and deformed. It can be seen that the eccentric specimen will produce bending moment during uniaxial tension, and the specimen cannot reach the maximum tensile load capacity, and finally the anchorage failure of the lower reinforcement of the specimen occurs, the minimum ultimate load is 67.4kN, the average load capacity is lower than other defective specimens, and the bond stress of the reinforcement when reaching the yield point is 16.91MPa, which is higher than that of the non-defect specimen.

The load-strain curves of the longitudinal grouting defect specimens varied considerably in the early stages. It is believed that the longitudinal grouting defects subject the grouted material to additional bending moments during tension and produce cracks early. The curve flattens out after the rebar enters the yielding stage and fails after a long platform section. The strain difference of the reinforcement is relatively stable. During the reinforcement stage, the reinforcement is effectively stretched. After the load exceeded the bond force, the reinforcement was pulled out, and the ultimate load was 72.47 kN, and the bond stress of the reinforcement when it reached the yield point was 18.58 MPa.

In this test, the bearing capacity of eccentric defect specimen is the lowest, which has also been verified in other research reports. Anchorage length defect and longitudinal grouting defect will affect the bearing capacity of the specimen too.

3.2 *Failure mode*

In the test, there were two forms of damage, namely reinforcement tensile failure and reinforcement pull-out failure, as shown in Figure 2, and the results are listed in Table 1. At the same tensile load, the smaller the anchorage length, the greater the bond stress per unit area of the reinforcement. Similarly, the greater the bond stress per unit area of grout material on the bond surface of the grout material. Many tests have shown that 7d is a safe anchorage length (d is the diameter of the reinforcement). When the anchorage length is 4d, the stress area is reduced by 40%, and most of the uniaxial tensile results are tensile failure of reinforcement.

Eccentric defect and longitudinal grouting defect belong to non-axisymmetric defects. For the eccentric defect specimen, in the process of unidirectional tension, the grouting material on the side of the steel bar close to the cylinder wall is pulled, and the grouting material on the side away from the cylinder wall is compressed, which leads to deeper damage of the grouting material on the one hand and lower bearing capacity of the specimen on the other hand. The horizontal longitudinal grouting defect leads to the reduction of the cross-sectional area of the grouting material. When the steel bar is not exposed, the performance of the specimen is weakened, but the steel bar can still undergo yield and strengthening for effective tension in the tensile process; When the steel bar is exposed, the bond stress increases further, the performance of the specimen is seriously weakened, and most of the steel bar is pulled out directly.

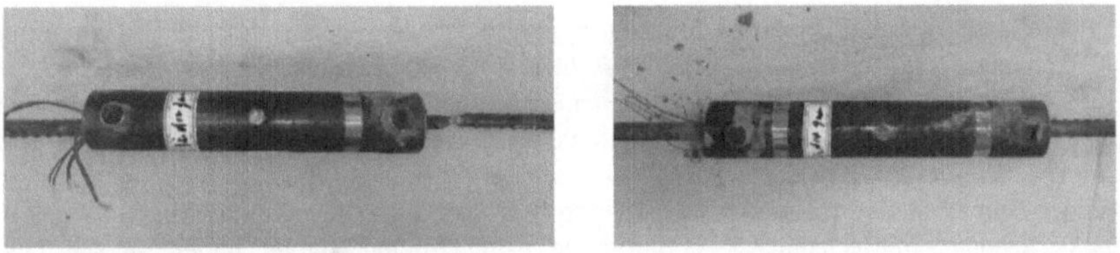

Figure 2. Failure mode test photos.

Table 1. Designed parameters and test results of specimens.

Specimen ID	Type of defects	Failure mode	Ultimate load (kN)	Bond slip (mm)	Bond stress at yield point (MPa)	Bond slip at yield point (mm)
BM-1	Full-grouted	TF	73.10	29.81	16.17	7.38
BM-2	Full-grouted	TF	73.13	29.24	16.01	7.81
MG-1	Anchor length defect	P	72.84	30.60	28.80	8.09
MG-2	Anchor length defect	P	72.53	26.33	28.98	7.56
PX-1	Eccentric defect	P	67.40	21.72	16.91	7.15
PX-2	Eccentric defect	P	72.67	36.47	16.70	7.19
XG-1	Longitudinal grouting defect	P	72.47	20.78	18.54	8.35
XG-2	Longitudinal grouting defect	P	72.22	39.92	16.58	8.90

Note: TF denotes the tensile fracture of rebars, P denotes the pull-out failure of rebar.

4 CONCLUSION

Based on the results and discussions presented above, the conclusions are obtained as below:

(1) Anchorage length defect, eccentricity defect, and longitudinal grouting defect all have some influence on the bearing capacity of the specimen. Among them, the eccentric defect has the greatest influence on the bearing capacity, and the bearing capacity decreases by 4%.

Anchorage length defect, longitudinal grouting defect bearing capacity decreased by 0.59%, 1.05% respectively.

(2) Defect resulted in elevated bond stress of the specimen. The bond stress of the anchor length defective specimen was significantly greater than that of the other defective specimens when it reached the yield point, elevated by 44% compared to the non-defective specimens, and the bond stress of the eccentric defective specimens and longitudinal grouting defective specimens was slightly elevated compared to the non-defective specimens, by about 7%.

(3) Defects resulted in elevated joint slip of the specimen when it reached the yield point, with no significant differences between individual defects.

ACKNOWLEDGMENTS

The authors would like to acknowledge financial support from the Nanjing Institute of Technology Foundation of China (No. CKJA201806) and Jiangsu Students' Platform for innovation and entrepreneurship training program (No. 202011276018Z).

REFERENCES

C.J. Hu, P.F. Yao, and Zhi Zhou. Bonding and anchoring performance of connecting reinforcement in precast assembled concrete bridges [J]. *Journal of Harbin Institute of Technology*, 2021, 53 (03): 82–93.

Chao Liu, Liufeng Pan, Hai Liu, Huawei Tong, Yubing Yang, Wei Chen. Experimental and numerical investigation on mechanical properties of grouted-sleeve splices[J]. *Construction and Building Materials*, 2020, 260:

Ghayeb Haider Hamad, Razak Hashim Abdul, Sulong N.H. Ramli,Mo Kim Hung, Abutaha Fuad, Gordan Meisam. Performance of Mechanical Steel Bar Splices using Grouted Couplers under Uniaxial Tension [J]. *Journal of Building Engineering*, 2020 (prepublish):

Jianwei Chen, Zhanwen Wang, Shilong Ju, Ziye Liu. Experimental study and numerical simulation on tensile behavior of eccentric connection of semi grouted sleeve steel bar [J] *Journal of building structure*, 2020, 41 (S2): 160–171 DOI:10.14006/j.jzjgxb.2020.S2.0019.

P.Poluraju, G.Appa Rao. Seismic Behaviour of Precast Reinforced Concrete Beam-Column Connections: A Literature Review[C]//Proceedings of 2013 2nd International Conference on Sustainable Construction Materials and Computer Engineering (ICSCMCE 2013 II)., 2013:16–20.

Qihao Han, Dayang Wang, Yongshan Zhang, Weijun Tao, Yong Zhu. Experimental investigation and simplified stiffness degradation model of precast concrete shear wall with steel connectors[J]. *Engineering Structures*, 2020, 220:

Qinglin Zheng, Ni Wang, Li Tao, Wenjie Xu. Experimental study on the influence of grouting defects on the performance of reinforced sleeve grouting connection specimens [J] *Architectural Science*, 2017, 33 (05): 61–68 DOI:10.13614/j.cnki.11-1962/tu.2017.05.010.

Quanwei Liu. Research on the seismic performance of assembled concrete shear walls with different connection methods[D]. Chang'an University, 2018.

Shun Xiao, Xiangmin Li, Qingfeng Xu. Research Progress on quality inspection and defect treatment of sleeve grouting in prefabricated integral concrete structure [J] *Building structure*, 2021,51 (05): 104116 DOI:10. 19701/j.jzjg.2021.05.016.

Wei Zeng. Research on high performance sleeve grouting material for prefabricated building [D] Shenyang Architecture University, 2016.

Xu Ma, Fangqi Yang, Jianxin Li, Xinxiang Wang, Jiangxiong Wei, Yini Li. Experimental study on the influence of grouting defects on reinforced sleeve grouting connection in prefabricated buildings [J] *Guangdong building materials*, 2020, 36 (11): 45–47.

Advances in Civil Engineering: Structural Seismic Resistance, Monitoring and Detection – Mohd Johari Mohd Yusof and Junwen Zhang (Eds)
© 2023 The Authors, ISBN 978-1-032-31491-4

Monitoring and analysis of underground space response during shield tunneling

Jie Yu
Hangzhou Jinghang Canal Second Channel Construction Investment Co., Ltd, Hangzhou, China

Mengmeng Liu
College of Harbour, Coastal and Offshore Engineering Hohai University Nanjing, China

Jun Cao & Yu Sun
Nanjing Hydraulic Research Institute Nanjing, China

Yang Sun
College of Harbour, Coastal and Offshore Engineering Hohai University Nanjing, China

ABSTRACT: Based on the shield construction close to the underground structure, this paper analyzes the impact of shield tunneling on the safety of underground space according to the field monitoring data. The results show that the adjacent construction caused the high earth pressure and high pore water pressure of the soil between the tunnel and the underground structure. The high pore water pressure led to the large deformation of the underground diaphragm wall, which seriously affected the safety of the underground structure. The instantaneous opening of the enclosed door upon shield arrival caused the collapse of the soil in front of the shaft well, which seriously affected the construction safety. By slowing down the shield tunneling speed and secondary grouting, the safety of underground space could be maintained.

1 INTRODUCTION

It is easy to cause large deformation of adjacent structures and soil during shield construction, which will have a serious impact on the safety of underground space. Chen 2017; Diao 2018; Sun 2021; Wang 2021 put forward specific measures for shield tunneling through complex structures to reduce the impact on the surrounding environment and maintain the safety of underground space. However, at present, relevant achievements are focused on the influence of shield tunneling on surrounding soil and underground structure in unconstrained underground space, and there is no precedent for the analysis of soil response characteristics during shield tunneling in semi-closed space in soft soil area. Therefore, based on the field monitoring data of the tunnel from Longyang station to Baiyang Station of Shanghai Metro Line 7, this paper studies the disturbance characteristics and mechanism of the surrounding soil adjacent to the built underground diaphragm wall during the shield construction process. It can provide suggestions for similar projects.

2 MATERIALS AND METHODS

To monitor the ground response, measuring points were arranged 35 m before the shield machine reached the arrival shaft. Figure 1 shows the layout of the measuring points. The monitored content of the surrounding soil included the ground settlement, earth pressure, pore pressure and lateral displacement. Lateral displacement was observed with an inclinometer tube at a depth of 29 m. The

DOI 10.1201/9781003310884-56

pore pressure was monitored by a piezometer, and the depths of K-1, K-2 and K-3 are 10.4 m, 9.9 m and 9.6 m, respectively, which corresponded to the depth of the centre of the tunnel. The earth pressure was observed with the earth pressure cell, and the depth of T-1 was 9.6 m, which was also the depth of the top of the tunnel. Furthermore, the ground settlement of the soil was measured by a level.

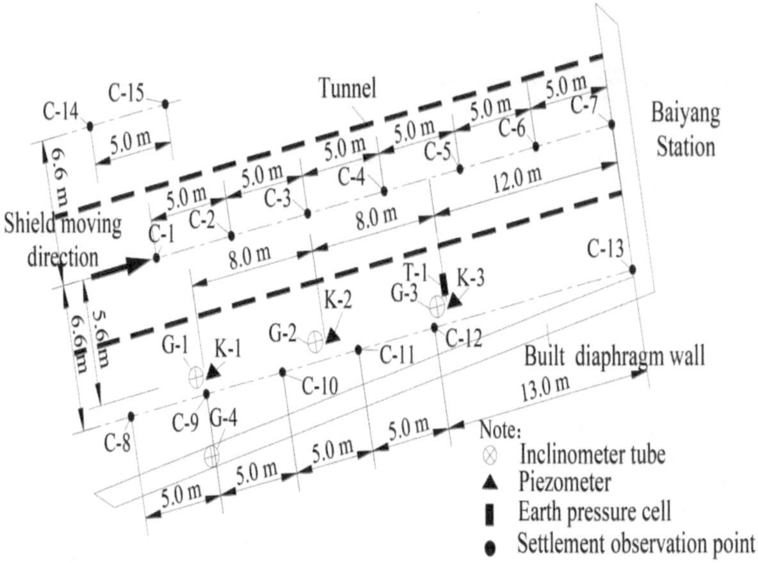

Figure 1. Layout of the measuring points.

3 RESULTS & DISCUSSION

3.1 *Analysis of earth pressure variation*

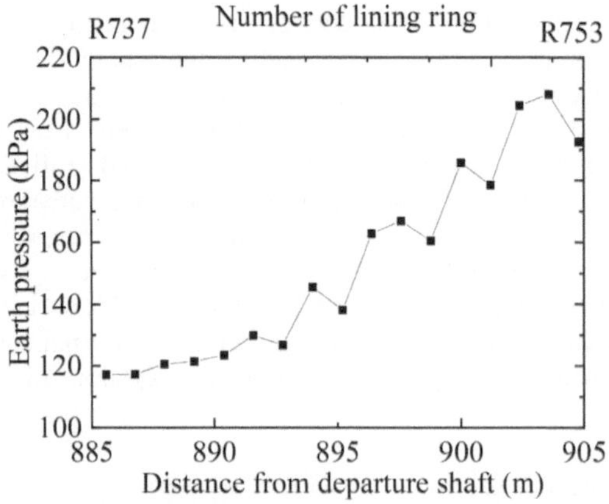

Figure 2. Earth pressure of T-1 at different distances.

Figure 2 illustrates the readings of earth pressure cell T-1 at different distances during shield driving, where the abscissa stands for the distance from the departure shaft. This figure shows that, between the segments R737 and R741 (from 886 to 890 m), the earth pressure of T-1 was approximately 120 kPa. The earth pressure then increased gradually as the distance to T-1 decreased. Accordingly, the maximum earth pressure was 208 kPa and was obtained from R753 (903 m), which was around the location of T-1. However, after the jacked distance reached R753, the earth pressure decreased as the distance to T-1 increased. Subsequently, 60 days later, after the end of tunnel construction, the earth pressure of T-1 stabilised at approximately 140 kPa. Furthermore, the earth pressure caused by the extrusion of shield driving could not be fully released because of the built diaphragm wall, so the final stability value was higher than the initial value of 116 kPa by approximately 20%.

The earth pressure cell T-1 was buried at a depth of 9.6 m. The groundwater level was approximately 2.0 m underground so that the effective vertical stress at T-1 was approximately 92 kPa. Therefore, the lateral earth pressure at rest was approximately 46 kPa with a coefficient of lateral earth pressure of 0.5 (K0). Thus, the total horizontal earth pressure at rest was approximately 122 kPa. According to the construction experiences of Shanghai, the designated chamber pressure for the EPB shield is approximately equal to the total horizontal earth pressure at rest. From this result, we can find that the existence of the nearby diaphragm wall caused a very high earth pressure between the shield and the diaphragm wall.

Therefore, due to the shield construction close to the diaphragm wall, the high earth pressure generated by the nearby soil seriously affected the safety of the diaphragm wall structure. It was necessary to take reasonable engineering measures to release the soil pressure.

3.2 *Correlation analysis between pore water pressure and diaphragm wall*

Figure 3 shows the readings of the piezometer K-1 at different distances during shield driving, where the abscissa stands for the distance from the departure shaft, and the ordinate denotes the excess pore pressure that decreased the initial pore pressure caused by shield driving.

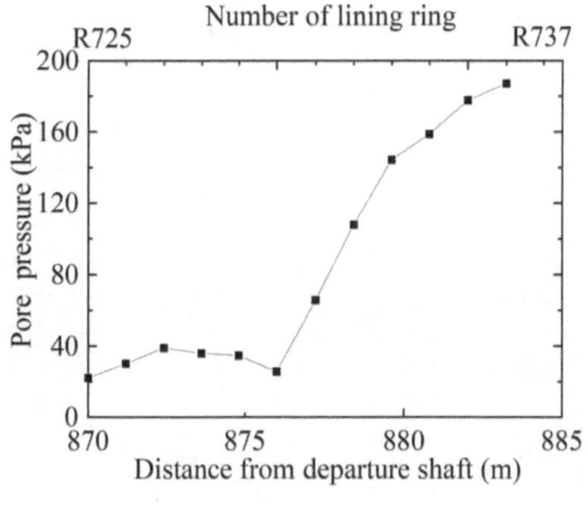

Figure 3. Pore pressure of K-1 at different distances.

As shown in Figure 3, from R725 to R730 (from 870 to 876 m), the pore water pressure varied from 20 to 40 kPa, and the amplitude was small. From R730 to R736 (from 876 to 883 m), the value of K-1 increased rapidly as the distance from the departure shaft increased. After the jacked distance reached R736 (883 m from the departure shaft), the value of K-1 increased dramatically to 187 kPa. After that, because of the rapid increase of the value beyond the range of K-1, damage to K-1 occurred.

Figure 3 indicates that the value of K-1 increased dramatically with tunnel excavation and that final damage to K-1 occurred. Furthermore, the measuring point C-9 near K-1 also experienced ground upheaval of 4.82 mm. The reason for the above phenomena was that the soil on the right side of the shield was constrained within a Γ-shaped semi-enclosed space surrounded by the adjacent diaphragm wall. Because of the existence of the diaphragm wall, the pore pressure dissipated at a very slow rate and resulted in the accumulation of pressure, which might affect the safety of the built diaphragm wall. In this case, shield driving was temporarily suspended. After the remaining pore pressure dissipated for a period of time, the advancement speed of the shield slowed down from 30 mm/min to 20 mm/min when the shield machine continued to drive forward.

In Figure 1, the inclinometer tube G-4 was pre-installed in diaphragm wall. Figure 4 shows the relationship between the pore pressure of K-1 and the lateral displacement of the diaphragm wall. When the jacked distance reached R720, the value of K-1 was 5.8 kPa. Meanwhile, the diaphragm wall produced a maximum displacement of 2.5 mm at a depth of 10 m. When the jacked distance reached R736, the pore water pressure at K-1 reached the maximum value of 187.2 kPa. Meanwhile, the diaphragm wall produced a maximum displacement of 8.5 mm at a depth of 10 m. The maximum displacement of the diaphragm wall corresponded to the depth of the tunnel axis, and the overall deformation of the diaphragm was arched. This figure shows that the readings of G-4 increased dramatically as the pore pressure increased. Therefore, it was important to control the pore pressure during tunnel construction for the safety of the adjacent diaphragm wall.

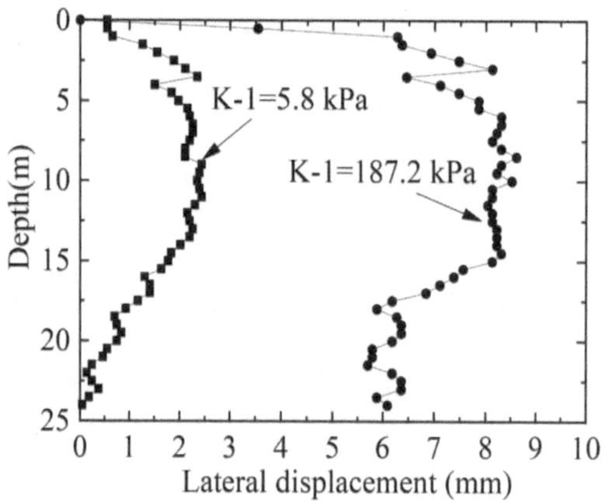

Figure 4. Lateral displacement of G-4 in different pore pressures.

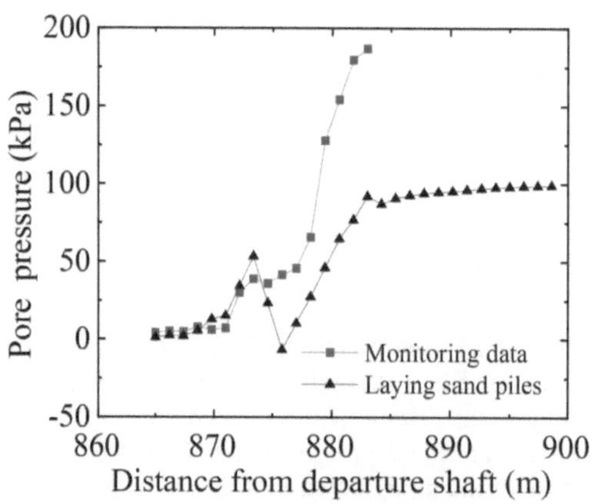

Figure 5. Variation of K-1 pore water pressure after laying sand piles.

In this paper, finite element analysis software PLAXIS 3D was used to establish a three-dimensional finite element model of soil-underground structure-tunnel to simulate the influence of shield tunneling on underground space. A row of drainage lines arranged at intervals of 6 m were set parallel to the tunnel axis at 6.6 m and adjacent to the diaphragm wall to simulate sand piles. The nearest sand pile to K-1 was 1 m.

Figure 5 shows that the variation range of pore water pressure at K-1 became smaller after sand pile placement. The maximum pore water pressure at K-1 decreased by 46.3%, and the reading was close to 90 kPa. As the sand pile changed the infiltration path of the nearby soil, when the shield tunneling continued to approach K-1, the pore water pressure at K-1 did not rise, but remained stable.

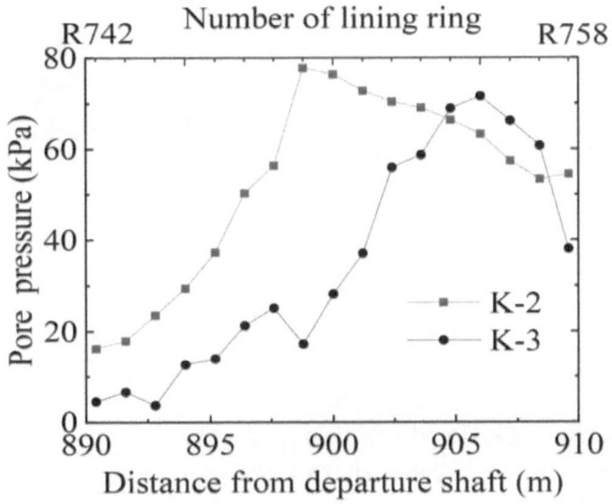

Figure 6. Pore pressure of K-2 and K-3 at different distance.

Figure 6 shows that the variation range of pore water pressure at K-2 and K-3 were small during excavation, and the decrease in the shield excavation velocity effectively alleviated the accumulation of the pore pressure. From R742 to R758 (from 890 to 910 m), the values of K-2 and K-3 changed significantly. The maximum value of K-2 was 77.8 kPa, recorded in R749 (899 m from the departure shaft) around the location of K-2. The value of K-2 was 35 kPa after the end of tunnel construction. Subsequently, 60 days later, the pore pressure almost dissipated and was approximately at 21 kPa. The maximum value of K-3 was 71.6 kPa, recorded in R755 (906 m from the departure shaft) around the location of K-3. The value of K-3 appeared to be stable at approximately 11 kPa 60 days after the end of tunnel excavation. However, the variation ranges of pore water pressure at K-2 and K-3 were much smaller than that at K-1. The reason was that considering the safety of diaphragm wall, the shield tunneling speed had been adjusted reasonably.

3.3 *Analysis of soil collapse in front of the arrival shaft*

Figure 7a shows that there was little change in the value of the ground displacement, which varied from -3 to +3 mm before the shield machine reached the arrival shaft. In the last stage of excavation (i.e., the shield machine reached the arrival shaft), C-1~C-3 appeared to be stable, while C-4~C-7 fluctuated sharply. The maximum ground displacement suddenly became -68 mm, as recorded by C-6. Simultaneously, a ground crack was found above the tunnel alignment close to the arrival shaft (see Figure 7b). Reasons for the aforementioned phenomena included the collapse of the ground over the tunnel when the enclosed door of the arrival shaft was opened. With the opening of the enclosed door, the soil over the shield collapsed so that a large displacement was measured over the centreline of the tunnel, which also caused the large lateral displacement of G-2 and G-3 (refer to Figure 8b). After the shield machine completely entered the shaft, secondary grouting was used to stabilise the ground, and the tracking measurement of C-4~C-7 gradually stabilised at around -2 mm 60 days later.

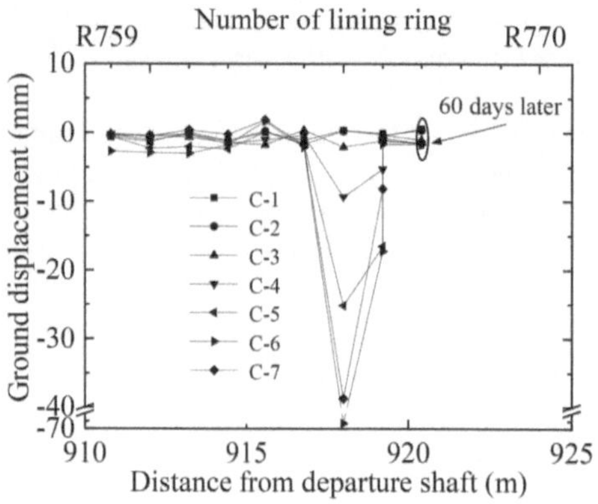

(a) Ground displacement of C-1~C-7 at different distances

(b) Ground crack above the tunnel axis

Figure 7. Soil collapse in front of the arrival shaft.

Figure 8a shows the readings of G-2 and G-3 5.6 m from the tunnel centreline when the jacked distance reached R754 (905 m from departure shaft). This figure shows that the slope of the observed displacement curve increased significantly within plus or minus 10 m of the tunnel centreline. As the distance from the arrival shaft became shorter, the surrounding soil was gradually extruded towards the direction of the built diaphragm wall. A maximum lateral displacement of +5 mm towards the direction of the built diaphragm wall at G-2 was recorded at a depth of 13.5 m, which corresponded to the depth of the centre of the tunnel. Similarly, G-3 showed a maximum reading of 11.5 mm at a depth of 12 m. Furthermore, the lateral displacement at G-3 was almost positive.

Figure 8b shows that, at R764 (917 m from the departure shaft, i.e., the shield machine just reached the arrival shaft), the lateral displacement abruptly turned to the new tunnel instead of the built diaphragm wall. And the lateral displacements of G-2 and G-3 fluctuated sharply from positive to negative (see Figure 8b). A maximum lateral displacement of -19 mm towards the direction of the new tunnel was recorded at a depth of 3 m. This phenomenon was consistent with the readings of C-4~C-7 in Figure 7a and was caused by the instantaneous exposure of the surrounding soil near the enclosed door of the arrival shaft. Furthermore, from R754 to R764 (from 905 to 917 m), the maximum difference between two readings was nearly 23 mm at G-3, which was relatively large compared to the maximum of previous value.

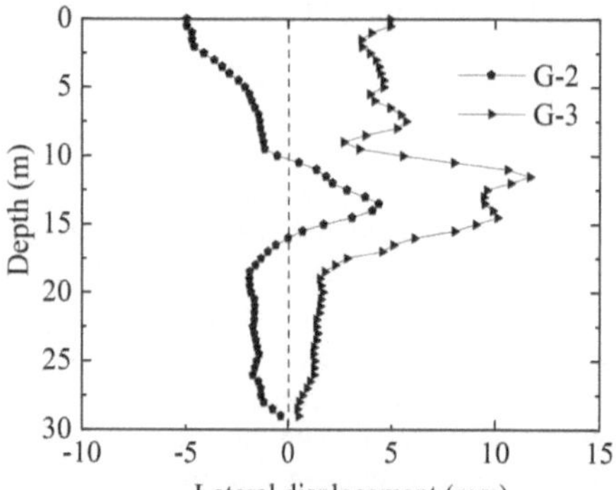

(a) Lateral displacements of G-2 and G-3 at R754

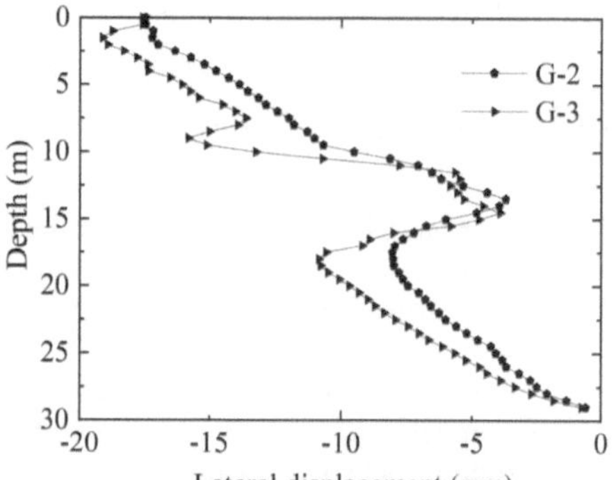

(b) Lateral displacements of G-2 and G-3 at R764

Figure 8. Lateral displacement change of soil.

4 CONCLUSIONS

(1) With the emergence of the peak value of excess pore water pressure in the soil between the tunnel and the diaphragm wall, the diaphragm wall correspondingly produced a large lateral displacement, and the rise and fall of excess pore water pressure would obviously affect the safety of the diaphragm wall structure. The arrangement of sand piles was conducive to the dissipation of pore water pressure.
(2) At the moment when the shield entered the arrival shaft and the enclosed door was opened, the soil in front of the arrival shaft would obviously crack and collapse, and the later deformation of the soil in front of the arrival shaft could be effectively controlled through secondary grouting.
(3) Reasonable control of shield tunneling speed could reduce the influence of shield construction on surrounding soil, effectively slow down the change of excess pore water pressure and maintain the safety of adjacent structures.

ACKNOWLEDGMENTS

The research work described herein was funded by Science and Technology Project of Zhejiang Provincial Department of Transportation (Grant Number: 2020014) and Fundamental Research Funds for the Central Universities (Grant Number: B210203068).

REFERENCES

D. C. Chen & J. P. Hu. (2017). Research on the influence of shield construction on the nearby shallow foundation frame structure. *Journal of Railway Science and Engineering*. vol. 14(03), 552–559.

L. Wang & J. Q. Zhong. (2021). Numerical Analysis of Ground Settlement Induced by Shield Tunnelling near TBM Launch Shaft. *Modern Tunnelling Technology*. 1–10.

P. C. Diao & P. Yang. (2018). Application and Measurement of Secondary Consolidation by Liquid Nitrogen Freezing to Small Shield Receiving in Water-rich and Soft-weak Strata. *Tunnel construction*. vol. 38(06), 1044–1051.

Q. T. Sun & X. Y. Wang. (2021). Shield Excavation Technology of Ultra-small Spacing Tunnel in Complex Environment. *Modern Tunnelling Technology*. 1–8.

Advances in Civil Engineering: Structural Seismic Resistance, Monitoring and Detection – Mohd Johari Mohd Yusof and Junwen Zhang (Eds)
© 2023 The Authors, ISBN 978-1-032-31491-4

Numerical analysis of tensile damage mode of expanded head anti-floating anchor in powdery clay

Xuefei Yuan
School of Civil Engineering and Architecture, University of Jinan, Jinan, China

Feng Ren
The Engineering Technology Research Center for Urban Underground Engineering Supporting and Risk Monitoring of Shandong Province, Jinan, China

Shuting Ren
Nanjing Hua Zhi Metallurgical Group Company Limited, Nanjing, China

ABSTRACT: In this paper, we use ABAQUS finite element numerical analysis software to theoretically study the damage model of the expanded head anti-floating anchor rod under tension in pulverized clay. The results show that: 1) the mode of anchor rod under tension damage is related to the burial depth ratio L_1/D_2, when $L_1/D_2 \leq 2$, the anchor rod undergoes shallow burial damage; when $L_1/D_2 \geq 6$, the anchor rod undergoes deep burial damage; when $2<L_1/D_2<6$, the anchor rod undergoes mixed damage. 2) When the burial depth of anchor rod L_1 is greater than 6m, the ellipse formed by the maximum principal stress cloud around the anchor rod is roughly $4D_1$ in height and $5D_1$ in width. 3) When the anchor is in the elastic stage, the load-displacement curves of shallowly buried anchor, transitional anchor, and deeply buried anchor are sloped straight lines with the same slope; after the anchor enters the plastic stage, the slope of the load-displacement curves decay at different rates for anchor rods with different burial depths. 4) The tensile damage of the deeply buried anchor rod roughly goes through the elastic stage, the plastic zone formation stage, the plastic zone soil strengthening, and the range expansion stage.

1 INTRODUCTION

To relieve the pressure of land resource shortage due to rapid economic development, China has started to develop underground space vigorously. When the buoyancy force generated by the groundwater is greater than the self-weight and permanent load of the building, the building will be displaced upward, which may cause the building destabilization and damage (Wang 2015), so the anti-floating design becomes pivotal. At present, China's commonly used anti-floating measures include the anti-floating plate method, weight pressing method, anchor bolt method, and anti-floating pile method (China Construction Southwest Investigation and Design Institute Co. 2019). With the advantages of simple force, flexible arrangement, convenient construction, short construction period, and low cost, the expanded head anti-floating anchor is widely used in the design of basement anti-floating (Cai 2014). Because the damage model of the expanded head anti-floating anchor in tension is complex and can easily cause irreversible engineering accidents if not designed properly, it is important to study the damage model of the expanded head anti-floating anchor.

Ghaly et al. (Ghaly 1991) studied helical anchors in dense, medium dense, and loose sandy soils under tension by scaled-down model tests, respectively, and concluded that the damage mode of helical anchors is related to the ratio of burial depth to anchor disc diameter. When the ratio of burial depth to anchor disc diameter is greater than or equal to 11, 9, and 7, respectively, the spiral anchor undergoes deep burial damage, when the ratio of burial depth to anchor disc diameter is less

 DOI 10.1201/9781003310884-57

than 11, 9 and 7, respectively, the spiral anchor undergoes shallow burial damage. Meanwhile, Yuan Chi et al. (Yuan 2019) used ABAQUS software to simulate the uplift damage of helical anchors in loose sand, and the conclusions reached were basically the same as those of the tests in loose sand by Ghaly et al. And Guo Gang et al. (Guo 2012) studied the tensile damage of expanded anchor rods in sandy soil by scaled-down model tests, and also concluded that the damage of anchor rods was related to the ratio of burial depth T to expanded head diameter D. When $T/D \geq 9$, deep burial damage occurred in anchor rods, when $5 < T/D < 9$, transition damage occurred in anchor rods, and when $T/D \leq 5$, shallow burial damage occurred in anchor rods. Therefore, this paper uses the finite element software ABAQUS to conduct a theoretical study on the tensile damage of expanded head floating anchors in powdered clay soil.

2 FORCE ANALYSIS OF ANTI-FLOATING ANCHORS WITH THE EXPANDED HEAD

2.1 *Force composition of the expanded head anti-floating anchor*

The enlarged head anti-floating anchor consists of enlarged head section L_1, common anchor section L_2, the anchor bar, etc. The specific construction is shown in Figure 1. Under the action of tension P, the anchor rod will be subjected to the frictional resistance $f_1$and f_2 generated by the grouted sidewall and soil, the resistance of the soil to the expanded head section T_1, the suction force T_2 generated at the bottom when the anchor rod is displaced upward, and the self-weight G of the anchor rod, and the force sketch of the anchor rod is shown in Figure 2. Since the suction force T_2 and self-weight G are small, they are generally neglected in practical engineering. Resistance T1 is the single most important force in the tensile resistance of the expanded head floating anchor.

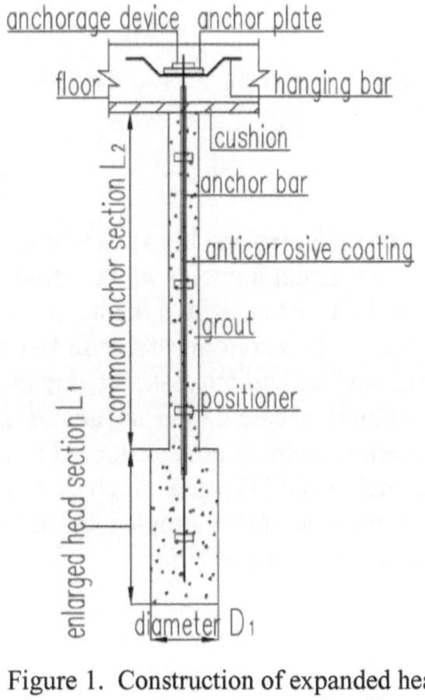

Figure 1. Construction of expanded head anti-floating anchor.

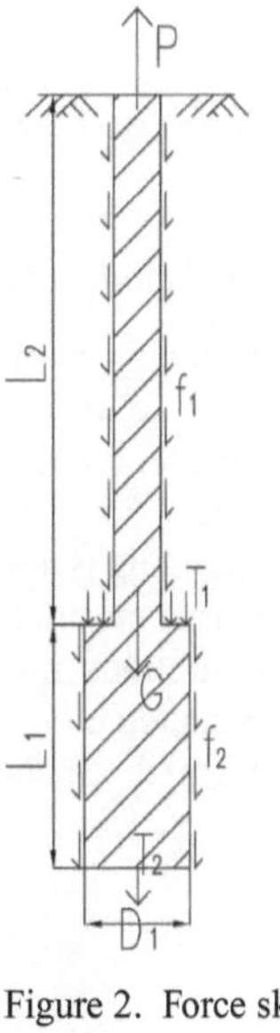

Figure 2. Force sketch.

2.2 *Damage types of expanded head anti-floating anchors*

When the expanded head anti-floating anchor is subjected to ultimate tension, the following types of damage to the anchor rod will occur roughly: 1) the anchor bar is pulled out from the expanded body; 2) the anchor bar breaks (or yields); 3) the anchor rod is pulled out with the surrounding soil in a conical shape, and the soil is damaged by shear as a whole.(Fu 2014), as shown in Figures 3, 4) the soil at the end of the expanded head is squeezed, and local shear damage occurs (Chen 2013), as shown in Figure 4.

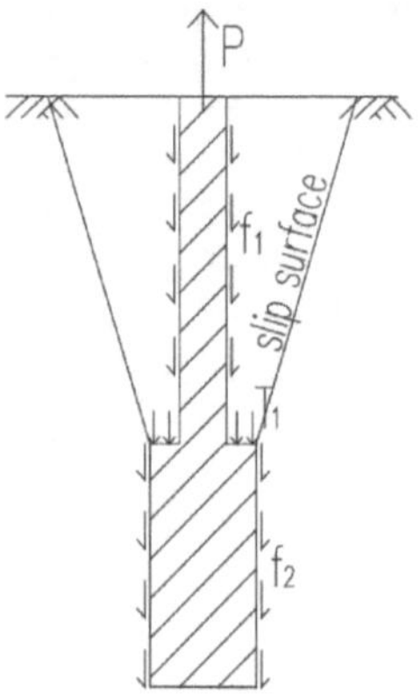

Figure 3. Overall shear damage.

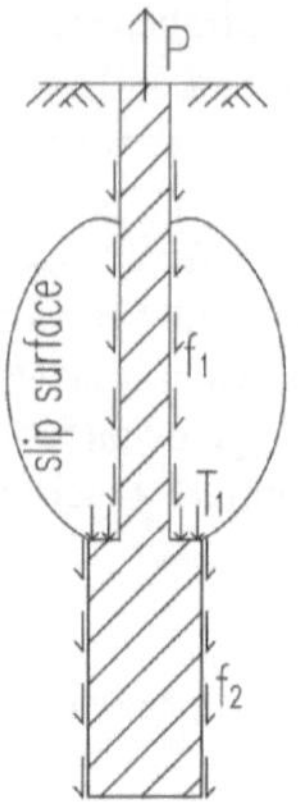

Figure 4. Local shear damage.

From engineering experience and basic damage tests of anchor rods, it can be concluded that most of the damage of expanded head anti-floating anchor rods are fracture or yielding of the anchor bars in the anchor rods and pulling out of the anchor bars from the expanded body, for which two kinds of damage can be solved in engineering by adjusting the material parameters of the anchor rods. This paper focuses on the latter two complex damage modes by numerical simulation. Combining the contents of the study by Guo Gang (Guo 2012) and others, the overall shear damage and local shear damage of the expanded head anti-floating anchor correspond to shallow and deep burial damage, respectively, and the burial depth ratio L_2/D_1 is used to distinguish shallow and deep burial damage.

3 NUMERICAL ANALYSIS OF ANTI-FLOATING ANCHORS WITH THE EXPANDED HEAD

3.1 *Computational model and basic assumptions*

In ABAQUS software, the numerical model of the single anchor rod under tension is an axisymmetric spatial model. Therefore, the calculation model can be 1/4 of the actual model, with the length and width of the model are taken as 10 times the diameter of the enlarged head, and the height of the model is taken as 2 times the total length of the anchor rod. The model's bottom is set with fixed restraint, the surrounding is normal restraint, the upper part is free surface without restraint, and the calculation model is shown in Figure 5. In order to facilitate the numerical calculation and analysis, the following assumptions are made: 1) the anchor rod and the reinforcement are considered as a whole and are simulated by solid units, and the strength of the anchor rod is large enough that no damage will occur; 2) the anchor rod and the soil are uniformly continuous and isotropic materials; 3) the soil material conforms to the Mohr-Coulomb principal structure relationship.

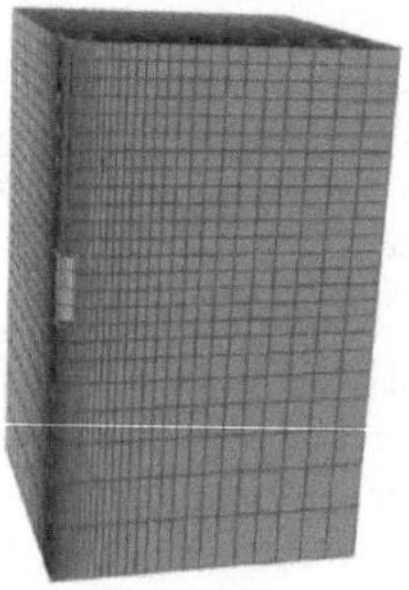

(a) Soil model

(b) Anchor model

Figure 5. Numerical calculation model.

3.2 *The value of material parameters and the setting of the test group*

The physical parameters of the pulverized clay were taken in combination with the geological survey data and engineering experience of a project in Nanjing. The Material parameters of anchor rods are taken according to engineering experience. The values are shown in Table 1. During the numerical analysis: The length of the enlarged head section L_l and the diameter of the enlarged head D_l were kept constant, and the burial depth of the enlarged head was changed by changing the length of the common anchorage section L_2, and tension was applied to the anchor rod until the anchor rod broke, and the relationship between the burial depth ratio L_2/D_l and the damage model of the anchor rod was studied and analyzed. The specific experimental setup is shown in Table 2.

Table 1. Parameters of anchor and soil layer.

γ(N/m^2)	E (Pa)	Poisson's ratio υ	φ(°)	C (Pa)	
Enlarged head anchors	2700*	30×10^9*	0.20*	–	–
Powdery clay	1850	30×10^6	0.30*	15.0	20000

* indicates the empirical value.

Table 2. Experimental group settings.

Group number	1	2	3	4	5	6	7
D_l(m)	0.5	0.5	0.5	0.5	0.5	0.5	0.5
L_l(m)	1.0	1.0	1.0	1.0	1.0	1.0	1.0
L_2(m)	1.0	2.0	3.0	4.0	5.0	6.0	7.0
L_2/D_l	2.0	4.0	6.0	8.0	10.0	12.0	14.0

3.3 *Numerical simulation results*

Because the numerical simulation of the single expanded head anti-floating anchor is spatially axisymmetric, the nephogram map of the calculation results is also spatially axisymmetric distribution. For the convenience of analysis, the calculation results are displayed in a mirror image 1/2 model.

By observing the maximum principal stress nephogram diagram of the numerical simulation results of the enlarged head anti-floating anchor rod under tension, it is found that the maximum principal stress nephogram diagram of the soil around the anchor rod has three shapes: 1) when the burial depth ratio $L_2/D_l \leq 2$, the maximum principal stress nephogram diagram of the soil is scattered in an inverted cone from near the end of the enlarged head and extends to the ground, as shown in Figure 6-(a), the soil stress within the inverted cone is divided into compressive stress except for the head of the enlarged head. The rest are tensile stresses, and the soil outside the conical range generates compressive stresses under the action of self-weight, so when the anchor rod is under tension, a conical slip crack surface will be generated and the soil will undergo overall shear damage, i.e. shallow burial damage; 2) When the burial depth ratio $L_2/D_l \geq 6$, the shape of the maximum principal stress nephogram diagram of the soil around the end of the enlarged head is roughly a closed ellipse, and the uppermost part of the ellipse has a certain depth from the ground, as shown in Figure 6-(c)(d) shows, it is not difficult to analyze the nephogram diagram to find that the soil near the expanded head end within a certain range produces a flames-like compressive stress zone, and the compressive stress zone is enveloped by a larger elliptical tensile stress zone, and the soil outside the elliptical range produces tensile stress, and the soil undergoes local shear damage, i.e., deep burial damage; 3) When the burial depth ratio L_2/D_l is between 2 and 6, the maximum principal stress nephograms around the end of the enlarged head is similar to the above case 2), and the soil undergoes local shear damage, but near the ground, the soil around the anchor rod generates a conical stress area and is larger than the stress of the surrounding soil, and small-scale conical shear damage occurs, and the stress nephogram is shown in Figure 6-(b),The damage occurring in the anchor rod has the characteristics of both deep burial damage and shallow burial damage, which is called transitional anchor rod, from a safety point of view, this damage is attributed to shallow burial damage. The above three cases are combined because the transition anchor is a shallow buried anchor. Therefore, there are two modes of damage of the expanded head anti-floating anchor in powder clay under tension. They are related to the burial depth of the expanded head: when the burial depth ratio $L_2/D_1<6$, shallow burial damage of anchor rod occurs; when the burial depth ratio $L_2/D_1 \geq 6$, deep burial damage of anchor rod occurs.

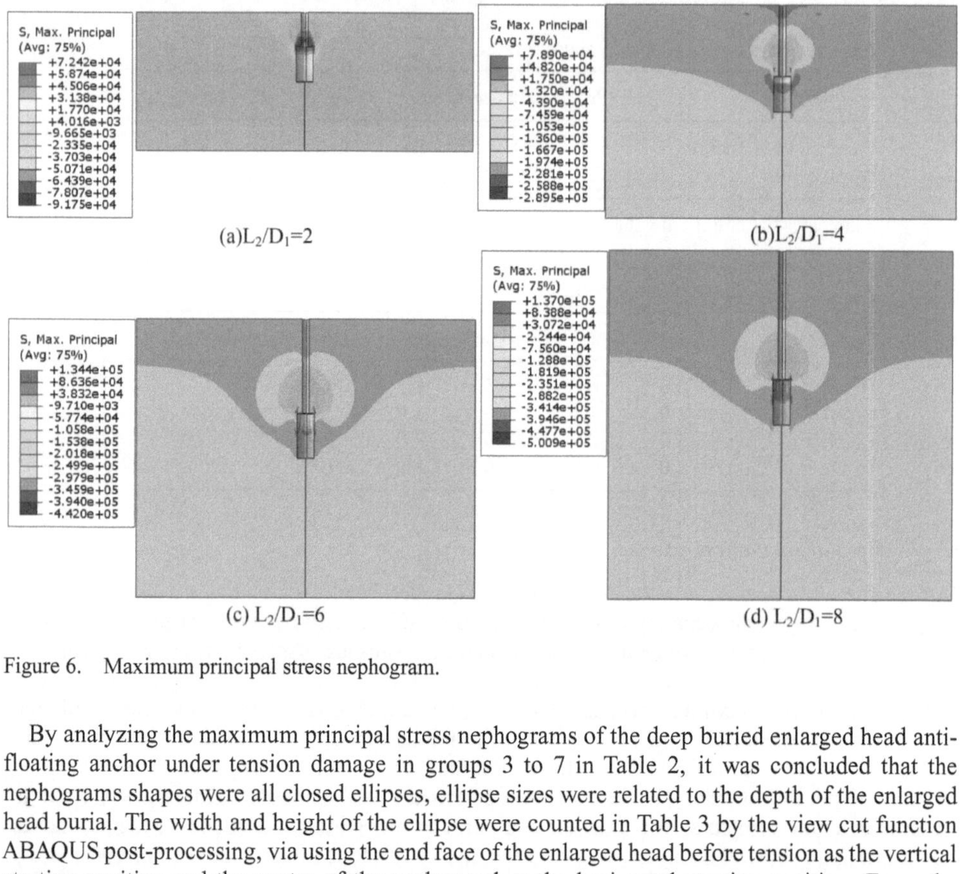

Figure 6. Maximum principal stress nephogram.

By analyzing the maximum principal stress nephograms of the deep buried enlarged head anti-floating anchor under tension damage in groups 3 to 7 in Table 2, it was concluded that the nephograms shapes were all closed ellipses, ellipse sizes were related to the depth of the enlarged head burial. The width and height of the ellipse were counted in Table 3 by the view cut function ABAQUS post-processing, via using the end face of the enlarged head before tension as the vertical starting position and the centre of the anchor rod as the horizontal starting position. From the analysis of the data in Table 3, it can be obtained that the size of the ellipse formed by the maximum principal stress nephogram gradually tends to be stable with a width of $5D_1$ and a height of roughly $4D_1$ after the anchor rod burial depth is greater than 6m.

Table 3. Ellipse height and width dimensions.

D_1(m)	0.5	0.5	0.5	0.5	0.5
L_2(m)	4.0	5.0	6.0	7.0	8.0
Elliptical width (m)	2.89	2.77	2.50	2.50	2.50
Elliptical height (m)	2.05	2.14	2.12	2.07	2.02

From the numerical simulation results of the tensile damage of anchor rods, the load-displacement curves corresponding to the damage of anchor rods with different burial depths are different. The load-displacement curves corresponding to the tensile damage of anchor rods are plotted in the same coordinate system with four different burial depths, to facilitate the analysis of the differences of load-displacement curves, as shown in Figure 7.

When the displacement is very small, the load-displacement curves corresponding to the four different buried anchor rods are sloping straight lines with the same slope and in the elastic stage. With the increase of displacement, the anchor rod enters the plastic stage: the slope of the load-displacement curve of shallowly buried anchor rod decays almost to 0, forming a straight line parallel to the horizontal coordinate, the bearing capacity of the anchor rod quickly reaches the limit value and instantaneous damage occurs, and shallowly buried anchor rod should be avoided in engineering. After the deeply buried anchor rod enters the plastic stage, the trend of the slope of the curve is the same. With the increase of displacement, the slope decays more slowly, and the slope is greater than 0. The anchor rod will not be damaged instantly. Under the same tensile force, the displacement of anchor rods with large burial depth is smaller than that of anchor rods with small burial depth, which indicates that increasing the burial depth can improve the pullout resistance of anchor rods and the depth of burial of anchor rods can be increased reasonably to improve the pullout resistance in engineering. After the transitional anchor rod enters the plastic stage, the curve characteristics are similar to those of the deeply buried anchor rod within a certain displacement range. Still, after the displacement exceeds this range, the slope decay rate of the curve is between the deeply buried anchor rod and the shallow buried anchor rod, but close to 0, similar to the damage characteristics of the shallow buried anchor rod, so the transitional anchor rod is classified as the shallow buried anchor rod.

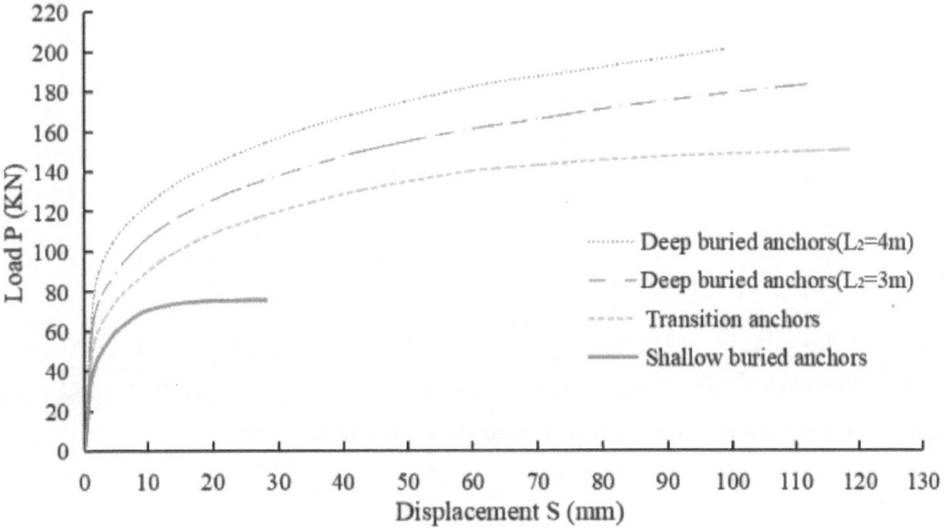

Figure 7. Load-Displacement curve.

3.4 *Deeply buried expanded head anti-floating anchor damage process*

Through numerical simulation of the tensile damage of the anchor rod with a deeply buried expanded head, it is concluded that the damage process of the anchor rod is similar to the findings of Zeng Qingyi et al. The soil around the anchor rod roughly goes through three stages: elastic stage, plastic zone formation stage, plastic zone soil strengthening and range expansion stage (Zeng 2010).

Elastic stage: When the tension force P is small, the frictional resistance f_1 and f_2 are generated between the sidewall of the grouted body of the anchor rod and the soil layer, at this time, the tension force P is balanced with the frictional resistance f_1 and f_2 and the self-weight G of the anchor rod, see Figure 8-(a), the anchor rod does not undergo upward displacement, and the soil body around the anchor rod is in the elastic stage. Plastic zone formation stage: When the tensile force P is greater than the ultimate frictional resistance between the sidewall of anchor grout and soil layer, the anchor will be displaced upward, and the soil at the front of the enlarged head will be squeezed, with the increase of displacement, the soil at the front of the enlarged head will gradually develop from elastic deformation to local plastic deformation, forming a local plastic zone, see Figure 8-(b). If the tension P continues to increase, the local plastic zone of the soil at the front of the enlarged head is connected into a closed whole, and the range increases, as in Figure 8-(c). Strengthening and expansion of the plastic zone: After the plastic zone of the soil at the front of the enlarged head is connected, with the increase of tension, the soil in the plastic zone at the front of the enlarged head is continuously compressed by the surrounding soil pressure and the extrusion of the enlarged head, and when the surrounding pressure is greater than the ultimate pressure that the soil can withstand in its initial state, the soil shears and shrinks, the strength of the soil is increased, and the pullout force of the anchor increases. If the tensile force continues to increase, when the surrounding soil pressure is less than the ultimate pressure that the reinforced soil can withstand, the soil in the plastic zone will shear and expand, the plastic zone will expand and form an ellipse, and local shear damage will occur in the soil around the anchor rod, as shown in Figure 8-(d).

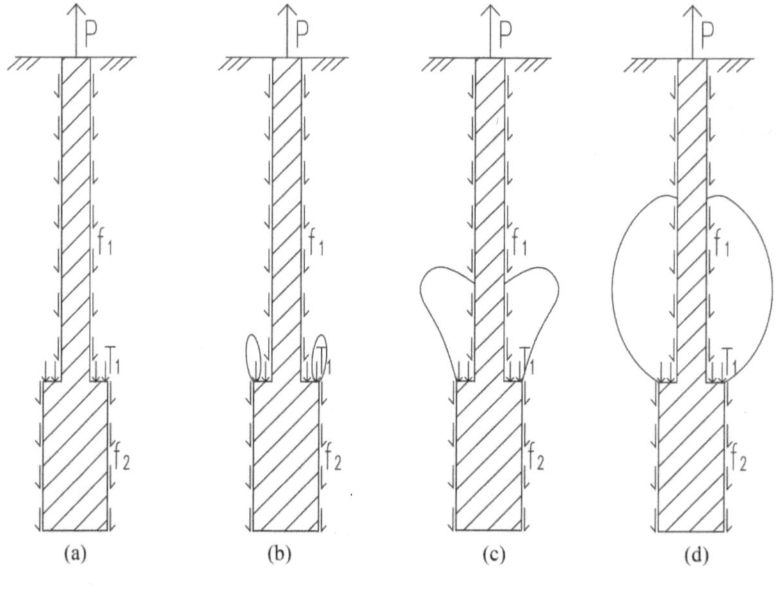

Figure 8. The tensile damage process of the expanded head anti-floating anchor.

3.5 *Discussions*

In this paper, based on the indoor test of the expanded head anchor by Guo Gang [6] et al. and the theoretical analysis of the tensile damage of the expanded head anti-floating anchor in powdered clay using the finite element software ABAQUS, the research results are at the theoretical stage, so that the research content of this paper can bring better guidance to the actual project, the next step is to conduct an in-depth study by combining the results of the theoretical analysis with field tests on the tensioning of the expanded head floating anchors in pulverized clay.

4 CONCLUSION

Based on the results of the above study, the following conclusions were drawn:

(1) There are two modes of tensile damage to the expanded head anti-floating anchor in powdery clay soil: when the burial depth ratio $L_2/D_1<6$, shallow soil burial damage occurs to the anchor

rod. When the burial depth ratio $L_2/D_1 \geq 6$, the anchor rod undergoes deep burial damage, combining with the findings of Guo Gang [6] et al. on the study of expanded body anchors subjected to tension in sandy soil, it is deduced that the damage modes of expanded head anti-floating anchors in different geotechnical layers correspond to different burial depth ratios and the reasonable burial depth should be determined according to the bearing layer of anchors in practical projects by combining geological survey data.

(2) In powdery clay soil, when the expanded head anti-floating anchor produces small displacement, the corresponding load-displacement curves of shallowly buried anchor, transitional anchor, and deeply buried anchor are all oblique straight lines with similar slope and are in the elastic stage, after the anchor enters the plastic stage, the slope of the load-displacement curve of shallowly buried anchor decays rapidly to 0. It is a straight line parallel to the horizontal coordinate, and the anchor is instantaneously damaged, so shallowly buried anchor should be avoided in engineering. The decay rate of the load-displacement curve slope of the deeply buried anchor is slow and greater than 0. The decay rate of the load-displacement curve slope of transition type anchor is between the deep-buried anchor and shallowly buried anchor, but close to 0, similar to shallow buried anchor, so the transition type anchor is classified as the shallow buried anchor.

(3) In the powder clay soil, when the burial depth of the expanded head anti-floating anchor is greater than 6m, the ellipse range formed by the maximum principal stress nephogram diagram of the soil at the end of the expanded head tends to stabilize, with an ellipse width of $5D_1$ and a height of roughly $4D_1$; the tensile damage of the deeply buried expanded head anti-floating anchor will roughly go through the elastic stage, the plastic zone formation stage and the plastic zone soil strengthening and range expansion stage.

ACKNOWLEDGMENTS

The author would like to thank you for the support from the National Fund Project (Project No. 51208228).

REFERENCES

Cai H.W. (2014). Application of anti-floating anchors in the design, construction and acceptance of an engineering project[J]. *Journal of Underground Space and Engineering*., 10:1926–1929.

Chen Shuai et al. (2013). Analysis and numerical simulation of uplift bearing capacity of floating anchors[J]. *Journal of Water Resources and Construction Engineering*., 11(5):39–44.

China Construction Southwest Investigation and Design Institute Co. *476JGJ-2019 technical standard for anti-floating of construction projects* [S]. Beijing: China Construction Industry Press, 2019:32–33.

Fu Man-kwong et al. (2014). Study on the stability verification formula of float-resistant anchor and anchor anti-float system[J]. *Journal of Geotechnical Engineering*., 36(11):1971–1982.

Ghaly A, Hanna A, Hanna M. (1991). Uplift Behaviour of Screw Anchors in Sand, I: Dry Sand[J]. *Journal of Geotechnical Engineering, ASCE*, 117 (GT5): 773–793.

Guo Gang et al. (2012). Experimental study on vertical pullout damage mode of expanded body anchors with different burial depth[J]. *Industrial Construction*., 42(1):123–128.

Wang Y.Y. & Wang N.W. et al. (2015). Design and testing of anti-floating anchors in a project[J]. *Journal of Water Resources and Construction Engineering*., 13(2):171–175.

Yuan Chi et al. (2019). Numerical analysis of uplift bearing capacity and damage mode of helical anchor in loose sand[J]. *Journal of Northeastern Electric Power University*., 39(6):50–57.

Zeng Q.Y. & Yang X.Y. & Yang C.Y. (2010). Mechanical mechanism and calculation method of expanded head anchors[J]. *Geotechnics*., 31(5):1359–1367.

Advances in Civil Engineering: Structural Seismic Resistance, Monitoring and Detection – Mohd Johari Mohd Yusof and Junwen Zhang (Eds)
© 2023 The Authors, ISBN 978-1-032-31491-4

The experimental study on the CO_2 erosion characteristic of concrete

Yufei Sun, Haiting Wang, Haodang Lu, Yong He & Lihao Fan
Powerchina Xibei Engineering Corporation Limited, Power China, Xi'an, China

ABSTRACT: In this paper, chemical analysis and microanalysis methods were employed for experimental research on mortar specimens with different water-to-binder ratios and different ages. The test results show that under the CO_2 erosion conditions, the cumulative dissolution ratio of CaO of the specimens continues to increase, and the erosion effect takes place from the outside to the inside, which only affects the surface area of the specimens in a short time, and reducing the water-to-binder ratio improves the resistance of concrete. The ability to decompose erosion is found to be advantageous.

1 INTRODUCTION

The durability of concrete is said to be the ability of concrete to resist damage caused by environmental effects, chemical erosion, abrasion, or other continuous effects (Bai & Wang 2004; Li & Cao 2001; Monteiro & Mehta 2006). Concrete materials encounter a series of complex and severe physical and chemical erosion effects during use, including soft water, erosion due to CO_2, sulfate, and chloride ion corrosion, etc.; CO_2 corrosion is considered the major factor in decreasing the durability of concrete (Du & Li 2004; Xu & Zhang 2017; Wang & Guo 2012; Zhao & Xing 2013).

Generally, water itself often contains a small amount of $Ca(HCO_3)_2$, and there must be a certain amount of free CO_2 in equilibrium. When the carbonic acid in the water increases, more $Ca(HCO_3)_2$ will be generated due to the erosion of CO_2. The chemical reaction formula is shown in formula 1 (Guo 2012). The corrosive environment of CO_2 further promotes physical $Ca(OH)_2$ leaching by changing the surface structure of cement-based materials. This double corrosion due to chemical attack and physical leaching makes it more significant to concrete damage (Yuan 1996).

$$CO_2 + H_2O + Ca\,(OH)_2 \rightarrow Ca\,(HCO_3)_2 \tag{1}$$

The hydration products in concrete can only exist stably in a highly alkaline environment. Due to CO_2 erosion, once the pH value in the concrete pore solution decreases, the originally stable hydration products begin to decompose; thereby, changing the composition and structure of the concrete material. It affects the stability of the system (Hu 2009), and the decrease in the pH value of the pore solution leads to formation of a passive film on the surface of the steel bar thus activating it, which may cause damage to the passive film, which will lead to corrosion of the steel bar (Andrade & Izquierdo 2005). Therefore, CO_2 decomposing erosion leads to a series of destructive effects on concrete.

Based on the experiment, this paper uses chemical analysis, scanning electron microscopy, and mercury intrusion method to study the impact of CO_2 decomposing erosion on concrete.

2 EXPERIMENTAL

2.1 *Materials*

The cement is Jidong 42.5 is an ordinary Portland cement. The fly ash is Grade I fly ash from Mudanjiang Second Power Plant. The aggregate is sand and gravel processing system of the lower

 DOI 10.1201/9781003310884-58

reservoir from a pumped storage power station. The CO_2 decomposable corrosive medium uses PY-04 softening water equipment, and continuously injects CO_2 into the soft water at a constant speed during the test to simulate CO_2 decomposing erosion conditions. Dissolution erosion medium uses soft water as the dissolution erosion medium to simulate dissolution erosion. The water quality analysis is shown in Table 1.

Table 1. Chemical index results of water erosion media in simulated decomposing erosion environment.

Type of erosion	Main performance index				
	Ca^{2+} (mmol/L)	Mg^{2+} (mmol/L)	pH value	Aggressive CO_2 concentration (mg/L)	HCO_3^- (mmol/L)
CO_2 decomposing erosion	0.03	0.01	5.13	150	/
Dissolution erosion	0.03	0.01	7.16	/	0.64

2.2 *Experiment method*

Specimen preparation: The test mixes two-stage normal concrete, as shown in Table 2, and removes small stones and middle stones through screening, forming 40 mm × 40 mm × 40 mm mortar specimen. During the test, the concrete forming and mortar specimen maintenance are in accordance with the *Hydraulic Concrete Test Regulations* (DL/T 5150-2001).

Table 2. Concrete mix ratio.

W/C	Gradation	Fly ash content (%)	Sand rate (%)	Slump (mm)	Air content
0.3	2	20	34	75	1.2
0.4		20	36	80	1.0
0.5		20	38	82	1.0
0.6		20	40	91	1.3

Erosion test: Put the specimens cured to a certain age into the CO_2 decomposing erosion test equipment. The test equipment uses a cylindrical container with constant speed stirring, and conducts dissolution type erosion test under the same conditions as the comparison. The test period is at regular intervals. Change the CO_2 erosion solution and the dissolution type erosion solution at regular intervals. The amount of solution to be replaced each time remains constant. While changing the water, samples of the erosion solution are collected, the ion concentration changes in the solution are measured, and the solution is analyzed in a certain erosion test period. At the same time, the microscopic analysis method is used to analyze the microscopic morphology and porosity of the specimen after decomposing erosion.

3 ANALYSIS

3.1 *Analysis of changes in cumulative dissolution ratio of CaO*

The calcium ion concentration is determined by EDTA titration with an accuracy of ±0.04 mmol/L, which is approximately equivalent to ±2 drops of EDTA solution.

The test measures the calcium ions precipitated in the erosion solution for different immersion times, and calculates the cumulative dissolution ratio of CaO in the test pieces with different water-to-binder ratios. The test results are shown in Figures 1 and 2.

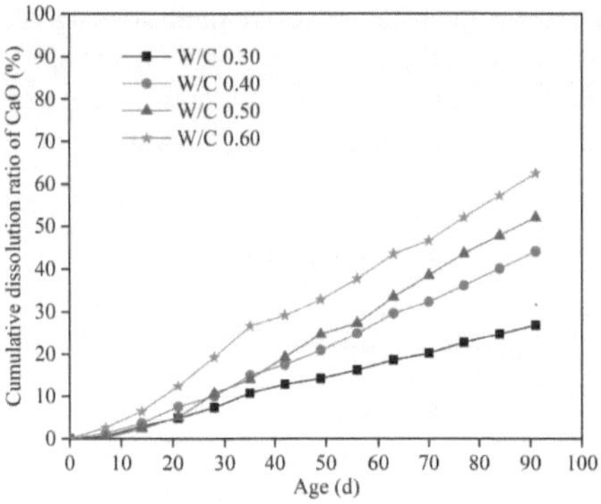

Figure 1. Changes of cumulative dissolution ratio of CaO in CO_2 decomposing erosion test with age.

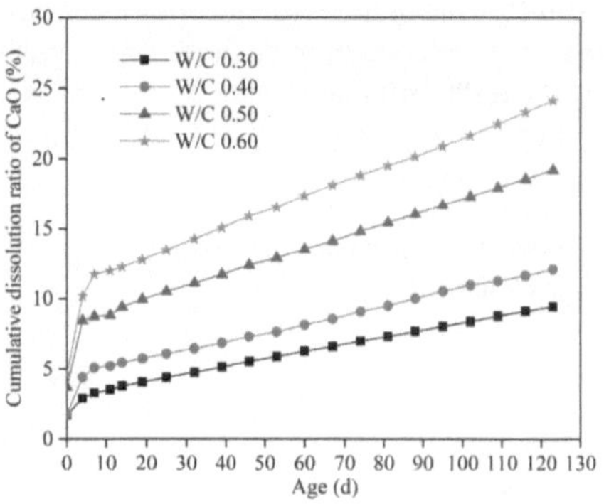

Figure 2. Changes of cumulative dissolution ratio of CaO in dissolution erosion test with age.

It can be seen from the test results that there is a significant difference in the accumulation and dissolution processes of CaO for decomposing erosion and dissolution erosion. In the early stage of the test, the cumulative dissolution rate of CaO for CO_2 decomposing erosion was low, but it increased in the middle and late stages of the test and maintained at a higher rate. While the rate of increase of the cumulative dissolution rate of CaO for dissolution erosion was higher in the early stage, it decreases in the mid-term, and then slightly increases, and maintains a certain rate of growth. The overall growth rate is significantly less than the CO_2 decomposing erosion.

Since the CO_2 decomposing erosion test medium adopts soft water to pass CO_2 to simulate the erosion process, the erosion process includes two parts: CO_2 decomposability and dissolution type. The cumulative dissolution ratio of CaO in the dissolution type erosion process varies with age. The comparison of period changes, the CO_2 decomposing erosion process is divided into two stages:

a. Early surface carbonization deposition stage
In this stage, because the specimen is in a state of hydrostatic curing before the start of the test, there is calcium hydroxide generated due to hydration in the pores of the specimen, and a highly alkaline environment is present in the pores. In this environment, the pore solution is decomposed and corroded. The main product should be insoluble calcium carbonate instead of soluble calcium bicarbonate (the reaction is shown in formulas 2, 3, and 4), so it may play a role in the early decomposing erosion. A buffering and hindering effect manifests as a slower growth rate of the cumulative dissolution ratio of CaO at the beginning of the experiment.
b. Continuous degradation stage
In this stage, as the Ca^{2+} in the pores produced insoluble $CaCO_3$ during the initial stage of the test, the alkalinity in the pores gradually decreased during the reaction, and the calcium hydroxide CH accumulated on the surface was quickly consumed, while the Ca^{2+} needed to be deeper in the specimen. At this time, CO_2 will further react with $CaCO_3$, covering the surface of the specimen (as shown in formula 5) to form soluble $Ca(HCO_3)_2$. The soluble $Ca(HCO_3)_2$ will corrode the replacement when environmental water is taken away, which further promotes the conversion of this insoluble matter to soluble matter, so that the insoluble matter covering the surface layer is gradually reduced and the pores of the specimen are exposed again, which is due to erosion. As the reaction progresses further, the porosity of the surface layer of the test piece will further increase, so that the decomposing erosion effect will further advance to the deeper level of the test piece, and the growth rate of the cumulative dissolution ratio of CaO will increase.

$$H_2CO_3 \rightarrow H^+ + HCO_3^- \tag{2}$$

$$2OH^- + H^+ + HCO_3^- \rightarrow CO_3^{2-} + 2H_2O \tag{3}$$

$$Ca^{2+} + CO_3^{2-} \rightarrow CaCO_3 \tag{4}$$

$$CaCO_3 + CO_2 + H_2O \rightarrow Ca^{2+} + HCO_3^- \tag{5}$$

During the CO_2 erosion test period, the increase in the cumulative dissolution ratio of CaO did not show a slowing trend, indicating that the CO_2 erosion continued during the test period, and the impact on the specimen was more obvious within a certain period.

During the CO_2 erosion process of the specimens with different water-to-binder ratios, the growth rate of the cumulative dissolution ratio of CaO of the specimens also has a certain difference. The larger W/C, the greater the growth rate of the cumulative dissolution ratio of CaO, indicating that the water is reduced. The glue ratio is beneficial to improve the CO_2 erosion resistance of concrete.

3.2 *Specimen microstructure analysis*

To compare and analyze the microscopic morphology of the specimens subjected to CO_2 erosion, while the corrosion test is carried out, the specimens formed under the same conditions will be subjected to standard curing, and the standard curing conditions and CO_2 erosion will be carried out. The specimens (W/C 0.3 and W/C 0.6, respectively, and the corrosion age of the specimens is 90d) are subjected to microscopic observation and analysis.

From the SEM analysis test results (shown in Figure 3), it can be seen that: For the test piece with W/C 0.6, after it is corroded by CO_2 decomposing, there are traces of erosion on the surface area of the test piece, the structure is relatively loose, and has a number of holes (Figure 3a), and the crystals inside the specimen are complete, the structure is dense, and the holes are relatively few (Figure 3b). The surface of the specimen cured under standard conditions (Figure 3c) did not show obvious signs of erosion, and the structure was relatively dense. The interior (Figure 3d) was similar to the inside of the eroded specimen, and the structure was denser. This shows that the role

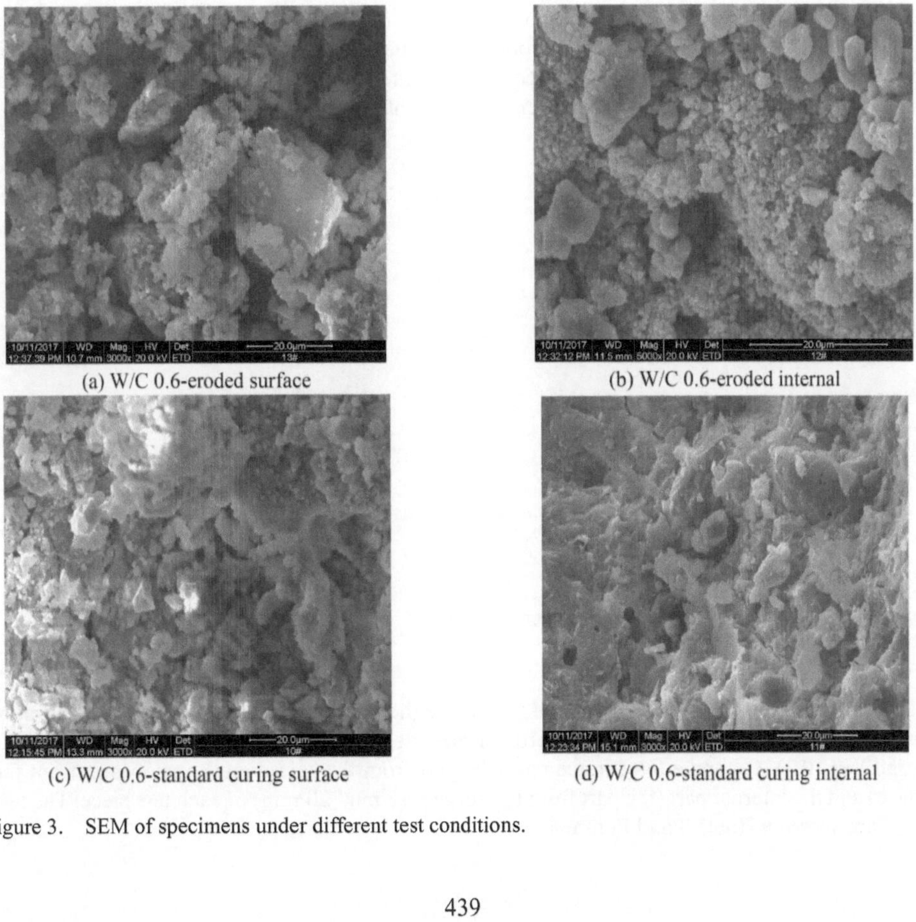

(a) W/C 0.6-eroded surface

(b) W/C 0.6-eroded internal

(c) W/C 0.6-standard curing surface

(d) W/C 0.6-standard curing internal

Figure 3. SEM of specimens under different test conditions.

(e) W/C 0.3-eroded surface

Figure 3. Continued.

of CO_2 decomposing erosion should be carried out from the outside to the inside, and will not affect the internal structure of the concrete in a short time. For the test piece with W/C 0.3, after it is corroded by CO_2 decomposing, the surface of the test piece (Figure 3e) is also partially eroded, but the degree of erosion is obviously smaller than W/C 0.6, which also shows that the smaller water-to-binder ratio, the stronger the ability of the concrete to resist CO_2 decomposing erosion.

The test performed energy spectrum analysis on the surface of the specimen with W/C 0.6 under CO_2 decomposable erosion conditions and under standard curing conditions (as shown in Figure 4). The test results show that: After CO_2 eroded, the value of Ca/Si on the surface is 0.53, which is significantly smaller than the value of Ca/Si (4.88) on the surface of the specimen under standard curing conditions. Therefore, there has been a significant loss of calcium.

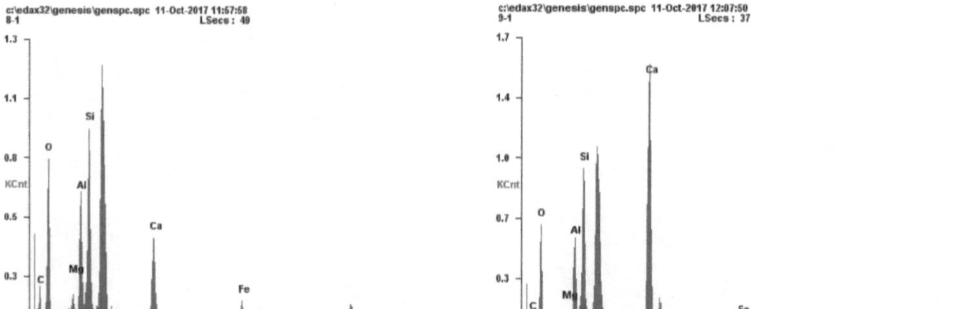

(a) Eroded Ca/Si=0.53

(b) Standard curing Ca/Si=4.88

Figure 4. Spectral analysis of specimens under different test conditions.

The test uses mercury intrusion method to analyze the pore characteristic parameters and pore size distribution of the specimen and the eroded specimen under standard curing conditions (W/C 0.6 and W/C 0.3). Compare the surface part (the part from the surface to 3 mm–5 mm from the surface) and the internal part (the part from the surface 15 mm–20 mm) of each test piece. The test results are shown in Table 3 and Figure 5.

Table 3. Pore characteristic parameters and pore size distribution of decomposable erosion specimens.

Serial number	W/C	Test conditions	Test site	of surface Porosity (%)	Growth rate Median porosity after dissolution (%)	pore size (nm)	Pore size (nm) distribution (%) <20	20~50	50~200	>200
1	0.6	Standard curing	Surface	14.92	0.00	31	34.47	26.63	13.81	25.09
2		Standard curing	Internal	15.55	—	38	34.50	27.10	12.75	25.65
3		Eroded	Surface	19.23	28.89	31	36.56	25.75	9.00	28.69
4		Eroded	Internal	15.65	—	31	39.13	22.69	12.69	25.49
5	0.3	Standard curing	Surface	13.02	0.00	37	37.01	25.39	14.68	22.92
6		Eroded	Surface	13.34	2.46	34	38.14	24.74	13.96	23.16

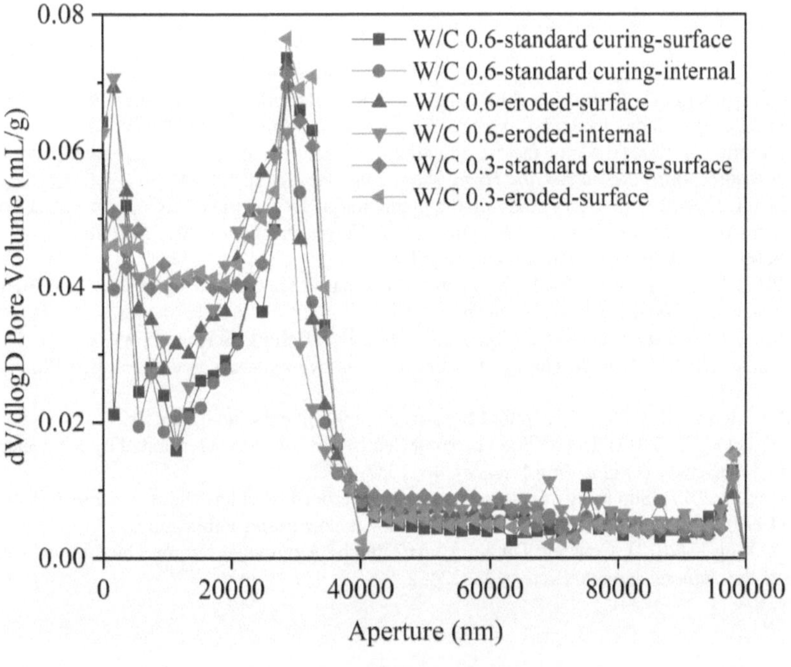

Figure 5. Analysis of porosity of specimens under different test conditions.

The surface porosity of the specimen with W/C 0.6 is significantly increased compared with the specimen cured under standard conditions after erosion, and harmful pores (aperture greater than 200 nm) increased, indicating that erosion is to a certain extent. The porosity of the surface layer of the specimen is increased, and some capillary channels and tiny pores are enlarged. However, the pore size distribution and porosity of the specimen under standard curing conditions are basically close to the pore size distribution inside the eroded specimen, indicating that the erosion of the specimen does not penetrate deep into the specimen, and only affects the surface area of the specimen.

The surface porosity of the sample surface of the sample with W/C 0.3 after erosion is not obvious compared with that of the sample surface cured under standard conditions, indicating that the decomposable erosion of the sample corrodes the surface part of the sample. Compared with the specimen with W/C 0.6, the porosity of the surface layer has a smaller change, which further shows that reducing water-to-binder ratio is beneficial to improve the decomposing and corrosion resistance of concrete.

4 CONCLUSION

The cumulative dissolution ratio of CaO precipitated at the initial stage of the concrete specimen subjected to CO_2 erosion has a slower growth rate, which has a certain inhibitory effect on the corrosion process, and then the growth rate increases and is basically stable, indicating the CO_2 erosion effect. Continue to proceed and aggravate the deterioration of the specimens and the increase in the cumulative dissolution ratio of CaO has not shown a trend of slowing down, indicating that CO_2 erosion will continue to affect the concrete specimens for a long time.

The greater water-to-binder ratio the greater the increasing rate of the cumulative dissolution ratio of CaO is also greater, indicating that reducing water-to-binder ratio is beneficial to improving the CO_2 erosion resistance of concrete.

The effect of CO_2 decomposing erosion should be carried out from the outside to the inside, and to a certain extent. The porosity of the surface part of the specimen is increased to expand the capillary channels and tiny pores of the concrete surface part.

REFERENCES

Andrade C. & Izquierdo D. (2005). Benchmarking through an algorithm of repair methods of reinforcement corrosion: The repair index method. *Cement & Concrete Composites*. 27(6), 727–733.

Bai B.Z. & Wang Y.R. (2004). Major factors affecting the durability of hydraulic structures and preventive counter measures. *Water Resources and Hydropower Engineering*. 10, 13–16.

Du Y.J. & Li Y.T. (2004). Preliminary study on the prediction of service life of high performance concrete in sulphate erosion resistance. *Journal of Northwest A&F University (Natural Science Edition)*. 32, 100–102.

DL/T 5150-2001 Test code for hydraulic concrete [S].

Guo G.F. (2012). Study on Corrosion Characters of Cement-based Materials in Aggressive CO_2 Environment. Guangzhou: *South China University of Technology*.

Hu S.G. (2009). *Advanced Cement-Based Composite Materials*. Beijing: Science Press.

Li J.Y. & Cao J.G. (2001). New development of the study on hydraulic concrete durability. *Water Power*. 4, 44–47.

Monteiro P. & Mehta P.K. (2006). Concrete: Microstructure. Properties and Materials.

Wang H.L. & Guo C.L. (2012). Degradation mesomechanism of concrete deteriorated by soft water. *Journal of Zhejiang University (Engineering Science)*. 46, 1887–1892.

Xu P. & Zhang Y. (2017). Summary of research on concrete eroded by chlorideions. *Concrete*. 9, 45–48.

Yuan R.Z. (1996). Cementitious Materials. Wuhan: National New Books Publications.

Zhao Z.Z. & Xing F. (2013). Concrete Durability Affected by Aggressive Carbon Dioxide in Groundwater. *Journal of the Chinese Ceramic Society*. 41, 224–229.

Advances in Civil Engineering: Structural Seismic Resistance, Monitoring and Detection – Mohd Johari Mohd Yusof and Junwen Zhang (Eds)
© 2023 The Authors, ISBN 978-1-032-31491-4

Research on portable falling weight deflectometer of testing subgrade quality

Jianyou Yu
Yanchong Construction Office of Hebei Expressway Group Limited, Zhangjiakou, Hebei, China

Yunfei Zhao*
School of Urban Geology and Engineering, Hebei GEO University, Shijiazhuang, Hebei, China

Weichao Liu
Department of Civil Engineering, Shijiazhuang Tiedao University, Shijiazhuang, China

Zhizhong Liu
Yanchong Construction Office of Hebei Expressway Group Limited, Zhangjiakou, Hebei, China

Guangqing Yang
Department of Civil Engineering, Shijiazhuang Tiedao University, Shijiazhuang, China

ABSTRACT: Subgrade quality inspection is a method of subgrade delivery acceptance. So the accuracy of the inspection directly affects the progress of the project and the quality of operations. Therefore, the selection of inspection methods is particularly important. The current construction standard of highway in China is based on the traditional quality inspection methods, which have problems like damage the roadbed or took too much time. A portable drop hammer deflector and Beckman beam method combined detection were used to test the quality of the road surface, and the correlation between the static and dynamic deflection values and the static and dynamic rebound modulus was obtained. The results show that the static and dynamic deflection has a good exponential correlation and the static and dynamic rebound modulus has a good logarithmic correlation. The results also indicated that quick test methods such as PFWD could do a good job on subgrade quality inspection as well.

1 INTRODUCTION

Subgrade is the foundation of highway, and having the quality of subgrade under control is the key to subgrade construction. As the physical index of subgrade process quality inspection, compaction can only judge whether the subgrade is compacted. However, it cannot tell whether the subgrade structure is stable. Deflection value and modulus of resilience can be the judgment criteria for the mechanical properties of subgrade structure, so it is the basis for the final roadbed quality judgment.

PFWD is the abbreviation of Portable Drop Weight Deflection Tester. It is a new type of rapid non-destructive testing equipment developed on the basis of FWD. It is a method to evaluate the bearing capacity of subgrades and foundations by measuring subgrade dynamic elastic modulus. In recent years, with the development of highway engineering, many experts and scholars have conducted research on PFWD and obtained research results, which provide a reference for rapid non-destructive testing of roadbeds in China. The joint methods of the PFWD method, the bearing plate method and the Beckman beam method were used to have the red sandstone roadbed detected, and have the correlation between the indicators analyzed, which obtained the value range of the red sandstone roadbed PFWD detection index (Yang 2010). The relationship between pressure and

*Corresponding Author: 89zyf@163.com

DOI 10.1201/9781003310884-59

deflection in the actual test of PFWD was studied, and the results show that the pressure of the instrument has a good linear relationship with deflection, and the relationship between modulus and pressure is a power function (Zha & Xiao 2011). The PFWD was used as a platform to predict the static deflection of Beckman beam for two different subgrade structures of gravel-bearing clay and gravel-bearing clay. The prediction results show that the BP neural network model is better than the multiple-regression-model (Sun et al. 2012). The rapid detection method of active subgrade in hot and humid areas have been studied, using sand filling method, bearing plate method, PFWD, DCP and other methods to have the subgrade detected, and have the physical mechanics such as deflection and elastic modulus established. The correlation between indicators and PFWD and DCP has been studied, and a method for PFWD to detect the performance of existing roadbeds is proposed (Tatsuoka 2021; Zhang et al. 2019).

The rapid non-destructive testing method is the requirement of highway subgrade construction and the future trend. However, the current construction standard of highway in China is based on the traditional quality inspection methods such as Beckman beam method and sand filling method. The traditional quality inspection methods used in the roadbed filling process and roadbed surface inspection have some problems such as damage to the roadbed structure, complicated operation process and long test time. Therefore, the non-destructive testing method such as PFWD has been proposed. But how to translate the results from PFWD into the Beckman beam method is the problem. So, based on the field test, regression analysis method is used to establish a correlation between the PFWD method and the Beckman beam method. So that it could provide a reference to the application in highway construction.

2 RESEARCH ON PFWD DETECTION METHOD

2.1 *Project overview*

The test project is based on the iron tailings roadbed of the expressway in the northern China. The project scale is large and construction period is short. It is very difficult to construct in such a harsh climate, especially for the construction of subgrade, and the quality of subgrade construction is the basic guarantee for the entire project. Subgrade inspection is an important way to inspect the construction quality. The Beckman beam method is used in the subgrade surface quality inspection of the project, that is, the standard deflection vehicle is used to inspect the measuring points after the subgrade is filled, but this method is used for inspection It takes a long time and requires more than three people to cooperate and complete. The PFWD instrument is also used in this project, and the instrument is directly placed on the detection point for detection. The detection time is short. It only requires two people to do it. Also, there is no standard value for the PFWD test. So it is necessary to set up a test section and use the PFWD method and determine the detection standard value of the instrument.

2.2 *Test section setting and joint tests*

The test section of the joint test was selected from K60+000 to K61+600 that was undergoing subgrade filling at the time, and test points were set every 20 m in the area. The test points were marked with spray paint or white gray, and the number was increased according to the mileage number. The layout of detection points is shown in Figure 1.

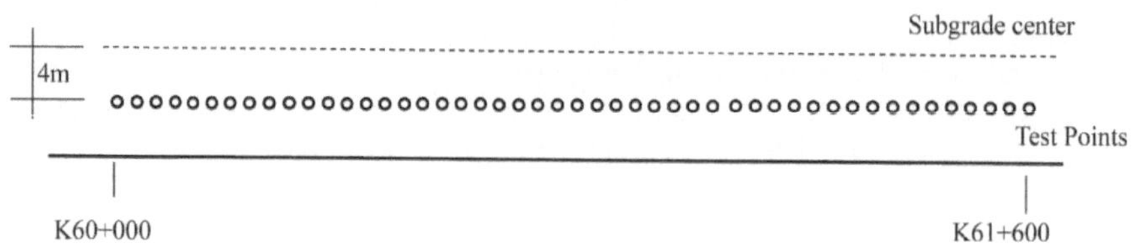

Figure 1. Layout of inspection points.

After the test section and measuring points are arranged, the joint inspection will be carried out. The joint inspection steps are as follows:

(1) Beckman beam method. Before testing the roadbed, inspect the instrument materials needed for testing, such as: standard car performance, accuracy of deflection gauge dial indicator, accuracy of thermometer, etc. During inspection, the standard car should move slowly forward based on the commander, and observe the dial indicator reading L1 during the driving process, record the dial indicator reading when it reaches the maximum value, and the car continues to move forward until it drives out of the deflection range (over 3 m) to stop, after the dial indicator turns stable, record the final reading L2, and then calculate the Beckman beam detection static deflection value LB.
(2) PFWD method. First, is to smooth the surface of the measuring point to remove the influence of uneven surface factors on the side measurement results. Then place the equipment on the surface of the corresponding roadbed measuring point, slightly rotate the instrument to make the instrument base and the roadbed surface contact well, turn on the instrument, and perform three pre-drop hammers on the test points before the formal test to discharge the impact of the un-compacted surface on the measurement results. Then perform five test drops for each test point, and record each falling hammer. The dynamic deflection value LP and the dynamic resilience modulus EP are the average dynamic deflection value and the average dynamic resilience modulus when processing the data.

2.3 *Determination of detection standard value*

After completing the test, analyze the test results of the two test methods. The Beckman beam method detects the static deflection value, and the static elastic modulus can be inversely calculated by the static deflection value; the PFWD method detects the dynamic deflection value and the dynamic elastic modulus. Two detection quantities are considered when determining the standard value of PFWD. The static deflection value and the dynamic deflection value can be used to calculate the dynamic deflection standard value, and the static and dynamic rebound modulus can be used to calculate the standard value of dynamic rebound modulus.

(1) Correlation analysis of static deflection value and dynamic deflection value. When processing the data, four models of exponential, linear, logarithmic and power are used to draw the regression curves of static deflection and dynamic deflection, as shown in Figures 3 to 6.

Figures 2 to 5 are the four regression curves of dynamic and static deflection regression analysis. Except for the logarithmic type, the other three models have good correlation. And the exponential correlation index R2 is 0.8807, which is the best, as it shown in formula (1):

$$L_P = 23.511e^{0.0165L_B} \tag{1}$$

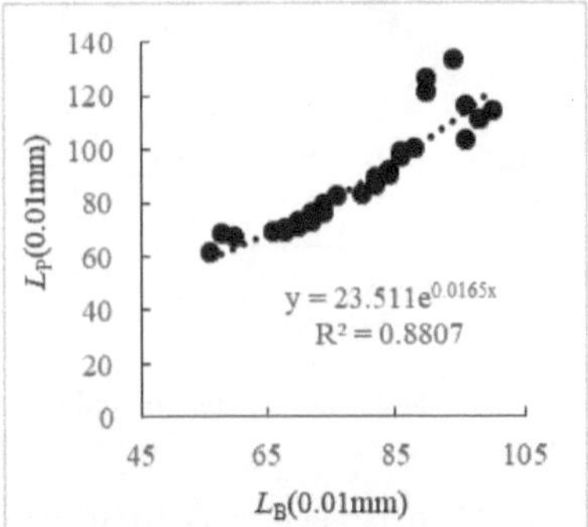

Figure 2. Dynamic and static deflection index regression curve.

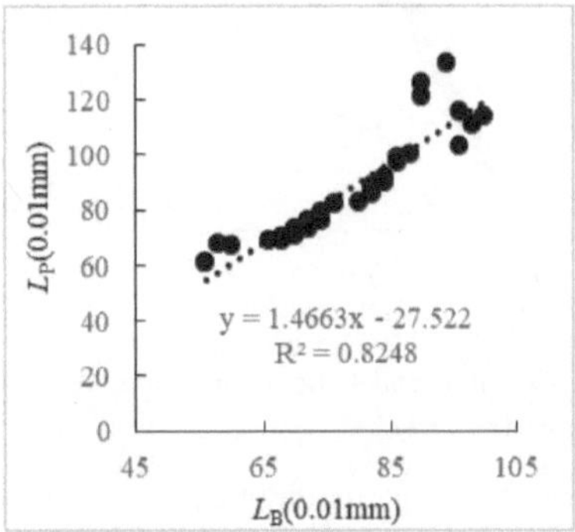

Figure 3. Dynamic and static deflection index regression curve.

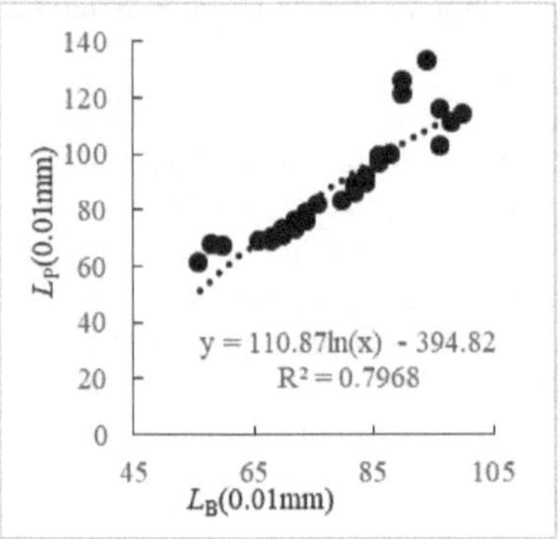

Figure 4. Dynamic and static deflection index regression curve.

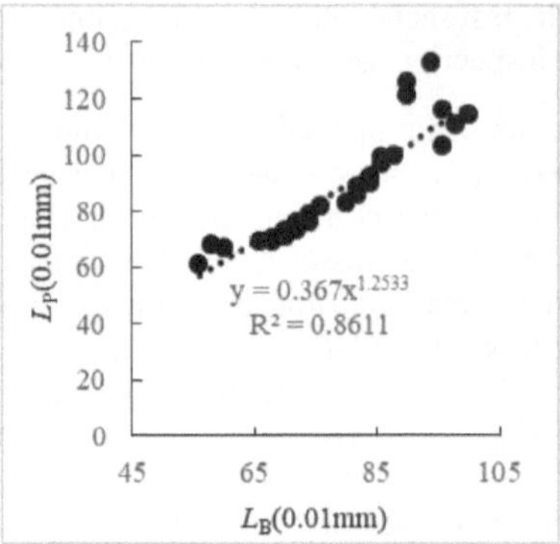

Figure 5. Dynamic and static deflection index regression curve.

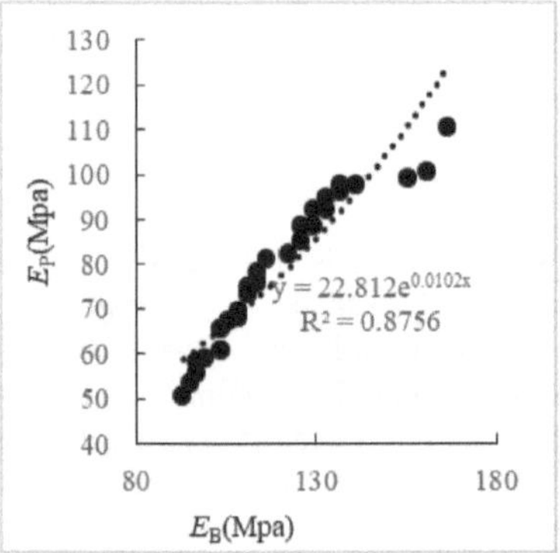

Figure 6. Dynamic and static rebound modulus index regression curve.

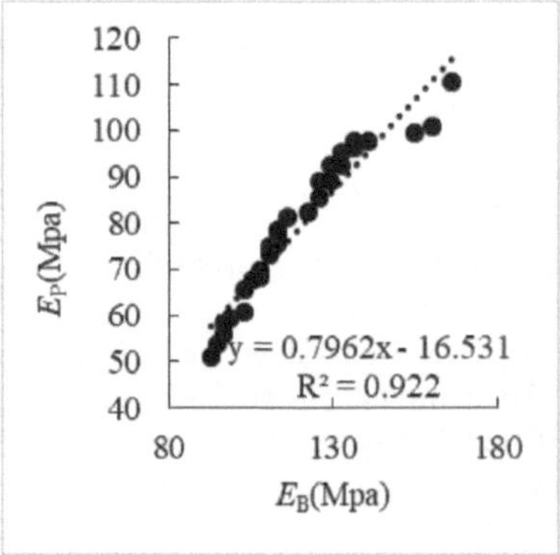

Figure 7. Dynamic and static rebound modulus index regression curve.

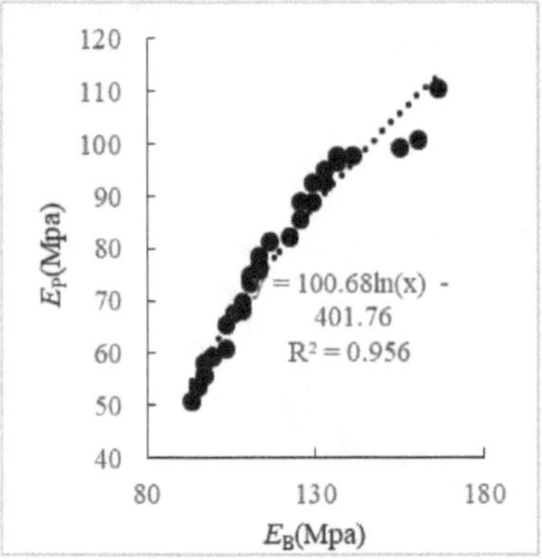

Figure 8. Dynamic and static rebound modulus index regression curve.

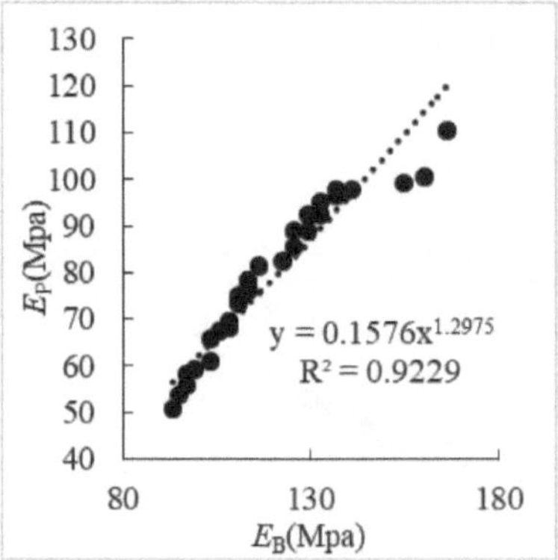

Figure 9. Dynamic and static rebound modulus index regression curve.

Based on the design requirements, the static deflection LB should not larger than 120 (0.01 mm) and the formula (1) can be calculated as the standard value of the dynamic deflection LP is 170 (0.01 mm).

(2) Correlation analysis of static and dynamic resilience modulus. Four models of exponential, linear, logarithmic and exponential are used to draw the regression curves of static and dynamic modulus of resilience, as shown in Figures 6–9.

Figures 6–9 show the regression curves of the four models of dynamic and static resilience modulus. The logarithmic correlation index R2 is 0.956, which is the best, as shown in the formula (2):

$$E_P = 100.68 \ln E_B - 401.76 \tag{2}$$

The standard value of static rebound modulus calculated according to the design deflection value EB is 77.6 Mpa, and the standard value of dynamic rebound modulus EP calculated by formula (2) is 36.4 Mpa.

3 CONCLUSIONS

To study the applicability of the PFWD in highway construction in China, a series of tests have been conducted on-site. The regression analysis method is used to establish a correlation between the two detection methods, the results are as follows.

The test section roadbed surface quality is jointly detected by the PFWD method and the Beckman beam method, and the dynamic deflection value and the dynamic backfill modulus control index of the test section can be obtained. When the static deflection value is 120 (0.01 mm), the standard value of dynamic deflection is 170 (0.01 mm); when the static elastic modulus is calculated back to 77.6 MPa, the standard value of dynamic elastic modulus is 36.4 MPa. When the test reaches this standard value, the subgrade filling quality meets the specification requirements.

REFERENCES

Fumio Tatsuoka, Takeshi Hashimoto, Kazuyoshi Tateyama. (2021) Soil stiffness as a function of dry density and the degree of saturation for compaction control. *Soils and Foundations*. 61(4): 989–1002.

Sun L, Wang D.Z, Zhang H.M. (2012). Predictive models of subgrade deflection using data from portable falling deflectometer. *Journal of Southeast University (Natural Science Edition)*. 42 (05): 970–974.

Yang H. (2010). The application of PFWD in the rapid detection of resilience modulus of red sandstone roadbed. *Heilongjiang Transportation Science and Technology*, 33 (07): 70–71+73.

Zha X.D, Xiao Q.M. (2011) Relation between Pressure and Deflection Measured with Portable Falling Weight Deflectometer. *Journal of Highway and Transportation Research and Development*, 28 (10): 24–28+53.

Zhang J.H, Deng Z.H, Liu J, etc. (2019) Research on the rapid detection method of existing roadbed in the hot and humid area of the south. *Journal of China & Foreign Highway*. 39(06): 6–10.

Advances in Civil Engineering: Structural Seismic Resistance, Monitoring and Detection – Mohd Johari Mohd Yusof and Junwen Zhang (Eds)
© 2023 The Authors, ISBN 978-1-032-31491-4

Influence of grouting pressure on segment displacement and internal force when the slurry shield passes through water rich sand layer

Shaopeng Wang
Sinohydro Bureau 11 Co., Ltd., Zhengzhou, China

Yilei Zhang
School of Human Settlements and Civil Engineering, Xi'an Jiaotong University, Xi'an, China

Wenhao Liu
Sinohydro Bureau 11 Co., Ltd., Zhengzhou, China

Jianxun Ma
School of Human Settlements and Civil Engineering, Xi'an Jiaotong University, Xi'an, China

Yangyang Bai
China Railway First Survey and Design Institute Group Co., Ltd., Xian, China

Dong Luo, Wenxiao Li & Hongxiang Yan
School of Human Settlements and Civil Engineering, Xi'an Jiaotong University, Xi'an, China

Chunqi Yang
China Power Construction (Xi'an) Rail Transit Construction Co., Ltd, Xian, China

Zengming Yue
Sinohydro Bureau 11 Co., Ltd., Zhengzhou, China

ABSTRACT: Taking the phase III project of Xi'an Metro Line 1 as the research background, the real construction process is simulated by establishing a three-dimensional finite element model, and the variation laws of segment displacement and internal force under different grouting pressure are analyzed by changing the value of grouting pressure. The results show that: with the increase of grouting pressure, the vertical displacement of completed segments increases; with the increase of grouting pressure, the internal force of the completed segment increases linearly.

1 INTRODUCTION

In recent ten years, shield technology has been more and more widely used in China, and has become an indispensable key technology for large scale river tunnel and urban subway construction. China's shield technology is constantly mature, and the construction volume of the project has ranked among the top in the world. According to the different stability system of the excavation surface and slag discharge system, there are two commonly used shield methods: earth pressure balance shield and slurry balance shield. Slurry balance shield has a wide range of application and can adapt to soft sand layer, gravel layer and soft rock layer, but the cost is slightly higher than earth pressure balance shield. Different shield machines have a certain scope of application to the stratum, and should be comprehensively compared and selected in terms of geological adaptability, safety, economy and impact on the environment.

 DOI 10.1201/9781003310884-60

At present, there are four main methods to analyze the deformation caused by shield construction: empirical method, analytical method, numerical simulation method and model test method.

Empirical method is based on a large number of measured data. Its parameter acquisition method is relatively simple and the prediction of general projects is usually accurate. Therefore, it is widely used in projects. Since peck proposed that the horizontal surface settlement trough caused by shield construction can be fitted with Gaussian distribution curve, scholars all over the world have done a lot of work and constantly revised the formula in combination with the engineering experience in this region, making this method the most commonly used method for predicting settlement. However, all empirical formulas including peck formula (Peck 1969) are based on a large amount of engineering experience. Under different construction methods and stratum properties, the prediction results of each empirical formula are quite different. Therefore, under the condition of less engineering experience, the accuracy of surface settlement prediction will be greatly reduced.

Analytical method deduces the calculation formula of stratum movement according to the corresponding stratum stress changes during tunnel construction. Sagaseta (1987) aimed at the problem that excessive ground deformation during shield construction seriously affects the normal passage of the ground, through the reverse fitting of field data, derived the three-dimensional calculation formula of ground deformation caused by shield construction. Mindlin (1936) regarded the tunnel as a cavity under the action of gravity and studied the stress distribution around the circular tunnel.

Numerical simulation method can simulate the complex characteristics of geotechnical media and the actual construction process. Now it has become a powerful tool to solve complex geotechnical problems. Sun Jun et al. (2002) based on the phase II project of Shanghai Rail Transit Pearl line, used ANSYS software to simulate the development and change law of soil displacement and structural internal force of overlapping tunnel in shield propulsion. Zhang Limin et al. (2004) used FLAC3D to analyze the shield construction process, and compared with the measured data to confirm the reliability of the numerical analysis method. Zhang Haibo et al. (2005) used three-dimensional nonlinear finite element software to study the influence relationship of soil layer properties, tunnel overburden thickness and relative position between tunnels.

Model test method refers to the establishment of a model and test according to the similarity theory for a specific project in the laboratory. Because the test can be controlled artificially and the test conditions can be changed, the influence law of single factor or multi factor can be determined. However, due to some differences between the sample and the in-situ stress field, the size effect of the model, approximate boundary conditions and initial conditions, the model test is still difficult in quantitative analysis and may be distorted. The model test is expensive and difficult to realize in large-diameter shield tunnel.

This study takes the phase III project of Xi'an Metro Line 1 as the research background. Due to the differences in geological conditions in different regions, the construction experience of similar subways at home and abroad can not solve the problems faced by Xi'an Metro. Therefore, it is of great significance to study the influence law of grouting pressure on the displacement and internal force of completed segments to ensure the safety of construction process.

2 PRINCIPLE OF THE SLURRY SHIELD

The main feature of slurry balance shield construction is that a diaphragm is set behind the cutterhead at the front end of the shield, a slurry pressure tank is formed between the cutterhead and the diaphragm, and the pressurized slurry is pumped to the slurry tank. When the mud tank is filled with mud with a certain pressure, a mud film will be formed on the excavation surface. The pressure of the mud tank acts on the soil on the excavation surface through the mud film, so as to maintain the stability of the excavation surface. Mud film formation and shield tunneling form a dynamic balance process. The cutterhead completed on the front of the shield rotates continuously to cut the soil containing mud film, and the mud film will be quickly reformed on the new excavation surface. After the soil cut by the cutterhead is mixed with the mud water in the mud tank, it is stirred by the mixing device to form high concentration mud water, which is then transported to the ground by the mud discharge pipeline.

In order to ensure the stability of the excavation surface, there is always a layer of mud film in the mud water balance shield whether in the excavation stage or segment assembly stage. When the mud film is cut by the cutter head of the cutter head, the new mud film will be formed rapidly and repeat again and again, so as to achieve the effect of maintaining the stability of the excavation surface. The structure and principle of the slurry shield are shown in Figures 1 and 2.

Figure 1. Schematic diagram of the slurry shield structure.

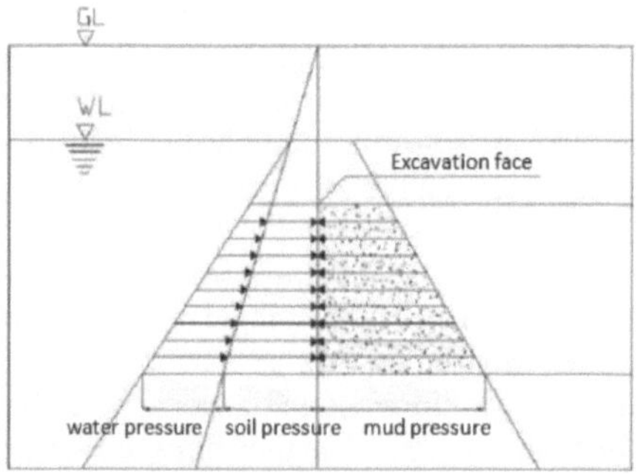

Figure 2. Geometrical dimensions of elements.

3 FINITE ELEMENT SIMULATION

The finite element model consistent with the actual project is established by Midas GTS NX software for analysis. According to the geological exploration report, the soil is simplified into four layers, and the modified Mohr Coulomb model is adopted. The soil parameters are shown in Table 1. The shield shell diameter is 6.28m, the segment diameter is 6m, the shield shell material is Q345 steel, the grouting layer and segment are C50 concrete, and the hybrid grid generator is used to divide the grid of the model. The boundary conditions are that the bottom surface is fixed, the normal displacement is restrained around, the top surface is free, the soil and grouting layer adopt solid element, the shield shell and segment adopt shell element, the tunneling pressure is simulated by applying surface load on the tunnel face, and the grouting pressure is simulated by applying surface load on the segment and soil surface, The shield process is simulated by continuously activating and passivating the load at the corresponding position.

The dynamic analysis is realized by activating and passivating the corresponding elements, loads and boundaries, and the simulation analysis closest to the real construction process is completed.

The dynamic construction process is mainly reflected in three aspects: the change of load position; transformation of material properties; activation and passivation of corresponding units. The finite element model is shown in Figure 3.

Table 1. Soil parameters.

Soil name	Poisson's ratio	Bulk density (g/cm^3)	Elastic modulus (MPa)	Cohesion (kPa)	Friction angle (°)	Dry density (g/cm^3)
Loess	0.30	1.85	8.70	33.00	25.30	1.53
Medium sand 1	0.35	2.20	30.00	0.00	34.50	1.77
Silty clay 1	0.30	2.00	7.00	30.80	18.20	1.64
Medium sand 2	0.25	2.00	30.00	5.60	32.87	1.74
Silty clay 2	0.29	2.02	7.00	47.00	27.30	1.67
Fine sand	0.28	2.00	20.00	0.00	30.48	1.74
Artificial fill	0.33	1.90	8.00	33.00	25.30	1.53

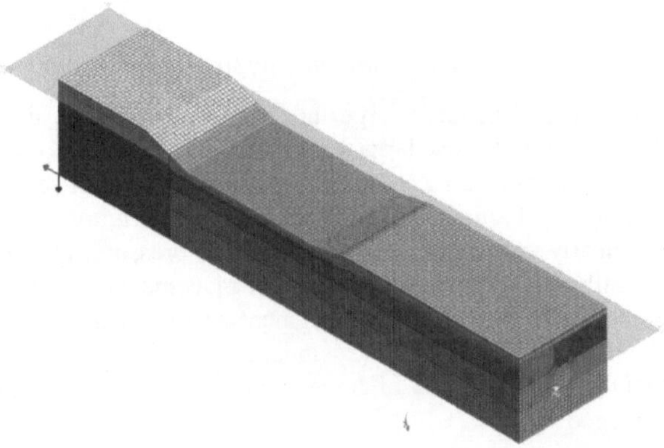

Figure 3. Finite element model.

4 ANALYSIS OF NUMERICAL SIMULATION RESULTS

Shield tail grouting is an important step in subway tunnel construction. When grouting to shield tail pores, a surface load will be formed on the segment surface, which is easy to deform or even destroy the segment. Therefore, it is of great significance to study the influence law of grouting pressure.

During the simulation process, other conditions remain unchanged. The grouting pressure is taken as the gradient of $50kN/m^2$ to study the variation law of segment displacement and internal force under the grouting pressure of 300, 350, 400, 450 and $500kN/m^2$.

4.1 *Influence law of different grouting pressure on vertical displacement of completed segments*

As the load directly loaded on the segment, the grouting pressure will have a great impact on the displacement of the segment. When the grouting pressure is too large, the segment will have a great displacement and have an adverse impact. The segment displacement under different grouting pressures is shown in Figure 4. It can be seen from the figure that with the increase of grouting pressure, the vertical displacement of the completed segment shows an increasing trend. The reason for this phenomenon is that the segment is made of reinforced concrete and prefabricated in the factory, so the segment quality is relatively consistent, so the response to deformation is relatively consistent, so it shows the deformation curve in the figure.

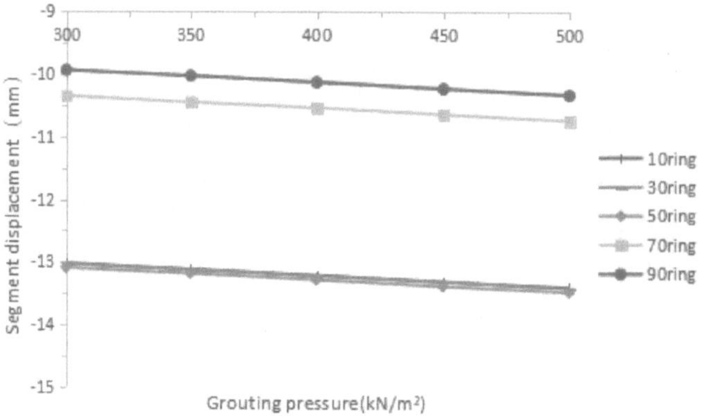

Figure 4. Relationship between grouting pressure and segment vertical displacement.

4.2 *Influence law of different grouting pressure on internal force of completed segments*

Too much grouting pressure will produce large internal force and even damage the segments, and too little grouting pressure will cause large settlement on the surface. Therefore, it is of great significance to select appropriate grouting pressure. The internal forces of segments under different grouting pressures are shown in Figure 5. It can be seen from the figure that the internal force of the segment increases linearly with the increase of grouting pressure, and the variation laws at different positions are similar. In combination with the displacement change trend of the segment, since the grouting pressure acts directly on the segment and the segment is a precast concrete product, the response to the load is similar to the standard linear elastic material. During the shield process, the displacement and internal force of the segment can be predicted according to the elastic material before it is damaged.

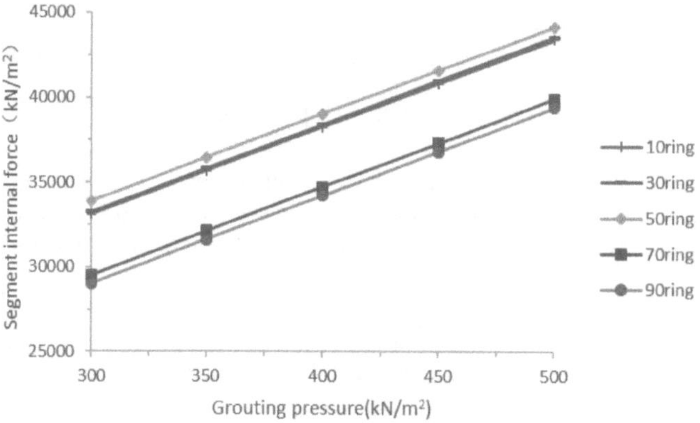

Figure 5. Relationship between grouting pressure and segment internal force.

5 CONCLUSION

In this paper, the finite element model is established by using Midas GTS NX software. By changing the value of grouting pressure, the influence law of different grouting pressure on segment displacement and internal force in the process of slurry shield is analyzed, and the following conclusions are obtained:

(1) With the increase of grouting pressure, the vertical displacement of the completed segment shows an increasing trend. The reason for this phenomenon is that the segment is made of reinforced concrete and prefabricated in the factory, so the segment quality is relatively consistent, so the response to deformation is also relatively consistent.
(2) The internal force of the segment increases linearly with the increase of grouting pressure, and the change laws of different positions are similar, indicating that the influence of grouting pressure on each part is relatively linear. In the construction process, the grouting pressure shall be reasonably controlled to ensure that the changes of various parameters in the construction process meet the specification requirements.

6 OUTLOOK

The main content of this paper is the analysis of the influence of grouting pressure on segment displacement and internal force when slurry balance shield passes through water rich sand layer. There are still some problems that can be further improved in the follow-up research:

(1) The node and load interval can be further refined. Due to the large research model in this paper, more nodes and loads cannot be selected for research. The load can be further refined and more nodes can be selected to verify and improve the universality of the law obtained in this study.
(2) The underpass process can be analyzed according to different water levels, different tunnel buried depths, different segment sizes and other factors, so as to further improve the research of relevant contents.
(3) Theoretical calculation and scale test can be used for comparative analysis to improve the theoretical research of shield process.

REFERENCES

Mindlin RD. Force at a point in the interior of a semi-infinite solid [J]. *Physics*, 1936, 7(5): 195–201.
Peck R B. Deep Excavations and Tunneling in Soft Ground [J]. *Proc. of 7th ICSMFE, Mexico*, 1969
Sagaseta C. Analysis of undrained soil deformation due to ground loss [J]. *Geotechnique*, 1987, 37(3): 301–320.
Sun Jun, Liu Hongzhou Three dimensional numerical simulation of soil deformation during shield construction of overlapping tunnel [J] *Journal of Tongji University* (NATURAL SCIENCE EDITION), 2002 (04): 379–385.
Zhang Haibo, Yin Zongze, Zhu jungao Numerical simulation of the influence of shield construction of close overlapping tunnel on old tunnel [J] *Geotechnical mechanics*, 2005 (02): 282–286.
Zhang Limin, Li Dayong Numerical simulation of soil deformation characteristics during shield tunneling [J] *Geotechnical mechanics*, 2004 (S1): 75–78.

Advances in Civil Engineering: Structural Seismic Resistance, Monitoring and Detection – Mohd Johari Mohd Yusof and Junwen Zhang (Eds)
© 2023 The Authors, ISBN 978-1-032-31491-4

Stress performance analysis of slitted reinforced concrete low shear wall with energy dissipation lead damper

Yuhang Mao & Chong Rong
Department of Civil Engineering Xi'an University of Architecture and Technology, Shaanxi, China

ABSTRACT: Low shear walls are often seen in high-rise buildings, which are prone to shear failure under strong earthquakes, have low ductility and energy consumption, and affect the seismic performance and safety performance of the overall structure. With regard to this problem, this paper proposes a new type of slotted low-rise shear wall with lead energy dissipation dampers. Based on the finite element software ABAQUS, a nonlinear finite element model of this type of reinforced concrete shear wall is established. Through nonlinear analysis of ordinary low shear walls, slotted low shear walls and slotted low shear walls with energy dissipation dampers, the hysteresis curves and failures of three different types of shear walls are compared—shape, bearing capacity, and ductility. Research shows that the new slit low shear wall with energy dissipation dampers has excellent energy dissipation capacity, and can effectively improve the wall stiffness and bearing capacity reduction caused by slits. This study has good research value and application prospects.

1 INTRODUCTION

With modernization in urban China, the number of high-rise buildings is increasing. At the same time, the buildings require higher individualization. Therefore, complex high-rise buildings are common, and the forms and stress states of their internal components are complicated. Among them, low-rise shear walls with an aspect ratio of less than 1 often appear in modern high-rise buildings due to design limitations. The low shear wall exhibits a shear failure form under the action of an earthquake, with poor ductility and poor seismic performance. Therefore, under limited conditions, the development of a low shear wall with good seismic performance is still a problem to be solved. Existing measures to improve the seismic performance of reinforced concrete low-rise shear walls in various countries can improve the seismic performance of low-rise shear walls in varying degrees

To improve the seismic performance of low-rise shear walls, Professor Kiyoshi Muto (Muto 1969) of Japan proposed the concept of shear walls with vertical joints in 1965, which includes setting up of a number of vertical joints in the wall and obtain them in actual projects. However, this slitting measure will cause the rigidity and bearing capacity of the shear wall to decrease. Dai (1992) proposed that a low-shear wall with horizontal short slits to form an x-shaped weakened shear wall (Dai 1992; Li 1992), that is, to open multiple stepped horizontal semi-through joints along the diagonal direction of the wall to improve the stiffness of the wall. Experimental and theoretical studies have shown that the ductility of this kind of wall can be increased by more than twice when the bearing capacity is reduced by 10%.. In 1992, Professor Lu (1995) at Southeast University proposed a slit shear wall with friction control device. A friction damper was used as a control device in the vertical seam. Experiments show that this type of shear wall has excellent seismic performance, obvious improvement in energy consumption, and is easy to repair and not easy to damage. In 2000, Professor Jiang (2000) at Tongji University proposed a slitted shear wall filled with neoprene tape. This kind of shear wall is a vertical slit in the middle of the wall, and the vertical slit is filled with neoprene tape. Experimental research shows that energy consumption and

 DOI 10.1201/9781003310884-61

damping increase significantly, but the bearing capacity decreases. In 2020, Professor Dang (2000) and others proposed a vertical seam fabricated shear wall with metal dampers on the basis of slitting, and the prefabricated wall panels were connected with x-shaped metal dampers. Experiments show that the rigidity bearing capacity and ductile energy dissipation capacity of the shear wall can be significantly improved, and the energy consumption value is close to the sum of the two. The above research shows that setting up vertical joints and installing energy-consuming devices in the joints is an effective measure to improve low-rise shear walls. Although the ductility and energy consumption of the slitted shear wall are considerably improved compared with the overall shear wall, the reduction in the bearing capacity and the weakened stiffness of the shear wall caused by the slit will limit its application in engineering.

This paper proposes a new type of shear wall by setting energy dissipation dampers, which not only improves the stiffness and bearing capacity reduction caused by the slit but also ensures excellent energy dissipation capacity. Based on the existing research, the ABAQUS finite element analysis software is used to perform numerical simulation analysis on the slotted reinforced concrete low-rise shear wall with energy-dissipating lead dampers, and a new type of low-rise shear wall structure is proposed. The feasibility of adding energy-dissipating lead dampers to the slit shear wall indicates a research direction for improving the seismic performance of low-rise shear walls

2 DEFORMATION OF SHEAR WALL AND FORCE MECHANISM OF LEAD DAMPER

2.1 *Working principle of damper*

Under horizontal load, the whole slitted shear wall will be bent and deformed. As shown in Figure 1(b)(c), the bending deformation of the wall between the joints cause the two points A and B to produce the angular displacement θ and the relative vertical displacement Δ in the same direction.

For a general slotted shear wall, the width of the wall is usually 400 mm, and the displacement angle of the vertex is 1/60. According to formula $\Delta = \theta b$, the displacement of points A and B is about 6.7 mm. It can be seen that the relative displacement between the wall column is small. When the horizontal load is small, the ordinary mild steel damper cannot enter the elastic-plastic energy dissipation stage, and cannot enter its optimal working state.

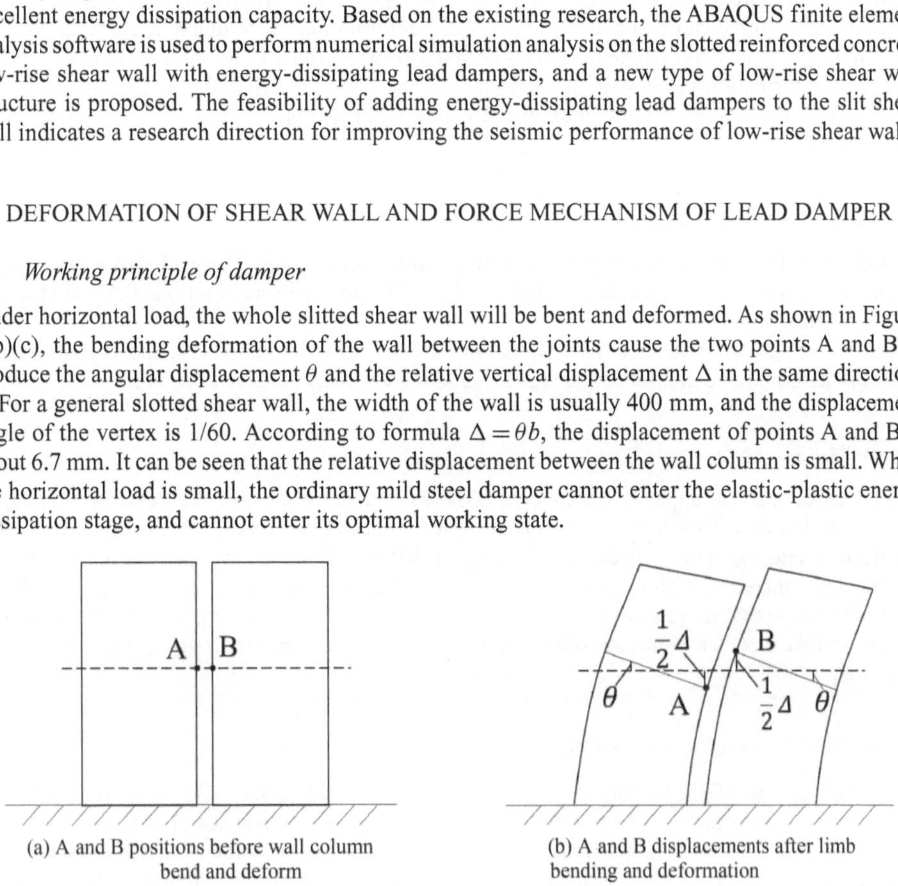

Figure 1. Shear wall deformation under horizontal load.

Therefore, based on the deformation characteristics of the abovementioned slotted shear wall, this paper sets up shear-type lead dampers to increase the rigidity of the shear wall while increasing its energy consumption capacity. When the horizontal displacement is small, the lead damper enters the working yield stage. At this time, the lead damper can be considered to only produce vertical shear deformation and transmit shear force to each wall limb. When the horizontal displacement is large, both sides of the wall limb experience large displacements. The horizontal displacements cannot be ignored. Bending deformation occurs between the wall columns. At this time, the dampers are also bent and deformed to transmit a certain horizontal force.

3 ESTABLISHMENT AND VERIFICATION OF FINITE ELEMENT MODEL OF SLOTTED SHEAR WALL

3.1 *The design concept of lead damper*

To overcome the large direct or indirect loss caused by the damage due to ductile member relying on the elastoplastic deformation of the structure itself to dissipate energy based on the ductile design concept. Based on the concept of structural vibration control, this article chooses lead as an energy-consuming device. Under small earthquakes, it gives shear wall a certain degree of integrity. During large earthquakes, lead dampers are used as energy-consuming components, and their shear transmission effect is limited, so they are transformed into shear walls with slits to obtain better ductility.

The salient feature of the lead shear damper is that it is very sensitive to small deformations, and a large damping force can be obtained under small deformations. The dynamic recrystallization function of lead makes it possible to cycle thousands of times under different deformations within the design deformation range without degradation, and keep the original mechanical properties unchanged. Its constitutive relation is similar to the ideal plastic body, so lead is a very good energy dissipation material.

3.2 *Model design and its parameters*

Use ABAQUS finite element analysis software to build model sw-3. The wall dimensions are H (height) × B (width) × t (thickness) = 800 × 750 × 75 mm. The aspect ratio is 0.67 and the slit width is 20 mm.

The strength grade of concrete is C60, the modulus of elasticity is 36 GPa, and Poisson's ratio is 0.2. In the model, the dilation angle is 36.3°, and the viscosity parameter is 0.05.

3.3 *Build a 3D model*

Separate modeling is adopted. Concrete uses C3D8R solid elements, and steel reinforcement uses T3D2 truss elements. To ensure calculation accuracy and calculation cost, the grid size of 50 mm is uniformly selected. The embedded technology in ABAQUS is used to make the reinforcement built into the concrete. This method can be used to simulate the contact relationship between them. But it is believed that no relative slip occurs at each node, so the bond slip between the concrete element and the steel bar is not considered in the model. Finally split the wall through the cutting command.

3.4 *Material constitutive relationship*

Concrete makes use of plastic damage model, which uses isotropic elastic damage combined with isotropic tensile and compressive plasticity to replace the inelastic behavior of concrete.

The stress and strain of concrete under uniaxial tension and compression are determined according to the "Code for seismic design of buildings" (GB50010-2010). The constitutive relationship of steel bars is a two-fold line constitutive. The concrete uniaxial tensile stress-strain curve equation can be determined according to the following equations (1) and (2).

$$\sigma = (1 - d_c)E_c\varepsilon \tag{1}$$

$$d_t = \begin{cases} 1 - \rho_t\left[1.2 - 0.2x^5\right]x \leq 1 \\ 1 - \dfrac{\rho_t}{a_t\,(x-1)^{1.7} + x} \quad x > 1 \end{cases} \tag{2}$$

Where a_t is the parameter values of the descending section of the concrete uniaxial tensile stress-strain curve are taken according to (GB50010-2010) Table C.2.3; $f_{t,r}$ the typical value of concrete uniaxial tensile strength, the value can be f_t,f_{tk} or f_{tm} according to the needs of actual structure analysis. $\varepsilon_{t.r}$ is peak compressive strain of concrete; d_t is damage Evolution Parameters of Concrete under Uniaxial Compression.

The concrete compressive stress-strain curve equation is determined by the following equations (3), (4), (5), and (6).

$$d_c = \begin{cases} 1 - \dfrac{\rho_c n}{n - 1 + x^n} & x \leq 1 \\ 1 - \dfrac{\rho_c}{a_c (x-1)^2 + x} & x > 1 \end{cases} \tag{3}$$

$$\rho_c = \frac{f_{cr}}{E_c \varepsilon_{cr}} \tag{4}$$

$$n = \frac{E_c \varepsilon_{cr}}{E_c \varepsilon_{cr} - f_{cr}} \tag{5}$$

$$x = \frac{\varepsilon}{\varepsilon_{cr}} \tag{6}$$

Where a_c is parameter value of the descending section of the stress-strain curve of concrete under uniaxial compression; $f_{c.r}$ is the typical value of concrete uniaxial compressive strength, the value can be A, B, or C according to the needs of actual structure analysis; $\varepsilon_{c.r}$ is the peak compressive strain of concrete; d_c is damage evolution parameters of concrete under uniaxial compression.

3.5 *Boundary and load settings*

In the experiment, the bottom of the wall was fixed, a distribution beam was used above the shear wall to load the vertical pressure on the top, and then a horizontal load was applied on the side. In the model, the bottom is completely fixed to constrain the degrees of freedom of all elements at the bottom.

The load conditions are applied in two steps: the first step is to apply a vertical load. To prevent stress concentration, a reference point should be set to couple with the top of the shear wall, and the vertical load should be applied to the coupling point. The magnitude of the vertical force is calculated according to the required axial compression ratio. The second step is to apply a horizontal load, adopt the displacement loading mode setting, and select the appropriate loading curve.

3.6 *Validation of the finite element model*

To check the reliability of the finite element simulation analysis, the test results of the sw3 slit low shear wall specimens in the literature[9] are compared with the finite element simulation analysis results. The comparison results are shown in Figure 2 as well as in Table 1.

The concrete strength design grade is C60. The shear wall panel is 800 mm × 750 mm. The board thickness t = 75 mm, and the vertical joint width is 20 mm. For detailed data see reference (Zuo 1997).

It can be seen from Table 1 and Figure 2 that compared with the skeleton curve of the experiment, the initial stiffness of the load-displacement curve obtained by the finite element analysis is slightly larger, which results in a smaller yield displacement. This is because the bond slip between steel and concrete is ignored in the simulation, but the error is within an acceptable range. The bearing capacity of each characteristic point of the shear wall differs from the test value by about 20%, which is in good agreement with the test curve. It can be seen that the skeleton curve obtained by the test and the finite element simulation analysis is also in good agreement. Explains the ABAQUS modeling described in this section methods. It is feasible to study the method of low shear wall with vertical joints.

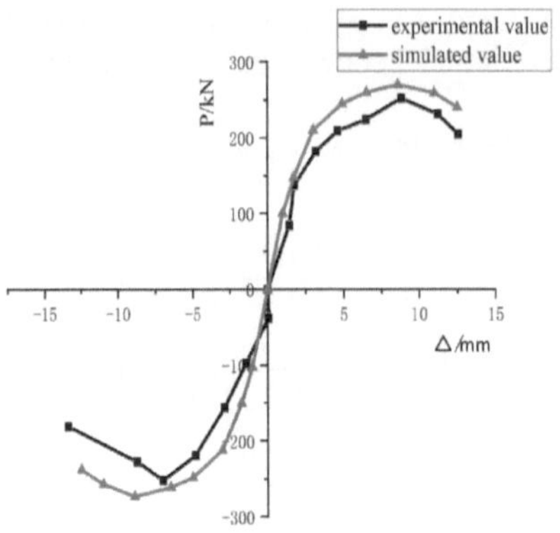

Figure 2. Comparison of load-displacement curves of specimens.

Table 1. Test value and simulation value of specimen displacement and load.

Specimen number	Project	Yield load point			Peak load point			Ultimate load point		
		Experiment	Simulation	Ratio	Experiment	Simulation	Ratio	Experiment	Simulation	Ratio
sw3	Strength	214.6	225.7	1.05	251.9	270.3	1.07	205.6	229.8	1.12
	Displacement	5.3	3.7	0.7	8.82	8.63	0.98	12.07	13.28	1.1

4 SEISMIC PERFORMANCE OF REINFORCED CONCRETE LOW-RISE SHEAR WALL WITH LEAD DAMPER

4.1 *Building the wall model*

To investigate the effect of lead dampers on the seismic performance of low shear walls, nonlinear finite element simulations of low shear walls with whole section (sw-0), low shear walls with vertical joints (sw-1), low shear walls with vertical joints, and lead dampers (sw-2) were performed.

4.1.1 *Slotting measures*

The external dimensions and reinforcement of each test piece are the same. The wall panel size is 1600 mm × 1000 mm × 110 mm, and the aspect ratio is 0.67. Use the cutting command to slit the wall and disconnect the horizontal steel bars. For the slit shear wall, set three vertical seam, where width of opening is 20 mm, height is 500 mm, spacing between seams is 400 mm, horizontal reinforcement is 8@150, vertical reinforcement is 10@100 concrete strength. The concrete strength grade is C30, the elastic modulus is 30 Gpa, and the Poisson's ratio is 0.2 as shown in Figures 3 and 4. For other modeling processes, refer to Section 2.

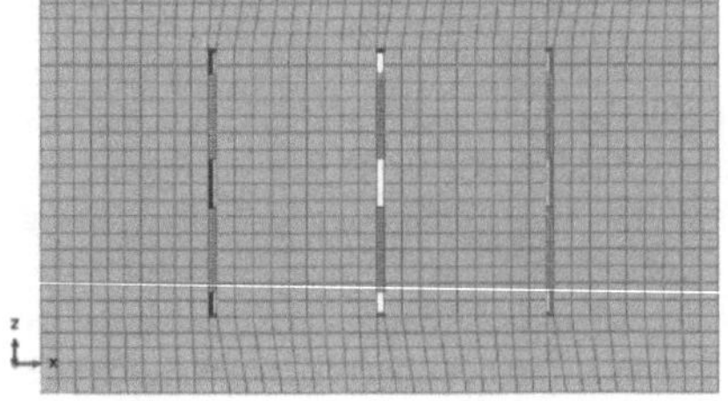

Figure 3. Reinforced concrete model of shear wall with vertical joints.

Figure 4. Rebar model of shear wall with vertical joints.

4.1.2 *Material selection of lead*

The size of the lead damper is 200 mm × 100 mm, the equivalent yield stress $\sigma_y = 19.8MPa$ elastic *modulus* $E = 16{,}460\ MPa$, the Poisson ratio is 0.42, and the solid element simulation is adopted. The mesh size is 5 mm to ensure the accuracy. Six groups of lead blocks are placed at the wall openings using the bind command. Lead will undergo shear yield when the stress is small, and the yield shear stress is y_τ, which can be regarded as an ideal elastoplastic solid, and the constitutive relationship can be expressed as (7)

$$\tau = \begin{cases} G\gamma & 0 \leq \gamma \leq \gamma_y \\ \tau_y & \gamma_y \leq \gamma \end{cases} \tag{7}$$

where G is the elastic shear modulus of lead; γ_y is the yield shear strain of lead.

4.2 *Damage characteristics*

Figure 5 shows the tensile damage clouds of concrete when the four specimen models reach the cracking load and the peak load. The concrete material is mainly damaged due to tension, and the distribution of damage factors on the damage clouds can reflect the damaged cracking of concrete in an image, and the damage of the models can be roughly understood by the animation playback. Through the analysis of the concrete tensile damage cloud diagram of each model, the following characteristics are mainly observed.

(1) It can be seen from the figure that when the cracking load Pcr = 518 kN and Δ = 1 mm, the diagonal cracks of the sw-0 shear wall appear at the feet of the two columns first and rapidly extend upwards; the cracks at the middle wall panel are roughly along the diagonal direction of the wall, and there is a local stress concentration phenomenon at the top of the corner. When Pcr = 420 kN and Δ = 2.25 mm, sw-1 appears diagonal cracks at the ends divided by the through joints. When Pcr = 549 kN and Δ = 2.73 mm, sw-2 first appears horizontal cracks between the joints.

(2) When the peak load is Pm = 647 kN and Δ = 3.53 mm, the wallboard in the middle of sw-0 shows a 45° diagonal crack along the diagonal direction of the wallboard, which is a typical shear failure, and when Pm = 463 kN, Δ = 3.65 mm. When compared with Pm = 547 kN and Δ = 3.6 mm, sw-1 and sw-2 are divided into several wall slab columns with greater aspect ratio due to the existence of vertical joints, and the cracks are distributed and integrated along the diagonal direction of the short wall slab column. Compared with the tensile damage cloud map, the cross-section wall shows an obvious bending failure form.

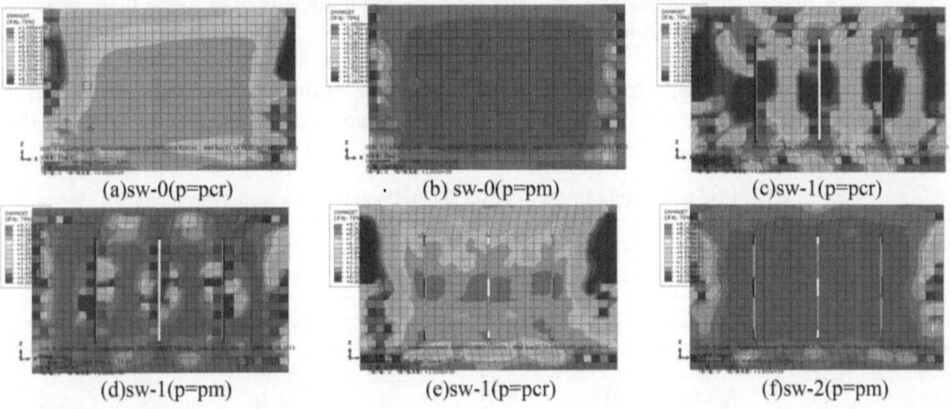

(a)sw-0(p=pcr) (b) sw-0(p=pm) (c)sw-1(p=pcr)

(d)sw-1(p=pm) (e)sw-1(p=pcr) (f)sw-2(p=pm)

Figure 5. DAMAGET tensile damage cloud diagram under each analysis model.

4.3 *Hysteresis characterization*

The hysteresis curves of various types of shear wall models are shown in Figure 6.

Through the comparison of each curve, it can be found that compared with the common low shear wall, the low shear walls with vertical joints and lead dampers have the following characteristics: the hysteresis is full and the energy consuming capacity is significantly increased; the displacement under ultimate load is significantly increased and the displacement ductility is significantly improved; and the load bearing capacity is reduced. At the same time, compared with the open-joint low-profile shear walls, the load-carrying capacity, stiffness, and energy dissipation capacity are significantly improved by adding energy dissipation dampers. In summary, it can be seen that this measure with lead energy dissipation dampers effectively improves the seismic performance of low shear walls.

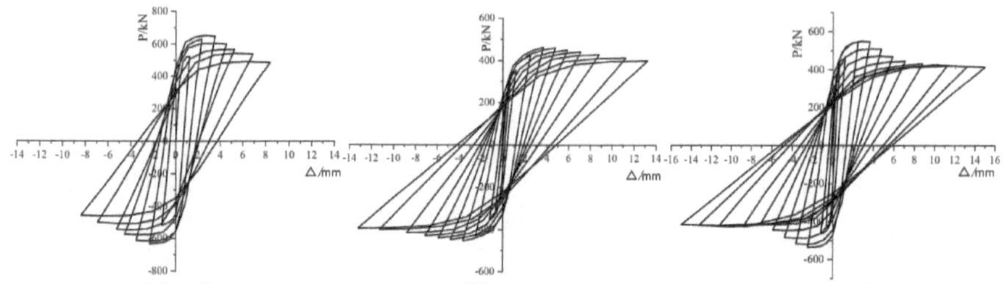

Figure 6. Hysteresis curves for each model.

4.4 *Model bearing capacity and ductility analysis*

Table 2 shows the yield load, peak load, ultimate load, and displacement ductility of each shear wall calculated by ABAQUS. The calculated forward skeleton curve of each test piece is shown in Figure 9; it can be seen from Table 2 and Figure 11:

(1) The three types of shear walls sw-0, sw-1, and sw-2 reach their respective peak loads near 3.6 mm. For yield load, peak load, and ultimate load, ordinary low-rise shear walls are the highest, with loss the slit reinforced concrete low shear wall with lead dampers is the second, and the ordinary vertical slit shear wall is the lowest.
(2) After sw-0 and sw-2 reach the peak load and before = 9 mm, the rate of decrease in the bearing capacity of the two is the same overall.
(3) The bearing capacity of sw-1 and sw-2 basically tends to be the same in the later stage. The reason is that the shear force transmitted by the lead damper is limited in the later period, which means that the slotted shear wall with lead damper is transmitted by the lead damper in the

Table 2. Comparison of load and displacement of characteristic points of each analysis model.

Specimen number	Yield load point		Peak load point		Ultimate load point		displacement ductility factor	
	p_y/kN	Δ_y/mm	p_m/kN	Δ_m/mm	p_u/kN	Δ_u/mm	μ	μ_i/μ_0
sw-0	578	1.82	647	3.53	550	6.20	3.41	1
sw-1	370	1.47	463	3.65	393	14.38	8.97	2.6
sw-2	450	1.01	547	3.6	464	6.0	5.87	1.7

later period. The shear force is limited, so its working condition is similar to that of ordinary slit shear wall.

4.5 *Analysis of stiffness attenuation process*

The curve of the attenuation process of the secant stiffness K with the displacement angle of each member obtained in the test is shown in Figure 8.

(1) The elastic stiffness of sw-2 is significantly higher than that of sw-1, indicating that adding lead dampers is a very effective measure to improve the stiffness of components.
(2) The stiffness degradation of sw-2 is slow, which indicates that the energy dissipation device can stabilize the later performance of the structure.
(3) The decrease trend of the later stiffness of sw-2 is consistent with that of sw-1, indicating that the decrease of the working performance of the damper in the later stage changes it into a working state similar to that of the slotted shear wall.

In summary, the addition of energy-consuming lead dampers can effectively improve the stiffness reduction caused by the opening of the overall shear wall, and the lead dampers are equivalent to a kind of fuse. In the early stage, energy consumption and initial rigidity are provided for the component to ensure that the component is in an elastic state, and the original performance of the component will not be affected after the withdrawal of work in the later stage.

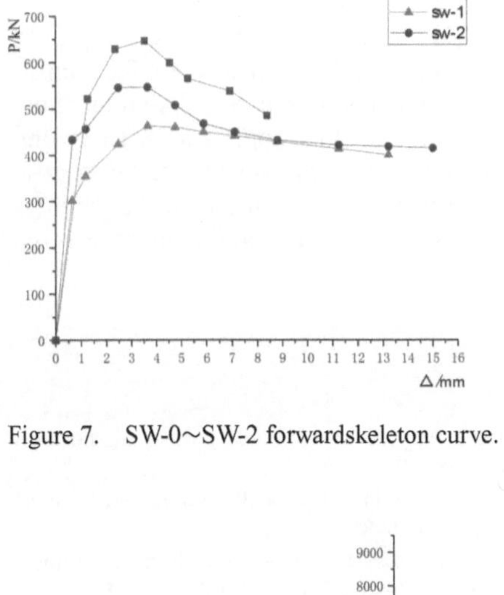

Figure 7. SW-0~SW-2 forwardskeleton curve.

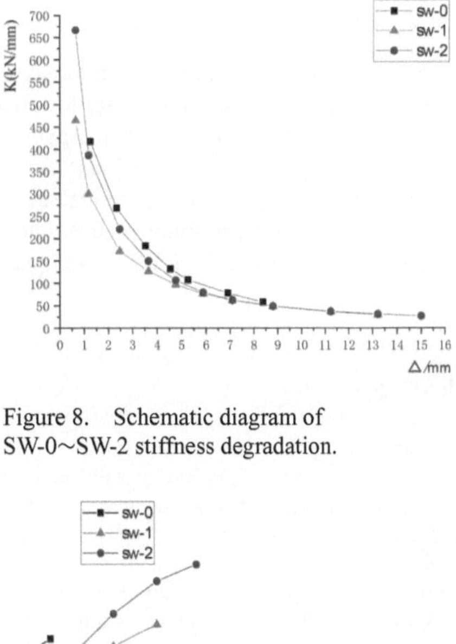

Figure 8. Schematic diagram of SW-0~SW-2 stiffness degradation.

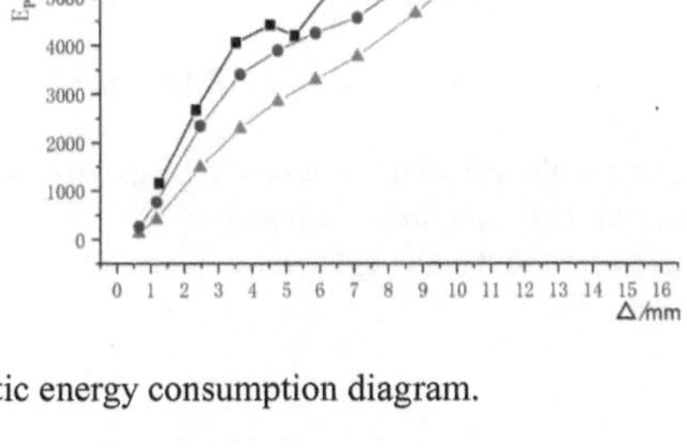

Figure 9. Single-turn hysteretic energy consumption diagram.

4.6 *Energy consumption capacity and analysis*

It can be seen from Figure 13 that the energy consumption capacity of the slotted low-rise shear wall sw-1 and the slotted low-rise shear wall sw-2 with energy dissipation lead dampers are, respectively, higher than that of the ordinary overall low-rise shear wall, that is, 12% and 54%. Compared with sw-0, sw-2, and sw-1 have weaker energy consumption than the overall shear wall under the same displacement. However, by adding energy-consuming lead dampers, the energy consumption of sw-2 was similar to that of sw-0 in the early stage. In summary, it shows that lead damper can effectively improve the energy-consuming capacity of slotted low-rise shear walls.

5 CONCLUSION AND OUTLOOK

Based on the results and discussions presented above, the conclusions obtained are as below:

(1) In the process of horizontal deformation, shear deformation occurs between the slits in the shear wall and the deformation is small. Lead dampers with small deformation and energy dissipation effect should be selected. Therefore, this paper proposes a slit with energy dissipation dampers in a low-rise shear wall.
(2) The installation of energy-consuming lead dampers between the joints can significantly improve the load-bearing capacity and stiffness of the slit shear wall, and effectively alleviate the problems of reduced load-bearing capacity and weakened stiffness after slitting.
(3) Compared with the traditional low-rise shear wall, the slotted low-rise shear wall with energy dissipation dampers can significantly improve the ductility and energy dissipation capacity of the components under the condition of ensuring the bearing capacity and rigidity.

In summary, the measures of slitting and installing inter-slit dampers have effectively improved the seismic performance of the traditional low-rise shear wall, which confirms the effectiveness of the structural measures proposed in this paper. However, this research is only based on finite element analysis and cannot provide a quantitative reference for practical applications. Therefore, further experimental research and theoretical analyses are required. At the same time, the specific structure, assembly method, number of installations, and arrangement positions of the dampers installed between the seams will all have an impact on the test results, and therefore more research is needed.

REFERENCES

Dai, H, & Chen, G (1992). Analysis of dynamic response and energy consumption of horizontal short-slit walls and low-shear walls. *J. Journal of Southeast University*, 58–65.

Dai, H & Ding, D J (1992). Shaking table test study of contrast between low shear wall with horizontal short slits and ordinary shear wall models. *Engineering mechanics*, 9(2) 76–85.

Dang, L J & Liang, S T (2020). Analysis on mechanical behavior of precast shear walls with metal dampers installing along vertical joints. *J. Journal of Southeast University* (Natural Science Edition), 50(5):8

Jiang, H J & Lu, X L (2000). Experimental study on a new type of energy dissipation shear wall model under cyclic loading. *World Information on Earthquake Engineering*. 16(2)63–67.

Kiyosh, Muto (1969). Reinforced Concrete Shear Walls for High-rise Building Structures SEAOC Convention of HAWAII. 10

Li, A Q (1992). Research on Anti-seismic Control and Control Device of Reinforced Concrete Shear Wall Structure. D. Southeast University.

Lu, X L & Meng, L (1995). Shaking table test study of a new type of seismic energy dissipation shear wall structure model. *World Information on Earthquake Engineering*, 29–34.

Zuo, X B (1997). Research on improving the seismic behavior of high strength concrete shear walls. D. Tianjin University.

Advances in Civil Engineering: Structural Seismic Resistance, Monitoring and Detection – Mohd Johari Mohd Yusof and Junwen Zhang (Eds)
© 2023 The Authors, ISBN 978-1-032-31491-4

Photo-aging performance of high-density polyethylene sheath in dry environment

B. Lin, C. Zhang & Chunxia Zhang
Research Institute of Highway Ministry of Transport, Beijing, China

ABSTRACT: The service life of the sling greatly affects the structural safety and durability of the cable bridge, and its protection mainly focuses on its own performance and external protection measures. As the anti-corrosion properties of high-strength steel wire materials become stronger and stronger, the replacement time of the sling will become relatively long. However, most cable sheathing made of high-density polyethylene (HDPE) will be aging and shorten their service life due to complex environment, which may lead to shorter service life than that of the sling replacement, affecting the service life of the sling. Therefore, we investigated the aging performance of HDPE sheath in dry environment and designed artificial daylight aging test and tensile test under different stress levels. The results showed that with the continuous aging of HDPE sheath, the damage forms such as stains, black spots and cracks appeared successively on the surface of HDPE sheath, and mechanical properties of HDPE sheath were also declining. The initial stress also affected the descent rate of mechanical properties of HDPE. These tests and results can provide reference for protecting sheath in dry places with little rain.

1 INTRODUCTION

Cable bridge, as a kind of bridge with long span, beautiful shape and convenient construction, has been widely used at present. But in recent years in the operating system of cable bridge inspection, the sling protective layer material was affected by environmental corrosion factors and its own expansion deformation, sunshine and temperature change because of the severe operating environment, causing the protective layer of the sling was damaged (Yang 2018; Ye 2003; Wu 2008). The damaged protective coating would cause wire rust, resulting in cable failure which had to replace the sling. Cable replacement seriously affects the structural safety and durability of the normal use of the bridge, and brings many unsafe factors to the normal operation of the bridge. At the same time, premature cable replacement causes the continuous increase of the operating cost of the bridge, and a very adverse social impact. Therefore, it is of great social significance to scientifically predict the service life of the external protective material of the sling in various environments, so that the it can be replaced before the its performance reaches the normal limit life, and then extend the service life of the sling.

The service life of the sling is closely related to its environment and protection system. Harsh environment is the direct factor that destroys the normal service life of the sling, among which atmospheric corrosion, water corrosion and soil corrosion are the three main corrosive environments. Especially for atmospheric corrosion, the researchers drew the regional division map of atmospheric corrosion in different regions of China. In view of these corrosive environments, various anti-corrosion protection systems and protection methods are put forward, such as winding wire, wrapping, dehumidification, sealing, etc., or a combination of various means. At the same time, a series of studies have been carried out on external protective materials, and some solutions and new designs have been proposed from the aspects of cracking mechanism and aging performance of external protective materials (Liu 2013; Parsons 2000).

DOI 10.1201/9781003310884-62

With the protection system of the sling is increasingly perfect, the anti-corrosion properties of high-strength steel wire materials themselves are getting stronger and stronger, where the replacement time of the sling will be relatively long (Zhang 2010). However, most of the external protective material of the sling is made of polyethylene material, which will produce aging and shorten the service life under the action of complex environment around China, and cannot reach the normal service life of the material (Zhang 2018). The service life of the external protection may be shorter than that of the sling replacement, so it needs to be replaced in advance. The study on aging properties and replacement life of the outer protective material of the sling will become a new trend of the sling protection research. Therefore, it is necessary to explore the aging performance and replacement life of the outer protective material of the sling, so as to better prolong the service life of the sling.

Here, we investigated the aging performance of HDPE sheath in dry environment and designed artificial daylight aging test and tensile test under different stress levels. First, the different initial strains were applied to the HDPE test pieces by the fixture, and then the processed test pieces were put into the artificial sunlight aging test chamber for aging test. After reaching the set time, the mechanical properties of the specimens after the test were measured. The results showed that with the continuous aging of HDPE sheath, the damage forms such as stains, black spots and cracks appeared successively on the surface of HDPE sheath, and mechanical properties of HDPE sheath were also declining. The initial stress also affected the descent rate of mechanical properties of HDPE. These tests and results can provide reference for protecting sheath in dry places with little rain.

2 CONSTRUCTION AND GEOMETRICAL DIMENSIONS OF SPECIMENS

2.1 *HDPE sample and test environment*

The aging test sample was made as the plate structure with length of 200 mm, width of 100 mm and thickness of 8 mm from raw materials of the KYLH-3 bridge cable HDPE sheath (Qingdao Keya Plastics Co., LTD, China). The test temperature was set at 60/50°C and humidity was set at 80% according the climate characteristics of dry and little rain. UVA-340 UV lamp (ATLAS Company, USA) was used to simulate the distribution of the violet outer light spectrum of the positive light. ZN-420 accelerated aging test chamber (Wuxi Colon Test Equipment Co., LTD, China) was used for aging tests.

2.2 *Experimental scheme and process*

Several HDPE samples were placed in a ZN-420 accelerated aging test chamber. UVA-340 UV lamp was used to simulate the ultraviolet spectrum distribution of sunlight. The experiment was firstly carried out for 8 hours at the ultraviolet irradiance of 9.8 $W \cdot m^{-2}$ and the temperature of 60°C, and then for 4 hours at the temperature of 50°C and the humidity of 80% without ultraviolet irradiation, which was regarded as a cycle test. Before the test began, the HDPE sheath could be fixed with a tensioning fixture and the initial strain was applied to it.

The samples were divided into three groups, and more detailed information of all samples could be seen in Table 1. The first group (a total of 4 pieces) was stored in a dry place as the control group, where 2 pieces were subjected to the initial strain of 5 MPa, 2 pieces were not subjected to initial strain. The second group (a total of 4 pieces) was subjected to light aging test from 0-960 hours in the accelerated aging test chamber, where 2 pieces were subjected to initial strain of 5 MPa, and 2 pieces were not subjected to initial strain. The third group (a total of 4 pieces) was subjected to light aging test from 0 to 1440 hours in the accelerated aging test chamber, where 2 pieces were applied the initial strain of 5 MPa, and 2 pieces were not applied the initial strain.

After reaching the specified aging time, the specimens were taken out and stored at ambient temperature for at least 16 hours. Then, the specimens were washed with organic solvent and compared with the control specimens in group 1.

Table 1. Test group.

Group	Temperature (°C)	Humidity (%)	Strain (MPa)	Aging time (h)
1A	/	/	0	960
1B	/	/	5	1440
2A	60, 50	80	0	960
2B	60, 50	80	5	1440
3A	60, 50	80	0	960
3B	60, 50	80	5	1440

2.3 *Tensile test*

After the artificial accelerated aging tests completed, these test samples were subjected to tensile tests to measure the mechanical properties after aging. The aged specimens were cut and processed into tensile test specimens. The size of the specimen was standard type 1B (ISO 2006).

The electronic universal testing machine (WDW-50) was used for tensile test, and the tensile speed was 10 mm /min. The force and strain data of the whole process were recorded by the computer, and processed into graphs and required data by the software of the testing machine.

3 TEST RESULTS AND DISCUSSIONS

After the aging test, the appearance of the test samples was observed first, and the differences in appearance of the three groups were compared and analysed. After that, the aging HDPE samples were cut and processed into tensile test specimens. Finally, the mechanical properties of these tensile specimens were tested. Meanwhile, the mechanical properties of the three groups of specimens after aging were compared and analysed.

3.1 *The appearance changes of HDPE samples after aging*

First, HDPE samples in the second and third groups were observed to record their appearance characteristics, and then compared with the samples without artificial accelerated aging test. Then, the HDPE samples were cut open to see the colour change inside, which would to explore the depth of the influence of light aging on the sample. The detailed results are as follows.

In the first group of four HDPE samples, the colour of two 1A samples remained basically unchanged without any other aging phenomena, while the colour of two 1B samples only became lighter without any other changes. In the second group of four samples that underwent artificial accelerated aging, the colour of two 2A samples became lighter than the four samples of Group 1, but no other changes, while the colour of two 2B samples became lighter than the four samples of Group 1 and two 2A samples with no cracks, warping and other aging phenomena. In the third group of four samples after artificial accelerated aging for a long time, the colour of 3A and 3B samples became very light with the stains appeared, but there were no cracks, warping and other aging phenomena in 3A and 3B samples.

Through comparison, it can be concluded that after aging, almost all HDPE samples have corrosion aging phenomenon, but it is a mild aging phenomenon, mainly in the sample material discoloration, without moderate corrosion and destructive corrosion phenomenon. Meanwhile, as the aging time increased, the aging phenomenon of HDPE surface became more obvious. The aging degree of HDPE was also affected by the presence or absence of initial stress, and the sample with initial stress became lighter in colour than that without initial stress.

As for the internal colour of the sample after cutting, we found that the aging specimen only showed fading phenomenon on its surface, while the interior colour of the specimen was still black.

This indicated that photoaging occurred only on the surface of the specimen, and the interior was not affected by lamp light.

3.2 *The mechanical property changes of HDPE samples after aging*

The mechanical properties of the treated standard type 1B aging specimens were tested by electronic universal testing machine. The results showed that the stress-strain curves of the three groups of specimens showed similar characteristics in different aging periods. Meanwhile, three parameters (Yield strength, elastic modulus, fracture strain) were used to explore the change of material properties after aging, which was shown in Table 2.

By comparison, it could be found that with the aging time increased, the yield strength, elastic modulus and elongation at break of the specimens all decreased. For the specimens with and without initial strain, the yield strength of the specimens with initial strain would decrease. After aging, the elastic modulus and elongation at break of the specimen with initial strain would decrease faster than that of the specimen without strain, indicating that the initial strain would strengthen the aging process of HDPE material to some extent.

Table 2. The mechanical properties of HDPE specimens.

Group	Yield strength (MPa)	Elastic modulus (MPa)	Elongation at break (%)
1A	29.76	582.42	68.76
1B	29.12	583.61	67.54
2A	24.85	568.58	54.48
2B	24.37	553.73	53.39
3A	22.31	536.25	45.63
3B	21.24	511.39	44.87

3.3 *Discussions*

The cable structure of cable-stayed bridge is the main load-bearing structure. In particular, the reliability of cable structures depends on the safety of cable body structure, durability and suitability of HDPE protection systems. The durability of HDPE protection system will directly affect the safety and applicability of cable structure, and then the safety and reliability of the entire bridge. Generally speaking, with the increase of the aging time of HDPE sample, the yield strength, elastic modulus and tensile elongation at break of the material will also decrease, and the results of this experiment also conform to this rule. For the effect of initial stress on the cracking of HDPE sheath, there is a critical value for the effect of initial stress. When the initial stress is less than the critical value, the stress on the one hand reduces the activation energy of chain fracture and accelerates the aging reaction, on the other hand, the stress will hinder the aging reaction by reducing the diffusion of oxygen (Chen 2004; Zhu 2008). Macroscopically, the stress has little influence on the aging performance. For this experiment, the initial stress applied did not reach the critical value, so the cracking of HDPE material was not caused. Therefore, the aging test under the condition of large initial stress should be carried out. On the other hand, the HDPE actually operates under cyclic stress and is exposed to even harsher conditions. In this experiment, a constant load is applied, so cyclic stress should also be applied to explore this aging process.

4 CONCLUSION

The performance of HDPE sheath in different environments will affect its own life. The photo-aging performance of HDPE in dry environment were studied in this paper. The mechanical properties and corrosion characteristics of HDPE were analyzed by loading different initial stress and artificial accelerated aging time. Based on the experimental results and discussions presented above, the conclusions are obtained as below:

(1) Initial stress is not the main cause of HDPE aging cracking.
(2) Environment and load will produce a certain degree of aging on the surface of HDPE specimen, but all belong to mild corrosion aging.
(3) With the increase of aging time, the yield strength, elastic modulus and elongation at break of HDPE samples all decrease to a certain extent, and the change rules of the three are consistent.

ACKNOWLEDGMENTS

This research was supported by the Special Fund Project for Basic Scientific Research Business Expenses of Central Public Welfare Scientific Research Institutes—Analysis and Evaluation of Aging Performance of Outer Protective Materials of Pull-Down Slings under Complicated Environmental Conditions (Approval No.: 2020-9056).

REFERENCES

Chen, R. & D. R. Tyler (2004). Origin of tensile stress-induced rate increases in the photochemical degradation of polymers. *Macromolecules* 37(14), 5430–5436.

ISO 527-2: 1993. (2006) *Plastics-Determination of tensile properties*. China: China Standards Press.

Liu, S. (2013). Damage mechanism and prevention countermeasures of HDPE protecting sheath of stayed cables. *Journal of Chongqing Jiaotong University* (Natural Science Edition) 32(S1), 888–893.

Parsons M. et al. (2000). Effect of strain rate on stepwise fatigue and creep slow crack growth in high density polyethylene. *J. Mater. Sci.* 35(8), 1857–1866.

Wu, W. & C. Li (2008). Discussion on anti-corrosion technology for stay cables. *Technology of highway and transport*, 4, 47–49.

Yang, S. et al. (2018). Damage mechanism of outer sheath of slings in service and life prediction method. *Journal of Chongqing Jiaotong University* (Natural Science Edition) 37(5), 12–17.

Ye, J. & J. Zhong (2003). Application and assessment of anti-corrosion technologies for stay-cable. *Corrosion & protection* 24(5), 221–223.

Zhang, B. & H. Zhang (2010). A study of lifetime of HDPE used in bridge cable. *Shanxi architecture* 36(10), 292–294.

Zhang, R. et al. (2018). The field repair process and its performance study of the cable-stayed HDPE cable. *Construction Quality* 36(11), 86–92.

Zhu, Z. et al. (2008). Research on the aging behaviors of polymer materials under stress. *Polyvinyl Chloride* 36(02), 6–9.

Advances in Civil Engineering: Structural Seismic Resistance, Monitoring and Detection – Mohd Johari Mohd Yusof and Junwen Zhang (Eds)
© 2023 The Authors, ISBN 978-1-032-31491-4

Reliability detection technology of hinge joint connection of hollow slab based on Ultrasonic

Peng Dong
Highway Research Institute of the Ministry of transport, Beijing, China

ABSTRACT: If the service life of the hollow slab is extended, the hinge joint is easy to break and fall off, thus affecting the operation safety and service life of the bridge. Through the comparison of the nondestructive testing methods of concrete, the ultrasonic method was selected as the hinge joint damage detection method. From the perspective of the possible failure characteristics and loading features of the hinge joint, an on-site operable reliability detection method of hinge joint connection of hollow slab based on the ultrasonic method was established, which would guide the actual engineering construction and the defect detection of existing hinge joints. In case of a further comprehensive improvement of the detection quality of the hinge joint, the detection cost would be saved.

1 INSTRUCTION

If the service life of the hollow slab is extended, the hinge joint is easy to break and fall off, thus affecting the operation safety and service life of the bridge. As a result, in-depth research on the damage identification of the hinge joint of the hollow slab bridge was carried out to explore an effective method that can timely and accurately identify the damage of the hinge joint in the early stage of the hinge joint damage, which was of great importance to the strengthening of the quality control of the hinge joint and the improvement the safety of bridge operation. Through the comparison of the nondestructive testing methods of concrete, the ultrasonic method was selected as the hinge joint damage detection method (Chen 2015; Li 2017; Wang 2015). From the perspective of the possible failure characteristics and loading features of the hinge joint, research on reliability detection method of hinge joint connection of hollow slab based on the ultrasonic method was carried out, which would be beneficial to the strengthening of the operability and convenience level of on-site detection of the hinge joint, the enhancement of the possibility of popularizing and applying hinge joint detection on the site on a large scale and the improvement of the quality level of the hinge joint as a wholeConstruction and Geometrical Dimensions of Specimens.

2 DETECTION PRINCIPLE

To realize the fast area array scanning of imaging targets, it is necessary to obtain holographic data through densely arranged transceiver units. However, the number of transceiver units will be too large to be realized in engineering according to the antenna layout requirements of single-transmitting and single-receiving millimeter-wave imaging half-wavelength. To significantly reduce the number of transmitting and receiving channels, a multi-transmitting and multi-receiving array with sparse antenna elements can be used for holographic imaging. With regard to the typical ultrasonic multi-transmitting and multi-receiving array imaging, the ultrasonic signal sent by the transmitting antenna irradiates the object, the reflected signal is received by all receiving antennas on the receiving plane at the same time. Then, all holographic data can be obtained by switching the transmitting unit in turn (Liu 2018; Qian 2018; Song 2006).

DOI 10.1201/9781003310884-63

The ultrasonic tomography instrument was used to detect the internal defects of hinge joints as shown in the figure.

Figure 1. Operation panel.

The principle of the array type ultrasonic transverse wave detection is mainly based on the ultrasonic pulse echo method and the imaging principle is provided with the signal processing technology of synthetic aperture focusing (SAFT). The transmission and reception were realized by transmitting and receiving the energy converter. One sensor transmitted the compressed wave pulse and the other sensor received the reflected pulse. After testing the time Δt from the beginning of the pulse to the arrival of the echo, the depth d of the reflecting surface (bottom interface or defect location) could be calculated when the wave velocity C was known.

The sensor was taken as both a transmitter and a receiver. The control unit in the antenna was used for exciting one row of the sensor and the other rows of the sensor were used for receiving pulses as a receiver. This process was repeated until each sensor of the other rows acted as a transmitter. The computer processed the transmission time of each reflected pulse. After obtaining the transmission time data at the test location, a signal processing technology called synthetic aperture focusing (SAFT) is used for reconstructing the 2D image inside the concrete component at the test location for the equipment.

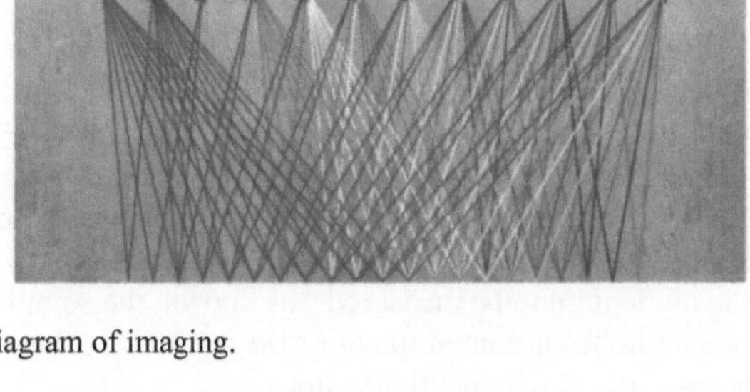

Figure 2. Schematic diagram of imaging.

3 DETECTION TECHNOLOGY METHODS

3.1 *Test area confirmation*

Before the test, the concrete situations of the hollow slab and the hinge joint were fully understood, including the concrete strength, thickness, internal form of structure and other information.

According to the apparent inspection results of hinge joint, the defect position of hinge joint is preliminarily determined. If the water seepage and other diseases exist among the hinge joint, it can be preliminarily determined that the hinge joint has internal defects such as cavity. In this case, the area is regarded as the test area.

It can be preliminarily determined that the hinge joint has internal defects such as cavity and damage. The grid was drawn in the test area and the distance size is confirmed according to step size settings.

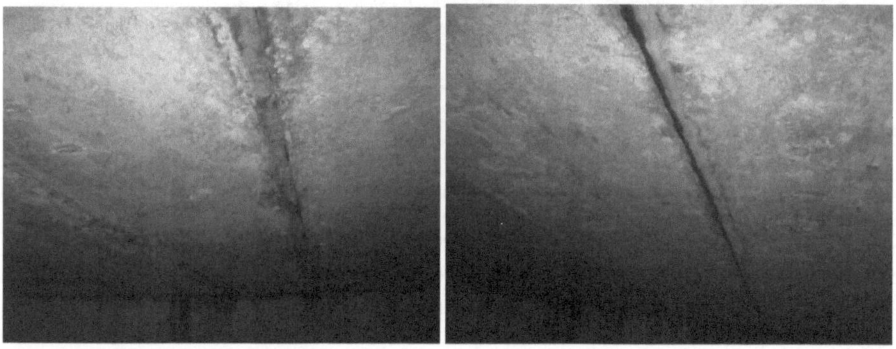

Figure 3. Water seepage at hinge joints.

3.2 *Test area cleaning*

The sandpaper and other tools were used to polish the bottom surface of the predetermined test area and remove the attachments on the concrete surface (such as dust, laitance, colloidal protective layer, unconsolidated layer, etc.) to keep the bottom surface flat and smooth.

3.3 *Wave velocity calibration*

To carry out correct image reconstruction and visualization, the correct sound velocity value was needed. The accuracy of the sound velocity value affected the size and depth of the test reflector. Before ultrasonic testing of hinged joints, the wave velocity needed to be calibrated. The wave velocity calibration needed to be carried out at the defect-free location of concrete. If there are defects in the calibrated wave velocity location or factors affecting the calibration, the test results would be affected. Before hinge joint detection, the wave velocity could be calibrated at the place where the hinge joint was free of defects or at the place where the surrounding hollow slab was free of defects.

(1) If the thickness of the hinge joint to be measured was unknown, the instrument would measure the sound velocity of the wave spreading on the inspection surface between DPC energy converters. The automatic measurement function of the equipment was turned on. 5-8 arbitrary points were selected for measuring the velocity and recording on the bottom surface of the hinge joint to be measured. The average value should be taken.
(2) If the thickness of the hinge joint to be measured was known, the sound velocity on the hinge joint to be measured should be calibrated to adjust the sound velocity value until the bottom echo indication appears at the correct depth location.

3.4 *Local test*

Before the continuous test of the MAP mode, a local test (under REVIEW mode) should be carried out to adjust the key parameters of the test equipment to the ideal state, which will be applied to the subsequent continuous test of the MAP mode.

Figure 4. Schematic diagram of elevation of hinge joint

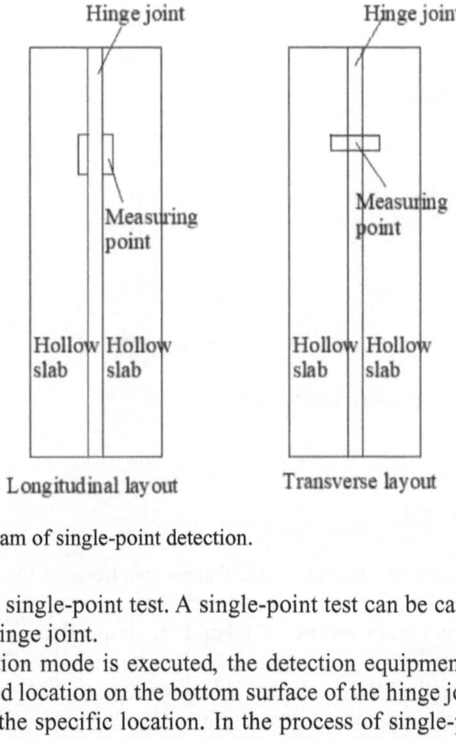

Figure 5. Schematic diagram of single-point detection.

The local test is also a single-point test. A single-point test can be carried out when detecting a point of concern of the hinge joint.

When the local detection mode is executed, the detection equipment is lifted up and in close contact with the measured location on the bottom surface of the hinge joint to analyze the internal structure and section of the specific location. In the process of single-point detection of internal defects of hinge joint, for the sake of ensuring the coupling effect between the detection equipment and the tested surface, it is necessary to reasonably select the longitudinal or transverse layout of the detection equipment according to the flatness of the bottom surface of hinge joint and whether the hollow slabs on both sides are staggered. If the flatness of the detected location is high, the detection equipment can be arranged horizontally to obtain the defect detection data on both sides of the hinge joint center line as far as possible. If the flatness of the tested location is poor or the hollow slabs on both sides of the hinge joint are staggered, the detection equipment can be arranged longitudinally along the edge of the hollow slab.

From the perspective of these results, the necessary information can be received immediately or the mobile device direction can be selected to receive more complete information. It is important to adjust the parameters before the continuous test of MAP mode to an appropriate value to obtain the best image quality and recheck a certain point of concern.

3.5 *Continuous test of the MAP test area*

After the test area has been confirmed and the local test parameters have been adjusted to the appropriate values, the longitudinal or transverse layout of the detection equipment should be reasonably selected according to the flatness of the bottom surface of the hinge joint and whether the hollow slabs on both sides are staggered to carry out a continuous test on the test area. According to the scope of the test area, it is needed to reasonably plan the test moving step size and direction before test setting. The data will be saved automatically after the test in the test area.

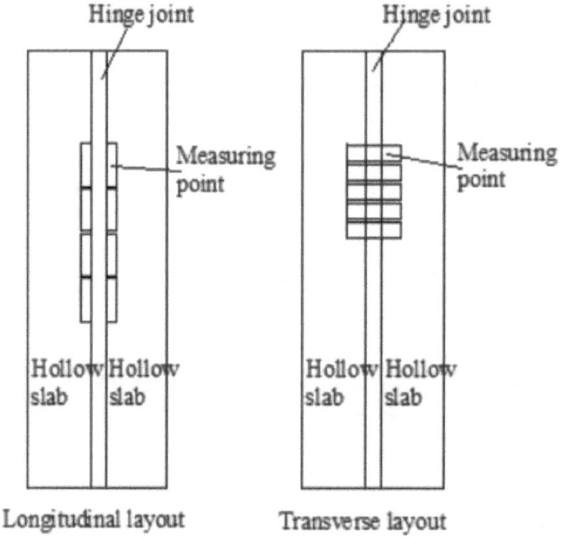

Figure 6. Schematic diagram of continuous detection.

4 RESULT DETERMINATION

Due to the different reflectivity of ultrasonic at different locations of the internal structure of the hinge joint, the reflectivity of each point of the tested hinge joint is coded. The reflected area is rendered as a visible spectrum from yellow to red and the non-reflected area is rendered as light blue and dark blue to obtain an image that can visually identify the internal structure of the concrete for a reasonable judgment on the internal defects of the structure. The details are as follows:

Blue in the image indicates an area where there is no ultrasonic reflection. Light blue and dark blue show that the light blue is mainly caused by slightly different material consistency but the overall sound wave velocity difference is very small (material characteristics of concrete). Red in the image indicates that the closer to the center of the reflector (defect, steel bar, crack, etc.) is, the more ultrasonic energy will fall on the receiving sensor of the antenna array, and the larger the corresponding signal amplitude is, the redder the color is, and vice versa. Yellow green in the image indicates that the lower the energy of the ultrasonic wave near the edge of the reflector is, the smaller the amplitude is.

Through the acquisition of ultrasonic signal, the redder the location in the image is closer to the center of the reflector (defect, steel bar, crack, etc.), indicating that the greater the possibility of defects in this location is. However, due to the complex internal structure of concrete, the ultrasonic

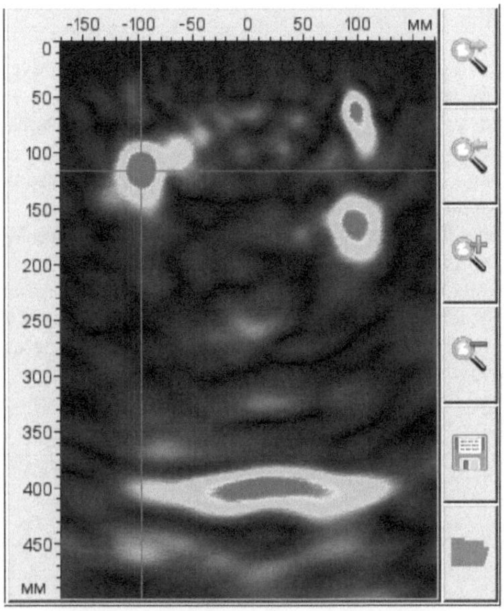

Figure 7. Schematic diagram of image information.

signal is reflected many times and there are many red areas in the image. So, the collected energy information may interfere with the determination of internal defects. Based on this, the color change of the image signal is observed by gradually adjusting the gain coefficient. If the gain coefficient is adjusted to a reasonable range, a reasonable judgment can be made on the internal defects of the hinged joint structure according to the color depth of the image.

If the gain coefficient decreases to a certain extent, the red area in the image is still obvious, so it can be considered that there are structural defects in this area. The size of the red area range is used for dividing the disease degree of internal defects of hinge joint into three grades, such as slight, medium and serious, which provides a basis for the treatment of internal defects and diseases of hinge joint.

5 CONCLUSION

On the one hand, the establishment of the reliability detection method of hinge joint connection of hollow slab based on the ultrasonic method can quickly locate the internal defects of hinge joint; on the other hand, it has good technical advantages such as easy operation and convenience of on-site detection, which will be beneficial to the improvement of the technical process of hinge joint damage detection, the strengthening of the quality and level of hinge joint damage detection and the reduction of the cost of bridge maintenance. The research results have certain guiding significance for the daily work of bridge management and maintenance organizations and have an extensive promotion prospect.

ACKNOWLEDGMENTS

This work was financially supported by Central Public-Interest Scientific Institution Basic Research Fund (2020-9055).

REFERENCES

Chen Guodong. (2015) Research on Application of Ultrasonic in Nondestructive Detection of Concrete Pile Foundation. *Wuhan University of Technology*.

Li Juanjuan. (2017) Research on Ultrasonic Testing Method for Internal Damage of Concrete Structures. *Nanchang Hangkong University*.

Liu Yanping and Wang Yinguan. (2018) Spectral Analysis Technique of Ultrasonic Pulse-Echo in Measuring Sound Velocity. *Technical Acoustics*. pp. 176–177.

Qian Chen. (2018) Research and Implementation of Concrete Defect Detection Method Based on Ultrasonic. *Nanjing University of Information Science and Technology*.

Song Huansheng, Zhao Xiangmo, Wang Guoqiang and Qi Xiuzhen. (2006) Ultrasonic Tomography System for Nondestructive Concrete Structure Testing. *Journal of Traffic and Transportation Engineering*. pp.73–77.

Wang Yu. (2015) Research on Ultrasonic Array Imaging Algorithm for Internal Exceptions of Concrete. *Nanchang Hangkong University*.

Advances in Civil Engineering: Structural Seismic Resistance, Monitoring and Detection – Mohd Johari Mohd Yusof and Junwen Zhang (Eds)
© 2023 The Authors, ISBN 978-1-032-31491-4

Study of fiber-reinforced cement matrix composites based on comprehensive analysis

YuFu Li*
School of Civil Engineering, Guangdong Construction Polytechnic, Guangzhou, Guangdong, China

YinBo Li
School of Architecture and Engineering, Guangdong Baiyun University, Guangzhou, Guangdong, China

ABSTRACT: Concrete is widely used in various fields of civil engineering, but the material has defects such as low tensile strength, poor toughness, and difficult to control crack width after cracking, which not only affect the service performance of the structure but also reduce the life of the project under environmental conditions, such as corrosion, and even cause safety accidents. The fiber-reinforced cementitious composite material (hereinafter referred to as X material) proposed in this paper is a new disordered short fiber-reinforced cementitious material with stress-strain hardening characteristics, high tensile strength, and excellent crack control ability. X material as a repair material to replace the damaged area of the structure has broad application prospects in increasing the structural load-bearing capacity and improving structural durability. Based on the study of the development background and research progress of X material, this paper further summarizes the advantageous characteristics of its mechanical properties, and conducts a study on the design of matching ratio and the prospect of practical engineering application of domestic X material. It aims to promote the development of research in this field to enhance the durability of concrete materials, improve the brittleness of concrete materials, and reduce the risk of cracking of concrete structures.

1 INTRODUCTION

Concrete is widely used in various fields of civil engineering; however it has several disadvantages such as low tensile strength, poor toughness, and difficult to control crack width after cracking, which not only affect the service performance of the structure, but also reduce the life of the project under environmental conditions such as corrosion, and even cause safety accidents. The fiber-reinforced cementitious composite material (hereinafter referred to as: X material) proposed in this paper is a new disordered short fiber-reinforced cementitious material with stress-strain hardening characteristics, high tensile strength and excellent crack control ability. X material as a repair material to replace the damaged area of the structure has a broad application prospect in increasing the structural load-bearing capacity and improving structural durability (Li et al. 2021). Based on the study of the development background and research progress of X material, this paper further summarizes the advantageous characteristics of its mechanical properties, and conducts a study on the design of matching ratio and the prospect of practical engineering application of domestic X material. It aims to promote the development of research in this field to enhance the durability of concrete materials, improve the brittleness of concrete materials, and reduce the risk of cracking of concrete structures.

*Corresponding Author: lyf999999@163.com

DOI 10.1201/9781003310884-64

2 DOMESTIC AND INTERNATIONAL RESEARCH PROGRESS

2.1 *Introduction of X material material*

X materials are usually made of cement, mineral admixture, quartz sand as the matrix, and fiber as the reinforcing material, and their ultimate tensile strain is usually in the range of 3%–7% when the fiber volume admixture is less than 2%. X materials have tensile strain hardening performance and multi-joint cracking characteristics, and the multi-joint cracking width in the saturated state is less than 0.1 mm. They have good compressive toughness and deformation coordination performance with reinforcement. Because of its excellent mechanical properties, X material provides a new solution for concrete structure to solve durability, engineering structure cracking, and earthquake prevention, as well as renovation and reinforcement of old structures. In recent decades, many researchers and research institutions at home and abroad have focused on the basic mechanical properties and durability of X material, such as direct tensile properties, uniaxial compressive properties, bending properties, shear properties, fracture properties, shrinkage creep properties, resistance to heat and humidity aging, fatigue impact properties, resistance to permeability, resistance to freeze-thaw cycles, etc., while in the published literature on the interface between X materials and concrete (Yuan et al. 2021). The number of studies on the bonding properties of X materials with concrete interface is limited. One of the objectives of the study on the interfacial bonding performance of X materials and existing concrete is to gain a comprehensive understanding of the interfacial stress characteristics, explore the interfacial bonding mechanism, and propose a test method for evaluating the bonding performance of X materials and existing concrete matrix. The other is to understand the shortcomings of interfacial bonding, propose improvement measures, and provide a basis for the reinforcement of concrete structures with X materials (Yuan et al. 2021).

2.2 *Study of basic mechanical properties of X materials*

X material is a high toughness cementitious composite material proposed by Professor Li V C of the University of Michigan in the early 1990s based on micromechanical principles. Since its inception, X material has been preferred and researched by domestic and foreign researchers for its flexible fabrication process, high strength, excellent crack resistance and crack control, good ductility and toughness, high shear resistance, and other good mechanical properties. Many types of fiber-reinforced cementitious composites have been developed on its basis, such as PE-X material formed by adding polyethylene fibers to the cement base, carbon fiber reinforced X material made by adding carbon fibers, steel fiber reinforced X material made by adding steel fibers, etc. (Wu et al. 2021). The most obvious characteristic of X material is the strain-hardening property under the action of axial tension, when it is subjected to tension, first, the cement base is pulled and cracked, while the fibers distributed in a disorderly manner still connect the cement base at both ends of the crack across the crack, playing the role of bridging and transferring the tensile stress, unlike ordinary concrete which is prone to cracking under the action of tensile stress and will break once cracked. The emergence of X material has fundamentally improved the shortcomings of concrete with insufficient tensile capacity. The strain-hardening properties are obvious in both tensile and bending loads.

2.3 *Research on the performance of structural members of X materials*

In engineering, the life and durability of reinforced concrete structure largely depends on the early appearance of macro cracks in the concrete and the rapid development of the crack width. As the concrete in the reinforced concrete structure wrapped in the external reinforcement, the role of protection for the reinforcement, once the surface of the concrete cracks and cracks reach a certain width, the internal reinforcement of the structure comes in direct contact with the air, the external air, acidic gas, or liquid and other aggressive media will enter the structure through the cracks and contact the reinforcement causing corrosion of the reinforcing steel (Yang et al. 2021). Once corrosion of the reinforcement takes place, not only will it reduce the effective bearing area of the reinforcement, but it also reduces the load-bearing capacity of the component. The rust phenomenon in the process of corrosion of the reinforcement will further intensify the development of cracks in the concrete protective layer, and in serious cases will cause the protective layer spalling. This is a serious threat to the safety and durability of the structure. X material with good crack control ability can prevent the concrete from easily cracking. Studies have shown that even under fatigue loading, the crack width of X material can be well maintained within a certain range. It is found that after 100,000 cycles of loading, the crack width of X material joint plate is about 50 um, while the crack width of concrete joint plate under the same condition is more than 0.6 mm. Thus, it can be seen that X material has more excellent crack width control performance compared with conventional concrete (Luo et al. 2021).

3 ADVANTAGES OF X MATERIALS

According to the research results of X materials at home and abroad, its main advantageous features can be summarized as follows:

- X material can produce multiple fine cracks and the ultimate crack width does not exceed 0.1 mm when under tension, and has significant strain hardening characteristics, and the ultimate tensile strain can be stable to more than 3%.
- The compressive strength of X material is close to that of ordinary concrete, but the compressive toughness and ductility of X material are much better than that of ordinary concrete.
- The shear resistance of X material members is greatly improved compared with ordinary concrete, which can reduce the use of transverse reinforcement.
- Material X members have excellent damage resistance, significant strain hardening phenomenon, and slow stiffness degradation.
- X material members have high ductility, the damage form is ductile damage, and because the deformation between X material and reinforcement is coordinated, no relative slip phenomenon will occur, thus avoiding splitting damage;
- X material members have superior energy dissipation capacity, due to its multiple and dense micro-cracking steady-state cracking characteristics, the reinforcement can yield only in a longer range, thus making fuller use of the characteristics of the reinforcement.
- Good durability of material X. Since material X does not contain coarse aggregates, its fine aggregates have good denseness, and the fine and dense micro-cracks can effectively prevent the infiltration of corrosive media when damaged by force, thus greatly extending the life of structural members.
- As X material is mixed with disorderly distributed fibers, its bridging effectively improves the cracking strength of the members, reduces the use of steel reinforcement, and saves costs.

4 STUDY ON THE FITMENT OF DOMESTIC X MATERIALS AND PRACTICAL ENGINEERING APPLICATIONS

4.1 *Analysis of the matching material performance of domestic X materials*

Cement: X material usually uses silicate cement, which should not be placed for too long because it sets and hardens quickly, and can be prepared before the test. To ensure the flow of X material, it is also necessary to require the fineness of cement to meet the requirements. There should not be agglomerates, which could affect the dispersion of fibers in the matrix. The total alkali content in cement should not be too high, as it will accelerate the loss of flow of the mix, while the alkali content also affects the compatibility between cement and water reducing agent. In view of the above factors, the cement used in this test is P.O52.5 grade ordinary silicate cement produced by a cement company, whose basic physical and mechanical indexes are shown in Table 1.

Table 1. Cement basic physical properties index table.

Fineness (0.08 mm sieve margin)	Initial setting time/min	Final setting time/min	Validity boiling method	Total alkali content %	Tensile strength (28d)/MPa	Compressive lightness (28d)/MPa
8.0	105	162	Qualification	0.4	12	52.5

Fly ash: Fly ash is a solid waste discharged from the flue gas after coal combustion. Due to the accelerated development of coal power industry, dust pollution is severe, among which the emission of fly ash is also increasing each year, which seriously pollutes the air, water, and the environment, and even threatens health and life of people. Therefore, reasonable treatment of fly ash can turn waste into treasure, reduce energy consumption, and meet the concept of green and sustainable development. Fly ash has low cost and is a type of cementitious material. When added to cement mortar matrix, it can significantly reduce the heat of hydration of the matrix and improve the working performance of X material. In addition, the interaction between fly ash and cement can play a dispersing role for cement particles, which can effectively improve the fluidity and compatibility of X material matrix. The fly ash matrix material also plays the role of water reduction, lubrication, and slow setting. After several test adaptations found that fly ash has a significant effect on the performance of multi-crack carry out. Especially in the direct tensile test and thin plate four-point bending test, the Class I fly ash performed better with a sieve residual of less than 5% in the 48 μm square pore sieve, the presence of a large number of smooth round bead particles, and very high strength. Therefore, the test fly ash in this paper uses the Class I fly ash produced by a power plant, and its performance index is shown in Table 2.

Table 2. Fly ash performance index table.

SiO_2%	Al_2O_3%	Fe_2O_3%	CaO%	MgO%	SO_3%	K_2O%	Na_2O%	Burning loss%
53.8	31.5	5.9	3.8	1.3	1.5	1.6	0.2	0.4

Quartz sand: Quartz sand is the most commonly used fine aggregate in X material. As X material does not use coarse aggregate, quartz sand can be used as the filler in cementitious material. It can be used to filling the gap between the matrix of X material, make the aggregate gradation continuous because the cement and fly ash lead to condensation and hardening of dry shrinkage and wet expansion, will cause the matrix volume address larger problems with the addition of quartz sand and cementitious material to form a basic skeleton. The addition of quartz sand, with the cementitious material to form the basic skeleton, effectively controls the change of the volume of the matrix. In addition, quartz sand also helps to give full play to the reinforcing and cutting effect of the fiber, and its basic performance index is shown in Table 3.

Table 3. Basic performance index of quartz sand table.

Sand type	Apparent density Kg/m^3	Bulk density Kg/m^3	Fineness modulus	Grain size
Quartz sand	1600	1300	5.0	100–150

PVA fiber: This type of fiber has great influence on the mechanical properties of X material. Ordinary fiber added to the concrete: Shi, although it can improve the material bending shear, it cannot improve the early crack resistance and deformation capacity of concrete. X materials are mainly used synthetic fibers. Among them, PVA fiber is a commonly used reinforcing and cutting materials, with the advantages of good dispersion, good adhesion with cement, good durability, etc. Due to its high tensile strength and elastic modulus, it can realize the excellent tensile strength of X material and the stable cracking of many microfine cracks, and improve the bending toughness of X material, which is the key factor to decide whether the many excellent properties of X material can be realized. The selection of PVA fiber is important; if the diameter is too small or the fiber is not surface treated, it will lead to difficulty in fiber dispersion, and the agglomeration phenomenon will easily appear during mixing. If the tensile strength and modulus of elasticity of the fibers do not meet the requirements, the fibers will be easily pulled off instead of being pulled out in the later X material tensile test, and the phenomenon of strain softening and less crack damage will occur, which will eventually affect the realization of the unique and excellent properties of X material. In this test, the K-II RECIS type PVA fiber produced by Kuraray of Japan was used, and its physical parameters are shown in Table 4.

Table 4. PVA fiber performance index.

Type	Length	Diameter/μm	Ultimate elongation/%	Tensile strength/MPa	Modulus of elasticity/GPa	Density g/cm^3
k-2	15	40	7.0	1650	43.5	1.5

Water: The water used in this test is ordinary tap water with a density of 1.0g/cm^3.

Additives: After the substrate is added to PVA fiber, the mobility is obviously reduced. To improve the fluidity of the mix and the dispersibility of PVA fiber, appropriate amount of water reducing agent and thickening agent should be added during the mixing of substrate material. The water reducing agent is added mainly to prevent the phenomenon of water secretion in X material, which has good dispersion effect on the cement particles inside the base material and a high water reduction rate. The amount of water reducing agent sometimes needs to be fine-tuned according to the water secretion of the mix in the mixing process, and record the accurate fine-tuning data, to ensure that under different matching ratios, we can obtain a new mix with good fluidity, no water secretion, and uniform fiber dispersion. In this paper, we choose polycarboxylic acid high efficiency water reducing agent, which is in powder form, light yellow in color and 1.05% density. Thickening agent mainly to thicken and water retention, in the high water-cement ratio of the ratio, thickening agent can prevent cement mortar segregation and mix water secretion, improve the dispersion of fibers, can increase the cohesion of the matrix, but also improve the homogeneity and hardening ability of the matrix material. In this paper, the powdered cellulose ether thickener, which is called "VMA" in foreign countries, is selected on the basis of the pre-matching, and the mix obtained has good fluidity and viscosity.

4.2 *Engineering application*

In the actual use of engineering structures, due to the poor crack resistance of ordinary concrete is often used with cracks, cracking form is generally a single cracking, crack width is large, is not conducive to the durability of steel protection, especially in the harsh use of the environment, such as marine chloride environment, chemical corrosion environment, the cracking of concrete so that the rapid invasion of aggressive substances, accelerating the corrosion of reinforcing steel, corrosion of reinforcing steel. The excellent crack control ability of X material can effectively prevent the erosive material from entering the structure and rusting the reinforcement, using X material as the protective layer of reinforced concrete structure, replacing the partially damaged concrete, strengthening the nodes of the structure, etc. It will effectively improve the service life and safety of the structure.

The first application scenario of X material is in the repair of irrigation channels, because X material does not contain coarse aggregate, has good confinement, and the cracks are fine and dense, so that the impermeability and corrosion resistance are guaranteed, and can be used to repair irrigation channels.

The second application scenario of X material is in the bridge deck seamless connection project, using X material matrix and steel deformation coordination ability, strong tensile strength and can reduce the number of configuration used to limit the width of the crack reinforcement and obtain good economic effect and other advantageous features. For example, the Meiyuan Bridge adopts X material deck slab, using X material to achieve a seamless connection of the deck slab, the overall amount of X material during the whole construction process is about 800 m, and the construction cost of the bridge is reduced accordingly, and the stress concentration phenomenon and fatigue resistance of the deck slab are significantly improved. In addition, due to the poor deformation ability and low tensile strength of concrete material, the deformation of adjacent deck slabs will be uncoordinated due to temperature change, shrinkage, and creep, which will easily lead to wide main cracks and affect normal use. Deformation inconsistency and poor crack resistance can be significantly improved, and the bridge deck can be seamlessly joined for a long time.

5 CONCLUSION

This paper shows that fiber-reinforced cementitious composites (X materials), with their excellent mechanical properties and crack control ability, provide a new solution to the problem of cracking and earthquake resistance and durability of engineering structures. PVA (high modulus polyvinyl alcohol fiber) is mainly used in X materials, which has superior performance and is harmless to humans, and is a green building material preferred by the construction industry today. In the future, we should also carry out research through the following aspects: axial tensile, straight shear, oblique shear test, to explore the interface roughness, existing concrete and X material strength, interfacial agent, and other factors on the X material existing concrete interface bonding performance. In this way, the microscopic mechanism of interfacial bonding is explored, the influencing factors are quantified through the test data, a strength calculation formula applicable to engineering practice is established, and an intrinsic relationship model describing the whole process of bonding–slip is built.

ACKNOWLEDGMENTS

This work was supported by Guangdong Science and Technology Innovation Strategy Special Fund (Science and Technology Innovation Cultivation for University Students, No. pdjh2020a0943.

REFERENCES

Li Ke, Wei Yaoxin, Jin Leilei, Wang Xinling. Compressive strength of high-strength stainless steel stranded wire mesh/ECC confined concrete [J/OL]. *Journal of Huazhong University of Science and Technology* (Natural Science Edition):1–6 [2021-11-03]. https://doi.org/10.13245/j.hust.211222.

Luo Yinjian, Li Xiudi, Wang Qifan, Yang Zhuanhong. Progress in the study of ECC dynamic properties and dynamic intrinsic structure relationship[J]. *Contemporary Chemical Industry*, 2021, 50(07):1690–1697.

Wu Tong, Shao YJ, Li GJ. Experimental study and damage analysis of composite torsional performance of reinforced PVA-ECC columns[J/OL]. *Building Structures*:1–9[2021-11-03]. http://kns.cnki.net/kcms/detail/11.2833.TU.20210818.1359.006.html.

Yang Z.H., Li X.D., Wang Q.F., Luo Yinjian. Research progress on dynamic mechanical properties and blast and impact resistance of engineered cement matrix composites[J]. *Silicate Bulletin*, 2021,40(08):2485–2496.

Yuan Chaoqing, Dai Xiaohui, Zhang Jie, Song Shuang. Study on the optimal ECC height of cement matrix composite double steel plate combination shear wall[J]. *Journal of Hebei Engineering University* (Natural Science Edition),2021,38(03):37–45.

Yuan Chengfang, Wang Di, Li Haofei, Zhang Zhe. Influence of fiber doping on the flow properties and mechanical properties of recycled brick powder ECC[J]. *Journal of Building Science and Engineering*, 2021,38(05):74–82.

Advances in Civil Engineering: Structural Seismic Resistance, Monitoring and Detection – Mohd Johari Mohd Yusof and Junwen Zhang (Eds)
© 2023 The Authors, ISBN 978-1-032-31491-4

Application of direct-cast environmental modifier in ultra-thin cover asphalt mixture

Xudong Yang
China Merchants Highway Network Technology Holdings Limited, Beijing, China

Jifen Wei
Guangxi Huatong Expressway Co., Ltd., Guilin Guangxi, China

Qiwei Zhou*
Chongqing Transportation Research and Design Institute Co., Ltd., Chongqing, China

Zhiyong Deng
Guangxi Huatong Expressway Co., Ltd., Guilin Guangxi, China

ABSTRACT: A direct-cast composite modified asphalt mixture was prepared by direct-cast environmental modifier, SBS and mineral materials and applied to ultra-thin overlay asphalt mixture. The effects of different dosage and type of direct casting modifier on high temperature performance, water stability and low temperature crack resistance of the mixture were studied. The results show that compared with SBS modified asphalt, when the SP-2 direct-input modifier content is 0.6 % of the mineral mass, the high-temperature stability increases by 111 %, and the freeze-thaw splitting strength ratio increases by 16 %. However, the low-temperature crack resistance decreases to a certain extent, and the optimal content is less than 0.4 % of the mineral mass. Through the test section paving verification, direct cast modified asphalt mixture has more convenient construction, more excellent road performance, and its performance meets the requirements of the current specification.

1 INTRODUCTION

Ultra-thin overlay is widely used in the preventive maintenance of expressways and national and provincial trunk lines due to its advantages of thin paving thickness, high driving comfort, and good durability, serving as a hot spot of preventive maintenance technology in China in recent years (Xu et al. 2020). Polymer modified asphalt is the core material of ultra-thin overlay. In recent years, it is mainly composed of polyolefin, SBS and SEBS, showing two kinds of modification modes. One is the common wet modified asphalt, which prepares modified asphalt after shearing the modifier and asphalt, and then mixes with aggregate to produce mixture. The other is dry modification, that is, the modifier and aggregate are first mixed and then added with matrix asphalt or modified asphalt to produce mixture (Cui 2016; Yao et al. 2018). Dry modification not only saves the production process of finished modified asphalt, reduces the emission of asphalt smoke, benzene soluble substances and benzopyrene, but also solves the problem of uneven dispersion of modifiers. It is a hot technology for modified asphalt mixture at home and abroad in recent

*Corresponding Author: zhouqiwei@cmhk.com

DOI 10.1201/9781003310884-65

years (Li 2014; Liu 2018). Domestic road engineers Zhou Zheng (2018) studied the direct-cast rubber powder and SBS composite modified asphalt mixture gradation design with the advantages of wet modified asphalt. It is concluded that the advantages of wet modified asphalt, and site construction is convenient, road performance and wet modification are equivalent; Xu Qiang (2020) studied the performance of direct-cast SBS modified asphalt SMA mixture, and concluded that direct-cast SBS modified asphalt SMA mixture has good construction workability, and has obvious advantages in high temperature stability and water stability. Liu Ning et al. (Liu et al. 2020) studied the dry mixing process and concluded that MEPE effectively enhanced the interface between aggregate and asphalt, thus more significantly improved the road performance of asphalt mixture. Qiu Guoping et al. (Qiu et al. 2020) studied the performance of direct-cast mixture and found that the high temperature rheology of asphalt was significantly improved after adding modifier. PG high temperature performance classification can be improved by 2-3 grades, and the improvement of high temperature sensitivity is not obvious. The high temperature stability and water stability of direct-cast modified asphalt mixture are better than those of SBS modified asphalt mixture. In summary, the direct-cast modified asphalt mixture not only has excellent mixing process and good road performance, but also is a green and environment-friendly modification technology.

In this paper, SBS, environment-friendly direct-cast modifier, and mineral aggregate were used to prepare high performance ultra-thin overlay asphalt mixture. The high temperature performance and water stability were analyzed, and the test section was paved at the same time to provide data reference for the application and promotion of direct-cast ultra-thin overlay.

2 RAW MATERIAL

2.1 *Asphalt*

SBS modified asphalt adopts the cost modified asphalt provided by an asphalt plant in Guangxi, the specific indicators are shown in Table 1.

Table 1. Technical indexes of SBS modified asphalt.

Test project		Technical standard	Test result
Needle penetration (25°C)/0.1 mm		40-60	55
Prolongation (5°C)/cm		≥20	37
Softening point/°C		≥60	68
In the wake of TFOT	Quality change/%	−1.0–1.0	0.2
	Penetration ratio (25°C)/%	≥65	75
	Prolongation at 5°C/cm	≥15	18

2.2 *Mineral material*

In this paper, the coarse and fine aggregates are diabase of Longsheng, Guangxi, and the mineral powder is limestone powder of Longsheng, Guangxi. All indexes of the material meet the requirements of relevant specifications.

2.3 *Direct injection modifier*

The direct casting modifiers are SP-1 and SP-2 modifiers developed independently. The main components are rubber and recycled PE. The specific indicators and dosage are shown in Tables 2 and 3.

Table 2. Technical index of direct injection modifier.

Test project	Technical standard	Test result SP-1	SP-2
Single particle mass/g	≤0.03	0.01	0.01
Density/(g/cm^3)	≤ 1.0	0.75	0.92
Melt index/(g/10 min)	≥2.0	4.2	4.2
Ash content/%	≤ 1.0	0.1	0.1

Table 3. Dosage of direct injection modifier.

Groups	Mineral material/part	SBS modified asphalt/part	SP-1 /part			SP-2 /part		
1	100	5.5	0.2	–	–	–	–	–
2	100	5.5	–	0.4	–	–	–	–
3	100	5.5	–	–	0.6	–	–	–
4	100	5.5	–	–	–	0.2	–	–
5	100	5.5	–		–	–	0.4	–
6	100	5.5	–	–	–	–	–	0.6
7	100	5.5	–	–	–	–	–	–

2.4 *Synthetic gradation of mixture*

The aggregate of 0–3 mm, 3–6 mm, 6–11 mm diabase and mineral powder were used for gradation design. The final design results are shown in Table 4.

Table 4. Results of mix proportion design.

Sieve aperture/mm	13.2	9.5	4.75	2.36	1.18	0.6	0.3	0.15	0.075
Synthetic gradation/%	100.0	95.4	37.6	28.4	23.4	18.3	15.2	12.3	8.5

Based on the above gradation, the Marshall compaction test was used to determine the optimum oil-stone ratio of 5.5 %.

2.5 *Preparation of direct-cast modified asphalt mixture*

Preparation process of high viscoelastic modified asphalt as shown in Figure 1.

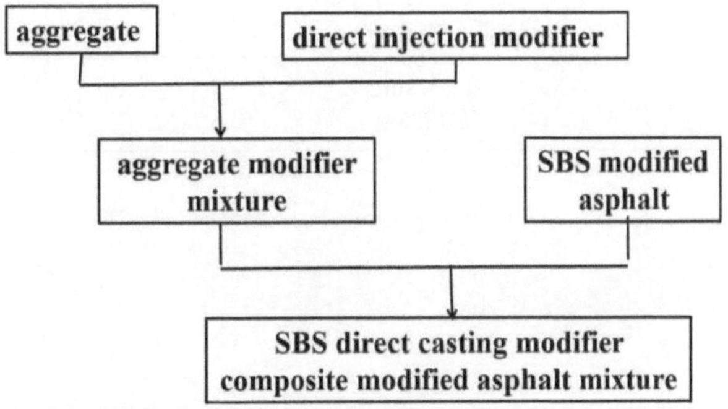

Figure 1. Preparation process of direct-casting modified asphalt mixture.

The preparation process of direct casting modified asphalt mixture in the laboratory is as follows: When the aggregate is heated at 180°C, the direct casting modifier is added to mix for 60 s, then the SBS modified asphalt is added to mix for 60 s, and finally the mineral powder is added to mix for 60 s.

3 ROAD PERFORMANCE ANALYSIS OF DIRECT CAST MODIFIED ASPHALT MIXTURE

3.1 *High-temperature stability*

The rutting test was used to evaluate the high temperature stability of the mixture. The parallel specimens were 2 in each group. The test results were averaged and are shown in Table 5.

Table 5. Wheel tracking test of direct-cast mixture.

Groups	Test results (times/mm)	Technical requirement (times/mm)
1	4535	≥3000
2	5748	
3	6412	
4	4987	
5	7852	
6	9035	
7	4268	

The following conclusions are drawn from Table 5: the anti-rutting performance of the direct-cast mixture increases with the increase of the direct-cast agent content, and the increase of SP-2 is more obvious; when the content of SP-2 is 0.6 % of the proportion of mineral aggregate, the dynamic stability of the mixture is greater than 9000 times/mm, which is 111 % higher than that of SBS modified asphalt mixture. The reason is that SP-2 contains a certain proportion of recycled PE, which not only improves the hardness of asphalt, but also further enhances the high temperature stability of the mixture.

3.2 *Water stability*

The freeze-thaw splitting test was used to evaluate the water stability of direct cast modified asphalt mixture. There were 8 parallel specimens in each group. The test results were averaged, and were shown in Table 6.

Table 6. Freeze-thaw splitting test results of mixture.

Groups	Test results TSR (%)	Technical requirement (%)
1	87.6	≥80
2	94.3	
3	97.8	
4	90.2	
5	96.3	
6	99.4	
7	85.2	

According to the analysis of Table 6, the water loss resistance of direct-cast mixture increases with the increase of direct-cast agent content, and the improvement of SP-2 type is more obvious. When the content of SP-2 is 0.6 % of the proportion of mineral aggregate, the dynamic TSR of the mixture is 99.4 %, which is 16.6 % higher than that of SBS modified asphalt mixture. The reason is that SP-2 contains a certain proportion of recycled PE and rubber, which improves the adhesion between asphalt and stone and further improves the water resistance of the mixture.

3.3 *Low temperature crack resistance*

The low temperature crack resistance of the mixture was evaluated by low temperature trabecular test. There were six parallel specimens in each group. The test results were averaged. The specific test results are shown in Table 7.

Table 7. Low temperature bending test results of mixture.

Groups	Average bending strain of test results ($\times 10^{-3}$)	Technical requirement ($\times 10^{-3}$)
1	3108	≥2500
2	2752	
3	1947	
4	2975	
5	2504	
6	1760	
7	3240	

From the test results in Table 7, it is concluded that the low temperature crack resistance of direct-cast mixture decreases with the increase of direct-cast agent content, and the decrease of SP-2 is more obvious. The reason is that SP-2 contains a certain proportion of recycled PE, which reduces the viscoelasticity of asphalt, makes asphalt brittle, and further reduces the low-temperature crack resistance of mixture.

Based on the above mixture indoor road performance comparison test, the best dosage of direct modifier is not more than 0.4 % of the proportion of mineral aggregate, and the comprehensive performance of SP-2 is better than that of SP-1, so SP-2 is selected to pave the test section.

4 ENGINEERING APPLICATION

4.1 *Paving of test section*

Combined with the above test data, the test section of ultra-thin overlay of direct cast modified asphalt mixture is paved on Guilin Guiyang Expressway in Guangxi, and the length of the test section is 1100 m.

The original pavement adopts fine milling process, and the milling depth is 1.8 cm. After milling, the high-pressure suction pump is used to remove dust on the pavement, and then the automatic spraying equipment is used to spray the fast-cracking high viscosity emulsified asphalt, with the dosage of 0.5 kg/m^2.

The paving temperature of the direct casting ultra-thin cover is controlled at 175°C and the paving speed is 9 meters/min. The steel-glue-steel combination is adopted, the preliminary rolling is carried out twice, the compound pressure is rubbed by rubber wheel for 4-6 times, and the steel wheel roller is re-static pressed for twice to ensure the compactness as shown in Figures 2–4.

Figure 2. Direct feeding bin.

Figure 3. Mixture paving.

Figure 4. Mixture rolling.

4.2 *Field test*

Water seepage, structural depth and thickness tests were conducted on the site after construction. Test and field test results are shown in Table 8.

Table 8. Field structural depth detection results.

Test items	Technical requirement	Test results
Texture depth (mm)	$\geq$0.5	0.95
Water permeability coefficient (mL/min)	$\leq$ 120	40
Thickness (mm)	18mm	18.3

From the field test results, the main technical indexes such as the structural depth, water permeability coefficient and thickness of the direct-cast modified asphalt mixture can meet the requirements of the design documents and the current technical specifications, and the road performance is good.

5 CONCLUSIONS

- Direct casting modifier can effectively improve the rutting resistance and water loss resistance of ultra-thin overlay mixture;
- The low-temperature crack resistance of the mixture was reduced to a certain extent by the direct-input modifier, and the dosage of the direct-input modifier was less than 0.4 % of the ore mass;
- Verified by the test section, the direct-input modifier can be used for ultra-thin cover.

REFERENCES

Cui H.B. (2016) Application of direct-cast asphalt mixture modification technology in asphalt concrete pavement construction. *Inner Mongolia Highway and Transportation*, (5): 6–9.

Li J.D. (2014) Quantitative research on emission reduction effect of asphalt mixture direct casting process. *Urban road and bridge and flood control*, (3): 149–151.

Liu N, Zhong H.Y, Lin Y.M, et al. (2020) Study on mixing process of direct cast modified asphalt mixture. *Road construction machinery and construction mechanization*, (37): 28–31.

Liu N. (2018) Direct cast asphalt mixture modifier and its mechanism. *Road construction machinery and construction mechanization*, (35): 50–52.

Qiu G.P, Yang Zh.W, Zhang J.SH. (2020) Road performance of direct-cast environmentally friendly asphalt modifier. *Road construction machinery and construction mechanization*, (35): 79–84.

Xu O.M, Cao Zh.F, Li M.Y, et al. (2020) Application and development of ultra-thin overlay asphalt concrete . *Chinese science and technology papers*,15 (4): 425–431.

Xu Q, Wang L.zh. (2020) Study on technical performance of direct-cast SBS modified asphalt SMA mixture. *Petroleum Asphalt*, 12: 28–32.

Yao H.R, Zh X.L, Xi M.Sh, et al. (2018)Research and application of direct-cast modified asphalt in Japan. *Petroleum Asphalt*, 12 (6): 7–11.

Zhou Zh. (2018) Gradation Design of Xuanqu high-speed direct-cast rubber powder and SBS composite modified asphalt mixture. *Highway*, 10(10): 67–70.

Advances in Civil Engineering: Structural Seismic Resistance, Monitoring and Detection – Mohd Johari Mohd Yusof and Junwen Zhang (Eds)
© 2023 The Authors, ISBN 978-1-032-31491-4

Experimental study on mechanical performance of slag-fly ash based geopolymer

Ying Wang, Lingfeng Yang, Meichun Zhu & Chongqi Fang*
School of Civil Engineering, Shanghai Normal University, Shanghai, P.R. China

ABSTRACT: In order to investigate the correlations between the mechanical performances and the components of slag-fly ash based geopolymer, a series of tests were performed including Portland cement mortar specimens and different proportion of slag for fly ash based geopolymer mortar ones. The test results indicate that the strength of pure fly ash geopolymer is lower than that of Portland cement. Furthermore, when the slag weight proportion is more than 20%, both the flexural strength and compressive strength of slag-fly ash based geopolymer are higher than that of Portland cement. It provides a reference to prepare geopolymer concrete.

1 INTRODUCTION

For Portland cement, the production process is highly energy intensive and leads to a lot of greenhouse gas emissions due to a quantitative of fossil fuel consumption. Nevertheless, for the geopolymer, the production differs widely from Portland cement. It's a polymerization process of alkali activation of industrial waste material such as fly ash and alkali activator such as the combination of sodium hydroxide and sodium silicate solution. In this respect, geopolymer provides an alternative binder to the Portland cement (Davidovits J 1991).

Since the geopolymer technology was introduced by professor Davidovits (1989) in 1970's, many researchers have carried out extensive and in-depth research on this kind of prospective construction material. For the geopolymer is obtained by polymerization reaction from serval kinds of material, geopolymer performances varies with the proportion of each kind of material, producing conditions, etc. Jing Wei et al. (2006) studied the optimum mix proportions of fly ash VS kaolin for given strength, and discussed some influencing factors of strength. Shang Jianli et al. (2011) studied the material mix proportion of the slag fly ash based geopolymer strength. He Tingshu et al. (2009) explored the effect of different activators on geopolymer strength.

In this paper, the flexural strength and compressive strength of slag-fly ash based geopolymer with different proportions (low content) of slag were tested. And comparison with Portland cement was carried out though experimental tests. Some useful conclusions were obtained for preparing geopolymer concrete and geopolymer members.

2 EXPERIMENTAL PROGRAM

2.1 *Materials*

The fly ash used in this study was class C fly ash of level II, and the chemical composition is shown in Table 1, the physical properties is shown in Table 2.

The slag used in this study was S95 grade slag powder, the main index is shown in Table 3.

*Corresponding Author: congqifang@msn.com

 DOI 10.1201/9781003310884-66

Table 1. Chemical composition of fly ash.

Chemical	SiO_2+Al_2O_3+Fe_2O_3	f-CaO	SO_3	H_2O	$CaSO_3 \cdot 0.5H_2O$
Component (Wt%)	81.37	2.0	2.0	0.2	2.0

Table 2. Physical property of fly ash.

Fineness (45μm sieve residue %)	water demand ratio (%)	Loss on ignition (%)	strength activity index (%)	Density (g/cm^3)
21.0	100	2.7	81.0	2.0

Table 3. The main index of slag.

Density (g/cm^3)	specific surface area (m^2/ kg)	Fluidity ratio (%)	SO_3 (Wt%)	Water content ratio (%)	7d activity index (%)	28activity index (%)
2.9	420	100	0.18	0.1	77	97

The sand used in this study was China ISO Standard Sand, and the particle size distribution was accordance to Chinese standards (The State Bureau of Quality and Technical Supervision. 1999). Alkaline activator used was the combinations of liquid sodium silicate (Na_2SiO_3) and solid sodium hydroxide (NaOH) with 96% purity. Sodium silicate used in this experiment had a weight percentage of 26.5% and 8.3% for SiO_2 and Na_2O respectively. Ordinary Portland cement with strength grade 42.5 (P.O42.5) was also selected in this study. The pictures of materials are shown in Figure 1.

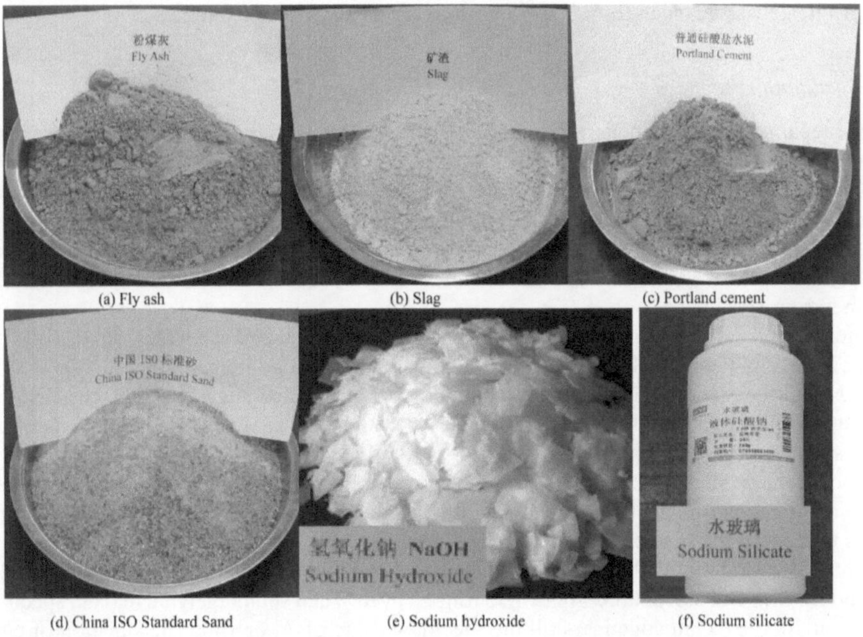

(a) Fly ash (b) Slag (c) Portland cement

(d) China ISO Standard Sand (e) Sodium hydroxide (f) Sodium silicate

Figure 1. Materials.

2.2 *Mix proportion*

Two kinds of mortar specimens were made in this experimental study. One was made of Ordinary Portland cement, the other was of geopolymer. The size of both was 40mm×40mm×160mm. For each specimen, the mix proportion by mass was one part of the binder, three parts of Standard Sand, and one half of water. For the cement specimens, the binder was the Portland cement, however, a mixture of slag powder and fly ash for the geopolymer ones. Normally, three prismatic specimens were prepared simultaneously, the amount of materials was 450gram for binder, 1350gram for Standard sand, and 225gram water.

For the geopolymer specimens, the fly ash was the base material for the binder, while the slag powder was the admixture to replace the fly ash by mass of 0%, 2%, 7%, 12% and even 17%. The details of mix proportions for three specimens are given in Table 4.

Table 4. Mix proportions of mortar specimens (3 specimens).

Sample	Portland cement (g)	Fly ash (g)	Slag (g)	Standard Sand (g)	Water (g)	Alkaline activator (g)
PM	450.0	0	0	1350.0	225.0	0
GM-FA		450.0	0		88.9	246.4
GM-FS2		441.0	9.0			
GM-FS7	0	418.5	31.5			
GM-FS12		396.0	54.0			
GM-FS17		373.5	76.5			

Through literature research (Han 2016; Mao et al. 2016), polymerization reaction was the best when the modulus of liquid sodium silicate was about 1.3. So in this study, the modulus of liquid sodium silicate was changed from 3.3 to 1.35 to serve as alkaline activator.

2.3 *Casting of specimens*

For the preparation of specimens, all the solid materials were put into the 105°C oven for 6h, then cooled down to room temperature.

Alkaline activator used in this study was the mixed solution by solid sodium hydroxide, liquid sodium silicate and pure water. The NaOH solution was mixed by solid NaOH and pure water at least 8h prior to mixing with liquid Na_2SiO_3 for heat releasing. The NaOH solution and liquid Na_2SiO_3 were mixed together to form one alkaline activator solution 24h before mixing with binder to form mortar specimens.

To mix each batch of mortar mechanically using the mixer. Poured the water into the bowl and added binder, and then the alkaline activator, then stared the mixer immediately for 30s with low speed. Added the Standard Sand steadily and let the mixer run for 30s with high speed. Stopped the mixer for 90s, removed the mixture in the bowel with a rubber scraper. Continued the mixing at high speed for 60s.

To mould the specimens immediately after the preparation of mortar. Put the mixture into each of the mould compartments by two layers, and compacted each layer with 60 jolts. Labeling the moulds to identify the specimens.

To move the mould with the mixture into the 60°C oven after standing for 60 min. Carried out demoulding with due precautions after 24h curing in oven. Then submerged the marked specimens in water at 20°C±1°C in containers till the specimens' age of 7d or 14d. The situation of curing specimens in water is shown as Figure 2.

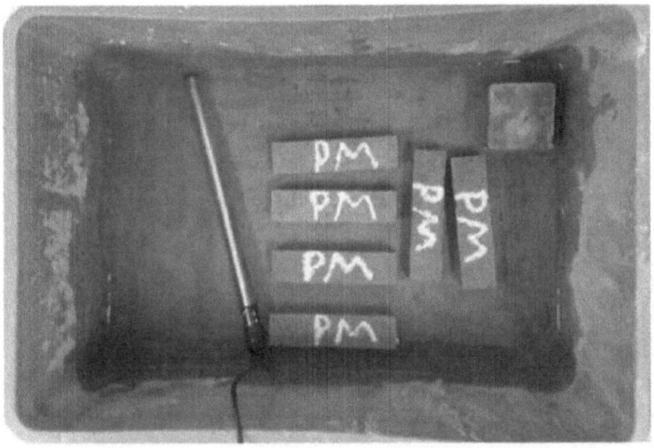

Figure 2. Specimens in water curing.

2.4 *Specimens testing*

Removed the specimens from the water not more than 15 min before the test is carried out. And covered the specimens with a wet fabric until tested.

Use the centre-point loading method to determine the flexural strength. Placed the prism in the test machine with one side face on the supporting rollers and with its longitudinal axis normal to the supports. Applied the load vertically and increase it smoothly at the rate of 50 N/s±10 N/s until fracture. Keep the prism halves damp until tested in compression.

The compressive strength determination should be carried out on the lateral of semi-section prismoid. Applied the load vertically at the rate of 2400N/s ±200 N/s till near the fracture load.

3 TEST RESULTS AND DISCUSSION

Both the flexural load and compressive load were tested for each specimen. According to the method of reference (The State Bureau of Quality and Technical Supervision. 1999), both the flexural strength and compressive strength were calculated for each specimen.

Flexural strength test result is defined as the mean value of the flexural strength determinations made on a set of three prisms. If one result within the three determinations varies by more than ±10% from the mean, then calculate the mean of the rest two as the result.

Compressive strength test result is defined as the arithmetic mean value of the six compressive strength determinations made on a set of six halves prisms. If one result within the six determinations varies by more than ±10% from the mean, then calculate the mean of the rest five as the result.

Both the flexural strength test results and compressive strength test results are given in Table 5.

Table 5. Test results summary.

		Flexural strength (MPa)		Compressive strength (MPa)	
Sample	Slag (Wt%)	7d	14d	7d	14d
PM	0	3.31	4.23	23.18	31.18
GM-FA	0	2.62	3.82	13.57	18.45
GM-FS2	2	2.40	3.93	17.02	28.17
GM-FS7	7	2.20	3.56	23.53	33.40
GM-FS12	12	1.99	4.32	27.87	37.15
GM-FS17	17	2.65	5.01	31.90	41.17

For convenience of comparison, Figures 3 and 4 show the conversion of test results in Table 5.

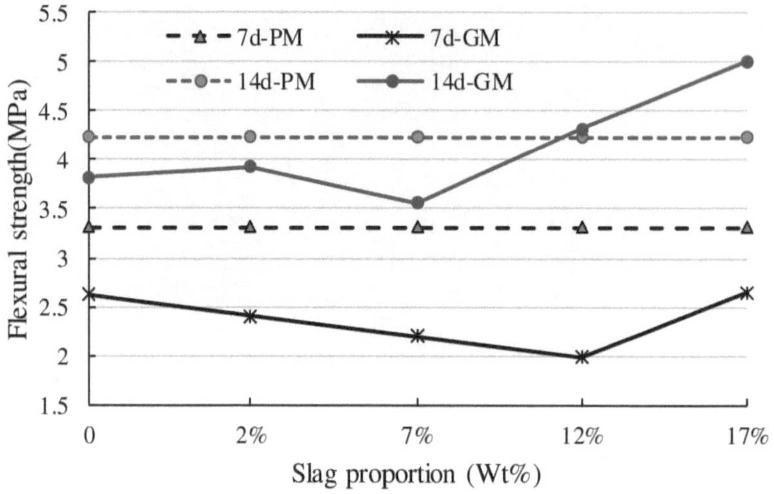

Figure 3. Flexural strength.

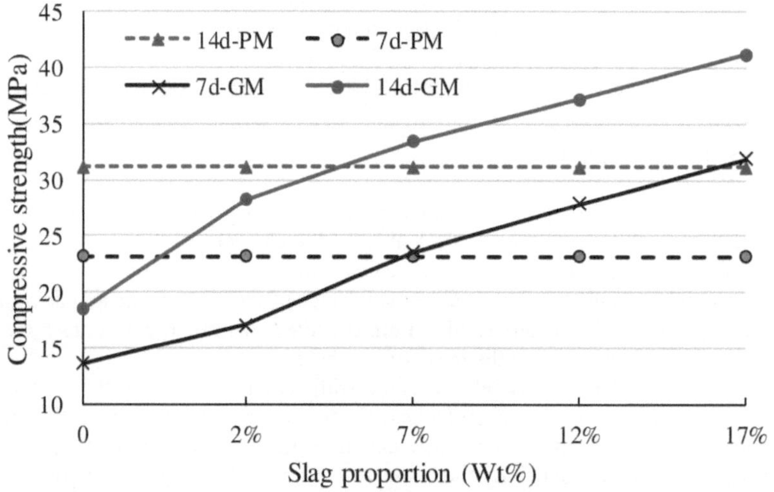

Figure 4. Compressive strength.

From Figures 3 and 4, it shows that both the flexural strength and compressive strength of pure fly ash geopolymer are lower than that of Portland cement.

From Figure 3, we can see that low slag content (<12%) has little effect on the flexural strength for fly ash based geopolymer. Based on the rising tendency of the curves, it's easy to predict that the flexural strength of fly ash based geopolymer with slag content more than 20% would higher than that of Portland cement at the same age.

From Figure 4, it shows that with the increasement of slag content, the compressive strength of geopolymer is becoming stronger and stronger. And eventually, the compressive strength of fly ash based geopolymer is stronger than that of Portland cement when the slag content is more than 17%.

4 CONCLUSION AND PROSPECT

In this paper, the experimental investigation of slag-fly ash based geopolymer has carried out. Both the flexural strength and compressive strength were tested for Portland cement mortar specimens and slag-fly ash based geopolymer ones. The authors focused on the effects on mechanical performances of the specimens with different amounts of slag content. Based on the research work mentioned above, it can be concluded that:

(1) The strength of pure fly ash geopolymer is lower than that of Portland cement. So, for the purpose of high strength structural members, the admixture is essential for fly ash based geopolymer structure.
(2) The strength of slag-fly ash based geopolymer is higher than that of Portland cement only when the mass weight of slag powder is more than 20%.

In this study, the authors only focused on the low content of admixture of fly ash based geopolymer, more research work in the future might carry out seeking for the best proportion of admixture, testing class F fly ash, and for the optimum modulus of sodium silicate.

ACKNOWLEDGMENTS

This paper was supported partially by the project from Shanghai Science and Technology Committee (NO. 18070502900).

REFERENCES

Davidovits J 1989 Geopolymers and geopolymeric materials. *Journal of Thermal Analysis*. **35**(2) 429–41.
Davidovits J 1991 Geopolymers: inorganic polymeric new materials. *Journal of Thermal Analysis*. **37** 1633–56.
Jianli SHANG and Lin LIU 2011 Study on preparation and mechanical properties of slag and fly ash based geopolymeric. *Bulletin of the Chinese Ceramic Society*. **30**(3) 741–44.
Mingjie MAO, Jingyang REN, Wenbo ZHANG and Wei CHEN 2016 Study on the mechanical properties of fly ash gropolymer concrete. *Concrete*. (5) 78–80.
Tanfyi HAN 2016 *Analysis of influencing factors on geopolymerzation process of alkali-activated ash*. (Changsha: Changsha University of Science & Technology).
The State Bureau of Quality and Technical Supervision. 1999 *Method of testing cements-Determination of strength: GB/T 17671-1999*. (Beijing: Standards Press of China).
Tingshu HE and Guoqiang WEI 2009 Effect of activators excited cement specimen with different amount of fly ash on strength. *Concrete*. (5) 62–64.
Wei JING, Jianhong LIU and Feng LIN 2006 Preparation and study on sly ash based gropolymer. *Journal of Science of Teachers' College and University*. **26**(4) 40–42.

Advances in Civil Engineering: Structural Seismic Resistance, Monitoring and Detection – Mohd Johari Mohd Yusof and Junwen Zhang (Eds)
© 2023 The Authors, ISBN 978-1-032-31491-4

Research on the technical design of maintainability and renewability of SI housing

Shenqi Gan*
School of Architecture and Urban Planning, Anhui Jianzhu University, Hefei, Anhui, China

Hong Zhang
School of Architecture, Southeast University, Nanjing, Jiangsu, China

ABSTRACT: According to the development status and construction technology of SI residence in developed countries and China, and through the early renovation cases in Holland and Japan, this paper focuses on the technical design method of maintenance and renewal of industrialized residence. The paper also studies the technical design methods that can be maintained and updated from the aspects of the standardized design of components, the connection design and the modulo-digital coordination system of the filling body, etc., the concrete design idea and construction flow that can be maintained and updated are put forward.

1 INTRODUCTION

SI house is the main architectural form of modern industrialized house (Gan 2019). The development of SI housing system in different countries has its own advantages and disadvantages, and the developed countries are represented by the Netherlands and Japan. Based on the development status and early renovation cases of Holland and Japan, as well as the construction situation and development advantages of CSI housing in China, this paper will deeply study the maintainable renewal nature and technical design methods of industrial residential buildings.

Since the 18th National Congress of the Communist Party of China, the implementation of construction industrialization and the development of green buildings have become an important agenda of national construction and development, and the requirement of "100-year residence" has emerged. In the case of SI houses, this can be achieved by making them maintainable and updated (Gan 2020). In this paper, the durability design of support body S is proposed to solve its maintenance and other related problems. Secondly, in view of the construction of functional and built-in update mainly embodied in the filling body I adaptability, and direct response to the demand change and personalized requirements of users, so the paper from the filling body of SI residential modular coordination system, the standardization of SI residential design as well as the installation of the filling body product positioning of the adaptability of filling body part component I can update sex has carried on the detailed research, finally proposes the design of the filling body maintenance updates I thought and construction process.

*Corresponding Author: Gansq@ahjzu.edu.cn

 DOI 10.1201/9781003310884-67

2 CONCEPTS AND FEATURES OF SI HOUSING SYSTEM

2.1 *Western SI housing system and its early renovation and renewal*

Holland and Japan are the representative pioneers of western construction industrialization. In the early 1960 S, the Dutch architect famous professor John. Habraken puts forward the theory of SAR to Support the body, the theory for the first time put forward the design and construction of the residence is divided into two parts, the residential development as the main structure to Support body S (Support), will develop the load bearing part of filling body part I (Infill), which is the source of the SI residential theory (Wang 2016). The core content of the design concept of SI house is to design and construct support body S and filling body I respectively. The supporting body S is constructed through on-site casting, and the filling body I is transported to the site for assembly after standardized production in the factory. This can obviously reduce the amount of wet work, improve work efficiency, and reduce the pollution to the environment.

On this basis, in the 1970s, professor Habraken puts forward the OB (Open Building) Open architecture theory, the theory emphasizes the residential design and construction work Open to the public, society and environment, thus creating a flexible rich residential, increasing the sustainability of housing, and therefore is considered to be the basic theory of the residential industrialization.

2.2 *CSI system in China*

China put forward the concept of CSI housing (China SI) in 2006, forming a new industrial housing construction system with Chinese characteristics. The main feature of CSI housing system is: according to the partition standard of housing support S part and housing filler I part, the housing filler is decompressed into various components, and the standard chemical plant production. The standardization of part components makes the on-site assembly and construction more convenient and easy, and can solve the problems such as excessive consumables, high energy consumption, heavy pollution and waste of secondary decoration in current residential construction in China (Figure 1). The important feature of CSI residential filling body is standardized design, and it has advantages in terms of adaptability and variability (Wu 2017).

It should be particularly pointed out that China began to implement the strategy of "Centennial Housing in China" in 2012. Strategy of "China's one hundred residential" integrated integration technology between the large space structure, the house, within the external wall thermal insulation, dry floor heating, ventilation in an all-round way and whole sanitary ware system, integral kitchen, pipeline and so on many technology, obtained the rapid development in recent years, Shanghai William residence 11 floor and Jinan "The home of tomorrow ii project" is among the more typical case of success.

Figure 1. Tomorrow house experimental room.

3 RESEARCH ON MAINTAINABLE AND RENEWABLE DESIGN OF SUPPORT BODY S

3.1 *Main body design of the structure*

The research on the maintainable and updated design of SI housing system is still in its initial stage. Therefore, based on the study of some excellent cases of SI houses at home and abroad, we study the construction characteristics, especially the maintainability and renewal properties of SI houses. The renovation project of Wahlberg in the Netherlands and the renovation project of Camp Housing in Koichi Prefecture in Japan mentioned above dates back to earlier times, and the main feature of which is that the inner filling body is completely removed, but the supporting body is still retained due to its strong durability. This gives us a strong inspiration that, in order to achieve the national "century-old residence" longevity requirements, we should focus on the durability design of the support S of SI residence, and then study the maintenance and quality assurance of the support and other related issues (Wang 2015).

The support body S shall meet the requirement of durability design life of 100 years, and shall be guaranteed and strengthened in strict accordance with the requirements of various architectural design codes and strengthening measures. For SI residence, the main structure of supporting body S (mainly refers to foundation, beam, column and floor slab) should be integrated into an integrated structure by integral cast-in-place method to improve its overall stability and durability. Second, it is important that the support body and the fill body be fully separated from each other. Because the two have different service life, only through their own independent design can ensure the long life of the century-old residential demand. This should be the basic requirement of CSI residential design (Li 2019).

3.2 *Research on structural connection design*

In order to ensure the maintainability and renewal of SI house, the connection of other parts other than the support body S should be easy to disassemble and reuse, so as to achieve the reversible connection, which is the intuitive technical requirement of maintainability and renewal. Therefore, the reversible connection of components should be an important application research content that can be maintained and updated. The use of reliable connections is essential to the robustness of prefabricated buildings. In terms of the design and connection of the supporting body S, a series of prefabricated buildings "micro-row house", "dream house" and "spring house" developed by Zheng Studio of Southeast University mainly have the following characteristics (Zhang 2017).

3.2.1 *Reversible connection*

In general, the reversible connection of the structure mainly pays attention to the following points:

(1) The main structure and members of the support body S can be transited through small concrete works, and the openings and embedded metal parts are reserved.
(2) The support body S and the inner parts can be connected through the reserved hole and the embedded parts.
(3) the decoration parts should all be connected by dry method between the components, in order to ensure that the construction is simple and convenient to disassemble.
(4) The bolt connection is a reversible connection structure because of its repeatable disassembly performance, which is not only convenient for disassembly and assembly, but also reusable for the disassembled material.

3.2.2 *Connection between structure and wall parts*

When the structural part of SI house is cast in situ, the connection of the outer wall is particularly important. Exterior wall should adopt the construction method of factory prefabrication, field assembly, at present, the widespread adoption of "after installation method", namely in the construction of the main structure construction is completed, then preformed PC panel as a bearing structure installed on the main structure (Zhan 2016), generally USES dry connection, including the main structure can be a steel structure, cast-in-situ concrete structure or precast concrete structure (Figure 2).

Figure 2. Back mounting method of PC wallboard.

"Post installation method" in the installation process is easy to produce error accumulation, so the main building construction precision and PC components of the production precision requirements are very high, which should cause us to attach great importance to.

3.2.3 *Connection between structure and ground parts*

The connection between the structure and the ground parts refers to the connection between the support body S and the ground floor slab. Dry connections shall be used for SI residences:

(1) Using cast-in-place floor slab;
(3) Reserved metal parts and reserved holes, so as to connect with other ground parts;
(3) For the double floor, the use of bolt feet and pressure plate combination system, and dry connection construction of overhead floor, not only equipment pipeline laying flexible, through the constraints of support, the operation is also relatively simple.

3.2.4 *Connection between structure and other parts*

The connection between the structure and the parts mainly includes the connection between the support body S and the ceiling parts, the prefabricated staircase, the vertical tube well and the household wall. The following technical requirements should be paid great attention to:

(1) The connection between the structure and the ceiling: that is, the connection between the support body S and the ceiling, the metal parts and the hole need to be reserved.
(2) The connection between the structure and the stair parts: China generally adopts the stair prefabrication, then it is necessary to reserve the connecting member on the structure body and the stair parts.
(3) The connection between the structure and the vertical pipe well: before the installation of the pipe, the caliber should be cut out at the predetermined position, and the diameter should be compacted with glue after the installation.
(4) The connection between the structure and the household wall: in the position of the household wall to do a small concrete structure, embedded metal members, through bolts and the main structure firmly connected (Figure 3).

Figure 3. Connection between structure and partition wall.

4 STUDY ON MAINTAINABLE AND UPDATED DESIGN OF FILLING BODY I

4.1 *Main design methods for maintainable and updated filling body S*

The maintainable and updated design of the prefabricated building is mainly reflected in the adaptability of its filling body I. The so-called adaptability means that the component components of the filling body I should be convenient for maintenance and replacement (Li 2017). A series of prefabricated buildings, such as "Micro-row House", "Dream House" and "Green House", developed by Zheng Studio of Southeast University, have solved this design problem well. Their components can not only be easily maintained and updated, but also be reused after replacement, with the reuse rate reaching 70% (Gan 2014). Now, the adaptive design method of filling body I is specifically proposed:

Firstly, the filling body I should be designed separately from the perspective of components. Different from the support body S, the part of the filler body I is more direct to respond to the user's demand changes and personalized requirements, and its service cycle will be as short as ten years or even the replacement of some parts will be more frequent. Therefore, the filler body I should be separated and designed separately compared with the support body S.

Secondly, standardized design is the basis of adaptive design. Only by adopting the standardized design scheme, it is possible to realize the partial production of the filling body parts and achieve the technical requirements of convenient maintenance and update.

Thirdly, the reversible connection of components is the main content of the adaptive design of filling body. Modular coordination and reversible connection techniques should be effective technical means to solve this problem.

4.2 *Modular coordination of SI residential filling body*

Building modular coordination refers to the application of modules and modular series in architectural design (Li 2019). Modular coordination is the basis of standardization in housing industry. Through modular coordination, the standardization design and size coordination of supporting body and filling body components can be solved, and the universality of filling body components can be further improved.

The modular design of filling parts can make its size more normalized and standardized. Through the size coordination between each part or combination, the versatility and mutual tolerance of filling parts can be effectively improved. At the design stage, part module design should first clear the function of each product component and size, the position of the part component in space, materials used for the production of parts and its utilization, part component and the size of the relationship between the space environment around, proportional relationship between each other, and many other issues, on the basis of setting up a product component of the filling body standardization system sizes.

At present, most of the implementation standards in China take the basic module 1 M=100 mm as the standard to standardize the size specification of building parts, when the spatial application dimension unit 3M=300 mm is decomposed using the fractional modulus cardinal numbers 1/10m, 1/5m and 1/2m, then

$$1/10 * 3M = 30mm \tag{1}$$

$$1/5 * 3M = 60mm \tag{2}$$

$$1/2 * 3M = 150mm \tag{3}$$

By now, the theory that Equation (1) is used as the module system to determine the components of the grading unit has been confirmed and widely applied in China. The 3M module can further unify the rules of residential plane layout design.

4.3 *Standardized design of SI house*

The standardized design of SI residence requires that parts and components should be designed according to modular coordination standards to form standardized and serialized products, which can not only improve the versatility of filling parts, but also make future maintenance and update convenient and feasible. This requires us to achieve part standardization, node standardization, product standardization and tolerance with the size.

(1) Standardization of parts
The components of SI house are independent components of a certain part of the building with specific functions. The degree of standardization of components directly determines the degree of industrialization of residential products. Standardized parts system should meet the unified modular coordination rules, the connection mode should be standardized, and as far as possible to use dry connection, installation should have compatibility and versatility.

(2) Node standardization
Residential buildings contain the connection of multiple nodes. In the design stage of filling body I, it is necessary to ensure the standardized connection of each component, and the port of residential equipment pipeline should be reserved according to the functional requirements, so as to ensure the close connection of indoor and outdoor pipeline equipment, as well as the close combination of indoor and outdoor nodes of all circles.

(3) Product standardization
Should fill body standardization design system of product components, including the modular coordination and reasonable parameters, size cooperate and tolerance parameters and connection way in construction and connection specification, etc., make the product components, standardization, generalization and serialization directory, such as design, production, installation, maintenance, update the stage can reach a unified standard.

(4) Tolerance fit size
In the design stage, the tolerances should be fully considered in order to control the tolerances within an acceptable range. On the premise of full consideration of various factors, through the relationship between the tolerance fit size and the size of parts (Figure 8), the difference between the actual size and the manufactured size of parts can be determined. As shown in the figure, the fabrication tolerance is TM, the installation tolerance is TE, the position tolerance is TS, and the connection space is ES. The allowable error between component datum level and installation datum level is d, then the size of the value of d directly determines

the minimum and maximum dimensions of the component parts. The relationship between d and the manufacturing tolerance Tm, installation tolerance Te, position tolerance TS, and connection space ES shall satisfy the following formula:

$$0 \leq es \leq d \leq es + tm + te + ts \tag{4}$$

4.4 *Positioning and installation of parts of filling body I*

In order to realize the maintainability and renewal of industrial residential buildings in the future, the filling parts must be standardized and serialized and installed correctly. The positioning methods of the filling components in the spatial grid include center line positioning method and interface positioning method, which have their advantages and disadvantages. The center line positioning method and interface positioning method are superimposed to form a unified modular grid, which can give full play to their respective advantages and enable the main structural components and built-in structural components to meet the positioning requirements of datum and modular size at the same time (Figure 4).

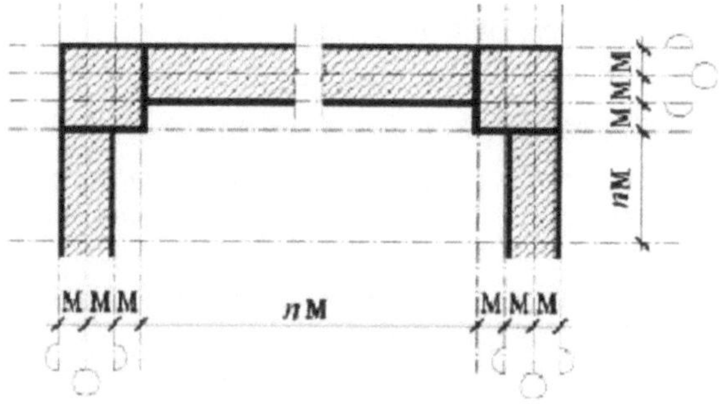

Figure 4. Center line positioning and interface positioning stack.

The partition parts of the filling body include the structural column, surface plate and keel in the partition wall, and the partition parts are positioned as a whole. The specific positioning method is divided into two situations: when the thickness direction of the plate parts is not combined with other parts or there is no requirement for the modulus space, the centerline positioning method is used; when a group of plate parts are assembled for installation, the interface positioning method should be used.

Considering the maintainability and renewal in the future, industrial residential buildings should also start from the construction process of filling inside, which is to manage the construction process sequentially and separately according to the project scale and characteristics. Through the case study of the construction and construction of a series of prefabricated buildings such as the "micro-row house" in the main studio of Southeast University, the process can be further summarized based on the basic construction process of maintainable and renewable interior decoration.

5 CONCLUSIONS

Based on the early residential renovation projects in Holland and Japan and a series of prefabricated buildings developed by "Zheng" Studio of Southeast University, this paper makes an in-depth study of the maintainability and renewal nature of SI house, and proposes to study its maintenance with the durability of the building as the target, and study its renewal with the functionality of the building and user needs as the target. Paper specific support body S durability design method is proposed, the filling body I adaptability design method and design method of connection, and from the SI residential modular coordination of filling body, the installation of the filling body product positioning, the standardization of SI residential design and the connection design of filling body and supporting body of SI residential renewable sex has carried on the detailed research, in this paper, the design of the corresponding technology and the design method, finally put forward in the specific case SI residential maintenance update connection installation and construction process diagram.

ACKNOWLEDGEMENTS

It is a project supported by the 2019 Talents Introduction (doctor) Scientific Research Starting fund Project from Anhui Jianzhu University (2019QDZ50). The corresponding author is Shenqi GAN.

REFERENCES

Gan Shenqi, Zhang Hong. Modern construction technology and development of "micro-row future house" [J]. *Building Technology*, 2014(10): 933–936

Gan Shenqi, Zhang Hong. Research on maintainability and renewal of industrial residential buildings [J]. *Housing Science and Technology*, 2020(09): 23–26+48

Gan Shenqi, Zhang Hong. Research on the concept and characteristics of western contemporary architecture renewal [J]. *Urban Architecture*, 2019 (02): 192–198

Li Min, Xia Haishan, Li Junjie. Modular coordination method and application research based on SI housing theory [J]. *Huazhong Architecture*, 2019(05): 47–52

Li Zhongfu, Han Xu. Study on Interface Type and Method of SI Residential System [J]. *Journal of Engineering Management*, 2017(03): 87–91

Li Zhongfu, JIN Yuge, Li Long. Research on Structure Selection and Construction Mode Selection of SI Residential Supports [J]. *Construction Economics*, 2019(06): 94–99

Wang Haining, Zhang Hong, Yin Jiang. Research and problem analysis on application of building information model [J]. *Construction Science and Technology*, 2015(23): 47–49

Wang Xiaomeng, Ma Tao. *SI Residence Design – Building Century-old House* [M]. Beijing: China Architecture and Building Press, 2016

Wu Xiuluan, Cheng Shuai. Study on Basic Theory and Structure Composition of CSI Residential Building [J]. *Architectural Design Management*, 2017(11): 88–90

Zhan Yaoyu, Tang Xuemei, Wang Qi, et al. Introduction to prefabricated exterior sandwich wall panels in Changtai pilot building of Fujian building industrialization [J]. *Concrete World*, 2016 (01): 56–64

Zhang Hong, Cong Meng, Zhang Ruizhe et al. Design, development, improvement and application of a pre-assembled house system: building product model and new architecture construction [J]. *New Architecture*, 2017 (02): 19–23

Advances in Civil Engineering: Structural Seismic Resistance, Monitoring and Detection – Mohd Johari Mohd Yusof and Junwen Zhang (Eds)
© 2023 The Authors, ISBN 978-1-032-31491-4

Kindergartens design and research of left-behind children in rural areas in Western China

Xinye Zhang
School of Urban Design, Wuhan University, Wuhan, China

Yu Zhang
New Institute, Gouda, The Netherlands

ABSTRACT: The problem of left-behind children has been a hot topic in China. Especially in Southwest China, the condition of the left-behind children is more serious. From the perspective of kindergarten, this study aims to provide solutions to left-behind children's life safety and psychological problems. The status quo and problems of left-behind children from the state's perspective, the elderly, children, parents, and staff are firstly analyzed. We propose solutions from four aspects: Site selection, function determination, space design, and material access. First of all, selecting a site with a simple and safe traffic environment and site should cooperate with other buildings around the site. Based on the original functions, we have added an archive, an auditorium, a staff dormitory, a home-school cooperation space, and separate psychological counseling rooms for children and staff. In terms of space design, we concentrated all children's outdoor activities into one piece, tried to enlarge its area, and made the surrounding buildings form a state of the enclosure. At the same time, we ensure that children's activity range does not leave the sight of teachers. Design a double dormitory for teachers and childcare workers, with public space for common use. According to the local architecture, the materials are selected while ensuring quality, minimizing money, and adapting the architectural style to the local culture. They are then choosing Puchang Village, Ganluo County, Liangshan Prefecture, Sichuan Province as an example. A Kindergarten is designed to express our thought for solving the problems of the left-behind children.

1 INTRODUCTION

Our country has not brought preschool education into the scope of public services, relying too much on social forces to run preschool education. Especially since the late 1990s, with the deepening development of China's economic transformation and social transformation, many kindergartens in cities have been gradually separated from state-owned enterprises and institutions. A large number of kindergartens run by original enterprises and institutions have been closed, resulting in a sharp decline in the total number of kindergartens. In preschool education, the trend of "private advance and state retreat" is very obvious. In recent years, under the background that "it is difficult and expensive to enroll in kindergartens" has been increasingly concerned by the whole society, the Chinese government has successively issued a series of policy documents to promote the development of preschool education. The Outline of the National Medium - and Long-Term Education Reform and Development Plan (2010-2020) emphasizes "basic universal preschool education", "clear government responsibilities," and "focus on the development of preschool education in rural areas" as the main task of development of preschool education. "The several opinions on the current development of pre-school education under the State Council for the" meet the demand of school-age children into the park and promote the scientific development of preschool education cause

 DOI 10.1201/9781003310884-68

put forward ten major measures. This series of policy documents show that children's pre-school education has entered the policy agenda.

When referring to the current situation of left-behind children in Southwest China. They can agree with this behavior, learning and can accurately understand the purpose of the study and significance, and in the learning process can adopt appropriate learning methods and strategies. However, most left-behind children learning attitude is not correct and cannot be in the study plan, time distribution, effective management strategy application, etc. The long separation from parents largely mean the left-behind children lack family education. At the same time, the physiological and psychological needs are not being met, often appear negative emotions and bad personality, academically usually show tired, play truant or grades at low levels. Therefore, strengthen communication with the left-behind children, parents to accompany children, regardless of the body and mind or academic is crucial.

Many research has been paid on preschool education regarding the parents and school itself. In terms of schools, Liu, Dan, and Ai put forward that the state should increase investment and strengthen teacher training to improve the quality of kindergartens. Some researchers have discussed left-behind children in ethnic minority areas. For example, Zhang studied Jinzhu Miao Village in Xiushan County, Chongqing, and Luo studied the Korean nationality in Heilongjiang. Zhang pointed out that the main difficulties encountered in children's education in ethnic villages in Jinzhu Miao Village included the lack of early family education, the absence of kindergarten education, the deficiency of village education, etc.

In this study, we try to solve the life safety and psychological problems of left-behind children. The status quo and problems of left-behind children from the state's perspective, the elderly, children, parents, and staff are firstly analyzed. Then a village in the rural area of Southwestern China in the Sichuan Province is chosen as an example. A Kindergarten is designed to express our thought for solving the problems of the left-behind children.

2 THE PLIGHT OF LEFT-BEHIND CHILDREN OF PRESCHOOL AGE IN RURAL AREAS

2.1 *Government*

In 2018, the State Council promulgated and implemented "the Opinions of the CPC Central Committee and the State Council on Deepening the Reform and Standardizing the Development of Preschool Education." This document has laid a certain foundation for the future legislation of preschool education law in China. Until September 7, 2020, the Ministry of Education has studied and formed the "Preschool Education Law of the People's Republic of China draft (draft for soliciting public opinions)" and solicited public opinions. This is a key breakthrough in China's preschool education law. However, at the moment, we still have no specific pre-school education law to follow, which poses a serious obstacle to government regulation. In general, the government meets the following challenges.

From the perspective of education investment in China at the present stage, due to the unbalanced economic development, the education investment in eastern China is significantly higher than that in western China. Coastal areas are significantly higher than those in inland areas, and urban areas are significantly higher than those in rural areas. The gap in education investment also reflects the huge differences in people's mass economy in different regions. Compared with urban areas, children in rural areas receive education in a more difficult environment and poor conditions. Specific manifestations include that the number of kindergartens in rural areas is very small. Still, the number of students is large, the infrastructure is not perfect, and the teacher team needs improvement.

2.2 *Elderly guardians*

Place the figure as close as possible after the point where it is first referenced in the text. If there are many figures and tables, it might be necessary to place some before their text citation. If a figure or table is too large to fit into one column, it can be centered across both columns at the top or the bottom of the page.

2.2.1 *Adverse effects of inter-generational rearing on the left-behind children*

Left-behind children are often entrusted to their grandparents because their parents work outside the city for many years, which is the condition of generational separation. Under the guardianship of grandparents, it is impossible to avoid the following problems, i.e., life safety, physical and mental health, communication, cultural knowledge. For life safety, inter-generational guardians attach great importance to the personal safety of left-behind children but cannot teach children safety first aid knowledge. The safety education of left-behind children is often limited to simple exhortation while neglecting to popularize the methods and common sense of saving themselves in the face of sudden danger. Moreover, they have no time and energy to tell the safety of left-behind children all the time.

2.2.2 *Architectural solutions*

There are different kinds of solutions to solve these problems. Only architecture-related solutions are mentioned here. Firstly, the enhancement of communication between left-behind children and their parents is really important. Left-behind children whose parents go out weakens their original emotional support from their parents, resulting in the "family hunger" of left-behind children for a long time, leading to the lack of parental education for left-behind children. Therefore, parents should first take the initiative to increase contact and communication with left-behind children by telephone, letter, video, and other means. Second, parents should try to choose cities close to home to work and increase the frequency of returning home. Parents who stay at home for a long time can give left-behind children a better growth environment and give timely help for the troubles and problems of left-behind children. Third, there should be more communication between parents and grandparents. Due to the different times and experiences between parents and grandparents, they have different opinions on the education of left-behind children. They need to communicate with each other to understand each other and give better education to children. Therefore, an archive could be created for communicating with their parents. This space will be detailed in the children section.

2.3 *Left-behind children*

2.3.1 *Current situation faced by left-behind children*

The problems faced by left-behind children can be roughly classified into the following three aspects: safety problems, psychological problems, and a weak ability to communicate. Firstly, when discussing the safety problems faced by left-behind children, we are usually divided into the following two categories: accidental injury and criminal injury. The main accidental injuries suffered by left-behind children include knife cuts/sharp instrument injuries and burns. The possibility of these injuries is exacerbated by the burden of family services on the elderly, and even some children begin to share the burden too early, as young and middle-aged family workers migrate. In addition, falls, sprains, and bump/crush injuries are also common. Lively and active is the nature of children. Most of the time, they are playing rough-and-tumble. In the process of playing, because guardians cannot always guard left-behind children, coupled with their lack of awareness of danger, they are easy to be hurt. Moreover, when the injury occurs, the left-behind children often cannot get timely and effective help due to the absence of their parents, which further aggravates the injury. Other unintentional injuries include animal bites, traffic accidents, drowning, electrocution, poisoning, etc. Most of the Inter-generational guardians are of low educational level. They lack awareness and ability of safety education, so they are not competent for safety education for left-behind children. Therefore, parents often place their hopes on kindergarten. However, In China's rural areas, especially in the Midwest, many rural schools, school conditions, teachers, and the restriction of the management system, there is no system. Because of the management measures of left-behind children, some schools made the safety management system, from the content point of view mostly too general. Also, it is difficult to implement in the actual teaching activities.

2.3.2 *Architectural solutions*

The architectural solutions for left-behind children including 3 aspects: site selection, function determination, and space design. First, for the site selection, a safe space should choose. Traffic around the site should be simpler, and it is best to avoid contact between children and motor vehicles.

Then, in terms of function determination, the child's psychotherapy room and an archive should be added to the function. Among them, the psychological counseling room can help children with psychological problems promptly. Because children's age and mind are not mature, they may not realize their psychological problems promptly, so it also requires the teacher to carefully observe the children's find out. After the discovery, the teacher can take the child to the psychological counseling room for counseling. As we mentioned in the inter-generational guardian section, The lack of parental care is the main reason for the psychological problems of left-behind children. To fundamentally solve this problem, we choose to add an archive in the kindergarten. It can be used to store letters sent from parents far away or even videos of video calls. For families that do not have the conditions to send letters or video calls, they can write letters or record videos in advance, save them in the archive, and send them to the children regularly by the teacher. If possible, children can also contact their parents here and send them some small cards.

Finally, space design should be more suitable for children. Naturally, it is the best classroom for children, which is also the rural environmental advantage. Therefore, children should be provided with outdoor venues as much as possible and pay attention to the safety of children in outdoor activities. It should be ensured that the teacher has no dead angle of sight and can fully observe the scope of the child's activities. At the same time, it is necessary to reduce the activities within the dangers, such as stairs, deep water, dangerous plants, etc. In addition to the outdoor space, the indoor space should also be integrated into the richer teaching instruments. In Montessori's educational theory, children come to kindergarten and are exposed to all kinds of development and promote their independent learning and exploration to the greatest extent. She suggested that the classroom should be divided into seven areas: living area, sensory area, language area, mathematics area, culture area, art area, and reading corner. All areas are equipped with various teaching aids for children to freely choose "work". We will incorporate this concept into the kindergarten design.

2.4 *Parents*

At present, most families in rural areas do not have a clear understanding of the importance of preschool education and do not pay enough attention to it. The main labor force in the family (parents) is of low educational level, poor educational level, and outdated concept of child-rearing. On the other hand, forced to make a living, they generally have the so-called "drag" education concept, failing to realize the vicious circle brought about by it and paying no attention to the growth and education of children. The other situation is the parents' excessive mentality of expecting their children to be successful, the emergence of the concept of young adults, strengthen intellectual education, completely treat children as young adults, give them a sense of competition, so that children on the back of the undue academic burden. From the Cognition of migrant workers' parents on left-behind children, it could be found that there are some differences between parents' cognition and children's actual situation. It is mainly reflected in four aspects: guardianship, living conditions, learning situation, and psychological status. Overall, compared with the understanding of the living conditions and learning conditions, migrant workers' parents know less about the emotional world of left-behind children. Generally speaking, parents of migrant workers pay attention to left-behind children's study, life, character, and psychology in order of importance. Parents of migrant workers pay the most attention to the study of left-behind children, and relatively speaking, the emotional world of left-behind children is seriously neglected. The solutions of the part of parents are similar to that of the inter-generational guardians and left-behind children. Because they can't accompany the children all year-round, in addition to strengthening their remote communication, we can only deal with it from the guardian and the left-behind children themselves.

2.5 *Kindergarten staff*

The kindergarten staff includes teachers, nurses, and logisticians. Among them, the rural kindergarten teachers have the problems of poor treatment and heavy pressure. First of all, the treatment of rural kindergarten teachers is not very good. Rural preschool teachers have problems such as identity is not recognized, a hard-working environment and heavy tasks, low wages and poor welfare treatment, few training opportunities, professional development difficulties, and no opportunity for promotion. Many problems also lead to some teachers often in a state of sub-health in body and mind.

Secondly, Preschool teachers in rural areas are under great pressure. For rural preschool teachers, in addition to teaching pressure, pressure from parents, and more security pressure, some of which are the preschool education itself, some are caused by the local special geographical environment and economic conditions.

For the nurses and logisticians, many rural kindergartens are short of them. In addition to completing children's education and care work, teachers also complete repairing desks and chairs, cleaning, painting walls, and renovating the playground. Teachers have a large amount of labor and heavy load. The class size of some rural kindergartens is seriously over the limit, which increases a lot of burden on teachers both in teaching and in safety management. For the kindergarten with them, the separation of education and health care is the biggest problem.

So, there are 2 architectural solutions for the kindergarten staff's problems. First of all, we should establish a good living environment for all employees and combine the public areas of teachers and caregivers to strengthen the communication between them. Then set up the psychological healing space for the staff, and let them release the pressure in time.

3 KINDERGARTEN DESIGN: AN EXAMPLE OF PUCHANG VILLAGE

3.1 *Site information*

Puchang Village, Ganluo County, Liangshan Prefecture, Sichuan Province, has a total population of 2,356, with 334 households and 376 left-behind children. The number of pre-school children among left-behind children is 83. The climate belongs to the subtropical monsoon climate zone, dry and wet, winter half of the sunshine is sufficient, little rain dries warm; The summer half-year cloud rain is more, the climate is cool. Agriculture to the cultivation of corn, rice, potato-based.

Figure 1. Map of puchang village.

The busy season is around June, July, and August. Mineral resources are mainly lead-zinc ore, iron sulfide, and so on.

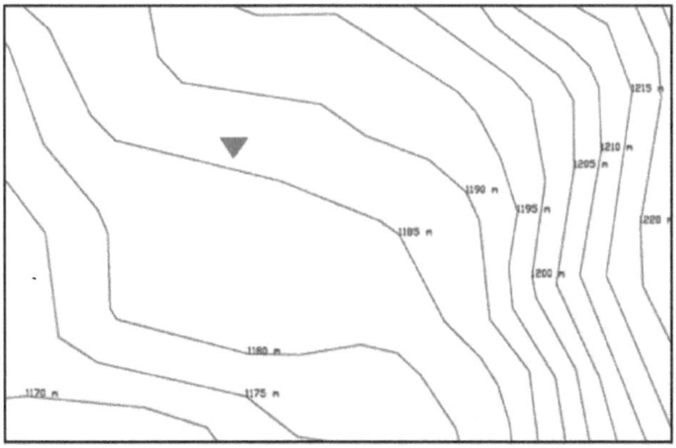

Figure 2. Contour line map of puchang village.

The site is located in the middle of the village, next to the middle school and primary school. The location helps parents pick up and drop off. At the same time, its terrain is relatively flat, suitable for children's activities.

3.2 *Location analysis*

Figure 3. Road network.

For the sake of children's safety and convenience, we chose a place far away from the car road and close to the pedestrian road. The site is accessible by two main roads, one parallel to and one perpendicular to the field's sideline. We consider setting the main entrance to pick up and drop off children on the vertical road on the south side, where the traffic flow is relatively simple, to protect the safety of children. The secondary entrance to the delivery is placed on the north side of the

Figure 4. Public service area.

Figure 5. Cultivated land.

parallel road. This road is wider, can be opened, the road conditions are more complex. It is not conducive for the child's safety but can be used as a delivery entrance.

The existing schools and hospitals in the venue form a major public service area. To cooperate with the existing buildings, we chose the kindergarten site next to this area. In this way, children of kindergarten, primary school, and junior high school in Puchang village are concentrated in this area, making it the education center. At the same time, the site is not far from the hospital, and in the same road section, if there is an emergency, it is very fast and convenient to send the doctor.

There are many cultivated lands around the site, including two terraces. It is worth mentioning that the terrace on the northeast side of the site is relatively high. The parents of workers and peasants can see the children's activities in the kindergarten below to a certain extent. At the same time, this arable land can also become a good place for children to learn.

3.3 *Concept*

We tend to integrate Montessori's educational theory into kindergarten design, take children as the center, cooperate with the good learning environment and rich teaching aids, let children self-issue active learning. Teachers play the role of assistant. They must have a deep understanding of the child's spiritual world and children's development to know the situation to provide children suitable, timely assistance and guidance. In Montessori's educational theory, children come to kindergarten and are exposed to all kinds of development and promote their independent learning and exploration to the greatest extent. She suggested that the classroom be divided into seven areas: living area, sensory area, language area, mathematics area, culture area, art area, and reading corner. All areas are equipped with various teaching aids for children to freely choose "work ". We will incorporate this concept into the kindergarten design.

3.4 *Functions*

According to "Code for architectural design of nurseries and kindergartens", the functional space of the kindergarten is mainly divided into living space and service space. The service space is mainly distributed on the first floor near the gate, the southwest and northwest of the site. The living space is mainly distributed in the southeast and northeast, far away from the gate, which helps to manage the children. On the second floor, there is only a service space, so children can't go up. To a certain extent, the danger that children may appear on the stairs is avoided. The area surrounded by the three main buildings serves as an outdoor activity area for children.

Figure 6. Plan partition and position annotation.

Then, we will explain each part in detail. The clear height of the indoor space in living rooms should not exceed the criteria in Table 1. So, the corresponding heights are 4m for music and physical playroom and 3m for others.

Table 1. Space height requirement(meters).

The name of rooms	Clear height(meters)
Activity room, dormitory, suckling chamber	2.80
Music and physical playroom	3.60

The built-up area required in the specification is shown in the figure below: Four teaching classes, two small classes, one middle class, and one large class. Two small classes share a bathroom, and the middle class and the big class share a bathroom. Among them, the activity room and dormitory of the small class are merged, and the middle class and large class are separated. All Cloakrooms are combined with the activity room. The music and sports activity center are on the southeast of the site.

Table 2. Space height requirement(meters).

Scale/name of living rooms	Large size	Medium size	Small size	Note
Activity room	50	50	50	Area of each class
Dormitory	50	50	50	Area of each class
Bathroom	15	15	15	Area of each class
Cloakroom	9	9	9	Area of each class
Music and sports activity center	120	120	90	The public area of the whole school

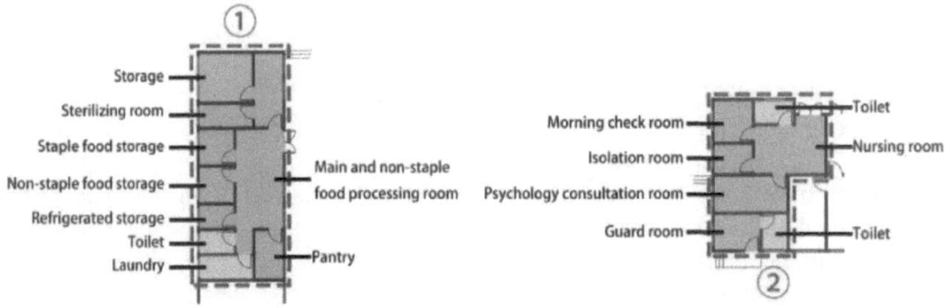

Figure 7. Detailed functional distribution.

The layout of functions in the service space is shown in Table 3 Figure 3.4.2. The effective area of serving rooms should not be less than the criteria in Table 4. The location of the room is shown in Figure 7.

Table 3. The minimum effective areas in serving rooms (m^2).

Scale/ name of rooms	Large size	Medium size	Small size
Nursing room	12	12	10
Isolation room	2×8	8	8
Morning checkroom	15	12	10

Table 4. The detailed area of the auxiliary space (m^2).

		Scale		
Kitchen	Name of rooms	Large size	Medium size	Small size
	Main and non-staple food processing room	45	36	30
	Staple food storage	15	10	15
	non-staple food storage	15	10	
	Refrigerated storage	8	6	4
	Pantry	18	15	10
Sterilizing room		12	10	8
Laundry		15	12	8

The clear width of the corridor in the main building should not be less than the criteria in Table 5. So, the width of the corridor we designed for children is 2 meters, and the narrowest part of the corridor for teachers is 1,5 meters.

Table 5. The detailed area of the auxiliary space (m^2).

	Arrange of rooms	
Name of rooms	Double-sided cloth room	single-sided cloth room or veranda
Living rooms	1.8	1.5
Serving rooms	1.5	1.3

Our new additional features are shown in Figure 8. Based on the original functions, we have added an archive, an auditorium, a staff dormitory, a home-school cooperation space, and separate psychological counseling rooms for children and staff. The teacher's psychological counseling room is placed on the second floor, and the environment is quiet and undisturbed. The archive is beside the auditorium, which is convenient for playing video and having some meetings. It can be used to store letters sent from parents far away or even videos of video calls. For families that do not have the conditions to send letters or video calls, they can write letters or record videos in advance, save them in the archive, and send them to the children regularly by the teacher. If possible, children can also contact their parents here and send them some small cards. A lecture on common sense can be held in the auditorium for both the children and the inter-generational guardians. It can also become a place for teacher training and team building. The home-school cooperation space is called the parents and teachers exchange center, a place for parents and teachers to exchange information about their children and communicate with the children together.

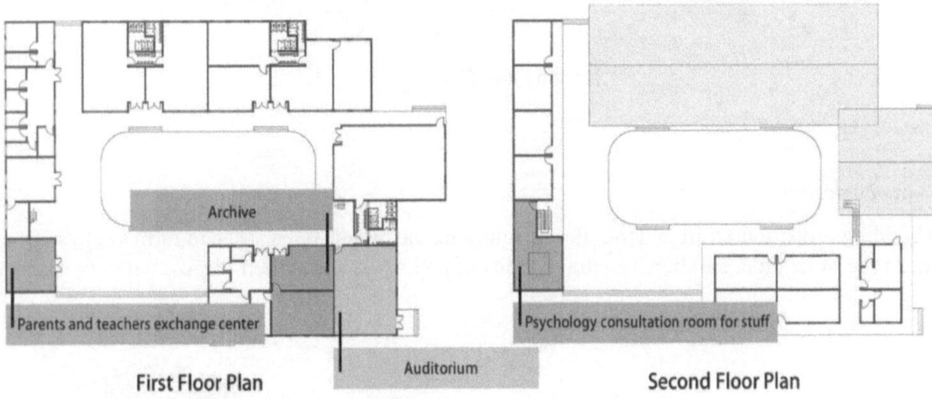

Figure 8. New added functional area.

The functions of children and staff living areas are shown in Figures 9 and 10. The children's activity rooms are divided into small sections. They have the corresponding teaching aids, which include: sensory area(S), language area(L), mathematics area(M), culture area(C), art area(A), and reading corner(R). The art space is combined with the music and sports activity center, the reading corner is combined with the archive. It is worth mentioning that the senior class students have the high cognitive ability and will connect with the primary schools, so they have their book corner. The staff dormitory is divided into a single dormitory and double dormitory, mainly for teachers and nurseries. The double dormitory on the south side is practical for class teachers and nurseries and strengthens their daily communication.

Figure 9. Children living space.

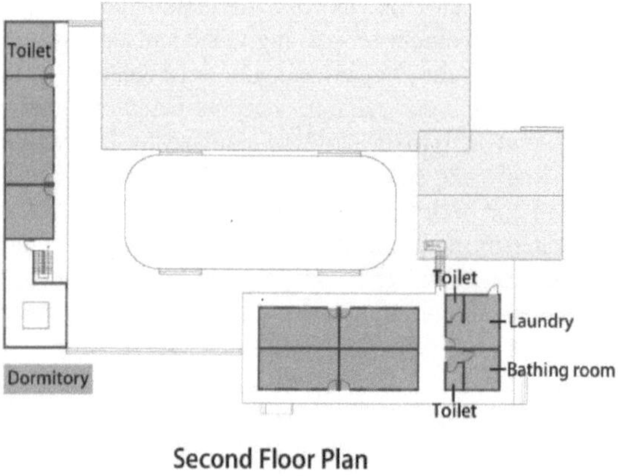

Figure 10. Stuff living space.

3.5 *Space design*

The building is distributed around the site, surrounding the middle open space to form a safe sports field. At the same time, children have three kinds of activity space at their disposal: indoor space,

Figure 11. Sections of sight analysis.

semi-outdoor space, and outdoor space. Children can explore and study in different functional areas of indoor space, run and play outdoors, and move under the eaves when it rains or is exposed to the sun. The transition of these spaces gives children different experiences.

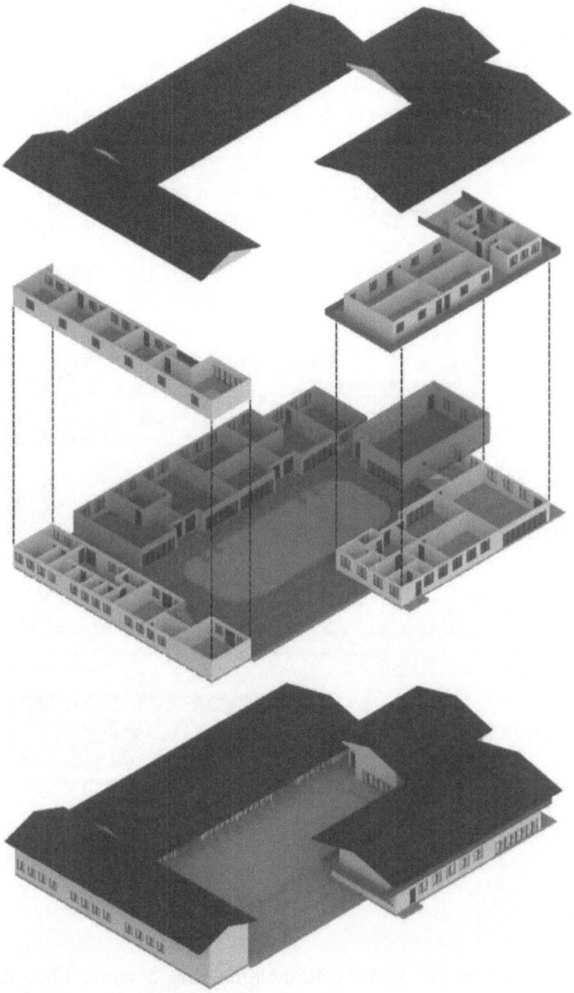

Figure 12. Explosion analysis.

For the second-floor teacher's living space, there are also indoor and semi-outdoor corridors. As shown in the section, the teachers' living areas on the second floor face the kindergarten with corridors inside. In the corridor, the teacher can see the children's activities in the middle field, grasp the overall situation, and provide protection for the safety of the children.

3.6 *Material selection*

In selecting architectural style, we selected the Yi tile-roofed houses with the same style as the local architectural style. The tiled house uses the local material to use the soil foundation, the tree as the material section to build the cover to use the wood structure to bear the load to take the soil foundation wall or the scholar wall as the maintenance structure, the house item is "人" shape slope surface. These materials are inexpensive and readily available locally.

Figure 13. Rendering with the surroundings.

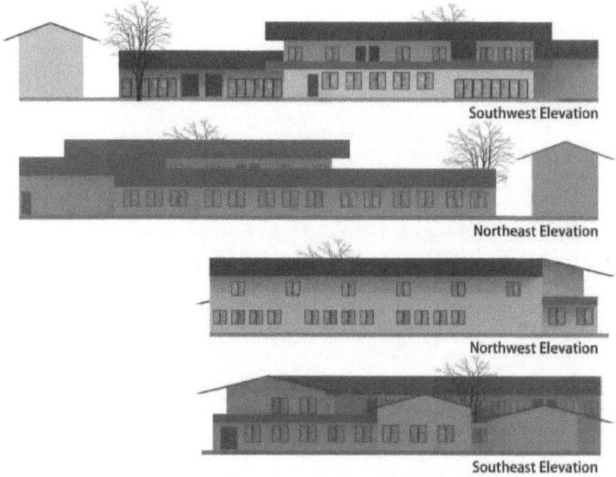

Figure 14. Elevation of four direction.

4 CONCLUSIONS

The left-behind children in rural areas of Western China have become a heated issue in China. In this paper, the key factors that influence the kindergarten design were carefully analyzed. Architectural solutions, i.e., location selection, function arrangement, space design, were presented. The following conclusions can be drawn:

Based on our design concept, the problems of Security, psychology, and communication can be solved to some extent for left-behind children. However, more policy support from the government is urgent. To solve these problems, firstly, the government should introduce relevant laws and regulations as soon as possible to make preschool education in rural areas have laws to follow. Secondly, The country needs to pay attention to the economic development of western and inland areas and increase the financial input of rural education appropriately. Thirdly, while ensuring their professional quality, their treatment should be appropriately improved as for kindergarten teachers. At the same time, regular assessment and training of teachers are carried out to ensure the maintenance and improvement of teachers' qualifications. Fourthly, to develop the local economy, increase farmers' income, reduce the number of left-behind children, and solve left-behind children's problem from the source. The government can optimize and upgrade its own industries to promote the healthy and sustainable development of the local economy. Finally, based on formulating and promulgating special preschool education law, we should establish an education supervision mechanism and rural areas to establish detailed supervision points.

(1) The overall cultural and ideological level of guardians of left-behind children needs to be improved. Although we have added the function of the lecture hall in kindergarten and held regular lectures to popularize common sense, this is far from enough. Besides policy support, improving the overall cultural and educational level in rural areas takes more time.
(2) In addition to the normal function design, the space design of left-behind children's kindergarten should also focus on the mental health of children and staff. Mental health problems have been gradually known in recent years, but many people do not pay attention to mental health problems in rural areas and do not even understand what mental health is. Therefore, rural mental health problems are worthy of attention, left-behind children and teachers under great pressure.
(3) Kindergartens for left-behind children should be a medium for communication among children, guardians, and teachers. Because of the lack of parental role, other care and companionship around left-behind children are particularly important. The most important are the teacher and the acting guardian, who have their own understanding of the different aspects of the child. The most important of them are the teacher and the acting guardian. They understand the different aspects of the children, and they all have their own places that they cannot find at ordinary times. Only when the two cooperate with each other can they understand the complete children. And make targeted education programs for children.

REFERENCES

Bing Wang. Investigation and thinking on the present situation of investment in preschool education of rural families in Henan Province. Educational Comments. 2015.

Chunmei Ai. Educational problems and countermeasures of left-behind preschool children in rural areas. *Modern Educational Science*. 2007, Vol 4: 121-122.

Congcong Zheng. Research on the preschool education needs of parents of left-behind children. Zhejiang Teachers College, 2014.

Dongquan Kou. Review the environmental problems faced by socialization of left-behind children in rural areas and repair measures – from the perspective of ecological system development view. *The journal of Teachers College in Guangxi*. 2007, No.1: 35-37.

Jianhua Wang, Tian Liu. Investigation on the study status of rural left-behind children in southwest China. *Journal of Hainan Normal University*. 2017, No.1, Vol 30: 119-127.

Li Zhang, Yuejian Cai. Problems and countermeasures of family education for rural left-behind children. The College of Education in Xinan University. 2008.

Liusong Dan. Popularization of Rural Pre-school Education: Challenge, Opportunity and Strategy. *Jixu Educational Research*. 2010, Vol 3: 51-53.

Weimin Zhang, Min Zhang. The Dilemma and Solution of Ethnic Preschool Education – Based on the Field Investigation of Jinzhu Miao Village in Xiushan County, Chongqing. Pre-education research. 2014.

Yan Liu. Present situation and thinking of preschool education for left-behind children in rural areas. The Educational College in Suihua College. 2012.

Zhengri Luo. Investigation report on the status quo of Korean preschool education in Heilongjiang province. Heilongjiang National periodicals. 2011.

Advances in Civil Engineering: Structural Seismic Resistance, Monitoring and Detection – Mohd Johari Mohd Yusof and Junwen Zhang (Eds)
© 2023 The Authors, ISBN 978-1-032-31491-4

Research status and prospect of architecture of bamboo

Yue Liu
College of Architectural Engineering, Kunming University of Science and Technology-Oxbridge College, Kunming, China

Yuan Yuan
Guangdong Yuanshun Architectural Design Co., Ltd. Jiangxi Branch, Ganzhou, China

Shaodong Li
Yunnan Design Institute Group Co., Ltd, Kunming, China

ABSTRACT: This paper focuses on the application prospects of bamboo, a traditional building material, under the background of deepening and standardization of green building construction. The author expounds the background of bamboo research, introduces the construction methods and limitations of traditional bamboo, and analyzes the research status of new bamboo buildings and the advantages and disadvantages of each kind of bamboo buildings by means of classification and statistics. Finally, the author suggests that we should adhere to the combination of low technology and high technology and continue to adhere to the research and application of new bamboo construction system.

1 INTRODUCTION

Construction is one of the oldest production activities of mankind and is necessary for human survival. With developments in human productivity, construction technology is constantly developing. Building materials are constantly updated with the development of building technology. With the development and expansion of cities, the energy consumption caused by construction activities is also increasing dramatically. The energy consumption of construction activities alone accounts for 40% of the world's resources. With the substantial use of non-degradable recycled materials, such as concrete, brick, glass, and steel, construction waste has also increased to 50% of the world's total waste, causing considerable damage to the environment.

In foreign countries, a large number of light wood structure houses are being built, which has been a big step in environmental protection. Timber is required to be planted in batches and forests, at the same time, corresponding standards and specifications are also formulated for logging. Industrialized timber production makes foreign forests effectively protected and reasonably utilized. However, large-scale wood production is not suitable for China, because China's logging is not standardized, and due to previous deforestation, China's forest coverage rate is only 12%, which is far lower than the international forest coverage rate of 22%.

Due to shortage of resources, China is also vigorously promoting prefabricated buildings, and China's rich bamboo resources make more scholars focus on this material. This paper analyzes the performance characteristics of bamboo, introduces the construction mode of traditional bamboo buildings, and analyzes its advantages and disadvantages. To further explore the mode of modern bamboo buildings, this paper sorts out the existing common construction modes of new bamboo buildings, points out that the research of new bamboo buildings should still organically combine

 DOI 10.1201/9781003310884-69

low technology with high technology, and advocates to explore more possibilities of new bamboo buildings along this route.

2 CHARACTERISTICS AND LIMITATIONS OF TRADITIONAL ARCHITECTURE OF BAMBOO

Bamboo has been used as a structural material since thousands of years in human architectural history. In the early stage, human beings used unprocessed raw bamboo to build houses and construct regional folk houses, such as the Diaojiao building in Western Hunan and the Ganlan bamboo building of Dai nationality in Yunnan. These buildings not only possess artistic beauty consistent with the local folk customs, but also use local materials, which can greatly save the construction cost. For a long time, the original bamboo structure architecture occupied an important position in history. However, due to limitations of bamboo-producing areas and the development of brick and tile materials, the dominant building materials have been wood and brick since long. Since modern times, with the emergence of steel and concrete, reinforced concrete has taken a leading role in the field of building materials. Compared with the advantages of reinforced concrete buildings, the limitations of traditional bamboo are also very obvious.

2.1 *The distribution of bamboo resources and its seismic performance complement each other*

The seismic performance of bamboo is excellent, and the strength to weight ratio of bamboo is about 0.017, second only to 0.020 of steel. The stiffness to weight ratio is about 33, which is better than steel, wood, concrete, and other materials. Yunnan witnesses the most abundant natural distribution of bamboo resources in China. Yunnan is one of the origins and modern distribution centers of bamboo plants in the world, with more than 200 species of 26 genera. It is known as "the hometown of bamboo in the world" and "the hometown of national bamboo culture".

At the same time, Yunnan is located in the east of the collision zone between the Indian Ocean plate and the Eurasian plate. It is one of the places where destructive earthquakes occur frequently in China, accounting for 20% of the average amount of destructive earthquakes in China, and the area where destructive earthquakes may occur accounts for 84% of the whole province. From 1970 to 2010, earthquakes with a magnitude of more than 7 will occur every four years, causing countless casualties and direct economic losses. Therefore, bamboo can become the preferred material for local construction in Yunnan, which is also in line with the principle of adjusting measures to local conditions.

2.2 *Contradiction between availability and durability of bamboo materials*

Traditional bamboo self-built houses have common defects: short service life, poor stability, easily to be damaged, and so on. For example, the Dai traditional bamboo house only goes through simple River rinsing and drying when taking and using bamboo, which cannot completely remove the sugar of bamboo so that bamboo is particularly easily to be damaged by insects. First, the traditional bamboo building has not undergone fire prevention treatment, so the safety of the traditional bamboo building is very poor; second, bamboo itself is hollow and thin after weaving. Heat insulation and sound insulation are great problems in hot and humid areas. Moreover, although the seismic performance of bamboo itself is excellent, there are potential safety hazards in the traditional bamboo building structure because the joints connected by traditional bamboo cannot meet the strength requirements and the anti-corrosion measures are not in place. Finally, the traditional bamboo buildings built by residents themselves have not been designed scientifically so that they cannot keep up with the requirements of residents' modern life in function, which makes residents think that only living in brick houses is a symbol of good life.

Therefore, to overcome the defects of traditional architecture of bamboo, we must find a modern bamboo structure system in line with modern production.

3 RESEARCH STATUS OF MODERN ARCHITECTURE OF BAMBOO

The situation of China's forest resources is grim and bamboo resources are relatively rich. On November 29, 2005, the general office of the State Council transmitted the opinions on accelerating the work of wood conservation and substitution jointly issued by 12 national departments, such as the National Development and Reform Commission, the Ministry of Science and Technology, the Ministry of Construction, the Ministry of Railways, the State Environmental Protection Administration, and the forestry bureau, which pointed out that China needs to find a suitable building material to replace wood. As early as the 1950s, many Chinese scholars studied bamboo, which laid a certain foundation for the study of modern bamboo structure.

At present, there are several kinds of bamboo buildings in terms of materials: bamboo architecture with round bamboo, bamboo architecture with round bamboo and steel, and bamboo architecture with bamboo plywood. In terms of building types, there are landscape sketch bamboo buildings, small bamboo buildings, and long-span bamboo buildings. Geographically, there are domestic bamboo architecture, Southeast Asian bamboo architecture, Japanese bamboo architecture, European Bamboo architecture, and American bamboo architecture.

3.1 *Bamboo architecture with round bamboo*

Bamboo architecture with round bamboo is particularly common in Southwestern China. Traditional Dai bamboo buildings are built by round bamboo. Not only Dai bamboo buildings, but also many temporary buildings in bamboo-producing areas in the southwest are made of local materials and round bamboo. Bamboo architecture with round bamboo is a kind of building with beam-column structure composed of raw bamboo collected for insect prevention, binding, and bucket crossing (Figure 1). In 2016, China Engineering Construction Association also formulated the corresponding technical specification for round bamboo structure construction (technical specification for round bamboo structure construction standard of China Engineering Construction Association).

Figure 1. Nodes of temporary bamboo buildings.

Figure 2. The wind pavilion.

Domestic architects have made some attempts at modern bamboo architecture, especially the practical attempts of college teachers and students, such as the wind pavilion designed and built by teachers and students at Tongji University in Anji (Figure 2), and the Huahai Bamboo Gallery designed and built by teachers and students at Beijing Forestry University in Anji (Figure 3). Functionally, they all belong to sketch buildings. Structurally, they make full use of the mechanical properties of bamboo and bend it to realize the designer's modeling intention. The main body of the building is supported by a simple arch structure and a transverse tie system.

Vietnamese architect Vo Trong nghia loves round bamboo buildings very much. Due to the unique mechanical properties of bamboo: high strength to weight ratio, high radial compressive, and tensile strength, which is similar to the physical properties of steel, it is very suitable for spatial grid structures. Many original bamboo architectural works of Vo Trong nghia give full

Figure 3. The Huahai bamboo gallery.

Figure 4. Vedana restaurant.

play to the performance of original bamboo. Vedana restaurant (Figure 4) and diamond island community center (Figure 5) both use round bamboo to build a huge dome structure, creating a spatial artistic conception like the Roman pantheon. To increase the structural strength, he tied the bamboo into bundle columns, and formed a spatial structural system through the interspersed and binding between the bamboos. Such bamboo buildings are not only bold and novel in materials, but also close to nature in texture, and their transparent architectural pattern also makes buildings organically integrate with nature and echo the local climate characteristics.

Figure 5. Diamond island community center.

Figure 6. Bamboo sports hall.

Thai architects also have many explorations on round bamboo buildings, such as bamboo sports hall for Panyaden International School designed by Chiang Mai life Architects (CLA) (Figure 6). The use of bamboo to make long-span buildings is a major feature of the stadium. The innovative structural design is based on newly developed prefabricated bamboo trusses with a span of over 17 m without steel reinforcement or connections. Second, the design and material enable a cool and pleasant climate all year round through natural exploitation and regulation. At the same time, the exposed bamboo structure is a feast for the eye and an exhibition of masterly handicraft (CLS 2017).

Round bamboo structure is the most natural bamboo building form. The bamboo after high-temperature deformation treatment, combined with traditional binding and screw connection, makes the round bamboo building pure and beautiful. However, round bamboo buildings also have a series of shortcomings: (1) Round bamboo buildings are more complex in the process of bamboo screening. The root of bamboo is thick and gradually becomes thinner from bottom to top, which also leads to the screening and splicing of bamboo with similar diameter in round bamboo buildings, which is not an easy work. (2) Because bamboo is a tubular structure and hollow, a large number of round bamboos need to be bound together when the round bamboo is subjected to lateral force, which not only wastes the bamboo, but also the binding nodes are not easy to make. (3) Because round bamboo buildings need a large number of bamboo materials to be screened and their quite complex node modes, the construction speed of round bamboo buildings is slow and difficult to implement. (4) The durability of bamboo is not high. For round bamboo buildings, it is difficult to replace the old with the new. (5) There is a gap between the original bamboo and the original bamboo, so it is difficult to achieve a complete and closed space. (6) Fire prevention of exposed raw bamboo cannot be guaranteed.

3.2 *Bamboo architecture with round bamboo and steel*

Different from the binding and interspersed method of round bamboo structure buildings, some architects try to connect bamboo components with other node treatment methods. Bamboo architecture with round bamboo and steel is mainly used to pour concrete at the end of round bamboo, then insert steel components, and make steel hoop fixed joints at the end of round bamboo (Figure 7). Colombian (South American) architect Simon Velez has always paid attention to the use of low-cost building materials. In several of his famous works, a large number of bamboo materials are used. Typically, his bamboo buildings are connected by concrete bolt bamboo.

German (European) architect Markus Heinsdorff has many bamboo architectural works. In November 2008, he built 21 bamboo exhibition booths at Guangzhou station, Avenue of the foreign ministries of Germany and China. He fixed the end of the bamboo with steel rings and steel plates. All the exhibition booths were overlapped on site and removed immediately after use. This is also fully reflected in the German Chinese peer house Pavilion of the 2010 Shanghai World Expo (Figure 8).

Figure 7. Nodes of steel and bamboo.

Figure 8. German Chinese peer house pavilion.

Figure 9. A experimental model.

Figure 10. The Bamboo and Rattan Pavilion.

3.3 *Bamboo architecture with bamboo plywood*

The glued bamboo glubam studied by the Institute of Modern Bamboo and Composite Structure led by Professor Xiao Yan at Hunan University has considerably alleviated the pressure of reconstruction work in the post Wenchuan disaster reconstruction, and has an obvious harmful effect on aftershocks.

Professor Xiao Yan's glubam plywood technology changes the shape of bamboo, industrializes and standardizes bamboo under the condition of ensuring mechanical properties, eliminates the difficulty of bamboo screening, and the buildings reconstructed after Wenchuan earthquake have good fire resistance. However, the production and processing of bamboo plywood have high requirements for instruments and equipment, and professional instruments and equipment are needed to produce glue injection. The processing price of integrated materials is high, so it is not practical to use them in poverty alleviation projects, and the need for poverty alleviation projects are villagers' self-construction and local materials. Therefore, the author believes that the application of plywood in poverty alleviation projects is not practical (Chen et al. 2011).

3.4 *Two-way orthogonal bamboo strip covered bamboo building with wood skeleton*

According to the advantages and disadvantages of round bamboo building, round bamboo steel member building and glubam bamboo plywood building analyzed above, the Green Local Research Institute of Kunming University of science and technology has developed a new bamboo building—wood skeleton two-way orthogonal bamboo strip covered bamboo building form (Figure 9). Due to the different diameters of round bamboo, the research institute chose to cut round bamboo into bamboo strips with uniform width and thickness; and the bamboo strips are fixed on the wood skeleton side by side, to form the wood skeleton bamboo strip covering structure. On this basis, this new bamboo architectural form is studied.

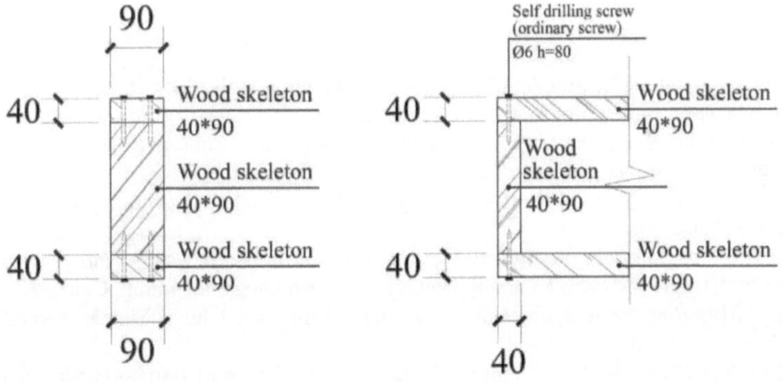

Figure 11. Bamboo beam

The two-way orthogonal bamboo strip cladding beam of wood skeleton is made of 40 mm * 90 mm wood, with a length of 80 mm, φ 6 mm self-drilling screws (or ordinary nails) are nailed to the wood skeleton (Figure 11), and then the bamboo strips are nailed to the wood skeleton with T50 air nails on both sides of the wood skeleton, with a total of two layers of bamboo strips. Cut off the excess bamboo strips and make a two-way orthogonal bamboo strip covered beam with wood skeleton.

3.5 *Non-structural bamboo building*

The Bamboo and Rattan Pavilion of the 10th China Flower Expo by ARCPLUS Shanghai Archi-Scientific Creation Center are depicted in Figure 10. With parametric design, they tried a three-dimensional weaving pattern of twisting from a plane along with surfaces with an interwoven element and a spatially woven surface is formed with changes of spatial depth and strong optical effects.

In this system, steel is used to make the framework of double-layer cable net structure, and bamboo is used to make double-layer two-way space weaving. This practice effectively created a new form of artistic expression of bamboo architecture (ARCPLUS 2021).

4 CONCLUSION—DEVELOPMENT PROSPECT OF BAMBOO ARCHITECTURE

Based on the results and discussions presented above, several conclusions are enlisted below:

(1) Bamboo is a perennial plant with short growth cycle, fast growth speed, and superior mechanical properties.
(2) In the construction process of bamboo buildings, the concept of low-tech construction is fully reflected, that is, the construction purpose is realized by the simplest construction method. Of course, through the above analysis, we also see that low technology does not mean poor quality and rough. Through the efforts of architects, high artistic achievements can still be achieved. It can be said that bamboo architecture is promising!
(3) If bamboo can meet the industrialization standard, the cost is relatively low, and the public can afford the cost of materials, it can reach the degree of self-construction. On the one hand, the construction period is shortened, and on the other hand, the labor cost and material cost are reduced.
(4) With development and progress in material science and technology, it will be a significant progress for green construction to study more practical and efficient bamboo building systems that meets the requirements of construction specifications.

REFERENCES

ARCPLUS, https://www.gooood.cn/the-bamboo-and-rattan-pavilion-of-the-10th-china- flower-expo-by-arcplus-shanghai-archi-scientific-creation-center. htm?lang=en.2021

Chen, Guo& Shan, Bo& Xiao, Yan. 2011. Experimental study on seismic performance of light bamboo structure buildings, *Vibration and Impact*, 2011;10.

CLS, https://www.gooood.cn/bamboo-sports-hall-for-panyaden-international- school-thailand-by-chiangmai-life-architect.htm?lang=en. 2017

Wang, Yajing. 2013. *Study on light roof renewal of Dai Folk Houses in Xishuangbanna with raw bamboo as the main material*, Degree-Master, Kunming University of technology, Kunming, China. 2013.

Yang, Baozhang.2014. *Bamboo and Bamboo Architecture.* Kunming, China: Yunnan People's Publishing House. 2014.

Yuan, Yuan& Liu, Yue.2018. The Development of Architecture with Short Bamboo Strips Nailed Web and The Research Status of The Architecture of Bamboo, *Journal of Kunming University of Technology (Social Science Edition).* 2018

Yunnan Design Institute. 1986. *Yunnan Folk Houses.* Beijing, China: China Construction Industry Press. 1986: 209–214.

Advances in Civil Engineering: Structural Seismic Resistance, Monitoring and Detection – Mohd Johari Mohd Yusof and Junwen Zhang (Eds)
© 2023 The Authors, ISBN 978-1-032-31491-4

Research on application and development of high-performance concrete at home and abroad

Ming Wei & Min Zhao
Kunming University of Science and Technology Oxbridge College, Kunming, China

ABSTRACT: High-performance concrete (HPC) is a new type of building material. This paper summarizes the application development, existing problems, and trends of HPC at home and abroad, which can be used as a reference for related engineers.

1 INTRODUCTION

High-performance concrete (HPC) is a new type of high-tech concrete, which is made based on modern concrete technology that involves improving the performance of ordinary concrete. It considers durability as the main index of design, and according to different application requirements, it guarantees the following key performance points: durability, workability, applicability, strength, volume stability, and economy.

2 THE DEVELOPMENT OF HIGH-PERFORMANCE CONCRETE

HPC is developed from high-strength concrete. However, it requires less technical performance than high-strength concrete. Generally, the configuration of HPC is characterized by adopting low water-binder ratio, selecting high-quality raw materials, and adding sufficient quantities of fine mineral admixtures and efficient admixtures.

2.1 *The definition of high performance concrete*

In May 1950, National Institute of Standards and Technology (NIST) and American Concrete Association (ACI) first proposed the concept of HPC. However, up to now, the requirements and meanings of HPC are completely different in different countries. American engineers and technicians believe that HPC is a kind of concrete that is easy to pour, tamp, and not separate. Concrete that can maintain high strength, toughness, and volume stability for a long time and has extended service life in harsh environments. According to American Concrete Association, this kind of concrete does not necessarily need high compressive strength, but it still needs to be above 55 MPa, and it needs high chemical corrosion resistance or other properties. Japanese engineers and technicians believe that HPC is a kind of concrete with high filling capacity, which can be poured easily without vibrating in the fresh mixing stage. In the early stage of hydration and hardening, cracks caused by hydration heat or drying shrinkage are rarely seen. It has sufficient strength and durability after hardening. According to Canadian engineers and technicians, HPC is a kind of concrete with high elastic modulus, high density, low permeability, and high corrosion resistance.

According to the requirement of HPC in various countries, it can be considered that HPC has high impermeability (the key property of high durability). High volume stability (low shrinkage, low creep, low temperature deformation, and high elastic modulus); high compressive strength; good workability (high fluidity, high cohesion, and self-compacting). In China's Technical Specification

DOI 10.1201/9781003310884-70

for Application of High Performance Concrete (CECS207-2006), HPC is defined as concrete produced by conventional materials and processes, with various mechanical properties required by concrete structures, high durability, high workability, and high volume stability.

3 TECHNICAL ROUTE OF HIGH PERFORMANCE CONCRETE

HPC is developed using high-strength concrete. However, HPC requires less technical performance than high-strength concrete. Generally, the development of HPC can be divided into three stages as follows.

3.1 *High-strength concrete formed by vibration and pressure-technological innovation*

Before the advent of superplasticizer, to obtain high-strength concrete, W/C (water-cement ratio) is generally reduced, and it is formed by strong vibration and pressure. That is, mechanical pressure is applied to concrete to squeeze out air and residual water in concrete and reduce porosity. However, this process is not suitable for field construction and is difficult for adoption. It is only widely used in the production of precast concrete slabs and precast piles. And used together with autoclaved curing.

3.2 *Mix high-efficiency concrete with high-efficiency water reducing agent—The fifth component innovation*

The appearance of superplasticizer in the late 1950s is a new development stage for high-strength concrete. The representations are naphthalene series, melamine series, and modified wood calcium series of superplasticizers, all of which are widely used at present.

High-strength concrete with high fluidity and compressive strength of 60~100 MPa can be obtained by common process, adding high-efficiency water reducing agent, and reducing water-cement ratio. However, the concrete prepared only with high efficiency water reducing agent is associated with the problem of large slump loss.

3.3 *Using mineral admixtures to prepare high performance concrete—The sixth component innovation*

In 1980s, use of mineral admixtures rose and developed into the sixth component of HPC, which complemented the fifth component and became an indispensable part of HPC. At present, the technical route of preparing HPC is to mix high-efficiency water reducing agent and mineral admixture into concrete at the same time. The mineral admixture of HPC is prepared using a micropowder-assisted cementing material with high specific surface area. For example, silica fume, fine ground slag powder, ultra-fine fly ash, etc., can form a micro-compact system through the interstitial filling effect of micropowder to improve the interface structure and interface bonding strength.

4 CHARACTERISTICS OF HIGH-PERFORMANCE CONCRETE

4.1 *Self-compactness*

HPC has low water consumption, good fluidity, and high segregation resistance, so it has excellent filling property. Therefore, proper HPC with high fluidity has better self-compactness.

4.2 *Volume stability*

HPC has high volume stability, which is characterized by high elastic modulus, low shrinkage & creep, and low temperature deformation. The elastic modulus of ordinary concrete is 20~25 GPa, and the elastic modulus of HPC with suitable materials and mix ratio can reach 40~45 GPa. Adopting coarse aggregate with high elastic modulus and high strength and reducing the content of cement paste in concrete, the dry shrinkage value of 90-day-old HPC prepared with reasonable mixture ratio is lower than 0.04%.

4.3 *Strength*

Compressive strength of HPC exceeds 200 MPa. At present, HPC with average strength of 100~120 MPa in 28 days has been applied in engineering. Compared with high-strength concrete, the tensile and compressive strengths of HPC are obviously increased, and the early strength of HPC develops faster, while the growth rate of the latter is lower than that of ordinary strength concrete.

4.4 *Heat of hydration*

Because of the low water-cement ratio of HPC, the hydration reaction is terminated earlier, so the hydration heat will be reduced accordingly.

4.5 *Shrinkage and creep*

The total shrinkage of HPC is inversely proportional to its strength, and the higher the strength the smaller is the total shrinkage. However, early shrinkage of HPC increases with the improvement of early strength. Relative humidity and ambient temperature are still the two main factors affecting the shrinkage performance of HPC. The creep deformation of HPC is significantly lower than that of ordinary concrete, compared with ordinary strength concrete, the total creep amount of HPC (sum of basic creep and dry creep) is significantly reduced. In the total amount of creep, the decrease of dry creep value is more significant, and the basic creep is only slightly decreased. However, the ratio of dry creep to basic creep decreases with the increase of concrete strength.

4.6 *Durability*

Except that the frost resistance and impermeability of HPC are obviously higher than those of ordinary concrete, the permeability of HPC is obviously lower than that of ordinary concrete. Because of its high compactness and impermeability, the chemical corrosion resistance of HPC is significantly better than that of ordinary strength concrete.

4.7 *Fire resistance*

Under the effect of high temperature, HPC will burst and peel off. Because of the high density of concrete, free water is not easy to be discharged from the pores quickly, and the steam pressure formed in the concrete can almost reach the saturated steam pressure when subjected to high temperature. At 300°C, the steam pressure can reach 8 MPa, while at 350°C, the steam pressure can reach 17 MPa. Such internal pressure can produce tensile stress of 5MPa in concrete, which will cause explosive denudation and shedding of concrete. Therefore, the high temperature resistance of HPC is a problem worthy of attention. In order to overcome this performance defect, organic fibers can be mixed into high-performance and high-strength concrete, and the fibers in concrete can melt and volatilize at high temperature. Many connected pores are formed, so that the steam pressure produced by high temperature can be released, thus improving the high temperature resistance of HPC.

To sum up, HPC is concrete that can better meet the structural functional requirements and construction technology requirements, and can maximize the service life of concrete structures and reduce the project cost.

5 THE POPULARIZATION AND APPLICATION OF HIGH PERFORMANCE CONCRETE

As a new building material, the durability of HPC is more than twice that of ordinary concrete, which can increase the safe service life of concrete structure and reduce the waste and construction waste caused by repair or demolition. A large number of industrial by-products and wastes can be used to minimize the consumption of natural resources and energy and reduce environmental pollution; Shrinkage is small, suitable for building efficient prestressed structures; HPC is suitable for the construction and repair of high-rise, long-span, large-volume, long-span bridges, undersea tunnels, expressways and structures used in harsh environments, such as nuclear reactors, offshore structures and structures in corrosive medium environments. Other intelligent HPC used for special purposes has its unique advantages which cannot be replaced by other concrete. Because HPC has many superior properties, it has shined brilliantly since it came into being, and the research and application momentum of HPC has been developing rapidly all over the world.

5.1 *The research and application status of HPC abroad*

From 1986 to 1993, French government organized 23 units, including government research institutions, universities and construction companies, to carry out the research project of "New Method of Concrete", carry out research on HPC, and set up demonstration projects. In 1996, the French Ministry of Public Works and the Ministry of Education and Research organized a four-year national research project "High Performance Concrete 2000". Invested 5.5 million dollars in research. Three HPC cable-stayed bridges built in France-Perti Bridge, and the recently built Loen River Bridge and Normandy Bridge are also made using HPC.

In 1994, 16 agencies of the US federal government jointly put forward a proposal on the application of HPC in infrastructure construction, and decided to invest 200 million US dollars in research and development within 10 years. More than 100 bridges with HPC decks have been built in New York State. In Washington State, the highway department is working out the standard of HPC beams.

At present, the current design code for concrete structures in Germany has reached C110 level, and its strength grade is the highest in the world today. The Science and Engineering Research Fund of the Royal Norwegian Institute of Science and Technology continuously supports the research of high-strength concrete and HPC. The Great Belt Project in Denmark is a large-scale tunnel-bridge connecting structure, and the specified design service life is 100 years. These foreign protests about the application of HPC have great enlightening reference value for us.

5.2 *Research and application status of HPC in China*

In recent years, the research and application of HPC in China have developed rapidly. China is a big country that produces and uses concrete, and the quality of concrete is constantly improving. It is only about 10 years to get involved in the research and application of HPC. With the continuous recognition of the superiority of HPC, the progress of concrete application technology and the acceleration of urban construction, HPC has been developed rapidly.

HPC has been widely used in practical projects, especially in high-rise buildings, long-span bridges, offshore oil production platforms, mine projects, seaport docks and other projects. For example: Shanghai Jinmao Tower (C60), Beijing Jing'an Center Building (C80), Liaoning Property Building (C80), Nanjing Hilton International Hotel (C30 and C50), Changchun International Trade City (C55), Guangzhou Humen Bridge (C50) and Shanghai Yangpu Bridge (C50) are all examples of application.

Many research units across the country have developed common pumping HPC, HPC with a large amount of fly ash, high-flow self-compacting HPC, fiber-reinforced HPC, lightweight aggregate HPC, underwater non-dispersible HPC, HPC for harbor engineering and marine engineering, and HPC with high throwing fiber. Developed C30-C80 HPC with various strength grades and complete concrete durability testing equipment, and mastered the complete set of construction technologies and various concrete durability testing technologies. Among them, C30 HPC with excellent durability will soon be used on a large scale in Shenzhen subway project with complex geological conditions. Recently, there is also a report that Shanghai has conquered the "life" of HPC, which can reach 100 years. This concrete has been used in Donghai Bridge. A new achievement, "Research on High Performance Preparation Technology for Salt Pollution Resistance", hosted by the Science and Technology Department of Guangzhou Province, has been scientifically demonstrated by experts in many aspects and passed the appraisal. In addition, in the technical specification of HPC, the concrete strength grade has reached C100.

5.3 *Problems in HPC research and application in China at present*

On the one hand, the research on properties: (1) self-shrinkage caused by self-drying; (2) brittleness; (3) Scientific classification and quality standard of mineral fine powder and its compatibility with concrete admixture; (4) Composite super-superposition effect of HPC multi-component composites; (5) Toughness of high-strength and HPC. In addition, the small test pieces produced in the laboratory according to the current specifications, the results of "durability" test under the conditions of collision and forced process are inconsistent with the performance of the concrete in the actual structure.

On the other hand, it is engineering application. If HPC is only used in long-span bridges, lower columns of high-rise buildings and some components and parts requiring high strength and high performance, it will not only greatly limit the expanded use of HPC, but also limit its many advantages and performances, especially the exertion of the green advantages greatly advocated now. This is not consistent with the need for the country to vigorously develop and popularize "energy saving and green building".

5.4 *Development trend of high performance concrete*

Firstly, improve the understanding of HPC, perfect norms and establish laws and regulations, and increase engineering application.

The development of HPC, but for more than ten years, people who are used to ordinary concrete don't know enough about it, which hinders the wide application of HPC. High-strength and HPC has been basically accepted, while low-strength and HPC has not been universally recognized by engineers, which brings great obstacles to the popularization of low-strength and HPC. The main reason is that the understanding of cost and performance is not in place. The literature points out that low-strength concrete has good economic benefits and good social benefits at the same time. At the same time, people should realize that "high-quality projects must have high performance". The proportion of concrete in the whole project cost is very small, considering the project quality, construction convenience, durability and so on. Even if the cost of HPC is slightly increased, it is worth it. To be a big account, it is necessary to comprehensively consider the economic and social benefits brought by HPC.

In order to adapt to the development of "energy-saving and green building" in China, the state should improve the norms and formulate corresponding mandatory laws and regulations to vigorously promote the application of HPC.

Secondly, increase the research on the performance of HPC.

(1) The constant pursuit of high workability of HPC. Good workability can speed up construction, improve labor intensity, and be beneficial to environmental protection. HPC should be fully mechanized, automated and modernized in management.
(2) Improve the durability design of concrete structure engineering. The construction industry consumes nearly 40% of the world's resources. If the life of buildings is doubled, the consumption of resources and energy and environmental pollution will be reduced by half. Extending the service life of concrete, improving the durability of concrete, reducing the repair of concrete structure and reducing the demand for cement concrete will improve the utilization rate

of resources and energy. Reduce economic costs, avoid economic losses, save resources and energy, and reduce the ecological environment load, which is in line with the sustainable development strategy.

(3) Developing towards ultra-HPC (UHPC). At present, ultra-HPC has appeared, and the more successful active fine-grained concrete, grouting fiber concrete and sub-dense reinforced composite materials are characterized by high strength and high compactness. A large number of fiber reinforcement is used to overcome the brittleness of concrete materials, and false toughness is used to prevent sudden fracture of concrete. The manufacturing cost is much higher than HPC, Generally, it can be used in some special projects such as military engineering and nuclear power plants. Canada has applied reactive concrete to a bridge, and France is preparing to study and apply UHPC in a nuclear waste tank. With the progress of science and technology, more UPHC structural projects will be produced in the future.

Thirdly, the development of HPC to green HPC (GHPC).

Today, with the increasing popularity of green environmental protection, whether concrete can be used as the most important engineering structural material for a long time depends on whether it can become a green building material, so GHPC will assume historical responsibility. GHPC can save more cement clinker, reduce environmental pollution more effectively, and at the same time reduce material consumption and energy consumption greatly. More fine additives mainly containing industrial waste residue can be added to save clinker, improve the environment and reduce secondary pollution. It can give full play to the advantages of HPC, minimize the consumption of cement and concrete, and achieve the purpose of saving resources, energy and improving the environment. Fly ash is the combustion waste residue of thermal power plants, and its annual emissions in China are over 140 million tons.Fly ash has pozzolanic activity, which can reduce the initial hydration heat and dry shrinkage, improve the workability of fresh concrete, increase the later strength of concrete and significantly improve the durability of concrete. Making full use of fly ash as admixture to develop HPC can save resources and energy and reduce secondary pollution. Achieving good economic and social benefits is a great contribution to the realization of China's sustainable development strategy.

6 CONCLUSION

HPC is the product of concrete technological progress, and its production needs qualified operators, perfect production equipment and high-level quality management control. To popularize HPC, we should constantly sum up experience, study the problems found in popularization and application, and solve them as soon as possible. We believe that with the development and increasing application of HPC technology, the overall level of China's construction industry will be greatly improved.

ACKNOWLEDGMENTS

This work was financially supported by fund of Yunnan Provincial Department of Education (2017ZDX258).

REFERENCES

Azmee, Norzaireen Mohd, and Nasir Shafiq. "Ultra-high performance concrete: From fundamental to applications." *Case Studies in Construction Materials* 9 (2018): e00197.

Chen, How-Ji, Yi-Lin Yu, and Chao-Wei Tang. "Mechanical properties of ultra-high performance concrete before and after exposure to high temperatures." *Materials* 13.3 (2020): 770.

JingjingZhu. Research on the problems and development of high performance concrete[J]. *Modern Trade Industry*, 2011, 23(014):279–279.

Li, Junquan, et al. "Durability of ultra-high performance concrete–A review." *Construction and Building Materials* 255 (2020): 119–296.

Salimi, Javid, Amir Mohammad Ramezanianpour, and Mohammad Javad Moradi. "Studying the effect of low reactivity metakaolin on free and restrained shrinkage of high performance concrete." *Journal of Building Engineering* 28 (2020): 101053.

Travush, V. I., D. V. Konin, and A. S. Krylov. "Strength of reinforced concrete beams of high-performance concrete and fiber reinforced concrete." *Magazine of Civil Engineering* 1 (77) (2018): 90-100.

Yoo, Doo-Yeol, and Min-Jae Kim. "High energy absorbent ultra-high-performance concrete with hybrid steel and polyethylene fibers." *Construction and Building Materials* 209 (2019): 354–363.

Yu, Yang, et al. "A novel optimised self-learning method for compressive strength prediction of high performance concrete." *Construction and Building Materials* 184 (2018): 229–247.

Zhang, Zi-Yu, et al. "Seismic performance of a novel interior precast concrete beam-column joint using ultra-high performance concrete." *Engineering Structures* 222 (2020): 111–145.

Advances in Civil Engineering: Structural Seismic Resistance, Monitoring and Detection – Mohd Johari Mohd Yusof and Junwen Zhang (Eds)
© 2023 The Authors, ISBN 978-1-032-31491-4

Demonstration of tunnel construction method for Guangzhou airport expansion project

Xuekui Gao & Zhijian Li
China Airport Planning & Design Institute Co., Ltd., Beijing, China

ABSTRACT: Guangzhou Baiyun airport plans to build two tunnels under three runways in service respectively. This paper introduces the comparison and selection process of tunnel construction methods. Based on the investigation of engineering cases and the adaptability analysis of different construction methods, this paper focuses on the comparison and demonstration of open excavation method, pipe curtain method and shield method. The demonstration results show that the shield method has the advantages of small impact on navigation, controllable risk and moderate investment. It is recommended to use the shield method to cross the operating runway. The research provides a strong support for the decision-making of project construction, and has reference significance for similar projects in airport projects.

1 INTRODUCTION

Guangzhou Baiyun Airport is one of the three major international hub airports in China. The expansion of airport is an important measure to enhance the competitiveness of Guangzhou international aviation hub. Two tunnels in the airport expansion project under cross three runways and several taxiways in server. The airport is currently very busy. Therefore, on the premise of not affecting the normal operation of the runway or controlling the impact within an acceptable range, selecting a reasonable construction method to ensure the operation safety of the airport is the key and difficult point of the project.

Most of tunnels are located in the airfield area, with a length of about 2100m and 1400m respectively. The project has the following characteristics: (1) the tunnel is used by ground service vehicles in the airport. Both ends are connected with ground roads. The line is relatively short, shallow buried depth and large section; (2) The runway surface settlement control standard is very strict; (3) The ground dynamic load is large, the maximum taxiing weight of a380f aircraft is about 592t, and the grounding load shall also be considered on the runway; (4) The airport is located in karst area, with complex geological conditions, abundant groundwater and soft stratum and hard stratum. The tunnel passes through the soil layer mainly including miscellaneous fill, silty clay layer, medium sand, coarse sand, silty clay layer, gravel layer, and locally distributed Carboniferous carbonaceous shale, argillaceous siltstone and limestone. It may be distributed with gray rock buds, and the shallow surface of the gently weathered limestone is about 21m deep. Soil caves are unevenly distributed in the stratum and karst caves are distributed in limestone. The buried depth of groundwater is about 1.8m ~ 4.2m. The grade of tunnel surrounding rock is considered as grade VI.

There are few engineering cases that can be referred to for the project and there is no mature experience. The research results provide strong support for project decision-making and lay a foundation for the smooth implementation of subsequent projects. This project will be the first application of shield construction method in similar airport professional projects.

2 CONSTRUCTION SCALE AND STANDARD

There are no corresponding standards and specifications for the project, so it is necessary to study and determine the control standards of main factors first.

 DOI 10.1201/9781003310884-71

(1) Main technical indexes of tunnel

The tunnel is constructed according to the standard of service Lane in the airfield area. According to the traffic demand analysis, there are four two-way lanes in total, and the section layout of one-way lane is 4.5m+3.5m; The live load at the top of the tunnel is based on the maximum aircraft load that may pass on it, A380 (592t) or b747-8 (440t); The seismic fortification intensity is 6 degrees.

(2) Pavement settlement control standard

The key of the project is runway surface settlement control. There is no unified regulation on the pavement settlement control standard. The runway pavement settlement control value can be divided into acceptable, allowable and excessive limit values. It is generally considered that there can be about 30mm differential settlement within 45m (ICAO 2018). Based on the requirements of relevant design specifications and safety management regulations, it is recommended that the pavement settlement control indicators are: the height difference within 3m of the pavement area shall be less than 10mm, the plate joint staggering shall be less than 5mm, and the total settlement shall be less than 30mm. At the same time, the differential settlement rate of runway shall not be greater than 1‰, and the differential settlement rate of taxiway shall not be greater than 1.5‰. In the project implementation stage, the construction settlement control index shall be combined with the current longitudinal and transverse slope and unevenness of the pavement to ensure that the runway and other facilities still meet the airworthiness standard after the pavement deformation caused by the project is superimposed.

(3) Non-suspension construction area and space restrictions

According to the non-suspension construction experience of airport projects in recent years, in order to strengthen safety management and ensure flight safety, during the use of the runway, the strictly restricted areas are within 105m on both sides of the runway centerline and within 300m at the end of the runway, and construction operations in the strictly restricted areas are not allowed. Outside the area, 24-hour all-day construction operation can be carried out, but the clearance requirements for moving objects in the internal transition surface shall be met.

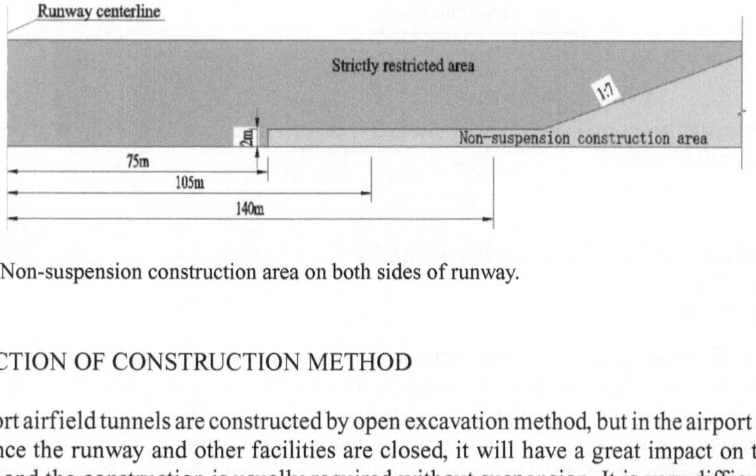

Figure 1. Non-suspension construction area on both sides of runway.

3 SELECTION OF CONSTRUCTION METHOD

Most airport airfield tunnels are constructed by open excavation method, but in the airport expansion project, once the runway and other facilities are closed, it will have a great impact on the airport operation, and the construction is usually required without suspension. It is very difficult to cross the existing runway without suspension. It is necessary to ensure its own construction safety and the operation safety of the airport flight area, so as to avoid a great impact on the normal operation of the airport. From the engineering experience at home and abroad, there are successful cases of open excavation method, shield method, pipe curtain method (Yuan 2004), mining method (Sun 2014) and so on.

The selection of construction method shall meet the requirements of construction operation safety, project enforceability, environmental impact during construction, construction period and cost, and construction management in flight area. Through comprehensive comparison and selection, it is considered that the shield method has little impact on navigation, controllable risk,

Table 1. Cases of airport underground crossing project.

Project name	Project introduction	Construction method
Guangzhou Baiyun Airport, North approach tunnel under the apron	The runway is closed, the tunnel width is 31.9m, and the maximum soil cover thickness is 3m	Open excavation method
Guiyang Airport, Service vehicle tunnel under the runway	The runway is closed, after the new second runway is put into use, close the first runway and turn over the construction	Open excavation method
Beijing Capital International Airport, t2-t3E tunnel under the middle runway	The tunnel is 23.2m wide, and the covering soil at the lower part of the middle runway is about 6m thick, with soil stratum	Pipe curtain method
Taipei Songshan Airport, Fuxing North Road Tunnel under the runway	The undercrossing length under the runway is 100m long, 5m deep and soil stratum	Pipe curtain method
Beijing Capital International Airport, the service vehicle tunnel under the taxiway	The undercrossing length is 142m, the minimum buried depth is 3.5m, and the soil stratum	Pipe roof + Box Culvert Jacking
Heathrow airport, car tunnel under the taxiway	The tunnel length is 1280m, the depth is 16m, the shield diameter is 8.8m, and the stratum is mainly clayey soil.	Shield method
Shanghai Hongqiao Airport, Xianxia west road tunnel under the detour taxiway	The tunnel shield interval is 1040m long, the tunnel depth is about 26m, shield diameter 11.58m, soil stratum.	Shield method
Addison airport, the car tunnel under the runway	The bench method is adopted for construction, with the covering thickness of about 8.5m and soft limestone stratum	Mining method

moderate construction period and investment, and it is recommended to use the shield method to cross the operation runway.

4 KEY POINTS OF SHIELD TUNNEL DESIGN

4.1 *Lining structure design*

According to the road gauge, the inner diameter of the tunnel segment is 10.3m, the fan can be suspended in the space above the building gauge, and the space under the road surface can be used as evacuation channel and cable channel. In combination with the design conditions of underground roads and stratum conditions, the upper soft and lower hard stratum shall be avoided as far as possible. The covering soil of the tunnel top shall be controlled at 6.0m ~ 8.5m, which belongs to ultra-shallow tunnel. After calculation and analysis, and considering the durability requirements of the structure, it is determined that the thickness of the segment is 500mm, the width of the lining ring should be 2.0m, staggered assembly, and the segment joint adopts the form of inclined bolt.

4.2 *Selection of shield machine*

The selection of shield machine shall comprehensively consider the geology, construction period requirements, cost and other factors of interval tunnel. For tunnel shield construction in Guangzhou, the following basic principles shall be followed (Zhu 2021):

Table 2. Comparison and selection of construction methods.

Construction method	Open excavation method	Pipe curtain method	Shield method
Formation adaptability	Various geological conditions	Mostly used in soil stratum	Various geological conditions
Construction period	Short	Long	Relatively long
Impact on airport operation	Large, runway alternately closed	Small, does not affect Runway Operation	Small, does not affect Runway Operation
Main risks	Water gushing and collapse of deep foundation pit	Excessive ground deformation, pipe curtain leakage, support instability, pipe curtain jacking construction obstruction, etc	Water leakage and slurry leakage of shield in and out of the tunnel, ground settlement or uplift, and difficult construction in Composite stratum
Settlement control	Foundation treatment can be carried out, and the post construction settlement can be controlled	The settlement during and after construction is controllable	Settlement control is relatively easy
Treatment of karst cave	Ground treatment	Combination of tunnel and ground treatment	Combination of tunnel and ground treatment
Tunnel operating conditions	The buried depth is shallow, and the ramps on both sides are short	The buried depth is deep and the ramps on both sides are long	The depth is relatively large, and the ramps on both sides are long
Project cost	Low	High	Moderate
Overall evaluation	Advantages: low investment, simple construction and low risk Disadvantages: the impact on airport operation is relatively large, and the runway needs to be closed	Advantages: it does not affect the runway operation and the risk is controllable Disadvantages: high investment, long construction period and impact on taxiway operation	Advantages: relatively little impact on airport operation; Mature technology and moderate investment Disadvantages: the buried depth is relatively large

(1) The selected shield machine shall meet the construction needs of the stratum crossed by the tunnel.
(2) The shield machine is required to have sufficient functions for controlling surface settlement and pavement area settlement and have good operation performance.
(3) Considering the difficulties in the investigation of the airport runway area and the possibility that it is impossible to find out the geological conditions, the shield machine shall have the function of advanced geological prediction.
(4) Considering that there is no ground reinforcement condition for the underpass and there may be weak stratum at the base, the shield machine shall have the function of advance grouting in the shield body. It shall also have the function of opening warehouse and changing tools under normal pressure without ground reinforcement.

Based on the geological conditions of the project and the strict control requirements for the settlement of the airport pavement area, the slurry balance shield machine is recommended in view of the good balance stability of the excavation surface of the slurry balance shield and the performance of controlling land settlement. The shield mainly crosses the clay layer and sand layer, but in the limestone area, the rock surface fluctuates greatly. Therefore, the cutting tool is mainly divided into cutting tool and scraper, and a certain number of hobs shall be equipped.

4.3 *Deformation control measures*

The influencing factors of ground deformation caused by shield tunnel construction include engineering geological conditions, shield diameter, buried depth and construction control. During shield construction, measures to reduce surface settlement include:

(1) Tunnel alignment optimization. The horizontal alignment shall be straight as far as possible, and the large slope uphill construction shall not be adopted for the vertical section of the line as far as possible.
(2) Lining form optimization. Waterproof Optimization: combine the elastic sealing gasket with water swelling materials, and make full use of the dual water sealing effect of elastic compaction water stop of sealing gasket and expansion water stop of swelling materials; One or two grouting holes shall be added to the segment under the runway surface to ensure the effectiveness of the grouting holes.
(3) Control the driving speed. In order to meet the safety of runway under aircraft dynamic load and ensure the stability of aircraft landing time area, the shield tunneling rate shall be controlled within 5mm / min.
(4) Keep the excavation surface stable. Control the excavation parameters such as soil bin pressure and excavated volume.
(5) Shield propelling attitude control should be corrected slowly.
(6) Information construction.

4.4 *Karst Treatment Measures*

Most of the project is located in the current airfield area, and the geological conditions are complex. It is difficult to find out the distribution of soil caves and karst caves in the survey stage. Therefore, it is necessary to configure advanced detection equipment in the shield machine (Qian 2017).

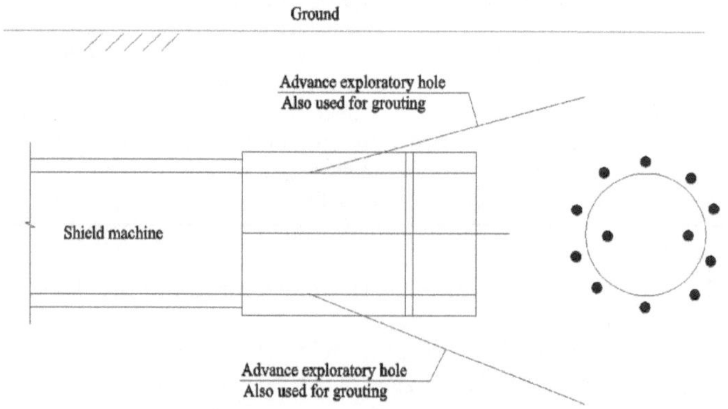

Figure 2. Schematic diagram of karst cave treatment in shield machine.

When the karst cave to be treated is found through advanced detection, the ground treatment method shall be preferred. When there are no ground treatment conditions, it can be treated in the tunnel. For Karst Treatment in shield tunnel, late grouting holes can be reserved on the tunnel segment, and the grouting hole diameter matches the segment hoisting hole. The cavity is mainly filled with water injection mud.

5 CONCLUSION

There are few engineering cases that can be referred to for the project and there is no mature experience. Through comparison and selection from the aspects of construction period, impact on navigation, risk, settlement control and investment, it is considered that the shield method has little impact on navigation, controllable risk, moderate construction period and investment, and it is recommended to use the shield method to cross the existing runway. This paper mainly introduces the construction method, the demonstration process of construction scheme and the key points of shield design technology. It is suggested to further study the key technologies such as aircraft load action, structural calculation and analysis, operational runway monitoring and rapid pavement repair. This project will be the first application of shield method in civil aviation professional engineering, which is of great significance.

REFERENCES

ICAO, annex 14 to the Convention on International Civil Aviation-airports, Volume I, airport design and operation (Eighth Edition), July 2018

Qian Qihu, Main Developments and Directions of Geological Prediction and Informatized Technology of Tunnel Construction [J], *Tunnel Construction*, (2017) 03-0251-13

Sun Jun, et al., *Research and application of key technologies in tunnel structure design* [M], Beijing: People's Communications Press Co., Ltd., December 2014

Yuan Jinrong, Chen Hong, A new method of using small diameter pipe jacking machine to build large section underground space-pipe curtain construction method [J], *Underground engineering and tunnel*, 2004 (1): 23-26

Zhu Weibin, Zhong Changping, et al., Challenges and Prospects of Construction Technology for Extra-large Diameter Composite Shields [J], *Modern Tunnelling Technology*. Vol.58, No.3 (Total No.398), Jun.2021

Advances in Civil Engineering: Structural Seismic Resistance, Monitoring and Detection – Mohd Johari Mohd Yusof and Junwen Zhang (Eds)
© 2023 The Authors, ISBN 978-1-032-31491-4

Analysis of influencing factors of Rayleigh surface wave exploration depth

Lipeng Yang*
Monitoring Company, Shenyang Metallurgical Inspection Engineering Co., Ltd, Liaoning, China

ABSTRACT: With developments in society and modernization, the scope of application of geophysics is expanding. Nowadays, the masses of society have gradually realized the importance of harmonious coexistence with the natural world, and at the same time, they are paying more and more attention to the shallow surface of the earth. As a result, surface engineering constructions are increasing; thus, shallow geophysical exploration methods have become one of the important development directions for applied geophysics in China. This article discusses the development of Rayleigh waves and the factors affecting exploration depth.

1 INTRODUCTION

Just like human civilization, the development of various disciplines will have advantages as well as disadvantages. After the test of time and the unremitting efforts of scientific researchers, their beautiful appearance can be displayed in front of everyone like today's modernization. The decades of surface wave development have also experienced a process that no one cares about, slow development, and now full of vitality. Although a part of the theory of surface waves can be applied to our daily production practice, there still are many characteristics that have not been discovered (Medina 2020). As an emerging shallow surface exploration method, the Rayleigh surface wave is often compared with conventional shallow seismic exploration methods. It has the advantages of non-destructive detection, high efficiency, wide application range, high detection accuracy, portable instruments, and small site restrictions. Due to the abovementioned advantages, it has been widely used in shallow surface engineering exploration, geological disasters, and other fields. Till now, the research on Rayleigh waves at home and abroad has made great progress (LEI 2019).

2 RESEARCH OVERVIEW

In the early research on elastic waves, concepts of body waves and surface waves were present. As early as 1887, the British scholar Lord Rayleigh had theoretically discovered and confirmed the existence of Rayleigh waves. At the same time, there are also Love surface waves and Stoney surface waves in the elastic body (Wang 2021). However, since the surface wave is the most destructive seismic wave in natural earthquakes, it causes a strong interference in conventional shallow seismic exploration. These characteristics made everyone's impression of it at that time. The surface wave is useless and has a strong interference stage. Therefore, before the early realization of more characteristics and applicability of surface waves, the research on surface waves was mainly carried out from the aspect of reducing or eliminating their influence. In the early 1950s, the scholar Press (Press, 1956) successfully obtained the phase velocity and dispersion curve of surface waves in

*Corresponding Author: ylp8246@163.com

 DOI 10.1201/9781003310884-72

the time domain. Since then, he has discovered the dispersion characteristics of Rayleigh surface waves, one of the most important characteristics of surface waves. Gradually, he began to use the dispersion characteristics of surface waves to study geological bodies and the internal structure of the earth (Zhang 2020).

In 1960, the Waterway Laboratory of the Mississippi Army Engineering Corps of the United States officially started the research on the engineering investigation method of surface waves, but it was not successful due to insufficient technical conditions at the time. Later, with the rise and development of computer technology, every subject field in geophysics was developing rapidly, which also included the technology of surface wave exploration.

In 1973, the study by FK Chang and others in the United States truly used transient Rayleigh waves to solve many problems in engineering geology, which attracted the attention of many geophysicists. The theory began with a more in-depth study. In 1982, Japan VC Co., Ltd., after years of research, launched the GR-810 Sato-type fully automatic survey machine, which mainly stimulates steady-state surface waves to conduct velocity exploration of various layers. This is also the first time that scholars have conducted impact surface surveys. Before studying the influence factors of wave exploration depth, we will analyze various aspects that affect its exploration depth before we can start our subsequent research and application.

This chapter will give a brief talk on theory, observation system establishment, data processing, and even geological situation. It will provide guidance and pavement for future research directions and the appearance of instruments that use steady-state excitation to generate Rayleigh waves. In 1983, Nazarian et al. proposed the SASW surface wave spectrum analysis for phase velocity calculation. This method uses the phase difference between two different records to establish a dispersion curve, and then calculates the shear wave velocity of the underground medium by inversion. In 1999, the University of Kansas proposed the MASW surface wave multi-channel analysis method based on predecessors, which significantly improved the accuracy and depth of Rayleigh wave exploration. Moreover, the beginning of Chinese research on shallow Rayleigh wave engineering exploration was also in the 1980s.

In 1987, to solve the problem of railway and highway roadbed exploration, China introduced the GR-810 steady-state surface wave survey system from Japan. In 1988, Wu Shiming and other experts from China began to use Rayleigh surface wave exploration to test the wave speed of various underground layers and published "Waves in Soil Medium" in 1996. The book mainly talks about Rayleigh surface waves, investigation methods, and technical issues. In 1989, Yang Chenglin and others used the TRE-RALOG shallow seismic instrument and cooperated with domestic controllable sources to form the first domestic steady-state Rayleigh wave exploration system, and then divided the Quaternary strata on this basis and the effect of foundation backfilling, and published "Reyleigh Wave Exploration Principles" in 1993, which introduced in depth the technology and principles of Rayleigh wave exploration methods.

The national industry standard has also been included in the geophysical prospecting regulations, and it has become a relatively hot technology in shallow exploration projects. Later, the development of surface wave exploration is mainly carried out in the aspects of forward and inverse algorithms, dispersion curve extraction, field data acquisition technology, and high-order surface wave characteristics. The main developments are as follows.

In 2007, in terms of the evolution of forward calculation, Xu Y, Miller RD and others proposed the use of dense spatial sampling, a smaller time step, and a second-order staggered grid, which is similar to the processing method of sponge boundary conditions. In combination, the forward simulation technology of Rayleigh surface wave finite difference was successfully realized. In his doctoral dissertation, Xiong Zhangqiang also achieved an algorithm for finite-difference forward modeling of Rayleigh waves in elastic media. These improved algorithms can effectively reduce or eliminate boundary reflections and numerical dispersion phenomena that were inevitable in previous forward simulations. This allows us to understand the propagation process of Rayleigh waves more clearly. This is also the case for Rayleigh waves. The research on surface wave inversion has laid a certain foundation for forward modeling.

In 2003, Song Xianhai et al. from China applied the improved $\tau-p$ transformation algorithm to extract the dispersion curve of transient Rayleigh waves, and the results were very effective and discussed the effectiveness of Rayleigh waves in low-velocity interlayers In the same year, Liu Jiangping used the $\tau-p$ transformation to separate the wave fields, and proposed a method for calculating the Rayleigh wave velocity of adjacent channels, which improved the resolution in the vertical and horizontal directions of the Rayleigh wave.

In 2008, Song Xianhai successfully applied the generalized pattern recognition algorithm to the nonlinear inversion of the Rayleigh wave dispersion curve for the first time, which greatly improved the calculation accuracy and time required for the Rayleigh wave inversion. Regarding the progress in the factors affecting the Rayleigh surface wave exploration, in 1997, Xia Yujing carried out theoretical research. Therefore, the research on the depth of surface wave exploration has not reached a certain depth so far. The effect of influencing factors has not been observed from the perspective of numerical simulation, and the actual influence in the field has not been observed from the natural experimental site of loess, or even the effect has not been observed. The amplitude spectrum of seismic records at different exploration depths is discussed to study the exploration depth.

3 BASIC THEORIES AFFECTING THE EXPLORATION DEPTH OF RAYLEIGH WAVES

3.1 *Depth of detection formula*

In the previous article on the attenuation characteristics of Rayleigh waves, we learned that in terms of energy, Rayleigh waves are only distributed near the free interface, and the energy of Rayleigh waves within a frequency wavelength can account for more than 90%, which also determines that with the current technical level, the Rayleigh wave detection depth can only be expressed as a distance within one wavelength, but as long as the artificial excitation frequency is large and the wavelength is large, the detection depth can still be considerable. Similarly, the amplitude energy of Rayleigh waves decays exponentially with the depth of propagation. Usually, we define the depth at which the amplitude ratio decays to 1/e as the penetration depth, but it can only be used as a reference depth coefficient, where it is the amplitude of the transverse vibration and the amplitude of the longitudinal vibration. According to different rock and soil media, we can calculate the relationship between penetration depth (expressed in multiples of wavelength) and amplitude energy, as shown in Table 1. In this way, we can determine the proportional relationship between depth and wavelength.

Table 1. The relationship between Poisson's ratio and penetration depth.

Poisson's ratio σ	0.10	0.20	0.25	0.30	0.35	0.40	0.45	0.48
u_0	0.665	0.722	0.735	0.785	0.819	0.857	0.898	0.926
u_0/e	0.245	0.266	0.277	0.289	0.301	0.315	0.330	0.341
Penetration depth H	0.55λ	0.62λ	0.65λ	0.70λ	0.75λ	0.79λ	0.84λ	0.87λ

3.2 *Average speed*

Due to the dispersion characteristics of Rayleigh waves, for a fixed frequency, its corresponding wave speed is constant, that is, different depths underground correspond to a stable average speed value. Therefore, the average speed of Rayleigh wave propagation is determined by the underground stratum and their combination. After we determine the geological body and its average velocity, the maximum exploration depth of Rayleigh waves can be calculated.

Generally, the greater the formation velocity, the greater the exploration depth. For example, assuming that the lowest frequency of the surface wave used is 5 Hz, in a software with a β value of 0.5 and a shear wave velocity ranging from 100 to 160 m/s, the exploration depth H = 10 to 16 m; and in a hard rock formation with a β value of 0.8, the average shear wave velocity can reach more than 1500, and its exploration depth can reach more than 240 m. The exploration depth of these two cases is very different, it can be seen that the average speed has a great influence on the exploration depth.

3.3 *Source frequency*

When the average speed remains unchanged, that is, the formation conditions remain unchanged, and the exploration depth is inversely proportional to the frequency of the Rayleigh wave used. In the process of Rayleigh surface wave exploration, the excitation mode of the transient surface wave source directly affects the frequency range, and source excitation has become one of the important links that restrict the depth of investigation. In other words, if the required exploration depth is small, then the main frequency of the seismic source should be higher; if the required exploration depth is larger, then the main frequency of the seismic source should be lower. We determine the best excitation method according to the exploration depth and the need to distinguish thin layers. In the surface wave engineering survey work, the commonly used seismic sources include hammering seismic source, falling weight seismic source, and explosive seismic source.

3.4 *Source energy*

For Rayleigh surface wave exploration, the most ideal pulse source is broadband, and the source energy is evenly distributed in each frequency. To achieve the ideal state of the seismic source, the magnitude of the seismic source energy is usually changed. However, the source energy should not be too small, because it determines the strength of the Rayleigh wave signal received by the geophone. We should ensure that a seismic source of sufficient intensity could be excited, so that the low-frequency wave can produce enough vibration in the medium within a wavelength depth range after the energy is attenuated, otherwise, the geophone will not receive the deep vibration information, which will affect the exploration depth.

4 IMPACT OF THE ESTABLISHMENT OF THE OBSERVING SYSTEM

4.1 *Offset*

The choice of offset is a factor that has a significant impact on the depth of surface wave exploration in the observation system. The reason is that during the excitation and propagation of the Rayleigh wave, its energy will have a process from generation to attenuation to disappearance. If we receive the signal in the area where the Rayleigh wave energy is most developed, the signal-to-noise ratio will increase. There is also a greater depth of exploration. On the contrary, if the offset is too close or too far, the exploration depth will be reduced. If the offset is set too close, it will be difficult to avoid noise interference containing sound waves, direct waves, reflected waves, refracted waves, high-order surface waves, etc. Only by increasing the offset can they be distinguished and processed. However, if the offset is set too far, taking into account the energy attenuation, there will be a possibility that the detector will not receive low-frequency energy. At the same time, due to the effect of the ground filter, the high-frequency components in the surface wave record will be filtered out, which will also affect the resolution of shallow formation exploration.

4.2 *Track spacing*

In the frequency wavenumber spectrum, the track spacing determines its maximum wavenumber limit, because the smallest track spacing at the distance D can only track the surface wave with the wavelength D. Therefore, in the frequency wavenumber domain, interference with a wavenumber greater than 1D can be directly cut off so that the signal-to-noise ratio of the data can be improved. In addition, it should be noted that when the exploration depth is large, the track spacing used must be correspondingly increased. However, because the surface wave exploration uses a two-dimensional Fourier transform, if the track spacing is selected too large, the processing result will be affected by the data. Insufficiency leads to the appearance of spatial aliasing, which affects the pick-up of the final dispersion curve, and reduces the resolution of the survey. If the track spacing is selected too small, the received signal will cause calculation errors due to insufficient phase difference. These all issues should be paid attention to obtain the best prospecting results.

4.3 *Arrangement length*

We know that an observation system with an arrangement length of L can only track a surface wave wavelength of 2L at most, and the reciprocal of this maximum wavelength 2/1L determines the minimum wavenumber limit in the frequency wavenumber spectrum. We can directly cut off the wavenumber less than 2/1L interference signal to improve the data signal-to-noise ratio. At the same time, the longer the acquisition arrangement of the observation system, the longer the wavelength of the surface wave that can be accurately reflected in the data, and the deeper formation aids in better distinguishing the information.

4.4 *Detector*

Choosing the right detector is a link that is easily overlooked in actual work, but it is very important to choose the right frequency detector. Because a general geophone has a natural frequency, it responds better to seismic sources whose main frequency is in its natural frequency range. In surface wave survey technology, the frequencies of the geophone and the seismic source are matched. When we are exploring the field, we must adopt measures to local conditions and choose the appropriate geophone according to the depth of the exploration target. Usually, if we need to detect deeper formations, the natural frequency of the geophone should be smaller.

5 INFLUENCE OF SURVEY AREA PROFILE

The influence of the geological profile of the survey area is mainly related to the basic theory of Rayleigh surface wave exploration depth. In actual work, the average wave velocity of the formation, the energy attenuation rate, and even the coupling degree of the seismic source and the surface will change with the change of the work area, so the exploration depth under different geological profiles is different.

5.1 *Overlay*

The thickness and softness of the overburden have a great influence on the exploration depth. The geological conditions of the overburden often determine the maximum exploration depth in this area, such as the loess overburden. Loess is a silty soil-like deposit formed through hundreds of thousands of years of weathering, erosion, and accumulation. The thickness is large, distributed from tens to hundreds of meters. Soft and loose loess not only has low Rayleigh wave speed but also has a strong absorption and attenuation effect on seismic waves. Therefore, the exploration depth of loess is generally within 10~20m, and it can reach 40~50m on some dense soils, but these are all up to less than the buried depth of most loess.

5.2 *Bedrock*

The propagation velocity of the Rayleigh wave in the bedrock is relatively high, so its exploration depth is relatively large. However, there are many types of bedrock, from sedimentary rock, and metamorphic rock to magmatic rock to their finer classification. There are obvious differences in hardness and compactness, which makes the propagation speed of Rayleigh waves have a great deal. The difference from the lowest 350-400 m/s to 2000-3000 m/s, their exploration depths range from 50-200 m or more. In addition to the hardness and compactness of the bedrock that can affect the depth of surface wave exploration, the integrity of the bedrock will also affect it. For example, the development and water content of fissures, fault fracture zone and artificial blasting, etc. All these conditions that lead to incomplete bedrock and water content will reduce the wave speed of the rock formation to a certain extent, resulting in a shallower final exploration depth.

5.3 *Energy attenuation and surface coupling*

Generally, for geological bodies with fast energy decay, the coupling between the excitation source and the surface is not good. Earlier, we introduced that geophones with too fast energy decay will not receive deep vibration information. We should also pay enough attention to this soft geological body. The requirements for good coupling are surface contact, flatness, uniformity, and firmness. Poor coupling conditions can cause many problems. First, the surface is irregular, and the miscellaneous soil and weeds will absorb a large amount of seismic source energy and reduce the exploration depth; second, the uneven surface, the more contact points between the seismic source and the ground surface, will generate many secondary waves, which will affect the accuracy of the signal; third, the coupling surface is soft, and the seismic source shifts and tilts during the excitation process, which will greatly reduce the exploration depth and accuracy. In practice, auxiliary devices can be used to improve coupling conditions.

6 IMPACT OF DATA PROCESSING METHODS

The data processing flow of the Rayleigh surface wave is the same as that of conventional shallow seismic exploration. It is necessary to remove noise interference, and then perform dispersion energy imaging of surface wave data, dispersion curve extraction, dispersion curve inversion, and finally inversion of the obtained shear wave velocity. The picture shows the results of Rayleigh surface wave exploration. Since the shear wave velocity map can provide shear wave velocity information in different strata, we can use this feature to solve shallow geological problems such as strata division, foundation reinforcement effect evaluation, and non-destructive testing of highways and airport runways. When removing noise interference, if the resolution of noise is not accurate enough, it will have a great impact on our data processing results. In the process of dispersive energy imaging, the different imaging methods used will also produce considerable changes in the extraction of the dispersion curve. At the same time, the inversion of the dispersion curve is also a very important step. Different methods are used, and the choice of penetration coefficient will affect the accuracy of the results. In this regard, we will choose its focus in the following chapters for research.

7 CONCLUSION

This article first discusses the development of Rayleigh waves and concludes that the most basic factors affecting the exploration depth are the shear wave velocity of the geological body, the source frequency, and energy. In terms of the establishment of the observation system, the offset, arrangement length, track spacing, and even the natural frequency of the geophone will have an impact on the exploration results; the influence of the survey area profile is mainly the embodiment of the theoretical formula, and the survey area has a great influence on the source coupling. Finally, a brief discussion on the data processing is also made.

REFERENCES

I. Medina, J. J. Hernández-Gómez, C. R. Torres-San Miguel et al. Prototype of a Computer Vision-Based CubeSat Detection System for Laser Communications[J]. *International Journal of Aeronautical and Space Sciences*, 2020 (prepublish)

Jiliang Zhang, Honglin Ran, Xiaojun Pan et al. Outage analysis of wireless-powered relaying FSO-RF systems with nonlinear energy harvesting[J]. *Optics Communications*, 2020, 477

Wang Yi, Wang Yaping Research on the performance of SIMO RC-OFDM system with exponential Weibull distribution in the downlink satellite–terrestrial laser communication[J]. *Optics Communications*, 2021, 491

Yao LEI, Xiaoming LI, Lizhong ZHANG Experimental Study on PAT System for Long-Distance Laser Communications Between Fixed-Wing Aircrafts[J]. *Photonic Sensors*, 2019, 9(2)

Advances in Civil Engineering: Structural Seismic Resistance, Monitoring and Detection – Mohd Johari Mohd Yusof and Junwen Zhang (Eds)
© 2023 The Authors, ISBN 978-1-032-31491-4

Research on designing scheme for the structure conversion of the reinforced concrete shaft tower

Feng Song
Rail Transportation Engineering Co., Ltd. of China Railway 19th Bureau Group Co., Ltd, Beijing, China

ABSTRACT: Based on the characteristics that the shaft tower frame column pile foundation construction is difficult and affects the construction period by using the freezing method, this paper puts forward a reasonable load-bearing scheme of transferring the inner column on one side of the shaft. In order to facilitate the construction of vertical shaft, two columns must be removed from the bottom layer to form a well tower with bottom supporting column conversion. Using PKPM / SATWE and MIDAS / Gen software, this paper establishes a variety of models by using the method of honeycomb gradually hollowing out solid web beam, and analyzes the effects of three conversion schemes of solid web beam, open web truss and truss on the stress and deformation of the structure. The results show that the well tower structure is more suitable for the truss conversion scheme; The method of gradually hollowing out the beam by using shell element is proposed, which provides a reasonable and effective idea and method for the analysis and design of transfer members in similar projects.

1 INTRODUCTION

With the development of diversified building functions, a high-rise building often needs a larger space in the lower part for halls, entertainment venues, etc., while the upper part may be used as offices, hotel rooms, etc. In this way, it will lead to uneven distribution of the upper and lower components of the structure. In the design, by setting up conversion floors on the floors where the upper and lower vertical structures are not continuous, the function of the upper vertical structure can be transferred to the lower vertical structure (Zhang 2011), thereby meeting people's requirements for building functions.

There are many domestic studies on the conversion of high-rise structures. Han xiao lei and others have conducted in-depth research on many problems of high-rise building beam conversion (Han 2004). Ma xue qiang and others used ABAQUS finite element software to carry out elastoplastic analysis of the complex nodes in the large-span truss transition layer (Ma 2017). Zhao jian min and Zhang jian ping analyzed the elastic buckling of the conversion structure, analyzed the coordinated work of the upper chord and the floor, and summarized the structural design points and design measures of the conversion truss (Zhao 2016).

The design concept of the transfer floor structure can also be extended and applied to the tower structure. In the construction of a mine shaft, the shaft tower is used as a platform for subsequent shaft excavation, usually with a built-in structure of 4 or 6 columns. When the freezing method is used for construction, the shaft tower has two frame columns located in the freezing circle. The construction is difficult and the period is long. In order to speed up the construction progress, the lower frame column can be adjusted to form the form of supporting column conversion. The beam conversion method is divided into three options: solid web type, open web truss type and ordinary truss type.

The beam conversion scheme is suitable for structures with small spans, and the truss conversion is suitable for structures with large spans and upper loads. Although the conversion of open-web

DOI 10.1201/9781003310884-73

truss conversion has a large span, the upper loads that can be borne are smaller. For the shaft tower structure, it is more suitable for use due to its characteristics such as high concentrated load, large live load on the floor, large opening on the horizontal floor, large first story, large vertical story height change, and large conversion span. For the truss conversion scheme, there are few researches on the force analysis of the truss conversion scheme of the shaft tower structure.

Based on specific engineering cases, this paper uses PKPM/SATWE and MIDAS/GEN software to analyze the three conversion schemes of solid web, open web truss and truss, and comprehensively considers that the truss conversion scheme is the most reasonable, and proposes that it can be used The shell element analyzes the method of gradual hollowing out of the beam, and at the same time puts forward relevant design suggestions for the design ideas and schemes of the conversion member.

2 ENGINEERING BACKGROUND

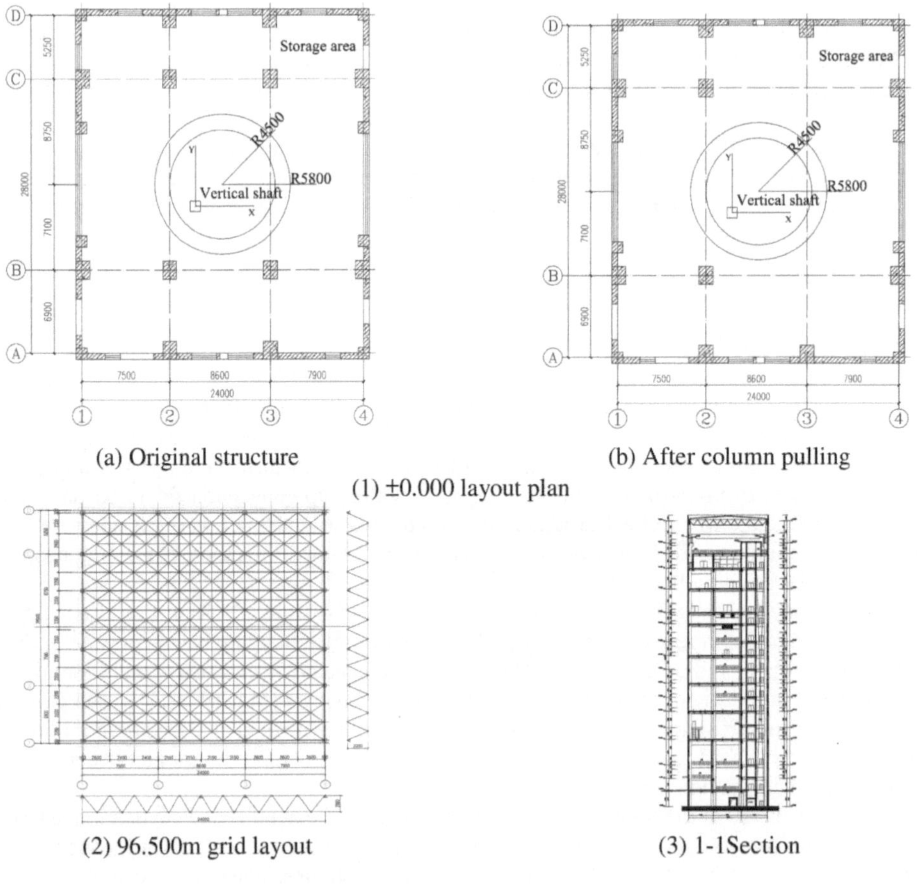

Figure 1. Main plane and section of shaft tower.

A certain well tower is a reinforced concrete frame shear wall structure with a structure length of 28.0m, a width of 24.0m, and a height of 96.5m. The plan and section diagrams of the tower are shown in Figure 1. The basic wind pressure in this area is w0=0.60kN/m^2, and the basic snow pressure is s0=0.40kN/m^2. The safety level is level 1, and the design service life is 95.5 years. The seismic fortification category is Class B (GB 50223-2008), the seismic fortification level is Class

II, the 3 to 10 floors after the columns are drawn are classified as Class II, and the bottom 1 to 2 floors are Class I (JGJ3-2010), the structural fortification intensity is 7 degrees, and the earthquake is divided into groups It is the first group, the site category is Category II, and the basic seismic acceleration value is 0.10g (GB 50011-2010).

3 MODEL PARAMETER

In order to facilitate the calculation and reflect the force characteristics of the structure, only 10 basic floors are reserved, and the heater room below the ground, the skirt room around the main structure of the shaft and the small mezzanine in the middle are not considered. The concrete grade is C40, the longitudinally stressed steel bar is HRB400, and the stirrup is HPB300. Because the structure underground is a rock-socketed piled raft, it can be considered that the upper structure and the underground foundation are rigidly connected; and when analyzing the well tower, only the working conditions of the two hoists working at the same time are considered, and only the calculation of the horizontal earthquake Loads, other loads are not calculated.

4 CONVERSION SCHEME ANALYSIS

4.1 *Proposed conversion scheme*

Extract one layer of ③/B column and ③/C column, and increase the beam span to 16.5m. Because the self-weight and the increase of the beam cross-section lead to a large internal force, the honeycomb type is adopted to gradually remove the concrete inside the beam for analysis. That is to say, the solid web beam conversion scheme (option 1) is adopted first, then the open-web truss conversion scheme (option 2) is adopted, and finally the truss conversion scheme (option 3) is adopted.

4.2 *Internal force calculation of conversion scheme*

4.2.1 *Beam solid abdomen*

Option 1: Gradually increase the height of the beam to study the changes in the internal force of the structure with different beam cross-sectional dimensions. The beam width remains unchanged at 900mm, and four sizes of beams with beam heights of 2500mm, 3000mm, 3500mm, and 4000mm are used. List the internal forces of the four sizes of beams under dead load and live load.

It can be seen from Tables 1 and 2 that the mid-span bending moment calculated by PKPM is about 10% larger than that of MIDAS. Under the action of dead load or live load, the negative bending moment of the transfer beam support decreases with the increase of the beam height, while the positive bending moment in the mid-span of the beam section gradually increases with the increase of the beam height, and the increase is greater When the beam height is 4000mm, the section bending moment is more than twice as large as the section bending moment of the beam height 2500mm. Therefore, when the beam height is increased, the internal force of the conversion beam will also increase, that is, only increase the size of the beam and design the conversion beam The plan is not feasible, and it must be hollowed out and optimized.

4.2.2 *Optimization scheme of beam hollowing*

When the beam height meets the requirements, the scheme of hollowing out the concrete in the beam shall be selected. In the analysis of elastic floor slab, considering the section effect of beam and column, the plate element is adopted to simulate the well tower transfer beam. The plate element and the column corresponding to the plate are divided along the height direction. The beam unit is divided every 0.1M in the height direction and 0.2m in the width direction, as shown in Figure 2. At the same time, the floor slab is also divided according to the beam division method, with length and width of 0.2m × 0.2m, as shown in Figure 3.

Table 1. bending comparison of structure ②-④/B axis transfer beam after column pulling.

Dimension (b×h)/mm×mm	Working condition/ internal force		Bending moment/(kN·m)			
			Transfer beam left	Secondary beam left	Secondary beam right	Transfer beam right
900×2500	Dead load	MIDAS	−6525.4	10519.9	10996.7	−5355.7
		PKPM	−6544.4	12450.5	12765.6	−5379.2
		Amplification	0.3%	18.4%	16.1%	0.4%
	Full live load	MIDAS	−2226.5	5691.1	6655.9	−3739.2
		PKPM	−2058.3	6788.8	7483.7	−3698.0
		Amplification	−7.6%	19.3%	12.4%	−1.1%
900×3000	Dead load	MIDAS	−6751.9	14493.7	15007.9	−5234.3
		PKPM	−6327.4	16691.8	17029.2	−5256.6
		Amplification	−6.3%	15.2%	13.5%	0.4%
	Full live load	MIDAS	−1955.5	7936.1	8914.5	−3755.6
		PKPM	−1568.0	9163.0	9859.3	−3690.7
		Amplification	−19.8%	15.5%	10.6%	−1.7%
900×3500	Dead load	MIDAS	−6220.6	18609.5	19153.0	−4856.9
		PKPM	−5484.9	20947.3	21297.0	−4968.2
		Amplification	−11.8%	12.6%	11.2%	2.3%
	Full live load	MIDAS	−1326.2	10257.3	11239.5	−3592.8
		PKPM	−806.1	11527.8	12220.8	−3560.9
		Amplification	−39.2%	12.4%	8.7%	−0.9%
900×4000	Dead load	MIDAS	−5214.0	22681.0	23239.7	−4384.2
		PKPM	−4321.8	25018.3	25371.2	−4632.1
		Amplification	−17.1%	10.3%	9.2%	5.7%
	Full live load	MIDAS	−500.5	12533.4	13512.4	−3354.0
		PKPM	63.2	13759.7	14446.3	−3386.4
		Amplification	−112.6%	9.8%	6.9%	1.0%

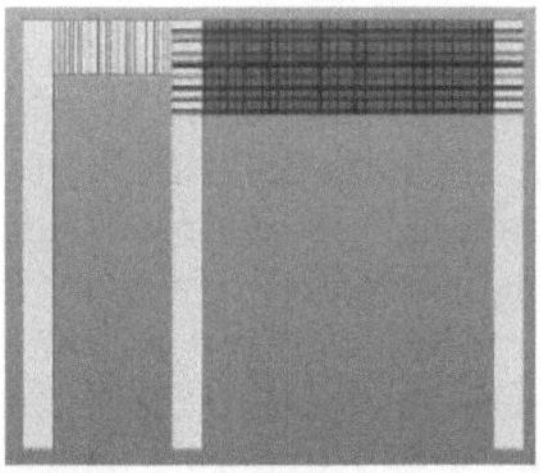

Figure 2. unit division of axis ②-④/B solid transfer beam.

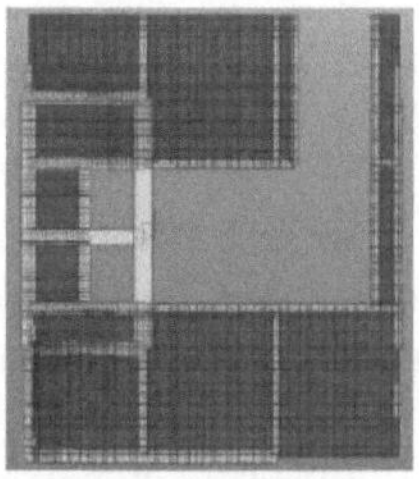

Figure 3. division of 18.200m floor unit.

The results show that the internal forces of shell element and beam element are basically the same. Then, considering the hollowing mode and multiple calculations, two feasible schemes are obtained, namely, Vierendeel Truss conversion scheme and truss conversion scheme, as shown in Figure 4.

4.3 *Influence of pumping column on overall calculation results of well tower structure*

In order to fully consider the influence of pumping column on the stress, deformation and other properties of well tower. The results of solid web transfer, vierendeel truss transfer and truss transfer are compared to study the influence of different methods on the structure.

Table 2. shear force of structure ②-④/B axis transfer beam after column pulling.

Dimension (b×h)/mm×mm	Working condition/ internal force		Shearing force/(kN) Transfer beam left	Secondary beam left	Secondary beam right	Transfer beam right
900×2500	Dead load	MIDAS	2291.3	1672.7	−1788.6	−2351.3
		PKPM	2537.6	1879.9	−1956.5	−2565.6
		Amplification	10.7%	12.4%	9.4%	9.1%
	Full live load	MIDAS	974.6	866.7	−1268.5	−1363.2
		PKPM	1082.9	974.7	−1344.7	−1442.0
		Amplification	11.1%	12.5%	6.0%	5.8%
900×3000	Dead load	MIDAS	2828.0	2112.8	−2236.5	−2888.1
		PKPM	3057.8	2295.6	−2423.8	−3130.4
		Amplification	8.1%	8.7%	8.4%	8.4%
	Full live load	MIDAS	1204.1	1096.3	−1556.5	−1651.1
		PKPM	1301.9	1193.7	−1639.9	−1737.1
		Amplification	8.1%	8.9%	5.4%	5.2%
900×3500	Dead load	MIDAS	3293.2	2481.2	−2669.0	−3409.4
		PKPM	3506.9	2640.2	−2870.9	−3675.0
		Amplification	6.5%	6.4%	7.6%	7.8%
	Full live load	MIDAS	1400.8	1293.0	−1830.2	−1924.8
		PKPM	1488.3	1380.1	−1918.0	−2015.2
		Amplification	6.2%	6.7%	4.8%	4.7%
900×4000	Dead load	MIDAS	3968.0	2789.2	−3082.0	−3911.4
		PKPM	3897.3	2926.1	−3287.9	−4189.6
		Amplification	−1.8%	4.9%	6.7%	7.1%
	Full live load	MIDAS	1569.5	1461.7	−2087.7	−2182.3
		PKPM	1646.8	1538.5	−2173.5	−2270.8
		Amplification	4.9%	5.3%	4.1%	4.1%

(a) Scheme 2: Vierendeel Truss conversion scheme

(b) Scheme 3: truss transfer scheme

Figure 4. optimization scheme for excavation of axis ②-④/B beam.

Table 3 compares the mass of well tower, lateral stiffness and natural vibration period of layer 1 before and after column pumping. It can be seen that after column extraction: the mass under live load remains unchanged, but the mass under dead load changes a little, which is the mass after taking out beam concrete; The stiffness in X and Y directions decreases slightly after column extraction; The three parameters have little change compared with those before column pumping. This is in line with the initial assumption of the hollowed out scheme, so the design method of using the conversion scheme of shell element gradually hollowed out beam is feasible.

Table 3. Comparison table of overall structure information before and after column extraction.

		Quality/(t)		Rigidity/(kN/m)		Natural vibration period/(s)
		Dead load	Live load	X direction	Y direction	T_1
Before column pumping		24093	7129	45751868.9	30306216.1	2.0778
After column extraction	Option 1	24117	7129	44589263.4	30015618.5	2.0758
	Option 2	23964	7129	45261099.7	30097828.5	2.0787
	Option 3	24009	7129	45751868.9	30306216.1	

5 STRESS ANALYSIS OF TRANSFER MEMBERS

When analyzing the internal force of the pumping column of the well tower, the size of the transfer beam can be determined according to the actual situation, the model can be established, and the calculation results of each component can be obtained (Zhao 2020); Then analyze the floors close to the transfer members, and determine the reinforcement according to the internal force.

5.1 *Internal force of Vierendeel Truss conversion scheme and truss conversion scheme under single working condition*

In Midas software, for the shell element, the internal force of each section is calculated by using the resultant force in each direction. The chord and web dimensions of the two truss schemes are taken as 1000×900, the end rod is the original column size, as shown in Figure 5.

Calculate the internal force of each member under various working conditions. It can be seen that the internal forces of members under the same working conditions of the two schemes are very different. For example, the bending moment of members 1 and 2 in scheme 3 is 30% less than that in scheme 2, because oblique web members are added in scheme 3. Therefore, the truss conversion scheme is recommended as the final design scheme.

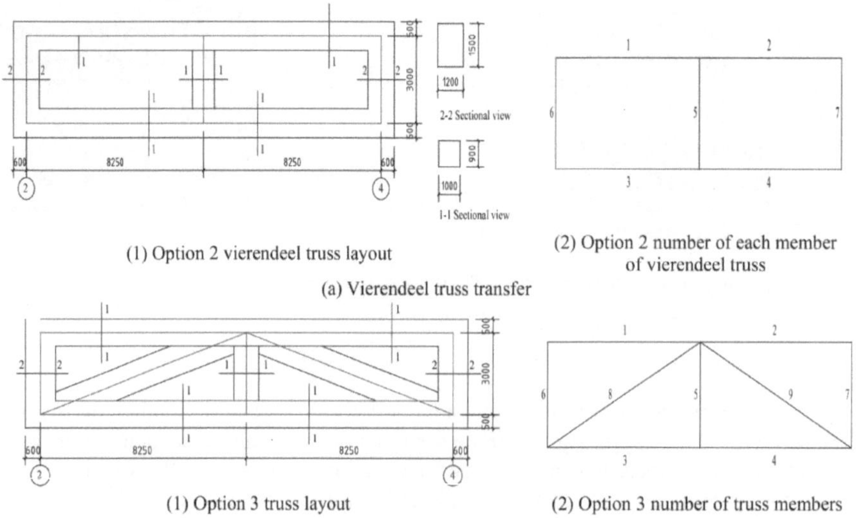

(1) Option 2 vierendeel truss layout

(2) Option 2 number of each member of vierendeel truss

(a) Vierendeel truss transfer

(1) Option 3 truss layout

(2) Option 3 number of truss members

(b) Truss conversion

Figure 5. section size and number of transfer member.

5.2 *Influence of height span ratio on mechanical and deformation performance of truss transfer scheme*

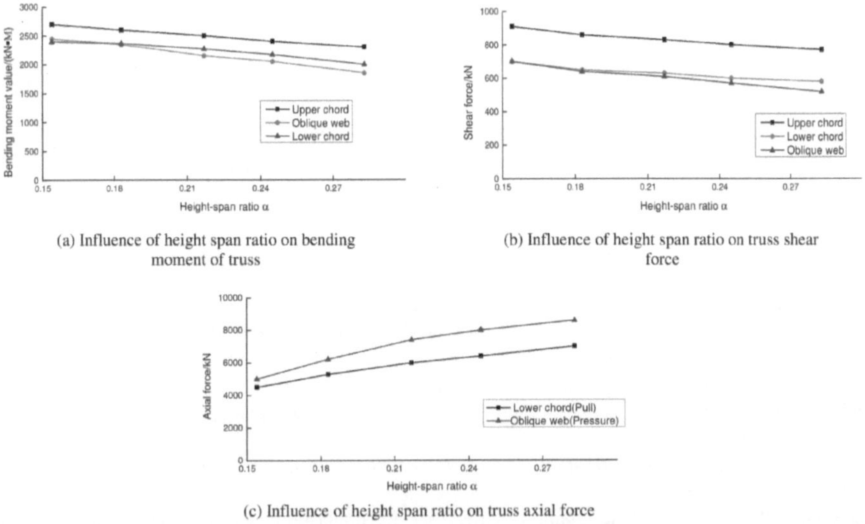

(a) Influence of height span ratio on bending moment of truss

(b) Influence of height span ratio on truss shear force

(c) Influence of height span ratio on truss axial force

Figure 6. Comparison of influence of height span ratio on truss internal force.

Truss height span ratio α refers to the ratio of total truss height H to total truss span L:$\alpha = H/L$. In truss structures, the height span ratio usually affects the stability of beams. In the previous part, the stress of the well tower of the three schemes is analyzed, and the design method of truss conversion scheme is determined. On this basis, the stress and deformation properties of truss members with height span ratio of 0.152, 0.182, 0.212, 0.242 and 0.273 are analyzed respectively.

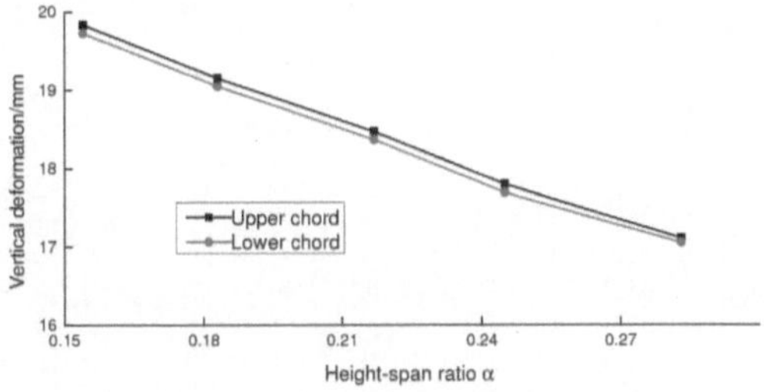

Figure 7. Comparison of influence of height span ratio on truss deformation.

In order to compare and analyze the influence of different height span ratio on the internal force of truss members under vertical load, the influence curves of height span ratio on the internal force and deformation of truss are drawn, as shown in Figures 6 and 7.

It can be seen from Figure 6 that with the increase of height span ratio, the bending moment M and shear force F of truss members are decreasing, but the axial force of diagonal web members gradually increases, indicating that the greater the height span ratio, the greater the adjustment effect of truss on internal force; The mid span vertical displacement of the upper and lower chord of the truss decreases gradually, but both can meet the requirements of the code.

In the actual project, limited by the building function, construction technology and other factors, the height span ratio of the truss can not be increased indefinitely. For the well tower structure, the truss with a height of 4.0m is acceptable. Therefore, the truss conversion design scheme with a height of 4.0m is adopted in this paper.

6 CONCLUSION

According to the results, the following conclusions are drawn:

(1) When analyzing the structural transformation scheme, the in-plane deformation of the floor must be considered, which cannot be calculated as a rigid floor.
(2) The shell element can be used to analyze and calculate the transfer members to reduce the influence of self weight.
(3) By gradually excavating the beam, the solid web, open web truss and truss conversion schemes are analyzed in turn, and the truss conversion scheme is recommended.

REFERENCES

GB 50011-2010. (2010). Code for seismic design of buildings [S]. Beijing: China Construction Industry Press.

GB 50223-2008. (2008). Classification standard for seismic fortification of building engineering [S]. Beijing: China Construction Industry Press.

Han Xiao Lei, Yang Kun, Zheng Yi, et al. (2004). Structural design of overrun high-rise buildings with beam transfer storey [J]. *Journal of Kunming University of science and Technology (Science and Engineering Edition)*, (06), 84–87.

JGJ3-2010. (2010). *Technical specification for concrete structures of high-rise buildings* [S]. Beijing: China Construction Industry Press.

Ma Xue Qiang, Wang Shu Le, Guo Qiang, et al. (2017). Elastoplastic analysis of large span truss transfer structure [J]. *Building structure*, 47 (S1), 507-513.

Zhang Bo. (2011). Tall buildings beam type transfer layers structure design principle and its application [D]. Changsha: Hunan University, 63-68.

Zhao Jian Min, Zhang Jian Ping, Wu Zhao Hua. (2016). Analysis and design of large span transfer truss structure [J]. *Building structure*, 46 (11), 21-26.

Zhao Xiang Dong. (2020). Study on ceiling transfer floor system of high speed railway station building [J]. *Railway construction technology*, (07), 36-38+47.

Advances in Civil Engineering: Structural Seismic Resistance, Monitoring and Detection – Mohd Johari Mohd Yusof and Junwen Zhang (Eds)
© 2023 The Authors, ISBN 978-1-032-31491-4

Growth law and dynamic response of shotcrete mechanical properties under mechanized full-section tunnel construction

Shuang Chen
Yunnan Construction Infrastructure Investment Co., Ltd., Kunming, Yunnan, China

Lianchao Ye
Chongqing Jiaotong University, Chongqing, China

Ke Li
China Merchants Chongqing Communications Technology Research & Design Institute Co., Ltd, Chongqing, China

Rong Xiang
Chongqing City Public Rental Housing Construction Co. LTD, Chongqing, China

ABSTRACT: Taking Caoguoshan tunnel as a research objective, the change law of macro mechanical properties of shotcrete with age was analyzed by field test, and the vibration law of shotcrete with distance of face was obtained by field monitoring under cyclic blasting load. The results show that the compressive strength of shotcrete prismatic body is 30.5% on 1d, 77.8% on 7d, and 94.5% on 14d, and the compressive strength of shotcrete increases significantly in the first 7 days. The elastic modulus of static compression in 1d is 36.6% of 28d. The elastic modulus of 7d is 59.3% of 28d. The elastic modulus of 14d was 78.2% of 28d. The vibration velocity of the shotcrete with increase in distance of the face is not strictly monotonous, but the general law is gradually reduced.

1 INTRODUCTION

During mechanized full-section construction, the shotcrete of the first branch of the tunnel bears the impact of the cyclic blasting load before it reaches the design strength under the cyclic blasting operation. This inevitably leads to the deterioration of the primary shotcrete, causing reduction in its macro mechanical properties (Qing 2021; Wu 2021), weakening its support ability for surrounding rock, and thus threatening the safety of the operators in the tunnel. Therefore, it is necessary to study the mechanical properties of primary shotcrete and vibration law of primary shotcrete under cyclic blasting condition.

Currently, studies on the impact of blasting vibration load on concrete mainly focus on mass concrete. Changes in mechanical parameters of each measuring point caused by blasting are monitored to determine the safe blasting vibration velocity under different age conditions, providing theoretical basis for construction (Lu 2000, 2001; Shu 1999; Zhu 1990). However, it is obviously not feasible to control the blasting effect only from the safe vibration velocity. The blasting effect is not only affected by the amount of charge but also is closely related to the borehole layout scheme, blast center distance, and explosive performance (Wu 1997). Li Ning et al. (Chen 1996; Li 1996) established model tests based on the field investigation data and performed in-depth studies on the influence of cyclic blasting on the mechanical properties of shotcrete at different ages of initial support after tunnel excavation. Zhang Lvsuo (Zhang 2005) and Hu Feng (Huang 1996) et al. studied the influence of shaft blasting on concrete based on model test, and theoretically explained the influence of blasting on mechanical properties of concrete at different ages. Ding Taishan et al.

DOI 10.1201/9781003310884-74

(Ding 2006) studied the influence of blasting air wave on cumulative damage of concrete at different ages in combination with damage theory and numerical simulation, and proposed the allowable damage threshold of concrete. Based on basic assumptions, such as isotropy and homogeneity, Chen Ming et al. (Chen 2008) introduced elastic wave to study the influence of elastic p-wave and s-wave on concrete failure forms and safe vibration velocity, providing a basis for the project.

Taking the blasting construction of Caoguoshan tunnel as the background, this study adopts field test to study the change in macro-mechanical properties of shotcrete with respect to age, and studies the vibration law of shotcrete under the action of cyclic blasting through field measurement to provide theoretical basis for the project.

2 ENGINEERING MONITORING

2.1 *Engineering situation*

This paper takes Caoguoshan tunnel as the research background, which is the longest separated tunnel on Jinping expressway. It is located in Adebo township and Jinhe Town. The entrance is located in Daijiazhai Village of Adebo Township, and the exit is located in Kuchahe Village of Jinhe Town. According to regional geological data and relevant in situ stress tests, the tunnel area has a large in situ stress, which makes it more likely for the rock mass of the tunnel wall to peel off and fall out during construction, and new cracks can easily be developed. At the same time, the tunnel is a high-stress rock burst area near the excavation face. The joints and fractures are well developed, the rock mass is relatively complete, and the surrounding rock stability is poor. The initial support adopts C25 shotcrete.

2.2 *Monitoring program*

The study involves blasting the vibration signal acquisition using TC-4850, collecting x, y, z directions of the blasting vibration velocity, filtering and measuring data, and extracting the peak particle velocity, the sensor pointing in the direction of tunnel excavation, y x direction pointing in the direction of tunnel radial, z direction perpendicular to the xy plane. To obtain the longitudinal distribution law of blasting vibration, and considering the convenience and safety of field testing, blasting vibration measurement points are set up in the reserved cavity near the side wall of the tunnel face. The blasting vibration law of each blasting operation cycle is collected following the construction progress on site until the completion of the second lining construction. The arrangement of measuring points is shown in Figure 1.

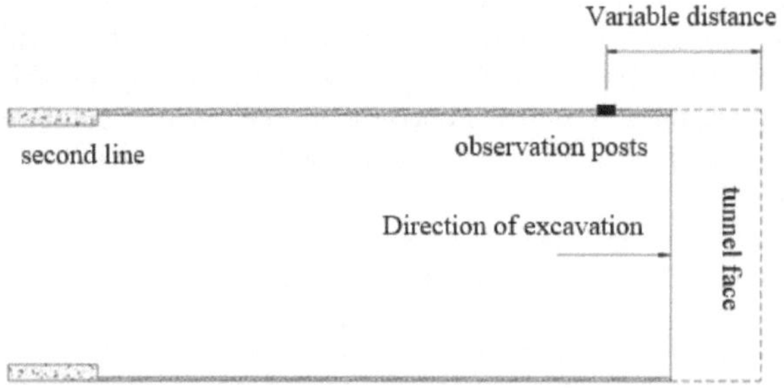

Figure 1. Layout of blasting vibration test points.

3 STUDY ON MECHANICAL PROPERTIES OF SHOTCRETE

This study relies on the second site laboratory of Manjin Expressway Project Department to carry out the test on the growth law of shotcrete mechanical properties. The compressive test and elastic modulus test of prisms are carried out according to the test rules of Cement and Cement Concrete for Highway Engineering jTGE30-2005.

3.1 *Test raw material*

Cement varieties from huaxin Cement Co., Ltd. produced ordinary Portland cement P.042.5 cement; coarse aggregate is the medium sand from the gravel field at the Manjiang end of Caoguoshan tunnel. Fine aggregate is 5–10 mm gravel from the gravel field at the end of the tunnel in Caoguoshan. The shotcrete fit is shown in Table 1.

Table 1. Mixture of shotcrete.

	Amount of various materials (kg)					
Project	Cement	Fine aggregate	Coarse aggregate	Water	Admixture	Additive
Each m^3	495	808	809	188	–	24.75

3.2 *Concept design*

The compressive strength tests and static compressive elastic modulus tests of prisms were carried out at 8h, 16h, 1d, 2D, 3D, 4D, 5d, 6d, 7d, 10d, 14d, 20d, and 28d, respectively. The growth law of shotcrete mechanical properties under 13 ages has been studied. There are 13 groups of prismatic compressive test and elastic modulus test, respectively, with 3 test blocks in each group, 78 pieces in total. The specimens of prismatic compression test and compression elastic modulus test were all 150 mm×150 mm×300 mm prismatic specimens. All specimens were placed in the standard curing room for curing.

3.3 *Outcome analysis*

Results of compressive strength and elastic modulus of shotcrete are summarized in Table 2. It can be seen from Table 2 that the compressive strength of the prism in 1d is 30.5% of that in 28d. The compressive strength of 7d reaches 77.8% of 28d. The compressive strength of 14d is 94.5% of 28d. It can be seen that the compressive strength of concrete increases considerably in the first 7d. The elastic modulus of static compression in 1d is 36.6% of 28d. The elastic modulus of 7d is 59.3% of 28d. The elastic modulus of 14d was 78.2% of 28d.

Figure 2 Failure mode of shotcrete elastic modulus specimen.

Through experimental research and theoretical analysis, the data of compressive strength and elastic modulus of shotcrete are fitted and analyzed. The fitting results have good correlation. The calculation model of compressive strength is the logarithmic function with correlation coefficient R2 = 0.934. The calculation model of elastic modulus is the logarithmic function with correlation coefficient R2 = 0.908. The following equations are the calculation models of compressive strength and elastic modulus, respectively.

$$f_{ck} = 4.119 - 5.316 \ln (t + 0.646) \tag{1}$$

$$E_{c,t} = -52.331 + 34.349 \ln (t + 8.28) \tag{2}$$

As the compressive strength and elastic modulus of concrete are both functions of time, the relationship between compressive strength and elastic modulus and time is shown in Figure 3. It is obvious from the figure that compressive strength increases rapidly in the early period and slowly in the later period. When the specimen time reaches 28d, the concrete strength can be considered to be stable. The elastic modulus increases rapidly in the early period and slows down in the late period. When the specimen time reaches 28d, the elastic modulus of concrete gradually tends to be stable. At the age of 4 days, the elastic modulus of shotcrete reaches 34.4 GPa, which is equivalent

Figure 2. Failure diagram of shotcrete elastic modulus specimen.

Table 2. Summary of compressive strength and elastic modulus of shotcrete.

Serial number	Age strength relation	Prismatic compressive strength/MPa	Elastic modulus of prismatic compressive strength/GPa
1	8h	4.8	24.0
2	16h	5.5	25.7
3	1d	6.2	25.0
4	2d	8.6	20.4
5	3d	8.8	32.2
6	4d	12.0	36.2
7	5d	12.1	30.6
8	6d	17.8	35.7
9	7d	15.8	40.5
10	10d	17.7	57.6
11	14d	19.2	53.4
12	20d	19.8	64.0
13	28d	20.3	68.3

to the design value of elastic modulus. At the same time, the compressive strength of shotcrete at the age of 4 days reaches 12.5 MPa, which reaches the design value of compressive strength.

4 ANALYSIS OF DYNAMIC RESPONSE CHARACTERISTICS

Through field measurements, the vibration law of blasting initial branch in tunnel was studied. Caoguoshan tunnel was set up with 6 blasting vibration test points, the pile numbers of which were K35+650, K35+750, K35+788, K35+810, K35+850, and K35+930, and the blasting vibration was tested 61 times following the construction progress cycle. The outliers of the measured data were removed and the overall vibration curve of the initial support is depicted in Figure 4. The measured data of the on-site K35+810 measurement point are shown in Table 3.

The results show that the amplitude of vibration velocity change after blasting in Y and Z directions is larger than that in X direction, and the vibration velocity of the initial blasting decreases gradually with increase in distance from the face of the hand, and the influence of blasting on the initial support decreases gradually. With regard to the primary support during the vibration velocity

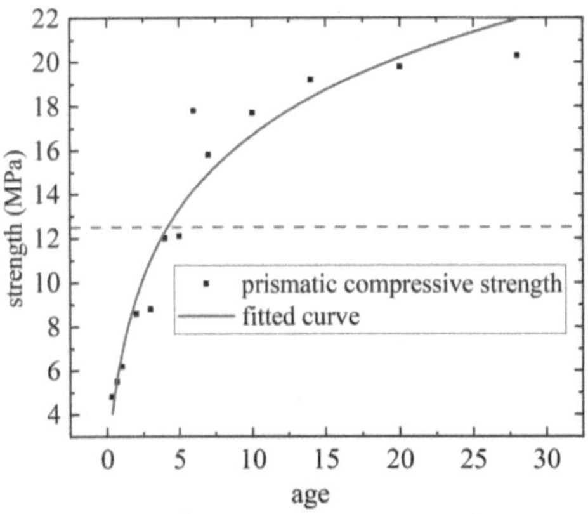

(a) Compressive strength

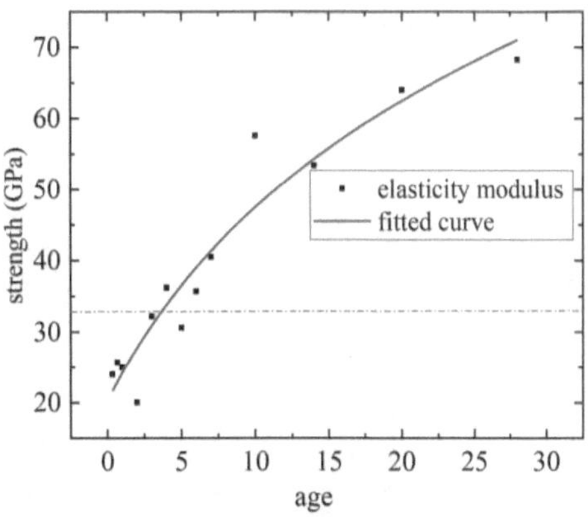

(b) Elastic modulus

Figure 3. Growth curve of shotcrete mechanical properties.

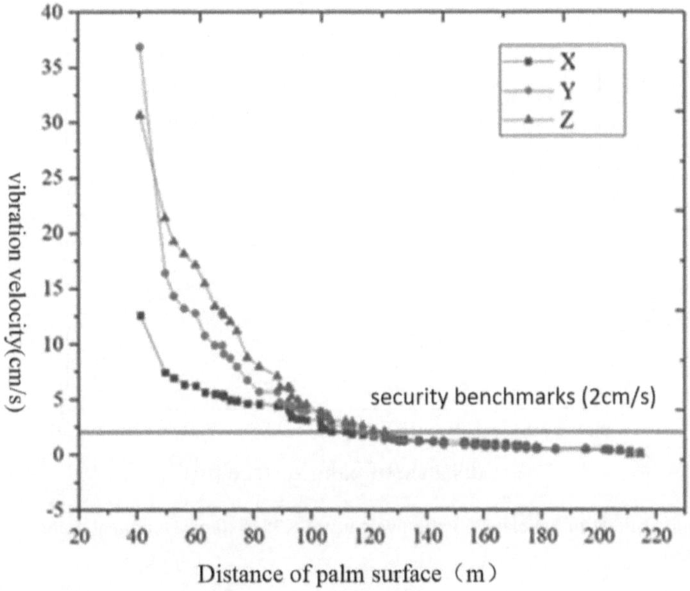

Figure 4. Variation rule of vibration velocity of primary blasting.

of the closer distance constraints, the higher the vibration velocity values, the higher is the change of amplitude. Through the study on the mechanical behavior of primary support above, we know that at the beginning the physical and mechanical properties and the curing ages have great correlation, where nearer the distance constraints, the shorter is the early support of curing age, thus not fully meeting the design strength. This also makes the blasting vibration velocity relatively larger in the places closer to the palm surface.

Table 3 consists of the data of initial blasting vibration measured by pile number at measuring point K35+810 and corresponding blasting parameters. Figure 4 depicts the curve of actual vibration velocity measured at measuring point K35+810. As can be seen from Figure 5, the blasting vibration velocity of the primary branch does not decrease monotonically with the increase in the distance between the face of the palm, but the overall rule is gradually decreasing. It can be seen from the

Table 3. Statistical table of test results of blasting vibration velocity at K35+810 measuring point.

Pile number of palm face	Station pile no	Dynamite (kg)	Distance between measuring point and palm surface (m)	Velocity of vibration(mm/s)			The actual footage
				X	Y	Z	
K35+903.2	K35+810	312	93.2	21.2	25.4	26.5	3.2
K35+906.4	K35+810	315	96.4	19.5	17.6	29.6	3.6
K35+913.2	K35+810	315	103.2	11.6	9	13.8	3.8
K35+917	K35+810	324	107	17.1	14.6	13.3	3.6
K35+924.2	K35+810	315	114.2	11.4	11.9	13	3.6
K35+927.8	K35+810	315	117.8	12.3	14.4	14.7	3.8
K35+931.6	K35+810	315	121.6	6.9	12.1	12.2	3.8
K35+935.4	K35+810	315	125.4	8.6	11.2	11.9	3.7

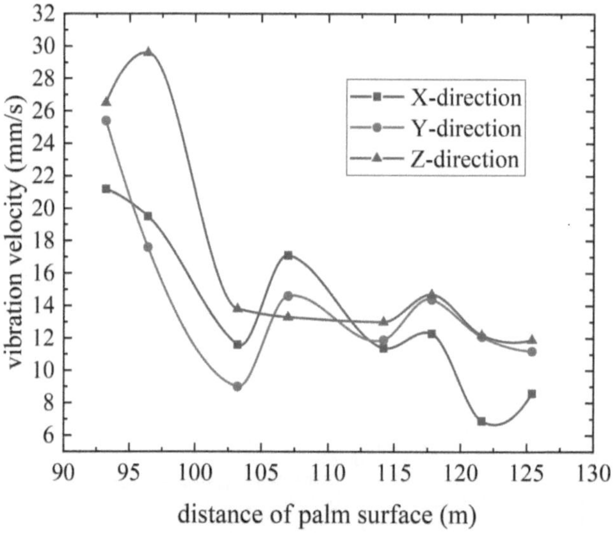

Figure 5. Variation rule of initial blast vibration velocity at K35+810 measurement point.

table that there are many factors affecting the vibration speed, such as the distance of the face of the hand, the amount of explosive, the actual driving distance, the temperature, humidity, and the grade of surrounding rock not listed in the table also affect the vibration speed.

5 CONCLUSION

(1) The experimental results show that the compressive strengths of shotcrete prismatic body are 30.5%, 77.8%, and 94.5% in 1d, 7d, and 14d, respectively, and the compressive strength of shotcrete increases considerably in the first 7 days. The elastic modulus of static compression in 1d is 36.6% of 28d. The elastic modulus of 7d is 59.3% of 28d. The elastic modulus of 14d is 78.2% of 28d.

(2) According to the regression curve analysis of test results, the compressive strength of shotcrete increases rapidly in the early stage and slows in the late stage with time. When the specimen time reaches 28d, the concrete strength can be considered stable. The elastic modulus increases

rapidly in the early period and slowly in the late period. When the specimen time reaches 28d, the elastic modulus of concrete gradually tends to be stable. At the age of 4 days, the elastic modulus of shotcrete reaches 34.4 GPa, which is equivalent to the design value of elastic modulus. At the same time, the compressive strength of shotcrete at the age of 4 days reaches 12.5 MPa, which reaches the design value of compressive strength.

(3) The blasting vibration velocity of the primary support gradually decreases with increase in distance from the face of the hand, and the impact of blasting on the initial support also gradually decreases. The closer the vibration velocity of initial support is to the face of the hand, the larger the value of vibration velocity is, the higher the amplitude of its change is, the greater is the influence of age on concrete.

(4) There are many factors that affect the vibration velocity of the primary support, such as the distance of the palm surface, the amount of explosive, the actual tunneling distance, the temperature, humidity, and the grade of surrounding rock not listed in the table also affect the vibration velocity of the primary support, which also influence the vibration velocity to not decrease monotonously with the distance of the palm surface, but the overall law gradually decreases.

ACKNOWLEDGMENTS

This paper is supported by the Science and Technology Project of The Department of Transportation of Guizhou Province (2020-122-031).

REFERENCES

Chen Yun-sheng, Li Ning, Lu Qiang. Experimental Study on Detonation Simulation of Concrete Lining [J]. *Journal of Xi'an University of Technology*, 1996 (1): 47–51.

Ding Taishan, Li Wanxi. Damage Analysis of blasting Construction on shotcrete [J]. *Journal of Underground Space and Engineering*, 2006 (5): 834–838.

Journal of China Coal Society, 1996 (5): 56-59. Huang Qi, Hu Feng. Experimental Study on Mechanical Properties of Concrete Under Explosion Load [J]. *Journal of China Coal Society*, 1996 (5): 56–59.

Li Ning, Chen Yunsheng, Han Bao.Experimental study on the effect of blasting vibration on lining damage characteristics [J]. *Engineering Blasting*, 1996 (2): 6–10.

Lu Wenbo, Lai Shiju, Zhu Chuanyun, et al. *Explosion and Shock Waves*, 2001 (1) : 67–71.

Lu Wenbo, Zhang Jianmin, Zhu Chuanyun, et al.Blasting vibration control of rock excavation adjacent to newly poured concrete. *Blasting Materials*, 2000 (6): 28–31.

Qin Y Wang L Su H M et al. Study on dynamic statistical damage constitutive relation of cretaceous frozen sandstone[J]. *Engineering Blasting*, 2021, 27(3): 42–50.

Rock and Soil Mechanics, 2008 (2): 455–459,464. (in Chinese)) (Chen Ming, Lu Wenbo.

Shu Daqiang. Influence and Control of Blasting Vibration on quality of Freshly poured Concrete [J]. *Journal of Wuhan University of Water Resources and Electric Power*, 1999 (4): 5–9.

Wu Delun, Ye Xiaoming. *Chinese Journal of Rock Mechanics and Engineering*, 1997 (3): 67–74.

Wu W Y, Lu W. Analysis of explosion damage factors in reinforeed concrete columns [J]. *Engineering Blasting*, 2021, 27(2): 58–65.

Zhang C, Hu F, Zou S. Effects of blast induced vibrations on the fresh concrete lining of a shaftf[J]. *Tunnelling & Underground Space Technology Incorporating Trenchless Technology Research*, 2005, 20(4):356–361.

Zhu Z G, Zhang Z Z, Tong J Y, et al. Influence of Blasting on freshly poured Concrete and its control standard [J]. *Blasting*, 1990 (3): 28–32.

Advances in Civil Engineering: Structural Seismic Resistance, Monitoring and Detection – Mohd Johari Mohd Yusof and Junwen Zhang (Eds)
© 2023 The Authors, ISBN 978-1-032-31491-4

Research on friction characteristics of reinforced-soil interface under static and dynamic load

Yunfei Zhao
School of Urban Geology and Engineering, Hebei GEO University, Shijiazhuang, Hebei, China

Guangqing Yang*
Department of Civil Engineering, Shijiazhuang Tiedao University, Shijiazhuang, China

Zhi Wang
China Railway Fifth Survey and Design Instuitute Group Co. Ltd., Beijing, China

Xunmei Liang
Taian Road Engineering Materials Co. Ltd., Taian, Shandong, China

ABSTRACT: In order to improve the stability of the embankment or prevent cracking on pavement, the reinforced materials such as geogrid, has been used on multiple areas. However, the proper way to use this kind of materials has become a problem. In order to understand the friction characteristic under static and dynamic load, laboratory tests were conducted. The test results indicated that the normal stress is the key factor to the shear strength of the reinforced soil, and shear strength of the interface develops with the displacement. Moreover, the pull-out force under dynamic load increased with normal load, the shear strength increased as the frequency and amplitude increased.

1 INTRODUCTION

Reinforced materials such as geosynthetics have been used in multiple areas, like highways, airports, etc. The inner stability design of this kind of structure was based on the friction mechanism between geosynthetics and soil.

The influence of interface normal stress, percentage of transverse ribs, grid width and drawing rate on the interaction mechanism of the reinforcement and soil interface and the force of the transverse ribs of the grid with the aid of the medium-scale pull-out test of geogrid and sand (Xiao 2019). The large-scale pull-out test equipment was used, and has the interface friction angle and the friction coefficient ratio analyzed under different normal pressures and grids (Wang 2011). The friction characteristics under different normal stresses were studied through pull-out tests (Chen 2011). The test found that when the normal stress and sand density are relatively low, as the stress continues to increase, the thickness of the reinforcement-soil interface decreases, and the two have a negative linear relationship. A series of pull-out test and direct shear test have been conducted on biaxial geogrid to study the interaction mechanism of the reinforcement (Moraci 2014; Cao 2019).

A large number of studies have been conducted on the interface between reinforcement and soil nowadays. However, most of them focused on friction characteristics on uniaxial geogrid, not the biaxial geogrid. To study the friction characteristics of biaxial geogrid-soil under static and dynamic load, a series of pull-out tests were carried out, so the rest results could provide a reference to the design work using biaxial geogrid in the field.

*Corresponding Author: yanggq@stdu.edu.cn

 DOI 10.1201/9781003310884-75

2 TEST SCHEME

2.1 *Material properties*

(1) Filling soil

Based on the laboratory test results, the grading particle curve of the soil was shown in Figure 1.

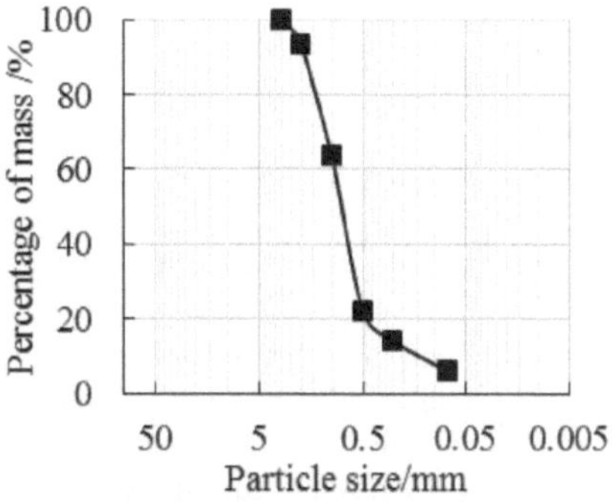

Figure 1. Curve of particle size distribution of filling soil.

(2) Geogrid

The specifications of the biaxial plastic tensile geogrids have been summarized in Tables 1 and 2.

Table 1. Specifications of High Density Polyethylene (HDPE) Geogrid.

Rib length/mm		Rib thickness /mm		Rib width/mm		
Cross direction	Machine direction	Cross direction	Machine direction	Cross direction	Machine direction	Crosspoint thickness/mm
32	33	1.7	1.5	2.0	2.0	4.5

Table 2. Technical Specifications of High Density Polyethylene (HDPE) Geogrid.

Items	Indexes
Tensile strength /(kN·m^{-1})	≥ 40
Tensile strength @ 2 % strain /(kN·m^{-1})	≥ 18.2
Tensile strength @ 5 % strain /(kN·m^{-1})	≥ 34.7
Peak strain /%	≤ 13

2.2 *Normal stress, pull-out rate*

Three normal stresses (25 kPa, 50 kPa and 75 kPa) are proposed to simulate the pull-out of geogrids at different depths in practical engineering. The pull-out velocity was set as 2 mm / min.

2.3 *Dynamic load frequency and amplitude*

In fact, in the scope of application of reinforced structures, not only the influence of static load, but also the live load applied on it that was necessary to be considered. The frequency from pull-out test under dynamic load studied before was for uniaxial geogrid (Jin 2020), from a high-speed train

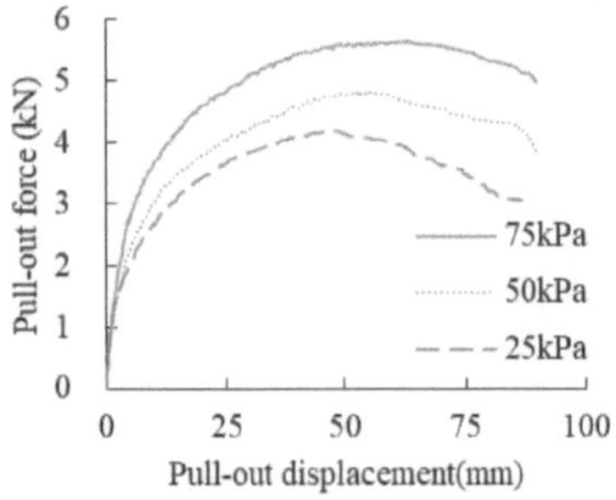

Figure 2. Pull-out force and displacement.

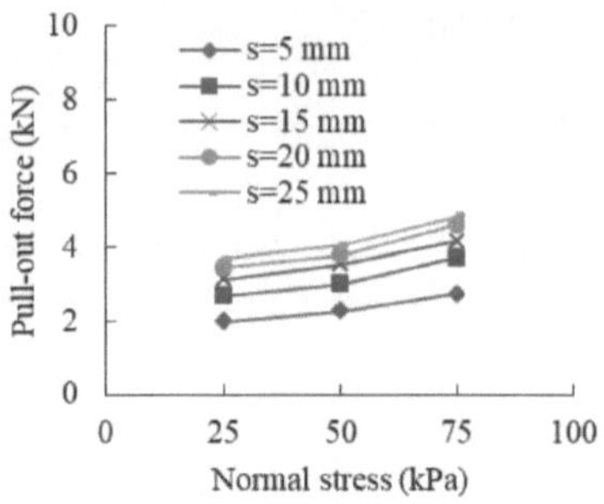

Figure 3. Pull-out force and stress under different displacements.

system, and it was too large for traffic load of highway engineering, so the frequency of dynamic load used in this paper was relatively small. Referring to its scheme, the sinusoidal waveform was used to impose cyclic dynamic load. Under the guidance of the relevant test results under the existing static load conditions, the influence of frequency and amplitude on the friction characteristics under dynamic load conditions was explored.

The initial normal load was set to 50 kPa, and the amplitude was set to 10 kPa, 20 kPa, and 30 kPa in the test scheme.

3 RESULTS AND DISCUSSION

3.1 *Influence of normal stress on friction characteristics of geogrid-soil*

The relationship between pull-out force and pull-out displacement (s) of geogrid-soil interface under different normal stress conditions was shown in Figure 2. According to the test results, the pull-out force showed a linear growth trend with the increase of the pull-out displacement under different normal stress conditions at the beginning, and then it slowed down on the increasing rate and turned to be flat after reached the peak. Overall, the curve showed strain softening under small load.

The break was set when the pull-out displacement reached at 90 mm, The pullout test with the corresponding normal stresses of 25 kPa, 50 kPa and 75 kPa, the maximum pull-out force is 4.197 kN, 4.811 kN and 5.649 kN, respectively, at 48.149 mm, 53.946 mm and 62.101 mm. The maximum pull-out force increases with the increase of normal stress.

The embedded effect of geogrid was not obvious under small normal stress and the friction between soil particles and geogrid-soil was small, resulting in the smaller maximum pull-out force in the pull-out test. Several points were randomly selected for analysis in the pull-out process, and the relationship curve between the pull-out force of geogrid and the normal stress was plotted, as shown in Figure 3. With the increase of normal stress, the pull-out force of geogrid increases at the same displacement. Under the relatively large normal stress, soil particles are continuously compacted and squeezed, and the embedded effect was fully activated, resulting in the pull-out force increases with the increase of normal stress at the same pull-out displacement. It can also be seen that the pull-out force increases with the increase of displacement.

The interfacial shear stress and the interfacial friction coefficient were calculated, and the relationship curves were carried out according to the Mohr-Coulomb strength theory, as shown in Figures 4 and 5. Calculated shear strength parameters c_{gs}= 9.538 kPa, φ_{gs} = 4.614° . According to other literatures, this result showed a good meet to the pattern. The cohesive of the soil increases, and friction angle decreases after being reinforced.

The results showed that the friction coefficient of reinforcement-soil interface decreases with the increase of normal stress, from 0.466 at 25 kPa to 0.209 at 75 kPa. The reason of it was that the

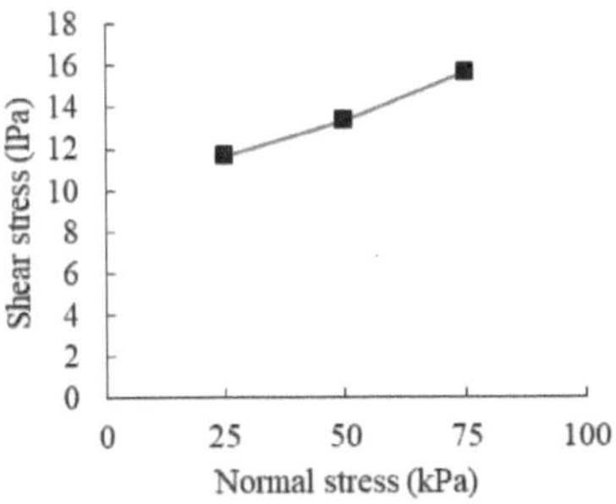

Figure 4. Shear and normal stress fitting curve.

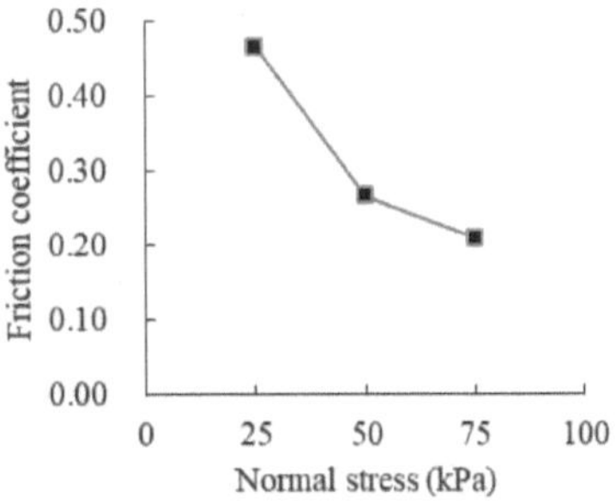

Figure 5. Friction coefficient under different normal stress.

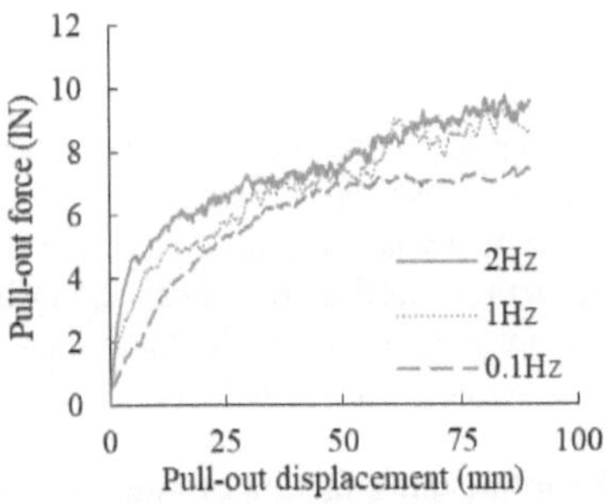

Figure 6. Pull-out force and displacement.

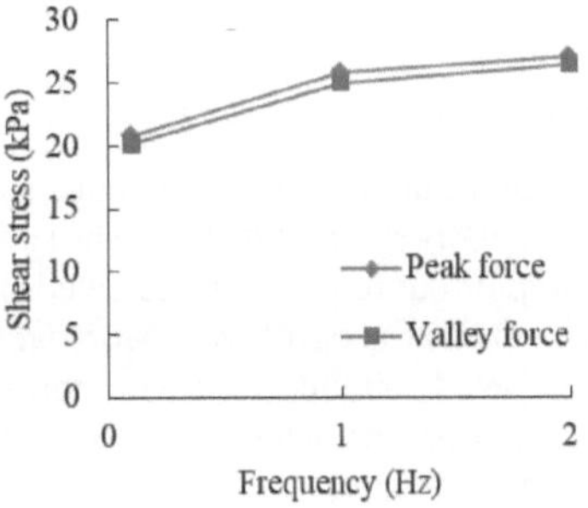

Figure 7. Shear stress under different frequencies.

dilatancy of the sandy soil. Under low normal stress, the geogrid could easily have the soil particles rotated in the process of test, which the dilatancy is obvious. With the increase of normal stress, the movement of soil particles was limited, and the soil particles could not easily rotate and move, so the dilatancy decreases. The interfacial friction coefficient was not a constant under different normal stresses.

3.2 *Influence of dynamic load frequency and amplitude on friction characteristics of geogrid – soil interface*

The normal load was applied to the soil by the bearing plate, and it was set as cyclic dynamic load. The sine wave shape is used at the frequencies of 0.1 Hz, 1 Hz, 2 Hz and amplitudes of 10 kPa, 20 kPa and 30 kPa.

It can be seen from the cyclic trend of the pull-out curve shown in Figure 6 that under the action of dynamic load, the relationship curve between pull-out force and pull-out displacement is similar to that under static load. The pull-out force increased rapidly at the beginning, and tends to be flat and stable in a certain range after reached a certain displacement, then showed strain softening.

When $f = 0.1$ Hz, the peak value of the maximum pull-out force was 7.487 kN, and the valley value was 7.246. When $f = 1$ Hz, the peak force was 9.302 kN, and the valley force was 9 kN. When $f = 2$ Hz, the peak force was 9.743 kN and the valley force was 9.523 kN. The maximum pull-out force increases with the increase of frequency. Figure 8 indicated that under the same frequency, the larger the amplitude was, the greater the maximum pull-out force was.

Due to the disturbance effect of dynamic load on the soil, the surface friction increased with the relative density of the filler increase, as shown in Figure 9. Therefore, as the frequency and amplitude increased, the shear strength increased as well.

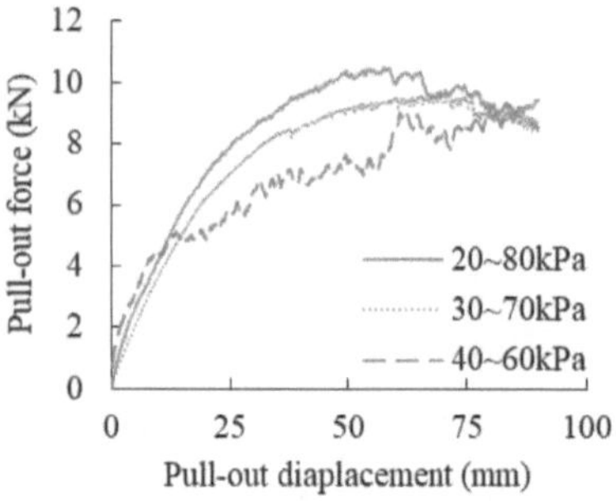

Figure 8. Pull-out force and displacement.

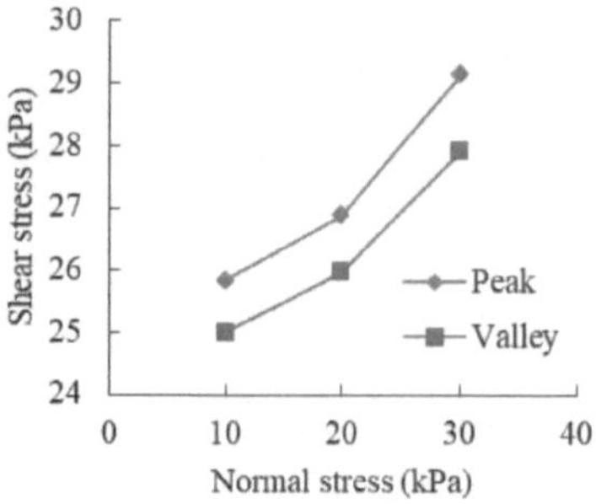

Figure 9. Shear stress under different frequencies.

4 CONCLUSIONS

The main conclusions are as follows:

(1) The pull-out force-displacement curve of reinforced soil interface is obviously nonlinear, which is generally strain-softening. The shear strength of the interface develops with the displacement.
(2) The pull-out test results indicated that the larger the normal stress was, the greater the pull-out force of geogrid was. Moreover, the friction coefficient as well as the shear strength of reinforced soil increased with normal stress. Hence, the normal stress was the main factor to the friction characteristics of geogrid-soil interface, and it should be considered during the design procedure.
(3) The frequency and amplitude were considered to be the influencing factors of interface characteristics under dynamic load. However, the test results indicated that the amplitude was the key factor to the pull-out force as well as the shear strength, and the frequency affected less.

REFERENCES

Cao Wenzhao, Zheng Junjie, Zhou Yanjun. (2019) Comparative Experimental Investigation of Geogrid-soil Interface Behavior of Biaxial and Triaxial Geogrid. *Joumal of Hunan University* (Natural Sciences) 46(01): 109–116.

Chen J., Li, H., Liu, J., Zhou, J. (2011) Mesoscopic study of interface properties of geogrid-reinforced soil. *Rock and Soil Mechanic*. 32(S1), 66–71.

Jin J. (2020) *Geogrid Creep Hehavior and Interaction Mechanism of Geogrid-Soil Interface*. Shijiazhuang Tiedao University. Shijiazhuang, China.

Nicola Moraci, Giuseppe Cardile, Domenico Gioffre,e t al. (2014) Soil Geosynthetic interaction: design parameters from experimental and theoretical analysis. *Transportation Infrastructure Geotechnology*, 1(2).

Wang, J., Zhou, J., Deng, Y., Huang, L. (2011) Pull-out test analysis of sand and geosynthetics. *Journal of Guangxi University: Nat Sci Ed*. 36(04), 659–663.

Xiao, C., Luo, Y., Wang, Z., Hu, H. (2019) Pull-out test of interface characteristics between geogrid and sand. *Journal of shenzhen university science and engineering*. 36 (03), 252–259.

Advances in Civil Engineering: Structural Seismic Resistance, Monitoring and Detection – Mohd Johari Mohd Yusof and Junwen Zhang (Eds)
© 2023 The Authors, ISBN 978-1-032-31491-4

Analysis and application of deployable truss bridge

Haofeng Ling†
Sydney University, Camperdown, Australia

Yixin Sun†
Chang'an University, Xi'an, China

Weijie Wang†
Dalhousie University, Halifax, Canada

ABSTRACT: Our group is paying attention to use the deployable structure to solve the problem of building a bridge rapidly. As a new structure, the deployable structure has plenty of advantages that traditional structure does not have, such as high utility rate of space and being produced forward in the factory. We are trying to use CAD and MATLAB to show the deployable structure sample and use the quantitative calculation to show the feasibility of this kind of structure. Also, we have shown the possible uses for this structure, especially in the disaster area supporting. At the end of the paper, we concluded our result at this stage, show the problem we now have, and put forward our vision of the future in this research.

1 INTRODUCTION

There is 71% surface of the earth covered with water, like ocean, sea, river, lake, etc. Therefore, the ancient people have paid plenty of attention to building bridges. It seems that the earliest bridge is the floating bridge which is combined with several boats. However, with the development and advancement of technology, the demand for the bridge in transportation had become larger than before. In this way, many different types of bridges have been created, such as the beam bridge, Truss bridge, Arch bridge, Cable bridge, etc. Nowadays, bridges have been used for helping vehicles, trains, pedestrians for many years to traverse an open space or gap. There is a very mature system all over the world for choosing a bridge. For the bridge with short to medium spans, people usually use the 'beam' structural form. Generally, it is simply supported with bearings at either end of each span or continuous across supports. Its foundations and bridge deck are all connected for the integral bridge, and no bearings and expansion joints, which can help to reduce the construction and maintenance costs. Prestressed concrete beams typically form this kind of bridge with an in-situ concrete deck slab, so its single spans can be up to 46 meters. However, to increase the structural efficiency, a Truss bridge has been created. This kind of bridge has relatively lightweight structures with the ability to span longer distances than beam bridges. However, a truss bridge has lots of connections, which can fatigue and require maintenance. If the truss ridge is too structurally 'efficient', the truss can sometimes lack 'redundancy', leading to a risk of sudden failure. The arch bridge is another good choice. In ancient times, people usually chose the arch bridge because they found that the arch bridge can change some pressure from vertical to horizontal, decreasing the demand for the material of building the bridge. Nowadays, arch bridge's span distances are more typically in the range of 60-150m but can up to 400/500m. The arch bridge is more structurally efficient than beam bridges and integral bridges. It acts in compression, so it is a structurally efficient solution for increasing the bearing capacity of the bridge. However, it needs to resist

†These authors contributed equally.

DOI 10.1201/9781003310884-76

horizontal thrust at supports, and the thermal stresses need to be carefully considered. Last but not least, the cable bridge is also a very famous type of bridge, such as the Akashi Kaikyo Bridge of Japan, whose length is 1991m and can resist the 8.5 magnitude earthquake. Cable bridge is used for spans where beam bridges are no longer feasible. Therefore, Cable-stayed bridges are usually an efficient solution when the total bridge length is 200-400m, and its maximum spans can achieve 1000m and so on. For the cable bridge, its cables support most of the deck weight in tension and then transfer the loads down through the towers, which means cables generally at an angle >25°, which means the towers are usually quite tall.

However, we have focused that with the frequent geological disaster such as earthquakes, floods, and debris flow, people need a quick and efficient bridge to span the big gap and turbulence (Yeh 2015). Many countries now are using floating bridges in emergencies (Sedlacek G & Trumpf H 2002). However, when facing the debris flow or the big gap, the floating bridge can't be used, so people usually use the rope to save the survivor, which is quite dangerous. Therefore, we are looking for a new way to solve this situation. The main challenge in building the bridge in the disaster area is that the bridge needs to be set up quickly and solid to fulfill the demand of saving the survivor and transport the relief material (Teixeira A 2014). However, build a bridge is a very complex process. It needs site investigation, the bridge design, the type of bridge, the material used on the bridge, the excavation, and even the health and safety should be considered carefully, especially in the disaster area. The site environment is very complex. It may have the risk of the secondary disaster (Chikahiro 2016). We have realized that it is difficult to build a new bridge in the disaster area. So, we are thinking about the possibility of transporting a whole bridge to the site because, in the nowadays bridge-building procedure, one of the commend methods in building the bridge in a short time is to produce the body of bridge advanced and transport it to the site and assemble it quickly (Nakazawa M 2012). To decrease the length of the bridge that we build advance to reduce the difficulty of transporting it. We decide to use the deployable structure in this paper. The deployable structure can help us save space during fold and provide a long span after deployment.

2 METHOD

The bridge uses a scissor-like linkage as the linkages. The scissor-like linkage is a linkage that can extend or set a beam by mechanical means. The terms 'scissor like' derive from a constructed mechanism with connected, folding supports in a crisscross bisect pattern (Chikahiro 2016).

As Figure 1. shows, the three circled adjacent triangular elements will be used as the main body of the bridge (Chikahiro 2017). In the front view, AD equals the BD and BC, which indicates the projections of the scissor-like structures as the right three diagrams show. CD represents the top surface, are fixed and is the angle within 0 to 90 degree describing structure opening status. Three types of scissor-like components meet the length relations as the three schemes demonstrate.

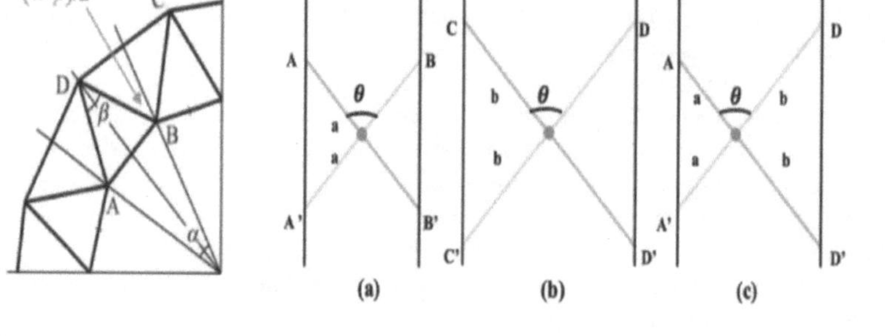

Figure 1. Front view and side details of interesting structure.

With the concept we obtained, the 3D scheme can be developed as Figure 2 (Chikahiro 2014). The folding process is achieved by stretching A and A' apart, and unfolding is achieved by contracting A and A' together. With structure deploying, the dark blue outframes trend to change their length, so we plan that using strings as out frames to connect the points, which not only can bear tension and also can constrain the opening range to make the bridge performs as designated (Thomas 2013). Strings cannot be defined as a rigid body and can be directly interpreted by the opening range, which will not affect the degree of freedom, so we would not consider strings into the analysis.

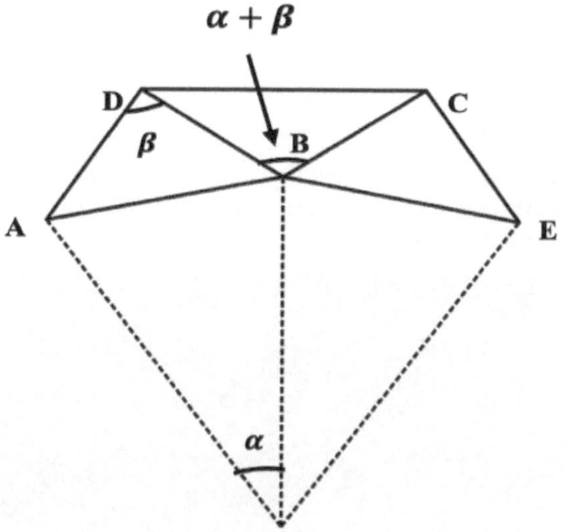

Figure 2. Geometric relations in the front view.

From Figure 2, we can conclude the geomatic relations (Luchsinger 2004). For further considerations on simulation, the relations are expressed as the coordination of A, B, C, D, E:

$$Ay = 0,\ Ax = 0 \tag{1}$$

$$Ey = 0,\ Ex = 2ABsin\left(\frac{\alpha+\beta}{2}\right) \tag{2}$$

$$Dy = ADcos\left[\beta - \left(\frac{\alpha+\beta}{2}\right)\right], Dx = ADsin\left[\beta - \left(\frac{\alpha+\beta}{2}\right)\right] \tag{3}$$

$$Cy = Ey + CEcos\left[\beta - \left(\frac{\alpha+\beta}{2}\right)\right], Cx = Ex - CEsin\left[\beta - \left(\frac{\alpha+\beta}{2}\right)\right] \tag{4}$$

$$By = Dx - BDcos\left(\frac{\alpha+\beta}{2}\right), Bx = \frac{Xc + Xd}{2} \tag{5}$$

From Figure 2, the equations (6) (7) and (8) can be obtained. From the three equations, equation (9) is derived, which will use to determine the particular beam lengths in simulation (Sedlacek 2004).

$$AD = BD = (a+b)\,sin\frac{\theta}{2} \tag{6}$$

$$CD = 2ADsin\frac{\alpha+\beta}{2} \tag{7}$$

$$AB = 2ADsin\frac{\beta}{2} \tag{8}$$

$$\frac{b}{a} = \frac{cos\frac{\alpha}{2} + sin\frac{\alpha}{2}\left(sin\frac{\alpha}{2} + \sqrt{1 + 2cos\frac{\alpha}{2}}\right)}{cos^2\frac{\alpha}{2}}, \beta = 2arcsin\frac{a}{a+b} \tag{9}$$

Based on the equations and coordination on the planar projection, with the α and β to be set up manually, the ratio of b/a is determined. So, the first step of my simulation is to set up the β=100deg, α=40deg, and b/a=0.7421.

Additionally, to demonstrate the motion trajectory clearly and avoid some unpredicted track, I fixed point A as zero points in the code, and confine point E can only move horizontally, which means Ey=0 always.

3 RESULT

3.1 *Link mechanism method*

The conceptual design of this kind of deployable truss bridge is showed in Figure 1. Several Isosceles trapezoids are made as the frames, and four equal scissor-like linkages connect every two frames. Cables connect the corners of the frames.

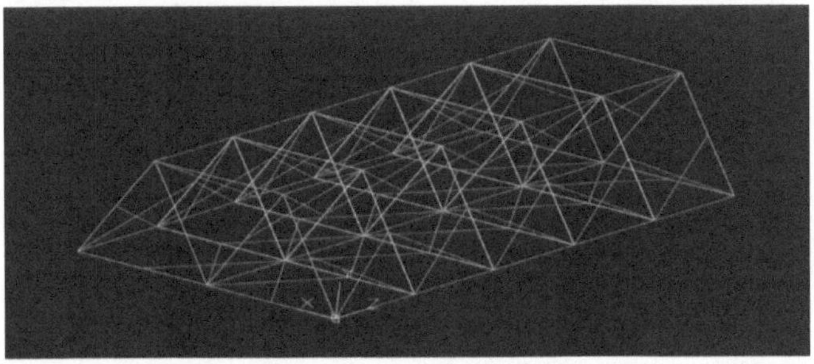

Figure 3. perspective drawing of the whole bridge.

The Isosceles trapezoids frames are strength by two beams. Thence, the Isosceles trapezoids are separated into three equilateral triangles, which is shown in Figure 3. The two triangles on both sides are positive, and the triangle in the middle is negative. These equilateral triangles not only strengthen the frame structure but also separate the frame into three individual parts. Two positive triangle parts are used for the vehicle to pass. And the middle negative triangle parts can be used for pedestrians and pipes and light systems.

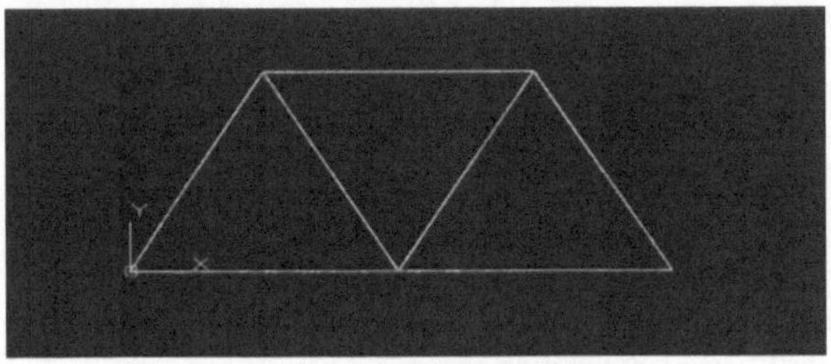

Figure 4. Front perspective drawing.

The linkages are set between each frame. Every two frames are connected by six similar linkages, which are shown in Figure 5. Two of the linkages are set at the bottom of the frame, shown in Figure 6. They are opposite to the middle axis and close to the axis. Another two linkages are setting on the side planes, which are also opposite to the middle axis, shown in Figure 7. The last two linkages are setting at the planes of the internal triangle, which is shown in Figure 8. So, their bottoms are connected.

The scissor-like linkages are similar, which means the joints are also similar. Figure 9 shows the different locations of different joints. Points A and C have rotated joints. They can rotate from 0 to

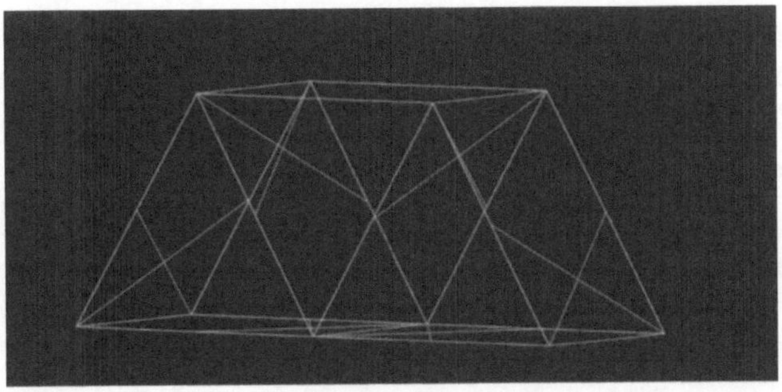

Figure 5. Perspective drawing of part of the bridge.

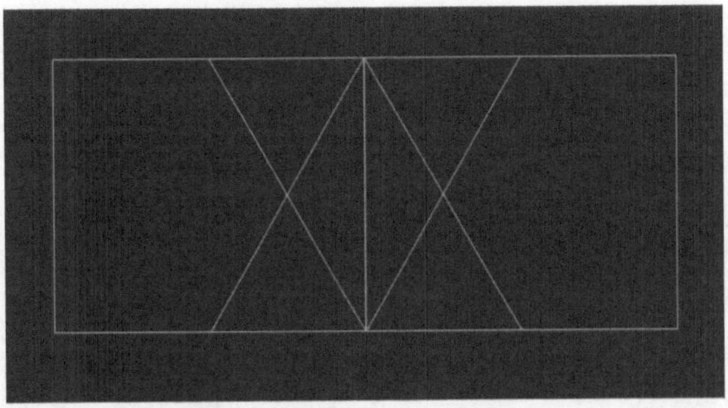

Figure 6. Bottom part of perspective drawing.

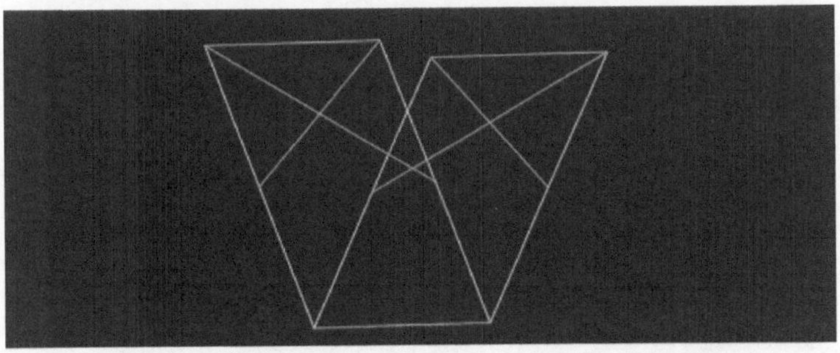

Figure 7. Middle part of perspective drawing.

60 degrees so that the scissor-like linkages can keep two opposite equilateral triangles (ΔABE and ΔCDE). Point E is a double rotate joint. It connects AD and BC, so it can rotate freely. The point B and E are rotational and sliding joints, which can slide along AF and CG. They can also rotate from 0 to 60 degrees.

The types of joints are shown in Figures 10, 11, and 12.

The road plates are designed to be set on the beam, which is shown in Figure 13. Three closed plates can connect each plate by concave-convex joints (Zhang 2020). Each plate has two concave joints and two convex joints. So, the plates can be connected by put convex joints on the concave joints. In addition, the plate also has concave joints beneath to fixed on the frames (Feo 2012). To strengthen, the fixed, the concaves are penetrated by a cylinder to fix the plate on the beam.

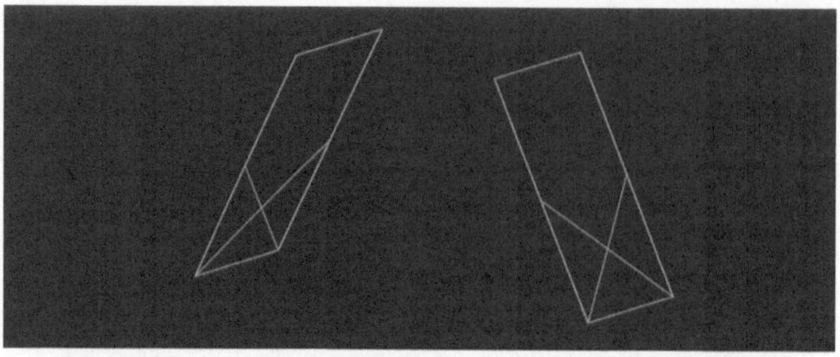

Figure 8. Side part of perspective drawing.

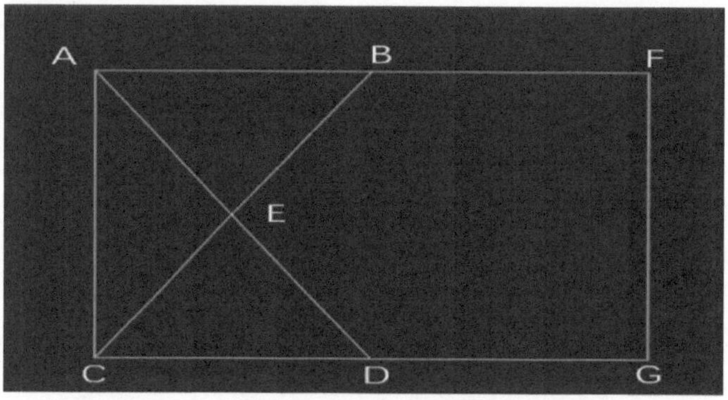

Figure 9. Scissor-like linkages.

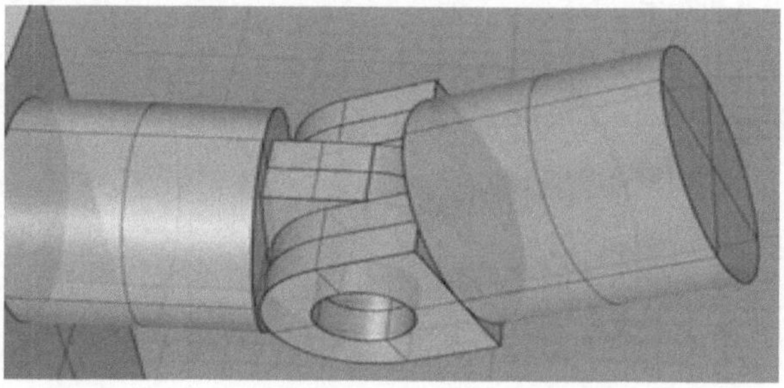

Figure 10. 0-60 degrees rotate joint.

Also, we have done some kinematic analysis for this system. Since all the linkages are the same, one of the linkage systems is shown in Figure 14.

The degrees of freedom of this kind of linkage is one, which includes 4 links (AB, BC, DC, and CB) and 5 joints (A, B, C, D, and E). DOF is showed in

$$n = 3 * (4 - 5 - 1) + 1 * 2 + 2 * 2 + 1 * 2 = 1 \tag{10}$$

The lengths of green linkages are half of the side length of a triangle of the frame. Hence, it can create equilateral triangles structure. AB=BE=AE=DE=CD=CE=1/2AF=1/2CG.

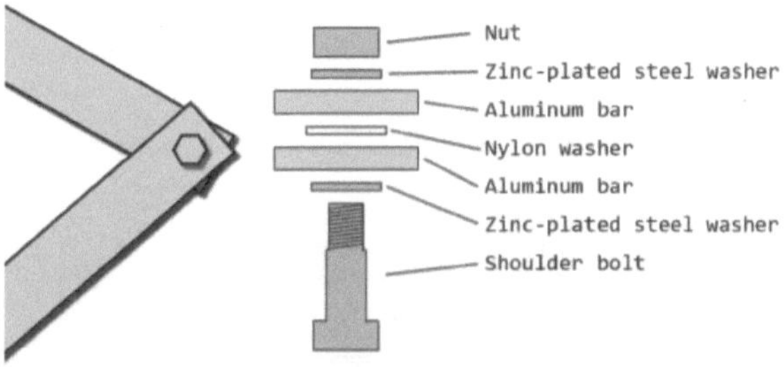

Figure 11. Pin joint connects two linkages.

Figure 12. Pin-in-Slot joint.

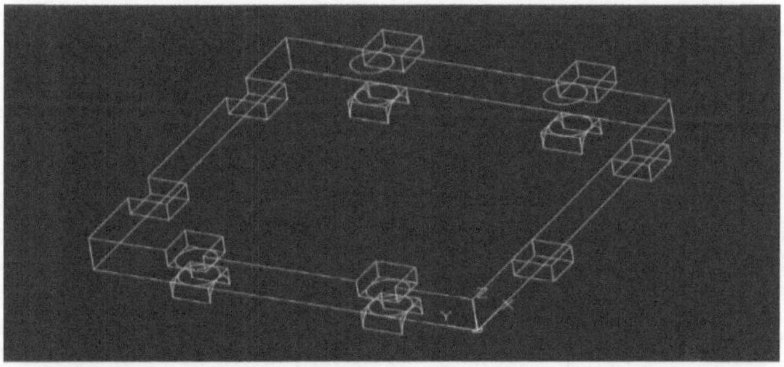

Figure 13. Road plate with concave and convex joints.

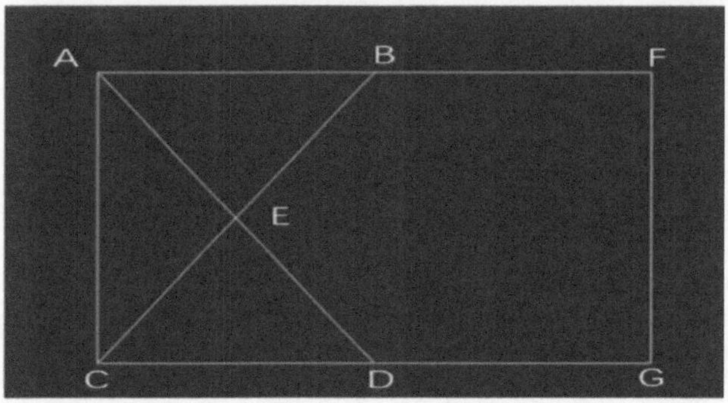

Figure 14. Scissor-like linkages.

The movement of point E depends on the movement of linkages of CB and AD. Since The trajectory of point E is part of the circle. The polar coordinate is used to indicate the position of point E.

$$CD = \frac{CG}{2} + vt \tag{11}$$

$$E = \left(\frac{CG}{2} * cos\theta, \frac{CG}{2} * sin\theta\right) \tag{12}$$

$$\theta = cos^{-1}\left(\frac{CG}{\frac{CG}{2} + \frac{vt}{2}}\right) \tag{13}$$

The movement begins with the scissor-like linkages is unfolded. Therefore, CD equals 1/2CG at first. Then, the D slide along with the CG with velocity v. Hence, then the length of CD is CG/2 +vt when time is t. Since CE equals 1/2CG, point E is moving on the part of a circle, whose centre is C and radius is CE. So, the trajectory of point E can be expressed. In addition, the θ (∠BCD) can be express by the length of CG and the movement of D.

3.2 *Overall extension method*

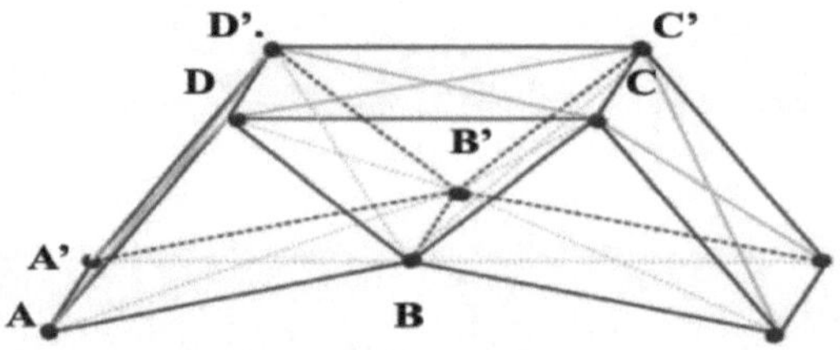

Figure 15. Scheme of the folding bridge.

As Figure 16 shows and the concepts discussed in the lessons, the angles between points are fixed, which implies the geometric Figures on the front and back sides will be contracted proportionally with structure folding. The simulation will be implemented through MATLAB by establishing rules of the geometric change.

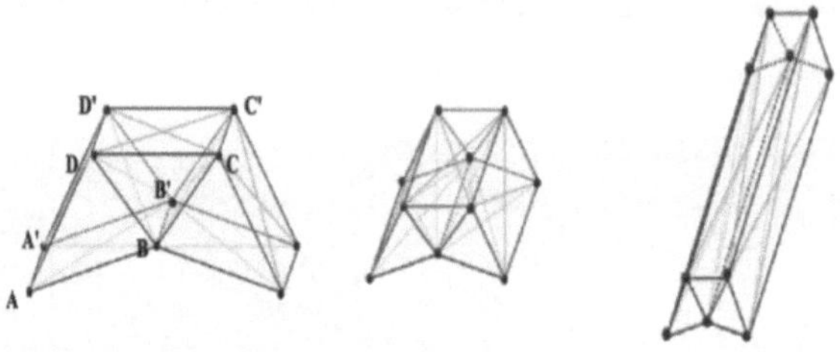

Figure 16. Scheme of the folding process.

With scissor-like structures instead of beams, when fixed the horizontal degree of freedom (DOF) at unfolding status, the expand-truss bridge is stable in structural considerations (A, M & Y&O 2013). The method of fixing horizontal DOF is that we exploit tenable cables to constraints the unfolding range with their length limit. The surface pads are to be placed on the top surfaces and two inclines surfaces after the structure is well deployed.

As Figure 17 produced by MATLAB, the motion trajectories of D, C, B, and E four points converge to point A with a folding process. It implies the entire structure will be folded in a pillar, and this form is exactly what we expected to obtain, that can be transported and hauled conveniently.

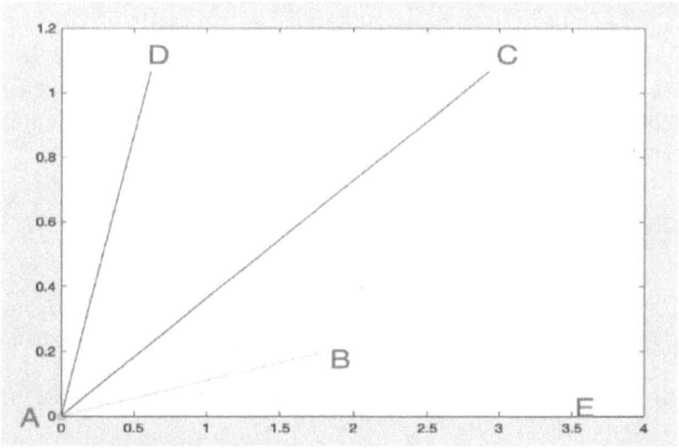

Figure 17. Simulation of motion trajectories.

4 CONCLUSION

4.1 *Summary*

We have shown our results with MATLAB and CAD in this paper. It is shown that it's possible to build a deployable truss bridge. This can help overcome the challenge of building a bridge in the disaster site and decrease the time of building. This will be valuable in saving the life and transporting relief material. Also, there are some other advantages to this deployable truss bridge. For example, it can save space and increase the utility rate of space, and this kind of bridge can be folded and reused after the rescue mission has finished, which can reduce the expenditure of support.

4.2 *Existing question*

There are some disadvantages to using this deployable truss bridge. First of all, we have to consider the stability and bearing capacity of this deployable truss bridge. We know that many bridges are composed of piers. These structures can help the bridge desk transmit the force acting on it to the column and decrease the pressure on both sides of the bridge. However, it is difficult to add the piers in the deployable structure because the real water line is not the same in the different cases. So, it is impossible to make the deployable bridge with piers in factory advanced. Therefore, without the pier's support, the bending capacity of deployable truss bridge will be decreased, and normally the vehicles used in rescue and relief work need a very high self-weight to keep their stability and not roll over. This will be a new challenge to this deployable truss bridge. What's more, the safety issue is what we also have to consider. We cannot transport the whole bridge to the site in one truck because the condition of the width of the road in the disaster area will not allow such a huge truck to move, but it is also impossible to make such a big carrier to carry such a bridge. So, we have to produce it in several main components in factory advanced and assemble it in the site. There are three main ways to assemble it: welding, bolts, and high strength bolts. In a welding process, the parts connected are heated to a sufficiently high temperature to fuse together with supplementary molten metal added to the joint. This method is good in the simplicity of the operator. However, the connection part made in this method is quite brittle and has to prevent residual stress and deformation. For bolt connection, it transmits the force and load by its screw. The high-strength bolt is the same as the bolt connection but adds some large prestress when tightening the bolts. We have mentioned before that without the pier, the bending pressure will all act on the connection. Whether it can bear this load is also an issue.

4.3 *Future expectations*

During the research, we have found many problems should be considered. We decide to make some new research to solve the bending capacity problems by the structural way, like finding the method to connect the components of the deployable truss bridge together in high strength, which can help decrease the requirement of the material.

REFERENCES

Chikahiro Y., Ario I., Nakazawa M., Ono S., Holnicki-Szulc J., Pawlowski P., Graczykowski C., Watson A. (2016) Experimental and numerical study of full-scale scissor type bridge, Automation in Construction, 71: 171–180.

Chikahiro Y., Ario I., and Nakazawa M. (2016) Theory and Design Study of a Full-Scale Scissors-Type Bridge, *Journal of Bridge Engineering*, 21: 401-605.

Chikahiro Y., Ario I., Pawlowski P., Graczykowski C., Nakazawa M., Holnicki-Szulc J., Ono S. (2017) Dynamics of the scissors-type Mobile Bridge, *Procedia Engineering*, 199: 2919–2924.

Chikahiro Y., Ario I., Nakazawa M., Ono S., Holnicki-Szulc J., Pawlowski P., Graczykowski C. (2014) An experimental study on the design method of a full-scale MobileBridge for a moving vehicle, The proceedings of 4th International Conference on Mobile, Adaptable and Rapidly Assembled Structures, 12: 94–103.

Feo L., Marra G., and Mosallam A. S. (2012) Stress analysis of multi-bolted joints for FRP pultruded composite structures, *Composite Structures*, 94: 3769–3780.

Luchsinger R. H., Pedretti A., Steingruber P. and Pedretti M. (2004) The new structural concept Tensairity: *Basic principles, Progress in structural engineering, mechanics and computation*, 32: 323-328.

Nakazawa M., Ario I. (2012) Structural characteristics of scissors type-emergency bridges, 5th Australian Small Bridges Conference, Australia, 66: 212-213

Sedlacek G., and Trumpf H. (2002) Mobile Light-Weight Fixed Bridges out of Pultruded Fibre-Reinforced Polymer Profiles, *Science and Engineering of Composite Materials*, 10: 383–396.

Sedlacek G., Trumpf H., and Castrischer U. (2004) Development of a Light-Weight Emergency Bridge, *Structural Engineering International*, 14: 282–287.

Teixeira A. M. A. J., Pfeil M. S., and R. C. (2014) Structural evaluation of a GFRP truss girder for a deployable bridge, *Composite Structures*, 110: 29–38.

Thomas G. R. and Sia B. J. (2013) Rapidly deployable bridge system, *Structures Congress*, 61: 656-667.

Yeh F.-Y., Chang K.-C., Sung Y.-C., Hung H.-H., and Chou C.-C. (2015) A novel composite bridge for emergency disaster relief: Concept and verification, *Composite Structures*, 127: 199–210.

Zhang J.- F., Zhao J.-J., Tian G.-F., Zhang P., Deng E.-F., He J.-M., Yu H.-X., and Fan Y.-F. (2020) Tensile behavior of the connection between nut-free high-strength bolt and endplate, *Journal of Constructional Steel Research*, 174: 106-301.

Advances in Civil Engineering: Structural Seismic Resistance, Monitoring and Detection – Mohd Johari Mohd Yusof and Junwen Zhang (Eds)
© 2023 The Authors, ISBN 978-1-032-31491-4

Study on optimization of initial support of tunnel in soft surrounding rock

W.K. Zhang, J.P. Liu, D. Wang, H.Y. You, H.X. Song & Z.J. Hu
Shandong University of Science and Technology, Qingdao City, Shandong, China

ABSTRACT: With rapid developments in traffic construction in China, tunnel engineering plays an important role in traffic construction, and the safety of tunnel construction is top priority. In this paper, numerical simulation and practical engineering are combined to study the optimization of the initial support of shallow buried tunnel in weak surrounding rock. The following conclusions are drawn: (1) when the NATM is used for the initial support of shallow buried soft rock tunnel, the earlier the support time is, the smaller is the deformation of tunnel surrounding rock; (2) different initial support time of tunnel will lead to different plastic zones of surrounding rock. It should be closed into a ring as soon as possible to improve the overall stability of surrounding rock; (3) the thickness of initial support structure shall be reasonably selected according to the actual situation.

1 INTRODUCTION

The deformation of the weak surrounding rock tunnel is large during the excavation process. Even after applying support, the deformation of the surrounding rock continues to increase at a relatively fast rate for a long time, which makes it difficult to control the deformation. The deformation of the dome settlement and horizontal convergence is also more obvious. After the deformation is maintained for a long time, the subsequent structural damage is prone to occur, which can also lead to tunnel collapse. As the most important safety guarantee in the process of tunnel excavation and construction, the initial support structure must be given sufficient attention. Scholars at home and abroad have conducted extensive research in this area.

Austrian engineer L.V. Rabcewicz proposed a new tunnel construction method called "New Austrian Method", which uses a surrounding rock as a bearing body and points out the direction for surrounding rock support (Han 1987). Feng et al. analyzed the optimal support time of the initial support structure under different footage cycles, that is, the initial support structure is applied every two cycles of footage to ensure safety and stability of the tunnel (Feng et al. 2019). Wang et al. established a tunnel support system suitable for squeezing formations, which is helpful for tunnel construction in squeezing formations. Garnes et al. proposed a support design that provided sufficient support force (Garnes et al. 2019). Wu et al. found that the pressure and displacement of tunnel support with a flexible layer could be improved compared with rigid support structures (Wu et al. 2019). Based on support optimization theory, Liu Huairong considers the surrounding rock displacement field and the force of the steel arch frame, and selects Midas/GTS to support parameters such as the length of the tunnel locking anchor pipe, the driving direction, the layout method, the initial support spacing, and the thickness of the second lining. Liu carried out optimization analysis (Liu 2016). Liu Junhui chose Midas/GTS finite element analysis software to optimize the supporting parameters such as the thickness and strength of the shotcrete, the type and spacing of the steel arch, and the length of the lock-foot anchor pipe according to the characteristics of the composite lining structure (Liu 2015).

A number of scholars have carried out research on the optimization of initial support parameters, but there are few studies on ultra-shallow buried tunnels, lack of optimization analysis of initial support construction under the new Austrian method of soft surrounding rock tunnels, and the combination of theoretical calculations and actual engineering is not in depth. The method of combining numerical simulation, theoretical calculation, and field practice is used to optimize the design of the initial support construction time and concrete thickness of the shallow-buried soft surrounding rock tunnel.

2 WORKING CONDITION OVERVIEW

The main line of the 410 m mine method tunnel section in the fourth work zone of the Guangfo Ring Road GFHD-2 standard is a double-hole single-line construction. The tunnel section starts at Guangshan 1st Road, Tianhe District, Guangzhou, and passes through Yinpai Mountain to the north. The surrounding environment of the tunnel is simple, the area passing through is hilly and low mountain landform, along the line are denuded mounds and small alluvial basins or valleys between mountains, and the terrain is very undulating. After on-site investigations, it was found that there are many mountains and few people in the section of the tunnel, no pipelines pass through, and a basketball court goes down.

The tunnel section is elliptical. According to the grade of surrounding rock, the tunnel has multiple excavation sections. The buried depth of the entrance section is 5.8 m, which is a shallow tunnel. The long axis of the tunnel section is 9.6 m, the short axis is 6.4 m, and the surrounding rock grade is VI. The rock quality is soft rock. Due to the large ups and downs of granite in local sections, the moderately weathered granite intruded into the tunnel, resulting in large differences in the strength of the driving face and difficulty in excavation.

3 WORKING CONDITION OVERVIEW

Determination method based on surrounding rock relaxation load angle:

The code method determines the boundary value of the deep and shallow burial of the tunnel. The "Code for Design of Railway Tunnels" stipulates that the boundary between the deep and shallow burial of the tunnel is as follows (JTG 3370.1-2018):

$$\begin{cases} h \lhd h_a & \text{Super shallow tunnel} \\ h_a \leq h \leq 2.5h_a & \text{Shallow tunnel} \\ h \rhd 2.5h_a & \text{Deep tunnel} \end{cases} \tag{1}$$

Where $h =$ Thickness of covering layer above tunnel vault (m); and $h_a =$ Calculated height of vertical load of the deep-buried tunnel (m).

The calculated height of the vertical load of the deep tunnel is:

$$h_a = 0.45 \times 2^{S-1}\omega \tag{2}$$

Where $\omega =$ Width influence coefficient; and S= Surrounding rock level.

The influence coefficient of surrounding rock width during tunnel excavation is

$$\omega = 1 + i\,(B + 5) \tag{3}$$

Where $B =$ Tunnel width; and $i =$ The rate of increase or decrease of surrounding rock pressure for every 1 m increase or decrease of B: when taking = 0.2; when taking = 0.1.

Bring into the formula to get:

$$\omega = 1 + i\,(B + 5) = 1 + 0.1 \times (6.4 - 5) = 1.14$$

$$h_a = 0.45 \times 2^{S-1}\omega = 0.45 \times 2^{6-1} \times 1.14 = 16.416m$$
$$h = 5.8m \lhd h_a = 16.416m$$

The tunnel studied in this paper is judged to be an ultra-shallow buried tunnel.

4 MECHANISM OF SUPPORTING STRUCTURE

4.1 *Supporting the function of shotcrete*

The purpose of shotcrete is to cover the surface of the newly excavated surrounding rock of the tunnel. Its aim is to prevent the weathering of the surrounding rock from long-term contact with air and to prevent liquefaction and large amount of soil loss due to contact with water. Insufficiently covered surrounding rock, long-term contact with the air will cause the surrounding rock to deteriorate gradually, which lowers the stability of the tunnel.

4.2 *The role of steel arch support*

During the construction of the weak and broken surrounding rock tunnel, the surrounding rock has poor self-stabilization ability and a short self-stabilization time after the tunnel is excavated. Ordinary spray-anchored support is not enough to provide effective support resistance. In this case, to control the plastic deformation of the surrounding rock, a steel arch with higher supporting strength and rigidity must be used to support the tunnel with spray-anchored support. To provide support for the tunnel in a short time after excavation, strong support of surrounding rock is needed (Siddall et al. 1992).

4.3 *Bolt supporting function*

When the bolt is supported, the bolt body and the surrounding rock are displaced with each other to generate anchoring force, which will prevent the surrounding rock from being damaged due to excessive deformation. Generally, the study of the interaction mechanism between bolts and surrounding rock can be carried out from the following three stages (Xu et al. 2019):

1) Early role of the anchor rod
   At this time, the role of the anchor rod is mainly to control the sliding, falling, and instability of the rock mass. The more timely the anchor rod is installed, the better the effect.
2) The mid-term role of the bolt
   This stage is mainly due to the rheological effect of the rock. With time, the strength of the rock mass continues to decrease, the roof sinks, and the force on the bolts increases gradually, and finally, a certain range of failure zones is formed.
3) The later role of the anchor
   When the support time is very long, the damage zone of the surrounding rock will be further enlarged, the deformation of the surrounding rock will increase sharply, and the force of the bolt will increase.

5 OPTIMIZATION ANALYSIS OF INITIAL SUPPORT DURING TUNNEL EXCAVATION

The tunnel body structure is designed according to the principles of New Austrian Methodology, that is, a composite lining that combines the initial support and secondary molding concrete with system bolts, shotcrete, steel mesh, and steel arches, and cooperates with advanced pre-grouting support. According to the engineering analogy and combined with the results of the structural check, the support parameters of the tunnel body lining are drawn up, and the reserved deformation is 10 cm.

5.1 *Principles of surrounding rock pressure calculation*

When $h_1 \leq h_a$ the tunnel is fully self-heavy and shallow,

Vertical surrounding rock pressure:

$$q = \sum_i \gamma_i h_i \tag{4}$$

Lateral pressure:

$$e_1 = \gamma h_1 \lambda \tag{5}$$

$$e_2 = \gamma h \lambda \tag{6}$$

Where: $q=$ Vertical surrounding rock pressure, KPa; $\gamma_i=$ The soil bulk density of each layer, KN/m^3; $h_i=$ The thickness of each layer, (m); e_1,$h_1=$ The side pressure of the cave roof and the depth of the cave roof,(m); e_2, $h=$ The pressure on the bottom side of the cave and the depth of the cave roof; and $\lambda=$ Side pressure coefficient, $\lambda = tg^2(45° - \phi_0/2)$, in, ϕ_0 Calculated friction angle of surrounding rock.

Bring in the formula to get:

$$q = \sum_i \gamma_i h_i = \gamma_1 h_1 + \gamma_2 h_2 = 18.9 \times 13.78 + 19.2 \times 1.62 = 291.546KPa$$

$$\lambda = tg^2(45° - \phi_0/2) = tg^2(45° - 37.01°/2) = 0.25$$

$$e_1 = \gamma h_1 \lambda = 18.9 \times 5.8 \times 0.25 = 27.405KPa$$

$$\lambda = tg^2(45° - \phi_0/2) = tg^2(45° - 40.26°/2) = 0.21$$

$$e_2 = \gamma h \lambda = 19.2 \times 15.4 \times 0.21 = 62.093KPa$$

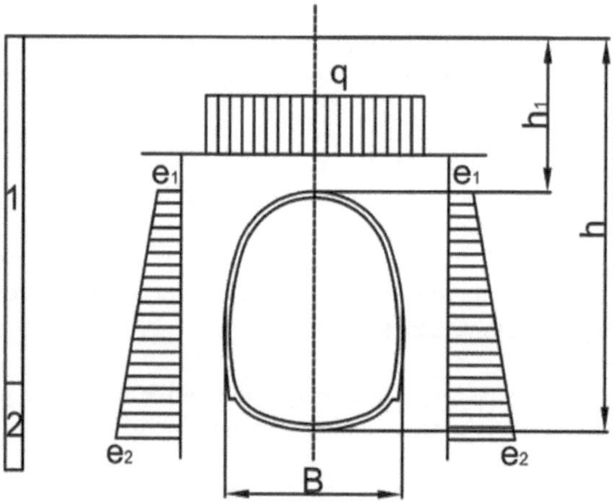

Figure 1. Tunnel force diagram.

5.2 *Model analysis and establishment*

The tunnel is shallowly buried and has a large cross-sectional area, so the two-step temporary invert method is adopted for the tunnel excavation. The two-step temporary invert method is used after the excavation of the upper step is completed, the construction of the upper step is temporarily inverted, and the upper step arch is closed into a ring to ensure the stability of the initial support. Then, after excavating the lower steps to form all closed loops, the temporary invert arch is demolished.

5.2.1 *Calculation parameters*

The entrance section of the tunnel is grade VI, and the surrounding rock is soft. The section that the tunnel passes through is mainly silty clay, and mechanical excavation is used to reduce soil disturbance.

Table 1. Engineering mechanical parameters of each soil layer.

Name	Severe (kN/m^3)	Elastic Modulus (MPa)	Poisson's ratio	Cohesion c(kPa)	Internal friction angle (°)	Expansion angle (°)
Silty clay	18.9	2e3	0.35	31.61	16.39	37.03
Fully weathered granite	19.2	29.823e3	0.25	31.46	24.03	40.26
Moderately weathered granite	26.1	35.266e3	0.30	50.00	42.5	45.10

Model establishment

This paper mainly studies the application time of initial support and concrete thickness: S40H25 indicates that the softening degree of elastic modulus is 40%, the thickness of shotcrete is 25 cm, and the softening degree of elastic modulus can be assumed as the initial application time.

Table 2. Model parameters.

Name	Modulus of elasticity softening degree (%)	Shotcrete thickness (mm)
S40H23	40	23
S50H23	50	23
S60H23	60	23
S40H25	40	25
S40H27	40	27

Timing analysis of initial support

1) Vertical displacement of surrounding rock

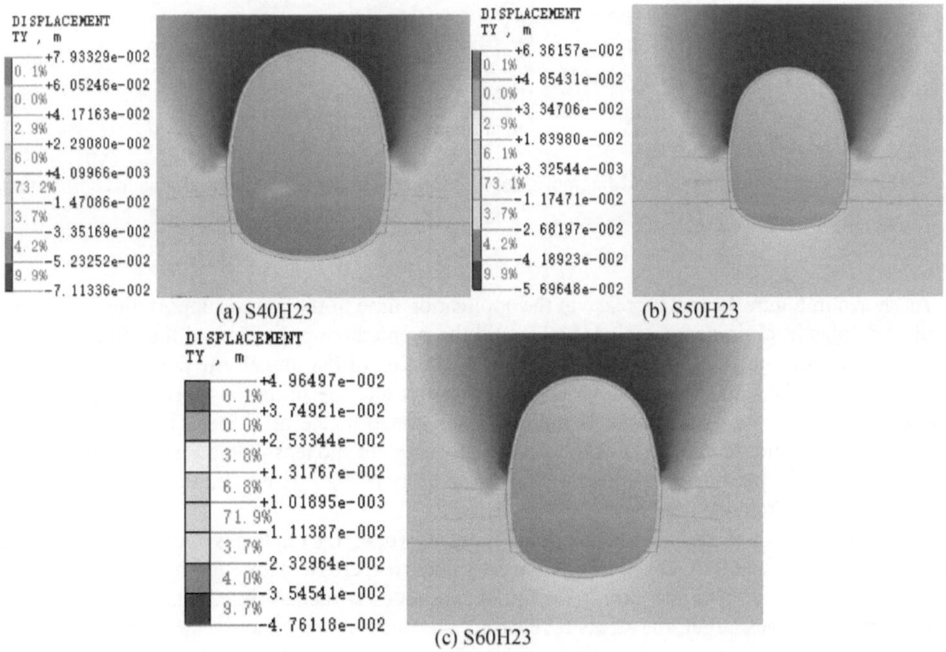

(a) S40H23 (b) S50H23 (c) S60H23

Figure 2. Vertical displacement cloud diagram of surrounding rock under different initial support construction time.

According to the analysis in Figure 2, with continuous increase in the initial support construction time, the vertical displacement of the surrounding rock shows an overall increasing trend. The maximum vertical displacements are 79.33 mm, 63.61 mm, and 49.64 mm, respectively, all of which occur in temporary inverts. The location of the application and the arch bottom is uplifted obviously, the deformation is greater than the settlement of the archtop, and the deformation of the tunnel is within the allowable limit of deformation.

2) Analysis of plastic zone of the surrounding rock

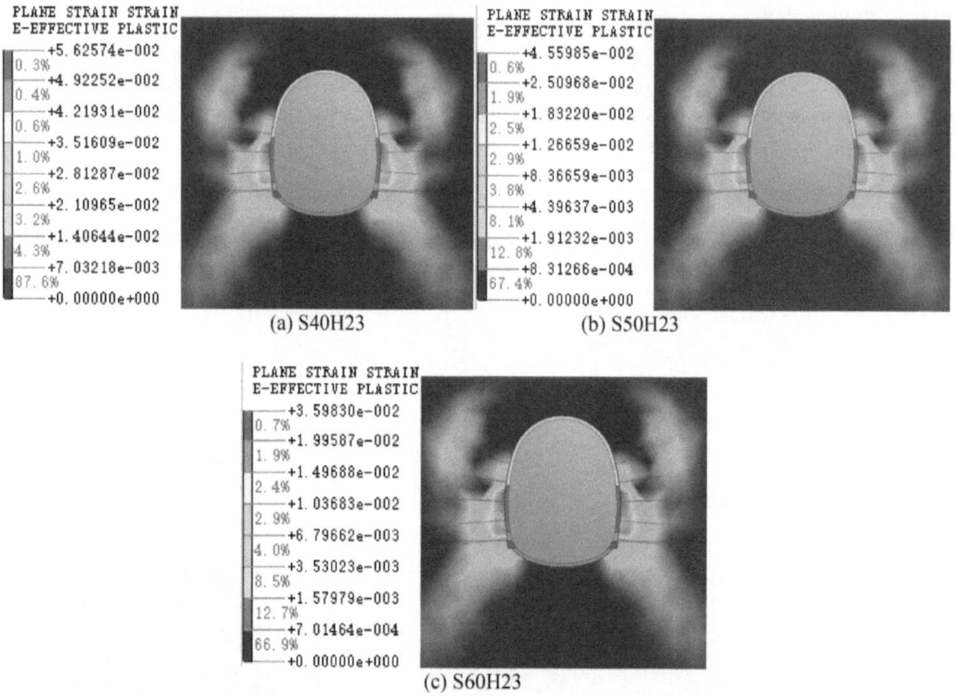

(a) S40H23 (b) S50H23

(c) S60H23

Figure 3. Vertical displacement cloud diagram of surrounding rock under different initial support construction time.

As shown in Figure 4, with increase in the application time of the initial support, the maximum equivalent plastic strain decreases gradually, and the premature application of the initial support makes the plastic zone larger. Moreover, during excavation of the tunnel, the plastic zone is first concentrated at the intersection of the temporary invert and the initial support. In general, the influence range of the plastic zone is mainly concentrated at the arch toe. Therefore, shallow tunnels should be closed into a ring in time to form a complete and reasonable force system, reduce the diffusion of the plastic zone, and maintain the stability of the surrounding rock.

3) Analysis of shotcrete thickness

According to the analysis in Figure 2(c) and Figure 4(a)(b), as the thickness of concrete increases, the vertical displacement of surrounding rock continues to decrease. However, it can be seen from the above analysis that with increase in the initial support construction time, the surrounding rock deformation becomes larger, the stress is released, the force on the concrete structure is reduced, and the deformation of the concrete is reduced. Therefore, the thickness of the concrete should be reduced. It should be based on the actual situation combined with economic benefits, etc., a reasonable choice.

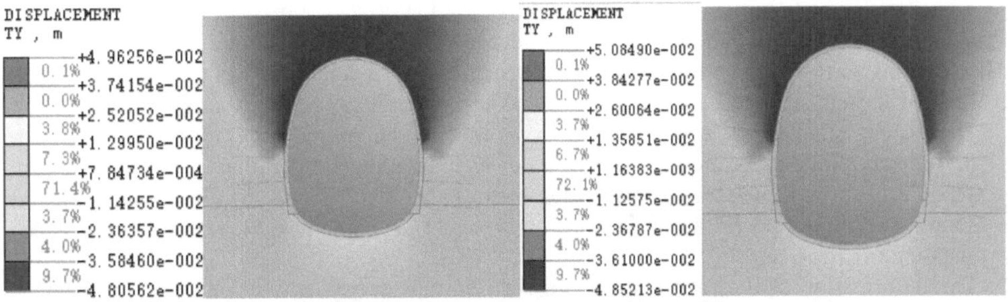

(a) S40H25 (b) S40H27

Figure 4. Vertical displacement cloud diagram of surrounding rock under different concrete thicknesses.

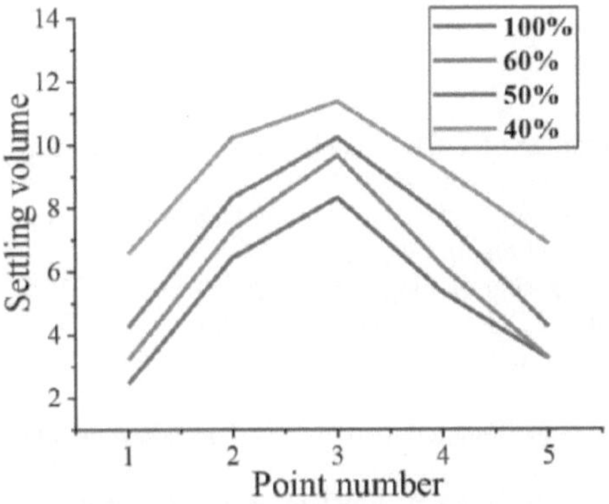

Figure 5. Surface settlement diagrams at different timings of the initial support construction.

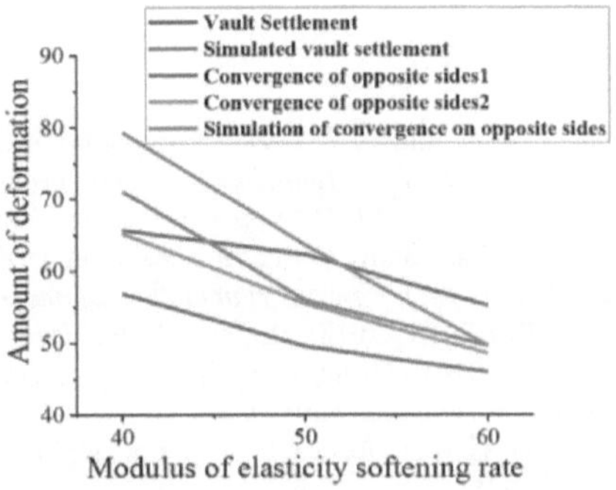

Figure 6. Deformation diagrams of surrounding rock at different timings of the initial support construction.

5.2.2 *On-site monitoring*

On the construction site, the inverted arch settlement of the tunnel, the side convergence, and the ground settlement were monitored. The tunnel was constructed for two-step temporary inverted arch construction. One vault settlement monitoring point and two pairs of opposite side settlement monitoring points were arranged respectively, and the surface settlement was monitored. They were arranged on the centerline, the sideline of the tunnel, and at the distance from the centerline twice the width of the tunnel, that is, a total of five points.

It can be seen from the figure that with increase in the initial support construction time, the surrounding rock and ground settlements increase, and the dome settlement and the convergence of the opposite edges are also in line with the simulation results. The research results have guiding significance in reality.

6 CONCLUSION

Taking Longda Tunnel as the background, this paper studies the optimization of initial support during the tunnel excavation process, and draws the following conclusions:

(1) When the New Austrian Tunneling Method is adopted for the initial support of the shallow-buried soft surrounding rock tunnel, the earlier the support time, the smaller is the deformation of the surrounding rock of the tunnel. Moreover, the uplift of the arch bottom of the tunnel is greater than the settlement of the vault top, followed by convergence around the tunnel. The

tunnel is shallowly buried, with less overlying soil, low vertical pressure, and slippage of the overlying soil, resulting in the small settlement of the vault and offset to both sides.

(2) According to the distribution of the plastic zone of the surrounding rock, the initial support time of the tunnel is too early, which results in large plastic zone of the surrounding rock. The plastic zone mainly appears at the excavation steps. As the excavation progresses, the plastic zone shifts to the arch toe, which is consistent with the larger uplift of the arch bottom. Therefore, for shallow tunnels, due to less disturbance, they should be closed into a ring as soon as possible to form a complete force system and maintain the overall stability of the surrounding rock.

(3) The deformation of the tunnel vault and arch bottom decreases with increase in shotcrete thickness, but the arch waist slightly increases. To give full play to the role of other supporting structures and consider economic benefits, the thickness of concrete should be appropriately reduced, and a reasonable choice should be made according to the actual situation.

REFERENCES

Feng, J. M. & Yan, C. W. (2019). Evaluation of installation timing of initial ground support for a large-span tunnel in hard rock. *Tunneling and Underground Space Technology*. pp. 93–103.

Garnes, R. S. et al. (2019). Support System for Tunnelling in Squeezing Ground of Qingling-Daba Mountainous Area: A Case Study from Soft Rock Tunnels. *Advances in Civil Engineering*. pp. 19–29.

Han, R. G. (1987). *New Austrian method of underground engineering*. Beijing. Science Press.

JTG 3370.1-2018. (2018). Highway Tunnel Design Code Volume One Civil Engineering. China.

Liu, H.R. (2016). Optimal analysis of support parameters for large deformation tunnels in carbonaceous shale. *Lanzhou Jiaotong University*.

Liu, J. H. (2015). Research on optimization of soft rock tunnel support parameters. *Lanzhou Jiaotong University.*

Siddall, R. G. & Gale, W. J. (1992). Strata Control-A New Science for An old Problem. *Mining Engineer.* pp. 341–356.

Wu, K. & Z. S. Shao, Z. S. (2019). Visco-Elastic Analysis on the Effect of Flexible Layer on Mechanical Behavior of Tunnels. *International Journal of Applied Mechanics.* pp. 11–26.

Xu, J.C. & T.D. Ni, T.D. (2019). Displacement ratio dichotomy back analysis of surrounding rock-initial support system of weathered rock tunnel. *Arabian Journal of Geosciences*. pp. 12–30.

Part 3. Intelligent model prediction and disaster risk assessment

Advances in Civil Engineering: Structural Seismic Resistance, Monitoring and Detection – Mohd Johari Mohd Yusof and Junwen Zhang (Eds)
© 2023 The Authors, ISBN 978-1-032-31491-4

Analysis on risk control of deep foundation pit construction of subway station

Cheng Zhang, Ziyu Meng & Jianwu Gong*
Wuhan University of Science and Technology, Wuhan, China

ABSTRACT: According to the deep foundation pit construction of a subway station project, the construction risk factors of foundation pit were identified from two aspects of engineering risk and environmental risk, and the occurrence mechanism was analyzed. Based on the fuzzy comprehensive evaluation method, the analysis model of construction risk has been established, and the weight values of risk factors were determined by analytic hierarchy process, then the risk during construction has been analyzed qualitatively and quantitatively. Results show that the construction of the deep foundation pit engineering has higher risk in the aspects of foundation pit collapse, formwork collapse, incline or crack of surrounding buildings, and damage of pipeline, and some necessary measures should be taken during the construction. According to relative risk factor, the control methods and suggestions have been put forward from some aspects of bored pile construction, suspending pile construction, anchor cable construction, drilling and blasting construction of foundation pit, formwork system construction, and foundation pit deformation control, surrounding pipelines and building protection.

1 INTRODUCTION

With the acceleration of urbanization, subway engineering has increasingly become the main direction of urban public transport development. However, the subway project has large investment, long construction period and complex technology, and involves a large number of deep foundation pit projects. The factors affecting the construction of deep foundation pit have the characteristics of uncertainty, randomness and fuzziness. It is particularly necessary to analyze and evaluate the risk of deep foundation pit construction.

At present, a lot of research work has been done on the risk analysis of deep foundation pit construction. The risk assessment methods are mainly qualitative and quantitative analysis based on probability statistics and fuzzy mathematics (MA 2020; DENG 2009). Based on the accident data of deep foundation pit engineering, a fuzzy comprehensive evaluation method of deep foundation pit risk was proposed based on Bayesian network (ZHOU 2009). The fuzzy comprehensive evaluation method for risk assessment was put forward according to the influencing factors of subway station engineering construction (CHEN 2006). According to practical engineering, the risk of deep foundation pit engineering during construction was evaluated by using expert investigation method and analytic hierarchy process, and the common methods of construction risk treatment was discussed (HUANG 2005). Aiming at the weight ranking of risk factors in analytic hierarchy process, The problem of solving matrix eigenvector was solved by using MATLAB program (OUYANG 2012). The weight of the model was determined using the coefficient of variation method, which overcomes the disadvantage of long investigation cycle when determining the weight of risk indicators by expert investigation method (ZHANG 2002). In addition, the combination of dynamic

*Corresponding Author: gongjianwu@wust.edu.cn

DOI 10.1201/9781003310884-78

monitoring and numerical simulation is commonly used in engineering to analyze the safety of deep foundation pit excavation process (SHI 2003; WANG 2013).

Combined with the practice of deep foundation pit engineering of subway station, the risk of deep foundation pit construction process is evaluated from two aspects of engineering structure risk and environmental risk by using multi-level fuzzy comprehensive evaluation method. Specific construction control measures are analyzed according to relevant risk factors, so as to provide a reference for similar engineering design and construction.

2 PROJECT OVERVIEW

The subway station project is located at the intersection of two urban trunk roads. The station is constructed by open cut method. The foundation pit is 175m long, 30m wide and 25m deep, the supporting structure adopts bored cast-in-situ piles and anchorage, and the groundwater is treated by pit stoppage and pit collection. The main structure of the station is a three-story, three-span frame structure. The top plate of the station is covered with soil about 2.7m ~ 3.5m.

The cross section of foundation pit and stratum geological conditions is shown in Figure 1.

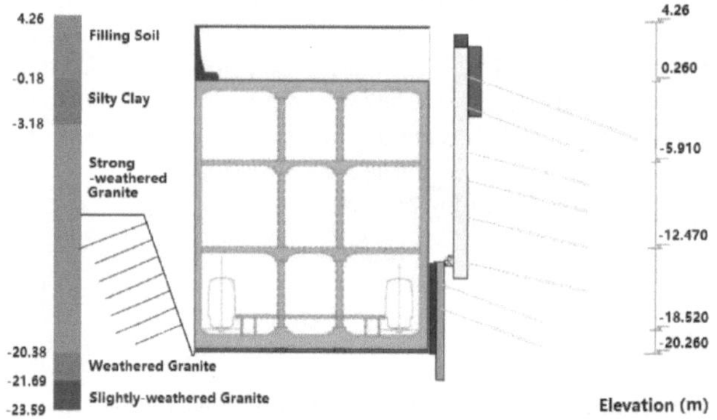

Figure 1. Cross section of foundation pit.

According to the pipeline data, there are mainly pipelines such as sewage, rainwater, drinking water, power supply and communication around the station. The buried depth is about 1.0~3.5m. There is a rainwater culvert across the station, the size is 4.5m×1.8m, the buried depth is 3.3m. And the buried depth of a DN500 sewage pipe is 4.4m and crosses the main structure of the station, which has a great impact on the project construction.

3 RISK SOURCE IDENTIFICATION AND ANALYSIS

3.1 *Analysis of engineering structure risk*

3.1.1 *Bored pile construction*

The foundation pit of the station is about 25m deep, and the strata are mixed fill, mucky silty clay, and coarse sand, strong lower rock layer, moderate and breezy tuff. The foundation pit support structure adopts bored cast-in-place pile.

During the construction of rotary drilling in soil and strongly weathered rock stratum, when encountering uneven soft and hard or boulder, the position deviation will occur, and the collapse of hole may occur in the drilling process. When the rotary drill is constructed in moderately weathered and slightly weathered rock, there is a risk that the bit cannot enter the rock.

3.1.2 *Construction of the sling pile*

The bottom of the foundation pit is located in the micro-weathered layer, and the hard rock face is relatively high, and the sling pile is used. The length of the bored pile is 8.2m, and the bottom of the pile is 7.55m away from the bottom of the foundation pit.

Due to the limited geotechnical drilling data and the different distribution of stratigraphic layers, there may be a large depth of micro-weathered rock in some areas. It will result in insufficient depth of the sling piles deep into the micro-weathered rock and cannot support the surrounding, resulting in the risk of pile foundation damage and foundation pit collapse in the process of foundation pit excavation.

3.1.3 *Prestressed anchor construction*

The system adopts suspension piles and anchor cables and rock anchors. The upper foundation pit adopts a sling pile with a diameter of 1m, the spacing is 1.5m, and the vertical spacing of anchor cable is 2.5m.

There is a risk of deviation between the drilling position and inclination of prestressed anchor cable, which will increase friction and cause prestress loss. If debris falls into the hole due to unclean hole-cleaning and insufficient borehole protection, it will lead to insufficient adhesion between mortar and rock mass, resulting in the risk of overturning of foundation pit retaining structure.

3.1.4 *Foundation pit drainage measures*

The main structure of the foundation pit is mainly located in silty clay, medium coarse sand and sticky clay gravel. If the drainage measures in the foundation pit are not in place, water will accumulate in the pit and affect the stability of the foundation pit.

3.1.5 *Construction of the main structure formwork*

As the materials do not meet the requirements, such as mismatch of jacking specifications, insufficient wall thickness of steel pipe, sliding wire of rotating fastener, non-standard operation and influence of natural factors, there is a risk of instability of formwork support system.

3.2 *Analysis of engineering environmental risk*

3.2.1 *Tilting or cracking of surrounding buildings*

One building is a frame-shearing structure with 22 floors above ground and one underground floor. The basic form is reinforced concrete independent column base and some rafts of waterproof board, which is 19.0m away from the main pit of the station. Another building has two floors above ground and one underground floor. The foundation is reinforced concrete independent column base and part of the raft of waterproof board. The closest distance to the main foundation pit is 40m.

These buildings have a long service life and poor foundation resistance to deformation, and are closer to the main pit. The foundation pit construction will cause the horizontal movement of the stratum horizontally. When the support strength is insufficient and the horizontal lateral displacement is too large, the surrounding buildings will be inclined or cracked.

3.2.2 *Deformation of underground pipelines*

The pipelines in the first phase of construction mainly include: DN600 rainwater pipe protected by suspension (buried depth 1.35m); DN300 sewage pipe protected by suspension (buried depth 1.30m); DN600*200 communication pipe suspended by protection (buried depth 1.35m); DN300 water supply pipe protected by suspension (buried depth 1.6m).

The formation deformation caused by the construction of the foundation pit of the station will cause additional internal force and deformation of the pipeline. When the deformation is too large, it is easy to cause problems such as cracking of the pipeline and opening of the non-rigid connection joint.

4 CONSTRUCTION RISK ASSESSMENT

Aiming at the complex environmental conditions of subway station, the multi-level fuzzy comprehensive evaluation method is used for quantitative risk assessment. The risk hierarchy of the deep foundation pit construction is made of three layers. The target layer (A) is construction risk of subway station, the criterion layer (B) is a potential risk accident, and the indicator layer(C) is an evaluation index. The risk hierarchy of deep foundation pit construction is shown in Figure 2.

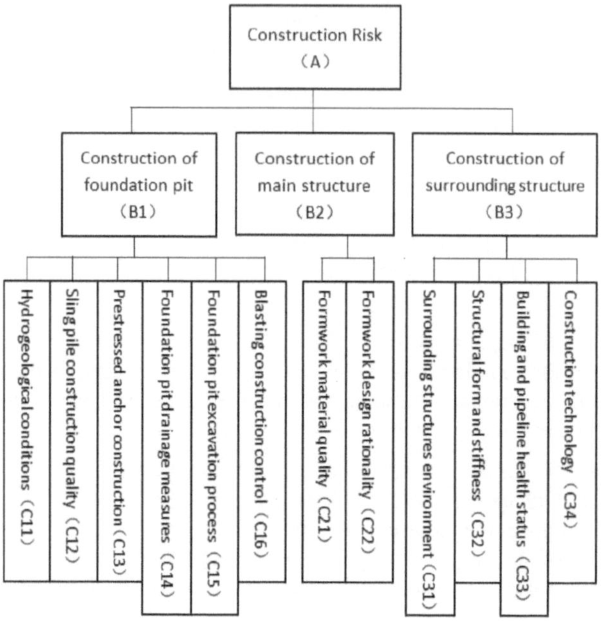

Figure 2. Risk hierarchy of deep foundation pit construction.

Taking the weight of the criterion layer B to the target layer A as an example, the judgment matrix is established as shown in Table 1.

Table 1. Judgment matrix of rule hierarchy B to target layer A.

A	B1	B2	B3
B1	1	4	7
B2	1/4	1	3
B3	1/7	1/3	1

According to the risk matrix recommended in the "code for risk management of underground works in urban rail transit" (GB50652-2011) in China, the risk occurrence level can be calculated, and the corresponding control plan can be formulated according to the risk acceptance criteria in the code.

In combination with the specific engineering conditions of the subway station, the risk factor levels and overall risk levels are obtained by using the above methods, as shown in Table 2.

Table 2. Risk factor and risk level of subway station.

Risk Factor	Indicator level risk level	Criteria layer risk level	Target level risk level
Hydrogeological conditions	IV	II	II
Sling pile construction quality	III		
Prestressed anchor construction	III		
Foundation pit drainage measures	III		
Foundation pit excavation process	III		
Blasting construction control	II		
Formwork material quality	IV	III	
Formwork design rationality	III		
Surrounding structures environment	III	III	
Structural form and stiffness	III		
Building and pipeline health status	IV		
Construction technology	III		

5 ANALYSIS OF CONSTRUCTION CONTROL MEASURES

5.1 *Structure risk control measures for foundation pit engineering*

5.1.1 *Bored pile rotary drilling construction*

According to the formation conditions, the appropriate drill bit is selected. It is strictly prohibited for the surrounding vehicles to disturb the soil layer during the hole forming process.

5.1.2 *Construction of the sling pile*

The length of the pile is adjusted according to the surrounding rock condition reflected by the hole drilling slag, and the pile foundation is ensured to be deep enough in the micro-weathered rock. After the foundation pit is excavated to the bottom of the sling pile, the lock anchor bolt and the steel support shall be constructed in time to prevent the hoisting pile from being in a cantilever state for a long time. When the blasting of the rock below the bottom of the pile is carried out, pre-split blasting is adopted at the excavation boundary to prevent over-excavation and the phenomenon of hanging of the sling pile.

5.1.3 *Prestressed anchor cable construction*

To ensure that the drilling quality meets the requirements, the tensioning equipment must be tested before it can be used. Before the cable is tensioned, the design axial tension value should be used for pre-tensioning, so that the parts are in close contact and the cable body is completely straight. After the anchor strength meets the design requirements, the tension can be tensioned and the tensioning process is strictly carried out according to the design requirements.

5.1.4 *Foundation pit blasting*

The millimeter-difference bench blasting is adopted. When the blasting depth is within 2m from the bottom of the sling pile, the single-row or double-row damper measures are adopted, and the low-power explosive is used to limit the maximum explosive quantity of a single blast.

5.1.5 *Main structure formwork construction*

According to the design requirements and the site conditions, select the appropriate template support system, and make a scientific and reasonable template erection scheme. The quality of the components should be checked before the template is set up and must meet the requirements of the corresponding standards. The erection operation of the formwork should be standardized and reasonable, and the stability of the high support form should be paid attention to during the process of pouring concrete.

5.1.6 *Deformation control of foundation pit*

Ensure the construction quality of the foundation pit support system, complete the pumping and drainage work, strengthen the monitoring during the construction process, the maximum settlement of the foundation pit shall not exceed 0.15%H, and the maximum horizontal displacement of the supporting structure shall not exceed 0.15%H. In case of any abnormal situation, it shall be reported in time and emergency measures shall be taken.

5.2 *Environmental risk control measures for foundation pit engineering*

5.2.1 *Pipeline protection*

The following measures should be taken to protect the pipeline during excavation of foundation pit.

(a) Further explore the environment condition of pipeline.
(b) The bottom soil excavation can be carried out only after the pipeline suspension structure is inspected and qualified by the pipeline management department.
(c) The soil shall be excavated manually at the cable position to prevent damage to the pipeline.

5.2.2 *Surrounding buildings*

In order to reduce the impact of foundation pit construction on surrounding buildings, the following measures are adopted.

(a) Ensure the construction quality of the foundation pit supporting structure and control the ground displacement caused by the foundation pit construction.
(b) Strengthen the monitoring work of the stratum and adjacent buildings, and take timely measures when the building is inclined or cracked.

6 CONCLUSION

Combined with the practice of deep foundation pit engineering of a subway station, the risk factors and control measures of the foundation pit construction process were analyzed from the aspects of engineering structure and environment, and the following conclusions were drawn.

(1) The risk factors of foundation pit collapse are mainly hydrogeological conditions, construction quality of sling piles, prestressed anchor, and foundation pit supporting structure, drainage measures for foundation pits, excavation process and blasting control. The construction risk of the main structure is mainly related to the quality of formwork material materials, formwork design rationality and construction quality.
(2) The deep foundation pit construction of the subway station has the risks of foundation pit collapse, collapse of the main structure formwork, inclined cracking of surrounding buildings, and pipeline damage. The results of fuzzy evaluation analysis show that the overall risk level of the foundation pit project is level II, and relevant control measures must be taken in the construction process
(3) According to the structural risk and environmental risk factors of the foundation pit construction process in the subway station, some control measures and suggestions are put forward, which can provide a reference for construction risk control of similar projects.

REFERENCES

Chen Shenlong, CHEN Longzhu, SONG Chunyu. (2006). Fuzzy synthetic evaluation on risk of subway station construction. *Chinese Journal of Underground Space and Engineering*. 2(1), 32–41.

Deng P. (2020). Overview of research status of foundation pit risk assessment. Sichuan Cement. (10), 325–326.

Huang Hongwei, BIAN Yihai. (2005). Risk management in deep foundation pit engineering construction. *Chinese Journal of Underground Space and Engineering*. 1(4), 611–614.

Ma Ningning. (2020). Risk analysis and control of deep foundation pit construction in subway station. Smart City. 6(11), 222–223.

Ouyang Jingwen, WANG Yueming, ZHANG Xianle, et al. (2012). Risk analysis and application of matlab based on ahp-fuzzy comprehension evaluation. *Sichuan Architecture*. 32(2), 280–283.

Shi Yalin, YANG Shunan. (2003). Safety monitoring in deep foundation pit retaining. *Journal of Safety and Environment*. 3(2), 32–35.

Wang Hongde, QIN Yubin. (2013). Process simulation and safety analysis for the metro station deep foundation excavation in case of weak geo-conditions. *Journal of Safety and Environment*. 13(2), 196–201.

Zhang Tienan, LI Jinglei. (2002). Application of multi-step fuzzy comprehensive evaluation. *Journal of Harbin Engineering University*. 23(3), 132–135.

Zhou Hongbo. (2009). Risk assessment of deep foundation pit by using fuzzy comprehensive evaluation method based on bayesian networks. *Journal of Shanghai Jiaotong University*. 43(9), 1473–1478.

Advances in Civil Engineering: Structural Seismic Resistance, Monitoring and Detection – Mohd Johari Mohd Yusof and Junwen Zhang (Eds)
© 2023 The Authors, ISBN 978-1-032-31491-4

The analysis report of the design research benefit of the light-duty anti-corrosion partially prestressed RPC transmission tower in the power grid

Xinmin Yu & Xiaolin Shi
State Grid Fujian Economic Research Institute, Fuzhou, China

ABSTRACT: In order to enhance the anti-corrosion ability and wind resistance of electric transmission tower lines, and greatly reduce the accidents of broken transmission towers, oblique transmission towers and inverted transmission towers, theoretical analysis, numerical simulation analysis, experimental research and other methods are used to analyze the effect of light-duty anti-corrosive partially prestressed reactive powder concrete (RPC) transmission towers in the power grid. Application for research. The lightweight anticorrosive part of the prestressed RPC transmission tower adopts a standardized design, which has the advantages of light weight, high strength, durability and strong corrosion resistance. It is suitable for distribution line projects of 10kV and below, and can provide effective solutions for the safe operation of distribution network lines. It can reduce the large amount of manpower and material resources spent in post-disaster reconstruction, greatly improve the operational safety and power supply reliability of the power grid, reduce power outage losses, and reflect major economic and social benefits.

1 INTRODUCTION

The lightweight anticorrosive partially prestressed RPC transmission tower is a new type of high-strength concrete transmission tower that uses hybrid fiber ultra-high performance concrete material and is combined with some prestressed steel bars. This product adopts a standardized design and has light weight, high strength and durability. And the advantages of strong corrosion resistance. It is suitable for distribution line projects of 10kV and below, and can provide effective solutions for the safe operation of distribution network lines (Dong and Chen 2019). The appearance of the RPC transmission towers is shown in Figure 1.

The main research content of this project is to study the application of light-duty anti-corrosion partially prestressed RPC transmission towers in power grids through theoretical analysis, numerical simulation analysis, and experimental research, and propose a RPC material mix suitable for transmission tower production, RPC transmission towers The manufacturing process, the reinforcement of some prestressed RPC transmission towers, the connection method and the new assembled foundation are designed. After the promotion and application of the technical results of this project, the corrosion resistance and wind resistance of the transmission tower lines can be significantly enhanced, transmission tower breakage, oblique transmission tower and transmission tower inverting accidents can be greatly reduced, the operation safety of the power grid and the reliability of power supply can be greatly improved, and the loss of power outage can be reduced. Which can reduce the large amount of manpower and material resources spent in post-disaster reconstruction and reflect major social and economic benefits.

 DOI 10.1201/9781003310884-79

Figure 1. The appearance of RPC transmission towers: (a) Appearance of the bottom. (b) Appearance of the top.

2 FEATURES OF LIGHTWEIGHT ANTICORROSIVE PARTIALLY PRESTRESSED RPC TRANSMISSION TOWER PRODUCTS

2.1 *Light weight*

The compressive strength of RPC concrete is more than twice that of ordinary concrete (Tepfers 1979). Under the same bearing capacity, the wall thickness of RPC transmission towers can be reduced by half compared with traditional transmission towers, and the weight is reduced by 30% ~ 50%.

2.2 *High strength*

RPC transmission towers are designed with composite high-strength concrete and partial prestressed high-strength reinforcements. Under the same specifications, the strength of RPC transmission towers can be increased by 10% ~ 20% compared with ordinary transmission towers, and the breaking strength can reach 2.6 times the bending moment of the standard inspection (Albermani, Kitipornchai et al. 2009).

2.3 *Strong durability*

RPC composite concrete has high impermeability, oxidation resistance and salt spray corrosion resistance. In addition, some prestressed tendons are used to effectively improve the resistance to longitudinal cracks of the transmission tower. The durability of RPC transmission towers is significantly improved compared with ordinary transmission towers (Albermani and Kitipornchai 2003).

3 TECHNOLOGY HIGHLIGHTS

Traditional non-prestressed transmission towers have problems such as heavy weight and longitudinal cracks. This product solves the above problems through research and adoption of new materials and new processes. The highlights are as follows:

A new type of hybrid fiber super high-strength concrete was developed. Through theoretical calculations and laboratory tests, basalt fibers and fibers commonly used in current ultra-high-strength transmission towers, and a new type of hybrid fiber high-strength concrete ratio is developed to improve the durability, corrosion resistance and corrosion resistance of transmission towers as a whole. Problems such as transverse cracks (Chen et al. 2010).

A high-strength electric transmission tower with partially prestressed reinforcement and hybrid fiber was developed. Through theoretical calculations, finite element numerical simulations, and actual transmission tower tests, we explored the stress mechanism of high-strength concrete and mixed reinforcements, and formed a set of calculation methods for the mechanical properties of partially prestressed reinforced hybrid fiber high-strength electrical transmission towers, based on reducing the overall weight. Improve transmission tower strength. The experimental sample is shown in Figure 2.

Compile a complete set of standardized atlas of light-duty anti-corrosion prestressed RPC transmission towers suitable for production and processing in Fujian Province, which is convenient for design, selection and application (Chen et al. 2015).

Figure 2. Experimental specimens of RPC transmission towers.

The compressive strength of RPC concrete transmission tower is 100 ~ 170 MPa, and the compressive stress at the most dangerous point of RPC concrete column is about 85Mpa, which is much smaller than the compressive strength of RPC concrete transmission tower, which meets the maximum actual working conditions. According to the load requirements, the transmission tower will not be damaged.

Numerical simulation results show that the various parts of the rigid flange only appear on the inner side of the transmission tower under bending and the welding point of the bottom of the plate. The maximum compressive stress of the stiffening plate is 117.4 MPa, which is less than the yield limit of Q345 steel, so the test design is obtained. Rigid flange, its strength meets the test requirements.

It is feasible to use rigid flanges to connect part of the prestressed RPC transmission towers with the foundation, and provide reference for the connection design of other types of large bending moment and high strength concrete transmission towers. Through theoretical research, the connection form of the RPC transmission tower is designed, the feasibility of the connection form is verified by numerical simulation, and the design method of the connection form of the node is revised, and a rigid flange design method suitable for RPC transmission tower is proposed.

4 ECONOMIC BENEFITS

4.1 *The promotion of this project can greatly improve the structural strength of the distribution network transmission towers, avoid accidents and reduce economic losses*

At present, most of the 10kV distribution network overhead lines in Fujian Province mainly use non-prestressed ordinary concrete transmission towers, and both prestressed and non-prestressed transmission towers are used for 35kV distribution network overhead lines. The number of steel bars inside the ordinary reinforced concrete transmission tower is limited by the geometric size and structure of the transmission tower, so the strength of the transmission tower is limited. Another defect is that it is easy to produce transverse cracks, so that the steel bars are exposed, which is easy to rust and affect the service life of the transmission tower. Fully prestressed reinforced concrete transmission towers increase the strength and quantity of steel bars, and at the same time increase the pressure on the concrete. Excessive compressive stress values can easily cause longitudinal cracks in the transmission towers, and also affect the strength of the transmission towers. Therefore, the increase in the strength of prestressed concrete transmission towers is also limited. In this project, both prestressed steel bars and non-prestressed steel bars are used in the reinforced concrete transmission towers, which can overcome the shortcomings of the above two types of transmission towers at the same time, and can increase the configuration of steel bars, thereby increasing the strength of the transmission towers. Part of the prestressed reactive powder concrete transmission towers are the rational selection of prestressed and non-prestressed steel bars and high-strength concrete to organically combine the three, thereby greatly improving the strength and crack resistance of the concrete transmission towers. Ensure that the transmission towers will not break when encountering strong typhoons, reducing direct economic losses.

4.2 *Utilize the excellent durability of RPC material to extend the service life of the transmission tower*

The RPC transmission tower has superior salt corrosion resistance, improves the durability of transmission towers, and extends the service life of transmission towers. The service life of RPC transmission towers is twice that of ordinary concrete transmission towers, and RPC transmission towers are maintenance-free in the future. The cost is reduced during the entire life cycle of the transmission tower in service. Taking the 230mm tip diameter and 12m long transmission tower as an example, the annual cost of each ordinary transmission tower during the full life cycle is 38.36 yuan, and the annual cost during the full life cycle of the RPC transmission tower It is 26.38 yuan, and the application of RPC transmission towers can save 31.2% of the cost of transmission towers. The experimental process is shown in Figure 3.

Figure 3. Experimental photo of RPC transmission tower.

4.3 *The proposed new fabricated foundation can reduce the investment of manpower and material resources*

When the anti-overturning ability of the new prefabricated foundation proposed in this project is the same as that of the traditional chuck, the weight of a single piece is 1/4 of that of the traditional chuck. It can be done by people, which reduces labor costs by 50%.

The new prefabricated foundation has the characteristics of light weight of a single piece, which reduces the transportation investment, can adapt to the application of mountain power grid lines, and reduces the investment of manpower and material resources.

The transmission towers can greatly improve the structural strength of the electric poles of the distribution network, avoid accidents and reduce economic losses.

5 SOCIAL BENEFITS

5.1 *Improve the strength of electric transmission towers, ensure the safe operation of the power grid, and strongly support social and economic development*

The promotion of this project can reduce power outages caused by typhoons, improve power supply reliability, and ensure that people's normal lives are not affected, thereby improving corporate image and having significant social benefits.

5.2 *Ensure continuous and safe power supply and maintain social stability*

Because RPC transmission towers are more than 30% lighter than ordinary transmission towers under the same strength, they can be used as emergency repair transmission towers and play a major role in typhoon and other disaster repairs. The use of RPC transmission towers can optimize the requirements for wind protection measures, reduce construction costs and improve engineering safety.

The promotion and application of the results of this project can avoid large-scale power outages in the power grid and provide power guarantee for post-disaster reconstruction. Avoiding the social impact and indirect losses caused by power emergencies to the greatest extent, maintaining social stability and the safety of people's lives and properties, has significant social benefits.

6 FUNCTIONS AND APPLICATION SCENARIOS

The light anti-corrosion partially prestressed RPC transmission tower has great technical and economic advantages. The functions and application scenarios are as follows:

Reduce the self-weight of the transmission tower, which is convenient for transportation and construction, and can be used as an emergency repair transmission tower to improve the efficiency of emergency repair.

Increase the strength of transmission towers and reduce the probability of transmission tower breakage and inverted transmission towers in extreme weather such as typhoons.

Strong durability and corrosion resistance, solve the problem of salt spray corrosion in coastal areas.

It is not necessary to pull the cable to solve the problem of land occupation and youth compensation for the distribution line project.

Large bending moment transmission towers can replace narrow base towers and steel pipe transmission towers, reducing project cost.

7 CONCLUSION

Based on the mechanical properties of RPC materials and referring to the relevant design specifications of some pre-stressed concrete transmission towers, the anti-corrosion and high-performance RPC was tested. Based on the mechanical properties of RPC, the optimized design of the structure of light-weight anti-corrosion partially pre-stressed RPC transmission towers was completed. The promotion of this project can greatly improve the structural strength and anti-corrosion characteristics of power transmission towers in the distribution network. In typhoon weather, it can reduce the occurrence of accidents such as broken transmission towers, broken transmission towers, and inverted transmission towers, and improve the safety and reliability of the power grid, which can avoid typhoons. A large number of emergency repair work brought about after the disaster saves a lot of manpower and material resources, reduces the workload of power emergency repair, and has significant economic benefits. It can also reduce power outages caused by typhoons, ensure that people's normal lives are not affected, and improve corporate image, which has significant social benefits.

REFERENCES

Albermani, F. & S. Kitipornchai (2003). Numerical Simulation of Structural Behaviour of Transmission Towers. *Thin-Walled Structures* 41(2–3): 167–177.

Albermani, F., et al. (2009). Failure analysis of transmission towers. *Engineering Failure Analysis* 16(6): 1922–1928.

Chen, L., et al. (2015). Application and design of composite materials in 220 kV power transmission line tower. *Hunan Electric Power* 42(3): 247–262.

Chen, Y., et al. (2010). Corrosion and Protection of Transmission Towers. *Electric Power Construction.*

Dong, C. & X. Chen (2019). Problems in Distribution Network and Safe Operation Management Measures of Distribution Lines. *Modern Industrial Economy and Informationization.*

Tepfers, R. (1979). Fatigue Strength of Plain, Ordinary, and Lightweight Concrete. *ACI Journal* 76.

Advances in Civil Engineering: Structural Seismic Resistance, Monitoring and Detection – Mohd Johari Mohd Yusof and Junwen Zhang (Eds)
© 2023 The Authors, ISBN 978-1-032-31491-4

Research on cross-sea bridge construction based on navigational risk analysis

Shengli Liu, Qinrong Li, Xinpei Hua, Xinrui Zheng & Min Dong
Transport Planning and Research Institute of the Ministry of Transport, Beijing, China

ABSTRACT: To support the construction of cross-sea bridge, the navigation risk in the waters after the completion of the cross-sea channel are required to be predicted. Based on the IWRAP basic theory recommended by the International Maritime Organization and less affected by subjective factors, taking the Zhoushan-Shanghai sea crossing bridge as an example, the navigation risk of the waters of the proposed sea crossing channel is calculated from the aspects of scene setting, channel terrain modeling, bridge area traffic flow fitting, navigation hole modeling, etc. According to the bridge navigation-setting standards of cross-sea bridge, the change of ship channel/route after bridge construction is analyzed. The navigation risks of four cross-sea channel construction schemes are calculated and compared, respectively, and the channel construction scheme with the smallest change in risk level is obtained, which provides technical support based on navigation risk considerations for supporting the construction of the cross-sea bridge.

1 INTRODUCTION

With rapid economic and social developments in the Yangtze River Delta, the Guangdong-Hong Kong-Macao Greater Bay Area, and the Beibu Gulf Urban Agglomeration, a number of cross-sea bridges, such as the Hong Kong-Zhuhai-Macao Bridge, ShenZhen-ZhongShan Bridge, East Sea Bridge, and Hangzhou Bay Bridge have been constructed in China in the estuaries and bays, such as the Yangtze Estuary, the Hangzhou Bay, and the Pearl River Estuary. In the waters where it is proposed to build cross-sea bridges, the navigation environment and situation will be considerably changed after the bridge construction. On the one hand, building bridges on the estuaries and bays, which are generally relevant to the industries development, functional adjustment, and shoreline utilization of related port areas, the construction of the bridge changed the collecting and dispatching system in these waters and the vessel traffic in these waters also changed correspondingly. On the other hand, the construction of cross-sea bridge will cause obvious changes to the traffic organization, by laying piers to form several or more navigable holes, which forces the reconstruction of the traffic organization of the waters in the whole bridge area. The bay area waters are usually flat with abundant water depth, where the ships are of free navigation. After the construction of the bridge, the ships are gradually merged into navigable holes and the route widths are considerably narrowed near the navigable holes. Part of the channels are affected by the investment and risk control, so no navigable holes are reserved in the existing routes, which leads to the disappearance of some existing routes. Other parts of channels are adopted by the scheme of bridge-tunnel combination, which lead to partial ships' preferring the tunnel navigation in the premise of no significant increase in voyages. Therefore, the construction of cross-sea bridges significantly influences the navigational risk of bridges' location in waters, which is necessary to predict and analyze the navigation situation according to the traffic organization scheme after the completion of bridges.

Usually, at the planning stage of cross-sea channels, there are many channel lines and schemes for being compared and selected. At the decision-making stage of a channel construction scheme,

 DOI 10.1201/9781003310884-80

a very important factor that needs to be taken into account is the navigational risk, which is caused by the scheme. Hence, in the preliminary work of cross-sea bridges construction, how to quantify the changes of navigational risk for the whole waters caused by different schemes is one of the major problems faced by engineering decision-makers.

The universally used risk analysis methods at present, including the analytical hierarchy process, fuzzy comprehensive evaluation method, Bayesian network method, grey relation analysis, Monte Carlo method, and artificial neural network method, are all based on the huge basic data and experienced personnel for weighting value judgment. Only limited information on various aspects after the bridge construction can be predicted by navigational risk analysis of cross-sea bridges construction at the decision-making stage. Moreover, few persons can scientifically weigh the risk factors after the construction of the bridge and it is difficult for an existing method to be adopted for large sea area risk assessment. Hence, there is urgency to develop a risk analysis method that is based on the extension of historical statistics law with minimal human factors as well as easy operation procedures. Many risk analysis methods have been previously reported. Chen et al. combined Bow-tie with fuzzy set theory to construct the vessel-bridge collision risk assessment model of the cross-sea bridge and disclosed the main factors of vessel-bridge collision by sensitivity analysis method (Chen 2018). Yang et al. proved the vessel-bridge model with HUGIN software and ranked the risk factors (Yang 2015). Gao et al. established a finite element model by Ansys/LS-Dyna to simulate the vessel collision at different water levels and impact angles, and analyzed the vessel-bridge collision risk with AASHTO standard (Gao 2020). Ke et al. utilized the IWRAP model to simulate and verify the vessel traffic flow data as well as screened out the best fitting distribution function and improved the accuracy of operation results with the IWRAP model (Ke 2018). The risk of vessel collision and grounding in the Yangtze Estuary was evaluated with the IWRAP model by Huang et al (Huang 2017). Various navigational risk assessment methods were comprehensively taken into consideration and based on IWRAP (IALA Waterway Risk Assessment Programme), which is recommended by international maritime organization, the navigational risks in different schemes before and after the construction of Zhoushan to Shanghai cross-sea channel were calculated in this work, which provided technical support to the decision-making of scheme and scientific foundation to the optimization of navigable holes setting in the cross-sea channels.

2 PROJECT OVERVIEW

Located outside the baymouth of Hangzhou Bay, Zhoushan-Shanghai cross-sea channel plays an important role in strengthening the traffic link between Zhoushan Archipelago New Area and Shanghai, which is of great significance to promote the coordinated development of Ningbo-Zhoushan-Shanghai, as well as accelerate the economic integration of Yangtze Delta region and the development process of excellent urban agglomerations. This project is responsible for the cross-sea channel from Daishan to Yangshan, which forms the necessary route for the ships to enter or leave Jiaxing Port and Hangzhou Bay Port area of Shanghai Port. The waters where the Zhoushan-Shanghai cross-sea channel is built mainly involve Caojing East Channel, Jinshan Channel, and Daishan North Channel. The navigable holes of the East Sea Bridge pass through five routes of Yuxingnao Island Route, Recommended Route of West Route, Customary Route of West Route, and East and West branches of small vessel route of Yushan Bridge. All of these channels and routes are natural waterways. Two channel lines and four construction schemes were proposed in the previous research. The orientation of channel line 1 and channel line 2 were shown in Figure 1. Among which, there are three construction schemes included in line 1, Scheme A is the full-bridge scheme, Scheme B is the "North Bridge South Tunnel" scheme where the existing west channel is maintained while Scheme C is the "North Bridge South Tunnel" scheme where the customary route of West Route is canceled. There is only one scheme included in line 2, which is the "North Tunnel South Bridge".

As shown in Figure 1, channel line 1 crosses the East Canojing Route, the Jinshan Route, the Customary Route of West Route, Daishan North Route, and the West branch of the small vessel

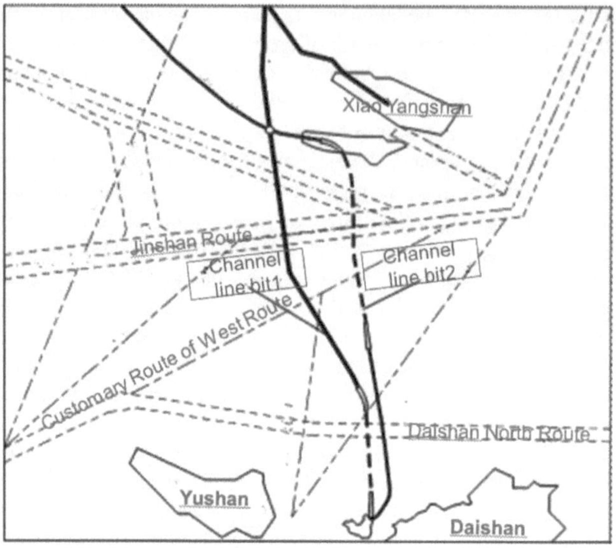

Figure 1. Line location and orientation in scheme.

route of Yushan Bridge and channel line 2 crosses the East Canojing Route, the Jinshan Route, the customary route of West Route, Daishan North Route, and the East branch of small vessel route of Yushan Bridge.

Scheme A is a full-bridge scheme where three channel bridges including the Jinshan Channel Bridge, the Customary Route Bridge of West Route, and the Daishan North Channel Bridge are constructed.

Scheme B is the "North Bridge South Tunnel" scheme where the Jinshan Channel Bridge, the Customary Route Bridge of West Route, and the Daishan North Tunnel are constructed.

Scheme C is also the "North Bridge South Tunnel" scheme where the Jinshan Channel Bridge and the Daishan North Tunnel are constructed.

Scheme D is the "North Tunnel South Bridge" scheme where the Jinshan Channel Tunnel, the Customary Route Tunnel of West Route, and the Daishan North Channel Bridge.

Table 1. Channel scheme setting.

	Scheme			
Channel/Route	A	B	C	D
Jinshan Route	Bridge	Bridge	Bridge	Tunnel
Customary Route of West Route	Bridge	Bridge	Canceled	Tunnel
Daishan North Route	Bridge	Tunnel	Tunnel	Bridge

3 NAVIGATIONAL RISK CALCULATION FOR CROSS-SEA CHANNELS

3.1 *Theoretical basis of risk calculation*

The navigational risk of cross-sea channels construction is closely related to the location and span of bridges in the channels, the clearance height, and clearance breadth of navigable holes formed by the bridge construction and the water depth, tide, vessel traffic flow, vessel behavior, and vessel trajectory distribution in the bridge construction area. A method of modeling treatment of the

Table 2. Navigating hole setting.

Route of navigable holes	Navigable vessel DWT/t	Arrangement of navigable holes	Clearance height/m	Clearance breadth/m
Jinshan Route	225000 ton cruise ship 100000 ton oil tanker 20000 ton oil tanker 2000 ton tanker	Two Holes: each single hole should allow the one-way navigation of 100000 ton oil tanker	69.6	670×2
Customary Route of West Route	20000 ton oil tanker	Single Hole with bidirection: allow the bidirectional navigation of 20000 ton oil tankers	50.2	960
North Daishan Route	100000 ton oil tanker 50,000 ton oil tanker	Two Holes: each single hole allows one-way navigation of 100000 ton oil tanker	55.9	560×2

traffic flow in the proposed bridge waters was presented in this work, by analyzing the cause types of vessel traffic accident, the vessel behavior and trajectory distribution in the bridge area were quantified, and the number of probable accidents in the area was calculated and forecasted. The navigational risk after cross-sea bridge construction mainly includes the collision risk of vessels and bridge (vessel-bridge collision risk) and the collision risk between vessels in the bridge waters (vessel-vessel collision risk). The theoretical basis is:

$$N = C \times P_G \times P_C \tag{1}$$

Where N is the number of collision, which means the risk level of cross-sea bridge construction; C is the traffic flow of proposed bridge area; P_G is the probability of head-on and P_C is the accident cause probability. To determine the quantity of collision accidents, the number of vessels with head-on probability in the waters should be calculated first.

3.1.1 *Calculation of the number of ships with head-on probability*

The head-on scenarios of vessels can be divided into four types, including head-on, overtaking, crossing encounter, and stranding, which are illustrated in Figure 2.

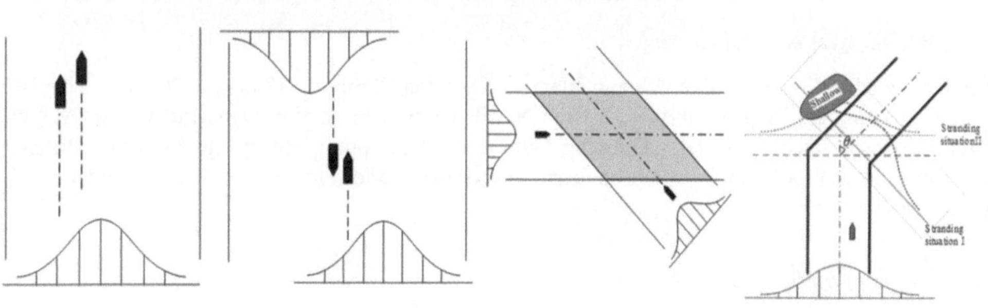

(a) Head-on (b) Overtaking (c) Crossing (d) Stranding

Figure 2. Four head-on scenarios.

The number of vessels with head-on probability can be calculated using the following functions:

$$N = L_C \sum_{ij} P_G \frac{V_{ij}}{V_i^{(1)} V_j^{(2)}} \left(Q_i^{(1)} Q_j^{(2)} \right) \tag{2}$$

$$P_G = \int_{-\infty}^{\infty} \int_{yj=-yi-\bar{B}}^{-yi+\bar{B}} f_{Y_i}(y_i) f_{Y_j}(y_j)\, dy_i dy_j \tag{3}$$

$$V_{ij} = \sqrt{(V_i^{(1)})^2 + (V_j^{(2)})^2 - 2V_i^{(1)}V_j^{(2)} \cos\theta} \tag{4}$$

Where N is the vessel quantity with head-on probability in the case of approaching. L_C is the channel length. P_G is the probability of vessel encounter. V_{ij} is the relatively speed. $V_i^{(1)}$ and $V_j^{(2)}$ represent the average speed of vessel type i and j in the direction of 1 and 2, respectively. $Q_i^{(1)}$ and $Q_j^{(2)}$ stand for the number of vessel type i and j in the direction of 1 and 2, respectively. Yi and yj are the horizontal scale of vessel type i and j respectively. $\bar{B}$ is the width of the vessel. $f_{Y_i}(y_i)$ and $f_{Y_j}(y_j)$ represent the probability density function of vessels coming and leaving, which follows normal distributions with parameters (μ_i, σ_i) and (μ_j, σ_j), respectively.

3.1.2 *The PC value of accident cause probability*

The P_C value of accident cause probability is related to various factors, such as human errors, configuration of navigation facilities, advancement of vessel system, quality of crew and environmental conditions. Based on the Bayesian network method and combining different researchers' theories around the world, the value of P_C proposed in this work is 9.0×10^{-5}.

3.2 *Scene setting*

In the bridge location waters, the main pier, auxiliary pier, and transition pier where the navigable holes are located are all obstacles for ship navigation, which is also one of the most direct factors of cross-sea bridges affecting navigation of bridge waters. In general, the main pier, auxiliary pier, and transition pier, which are within three times the ship's designed length on both sides of the navigable holes, may cause a threat of ship collision. In the four listed head-on scenarios, the ship-bridge accident corresponds to stranding, which means the bridge can be regarded as a shoal obstacle and the collision caused by ship maneuvering and bridge pier may lead to two kinds of stranding, that is, no steering action was taken before the ship collided with bridge pier and the steering action have been taken but still did not avoid colliding with the bridge pier. However, the other three situations of ship collision accidents correspond to encounter, overtaking, and crossing, respectively.

3.3 *Modeling for channel terrain*

Sea maps such as "Nanhuizui to Volcano Islands" (Drawing Number: 13339) and "Craggy Islands" (Drawing Number: 13341) published by the Naval Aviation Protection Department and "Ma'an Archipelago to Daishan Island" (Drawing Number: 52001) published by the Maritime Bureau were selected to proofread the water environment for depth and reference so that the modeling of waters can be completed.

3.4 *Fitting of traffic flow*

3.4.1 *Setting up of AIS ship flow observation line*

Selecting the AIS ship trajectory distribution map in the bridge waters, setting ship flow observation line, and determining the channel/route distribution situation, as shown in Figure 3, 25 AIS ship flow observation lines were set in this work.

3.4.2 *Fitting of traffic flow distribution function*

Based on the five basic mathematical distribution laws of Normal Distribution, Logarithmic Distribution, Uniform Distribution, Weber Distribution, and Gumbel Distribution, the distribution and

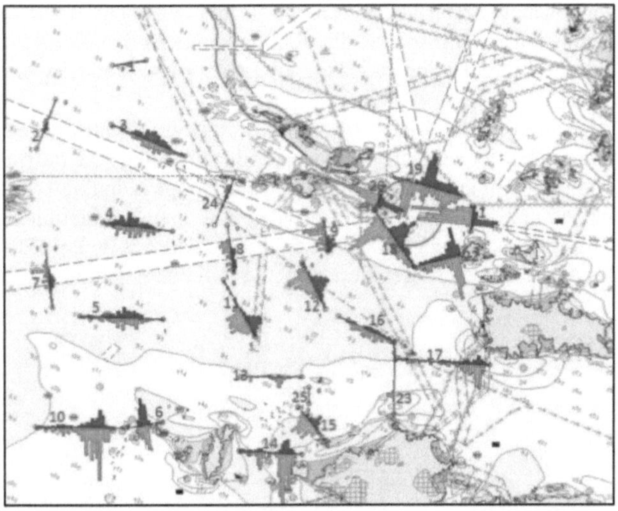

Figure 3. Distribution of AIS ship flow observation lines in the waters of Zhoushan to Shanghai cross-sea channel.

its parameters were determined by mathematically fitting sectional transverse distribution function cylinder diagram, as showed in Figure 4.

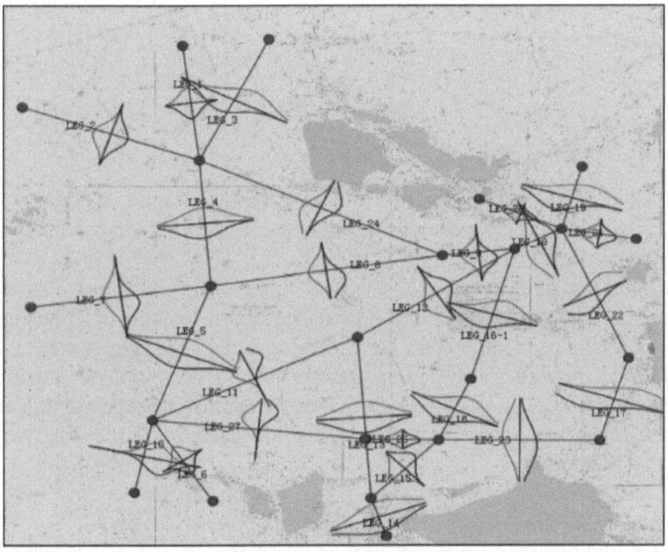

Figure 4. Setting diagram of traffic flow distribution function in the research water area.

3.5 *Modeling of navigable holes*

For these four channel construction schemes, it is necessary to fit the distribution function of the traffic flow in navigable holes waters. As the construction of bridges in the schemes is not finished, the specific traffic flow distribution after their completion is unknown. Therefore, according to the bridge navigation scheme, the ship traffic flow distribution law in the navigable holes in the waters or the near navigable holes of similar bridge can be selected in analogy to the proposed bridge to determine the character values of traffic flow distribution functions like average value and standard deviation after the construction of bridge. And the long-term forecasted flow of channel and route is adaptability adjusted.

The Donghai Bridge and Pingtan Strait Bridge Highway Bridge, which have similar conditions like the proposed bridge, are selected as the reference bridge to analogy the traffic flow distribution. The main navigable holes of Pingtan Strait Bridge Highway Bridge is double-hole and bi-directional navigation whose navigation width is 2 x 180 m. While the main navigation of Donghai Bridge is single hole with bi-direction navigation whose navigable width is 300 m. The AIS data for the main navigable holes of these two bridges were selected to function fitting, whose result is shown in Figure 5.

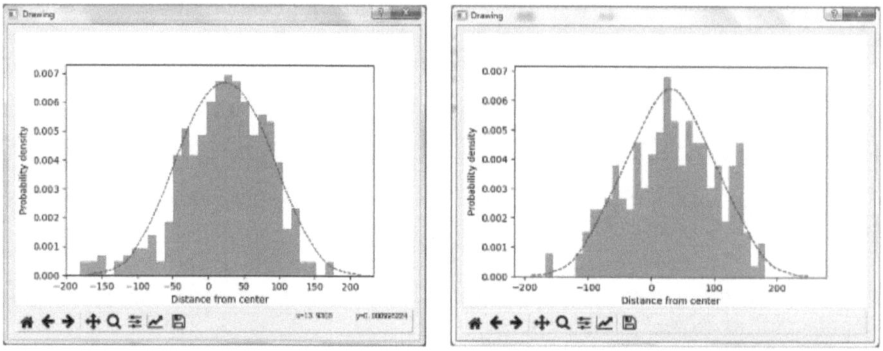

(a) Pingtan Strait Bridge Highway Bridge (b) Donghai Bridge

Figure 5. AIS data function fitting diagram for the above two bridges.

3.6 *Risk Calculation result*

Comparing the risk level after completion of channel construction with no construction of channel and no construction of channel waterway/route adjustment, the following can be obtained:

3.6.1 *Model reliability verification*

According to the observed data of AIS ship flow in the channel waters in Zhoushan to Shanghai, the total number of accidents in this waters was predicted to be 6.584. The statistics from Zhoushan Maritime Bureau and Yangshan Maritime Bureau state that the total number of accidents from 2008 to 2016 occurred in this water was 59, which means the annual average number was 6.556. The predicted number was in good agreement with actual average annual accident, which proved the reliability of model prediction.

3.6.2 *Navigational risk changes before and after the channel construction*

The navigational risk of the waters before and after the channel construction can be calculated by the model. The risk assessment cloud chart in the bridge waters before the channel construction was shown in Figure 6 and the risk assessment cloud chart in the waters after the bridge construction was taken from Scheme B as an example, which is shown in Figure 7.

The calculation results of navigational risk and the risk changes of the above four channel construction schemes are illustrated in Table 3.

3.6.3 *Risk assessment result*

The risk value of no channel construction is 10.729 per year by calculations. Comparing the four channel construction schemes, the navigation safety risk value of Scheme A is the largest, which

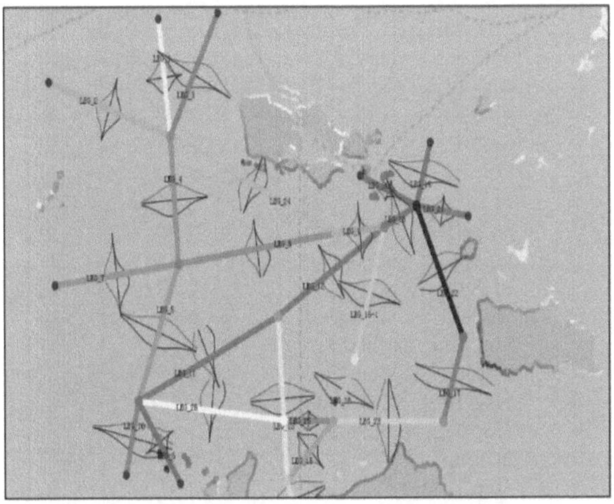

Figure 6. Cloud chart of navigational risk assessment before bridge construction in the study water area.

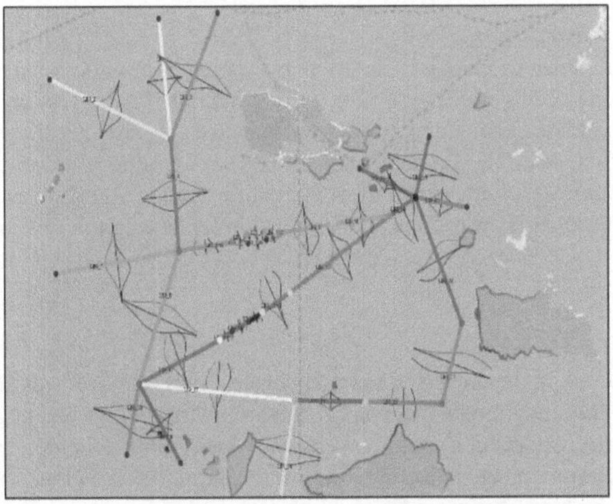

Figure 7. Cloud chart of navigational risk assessment after bridge construction in the study water area.

was 12.639 per year increasing by 17.8% compared with no channel construction. However, the value of Scheme C is the lowest, which was 10.842 per year increasing by 1.1% compared with no channel construction. Moreover, Schemes B and D were in medium changes of navigation safety value.

In the waters near the navigable holes, the risk caused by bridge construction were mainly from two aspects: first, due to the narrowing of the channel, the ships density was increased, which led to ship collision risk between ships due to sheering. Second, the ship-bridge collision risk increased caused by the increased obstacles like piers, which was introduced during the bridge construction in the water. Overall, the risk of ship-ship collision was much higher than that of ship-bridge collision. Therefore, in the later stage of bridge construction, when paying attention to the protection of bridge itself, we should also be concerned about the strengthening of ship traffic organization in the bridge area.

Table 3. Navigational risk calculation results and risk changes of four bridge construction schemes.

Bridge construction scheme	Risk type	Possibility of accident (per year)	Risk value change (%)
No channel constructed	Ship-Bridge collision	2.679	0
	Ship-Ship Collision	8.050	
	In Total	10.729	
Scheme A	Ship-Bridge Collision	2.694	17.8
	Ship-Ship Collision	9.945	
	In Total	12.639	
Scheme B	Ship-Bridge Collision	2.680	13.5
	Ship-Ship Collision	9.535	
	In Total	12.173	
Scheme C	Ship-Bridge Collision	2.377	1.1
	Ship-Ship Collision	8.465	
	In Total	10.842	
Scheme D	Ship-Bridge Collision	2.501	3.9
	Ship-Ship Collision	8.643	
	In Total	11.144	

3.6.4 *Risk control measures*

According to the experience of channel's safety management from home and abroad, it was suggested to implementing ship's routing system, strengthening regional safety management, and establishing bridge collision avoidance system to decrease the channel waters risk. Depending on the experience of international bridges, such as Fermain Bridge, the waters safety risk could be decreased by around 60% with series of combination management measures. All in all, the navigational risk of all schemes was controllable by complementing the risk mitigation measures.

4 CONCLUSION

The navigation safety risk in the cross-sea channel construction schemes was analyzed and quantified in this work, and the navigational risk changes after construction were using different schemes, which provided support for decision-making of actual construction scheme.

Due to traffic environment changes, such as route/channel narrowing, the risk of ship-ship collision increased, which was the main factor of navigation safety risk changing caused by bridge construction.

In the premise of implementing risk control mitigation measures, it was all controllable for the risk changes in the waters, which were caused by the combination of the entire bridge, the north bridge south tunnel, and the north tunnel south bridge proposed by the construction of Zhoushan-Shanghai cross-sea channel.

REFERENCES

Chen Weijiong; Lu Yining; Zhang Shanjie; Xie Qimiao; Xu Xunqi; Guo Jinchun. A method for assessing risk of ship sea-bridge collision based on fuzzy-Bow-tie [J]. *China Safety Science Journal*. 2018,28(01):87–92. (Full-text in Chinese)

Chen Xingwei; Hu Shenping; Fang Cheng. Analysis of Factors Affecting the Severity of Maritime Traffic Accidents[J]. *Journal of Wuhan University of Technology* (Transportation Science & Engineering). 2019(06):93–97. (Full-text in Chinese)

Gao Xiao; Zhao Hongyuan; Yao Haozhen. Risk analysis of ship-bridge collision in urban inland waterway based on AASHTO specification [J]. 2020,57(S1):116–119. (Full-text in Chinese)

Gen Bo; Shang Junnian; Wang Hong and Luo Qiang. Research on Ship Collision Protection Standard of Xiazhang Sea-crossing Bridge Based on Risk[C]. China Conferences. 2010. (Full-text in Chinese)

Huang Chun; Chen Guowei; Chai Tian. IWRAP MK || Software Application in Ship Collision and Grounding Risk Assessment in Yangtze River Estuary Area[J]. Navigation of China. 2017 (01): 79–82. (Full-text in Chinese)

Ke Ranxuan; Hu Xuzhen; Chen Yi. Ship Traffic Flow Data Processing for IWRAP [J]. *Navigation of China*. 2018, 41(04):77–82. (Full-text in Chinese)

Wang Qunpeng. Study on the Model of Dynamic Risk Evaluation on Navigation Safety at Construction Period of Complex Maritime Engineering[D]. *Chinese Master's Theses Full-text Database*. 2014. (Full-text in Chinese)

Wu Jianjun; Gao Shenghui; Wan Yaoyi. Study on navigational risk and Emergency Transportation Route of Tanker in Shanghai Port[J]. *Navigation*. 2021(02):26–30. (Full-text in Chinese)

Yang Xiangrui. Research of Risk Assessment for Ship-Bridge Collision Based on Bayesian Network [D]. *Chinese Master's Theses Full-text Database*. 2015. (Full-text in Chinese)

Advances in Civil Engineering: Structural Seismic Resistance, Monitoring and Detection – Mohd Johari Mohd Yusof and Junwen Zhang (Eds)
© 2023 The Authors, *ISBN 978-1-032-31491-4*

Global earthquake disaster and emergency response in 2021

Yanyun Nan, Yigang Li*, Kang Liu, Boyang Dai, Junyan Lai & Yuan Zhang
National Earthquake Response Support Service, Beijing, China

ABSTRACT: Based on the global earthquake catalog and disaster data in 2021, statistics were obtained on the data of catastrophic earthquakes with Ms $\geq$ 6.0, including those abroad as well as all disastrous earthquakes in China in 2021. Combined with typical earthquake cases, this paper analyzes the characteristics of earthquake damage and emergency work at home and abroad in 2021, as well as the matters needing most attention in earthquake emergency response, especially cross-border rescue under the normalization of the epidemic situation, hoping to provide a reference for future earthquake trend analysis and emergency rescue. The results show that the frequency of global earthquake disasters in 2021 is relatively low and the distribution is relatively concentrated; however, the earthquake damage is severe. Poor earthquake resistance of buildings is still the main cause of casualties. In addition, earthquake emergency response will be faced up with huge challenges under epidemic prevention and control for a long time in the future.

1 INTRODUCTION

In 2021, 112 earthquakes with Ms $\geq$ 6.0 were reported across the world, including 93 earthquakes with magnitude from 6.0 to 6.9, 18 earthquakes from 7.0 to 7.9, and 1 earthquake from 8.0 to 8.9. The frequency of earthquakes with Ms $\geq$ 6.0 is slightly lower than the average level of 115.6 times per year in the past decade (Feng 2012, 2015, 2016, 2017; Li 2019; Nan 2019; Sun 2018), and the largest earthquake with a magnitude of 8.1 occurred on July 29 in the waters in the south of Alaska. The earthquake with the largest number of casualties occurred in Haiti with the Ms7.3 on August 14, which killed 2,577 people (including 329 missing cases) and injured 12,763 people.

2 EARTHQUAKE DISASTERS

2.1 *Foreign earthquake disasters*

In 2021, there were 106 earthquakes with Ms $\geq$ 6.0 abroad, resulting in 2,707 deaths (including 332 missing cases) and more than 19,525 injuries (Table 1). Among them, the Ms7.3 earthquake in Haiti on August 14 killed 2,577 people (including 329 missing cases) and injured 12,763 people, accounting for 95.2% and 65.38% of the total number of casualties with Ms $\geq$ 6.0 abroad in 2021, respectively. It is the earthquake with the most casualties this year. Compared with the casualties caused by earthquakes with Ms $\geq$ 6.0 in the past 10 years (Feng 2012, 2015, 2016, 2017; Li 2019; Nan 2019; Sun 2018), the overall casualties this year is relatively serious (Table 2).

*Corresponding Author: 86219908@qq.com

 DOI 10.1201/9781003310884-81

Table 1. Statistics of earthquake disasters with Ms ≥ 6.0 abroad in 2021.

Date	Beijing Time	Location	Magnitude (M)	Casualties		
				Deaths	Missing	Injuries
01-15	02:28	Sulawesi, Indonesia	6.2	105	3	6489
01-19	10:46	Argentina	6.6			14
02-13	01:01	Tajikistan	6.1	1		
02-13	22:07	Off the east coast of Honshu, Japan	7.1	1		186
04-10	15:00	Sea area south of Java island, Indonesia	6.0	8		39
07-10	10:14	Tajikistan	6.4	5		
07-31	01:10	Peru	6.3			34
08-12	01:46	Waters near Mindanao	6.5	1		
08-14	20:29	Nips Province, Western Haiti	7.3	2248	329	12763
09-08	09:47	Mexico	7.1	3		
11-14	20:08	Hormuzgan Province in southern Iran	6.3	2		
11-28	18:52	Peru	7.3	1		
12-14	11:20	Flores sea	7.3			

Table 2. Comparison of casualties with Ms ≥ 6.0 in foreign and domestic disaster earthquake from 2012 to 2021.

Year	Earthquake with Ms ≥ 6.0 in foreign		Disaster earthquake in China	
	Deaths (including missing cases)	Injuries	Deaths (including missing cases)	Injuries
2012	400+	Thousands	86	1331
2013	800+	2000+	294	15671
2014	19	hundreds	736	3688
2015	9529	Nearly 30000	6	357
2016	1143	20,000+	2	103
2017	1126	15,000+	38	638
2018	3631	20,000+	0	85
2019	212	4336+	17	425
2020	182	2912	5	30
2021	2707	19,525+	9	214

(1) The Ms6.2 earthquake in Sulawesi, Indonesia

At 2:28 on January 15, 2021, an earthquake of magnitude 6.2 occurred in Sulawesi, Indonesia. The epicenter was located at 3.04°S and 118.99°E, with a focal depth of 10 km. The earthquake killed 105 people, left 3 missing, injured 6,489, destroyed 1,150 buildings, damaged 7,683 buildings, and displaced more than 27,000 people. Local power supply was cut off, communications were intermittent, and the roads connecting the disaster-stricken areas were disrupted by multiple landslides.

Indonesia's National Disaster Management Agency coordinated emergency rescue operations after the earthquake. Local rescuers actively treated the wounded, built emergency shelters, evacuated affected people, and counted casualties and property losses. Indonesia's Meteorology, Climatology, and Geophysics Agency reminded residents near the epicenter to be vigilant to prevent aftershocks.

(2) The Ms7.1 earthquake off the east coast of Honshu, Japan

At 22:7 on February 13, an Ms7.1 earthquake occurred off the east coast of Honshu, Japan (the waters near Fukushima), with the epicenter located at 37.70°N, 141.80°E, and a focal depth of 50 km. The earthquake killed 1 person, injured 186 people, damaged 9,137 buildings, and caused minor damages to gas pipelines, roads, port facilities, and railway facilities in Miyagi and Fukushima prefectures. Some of the nuclear sewage tanks in the Fukushima Daiichi nuclear power plant were displaced, however no leakage occurred. In addition, earthquakes cause secondary disasters such as mountain collapses and avalanches.

The Crisis Management Center of the Prime Minister's Residence of the Japanese Government set up a response room for the official residence, the National Police Agency established a Disaster Preparedness Headquarters, and Fukushima and Miyagi Prefectures established Disaster Response Headquarters to carry out decision-making and deployment. The Prime Minister of Japan instructed to collect the data on disaster situation as soon as possible, and the Minister of Defense asked the Self-Defense Force to provide optimum disaster response service and carry out emergency response work urgently. Tokyo Electric Power Company organized personnel to conduct risk inspections on nuclear power plants. Experts from the University of Tokyo reminded people to be alert from earthquakes of the same size in the coming week.

(3) The Ms6.0 earthquake in the waters south of Java Island, Indonesia

At 15:00 on April 10, an earthquake of magnitude 6.0 occurred in the waters south of Java Island, Indonesia. The epicenter was located at 8.85°S and 112.5°E, with a focal depth of 10 km. The earthquake killed 8 people, injured 39, destroyed 85 buildings, and damaged 1,104 buildings. Public buildings and settings in many counties and cities in East Java were damaged in the earthquake, and people panicked and fled to open areas for safety.

Various disaster relief agencies in Indonesia set up emergency shelters to accommodate the homeless victims, and opened up public kitchens to provide food. Rice, vegetables, instant noodles, blankets, tarpaulins, masks, hand sanitizers, and other basic household items were also provided.

(4) The Ms7.3 earthquake in Haiti

At 20:29 on August 14, 2021, an Ms7.3 earthquake occurred in the province of Nips in western Haiti. The epicenter was located at 18.35°N, 73.45°W, with a focal depth of 10 km. The earthquake killed 2,248 people, left 329 missing, injured 12,763, destroyed 53,815 buildings, and damaged 83,770. After the earthquake, seawater found way into the coastal cities, such as Laika, causing panic among the people. Some people took refuge in the mountains. From August 16 to 17, tropical storm "Grace" brought heavy rains to the disaster area, triggering hundreds of landslides.

Due to serious damages to roads and bridges, the emergency rescue after the disaster was mainly carried out by air. The U.S. Southern Command, the Coast Guard, the Bahamas, and other countries dispatched helicopters and other aircraft to transport rescue personnel and supplies to the disaster area for emergency response. The Caribbean Disaster Emergency Management Agency (CDEMA) worked with local officials in Haiti to develop a rescue plan, sending vaccinated personnel to work in the disaster area. Venezuela, Chile, Mexico, Panama, and other Latin American countries, as well as Taiwan Province of China donated masks, protective clothing, oxygen generators, sleeping bags, dry food, other anti-epidemic utilities, as well as relief materials to Haiti.

The European Union, the United Nations, the United Kingdom, and others allocated funds for Haiti to provide basic health care, clean water, emergency shelter, and sanitation. Israeli humanitarian organization SmartAID and international aid organizations sent volunteers to Les Cayes, Jeremie, and other areas to carry out disaster assessment work.

2.2 *Domestic earthquake disasters*

In 2021, there were 6 disastrous earthquakes in China, resulting in 9 deaths, 214 injuries, and direct economic losses of 9,953.13 million yuan (Table 3). Among them, the Ms6.0 earthquake in Luxian County, Sichuan on September 16 killed 3 people and injured 159 people, accounting for 33.33% and 74.3% of the total number of casualties in domestic disastrous earthquakes in 2021, respectively. It was the earthquake with the most serious casualties in China this year. Compared with the casualties caused by earthquakes in China in the past 10 years (Chen 2016; Feng 2017; Li 2019; Nan 2019; Sun 2018; Wen 2018; Zheng 2014, 2015), the casualties this year were generally relatively less (Table 3).

Table 3. Statistics of earthquake disasters in China in 2021.

Date	Beijing Time	Magnitude (*M*)	Location	Casualties		Direct economic loss (10,000 yuan)
				Deaths	Injuries	
03-24	05:16	5.4	Baicheng County, Aksu Prefecture, Xinjiang	3	0	7153
05-21	21:48	6.4	Yangbi County, Dali Prefecture, Yunnan Province	3	34	324,000
05-22	02:04	7.4	Maduo County, Golog Prefecture, Qinghai	0	19	410,000
06-10	19:46	5.1	Shuangbai County, Chuxiong Prefecture, Yunnan	0	2	3440
06-12	18:00	5.0	Baicheng County, Aksu Prefecture, Xinjiang	0	0	3720
09-16	04:33	6.4	Lu County, Luzhou City, Sichuan	3	159	247,000

(1) The Ms5.4 earthquake in Baicheng County, Aksu Prefecture, Xinjiang

At 5:14 on March 24, an earthquake of magnitude 5.4 occurred in Baicheng County, Aksu Prefecture, Xinjiang. The epicenter was located at 41.70°N latitude, 81.11°E longitude, and the focal depth was 10 km. The earthquake killed 3 people and caused a direct economic loss of 71.53 million yuan.

The party committee and government of the autonomous region launched emergency plans, organized forces to verify the disaster situation, and relocated and resettled the affected. The emergency, earthquake, firefighting, and other departments of the autonomous region and the local government organized forces to rush to the disaster area to dispatch disaster relief materials, carry out aftershock monitoring, strengthen the investigation of hidden dangers in key areas, implement various preventive measures, and do their best to reduce casualties and property losses.

(2) The Ms6.4 earthquake in Yangbi County, Yunnan

At 21:48 on May 21, an earthquake of magnitude 6.4 occurred in Yangbi County, Yunnan Province. The epicenter was located at 25.67°N, 99.87°E, and the focal depth was 8 km. The earthquake caused 3 deaths and 34 injuries, with a total direct economic loss of 3,188.1 million yuan.

The Ministry of Emergency Management launched a three-level earthquake emergency response, and the Secretary of the Party Committee and Minister of the Ministry gave important instructions. The Yunnan Provincial People's Government launched an emergency response, and the government leaders led a working group to the earthquake-stricken area to guide the earthquake relief work. The national, provincial, municipal, and county earthquake departments formed a joint earthquake site task force to carry out emergency response work, such as earthquake monitoring, trend research and judgment, intensity assessment, and disaster loss assessment.

(3) The Ms7.4 earthquake in Maduo, Qinghai

At 2:4 on May 22, an earthquake of magnitude 7.4 occurred in Maduo County, Guoluo Prefecture, Qinghai Province. The epicenter was located at 34.59°N and 98.34°E, with a focal depth of 17 km. The earthquake caused 19 injuries, many houses were damaged, roads and bridges were severely damaged, and the direct economic loss was 4.1 trillion yuan.

The Qinghai Provincial Party Committee and the Provincial Government launched the second-level emergency response to the earthquake and dispatched a working group to the front line overnight to guide and assist local areas in carrying out earthquake relief work. Provincial emergency, earthquake, fire, and other departments organized forces to rush to the disaster areas to carry out earthquake damage investigation, aftershock monitoring, search and rescue, and other emergency response work.

(4) The Ms6.0 earthquake in Lu County, Sichuan

At 4:33 on September 16, an earthquake of magnitude 6.0 occurred in Lu County, Luzhou City, Sichuan Province. The epicenter was located at 29.2°N, 105.34°E, with a focal depth of 10 km. The earthquake killed 3 people and injured 159 people. Around 77,786 people were urgently transferred to avoid danger, 27,974 people were transferred and resettled, and the direct economic loss was 2.47 billion yuan.

The State Council Earthquake Relief Headquarters Office and the Emergency Management Department launched a three-level earthquake emergency response and dispatched a working group to the scene to guide the earthquake relief work. The Sichuan Provincial Party Committee and the Provincial Government launched the second-level emergency response to the earthquake emergency and held a special meeting to study and deploy the earthquake relief work. Provinces, cities, and counties formed joint headquarters on-site to organize all forces to fully participate in the earthquake relief work.

3 CHARACTERISTICS OF EARTHQUAKE DAMAGE AND EMERGENCY RESPONSE

3.1 *Characteristics of earthquake damage*

(1) Characteristics of earthquake magnitude and frequency

The frequency of global earthquake disasters is slightly lower than the level of the past decade, but the degree of damage is relatively high. In 2021, there were 14 earthquakes with Ms $\geq$ 6.0 in the world, which had casualties, including 10 earthquakes with a magnitude from 6.0 to 6.9, causing 128 deaths, 3 missing, and 6769 injuries; 5 earthquakes with a magnitude from 7.0 to 7.9, causing 2253 deaths, 329 people were missing, and 12,968 were injured.

(2) Distribution characteristics

Earthquakes with Ms $\geq$ 7.0 are concentrated in space and distributed on the northeastern boundary of the Australian plate. Foreign catastrophic earthquakes are mainly concentrated in three countries including Indonesia, Japan, and Haiti, and domestic catastrophic earthquakes are mainly concentrated in four provinces (autonomous regions), including Xinjiang, Yunnan, Qinghai, and Sichuan.

(3) Disaster-causing characteristics

The earthquake with high magnitude caused serious damages, such as the Ms7.3 earthquake in Haiti, which killed 2,248 people, left 329 people missing, and injured 12,763 people. Small earthquakes caused disasters, such as the Ms5.8 earthquake in Pakistan on October 7, which killed 42 people and injured 229 people; and Ms5.4 magnitude earthquake in Baicheng, Xinjiang on March 24, which killed 3 people.

(4) Causes of casualties

The main cause of casualties was collapse of houses. The main reason for the large number of casualties caused by the earthquake in Haiti abroad is the poor quality of buildings and their ability to resist earthquakes. In recent years, with the progress of housing security in poverty alleviation in China, the earthquake resistance of houses and other infrastructure has been gradually improved. However, in some areas such as Yunnan, some casualties have been caused due to incomplete demolition of old houses.

3.2 *Emergency features*

After the earthquake, many foreigners entered the disaster area, and the challenge of epidemic prevention and control was even higher. To prevent the superimposition of the disaster and the epidemic, and to coordinate the earthquake relief and epidemic prevention and control, the government in the disaster-stricken area should take measures to prevent earthquakes by timely formulating epidemic prevention and control work plans, strengthening the sanitation and killing of mass resettlement sites, nucleic acid testing, vaccination, and drinking water monitoring of key populations.

Disaster relief may lead to the importation and spread of the epidemic. Resolutely adhere to the bottom line of "no major epidemic after a major disaster" to ensure that the epidemic prevention and control situation in the disaster area continues to be stable and the earthquake relief work is carried out in an orderly manner. Rescue countries and provinces should take personal protection when carrying out cross-border and provincial rescues to ensure the absolute safety of personnel and materials, and strictly prevent the spread of the epidemic.

4 CONCLUSION

Based on the results and discussions presented above, the conclusions are obtained as below:

(1) The frequency of global earthquake disasters in 2021 is relatively low and the distribution is relatively concentrated, but the earthquake damage was serious.
(2) Poor earthquake resistance of buildings is still the main cause of casualties. With the progress of housing security, the earthquake resistance of houses and other infrastructure has been gradually improved. However, some old houses were not completely demolished and used for other purposes, which became one of the important reasons of casualties.
(3) Rescue countries and provinces should take personal protection when carrying out cross-border and intra-provincial rescues, ensure absolute safety of personnel and materials, and strictly prevent the spread of the epidemic.

This paper only conducts a statistical analysis of the catastrophic earthquakes with Ms ≥ 6.0 in foreign countries and domestic catastrophic earthquakes in 2021. The next step will be to conduct a comprehensive analysis and statistics of all disastrous earthquake data in the world in 2021 to provide a better reference for earthquake emergency personnel.

REFERENCES

Chen, T., Zheng, T.Y. (2016). Review of earthquake damage losses in mainland China in 2015. *Journal of catastrophology*, 31, 133–137.

Feng, W., Li, W., Chen, T., et al. (2012). Overview of worldwide earthquake disasters in 2012. *Journal of catastrophology*, 28, 133–137.

Feng, W., Zhu, L., Hou, J.S., et al. (2016). Overview of worldwide earthquake disasters in 2014. *Technology for Earthquake Disaster Prevention*, 11: 420–426.

Feng, W., Zhu, L., Zhao, M.S. (2017). Overview of worldwide earthquake disasters in 2016. *Recent Developments in World Seismology*, 29–32.

Feng, W., Zhu, L., Zhao, R.G. (2015). Overview of worldwide earthquake disasters in 2013. *Recent Developments in World Seismology*, 37–40.

Feng, W., Zhu, L., Zhao, R.G. (2016). Overview of worldwide earthquake disasters in 2015. *Recent Developments in World Seismology*, 22–27.

Li, Y.G., Nan, Y.Y., Liu, K., et al. (2019). Earthquake disasters and emergency response in 2018. *China Emergency Rescue*, 4–9.

Nan, Y.Y., Li, Y.G., Liu, K., et al. (2020). Earthquake disasters and emergency response in 2019. *China Emergency Rescue*, 4–9.

Sun, G., Li, Y.G., Du, X.X., et al. (2018). Overview of earthquake disasters and emergency response in 2017. *China Emergency Rescue*, 8–14.

Wen, X.T., Zheng, T.Y. (2018). Review of earthquake damage losses in mainland China in 2016. *Journal of catastrophology*, 23, 141–144.

Zheng, T.Y., Feng, W., Zheng, Y. (2015). Review of earthquake damage losses in mainland China in 2014. *World Earthquake Engingeering*, 31, 202–208.

Zheng, T.Y., Zheng, Y. (2014). Review of earthquake damage losses in mainland China in 2012. *Journal of Natural Disasters*, 23, 166–170.

Zheng, T.Y., Zheng, Y. (2015). Review of earthquake damage losses in mainland China in 2013. *Journal of Natural Disasters*, 24, 239–246.

Advances in Civil Engineering: Structural Seismic Resistance, Monitoring and Detection – Mohd Johari Mohd Yusof and Junwen Zhang (Eds)
© 2023 The Authors, ISBN 978-1-032-31491-4

The safety impact of subway underpass on the existing airport expressway

Shiping Jia & Yufu Li
Guangdong Construction Polytechnic, Guangzhou, China

Yinbo Li
Guangdong Baiyun University, Guangzhou, China

ABSTRACT: The paper analyzes the safety impact of subway underpass construction on the bridge and subgrade of the existing airport expressway. First according to various specifications and engineering experience, determined the impact safety standards. Second, used the finite element simulation the construction to analyze the impact. Then, based on the established evaluation standards, use the analysis results to make a reasonable assessment of the road safety status. Finally, the monitoring data of the construction process are closely tracked and timely compared with the analysis results to ensure the smooth completion of the underpass construction, thus verifying the reliability of the evaluation system.

1 INTRODUCTION

With the acceleration of urbanization, it often happens that the subway outside the city passes through the existing expressway. This paper studies the safety influence of subway underpass construction on existing expressway bridges and roadbed with engineering examples. Work idea: establish evaluation standard; three-dimensional finite element construction simulation analysis; carry out evaluation; monitor comparative demonstration; complete evaluation, so as to form experience with reference value.

There is no standard agreement on the safety standard of subgrade for subway undercrossing expressway, so it is necessary to refer to relevant contents in various specifications (JTG D30—2015; JTG F80/1—2017; JTG/T3610—2019) and determine it based on similar engineering experience. At present, scholars mainly use theory combined with numerical analysis to carry out research, and control pavement settlement through reasonable advanced support and optimized construction method combined with dynamic monitoring data in the construction process (Chen & Bai 2011; Mao et al. 2013). Some set the settlement control standard for the undercrossing road of shallow buried tunnel of railway line within 20mm, and compared and optimized the three construction methods (Zhao et al. 2020); Someone proposed to use driving comfort to determine the control benchmark of pavement settlement caused by tunnel excavation from the perspective of meeting pavement functionality (Zhang 2008).

Similarly, subway undercrossing highway bridges also need to refer to various codes to determine safety standards (JTG 3363—2019; JTG/T H21-2011). Some people established an integral 3D finite element model to study the influence of shield tunnel construction on adjacent overpass pile foundation (Zhang & He 2003), the results show that the construction will not only cause the existing pile foundation to deviate from the horizontal direction of the tunnel, but also adversely affect the side wall of the new tunnel adjacent to the pile foundation, which requires reinforcement measures. some simulated the stress and deformation of abutment pile foundation caused by shield tunneling by using three-dimensional nonlinear finite element method (Zhang et al. 2007), the

 DOI 10.1201/9781003310884-82

results show that in the process of shield tunneling, flexural deformation will occur in the direction of parallel and vertical shield axes of adjacent pile foundation, resulting in significant additional axial force and bending moment in the pile foundation, which reach the maximum value when the tail of shield passes through the plane of pile foundation. In view of the design and construction scheme of the underground tunnel in Xicun Station of Guangzhou Metro Line 5, some studied the stress characteristics and displacement variation rules of the existing pile foundation during tunnel excavation by using the finite difference method (Li et al. 2006), the pile side presents negative friction resistance during construction, and the bearing capacity of pile foundation is reduced by 40%. Others used the finite difference program FLAC 3D to predict and analyze the ground settlement caused by tunnel excavation and its influence on adjacent bridge piles (Gao et al. 2012; Zhang et al. 2011), the research shows that the maximum value of surface settlement is not more than 1cm, the deformation of bridge pile foundation is mainly horizontal deformation, and the pile foundation is slightly inclined due to underpass, and the tilt will become larger and larger with the tunneling of shield machine, and the maximum value will be reached when the tunnel is connected.

2 SAFETY CONTROL STANDARD FOR UNDERPASS CONSTRUCTION

2.1 *Safety control standard for subgrade of airport expressway under subway*

Considering the importance of the airport expressway, the maximum settlement value of the road subgrade caused by the construction of the subway crossing the airport expressway shall not exceed 20mm, the maximum uplift value shall not exceed 10mm, and the road slope control value shall not exceed 0.50% as the safety control standard of the undercrossing construction subgrade.

2.2 *Safety control standard for bridge under subway crossing airport expressway*

The safety control standards for the subway crossing the bridge section of the airport expressway are determined as follows:

(1) Longitudinal adjacent pier uneven settlement difference: bridge deck can not form more than 0.20% of the additional longitudinal slope, span≤70mm, side span≤50mm (span 35m, side span 25m);
(2) The differential value of uneven settlement between transverse bridge piers should meet the stress requirements of the upper cover beam, generally not more than 5mm;
(3) The lateral deformation of bridge pile causes the change of internal force of pile foundation, which should meet the stress requirement of pile foundation.

3 UNDERPASS FINITE ELEMENT MODEL

3.1 *Modelling*

A certain subway section in Guangzhou passes diagonally under a middle bridge of an expressway, the distance between the middle lines is about 24 ~ 26m, the buried depth of the subway shield is about 8m, and the buried depth of the river under the bridge is about 4m.The oblique Angle between the left line and the middle line of the highway is about 73.25°, the intersection Angle of the right line is about 76.41°, and the length of the oblique is about 33.8 meters. The right line of subway is close to pier 1#, and the closest distance of shield tunneling on the right line is only 2m outside the pile foundation under the pier.

Considering the complex formation distribution and the geological conditions revealed by typical boreholes, the formation is reasonably simplified into five layers according to the equivalent modulus, and the Moore-Coulomb constitutive model is selected and simulated by solid element. Roadbed, bridge structure and pile foundation are simulated by solid element and isotropic elastic constitutive model is adopted. Slab element simulation of tunnel support structure. In addition to the surface, boundary constraint conditions in corresponding directions were applied around the model. The slurry pressure applied on the excavation surface of shield tunneling machine was controlled at 50~100kPa, and the minimum value was used for the underside of river surge. The solid model was established as shown in Figure 1. The side pier (pile) and adjacent pier (pile) that are close to the two ends of the bridge pier and below the left and right lines of the tunnel are taken as research objects, as shown in Figure 2. The pile foundation near platform 0# and pier 1# has a fine mesh size of 0.75m, while the pile foundation far from the subway has a coarse mesh size of 1m. The model has a total of 411,316 units.

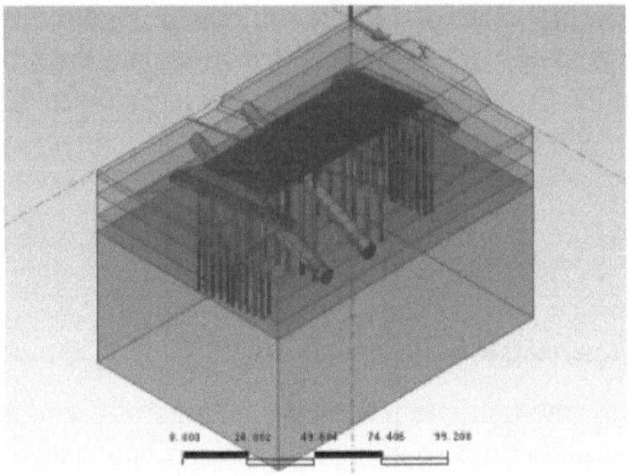

Figure 1. Axonometric drawing.

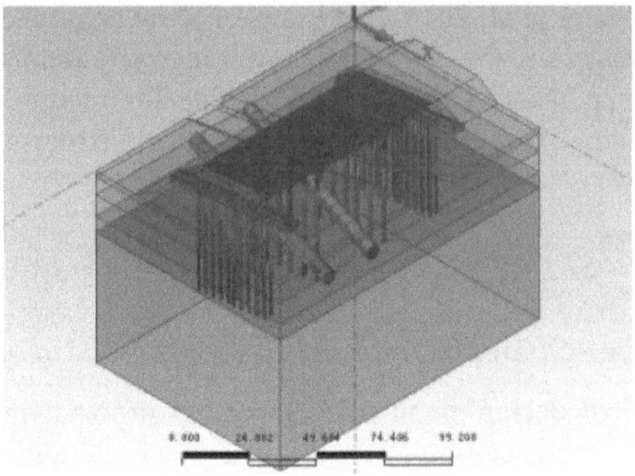

Figure 2. Distribution of research objects of bridge piers and abutments.

3.2 *Construction steps and calculation*

According to the construction steps, simulate the whole process of shield construction in the process of undercrossing the airport expressway:

(1) Initial stress analysis, displacement clearance;
(2) Control the excavation sequence of the left and right lines, the right line is excavated first and then the left line, and the distance between the left and right lines is controlled at 100m;
(3) After each excavation, the segment lining was constructed in time and a thrust force of 50~100kPa was applied at the face of the face to simulate the mud pressure of the shield machine head. The pressure of 50kPa was applied at the downstream upwelling section and 100kPa at the rest sections. In the next construction step, the thrust of the previous step is cancelled and the new thrust is added.

4 INFLUENCE ANALYSIS OF UNDERPASS CONSTRUCTION

4.1 *Analysis on the influence of highway subgrade*

The right line of subway is constructed first. Since the length of underpass section is less than 100m, and the interval between the excavation faces of left and right lines is 100m, the construction of left line begins after the completion of underpass construction. The calculation results of the two typical construction steps are as follows:

In the 17th working condition, the right line is finished, and the left line is just finished. The results show that when the left subgrade is under the construction of the right line subway, the subgrade presents a trend of uplift as a whole due to the influence of the thrust force of the palm surface, and the uplift value increases gradually with the construction of underpass. After the completion of underpass, the maximum uplift value of the pavement edge of the subgrade near the subway side is 2.20mm, and the uplift value of the middle line of the road surface is 1.78mm. As the subgrade on the right side is far from the subway line, it is not affected by the thrust. The uplift or settlement of the subgrade on the right side is less than 1mm, so the influence can be ignored.

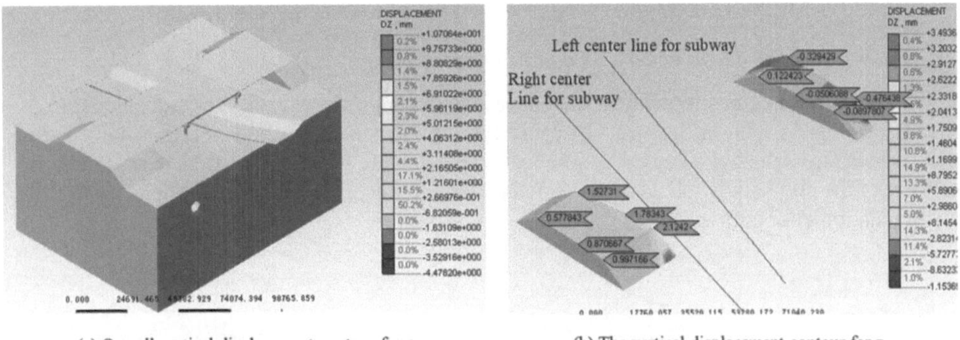

(a) Overall vertical displacement contour for z (b) The vertical displacement contour for z

Figure 3. Condition 17 vertical displacement contour.

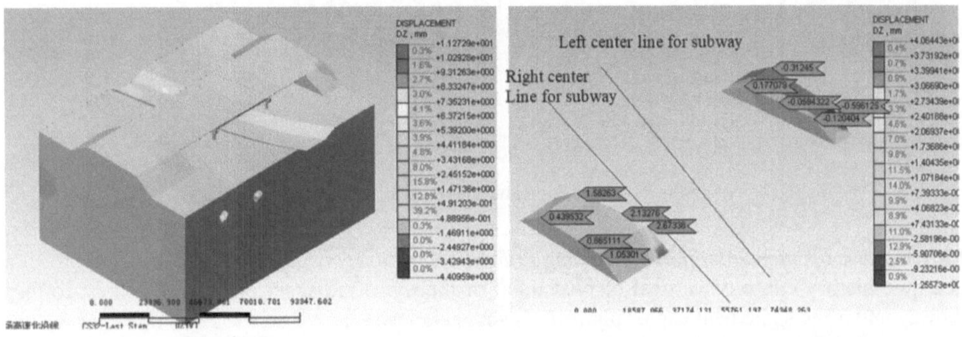

(a) Overall vertical displacement contour for z (b) The vertical displacement contour for z

Figure 4. Condition 39 vertical displacement contour.

In the 39th working condition, both the left and right lines were completed. The analysis shows that, after the completion of the left line, the maximum uplift value of the roadbed edge near the subway side increased from 2.10mm to 2.70mm, and the uplift value of the middle line of the road surface also increased from 1.78mm to 2.13mm. Because the subgrade on the right side is far from the subway line, the influence is still small, and the deformation is less than 1mm, so the influence can be ignored.

The deformation of subgrade structure in the whole construction process is shown in Table 1.

Table 1. Summary of calculation results of each construction step.

	The subgrade structure		
Location	Maximum settlement or uplift of subgrade/(mm)	Tilt value (longitudinal)	Tilt value (transverse)
Subgrade near 0# abutment	+2.70	0.01%	0.03‰
Near 1# abutment subgrade	-0.60	0.01%	0.01‰

After the construction of the airport expressway under the subway is completed, the maximum uplift value of the roadbed top is +2.70mm without considering the road surface load, which is less than the control standard of +10mm. The maximum tilt value of highway subgrade structure is 0.01%, which is also less than 0.50% of the control standard.

4.2 *Analysis of influence of underpass construction on bridge*

Based on the above control standards, the uneven settlement difference of longitudinal and transverse adjacent piers of the bridge structure caused by subway penetration and the lateral deformation of the bridge pile foundation caused by soil lateral deformation may lead to the change of the internal force of the pile foundation. Figure 5 Cloud diagram of displacement of rear bridge and bottom pile foundation after underpass.

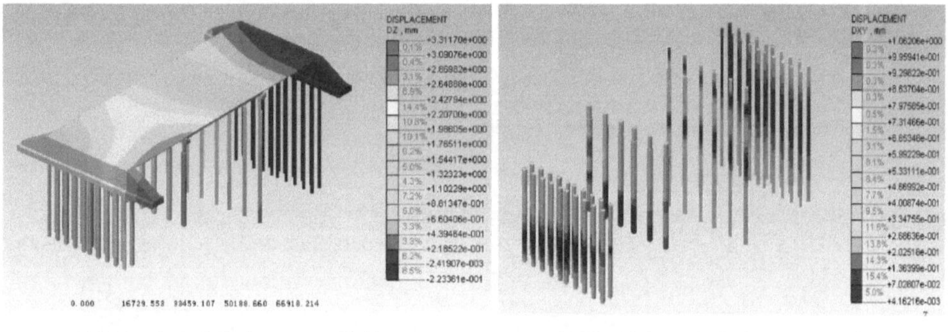

(a) The vertical displacement of bridge (b) The horizontal displacement of pile

Figure 5. The displacement contour of the bridge and the lower pile after the underpass.

4.2.1 *Uneven settlement difference of longitudinal and transverse adjacent pier*

The time-history curve of vertical displacement of the top of adjacent pier (pile) nearest to the subway under the same bent cap or platform is shown below.

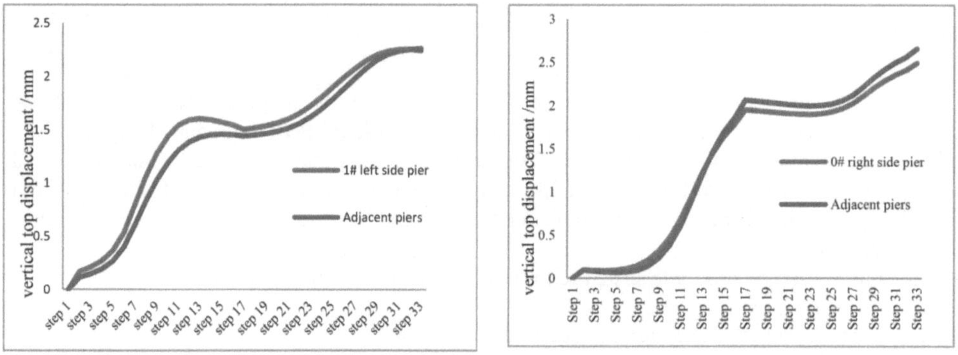

(a) Time-history curve of vertical displacement of top of adjacent pier 1#

(b) Time history curve of vertical displacement of adjacent pile tops under 0# platform

Figure 6. Time-history curve of vertical displacement at the top of adjacent pier or pile (nearest to subway) under the same bent cap or platform.

The maximum differential settlement results at the top of adjacent piers (piles) under the same bent cap or platform are shown in Table 2.

Table 2. Maximum lateral differential settlement of bridge pier (abutment).

location	vertical displacement		
	Side pier (pile) top/(mm)	Top of adjacent pier(pile)/(mm)	Maximum differential settlement/ (mm)
0 # bridge abutment	2.68	2.49	0.19
1 # pier	1.55	1.31	0.24
2 # pier	0.58	0.41	0.17
1 # bridge abutment	0.37	0.28	0.09

In the same way, the maximum differential settlement results of the longitudinal adjacent piers of the bridge caused by subway underpass are shown in Table 3.

Table 3. Maximum longitudinal differential settlement of adjacent pier.

location	Maximum differential settlement of adjacent pier /(mm)		
	The 1 # pier	The 2 # pier	The 1 # bridge abutment
0 # bridge abutment	0.62	—	—
1 # pier	—	1.70	—
2 # pier	—	—	0.89

4.2.2 *Analysis of horizontal displacement and internal force of pile foundation near subway*

The research shows that the maximum lateral displacement of the pile foundation closest to the subway occurs at the height of the middle line of the shield tunnel about 1mm in the depth range of the subway structure after underpass construction, and the lateral displacement of the other adjacent piles gradually decreases with the increase of the distance from the subway.

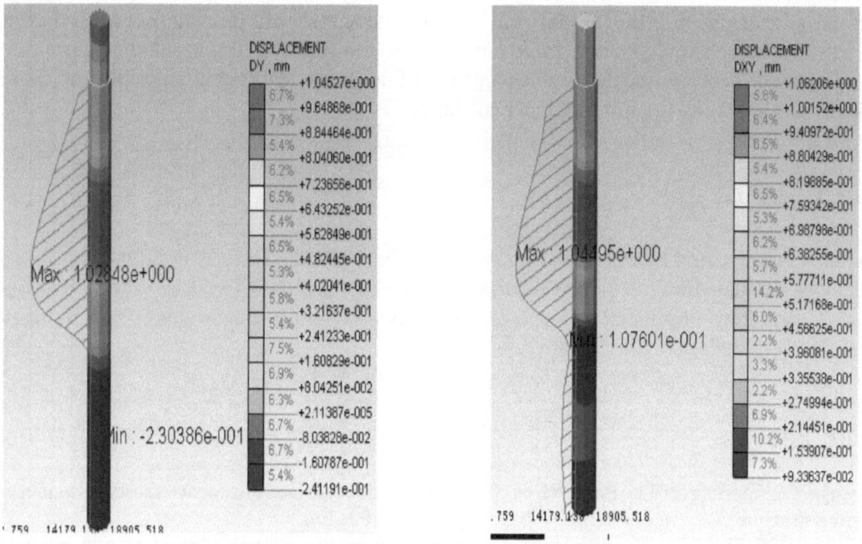

(a) Horizontal displacement along the bridge (b) Total horizontal displacement

Figure 7. Horizontal displacement of the pile closest to the subway.

4.2.3 *Analysis of influence of underpass construction on bridge*

(1) The non-uniform settlement value of bridge pier caused by construction is very small, and the maximum differential settlement is only 0.24mm, which will not cause the big stress change of the cap beam and can meet the structural stress requirements.

(2) As a result, the longitudinal differential settlement value of the adjacent pier of the bridge is larger than that of the lateral differential settlement, and the maximum differential settlement is 1.70mm when the construction of the left and right lines of the middle-span subway is completed, which meets the safety control standard of the middle-span differential settlement ≤70mm.

(3) The horizontal and lateral displacement of the pile foundation closest to the subway occurred, and the maximum displacement occurred at the height of the middle line of the shield tunnel was about 1mm. The lateral displacement of the other adjacent piles was small and had little influence on the stress of the pile foundation, so the internal force analysis of the pile foundation caused by this was not required.

5 CONCLUSION

(1) The construction of the project has been completed, the third-party monitoring data feedback from the site shows that the monitoring data is consistent with the calculation results, and there is no warning in the whole construction process, and the underpass construction is successfully completed, which verifies the reliability of the evaluation system.

(2) The slurry shield machine is a low settlement and safe construction method, and the slurry pressure of 50∼100kpa is reasonable, which can effectively control the deformation of excavation face and foundation settlement.

6 THE LIMITATIONS AND NEXT RESEARCH PLAN

Groundwater is a constant hydrostatic pressure, which can not reflect the influence of water level change, and the feedback of construction monitoring and measurement results is not timely. Therefore, advance geological forecast and dynamic design should be made, and the design and construction scheme should be optimized and adjusted in time. The longitudinal distance of excavation surface of subway double line has a great influence on settlement and uplift of subgrade. The next step is to study the minimum safe distance.

ACKNOWLEDGMENTS

The work was supported by the following two projects: 2020 Characteristic Innovation Project of Universities in Guangdong Province (Natural Science) (No. 2020KTSCX235) and 2020 Special Fund of Guangdong Provincial Science and Technology Innovation Strategy ("Climbing Plan" Special Fund) (No. Pdjh2020a0943).

REFERENCES

Chen Xingxin & Bai Bing (2011). Research on control standards of surface subsidence caused by tunnel sunder existing structure. *Journal of Engineering Geology* 19(1):103–108.

Gao Wei et al. (2012). Analysis of the influence of shallow buried tunnels on bridge foundation. *Highway* 11: 123–128.

JTG D30—2015, Specifications for Design of Highway Subgrades. Beijng: China Communications Press Co., Ltd.

JTG F80/1—2017, Quality Inspection and Evaluation Standards for Highway Engineering. Beijng: China Communications Press Co., Ltd.
JTG/T H21-2011, Standards for Technical Condition Evaluation of Highway Bridges. Beijng: China Communications Press Co., Ltd.
JTG/T3610-2019, Technical Specification for Construction of Highway Subgrades. Beijng: China Communications Press Co., Ltd.
JTG 3363—2019, Specifications for Design of Foundation of Highway Bridges and Culverts. Beijng: China Communications Press Co., Ltd.
Li Qiang et al. (2006). The influence of underground excavated tunnel construction in subway station on the existing pile foundation. *Chinese Journal of Rock Mechanics and Engineering* 25(1): 184–190.
Mao Yuanfeng et al. (2013). Characteristic analysis of pavement deformation caused by subway shield tunneling under the expressway. *Railway Standard Design* 8: 84–88.
Zhao Lin. et al. (2020). Research on settlement and control of super shallow tunnel underneath the expressway. *Journal of Henan University* (Natural Science Edition) 50(1): 118–126.
Zhang Peng (2008). Criteria for road subsidence control caused by shallow tunnels underneath the highway. *Journal of Beijing Jiaotong University* 32(4): 137–140.
Zhang Heng et al. (2011). Analysis of the influence of shield construction on the ground surface and bridge pile foundation. *Chinese Journal of Underground Space and Engineering* 7(3):552–557.
Zhang Haibo et al. (2007). Analysis of the impact of shield tunneling on bridge pile foundations. *Railway Construction* 8: 37–40.
Zhang Zhiqiang & He Chuan (2003). Research on mechanical behavior of adjacent pile foundation construction of Shenzhen Metro tunnel. *Chinese Journal of Geotechnical Engineering* 25(2): 204–207.

Advances in Civil Engineering: Structural Seismic Resistance, Monitoring and Detection – Mohd Johari Mohd Yusof and Junwen Zhang (Eds)
© 2023 The Authors, ISBN 978-1-032-31491-4

Cause analysis of coal mine water disaster accident using Bayesian networks

Shaoyu Li, Wei Yan & Dan Yu
Shandong University of Science and Technology, Qingdao, China

ABSTRACT: Coal mine water disaster is a complicated phenomenon involving many unknown causes and complex relationships. It may lead to severe economic losses and casualties if occurring. In this study, the reports on coal mine water disaster accidents were manually collected and analyzed to acquire the keyword frequency. The accident impact factors (AIFs) of coal mine water disasters were identified under cause theory. The Bayes network model was established using a data-driven method to input 139 coal mine water disaster cases to study and analyze the impact relationship among AIFs under different accident levels. The study discovered some key factors impacting coal mine flood accidents, set the different states of the elements in scenario settings to analyze the evaluation relationship among AIFs quantitatively, and finally performed scenario simulation to understand the difference of AIF status in accidents at different levels. The water hazard problems in the accident reports were summarized. Serving as a reference for coal mining companies to investigate and rectify water hazards, this study can help coal companies develop preventive measures to reduce accident hazards.

1 INTRODUCTION

China is one of the countries with the most severe coal mine water disaster in the world(Wu. 2014). Coal mine water disaster has the characteristics of strong suddenness and high harmfulness. Once it occurs, its consequences are immeasurable. The most severe coal mine water disaster in history occurred on August 17, 2007. In the Huayuan coal mine of Xintai coal mining area in North China, the river broke through the abandoned shaft and caused water seepage. A total of 12.6 million cubic meters of water and sand penetrated the mine, causing 181 deaths, which caused enormous casualties and severe economic losses. Coal mine water disaster is a relatively complex adaptive phenomenon (Gong et al. 2011). When many unknown factors are involved, it will be challenging to assess disasters due to the lack of accident information and experience. Therefore, it is of great significance to deeply understand the causes and influencing factors of coal mine water disaster for strengthening and improving water disaster prevention and control measures.

In the past, empirical and statistical methods were often used to study the problem of coal mine water disasters (Wu et al.2008). Generally, the research idea was to understand the change law of water inrush index nodes in the field observation, and then realize the mechanism research on the occurrence process of coal mine water disasters. Commonly used research methods include index evaluation method, such as vulnerability index method(Wu et al. 2007), attribute comprehensive evaluation method(Li et al. 2015), fuzzy comprehensive evaluation method(Wang et al. 2012); Numerical simulation, such as multi-agent simulation(Gong et al. 2011), geographic information system (GIS) (Ding et al. 2013), dynamic model(Li et al. 2013); Physical experiment method(Wu et al. 2004; Zhang et al. 2013b); Machine learning, such as extreme learning machine(Zhao et al. 2014), support vector machine(Yan et al. 2008).

The above research has established the research ideas and methods of coal mine water disaster. Still, it needs complete field data information and expert identification ability, which is difficult

 DOI 10.1201/9781003310884-83

to obtain in the coal mine field operation environment. At the same time, only a small number of field cases can not well simulate the scene of flood accidents. It is necessary to find a more common method to count the occurrence law of flood disasters to meet the requirements of disaster prevention. Therefore, given the uncertainty of water disasters, this paper puts forward the idea to study water disasters in the theoretical framework of accident causes. The accident cause theory takes the accident as the research object and realizes the purpose of preventing the accident from the grounds and laws, which is widely used in the field of mining safety management (Liu et al. 2016; Wang et al. 2016; Zhang et al. 2011). Cause analysis helps analyze the mechanism of water hazards and reveal the process and causes of water disaster factors. As an uncertainty analysis tool, the Bayesian network (BN) has a decisive advantage in the analysis and research of water disasters t (Dong et al. 2012; Wu et al. 2016), which can deal with complex node dependence and evolve. However, the traditional BN method is challenging to transform the actual problem into probability to express the uncertainty. It isn't easy to obtain accurate information in the coal mine production system environment (Liu et al. 2020). Combining BN with accident cause theory can solve this problem well. Firstly, BN can construct multi-state nodes based on the water disaster factors from accident cause analysis, which is suitable for water disaster scenario analysis; Secondly, BN can transform the process and cause analysis of water disaster by accident cause theory into the causal relationship among nodes. In addition, the node description of each accident through cause analysis can be used as the input evidence to form conditional probability, which avoids the bias effect caused by subjective consciousness. At present, there are few studies from the perspective of accident analysis among a large number of coal mine water disaster analysis papers published in academic journals. Therefore, based on this research, the Bayesian network is constructed using the theoretical framework of the cause of accidents to identify and analyze the causes and effects of water disasters.

This paper first proposed a recognition mode of keywords of coal mine water disaster from the accident text reports. It focused on the use of accident cause theory to identify the keywords to extract AIFs from the accidents. Using expert judgment and data-driven method for structure learning and parameter learning to construct BN model, and ensure the objectivity and universality of the model through sensitivity analysis and historical accident verification. On this basis, the key factors that affect coal mine water disasters were established. The relationship among AIFs was discussed in the case of a single element or a combination. This framework model effectively reflected the objective problems existing in historical coal mine water disasters and the potential relationship among the causes. Relevant departments can form references and strengthen coal mine water disaster prevention and control work more effectively.

2 IDENTIFICATION OF AIFS AND ESTABLISHMENT OF BN NODE

2.1 *Identification of AIFs*

To effectively identify and select AIFs, it is necessary to obtain and screen reports manually. The primary sources of accident reports include: (1) National Coal Safety Production Network (http://www.mkaq.org/); (2) Ministry of Emergency Management of the People's Republic of China (https://www.mem.gov.cn/). The latest report of accident report up to September 2020 collected 592 data of coal mine water disaster accidents, among which 139 accident reports provided four conditions including 'person', 'management', 'equipment' and 'environment' to describe and clarify the cause of the accident. To determine the closely related factors, keyword screening and artificial statistics from the detailed report were carried out to summarize the distribution of keywords related to water disasters. Figure 1 summarizes the visual display of natural language data obtained through word cloud technology and the probability distribution of 32 keywords in the processed report. WC is a visual display tool based on Python software, in which the frequency of each word is reflected in the size of the font (the larger the font, the higher the frequency). With the help of WC visualization technology, managers can better identify and screen the influencing factors, and find out the potential factors that may exist.

Specifically, the fundamental hazard of man-made conditions refers to employees' unsafe behaviors and hazard handling procedures, such as illegal mining and emergency measures. The fundamental hazard of the environment refers to the working environment and the natural environment, such as groundwater, gas, water pressure. The fundamental hazard of machines refers to drainage and ventilation systems. The fundamental hazard of management refers to the organization's management system, including on-site management and technical management. Referring to the previous analysis and research on indicators and factors of coal mine disaster sources (Yun et al. 2012), the influencing factors of coal mine water disasters reflected in the accident report are described as follows: illegal command, illegal mining, exploration and drainage measures, rescue response, precipitation, geological structure, water inflow, toxic gas, safety specifications and waterproof and drainage system.

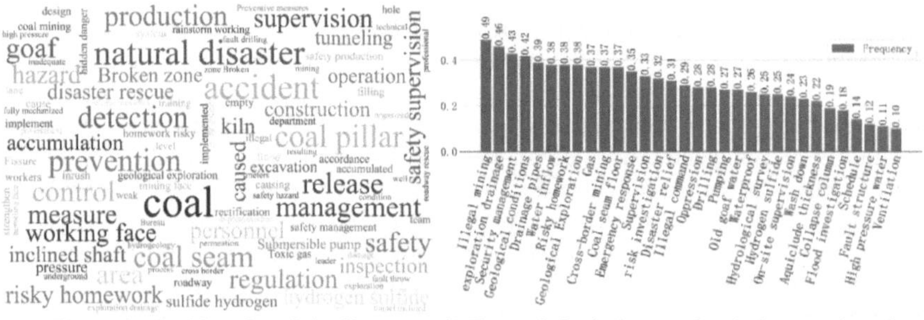

(a) Keyword recognition of word cloud image (b) Keywords for the impact of coal mine water hazards

Figure 1. Text mining word frequency technology statistics.

The status of all node indicators is derived from the scene of the accident site report. Node status 1 indicates a bad status, intermediate status indicates a general status or not reflected in the report, and node status 3 indicates a positive status. BN node and description are shown in Table 1.

Table 1. BN node and description.

Category	Factor	Describe	State
Environment condition	Precipitation	The report recorded heavy rainfall or snowfall at the time of the accident	Pr_1 *(rich)* Pr_2 *(general)* Pr_3 *(little)*
	Geological structure condition	The complexity of mine hydrogeological types and the development of fault joints	Gs_1 *(complex)* Gs_2 *(middle)* Gs_3 *(simple)*
	Water inflow	According to the records of water inflow at the accident site	Wi_1 *(high)* Wi_2 *(general)* Wi_3 *(litter)*
	Hydrogen sulfide	The accident report records the absolute emission and relative emission of hydrogen sulfide gas in the coal mine	Hs_1 *(high)* Hs_2 *(general)* Hs_3 *(litter)*
Artificial condition	Illegal command	Improper command of relevant responsible person leads to irregular safety behavior of workers, abuse of power, violation of orders or forcing workers to violate rules and regulations and take risks in mining operation	Lc_1 *(yes)* Lc_2 *(general)* Lc_3 *(no)*

(continued)

Table 1. BN node and description.

Category	Factor	Describe	State
	Rescue measure	Including coal miners, management, mine manager, etc. in the face of danger response measures, emergency state, clear contact personnel, withdrawal procedures are timely	Rm_1 (*nottimely*) Rm_2 (*general*) Rm_3 (*timely*)
	Exploration and drainage measure	Through reasonable exploration methods to find out the water-bearing structure layer, and to release the water and gas in the water	Ed_1 (*invalid*) Ed_2 (*general*) Ed_3 (*effective*)
	Illegal mining	The mining method is unreasonable, not following the rules and regulations of mining, risk recklessly, digging indiscriminately	Im_1 (*yes*) Im_2 (*general*) Im_3 (*no*)
Management condition	Safety specification	The quality of the management standard of the safety management organization	Ss_1 (*bad*) Ss_2 (*general*) Ss_3 (*good*)
Equipment condition	Waterproof and drainage system	The report records the layout and use of drainage system equipped with waterproof and waterproof coal pillar, waterproof and drainage pipeline, and sump in the mine	Wd_1 (*bad*) Wd_2 (*general*) Wd_3 (*good*)
Accident results	Accident type	According to the type of water source, the accidents can be divided into goaf water disaster, surface water disaster, underground water disaster, and other water disasters	T_1 (*s_o*) T_2 (*s_g*) T_3 (*s_k*) T_4 (*s_t*)
	Accident level	Casualties and economic losses were caused by accident. Generally, it can be divided into the ordinary, large, grave, and extremely grave accidents	R_1 (*r*_1) R_2 (*r*_2) R_3 (*r*_3) R_4 (*r*_4)

2.2 *BN structure learning*

Structural learning is used to mine the causal relationship among factors and express it with the directed edge. In the process of structural learning, the use of prior knowledge is crucial, which can make up for the fitting problem caused by simple data (Li et al. 2020). Therefore, the structure and correlation in BN can be obtained by combining historical data and related domain knowledge. Figure 2 shows the establishment of the BN structure and explains the dependence among some nodes.

(1) 'Precipitation' can affect 'water inflow' and 'accident type'.
According to experts engaged in related fields, the source of coal mine water inflow is mainly infiltration and inflow of atmospheric precipitation, river supply around the coal mine, and water accumulation in the old goaf. Surface water disaster is more severe in the rainy season.
(2) 'Geological structure condition' can affect 'hydrogen sulfide' and 'water inflow'.
The complexity of the geological type of the mine itself determines the number, scope, and water accumulation of the old goaf. Generally, ponding is easy to occur under complex geological conditions, and the water circulation is poor, which may react with rocks to generate toxic gases such as hydrogen sulfide.
(3) 'Safety specification' can affect 'illegal mining' and 'illegal command'.
Experts in the coal production industry said illegal production in most of the coal mines involved in the accident. The managers did not restrict the construction personnel, the rules and regulations were chaotic, and the safety education was ineffective.

(4) 'Waterproof and drainage system' can affect 'water inflow'.
Properly strengthening the drainage capacity, leaving enough waterproof coal pillars to prevent the collapse zone from penetrating the aquifer can reduce the mine water inflow.

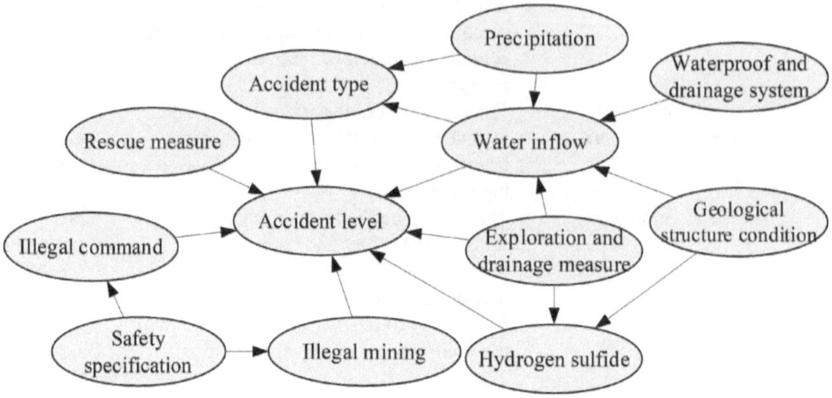

Figure 2. Establishment of BN accident model.

3 TEST RESULTS AND DISCUSSIONS

3.1 *Quantitative research of AIFs*

Accident diagnosis is to analyze the probability of each state under the condition of the accident through reverse inference and analysis of the condition of the known accident result and put forward prevention and control measures. In the AIFs defined in Table 1, the accident level is defined as the final result variable as the final direction of the arrow, which represents the severity of the accident result. The 139 events evaluated manually are used as the initial node state input to complete the construction of the entire network, and the software netica is used to calculate the CPT value of all relational nodes. netica can be used for deep learning networks and data updates. After obtaining the table, it can perform accident diagnosis and accident prevention by calculating the posterior probability. The calculation results are displayed intuitively and smoothly through the graphics package, as shown in Figure 3.

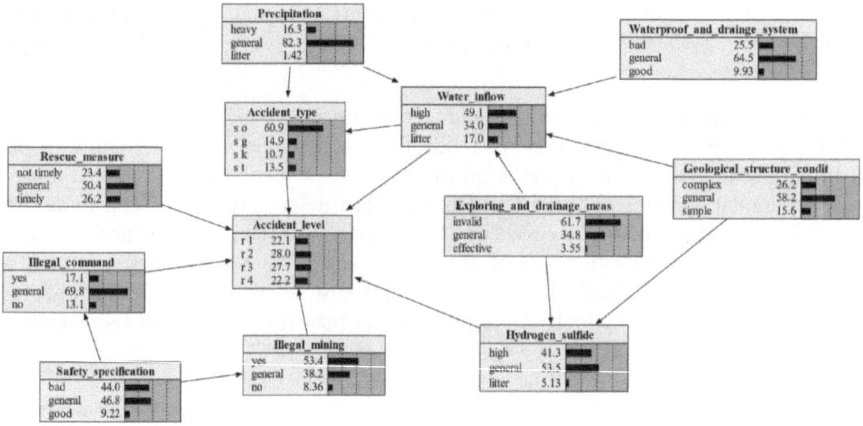

Figure 3. The possibility of BN under different accident levels.

3.2 *Sensitivity analysis*

Evidence sensitivity analysis is to analyze the sensitivity of the query node to the target result node, and then determine the node value that has a more significant impact on the target node. In order to find the nodes that have a significant influence on the accident results, it is necessary to select 'accident level' as the target for sensitivity analysis. The sensitivity analysis of AIFS with 'accident level' as the node is shown in Table 2, shows that the effect of 'hydrogen sulfide' is largest, indicating that reducing the hydrogen sulfide content in the mine can effectively reduce the casualties and economic losses caused by accident. The management department should take necessary measures to reduce the concentration of toxic gases in the mine. To obtain the influence level of AIFs on the 'accident level' and carry out the next round of diagnostic analysis, all AIFs are compared and sorted. The three most critical influencing variables on the 'accident level' are selected: 'hydrogen sulfide release'> 'rescue measures'> 'water inflow'.

Table 2. Sensitivity analysis of AIFS with 'accident level' as the node.

Node	Mutual Info	Percent (%)	Variance of Beliefs
Accident level	1.99636	100	0.5609256
Hydrogen_sulfide	0.00606	0.303	0.0005958
Rescue_measure	0.00237	0.119	0.0002081
Water inflow	0.00178	0.089	0.0001565
Accident type	0.00140	0.0702	0.0001207
Illegal_command	0.00092	0.0459	0.0000792
Illegal_mining	0.00072	0.0359	0.0000624
Geological_structure_condition	0.00061	0.0306	0.0000576
Exploring_and_drainage_measure	0.00039	0.0197	0.0000344
Safety_specification	0.00032	0.0158	0.0000276
Precipitation	0.00008	0.00385	0.0000066
Waterproof_and_drainge_system	0.00004	0.00202	0.0000037

3.3 *Scenario analysis*

This paper incorporates scenario analysis into coal mine water hazard analysis, which helps understand the cause process and explain specific situations. Select the 'water inflow' and 'hydrogen sulfide' containing the parent node as the two nodes of the scenario analysis, and observe its impact by changing the parent node's state.

3.3.1 *Scenario analysis 1: water inflow*

Water inflow is the research focus of mine prevention and control and the basis of technical design, so it is of great significance to conduct scenario analysis on it. Under this framework, the parent nodes of 'water inflow' are 'precipitation', 'waterproof and drainage system', 'geological structure conditions', and 'exploration drainage measures'. Figure 4 shows the scenario inference results of 'water inflow'. This shows that 'precipitation', 'waterproof and drainage system' and 'geological structural conditions' have a greater impact on 'water inflow', but the node with the greatest degree of relative change is 'geological structural conditions'. This is manifested in the fact that when the 'geological structure conditions' change from 'simple' to 'complex', the probability of the 'water inflow' being 'high' has more than tripled (increased from 16.7% to 60.0%). Geological structure conditions determine the development level of folds and faults and the water richness of the aquifer. Simple geological structures are not easy to accumulate water sources. Even if the precipitation and rainfall conditions are serious, the water inflow will not be very high (only 16.7% increased to 33.5%). Figure 4 also shows that when the 'precipitation' is 'severe', the probability of 'water inflow' being 'high' is the largest (80.0%). As the main source of water supply for coal mines, precipitation is often at the same time as the disaster-causing time or slightly later than atmospheric precipitation. Therefore, in the rainy season, attention should be paid to water inrush monitoring and hidden danger investigation, and more waterproof measures should be taken if necessary.

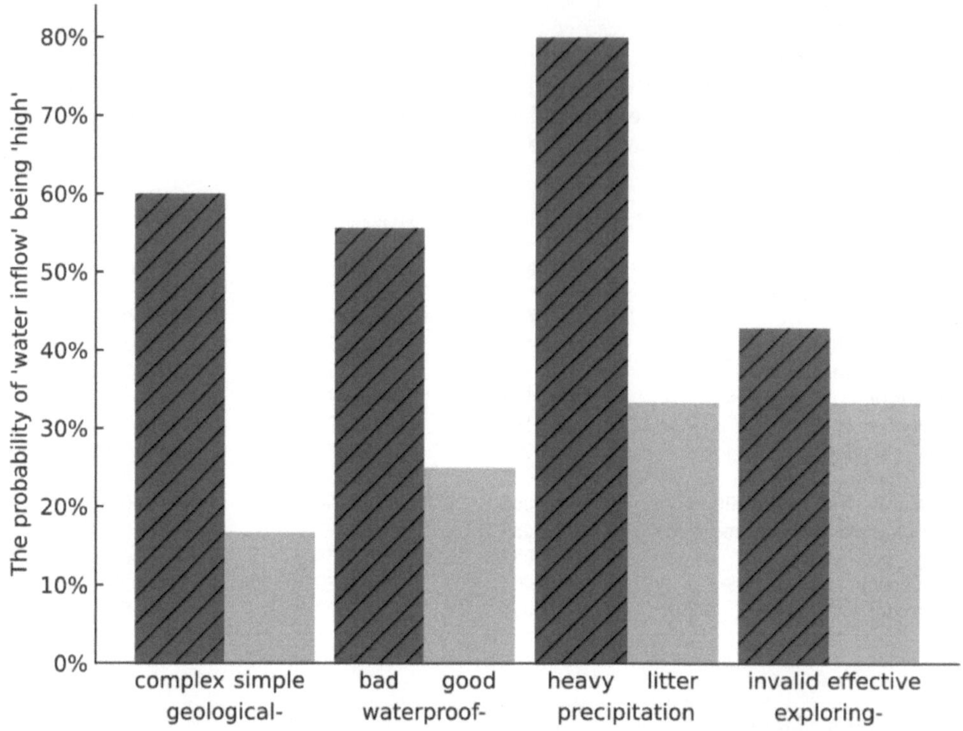

Figure 4. The probability value of 'water inflow' being 'high' when the state of the parent node changes.

3.3.2 *Scenario analysis 2: hydrogen sulfide*

In most cases, hydrogen sulfide gushes out along with water penetration, which is highly concealed and lethal. Once the number of people underground is large, the impact of the accident will be catastrophic. Fig.5 shows the probability inference results of 'hydrogen sulfide' under the combination of 'exploration drainage measures' and 'geological conditions'. It shows that the probability of a high hydrogen sulfide concentration caused by ineffective exploration and drainage measures is as high as 60.6%. In the case of complex geological structure conditions, effective exploration and drainage measures will reduce the probability to 33.3%. Under the condition of a simple geological structure, the probability of high hydrogen sulfide release is much lower than that of the former, and the influence of exploration drainage design is not so significant. According to the scenario analysis, compared with the exploration and drainage measures, the geological structure has a more significant impact on hydrogen sulfide. Under the complex geological structure conditions, effective exploration and drainage measures can reduce the release of hydrogen sulfide. Therefore, it is necessary to find out the hydrogeological conditions around the mine before mining. On this basis, the geological report data to meet the needs of production are designed and submitted by professionals.

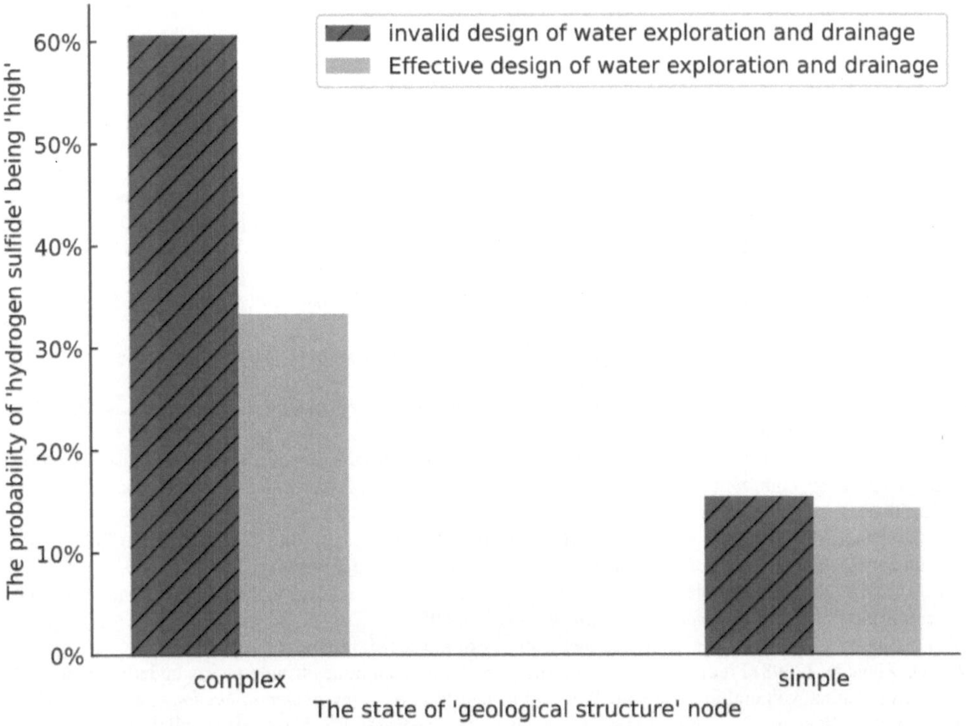

Figure 5. Probability of 'high' state of 'hydrogen sulfide' under combinatorial analysis.

4 CONCLUSION

Based on the scenario analysis, the conclusions are obtained as below:

(1) Geological structure condition is an important influencing variable of water inflow and hydrogen sulfide emission. Under the condition of high water inflow and hydrogen sulfide emission, it is easy to lead to grave accidents and especially grave accidents.
(2) Precipitation is closely related to water inflow, which shows that when the precipitation conditions are severe, the probability value of high water inflow will become very high (80%).
(3) In the case of complex geological conditions, if effective water exploration and drainage design is carried out, the release of hydrogen sulfide in the mine can be effectively reduced.

ACKNOWLEDGMENTS

This work was partially supported by the State Administration of Work Safety, Key Technology and Technology Project for the Prevention and Control of Major Accidents in Work Safety (Gaoxiao-0008-2017AQ, Shangdong-0016-2018AQ) and partially supported by the National Natural Resources Project (51509149)

REFERENCES

Ding, L. (2013) Early Warning System for Coal Mine Water Hazards Based on GIS. *Mining Safety & Environmental Protection* 40, 46–48+51.

Donglin, D., Wenjie, S., Sha, X. (2012) Water-inrush Assessment Using a GIS-based Bayesian Network for the 12-2 Coal Seam of the Kailuan Donghuantuo Coal Mine in China. *Mine Water and the Environment* 31, 138–146.

Gong, C., Li, L., Zhu, K., Gao, Y. (2011) Evolutionary Model of Coal Mine Water Hazards Based on Multi-Agent. *Systems Engineering Procedia* 2, 358–365.

Li, C., Li, J., Li, Z., Hou, D. (2013) Establishment of spatiotemporal dynamic model for water inrush spreading processes in underground mining operations. *Safety Science* 55, 45–52.

Li, L.P., Zhou, Z.Q., Li, S.C., Xue, Y.G., Xu, Z.H., Shi, S.S. (2015) An Attribute Synthetic Evaluation System for Risk Assessment of Floor Water Inrush in Coal Mines. *Mine Water and the Environment* 34, 288–294.

Li, M., Wang, H., Wang, D., Shao, Z., He, S. (2020) Risk assessment of gas explosion in coal mines based on fuzzy AHP and bayesian network. *Process Safety and Environmental Protection* 135, 207–218.

Liu, Q., Meng, X., Hassall, M., Li, X. (2016) Accident-causing mechanism in coal mines based on hazards and polarized management. *Safety Science* 85, 276–281.

Liu, Q., Dou, F., Meng, X. (2020) Building risk precontrol management systems for safety in China's underground coal mines. *Resources Policy*.

Wang, Q., Wang, H., Qi, Z. (2016) An application of nonlinear fuzzy analytic hierarchy process in safety evaluation of coal mine. *Safety Science* 86, 78–87.

Wang, Y., Yang, W., Li, M., Liu, X. (2012) Risk assessment of floor water inrush in coal mines based on secondary fuzzy comprehensive evaluation. *International Journal of Rock Mechanics and Mining Sciences* 52, 50–55.

Wu, Q., Wang, M., Wu, X. (2004) Investigations of groundwater bursting into coal mine seam floors from fault zones. *International Journal of Rock Mechanics and Mining Sciences* 41, 557–571.

Wu, Q. (2014) Progress, problems and prospects of prevention and control technology of mine water and reutilization in China. *Journal of coal industry* 39, 795–805.

Wu Q, Zhilong, Z. (2007) A new practical methodology of the coal floor water bursting

Wu, Q., Zhou, W. (2008) Prediction of groundwater inrush into coal mines from aquifers underlying the coal seams in China: vulnerability index method and its construction. *Environmental Geology* 56, 245–254.

Wu, J., Xu, S., Zhou, R., Qin, Y. (2016) Scenario analysis of mine water inrush hazard using Bayesian networks. *Safety Science* 89, 231–239.

Yan, Z., BAi, H., Zhang, H. (2008) A Novel SVM Model for the Analysis and Prediction of Water Inrush from Coal Mine. *China Safety Science Journal*, 166-170.

Yunxiao, F., Ming, L. (2012) Design of Hazards List based on Hazard Components for Chinese Coal-mine. *Procedia Engineering* 45, 264–270.

Zhang, R., Jiang, Z., Zhou, H., Yang, C., Xiao, S. (2013b) Groundwater outbursts from faults above a confined aquifer in the coal mining. *Natural Hazards* 71, 1861–1872.

Zhang, Y., Wang, K., Zhang, R. (2011) Theoretical research on hazards and accident prevention. *Procedia Engineering* 26, 16–24.

Zhao, Z., Hu, M. (2014) Multi-Level Forecasting Model of Coal Mine Water Inrush based on Self-Adaptive Evolutionary Extreme Learning Machine. *Applied Mathematics & Information Sciences Letters* 2, 103–110.

Advances in Civil Engineering: Structural Seismic Resistance, Monitoring and Detection – Mohd Johari Mohd Yusof and Junwen Zhang (Eds)
© 2023 The Authors, ISBN 978-1-032-31491-4

Survey of earthquake casualties assessment

Chao Deng, Junqi Lin & Jinlong Liu
Disaster Prevention and Mitigation Engineering, Institute of Engineering Mechanics, China Earthquake Administration, Harbin, Heilongjiang Province, China

ABSTRACT: China is a country with frequent earthquakes. Earthquakes with a magnitude of more than 5 occur dozens of times every year, and people lose their lives because of earthquakes every year. The Wenchuan earthquake and the Tangshan earthquake took tens of thousands of lives instantly. Therefore, earthquake casualty assessment plays a decisive role in the early stage of an earthquake without accurate data. It will affect disaster relief materials, budget, and disaster relief priorities in each region. Therefore, earthquake people in China have studied the evaluation methods of earthquake casualties from generation to generation. This paper analyzes the causes of earthquake casualties and then summarizes the evaluation methods of earthquake casualties based on intensity, house vulnerability, population, GDP, and multi influencing factor indicators. Then, the advantages, disadvantages, and application scope of these methods are analyzed, and the future is prospected.

1 INTRODUCTION

China is located between the two major seismic belts globally - the circum Pacific seismic belt and the Eurasian seismic belt. The seismic fault zone is very active due to the compression of the Pacific plate, the Indian plate, and the Philippine Sea plate. Earthquakes in China are mainly distributed in five regions: Taiwan Province, Southwest China, Northwest China, North China, southeast coastal areas (Yang 2021).

China has severe earthquake disasters because of its high frequency, high intensity, shallow focal point, and wide distribution. Since 1900, 550000 people have died in earthquakes in China, accounting for 53% of the global earthquake deaths; Since 1949, more than 100 destructive earthquakes have hit 22 provinces (autonomous regions and municipalities directly under the central government), including 14 provinces in the eastern region, killing more than 270000 people, accounting for 54% of the death toll of various disasters in China. The earthquake affected an area of more than 300000 square kilometers, and 7 million houses collapsed (Li 2014). The seriousness of earthquakes and other natural disasters constitutes one of China's primary national conditions. Therefore, China pays special attention to earthquake casualty assessment. Since the reform and opening up, there have been hundreds of methods to evaluate earthquake casualties.

2 CAUSES OF EARTHQUAKE CASUALTIES

Scholars have studied the causes of casualties caused by fatal earthquakes. Many factors affect the degree of the earthquake disaster. These factors are divided into intensity factors that determine the destructive strength of the earthquake, vulnerability factors of building structure that determine

DOI 10.1201/9781003310884-84

the anti collapse ability of buildings, population factors that determine the population density of the region, and GDP factors that determine the economy of the region.

Earthquake intensity is the reflection of the intensity of ground vibration and its influence caused by earthquake. It can be seen from its definition that the intensity is not only the embodiment of the strength of the ground motion itself but also the strength of the impact caused by the earthquake, especially the damage degree of earth construction facilities such as houses and buildings. It is a description of the ground damage capacity of the earthquake. The greater the intensity, the more casualties.

Research shows that 75% of the deaths in earthquakes come from the destruction and collapse of buildings. Seismic waves are used in buildings, resulting in varying degrees of damage. Those who fail to escape in the buildings will directly die or be trapped, and the trapped people will eventually die due to untimely rescue. The higher the seismic capacity of the building, the lower the vulnerability, the lighter the damage degree in the earthquake, and the fewer casualties. On the contrary, the worse the seismic capacity of the building, the higher the damage degree and the more serious casualties.

The characteristic of earthquakes in China is that there are more earthquakes in the West and less earthquakes in the East. The population is large in the East and small in the West. In the case of frequent earthquakes in the west, there are no more earthquake casualties than in the East, because the population density in the west is lower than that in the East. According to the statistics of casualties caused by historical earthquakes, we can find that under the same conditions of earthquake intensity, economic development level, and geographical environment, the higher the population density in the disaster area, the greater the casualties.

The local GDP not only directly reflects the reserve and supply level of disaster relief materials but also indirectly reflects the local construction quality and earthquake emergency response capacity (Tang 2017). Generally, the higher the level of local economic development, the lower the injury risk of earthquake disaster personnel. Once, when GDP was underdeveloped, the huge number of earthquake casualties was because houses were in the stage of no seismic fortification. With the progress of the times, the houses designed in the last decade should meet the seismic code. The guiding ideology of building seismic design is Sanshui quasi two-stage design. However, in areas with high GDP, its seismic fortification capacity will also be improved, so it is a research rising in recent years to test the seismic capacity with GDP.

3 EVALUATION METHOD OF EARTHQUAKE CASUALTIES

3.1 *Evaluation of earthquake casualties based on intensity*

Intensity describes the damage degree of local ground level buildings affected by earthquake. The intensity is judged according to people's feeling, the vibration of furniture and articles, the damage degree of houses and buildings, and the damage phenomenon on the ground. The mainstream of intensity based casualty assessment method is to use a large number of statistical seismic data, generally at least 80 seismic data, and then carry out least square regression on the data to obtain the preliminary curve formula of magnitude or intensity. Then, other auxiliary parameters such as population density, earthquake occurrence time, fortification intensity, prediction level and focal depth are used as several coefficients for amplitude modulation. For example, in Liu Jinlong's method, he first found the relationship between intensity and death through formula fitting (Liu 2012).

$$\ln(D_m) = 12.2e^{-(\ln(\ln t)-2.445)^2/0.3^2} \tag{1}$$

$$Mag_m = 0.533Int + 2 \tag{2}$$

Then, in Jinlong Liu's method, he used the magnitude correction factor α_m. Its physical meaning is that when the epicenter intensity is the same and the actual magnitude is greater than the average magnitude corresponding to the epicenter intensity, the proportion of its high intensity area is relatively high. Similarly, the correction coefficient of population density is obtaine. Finally, the estimation model of earthquake deaths is obtained.

$$\alpha_m = \frac{\ln(D)}{\ln(D_m)} = \frac{Mag - 5}{Mag_m - 5} \tag{3}$$

$$D = \alpha_m \alpha_{den} D_m = e^{12.2\alpha_{den}\alpha_m e - (\ln(Int) - 2.445)^2/0.3^2} \tag{4}$$

Mengxin Li used SPSS software commonly used in statistics when regression based on intensity formula. She mainly studied all earthquakes with intensity above 6 in Yunnan since 1995. She inputs all seismic data into the software, carries out multiple nonlinear regression, and fits the formula based on intensity (Li 2018). The biggest advantage of her method is that the basic data are from Yunnan, so it is particularly applicable to Yunnan. Moreover, most of the earthquakes he used were after 2000, and most of the house structures were still the type of house structures at that time, so it is of great utility to evaluate the casualties of earthquakes now.

Table 1. Parameter estimate.

parameter	estimate	Standard error	confidence interval	
			lower limit	upper limit
b_0	−11.00	43.98	−99.373	77.373
b_1	2.000	11.078	−20.285	24.268
b_2	0.00004464	0.698	−1.403	1.403
b_3	0.019	0.013	−0.006	0.045
b_4	−0.00005184	0.0019	0.00036	0.00003529
b_5	0.000003062	0.0056	0.00002791	−0.00003404

Xiaojie Li adopted the global earthquake rapid assessment system for earthquake emergency released by USGS. Parameters for calculating model requirements θ and β. Firstly, to minimize the residual between the estimated personnel loss value and the actual personnel loss value, it is necessary to construct the residual objective function. Then the model parameters are calculated by numerical optimization algorithm with MATLAB and preprocessed data θ and β. Finally, the prediction model of personnel loss rate is obtained. By comparing the residual frequency histogram, he found that it is more accurate to calculate the East and West separately than to integrate them together. Then she superimposed the population data with the intensity map to obtain the population in each intensity area and then calculated the number of population deaths through the previous model (Li 2012). The advantages of this method are: the model has high sensitivity to earthquake events with high number of personnel losses; Therefore, when selecting earthquake cases, we should ensure that there are a sufficient number of earthquake cases with high intensity and high number of personnel losses.

In foreign countries, P.A. page combines the intensity with population density to regress the number of earthquake casualties during the day and at night (Xiao 1986).

$$\ln(N_D) = -22.73 + 10.6\ln(I) + 0.34\ln(\rho) \tag{5}$$

$$\ln(N_N) = -11.35 + 5.77\ln(I) + 0.36\ln(\rho) \tag{6}$$

Weihua Shi used 49 earthquakes in Yunnan and 6 major earthquakes in other regions as basic data, and fitted the earthquake casualties formula based on magnitude and the earthquake casualties formula based on intensity, respectively. According to the comparison, it is found that the fitting data of the method based on magnitude has poor stability and large dispersion, while the method based on intensity has good stability and dispersion, so the method based on intensity may be more accurate in predicting earthquake casualties (Shi 2012).

Instrument intensity refers to the seismic intensity calculated according to the seismic observation records. It can be quickly obtained by using the observation data of seismic stations after an earthquake according to a certain calculation method. The instrument seismic intensity is intuitive, simple, and fast, which can provide an important basis for rapid assessment of post earthquake disaster and decision-making of emergency rescue. Therefore, it has been paid attention by researchers and widely used in earthquake emergency work in recent years. Yongli Zhu put forward an earthquake casualty assessment method based on the combination of life loss function and instrument intensity probability density function in the rapid assessment method of earthquake life loss based on strong motion records (Zhu 2017). The main idea of this method is that after a destructive earthquake, the mean and standard deviation of the instrument intensity probability density function can be calculated according to the rapidly obtained strong motion record data.

$$\mu_s = q_1 I_1 + q_2 I_2 + q_3 I_3 + q_{11} I_{11} \tag{7}$$

$$\sigma_s = \sqrt{(q_1 I_1^2 + \sigma_1^2) + (q_2 I_2^2 + \sigma_2^2) + (q_{11} I_{11}^2 + \sigma_{11}^2)} \tag{8}$$

So as to determine the instrument intensity probability density.

$$f(I|\mu_s, \sigma_s) = \frac{1}{\sqrt{2\pi}\sigma_s} \exp\left[-\frac{(I - \mu_s)^2}{2\sigma_s^2}\right] \tag{9}$$

$$L = \int_5^{12} R_n(I) f(I|\mu_s, \sigma_s) dI \tag{10}$$

Now the average accuracy of the instrument intensity is 0.3 degrees. With the progress of science and technology, its accuracy will further increase (Tian 2020). The evaluation of earthquake casualties based on instrument intensity will be an indispensable part of earthquake casualties evaluation in the future. Because he will no longer rely on historical empirical data, he quickly evaluates the number of earthquake casualties through first-hand data, which provides more accurate and rapid data support for the earthquake relief of disaster relief agencies, so that disaster relief agencies can reasonably allocate materials, so as to increase the efficiency of disaster relief.

Intensity based casualty assessment is an empirical statistical method. Its disadvantage is that it uses all historical data. However, with the development of the times, in the past, the M 5.0 earthquake may have killed dozens of people, and now the M 6.4 earthquake in Yangbi, Yunnan Province has only killed three people. It is obviously unreasonable to fit the curve to evaluate the current casualties with the data 30 or even 50 years ago. Therefore, we should use machine learning to establish a set of algorithms, which can be used to evaluate the latest casualties by fitting the data of nearly 15 years, then updating the database every year, and constantly fitting the curve of nearly 15 years by updating the database once a year. Only in this way can the intensity based casualty assessment be more accurate and keep up with the times.

3.2 *Evaluation of earthquake casualties based on building vulnerability*

Building vulnerability is the most extensive method to evaluate earthquakes besides intensity. It can be divided into two categories: one is the seismic damage matrix method based on historical data and expert experience, and the other is the semi empirical and semi probabilistic method based on the consideration of structural seismic capacity and seismic risk (performance-based vulnerability method). The core of the former is to establish the building earthquake damage matrix, and obtain the probability of each structural form in intact, slight damage, medium damage, serious damage, and damage through statistics and analysis.

As an important founder of vulnerability, Zhiqian Yi proposed to fit the human death ratio D with the house damage ratio A after establishing the house earthquake damage matrix (Yin 1991).

$$\log^D = 12.479A^{0.1} - 13.3 \tag{11}$$

Lili Xie and Yuhong Ma's team used scientific methods to explore the evaluation methods of earthquake casualties. They summarized the previous methods, input the seismic data into each method, get the evaluation casualties of each method and compare them with the actual casualties, and finally determine the formula method based on the house collapse rate (Ma 2000). RD is the mortality rate; RB is the collapse rate of the house.

$$log_{10}^{RD} = 9.0RB^{0.1} - 10.07 \tag{12}$$

$$\mathrm{ND} = \mathrm{f_t} \cdot \mathrm{f} \cdot \mathrm{RD} \cdot \mathrm{M} \tag{13}$$

Qijia Zou uses the Tangshan earthquake as the basic data, obtains the failure probability matrix of houses with different structural types and the probability of the site suffering from different intensities through the vulnerability curve, then iteratively fits the casualty coefficient through the data of Tangshan earthquake, and then obtains the results of personnel injury and death (Zou 1995).

The vulnerability is based on the earthquake damage type of each house type, but the day and night use of the house type is different, only saying the vulnerability may lead to relatively large errors. Firstly, Xu Liao proposed to calculate the resident number of each building per hour according to the day and night use of the building (Liao 1999). Then, the casualty probability of each damage degree is counted according to the vulnerability matrix. Finally, the day and night variation is counted, and the day and night are weighted respectively to obtain the number of earthquake casualties.

The energy of an earthquake increases exponentially. If the magnitude increases by one magnitude, the energy of an earthquake will increase 32 times. Therefore, most of our existing seismic methods can only accurately evaluate the casualties of earthquakes below magnitude 7.0 With the increase of magnitude, the error becomes larger and larger. Based on the Wenchuan earthquake, Jie Zhang made a vulnerability based earthquake casualty assessment method that is only applicable to magnitude 7.0 or above. He established the earthquake casualty assessment with each damage degree and area of the house as the dominant factor and then used the time factor and population density factor for amplitude modulation (Zhang 2004). This method is used to evaluate Yushu earthquake, and the results are basically similar.

Most of the existing vulnerability assessment is to assess the total number of casualties caused by the last earthquake, while Suqin Zhou proposed a dynamic casualty assessment method based on vulnerability so that the casualty assessment can be accurate to the number of casualties per day, which provides better data support for the seismic headquarters. She calculates the structural damage matrix A according to statistics and determines the initial personnel disability probability distribution matrix C of the structure by using the existing seismic damage data and statistical distribution. Then, she can calculate the personnel casualty probability distribution matrix M of the structure at the initial time t_0 after the earthquake, and then calculate the probability of each dynamic disability level according to the disability state function, and then calculate the number of people at each disability level (Zhou 2004).

In the second method of vulnerability, that is, the semi empirical and semi probabilistic method based on the consideration of structural seismic capacity and seismic risk, Mingzhe He establishes the structural model based on the design drawings of building structure and HAZUS-MR manual; Then pushover analysis is carried out on the model of the structure, and the relationship curve between base shear and vertex displacement is obtained; Then, through coordinate transformation, the capacity curve obtained from Pushover analysis is transformed into A-D response spectrum curve. Combined with the site type of the structure and other factors, the elastic response spectrum curve for the model is established, the intersection of the elastic-plastic response spectrum curve and the capacity spectrum curve is obtained, the damage index DMI of the component is calculated, and then the overall damage index of the structure is calculated, Make the relationship curve between the overall damage index of the structure and the vertex target displacement. Finally, make the damage median through the curve, and then obtain the transcendence probability of the structure in a certain damage state, and then obtain the probability of the structure in a certain damage state. Substitute it into the HAZUS casualty model to obtain the casualty probability (He 2011).

Because this method uses modeling, the curve obtained is more accurate than others. Because this method needs to obtain more house architectural drawings, coupled with its high accuracy, it is more suitable for research and prediction of accurate earthquake casualties in a small area with known structural architectural drawings.

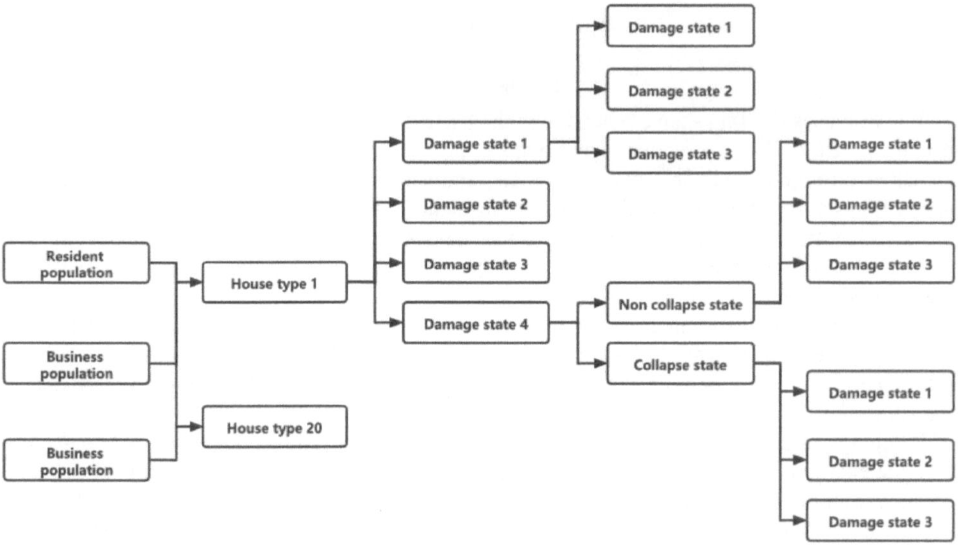

Figure 1. Event tree model for indoor casualty prediction.

The core of vulnerability based earthquake casualty assessment is to establish a relationship between casualties and house collapse rate, damage rate, and house type. The disadvantage of this method is that the classification of buildings is relatively rough. Due to local historical conditions and limitations, the accuracy and reliability of the building earthquake damage matrix may not be so accurate. Moreover, for areas with large disaster area, the building earthquake damage matrix often only counts several typical areas and estimates the overall building earthquake damage matrix according to the typical areas. Now, with the development of UAV technology, we can comprehensively evaluate the overall house damage in this area through aerial photography and high-precision HD camera. Then, through the high-definition map recognition function of the computer, the earthquake damage matrix of all houses in the area is established by the computer. In addition, with the implementation of rural urbanization and housing construction standards, residential houses have unified standards, and the difficulty of building classification has been greatly reduced. Therefore, the earthquake casualty assessment based on house vulnerability has a considerable development space.

3.3 *Population based earthquake casualty assessment*

In most studies, population is only an important factor, and most of them express the impact of population on earthquakes in the form of a coefficient. However, with the deepening of current research, more and more studies are based on population as the dominant factor.

Among many researchers, Jiayu Cheng uses the average family population to obtain the probability distribution of the number of families and then obtains the probability of working people in n-member families. At the same time, through statistics, she obtains the indoor probability of urban and rural workers and freelancers at different times of the working day. Finally, the probability of casualties can be calculated through these two data (Cheng 1993).

Not only Jiayu Cheng, but also Weijia Zhang, who advocated the use of population dynamics correction model to deal with population factors (Zhang 2021). He used the distribution law of population mobility to calculate the number of migrant workers, then calculated the population of Sichuan and Yunnan in the peak tourism season by using the number of people in the peak tourism season and then substituted the processed population data into the formula of Zhiqian Yi.

$$\mathrm{lgd} = \mathrm{aC} - \mathrm{b} \tag{14}$$

C is the house destruction rate and D is the personnel mortality rate. Add the statistical house data to calculate a and B. then apply Xianfu Bai's research to calculate the landslide coefficient d_{all}, and add the three coefficients a, B and landslide coefficient d_{all} to the rapid evaluation results, which can make the results very accurate.

Among the demographic factors, Jinglai Wang mainly believes that the outdoor population rate should be used to express the demographic factor. Through statistics, the outdoor population rate of rural population in Yunnan at different times is obtained. According to different outdoor population rate, the regional index of different regions is obtained. Then he divided Yunnan into five regions. Each region uses different regional indexes to express the degree of population casualties when the earthquake comes. The lower the regional index, the greater the casualties. Then the formula of population death and population index and intensity is obtained by formula fitting (Wang 1994).

Weihua Shi believes that the population density of the earthquake area has a great relationship with casualties. However, the population density changes every year, so the value of the ratio of the population density of the earthquake area to the population density of Yunnan Province should be calculated from the seismic statistical yearbook of Yunnan Province. (Shi 2012) When the statistics cannot be obtained, it should be obtained by multiplying the population density of Yunnan Province and the earthquake site a few years ago by the annual growth rate. The population density obtained in this way is the accurate population density.

3.4 *Evaluation of earthquake casualties based on GDP*

The GDP of a region is an important indicator to measure the development and underdevelopment of the region. In developed regions, houses are designed according to the Behavior-based seismic design idea, while in general regions, houses are designed according to the single degree of freedom response spectrum theory. In underdeveloped regions and even poor counties, houses are designed as Adobe houses, Masonry structure has no seismic structural design and seismic mechanical calculation. Moreover, GDP not only affects the earthquake resistance of houses in the region but also affects the disaster relief capacity of the region when the earthquake comes. If the disaster relief team can come earlier and the first batch of disaster relief teams can have good disaster relief equipment, the corresponding casualties of the earthquake will be smaller.

Jianhong Kang proposed that the weight values of engineering seismic capacity (seismic fortification capacity, secondary disaster defense), socio-economic disaster prevention capacity (residence type, per capita GDP), emergency rescue and recovery capacity (earthquake prediction, engineering prediction, medical rescue, and material reserve) should be obtained by analytic hierarchy process, so as to know the seismic capacity index of each region (Kang 2015). Then, different weight intervals are fitted according to the historical earthquake regression, and the relationship between mortality and intensity is regressed respectively. In this way, a formula group is obtained. Then, the formula is substituted with the township as the administrative unit to calculate the number of deaths in each township, which is added to obtain the number of deaths in the whole region.

Jianglin Li believes that the correlation function of influencing factors of regional earthquake casualties should be constructed (Li 2011).

$$Y = f(GDP, G, R, D, K, J) \tag{15}$$

Y is the value of correlation function; GDP is the per capita GDP value of the earthquake disaster risk analysis area; G is the urban function type factor in the seismic disaster risk analysis region; F is the regional urban economic development factor of earthquake disaster risk analysis; D is the regional geological environment factor for seismic disaster risk analysis; K is the regional disaster resistance factor of earthquake disaster risk analysis.

$$Y = 0.21G + 0.24R + 0.2D \tag{16}$$

It fully considers the multi index influencing factors of regional earthquake casualties, puts forward the evaluation model of regional earthquake casualties, then obtains the weight of each variable in the correlation function of regional earthquake casualties through the PCA analysis of the index system, and finally inputs the seismic data to obtain the ratio between per capita earthquake disaster casualties and per capita GDP. MDF is the ratio of earthquake casualties per capita to GDP per capita.

$$MDF = e^{-7.93} Y^{0.141} \left(\frac{I}{i}\right)^{2.81} \tag{17}$$

Jifu Liu also believes that GDP is a very important link in modern earthquake assessment (Liu 2009). He proposed the registration and digitization of seismic Isoseismal map. He scanned the selected seismic Isoseismal map, registered the Isoseismal map under the MapInfo environment, and digitized the seismic Isoseismal map above grade VI, so as to prepare for the later GDP distribution. Then, according to the collected population and GDP land area of each district and county involved in each intensity area at the time of the earthquake, the population and GDP values of each intensity area are converted according to the area ratio of each intensity area to each district and county. The number of deaths in each intensity area is redistributed according to the actual earthquake disaster assessment data, The life loss ratio in the intensity area is obtained by calculating the intensity zoning of each earthquake. The per capita GDP expressed by the constant price in 2000 is used as the classification index of earthquake life vulnerability. Taking the per capita GDP of 2700 yuan as the classification threshold, all statistical data are divided into two groups for regression statistics according to the following relationship formula to determine the relationship between life loss rate and earthquake intensity. R_{MDF} represents life loss rate, I is seismic intensity, and a, B, and C are coefficients.

$$R_{MDF} = CAI^{B} \tag{18}$$

3.5 *Evaluation of earthquake casualties based on multi factor indexes*

The above methods have a common feature. They are dominated by a single factor, and other factors are evaluated in the form of coefficients. However, we understand that the casualties caused by earthquakes should never be dominated by one factor. The casualties caused by earthquakes may be due to the high intensity, the high population density in the region, the high rate of house damage, the low GDP, the people sleeping at night, and the extreme meteorological disasters caused by earthquakes. These are very likely. The previous research based on single factor may overestimate the dominant factors, resulting in inaccurate casualty assessment. Now, with the deepening of research, there is an earthquake casualty assessment based on multi factor indicators.

This method mainly uses analytic hierarchy process as a means to solve the weight probability of different factors when there are many factors, and all factors can be gathered into the judgment matrix, such as earthquake magnitude, earthquake intensity, focal depth, occurrence time, focal depth, topography, polar weather, population density, material reserves, GDP of the region and other factors (Zhang 2017). Then, the weight of each factor is calculated through the judgment matrix. Finally, the obtained weight is the relative coefficient of each factor and then brought back to the previously summarized method to obtain a new formula.

The key of this method is analytic hierarchy process (AHP). It decomposes the elements related to decision analysis into objectives, criteria, indicators and other levels, and carries out quantitative and qualitative analysis on this basis. It can use less quantitative information to mathematicize the thinking process of decision-making. It is most suitable for the situation where the element structure is complex and the indicators need to be quantified on the basis of empirical judgment.

The analytic hierarchy process should first construct a judgment matrix, assuming that the criterion layer is A_1, then the index layer of A_1 is $B_1, B_2 \cdots B_N$, and then scale each index according to the relative importance proportion according to the investigation and expert suggestions to form a judgment matrix:

$$\begin{matrix} & B_1, B_1, B_1, B_n \\ B_1 & \\ B_2 & \\ B_3 & \\ B_n & \end{matrix} \begin{bmatrix} b_{11} & b_{12} & b_{13} & b_{1n} \\ b_{21} & b_{22} & b_{23} & b_{2n} \\ b_{31} & b_{32} & b_{33} & b_{3n} \\ b_{n1} & b_{n2} & b_{n3} & b_{nn} \end{bmatrix} \tag{19}$$

When the importance of B_i relative to B_j is expressed numerically, the nine order scaling method is usually used, and bij is taken as $1, 2, 3 \cdots 9$ or their reciprocal.

Then find the maximum eigenvalue and eigenvector of the judgment matrix. Then the eigenvalue of matrix B and its eigenvector can be expressed as $B\alpha = \lambda\alpha$. Then the maximum characteristic root of B can be calculated λ_{Max}, and its corresponding eigenvector α. Then yes α Normalize.

(1) Calculate the product M_i of each row element in the judgment matrix respectively:

$$m_i = \prod_{i=1}^{n} b_{ij} 1 \leq i \leq n \tag{20}$$

(2) Calculate the n-th root of M_I:

$$\alpha = \sqrt[n]{m_i} \tag{21}$$

(3) Pair vector α_1 standardization:

$$\alpha_i = \frac{\alpha_l}{\sum_{j=1}^{n} \alpha_j} \tag{22}$$

(4) Calculate the maximum eigenvalue of matrix B according to the eigenvector λ_{max}:

$$\lambda_{max} = \frac{1}{n} \sum_{i=1}^{n} \frac{(B\alpha)}{\alpha_i}, 1 \leq i \leq n \tag{23}$$

(5) Due to the complexity of the objective things studied by analytic hierarchy process, it is also necessary to test the randomness and consistency of the judgment matrix to judge the rationality of the index weight and test formula:

$$CR = \frac{CI}{RI} \tag{24}$$

$$CI = \frac{\lambda_{max} - m}{m - 1} \tag{25}$$

(6) Calculated results λ Is the weight corresponding to each factor.

Among the multiple influencing factors, using analytic hierarchy process to deal with each factor is a simple method. In the absence of computer, it can also be processed by hand. Today, with the development of computer and statistics, SPSS can be used to process the data, eliminate the redundant data that does not meet the expected value, and then use the built-in program of Amos to set the potential variables and observation variables, The structural equation model of influencing factors of earthquake casualties is established, the program parameters are defined, and then the influence coefficients of each of these influencing factors are obtained (Wu 2013).

Table 2. The relationship of variable definition.

	Exogenous variable		Endogenous variable
Potential variable	ζ_1 : Seismic destructive force level	ζ_3 : House damage	η_1 : Earthquake casualties
	ζ_2 : Regional fortification level	ζ_4 : Regional population	
		ζ_5 : Seismic destructive force level	
Observation variable	X_{11}: magnitude	X_{23}: House collapse area	Y_1: death toll
	X_{12}: Epicenter intensity	X_{31}: Damaged area of house	
	X_{13}: Area above grade V area	X_{32}: Gender ratio	
	X_{14}: Regional seismic fortification intensity	X_{33}: Population density	Y_2: Total casualties
	X_{21}: Design basic seismic acceleration value	X_{41}: Earthquake time	
	X_{22}: Earthquake prediction	X_{42}: Focal depth	

Because this method involves many factors, it can more accurately evaluate earthquake casualties. However, because too much accurate data is required, it may not work when the earthquake just occurs.

Figure 2. Structural equation model of earthquake casualties influencing factors.

This method can play its role only after certain data statistics are available. However, due to its balance, accuracy and many factors involved, I think it has a better prospect for its development in the future and will be an important method in the study of earthquake casualty assessment in the future.

4 CONCLUSION

The seismic evaluation method with intensity as the main factor focuses on statistical regression and big data evaluation, but everyone's focus is different. Jinlong Liu takes magnitude and population density as auxiliary factors, which is better applicable to cities and towns with large population density in the East and West. Mengxin Li uses all the recent seismic data of Yunnan, So his method is more suitable for Yunnan Earthquake. Xiaojie Li adopts the global earthquake rapid assessment system pgaer, so her method is highly sensitive to earthquake events with high number of personnel losses, P.a.page takes the day and night factors into account and has more accuracy for night earthquakes. Weihua Shi compares the magnitude with the intensity and concludes that using the intensity as the dominant factor is more accurate than using the magnitude. Yongli Zhi takes the instrument intensity as the method to evaluate earthquake casualties, which provides a method to quickly evaluate casualties directly through the data of seismic stations. With the rapid development of science and technology, magnitude based statistical regression methods will be fused and bound with big data, machine learning, and neural network to provide new blood and vitality for traditional evaluation methods.

The essence of vulnerability based method is to study the building damage matrix. Zhiqian Yi team, Lili Xie, and Yuhong Ma team first studied the relationship between house collapse rate and earthquake casualties. Then Qijia Zou studied the failure probability matrix of houses with different structural types and the probability of sites suffering from different intensities. Xu Liao also studied the occupancy rate of the house in different time periods to improve the accuracy of vulnerability. Jie Zhang also made a deep analysis of the large earthquake, so that the earthquake with 7.0 can have better adaptability with her method. Suqin Zhou also proposed a dynamic casualty assessment method based on vulnerability, so that the casualty assessment can be accurate to the number of casualties per day. Mingzhe He's semi empirical and semi probabilistic method based on the consideration of structural seismic capacity and seismic risk can enable researchers to more accurately predict earthquake casualties in a small area of known structural building drawings in the laboratory. However, in actual earthquakes, it is often impossible to get the damage data of houses at the first time. Therefore, I think vulnerability will be combined with high-definition aerial photography in the future to obtain house damage data through aerial photography. Coupled with the current standardization of residential houses, there will be considerable room for development of earthquake casualty assessment based on house vulnerability.

As for the main factor of population, the previous earthquake statistics were only imported directly based on the existing data, and the population data were not refined. Now the research is more and more careful, and the population data began to be dynamically revised. In non census years, the population data should be multiplied by the population growth rate according to the previous population data, This makes the basic data of seismic evaluation closer and closer to reality, so that the evaluation is closer and closer to the real value.

In the evaluation of earthquake casualties based on GDP, Jianhong Kang and Jianglin Li both believed that the weight of each factor was obtained by combining the engineering seismic capacity, socio-economic disaster prevention capacity, emergency rescue, and recovery capacity, so as to evaluate the earthquake casualties in the region as a whole, while Jifu Liu combined GDP with isoseismal line, taking GDP into account, This method is more accurate in practical application.

In the multi factor index of earthquake, it emphasizes the importance of multi factor joint influence. The casualties caused by earthquake are never the result of the influence of a single factor, and the influence of multiple factors is just in line with the characteristics of earthquake casualties, which can make the structure more accurate. The advantage of analytic hierarchy process is that it is simple to operate and easy to start. It is suitable for use when simple estimation or fast operation is required. Using Amos requires setting many parameters, which is difficult to operate. However, because it is computer analysis, the accuracy is more accurate than analytic hierarchy process.

At present, there are many post earthquake casualty assessment models, but the estimation results of different models are quite different. On the one hand, the difference comes from the collection of earthquake disaster data, on the other hand, it comes from the evaluation model itself. Therefore, we should compare each method, so as to select the most appropriate method for evaluation when the earthquake comes, so as to accurately evaluate earthquake casualties.

ACKNOWLEDGMENTS

This work was financially supported by National Key R&D Projects Of China (2018YFC1504503).

REFERENCES

Cheng, J.Y. (1993) Probability of impact of earthquake occurrence time on casualties. *Disaster science*, (02):13–16.

He, M.Z., Zhou, W.S., (2011) Prediction method of earthquake casualties based on earthquake damage index. *Journal of Harbin Institute of Technology*, 43(04):23–27.

Kang, J.H., Li, Z., An, J.F. (2015) Rapid assessment method of earthquake life loss based on macro earthquake disaster reduction ability. *Journal of disaster prevention and reduction*, 31(02):80–83. DOI:10.13693/j.cnki.cn21-1573.2015.02.014.

Li, J.l., Su, J.Y., Li, X.Z. (2011). Study on casualty assessment model of regional earthquake disaster. *Henan Science*, 29(07): 869–872. DOI:10.13537/j.issn.1004-3918.2011.07.032.

Li, X.M. (2018) Journal of natural disasters rapid estimation of earthquake injury based on SPSS analysis. *Scientific and technological innovation*, 2018(03):57–58.

Li, Y.Y., Su, G.F., Weng, W.G., Yuan, H.Y. (2014) Study on evaluation method of earthquake casualties. *Disaster science*, 29(02):223–227.

Liao, X., Gao, C.B., Han, S.X. (1999) Prediction of direct casualties in earthquakes at different times. *Journal of natural disasters* (04):92–96.

Liu, J.F., Chen, Y., Shi, P.J., Chen, J. (2009) Study on seismic risk analysis model in Chinese mainland (II): vulnerability model for life. *Journal of Beijing Normal University* (NATURAL SCIENCE EDITION), 45(04):404–407.

Liu, J.L., Lin, j.Q. (2012) Study on evaluation method of earthquake casualties based on epicenter intensity. *Journal of natural disasters*, 21(05): 113–119. DOI:10.13577/j.jnd.2012.0516.

Ma, Y.H., Xie, L.L. Study on estimation method of earthquake casualties. (2000) *Earthquake engineering and Engineering Vibration*, (04):140–147. DOI:10.13197/j.eeev.2000.04.021.

Shi, W.H., Chen, K.H., Xie, Y.Q., Li, X., Pang, W.D., Dang, B.Y. (2012) Study on prediction method of casualties in earthquake disasters in Yunnan. *Seismic research*, 35(03):387–392.

Tang, B.H. (2017) Study on the occurrence law of earthquake casualties and the physical and psychological trauma of victims. *Second Military Medical University*.

Tian, X.F., Zhang, W.D., Yuan, J., Zhu, Z.J. (2020) Comparative analysis of instrument intensity and macro intensity of Wenchuan Ms 8.0 earthquake. *Journal of Earthquake Engineering*, 42(05):1225–1231.

Wang, J.L. (1994) Quantitative estimation of earthquake deaths in Yunnan. *Disaster science*, (04):55–58.

Wu, J.F., Wang, H.X. (2013) Analysis of influencing factors of earthquake casualties based on Amos. *Earthquake engineering and Engineering Vibration*, 33(02):221–228.

Xiao, G.X. (1986) Economic problems in Earthquake Engineering. *World Earthquake Engineering*, (01):3–10+19.

Yang, F., Chen, H.L., Cai, J. A. (2021) Machine learning of the relationship between earthquakes in Chinese mainland and surrounding seismic belts. *Information Technology Committee of China Geophysical Society*, 026487.

Yin, Z.Q. (1991) Study on earthquake disaster loss prediction. *Earthquake engineering and Engineering Vibration*, (04): 87–96. DOI:10.13197/j.eeev.1991.04.010.

Zhang, J., Gao, H.Y., Liu, Q. (2011) Research on prediction model of earthquake casualties based on Wenchuan earthquake. *Chinese Journal of Safety Science*, 21(03): 59–64. DOI:10.16265/j.cnki.issn1003-3033.2011.03.002.

Zhang, J.W., Yang, T.Q., Bai, X.F., Chen, Y.H., Jiang, L.X. (2021) Study on dynamic evaluation method of earthquake casualties in Sichuan Yunnan region based on regional difference correction factors. *China Earthquake*, 37(01):197–205.

Zhang, Y., Yin, W.G., Guo, W.G., He, Y.F. (2017) Construction of index system of key influencing factors of casualties in earthquake disasters, 33(04): 87–94. DOI:10.13693/j.cnki.cn21-1573.2017.04.016.

Zhou, S.Y., Guo, Z.Y., (2004), Dynamic evaluation of earthquake casualties. *Journal of Huaqiao University* (NATURAL SCIENCE EDITION), (01):54–57.

Zhu, Y.L., Li, D.H.,Zhu, J.G. (2017) Fast evaluation method of earthquake life loss based on strong motion records. *Journal of Seismology*, 39(01):143–154.

Zou, j.Q., Mao, G.M., Sun, Z.K., Ma, C.Q., Ma, G.M. (1995) Study on vulnerability of earthquake casualties. *Journal of natural disasters*, (03):60–68.

Advances in Civil Engineering: Structural Seismic Resistance, Monitoring and Detection – Mohd Johari Mohd Yusof and Junwen Zhang (Eds)
© 2023 The Authors, ISBN 978-1-032-31491-4

Prediction of pile bearing capacity based on GM (1,3) model in karst area

Jingbin He
Power China Northwest Engineering Coporation Limited, Shaanxi Xian, China

Yunxiu Dong
School of Civil Engineering, Longdong University, Qingyang Gansu, China

ABSTRACT: Combined with the results of numerical simulation, based on the characteristics of grey theory was used to study the influence of cave height and span on the ultimate bearing capacity of pile foundation. The results show that the ultimate bearing capacity of pile foundation decreases with the increase of cavern height and span, and the influence of cavern height is greater than that of cavern span. The grey theory can better reflect the change rule of ultimate bearing capacity of pile foundation under the height and span of karst cave, and the error between the calculation result and the numerical simulation result is 4.69%, indicating that the grey theory can be better used in the prediction of ultimate bearing capacity of pile foundation.

1 INTRODUCTION

When constructing roads in areas where karst is widely distributed, it is inevitable to pass through areas where karst is developed. Because of the strong concealment of karst, when the pile foundation is built in the karst development area, it is easy to cause the foundation damage and the bearing capacity of the pile foundation to decline because the pile is worn through karst caves or there are karst caves around the pile.

Aiming at the ultimate bearing capacity of pile foundation in karst development area, Xia Lianand Yin Kaili (Xia 2018; Yin2015) analyzed the influence of different cave height, span, roof thickness and rock strength on pile foundation bearing capacity based on ANSYS finite element calculation software. Gong Chengzhong (Gong 2006) carried out indoor tests, field static load tests, theoretical analysis and numerical simulation, etc., and believed that the stability of karst roof should be checked when there are caves or cracks within the safe thickness range of karst roof. According to the basic principle of upper limit finite element method, Li Chun (Li 2019) introduced parameter k to characterize the reduction degree of ultimate bearing capacity of pile foundation by the existence of karst cavern, and discussed the influence of soil weight, rock-socketed depth, karst cavern radius, horizontal and vertical distance between pile and karst cavern on the upper bearing capacity of pile foundation. Zhao Jianbin (Zhao 2011) obtained the sequence of factors affecting the bearing capacity of the single pile of the project and carried out the bearing capacity prediction analysis through the grey relational degree analysis of the survey data, construction records and field load test data of the project. Based on the grey system theory, Wang Chenghua (Wang 2013) studied the characteristics of load-settlement relationship curve of bored pile and proposed an improvement method combining with engineering practice.

This paper analyzes the influence of different cave heights and spans on the ultimate bearing capacity of pile foundations based on finite element software. This research can provide a reference for accurately calculating and predicting the ultimate bearing capacity of pile foundations in karst areas, and guarantee the safety of pile foundation construction in karst areas.

 DOI 10.1201/9781003310884-85

2 FINITE ELEMENT MODEL ESTABLISHMENT

2.1 *Model establishment*

This paper establishes a karst-pile-soil interaction model based on finite element software, as shown in Figure 1. The ideal elastic constitutive model is used for pile foundation. The foundation soil includes overburden, bearing limestone and karst cave backfill, the Morh-Coulomb yield criterion is used for analysis. In this model, the pile diameter D is 1.8m, and the pile length L is 20m. The foundation soil is composed of silty clay and medium-weathered limestone from top to bottom, and the thickness is 3m and 37m respectively. The bottom of karst cave is 5m above the pile tip. The range of foundation soil is taken 14D around the pile, that is 25.2m. Take L from the pile bottom to the boundary, i.e. 20m. The boundary conditions are fixed x and y displacements on the side of foundation soil, and x, y and z displacements on the bottom.

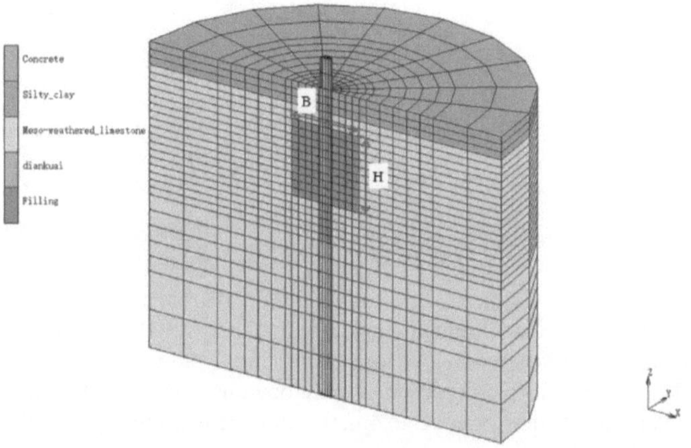

Figure 1. Karst-pile-soil interaction model.

2.2 *Parameter selection and calculation scheme design*

Starting from the influence factors of the span and height size of the cave, the influence of the size of the karst cave on the ultimate bearing capacity of the pile foundation is analyzed. The calculation scheme is shown in Table 1. The calculation parameters of the finite element model are shown in Table 2.

Table 1. Calculation scheme.

Conditions change	Pile length/m	Pile diameter/m	Cave Height H/m	Cave span B/m
Cave height change, cave span change	20	1.8	0, 2, 4, 6, 8, 10	0, 2, 4, 6, 8, 10

Table 2. Model calculation parameters.

Material	Modulus of elasticity/Pa	Poisson's ratio	Cohesive force/kPa	Angle of internal friction/°	Severe /(kN·m^{-3})
Pile Foundation	3×10^{10}	0.2	–	–	25
Silty Clay	3×10^{7}	0.35	17	18	19
Medium-weathered Limestone	2×10^{8}	0.25	110	25	23
Backfill Soil	1×10^{7}	0.3	10	5	16
Rigid Block	3×10^{11}	0.2	–	–	25

3 INFLUENCE OF KARST DIMENSION ON PILE FOUNDATION

3.1 *Analysis of influence of cavern height change on pile bearing capacity*

The $P - S$ curve when the height of the cave increases from 0m to 12m is shown in Figure 2. The influence of hole height change on ultimate bearing capacity of pile foundation and its attenuation is shown in Figure 3. According to code for Design of Building Foundation, the bearing capacity corresponding to pile top settlement reaching 40mm is the ultimate bearing capacity, and the reduction is the percentage of the reduction in the ultimate bearing capacity of the i-th working condition relative to the i-1th working condition.

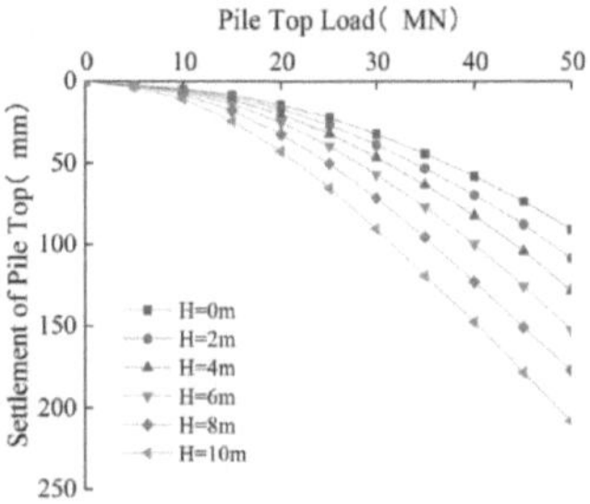

Figure 2. $P - S$.

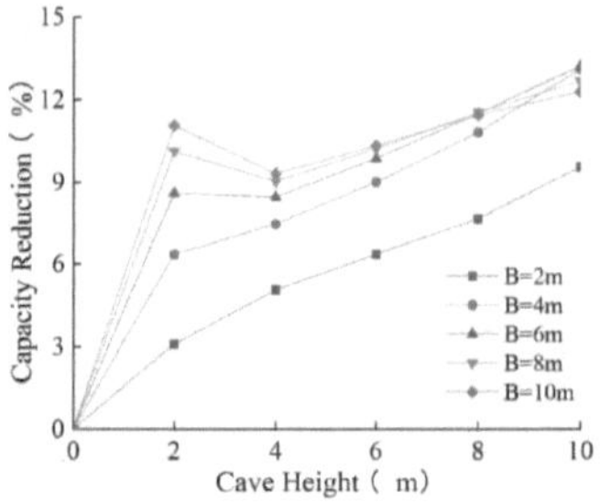

Figure 3. Ultimate bearing capacity decreases.

It can be seen from Figures 2 and 3 that, with a certain span of the cavern, the settlement of pile foundation under vertical load gradually increases and the ultimate bearing capacity gradually decreases with the increase of cavern height. The decreasing amplitude tends to be gentle with the increase of cave height, and this phenomenon is more obvious when the cave span is large. When the span of cave span B is less than or equal to 4m, the ultimate bearing capacity decreases with the increase of cave height. When the span of the cave is larger than 4m, the bearing capacity of pile foundation is significantly affected by the existence of karst cave. When karst cave appears, the bearing capacity decreases significantly, and the decreasing amplitude decreases first and then increases gradually with the height of the cave. The reason for the above phenomenon is that when the cave span is constant, the contact area between the pile side and the rock mass decreases with the increase of the cave height. Therefore, it can provide lower side friction resistance. Therefore, the higher the cavern, the lower the side friction resistance of the pile and the lower the ultimate bearing capacity.

3.2 *Analysis of the influence of karst cave span changes on the bearing capacity of pile foundations*

The P-s curve when the height span of the karst cave increases from 0m to 10m is shown in Figure 4, and the influence of the variation of the cave span on the reduction of the ultimate bearing capacity of the pile foundation is shown in Figure 5.

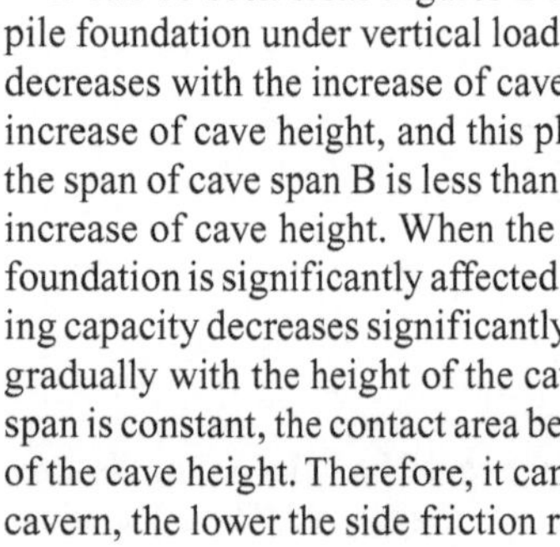

Figure 4. $P - s$.

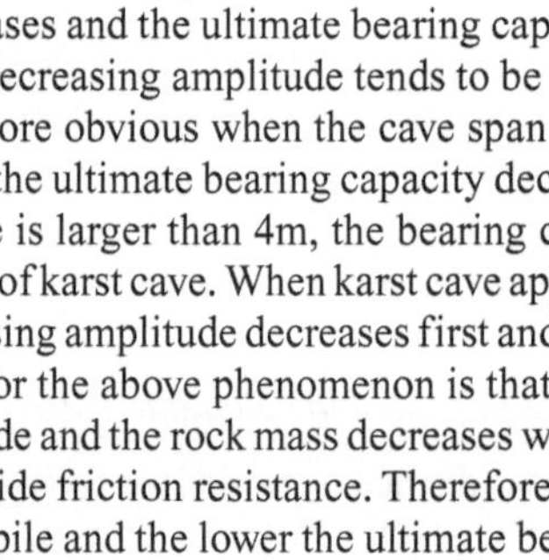

Figure 5. Decrease of ultimate bearing capacity.

It can be seen from Figure 4 and Figure 5 that when the cave height is constant, the ultimate bearing capacity of the pile foundation gradually decreases with the increase of the cave span, and the reduction first increases and then gradually decreases. It can be seen from Figure 5 that as the cave span gradually increases, the presence or absence of karst caves has a very significant impact on the bearing capacity of the pile foundation. When the cave appears, the influence of the gradual increase of the cave span on the bearing capacity of the pile foundation gradually decreases. The reason for the above phenomenon is that when the height of the cave is constant, when the span of the cave gradually increases, the downward force on the rock mass at the side of the pile when the pile is under load is equivalent to the force acting on the slab structure, and the greater the span, the greater the deflection and deformation of the board, so when a cave appears (the span increases from 0m to 2m), the bearing capacity decreases significantly. When the span continues to increase (4m~10m), the settlement of the pile body depends not only on the flexural deformation of the upper and lower rock layers of the karst cave section, but also on the side friction and pile tip resistance provided by the rock layers. Therefore, its influence gradually decreases as the span continues to increase.

In summary, when the pile passes through the karst cave, the influence of the cave height on the ultimate bearing capacity of the pile is greater than the influence of the cave span on the ultimate bearing capacity of the pile, and B>4m, the influence gradually decreases.

4 CALCULATE THE ULTIMATE BEARING CAPACITY OF PILE FOUNDATION BASED ON GM (1,3) MODEL

4.1 *Establishment method of GM (1, N) model*

The GM (1, N) model is a first-order gray model with N variables. If there are n variables such as $x_1, x_2, ..., x_n$, there are n number sequences.

When the cumulative generation method is used for $x_i^{(0)}$, we can establish grey differential equation:

$$\frac{dx_1^{(1)}}{dt} + ax_1^{(1)} = b_1x_2^{(1)} + b_2x_3^{(1)} + ... + b_{n-1}x_n^{(1)}$$

The parameter column is $\hat{a} = [a, b_1, b_2, ..., b_{i-1}]^T$. According to the least square method, $\hat{a}$ can be found as $\hat{a} = (B^TB)^{-1}B^Ty_n$, $y_n = [x_1^{(0)}(2), x_1^{(0)}(3), \cdots, x_1^{(0)}(n)]^T$, in where

$$B = \begin{bmatrix} -\frac{1}{2}(x_1^{(1)}(1) + x_1^{(1)}(2)) & x_2^{(1)}(2) & \cdots & x_n^{(1)}(2) \\ -\frac{1}{2}(x_1^{(1)}(2) + x_1^{(1)}(3)) & x_2^{(1)}(3) & \cdots & x_n^{(1)}(3) \\ \vdots & \vdots & & \vdots \\ -\frac{1}{2}(x_1^{(1)}(n-1) + x_1^{(1)}(n)) & x_2^{(1)}(n) & \cdots & x_n^{(1)}(n) \end{bmatrix} \quad (1)$$

4.2 *Calculation of ultimate bearing capacity of pile foundation based on GM (1,3) model*

When using the GM (1,3) model for analysis, the variables mainly include ultimate bearing capacity P, cave height H, and cave span B, which are $x_1^{(0)}, x_2^{(0)}, x_3^{(0)}$, and the variable curve is shown in Figure 6. In order to improve the accuracy and reduce the error, it is advisable to increase the ultimate bearing capacity. Therefore, the data in $x_1^{(0)}$ is uniformly processed as 35-$x_1^{(0)}(k)$. The data curve is shown in Figure 7.

Accumulate and generate $x_i^{(0)} = (x_i^{(1)}(1), x_i^{(1)}(2), ..., x_i^{(1)}(26))$ once, and the accumulative curve is shown in Figure 8.

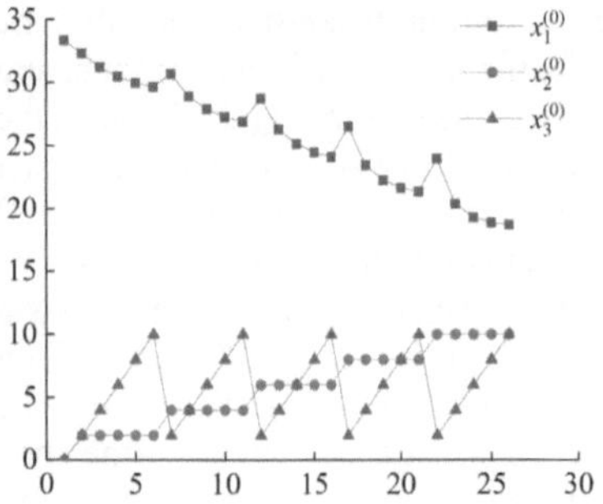

Figure 6. The original variable curve.

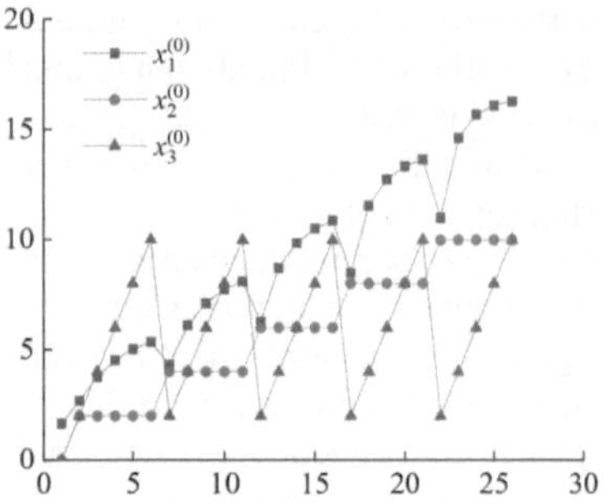

Figure 7. The processed variable curve.

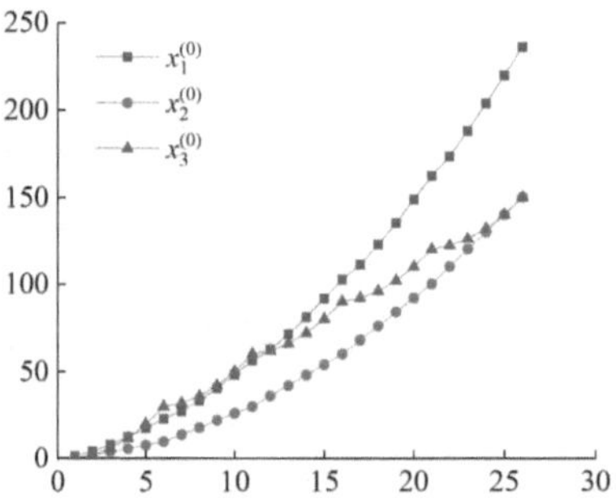

Figure 8. Accumulation curve.

According to formula (1), the matrix B can be solved, and the least square method can be used to obtain:

$$\hat{a} = (B^T B)^{-1} B^T y_n = [a, b_1, b_2]^T = [-0.03576, -0.1059, 0.16005]$$

The extended mode of GM (1, N) is GM (1, N, $x^{(0)}$) (Shen 2010), and the extended model formula is

$$x_1^{(0)}(k) = \sum_{i=2}^{N} \beta x_i^{(0)}(k) + (1-\alpha)x_1^{(0)}(k-1) \tag{2}$$

Among this, $\alpha = \frac{a}{1+0.5a}$, $\beta = \frac{b}{1+0.5a}$
Solved: α=-0.0364, β_1=-0.1078, β_2=0.1630, $1-\alpha$=1.0364.
The calculated formula for ultimate bearing capacity of pile foundation is

$$x_1^{(0)}(k) = -0.1078x_2^{(0)}(k) + 0.1630x_3^{(0)}(k) + 1.0364x_1^{(0)}(k-1) \tag{3}$$

Formula (3) is the formula for determining the bearing capacity derived from the gray theory. Finally, subtract 35 from the calculated result is the ultimate bearing capacity. Bring the data into the calculation formula for carrying capacity inspection. The comparison between the predicted value and the actual value is shown in Figure 9.

It can be seen from Table 4 and Figure 9 that the predicted value calculated by the GM (1, 3) model established by gray theory is consistent with the actual value, and the average error is 4.69%, which meets the error requirements of gray theory for predictive analysis. Therefore, the above calculation The formula can be used as a calculation formula for the ultimate bearing capacity of pile foundation under the influence of hole height and hole span. And in the formula, it can be seen that |b2|>|b1|, that is, the influence of the hole height on the bearing capacity is greater than the influence of the hole span on the bearing capacity, which is consistent with the law obtained in the previous section.

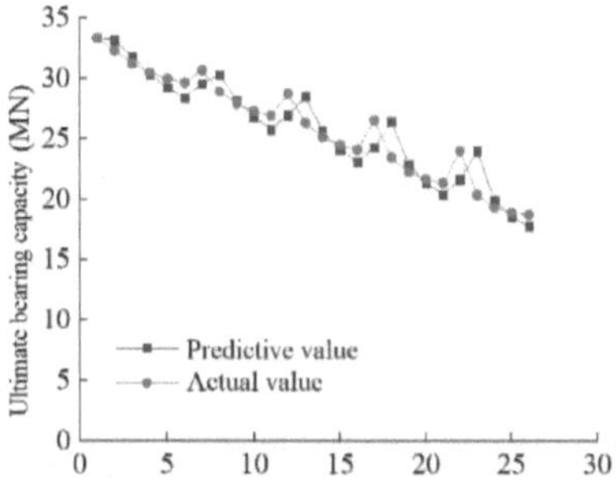

Figure 9. Comparison of predicted value and actual value.

5 CONCLUSION

Based on the finite element and grey theory analysis method, the influence of the hole height and the hole span on the ultimate bearing capacity of the pile foundation is analyzed, and the main conclusions are as follows:

(1) The change of height and span has a significant impact on the ultimate bearing capacity of pile foundation. With the increase of hole height and span, the ultimate bearing capacity of pile foundation decreases, and the influence of the change of hole height is greater than the change of hole span.
(2) The GM (1, 3) model established based on gray theory derives the calculation formula of the relationship between the ultimate bearing capacity of pile foundation and the size of karst cave. The average error between the calculation results and the numerical simulation results is 4.69%, which can better reflect and Predict the ultimate bearing capacity of pile foundation under the influence of hole height and hole span.
(3) In addition, the influence of cave location, rock strata strength and number of caves on bearing capacity can also be considered.

REFERENCES

Gong, CZ. & He, CL. (2006). Research on vertical bearing capacity of pile foundation in karst area. *J. Water Transport Engineering*: 92–96.

Li, C. (2019). Finite element upper limit analysis of ultimate bearing capacity of pile foundation in karst area. *J. China and Foreign Highway*, 39(02): 28–32.

Shen, ZT. (2010). Analysis of commodity housing prices based on grey system prediction theory. *J. Dalian University of Technology*.

Wang, CH. & Zhang, MN. (2013). Grey theory revision prediction method of bored pile load-settlement curve. *Journal of Shenyang Jianzhu University* (Natural Science Edition), 29(02): 270–276.

Xia, L. (2018). Research on influencing factors of bearing characteristics of pile foundation in karst area. *J. Highway*, 63(11): 168–171.

Yin, KL. & Hao, Sl. (2015). Numerical simulation study on the influence of karst on the bearing capacity of pile foundation in Wuhan area. *J. Hydrogeology and Engineering Geology*, 42(06): 96–102.

Zhao, JB., Shi, YQ., & Yang, J. (2011). Analysis of influencing factors on the bearing capacity of static pressure pipe piles based on grey theory. *Chinese Journal of Geotechnical Engineering*, 33(S1): 401–405.

Advances in Civil Engineering: Structural Seismic Resistance, Monitoring and Detection – Mohd Johari Mohd Yusof and Junwen Zhang (Eds)
© 2023 The Authors, ISBN 978-1-032-31491-4

Numerical analysis of slope stability

Jinhua Xu*
Wenhua College, Wuhan, China

Li Chen* & Yafen Song
School of Urban Construction, Hubei Business College, Wuhan, China

Chao Wang
CCCC Second Harbour Engineering CO., Ltd, Wuhan, China

ABSTRACT: Taking Shiman expressway as the background, based on Coulomb earth pressure theory, the numerical modeling of the slope behind multistage retaining wall is carried out by using finite element software, and the stress nephogram and displacement nephogram in different directions are obtained. In the project, the measured stress data and settlement of slope points in different directions are measured, and the measured stress curve and settlement curve are drawn. By comparing the measured value of the slope behind the retaining wall at all levels with the calculated value of the model, the reliability of the numerical simulation is verified, and the relevant laws of the slope stability behind the retaining wall are put forward, which provides the necessary theoretical basis for the prevention of the instability of the retaining wall and plays a certain guiding role for the engineering practice. The research results can provide reference for the design and analysis of similar multistage retaining walls.

1 GENERAL INSTRUCTIONS

In recent years, with the gradual deepening of China's western development, more and more expressways have extended into the western mountains (Shreya & Taku 2021). Due to the limitation of many conditions such as topography, landform, hydrology and geology, various high fill embankment retaining walls are applied more and more (Magnusson & Johansson 2019). In order to ensure the operation safety, it is necessary to observe and study the earth pressure and displacement of complex retaining walls during the construction stage. It mainly satisfies the following requirements: first, to check the construction quality; second, to observe the earth pressure and overall deformation characteristics of the retaining wall project, to provide information for further improvement of design theory and construction technology. The influence of earth pressure and displacement of multi-stage retaining wall on time and wall height provides a new subject for our research on retaining wall (OuYang et al. 2021).

At present, the research on the earth pressure and displacement of retaining wall is more and more in-depth in civil engineering. Various model tests and field tests focus on the change of the earth pressure and displacement of retaining wall (Wang et al. 2021). However, as a common retaining structure for High Fill Subgrade in mountainous area, multi-stage retaining wall (Zhang et al. 2019) is used. The research on earth pressure and displacement is rare. Is the distribution of the earth pressure and displacement with the height of the multi-stage retaining wall and the common single-stage retaining wall the same? What are the regularities of the displacement of each wall position and the magnitude and change of the earth pressure of the multi-stage retaining wall? How does the earth pressure and displacement of the multi-stage retaining wall change with time

*Corresponding Authors: 165688344@qq.com and 251076308@qq.com

 DOI 10.1201/9781003310884-86

and filling height? These problems will be analyzed and discussed at a certain level on the basis of field test of the subject. Firstly, some laws and experiences are summarized through research, which can be used for reference and improvement in the actual design and construction in the future. Secondly, do some useful exploration for deeper analysis and improvement, so that our research can constantly adapt to new situations and problems, and promote production and practice.

2 PROJECT OVERVIEW

Project overview k32 + 570 ~ k32 + 690 section of Shiman Expressway belongs to the high filling section of subgrade, one side is close to the mountain, and a multi-stage retaining wall is designed for the subgrade retaining wall on the right side of the route, in which the foundation is rubble concrete foundation, Above it are gravity retaining walls made of 5-level rubble materials (The number of retaining wall is the fifth, fourth, third, second and first level retaining wall from the foundation) and 4-level subgrade retaining walls. The height of subgrade retaining wall of each stage is 8m, the slope ratio is 1:2, and a 2m wide staggered abutment is set between each stage. The total filling height of this subgrade high filling section is up to 70m (the photo of its front wall is shown in Figure 1).

Figure 1. Photos of front wall of multistage retaining wall.

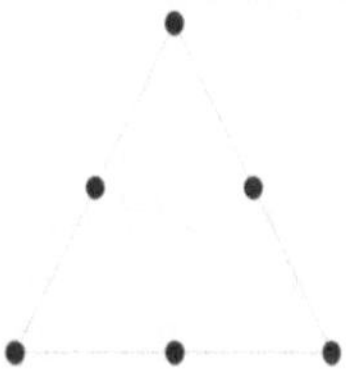

Figure 2. Plane2 unit.

3 NUMERICAL ANALYSIS OF BACK SLOPE OF RETAINING WALL

3.1 *Material model selection*

The conventional model of soil (DP model) is used as the material constitutive relationship model, which is suitable for granular materials such as concrete, rock and soil.

3.2 *Unit selection*

The elements selected in this paper are 2-dimensional 6-node triangular solid high-order elements, as shown in Figure 2.

3.3 *Boundary condition*

The chain bar support at the bottom of the model restricts its displacement in the Y direction; The earth filling boundary on both sides is supported by chain rods to restrict its displacement in X direction. Other boundaries are free boundaries.

3.4 *Finite element model and analysis*

3.4.1 *Finite element model and material parameters*

Figure 3 is the finite element model diagram of the slope behind the retaining wall established in Ansys. When modeling, the lower purple-black soil layer represents the undisturbed soil layer and the upper green-black soil layer represents the filling layer, i.e. the filling layer behind the retaining wall.

Table 1. Value of basic calculation material parameters.

Material type	Density /(kg.m^{-3})	Cohesion	Compression modulus/MPa	Poisson's ratio	Internal friction angle/(°)
Filling	2186	47	30	0.35	31.5
Foundation soil	2221	49	32	0.35	29

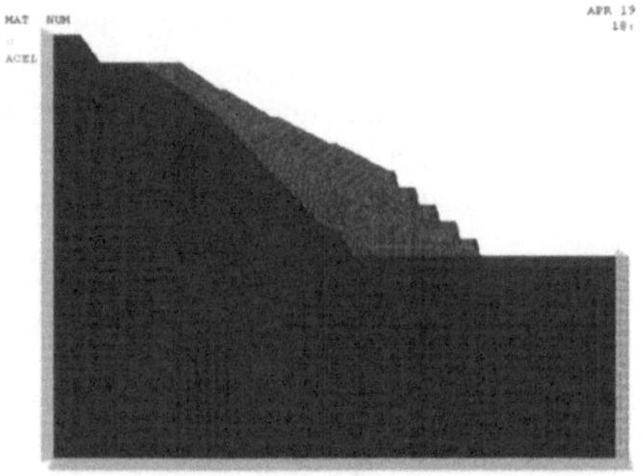

Figure 3. Finite element model established in Ansys.

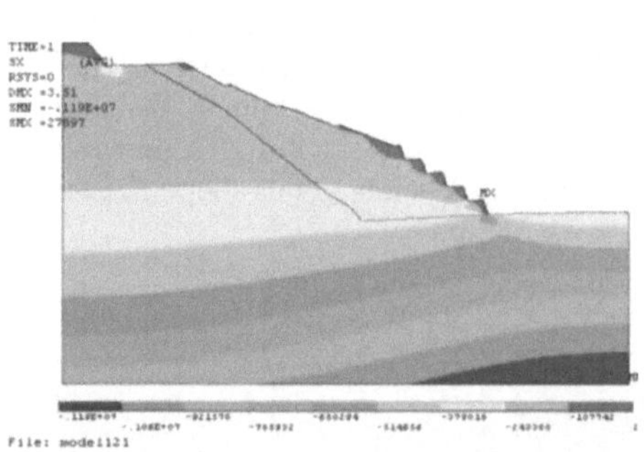

Figure 4. X-direction stress nephogram.

3.4.2 *Finite element model analysis*

1. Stress analysis of slope behind retaining wall

As can be seen from Figure 4, the stress along the X direction is positive at the top of the retaining wall, representing tensile stress, with a maximum value exceeding 25 Kpa. All below the initial top are negative. There is a high compressive stress in the slope area behind the retaining wall, and it is also very concentrated at the bottom of the retaining wall. The compressive stress in this area basically exceeds 240 Kpa.

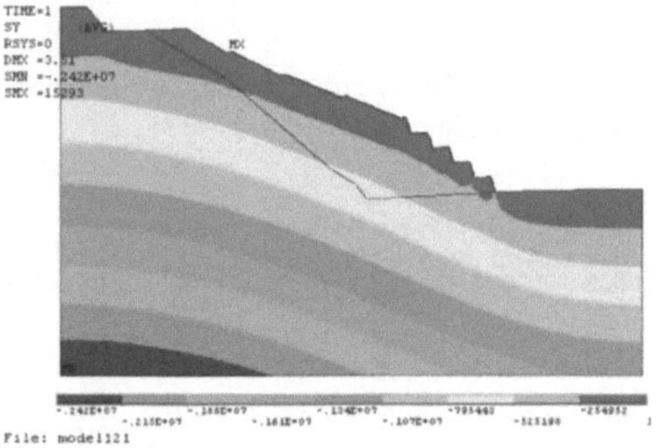

Figure 5. Y-direction stress cloud diagram.

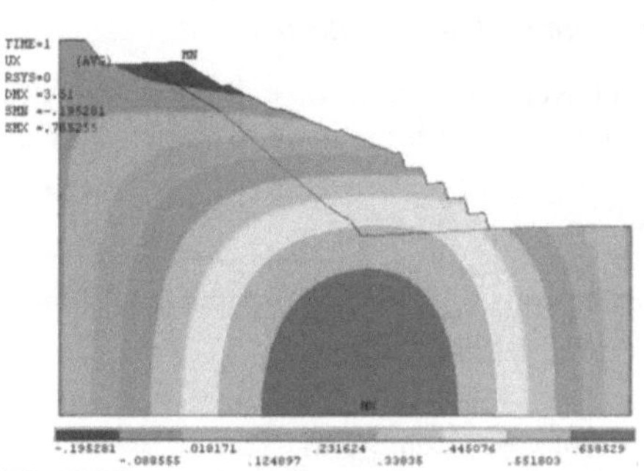

Figure 6. X-direction Displacement Cloud Map.

From Figure 5, it can be seen that the stress along the Y direction is positive at the top of the slope behind the retaining wall, representing tensile stress, but its value is relatively small, only about 15Kpa. Generally, its safety requirements can be met by basic structural measures.

2. Displacement analysis of slope behind retaining wall

Figure 6 shows a displacement cloud along the X direction. It is easy to see that there is a certain amount of displacement in the X direction (transverse direction). However, the value is small, which indicates that the geotechnical grille supported by the filling boundary link plays a very important role in limiting the horizontal displacement.

3. Distribution of displacement and settlement of slope behind retaining wall

From Figure 7, we can see that the settlement of the filling from high to low. The whole curve approximates a parabola, with the lowest settlement in the middle between the two sides and the maximum settlement in the middle of the filling up to over 40 cm.

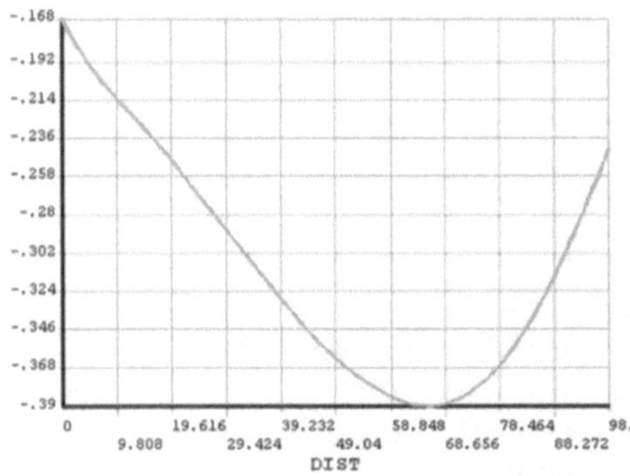

Figure 7. Settlement curve of each point in Ansys.

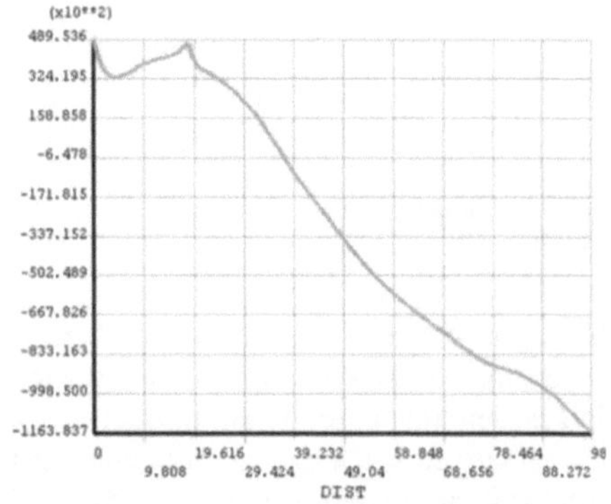

Figure 8. X-direction stress curve at each point.

Figure 8 shows the stress curve of each node along the X direction. Figure 8 shows the stress curve of each node along the Y direction. Figure 8 shows the stress curve of each node along the Z direction. Comparing with Figures 8–10, we can see that the trend of stress curves in these three maps is similar, the stress along the Y direction is positive at the top, which is shown as tensile stress, while the other areas are all compressive stress. The stress changes along the X and Z directions are similar, both of which gradually decrease from upper tensile stress to increase compressive stress.

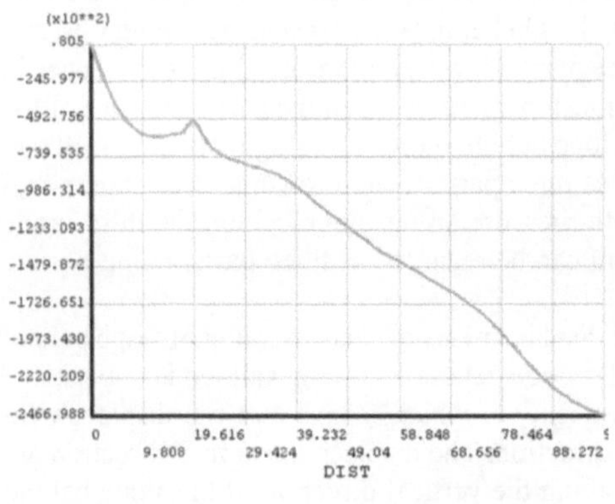

Figure 9. Y-direction stress curve at each point.

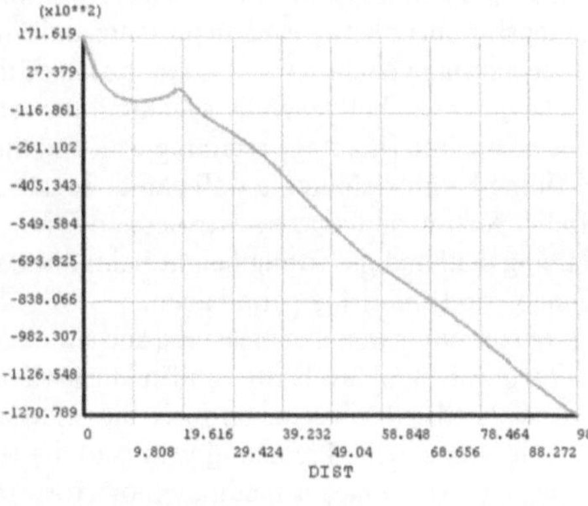

Figure 10. Z-direction stress curve at each point.

4 MONITORING RESULTS AND COMPARATIVE ANALYSIS

Figure 11 is a broken line comparing the measured value of earth pressure at each measuring point with the calculated value of the model after the construction of filling subgrade at the top of the wall (208 days) for the slope behind the first, second and third retaining walls. It can be seen from Figure 11 that the calculated earth pressure by the model is not significantly different from the measured earth pressure of the slope behind each retaining wall. The displacement calculated by the model is basically in agreement with the measured displacement value of the slope behind each retaining wall. That is to say, considering the influence of errors, the simulated value of the established model is basically in agreement with the actual situation, which shows tha t the calculated value of the finite element model is reliable to some extent.

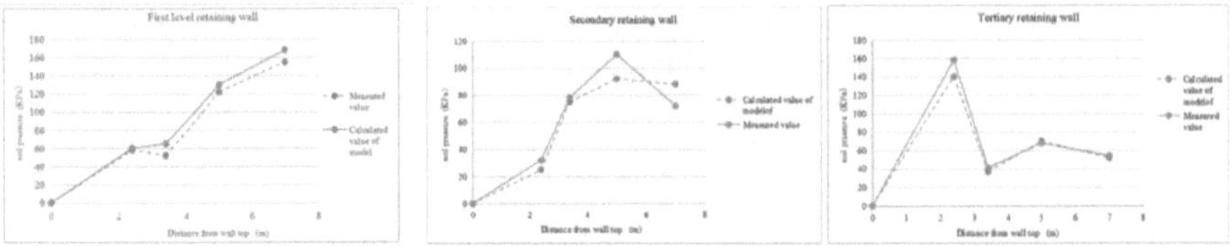

(a) First stage retaining wall (b) Secondary retaining wall (c) Tertiary retaining wall

Figure 11. Comparison between measured values and model calculated values of slope behind retaining walls at all levels.

During the model calculation of high filling section by using analysis software Ansys, the weak positions of the section, including the location with high tensile stress and the location with the largest settlement, are found through stress cloud diagram, displacement cloud diagram and settlement curve, which are basically consistent with the actual observed data and can prove that the results of the simulation numerical calculation are reliable.

5 CONCLUSION

Relying on the research project of slope stability behind retaining wall of Shiman Expressway, combining theory with practice, and combining analysis software Ansys to carry out numerical simulation, the simulation results are compared with the measured data, and the following conclusions are drawn:

When the filling height behind the retaining wall exceeds the wall, the earth pressure on the slope behind the retaining wall of different levels varies with the filling height of the upper part. The relationship curve between earth pressure and filling height behind the first stage retaining wall and the second stage retaining wall is composed of three parts. However, there are slight differences in the change trend, but generally speaking, the relationship curve between the earth pressure of the slope behind the first stage retaining wall and the slope behind the second stage retaining wall and the filling height behind the wall can be regarded as three parts: rapid growth section, horizontal stability section and slow growth section. The earth pressure on the slope behind the third grade retaining wall and the filling height behind the wall can be regarded as three parts: rising period, falling period and rising period again.

From the comparison of measured and simulated displacements of each direction of slope behind retaining wall at all levels, we can find that when the slope behind retaining wall tends to be stable as a whole, the displacements in X direction are negative, which means that the displacement along the slope behind retaining wall is to the left direction. The displacements in Y direction are all positive, which means that they move forward along the vertical direction of the slope behind the retaining wall. Displacement in Z direction is negative, which means that the displacement is downward along the vertical direction of the slope behind the retaining wall. The displacement curve in each direction is nonlinear. The deformation of the slope behind the retaining wall in all directions is relatively large.

REFERENCES

OuYang JiaYi; Liu Yong; Yao Kai; Yang ChenJun; Niu HuiFeng.Model Updating of Slope Stability Analysis Using 3D Conditional Random Fields[J]ASCE - ASME Journal of Risk and Uncertainty in Engineering Systems, Part A: Civil EngineeringVolume 7, Issue 3. 2021.

Simon Magnusson, Maria Johansson, Sandra Frosth et al. Coordinating soil and rock material in urban construction – Scenario analysis of material flows and greenhouse gas emissions[J] *Journal of Cleaner Production*, 2019, 241(C).
Thusoo Shreya, Obara Taku, Kono Susumu et al. Design models for steel encased high-strength precast concrete piles under axial-flexural loads[J] *Engineering Structures*, 2021, 228.
Tan Zhang, Hong Zheng, Cong Sun. Global method for stability analysis of anchored slopes[J] *International Journal for Numerical and Analytical Methods in Geomechanics* Volume 43, Issue 1. 2019. PP 124–137.
Wang Haojie, Sun Ping, Wang Gang, Wu Lizhou. Experimental and numerical study of shallow loess slope failure induced by irrigation[J]*Catena* Volume 206, 2021.

Advances in Civil Engineering: Structural Seismic Resistance, Monitoring and Detection – Mohd Johari Mohd Yusof and Junwen Zhang (Eds)
© 2023 The Authors, ISBN 978-1-032-31491-4

Influence analysis of blasting nearby the dam on the safety of earth-rock dam

Liang Hu
Zhejiang Institute of Hydraulics and Estuary (Zhejiang Institute of Marine Planning and Design), Hangzhou, China
Zhejiang Provincial Key Laboratory of Hydraulic Disaster Prevention and Mitigation, Hangzhou, China

Quan-hua Jin & Wei Jiang
Zhejiang Institute of Hydraulics and Estuary (Zhejiang Institute of Marine Planning and Design), Hangzhou, China

ABSTRACT: The blasting vibration produced by the blasting construction near the dam will have a certain impact on the earth-rock dam. Combining blasting theory and engineering practice, the dam finite element analysis model was established through Geo-Studio, and the impact of blasting vibration on the dam was analyzed from the aspects of dam slope stability and dam foundation liquefaction, and the blasting plan and dam safety monitoring during construction, the results show that under the action of blasting load, the safety factor of dam slope anti-sliding stability has been reduced, but it still meets the requirements of the specification, and the dam foundation will produce excessive pore water pressure and possible liquefaction zone. The facility protection puts forward reasonable suggestions, and the calculation and analysis results provide a reference for the analysis of the impact of blasting near the dam on the safety of the earth-rock dam.

1 INTRODUCTION

Earth-rock dam is one of the most widely used dam types in my country, and its filling technology has been rapidly developed since the 1950s. Many projects have faced problems such as expansion and blasting of mountains in the reservoir area, bridges or tunnels near the reservoir, and demolition and reconstruction of old dams. As an effective and economical method, blasting excavation is widely used. The harmful effects caused by blasting have attracted more and more attention. Heller L.W. used a full-scale nuclear blast to impact a rolling earth-rock dam, and verified the applicability of the seismic design procedure in the design of the earth-rock dam by observing the movement of the foundation and the dam. Some scholars at home and abroad have studied the dynamic response, anti-seismic measures and reinforcement scheme of dam body under earthquake conditions, and obtained some meaningful results (Heller 1973). Linsbauer simulated the explosion load at the bottom of the concrete gravity dam reservoir with preset cracks by shocking, and discussed the stability of the dam and the mechanism of crack expansion and failure (Linsbauer 2009). Ke-xu Zhang and Pan Jian-ping studied the seismic design method of tailings dam, and proposed the seismic measures that should be taken for the dam body (Pan 2007; Zhang 1988). Zhi-ying Xu used different methods to calculate and analyze the seismic liquefaction problem of high dams, and made a safety evaluation of the stability of the dam body (Xu 1983). However, the research on the safety control standards of blasting, the law of vibration propagation in different media, the impact on the surrounding buildings of the dam, and the safety criteria are still not enough (Jin 2008; Lu 2012). This paper analyzes the impact of a mountain blasting project in the dam body of the earth-rock dam, and puts forward corresponding suggestions and control measures.

 DOI 10.1201/9781003310884-87

2 PROJECT OVERVIEW

2.1 *Blasting engineering*

The project is a small reservoir mainly for irrigation and water supply. The total storage capacity is 210,000 cubic meter. The pivotal buildings include dams, spillways and water delivery culverts. The exceptional flood level is 72 meter. The dam type is a clay core dam with crest elevation 73.0m, the maximum dam height of 17m, crest length of 115.0m. The impervious body of the dam is a silty clay core wall, the shell soil is mainly cohesive soil and sandy soil, the dam foundation is a sandy silty clay layer with a thickness of 1.2 to 5.6 m.

The blasting project is located in the left side of the dam within 200m. The blasting area is divided into two areas with a linear distance of 100m~150m from the dam body of the reservoir, namely Zone A (more than 150m from the dam body boundary) and Zone B (distance from the dam body boundary is 100~150m). The blasting plan is deep-hole bench blasting and the construction follows the top-down sequence.

2.2 *The problem*

In addition to the rock breaking required by the project, the blasting construction will cause a strong disturbance to the rock mass to spread from near to far. The blasting seismic wave propagates in the rock body through the rock and soil medium and reaches the surface to form ground vibration. The blasting vibration is gradually attenuated from near to far. Since the blasting location is only within 200m of the dam, it is bound to have a certain impact on the dam. If the blasting strength exceeds a certain level, it may cause serious impact or damage.

Whether blasting vibration constitutes a safety impact on the surrounding environment is closely related to the three elements of the explosion source, and the seismic performance of the specific protection. Relevant norms control the impact of blasting generally by using allowable vibration speed of the building mass point, but there are no clear regulations on earth-rock dams, and the mechanics of dam body filling mechanics are complex, and single control parameter cannot accurately reflect blasting. It is necessary to analyze and evaluate the impact of blasting construction based on the actual situation of the engineering dam.

3 BLASTING IMPACT ANALYSIS

3.1 *Blasting load analysis*

The blasting impact load is similar to a natural earthquake. It is an energy wave, but the blasting impact load has the characteristics of shallow source, high intensity of load, high vibration frequency, and short duration. The paper will analyze and evaluate the blasting load, dam slope stability and dam foundation liquefaction using the SEEP, SLOPE and QUAKE modules in the Geo-Studio.

According to the above, the maximum particle vibration velocity is 0.80 cm/s in the A zone, and 0.92 cm/s in the B zone. When the peak vibration velocity of the particle is 1.0cm/s, the corresponding seismic acceleration is 0.01g. Considering the large amount of blasting and the number of times, this model appropriately amplifies the load effect, the corresponding seismic acceleration is 0.02g, the duration is 0.8, and the peak appears at 0.3s.

3.2 *Dam slope stability analysis*

The effect of blasting vibration is simulated by equivalent seismic conditions. The horizontal seismic acceleration is used to reflect the magnitude of the vibration, so as to obtain the influence on the stability of the dam slope. The influence on the stability of the dam slope is mainly manifested in the blasting vibration load, the pore water pressure in the soil increases, which leads to a decrease in the effective stress, and the strength of the soil is weakened, which may cause cracks in the dam body and instability of the dam slope.

According to the possible operating conditions of the reservoir, and considering the need for analysis and comparison, this modeling mainly considers three operating conditions, as shown in Table 1, and the results are shown in Figures 1 to 4 and Table 2.

Table 1. Analysis and calculation conditions(m).

Number	water level	Performance	Upstream water level	Downstream water level
1	Normal water level	Non-blasting load	70.50	53.00
2	Normal water level	Blasting load	70.50	53.00
3	1/3 reservoir water level	Non-blasting load	64.00	53.00
4	1/3 reservoir water level	Blasting load	64.00	53.00
5	Empty reservoir	Non-blasting load	60.00	53.00
6	Empty reservoir	Blasting load	60.00	53.00

The results show that under the action of blasting load, the safety factors of the anti-sliding stability of the upstream and downstream dam slopes corresponding are reduced: (1) The safety factor of the downstream slope of the normal water storage level is reduced from 1.559 to 1.409; (2) 1/ 3 The safety factor of the upstream slope of the reservoir water level was reduced from 1.743 to 1.512; (3) The safety factor of the upstream slope was reduced from 1.791 to 1.553 when the reservoir was empty, and the safety factor of the downstream slope was reduced from 1.613 to 1.486. But both were greater than the 1.25 required by the code. The dam slope is relatively stable indicating the blasting load, and the lower the reservoir water level, the higher the safety factor of the dam slope.

As the surface of the dam slope and the shallow layer may have defects such as loosening or hollowing out, during blasting vibration, collapse and sliding may occur in a small local area.

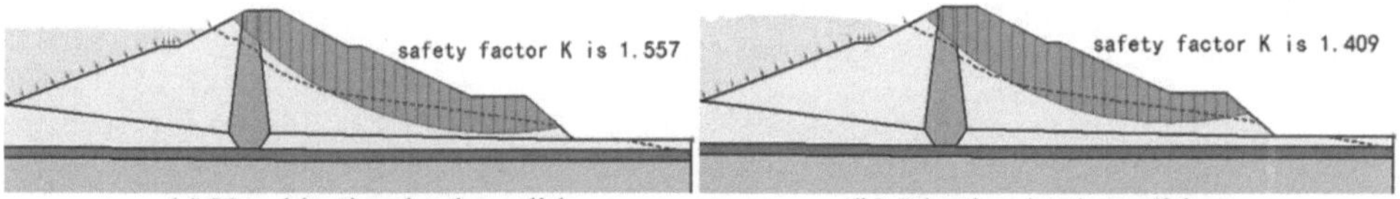

(a) Non-blasting load conditions (b) Blasting load conditions

Figure 1. Anti-sliding stability diagram of downstream slope at normal water storage level.

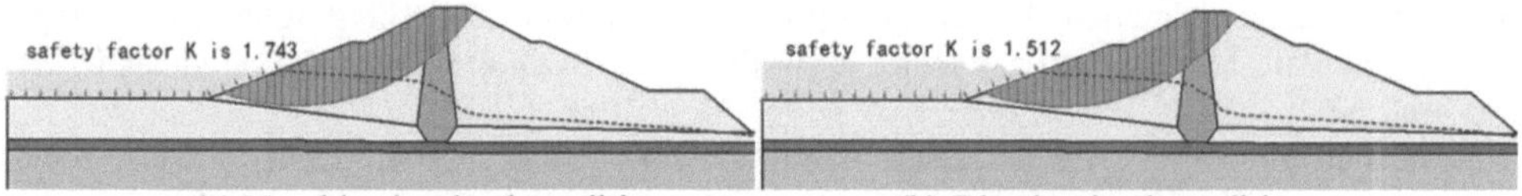

(a) Non-blasting load conditions (b) Blasting load conditions

Figure 2. Anti-sliding stability diagram of upstream slope of 1/3 reservoir water level.

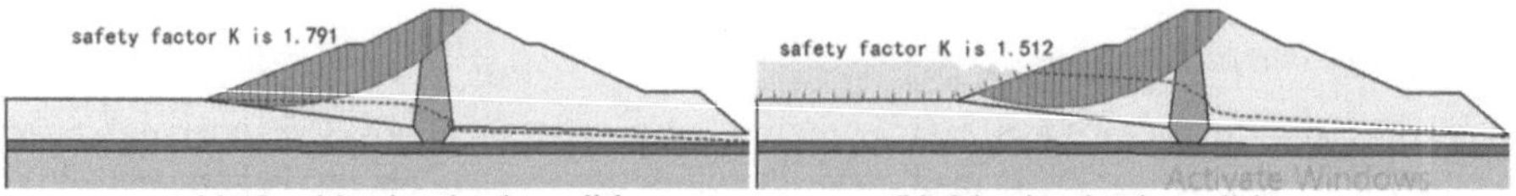

(a) Non-blasting load conditions (b) Blasting load conditions

Figure 3. Anti-sliding stability diagram of the upstream slope under empty storage.

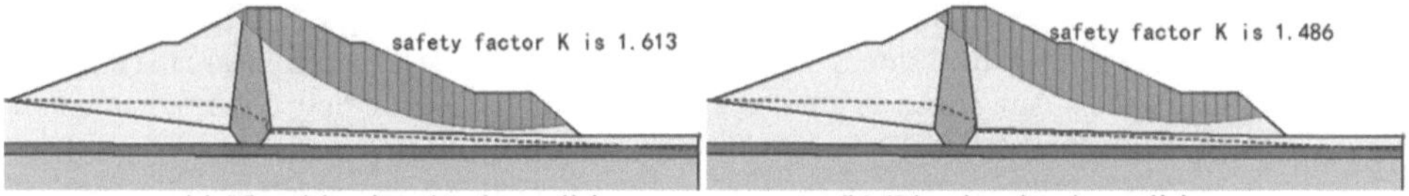

(a) Non-blasting load conditions (b) Blasting load conditions

Figure 4. Anti-sliding stability diagram of downstream slope under empty warehouse.

Table 2. Analysis table of dam slope stability results.

			Safety factor *K*	
Number	water level	Calculate location	Non-blasting load	Blasting load
1	Normal water level	Back water slope	1.559	1.409
2	1/3 reservoir water level	Waterfront slope	1.743	1.512
3	Empty reservoir	Waterfront slope	1.791	1.553
4	Empty reservoir	Back water slope	1.613	1.486

3.3 *Dam foundation liquefaction analysis*

It can be seen from Figure 5 that under the action of blasting load, there will be a certain area of excess pore water pressure in the upstream reservoir area of the dam, and the higher the reservoir water level, the larger the range of excess pore water pressure and the higher the maximum value.

The dam foundation fill is sand-containing silty clay. According to the seismic liquefaction judgment method, the foundation soil of this project has a particle size $d \leq 5$mm with a particle content exceeding 30%, and a particle size $d \leq 0.005$mm with a particle content of 9.9% and less than 16%. Therefore, there is the possibility of liquefaction of the dam foundation fill. Figure 6 is a schematic diagram of the possible liquefaction area of the reservoir area and the upstream dam slope obtained under different working conditions. It can be seen that the higher the reservoir water level, the larger the possible liquefaction area, and there is basically no liquefaction area at the dam foundation location when the reservoir is empty.

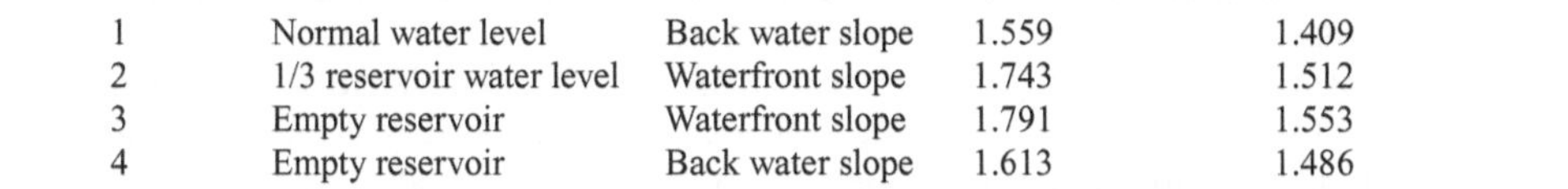
(a) Normal storage level conditions (b) 1/3 reservoir water level

Figure 5. Schematic diagram of the distribution of excess pore water pressure in the reservoir area.

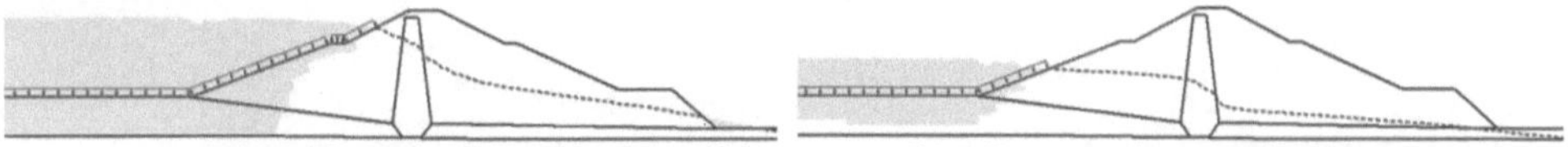
(a) Normal storage level conditions (b) 1/3 reservoir water level

Figure 6. Schematic diagram of the dam foundation liquefaction zone scope.

4 ENGINEERING SUGGESTION

4.1 *Reservoir operation*

Before construction, the reservoir was identified as a dangerous reservoir. According to the "Guidelines for Safety Evaluation of Reservoir Dams", the reservoir must be emptied before reinforcement.

According to the analysis result, the lower the reservoir water level, the higher the safety factor of the upstream and downstream dam slopes, and the smaller the possible liquefaction area of the dam foundation and upstream dam slope. Therefore, the reservoir should be emptied during mountain blasting construction near the dam.

4.2 *Monitoring and protection facilities*

According to relevant geological prospecting data, the soil and soil quality of the dam core wall is uneven; the two dam heads and dam foundation are not completely cleaned, and there is a hidden danger of leakage; there are leakage channels in the contact parts of the dam body, spillway and dam foundation; abnormal deformation of the dam surface may occur locally. Therefore, during the blasting construction, monitoring of the particle vibration velocity, dam surface deformation, dam seepage and other items should be real-time monitoring. And the dam should be emptied before blasting construction.

During the blasting construction, the reservoir management facilities should be covered, isolated and other protective facilities to avoid damage by flying rocks, and people within the blasting safety range should be notified to evacuate in time. Reservoir management units should prepare emergency plans.

5 CONCLUSION

(1) The blasting vibration wave is similar to the seismic load, which will bring harmful effects to nearby buildings. According to the explosive quantity of a single blasting and the characteristics of the bedrock, the vibration effect of blasting needs to be considered as an earthquake with an acceleration of 0.02g. Therefore, it is necessary to evaluate the impact of blasting near the dam on the safety of the reservoir dam.

(2) The impact of blasting vibration on the safety of earth-rock dams is mainly reflected in two aspects: dam slope stability and dam foundation liquefaction. By establishing a calculation model, apply horizontal seismic acceleration to simulate blasting vibration load. The results show that under the action of blasting load, the safety factor of dam slope anti-sliding stability has been reduced, but it still meets the requirements of the specification, and the dam foundation will produce excessive pore water pressure and possible liquefaction zone. The higher the reservoir water level, the lower the anti-sliding stability safety factor, the larger the value and range of excess pore water pressure, and the larger the range of possible liquefaction zones.

(3) Due to the complex causes of the blasting vibration effect, it is recommended that the blasting construction be carried out in an empty reservoir. At the same time, the safety of the dam should be comprehensively evaluated by combining the dam mass point vibration velocity and acceleration, dam deformation and seepage monitoring data.

6 OUTLOOK

(1) In the analysis and calculation, the characteristics of the dam foundation and the existing weak interlayers such as faults and fissures should be further considered, and the influence of the expansion of the fissures on the dynamic response under the blasting load should be studied, so as to make the safety protection of the dam more reasonable.

(2) Seepage has a great influence on the simulation of earth-rock dams. In the process of model, the analysis of the impact of blasting load on the seepage of earth-rock dams is insufficient, and the influence of seepage field on the dynamic response of the structure is not fully considered. This will be the focus of future research.

REFERENCES

Heller L W. Seismic effects on earth dam from explosion [J]. *Journal of Soil Mechanics &Foundations Div* 197399

Jian-ping Pan. Study on anti-seismic design method and anti-seismic measures of tailings dam [D] Dalian: Dalian University of Technology 2007

Ke-xu Zhang. Study on anti-seismic design and of tai-lings dam [J] *World Earthquake Engineering*, 1988 (1): 13–18

Linsbauer H. Damage potential of an upstream-side crack in a gravity dam subjected to an impact loading in the reservoir[C]. Proceeding of the 2nd international Conference on long term behaviour of dams, Graz, 2009: 817–822.

Yun-xiang Lu. The Influence Analysis of Dam Regulation Project Blasting Vibration on the Safety of Dam [D]. Da Lian: Da Lian University of Technology, 2012: 2–5.

Zhi-ying Xu, Zhu-jiang Shen. Earthquake response and liquefaction analysis of high tailing dam by various methods[J]. *Journal of Hydraulic Engineering*, 1983, (5):30–39.

Zhi-yu Jin. Preliminary study on the influence of rock mass blasting to dam safety [J]. *Dam and safety*, 2008(2): 18–20.

Advances in Civil Engineering: Structural Seismic Resistance, Monitoring and Detection – Mohd Johari Mohd Yusof and Junwen Zhang (Eds)
© 2023 The Authors, ISBN 978-1-032-31491-4

Research on application technology of BIM standard component modeling in highway engineering

Jin Jin Su*
Intelligent Engineering College, Guangzhou Huashang Vocational College, Guangzhou, China
Graduate School of Business, Universiti Tun Abdul Razak (UNIRAZAK), Malaysia

Benjamin Chan Yin Fah*
Graduate School of Business, Universiti Tun Abdul Razak (UNIRAZAK), Malaysia

ABSTRACT: BIM (Building Information Modeling) is a new engineering technology method that emerged at the beginning of this century. It builds models based on various relevant information data of engineering projects and simulates the real information of projects through digital information simulation technology. Compared with the traditional construction mode, BIM technology and management are more standard, standardized and more efficient. This paper focuses on the construction of BIM modeling technology and its application in highway engineering.

1 INTRODUCTION

From the conception of BIM to its development so far, the existing research is mostly in academic theory, improving the concept construction of BIM from a macro level and its advantages and necessity, and more attention is paid to BIM technology in the housing construction industry and transportation industry. Management applied research is less.

In traditional highway projects, there are many problems in the construction period, quality, cost, safety and other aspects of the project due to non-standard management. The level is getting deeper and deeper, and the application of BIM standardization technology in highway engineering is not perfect, and the problem of BIM standardization is becoming more and more prominent.

Therefore, the research content of this paper is the modeling technology and standardization of highway engineering BIM, and the application of standardized system in bridge engineering management, in order to supplement the research field of BIM technology management and improve the management level of highway engineering projects.

1.1 *BIM modeling technology concept and characteristics*

BIM technology, namely building information modeling technology. Dr. Chuck Eastman first proposed the concept of building information model in 1975. He believed that building information model should include all information, functions and performance requirements of different disciplines, and should integrate all the information of the whole life cycle of buildings in a project. into a 3D model.

1.2 *Research status at home and abroad*

1.2.1 *Current status of foreign research*

In 1992, Van Nederveen perfected "Building Modeling" and BIM modeling technology has since evolved. In 1999, Professor Tolnan proposed the concept of "Building Information Modeling" (Lesslin & wang 2011), but limited by the level of computer and network development, there was

*Corresponding Authors: Sujinjin@gzhsvc.edu.cn and 369639207@qq.com

 DOI 10.1201/9781003310884-88

no breakthrough in informatization research. In 2007, the American Academy of Building Sciences founded the Building SMART Alliance (U. S Army Crop of Engineers, Building Sciences 2006) focusing on the research work on information transmission and interaction in BIM technology. In the same year, the US National BIM Standards Program Committee, NBIMS) promulgated the US federal unified BIM standard (Azhar S et al. 2019).

Most of the foreign research is at the macro level, and the content of the research is mostly the advantages and necessity of BIM in engineering projects, and there is less research on the application of BIM technology in practical engineering.

1.2.2 *Research state in China*

my country's research on BIM technology focuses more on its application in practical engineering. Yu Fangqiang and Zhang Jianping of Tsinghua University (Yu et al. 2014) proposed a semi-automatic generation method of BIM sub-model views based on IFC, which provides technical support for BIM implementation and application for the whole life cycle of buildings. Li Ye of Wuhan University of Technology (Li 2014) proposed the application framework of BIM technology in the DB model, focusing on the issue of workflow and benefit distribution.

1.3 *BIM Standardization*

Although the China BIM Development Alliance, the state and local governments have released dozens of BIM standards for information exchange technical standards for various majors, most of them are independent, not applicable to each other, or even contradict each other. After exploration, Tsinghua University proposed that CBIMS divide BIM standards into three levels: national, local, industry, and enterprise, so that national standards guide the implementation of industry and enterprise standards, and ultimately rely on enterprise implementation standards to promote the implementation of BIM standards. In this paper, a key highway project in Sichuan Province is used as a case to introduce the modeling application technology of BIM standard components.

2 COMPONENT STANDARDIZED APPLICATION PROCESS

The standardization of BIM components in highway engineering is also a part of my country's BIM standard framework system. By formulating BIM standards for highway engineering, it can fill the gap in the application standard of my country's BIM technology in the field of highways and enrich the BIM standardization system. Moreover, the created highway professional component library can be used as a data resource library, which can be reused in subsequent practical projects to improve the modeling efficiency.

Component is the smallest unit of BIM model, and BIM component standardization is the smallest unit of BIM standardization. The BIM component is a replaceable entity component in the entire BIM model. It has irreplaceable specific functions, conforms to the interface standard of the entire BIM model, and can be assembled with other components, and finally the BIM model is composed of different components.

2.1 *Installation environment*

The main application of this component is to load it into the Civil 3D software of the Autodesk platform, and create a 3D road model with engineering properties through the "Road" function of the Civil 3D software. The road model can be used for engineering design, engineering drawing and model display, etc.Load the road model (*.dwg file) of Civil 3D through the mainstream model integration platform InfraWorks,[3] which can present the 3D model more intuitively.

2.2 *Component code*

(1) Clarify the actual requirements of the component. Before starting to write the component, set the necessary parameters and offset target of the component.

Figure 1. Civil 3D software model style.

(2) Custom widget framework. According to the principle from top to bottom, a total of 7 kinds of "internal parameters", "offset target point", "AP point line", "code setting", "structure point line", "output point" and "output value" are formulated Component composition.
(3) AP auxiliary point preparation. In this component, all auxiliary points on the section structure are marked with two digits, of which the single digits are "1" and "2", which respectively indicate the right and left sides of the section, and the ten digits indicate the position of the structure point. and sequence.
(4) Setting of dotted line code. You can directly assign values to the structure point line codes under different sections (type is String) through the variable name of "road width form".
(5) Configuration point\connection settings. After the AP auxiliary point is written, just attach the structure point Point to the corresponding auxiliary point Auxiliary Point, form the structure line through the connection of Link, and then write the code variables that have been set to form the structure point line of the final component.

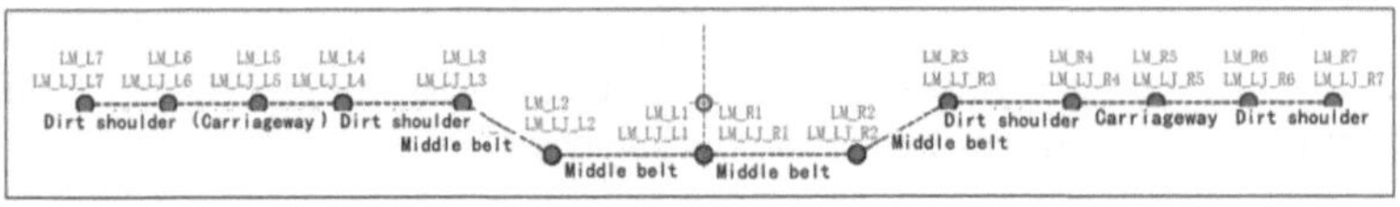

Figure 2. Code composition of full-width subgrade section (schematic).

(6) Output point and output value. When compiling BIM standard components of highway engineering, it is combined with road characteristics and component editor to compile.

2.3 *Road Modeling and Integration*

Leverage the Corridor feature in Autodesk Civil 3D to accurately model corridors from created corridor terrain surfaces and design alignments. And save it as a road *.dwg file, and then integrate the road model in the InfraWorks software.

2.4 *Basic parameters*

The basic parameters are roughly divided into three categories, which are input parameters of engineering elements, mark point parameters and component transfer parameters, and their types are usually Double.

Table 1. Input parameter information table (part).

serial number	member	Parameter name	parameter Type	parameter default value	source component
1	main road surface	no	no	no	no
2	road interchange	no	no	no	no
3	shoulder wall	Subgrade judgment	Double	0	pavement components

(continued)

Table 1. Input parameter information table (part).

serial number	member	Parameter name	parameter Type	parameter default value	source component
4	Fill	Subgrade existence	Double	0	pavement components
		Retaining Wall Existence	Double	0	shoulder wall components
		embankment wall distance	Double	0	shoulder wall components
		shear line existence	Double	0	shoulder wall components
		separate existence	Double	1	pavement components
		Whether the slope continues	Double	1	Superior Fill Slope Component

3 INSTANCE APPLICATION OF COMPONENTS

Taking the formulation of components of a key highway project as an example, the geological structure along the project is complex, the terrain fluctuates greatly, and the subgrade cross-section forms are diverse. By setting sampling lines, cross-sectional views are generated, and finally the road model is imported into InfraWorks to display the model effect.

3.1 *Workflow and Ideas*

First open Civil 3D, create a new DWG file, and assemble the assembly. Created in AutoCAD Civil 3D with reference to the terrain surface of the highway.

Second, refer the alignments containing the horizontal and vertical information to the drawing, and check whether the alignments are all within the surface range.

Thirdly, draw the offset line according to the design data to provide the basis for the subsequent creation of the road assembly to pick the offset target.

Finally, create a road and generate a main line road model of a road along the river. The generated model can be directly viewed in the software.

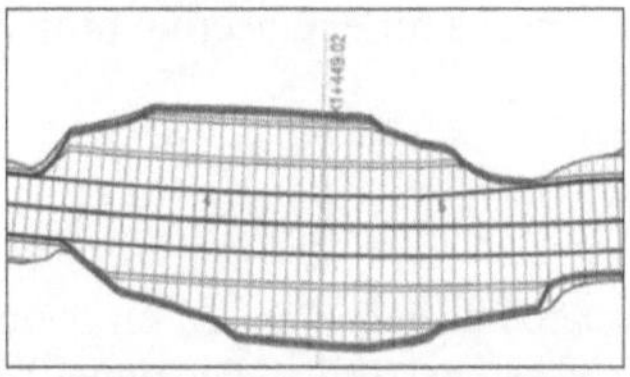

Figure 3. Interchange main road.

After the road model is made, it is imported into InfraWorks for rendering display. By setting the style rules of InfraWorks, various codes are given different materials to achieve the ideal model display effect.

3.2 *InfraWorks Model Schematic*

After the refresh is completed, by assigning different materials to the code, an intuitive road model effect is presented. The following is the display effect of the road model in InfraWorks.

Figure 4. Integral roadbed model diagram.

4 CONCLUSION

On the whole, BIM standardization technology has the following functions in the transportation industry:

(1) The three-dimensional rendering effect is good. Compared with the traditional 2D model, the advantages of the 3D digital model created by BIM are highlighted in three-dimensionality, accuracy and synergy.
(2) Promote the industrialization of the industry. Apply BIM technology in the management of the whole life cycle.
(3) Promote the transformation of industry models. BIM technology can effectively improve the productivity level of my country's transportation construction industry.

5 LIMITATIONS AND FUTURE RESEARCH PLANS

5.1 *Study Limitations*

Because the BIM standardization application technology has a certain threshold, and the expressway is a large-scale national infrastructure facility and cannot widely represent small and medium-sized projects, the samples selected in this paper have limitations.

5.2 *Future research plans*

(1) Expand the research field. In-depth study of the application of BIM standardization technology such as bridge engineering and railway engineering makes the research field of BIM standardization technology more extensive and representative.
(2) Improve research methods. In the future, the coordinated development of BIM standardization technology in different engineering fields can be further studied, in order to lower the threshold of BIM use and achieve better benefits for engineering projects.

FUNDED PROJECT

The school-level key scientific research project, "Based on the construction of the "teacher studio" training mode Construction Professional Skilled Personnel Training and Green Energy-saving Technology Innovation Research" (Project No. KY2022003).

REFERENCES

Azhar S, Hein M, Sketo B. Building information modeling (BIM): *Benefits, risks*, 2019.
Jerry Lesslin, Wang Xin. History of BIM [J]. *Architectural Creation*, 2011(6):148–152
Li Ye. Research on the application of BIM technology in DB mode [D]. *Wuhan University of Technology*, 2014.
The website of the building SMART alliance, a subsidiary of the American Academy of Building Sciences [[EB/OL].
U. S Army Crop of Engineers, Building Information Model-A Roadmap for Implementation, 2006.
Yu Fangqiang, Zhang Jianping, Liu Qiang. Semi-automatic generation of BIM sub-model views based on IFC [J]. *Journal of Tsinghua University* (Natural Science) Edition). 2014 (08): 987–992.

Advances in Civil Engineering: Structural Seismic Resistance, Monitoring and Detection – Mohd Johari Mohd Yusof and Junwen Zhang (Eds)
© 2023 The Authors, ISBN 978-1-032-31491-4

A new method based on 1D visco-hypoplasticity constitutive model for predicting consolidation and creep

Xiangming Ge
PowerChina Huadong Engineering Corporation Limited, Hangzhou, Zhejiang, China
HydroChina Itasca R&D Center, Hangzhou, Zhejiang, China

Andrés Alfonso Pena Olarte & Roberto Cudmani*
Chair of Soil Mechanics and Foundation Engineering, Rock Mechanics and Tunneling, Technical University of Munich, Baumbachstraße 7, Munich, Germany

ABSTRACT: The 1D-visco-hypoplastic model based on Hypothesis B was implemented within the framework of Terzaghi's consolidation theory to estimate time-dependent soil behavior. The model parameters are obtained from constant rate of strain tests. In this model, the creep was modified for a better prediction. The first loading relates to the strain rate. The over-consolidation ratio is taken into account in describing the yield surface. This method is verified by an isotropic compression test by Krieg (2005) and a well-recorded project by Soumaya (2005). The comparisons illustrate the good feasibility of this 1D-visco-hypoplastic model in geotechnical design.

1 INTRODUCTION

Terzaghi (1925) defined consolidation as a process that involves a decrease in water content of a saturated soil without replacement of water by air and creep as a continuous deformation under constant effective stress (Terzaghi & Karl 1943). Multiple methods were proposed for predicting this time-dependent soil behavior, i.e., the creep strain, in history. These methods utilized either Hypothesis A or B. In contrast to Hypothesis A, it was assumed in Hypothesis B that the time-dependent soil behavior occurred during the whole consolidation process (Graham & Yin 1996). Degago et al. (2009, 2011) illustrated that the observed time-dependent behavior of clays can be explained by Hypothesis B (Degago et al. 2009; Degago et al. 2011). A phenomenological model was proposed by Campanella and Vaid (1974) to describe the creep settlements which connects the strain to time (Campanella & Vaid 1974). Another phenomenological model was built by Hunter and Khalili (2000) to describe the creep failure which connects the failure potential to the stress peak (Hunter, Khalili 2000). Yin and Graham (1989, 1994) compared the prediction of creep to the test data and found that the 1D elastic visco-plastic model based on Hypothesis B predicted the soil behavior better than Hypothesis A (Yin 1989; Yin & Graham 1994). Gang Bi (2019) proposed an approach Creep as Rate of Loading and developed a power law model to describe consolidation and creep (Gang et al. 2019). This method required a few model parameters and was verified by oedometer tests and a historical case.

Nimunis and Krieg (2005) proposed a numerical model under oedometric conditions to predict viscous behavior (Niemunis & Krieg 1996). In the present study, the 1D-visco-hypoplastic model based on the descriptions of Nimunis (2005) developed by authors was implemented within the framework ofTerzaghi's consolidation theory. Furthermore, this method was verified by an isotropic compression test by Krieg (2005) and a well-recorded project by Soumaya (2005).

*Corresponding Author: roberto.cudmani@tum.de

DOI 10.1201/9781003310884-89

2 SEMI-ANALYTICAL 1D-VISCO-HYPOPLASTIC MODEL

2.1 *Description and assumptions*

This 1D rheological model was developed by Nimunis (1995) for predicting the differential settlement of Schweriner Castle in Germany. Figure 1 represents the soil behavior during the first loading, unloading, reloading, and the creep in logarithmic space. In comparison to the other methods, the viscous strain was taken into account during the primary as well as secondary consolidation. Thus, the total strain rate $\dot{\varepsilon}$ is the sum of the elastic strain rate and the viscous strain rate during the primary consolidation, or rather, the first loading phase. Note that the viscous strain disappears during unloading and reloading. The soil behavior during the first loading, unloading, and reloading is controlled by the governing equation (1) and (2). A yield surface is defined, which merely destinates the specific reference strain rate $\dot{\gamma}$. As shown in Figure 2, the yield surface is reduced to the Hvorslev's equivalent stress σ_e in the 1D-visco-hypoplastic model. The $\dot{\varepsilon}^{vis}$ is larger than the $\dot{\sigma}$, if the strain rate $\dot{\varepsilon}$ under loading is faster than the reference value, i.e., OCR<1, and vice versa. The governing equation (3) demonstrates that the $\dot{\varepsilon}^{vis}$ relates to the reference creep rate $\dot{\gamma}$ and the over-consolidation ratio OCR.

$$\dot{\sigma} = \frac{\sigma(1+e_0)}{\kappa} \cdot (\dot{\varepsilon} - \dot{\varepsilon}^{vis}) \tag{1}$$

$$\dot{\sigma}_e = \frac{\sigma_e(1+e_0)}{\lambda} \cdot \dot{\varepsilon} \tag{2}$$

$$\dot{\varepsilon}^{vis} = \dot{\gamma}\left[\frac{\sigma}{\sigma_e}\right]^n \tag{3}$$

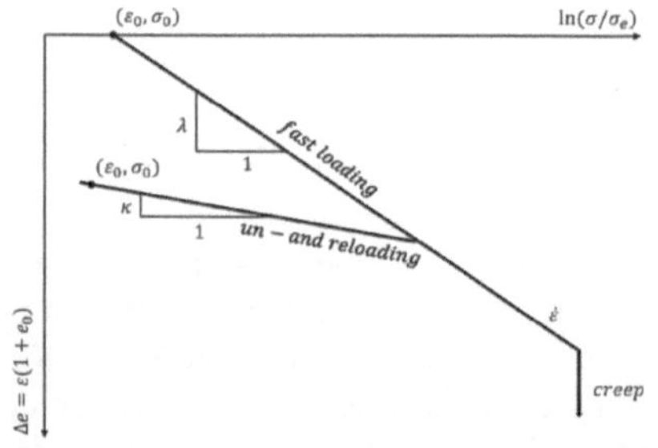

Figure 1. Schematic description of the first loading, unloading, and reloading as well as the creep in logarithmic space.

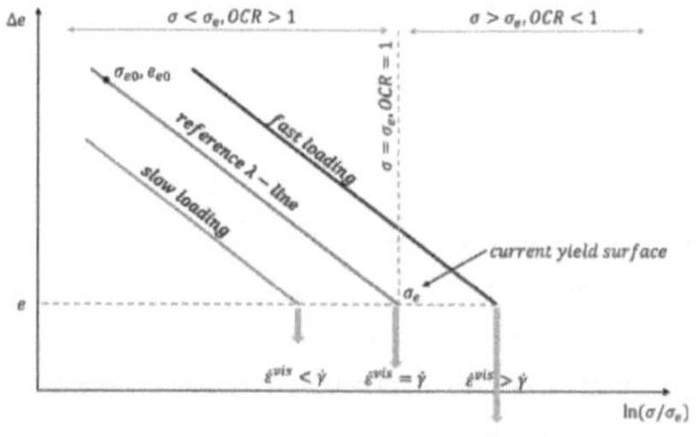

Figure 2. Compression lines and yield surface of the 1D-visco-hypoplastic model.

2.2 *Figure Model parameters and initial state*

All parameters of the 1D-visco-hypoplastic model are listed in Table 1. The compression and swelling indices are represented by λ and κ, the reciprocal of the viscosity index I_v by n, the product of compression index and the viscosity index by ψ, and a material parameter relating to

Table 1. Parameters for 1D-visco-hypoplastic model.

Type	Material constants	Reference values	State variables	Initial values
Basic	$\lambda, \kappa, n, \Gamma$	$e_{e0}, \sigma_{eo}, \dot{\gamma}$	e, σ	e_0, σ_0
Derived	ψ, I_v	ε_{e0}	$OCR, \sigma_e, \dot{\varepsilon}^{vis}, \varepsilon, t_0$	–

the reference λ-line by Γ. The reference values e_{e0}, σ_{e0}, $\dot{\gamma}$ lie on the reference λ -line. The $\dot{\gamma}$ is generally adopted as 1%/h, and the initial values can be used as the reference values e_{e0}, σ_{e0}. The state variables describe the current state of the soil and change themselves with time. The model parameters can be obtained from constant rate of strain (CRSN) tests.

2.3 *Analytical solution of 1D vertical consolidation*

The vertical consolidation is quantitatively expressed by the vertical consolidation coefficient c_v, which relates to the soil permeability and the effective stress state. It is adopted for solving this 1D problem:

- Since the pore water and soil particle are incompressible, the drainage velocity is assumed to be identical to the rate of settlement, but in the opposite direction, which can be expressed by equation (4).
- The total strain rate at the same position is constant with changing time, which can be expressed by equation (5).
- Depending on Terzaghi's consolidation theory, the drainage velocity is directly related to the excess pore water pressure, which can be expressed by equation (6).

$$v_w = -v_s \tag{4}$$

$$\dot{\varepsilon}(x, t) = const \tag{5}$$

$$v_w = -\frac{ku'(x)}{\gamma_w} \tag{6}$$

For solving this 1D coupling problem, some boundary conditions and initial conditions must be employed. For $H_{eff} = H$: $u(H) = 0$, $u(0) = u_{\max}$ and $u'(0) = 0$; for $H_{eff} = H/2$: $u(0) = u(H) = 0$, $u(H/2) = u_{\max}$ and $u'(H/2) = 0$. After the excess pore water pressure is dissipated, i.e., the effective stress is constant, merely the viscous strain occurs, which means $\dot{s}^{vis} = -\dot{\varepsilon}H = -\dot{\varepsilon}^{vis}H$.

3 VERIFICATION OF THE SEMI-ANALYTICAL 1D-VISCO-HYPOPLASTIC MODEL

3.1 *Consolidation and creep under isotropic compression*

This method was verified by Nimunis (1996) using an odometric test (Niemunis & Krieg 1996). In this study, another odometric test conducted by Krieg (2000) was employed for verifying the capability of this method. The soil samples are produced with d/h=100/20 mm (Krieg 2000). The soil sample was loaded as described in Krieg (2000), e.g. first loading, unloading and reloading, and the creep phase. The measured and derived parameters are summarized in Table 2. The compression parameter C_c and swelling parameter C_s were available instead of the compression index λ and swelling index κ in this odometric test. Due to different logarithmic space, the compression index λ and swelling index κ were identical to the values of the compression parameter C_c and swelling parameter C_s divided by 2.3. The reference value pairs σ_{e0} and e_{e0} can be selected on the reference λ -line at random. Here, the σ_{e0} and $e_{e0}$was selected as 610 kPa and 1.13, respectively. The viscous index I_v was defined by Krieg (2000) as a ratio of the creep parameter C_α to the compression parameter C_c(Krieg 2000). Therefore, the I_v was 0.028, which was also within the range of 0.01~0.04 suggested by Soumaya (Kempfert & Soumaya 2004). OCR defined as $OCR = \sigma_e/\sigma$ had a value of 6.1 for the sample prepared in the laboratory. Based on equation (3), the creep rate $\dot{\varepsilon}^{vis}$ was much lower than the expected value in the range of 10^{-8}-10^{-7}. Instead, an empirical creep rate of about 3.7x10^{-7} /s was used. Figure 3 represents the measured and simulated $e - \sigma$ relationships. This semi-analytical 1D-visco-hypoplastic model reproducs the soil behaviors including the creep during the first loading, unloading, and reloading stages.

Table 2. Parameters related to compression and viscosity (Krieg (2000)).

Measured parameters	C_c [-]	C_s [-]	C_α [-]	c_v [m^2/s]	δh [m]	γ [kN/m^3]	σ_{v0} [kPa]
	0.5	0.069	0.014	$3 \cdot 10^{-9}$	0.02	14.31	9.10
Derived parameters	σ_{e0} [kPa]	e_{e0} [-]	λ [-]	κ [-]	$\dot{\varepsilon}^{vis}$ [-]	I_v [-]	OCR [-]
	610	1.13	0.1938	0.0307	3.7×10^{-7}	0.028	6.1

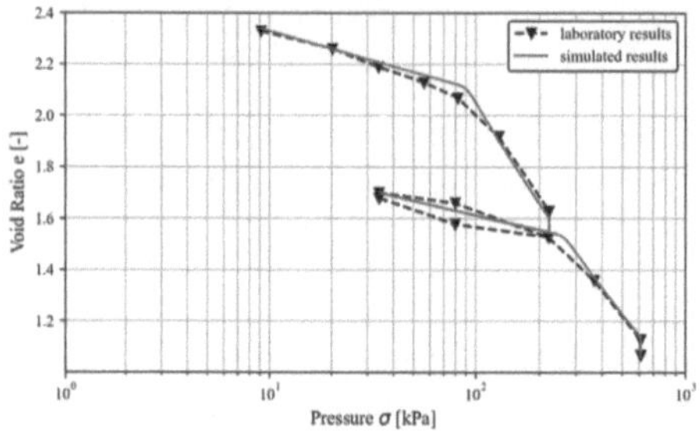

Figure 3. Comparison of void ratio-stress relationship obtained from laboratory experiments and simulations.

3.2 *Case study of Student Hostel in Constance by Soumaya (2005)*

Soumaya (2005) compared the measured and numerically calculated settlements of a student hostel in Constance and found that the numerically calculated settlements were about 50% larger than the measured onsite. Hence, a correction factor $\kappa \approx 0.7$ was proposed in further works (Soumaya-Ing 2006). Table 3 lists all soil properties required for calculating the settlements of the student hostel. The underlying clay with low permeability prevented the drainage downwards. According to the 20%-criteria, the effective height H of soil was determined as 7.4m (Werkle & Slongo 2018). Similarly, the compression index λ and swelling index κ can be determined based on the compression parameter C_c and the swelling parameter C_s. The viscosity index I_v was 0.03 as specified by Soumaya (2005). With an empirical adoption of $\dot{\varepsilon}^{vis} = 1.0 \times 10^{-7}$ /s, OCR=1.18 was determined depending on equation (3). The reference values σ_{e0} and e_{e0} were 61.8kPa and 0.88 respectively. The σ_{e0} was determined at the centre of theeffective height. The consolidation coefficient c_v was selected as 4 m^2/a in the range of 4 m^2/a to 6 m^2/a (Soumaya-Ing 2006).

Figure 4 represents the measured settlements at the characteristic points and the numerically calculated settlements using the 1D-visco-hypoplastic model. The measured settlements were different along the foundation causing an inclination and were more catastrophic than large settlements in geotechnical engineering. The green dashed line represents the average measured settlements at the characteristic points. The final average settlements were 5.60cm, which was smaller than 6.15cm determined using Asaoka's method (Kempfert, Soumaya 2004). The blue solid line represents the numerically calculated settlements using the 1D-visco-hypoplastic model. It can be observed that the numerically calculated settlements showed a satisfying agreement with the measured field settlements. The primary consolidation was well simulated, whereas the secondary settlements were overestimated. A possible reason was the overestimated viscosity index.

Table 3. Soil properties of marine clay (Soumaya (2005)).

Soil	γ [kN/m^3]	δh [m]	λ [-]	κ [-]	σ_{e0} [kPa]	e_{e0} [-]	$\dot{\varepsilon}^{vis}$ [-]	I_v [-]	OCR [-]	σ_{v0} [kPa]	c_v [m^2/a]
marine clay	19	7.4	0.0783	0.0174	61.8	0.88	1×10^{-8}	0.03	1.18	36	4

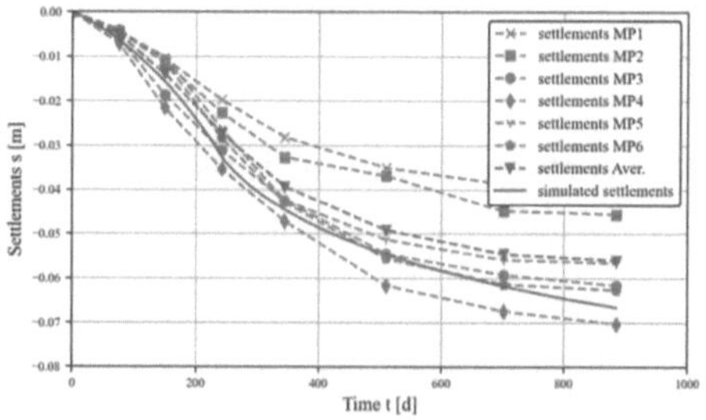

Figure 4. Simulated results in comparison to the measured settlements onsite.

4 CONCLUSIONS

This method combining the 1D-visco-hypoplastic model and the analytical solution by Terzaghi for the 1D consolidation was verified by an odometric test by Krieg (2005) and a well-recorded project by Soumaya (2005). This method was able to simulate the stress-strain relationships of soil under isotropic conditions. The back analysis for the long-term settlements of the hostel in Constance illustrated the capability of this method in geotechnical engineering. The application of this new method on multi-layer soil has to be verified in future works.

REFERENCES

Campanella R G and Vaid Y P 1974 Triaxial and Plane Strain Creep Rupture of an Undisturbed Clay *Canadian Geotechnical Journal* **11**(1):1–10.

Degago S A, Grimsta D G and Jostad H P et al. 2011 Use and misuse of the isotache concept with respect to creep hypotheses A and B *Géotechnique* **61**(10):897–908.

Degago S A, Grimstad G and Jostad H P et al. 2009 The non-uniqueness of the end-of-primary (EOP) void ratio-effective stress relationship *International Conference on Soil Mechanics & Geotechnical Engineering*.

Gang Bi and Jean-Louis et al. 2019 Power Law Model to Predict Creep Movement and Creep Failure *Journal of Geotechnical and Geoenvironmental Engineering* **145**(9).

Graham J and Yin J H 1996 Elastic visco-plastic modelling of one-dimensional consolidation *Géotechnique*. **46**(3):515–527.

Hunter G J and Khalili N 2000 A simple criterion for creep induced failure of over-consolidated clays.

J H Yin J G 1989 Viscous-elastic-plastic modelling of one-dimensional time-dependent behaviour of clays *Canadian Geotechnical Journal* **27**.

Kempfert H G and Soumaya B 2004 Settlement Back-Analysis of Buildings on Soft Soil in Southern Germany *Fifth International Conference on Case Histories in Geotechnical Engineering*.

Krieg S 2000 Viskoses Bodenverhalten von Mudden, Seeton und Klei *Veroff. Inst. Boden-u. Felsm* **150**.

Niemunis A and Krieg S 1996 Viscous behaviour of soil under oedometric conditions *Canadian Geotechnical Journal* **33**(1):159–168.

Soumaya-Ing B 2006 Bewertung von Setzungsmessungen flachgegründeter Gebäude in weichen Böden.

Terzaghi, and Karl 1943 *Theoretical soil mechanics* J. Wiley and Sons, inc.

Werkle H and Slongo L 2018 Modellierung des Baugrunds bei der Finite - Element - Berechnung von Bodenplatten *Bautechnik* **95**(9): 607–619.

Yin J H and Graham J 1994 Equivalent times and one-dimensional elastic viscoplastic modelling of time-dependent stress-strain behaviour of clays *Canadian Geotechnical Journal* **31**(1):42–52.

Advances in Civil Engineering: Structural Seismic Resistance, Monitoring and Detection – Mohd Johari Mohd Yusof and Junwen Zhang (Eds)
© 2023 The Authors, ISBN 978-1-032-31491-4

Analysis on construction index and standard of water-saving units of public institutions in southern water-rich region

Haibo Xu & Yuehua Wang
Zhejiang Institute of Hydraulics & Estuary (Zhejiang Institute of Marine Planning & Design), Zhejiang, China

ABSTRACT: Public institutions take the lead in saving water resources, which not only has direct practical significance for controlling and reducing water resources consumption, but also plays a positive demonstration effect and guiding role in guiding and promoting the whole society to save resources. The rapid development of urbanization puts forward higher requirements for the functions of public institutions. It is very necessary to strengthen the construction of water-saving units in public institutions. Combined with the current situation and characteristics of water use by public institutions in southern water-rich region, this paper deeply discusses the construction indicators and standards of public institutions in Zhejiang, Jiangsu, Guangdong and Shanghai, and puts forward scientific and reasonable suggestions combined with examples. It will play a positive role in promoting the construction of water-saving units in public institutions and sustainable social and economic development in southern water-rich region.

1 GENERAL INSTRUCTIONS

Water control is not only an important embodiment of national governance capacity, but also an inevitable requirement for the construction of ecological civilization. Water is the source of life, the essence of production and the basis of ecology, providing solid support for economic and social development. In 2014, general secretary clearly put forward the idea of "water saving priority, space balance, system management, power generated by government and market mechanism", emphasizing the importance of water saving and water security protection, emphasizing that water conservation should be given priority in terms of ideas, consciousness, measures and so on, providing scientific guidance and fundamental compliance for water saving and water management. The Fifth Plenary Session of the 19th CPC Central Committee clearly proposed to implement the national water-saving action, establish a rigid constraint system on water resources, and vigorously promote water-saving in agriculture, industry, cities and towns. The theme of "World Water Day" and "China Water Week" in 2021 is "deeply implement the new development concept and promote the intensive and safe utilization of water resources".

According to the regulations on energy conservation of public institutions (Order of the State Council No. 531), public institutions refer to state organs, institutions and organizations that use all or part of financial funds. Among them, organs and institutions, schools and hospitals are an important part of urban public water use. Water saving and sewage recycling in public institutions play an exemplary and exemplary role in improving the water-saving awareness of the whole society, and are an important part of the construction of water-saving society and water-saving city.

According to the world water development report 2021, the role of water in homes, schools, workplaces and health care facilities is often ignored or not given value equivalent to other uses. Water is the basic demand of human beings and the necessary resource to ensure drinking, maintain sanitation and cleanliness, and maintain life and health. The access to water and sanitation facilities is a human right. In 2020, the spread of New Coronavirus pneumonia was widespread worldwide.

 DOI 10.1201/9781003310884-90

The most vulnerable group was the most vulnerable group in the world, many of whom lived in informal settlements and city slums. Hand cleaning is very important for preventing the spread of COVID-19. Globally, more than 3 billion people and 2 / 5 of medical institutions lack adequate hand cleaning facilities.

Therefore, water conservation is an important measure to implement the sustainable development strategy. To build a number of water-saving units of public institutions, we need to improve the construction indicators and standards, constantly promote all kinds of public institutions at all levels to strengthen water-saving management and technological transformation, and guide the whole society to improve water-saving awareness.

2 NECESSITY OF WATER SAVING IN WATER-RICH AREAS AND ARACTERISTICS OF WATER USE OF PUBLIC INSTITUTIONS

2.1 *Necessity*

Although water resources are relatively abundant in water-rich areas, with the rapid development of economy and society, there are water shortages of water quality, resource and engineering types to varying degrees. Therefore, water conservation also needs to be encouraged in water-rich areas. The temporal and spatial distribution of water resources in water-rich areas is uneven, and the rainfall varies greatly throughout the year, showing the characteristics of abundant water in summer and low water in winter. The rainfall runoff from April to September in the flood season accounts for about 80% of the whole year, mainly in the form of flood, which is difficult to use water resources. The problems of low allocation rate of water-saving appliances and low water reuse rate in public institutions are prominent, and the way of water use is relatively extensive. Therefore, it is very necessary to vigorously advocate water-saving in water-rich areas to promote the construction of water-saving society. Water saving in public institutions also plays a leading role in water-saving for the whole society (Kong 2020; Liao 2005).

In August 2019, the office of the State Administration, the general office of the national development and Reform Commission and the general office of the Ministry of water resources jointly issued the code for water conservation management of public institutions, which clearly stipulates the operation and management of water conservation of public institutions in terms of water use planning and design, water intake quota, maintenance, metrological and statistical analysis, water quality and water treatment, water use system and performance evaluation. It provides an important basis for public institutions to analyze and evaluate their own water management performance.

2.2 *Characteristics of water use*

Water resources are indispensable for the normal operation of public institutions, and the fields, industries and industries involved in public institutions are different, so the water consumption of public institutions is also quite different. However, generally speaking, China's public institutions mainly have the following characteristics in terms of water consumption: (1) general organs and units. Their main water consumption departments are restaurants and public health regional offices. The main purpose of water is catering and maintaining the sanitary environment of office places. In terms of total water consumption, this part of water accounts for more than 65% of the total water consumption. (2) For the water consumption of special units, such as hospital institutions and schools, the main water consumption departments of the hospital are outpatient and ward, and the proportion of water consumption is about 60% of the total water consumption. Schools mainly use water in dormitories, teaching offices, canteens and other areas, and the proportion of water can reach about 70%. Of course, there are some differences among university, middle and primary schools. (3) There are great differences in water use among universities, middle schools and primary schools. Among them, universities, mainly dormitories and teaching office areas, need to use a lot of water resources, and the proportion of water can reach about 75%. In addition to dormitories, the main water departments of primary and secondary school campuses also have classrooms and canteens, and the total water consumption ratio can reach about 74% (Liu 2019; Qin 2017).

3 REQUIREMENTS OF RELEVANT STANDARDS

The construction of water-saving units in public institutions must have scientific construction indicators and standards in order to ensure the implementation and effectiveness of the construction of water-saving units in public institutions (Xu 2019).

In 2013, the Ministry of water resources, together with the office affairs administration of the State Council and the national water conservation office, issued the notice on carrying out the construction of water-saving units in public institutions, put forward relevant assessment standards and requirements, and took the construction of water-saving units as an important content of water-saving work in public institutions.

In February 2014, Zhejiang provincial bureau of organ affairs, Zhejiang provincial department of water resources and Zhejiang provincial water conservation office jointly issued the standard for the construction of water-saving units in provincial public institutions in Zhejiang Province, which carried out the creation of water-saving units in provincial public institutions in the province, and local governments at all levels carried out the creation of water-saving units at the city and county levels in accordance with the provincial standard. Jiangsu Province, Guangdong Province and Shanghai also issued their own construction standards.

In March 2019, the general office of the Ministry of water resources issued the notice on the construction of water conservation organs in the water conservancy industry, requiring that the organs of the Ministry of water resources, the organs of all directly affiliated units, and the water conservancy (water affairs) departments (bureaus) of all provinces (autonomous regions and municipalities directly under the central government) should be established as water conservation organs by the end of 2019. By the end of 2020, water conservancy bureaus at the provincial (municipal) and county levels will be established as water-saving organs.

In May 2020, the State Administration, the national development and Reform Commission and the Ministry of water resources jointly issued the implementation plan for leading action of water efficiency leaders in public institutions, carried out leading action of water efficiency leaders in public institutions on the basis of the construction of water-saving units in public institutions, accelerated the formation of a water-saving work pattern of "unit initiative, industry linkage and multi-party action", and promoted the construction of a water-saving society.

Generally speaking, both national and local construction indicators and standards basically require the construction of public institutions to meet the water-saving technology evaluation indicators such as water metering rate, popularization rate of water-saving appliances, per capita water consumption, leakage rate of water appliances and water supply pipe network, as well as rules and regulations, measurement statistics, management and maintenance, unconventional water utilization Water saving publicity and other water-saving management evaluation indicators. Evaluate according to these construction indicators and standards to judge whether it meets the requirements of water-saving units of public institutions.

4 MAIN CREATION CONTENTS

There are two main types of water-saving units created by public institutions: one is independent office, and the construction unit has independent water metering facilities, which can distinguish the water consumption of each functional area. Second, in the comprehensive building with centralized office of multiple departments, there are multiple units in one building or one floor building, which share public facilities such as toilet, tea room and canteen. It is difficult to distinguish the independent water consumption of each unit. Therefore, the number of on-the-job people of each unit is used to share the water consumption.

4.1 *Water metering rate*

The construction of water metering facilities is not only an important link to ensure the completion of water metering rate, but also the premise and foundation of water balance test. According to the acceptance standard, the allocation rate of water measuring instruments for primary water unit shall reach 100%, and points shall be deducted if secondary and tertiary water are not measured. Therefore, an important measure of public institutions in the process of building water-saving units is to install water meters in boiler rooms, central air conditioning, greening water, office buildings and canteens, improve the measurement system at all levels and realize zoning and grading measurement.

The allocation rate of primary and secondary water measuring instruments in Zhejiang Province shall reach 100% and 95%. If the primary water measuring rate does not meet the standard, no score will be given. For every decrease in secondary water measuring rate by 1%, 1 point will be deducted until it is completely deducted.

The metering rates of class I, II and III water meters in Jiangsu Province are required to reach 100%, 95% and 85% respectively. If the metering rate of class I fails to meet the standard, no score will be given; Deduct 0.5 points for every 1 percentage point lower than the metering rate of secondary water meter; If the metering rate of level III fails to meet the standard, points shall be deducted as appropriate until all points are deducted.

If the primary water metering rate of Guangdong Province reaches 100% and the secondary water metering rate reaches 95%, full marks will be given if they meet the requirements respectively, and 0 point will be given if one item fails to meet the standard.

Shanghai's primary water metering rate shall reach 100%, no score if it fails, and the secondary water metering rate shall reach 97.5%, full score if it all reaches. Deduct 2 points for every 0.5 percentage point lower than the secondary measurement rate, and no score for lower than 95%.

It can be seen from the table that Zhejiang, Guangdong and Shanghai all set up two-level measurement, and there is little difference in score. Guangdong Province and Shanghai have relatively strict requirements on water metering rate, and no score will be given if the metering rate of secondary water units does not meet the standard.

Table 1. Water metering rate.

Province/ City	Primary metering rate	Secondary metering rate	Tertiary metering rate	Remarks Remarks
Zhejiang	100%	95%	/	The full score is 9 points. If the primary measurement rate does not meet the standard, no score will be given
Jiangsu	100%	95%	85%	The full score is 8 points. If the primary measurement rate does not meet the standard, no score will be given
Guangdong	100%	95%	/	The full score is 8 points. If one item fails to meet the standard, no score will be given
Shanghai	100%	97.5%	/	The full score is 10 points, and no score will be given if the secondary measurement rate is lower than 95%

4.2 *Popularization rate of water-saving appliances*

In the popularization rate of water-saving appliances, water-saving appliances are the general name of mechanical equipment and storage equipment to improve water efficiency and reduce water consumption, such as water-saving faucets, toilets, showers and washing machines.

If the popularization rate of water-saving appliances reaches 100% in Zhejiang Province, 9 points will be obtained, and 1 point will be deducted for every decrease by 1% until it is completely deducted, which is equivalent to no score if it is lower than 92%.

If the popularization rate of water-saving appliances reaches 100% in Jiangsu Province, 8 points will be obtained. If it is lower than 96%, no point will be obtained, which is equivalent to deducting 2 points for every decrease by 1% until it is completely deducted.

If the popularization rate of water-saving appliances reaches 100% in Guangdong Province, 8 points will be given. If it is lower than 96%, no point will be given. At the same time, the value is assigned between partitions, and the change range less than 1% is allowed.

If the popularization rate of water-saving appliances reaches 100% in Shanghai, 6 points will be obtained. If it is lower than 98%, no point will be obtained, which is equivalent to deducting 3 points for every decrease by 1% until it is completely deducted.

It can be seen that the requirement for the popularization rate of water-saving appliances in three provinces and one city is 100%, and Shanghai has the lowest score and the most stringent requirement. Jiangsu, Guangdong and Shanghai have a minimum standard, with no score below 96% and 98%, while Zhejiang will gradually deduct points according to the penetration rate until it is all deducted. It can be seen that Zhejiang has the loosest requirements. Guangdong score is more detailed.

Table 2. Popularization rate of water saving appliances.

Province/ City	Popularization rate	Remarks
Zhejiang	100%	The full score is 9 points, and 1 point will be deducted for every 1% lower
Jiangsu	100%	The full score is 8 points, and those below 96% will not be scored
Guangdong	100%	The full score is 8 points, $P < 96\%$, 0 point; $96\% \leq P < 97\%$, 2 points; $97\% \leq P < 98\%$, 4 points; $98\% \leq P < 99\%$, 6 points;$99\% \leq P$: 8 points
Shanghai	100%	The full score is 6 points, and those below 98% will not be scored

4.3 *Per capita water consumption*

The per capita water consumption in Zhejiang Province refers to the ratio between the water consumption and the number of employees in employment. Units of the same type refer to units with similar properties, similar water habits and similar water uses. It is difficult to implement the average water consumption of the same type in the province in the previous natural year.

The per capita water consumption in Jiangsu Province is the ratio of the annual water consumption (excluding temporary water for infrastructure construction, water for family areas and external water supply) to the number of employees. In recent years, it is difficult to implement the statistics of the average water consumption of the same type in this region.

Considering that it is difficult to make effective statistics on the total water consumption of each unit, Guangdong temporarily takes the average daily water consumption per capita of organs (Institutions) in the local standard water quota of Guangdong Province to be issued as the base, multiplied by the total number of days in the calendar year of the current year as the average water consumption per capita of public institutions in Guangdong Province.

The unit of daily domestic water consumption per capita adopted in Shanghai is the ratio of daily domestic water consumption to relevant water users. Based on classified calculation, the per capita domestic daily water consumption is compared with the water quota value of corresponding industries in Shanghai water quota (Trial).

It can be seen that among the three provinces and one city, Zhejiang Province has the highest score; Guangdong and Shanghai calculate the standard number more reasonably, considering the number of people in and out of the staff; It is relatively difficult to adopt the average value of unit water consumption of the same type in Zhejiang and Jiangsu; Guangdong and Shanghai bureaus directly adopt quota indicators, which is easy to operate; Shanghai has not refined the zoning assignment, while Zhejiang, Jiangsu and Guangdong are relatively more refined.

Table 3. Per capita water consumption.

Province/City	Computing method	Remarks
Zhejiang	Annual water consumption / number of employees	It is determined according to the average water consumption of the same type of units in the province in the last natural year: per capita water consumption ≤ 0.9×Average value: 10 points; 0.9× average value < water consumption per capita ≤ average value, 8 points; Average value < water consumption per capita ≤ 1.1×Average value: 5 points; 1.1×Average value < water consumption per capita ≤ 1.2×Average value: 3 points; 1.2×average value< water consumption per capita, 0 point.
Jiangsu	Annual water consumption / number of employees	According to the average water consumption per unit of the same type in the region in recent years, the judgment score is: per capita water consumption ≤ 0.9×Average value: 8 points; 0.9×the average value <per capita water consumption ≤ average value, 6 points; Average value <per capita water consumption ≤ 1.1×Average value: 3 points; 1.1×Average value <per capita water consumption ≤ 1.2×Average value: 1 point; 1.2×Average value <per capita water consumption, no score. Other water users can verify the water consumption standard according to the relevant water consumption quota formulated by the province, and determine the score with reference to the above principles.
Guangdong	Annual water consumption / number of on-the-job employees	WR≤0.9×WP, 8 points; 0.9×WP < WR ≤ WP, 6 points; WP<WR≤1.1×WP, 3 points; 1.1×WP<WR≤1.2×WP, 1 point; 1.2×WP < WR, 0 point.
Shanghai	Average daily domestic water consumption per unit / total number of people using water	per capita daily water consumption≤ the quota value, 8 points will be obtained, and no point will be obtained if it is higher than the quota value.

5 EXISTING PROBLEMS AND COUNTERMEASURES

5.1 *Existing problems*

(1) Construction indicators and standards are not unified. The water-saving carriers applying for different competent departments (water conservancy, housing construction, etc.) often adopt different indicators and standards. Although the indicators of different competent departments basically contain the same content, they also increase the reporting burden of water-saving units.

(2) The incentive mechanism for water saving is not perfect. The establishment of water-saving units requires water balance test, water-saving technical transformation, water-saving facilities construction and other measures. The investment is large and the economic benefit is not obvious, which leads most units to ignore the ecological and social benefits of water-saving. In addition, the imperfect incentive mechanism leads to the low enthusiasm of units.

5.2 *Countermeasures and suggestions*

(1) Revise unified assessment standards

In 2015, in order to standardize urban water conservation and comprehensively improve urban water use efficiency, the Ministry of housing and urban rural development issued and implemented the new urban water saving evaluation standard (GB / T51083-2015), which puts forward higher requirements for the establishment of water-saving units. At the same time, in recent years, all localities have also issued normative documents, formulated the creation standards of water-saving schools, organs, benchmarks and water efficiency leaders, and widely carried out the creation activities of water-saving units. These standards provide a good reference for revising and perfecting the standards of water-saving units.

Example 1: Jinhua City integrates the relevant standards in the notice on printing and distributing the implementation plan of leading action of water efficiency leaders of public institutions and the notice of Zhejiang Water Conservation Office of Zhejiang Provincial Department of water resources of Zhejiang Provincial Bureau of institutional affairs on carrying out the establishment activities of water-saving units, so as to form construction indicators and standards in line with the actual situation of Jinhua City, and promote the establishment of water-saving units in municipal public institutions. For example, the per capita water consumption is determined by the average water consumption of the same type of units in the city in the previous natural year. It is suggested that the water quota is more practical and more operable. "Leakage rate of water appliances" should be literally understood as "the ratio of water leakage of water appliances to total public water consumption", but in order to increase operability, "the ratio of the number of leaking appliances to the total number of pieces" is adopted. Encouraging indicators reflect local characteristics, encourage innovative characteristics, encourage advanced water-saving experience, practices and characteristic water-saving publicity, etc.

Example 2: There is no separate construction standard for water-saving schools. As the main part of public institutions, the construction standard of water-saving schools generally adopts the construction standard of water-saving units of provincial public institutions in Zhejiang Province issued by Zhejiang provincial organ affairs administration, Zhejiang Provincial Department of water resources and Zhejiang Provincial Water Conservation Office. In 2020, the notice of seven departments including Zhejiang Provincial Department of water resources on carrying out water-saving benchmark leading action made clear requirements for school water-saving benchmark. This time, Jinhua water-saving schools suggest to adopt the water-saving benchmarking model, which is divided into colleges and universities and non-university, and mainly supplement and improve the construction standards of non-university and universities. Colleges and universities refer to the evaluation standards of water-saving colleges and universities, which is also to better connect the water-saving benchmarking construction of Zhejiang Province. The construction of non-university water-saving carrier is mainly based on the non-university water-saving benchmark, and the technical indicators include "water meter measurement rate" and "per capita comprehensive water consumption". The "unconventional water resources substitution rate" and "information management" have been added to the adjustment of incentive indicators, and the assessment scores have been partially adjusted. In addition, in order to better connect the actual water-saving schools in Jinhua with provincial water-saving units in the future, this standard is also formulated with reference to the construction standards of water-saving units in public institutions in Zhejiang Province, and the technical indicators are added with "cooling water make-up rate of central air conditioning and recovery rate of boiler condensate".

(2) Strengthen organizational leadership and review

In order to promote the establishment of water-saving units, assessment methods should be issued as soon as possible. Water saving management departments at all levels shall strengthen the overall planning and coordination with the competent departments of the industry, establish a joint meeting system for water-saving units, and make it clear that the competent departments of the industry have the guiding responsibility for the water-saving units within their systems. After the establishment of water-saving units, dynamic management should be implemented, review work should be strengthened, and the current working state of emphasizing creation over supervision should be changed.

Example 3: Jinhua City, Zhejiang Province has formulated the supervision and assessment measures for the establishment of water-saving carriers in Jinhua City to further clarify the contents of supervision and inspection, and recheck once every five years. Under special circumstances, it can be postponed for up to seven years. The object of each review is the municipal water-saving carrier named by the municipal water-saving competent department and other relevant departments. In order to simplify the review process, the review standard shall be implemented with reference to the construction standard of water-saving carrier. The water-saving carrier unit shall fill in the review self-assessment form, and the water-saving competent unit shall fill in the review summary form. The daily work shall be coordinated by the Municipal Water Resources Bureau, which shall guide and supervise the water conservation work within its administrative area. The competent departments of relevant industries are responsible for water conservation in their respective industries. The review of water-saving carriers shall be in the charge of the municipal water-saving department. The reexamination shall be carried out by listening to the report, checking the data and looking at the site, and the reexamination opinions shall be formed one by one. Those that pass the reexamination shall be re announced by the Municipal Water Conservation Department. If it fails to pass the reexamination, the title of water-saving carrier shall be revoked.

(3) Adhere to demonstration guidance and policy incentives

Water saving departments should strengthen the publicity of water-saving units and make use of important periods such as "World Water Day", "China Water Week" and "urban water-saving Publicity Week". Water saving departments shall organize thematic publicity activities. Give full play to the industrial advantages of radio and television, press and publication, tourism, transportation, municipal appearance and other departments. The water-saving department shall use television, radio, network, newspaper, carriage, platform and LED display screen to carry out various forms of water-saving unit publicity around water-saving models and water-saving measures. Water saving departments should establish water-saving benchmarks, implement water efficiency leader actions in the same industry, establish water-saving models in all districts and industries, and drive the water-saving work of the whole society. Water saving departments should increase financial investment and innovate incentive mode (Zhang2018).

Example 4: In 2020, Jinhua Water Resources Bureau stick to the bottom line of epidemic prevention, closely focus on the theme of "giving priority to water conservation and building happy rivers and lakes", put forward the concept of "wandering", create a strong water week atmosphere, and let the whole people share the achievements of water conservation and water protection. Jinhua rolling broadcast the World Water Day public service advertisement on the platform of 942 traffic radio and agricultural radio, calling on the audience to join the water-saving team. Experts are invited to explain the current situation of water quality and quantity of drinking water sources and the development of water conservancy work in Jinhua, so as to attract more citizens to join Jinhua water-saving action. Jinhua and Yiwu City issued the reward measures for water conservation in Yiwu City to implement the idea of giving priority to water conservation, promote the conservation, protection and rational utilization of water resources, mobilize the enthusiasm of the whole society for water conservation and promote water-saving management.

6 CONCLUSIONS

(1) The specific situation of public institutions in southern water-rich areas is different. After a period of practice, the construction of water-saving units has achieved practical results by formulating construction indicators and standards, clarifying creation requirements, establishing normalized management mechanism and strengthening safeguard measures.

(2) The construction of water-saving units in public institutions can drive public institutions at all levels to strengthen water-saving technological transformation and management system innovation, guide the whole society to improve water-saving awareness, promote water conservation and contribute to the construction of a water-saving society.

ACKNOWLEDGMENTS

The research was supported by National Key R&D Program of China (2019YFC0408804).

REFERENCES

Kong Qingjie, Wu Qiang (2020). Discussion on the construction of water-saving units in public institutions. *J. Low Carbon World*. 4, 90–91.

Liao Hongxi, Ge Xiaopeng (2005). Economic analysis of water-saving society in water-rich areas. *J. China Water Resources*. 7, 47–49.

Liu Youping (2019). Experience and Rethinking on the establishment of water-saving units of public institutions in water-rich areas in Southern China. *J. Pearl River*. 40(7), 122–127.

Qin Xiaoxia, Liu Hezhu (2017). How to promote the construction of county water-saving carrier. *J. Technology and Market*. 7, 400, 402.

Xu Xingyi, Hu Chunyan (2019). Construction status and work thinking of water-saving units in public institutions in Hunan Province. *J. Hunan Hydro & Power*. 1, 77–79.

Zhang quanmei (2018). Analysis on the management and implementation process of water-saving units. *J. China Market*. 30, 110, 112.

Advances in Civil Engineering: Structural Seismic Resistance, Monitoring and Detection – Mohd Johari Mohd Yusof and Junwen Zhang (Eds)
© 2023 The Authors, ISBN 978-1-032-31491-4

The challenges and suggestions for BIM implementation in China: Project-based statistical analysis

Jiang Lin & Zhou Kexiang
School of Economics and Management, Tianjin Chengjian University, Tianjin, China

Liu Rui
M.E. Rinker, Sr. School of Construction Management, University of Florida, Gainesville, FL, USA

Ji Ye
Hebei Jiaotong Vocational and Technical College, Shijiazhuang, China

ABSTRACT: There are more and more developing countries worldwide that need BIM technology to promote their AECO industry. It is important and urgent to understand the challenges and problems of BIM implementation in order to apply this technology more efficiently and effectively. We designed a questionnaire to investigate the current state of BIM implementation in construction industry in China. Then, the problems and challenges were summarized from the survey results by statistical analysis. This resulted in proposed suggestions for government policies to effectively BIM development and implementation within China. This result will provide a reference for BIM adoptions in other developing countries.

1 INTRODUCTION

With the rapid development of computer and information technology, BIM (Building Information Modeling) has become the technology trend to advance AECO (Architecture, Engineering, Construction, and Operations) industry. Many developed countries are vigorously developing BIM technology, so BIM implementation strategies are increasingly driving the application of this technology within the AECO industry on a global scale. For example, all public procurement projects in the UK have been mandated to use BIM since 2016, and BIM technology was required to be introduced for all public buildings costing over $27.6m in South Korea by 2016(Hill 2014).

In recent years the AECO industry has developed rapidly in the developing countries. For example, in China, the GDP of construction was about 3.5 trillion USD in 2019, which accounts for 25% of total GDP (National Bureau of Statistics of China 2020). The Ministry of Housing and Urban-Rural Development of the People's Republic of China (MOHURD) issued "The Outline of Development Program for Construction Industry Informationalization for 2003-2008" as early as in November 2003. However, the BIM implementation in AECO industry in China has not achieved satisfactory results even with the mandated policy by the government. Given China's unique institutional and cultural context, government-led promotion policies are the preconditions for BIM implementation. By March 2020, 63 BIM promotion policies and 8 relevant standards have been issued at the national level in China. Local governments have issued 320 BIM promotion policies and 41 BIM standards. (Figure 1).

Chinese scholars began to pay attention to BIM technology research in 2006. China National Knowledge Infrastructure (CNKI) database, the largest integrated knowledge resources database in China, showed that the number of BIM-related research papers had exploded in recent five years (Figure 2), and was about 3700 in 2019. However, there is a lack of systematic research on strategic approaches appropriately according to the current situation of BIM implementation

DOI 10.1201/9781003310884-91

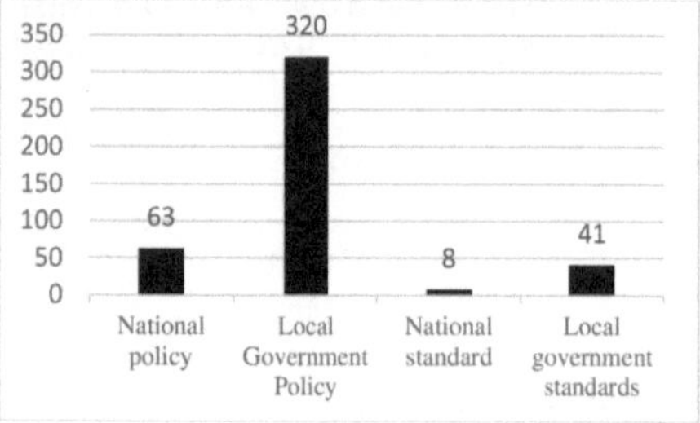

Figure 1. Statistics of BIM policies and standards at national and local government levels in China. (Unit: item)

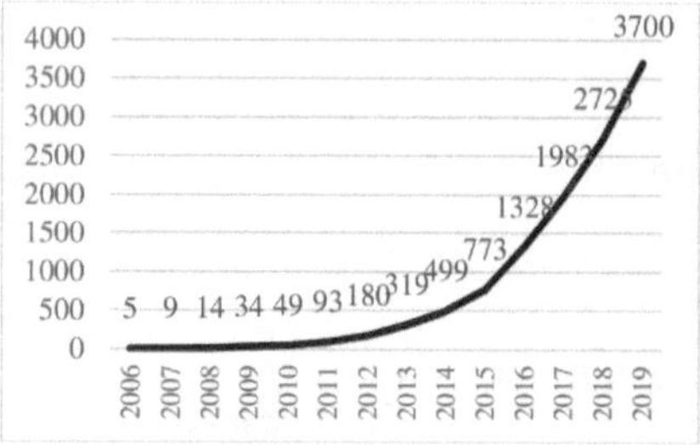

Figure 2. The tendency of the Research about BIM in China. (Unit: article)

within the Chinese AECO industry (Zhang 2016). Although some studies have proposed suggestions for establishing a unified workflow, improving participants' perception (Ding 2015; Jin 2017), few suggestions have been offered to connect the project performance and the government policies. Therefore, this study is aimed at filling this gap and providing recommendations regarding appropriate strategies for promoting the development and implementation of BIM by surveying the current situations on a project level in China and the experiences in other countries.

2 RESEARCH METHODS AND OBJECTS

2.1 *Research Methods*

This study follows a three-step research design to achieve the objectives of this research. First, a preliminary identification of problems or challenges that could affect BIM implementation in China's AECO industry was prepared by a systematic literature review. Second, a questionnaire survey with industrial practitioners was carried out in order to identify the root cause of the problems with BIM implementations. The last step is to process and analyze the data based on the questionnaire survey, and put forward some suggestions to solve the problems by learning the strategies and policies successfully implemented in other countries.

2.2 *Research Objects*

There are 18 questions in this survey, 100 questionnaires have been issued in total, 77 valid questionnaires have been recovered, with an effective rate of 77%. The objects of the survey are the projects under construction using BIM technology. This study included 75 projects located in 19 provinces in China and 2 overseas projects constructed by Chinese construction enterprises (located in the UK and Australia). The project types involved in civil infrastructures, public buildings, industrial factory building, bridge, underground utilities, roads, water conservancy projects, and a variety of other project types. All 77 projects investigated were completed within the last five years.

3 DATA ANALYSIS

3.1 *The Status of BIM Implementation*

BIM platform of a construction project may be established or applied by the owner, designer, or contractors. The survey reflects that there are only 33.8% BIM platform established by designers (Table 1). That means two-thirds of designers did not adopt BIM technology at their stages of projects, in which the 3D BIM model is established on the completed 2D drawings, namely, Reverse Modeling. After the completion of the project, 26.0% of the projects do not have BIM

Table 1. Statistics of questionnaire survey results.

Type	Questionnaire Questions	survey result
A. BIM implementation status	Q1 Subject of establishing and managing BIM platform	Designer (33.8%) Owner (37.7%) Constructer (49.4%), Others (11.7%).
	Q2 deliver to the owner or operator	Yes (74.0%), No (26.0%).
	Q3 How many management fields have adopted BIM	1 aspects (24.7%),2 aspects (20.8%), 3 aspects (23.4%),4 aspects (15.6%), 5 aspects (9.1%),6 aspects (6.5%).
	Q4 The frequency of BIM usage in daily work	Almost never (13.0%), Sometimes (33.8%), Often (41.6%), Frequently (7.8%), Almost everywhere (3.9%).
B. Management performance	Q5 The impact of BIM technology on the total project cost	Decreased significantly (13%), Decreased slightly (45.5%), Almost unchanged (32.5%), Increased slightly (5.2%), Increased significantly (3.9%).
	Q6 The impact of BIM technology on project work efficiency	Decreased significantly (3.9%), Decreased slightly (11.7%), Almost unchanged (18.2%), Increased slightly (48.1%), Increased significantly (18.2%).
C. Personnel competence	Q7 respondents' operational level of BIM software or platform	Almost incapable (20.8%), Simple (29.9%), General (24.7%), Familiar (19.5%), Proficient (5.2%).
	Q8 Learning difficulty of BIM software or platform	Very difficult (0%), Difficult (15.6%), General (62.3%), Easy (20.8%), Very easy (1.3%).
	Q9 Difficulty of data acquisition and integration when using BIM software or platform	Very difficult (0%), Difficult (22.1%), General (50.6%), Easy (22.1%), Very easy (5.2%).
	Q10 The degree of satisfaction of personnel	Very unsatisfied (0%), Unsatisfied (1.3%), General (41.6%), Satisfied (48.1%), Very satisfied (9.1%).
D. Horizontal collaboration	Q11 the proportion of traditional 2D design and 3D BIM model design in the project design stage	All 2D (15.6%), A few 3D(40.3%), Half 3D(18.2%), Most 3D (18.2%), All 3D(7.8%)
	Q12 The coherence and coordination of BIM software or platform application between the project design stage and construction stage	Totally uncoordinated (13.0%), Some uncoordinated (18.2%), Some coordinated (49.4%), Very coordinated (18.2%), Excellently coordinated (1.3%).
	Q13 The coherence and coordination of BIM software or platform application between the project construction stage and operation stage	Totally uncoordinated (9.1%), Some uncoordinated (14.3%), Some coordinated (51.9%), Very coordinated (19.5%), Excellently coordinated (5.2%).
	Q14 The difficulty of coordination with other project participators in BIM software or platform	Very difficult (2.6%), Difficult (6.5%), General (61.0%), Easy (24.7%), Very easy (5.2%).
	Q15 BIM system interface uniformity and system compatibility with other management systems	Very poor (3.9%), Poor (7.8%), General (46.8%), Good (37.7%), Very good (3.9%).
E. Vertical support	Q16 The superior's attitude towards the application of BIM technology in project management	Strongly disagree (0%), Disagree (0%), Not care (19.5%), Support (51.9%), Strongly support (28.6%).
	Q17 The attitude of governments at all levels to encourage and support the application of BIM in this project	Strongly disagree (0%), Disagree (1.3%), Not care (26.0%), Support (51.9%), Strongly support (20.8%).
	Q18 The actual policies or behaviors from governments at all levels to encourage or support the application of BIM in the project	Never (79.2%), Ever (20.8%).

deliverables to the owners or the facility managers (Q2). One reason is that the receiving parties had no corresponding department or personnel to take over. The other reason is the BIM platforms are not comprehensive and integrated.

It is only 5 projects among all 77 surveyed projects that BIM technology was applied throughout the life cycle of the project in different aspects including time control, quality control, cost management, safety monitoring, site planning, and contract management. Furthermore, there are 24.7% surveyed projects applied BIM technology in only one aspect of construction management (Q3). It shows that BIM has not been used fully in project management. On the other hand, in terms of the applying frequency of BIM platform, 13.0% of the projects set up BIM platforms but hardly used it, and 33.8% of the projects only used it sometimes (Q4). The reasons are concluded by inquiries as follows: 1) The limitation of BIM platform designed. The establishment of the platform was only for management in a certain aspect, not suitable for all the aspects, so most of the other BIM related work cannot be performed on the designated BIM platform.2) The silo working style of different stakeholders. 3) The attitude of the staff. Some people are resistant to the adoption of BIM as they are comfortable working in their traditional way, but not in the BIM process.

3.2 *Management Performance*

In terms of cost management, 58.4% of the project cost has been reduced after the application of BIM platform, 9.1% of the project cost has been increased, and 32.5% of the project cost did not change significantly (Q5). The application of BIM software and the establishment of the platform need additional investment in the early phase of the project. Only if it were made full use of in the later phrases, the BIM platform could achieve cost savings.

In terms of work efficiency (the different time needed for completing the same work), it was improved in 66.2% of projects (Q6) compared with the projects without BIM technology. It illustrated that BIM played an important role in increasing the efficiency of construction management. Certainly, it is well-known and needs no more discussion (Doumbouya 2016). However, the work efficiency of 15.6% of projects declined. That indicated that although BIM is recognized as an advanced management tool, it will still harm the project performance if it is not applied appropriately.

3.3 *Personnel Competence*

The management performance of BIM implementation is strongly associated with personnel competence and the operational level. 20.8% of the respondents said they have little knowledge of BIM, 29.9% said they can only make some simple operations, and 49.3% said they are skillful or mastered various operations (Figure 3). Many people were not familiar with the BIM platform, the main reason is that most of the BIM tools are complicated and has a long learning curve. Only 22% of respondents thought it is easy to learn the BIM platform and 78% think it is normal or difficult (Figure 4). Only 27.3% said it was easy to collect data when using the BIM platform for management (Q9).

An interesting result is that 98% of the respondents in the survey have a bachelor's degree or above, and 41.56% have worked in the construction industry for more than 10 years. They were engaged in relevant management in the AECO industry and should be the main customers of the BIM platform. If such persons think it is not easy to learn BIM, it is important to remind senior managers that BIM is not a simple management tool, which requires some certain professional training before implementing the BIM platform.

In terms of employee satisfaction, only1.3% of respondents felt unsatisfied (Q10). Although many could not operate the BIM platform skillfully, almost all the respondents are still satisfied with the situation. The reason for such satisfaction can be from two aspects. First, these projects proceeded smoothly and encounter few difficulties. Second, the professionals involved in these projects were not forced to use BIM tools.

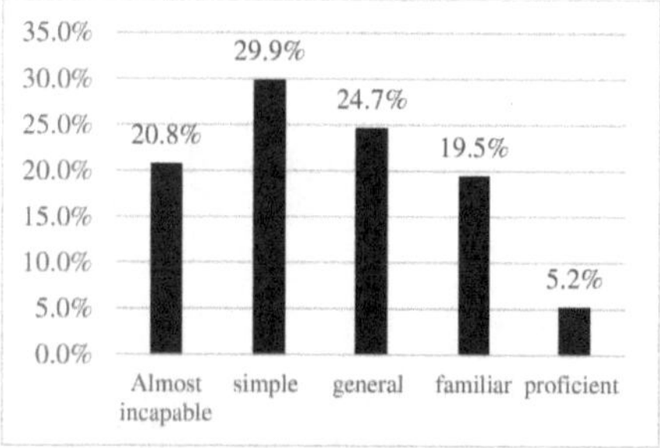

Figure 3. Respondents' operational level of BIM software or platform (Q7).

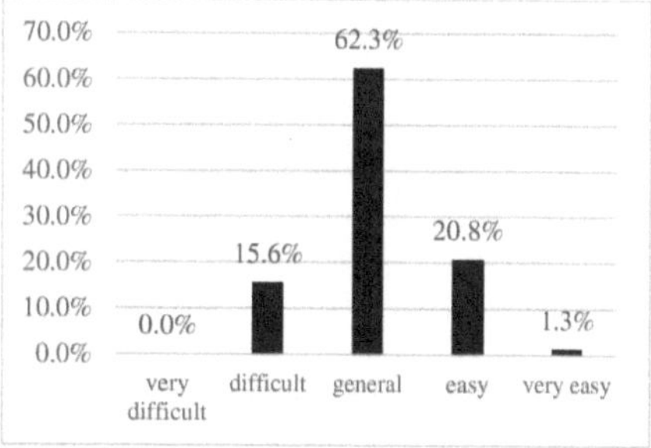

Figure 4. Learning difficulty of BIM software or platform (Q8).

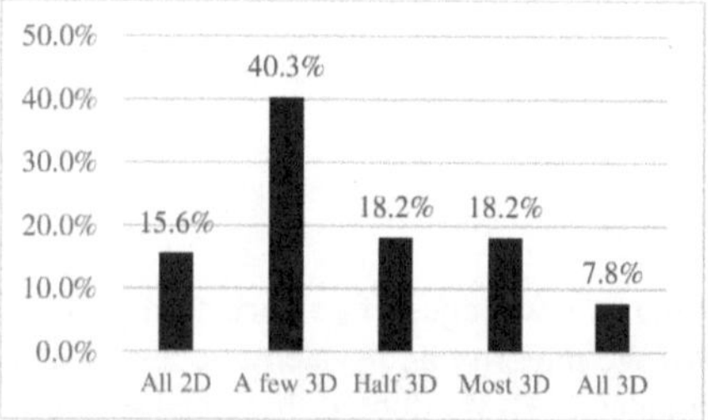

Figure 5. The proportion of traditional 2D design and 3D BIM model design in the project design stage (Q11).

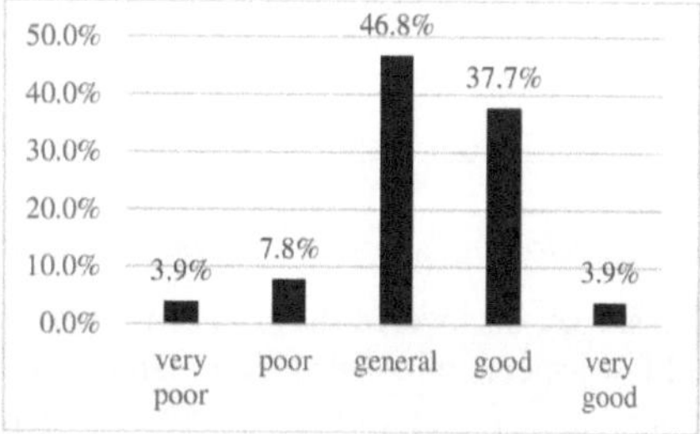

Figure 6. BIM system interface uniformity and system compatibility with other management systems (Q15).

3.4 *Horizontal Collaboration*

In the design stage, 15.6% projects are using all 2D drawings (Figure 5), 40.3% projects in which the 3D model is a minor part (less than 25%) of the design content, and 18.2% projects have 3D models as a major part of the total design content. There are 74.1% of the projects in which 3D models accounted for less than half of the total design content. That means there are many projects without BIM implementation in the design stage. The situation brings great obstacles to the establishment of the BIM platforms.

When operating the BIM platforms in the project life cycle, the survey showed that design was disjointed from construction in 31.2% projects (Q12), construction was disjointed in 23.4% projects (Q13). When considering the BIM platform with other project participants, the survey showed that the BIM platform of 41.6% projects demonstrate a good BIM software interface with others (Figure 6), and only 30% demonstrated easy coordination with others (Q14). This shows that BIM platforms exhibit poor coordination and consistency. Therefore, it is difficult to give full play to the advantages of BIM implementation, which will undoubtedly reduce the efficiency of project execution.

3.5 *Vertical Support*

The survey showed there are 80.5% of managers in the projects supported the BIM implementation (Figure 7), and 72.7% of all levels government departments supported, too (Figure 8). However, there did not seem to be much special support policies and measures in the implementation. 79% of respondents indicated that they did not receive any support (Q18). The supporting policies that have been accepted mainly included spiritual encouragement, such as leaders' observation and inspection, participation in awards, and other honors and awards. Among the projects that received

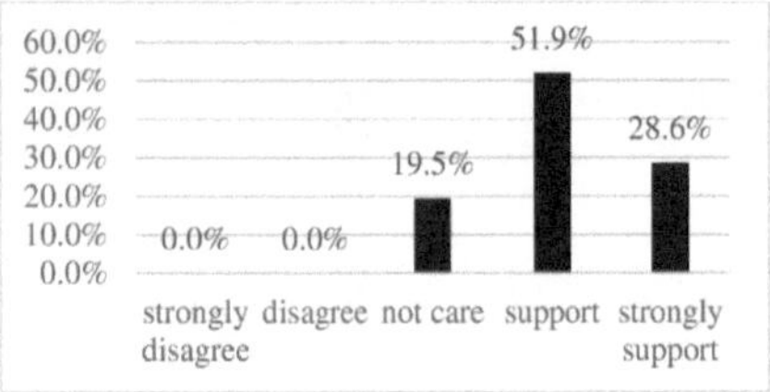

Figure 7. The superior's attitude towards the application of BIM technology in project management (Q16).

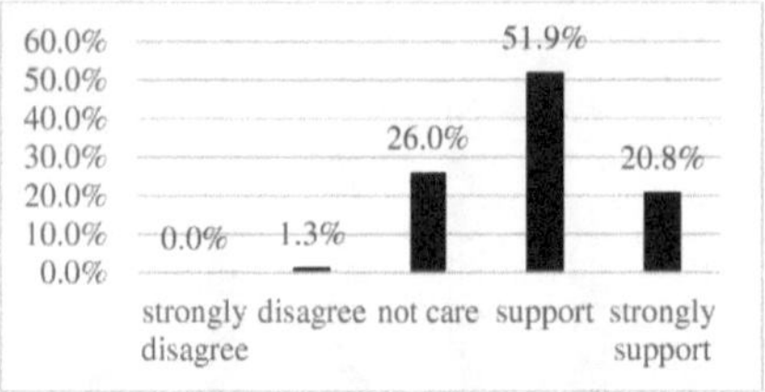

Figure 8. The attitude of governments at all levels to encourage and support the application of BIM in this project (Q17).

encouragement and support, only 7.8% of the projects received preferential policies or financial support.

4 CHALLENGES AND PROPOSALS

4.1 *Developing Lifecycle BIM Implementation* .

This survey discovered that BIM implementation in China in few stages of construction projects and few aspects of construction management was lack of the integrity and systematicness, which limited the applying frequency and management performance of BIM adopted. In the BIM implementation life cycle, ownership is the first problem. If the owners of the BIM deliverable are the clients, they may claim ownership of the data and documents, as they are paying for the design. Conversely, designers will claim that their designs remain their intellectual property (Eadie 2014). Therefore, the public construction projects owned by the government sectors might be a good chance for the lifecycle BIM implementation. In China, more than 60 strategy documents about BIM implementation have been issued by various departments at the national level, some of these documents were development guidelines, which guided the development direction of China's AECO industry and introduced BIM into the construction projects of various industries in China. Some of them are supporting policies for BIM implementation in China's ACEO industry, but almost these policies were not about encouraging for lifecycle BIM adoption or public construction projects. Consequently, more strategies are necessary to be issued for promoting BIM adopted integrally and systematically in the lifecycle of public projects led by the governments.

4.2 *Improving Personnel Training*

This survey shows that many people had low proficiency in BIM implementation, which is due to a lack of personal learning or lack of operability of BIM platforms. Although BIM training classes were held by MOHURD every year from 2012 to 2015, it was short-term (usually 2-3 days) training for basic knowledge of BIM. It was an urgent demand for continuous training in practical skills, knowledge updating, and technical assistance. In the United States, the public sector launched BIM programs, established BIM committees, and held various BIM conferences and training courses. As early as 2003, in the United States, the General Services Administration (GSA) Public Buildings Service (PBS) Office of the Chief Architect (OCA) established the National 3D-4D-BIM Program. They implemented BIM in over 200 active projects for the program, which was valued at over $12 billion (Yee 2009). With the support from industry technology leaders, the GSA BIM team drafted eight BIM Guide Series 01 to 08 over the past decade. For that reason, the governments in China should also consider setting up a special BIM promotion department following the examples of other countries, which can provide long-term training and technical assistance for BIM implementation, and promote the formulation of BIM standards or procedures by accepting the feedback of actual staff.

4.3 *Promoting Standard for BIM Implementation*

There are many barriers in the coherence between the sequential stages of the projects, the interfaces between different systems, and the coordination between the various participants. BIM technology received attention early on in China but the standards on BIM technology adoption had been issued later. Most of the standards had been issued and executed in the last three years. The earliest application standard of BIM technology in the AECO industry in China was the "Design Standard of Civil Building Information Model" issued by the Beijing Municipal Planning Commission in February 2014. The first BIM standard from the national governments was the "Unified Standard for Building Information Model Application" issued by the MOHURD in December 2016. Up to now, there are only 7 national standards that had been executed. In addition, the "Standard for storage of building information model" is also being developed. Almost these standards were executed in last two years. The projects investigated are projects completed in the last five years, many of which had been completed before the standards executed so the severe shortage of BIM standards was indicated in this survey. It can be predicted that the condition will be improved after all these standards executed, but there is still a certain gap between China and other countries with advanced BIM implementation. In short, the standards for BIM implementation still needs to be promoted in the future.

4.4 *Increasing More Practical Policies*

It is found in the survey that BIM implementation was supported both by superiors in the companies and by governments at all levels. However, there are much formal guidance and few practical policies received by the construction projects that were investigated. Particularly, there is a shortage of national incentive strategies, among which financial subsidies may be a feasible choice. In China there are not only more than 60 strategy documents at the national level, but also more than 300 from local governments. Some local governments had started to subsidize the projects with BIM implementation. In Shanghai, "the Notification on Further Strengthening the Popularization and Application of the Building Information Model Technology in Shanghai" was issued in 2016, which specified that the construction project with BIM implementation led by the owner in the design and construction two stages would be awarded for 20 CNY per square meter (maximum 300,000 CNY), or in design, construction and operation three stages would be awarded for 30 CNY per square meter (maximum 500,000 CNY). This policy also reflects the encouragement of the BIM implementation throughout the lifecycle of construction projects. It is hoped that more similar strategies and policies will be issued by national or local governments in the future in China.

5 CONCLUSION

The primary objective of this study was to facilitate the formulation of appropriate policies by investigating the challenges of BIM implementation in China, so as to provide reference for other developing countries to develop BIM implementation. The main method of investigation is a questionnaire survey, and the questions were divided into the following five categories: the status of BIM implementation, management performance, personnel competence, horizontal collaboration, and vertical support. Based on the data analysis from the survey, four main challenges were found, and then corresponding proposals were suggested finally, including developing lifecycle BIM implementation, improving personnel training, promoting standards for BIM implementation, and increasing more practical policies.

There are several limitations of this study. First, only 18 survey questions were selected to ensure that the questionnaire length is appropriate to get serious answers. Due to the limited questions asked, it may have some hidden problems in the BIM implementation but not discovered in this study. Second, there are only 77 valid questionnaires were received, and the survey did not cover all the provinces of China. So, the data may not be comprehensive enough. A further study is necessary to understand the geographic implementation differences in China. In addition, although the solution proposals to the challenges refer to the practices of other developing countries, whether they are feasible and effective in China or other developing countries need to be tested in future research.

REFERENCES

Ding, Z., Zuo, J., Wu, J., Wang, J.Y., 2015. Key factors for the BIM adoption by architects: A China study. Engineering, Construction and Architectural Management. 22 (6), 732–748.

Doumbouya, L., Gao, G., Guan, C., 2016. Adoption of the Building Information Modeling (BIM) for Construction Project Effectiveness: The Review of BIM Benefits. *American Journal of Civil Engineering and Architecture*, 4(3), 74–79.

Eadie, R., Odeyinka, H., Browne, M., McKeown, C., Yohanis, M., 2014. Building information modelling adoption: an analysis of the barriers to implementation. *J. Eng. Architect.* 2 (1), 77–101.

Hill, M., 2014. *The Business Value of BIM for Construction in Global Markets*. McGraw-Hill Construction, Bedford, MA.

Jin, R., Hancock, C., Tang, L., Chen, C., Wanatowski, D., Yang, L., 2017. Empirical Study of BIM Implementation - Based Perceptions among Chinese Practitioners. *Journal of Management in Engineering*. 33 (5), 04017025.

National Bureau of Statistics of China (NBSC). 2020. Statistical Communique of the People's Republic of China on the 2019 National Economic and Social Development. http://www.stats.gov.cn/tjsj/zxfb/202002/t20200228_1728913.html (in Chinese)

Yee, P., Matta, C., Kam, C., Hagan, S., and Valdimarsson, O., 2009. The GSA BIM story. *Journal of Building Information Modeling*, the Spring 2009 edition.

Zhang, J., Long, Y., Lv, S. and Xiang, Y., 2016. BIM-enabled modular and industrialized construction in China. *Procedia Engineering*, 145, 1456–1461.

Advances in Civil Engineering: Structural Seismic Resistance, Monitoring and Detection – Mohd Johari Mohd Yusof and Junwen Zhang (Eds)
© 2023 The Authors, ISBN 978-1-032-31491-4

Lishui city Liandu district Bihu plain water system connection project (Yuxi reservoir water diversion) water resources demonstration example analysis

Chenyang Zhang, Qi Chen & Ce Chen
Zhejiang Guangchuan Engineering Consultation Co. Ltd., Hangzhou, Zhejiang, China
Zhejiang Provincial Key Laboratory of Hydraulic Disaster Prevention and Mitigation, Hangzhou, Zhejiang, China

ABSTRACT: Bihu Plain, located in Bihu Town, Liandu District, has the oldest large-scale water conservancy project in Zhejiang Province. In recent years, the local tourism economy has developed rapidly, leading to increase in regional water use. Therefore, it is particularly important to attract Yuxi Reservoir water to supplement through Tongji Weir. This paper analyzes the demonstration of water resources in Bihu Plain, studies the necessity and rationality of water diversion (JI 2014), and concludes that the construction of Yuxi Reservoir diversion project can make the annual guarantee rate of water supply in Bihu Town reach more than 95%, and improve the daily guarantee rate of ecological environment water supply to more than 95%.

1 INTRODUCTION

Bihu Plain is located in the southwest of Liandu District, in the middle reaches of the Ou River, where two provincial roads Longli Line and Lipu Line meet 20 km away from the urban area. The plain slopes are located across southwest to northeast, high southwest and northeast, high east and west, and low in the middle.

Bihu Town, Liandu District is located in Bihu Plain, which is a water conservancy project area. Due to rapid development of the local tourism, which leads to rapid growth of permanent resident population and tourism population, the existing water supply plant can no longer meet the water needs of Bihu Town. The current irrigation area of Bihu Plain is about 4146 m^2. Due to uneven water abundance and contraction and the control of water diversion flow of Tongji Weir, both irrigation and ecological environment water cannot meet the demand. To address the domestic water supply needs of Bihu Town, while taking into account the Bihu Plain irrigation and ecological environment water demand, planning and construction of water diversion project is considered.

Bihu Plain water system connection project task is given priority for flood control and drainage (Xiao 2006), both water resources guarantee water landscape improvement comprehensive utilization, including a project for Yuxi Reservoir diversion, mainly used to guarantee water conservancy scenic area urban and rural residents living water, Tongji Weir irrigation area irrigation water, and ecological landscape water guarantee rate; thereby, promoting the development of local tourism economy.

Lishui Yuxi Water Conservancy Project is located on the mainstream of Ou River in Zhejiang Province, next to Lipu line of provincial road, 30 km away from Lishui city. It is a water conservancy project with comprehensive benefits, such as power generation, irrigation, shipping, and rafting.

Yuxi Reservoir diversion project diversion water flow 3 m^3/s. The starting point of the project is located at the left bank of the upstream of Daxi Yuxi Water Conservancy Project, and the water is diverted to Bihu Plain by the reserved inlet of Yuxi Reservoir, and the endpoint is located at the right bank of the upstream of Songyin Creek Tongji Weir. The tunnel goes through S33 Longliwen

DOI 10.1201/9781003310884-92

Expressway and the length of the water transmission line is 3.6km. The project plans to lead Yuxi Reservoir in Daxi River basin to Tongji Weir irrigation area in Songyin Creek downstream. The main buildings of the water diversion project include water inlet, water tunnel, water outlet, etc.

According to the Regulations of Water Resources Demonstration Management Measures for Construction Projects of the Ministry of Water Resources, water resources demonstration work has been carried out for water diversion projects (Feng 2016; Liu 2009).

2 WATER RESOURCES DEMONSTRATION WORK GRADE

According to the Guidelines for Water Resources Demonstration of Water Resources Construction Projects (SL525-2011), as per the principle of determining the highest grade (Sun 2011), the work grade of water resources demonstration of the project is grade 1. See Table 1 for specifics.

Table 1. Classification table of water resources demonstration.

Classification	Classification index	Basis for discrimination	This evaluation Work grade
Surface water intake	Domestic water intake volume	Water supply scale of 80,000 m^3/d, annual water intake of 28.38 million m^3,5~100,000 m^3/d	2
	Irrigation design flow rate	3.23 m^3/s,$\leq$10 m^3/s.	3
	Water diversion works Design flow	The 3 m^3/s, is located at 1 m^3/s ~5 m^3/s	2
	Irrigation area	Irrigation area of 42,000 mu, small irrigation area	3
Impact of water intake and water withdrawal	Water resources utilization	The discharge of Yuxi Reservoir of 3m^3/s, accounts for 2.7% of its average annual average flow, with a slight impact on water resources and utilization in Daxi Basin	3
	Ecology	There is no sensitive ecological problem, which has a slight impact on the ecological water in Daxi Basin and is conducive to the ecological water in Bihu Plain	3
	Water area management requirements	Six water functional zones are involved	1

3 WATER INTAKE AND WATER SOURCE DEMONSTRATION

The new Yuxi Reservoir diversion project is expected to solve the problems of insufficient water supply for Bihu Town as well as irrigation needs in Bihu Plain. The water diversion line enters Songyin Creek (upstream of Tongji Weir) from Yuxi Reservoir through a tunnel, and then enters Bihu Plain through the original water diversion channel of Tongji Weir. The rationality of Yuxi diversion flow and the necessity of the diversion project were analyzed by calculating whether the daily inflow flow of Tongji Weir combined with the diversion project from 1965 to 2015 could meet the sum of daily domestic water supply, irrigation, and ecological flow of the water supply area.

3.1 *Calculation of water demand*

3.1.1 *Analysis of water demand in Bihu Town*

The water supply planning area of Bihu Plain Water Conservancy Scenic Area is the Tongji Weir Water Conservancy Scenic Area. According to the water demand analysis and calculation of Bihu

Town, the maximum daily water consumption of urban and rural life is 109,000 m^3/d, and by tourist attractions is 12,300 m^3/d. The total maximum daily water consumption is 121,300 m^3/d.

Considering the water capacity of Bihu no. 1 water plant is 50,000 m^3/d, after the original Nanshan water plant no longer supplies water, at the end of the planning period, there will still be a water supply gap of 71,300 m^3/d in the planning area in 2025. Therefore, it is reasonable to set the daily water supply scale of the new water plant at 80,000 t/d, and the water diversion project combined with the inflow from Tongji Weir shall provide an average annual domestic water supply of 28.38 million m^3 to the new water plant in Bihu Town.

3.1.2 *Analysis of water demand for agricultural supplementary irrigation*

After completion of the diversion project, when the incoming water of Bihu Plain cannot meet the irrigation water demand, diversion of water should be carried out. According to the irrigation quota and planting area of various crops, the average water demand for supplementary irrigation in the irrigation area for many years is 20.51 million m^3, and the maximum daily irrigation flow is 3.23 m^3/s.

3.1.3 *Analysis of ecological environment water demand*

Ecological water demand in river course refers to the amount of water (flow) needed to maintain a certain form and function of the river ecosystem, which is related to the development and utilization of local water resources, the level of economic and social development, the evolution of water resources, ecological environment protection, rehabilitation, and construction goals [1]. According to the Guidelines for the Demonstration of Water Resources for Water Conservancy and Hydropower Construction Projects, the ecological water consumption is determined by 10%–20% of the annual average flow in principle. According to the comprehensive Management of Xuanping River Basin and Water System Connectivity Planning of Bihu Plain in Liandu District, Lishui City, the ecological base flow of Bihu Plain is 0.36 m^3/s, 10% of the annual average flow.

Ecological water in Bihu Plain is considered for water diversion from Tongji Weir. During calculation, priority should be given to ensure the ecological flow under Tongji Weir, and then ensure that the ecological environment is the best to meet the landscape needs of local tourism development. According to the Code for Calculation of Water Demand for Ecological Environment of Rivers and Lakes, if the best ecological environment condition in the river needs to be reached, the ecological environmental flow should account for 60%–100% of the annual natural flow (Zhao 2011). The ecological environmental flow of Bihu Plain is calculated at 60% of the average annual flow of 2.1 m^3/s.

The ecological flow under Tongji Weir takes 10% of the annual average flow of the Weir site, which is 5.8 m^3/s.

Therefore, the average annual ecological and environmental water demand in Bihu Plain is about 63.74 million m^3 and the water demand of the ecological environment under Tongji Weir is about 170.19 million m^3.

3.2 *Available water supply calculation*

Through analysis and calculation of the daily runoff of Yuxi Reservoir and 1954–2015, the annual average runoff flow of Yuxi Reservoir is 110.8 m^3/s, annual average runoff is 3.51 billion m^3, and annual average runoff depth is 1031 mm. The average annual runoff flow is 58.4 m^3/s, the average annual runoff is 1.84 billion m^3, and the annual average runoff depth is 912.8 mm.

Tongji Weir has no adjustment capacity. For daily average flow series from 1954 to 2015, for the long series, the flow frequency results of Tongji Weir section are shown in Table 2.

The water supply source of Tongji Weir includes the inflow of Songyin Creek and the inflow of Yuxi Reservoir, and the amount of available water supply is the minimum of the inflow and demand of the two sources.

To ensure normal water supply to the newly built water plant, Tongji Weir provides fixed water diversion of 0.9 m^3/s to the water plant every day. When the water inflow from Tongji Weir can

Table 2. Average daily flow results of Tongji Weir.

Location	Mean value (m^3/s)	Representative daily flow (m^3/s)			
		50%	75%	80%	90%
Tongji Weir	58.4	21	8.5	6.8	3.7

meet the downstream ecological flow in the wet season, there is no need for the ecological diversion project to provide water diversion. When the water inflow from Tongji Weir is not enough in the dry season, the water diversion project will remove the 0.9 m^3/s of water diversion provided to the water plant. In addition, the ecological flow was increased by 2.1 m^3/s, and the total water diversion was 3.0 m^3/s. During the irrigation period, the maximum water diversion of Tongji Weir is 3.23 m^3/s, which can meet the irrigation needs.

After the daily flow balance, the average water diversion of Yuxi Reservoir is 22.05 million m^3. The average annual flow of Tongji Weir is 58.4 m^3/s. The average annual amount of water is 1.84 billion m^3.

3.3 *Analysis of supply and demand flow balance before and after the completion of the Bihu Plain Diversion Project*

The regulation principle is "do not reduce the base flow of Songyin Creek, use river water first and reservoir water later, give priority to domestic water supply, combine ecological water with irrigation water, and the maximum flow of tongjiyan slugate shall not exceed 3.23 m^3/s". According to the daily flow balance adjustment calculation from 1965 to 2015, the annual water supply guarantee rate of the current ecological environment in Bihu Plain is very low. Due to small water flow from Tongji Weir during dry season, there are water shortages every year, so it is difficult to ensure the ecological environment and agricultural water in Bihu Plain. If measured by day, before the completion of the project, from 1965 to 2015, water supply flow assurance rate was about 77%. After the construction of the water diversion project, Yuxi Reservoir water source was introduced in the dry season, and the maximum diversion flow was 3 m^3/s, which improved the water supply guarantee rate of ecological water supply, and the daily guarantee rate of water supply flow could reach 99%. After the completion of the diversion project, Bihu Town's annual water assurance rate is expected to reach more than 95%.

3.4 *Analysis of rationality and reliability of water intake*

The main buildings of Bihu water diversion project include water inlet, water transmission tunnel, water outlet (no longer within the protection scope of drinking water source, allowing construction), etc. The water intake of the diversion tunnel is reserved for Yuxi Reservoir. The bottom sill elevation at 73.0 m is 4.6 m, below the designed minimum power generation level of the reservoir, which can meet the water intake requirements and the water intake setting is reasonable.

Yuxi Reservoir catchment area is 3407 km^2, annual average flow is 110.8 m^3/s, average annual runoff is 3.51 billion m^3, and annual average runoff depth is 1031 mm. Tongji Weir catchment area is 1969 km^2, multi-annual average flow of 58.4 m^3/s, multi-annual average runoff is 1.84 billion m^3, multi-annual average runoff depth of 912.8 mm. After the completion of the Bihu water diversion project, the water diversion flow of 3.0 m^3/s can be the average annual water diversion from the Yuxi Reservoir area of 22.05 million m^3. After years of daily flow balance adjustment calculation, the construction of water diversion project, the maximum flow of 3.0 m^3/s, the annual water supply guarantee rate of more than 95% improve the daily guarantee rate of ecological environment water supply to 95%, provide annual irrigation water of 20.51 million m^3, average ecological environment water 63.74 million m^3, and average ecological environment water under Weir of 170.19 million m^3.

4 CONCLUSIONS

Bihu Plain diversion project aims to provide domestic water to Tongji Weir irrigation water supply and Bihu Plain ecological environment. During the dry period, the upstream water cannot meet the local living water needs and ecological environment water needs. The maximum water flow from the upstream of Yuxi Reservoir to Tongji Weir is 3.0 m^3/s, which solves Bihu Town water requirement with water supply flow of 0.9 m^3/s and Bihu Plain ecological environment water with flow of 2.1 m^3/s. Combined with Songyin Creek water, domestic water supply of 28.38 million m^3 is provided to Tongji Weir irrigation area with an annual average irrigation water of 20.51 million m^3, and to provide annual ecological environment water of 63.74 million m^3. Through this analysis and demonstration, after the construction of the Yuxi Reservoir water diversion project, the annual guarantee rate of domestic water supply in Bihu Town is expected to reach more than 95%, and improve the daily guarantee rate of water supply of ecological environment to more than 95%. Therefore, the water diversion project is necessary and reasonable.

REFERENCES

Baohua Zhao. & Li Li. (2011). Water Resources Demonstration of Hanjiang Shuhe Hydropower Station Construction Project [J]. *Water Conservancy Technology and Economy*, (04):83–86.

Hongxiang Ji. &Jing Qiu (2014). Rational Analysis of Urban Water Supply Project [J]. *Guangdong Water Resources and Hydropower*, No. 219 05: 30–33.

Qinghua Feng. (2016). Water Resources Demonstration of Nanling County Urban Water Supply Project [J]. *China Water Transport* (second half), (02): 149–150.

Wangen Liu. (2009). On the Role of Construction Project Water Resources Demonstration in Water Resources Management [J]. *People Pearl River*, v. 30, No. 172, (03):8–9.

Zhaodong Sun. & Yongfeng Liu. (2011). Qinghai Water Canal Project Water Diversion Project [J]. *China Water Resources*, No.683 17:48–51.

Zhijuan Xiao. & Jiancang Xie. (2006). Case analysis of water resources demonstration of construction project [J]. *China Rural Water and Hydropower*, 02:8–11.

Advances in Civil Engineering: Structural Seismic Resistance,
Monitoring and Detection – Mohd Johari
Mohd Yusof and Junwen Zhang (Eds)
© 2023 The Authors, ISBN 978-1-032-31491-4

Orientation and exploration: Taking green city as the goal to develop new forms of housing development

Yuguan Hua†
School of Nanjing University, Jiangsu, Nanjing, China

Yiran Xu†
School of Beijing University of Technology, Beijing, Haidian, China

ABSTRACT: Nowadays, the pressure on resources and increased demand on local and global environments, coupled with the need and expectation to improve residents' overall quality of life, is high. How to respond to these challenges with local, regional, and international reach becomes important for the city and the local governments. Green building thus became the center of attention. Because of its unclear concept and wide range, the research group focused on its evaluation criterion—active house. To explore the potential factors and profound significance of active housing as a new direction of future housing development, the project summarized the existing problems of active housing through qualitative analysis, literature review, and case analysis and proposed the evaluation system of active housing based on environmental protection, energy consumption, and people-oriented concern. Nowadays, there are three main blocks to promote active housing: short development time, low popularity, low high-tech level, and high construction cost. Although these problems are inevitable, active housing is still the direction of future housing development plans.

1 INTRODUCTION

With global energy shortage crisis and gradual establishment of concepts of green environmental protection in the process of social development, the development of green city has become the goal in urban settings. Based on the "National New Town Planning (2014-2020)" put forward to promote the construction of the green city, it needs to accelerate the design and construction of green buildings (The Central People's Government of the People's Republic of China, 2014–2020). Therefore, green building has become the direction of architectural design pursuit. However, the evaluation criteria for green building are various, such as passive building and active building. As an evaluation standard for green buildings, active buildings can be evaluated more completely and accurately (Architectural Research Centre of Shanghai Institute, 2018). Therefore, this study takes active buildings as the research object. In the context of environmental protection, low energy consumption, and living comfort, this paper explores the possibility of active buildings as a new form of residential development and tries to verify that active houses will be the main direction of residential development in the future through the lack of humanism in architectural design.

†These authors contributed equally.

 DOI 10.1201/9781003310884-93

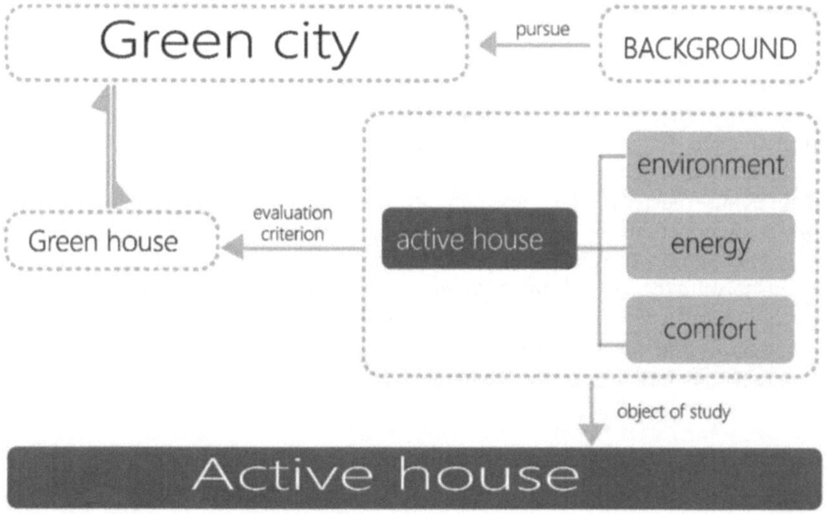

Figure 1. Mind map.

2 METHODOLOGY

2.1 *Qualitative analysis*

This study qualitatively analyzes the basic meaning of active housing in the context of architectural design. "Active" means to achieve a certain goal through self-control without relying on the external drive. Meanwhile, the opposite context of "passive" is defined as to achieve a certain goal must be driven by external forces to achieve. Otherwise, it will not be achieved (Eriksen & Song 2016). The residence is inseparable from people. Active residence does a good job of starting from the user's sense of experience, following the concept of people-oriented, taking comfort and health as the design goal, and can make more intelligent self-adjustment according to the changes of the environment through self-perception. It is more inclined to actively provide energy to the house through energy equipment, such as wind power and solar power (Ren 2021). However, passive houses mainly rely on the external environment and achieve energy saving through their construction practices while ignoring the building's own regulation and user's comfort.

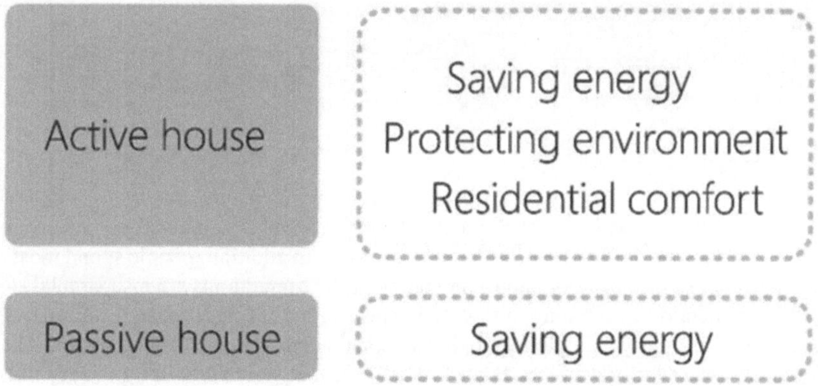

Figure 2. Qualitative analysis.

2.2 *Overview*

First, we focused on the green city by following up on the actual issues and narrowed the scope of our attention by referring to the relevant materials of a green city. In the end, we select the active house, one of the evaluation criteria of green building. Second, according to the research object selected in the first step—active residence, through relevant research reports, articles, and data were collected through websites such as CNKI to find the characteristics of the object and the existing pain points to determine the research direction. Finally, the most humanist characteristic of this object—indoor comfort is determined as the final research direction. By finding the factors that affect indoor comfort and analyzing foreign cases, summarizing and recombining the factors, this paper puts forward the methods to improve the related factors and analyzes the potential factors and far-reaching significance of active housing as new direction of future residential development. Finally, we carry on the dialectical thinking to the research content, carries on the deeper level imagination to put forward the prospect to the future.

3 MAIN BODY

3.1 *Overview*

To explore the possibility of active housing as the main form of residential development in the future, the project puts forward the idea that people should focus on the design of active housing from three aspects: ecological environment, energy consumption, and people-centered concern.

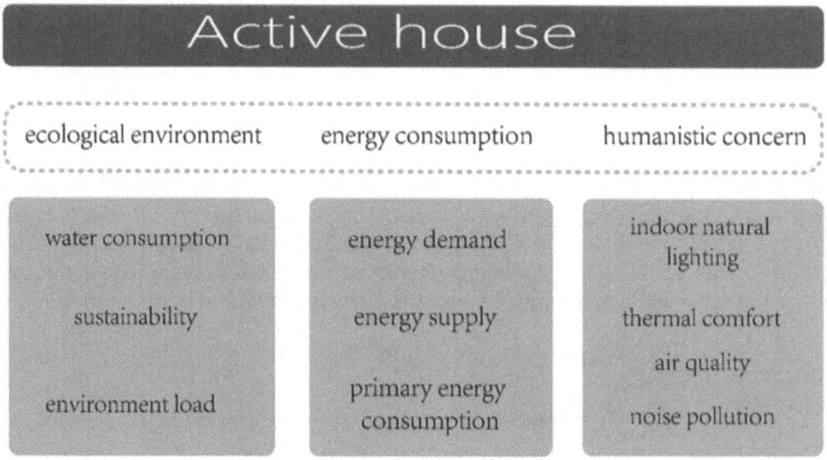

Figure 3. Three aspects of active house.

According to the survey, with the development of society, people's demand for comfort and health has increased, making health and comfort more of an issue. In addition, people spend most of their time in their homes, so the health of people and the "health" of buildings are inseparable. However, there is lack of attention to the human-oriented aspect in architectural design nowadays, which tends to ignore people's psychological and physiological needs. Therefore, this project selects the human-oriented aspect for in-depth research and believes that residential design needs to be more humanized from the perspective of human well-being. It is proposed that we should pay attention to four aspects: indoor natural lighting, thermal comfort, air quality, and noise pollution in the design.

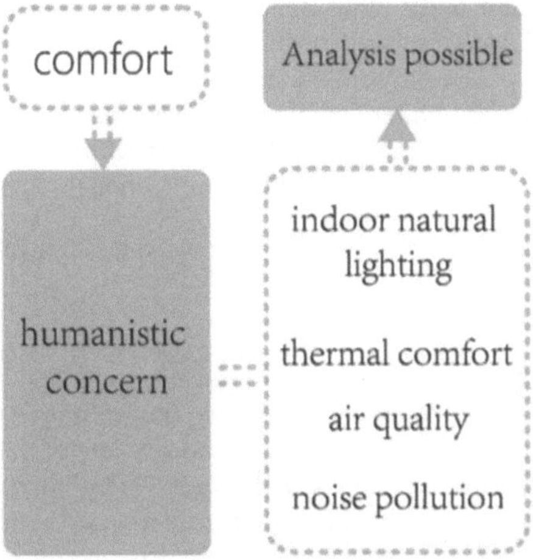

Figure 4. Analysis of humanistic concern.

3.2 *Natural lighting*

Indoor natural lighting is very important as it directly affects people's psychology and body health. The natural lighting of the residence is related to the space layout of the house, the window layout of the residence, and the selection of indoor partition materials. For example, the active building, the Willux Office Building, utilizes several of these aspects to ensure adequate and comfortable daylight in both summer and winter. The windows and shading of the building's facade are designed to absorb natural light in the winter and block excessive natural light in the summer. Inverted facades are designed to capture natural light more easily, and the office's transparent partition walls ensure plenty of light in the interior space.

Figure 5. Willux Office Building [3].

In terms of aspects of door space layout, one should choose the house that faces north and south by shortening the depth of the house as much as possible to ensure that the light enters the house. However, this study focuses on improving indoor natural lighting through window design and interior partition materials based on the existing building.

First, reflectors coated with highly reflective materials are used to reflect light into the room, increasing the indoor natural light and reducing glare (Wang 2019).

Meanwhile, the window shading design is also very important. It can be considered to add a shading system that can be automatically controlled inside to cope with the external environment in different seasons and different lighting conditions to ensure the right amount of natural lighting.

In addition, the partition material of the interior is also an important reason to affect the indoor natural lighting. Due to rapid changes in family structure, multi-generation residential buildings have become mainstream residential buildings. Therefore, many residential buildings are divided by adding or removing partition walls to change users' functional needs. The new type of light-transmitting material can replace the hard partition wall and play an important role in blocking and lighting to avoid forming a black room and prevent excessive building energy consumption. It not only can ensure the indoor lighting but also the privacy of the residence. Ground glass and toughened glass, or some other materials, are always used as partition material. It can increase indoor brightness through the wall and increase diffuse reflection through its uneven surface. It can make indoor light more natural and downy.

3.3 *Indoor thermal comfort*

Good ventilation is the most basic and efficient way to ensuring indoor thermal comfort. In addition, indoor thermal comfort is also related to heat recovery and utilization of buildings and building envelope structures (Wang et al. 2019). From the point of view of solving the problem of heat insulation in summer and heat insulation in winter, the active building adopts more efficient and energy-saving thermal insulation construction practices and new thermal insulation materials to optimize indoor thermal environment, reduce indoor heat loss, and reduce the energy consumption of the building (Zhou et al. 2009).

First is the heat recovery and utilization of buildings. The way of heat recovery can save resources and achieve the reuse of energy. The building can use the auxiliary heat source to adjust the indoor heat and absorb the heat for storage when the indoor heat is too much, and release it when required. For example, the active building Velux office building uses the ground source heat pump system to release the indoor heat into the soil in summer and release heat from the soil in winter to achieve summer cooling and winter heating (Zhang 2017).

Second, the active building envelope is one of the core technologies of this case. The thickened exterior walls provide complete separation between the interior and the external contact surfaces, reducing the impact of the external environment on the indoor temperature. In addition, the building envelope strategy also includes two ways. The first is constructing a double-layer hollow wall structure to reduce the impact of the external environment on indoor temperature. And then is to use the characteristics of wall materials to actively adjust the room temperature (Guo 2020).

3.4 *Indoor air quality*

To residents, the indoor air quality of the active house is as important as thermal comfort because it can affect cardiovascular issues or respiratory diseases. The factor is relevant to many specific issues, such as formation of mold or dampness (Feifer et al. 2018). These issues lower the user comfort and, in the long term, affect the occupants' health. Carbon-dioxide concentration, a part of the air quality, may even affect residents' brain (Bain & 1948). Therefore, active houses should provide good indoor air quality.

Table 1. Indoor air quality quantitative criteria and active house evaluation of standard fresh air supply (Feifer et al. 2018).

Area	Parameter	Criteria
Comfort	Standard fresh air supply	The fresh air supply shall be established according to the below limit values for indoor CO_2 concentration in living rooms, bedrooms, study rooms, and other rooms with people as the dominant source and that are occupied for prolonged periods: 500 ppm above outdoor CO_2 concentration 750 ppm above outdoor CO_2 concentration 1000 ppm above outdoor CO_2 concentration 1200 ppm above outdoor CO_2 concentration

Apart from CO2 concentration, some other factors impact the user's comfort in terms of air quality. A tool named IEQ-compass, which can holistically assess the indoor environment, both the building and the user, is considered.

Table 2. Parameters, criteria, and their weights for the assessment of acoustic IEQ (Tine Steen Larsen et al. 2020).

Parameter		Parameter weights	Criteria		Criterion weights
IAQ1	Impact of outdoor air	15%	1.1	Outdoor air quality and filtration	100%
IAQ2	Building ventilation and materials	35%	2.1a	Mechanical ventilation (and commissioning)	70%
			2.3	Emission from materials	30%
			2.1b	Natural ventilation (potential ACR)	35%
			2.2	Bathroom exhaust fan	35%
			2.3	Emission from materials	30%
IAQ3	Impact of household activities	30%	3.1	Options for drying clothes	3%
			3.2	Stove exhaust hood	5%
			3.3	Stove type (electricity or gas)	2%
IAQ4	Occupants' possibilities to adjust the IAQ	20%	4.1	Ventilation boost natural ventilation	35%
			4.2	Ventilation boost, mechanical ventilation	3%
			4.3	Automatic control of ventilation rate	35%

Since there have been several criteria that can effectively give researchers a view on the indoor air quality, it is feasible to set a requirement that should be satisfied by active houses.

Previous research under the northern climate shows that current houses have a high risk of reaching a high level of indoor CO2 concentration, affecting occupants' health, especially in the relatively hot seasons (Theofanis et al. 2021). Humidity can also be a problem for germs may thrive under high humidity circumstances, resulting in food spoilage or even contagion. The humidity problem is most severe in the kitchen, bathrooms, and washrooms, as these rooms are often the most humid in a house.

Limited by the energy efficiency, active houses often use passive strategies, or sometimes hybrid strategies, which combine passive ones with mechanic ones to cope with the air quality problem. This paper shows some examples of the applications. In the Federal University of Bahia, Brazil, the buildings applied floor ventilation grating and walls with an open structure that extend beyond the floor slap to favor the circulation. The windows also have gratings below an axis, enabling them to open to about 90°. Because of the hot and humid weather in that area, the design uses many techniques to improve the openness of the building. Instead of simply trying to lower the humidity level, it takes advantage of the wet air to take heat away from the building through natural ventilation, coping with thermal comfort and indoor air quality at the same time. In another experiment named active NetZero house, designers applied integrated apparatus to promote natural ventilation, an example of a hybrid strategy. In its design for ventilation during daytime in summer months (Ellis et al. 2017), like the former case, this building applies the open space concept as well. The orientation of the windows is properly designed to balance ventilation with heat capturing. The window utilizes the prevailing wind and adapts to the heating climate of the site by creating indoor heat circulation to increase airflow.

Research concerning the building design strategies in post-pandemic architecture offers a systematic idea (Qiao et al. 2020). The application of UV-based technology and biofiltration technology can efficiently prevent germs from infecting people.

3.5 *Noise pollution*

Noise pollution is now one of the main issues to be solved in modern cities due to health risks people are facing by excessive noise exposure (Wongeun Oh 2017). As sustainable architecture,

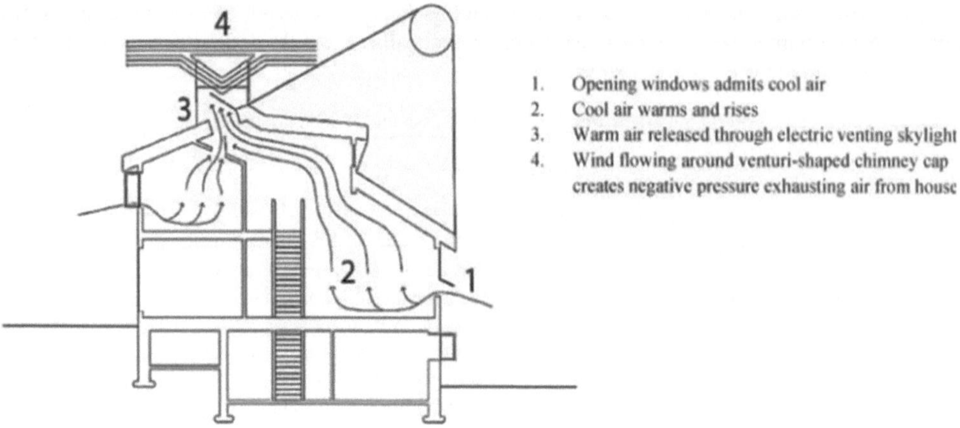

Figure 6. Flowchart of common solutions for enhancing IAQ through air purification and disinfection.

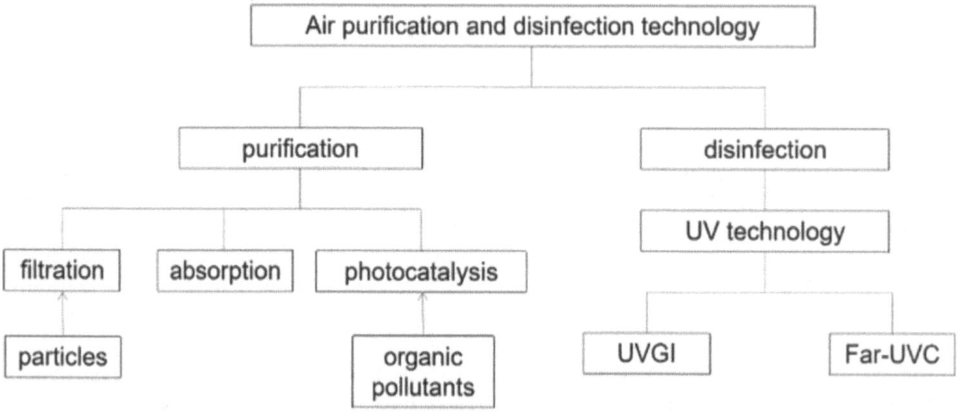

Figure 7. Flowchart of common solutions for enhancing IAQ through air purification and disinfection (Naglaa & Ehab 2021).

active houses should ensure residents' auditory health by mitigating noise. The AHA has announced a set of criteria in evaluating the houses' noise pollution level, and a system of lowering that level is being established.

In acoustic IEQ (ACOU), comprehensive factors from both the outside and the inside of the buildings are considered (Tine Steen Larsen et al. 2020). As can be seen in the table, noise from outside of the house is the most important factor influencing the acoustic IEQ. At the same time, occupants' possibility to adjust it shares minor weight, indicating that, unlike illumination or ventilation, which can be actively adjusted by the residents, ACOU adjustment relies heavily on the controlling noise system of the building itself.

One of the major problems concerning noise pollution control is that the sound figures are so variable that it is challenging to mitigate all types of noise.

One of the most commonly applied technologies to lower noise levels is the ANC (active noise control) system, which can produce reversed sound to counter the original noise. The current situation is that traditional ANC, such as Single-Channel Duct-Type Open-Window ones (Wongeun Oh, 2017), performs badly in mitigating low-frequency noise. It is most efficient under the frequency of 400 Hz to 1800 Hz. Another ANC window design that applies micro-perforated absorbers performs

Table 3. Parameters, criteria, and their weights for the assessment of acoustic IEQ (ACOU) (Tine Steen Larsen et al. 2020).

Parameter		Parameter weights	Criteria		Criterion weights
ACOU1	Noise from surroundings	35%	1.1	Low impact of external noise (e.g., traffic noise, industry	80%
			1.2	Possibility to open windows toward a silent side	20%
ACOU2	Noise from neighboring dwellings	35%	2.1	Low impact of noise from other dwellings – airborne noise	50%
			2.2	Low impact of noise from other dwellings – impact noise	50%
ACOU3	Noise from within the dwelling	25%	3.1	Technical installations	60%
			3.2	Reverberation time	40%
ACOU4	Occupants' possibilities to adjust the acoustic IEQ	5%	4.1	Possibility to open windows in multiple directions	100%

well when the sound is between 500 Hz and 800 Hz but makes little difference to noise with low frequency (Lam et al. 2020).

Fortunately, active control strategies against the low-frequency noise are being developed, and the primitive results are optimistic (Kang & Brocklesby 2005). In a previous study, researchers applied a top-hung window that is fully open and found out that the device outperformed other ANC apparatus using closed windows. Another experiment that put a natural ventilation unit over an ordinary window shows a relatively well effect on lowering noise pollution level under the frequency between 100 Hz and 400 Hz. (Natural Ventilation and Sound Insulation Windows). As material technology develops, the passive parts of the devices will perform better, so noise level of low frequency is highly possible to be well reduced in the future.

3.6 *Possibility analysis*

The above research demonstrates that the active residence has great advantages in environmental protection, energy saving, and user experience. It can not only cause a negative impact on the environment through the use of renewable energy but also can improve the comfort of the indoor environment to meet the needs of people's healthy lives.

However, it is not common now. According to the people-oriented concern, the feasibility analysis of active housing is carried out. Three aspects affect the feasibility: short development time with low popularity, insufficient high and new technology level, and high construction cost.

First, let's consider the short time of development and its imperfect situation. Compared with the passive house, because they are in the same pursuit of low energy consumption, it is the first step in meeting passive building standards (Eriksen & Song 2016). Therefore, the research on an active house is not in-depth, and its design method is difficult to effectively use by designers. It also lacks publicity among the public and is ignored by the public.

Second, the technology level not being high is also a major problem. What "active" really means is an integrated solution, a creative strategy that takes advantage of pure technology (Eriksen & Song 2016). To satisfy users' comfort needs, the design of active houses needs to pay special attention to natural lighting, air quality, and other aspects. The materials used in the early stage of construction and the implementation process need to be supported by research and the development of new technologies. For the continuous optimization in the later stage, innovative strategies are also needed to improve. Therefore, having high and new technologies is a necessary condition for the design of active houses. In addition, it is also a challenge to achieve energy conservation, environmental protection, and comfort while ensuring beautiful architectural design.

The final problem is the high cost of building active homes. High cost will lead to high housing prices, and high housing prices will bring cost-effectiveness to the society, which will reduce the performance of enterprises, distort managers' efforts, and then reduce their sensitivity to pay and perform (Liu & Zhuang 2021). For citizens, it will bring more life pressure and living burden to people. Therefore, even if the above two major problems are solved, the high cost of active housing using high technology is still a hurdle for promoting active housing.

4 CONCLUSION

Through analysis and description of the environmental protection, energy consumption, and people's concern for active building, the general advantages of environmental protection, low energy consumption, and the outstanding advantages of high comfort level of the active building are summarized. At the same time, this paper disclosed difficulties and problems that the active house still faces in the domestic promotion and puts forward feasible references. With the promotion of the advantages of active building, continuous improvement in new technology in the construction industry, and the constant coordination between cost and new technology, the active house is likely to become the main form of residential development in the future. The relationship between the city, architecture, nature, and humans has always been contradictory, which has always been the main problem that designers want to solve. It is hoped that in the future development of the construction industry, the boundary between architecture and nature can be blurred, and the relationship between humans, architecture, and nature can be explored. Therefore, they can coexist harmoniously, achieve urban low-carbon, and accelerate the development of green cities. This study is universal and can be read not only by architectural designers and designers of related fields but also by ecologists who care about ecological protection, investors concerned about housing development trends, and everyone who needs a comfortable home.

REFERENCES

Architectural Research Centre of Shanghai Institute. (2018) Active Building – Adaptive Development of Green Building Evaluation Criteria. Unpublished intranet document, Architectural Research Centre of Shanghai Institute.

B. Lam; D.Y. Shi; V. Belyi; S.L. Wen; W.S. Gan; Kelvin L.; Irene L. (2020) Active Control of Low-Frequency Noise through a Single Top-Hung Window in a Full-Sized Room, *Applied Sciences*, Vol.10, No.6817, ISSN 2076-3417.

C.Z. Qiao, Kai Ji, Zhen Zhang. (2020) *Research progress of indoor air purification technology*, School of Civil Engineering, North China University of Technology, Beijing, 100144, China.

Eriksen K.E., Y.H. Song. (2016) Interviews: Architects and users of sustainable design. *Journal of Architecture*, no.05:113-117.

Feifer, L., Imperadori, M., Salvalai, G., Brambilla, A., Brunone, F. (2018) Active House: Smart nearly zero energy buildings, ISBN 978-3-319-90814-4.

Guo J.N. (2020) Study on thermal comfort optimization design of rural residential for the aged in Rizhao area. Unpublished internal document, Shandong University of Architecture.

J. Ellis, J. Schwartz & R. Mora. (2017) Assessment of natural ventilation effectiveness foran active NetZero energy house, *International Journal of Ventilation*, 16:2, 134–151.

J. Kang; M.W. Brocklesby. (2005) Feasibility of applying micro-perforated absorbers in acoustic window systems, *Applied Acoustics*, Volume 66, Issue 6: 669–689.

JA Bain, JR Klein. (1948) Effect of carbon dioxide concentration on brain lactate.

Liu X.H., X.H. Zhuang. (2021) Real estate price, cost effect and executive. *Economic Problems*, no.02:124-128

Naglaa A.M., Ehab M.G. (2021) Indoor Air Quality: Rethinking rules of building design strategies in post-pandemic architecture, *Environmental Research*, Volume 193, February 2021, 110471

Ren Z.N. (2021) the application of passive energy saving in building design. Smart Buildings and Smart Cities, no.02:86-92

The Central People's Government of the People's Republic of China, 2014-2020. National New-Type Urban Planning. Beijing: CPC Central Committee and the State Council

Theofanis P., Despoina T., Sarka L., Paula W., Pawel W. Indoor humidity of dwellings and association with building characteristics, behaviours and health in a northern climate, *Building and Environment*, Volume 198, July 2021, 107885

Tine Steen Larsen, Lasse Rohde, Kim Trangbæk Jønsson, Birgit Rasmusse, Rasmus Lund Jensen, Henrik N. Knudsen, Thomas Witterseh, Gabriel Bek€o. (2020) IEQ-Compass – A tool for holistic evaluation of potential indoor environmental quality, *Building and Environment*, Volume 172, April 2020, 106707

Wang S.Y., Y.Y.Wang. (2019) Optimization of natural lighting and energy saving for rural residential buildings in eastern southern Hebei Province. Energy Conservation in the Building Industry, no.09:36-37

Wang W.J., T.Y. Fang, Ding R., S.G. Zhu. (2019) Study on Indoor Thermal Environment and Energy Saving Optimization of Existing Residential Buildings in Hot Summer and Cold Winter Area. *Journal of West Wanxi University*, 35(05):91-96

Wongeun Oh. 2017. Open-Window Active Noise Control for Environmental Noise, Suncheon National University, Korea. ISBN 978-93-84422-76-9

Zhang B.X. (2017) Study on the suitability of active buildings. Unpublished internal document, Shandong University of Architecture.

Zhou S.P., Tao Jing, Y.G. Ren. (2009) Qualitative and quantitative exploratory analytical modelling. *Computer Simulation*, 26(10):87-89

Advances in Civil Engineering: Structural Seismic Resistance, Monitoring and Detection – Mohd Johari Mohd Yusof and Junwen Zhang (Eds)
© 2023 The Authors, ISBN 978-1-032-31491-4

Risk assessment of construction safety of sea crossing cable-stayed bridge based on AHP

Lei Chen
Research Institute of Highway MOT, Beijing, China

ABSTRACT: In order to strengthen the construction safety management of highway bridge and tunnel engineering, optimize the construction organization and improve the effectiveness of safety pre-control on the construction site, the Ministry of Transport issued the guideline for safety risk assessment of highway bridge and tunnel engineering construction in 2011. The bridge engineering in this guideline is mainly aimed at land bridges [1-3]. Combined with the actual situation of the project, this paper enriches the indicators related to sea area construction safety, and uses analytic hierarchy process for analysis. The research results have reference significance for the risk pre assessment of similar projects.

1 INTRODUCTION

Affected by the complex offshore construction environment such as tide, wind, wave, strong light, seabed quicksand and marine traffic, offshore construction is difficult and requires high safety management. Offshore construction includes offshore platform, offshore trestle, offshore floating crane hoisting, offshore transportation, offshore mixing plant, etc.

Based on the weighted average method a fuzzy analytic hierarchy process (FAHP) for bridge construction safety risk assessment is constructed by ZHANG Yin-tao (2019); Lin Jin-teng and Chen Chang-song consider the combination of analytic hierarchy process can more accurately and comprehensively obtain the overall level of construction risk of cable-stayed bridge (Chen 2017; Lin 2019).

The index system method is recommended in the published guidelines for construction safety risk assessment. The relevant indicators in this method are mainly for land bridge construction, and lack of relevant indicators such as meteorology, tide, tidal range, anchorage and navigation grade with sea area characteristics. Combined with the construction safety impact analysis of sea crossing cable-stayed bridge, this paper studies and establishes the construction safety risk assessment index system of sea area cable-stayed bridge.

2 BASIC INFORMATION OF THE PROJECT

The bridge is a three tower steel box girder cable-stayed bridge (550m), and the substructure adopts diamond cable tower, wall pier and bored pile foundation. The upper part is covered with muddy clay and silty clay, and bedrock is crystalline tuff. The bridge crosses the channel and passes through 50000 ton oil tanker, 40000 ton general cargo ship and 30000 ton container. The sea fog day is about 30 days in the bridge project. The project area is affected by tropical cyclones, with an average of 2.2 per year. The tidal type in the project area belongs to irregular semi diurnal shallow water tidal type, and the diurnal inequality of tides is significant. In the construction sea area, there are important optical cables, drinking water pipes, multi terminal flexible DC-cables and skew of bridge main line, which greatly interfere with the construction of bridge foundation, limit the anchoring of construction ships, and have great potential safety hazards.

 DOI 10.1201/9781003310884-94

3 ESTABLISH RISK FACTOR SET

3.1 *Marine environment*

The construction index system of marine environment is established mainly according to the impact of factors such as whether the bridge site is a navigable area and anchorage on construction safety.

Table 1. Marine environment index system.

Risk category		Grading index	Recommended value
Marine environment A	Grading of waterway A1	I ~ II grade	$75 < R_{ij} \le 100$
		III ~ IV grade	$50 < R_{ij} \le 75$
		V ~ VI grade	$25 < R_{ij} \le 50$
		VII grade and beyond	$0 < R_{ij} \le 25$
	Shelter anchorage A2	The wind shelter anchorage is poor and far from the construction area	$75 < R_{ij} \le 100$
		The wind shelter anchorage is good and far from the construction area	$50 < R_{ij} \le 75$
		The wind shelter anchorage is poor and close to the construction area	$25 < R_{ij} \le 50$
		The wind shelter anchorage is good and close to the construction area	$0 < R_{ij} \le 25$
		greater than 6m	$75 < R_{ij} \le 100$
	Tidal range A3	4.5-6m	$50 < R_{ij} \le 75$
		3-4.5m	$25 < R_{ij} \le 50$
	range A3	less than 3m	$0 < R_{ij} \le 25$

3.2 *Meteorological environment*

Mainly consider disastrous weather, such as typhoon, squall line, fog and other bad weather, and establish an index system.

Table 2. Meteorological environment index system.

Risk category		Grading index	Recommended value
Meteorological environment B	Monsoon B1	greater than 60d	$75 < R_{ij} \le 100$
		40 < 60d	$50 < R_{ij} \le 75$
		20 < 40d	$25 < R_{ij} \le 50$
		less than 20d	$0 < R_{ij} \le 25$
	Annual average typhoon B2	greater than 3 times	$75 < R_{ij} \le 100$
		2 < 3 times	$50 < R_{ij} \le 75$
		1 < 2 times	$25 < R_{ij} \le 50$
		less than 1time	$0 < R_{ij} \le 25$
	Foggy days B3	greater than 50d	$75 < R_{ij} \le 100$
		30-50d	$50 < R_{ij} \le 75$
		15-30d	$25 < R_{ij} \le 50$
		less than 15d	$0 < R_{ij} \le 25$
	Annual rainfall B4	greater than 1600mm	$75 < R_{ij} \le 100$
		1000 < 1600mm	$50 < R_{ij} \le 75$
		500 < 1000mm	$25 < R_{ij} \le 50$
		less than 500mm	$0 < R_{ij} \le 25$

3.3 *Construction scale of cable-stayed bridge*

According to the height span ratio and side mid span ratio specified in 4.2.4 of *Guidelines for Design of Highway Cable-Stayed Bridge* (JTG/T D65-01-2007), the index system of cable-stayed bridge construction scale is established.

Table 3. Construction scale index system of cable-stayed bridge.

Risk category		Grading index	Recommended value
Construction scale index system of cable-stayed bridge C	Maximum single hole span C1	Exceeding or reaching similar bridge types at home and abroad	$75 < R_{ij} \leq 100$
		Lk≥800m	$50 < R_{ij} \leq 75$
		400m≤Lk < 800m	$25 < R_{ij} \leq 50$
		Lk < 400m	$0 < R_{ij} \leq 25$
	Tower height C2	H≥300m	$75 < R_{ij} \leq 100$
		200m≤10 < 300m	$50 < R_{ij} \leq 75$
		100m≤10 < 200m	$25 < R_{ij} \leq 50$
		H < 100m	$0 < R_{ij} \leq 25$
	Total length of bridge C3	Exceeding or reaching the maximum total length of similar bridge types at home and abroad	$75 < R_{ij} \leq 100$
		L≥2000m	$50 < R_{ij} \leq 75$
		800m≤L < 2000m	$25 < R_{ij} \leq 50$
		L < 800m	$0 < R_{ij} \leq 25$

3.4 *Pile group construction on water*

Comprehensively consider the impact of sea area geological conditions and hydrological conditions on construction safety, and establish an index system.

Table 4. Construction index system of pile groups on water.

Risk category		Grading index	Recommended value
Construction index system of pile groups on water D	Pile length D1	L≥100m	$75 < R_{ij} \leq 100$
		60m≤L < 100m	$50 < R_{ij} \leq 75$
		30m≤L < 60m	$25 < R_{ij} \leq 50$
		L < 30m	$0 < R_{ij} \leq 25$
	Geological conditions D2	Areas prone to adverse geological disasters	$75 < R_{ij} \leq 100$
		Soft soil stratum or gravel stratum, affecting construction safety and progress	$50 < R_{ij} \leq 75$
		There are unfavorable geology, but it does not occur frequently or there are special rocks and soils, which affect the construction safety and progress	$25 < R_{ij} \leq 50$
		The geological conditions are good, which basically does not affect the construction safety and progress	$0 < R_{ij} \leq 25$
	Hydrologic condition D3	Poor hydrological conditions, large scouring and large water level change	$50 < R_{ij} \leq 100$
		The hydrological conditions are good and the scouring is small, which basically has no impact on the construction safety	$0 < R_{ij} \leq 50$

(*continued*)

Table 4. Continued

Risk category		Grading index	Recommended value
	Construction period D4	Experience typhoon season, flood season, high temperature	$50 < R_{ij} \leq 100$
		Suitable construction period	$0 < R_{ij} \leq 50$
	Drilling method D5	The selected drilling equipment does not adapt to geology, hydrology and drilling platform	$50 < R_{ij} \leq 100$
		The selected drilling equipment is suitable for geology, hydrology and drilling platform	$0 < R_{ij} \leq 50$
	Construction area D6	at sea	$50 < R_{ij} \leq 100$
		tidal-flat are	$0 < R_{ij} \leq 50$
	Temporary structure design D7	Design scheme based on previous experience	$50 < R_{ij} \leq 100$
		Professional design verification scheme shall be adopted and made by enterprises with relevant qualifications	$0 < R_{ij} \leq 50$
	Oil and gas pipeline D8	The steel pipe pile and pile foundation of the operation platform are within 5 ~ 50m of the oil and gas pipeline	$50 < R_{ij} \leq 100$
		The steel pipe pile and pile foundation of the operation platform are within 50m of the oil and gas pipeline	$0 < R_{ij} \leq 50$
	Other uncertainties D9	Other uncertain factors due to different construction conditions	$0 < R_{ij} \leq 100$

3.5 *Cofferdam construction*

Comprehensively consider the impact of cofferdam design, tidal range and navigation environment on cofferdam construction safety, and establish an index system.

Table 5. Cofferdam construction index system.

Risk category		Grading index	Recommended value
Cofferdam construction index system E	Special design of steel cofferdam E1	Design scheme based on previous experience	$75 < R_{ij} \leq 100$
		Special design and safety check calculation were carried out, but the opinions of the original bridge design unit were not solicited and the corresponding design qualification unit was entrusted to issue the review and check calculation report	$50 < R_{ij} \leq 75$
		Carry out special design, carry out safety check calculation, solicit the opinions of the original bridge design unit, and entrust the corresponding design qualification unit to issue the review and check calculation report	$0 < R_{ij} \leq 50$
	Grading of waterway E2	I ~ II grade	$75 < R_{ij} \leq 100$
		III ~ IV grade	$50 < R_{ij} \leq 75$
		V ~ VI grade	$25 < R_{ij} \leq 50$
		VII grade and beyond	$0 < R_{ij} \leq 25$

(continued)

Table 5. Continued

Risk category	Grading index	Recommended value
Geological conditions E3	Areas prone to adverse geological disasters	$75 < R_{ij} \leq 100$
	Soft soil stratum or gravel stratum, affecting construction safety and progress	$50 < R_{ij} \leq 75$
	There are unfavorable geology, but it does not occur frequently or there are special rocks and soils, which affect the construction safety and progress	$25 < R_{ij} \leq 50$
	The geological conditions are good, which basically does not affect the construction safety and progress	$0 < R_{ij} \leq 25$
Tidal range E4	greater than 6m	$75 < R_{ij} \leq 100$
	4.5-6m	$50 < R_{ij} \leq 75$
	3-4.5m	$25 < R_{ij} \leq 50$
	less than 3m	$0 < R_{ij} \leq 25$
Tide E5	greater than 2.0m/s	$75 < R_{ij} \leq 100$
	1.0-2.0m/s	$50 < R_{ij} \leq 75$
	1.0-2.0m/s	$25 < R_{ij} \leq 50$
	less than 0.6m/s	$0 < R_{ij} \leq 25$
Climatic and environmental conditions E6	Extreme weather (strong wind, rainstorm, lightning strike) event prone area	$50 < R_{ij} \leq 100$
	The climate and environmental conditions are general and have no significant impact on construction safety	$0 < R_{ij} \leq 50$
Construction period E7	Experience typhoon season, flood season, high temperature	$50 < R_{ij} \leq 100$
	Suitable construction period	$0 < R_{ij} \leq 50$
Construction area E8	at sea	$50 < R_{ij} \leq 100$
	tidal-flat are	$0 < R_{ij} \leq 50$
Assemble and sink construction E9	Gantry crane	$50 < R_{ij} \leq 100$
	Floating crane	$0 < R_{ij} \leq 50$
Other uncertainties E10	Other uncertain factors due to different construction conditions	$0 < R_{ij} \leq 100$

3.6 *Pier (tower) column construction*

The evaluation index system is mainly established based on the collapse accident of concrete engineering.

Table 6. Construction index system of Pier (tower) column.

Risk category		Grading index	Recommended value
Construction index system of Pier (tower) column F	Height F1	$H \geq 80m$	$75 < R_{ij} \leq 100$
		$40m \leq H < 80m$	$50 < R_{ij} \leq 75$
		$15m \leq H < 40m$	$25 < R_{ij} \leq 50$
		$H < 15m$	$0 < R_{ij} \leq 25$
	Construction period F2	Experience typhoon season, flood season, high temperature	$75 < R_{ij} \leq 100$
		Suitable construction period	$50 < R_{ij} \leq 75$

(continued)

Table 6. Continued

Risk category		Grading index	Recommended value
	Wind conditions F3	greater than 60d	$75 < R_{ij} \leq 100$
		40 < 60d	$50 < R_{ij} \leq 75$
		20 < 40d	$25 < R_{ij} \leq 50$
		less than 20d	$0 < R_{ij} \leq 25$
	Climatic and environmental conditions F4	Extreme weather (strong wind, rainstorm, lightning strike) event prone area	$50 < R_{ij} \leq 100$
		The climate and environmental conditions are general and have no significant impact on construction safety	$0 < R_{ij} \leq 50$
	Template form F5	Support formwork method	$75 < R_{ij} \leq 100$
		Formwork turnover construction	$50 < R_{ij} \leq 75$
		Climbing formwork construction	$0 < R_{ij} \leq 50$
	Upper and lower channels F6	No upper and lower channels or simple channels	$75 < R_{ij} \leq 100$
		Scaffold erection of upper and lower passages	$50 < R_{ij} \leq 75$
		Qualified products or professional design, and safety checking calculation	$0 < R_{ij} \leq 50$
	Other uncertainties F7	Other uncertain factors due to different construction conditions	$0 < R_{ij} \leq 100$

3.7 *Cantilever assembly construction*

The index system shall be established by comprehensively considering meteorological conditions, construction equipment and other factors.

Table 7. Cantilever assembly construction index system.

Risk category		Grading index	Recommended value
Cantilever assembly construction index system G	Spreader anchorage G1	Adopt empirical design scheme	$75 < R_{ij} \leq 100$
		Adopt professional design verification scheme	$50 < R_{ij} \leq 75$
		Adopt qualified and reliable products	$0 < R_{ij} \leq 50$
	Hoisting mode G2	Hoisting with winch	$50 < R_{ij} \leq 100$
		Floating hoisting is adopted	$0 < R_{ij} \leq 50$
	Wind conditions G3	greater than 60d	$75 < R_{ij} \leq 100$
		40 < 60d	$50 < R_{ij} \leq 75$
		20 < 40d	$25 < R_{ij} \leq 50$
		less than 20d	$0 < R_{ij} \leq 25$
	Annual average typhoon G4	greater than 3 times	$75 < R_{ij} \leq 100$
		2 < 3 times	$50 < R_{ij} \leq 75$
		1 < 2 times	$25 < R_{ij} \leq 50$
		less than 1 time	$0 < R_{ij} \leq 25$
	Grading of waterway G5	I ~ II grade	$75 < R_{ij} \leq 100$
		III ~ IV grade	$50 < R_{ij} \leq 75$
		V ~ VI grade	$25 < R_{ij} \leq 50$
		VII grade and beyond	$0 < R_{ij} \leq 25$
	Construction area G6	at sea	$50 < R_{ij} \leq 100$
		tidal-flat are	$0 < R_{ij} \leq 50$
	Other uncertainties G7	Other uncertain factors due to different construction conditions	$0 < R_{ij} \leq 100$

4 ESTABLISH ANALYTIC HIERARCHY PROCESS STRUCTURE MODEL

4.1 *Importance ranking and weight of middle tier*

According to the influence degree of the middle layer index on the construction safety of the sea crossing bridge, the importance of the index system is sorted by using the expert investigation method to determine the weight γ.

Table 8. Importance ranking and weight of middle tier.

Serial number	middle tier	Importance ranking	Weight γ
1	Marine environment A	1	0.27
2	Meteorological environment B	3	0.18
3	Construction scale of cable-stayed bridge C	2	0.22
4	Pile group construction on water D	4	0.14
5	Cofferdam construction E	5	0.10
6	Pier (tower) column construction F	5	0.06
7	Cantilever assembly construction G	7	0.03

4.2 *Importance ranking and weight of the lowest layer*

According to the influence degree of the lowest level index on the construction safety of sea crossing bridge, the importance of the index system is sorted by expert investigation method to determine the weight.

Table 9. Importance ranking and weight of the lowest layer of marine environment.

Serial number	middle tier	lowest layer	Importance ranking	Weight γ
1	marine environment A	Grading of waterway A1	1	0.56
2		Shelter anchorage A2	3	0.11
3		Tidal range A3	2	0.33

Table 10. Importance ranking and weight of the lowest layer of meteorological environment.

Serial number	middle tier	lowest layer	Importance ranking	Weight γ
1	Meteorological environment B	Monsoon B1	1	0.44
2		Annual average typhoon B2	3	0.19
3		Foggy days B3	2	0.31
4		Annual rainfall B4	4	0.06

Table 11. Importance ranking and weight of the lowest layer of cable-stayed bridge.

Serial number	middle tier	lowest layer	Importance ranking	Weight γ
1	Construction scale index system of cable-stayed bridge C	Maximum single hole span C1	1	0.56
2		Tower height C2	2	0.33
3		Total length of bridge C3	3	0.11

Table 12. Importance ranking and weight of the lowest layer of pile groups on water.

Serial number	middle tier	lowest layer	Importance ranking	Weight γ
1	Construction index system of pile groups on water D	Pile length D1	8	0.03
2		Geological conditions D2	1	0.21
3		Hydrologic condition D3	3	0.16
4		Construction period D4	7	0.06
5		Drilling method D5	6	0.09
6		Construction area D6	2	0.19
7		Temporary structure design D7	5	0.11
8		Oil and gas pipeline D8	4	0.14
9		Other uncertainties D9	9	0.01

Table 13. Importance ranking and weight of the lowest layer of cofferdam construction.

Serial number	middle tier	lowest layer	Importance ranking	Weight γ
1	Cofferdam construction index system E	Special design of steel cofferdam E1	5	0.11
2		Grading of waterway E2	1	0.19
3		Geological conditions E3	6	0.09
4		Tidal range E4	2	0.17
5		Tide E5	3	0.15
6		Climatic and environmental conditions E6	7	0.07
7		Construction period E7	8	0.05
8		Construction area E8	4	0.13
9		Assemble and sink construction E9	9	0.03
10		Other uncertainties E10	10	0.01

5 RISK ASSESSMENT OF CONSTRUCTION SAFETY OF SEA CROSSING CABLE-STAYED BRIDGE

5.1 *Establishment of evaluation model*

Assessment model of construction safety accident possibility of sea crossing cable-stayed bridge: $R=\{A_i, B_i, C_i, D_i, E_i, F_i, G_i\}\cdot\gamma$, A_i-G_i value shall be taken in combination with the actual situation of the project.

Table 14. Importance ranking and weight of the lowest layer of pier (tower) column.

Serial number	middle tier	lowest layer	Importance ranking	Weight γ
1	Construction index system of Pier (tower) column F	Height F1	1	0.27
2		Construction period F2	3	0.18
3		Wind conditions F3	4	0.14
4		Climatic and environmental conditions F4	5	0.10
5		Template form F5	2	0.22
6		Upper and lower channels F6	6	0.06
7		Other uncertainties F7	7	0.03

Table 15. Importance ranking and weight of the lowest layer of cantilever assembly.

Serial number	middle tier	lowest layer	Importance ranking	Weight γ
1	Cantilever assembly construction index system G	Spreader anchorage G1	3	0.18
2		Hoisting mode G2	2	0.22
3		Wind conditions G3	5	0.10
4		Annual average typhoon G4	4	0.14
5		Grading of waterway G5	1	0.27
6		Construction area G6	6	0.06
7		Other uncertainties G7	7	0.03

5.2 *Estimation of accident possibility*

According to the accident possibility assessment model (Cai 2008, Liu 2008), $R=\{A_i, B_i, C_i, D_i, E_i, F_i, G_i\}\cdot\gamma=0.27\cdot\Sigma A\cdot\gamma+0.18\cdot\Sigma B\cdot\gamma+0.22\cdot\Sigma C\cdot\gamma+0.14\cdot\Sigma D\cdot\gamma+0.1\cdot\Sigma E\cdot\gamma+0.06\cdot\Sigma F\cdot\gamma+0.03\cdot\Sigma G\cdot\gamma=54.7$.

Among them, in the sea area construction environment, the relative score of navigation conditions is high (high construction risk). Therefore, it is recommended to take necessary navigation closure measures for key construction processes to ensure the safety of bridge construction.

6 CONCLUSION

(1) According to the construction characteristics of sea crossing bridge, this paper establishes the risk index system of sea crossing bridge; The analytic hierarchy process is used for comprehensive judgment. The calculation method can more accurately and comprehensively couple various influencing factors, and provide reference for construction safety risk assessment.

(2) Although the pre evaluation method based on analytic hierarchy process is proposed in this paper, and the assignment is mainly combined with the actual situation of the site, the actual factors and characteristics affecting the bridge engineering are different. In the use of this method, the pre-evaluation shall be carried out in combination with the actual situation of the bridge project, on the basis of improving the index factors affecting the sea crossing bridge, and in combination with the field data as far as possible.

REFERENCES

Chang-song Chen, Yi Wei. (2017) Application of fuzzy comprehensive evaluation method in construction risk assessment of cable-stayed bridges [J]. *Traffic science and engineering*, V33; No.126(03):43–49.

Jing-liang Liu, etc. (2008) *Safety Management*[M]. Beijing: Chemical Industry Press.

Jin-teng Lin. (2019) Construction risk assessment of cable-stayed bridges based on grey fuzzy analytic hierarchy process theory [J]. *Urban roads and bridges and flood control*, No.244(08):178–182+23.

Yin-tao Zhang, Fang-quan Zhang. (2019) Research on construction safety risk assessment of steel truss cable-stayed bridge based on FAHP [J]. *Northern transportation*, No.310(02):1–4+8.

Zhuang-hong Cai, Chong-xi He, etc. (2008) *Safety Assessment Technology*[M]. Beijing: Chemical Industry Press.

Advances in Civil Engineering: Structural Seismic Resistance, Monitoring and Detection – Mohd Johari Mohd Yusof and Junwen Zhang (Eds)
© 2023 The Authors, ISBN 978-1-032-31491-4

Safety evaluation method of cascade dams system based on fuzzy extension hierarchy analysis

Lifu Xu & Qianyun Hu
Changjiang Survey, Planning, Design and Research Co., Ltd. Wuhan, Hubei, China

Jiafeng Shi*
College of Water Conservancy and Hydropower Engineering, Hohai University, Nanjing, Jiangsu, China

ABSTRACT: The safety evaluation of cascade dams system is of great significance to ensure operation of cascade dams and the safety of surrounding areas. To improve the accuracy of safety evaluation, aiming at the characteristics of complex relationships, large scale and uncertain evaluation of cascade dams system, the safety evaluation index system of cascade dams system is established by using fuzzy theory and extension analytic hierarchy process (EAHP), combined with the engineering safety of dams. The extension interval judgment matrix is established by expert scoring, and the weights of factors at all levels are determined. The fuzzy membership function is used to fuzzify the expert scoring, and the final safety evaluation results are obtained by comprehensive calculation. The safety evaluation of engineering examples verifies the scientific authenticity and accuracy of the application of fuzzy extension analytic hierarchy process in the evaluation of cascade dams.

1 INTRODUCTION

Since the 1980s, China began building cascade dams on a large scale, and has become the country with the largest number of cascade dams in the world (Xiong, Ren et al. 2015). The cascade development is characterized by cascade rolling of the river basin, which greatly improves the utilization rate of hydropower resources. However, while improving the comprehensive operation efficiency, the safety risk of the dams system cannot be ignored. Due to the interconnected influence of cascade dams in the basin, the system scale is large and the influencing factors are complex. Once dam failure is detected, the consequences can be unimaginable. It may have a serious impact on the environment, economy, and population of the surrounding areas. The system safety evaluation of cascade dams system can control the operation state of dams and provide a basis for safety management to carry out reasonable command and dispatch. There are few studies on the safety evaluation of cascade dams in China. Due to the large subjectivity, the traditional analytic hierarchy process cannot objectively reflect its safety status. Some scholars use fuzzy comprehensive evaluation method, entropy weight method, grey correlation degree, and other methods to optimize this (Jiang et al. 2020; Yang 2013); however, there are still problems of insufficient qualitative and quantitative considerations, hence necessitating further research. In this paper, combined with the extension analytic hierarchy process (He et al. 2008; Wang et al. 2010) and fuzzy comprehensive evaluation (Gao & Sheng 2002; Liu et al. 2007), the cascade dams are taken as the research object. From the perspective of dams engineering safety, the safety evaluation index of cascade dams system is coupled to improve the accuracy of safety evaluation of cascade dams system.

*Corresponding Author: 958764170@qq.com

DOI 10.1201/9781003310884-95

2 METHOD

The traditional reservoir dam safety evaluation method is mainly a qualitative method, which has disadvantages of strong subjectivity and large limitations. At the same time, the dam system of cascade reservoirs has large scale, influencing factors and complex structural relationship. Its safety evaluation index has the characteristics of both qualitative and quantitative factors and has strong uncertainty, which is not suitable for the safety evaluation by traditional methods. In this paper, using extension and fuzzy theory, fuzzy extension analytic hierarchy process (FAHP) is proposed to evaluate the safety of cascade reservoirs and dams. This method includes the characteristics of refinement, quantification, and mathematicalization, and thus improves the accuracy of cascade system safety evaluation (Chen & Zeng 2017; Liu et al. 2021; Wang et al. 2020).

2.1 *Building hierarchical structure model*

The engineering safety evaluation of cascade dams is a multi-level and multi-factor comprehensive evaluation problem. According to the location of the dam in the cascade dams system, from top to bottom, dams are classified as dam 1, dam 2, and dam n.

According to the "Reservoir and Dam Safety Evaluation Guidelines" compiled and released by the Ministry of Water Resources of 2017 (Ministry of Water Resources of China 2017). The engineering safety of dams in cascade dams can be divided into seven aspects, namely engineering quality evaluation, flood control standard review, structural safety evaluation, seepage safety evaluation, seismic safety review, metal structure safety, and operation management evaluation; basically the criterion layer in the system. The seven aspects themselves include a subset of their influencing factors, which can be further subdivided. For example, the engineering quality evaluation includes three parts: engineering hydrogeological conditions, engineering construction quality, and dam foundation bank treatment. As shown in Figure 1, this level is called the index layer.

Since many factors affect the weight of the dam in the system, the importance of different dam individuals is not convenient for pairwise comparison. Therefore, five factors that have great influence on the weight of the dam, namely dam type, dam height, reservoir capacity, age of dam, and engineering level, are selected to establish the hierarchical relationship, and the weight of each dam in the system is obtained by the extension hierarchy method.

2.2 *Evaluation Steps*

Step 1: Determine the weight of each dam

In the cascade dams system, the importance of each dam is difficult to reflect directly. As shown in Figure 1, this paper selects five factors "dam type, dam height, storage capacity, dam age, project level", which have a great influence on the dam as the index, and determines the weight of each dam by extension analytic hierarchy process.

Experts compare the importance degree of five indexes leading to dam failure, establish an extension judgment matrix, calculate the relative weight between indexes, and obtain the weight value of each factor after single value processing. Then, the extension judgment matrix for the mutual comparison of the safety of each dam under different factors in the cascade system is established, and its weight vector is obtained. Combined with the single factor weight, the safety evaluation weight value of each dam in the cascade system is obtained.

Step 2: Taking the engineering safety of cascade dams as the standard, the experts compare the importance of each index at the same level to obtain the extension judgment matrix at different levels, and calculate the weight value of each index after normalization.

Step 3: Experts evaluate and score the 18 engineering safety factor indexes of each dam in the dams system through on-site inspection, tests, and calculation reviews.

Step 4: The expert score is fuzzified to obtain the fuzzy membership vector of each index.

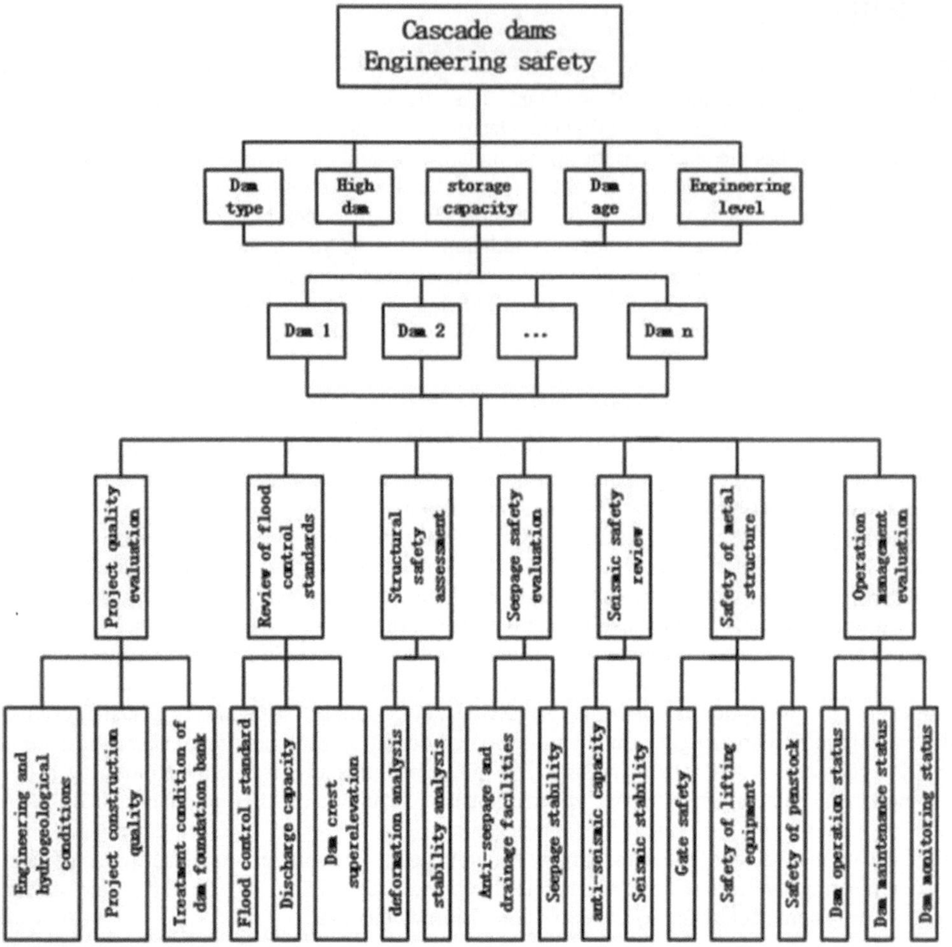

Figure 1. Safety evaluation index system of cascade dams system.

Step 5: The weight value of each index at the same level is coupled with the fuzzy membership vector calculation, and the obtained is the safety evaluation vector of the index in the previous level criterion layer.

Step 6: Repeat steps 3–5 according to the hierarchical relationship of the engineering safety evaluation index system of cascade dams until the fuzzy membership matrix of each dam safety evaluation is obtained.

Step 7: The weight value of each dam in the dams system is coupled with the calculation of fuzzy membership vector to obtain the engineering safety evaluation vector of the dams.

Step 8: According to the principle of maximum membership degree and the standard of safety evaluation table, the engineering safety evaluation level of cascade dams is obtained.

3 PROJECT PROFILE

Three reservoir dams, A, B, and C, were built along the river in a certain basin. Dam A is located in the eastern tributary of the basin. Its construction started in 1952 and completed in 1954. It is a comprehensive water conservancy project with flood control as the main part and irrigation, power generation, and tourism as a whole. It has a total reservoir capacity of 496 million m^3, flood control standards for the millennium. The dam is a concrete multi-arch dam with 20 dam stacks, 21 arches, and gravity dams at both ends. The maximum dam height is 75.9 m, the dam crest elevation is 129.96 m, and the installed capacity of the hydropower station is 310,000 kW. The dam has a very important geographical location, 17 km from the county, 60 km from the city, 65 km downstream roads, railways, high-speed rail, and other important civil and military communications facilities. Dam B is located in the eastern river of the tributaries of the basin. The dam is 25 km away from dam A, and is located in the upstream of the dam A. The two dams constitute a cascade hub, which is a water conservancy project based on flood control, combined with irrigation and power generation. It was started in 1956 and completed in 1968. The dam is a large concrete head dam, with 12 double piers, 3 single piers on both sides, and gravity dams at the turning section. The dam is 82 m high and the total storage capacity is 337 million m^3. The flood control standard is once in 5000 years. After the completion of the dam, the runoff was regulated jointly with dam A to give full play to the comprehensive utilization benefits. Dam C is located in the western tributaries of the basin, located in the upstream of dam A, 26 km apart. The two dams constitute a cascade hub, which is a water conservancy project with flood control as the main, water supply, power generation, and irrigation as the auxiliary. Its construction started in 2005 and completed in 2014. The dam is an RCC arch dam with maximum dam height of 104 m, a total reservoir capacity of 460 million m^3, and the total installed capacity of 2×25 MW. The construction of dam C raises the flood control standard of downstream dam A to 5000 years, which solves the problem of insufficient flood control standards.

4 EVALUATION RESULTS

According to the specific engineering safety index system and specific steps, experts are invited to evaluate and score. The five factors that affect the dam are compared with each other. After scoring by experts, the extension judgment matrix is established, and the following Table 1 is obtained:

Table 1. Judgment matrix of dam influence factors.

	Dam Type	Dam Height	Reservoir Capacity	Dam Age	Engineering Level
Dam Type	<1,1>	<0.29,0.4>	<1.5,2.5>	<0.4,0.67>	<2.5,3.5>
Dam Height	<2.5,3.5>	<1,1>	<1.5,2.5>	<1.5,2.5>	<5.5,6.5>
Reservoir Capacity	<0.4,0.67>	<0.4,0.67>	<1,1>	<0.4,0.67>	<0.67,2>
Dam Age	<1.5,2.5>	<0.4,0.67>	<1.5,2.5>	<1,1>	<4.5,5.5>
Engineering Level	<0.29,0.4>	<0.15,0.18>	<0.5,1.5>	<0.18,0.22>	<1,1>

Combined with Table 1 to calculate the weight coefficient $k = 0.9049$, $m = 1.0709$, and meet $0 \leq k \leq 1 \leq m$, the judgment matrix consistency is found to be ideal. Calculating the interval weight vector of each factor index $W = \begin{pmatrix} \langle 0.2866, 0.3519\rangle, \langle 0.6938, 0.7948\rangle, \langle 0.1953, 0.2677\rangle, \\ \langle 0.4526, 0.5449\rangle, \langle 0.1111, 0.1512\rangle \end{pmatrix}^T$. After getting the weight vector of each factor W, the interval vector is single-valued, and the important comparison weight of each factor affecting the dam is $V = (0.1857, 0.3939, 0.1132, 0.2665, 0.0407)$.

Similarly, the extension judgment matrix of each dam to five influencing factors in the dams system is established, respectively, and the corresponding weight value after a single value is calculated. The comprehensive calculation shows that the weight set of each dam in the dams system is $v = (0.5621, 0.1978, 0.2401)$.

Similarly, the importance of factors at the same level in the safety index layer of the cascade dams system project is compared. The interval number is given by expert scoring, and the extension judgment matrix of each index at the same level is established, respectively. The weight vectors of each index at the same level are calculated as $V_1 = (0.0226, 0.4171, 0.5603)$, $V_2 = (0.5488, 0.4106, 0.0427)$, $V_3 = (0.8678, 0.1322)$, $V_4 = (0.0317, 0.9683)$, $V_5 = (0.0317, 0.9683)$, $V_6 = (0.0226, 0.5589, 0.4186)$, and $V_7 = (0.6755, 0.2585, 0.0334)$. Then, the extension judgment matrix of each index in the criterion layer is established, and the weight set of each factor index in the criterion layer is calculated as $v_1 = (0.0226, 0.4171, 0.5603)$.

By using the ridge distribution membership function, the experts' scores of 18 index factors of three dams are fuzzified, and the membership degrees of each evaluation index are calculated. The weight value of each factor in each dam index layer is coupled with its membership degree to obtain the safety evaluation matrix of the index layer. The matrix is the membership matrix of each index in the criterion layer, and the matrix is coupled with the calculation of the weight set of the criterion layer to obtain the fuzzy membership matrix of the comprehensive evaluation of the dam's system engineering safety:

$$E_A = \begin{bmatrix} 0 & 0 & 0.0006 & 0.0188 & 0.1657 \\ 0 & 0 & 0 & 0 & 0.0398 \\ 0 & 0 & 0 & 0.1260 & 0.1644 \\ 0 & 0 & 0.0012 & 0.1372 & 0.1300 \\ 0 & 0 & 0 & 0.0788 & 0.0788 \\ 0 & 0 & 0 & 0 & 0.0254 \\ 0 & 0 & 0 & 0.0085 & 0.0249 \end{bmatrix} \tag{1}$$

$$E_C = \begin{bmatrix} 0 & 0 & 0 & 0.0021 & 0.1830 \\ 0 & 0 & 0 & 0 & 0.0398 \\ 0 & 0 & 0 & 0 & 0.2904 \\ 0 & 0 & 0 & 0.0423 & 0.2261 \\ 0 & 0 & 0 & 0 & 0.1576 \\ 0 & 0 & 0 & 0.0024 & 0.0230 \\ 0 & 0 & 0 & 0 & 0.0334 \end{bmatrix} \tag{2}$$

$$E_B = \begin{bmatrix} 0 & 0 & 0 & 0.0631 & 0.1219 \\ 0 & 0 & 0.0140 & 0.0038 & 0.0221 \\ 0 & 0 & 0 & 0.1452 & 0.1452 \\ 0 & 0 & 0 & 0.0073 & 0.2612 \\ 0 & 0 & 0 & 0.0763 & 0.0813 \\ 0 & 0 & 0 & 0.0076 & 0.0178 \\ 0 & 0 & 0 & 0.0113 & 0.0221 \end{bmatrix} \tag{3}$$

According to the above results, the engineering safety evaluation results of the cascade dams system and the dam safety are given in Table 2.

Table 2. Evaluation results.

	Evaluation Sets				
Dam	Very Unsafe	Insecurity	Less Secure	Basic Security	Security
A	0	0	0.0019	0.3692	0.6289
B	0	0	0.0014	0.3146	0.6715
C	0	0	0	0.0468	0.9532
Cascade Dams			0.0038	0.2810	0.7152

According to the comparison evaluation set of the results in Table 2 and the maximum membership principle, the engineering state of the three dams in the cascade dams is safe, and there are some safety risks in dam A and dam B. The whole cascade system is also in an engineering safety state.

5 CONCLUSION

Because of the shortcomings of the traditional analytic hierarchy process (AHP), combined with the characteristics of large scale, many influencing factors, and complex structural relationship of cascade dams, a fuzzy extension of comprehensive AHP was proposed. First, based on the engineering safety of cascade dams, the safety evaluation system of a single reservoir is improved, and the complex structure affecting the engineering safety of cascade system is comprehensively considered. The engineering safety index system of cascade dams is established, which makes the evaluation index system closer to the actual situation. The extension interval number optimization judgment matrix is introduced to improve the determination of index weight. The consistency is tested in the calculation process, which is more convenient and accurate than the traditional weight calculation. Through the fuzzy membership function of ridge distribution, the expert scoring is fuzzified and the boundaries of different evaluation levels are clarified, which improves the rationality and accuracy of system safety evaluation and makes the overall evaluation more objective. The rationality and accuracy of the fuzzy extension AHP evaluation method for the engineering safety evaluation of dams are verified by the engineering examples of three-stage series reservoirs. This method improves the accuracy of mastering the safety state of cascade dam projects, and provides a scientific basis for the management department to manage the cascade dams reasonably.

ACKNOWLEDGMENTS

This work was supported by National Natural Science Foundation of China (U2040223), Open Foundation of Changjiang survey, planning, design and Research Co., Ltd (CX2019K01).

REFERENCES

Chen X, Zeng Y. (2017) Comprehensive evaluation of reservoir dam safety based on analytic hierarchy process and fuzzy theory. *Journal of Water Conservancy and Construction Engineering*, 15(06).

Gao J, Sheng Z. (2002) Extension Analytic Hierarchy Process. *System Engineering*, 05.

He J, Liao W, Shi Y. (2008) Comprehensive dam safety evaluation method based on extenics. *Journal of Wuhan University* (Engineering Edition), 02.

Jiang F, Wu H, Liu Y, et al. (2020) Research on safety evaluation of uranium tailings dam group under earthquake based on grey correlation degree and variable weight theory. *Journal of South China University* (Natural Science Edition), 34(01).

Liu X, Wang S, Yu H, et al. (2021) Safety evaluation method of robot system based on extension analytic hierarchy process. *Combined machine tool and automatic machining technology*, 03.

Liu Y, Wang L, Shen L, et al. (2007) Fuzzy hierarchical comprehensive model for dam safety risk assessment. *Water science and engineering technology*, 01.

Ministry of Water Resources of China. (2017) Reservoir and dam safety evaluation guidelines, SL258-2017.

Wang G, Jia A, Zhou G. (2010) Evaluation method and application of green degree of CNC machine tools based on fuzzy extension analytic hierarchy process. *Mechanical Engineering Journal*, 46(03).

Wang N, Liu W, Gao H. (2020) Solving Optimal Spanning Tree Based on Extension Analytic Hierarchy Process. *Computer Application*, 40(S1).

Xiong Y, Ren Q, Tian Y, et al. (2015) Application of fuzzy comprehensive evaluation method in safety evaluation of cascade reservoirs. *Hydroelectric energy science*, 33(12).

Yang H. (2013) Fuzzy evaluation of earth-rock dam safety based on AHP-entropy combination weighting. People's Yellow River, 35(06).

Advances in Civil Engineering: Structural Seismic Resistance, Monitoring and Detection – Mohd Johari Mohd Yusof and Junwen Zhang (Eds)
© 2023 The Authors, ISBN 978-1-032-31491-4

Analysis and evaluation of mountain flood disaster in Yuyao

Yi-zhen Wenren, Yao Zhou & Ji-jiang Chen
Yuyao Water Resources Bureau, Zhejiang, China

Zheng Shi
Zhejiang Institute of Hydraulics and Estuary, Zhejiang, China

ABSTRACT: Mountain flood disaster analysis and evaluation is an important content and key technology of the mountain flood disaster prevention capacity improvement project, including flood control capacity of small watersheds in massif areas, early warning indicators of important villages, classification of crisis zone and drawing of crisis zone map. Taking the mountain flood disaster prevention capacity improvement project of Yuyao in 2020 as an example, through the calculation of design flood and rainstorm, the crisis zone is classified by using flood grade, the early warning threshold of important villages is analyzed and calculated, and the map of danger zones is drawn. It provides detailed and comprehensive basic technical support for mountain flood disaster early warning, plan preparation, transfer and resettlement, group survey and group prevention, mountain flood gully flood control, etc. It shows that the mountain flood disaster prevention ability of Yuyao has been improved and the defense system is relatively complete.

1 INTRODUCTION

The special climatic conditions and geographical environment of Zhejiang Province determine the prone and frequent characteristics of mountain torrent disasters. In recent years, with the global climate change and the increase of extreme weather events, there has been an obvious trend of heavy rainfall, especially localized heavy rainfall. The record of rainfall intensity has been constantly refreshed, and mountain torrents have tended to occur frequently. The casualties caused by mountain torrents and geologic disasters in small watersheds have accounted for more than 80% of the casualties caused by flood and waterlogging disasters. Mountain flood disaster has become the focus and difficulty of flood and typhoon prevention.

Yuyao is located in the east of Zhejiang Province. It has a subtropical monsoon climate. The feature is high in the South and low in the north, with a slight depression in the middle. In the south is Siming Mountain area. There are many small watersheds in the city, and most of them belong to mountain and stream rivers, with short source and rapid flow. During the flood season, the peak flow is large, fluctuates rapidly, and the flood retention is short. The floods surge and fall, which is easy to cause river flood, mountain landslide, debris flow and other mountain flood disasters and geologic disasters, and the threat of mountain flood disaster is great. There are 12 small watersheds vulnerable to mountain torrents in Yuyao. Since 2010, Yuyao has successively implemented the second phase of mountain torrent disaster prevention and control project, involving 11 townships and 43 Administrative Villages in the city. However, there are still shortcomings and weaknesses such as unclear responsibilities for mountain flood disaster prevention, incomplete monitoring and early warning, and inadequate early warning low precision. In 2019, according to the unified deployment of the Ministry of Water Resources and the Provincial Department of Water Resources, realizing the globalization of mountain flood disaster prevention objects, the inventory of risk management and control, the refined monitoring and early warning, and the standardization of defense work are the overall goals. Yuyao has carried out the construction and implementation

 DOI 10.1201/9781003310884-96

of the mountain flood disaster prevention capacity improvement project in 2020. The mountain flood disaster analysis and evaluation includes the flood control capacity of small watersheds in massif areas, early warning indicators of important villages, delimiting the grade of crisis zones and drawing the map of crisis zones. It is important content and key technology of the mountain flood disaster prevention capacity improvement project.

2 TECHNIACAL ROUTE

Based on the results of basic data processing and mountain flood disaster investigation, mountain flood disaster analysis and evaluation is carried out for specific disaster prevention objects such as important villages along the river. It is carried out in four stages: work preparation, rainstorm flood calculation, analysis and evaluation and results sorting. See Figure 1 for the technical route.

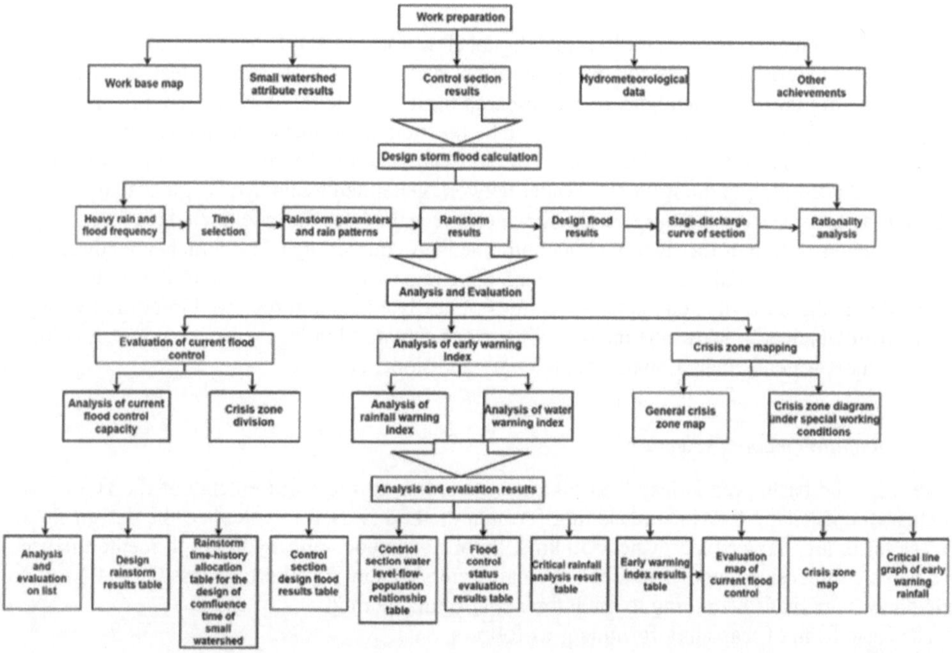

Figure 1. Technology roadmap.

3 DESIGN RAINSTORM FLOOD CALCULATION

Designing rainstorm flood calculation is the key step of mountain flood disaster analysis and evaluation. Analyzing the rainstorm flood characteristics of small watersheds in the mountain flood disaster prevention and control area and calculating the design rainstorm characteristic value and the corresponding design flood characteristic value are the basis for the evaluation of flood control capacity, the division of crisis areas and the determination of early warning indicators in the prevention and control area.

3.1 *Selection of watershed characteristic value*

The extraction of watershed characteristic value is an important basic work of analysis and evaluation. This time the extraction of watershed characteristic value uses topographic map volume algorithm and ArcGIS analysis method, and the comparative analysis of the results extracted by the two methods is carried out to obtain scientific and reasonable watershed characteristic values.

(1) Map volume algorithm

The rain-gathering area above each important village is regarded as a basin calculation unit. According to the 1:10000 (F≤200km^2) topographic map of the basin, the basin area F, the main ditch length L, and the main river ditch longitudinal slope J are respectively measured. The value of J is calculated as follows:

$$J = \frac{P_1L_1 + (P_1 + P_2)L_2 + (P_2 + P_3)L_3 + \ldots}{L^2} \tag{1}$$

In the formula: P_i-the elevation difference between a certain point and the exit section;

L_i——Segment distance;

L——The total length of the main ditch.

(2) ArcGIS analysis method

ArcGIS hydrological analysis toolbox and digital elevation model (DEM) are used for watershed hydrological analysis. There are two most important parts. The first is surface feature extraction. Through the digital terrain analysis of the research basin, the slope, slope direction, change rate and elevation of each point in the basin required by the model, the surface roughness of each DEM grid unit and the surface area of the basin can be obtained, and the river surface properties can be extracted according to the watershed water network generated by its hydrological analysis; The second is aqueous system feature extraction, which first calculates the catchment area on each grid unit, and then determines the river network combined with the setting of catchment area threshold.

According to the calculations above, through the comparison and selection of the results of the two methods, the watershed characteristic values of newly added and reviewed important villages in Yuyao are obtained. The results include: village name (natural village), section number, latitude and longitude, watershed area, mainstream length and slope, etc.

3.2 *Calculation method selection*

In general, the basin area is less than 50 km^2, according to the requirements of the Hydrology Manual, the Zhejiang Province reasoning formula method is used to calculate the design flood; for the basin area larger than or near 50 km^2, the instantaneous unit hydrograph method is used to calculate the design flood. The areas of small watersheds in Yuyao are all less than 50 km^2, so calculations are made according to the inference formula method.

The basic form of reasoning formula is as follows:

$$Q_{mp} = 0.278\frac{h_p}{\tau_p}F \tag{2}$$

F——Control the rainwater collection area above the section;

h_p——Net rainfall during the period (mm);

τ ——*Concentration time (H).*

3.3 *Analysis of concentration time*

Watershed concentration time is the most important parameter reflecting the characteristics of runoff generation and concentration in a small watershed, and it can be used as the longest duration that needs to be considered in the calculation of storm rainfall in the design of a small watershed. When determining the watershed concentration time, based on the results of the previous basic work, the inference formula is used to reverse the concentration time, combined with the rainstorm and the underlying surface conditions in the basin, and based on the trial calculation results, comprehensive analysis and determination of the watershed concentration time.

3.4 *Design rainstorm calculation*

This time, 10 minutes, 1 hour, 3 hours, 6 hours and 24 hours are selected as the typical duration, and the typical frequencies are 5-year return period, 10-year return period, 20-year return period, 50-year return period and 100-year return period.

Taking the villages along the river in the mountain flood disaster prevention zone of Yuyao as a unit, the average point rainfall and the Cv value of 10min, 1h, 6h, 24h, etc. were obtained by checking the map on the geographical location.

Then, the Kp value is obtained by checking the map from the line-fitting parameters (Cs/Cv=3.5) of the rainstorm duration, so as to obtain the design rainstorm for each duration and frequency of the small watershed where the villages along the river are located. Each small watershed is based on the design rainstorm value of each time and frequency of the village along the river to which it belongs, and then multiplied by the point-to-surface reduction coefficient (the point-to-surface coefficient is calculated based on the area of the watershed) to obtain the design rainstorm of each small watershed with each typical frequency.

The design rainfall in each period is calculated according to the rainstorm formula. The calculation formulas of design rainfall and rainstorm attenuation index in each period are as follows:

(1) Distribution of rainstorm process

When ti is between 10 and 60 minutes:

$$H_i = H_{10}(t_i/60)^{1-n_{10,60}} \tag{3}$$

$$n_{10,60} = 1 + 1.2851g(H_{10}/H_{60}) \tag{4}$$

When ti is between 1 and 6 hours:

$$H_i = H_6(t_i/6)^{1-n_{1,6}} \tag{5}$$

$$n_{1,6} = 1 + 1.2851g(H_1/H_6) \tag{6}$$

When ti is between 6 and 24 hours:

$$H_i = H_{24}(t_i/24)^{1-n_{6,24}} \tag{7}$$

$$n_{6,24} = 1 + 1.6611g(H_6/H_{24}) \tag{8}$$

(2) 24-hour rain pattern generalization

The difference of adjacent duration (ti) rainfall is the interval rainfall arranged from large to small. The maximum daily precipitation and 24-hour rain patterns are arranged according to the following rules: the last time of the largest item of rainfall in the period is at 18:00, and the second item of rainfall in the period is on the left of the largest item; The rest items are arranged from large to small, the odd items are arranged on the left, and the even items are arranged on the right. When the right row is full at 24:00, the rainfall in the remaining periods is arranged on the left from large to small.

According to the design rainstorm of each duration obtained by the chart search method, the rainstorm formula in the atlas of short duration rainstorm in Zhejiang Province is used to deduce the n value of each small watershed.

3.5 *Design flood calculation*

In this runoff generation part, the saturation-excess runoff model is used to calculate the net rainfall, meanwhile, the characteristics of mountain torrents in small watersheds are taken into account, and the deduction of evaporation is temporarily not considered in the calculation process. The saturation-excess runoff model is widely used in humid and semi-humid areas with abundant rainfall, and its calculation formula is as follows:

$$W_{mm} = \frac{1+B}{1-IMP}W_m \tag{9}$$

$$A = W_{mm}\left[1 - (1 - (1 - \frac{W_0}{W_m}))^{\frac{1}{1+B}}\right] \tag{10}$$

When $P \geq 0, R = 0$;

When $P > 0$ and $P + A < W_{mm}$; $R = R - (W_m - W_0) + W_m\left[1 - \frac{P+A}{W_{mm}}\right]^{B+!}$;

When $P > 0$ and $P + A \geq W_{mm}$, $R = P - (W_m - W_0)$.

Among them, P is the precipitation during the period; R is the total runoff during the period; Wm is the average water storage capacity of the basin, which is calculated as 100mm in this calculation; W0 is the initial average water storage of the basin. Under design conditions, Zhejiang Province generally takes the value of 75mm; Wmm is the maximum water storage content in the basin, and the value is 120mm through calculation; B is the parabolic index of water storage content, the medium watershed area (within 300km^2) takes the value 0.2~0.3, and the value of this estimate is 0.2; IMP is the ratio of the impervious area of the basin to the area of the entire basin, with a value of 0; A is the maximum point of water storage capacity in the previous period.

According to the assumption of the same frequency of rainstorm and flood, the design flood frequency of important riverside villages in Yuyao is the design rainstorm frequency, which is once in 5 years, once in 10 years, once in 20 years, once in 50 years, and once in 100 years.

3.6 *Calculation of the stage-dischrage relationship*

In the analysis and evaluation of mountain flood disasters, one of the important purposes of designing storm flood analysis and calculation is to obtain the stage-discharge curve of the control section where disaster prevention objects such as villages along the river are located.

(1) Using stage-discharge relationship or Manning formula and other methods to analyze the stage-discharge relationship, convert the design flood peak discharge of the control section of the villages along the river into the corresponding water level, and draw the stage-discharge curve.

(2) Parameter determination

In the conversion of water level and flow, the gradient and the roughness are two very important key parameters. The parameter values of the two are correct or not has an important influence on the calculation results of the design flood.

$$Q = \frac{1}{n} \times A \times R^{2/3} \times J^{1/2} \tag{11}$$

In the formula: Q—flow rate; n—roughness; R—hydraulic radius; J—hydraulic gradient.

1) Determination of the slope

Manning formula is used to analyze the the stage-discharge relationship.

The gradient refers to the gradient of the flood water surface, but when the mountain flood actually occurs, the water surface gradient is difficult to measure. Therefore, if there are data on the distribution along the route of historical flood traces, flood water surface curves that occurred at flood levels in recent years, or measured water surface lines when small and medium floods occurred, they are used as the gradient in the water level and flow conversion; If there is no corresponding water surface curves information, the riverbed gradient of the villages along the river will be used. In order to make the results of the analysis as reasonable as possible, this evaluation and calculation mainly focus on the gradient of the historical flood mark water surface curve. It also uses the measured water surface gradient and the riverbed gradient of mountain torrent disasters as a reference, and compares the three gradients. When there is little difference between the three, the flood mark water surface gradient is used as the final gradient. When there is a great difference between the three, the riverbed gradient is used as the final gradient, and through trial calculation and rationality analysis.

2) Determination of roughness

Roughness are determined according to the longitudinal and transversal photos of the river channel taken during the field survey, and the roughness in the conversion of water level and flow is comprehensively determined by referring to the roughness of the typical types and characteristics of natural or artificial river channels.

4 CONFIRMATION OF EARLY WARNING INDICATORS

4.1 *Analysis of current flood control capacity*

Based on the calculation and analysis of design flood, the evaluation of flood control status is to obtain the stage-discharge curve at the control section of disaster prevention objects such as villages along the river, and then to analyze the current flood control capacity of disaster prevention objects, classify mountain flood disaster risk areas, and provide support for the preparation of mountain flood disaster prevention plan, personnel transfer, temporary resettlement, etc. The main content of the current flood control capacity analysis is the frequency analysis of the flood peak flow corresponding to the disaster water level of villages along the river, the statistics to determine the disaster water level, and the comprehensive evaluation of the current flood control capacity. The current flood control capacity is expressed by the frequency of flow corresponding to the disaster water level, and the disaster water level is investigated on site. According to the stage-relationship of the control section, the flood peak discharge corresponding to the disaster water level is calculated, and the flood frequency corresponding to the flow is determined by analyzing the flow interpolation corresponding to each typical frequency, that is, the flood frequency corresponding to the disaster water level.

The flood frequency index is used to classify the crisis area, and the specific standards are shown in Table 1.

Table 1. Classification standard of danger area.

Crisis zone level	Flood return period (year)	Explanation
Extremely high crisis zone	Less than once in 5 years	Higher frequency
High crisis zone	Greater than or equal to once in 5 years, less than once in 20 years	Moderate frequency
Crisis zone	Greater than or equal to once in 20 years to the highest flood level in history	Rare occurrence frequency

According to the evaluation and analysis, 53 administrative villages and 97 natural villages in Yuyao are important villages for mountain flood disaster. The results are as follows: 39 extremely high crisis areas, 91 high crisis areas and 97 crisis areas.

4.2 *Determination of early warning indicators*

The early warning indicators include rainfall early warning indicators and water level early warning indicators, which are divided into two early warning levels: "ready to transfer" and "immediate transfer".

The analysis content of the rainfall early warning index includes the critical rainfall of each early warning period, and the early warning period is 1 hour, 3 hours and 6 hours. The critical rainfall is calculated by water level/flow inverse method, considering the water content of the watershed in general and wetter conditions, threshold analysis is carried out according to the calculated rainfall critical value, and the variation amplitude of the critical rainfall is reasonably determined. In order to allow a certain amount of time for the transfer of people in the crisis zone, this time, for small watersheds with a concentration time $\leq$ 2 hours, the immediate transfer index and the ready-to-transfer index will be reduced on the basis of the critical rainfall. The method of reducing the early warning index for immediate transfer is as follows: According to the design flood hydrograph of the control section, the corresponding flow 20-30 minutes before the occurrence of the flood peak is obtained, the corresponding water level from the water level flow relationship curve is

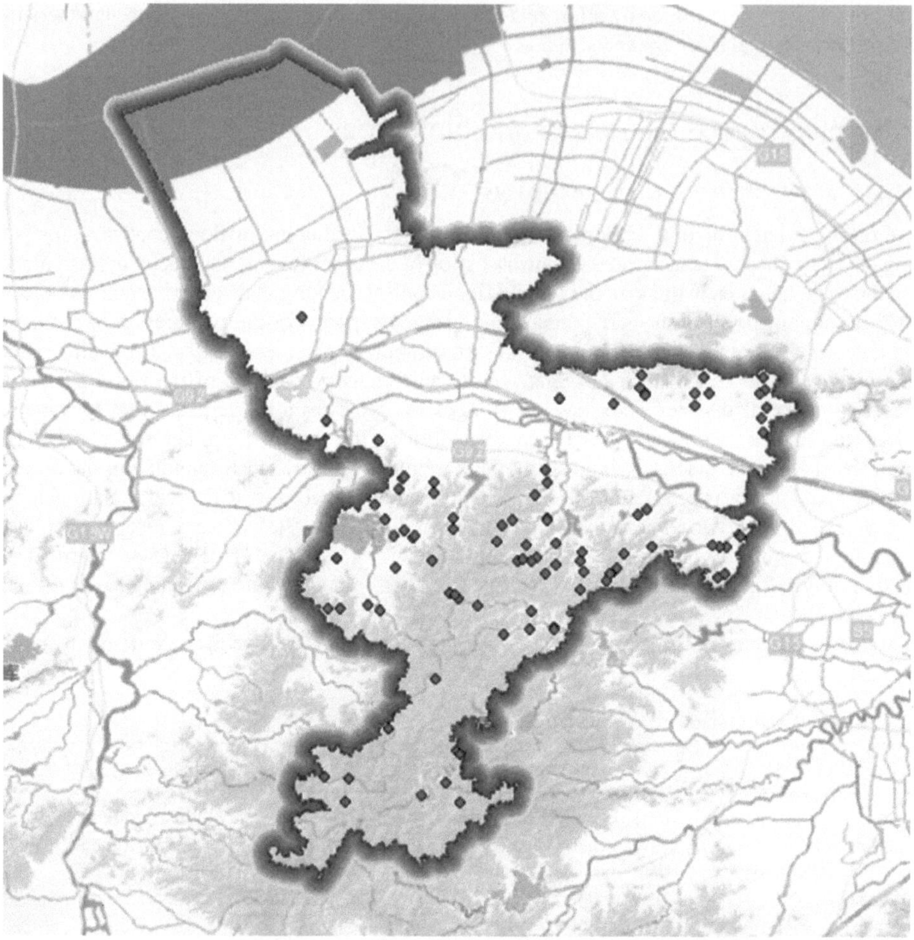

Figure 2. Distribution of important villages for mountain flood prevention in Yuyao.

obtained, and then the early warning index is deduced from the water level inversely; On the basis of the immediate transfer index, the ready-to-transfer index is pushed forward for 20 to 30 minutes. For small watersheds with a confluence time> 2 hours, the immediate transfer index is equal to the critical rainfall, and the ready-to-transfer index is reduced on the basis of the critical rainfall according to the above method. The forward push time is generally 20 to 30 minutes.

The water level early warning index analyzes the water level threshold under different early warning levels, and the critical water level ready for transfer and immediate transfer is determined through the historical mountain flood disaster records, historical flood traces and the actual situation of the dangerous area. Similarly, considering the reserved transfer time, the critical water level is reduced to obtain the water level early warning index, and the forward calculation time is not less than 30 minutes.

5 DRAWING OF DANGER ZONE MAP

The crisis zone map is on the base map of the investigation and evaluation of mountain flood disasters (or a larger scale base map), the relevant results of the evaluation of the current flood control situation are visually displayed on the drawings to provide support for the early warning of mountain floods, the preparation of pre-plans, the transfer of personnel, and temporary resettlement.

(1) Draw according to the design storm flood inundation range corresponding to the frequency of the crisis zone level;
(2) If there are weir ponds, small reservoirs, dikes, bridges and culverts in the upstream and downstream of the disaster prevention object, and there is a possibility of failure or blockage of floods, additional maps of the crisis zone under special working conditions should be drawn.

6 OVERALL EVALUATION

Through the first and second phases of mountain flood disaster prevention and control projects and the construction of capacity improvement projects in 2020, Yuyao has conducted a comprehensive investigation and evaluation of the key areas of the city's mountain torrent disaster threats. It also upgraded the mountain torrent monitoring and early warning platform, and supplemented and improved the monitoring system. A small watershed mountain torrent disaster prevention system has been established that integrates a responsible organization structure, a group strategy and group defense system, and a monitoring and early warning system platform. Yuyao has analyzed and evaluated the design rainstorms in the small watersheds where 97 important villages are located for flash flood disasters, and the design floods at each important control section. It also analyzed and determined the early warning indicators of 97 important villages, and drawn 97 flash flood disaster crisis area maps. More detailed and comprehensive basic technical support for mountain torrent disaster early warning, preparation of plans, transfer routes, temporary resettlement, popularization of disaster prevention awareness, mass monitoring and prevention, and flood control of mountain torrents are provided.

7 CONCLUSION

The prevention of mountain torrent disasters in small watersheds mainly relies on monitoring and early warning. A successful mountain torrent monitoring and early warning system includes organization and management system, monitoring elements, early warning thresholds, monitoring equipment and stations. The analysis and evaluation of mountain torrent disasters is the basis of defense work, and it is also the key technology of defense work. The determination of the threshold of monitoring and early warning is the core issue, which is directly related to the scientificity and accuracy of early warning. The calculation accuracy of rainfall early warning indicators, the determination of thresholds, and the rationality of the division of the danger zone range are highly professional, difficult, and have many influencing factors. With the improvement of various engineering measures, the adjustment of industrial structure, and the change of living conditions along the river, the technical indicators of its defense are also a process of dynamic changes. Only the review of early warning indicators is continuously carried out, and the normal management of dynamic updates and corrections is established, can we continuously improve the accuracy and reliability of mountain torrent disaster forecasting and early warning, continuously improve the defense capabilities, and better serve the protection of people's lives and property and the maintenance of social stability.

ACKNOWLEDGEMENTS

This research was financially supported by Ningbo Hydraulic Science and Technology Research Project (Grant No. NSKA202002).

REFERENCES

National Mountain Flood Disaster Prevention and Control Project Team. Technical requirements for analysis and evaluation of mountain torrent disasters[S]. 2014.

Water Resources Department of Zhejiang Province. Guidelines for Investigation of Mountain Flood Disasters in Zhejiang Province[S]. Hangzhou: 2019.

Zhejiang Hydrological Survey Bureau. Short-duration rainstorm in Zhejiang Province [G]. Hangzhou: Zhejiang Hydrological Survey Bureau, 2003.

Zhejiang Provincial Department of Water Resources. Notice of Zhejiang Provincial Department of Water Resources on Accelerating the Improvement of Mountain Flood Disaster Prevention Capability[Z]. 2019.

Zhejiang Water Conservancy and Hydropower Survey and Design Institute. Design Storm and Flood Atlas of Small and Medium-sized Watersheds in Zhejiang Province[M]. Hangzhou: Zhejiang Water Conservancy and Hydropower Survey and Design Institute, 1984.

Advances in Civil Engineering: Structural Seismic Resistance, Monitoring and Detection – Mohd Johari Mohd Yusof and Junwen Zhang (Eds)
© 2023 The Authors, ISBN 978-1-032-31491-4

Analysis on the seismogenic mechanism of Changdao earthquake swarm in Shandong province

Rui Zhao, JianChang Zheng, HuaWei Cui & DongPu Ji
Shandong Earthquake Agency, Jinan, Shandong, China

ABSTRACT: The Changdao earthquake swarm in 2017-2018 was relocated by using the data of the seismic observation report of Shandong Seismological Network, and the focal mechanism solutions of the significant earthquakes in the swarm were calculated by FOCMEC method. The results show that the overall distribution of the Changdao earthquake swarm is NWW. The epicentral location of the north earthquake swarm is obviously distributed along the Dazhushan-Weihai North Fault, which is a nearly upright high-angle fault, inclining to NNE, with a dip angle of about 80°, and a strike-slip type. The focal mechanism solutions of five earthquakes with ML4.0 or higher show that, the northern earthquake swarm was formed by the activity of the branch faults of Dazhudao-Weihai North Fault under the action of local tectonic stress field, and the southern earthquake swarm was related to Shendaokou Fault and Changdao-Zhifudao Fault. The data above ML 2.0 in the earthquake swarm were selected, using the double difference wave velocity ratio method to calculate the wave velocity ratio in Changdao area. The average values of wave velocity ratio of North earthquake swarm and South earthquake swarm in Changdao are 1.7301 and 1.7217, respectively, and the fluctuation range is between 1.69 and 1.78. During the active period of the earthquake swarm or before the occurrence of a large earthquake, the ratio of wave velocity increased gradually and decreased after the occurrence of the earthquake. The calculation of the source spectrum parameters shows that the high frequency attenuation coefficient γ is generally higher than 2, which shows that the seismogenic mechanism of the Changdao earthquake swarm is not simply the fault rupture, the high-pressure fluid intrusion may play an important role, which is in line with the process of expansion-saturation-occurrence. The average value of stress drop in this region is lower than the average level of Jiaodong Peninsula, and the faults are densely distributed and broken, which makes it difficult to accumulate large stress. The continuous activity of earthquake swarm belongs to the process of low energy rupture, which forms the activity characteristics of more moderate and small earthquakes but less moderate and strong earthquakes in this region, and has no background of strong earthquakes in the short term.

1 INTRODUCTION

As one of the precursors of seismicity, earthquake window is generally defined as aftershock activity or earthquake swarm activity in a specific tectonic area, which is sensitive to the change of tectonic stress field. The Changdao earthquake window is considered as an important precursor index of seismicity in eastern China (Zhao et al. 2009; Wang et al. 2013). Since 1970, there have been many window-opening activities in Changdao area, and many major earthquakes have occurred in subsequent neighboring areas. For example, in 1976, the Heishan earthquake swarm occurred on February, and Tuoji earthquake swarm occurred on March in Changdao area, followed by the Tangshan M7.8 earthquake on July 28th (Zhao et al. 1992). Earthquake statistics show that within one year after earthquake swarm activity occurred in Changdao area from 1970 to 2017, the proportion of earthquakes with M5.0 or higher occurred in eastern China reached 63.2%. From February 14, 2017, a new intensive activity of small earthquakes occurred in Changdao

DOI 10.1201/9781003310884-97

area, which reached the index of "earthquake window". Then, on March 3, 2017, the frequency of small earthquakes suddenly increased significantly, forming an explosive earthquake swarm activity, both in frequency and intensity significantly exceeding the historical activity level, but up to now, there is no correspondence between moderate and strong earthquakes. Therefore, what is the seismogenic mechanism of this Changdao earthquake swarm activity? It is highly concerned by seismologists and local governments in China.

Changdao area is controlled by the structures in NNE-Tanlu fault zone and its branches, and NW-Zhangjiakou-Penglai fault. The complex nature of fault movement will lead to the fracture of fine rocks. Because Changdao area is located offshore, is the seismogenic mechanism related to fluid intrusion? According to the mechanism of reservoir earthquake, the tectonic stress field in the upper crust of the reservoir area will be disturbed when storing water, especially the adjustment of local stress state in the potential source area. Li Yalin and other ultrasonic experiments show that when the rock pore or fracture system is completely saturated, the P-wave velocity (V_P) will obviously increase, while the S-wave velocity (V_S) will basically remain unchanged or slightly decrease, and the wave velocity ratio will obviously increase (Li et al. 1999). Cai Xinghui's research also shows that with the increase of water level, the permeability of rocks in the shallow surface increases sharply, and the water in fracture zone or fissure becomes saturated gradually. When the fluid pressure is greater than the critical value of rock pressure, the shallow fault expands to the ground under the action of surface force, thus inducing shallow earthquakes, showing an obvious process of expansion-saturation-earthquake generation (Cai 2013). The spectral high-frequency attenuation coefficient γ in the source parameters reflects the kinematic properties of earthquake rupture, reveals the overall geometry of the fault zone, and reflects the complexity of fault rupture. The general period is $0 < \gamma \leq 3$. The γ value of the traditional Brune disk rupture model is generally about 2, that is, the plane rupture of the fault occurs. However, when microseisms are induced by fluid intrusion in volcanic areas, the γ value is about 2, which is closer to 3 (Patane et al. 1994).

In this paper, the Changdao earthquake swarm is taken as the research object, the double-difference seismic positioning method is used for precise positioning, and the FOCMEC method, which is used to jointly calculate the focal mechanism solution of the initial motion and amplitude ratio of P-wave, SV-wave and SH-wave, is used to calculate the focal mechanism solution of five significant earthquakes in the swarm. Selecting the waveform and phase data of earthquakes above ML2.0, the source parameters of Changdao earthquake swarm are calculated, and the wave velocity ratio is calculated by using the method of "double difference wave velocity ratio". By analyzing the variation of high frequency attenuation coefficient and wave velocity ratio, seismicity and stress accumulation, the seismogenic mechanism of Changdao earthquake swarm is discussed.

2 DATA INFORMATION

Changdao area is located in Bohai Rim region, while Bohai Strait is located on Jiaoliao uplift belt, with complex geological structure, bounded by Yingkou-Weifang section of Tanlu fault zone in the west, adjacent to Bohai Bay Basin (Li et al. 2013), which is mainly controlled by structures in NNE-trending Tanlu fault zone and its branches, and NW-trending Zhangjiakou-Penglai fault (Wang et al. 2016), as shown in Figure 1 (frepresents NNE-trending fault, and F represents NW-trending fault.

According to the epicentral position of Changdao earthquake swarm in 2017–2018,. it can be divided into Changdao North Earthquake Swarm and Changdao South Earthquake Swarm. The red box in Figure 1 shows the study area of this paper, ranging from 120.6°–120.9°E and from 37.9° to 38.2°N. The black dots in the box show two earthquake swarms in Changdao. Up to September 5, 2018, the small earthquake activity in the two earthquake swarms basically tended to the level of earthquake background activity, with 2,499 localizable earthquakes recorded in the North Earthquake Swarm of Changdao and 308 localizable earthquakes recorded in the South Earthquake Swarm of Changdao.

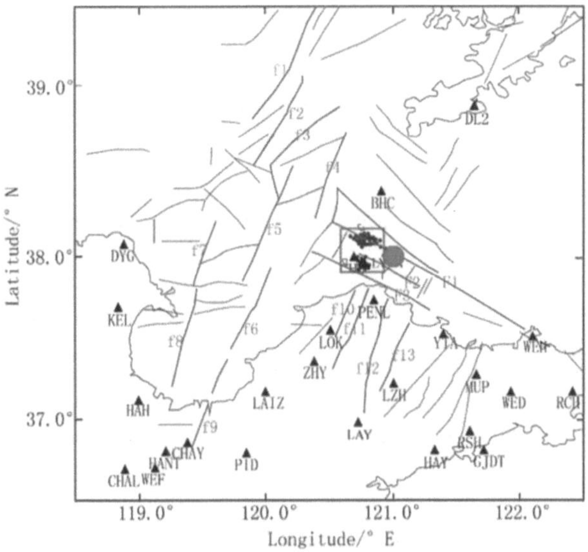

Figure 1. Distribution of faults, stations and earthquake swarms in Changdao area.

(Black triangle in the picture: Seismic stations; Red dots: The epicentre of the M7.0 earthquake near Changdao area in 1548; Red box: Location of study area; Black dot in the red box: Epicentre position of Changdao earthquake swarm; f: NNE trending fault; F: NW trending fault; f1: Fault of eastern Bohai; f2: fault of southeast Bohai; f3: Fault of Miaoxi No.2; f4: Fault of Miaoxi No.3; f5: Western fault of Yingwei east branch ; f6: Eastern fault of Yingwei east branch; f7: BZ34 fracture; f8: KL3 fracture; f9: Anqiu-Juxian fault; f10: Beigou-Linglong fault; f11: Fengyidian fracture; f12: Qixia fault; f13: Octagonal Shewopo fault; F1: Dazhudao-Weihai North Fault; F2: Shendaokou fault; F3: Changdao-Zhifu Island Fault)

3 PRECISE EARTHQUAKE LOCATION AND FOCAL MECHANISM SOLUTION

As shown in Figure 1, Changdao is located in the Bohai Rim region, with dense active faults and complex geological structure. Wang Zhicai et al. (Wang et al. 2006) completed acoustic detection of 11 shallow stratum profiles in Penglai-Weihai fault zone, and the results show that most faults are dominated by normal fault and strike-slip movement, and some faults have thrust movement, among which the characteristics of Dazhuzhu-Weihai North fault are different on different survey lines, and there are normal fault characteristics inclined to SW and thrust movement characteristics pointing to SW. In 1548, an earthquake with M7.0 occurred near Changdao area, and its epicenter was located at 38.0N and 121.0E, as shown in the red origin in Figure 1. Some scholars have shown that this M7.0 earthquake is related to the intersection of NNE northbound Gully-Linglong fault and NW-trending Zhangjiakou-Penglai fault, and the earthquake-generating position is probably located in Zhangjiakou-Penglai fault (Ge et al. 2010)· Since 1970, earthquakes of magnitude above 4 have frequently occurred in Jiaodong Peninsula and its offshore area, mainly distributed along NNE and NW faults, and the earthquakes occurred most frequently at the intersection of the two sets of faults, indicating that the dominant structure is still these two sets of faults, which are still active (Wan et al. 2013).

In view of the earthquake swarm activity in Changdao, a mobile station was set up in Dazhushan Island in the southeast of Changdao, which can obtain waveform data near the epicenter. In order to explore the seismogenic structure of the Changdao earthquake swarm in detail, this paper selects more than 1,600 earthquake cases with clear phases recorded by at least four stations from more than

2,400 earthquakes recorded by the Changdao earthquake swarm, and uses the double-difference seismic positioning method for precise positioning. The one-dimensional velocity model in positioning is shown in Table 1 (Li et al. 2012), and the wave velocity ratio is set to 1.732. The S-wave velocity is converted according to P-wave velocity and wave velocity ratio. The FOCMEC method, in which the initial motion and amplitude ratio of p-wave, SV-wave and SH-wave are used to jointly calculate the focal mechanism solutions, is used to calculate the focal mechanism solutions of five significant earthquakes in the earthquake swarm, namely, the earthquakes of magnitude 4.5 on March 3, magnitude 4.1 on March 21, magnitude 4.3 on March 27, magnitude 4.0 on April 8 and magnitude 4.1 on June 18, 2017. The results are shown in Figure 2 and Table 2.

Table 1. One-dimensional velocity structure used in double difference positioning calculation.

Depth (km)	0	8	15	20	30	35
P wave velocity(km/s)	2.89	4.44	6.22	6.36	7.07	8.00

Table 2. Results of focal mechanism solutions of 5 earthquakes with M ≥ 4.0 in Changdao earthquake swarm.

			Nodal surface 1			Nodal surface 2		
Earthquake cases	Date of earthquake occurrence	Magnitude/ ML	Trend/ °	Inclination angle/°	Sliding angle/°	Trend/ °	Inclination angle/°	Sliding angle/°
1	20170303	4.5	320.0	57.0	12.0	223.4	80.0	146.4
2	20170321	4.1	321.0	53.0	−10.0	57.1	82.0	−142.6
3	20170327	4.3	321.0	50.0	161.0	63.5	75.6	41.6
4	20170408	4.0	351.0	60.0	−57.0	118.6	43.4	−133.3
5	20170618	4.1	142.0	30.0	25.0	30.0	77.8	117.6

In Figure 2(a), *AA′* section is made along NWW to F1 dazhu island-weihai north fault, and *BB′*, *CC′* and *DD′* sections are made in NEE direction in the south, middle and north sections, respectively, and vertical profiles are obtained, as shown in Figure 2(b), (c), (d) and (e). The result shows: The focal depth of Changdao North Earthquake Group is concentrated in the range of 3–18km (AA′ profile), and the epicenters are concentrated on both sides of fault F1 (BB′ profile, CC′ profile and DD′ profile). The horizontal connection of the focal positions of different depths is almost upright, and the distribution characteristics along F1 fault plane are obvious. This fault is an almost vertical high-angle fault, which inclines towards NNE with an inclination of about 80°. The focal rupture mechanism is mainly strike-slip. However, the focal mechanism solutions of five earthquakes with ML≥ 4.0 show that the strike of nodal plane 1 in four cases is NW or NNW, and the strike of one case is quite different from that of F1, and its dip angle is obviously different from that of F1 fault. It shows that the epicentral location distribution of the earthquake swarm is consistent with the location of fault F1, and the sources are concentrated in the vicinity of fault plane F1, but it is not the direct dislocation of the fault that caused the earthquake. It is speculated that the earthquake swarm was formed by the activity of its branch faults under the action of local tectonic stress field. There is an earthquake gap between the southern earthquake swarm and the northern earthquake swarm in Changdao. From the horizontal direction, the earthquake is located between the Shendaokou fault F2 and the Changdao-Zhifudao fault F3. There should be more connections between them in the seismogenic structure, so the information is small and it is not convenient to speculate.

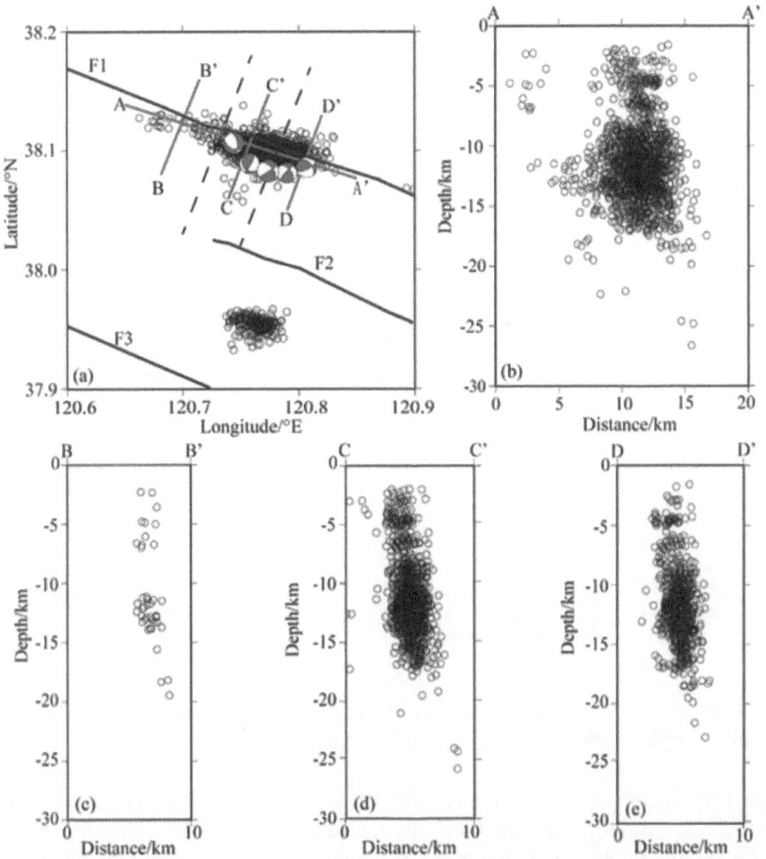

Figure 2. Fine positioning results of Changdao earthquake swarm. (a) Epicenter distribution map of Changdao earthquake swarm; (b) Vertical distribution of profile AA′; (c) Vertical distribution map of BB′ profile; (d) Vertical distribution map of CC′ profile; (e) Vertical distribution of DD′ profile. (F1: Dazhudao-Weihai North Fault; F2: Shendaokou fault; F3: Changdao-Zhifu Island Fault).

4 VARIATION CHARACTERISTICS OF WAVE VELOCITY RATIO

The calculation method of double difference wave velocity ratio is to group two earthquakes close to the same station in a small earthquake cluster, and deduct the same path of seismic rays based on the double difference of arrival time of P wave and S wave, so as to obtain the ideal result of wave velocity ratio in earthquake cluster seismicity area (Jia et al. 2017; Wang et al. 2011; Zhang & Lin 2014). The earthquake events above ML2.0 is selected, with epicentral distance of 50-130km, and at least 4 stations have clearly recorded P-wave and S-wave phases, and check them in advance by using the Heda line to reduce the arrival error of seismic wave phases. The diameter of the earthquake concentrated area of Changdao North earthquake swarm is 8 km, and that of South earthquake swarm is 4 km. The wave velocity ratio of this area is calculated by the method of "double difference wave velocity ratio", and the number of eligible earthquake events is 266 in North earthquake swarm and 20 in South earthquake swarm respectively. In Changdao North Earthquake Swarm, sliding window is carried out under the conditions of step length of a week, window length of a month and at least five earthquake cases. The data of Changdao South Earthquake Swarm is small, and the window sliding is carried out under the conditions of two earthquake cases as step length and eight earthquake cases as window length. The variation of wave velocity ratio with time is shown in Figure 3.

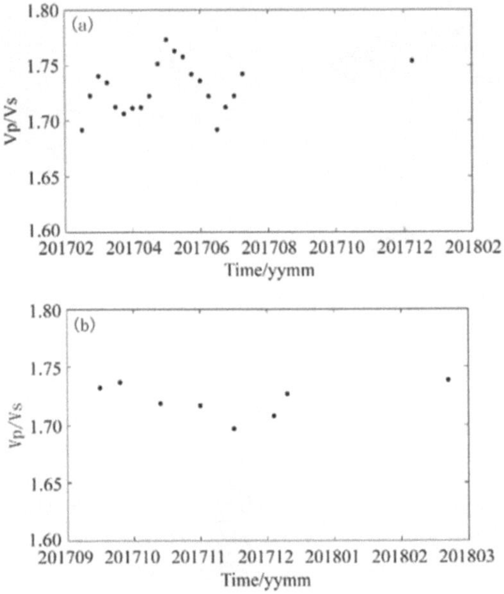

Figure 3. Calculation results of wave velocity ratio of North earthquake swarm and South earthquake swarm in Changdao. (a) Changdao north earthquake swarm, (b) Changdao Nanzhen Group.

The results show that the average wave velocity ratios of the north and south earthquake swarms in Changdao are 1.730 and 1.722, respectively, and the wave velocity ratios fluctuate between 1.69 and 1.78, and the fluctuation range of the north earthquake swarm is larger. According to the significance of wave velocity ratio in earthquake prediction, in North China, the anomaly of low value of wave velocity ratio may occur before the occurrence of moderate and strong earthquakes. Generally speaking, at the initial stage of rock compression, the wave velocity ratio increases with the increase of pressure, but when the pressure exceeds 50% of rock compressive strength before rock fracture, the wave velocity ratio obviously decreases with the increase of pressure. It can be seen from Figure 3 that the wave velocity ratio is in a low value state before the earthquakes of magnitude 4.5 on March 3, magnitude 4.1 on March 21, magnitude 4.3 on March 27, magnitude 4.0 on April 8 and magnitude 4.1 on June 18, 2017. On November 20, 2017, when the Southern Seismic Group of Changdao had an earthquake with ML3.0, the wave velocity ratio also showed a low value. Since February, 2018, the number of earthquake cases has obviously decreased, and the wave velocity ratio of the two earthquake swarms has slightly increased, which suggests that the rock pressure has not reached the critical point. From 2017 to 2018, the change of the wave velocity ratio of Changdao earthquake swarm is closely related to the regional seismicity process. When the fluid pressure is greater than the rock pressure, the shallow fault expands underground under the action of the surface force, thus inducing shallow earthquakes. The wave velocity ratio decreases obviously with the increase of pressure, and when it reaches saturation, it induces shallow earthquakes, showing an obvious process of expansion-saturation-earthquake generation.

5 SOURCE PROPERTY

The high frequency attenuation coefficient γ of the source spectrum in the source parameters reflects the kinematic properties of earthquake rupture, which can reveal the overall geometry of the fault zone. The magnitude of γ value can be used to judge whether the earthquake mechanism of Changdao earthquake group is related to fluid intrusion. In this paper, based on the data of earthquakes with magnitude of ML2.0 or higher in Changdao earthquake swarm, firstly, the observed waveform of the station is preprocessed, and the velocity spectrum is obtained by Fourier transform, and the theoretical source spectrum is solved by two-step inversion method (Zheng et al. 2016), and the high-frequency attenuation coefficient γ and stress drop of the source parameters are obtained. The results are shown in Figure 4.

The results show that the γ values of Changdao earthquake swarm are generally higher than 2, and fluctuate greatly at the initial stage of earthquake swarm activity, which further verifies that the seismogenic mechanism of Changdao earthquake swarm is not simply the fault rupture, but the fluid intrusion may play an important role. The average level of stress drop in the world is about 3 MPa, while the level of stress drop in Jiaodong Peninsula is relatively low, and the average stress drop of small earthquakes is about 1 MPa(Wang et al. 2013). The average stress drop of this Changdao earthquake swarm is 0.736 MPa, which is lower than the average level in Jiaodong Peninsula and should be related to the geological structure of Changdao, which leads to more moderate and strong earthquakes, but less moderate and strong earthquakes in history, which also enable this area to be used as an "earthquake window", reflecting the stress changes in eastern Shandong (Zhao et al. 2009).

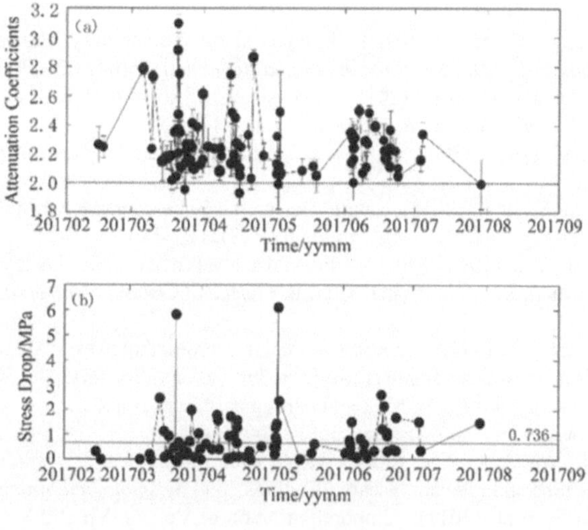

Figure 4. Source parameter diagram of Changdao earthquake swarm.

6 CONCLUSION

1. The Changdao earthquake swarm is generally distributed in NWW direction, and the epicentre of the North earthquake swarm is obviously distributed along the Dazhudao-Weihai North fault, which is a nearly upright high-angle fault with NNE inclination and dip angle of about 80°. The focal rupture mechanism is mainly strike-slip type. The results of focal mechanism solutions of five earthquakes with ML $\geq$ 4.0 show that the Northern Earthquake Swarm was formed by the activity of the branch faults of Dazhudao-Weihai North Fault under the action of local tectonic stress field. South earthquake swarm is related to Shendaokou fault and Changdao-Zhifudao fault.
2. The average values of wave velocity ratios of the North Earthquake Swarm and the South Earthquake Swarm in Changdao are 1.730 and 1.722, respectively. The North Earthquake Swarm on March, April and June, 2017 when the Swarm was active, and South Earthquake Swarm before

the earthquake with ML3.0 occurred on November, 2017, the wave velocity ratio increased gradually, but decreased when the earthquake occurred. In addition, the high-frequency attenuation coefficient γ of the source spectrum of the Changdao earthquake swarm was generally higher than 2, and the average stress drop was lower than the average level of Jiaodong Peninsula. It is considered that the seismogenic mechanism of the Changdao earthquake swarm is not simply the fault rupture, but the fluid intrusion may play an important role in the occurrence of earthquake swarm, and the change of wave velocity ratio accords with the process of expansion-saturation-seismogenic.

ACKNOWLEDGMENT

This paper is supported by the International Cooperative Research Project on Earthquake Predictability of China Earthquake Science Experimental Site (2018YFE0109700).

REFERENCES

Cai X. H. (2013). The measurement and research on the average wave velocity ratio in Fujian Province. *Plateau earthquake research*. 25(4), 31–34.

Ge F. G., Wang Z. C., Wang J. Q., et al. (2010). Preliminary study of quaternary faults on Changdao island of Shandong province. *Journal of Disaster Prevention and Reduction*. 26(04), 13–21.

Jia L. Z., Wang Z. S., Zhang Y. L., et al. (2017). Analysis of the 2014-2015 Jinzhai earthquake swarm by double-difference Vp/Vs ratio method. *Earthquake*. 37(1), 112–120.

Li B., Cui X., Ye Q. D., et al. (2012). Study of 1D velocity model in Shandong area. *North China Earthquake Sciences*. 30(04), 1–6.

Li H. K., Yang Y. B., Geng K., et al. (2013). Subdivision of the Bohai landmass from the north China Block and its implications. *Geological Survey and Research*. 36(3), 173–180.

Li Y. L., Xie X. P., He Z. H., et al. (1999). The experimental analysis of dynamic elastic parameters of exposure rock samples under formation conditions in CX basin. *Journal of Chengdu University of Technology*. 26(2), 161–167.

Patane D., Ferrucci F., Cresta S. (1994). Spectral features of microearthquakes in volcanic areas: attenuation in the crust and amplitude response of site at Mt. Etna, Italy. *Bull. Seism. Soc. Am.* 84(6), 1842–1860.

Wan B., Jia L. H., Dai Y. L., et al. (2013). Moderate-strong earthquakes and their tectonic correlation in the Liaodong peninsula. *Seismology and Geology*. 35(2), 300–314.

Wang C. Y., Duan Y. H., Wu Q. J., et al. (2016). Exploration on the deep tectonic environment of strong earthquake in North China and relevant research findings. *Acta Seismologica Sinica*. 38(4), 511–549.

Wang L. Y., Li Y. E., Li F., et al. (2011). Temporal variation of Vp/ Vs, Vp and Vs before and after 2008 Wenchuan earthquake. *Acta Seismologica Sinica*. 33(1), 1–14.

Wang P., Zheng J. C., Zhao J. H. (2013). Status of Seismic Source Spectral Parameters in the Jiaodong Peninsula. *China Earthquake Engineering Journal*. 35(2), 360–365.

Wang Z. C., Deng Q. D., Chao H. T., et al. (2006). Shallow-depth sonic reflection profiling studies on the active Penglai-Weihai fault zone offshore of the northern Shandong peninsula. *Chinese J. Geophys*. 49(4), 1092–1101.

Zhang Q. & Lin G. (2014). Three-dimensional Vp and Vp/Vs models in the Coso geothermal area, California: Seismic characterization of the magmatic system. *Journal of Geophysical Research Solid Earth.* 119(6), 4907–4922.

Zhao J. H., Lu H. P., Wu D. T., et al. (2009). Analysis on recent trend of seismic risks and seismic activity characteristics in Shandong peninsula as well as its nearby waters. *Seismological Research of Northeast China*. 25(2), 17–22.

Zhao X. L., Li Y. S., Hua A. J., et al. (1992). Analyzing earthquake hazard in Shandong province and surroundings by swam information entropy. *Seismological and Geomagnetic Observation and Research*. 13(4), 1–8.

Zheng J. C., Wang P., Xu C. P., et al. (2016). Robust inversion of seismic-source spectral parameters for the 2013–2015 Rushan swarm. *Chinese J. Geophys*. 59(11), 4100–4112.

Advances in Civil Engineering: Structural Seismic Resistance, Monitoring and Detection – Mohd Johari Mohd Yusof and Junwen Zhang (Eds)
© 2023 The Authors, ISBN 978-1-032-31491-4

Suitability evaluation of earthquake emergency shelter site selection

B. Zheng, J. Liu & H. Zhang
School of Architecture and Art, Central South University, Changsha, Hunan, China

W. Zhou
School of Management, Guangdong University of Technology, Guangzhou, Guangdong, China

ABSTRACT: In the process of evaluating the suitability of emergency shelters, researchers seldom explore future uncertainties and environment complexity through scenario planning. This paper explores the method of combining scenario planning and multi-criteria analysis that can be used to evaluate the site suitability of earthquake emergency shelters. The study determines 14 scenarios and 27 evaluation criteria of earthquake emergency shelters through relevant literature analysis and field investigation. The study adopts the developed method to evaluate the site suitability of three earthquake emergency shelters in Chenggong District, Kunming. It is found that the scoring results of site suitability of earthquake emergency shelters under different scenarios are significantly different, which indicates that the method can be well applied to the evaluation of earthquake emergency shelters.

1 INTRODUCTION

Earthquake disasters can cause great economic losses and heavy casualties. The seismic risk is the main source of the earthquake disaster risk. Ground motion, seismic risk, and the vulnerability of building groups in the ground motion have uncertainty (Wang & Gao 2019). In most earthquakes and the aftermath such as landslides, floods, fires, and tsunamis, the majority of casualties are caused by the collapse of buildings (Ketsap et al. 2019). In the process of urban development, it is very important to strengthen the comprehensive ability of urban earthquake prevention and disaster mitigation. Earthquake prevention and evacuation planning is an important part of urban earthquake disaster prevention planning (Dan et al. 2018). Here, earthquake evacuation refers to the rapid and organized evacuation of people in the hazardous area to a shelter within the time allowed for evacuation (Liu et al. 2015), and earthquake shelters are the core of urban earthquake disaster prevention planning, and are an effective platform to carry out post-disaster emergency responses (Zhang et al. 2021).

To determine an earthquake shelter, we need to comprehensively consider many different factors. The multi-criteria analysis (MCA) method has been widely used in the determination of earthquake shelters. According to relevant international and Chinese standards, Lai et al. (2015) compiled the minimum requirements for shelters for living purposes. Yin et al. (2020) constructed an evaluation index system for the suitability of emergency shelters site selection, established a spatial analysis model using ArcGIS, and evaluated the suitability of 27 shelters in Tianjin, China. Based on multiple different standards and geographic information systems (GIS), Xu et al. (2016) determined emergency shelters from the perspective of urban planning. Based on the analytic hierarchy process in the MCA Method, Li et al. (2020) evaluated the disaster mitigation capability of earthquake emergency shelters for primary and secondary schools in southern Ningxia, China. Based on appropriate scale, accessibility standard, and site selection standard, Şenik & Uzun (2020) evaluated the suitability of the open green space in the downtown Düzce as an emergency shelter and included earthquake disaster response strategies in the green space planning of downtown Düzce. Based on expert experience and extensive reading of related literature, Soltani et al (2015) built a multiple-criteria analysis frame system to determine emergency shelters after earthquakes. Tsioulou et al (2021) constructed a MCA method system to compare the relative suitability of different school buildings as emergency shelters.

The MCA method allows us to consider the advantages and disadvantages of the selected emergency shelter from several different perspectives. The uncertainty of ground motion, seismic risk, and the vulnerability of ground motion of building groups means that the situation after the earthquake has a higher degree of uncertainty and complexity. By analyzing the earthquake shelter with the MCA method, we cannot predict certain emergencies after the earthquake and take targeted measures in advance. Through scenario planning, we can build several different possible futures, which is conducive to exploring future risks, uncertainties, and complexity. Some researchers combined scenario planning and MCA method to form the MCA method guided by scenario planning (Schroder & Lambert 2011). Through the constructed scenario, we can explore future uncertainties. The combination of MCA method and the constructed scenario allows the evaluator to evaluate based on a number of different criteria in each scenario. In this study, the MCA framework guided by scenario planning was constructed to evaluate the suitability of earthquake emergency shelters in a variety of different scenarios.

2 RESEARCH OVERVIEW

Daniell et al (2017) pointed out that throughout the 100 severe earthquakes since 1900, the secondary disasters of earthquakes such as tsunami, fires, landslides, liquefaction, and fault fracture have caused huge economic losses and heavy casualties like earthquakes. Seismic events are quite random in time, space, and intensity distribution, and the vulnerability of building groups to the ground motion is uncertain (Wang & Gao 2019). Zhang et al. (2021) pointed out that extreme weather and climate events frequently occurred in recent years, and they are having an increasingly extensive impact. Therefore, there is an urgent need for meteorological services to provide more targeted and accurate predictions for earthquakes. Li & Yuan (2018) pointed out that after an earthquake, natural geological disasters such as collapses, landslides, and debris flows might frequently occur in the severely afflicted areas. In view of this, Karpouza et al (2021) proposed a method that allows mapping of the hazard zone of earthquake-induced secondary effects. The earthquake-induced secondary effects in this study should include landslides and soil liquefaction (Karpouza et al. 2021). Li & Geng (2012), and Liu and Wang (2017) pointed out that secondary fires are likely to occur after destructive earthquakes due to the complex power grids, pressure pipelines, and gas pipelines in cities. Suboyama-Kasaoka et al. (2014) believed that shortage of supplies is a common problem after the earthquakes. In addition, seeing as people often stay indoors at night, there will be an increase in casualties in case of earthquakes at night, especially during the time when most people are fast asleep. People who are fast asleep cannot respond to sudden earthquakes (Park et al. 2016).

Urban earthquake shelters are safe places that provide evacuation protection and basic life support services after an earthquake. Some researchers have established the evaluation index system for earthquakes evacuation shelters. Liu et al. (2010) established hierarchical principles and the model for site selection of emergency shelters for catastrophic earthquake disasters through comprehensive risk analysis and evaluation of earthquake disasters according to field investigation data and analysis data. Xu et al. (2016) summarized the principles of the site management of earthquakes evacuation shelters from the perspective of urban planning, which are the principle of safety, the principle of land use control, the principle of evacuation to the nearest place, the principle of economy, the principle of proper evacuation time and distance, the principle of maximum coverage, and the principle of population capacity. Based on these principles, a multi-criteria evaluation model was proposed to determine the location of earthquake shelters. Yao et al. (2021) mainly evaluated the suitability of the site of earthquake emergency evacuation shelters based on safety, accessibility, and availability. Xiong et al. (2014) pointed out that factors affecting the disaster mitigation and resilience capabilities of earthquake emergency shelters can be divided into environment support capability, emergency resource protection capability, and social control capability. Zhao et al. (2017) pointed out that evacuation time and construction costs are two important factors in the site selection of earthquake emergency shelters. Şenik & Uzun (2020) selected earthquake evacuation shelters based on appropriate scale, accessibility, and certain suitability criteria. Anhorn & Khazai (2015) proposed a method for considering the ability of open spaces to be used as earthquake emergency shelters. This method also considers the extent to which the site is suitable for being used as an emergency shelter and the accessibility of the site. Trivedi & Singh (2017) selected emergency shelters through a hybrid group decision support method. The factors related to site selection were determined by consulting some disaster management experts and extensively reviewing relevant literature[28].

A scenario is a set of future hypothetical events constructed to illustrate the possible chain of causal events and their decision points (Amer et al. 2013). Scenario planning provides an opportunity to imagine a reasonable future state, thereby helping to reduce risks, making use of opportunities, and avoiding potential threats in the process of strategy making Ramírez and Selin). Scenarios can also be combined with MCA to determine scores, weights, and standards of different criteria in different scenarios (Ward et al. 2016). Montibeller et al. (2006) and Schroder & Lambert (2011) bound a series of events in the future with scenarios and adjusted the weights of the evaluative criteria according to the scenarios.

3 RESEARCH METHOD

This section overviews the evaluation method based on the scenario in this study and focuses on the multiple-criteria comparison of the suitability of earthquake emergency shelters. This method was developed by reference to the papers of Schroder & Lambert (2011), Schroeder (2009), and Schoemaker (1993), and is suitable for evaluating the suitability of several different earthquake emergency shelters in different scenarios.

The first step in this study is to determine the relevant scenarios after the earthquake as well as the evaluation and decision-making criteria. The scenarios and decision-making criteria in this study were determined through guidance documents and academic literature of relevant government and non-government organizations. We believe that these scenarios and analytical decision-making criteria are widely applicable to the suitability evaluation of different earthquake emergency shelters, and we defined a group of scenarios S, where S_n represents the scenario as planned.

$$S = \{S_n : n \in \{0, \ldots N\}\} \tag{1}$$

Next, this study defined a group of evaluation and decision-making criteria C, where C_n represents the nth evaluation and decision-making criterion.

$$C = \{C_n : n \in \{0, \ldots N\}\} \tag{2}$$

Next, we defined the weight of the default S_0 evaluation criterion for the scenario in which there was no other emergency after the earthquake. In this study, we defined the weights of a group of default evaluation criteria as W_n^0. Based on this group of evaluation and decision-making criteria C and the scenario in which there was no other emergency after the earthquake S_0,

$$W_n^0 = \{W_1^0, W_2^0, \ldots, W_n^0\}$$

Where

$$\sum_{k=1}^{n} W_n^0 = 1\ 0 \leq W_n^0 \leq 1 \tag{3}$$

The weights are determined based on the rank sum method (Zardari et al. 2015). The researchers first ranked their importance according to the criteria in the default scenario S_0 in which there was no other emergency after the earthquake, where r represents the ranking of importance of this criterion in all the criteria. Its weight W_r^0 can be calculated by the following formula. Many experts determined the average weight of each criterion by averaging the weighted values given by ranking.

$$W_r^0 = \frac{2n - 2r + 2}{n^2 + n} \tag{4}$$

Next, this study generates the influence scores for each earthquake emergency shelter E_n according to the criteria and obtains the sum of the influence scores for each earthquake emergency shelter E_n according to the criteria. We defined the scoring function as Z_{jk}, where

$$Z_{jk} = \begin{cases} 1 & \text{if the earthquake emergency shelter } \mathrm{E_j} \text{ fully meets the criterion } \mathrm{C_k} \\ 0.5 & \text{if the earthquake emergency shelter } \mathrm{E_j} \text{ partially meets the criterion } \mathrm{C_k} \\ 0 & \text{if the earthquake emergency shelter } \mathrm{E_j} \text{ does not meet the criterion } \mathrm{C_k} \end{cases} \tag{5}$$

In this study, Z_n^0 was defined as the total score of the suitability of each earthquake emergency shelter E_n in the default scenario S0.

$$Z_n^0 = \sum_{n=1}^{n} W_n^0 Z_{jk} \tag{6}$$

The fourth step of this study is to redefine the weight of the evaluation criterion according to scenario Sn, and the weight re-determined is defined as W_k^n. This study defines that, if the importance of a criterion in scenario Sn is greatly increased or slightly increased than that in scenario S0, the weight is multiplied by three-halves and five-fourths respectively on the original basis, and if the importance of a decision criterion in scenario Sn is significantly deceased or slightly decreased than that in scenario S0, the weight is multiplied by two-thirds and four-fifths respectively on the original basis. If the importance of a decision criterion remains the same in the situational Sn, then the weight is multiplied by 1 on the original basis. When all the evaluation criteria in a certain category have completed the relevant scaling, it is necessary to adjust all the weights proportionally according to the principle that the sum of the weights in each category is 1.

$$W_k^n \propto = \begin{cases} W_k^0(\delta) & \text{if the importance of the criterion is greatly increased in scenario } \mathrm{S_k} \\ W_k^0(\varepsilon) & \text{if the importance of the criterion increases slightly in scenario } \mathrm{S_k} \\ W_k^0 & \text{if the importance of the criterion does not change in scenario } \mathrm{S_k} \\ W_k^0\left(\frac{1}{\delta}\right) & \text{if the importance of the criterion decreases slightly in scenario } \mathrm{S_k} \\ W_k^0\left(\frac{1}{\varepsilon}\right) & \text{if the importance of the criterion is greatly decreased in scenario } \mathrm{S_k} \end{cases} \tag{7}$$

where

$$\sum_{k=1}^{n} W_k^n = 1 \; 0 \leq W_k^n \leq 1 \tag{8}$$

The fifth step is to calculate the total suitability score of each earthquake emergency shelter E_n in the scenario Sn according to the redefined evaluation weight W_k^n in each scenario S_n.

$$Z_j^n = \sum_{k=1}^{k} W_k^n Z_{jk}$$

4 DETERMINATION OF SCENARIOS AND CRITERIA

Based on relevant literature analysis and field investigations, this study established relevant emergency scenarios (Table 1) and the evaluation index system for the suitability of the site of earthquake evacuation shelters (Table 2). In the research review of this paper, some relevant references used to develop scenarios and the evaluation index system are introduced. The field investigation was mainly made on some earthquake emergency shelters and the surrounding environment. In the process of the development of scenarios, ground motion, seismic risk, the vulnerability of building groups to the ground motion, and some common emergencies during and after the earthquake were mainly considered. The criteria were mainly determined based on three different aspects, which are the safety assurance capability, accessibility, and suitability of the site. The relevant comprehensive evaluation indicators for each aspect are summarized in detail.

Table 1. Emergency scenario Sn is used to reweight the criteria.

No.	Contents
S1	Flood after the earthquake
S2	Debris flow after the earthquake
S3	Landslide after the earthquake
S4	Collapse after the earthquake
S5	Fire after the earthquake
S6	Aftershock after the earthquake
S7	Sudden outbreak of epidemic after the earthquake
S8	Hurricane after the earthquake
S9	Tsunami after the earthquake
S10	Frequent extreme weather after the earthquake
S11	earthquake in wee hours
S12	Critical shortage of emergency supplies
S13	Severe earthquake
S14	Extremely serious damage to buildings

5 CASE STUDY

Kunming is the provincial capital of Yunnan Province and one of the central cities in southwest China. Its geographical coordinates are 102°10′ to 103°40′E and 24° 23′ to 26°33′ N. In 2020, Kunming's permanent population was 8.46 million (Kunming Municipal People's Government 2021). As Kunming is located between the Xiaojiang fault zone and the Puduhe fault zone, it is prone to earthquakes (Li et al. 2021). Kunming Municipal People's Government assigns great importance to the setting up of earthquake emergency shelters and has set up such shelters throughout the city (Kunming Earthquake Administration 2018a). The municipal government has classified earthquake emergency shelters into three categories based on the layout of the facility and the number of days for which it can accommodate people. Category A has the complete facility configuration and can shelter people for the longest time, i.e. minimum of 30 days. Category C has the simplest configuration and shelters people for the shortest period, i.e. maximum of 10 days (Kunming Earthquake Administration 2018b). This study selected three Category C shelters in Chenggong District of Kunming City as the research objects. They are Luolong Park, No.3 Middle School in Kunming and the Chenggong Campus of Yunnan Normal University (Figure 1) (Kunming Earthquake Administration 2018a).

The site suitability of the three shelters was evaluated using an established evaluation system for earthquake emergency shelters. In Table 3 the sequence numbers of the criteria are used to substitute for each criterion, and the specific content of the criteria can be found in Table 2 according to the sequence numbers in Table 3. The marking standard is described explicitly in the research methods section. Five experts were given questionnaires, in which they ranked the importance of each

Table 2. Evaluation indicator system of the suitability of the site of earthquake evacuation shelters.

Category	No.	Criterion
Safety assurance capability of the site	C1	Flat ground
	C2	Avoid places prone to natural disasters
	C3	Outside the range of the collapse of high-rise buildings
	C4	Convenient transportation
	C5	There are several roads for evacuation
	C6	Away from inflammable and explosive chemical plants
Accessibility of the site	C7	Appropriate evacuation time
	C8	Appropriate evacuation distance
	C9	Follow the principle of evacuation to the nearest place
	C10	First aid response speed
	C11	Close to residential areas
	C12	Evacuation capacity of the road
	C13	Whether people know the emergency shelter
	C14	Consider possible problems of women, children, and the disabled in the shelter
	C15	Set the emergency shelter close to hospitals, elementary schools, etc.
	C16	Varieties and supply of reserved substances
Suitability of the site	C17	Take care of vulnerable groups
	C18	Completeness of infrastructure
	C19	Economical site that avoid waste
	C20	Follow the principle of disaster prevention. Use as schools, hospitals, sports centers, and others in ordinary times. Convert into shelters after the earthquakes
	C21	Land use control
	C22	The emergency shelters can be easily maintained by victims
	C23	Population capacity
	C24	Maximum coverage
	C25	Emergency resource guarantee capacity
	C26	Consider the protection of historical heritages
	C27	Drainage Capacity

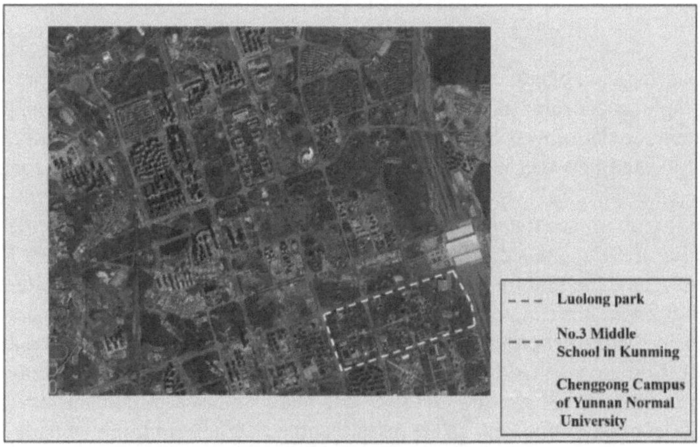

Figure 1. Geographical Location Map of Three Earthquake Emergency Shelters.

criterion in each category. The formula in research method section was used to calculate and convert ranks to weights. The final weight of each criterion was obtained by calculating the mean of the weights given by the five experts. The weights and evaluation results for each criterion are listed in Table 3.

Table 3. Weight of each criterion and evaluation results of the three shelters without any other emergency (S_0).

Category	Criterion SN	Weight	Luolong Park	No.3 Middle School in Kunming	Chenggong Campus of Yunnan Normal University
Site safety guarantee capability	C1	11.43%	0.5	0.5	0.5
	C2	24.76%	1	1	1
	C3	24.76%	1	1	0.5
	C4	8.57%	1	1	1
	C5	14.29%	1	1	1
	C6	16.19%	1	1	1
Site accessibility	C7	13.09%	1	1	1
	C8	12.73%	1	1	1
	C9	14.55%	1	1	1
	C10	9.82%	0.5	1	0.5
	C11	10.55%	1	0.5	1
	C12	8.72%	1	0.5	0.5
	C13	11.64%	0.5	0.5	0.5
	C14	7.27%	0.5	0.5	0.5
	C15	8%	0.5	1	0.5
	C16	3.64%	0.5	0.5	1
Site suitability	C17	7.58%	0.5	1	0.5
	C18	12.73%	0.5	0.5	1
	C19	3.64%	0.5	0.5	0.5
	C20	12.73%	0.5	1	1
	C21	6.36%	0.5	0.5	0.5
	C22	8.48%	0.5	0.5	0.5
	C23	11.82%	0.5	0.5	0.5
	C24	11.82%	0.5	0.5	1
	C25	13.03%	0.5	0.5	1
	C26	7.58%	1	1	1
	C27	4.24%	0.5	0.5	1

Further, three scenarios, namely, S5 (post-earthquake fire), S11 (earthquake in wee hours), and S12 (severe shortage of emergency supplies), were selected to redefine the weights of the criteria in different scenarios. The details of the methods of redefining the weights of the criteria under different scenarios are summarised in the research methods section. Table 4 lists the weights in the three emergencies for each criterion. The total score of each shelter under each scenario was calculated by the methods introduced in the research methods section. The specific scores are listed in Table 5.

It can be seen from Table 5 that in the absence of other emergencies after an earthquake (scenario S0), the total score of the site suitability of the emergency shelter in the Chenggong Campus of Yunnan Normal University is higher than that in No.3 Middle School in Kunming. The total scores of these two earthquake emergency shelters are significantly higher than that of Luolong Park. Relevant researchers can further analyze the evaluation score of each emergency shelter and the weight of each evaluation criterion of the emergency shelter in scenario S0 according to Table 3, and develop strategies for improvement. In scenario S5, where there is a fire after the earthquake, the scores of the three emergency shelters are higher than those in scenario S0, indicating that these earthquake emergency shelters with certain favorable conditions can be used to respond to the sudden fire after an earthquake. In the case of an occurrence of an earthquake in the wee hours, the total scores of the emergency shelters in Luolong Park and Chenggong Campus of Yunnan Normal University are slightly lower than those in scenario S0, while the total score of No.3 Middle School in Kunming is significantly lower, which also indicates that these three earthquake emergency shelters should take some measures to improve their ability to respond to earthquakes in the wee

Table 4. Redefined weights of the criteria under different situations.

Item	Criterion SN	S5 Post-earthquake fire	S11 Earthquake in wee hours	S12 Severe shortage of emergency supplies
Site safety guarantee capability	C1	9.56%	13.19%	10.26%
	C2	20.72%	28.57%	22.22%
	C3	20.72%	23.81%	22.22%
	C4	10.76%	8.24%	11.54%
	C5	17.93%	13.74%	19.23%
	C6	20.32%	12.45%	14.53%
Site accessibility	C7	15.53%	14.11%	11.21%
	C8	15.10%	13.71%	10.9%
	C9	11.50%	12.54%	12.46%
	C10	9.71%	8.46%	12.62%
	C11	8.34%	11.36%	9.03%
	C12	10.35%	9.4%	11.21%
	C13	11.50%	12.53%	9.97%
	C14	7.20%	7.84%	9.35%
	C15	7.91%	6.9%	8.57%
	C16	2.88%	3.13%	4.67%
Site suitability	C17	8.13%	9.13%	9.83%
	C18	10.93%	12.28%	13.76%
	C19	3.12%	3.51%	3.15%
	C20	10.93%	9.82%	11.01%
	C21	5.47%	6.14%	5.5%
	C22	10.93%	8.18%	9.17%
	C23	10.15%	11.4%	10.22%
	C24	10.15%	11.4%	10.22%
	C25	16.79%	15.71%	16.91%
	C26	9.76%	7.31%	6.55%
	C27	3.64%	5.12%	3.67%

Table 5. Summary of the site suitability for the three shelters under three scenarios.

	Luolong Park	No. 3 Middle School in Kunming	Chenggong Campus of Yunnan Normal University
S0 default scenario	2.279	2.373	2.403
S5 Post-earthquake fire	2.305	2.395	2.426
S11 Earthquake in wee hours	2.276	2.344	2.397
S12 Severe shortage of emergency supplies	2.256	2.364	2.390

hours. In the case of a critical shortage of emergency supplies after an earthquake, the site suitability of the emergency shelter in Kunming No. 3 Middle School and Chenggong Campus of Yunnan Normal University decrease slightly and the site suitability of the emergency shelter in Luolong Park decreases significantly. In general, the Chenggong Campus of Yunnan Normal University has the highest total score in this scenario.

6 CONCLUSION

Through scenario planning, we can create a variety of possible futures and explore future uncertainties and environmental complexity. By evaluating the suitability of earthquake emergency shelters through scenario planning and MCA method, we can comprehensively evaluate earthquake emergency shelters under a variety of possible emergency scenarios, which allows us to better predict the site suitability of earthquake emergency shelters in different possible future situations in advance. In this study, it is believed that most of the previous studies on the evaluation of the site suitability of earthquake emergency shelters mainly focused on the evaluation of the site suitability of earthquake emergency shelters based on different evaluation criteria, but neglected to explore the future uncertainties and environmental complexity through methods other than MCA. This study proves that this method can be used in the evaluation of earthquake emergency shelters. In this study, it is believed that more research can be carried out to explore the suitability of earthquake emergency shelters through the use of more methods combining scenario planning and MCA. Through these studies, we can better establish earthquake emergency shelters to minimize economic losses and casualties after earthquakes.

REFERENCES

Amer, M., Daim, TU., Jetter, A. (2013) A review of scenario planning. *Futures*. 46: 23–40.

Anhorn, J. & Khazai, B. (2015) Open space suitability analysis for emergency shelter after an earthquake. *Nat. Hazards Earth Syst. Sci.* 15(4):789–803.

Dan, B., Qiu, Y., Fan, K., Wang, K. (2018) Optimizing Allocation of Urban Seismic Shelter for Evacuation Based on GIS and Huff Model. *Geo. Geo Inf. Sci.* 34(2): 22–26.

Daniell, J.E., Schaefer, A.M., Wenzel, F. (2017) Losses Associated with Secondary Effects in Earthquakes. *Front. Built. Environ.* 3:1–14.

John, W. E., Harry, D.T., Marco, D. (2016) Theory and Background of Multi-Criteria Analysis: Toward a policy-led MCA for megaproject transport infrastructure appraisal. *Res. Transp. Econ.* 58(C):21–45.

Karpouza, M., Chousianitis, K., Bathrellos, G.D., Skilodimou, H.D., Kaviris, G., AntoNara-Ketsap, A., Hansapinyo, C., Kronprasert, N., Limkatanyu, S. (2019) Uncertainty and Fuzzy Decision in Earthquake Risk Evaluation of Buildings., *Eng. J.,* 23(5): 90–105.

kou,A. (2021) Hazard zonation mapping of earthquake-induced secondary effects using spatial multi-criteria analysis. *Nat. Hazards.* 109(1):637–669.

Kunming Earthquake Administration (2018 a) *Notice on the Results of the Second Classified Evaluation of Earthquake Emergency Shelters in Kunming*. http://fzjz.km.gov.cn/c/2018-11-14/3352306.shtml.

Kunming Earthquake Administration (2018b) *Classification and Evaluation Standard of Earthquake Emergency Shelters in Kunming*. http://fzjz.km.gov.cn/c/2018-07-13/2690913.shtml.

Kunming Municipal People's Government. (2021) The basic situation of Kunming. http://www.km.gov.cn/c/2021-07-14/4002080.shtml

Lai, Y., Li, Y., Du, X., Liu, C., Xu, J. (2015) Earthquake Shelter Planning of Newly-Built Communities. *Technol.* Earthquake Disaster Prev. 10(3): 629–638.

Li, T. & Geng,F. (2012) Study on Hazard Analysis Method of Earthquake Secondary Fire Disaster in Urban Area Based on GIS. *Prog. Ind. Civ. Eng.* 204–208:3457.

Li, C., Wang,X., Hu,B., Wang,S., Zhao,Y. (2021) Study on the Probability of Secondary Fire Based on Earthquake Damage. *Sci. Tech. Eng.* 21(18):7711-7716.

Li, T. & Yuan, Y. (2018) Risk Assessment of Secondary Geological Disasters Induced in an Earthquake-Stricken Area. *China Earthq. Eng. J.* 40(1):111–115.

Li,Z., Ma,Z., Li,G., Yu,S. (2020) Evaluation of Disaster Reduction Capability of Earthquake Emergency Shelters in Middle and Primary Schools in Southern Ningxia Based on AHP. Technol. *Earthquake Disaster Prev.* 15(3): 537–547.

Liu, Q., Ruan, X., Fu, B. (2010) A Study of Principles on Emergency Shelter Site Selection and Modeling Under Disastrous Earthquake. *J. Ocean Univ. China (Sci. Technol.)*. 40(8): 129–135.

Liu, X. & Wang, S. (2017) Potential Risk Analysis of Urban Post-earthquake Fire Based on Attribute Interval Recognition Theory. *J. Catastrophology.* 32(4):43–50.

Liu, X., Su, J., Wang, W., Ma, D. (2015) A theoretical model for roads' emergency earthquake evacuating in key area. *Syst. Eng. Theory Pract.* 35(1): 205–215.

Montibeller, G., Gummer, H., Tumidei, D. (2006). Combining Scenario Planning and Multi-Criteria Decision Analysis in Practice. *J. Multi-Criteria Decis. Anal.* 14: 5–20.

Park, J., Shin, M., Cho, G. (2016). A dynamic estimation of casualties from an earthquake based on a time-use survey: applying HAZUS-MH software to Ulsan, Korea. *Nat. Hazards.* 81(1): 289–306.

Ramírez, R., Selin, C. L. (2014) Plausibility and probability in scenario planning. *Foresight.* 16 (1): 54–74.
Schoemaker, P.J.H. (1993). Multiple scenario development: Its conceptual and behavioral foundation. *Strateg. Manag. J.* 14: 193–213.
Schroeder, M.J. (2009) Scenario based reweighting of an additive value function: Application to transportation planning. Master's thesis, University of Virginia.
Schroeder, M.J. & Lambert, J.H. (2011) Scenario-based multiple criteria analysis for infrastructure policy impacts and planning. *J. Risk Res.* 14(2): 191–214.
Şenik, B. & Uzun, O. (2020) An assessment on size and site selection of emergency assembly points and temporary shelter areas in Düzce. *Nat. Hazards.* 105(2):1587–1602.
Soltani, A., Ardalan, A., Boloorani, D.A., Haghdoost, A, Hosseinzadeh-Attar, M.J. (2015) Criteria for Site Selection of Temporary Shelters after Earthquakes: a Delphi Panel. *PLoS Curr.* 7: 1–11.
Trivedi, A. & Singh, A. (2017) Prioritizing emergency shelter areas using hybrid multi-criteria decision approach: A case study. *J. Multi-Criteria Decis. Anal.* 24(3–4): 133–145.
Tsioulou, A., Walker, J.F., Lo, D.,S., Yore, R. (2021) A method for determining the suitability of schools as evacuation shelters and aid distribution hubs following disasters: case study from Cagayan de Oro, Philippines. *Natural Hazards.* 105(2): 1835–1859.
Tsuboyama-Kasaoka, N., Hoshi, Y., Onodera, K., Mizuno, S., Sako, K. (2014) What factors were important for dietary improvement in emergency shelters after the Great East Japan Earthquake? *Asia Pac. J. Clin. Nutr.* 23(1):159–166.
Wang, D. & Gao, Y. (2019) Study on the probabilistic seismic disaster risk assessment of urban building complex. *Eng. Mech.,* 36(7):165–173.
Xiong, Y., Liang, F., Qiao, Y., Bo, T. (2014) Mitigation Capabilities Evaluation System of Beijing Earthquake Emergency Shelters. Technol. *Earthquake Disaster Prev.* 9 (4):921–931.
Xu, J., Yin, X., Chen, D., An, J., Nie, G. (2016) Multi-criteria location model of earthquake evacuation shelters to aid in urban planning. *Int. J. Disaster Risk Reduct.* 20:51–62.
Xu, J., Yin, X., Chen, D., An, J., Nie, G. (2016) Multi-criteria location model of earthquake evacuation shelters to aid in urban planning. *Int. J. Disaster Risk Reduct.* 20:51–62.
Yao, Y., Zhang, Y., Yao, T., Wong, K., Tsou, J.Y., Zhang, Y.(2021) A GIS-Based System for Spatial–Temporal Availability Evaluation of the Open Spaces Used as Emergency Shelters: The Case of Victoria, British Columbia, Canada. *ISPRS Int. J. Geo-Inf.* 10(63):63.
Yin, Z., Li, L., Wang, X., Huo, P. (2020) Location Selection of the Emergency Shelters in Tianjin: An Evaluation Based on GIS. Technol. *Earthquake Disaster Prev.* 15(3): 571–580.
Zardari, N.H., Ahmed, K., Shirazi, S.M., Yusop, Z.B. (2015) *Weighting methods and their effects on multi criteria decision makingmodel outcomes in water resources management.* Springer. Skudai.
Zhang, J., Chen, T., Huang, L., Su, G., Sun, Z., Chen, J. (2021) Community evacuation scheme based on multi-objective programming and agent simulations. *J. Tsinghua Univ. (Sci. Technol.).* 61(6): 536–542.
Zhang, T., Shu, B., Xie, L. (2021) Construction of On-site Emergency Meteorological Service System to Sudden-onset Disasters in Yunnan Province. *J. Catastrophology.* 36(2):181–186.
Zhao, X., Xu, W., Ma, Y., Qin, L., Zhang, J., Wang, Y. (2017) Relationships Between Evacuation Population Size, Earthquake Emergency Shelter Capacity, and Evacuation Time. *Int. J. Disaster Risk Sci.* 8(4):457–470.

Advances in Civil Engineering: Structural Seismic Resistance, Monitoring and Detection – Mohd Johari Mohd Yusof and Junwen Zhang (Eds)
© 2023 The Authors, ISBN 978-1-032-31491-4

The seismic damage and analysis method system for earthquake vulnerability of group culverts

Peng Du
Guangdong Earthquake Agency, Guangzhou, China

Lixin Wang
Shenzhen Academy of Disaster Prevention and Reduction, Shenzhen, China

Xiao Tan
Central Research Institute of Building and Construction (Shenzhen) Co., Ltd, Shenzhen, China

Jiajian Zhu & Yi Zhang
Guangdong Earthquake Agency, Guangzhou, China

ABSTRACT: The group culverts as one of the key nodes of seismic vulnerability analysis of traffic system must participate in the connection reliability analysis of traffic system directly. Based on the investigation and data analysis of the seismic damage of the culverts belonging to Longnan highway section after the Wenchuan earthquake in Sichuan Province, and the comparative study of the seismic damage phenomenon of the culverts after the Tangshan earthquake, five seismic damage grades and descriptions of the culvert are obtained. According to the phenomenon and cause of seismic damage, six main factors of the seismic vulnerability analysis method are selected, which are earthquake intensity, site type, structure type, portal building type, culvert span and culvert length. Using 100 actual group culvert cases among them with relatively uniform attribute distribution, the statistical analysis method, the support vector machine method and the earthquake damage degree evaluation method are studied respectively. The research results are the method system of culvert seismic vulnerability analysis. It is not only a supplement to the earthquake damage prediction of transportation system, but also can be applied to earthquake disaster risk assessment, the research of culvert seismic design method and the scientific management of culvert engineering.

1 INTRODUCTION

The post-earthquake traffic system is the lifeline project that determines the life and death speed of rescue and disaster relief. The road is the life channel that must be opened up at the first time for emergency rescue. It is very important to ensure smooth traffic as much as possible. It is of great practical significance to carry out the earthquake damage prediction of transportation system before the earthquake. It can predict the earthquake damage level and connectivity probability including various infrastructure, provide reference for maintenance and reinforcement, reduce the loss degree of earthquake disaster, and realize the scientific management of culvert engineering from the perspective of earthquake resistance.

In the past, when the vulnerability analysis of urban lifeline system was carried out, the consideration of traffic system was the network system including roads, bridges, tunnels, railways and other facilities. However, the culvert works of highway traffic system accounted for a considerable proportion. According to statistics, the number of culvert accounts for about 60%–70% of the total

DOI 10.1201/9781003310884-99

number of bridges and culverts. In plain areas, there are about 1–3 culverts per kilometer. In mountainous and hilly areas, there are about 4–6 culverts per kilometer on average. The cost of culvert accounts for about 40% of the total amount of bridges (Zhao 2010). Therefore, it is proposed to take the culvert works as the seismic damage influencing factor of traffic system and participate in the network connectivity analysis of traffic system.

The study area (or working area) earthquake disaster prediction is based on the theoretical basis of earthquake disaster and loss prediction method (Yin 1996), according to the characteristics of prediction object. At present, it has formed the working basis of "Code for earthquake loss estimation and its information management system". The research object covers the infrastructure such as buildings (structures), lifeline engineering system, secondary disaster sources and so on, which maintain the normal operation of the area.

The primary unit of lifeline engineering system is network system, and the secondary unit includes key nodes, effective segments, etc. The traffic system includes urban road system and highway network system. The proportion of culvert as the key node cannot be ignored. Therefore, the culvert, road, bridge, tunnel and so on should be the second level unit of the traffic system and participate in the connectivity reliability analysis of the traffic system network as the key node. The analysis method system of seismic vulnerability for group culverts is a supplement to the earthquake damage prediction method of traffic system. It can also be applied to the risk assessment of earthquake disaster, the research of culvert seismic design method and the scientific management of culvert engineering (He 2010), etc.

2 CULVERT ENGINEERING

The culvert is a drainage structure crossing the roadbed, which is composed of the culvert body, the portal building and the auxiliary facilities. The culvert body is the main part of water passage, which is usually composed of bearing structure, culvert abutment, culvert pier, waterproof layer, expansion joint and other parts. The portal building is a connecting structure of culvert body, roadbed and river course, which is composed of three parts: water inlet, water outlet and ditch bed reinforcement. Its type includes splay type, one-word type (end wall type), twist slope type, flat head type, corridor type, streamline type and drop well type. Its function is connecting the river course and stabilizing the roadbed slope.

According to the structural type, culverts are divided into arch culvert, slab culvert, circular pipe culvert, box culvert, inverted siphon culvert and steel corrugated pipe culvert, etc.; According to the building materials, it can be divided into reinforced concrete culvert, prestressed reinforced concrete culvert, stone culvert, wood culvert and steel corrugated pipe culvert, etc. The culverts of various structural types have their adaptability and demands for the construction environment:

Arch culvert: The tunnel body is arched, including stone arch culvert, concrete arch culvert and wood arch culvert. It is widely used because of its easy masonry, large bearing capacity, deep ditch and high embankment.

Slab culvert: It is composed of slab, abutment and foundation, including stone slab, reinforced concrete slab, etc. It has clear stress, simple structure. It is similar to the simple supported beam bridge with small span and suitable for the case of less stones and larger water inflow.

Circular culvert: It is composed of pipe body, foundation and joint. The pipe body is usually reinforced concrete structure, which has good adaptability and stress. It can be widely built on soft foundation, clay or sand foundation and rock foundation.

Box culvert: The whole reinforced concrete closed structure consists of cover plate, culvert body and foundation, with square or rectangular section. Because of its strong integrity, it is suitable for soft soil foundation except high cost.

Inverted siphon culvert: It is mainly composed of inlet section, horizontal section and outlet section. The inlet section is composed of inlet ditch, sedimentation tank and joint. The horizontal section is main body. The outlet section is composed of outlet well and outlet ditch.

3 SEISMIC DAMAGE OF CULVERT

3.1 *Statistics of seismic damage data*

In the Tangshan earthquake, 80 culverts of all kinds suffered from different degrees of damage, which had been divided into four categories (Liu 1986):

(1) The culvert abutment and pier were broken and collapsed, making the fracture and arch fall of culvert cover plate;
(2) The inclination and cracking of culvert abutment and pier caused serious cracking of culvert roof or arch ring;
(3) The splayed wall and wing wall collapsed, and the ancillary works such as paving were seriously damaged, but the main structure had less seismic damage;
(3) The splayed wall, wing wall and pavement are damaged.

Table 1. Tangshan seismic damage statistical classification of culvert.

Classification	I	II	III	IV
Number	18	13	35	14
Percentage	22.5%	16.2%	43.8%	17.5%

In Wenchuan earthquake, the seismic damage of 275 culverts is as follows:

Slab culvert: cracking and breaking of slab, culvert body and culvert bottom; concrete cavity, exposed reinforcement and corrosion of reinforcement, etc.;

Arch culvert: main arch ring cracking, exposed reinforcement; culvert body cracking;

Pipe culvert: main body damage; paving stone defect;

Portal buildings: the splayed wall and end wall were cracked, inclined, displaced, suspended and disconnected from the main body of the culvert;

The water dropping well was damaged and the rapid flow was broken;

Foundation: cracking, deformation, inclination, sinking and collapse of culvert abutment and pier; abutment cap fracture, foundation suspension; loosing and falling of concrete protective layer, exposed reinforcement;

Others: mountain sliding, subgrade collapse, slope subsidence, side ditch damage, serious siltation; bottom hollowing, pavement damage; embedded crack, side block, edge stone damage, etc.

On the basis of the above five categories, one type of impact of earthquake geological disasters is added as follows:

Table 2. The earthquake in Wenchuan Longnan road culvert classification statistics.

Classification	I	II	III	IV	V
Number	126	72	40	21	16
Percentage	45.8%	26.2%	14.5%	7.6%	5.8%

3.2 *Causes and commonness analysis*

The seismic damage of culvert is mainly caused by the following reasons:

(1) Large magnitude and high intensity;
(2) The actual strength of the structural materials is not up to the standard, and the embedded depth of the foundation is not enough, which leads to the lack of earthquake resistance;

(3) Only the main structure design and the portal structural design are considered for earthquake resistance, but the structural joints are not considered;
(4) The influence of seismic geological disasters, such as landslides and collapses;
(5) The disease appeared before the earthquake (Long 2006). For example, foundation settlement, culvert body cracks (Ou 2006), local carbonation of components, pitted surface, honeycomb, cavity, peeling off and even exposed reinforcement; the portal building has cracks, extroversion and so on. The seismic damage is more serious in these cases.

4 CLASSIFICATION OF SEISMIC DAMAGE

According to the summary of seismic damage data of culvert and the classification of seismic damage of bridges (Du 2010), five seismic damage levels of the culvert are divided as follows:

Basically intact: there is no earthquake damage, or there are only minor cracks in the culvert body, arch ring, culvert pier and culvert abutment, which are in normal use;

Slight damage: there are slight cracks in the culvert body, arch ring, culvert pier and culvert abutment, the width of that is not more than 0.5mm. The main structural, components and functions are intact, and they can be used without repair or simple short-term repair;

Moderate damage: there are obvious cracks in the culvert body, arch ring, pier, abutment and socket, with the width between 0.5mm and 1.0mm. The main structure and components have cavities and slightly exposed reinforcement. The culvert pier and culvert abutment are inclined and deformed, and the overlying subgrade soil sinks. The portal buildings are deformed and displaced. The main structure and function are affected to some extent, so it needs to be used after maintenance and reinforcement measures are taken;

Serious damage: there are serious cracks in the culvert body, arch ring, pier, abutment and socket, with the width greater than 1.0mm. The main components are exposed, the reinforcement is deformed and rusted, or even partially broken. The foundation is suspended and the abutment body is broken and sunk. The portal buildings have moved and toppled, even collapsed. The chute is broken. The main body of the culvert has not collapsed and its function has not been completely lost, but it can only be used after a long time and a large amount of work;

Destruction: the main body of culvert collapses and damages, the portal buildings collapses, or the main body and functions are completely damaged due to the earthquake and geological disasters, so they cannot be used anymore and need to be demolished and rebuilt.

The damage degree corresponding to these five damage grades is defined as 1.0, 2.0, 3.0, 4.0 and 5.0 respectively.

5 FACTORS AFFECTING SEISMIC DAMAGE

According to the seismic damage phenomena and causes of culvert samples, the influencing factors are selected as follows:

(1) Seismic intensity: mainly distributed in VII, VIII and IX degree areas.
(2) Site category: most samples are located in class II sites. Class I, III and IV sites are recommended coefficient values based on the statistical experience of related structures.
(3) The portal building type: the damage degree of splayed wall opening and end wall opening is equivalent, and they are classified into one category. The splayed wall/end wall opening and drop well opening are selected as factor classification.
(4) Culvert Type: mostly slab culvert, arch culvert, circular pipe culvert, box culvert, etc.
(5) Culvert span: according to relevant specifications and actual earthquake damage degree, 3m is selected as the span boundary value.
(6) Culvert length: three length divisions are defined with the boundary of 15m and 30m.
(7) Structural materials: used for the adjustment of seismic damage index value.
(8) Sand liquefaction and slope instability: refer to the relationship between site category, seismic intensity, liquefaction and slope instability in bridge seismic damage assessment.

6 SEISMIC VULNERABILITY ANALYSIS METHOD

6.1 *Statistical analysis method*

The statistical analysis method is quadratic discrimination method (Du 2015): The first discrimination is based on the influence factors of seismic damage and empirical statistical formula to calculate the index value of seismic damage; The secondary discrimination is based on the adjustment of the factors such as the failure degree of the foundation, the seismic damage of the auxiliary facilities and the seismic geological background to get the index value of the resulting damage.

Six seismic damage factors, including seismic intensity, site category, culvert type, portal type, culvert span and culvert length are taken as first discriminant calculation influence factors. Sand liquefaction and slope instability are considered as the influence factors of the second discriminant calculation. Other auxiliary facilities, such as subgrade slope, curb block and complex geological environment factors, are used as the empirical influence factors of secondary discrimination.

6.1.1 *First discrimination*

Most of the sample culverts that were built in the 1980s and 1990s had not any seismic fortification or measures. 100 of them are selected as the research object, and the seismic damage index is assumed to be the product of the coefficients of various influencing factors. The relationship mode is as follows:

$$\mathrm{Y_i} = \alpha \prod_{j=1}^{n} \prod_{k=1}^{r_j} \beta_{jk}^{x_{jk}} \xrightarrow{\text{Logarithm}} y_i = \phi + \sum_{j=1}^{n} \sum_{k=1}^{r_j} \lambda_{jk} x_{ijk} \xrightarrow{\text{least square method}} \hat{Y}_i = \alpha_0 \prod_{j=1}^{5} \prod_{k=1}^{2} \beta_{0jk}^{X_{jk}} \tag{1}$$

In formula (1), $i = 1, 2, \cdots, m$, m is the number of samples; $j = 1, 2, \ldots, n$, n is the number of influencing factors of bridge seismic damage; $k = 1, 2, \cdots, r_j$, r_j is the classification number of factor j; Y_i is the seismic damage degree of the ith sample bridge; x_{ijk} is the 16 observation numbers of each bridge, and its value is 0 or 1.

Order $y_i = \ln Y_i$ and take 100 sets of sample data pairs $(\ln Y_i, X_{ijk})$ into the second item of formula (1). Considering that the least square method is a mathematical optimization technique to find the best function matching of data by minimizing the square sum of errors, the curve is sought with the least square sum of the given point (x_{ijk}, y_i) distance, and use the least square method to get the category, regression value and correction value corresponding to the six seismic damage factors, as table 3 shows. By rechecking the seismic damage index of 100 culverts, the residual standard deviation is 0.451, the empirical correlation number is 0.948, and the error limit of the prediction model is 1.12 when the confidence level is 95%. 36% of the errors between the predicted value and the statistical value are within a quarter level, 86% within a half level and 95% within 1 level.

Table 3. Regression values and recommended values of seismic influencing factor coefficient.

j	Item	*k*	Category	Regression coefficient value	Suggested factor value
		1	VII degree	1.00	1.00
1	seismic intensity	2	VIII degree	1.93	1.95
		3	IX degree	3.00	3.00
2	site category	1	I		0.80*
		2	II	1.0	1.00
		3	III, IV		1.50*

(*continued*)

Table 3. Continued.

j	Item	k	Category	Regression coefficient value	Suggested factor value
		1	slab culvert	1.00	1.00
3	culvert type	2	arch culvert	1.16	1.15
		3	pipe culvert	1.24	1.20
4	portal type	1	splayed wall / end wall	1.00	1.00
		2	water dropping well	1.30	1.30
5	culvert span	1	≤ 3m	1.00	1.00
		2	>3m	1.05	1.10
6	culvert length	1	≤ 15m	1.00	1.00
		2	15~30m	1.03	1.15
		3	>30m	1.34	1.35
	α_0			1.20	1.18

*is the suggested coefficient value of supplementary category based on statistical experience

6.1.2 *Second discrimination*

(1) Coefficient correction

In view of the consistency of foundation failure as a site influence factor, the proportion relationship between different site categories, seismic intensity and the degree of foundation failure obtained by the improved seismic damage assessment method of beam highway bridge (Guo 2015), as Table 4 shows:

Table 4. Suggested value for foundation failure degree.

Seismic intensity	Site category II		Site category III		Site category IV	
≤ VII degree	few happened	1.0	no or slight happened	1.5	mostly happened	1.5
VIII degree	few happened	1.0	mostly happened	1.5	more serious failure degree	1.8
≥ IX degree	no or slight happened	1.5	more serious failure degree	1.5	more serious failure degree	1.8

(2) Experience correction

For two cases: the structural characteristic parameters are similar or consistent, but the damage degree is different; the deviation between the theoretical value and the actual seismic damage is large, and the index correction is shown in Table 5.

Table 5. Reference correction vulnerability index value of culvert.

NO.	Other seismic damage factors	Affirmative method
1	Reinforced	Reduce 0.5~1.0
2	Diseases	Increase 0.5
3	Seismic damage of culvert pier and abutment	Increase 0.5~1.0
4	Long time construction no seismic design or fortification measures	Increase 0.5~1.0
5	Seismic damage of auxiliary facilities (side ditch, side block, curb, etc.)	Increase 0.25
6	Incomplete basic data	Increase 0.25~0.5

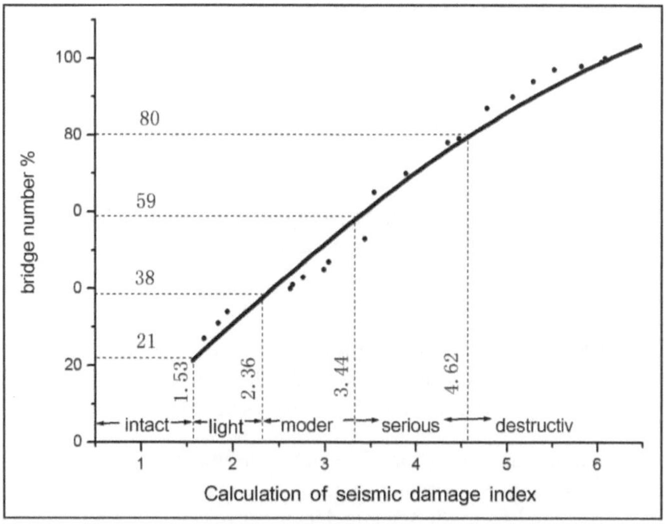

Figure 1. Frequency accumulated forecast curve of railway bridge earthquake damage.

The curve is fitted with quadratic polynomials, and the cumulative distribution of frequency of predicted seismic damage index of culvert samples is summarized. The curve is drawn as shown in Figure 1. According to the actual seismic damage rate, the boundary values of grade are determined again, which are 1.53, 2.36, 3.44 and 4.62 respectively.

6.2 *Support vector machine method (SVM)*

Based on the VC dimension theory of statistical learning theory (Vapnik 1995) and the principle of structural risk minimization, the method can distinguish different sample categories by constructing classification function, which is suitable for solving the problems of non-linear and high-dimensional space recognition.

6.2.1 *SVM principle*

Suppose a data sample set (x_i, y_i), $i = 1, 2, \cdots, n$, $x_i \in R^d$, $y_i \in \{-1, 1\}$ is the label category. By finding the optimal classification surface, the two kinds of sample data are separated, and the geometric interval to the surface is the smallest, and the training error rate is the lowest. The discriminant function is as follows:

$$f(x) = sgn\,(w \cdot x + b) \tag{2}$$

When $w \cdot x + b > 0$, then $y_i = 1$; when $w \cdot x + b < 0$, then $y_i = -1$.. If a given sample is approximately linearly separable, relaxation factor is introduced to allow the existence of misclassified samples; If the samples are nonlinear separable, the data samples are mapped from low dimension feature space to high dimension feature space by nonlinear mapping. In order to obtain the optimal values of w and b, the problem of classification can be transformed into a problem of finding the minimum value with constraints, and its mathematical expression is as follows:

$$\begin{cases} \min \frac{1}{2}\|w\|^2 + C\sum_{i=1}^{n} \xi_i \\ y_i[(wx_i) + b] - 1 \geq 0 \end{cases} (i = 1, 2, ...l) \tag{3}$$

In the formula, ζ is the relaxation factor; C is the penalty factor, and the solution to the optimal problem is the classification function (Platt 1999):

$$f(x) = sgn\left(\sum_{i=1}^{l} \alpha_i^* y_i K(x, x_i) + b^*\right) \tag{4}$$

Among them, $K(x, x_i)$ is the kernel function, which can realize mapping from low dimension space to high dimension space and solve nonlinear problems.

6.2.2 *Model establishment*

Based on six seismic damage factors and level information of each culvert sample, SVM model for vulnerability analysis of the group culverts is established (Tan 2016), MATLAB software and LibSVM toolbox are used to program the model. The steps are as follows:

(1) Determines the input and output vectors. The input parameter is the classification value corresponding to the seismic damage factor in Table 2, and the output parameter is the seismic damage degree corresponding to the seismic damage level.
(2) Select the kernel function. The commonly kernel functions are polynomial function, radial basis function and sigmoid function. The common selects radial basis kernel function is selected for this model.
(3) Determination of kernel function parameters and penalty factors. The experimental results are 10 and 1 respectively.
(4) SVM prediction model is obtained by training samples.

6.2.3 *Actual earthquake cases*

Among the above 100 sample culverts, 21 are basically intact, 18 are slightly damaged, 20 are moderately damaged, 21 are severely damaged and 20 are damaged. 82 culverts are randomly selected as training data, and the remaining 18 channels are tested for three times, with the accuracy of 78%, 94% and 100% respectively.

6.3 *Evaluation method of seismic damage degree*

The detailed size of culvert structure is ignored in this method. According to the field survey and earthquake damage assessment, the degree of seismic damage can be evaluated directly as a comparison or supplement of the above methods.

6.3.1 *Evaluation method*

The scoring method is the percentage system and the detailed method is as follows:

Table 6. Seismic damage evaluation method of culvert.

Category	Assessment project	Score calculation
1	The main structure is buried or even collapsed under the influence of earthquake and geological disasters.	Reduce 100 points
2	It is not or slightly affected by the earthquake geological disaster, the main structure is slightly damaged but the function is lost, such as the road surface is covered and damaged due to silting, etc.	Reduce 100 points

(*continued*)

Table 6. Seismic damage evaluation method of culvert.

Category	Assessment project		Score calculation
3	Seismic damage evaluation of culvert body components	there is no earthquake damage, or there are only minor cracks in the culvert body, arch ring, culvert pier and culvert abutment, which are in normal use.	Reduce 5 points
		there are slight cracks in the culvert body, arch ring, culvert pier and culvert abutment, the width of that is not more than 0.5mm. The main structural, components and functions are intact.	Reduce 10 points
		there are obvious cracks in the culvert body, arch ring, pier, abutment and socket, with the width between 0.5mm and 1.0mm. The main structure and components have cavities and slightly exposed reinforcement. The culvert pier and culvert abutment are inclined and deformed. The portal buildings are deformed and displaced. The main structure and function are affected to some extent.	Reduce 30 points
		there are serious cracks in the culvert body, arch ring, pier, abutment and socket, with the width greater than 1.0mm. The main components are exposed, the reinforcement is deformed and rusted, or even partially broken. The foundation is suspended and the abutment body is broken and sunk. The portal buildings have moved and toppled, even collapsed. The chute is broken. The main body of the culvert has not collapsed and its function has not been completely lost.	Reduce 60 points
		The main body of the culvert collapses, the structure is damaged and the function is lost.	Reduce 100 points
4	Seismic damage assessment of culvert auxiliary buildings and facilities	The portal buildings are slightly deformed and inclined; the auxiliary facilities are not damaged or slightly damaged.	Reduce 5 points
		The portal buildings are deformed and displaced; the edge stone, edge block and side ditch are slightly damaged.	Reduce 10 points
		Large deformation, displacement and even toppling of the portal buildings, breakage of the edge stone and side block of the chute, and damage of the side ditch.	Reduce 15 points
		Collapse of portal buildings and damage of auxiliary facilities	Reduce 20 points
5	Earthquake damage evaluation of foundation failure (liquefaction, slope instability, etc.)	No or slight appearance	Reduce 5 points
		Most appearance	Reduce 10 points
		Increased failure	Reduce 20 points
6	Structure status determination	Diseases	Reduce 15 points
		Reinforced	Add 10 points

In the same way, 100 sample culverts used to establish the seismic vulnerability analysis model of culverts are selected as the basic data for the definition of score grade in the seismic damage evaluation method, and the following classification criteria are deduced:

Table 7. Culvert damage grade and score interval.

Classification	Interval of value	Seismic damage grade
I	0~20 points (containing 20 points)	Slight damage
II	20~50 points (containing 50 points)	Moderate damage
III	50~80 points (containing 80 points)	Serious damage
IV	80~100 points (containing 100 points)	Destroy

6.3.2 *Actual earthquake cases*

The 64 culverts which are not used to establish vulnerability analysis model of group culverts are selected for seismic damage evaluation. The evaluation results are compared with the vulnerability analysis results. The evaluation seismic damage is the actual seismic damage in the corresponding seismic vulnerability analysis of group culvert. The comparison results of seismic damage show good consistency.

Table 8. Result of seismic damage contrast.

Method	Slight damage	Moderate damage	Serious damage	Destroy
Evaluation	6/64	19/64	21/64	16/64
Calculation	7/64	20/64	20/64	15/64

7 CONCLUSIONS

(1) In the analysis method of seismic vulnerability of culvert, the empirical statistical method is applicable to evaluate the probability level of overall failure of samples, rather than the calculated value of seismic damage level of each culvert. The prediction method of culvert seismic damage based on support vector machine is compared with the results of empirical statistics method and neural network method to verify the accuracy of its output of seismic damage level. The group culvert seismic damage evaluation method is easy to operate, which can be used for quick post-earthquake site evaluation and the seismic damage evaluation result can be given quickly, but there is a certain subjective disturbance. This method can be regarded as a supplement to the theoretical algorithm. Based on the above research results, the seismic vulnerability analysis method of group culvert has been given.

(2) The seismic vulnerability analysis method system of group culverts can be used for post-earthquake site assessment, seismic damage prediction, earthquake disaster risk prevention and control, planning of earthquake prevention and disaster reduction. The method derivation not only considers the compatibility with roads, bridges and tunnels when applied to the reliability analysis of traffic system connectivity, but also absorbs the new ideas of post-earthquake related research, which is easy to operate and constrained by the relevant industry specifications.

(3) The sample data used to analyze the seismic damage of culvert still has some limitations. The culvert samples are mainly distributed in the Longnan highway section, which is influenced by objective factors such as construction age, structural type, statistical quantity, uneven distribution of geographical location, design and construction, etc. The influence weight of site and type difference is weakened on account of the relatively centralized sample distribution. The basic data and destruction of samples show high similarity, which restricts the comprehensiveness of the method to some extent. The method system should be improved by adding sample types and modifying impact item categories.

ACKNOWLEDGMENTS

This work was financially supported by Guangdong Science and Technology Plan Project (2018B020208003), Guangdong Science and Technology Plan Project (2018B020207011).

REFERENCES

Du Peng and Jiang Hui (2010), "Seismic damage and vulnerability analysis of bridges at Longnan highway", *Journal of Earthquake Engineering and Engineering Vibration*. 30(3):107. (in Chinese)

Du Peng and Tan Xiao (2015), "Study on analysis method seismic vulnerability of group culverts", *China Civil Engineering Journal*. 48(5):71–75. (in Chinese)

Guo Endong and Zhao zhao (2013), "Improved methodology for seismic estimation of highway beam bridges", *Journal of Beijing University of Technology*, 39(1): 193–197. (in Chinese)

He Yi (2010), "Culvert management measures", *Modern agricultural science and technology*, 9:262. (in Chinese)

JC Platt (1999), editors, Using Analytic QP and Sparseness to Speed Training of Support Vector Machines. MIT Press.

Liu Huixian, editors, *Tangshan earthquake damage (three)*. Chapter 1–3. (in Chinese)

Long Yonggang(2006), "Disease and prevention of culvert in Expressway", *China Science and Technology Information*. 7:148–149. (in Chinese)

Ou Weiqiang (2006), "Analysis of cracks in top slab of culvert engineering", *Science& Technology Information*. 20:41–42. (in Chinese)

Tan Xiao and DuPeng(2016), "Application of SVM method in the seismic damage prediction of culvert", *World Earthquake Engineering*. 32(2): 243–245. (in Chinese)

VN Vapnik (1995), editors, *The Nature of Statistical Learning Theory*. Springer Verlag.

Yin Zhiqian, editors, *Earthquake Disaster and Loss Prediction Method*, Chapter 1. (in Chinese)

Zhao Jianghui (2010), "On culvert design for expressways", *Shanxi architecture*, 36(8): 283–284. (in Chinese)

Advances in Civil Engineering: Structural Seismic Resistance, Monitoring and Detection – Mohd Johari Mohd Yusof and Junwen Zhang (Eds)
© 2023 The Authors, ISBN 978-1-032-31491-4

The construction of the Hong Kong-Zhuhai-Macau Bridge (HZMB)

Yunxin Xu*
School of Civil Engineering, Beijing Jiaotong University Weihai Campus, Weihai, Shandong, China

ABSTRACT: Hong Kong-Zhuhai-Macau Bridge (HZMB) is one of the most significant fixed links in Chinese and even the world's history. It overcomes plenty of challenges, thus achieves impossible tasks. HZMB is divided into three navigation span bridges and a tunnel. Each part of the bridge wall varies inside. In order to maintain a stable structure, statics of bridge is not the only influential factor. Sea crossing bridges of considerable lengths have to set up artificial islands to ensure the connection between the bridge and the undersea tunnel. At the same time exchange traditional methods to reduce the water resistance. Despite these, HZMB still has other advanced designs that are worth exploring and they will all be introduced in this paper in general. This article explains both the overall structure of HZMB and moreover reveals details worth recommending. Research elaborates that both the structure and the construction processes include various ingenuity to maintain the stability and durability of HZMB.

1 INTRODUCTION

The construction of bridges is one of the most attractive topics since ancient time. Bridges not only serve as the shortcut for transportation which offers convenience for both economy and tourism but also represents the distinct literary flavor of a certain area. Especially in China, the history and development of bridges are extremely complex and engaging, from the basic wooden bridge to the stone arch bridge. In recent decades, iron and steel bridges occupy the major position until reinforced concrete bridges came into reality after the Second World War. The competition between bridges is as well a form of national strength contest. However, the construction of bridges, especially the development of its technique is an essential problem for governments. While designing a bridge, numerous factors should be taken into consideration. The structure should satisfy the social command and at the same time manage to sustain both internal and external loads such as the wind load, the traffic load, and the erosion of tides and waves. These influences accumulate inside the bridge's structure and when they reach a certain degree, the bridge will collapse under fatigue damage. Aims to solve the problem, designers of Hong Kong-Zhuhai-Macau Bridge (HZMB) set up several methods and thus the longest link between Hong Kong, Zhuhai and Macau came into reality. Since 2009 when the construction started, it soon gained the attention of people around the world. This article mainly uses Hong Kong-Zhuhai-Macau Bridge as an example to explain the basic structure of a sea-crossing bridge which combines a beam bridge, cable-stayed bridge and even the undersea tunnel as an entirety. This article will be divided into four parts to introduce this bridge elaborately.

Generally, this paper focuses on the innovation and other specific characteristics of HZMB through (1) Clarifying the aesthetic and cultural meaning and the economic benefit it brings to the adjacent area. (2) Split the whole structure as simple units according to the length of spans and the internal structure of decks. Analyze the separated parts, especially the technique for steel box

*Corresponding Author: xuyunxin332021@163.com

 DOI 10.1201/9781003310884-100

girders and piers which are differentiated from normal conditions. (3) Other factors are also taken into consideration including the corrosion or the assembling method.

1.1 *Background information*

Hong Kong-Zhuhai-Macau Bridge (HZMB) is the essential element of the transportation system in the Pearl Delta area. The Hong Kong-Zhuhai-Macao Bridge (HZMB) has a total length of 55-km, which covers 12-km Hong Kong Link Road, 29.6-km Main Bridge and 13.4-km Zhuhai Link Road. Thus, it becomes the longest bridge-cum-tunnel sea crossing in the world. With this link, the traveling time across the delta will be reduced from over 5 hours through road and by fast ferries to approximately 30 minutes (N. Hussain 2011). Moreover, it provides a safe and fast link to Hong Kong International Airport and Hong Kong Shipping Container Terminals which will manage to solve the problem of uneven development between the east and the west and further motivate the growth of the economy.

The construction of the Hong Kong-Zhuhai-Macau Bridge took about nine years from December 15th 2009 to October 24th 2018. As soon as the concept of Hong Kong-Zhuhai-Macau Bridge was proposed, it immediately led to widespread concern from people around the world. Since it is both the longest seacrossing ("China opens longest sea-crossing bridge". BBC News 2018) and the longest open-sea fixed link in the world, specialists from various fields are all wondering about the design and technics. As a matter of fact, to achieve the practicability and the needs for environmental protection, this bridge proposed many original structures and construction methods which are worth researching. In this report, three parts will be included to explain the construction of the Hong Kong-Zhuhai-Macau Bridge. The general structure is the first part, followed by the introduction of specific elements such as the deck and its box girders. At the end of the main body of this report, other related construction conditions and their matched solutions will be listed. The whole report will end up with the conclusion including some further expectations for deeper exploration.

1.2 *General Structure*

Despite infrastructures for boundary-crossing examinations, the general structure of the Hong Kong-Zhuhai-Macau Bridge (HZMB) also includes two artificial islands that serves as both linkages and supports for a 6-km-long undersea tunnel, marine viaducts, and three navigation span bridges (Hu 2015). This bridge originates from Hong Kong and first attaches to the east artificial island. Crossing the undersea tunnel which is approximately 6 km, the vehicles then arrive at the west artificial island and climb up to the bridge above. Along the bridge above there exist three sections of cable-stayed bridges/navigation span bridges. They are, from the west side to the east side, the Jiuzhou navigation span bridge, Jianghai navigation span bridge, and Qingzhou navigation span bridge. In the following report, they will be introduced separately.

Cable-stayed bridge consists of three main parts: towers (pylons), girders, and cables. The basic structure is shown in Figure 1 above. Among all these components, the cables which mainly sustain the tension between the deck and the pylon thus a triangular structure appears. The girder and towers of the structure endure the compression from both sides. Towers or pylons of the bridge are classified by their appearance. The Qingzhou Channel Bridge is one of the three navigable bridges in the main project of the Hong Kong-Zhuhai-Macao Bridge with a total length of 1150 m. It's a steel box girder cable-stayed bridge with two H-shaped pylons and double cable planes. The lengths of spans are 110 m, 236 m, 458 m, 236 m, and 110 m. During the information collection, Guruprasad (2016) even comes up with a model of three planes cable-stayed bridge and the comparison between a two planes bridge and a three planes bridge as the Figure 2 and Figure 3 below shows (Guruprasad D 2016).

Students from Far Eastern University ever applied software to imitate the realistic situation. They have taken the dead load, live load (the traffic load), wind load, and even the earthquake into consideration. Under these conditions, the members of the bridge react variously. Obviously, the

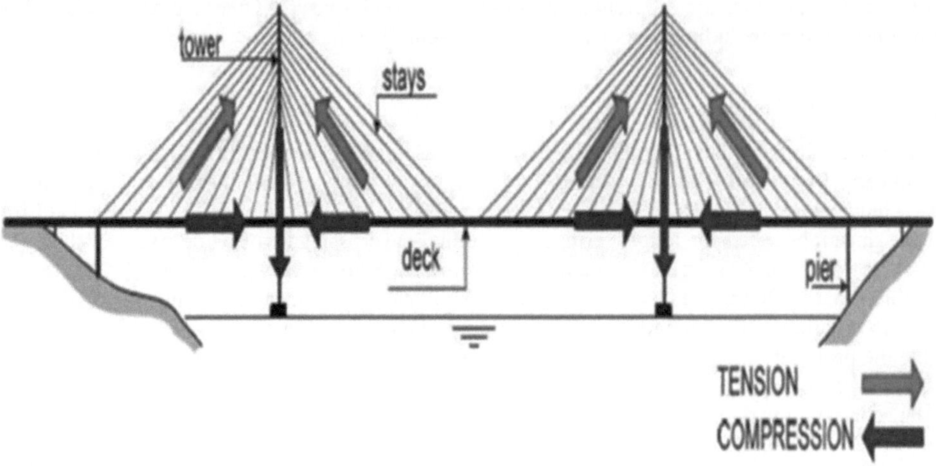

Figure 1. Structure of cable-stayed bridge and its force analysis on piers and cables (Alberto, 2020).

Figure 2. The model of two planes cable-stayed bridge (Guruprasad D 2016).

elements of the three planned cable-stayed bridges undergo a relatively smaller magnitude of axial forces. At the same time, the increasing number of planes represents the reduction of cross-sectional areas and hence reduces the consumption of steel. The price of steel in the market is much more expensive than concrete which always is applied as a substitute in engineering.

Table 1. Maximum forces on elements (Guruprasad D 2016).

	Cables	Cross beams	Longitudinal beams	Pylon
Two planes	6460	713	25076	104000
Three planes	3445	210	13462	6340

Figure 3. The model of three planes cable-stayed bridge (Guruprasad D 2016).

Table 2. Quantity of steel (Guruprasad D 2016).

	Cables	Transvers beams	Longitudinal beams	Pylon
Two planes	3590	3074	1366	8602
Three planes	2393	1648	1455	8042

Despite the Qingzhou navigation span bridge, Jianghai is a three towered cable-stayed bridge with the length of spans 129 m-258 m-258m-129m and Jiuzhou spans 85 m-127.5 m-268m-9127.5m-85m. The tower of these bridges not only consider practicability but also incorporate the Chinese culture into the appearance of the structure. The tower of Jiuzhou cable stayed bridge looks like the sail which means everything goes smooth in Chinese culture.

1.3 *Specific design*

The main composition of the Hong Kong-Zhuhai-Macau Bridge (HZMB) is a beam bridge instead of a navigation cable-stayed bridge. According to Su (2016), the bridge section has a total length of 22.9 km, including 3 navigation span cable-stayed bridges. The non-navigation span bridge uses a 110 m steel box continuous girder and 85 m steel box composite continuous girder (SU 2016). The Zhuhai and Macao area has comparatively shallow water for viaducts to span over while they are elongating towards the Lantau area, the depth of water increases, and thus the span varies. Among the Zhuhai and Macau region, the length of span fluctuates between 60 m and 80 m to suit shallow water. Over deeper water which is in proximity to the Lantau area, the spans investigated ranged from 90 to 120m. Based on the length of the span, the material and the structure of the box girder also vary. The main beam of Hong Kong-Zhuhai-Macao Bridge combination continuous bridge adopts the structure of a single-box, single-room with the same beam height. A single bridge's width is 16.3m and the height of the beam is 4.3 m.

The continuous combination of beam bridge steel main beams is in the form of an inverted trapezoidal groove and the material is Q235 steel whose modulus is 206,000 MPa, as shown in Figure 4. Shear studs are applied between the connected surface of the concrete surface and steel box. The concrete surface is made of concrete C60 whose modulus is 36,000 MPa. The combination of these two materials reduces the risk of crashing down and improves durability and stability (Li 2013).

In deeper water, the span will be prolonged and thus the structure of the box girder will change (Li 2013). An enormous cantilever structure is selected for steel box girders of HZMB. The thickness

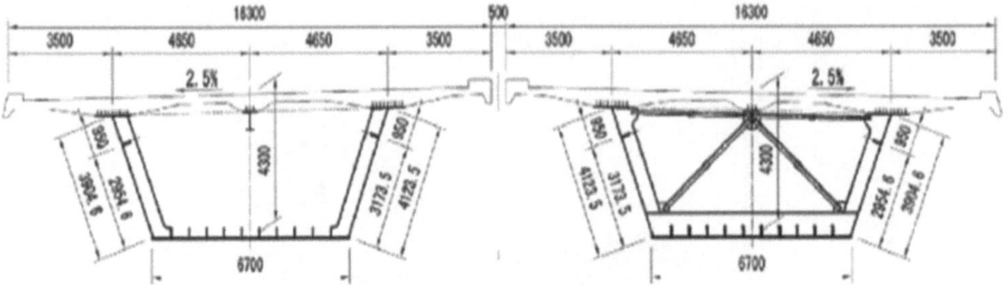

Figure 4. Split main beam section (unit: cm) (Li 2013).

of the panels on the top is unevenly distributed. The thinnest point is approximately 18 mm. U-rib is a crucial part of the structure. It locates right under the surface of the deck, responsible for transferring the load and thus improving the efficiency of the whole structure. There also exists a hollow structure and some special components for vibration damping. This will be extremely useful especially when the earthquake or other crustal movements happen.

Tuned mass dampers (TMDs) are usually made of concrete and their function is to reduce the vibration. For box girders, those cement blocks are hung in cells and connect to the walls with spring. While the vibration of the bridge, the blocks will wave in the opposite direction and partly eliminate vibration at the same time reducing the risk of bridges.

2 OTHER CONSIDERATIONS

According to Chen (2016), the following conditions are also considered:

(a) To satisfy the restriction for food quality and the regulations for silt control, the constrictivity of sea flow caused by the HZMB must be less than 10%. The pile foundation is buried beneath the seabed to avoid the resistance caused by it. Generally, the direction of the route has been kept perpendicular to the water flow to minimize the obstruction of the water flow and a horizontal curve is introduced to provide an interesting view of the main bridge as seen by the driver, passengers or visitors on the vehicles. Moreover, the whole section of the bridge has an S-shaped curve.

(b) Limited by the nearby airports, the height of auxiliary infrastructures during the construction of pylons should be restricted. Due to the limitation of the aviation height of Macau Airport, the steel tower of Jiuzhou Waterway Bridge adopts an innovative installation scheme, that is, "first hoist the entire bridge to the bridge deck, and then rotate it vertically on the bridge deck". During the installation, a large floating crane is used to hoist a steel tower about 70 meters in length and 1,000 tons in weight onto the slide on the bridge deck. Then the vertical rotation will be completed by two temporary hangers from two opposite edges of the deck.

(c) The middle area of the HZMB still serves as the important national reserve for endangered white dolphins. During the construction, the environment should be protected and the influence on habitat should be minimized. The National reserve of white dolphins is in the middle area of the bridge and during that section, the span will be prolonged as much as it could. The reduction of obstruction rate of the water flow as well protect wildlife to some degree.

(d) The marine environment provides extreme challenges for the sustainability and durability of HZMB. Height concentration of sodium ions, chloride ions, and the dense water vapor is more likely to corrode the piles and other metal structures of HZMB. Since the link is designed to serve the nation for more than a century, specialists are emerging to solve these problems. The bridge locates on the estuary where the corrosion from seawater is much more severe than the corrosion of river and streams. Plenty of researchers have focused on the chloride irons and their effects on both concrete and steel structure. The specialists managed to invent unique coating which could protect the bridge and improve the durability. This material could be recycled and it's environmentally friendly.

3 CONCLUSION

The construction of Hong Kong-Zhuhai-Macau Bridge (HZMB) sufficiently considers various conditions including its durability, stability, and the appearance. It's undeniable that specialist from different areas in China devote their efforts to set up this charming structure. This article covers all aspects of bridge construction from the structure design to the selection of materials. Among three navigation span bridges, the number of planes varies based on the requirement. The inverted trapezoid girder structure of the deck also considers both the stability and cost to construct the integration. Meanwhile Hong Kong-Zhuhai-Macau Bridge avoids maximum negative effects towards surrounding environment and nearby infrastructures. Despite the content above, there still are more things waiting for people to explore such as the structure of piers and the connection between decks. Numerous figures and articles are available on the internet, which reveal that the pier has unique U-shape structure for connection. This is a new technique for further research. Hoping that in the future, more chances will be provided to promote the exploration of the Hong Kong-Zhuhai-Macau Bridge (HZMB).

REFERENCES

Alberto M.B. Martins, Luís M.C. Simões, João H.J.O. Negrão. (2020). Optimization of cable-stayed bridges: A literature survey, Advances in Engineering Software, Volume 149, ISSN 0965-9978, https://doi.org/10.1016/j.advengsoft.2020.102829.

"China opens longest sea-crossing bridge". BBC News. 23 October 2018. Retrieved 28 October 2018.

CHAN, Man & POON, Wai & LEUNG, Yew & CHAN, Donald & PREMAUD, Valery & RIALLAND, Yves. (2016). Challenges in Hong Kong – Zhuhai – Macao Bridge (HZMB) Hong Kong Link Road Project. IABSE Symposium Report. 106. 797–804.

Guruprasad D, *Comparison of Two Planes and Three Planes Cable Configuration Of Cable Stayed Bridge*, Volume 03, 2016, Pages 1029-1031, e-ISSN 2395-0056, https://www.irjet.net/archives/V3/i9/IRJET-V3I9185.pdf

Hu, Z. N., Xie, Y. L., & Wang, J. (2015). Challenges and strategies involved in designing and constructing a 6 km immersed tunnel: A case study of the Hong Kong–Zhuhai–Macao Bridge. *Tunnelling and Underground Space Technology*, 50, 171–177. https://doi.org/10.1016/j.tust.2015.07.011.

Li P., 2013, The limited element analysis of the fatigue characteristics of the steel-mixed continuous beam bridge of the Hong Kong-Zhuhai-Macao Bridge. *Wuhan University of Technology*.

N. Hussain, Carlos Wong, Matt Carter, Samuel Kwan, Tak-Wah Mak,Hong Kong Zhuhai Macao Link, *Procedia Engineering*, Volume 14, 2011, Pages 1485–1492 https://doi.org/10.1016/j.proeng.2011.07.187

SU Quan-ke, XIE Hong-bing. Summary of Steel Bridge Construction of Hong Kong-Zhuhai-Macao Bridge[J]. *China Journal of Highway and Transport*, 2016, 29(12): 1–9.

Advances in Civil Engineering: Structural Seismic Resistance, Monitoring and Detection – Mohd Johari Mohd Yusof and Junwen Zhang (Eds)
© 2023 The Authors, ISBN 978-1-032-31491-4

Research on the rainfall-induced damage to railroad embankment slope: A case study from the railroad in Henan, China

Chuan Yin
Yunnan Construction and Investment Holding Group Co., Yunnan Kunming, China

Jiawei Fan
China Academy of Railway Sciences, Beijing, China

Yufang Zhang & Wenjiao Zhou
Railway Engineering Research Institute, China Academy of Railway Sciences Co. Ltd, Beijing, China

ABSTRACT: Affected by the severe rainfall action in Henan province, a huge range of railroad loess embankment was damaged under the action of rainwater infiltration and soakage. In order to solve the problem of loess embankment collapse and slippage, the methods of on-site investigation and numerical simulation were used. The research results indicated that: affected by the concentrated surface water erosion and infiltration, the railroad shoulder collapsed. To deal with the embankment damage, the prevention countermeasures of vertical steel flower tube + oblique steel anchor tube + slope toe retaining wall were proposed. The research results provide a reference for the prevention and control of existing railroad embankment damage in loess regions.

1 INTRODUCTION

Loess is a kind of collapsible soil which is widely distributed in China (Wang et al. 2017; Guo et al. 2019). The loess has poor water resistance. When it is exposed to water, the natural structure of loess is damaged so that strength of loess is decreased, and the slope stability of loess embankment is decreased as well (Cheng et al. 2017, 2018; Zuo et al. 2020). A series of research has been conducted to investigate the influence on rainfall to the strength and stability of loess slope (Chang et al. 2021; Hou et al. 2021; Li et al. 2020; Shen et al. 2019; Tang et al. 2015).

However, few studies were aimed at investigating the loess in Henan province because the rainfall in Henan province was not severe previously. Since July 2021, the northern region of China has suffered from heavy rainfalls. Henan province has suffered from the most severe rainfall in the last hundreds of years. From July 18th to July 20th, 2021, the maximum rainfall in the study site reached 438.7 mm, and the rainfall in two days was basically the same as the annual rainfall in the area. As a result, the railroad loess embankment in the study site was damaged by the heavy rainfall, inducing long-term traffic interruptions. Therefore, a case study was conducted to investigate a typical site within the loess embankment of the railroad in Henan province. By adopting field investigation and numerical simulation, the slope stability of loess embankment was analyzed, and corresponding prevention countermeasures were proposed.

2 THE STUDY SITE

According to field investigations, the stratum of this section is the Quaternary Upper Pleistocene alluvial strata which is mainly composed of light brown thick loess. The loess in the study site is lightly dense-medium dense, with a small particle size.

 DOI 10.1201/9781003310884-101

The average altitude in the study site is 110 meters with an annual average temperature of 11°C. The frost-free period is about 185 days, and the annual precipitation is 500–600mm.

During the on-site investigation, no regular surface water was discovered on the slope, and no groundwater was discovered in the cut slope. The groundwater is dominated by Quaternary pore water and bedrock fissure water. Quaternary pore water mainly exists in a river alluvial plain in the study site, with a uniform phreatic level and continuous distribution. Bedrock fissure water is the most widely distributed groundwater in the region, which exists in sandstone structural fissures and weathered fissures.

The layout of railroad was shown in Figure 1.

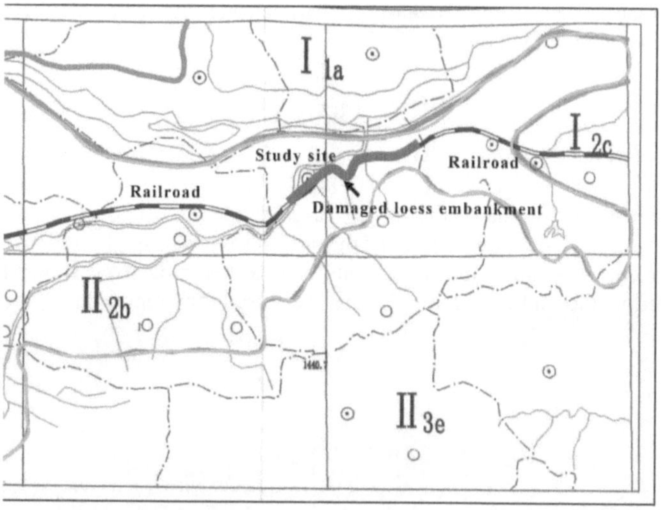

Figure 1. The layout of railroad.

3 RAILROAD LOESS EMBANKMENT SLOPE RAINFALL-INDUCED DAMAGE FIELD INVESTIGATION

The study site on the railroad were the embankment section. The slope height was 9 m. Influenced by the concentrated surface water erosion and infiltration, the study site of the railroad shoulder collapsed, forming a gap of 15 m in length, 3 m in width, and 2 m in depth. Under the continuous erosion of surface water, the gap was expanded. After the occurrence of the damage, the temporary reinforcement measures of cement sandbag backfilling were adopted, which was shown in Figure 2.

Figure 2. Temporary reinforcement measures of cement sandbag backfilling.

Due to the lack of protective measures on the original slope, the shallow deposits on the slope were relatively loose. Under the influence of severe rainfall and surface water infiltration, the soil of embankment slope was washed away by the rainwater, forming erosion grooves and local collapse, threatening the safe operation of the railroad. Thus, it was recommended that the embankment should be reinforced immediately.

4 STABILITY ANALYSIS OF THE EMBANKMENT SLOPE

To further investigate the stability of the railroad embankment after the rainfall, a finite element model has been constructed to simulate and calculate the stability factors of the slope under the natural condition and rainfall condition.

4.1 *Model setup*

The model was constructed through Plaxis2D software. A typical section was selected as the simulation section since it can display the most unfavourable condition within the damaged railroad embankment segment. According to the on-site measurement, a sectional CAD drawing of the railroad embankment slope was completed, and the model was established in the Plaxis2D software. The established model was shown in Figure 3, and the meshed model was shown in Figure 4. As for the yield criterion, the Mohr Coulomb criterion was used for strength calculation.

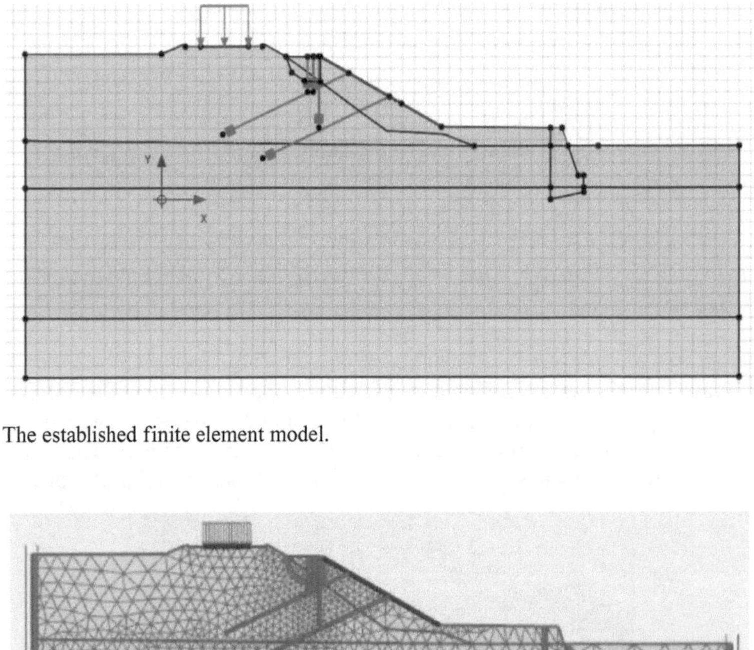

Figure 3. The established finite element model.

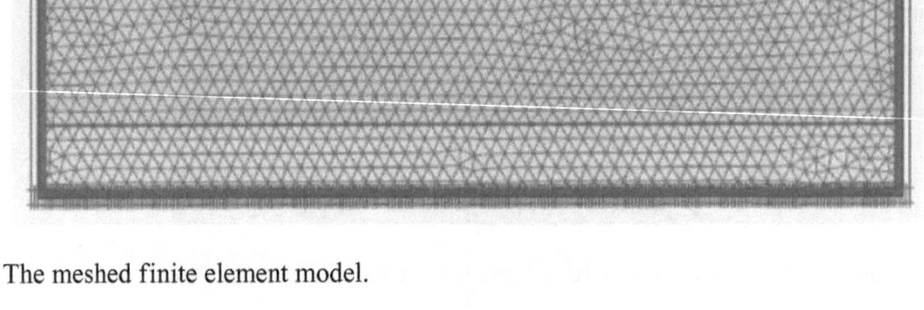

Figure 4. The meshed finite element model.

4.2 *Calculation parameter determination*

By on-site boring sample and conducting laboratory soil physical and strength tests, the physical parameters and strength parameters of the loess were obtained. The parameters used in the numerical simulations were displayed in Table 1.

Table 1. Parameters used in the numerical simulations.

Parameter	Bulk density (KN/m3)	Cohesion (kPa)	Friction (°)
Loess under natural condition	17	13.9	25
Loess under rainfall condition	19	9.4	11
Backfill soil	19	20	25
Coarse sand	21	10	27
Gravel soil	23	5	40

4.3 *Simulation results and analysis*

The slope stability analyzed using Plaxis2D was displayed by slope displacements strain increment. As shown in the model, the train and track load were a linear load of 68 KN/m which was evenly distributed within a range of 3.7 m.

4.3.1 *Natural condition*

The slope stability was analyzed by the limit equilibrium method, and the safety factor of the slope under natural condition were calculated. The calculated maximum shear strain increment cloud diagram was shown in Figure 5. The calculation result indicated that the safety factor of the slope in its natural condition was 1.753.

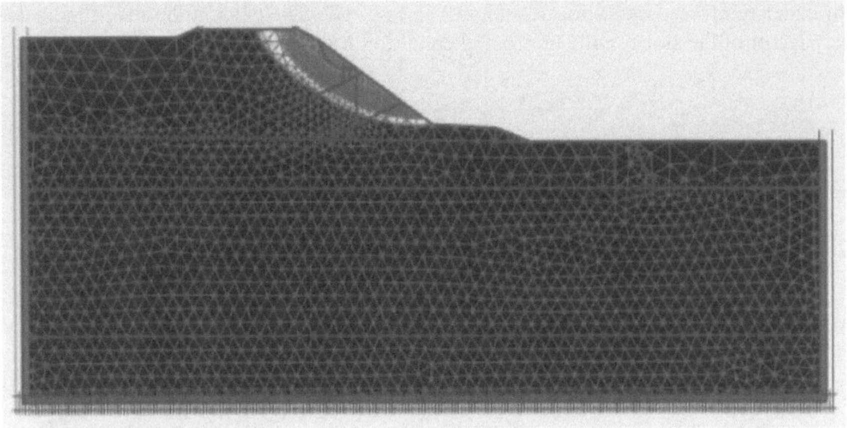

Figure 5. Cloud diagram of maximum shear strain increment of slope under natural condition.

4.3.2 *Rainfall condition*

The finite element calculation results of Plaxis2D indicated that the failure mode of the slope was a circular arc sliding under rainfall condition. The calculated maximum shear strain increment cloud diagram under rainfall condition was shown in Figure 6. The calculation result indicated that the safety factor of the slope in the rainfall condition was 0.977.

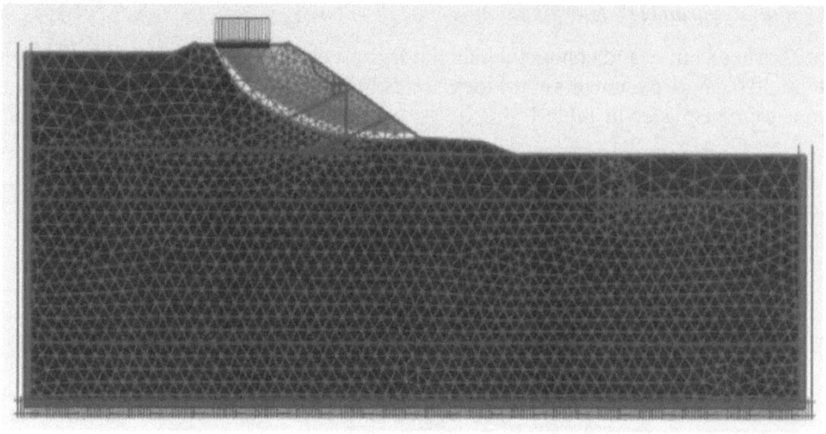

Figure 6. Cloud diagram of maximum shear strain increment of slope under rainfall condition.

4.4 *Prevention countermeasures*

The prevention countermeasures for the slope at the study site were described as follows. Firstly, set up a retaining wall at the toe of the slope. The top width of the retaining wall is 1 m, and 1.5 m of the retaining wall was exposed while 4.5 m of the retaining wall was buried. Secondly, two rows of oblique steel anchor pipes were constructed on the slope. The oblique steel anchor pipes were 12 m long, with a horizontal spacing (along the rail line) of 2.5 m and a vertical spacing of 2.5 m. Three rows of vertical steel floral tubes were constructed on the slope. One row of vertical steel floral tubes was 12 m long, and two rows of vertical steel floral tubes were 6 m long, with a horizontal spacing (along the rail line) of 0.5 m and a vertical spacing of 0.5 m.

In order to verify the effective of the prevention countermeasure design, use Plaxis2D to simulate the slope stability after reinforcement. The calculated maximum shear strain increment cloud diagram under reinforced condition was shown in Figure 7. The calculation result indicated that the safety factor of the slope in the reinforced condition was 1.793.

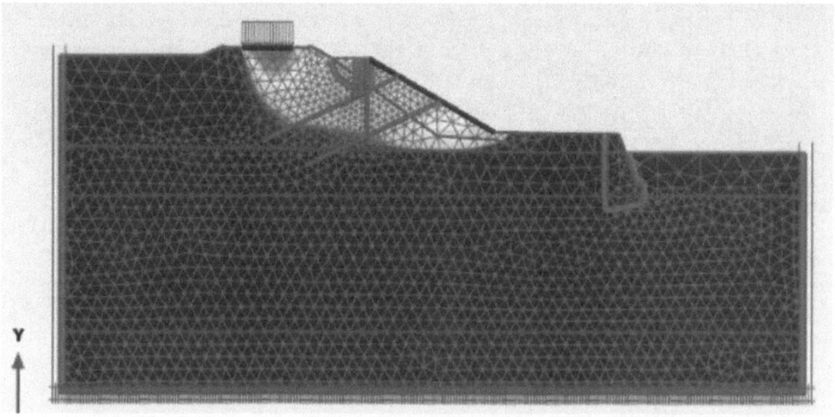

Figure 7. Cloud diagram of maximum shear strain increment of slope under reinforced condition

The safety factors calculated by Plaxis2D under three different conditions were displayed in the Table 2.

Table 2. Safety factors under three different conditions.Safety factors under three different conditions.

Conditions	Natural condition	Rainfall condition	Reinforced condition
Safety factors	1.753	0.977	1.793

5 CONCLUSION

By using field investigation and numerical simulation methods to study the slope stability of railroad loess embankment within the study site in Henan province, the following conclusions are obtained:

(1) Both field investigation and numerical simulation methods indicated that the heavy rainfall indeed induced damage to the railroad loess embankment. Cracks were discovered on the railroad embankment shoulder from field investigation, and numerical simulation result indicated that the safety factor of the slope under rainfall condition was less than 1.
(2) Oblique steel anchor pipes and vertical steel floral tubes were firstly adopted to dealing with railroad loess embankment stability problems in Henan region, which provides a new technology to reinforce railroad loess embankment.
(3) The effectiveness of the prevention countermeasures of utilizing oblique steel anchor pipes and vertical steel floral tubes were validated. It can be inferred from the numerical simulation results that the safety factor of the slope after reinforced condition was enhanced from 0.977 to 1.793.

ACKNOWLEDGMENTS

The authors acknowledge to the financial support from Science and Technology Plan Project of Yunnan Provincial Department of Transportation of China under grant number Yunjiao Science and Education 2018-34 and Yunjiao Science and Education 2018-35.

REFERENCES

Chang, Z., Huang, F., Huang, J., Jiang, S. H., Zhou, C., & Zhu, L. (2021). Experimental study of the failure mode and mechanism of loess fill slopes induced by rainfall. *Engineering Geology*, 280, 105941.

Chen, G., Meng, X., Qiao, L., Zhang, Y., & Wang, S. (2018). Response of a loess landslide to rainfall: observations from a field artificial rainfall experiment in Bailong River Basin, China. *Landslides*, 15(5), 895–911.

Cheng, S., Li, Z., Xu, G., Li, P., Zhang, T., & Cheng, Y. (2017). Temporal stability of soil water storage and its influencing factors on a forestland hillslope during the rainy season in China's Loess Plateau. *Environmental Earth Sciences*, 76(15), 1–10.

Guo, W. Z., Luo, L., Wang, W. L., Liu, Z. Y., Chen, Z. X., Kang, H. L., & Yang, B. (2019). Sensitivity of rainstorm-triggered shallow mass movements on gully slopes to topographical factors on the Chinese Loess Plateau. *Geomorphology*, 337, 69–78.

Hou, X., Li, T., Qi, S., Guo, S., Li, P., Xi, Y., & Xing, X. (2021). Investigation of the cumulative influence of infiltration on the slope stability with a thick unsaturated zone. *Bulletin of Engineering Geology and the Environment*, 1–14.

Li, M., Zhang, X., Yang, Z., Yang, T., & Pei, X. (2020). The rainfall erosion mechanism of high and steep slopes in loess tablelands based on experimental methods and optimized control measures. *Bulletin of Engineering Geology and the Environment*, 79, 4671–4681.

Shen, H. O., Zheng, F. L., Wang, L., & Wen, L. L. (2019). Effects of rainfall intensity and topography on rill development and rill characteristics on loessial hillslopes in China. *Journal of Mountain Science*, 16(10), 2299–2307.

Tang, Y. M., Xue, Q., Li, Z. G., & Feng, W. (2015). Three modes of rainfall infiltration inducing loess landslide. *Natural Hazards*, 79(1), 137–150.

Wang, L., Dalabay, N., Lu, P., & Wu, F. (2017). Effects of tillage practices and slope on runoff and erosion of soil from the Loess Plateau, China, subjected to simulated rainfall. *Soil and Tillage Research*, 166, 147–156.

Zuo, L., Xu, L., Baudet, B. A., Gao, C., & Huang, C. (2020). The structure degradation of a silty loess induced by long-term water seepage. *Engineering Geology*, 272, 105634.

Advances in Civil Engineering: Structural Seismic Resistance, Monitoring and Detection – Mohd Johari Mohd Yusof and Junwen Zhang (Eds)
© 2023 The Editors and Contributors, ISBN 978-1-032-31491-4

Author index